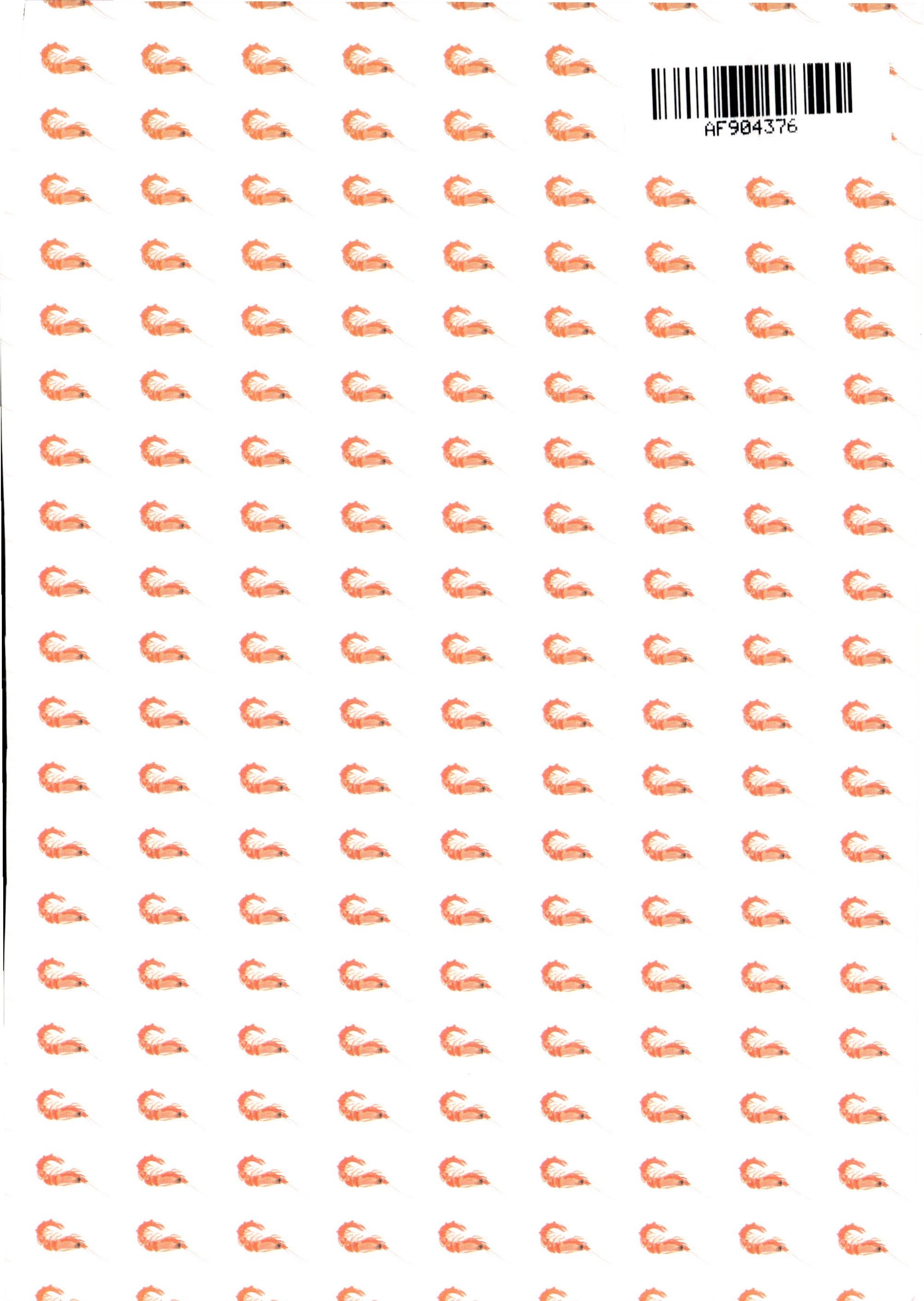

AF904376

THE SHRIMP BOOK II

THE SHRIMP BOOK II

EDITED BY VICTORIA ALDAY-SANZ

5m Books

First published 2022

Copyright © 5m Books Ltd 2022

All rights reserved. No part of this publication may be reproduced, stored in a retrieval system, or transmitted, in any form or by any means, electronic, mechanical, photocopying, recording or otherwise, without prior permission of the copyright holder.

Published by
5M Books Ltd,
Lings, Great Easton,
Essex CM6 2HH, UK,
Tel: +44 (0)330 1333 580
www.5mbooks.com

A Catalogue record for this book is available from the British Library

ISBN 9781789181043
eISBN 9781789181968
DOI 10.52517.9781789181968

Book layout by KSPM, Codsall, Wolverhampton

Printed by Ömur Printing & Binding, Istanbul, Turkey

Photos and illustrations by the authors unless otherwise indicated

Cover photograph by Jordi Bernadó, photographer (www.jordibernado.com)

GRACIAS PADRE!!

CONTENTS

PART III – PRODUCTION SYSTEMS

PART IV – FEEDS AND FEEDING

(SUB-EDITOR D. A. DAVIS)

PART V – BIOSECURITY AND HEALTH MANAGEMENT

PREFACE

First and above all, I and the 67 contributing authors would like to thank the readers of *The Shrimp Book* for their warm reception of that book. It is the dream of an author to be read, to be listened to, to share, even to be challenged and that was widely achieved. Thank you. We also thank the reviewers for their kind comments. These responses exceeded all our expectations and led to the request to prepare this new volume. We hope they will also be satisfied with this new volume.

Twelve years have passed since the publication of the first volume of *The Shrimp Book*. While the content of the first volume remains valid and useful, I felt a new volume was needed to capture the developments that have taken place over these years. Again, my aim was to try to produce the book I would like to have on my shelves, the book I could use as a reference. The first volume was conceptualized from the perspective of my academic years of experience looking into shrimp farming. However, I have spent these 12 years as a 'shrimp farmer' at industrial scale and this new volume is informed by my farming experience while still looking for the integration of academia and industry. *The Shrimp Book II* is complementary to the first volume, but it has a different perspective that emerges from the selection and combination of authors. The efforts of this new generation of authors together with our venerated ones bring together new concepts, new knowledge, new technologies and identify new challenges.

The content is structured in six parts with 27 chapters. Part I is general covering sustainability and insurance, then Part II covers the shrimp as an animal: functional anatomy, its genome and genetics. A third part on production systems: microbial, water and oxygen management, then Part IV on feeds and feeding including feeding behavior and automatic feeders. The fifth part is on biosecurity covering the response of shrimp to viruses, epigenetic management, new and old pathogens and the uses of disinfectants. Finally, Part VI addresses post-harvest issues discusses trade and quality control issues.

The book is addressed to a diverse readership. Its aim is to be of value to shrimp producers, researchers, service providers and students in any field of shrimp culture and to those involved in the regulation of the industry. As a deliberate editorial policy, there has been no attempt to uniformize the structure, the approach or the style of the chapters. The personal backgrounds and experience of the authors is reflected in their text which we hope will bring more immediacy to the reader. This level of 'freedom' had to be extended to the nomenclature of the white shrimp as some of the authors had strong views on whether *Penaeus* or *Litopenaeus* genus should be used, so the reader will find both nomenclatures used along the chapters.

This book could not have been produced without the inputs from a large number of people around the world. I am deeply in debt to the many colleagues that have guided me in the development of the book structure. I am particularly in debt to the authors for their generous effort and dedication particularly during the difficult period of the pandemic and the complications it added to our routine lives. I am particularly in debt to 5M Books who have trusted me with this project and supported throughout the process.

Of course, my deepest debt is for those who actually produce shrimp, all those who have patiently taught me over the years and have shared both their deep insight as well as their intuitive knowledge. My admiration for them has continued to bloom. In these last 12 years, adoption of technology has brought great improvements in production and we have seen the shift in trends particularly in the West. We are now at the edge of a new era of developments thanks to the molecular tools which will provide a completely different level of understanding of the production ecosystems, the shrimp physiology and the pathogens. The new approaches in *The Shrimp Book II*, will address novel, emerging questions and will question the *status quo* of orthodox beliefs.

Finally, the photograph of the cover requires an explanation. This is again a photograph of the great artist Jordi Bernado, one of best photographers of our generation and a troubadour through images. The photograph reflects the value of the family, the home, the land and the new generations with their defiant attitudes and their overwhelming strength. This new generation that must face new problems and are again bound to question the *status quo*.

Victoria Alday-Sanz
New Orleans
United States

CONTRIBUTORS

Diva January Aldama-Cano, Centex Shrimp, Faculty of Science, Mahidol University, Bangkok, Thailand, dr.divacano@gmail.com

Victoria Alday-Sanz, Director of Biosecurity, Breeding Programs and Research & Development, National Aquaculture Group, Naqua, Saudi Arabia, victoria_alday@yahoo.com

Edgar Amar, Southeast Asian Fisheries Development Center (SEAFDEC) Aquacualture Department, Iloilo, Philippines, eamar@seafdec.org.ph

Sopacha Arayamethakorn, National Center for Genetics Engineering and Biotechnology (BIOTEC), National Science and Technology Development Agency, Pathum Thani, Thailand, sopacha.ara@biotec.or.th

Maria de Fatima Arruda, Department of Physiology and Behavior, Center of Bioscience, Universidade Federal do Rio Grande do Norte, Natal, Brazil, mariadefarruda@gmail.com

Narong Arunrut, National Center for Genetics Engineering and Biotechnology (BIOTEC), National Science and Technology Development Agency, Pathum Thani, Thailand, narong.aru@biotec.or.th

Guillermo Bardera, Institute of Biomedical and Environmental Health Research (IBEHR), School of Health & Life Sciences, University of the West of Scotland, Paisley, PA1 2BE, UK, Guillermo.Bardera@uws.ac.uk

Kartik Baruah, Department of Animal Nutrition & Management, Faculty of Veterinary Medicine & Animal Sciences, Swedish University of Agricultural Sciences, Uppsala, Sweden, kartik.baruah@slu.se

Hao Bin, Fisheries Division, Food and Agriculture Organization (FAO) of the United Nations, Rome, Italy, Bin.Hao@fao.org

Peter Bossier, Lab Aquaculture & Artemia Reference Center, Faculty Bioscience Engineering, Ghent University, Gent, Belgium, Peter.Bossier@UGent.be

Claude E. Boyd, School of Fisheries, Aquaculture, and Aquatic Science, Auburn University, Auburn, Alabama, USA, boydce1@auburn.edu

David E. Brune, University of Missouri, Columbia, Missouri, USA, bruned@missouri.edu

Peter Bossier, Lab of Aquaculture & Artemia Reference Center, Department of Animal Sciences and Aquatic Ecology, Faculty of Bioscience Engineering, Ghent, Belgium, Peter.Bossier@UGent.be

Marcio Castro de Souza, Fisheries Division, Food and Agriculture Organization (FAO) of the United Nations, Rome, Italy, Marcio.CastroDeSouza@fao.org

Thawatchai Chaijarasphong, Center of Excellence for Shrimp Molecular Biology and Biotechnology (Centex Shrimp), Faculty of Science, Mahidol University, Bangkok, Thailand and Department of Biotechnology, Faculty of Science, Mahidol University, Bangkok, Thailand, thawatchai.chi@mahidol.edu

Sage Chaiyapechara, National Center for Genetics Engineering and Biotechnology (BIOTEC), National Science and Technology Development Agency, Pathum Thani, Thailand, sage.cha@biotec.or.th

Feiyang Chen, The Key Laboratory of Aquaculture Nutrition and Feeds Fisheries (Ministry of Agriculture and Rural Affairs), The Key Laboratory of Mariculture (Ministry of Education), Fisheries College, Ocean University of China, Qingdao, PR China, 1176789697@qq.com

Katherine A. Clayton, Biosciences, College of Life and Environmental Sciences, Centre for Sustainable Aquaculture Futures, University of Exeter, Exeter, UK, kc530@exeter.ac.uk

Mathias Corteel, IMAQUA, Lochristi, Belgium, mathias.corteel@imaqua.eu

Madeline Craig, World Wildlife Fund, Washington, DC, USA, Madeline.Craig@wwfus.org

Jean-Benoît Darodes de Tailly, Institute of Biomedical and Environmental Health Research (IBEHR), School of Health & Life Sciences, University of the West of Scotland, Paisley, PA1 2BE, UK, Jean-Benoit.deTailly@uws.ac.uk

D. Allen Davis, School of Fisheries, Aquaculture and Aquatic Sciences, Auburn University, Auburn, Alabama, USA, davisda@auburn.edu

Gaëtan De Gryse, Laboratory for Virology, Faculty of Veterinary Medicine, Ghent University, Ghent, Belgium, gaetanmadegryse@gmail.com

Tom Defoirdt, Center for Microbial Ecology and Technology, Department of Biotechnology, Faculty of Bioscience Engineering, Ghent University, Gent, Belgium, Tom.Defoirdt@UGent.be

Felix Dent, Fisheries Division, Food and Agriculture Organization (FAO) of the United Nations, Rome, Italy, Felix.Dent@fao.org

Xuan Dong, Yellow Sea Fisheries Research Institute, Chinese Academy of Fishery Sciences; Laboratory for Marine Fisheries Science and Food Production Processes, Pilot National Laboratory for Marine Science and Technology (Qingdao); Key Laboratory of Maricultural Organism Disease Control, Ministry of Agriculture and Rural Affairs; Qingdao Key Laboratory of Mariculture Epidemiology and Biosecurity, Qingdao City, Shandong Province, PR China, dongxuan@ysfri.ac.cn

Peter De Schryver, INVE Technologies NV, Dendermonde, Belgium, p.deschryver@inveaquaculture.com

Robert P. Ellis, Biosciences, College of Life and Environmental Sciences, Centre for Sustainable Aquaculture Futures, University of Exeter, Exeter, UK, r.p.ellis@exeter.ac.uk

Cesar M. Escobedo-Bonilla, Instituto Politécnico Nacional-CIIDIR Unidad Sinaloa, Guasave, Sinaloa, cesar_escobedomx@yahoo.com

Albert Ferrer, National Aquaculture Group (Naqua), Al Lith, Kingdom of Saudi Arabia, aferrerlladosa@gmail.com

Fahma Fiqhiyyah Nur Azizah, Graduate School of Agricultural and Life Sciences, The University of Tokyo, Tokyo, Japan, azizah-fahma@g.ecc.u-tokyo.ac.jp

Timothy W. Flegel, Center of Excellence for Shrimp Molecular Biology and Biotechnology (Centex Shrimp), Faculty of Science, Mahidol University, Bangkok, Thailand and National Center for Genetic Engineering and Biotechnology (BIOTEC), National Science and Technology Development Agency (NSTDA), Bangkok, Thailand, tim.flegel@gmail.com

Esther Garrido Gamarro, Fisheries Division, Food and Agriculture Organization (FAO) of the United Nations, Rome, Italy, Esther.GarridoGamarro@fao.org

Huaicheng Ge, The Key Laboratory of Aquaculture Nutrition and Feeds Fisheries (Ministry of Agriculture and Rural Affairs), The Key Laboratory of Mariculture (Ministry of Education), Fisheries College, Ocean University of China, Qingdao, PR China, 2711121655@qq.com

Jian Ge, Key Laboratory of Mariculture (Ocean University of China) Ministry of Education Fisheries College, Ocean University of China, Qingdao, PR China, gejian@stu.ouc.edu.cn

Yanlin Guo, The Key Laboratory of Aquaculture Nutrition and Feeds Fisheries (Ministry of Agriculture and Rural Affairs), The Key Laboratory of Mariculture (Ministry of Education), Fisheries College, Ocean University of China, Qingdao, PR China, wolin0521@vip.qq.com

John A. Hargreaves, Austin, Texas, USA, jhargreaves01@yahoo.com

Barbara Hostins, INVE, Dendermonde, Belgium, b.hostins@inveaquaculture.com

Dong Huang, The Key Laboratory of Aquaculture Nutrition and Feeds Fisheries (Ministry of Agriculture and Rural Affairs), The Key Laboratory of Mariculture (Ministry of Education), Fisheries College, Ocean University of China, Qingdao, PR China, 542725052@qq.com

Jie Huang, Network of Aquaculture Centres in Asia-Pacific, Bangkok, Thailand, jie.huang@enaca.org

Ming Huang, Key Laboratory of Mariculture (Ocean University of China) Ministry of Education Fisheries College, Ocean University of China, Qingdao, PR China, huangming@stu.ouc.edu.cn

Ornchuma Itsathitphaisarn, Center of Excellence for Shrimp Molecular Biology and Biotechnology (Centex Shrimp), Faculty of Science, Mahidol University, Bangkok, Thailand and Department of Biochemistry, Faculty of Science, Mahidol University, Bangkok, Thailand, ornchuma.its@mahidol.edu

Huang Jie, Yellow Sea Fisheries Research Institute, Chinese Academy of Fishery Sciences; Laboratory for Marine Fisheries Science and Food Production Processes, Pilot National Laboratory for Marine Science and Technology (Qingdao); Key Laboratory of Maricultural Organism Disease Control, Ministry of Agriculture and Rural Affairs; Qingdao Key Laboratory of Mariculture Epidemiology and Biosecurity, Qingdao City, Shandong Province, PR China, huangjie@ysfri.ac.cn

Sarocha Jitrakorn, National Center for Genetics Engineering and Biotechnology (BIOTEC), National Science and Technology Development Agency, Pathum Thani, Thailand, sarocha.jit@biotec.or.th

Lorenzo M. Juárez, Sea Products Development, Rockport, Texas, USA, lorenzojuarez@yahoo.com

Iddya Karunagar, Subba Meena, Mangalore, India, iddya.karunasagar@gmail.com

Indrani Karunasagar, Subba Meena, Mangalore, India, karuna8sagar@yahoo.com

Thuong Khuong Van, Ria2, Hungyen City, Vietnam, thuongkhuong03@gmail.com

Wansika Kiatpathomchai, National Center for Genetics Engineering and Biotechnology (BIOTEC), National Science and Technology Development Agency, Pathum Thani, Thailand, wansika@biotec.or.th

Keke Lei, The Key Laboratory of Aquaculture Nutrition and Feeds Fisheries (Ministry of Agriculture and Rural Affairs), The Key Laboratory of Mariculture (Ministry of Education), Fisheries College, Ocean University of China, Qingdao, PR China, 704731636@qq.com

Fuhua Li, Institute of Oceanology Chinese Academy of Sciences, Qingdao, Shandong, PR China, fhli@qdio.ac.cn

Xinxin Li, The Key Laboratory of Aquaculture Nutrition and Feeds Fisheries (Ministry of Agriculture and Rural Affairs), The Key Laboratory of Mariculture (Ministry of Education), Fisheries College, Ocean University of China, Qingdao, PR China, 1174469106@qq.com

Dandan Liu, Ocean University of China (OUC Laoshan Campus), Qingdao, PR China, liudandan0921@qq.com

Guilia Loi, Fisheries Division, Food and Agriculture Organization (FAO) of the United Nations, Rome, Italy, Giulia.Loi@fao.org

Roland Lulijwa, National Agricultural Research Organisation (NARO), Rwebitaba Zonal Agricultural Research and Development Institute (Rwebitaba-ZARDI), Fort Portal, Uganda, rlulijwa@gmail.com

Brett MacKinnon, Canadian Food Inspection Agency, Ottawa, Canada, mackinnonbrett@gmail.com

Melissa McLaws, Canadian Food Inspection Agency, Ottawa, Canada, melissa.mclaws@gmail.com

Aaron McNevin, World Wildlife Fund, Washington, DC, USA, Aaron.McNevin@wwfus.org

Shuoli Ma, The Key Laboratory of Aquaculture Nutrition and Feeds Fisheries (Ministry of Agriculture and Rural Affairs), The Key Laboratory of Mariculture (Ministry of Education), Fisheries College, Ocean University of China, Qingdao, PR China, 312554386@qq.com

Hugo H. Montaldo, Facultad de Medicina Veterinaria y Zootecnia, Universidad Nacional Autónoma de México, Coyoacán, Mexico, montaldo@unam.mx

Sergio Nates, Praire Aquatech, Brookings, South Dakota, USA, Sergio@prairieaquatech.com

Hans J. Nauwynck, D.V.M., Laboratory for Virology, Faculty of Veterinairy Medicine, Ghent University, Ghent, Belgium, Hans.Nauwynck@ugent.be

Stephen G. Newman, Aquaintech Inc., Lynnwood, Washington, USA, sgnewm@aqua-in-tech.com

Parisa Norouzitallab, Laboratory for Immunology & Animal Biotechnology, Ghent University, Laboratory of Aquaculture & Artemia Reference Center, Ghent University, Ghent, Belgium, Parisa.Norouzitallab@UGent.be

Sarah Penir, National Institute of Molecular Biology and Biotechnology and Philippine Genome Center, University of the Philippines, Quezon City, Philippines, smupenir@gmail.com

Leobert de la Pena, Southeast Asian Fisheries Development Center (SEAFDEC) Aquacualture Department, Iloilo, Philippines, leobertd@seafdec.org.ph

Manuel Poulain, INVE, Guarnizo, Spain, m.poulain@inveaquaculture.com

Anuphap Prachumwat, Aquatic Animal Health Research Team (AQHT), Integrative Aquaculture, Biotechnology, National Center for Genetic Engineering and Biotechnology (BIOTEC), National Science and Technology Development Agency (NSTDA), Bangkok, Thailand, anuphap@gmail.com

Juan C. Quintana-Casares, Maricultura del Pacifico, Mazatlan, Sinaloa, Mexico, juancarlos@maricultura.com.mx

Herbert E. Quintero, School of Agriculture, University of the Virgin Islands, St. Croix, US Virgin Islands, qfherbert@gmail.com

Liang Qiu, Yellow Sea Fisheries Research Institute, Chinese Academy of Fishery Sciences; Laboratory for Marine Fisheries Science and Food Production Processes, Pilot National Laboratory for Marine Science and Technology (Qingdao); Key Laboratory of Maricultural Organism Disease Control, Ministry of Agriculture and Rural Affairs; Qingdao Key Laboratory of Mariculture Epidemiology and Biosecurity, Qingdao City, Shandong Province, PR China, qiuliang@ysfri.ac.cn

Cesar Real R. Legisima, RMB Insurance Brokers, Vigo, Spain, cesar.real@rmb.es

Melba Reantaso, Fisheries Division, Food and Agriculture Organization (FAO) of the United Nations, Rome, Italy, Melba.Reantaso@fao.org

João Reis, School of Fisheries, Aquaculture and Aquatic Sciences, Auburn University, Auburn, Alabama, USA, jzt0062@tigermail.auburn.edu

Luke A. Roy, School of Fisheries, Aquaculture & Aquatic Sciences, Auburn University, Auburn, Alabama, USA, royluke@auburn.edu

Wanilada Rungrassamee, National Center for Genetics Engineering and Biotechnology (BIOTEC), National Science and Technology Development Agency, Pathum Thani, Thailand, wanilada.run@biotec.or.th

Francisco Saraiva Gomez, Pontos Aqua Holding LLC, New York, USA, f.saraivagomes@pontosaqua.com

Jason Scott, Scott Aquaculture Risk Management Ltd., Herriard, Hampshire, UK, jason.scott@scottaqua.com

Cibele Soares Pontes, Unit Specialized in Agrarian Sciences (Jundiaí Agricultural School), Universidade Federal do Rio Grande do Norte, Macaíba, Brazil, cibelepontes.ufrn@yahoo.com.br

Vanvimon Saksmerprome, National Center for Genetics Engineering and Biotechnology (BIOTEC), National Science and Technology Development Agency, Pathum Thani, Thailand, vanvimon.sak@biotec.or.th

Cynthia Palmes Saloma, National Institute of Molecular Biology and Biotechnology, and Philippine Genome Center, University of the Philippines, Quezon City, Philippines, cpsaloma@up.edu.ph

Patrick Sorgeloos, Laboratory of Aquaculture & Artemia Reference Center, Deapartment of Animal Sciences and Aquatic Ecology, Faculty of Bioscience Engineering, Ghent University, Gent, Belgium, Patrick.Sorgeloos@UGent.be

Kallaya Sritunyalucksana, Aquatic Animal Health Research Team (AQHT), Integrative Aquaculture, Biotechnology, National Center for Genetic Engineering and Biotechnology (BIOTEC), National Science and Technology Development Agency (NSTDA), Bangkok, Thailand, kallaya@biotec.or.th

Grant D. Stentiford, Centre for Environment, Fisheries and Aquaculture Science (Cefas), Weymouth, Dorset, UK, Grant.stentiford@cefas.co.uk

Suparat Taengchaiyaphum, Aquatic Animal Health Research Team (AQHT), Integrative Aquaculture, Biotechnology, National Center for Genetic Engineering and Biotechnology (BIOTEC), National Science and Technology Development Agency (NSTDA), Bangkok, Thailand, suparat.tae@biotec.or.th

Kwanta Thansa, Aquatic Animal Health Research Team (AQHT), Integrative Aquaculture, Biotechnology, National Center for Genetic Engineering and Biotechnology (BIOTEC), National Science and Technology Development Agency (NSTDA), Bangkok, Thailand, kwanta.tha@biotec.or.th

Olav Vadstein, Department of Biotechnology and Food Science, NTNU Norwegian University of Science and Technology, Trondheim, Norway, olav.vadstein@ntnu.no

Daisy Vanrompay, Laboratory of Immunology and Animal Biotechnology, Department of Animal Sciences and Aquatic Ecology, Ghent University, Ghent, Belgium, Daisy.Vanrompay@UGent.be

Weiwei Wang, Fisheries Division, Food and Agriculture Organization (FAO) of the United Nations, Rome, Italy,WeiWei.Wang@fao.org

Bryony A. P. Williams, Biosciences, College of Life and Environmental Sciences, Exeter, UK, B.A.P.Williams@exeter.ac.uk

Rod W. Wilson, Biosciences, College of Life and Environmental Sciences, Centre for Sustainable Aquaculture Futures, University of Exeter, Exeter, UK, R.W.Wilson@exeter.ac.uk

Dominic Wiredu-Boakye, College of Life and Environmental Sciences, Living Systems Institute, University of Exeter, Exeter, UK, D.Wiredu-Boakye@exeter.ac.uk

Yanjiao Wu, The Key Laboratory of Aquaculture Nutrition and Feeds Fisheries (Ministry of Agriculture and Rural Affairs), The Key Laboratory of Mariculture (Ministry of Education), Fisheries College, Ocean University of China, Qingdao, PR China, 825963735@qq.com

Jianhai Xiang, Institute of Oceanology Chinese Academy of Sciences, Qingdao, Shandong, PR China, jhxiang@qdio.ac.cn

Jianbo Yuan, Institute of Oceanology Chinese Academy of Sciences, Qingdao, Shandong, PR China, yuanjb@qdio.ac.cn

Yang Yu, Institute of Oceanology Chinese Academy of Sciences, Qingdao, Shandong, PR China, yuyang@qdio.ac.cn

Qingli Zhang, Yellow Sea Fisheries Research Institute, Chinese Academy of Fishery Sciences; Laboratory for Marine Fisheries Science and Food Production Processes, Pilot National Laboratory for Marine Science and Technology (Qingdao); Key Laboratory of Maricultural Organism Disease Control, Ministry of Agriculture and Rural Affairs; Qingdao Key Laboratory of Mariculture Epidemiology and Biosecurity, Qingdao City, Shandong Province, PR China, zhangql@ysfri.ac.cn

Wenbing Zhang, The Key Laboratory of Aquaculture Nutrition and Feeds Fisheries (Ministry of Agriculture and Rural Affairs), The Key Laboratory of Mariculture (Ministry of Education), Fisheries College, Ocean University of China, Qingdao, PR China, wzhang@ouc.edu.cn

Xiaojun Zhang, Institute of Oceanology Chinese Academy of Sciences, Qingdao, Shandong, PR China, xjzhang@qdio.ac.cn

Yangen Zhou, Key Laboratory of Mariculture (Ocean University of China) Ministry of Education Fisheries College, Ocean University of China, Qingdao, PR China, zhouyg@ouc.edu.cn

Chapter 1

SUSTAINABILITY IN SHRIMP AQUACULTURE

LORENZO M. JUÁREZ, CLAUDE E. BOYD, AARON MCNEVIN, MADELINE CRAIG AND SERGIO NATES

Summary

Shrimp farming and its associated technologies have sustained tremendous growth since their beginnings in the 1970s. With this growth also came concerns about the environmental sustainability of the new industry. In this chapter we examine the operational concept of sustainability as applied to shrimp farming and we frame the most useful and quantifiable approaches to efficiency in the use of resources such as water, land, feed, fishmeal, manpower, and agrochemicals. Also discussed are shrimp farming's effects on sensitive habitats such as mangrove forests, the effects of farm effluents on the receiving bodies of water, and impacts from the translocation of shrimp species via animal movements and associated shrimp diseases. In general, shrimp aquaculture has come a long way from its beginnings with regard to sustainability. It is envisioned that it will continue to evolve and follow the general patterns of other agribusinesses towards increasing efficiency in its use of resources, with advanced technologies and with consumer demand for ecologically friendly foods incrementally moving the industry towards better environmental performance.

Introduction

Aquaculture is one of the world's fastest-growing food production technologies, having more than tripled output between 2000 and 2019 (FAO, 2019a). The global supply of fish is projected to be 196.3 Mt by 2028, a 14% increase relative to 2018, and the majority of this growth will come from the aquaculture sector, where output is projected to increase by an average of 2 Mt/y to 102.2 Mt in 2028, an increase of 28% over the output in 2018 (OECD/FAO, 2019). These increases in production are required to keep up with the needs of a growing population, but our planet has limits, and rapid growth and competition can put the environment at risk.

Shrimp farming has been associated with negative environmental impacts such as destruction of habitats, water pollution, translocation of shrimp species and diseases, and excessive use of fishmeal and fish oil in feeds (Moss et al., 2001). Over the roughly 50 years of shrimp farming's existence, environmental sustainability has evolved from a blurry idea to concrete progress on making more efficient use of resources such as land, water, energy, manpower, and feed. The use of selectively bred and specific-pathogen-free stocks has, together with biosecurity, allowed better management of disease risks. Importers and consumers are ever more demanding of sustainable production methods, and good environmental practices are required of aquaculture firms by certifying bodies and government authorities. In today's world, there is no debate over whether aquaculture should be managed in an environmentally responsible and sustainable fashion (Engle, 2019; Engle and D'Abramo, 2016).

Shrimp farming technology is not well understood by the public. The realization that warmwater marine shrimp could be farmed can be related to the study of the life cycle of shrimp (Fig. 1.1). In nature, adult shrimp mature sexually, mate, and spawn in the open ocean. The eggs hatch and undergo metamorphosis through three basic larval stages: nauplius, protozoea, and mysis. The larvae then metamorphose to postlarvae and grow into early juvenile stages, which are carried by the currents into coastal marshes and estuaries. The juvenile stages of shrimp thrive in estuarine environments, where they grow on the plentiful food sources characteristic of these transitional ecosystems. Tides and currents can propel the juveniles into shallow water pools, where they may be cut off from the estuary. Shrimp aquaculture thus had its remote origins in juveniles that became trapped in the tidal ponds and marshes

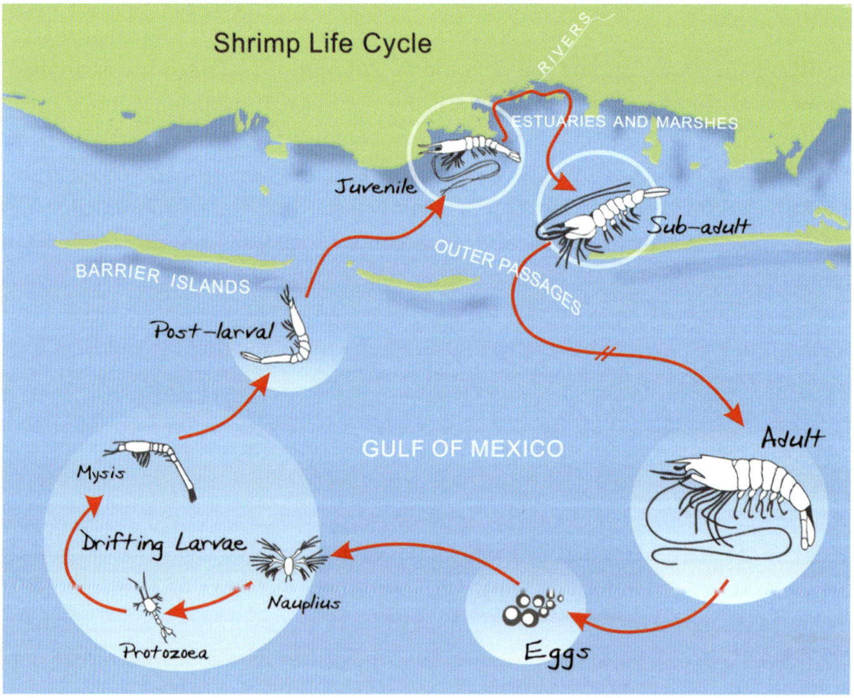

FIGURE 1.1 The life cycle of penaeid shrimp. Credit: Diagram by J.R. Allen in Spies et al. (2016). With permission and thanks to *Gulf of Mexico Science*, The Ocean Conservancy, and J.R. Allen.

of Southeast Asia and the Americas (Chamberlain, 2010; Hirono and Leslie, 1992; Lopez-Alvarado et al., 2016; Stickney et al., 2000). Soon people learned to harvest the young postlarvae from the sea (Motoh et al., 1994) and to transfer them into artisanal coastal ponds. Rudimentary fertilization, feeding, and water exchange management followed, heralding a path towards intensification that continues to this day.

While shrimp farming has been practised at an artisanal level in Southeast Asia and the Americas for a long time, modern shrimp farming only began when scientists developed methods to produce postlarvae in hatcheries. These efforts were pioneered in Japan in the 1930s by Dr Motosaku Fujinaga (also known as Hudinaga), who first closed the life cycle of the Japanese shrimp *Marsupenaeus japonicus,* allowing the development of a local modest-sized shrimp aquaculture industry (Hudinaga, 1935). One of Fujinaga's students, I. Chiu Liao, went on to refine culture techniques for shrimp in Taiwan, studying the comparative physiology of several species of shrimp and selecting the native prawn, *Penaeus monodon*, as the most suitable culture species for that country. These achievements, together with advances in phytoplankton cultivation, later catalysed progress in the Americas, with researchers at the US Bureau of Commercial Fisheries' Galveston Laboratory developing a more reliable method for larval culture that eventually became known as the 'Galveston Method' (Cook and Murphy, 1966). This ability to predictably produce shrimp postlarvae ('seed') and the interest generated in farming encouraged research and pilot projects in Florida, Texas, and Mississippi, which eventually spread into commercial projects in Honduras (1968), Ecuador (1969), Panama (1973), the rest of Latin America, and eventually Southeast Asia and many other warm-weather regions of the world. The interested reader is directed to the history of shrimp farming by Chamberlain (2010).

In its beginnings, shrimp production by aquaculture was modest and grew slowly (Fig. 1.2). For example, in 1970 the world's farms produced under 9000 t of shrimp, but by the 1980s growth started to accelerate, reaching more than 5 Mt in 2017, valued at about US$32 billion (Tacon, 2020). Clearly, shrimp farming has become an important food production technology, but growth has not always been sustainable, as evidenced by epizootics that crippled whole farming regions and countries. The development of shrimp rearing technology can be viewed as one of increasing intensity with lagging but improved control of the culture environment. As production and intensity increased, challenges emerged

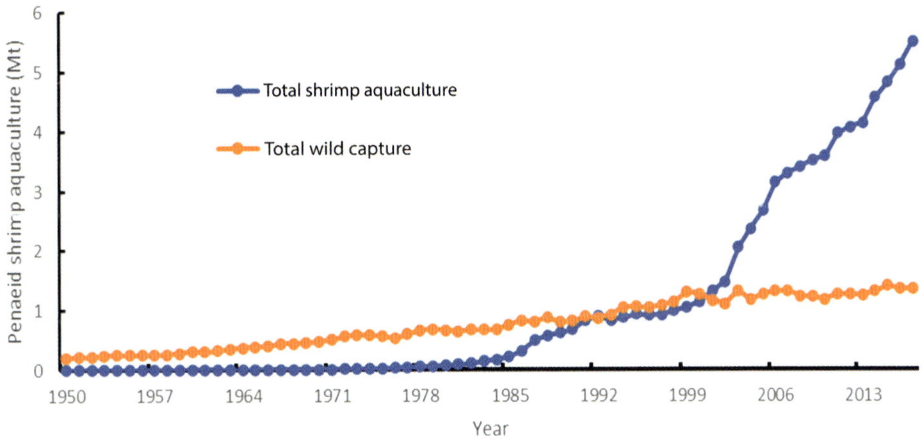

FIGURE 1.2 Shrimp supply from aquaculture and captures, 1950–2017. Data from FAO (2019a, b).

that threatened the sustainability of the industry and were subsequently addressed with improved management practices and technical innovations. The surge in shrimp production since the early 2000s primarily came from the introduction of Pacific whiteleg shrimp, *Litopenaeus vannamei*, to Southeast Asia, especially into China, India, Vietnam, Indonesia, and Thailand. Requiring less protein in its diet than *P. monodon* and being better domesticated to farming, *L. vannamei* has become the preferred species to drive intensification and increase output.

The global aquaculture community has been on a steep learning curve over the past 50 years, during which environmental considerations have evolved from a side-line concept into an indispensable way to conduct business. In this chapter we examine the different operational definitions of sustainability as they apply to shrimp farming. We also discuss the main interactions between shrimp farming and the environment, presenting quantitative measurements of resource efficiency and highlighting recent progress.

Towards a useful understanding of sustainability

Because the terms 'sustainable' and 'sustainability' have been overused and lack clear definitions, we prefer terms such as 'environmental stewardship' or 'environmental responsibility', or, even better, to refer directly to measurable indicators of resource efficiency. Nevertheless, a discussion of the different meanings and definitions of sustainability is presented below.

Despite its current relevance, sustainability as applied to aquaculture is not a clearly defined concept, and its use often results in confusion because it can mean different things under distinct disciplines. The public understands sustainability as the capacity of human activities to persist in time while maintaining a healthy environment. The most-cited definition is that of the United Nations' World Commission on Environment and Development, better known as the Bruntland Commission, which defined sustainable development as use of the environment and resources that meets the needs of the present without compromising the ability of future generations to meet their own needs (World Commission on Environment and Development, 1987). The Bruntland approach integrates environmental stewardship with social responsibility and economic gain, to reflect an understanding that focusing solely on economic growth ignores and impedes social development and environmental protection (Hove, 2004). The Bruntland definition has been criticized for its lack of specificity and usefulness, and for leading to the false impression that development can proceed with only minor changes (Stefanovich, 2000). Regrettably, sustainability has become more of a catchphrase and a marketing tool than a clearly defined and useful concept.

Economic theory provides a more useful definition of sustainable development, as economic systems where the productive base, or total capital, remains constant over time (Chapin et al., 2009). In this definition 'capital' refers to the comprehensive wealth of the system, including human, environmental, and economic components. This economic definition implies the impossibility of unlimited growth given the planet's limited resources, and presents the practical problem of assigning limits to the system and of quantifying the consequences experienced by unrelated third parties, the so-called externalities. In a sustainable aquaculture system, the overall value is maximized while environmental damage is minimized.

Social sustainability is related to the promotion of well-being within an organization's members. It focuses primarily on fair labour practices, but includes the overarching goals of reducing social inequality, increasing quality of life, and defending human rights. Social sustainability issues in aquaculture have focused on the more pressing needs to eliminate abject poverty, child labour, modern slavery, and other unethical practices that unfortunately still exist in some regions.

In ecology, sustainable biological systems are characterized as diverse, adaptable, resilient, and biologically productive over time (Chapin et al., 2009). Examples of ecologically sustainable systems are healthy oceans, lakes, rivers, wetlands, and forests. This ecological definition is not applicable to monoculture food production systems such as most forms of industrial agriculture and aquaculture.

Most countries have sensible laws and regulations to protect the environment. Aquaculture is often regulated by multiple government agencies that mainly deal with capture fisheries, agriculture, coastal management, water resources, and environmental protection. Although these compound bodies of regulation are often flawed, they represent the country's legal interpretation of environmental sustainability.

Business owners and managers often see sustainability as the strategic long-term management of resources, leading to the capacity of the business to thrive and persist in time. There is no sustainability in a system that is not financially rational. Central to this concept is the idea that business results should be predictable and profitable. Although the health of the environment is not the primary motivator of businesses, sensible business leaders recognize that a healthy environment is required for successful operations. Seafood buyers and consumers increasingly play a role, demanding products that have been produced in a manner taking social and environmental responsibility into consideration. These requirements have resulted in increased awareness and continuous improvements designed to manage the environmental footprint of aquaculture. A related idea is that sustainability and productivity go hand in hand. Productivity implies transactions that increase value, that is, producing more shrimp with less resources while making more money (Anthony Schuur, Galt, California, 2020, personal communication).

In practice, environmental sustainability becomes an economic question of choosing the best options among different possible practices and procedures. Typically, when faced with an operating or investment decision, an aquaculture manager can choose solutions with different environmental consequences. Ideally, the best technologies allow producers to increase productivity and reduce the environmental footprint. Engle (2019) summarized this approach in an editorial: 'perhaps it is time to bring the sustainability discussions down to earth and focus on more pragmatic, empirical, outcomes-based approaches to choices related to species/production systems, their sustainability, and their contribution to human food security'. Industry groups, companies, organizations, and governmental agencies have developed sets of good management practices that aid in making these choices. Additionally, 'green label' programmes sometimes help cover the cost of more expensive practices by providing price premiums or access to speciality markets for aquaculture products raised under socially and environmentally responsible practices.

Central to the idea of sustainability is making efficient use of resources such as energy, water, land, feed, and fertilizer. In this context, aquaculture is generally more efficient than traditional farming of terrestrial livestock (Hall et al., 2011). Reducing the use of scarce resources is key to providing healthy food for a growing population in an environmentally responsible manner. Because such use can be quantified, benchmarked, and compared, this approach eliminates the subjectivity of stakeholders and turns it into a more objective endeavour. Boyd and McNevin (2015) presented a detailed review of this approach. Naturally, there are differences between farming systems and species. For example, beef production requires no fishmeal or fish oil, while salmon farming requires large amounts of fish oil in comparison to other farmed fish. Unlike beef farming, salmon aquaculture does not require much land. These differences result in a host of subjective value judgements regarding the impacts of different agribusinesses.

The increased emphasis on the efficient use of natural resources, both on-farm and 'embodied' in feed (i.e. resources used for the production of the feed), has arisen because of the direct economic implications of resource inefficiencies. While this is not a novel management approach, the tightening of available resources and a more professionalized industry have sought improved precision in the production of shrimp. While land, brackish water, and energy are utilized on-farm, embodied land, freshwater, energy, and wild fish are needed to produce shrimp feed and other farming inputs. Thus, with increased intensification, the shrimp aquaculture industry has put itself in competition with other forms of animal husbandry for some of the same resources. For example, ideal shrimp farm sites are limited, and producers endeavour to minimize the use of land to maximize profits. This is also a result of the commoditization of shrimp on the global market. Downward price trends demand greater efficiency and will result in further consolidation of the sector. While shrimp farming has had substantial impact on its own sustainability through the degradation of habitat and production inefficiencies, its practices are changing for the better and can be considered part of the learning curve in the broader aquaculture sector.

Sustainability and culture intensity

Shrimp farming has been classified into categories of intensity according to its use of resources such as feed, aeration, energy, land, and water. Table 1.1 shows a summary of resource use estimates in five major shrimp-producing countries and in an indoor, hyper-intensive farm in the USA. The systems differ greatly in their inputs and outputs. Semi-intensive systems, like those common in Ecuador and Central America, have embraced a less risky and resource-intensive system of production to reduce stress and prevent disease outbreaks. These systems sometimes benefit from specific subsidies that translate into system externalities, where the producer does not have to bear the full costs (e.g. availability of inexpensive land concessions or leases, simplified discharge permits, or subsidized fuels), or from price premiums in niche markets specific to lower-intensity shrimp. At the other end of the spectrum are fully enclosed hyper-intensive systems popular in North America and Europe, which are highly biosecure and have virtually no direct interaction with the environment. More intensive systems are most akin to precision agriculture and friendlier to the environment due to their more efficient use of resources per unit weight of shrimp produced (Boyd et al., 2017, 2018).

In the past, some environmentalists have argued that low-density aquaculture is more sustainable than the intensive form. This was due to an association between extensive farming and natural productivity and/or wild capture fisheries. This approach makes sense until one considers and quantifies the use of finite resources by both forms of aquaculture. For example, Boyd and McNevin (2018) estimated that there are approximately 2.4 million ha of shrimp ponds in the world, but that approximately 94% of the farmed shrimp is produced in only 1.4 million ha, or 58% of the total. In other words, only 6% of the shrimp is produced in 42% of the total shrimp pond area. Land for ponds is often taken from wetlands and other valuable ecosystems, so making efficient use of land is highly desirable. Shrimp production systems need to become more efficient in their use of natural resources. More efficient production has a

TABLE 1.1 Summary of resource use estimates per tonne of shrimp (*L. vannamei*) produced by aquaculture in five major producing countries and in an indoor hyper-intensive system in the USA.

Variable	*Ecuador*	*India*	*Vietnam*	*Thailand*	*Indonesia*	*USA*[1]
Stocking rate (postlarvae/m²)	22.5	38.3	57.3	76.5	92.5	150
Survival (%)	»60	75.8	73.9	70.8	72.5	85
Number crops/y	3.6	1.97	1.87	2.25	2.5	3.9
Crop duration (d)	91	115	90	90	100	84
Aeration rate (hp/ha)	9	18	24	43	?	110
Yield (t/ha pond surface/y)	7.03	7.85	9.3	17.7	22.7	199
Yield (t/ha farm area/y)	5.37	6.54	5	8.34	12.97	49.75
Feed conversion ratio	1.32	1.48	1.33	1.49	1.4	1.3
Saline water use (intake):						
Water exchange (m³/t)	76,817	126,498	15,100	–	33,475	1,145
No water exchange	9862	6598	–	5400	4459	235
Effluent discharge (m³/t)						
Water exchange	75,317	123,163	14,077	–	33,046	286
No water exchange	6187	5087	–	4856	4030	–
Land use:						
Farm (ha/t)	0.19	0.152	0.196	0.12	0.077	0.020
Embodied in feed (ha/t)	0.33	0.369	0.331	0.371	0.402	0.324
Embodied in sugar or molasses (ha/t)	0.02	Rarely used	Not used	0.02	–	Not used
Total (ha/t)	0.54	0.521	0.527	0.511	0.479	0.344
Direct energy use:						
Construction and repair (GJ/t)	0.9	5.2	10.7	6.2	5.8	1.8
Farm operations (GJ/t)	23.3	56.9	26.4	32.6	37	35.5
Total (GJ/t)	24.2	62.1	37.1	38.8	42.8	37.2
Embodied energy:						
Fuel (GJ/t)	21.1	80.5	72.4	47.3	48.8	102.9
Feed (GJ/t)	12.4	12.7	10.9	11.9	11.7	11.2
Fertilizers[b] (GJ/t)	2.5	1.71	1.17	0.67	–	Not used
Disinfectants[b] (GJ/t)	0.1	3.67	4	4.18	–	0.11
Other[c] (GJ/t)	0.9	0.57	0.12	0.16	–	–
Total embodied energy (GJ/t)	37.0	99.2	88.6	64.2	60.5	114.2
Total energy (GJ/t)	61.2	161.3	125.7	103	103.3	151.4
Embodied freshwater use:						
Fuel (m³/t)	100	383	103	205	208	5.14
Feed (m³/t)	2112	1651	1484	1663	2003	1282
Fertilizers[a] (m³/t)	12	35	26	23	–	Not used
Disinfectants[b] (m³/t)	2	2	3	3	–	–
Other[d] (m³/t)	50	2	2	3.92	–	–
Total freshwater use (m³/t)	2276	2073	1618	2286	2211	1287
Wild fish use in feed (t/t)	0.89	0.66	0.63	0.67	0.61	1.69

a Data for a single, hyper-intensive, indoor production system in Texas, USA. Data provided by Thomas James, Sea Products Development (personal communication).

b Liming materials, fertilizers, and mineral amendments.

c Chlorine and iodine compounds, peroxides, copper sulphate, etc.

d Sugar, molasses, etc.

clear value proposition for producers: less land, water, energy, and/or feed use can result in higher profits (Engle et al., 2017).

The intensity of the culture system is also related to shrimp health and risk management. There are two – somewhat opposed – approaches to health management in relation to culture intensity. In the less intensive approach characteristic of traditional farms in Ecuador, stress is reduced by low culture densities and good water quality, which is maintained by exchanging pond water. Shrimp stocks in these systems have been naturally selected through many generations of exposure to the pathogens present in the environment and have become somewhat tolerant to those diseases. The second approach, which can be called 'controlled intensification', is a more fine-tuned form of farming that aims to increase control over the inputs and variables that affect production. In this approach, externalities are limited through greater containment of the production system and less dependency on the environment. An example is the type of system prevalent in Thailand, where diseases are excluded by using specific-pathogen-free stocks, disinfecting the water and reducing water exchanges, minimizing contact with pathogen vectors, using biosecure feeds, and enforcing other biosecurity practices. At the same time, the shrimp's natural ability to limit disease is boosted by stress-reduction practices such as providing good nutrition, stable water quality, and a healthy and diverse bacterial community, often aided by the addition of probiotics.

As in other agribusinesses, shrimp aquaculture has gone through transformations trending towards greater intensity of production. These transformations are typically accompanied by growing pains from pushing limits, which can stress the species under cultivation and contribute to outbreaks of disease or other causes of crop failure. The intensification of shrimp farming is primarily a result of: (1) better-formulated feeds, which provide better nutrient delivery; (2) improved mechanical aeration, which allows for greater culture density through improved oxygenation; (3) use of domesticated stocks, resulting in faster growth and increased disease tolerance; and (4) improved biosecurity and control of culture variables.

Use of freshwater

In the early days of *P. monodon* culture in Taiwan it was a common practice to use freshwater to dilute seawater to reduce the salinity to 15–20 ppt. The removal of groundwater required for this practice caused instances of aquifer depletion that resulted in land subsidence. Nowadays, the practice of diluting seawater for shrimp farming has been largely abandoned (Tang and Tang, 2006). Most shrimp farms today do not use freshwater directly for anything other than the personal needs of workers. The main exception is the case of inland farms that grow shrimp in freshwater that has been supplemented with salts (Roy et al., 2010). There is scarce information on the extent of this practice, but it is not common.

In some areas, seepage from shrimp ponds built on sandy soils has resulted in salinization of freshwater bodies and aquifers. Salinization has also resulted from the discharge of effluents. Sediment removed from ponds has also been disposed on vacant land, and rainfall on to these spoil piles resulted in saline runoff into surface waters. Communities around shrimp farms in Taiwan, China, Thailand, and India have had issues with salinization leading to lack of potable water (Cao et al., 2007; Dierberg and Kiattisimkul, 1996; Reddy et al., 2004; Teng, 2008). Observance of discharge regulations and good management practices help prevent the salinization of freshwater bodies by shrimp farm effluents.

The primary uses of freshwater in shrimp farming are indirect; that is, to irrigate the crops used to produce feed and to produce other inputs used in farming (Boyd et al., 2017, 2018; Chatvijitkul et al., 2017).

Much freshwater is used in the production of legumes and cereals and in the processing of seeds into meals and oils that constitute shrimp feed ingredients (Mekonnen and Hoekstra, 2011). Freshwater is also used in the manufacture of fishmeal and fish oil used in shrimp feeds. Other sources of embodied water include that used in the production of electricity, diesel fuel, gasoline, fertilizers, liming materials, molasses, and sugar. Chatvijitkul et al. (2017) estimated that feeds for *L. vannamei* contained 819–1666 (average = 1116) m^3/t and that *P. monodon* feeds had 809–1542 (average = 1130) m^3/t of embodied freshwater.

Tables 1.1 and 1.2 show the average values for freshwater embodied in feed for the production of *L. vannamei* and *P. monodon* in Ecuador, India, Indonesia, Thailand, and Vietnam. The *L. vannamei* countries in Table 1.1 used feed and aeration. In Ecuador all farms used feed, but only about half applied aeration (Boyd et al., 2021). The *P. monodon* farms assessed in India and Vietnam (Table 1.2) also used feed and aeration, but those in Thailand and Indonesia used extensive practices (no aeration and little or no added feed). In Vietnam, a huge area is devoted to extensive production of *P. monodon*, and there is also considerable extensive culture of this species in India. The production intensity at *P. monodon* farms in Vietnam and India (Boyd and McNevin, 2018) is similar to that reported for Indonesia and Thailand (Table 1.2). Table 1.3 shows the resources embodied in shrimp feed and sugar cane molasses. Average freshwater use for the production of *L. vannamei* and *P. monodon* in feed-based culture in Asia was 2047 m^3/t of shrimp and 2014 m^3/t of shrimp, respectively. The main contributor was the feed, accounting for on average 88.8% of the embodied freshwater; the next most important contributor was fuel (8.6%). There was no embodied freshwater used for extensive culture of *P. monodon* because no feed was used. For comparison, and using data from Mekonnen and Hoekstra (2011) and dress-out percentages of 60%, 74%, and 71% for beef, pig, and chicken, respectively, water use was 9249, 4431, and 3071 m^3/t for live weight production of beef, pig, and chicken, respectively. As with shrimp, most of the water used for production of these three meats was embodied in the feed. Shrimp farming requires less freshwater per unit of production than the three meats with which it was compared.

TABLE 1.2 Average resources used to produce 1 t of shrimp (*P. monodon*) in four Asian countries.

Item	*India*	*Indonesia*	*Vietnam*	*Thailand*
Stocking rate (postlarvae/m^2)	29.6	8.7	24.2	?
Survival (%)	74.0	14.5	58.4	?
Crops (no./y)	1.7	2.9	1.6	1
Crop duration (days)	156	116	188	–
Yield (t/ha pond surface/y)	3.88	0.56	3.60	0.30
Feed conversion ratio	1.36	No feed	1.71	–
Wild fish use (t/t)	1.77	0.0	2.23	–
Land use (ha/t)				
Shrimp farm	0.43	2.35	0.67	3.6
Embodied in feed	0.36	0	0.40	–
Total	0.79	2.35	1.07	3.6
Saline water use (m^3/t)				
With water exchange	191,500	Tidal	19,000	Tidal
Without water exchange	–	–	–	–
Effluent volume (m^3/t)				
With water exchange	187,308	Tidal	–	Tidal

TABLE 1.2 *continued*

Item	*India*	*Indonesia*	*Vietnam*	*Thailand*
Without water exchange	5336	–	–	–
Energy use (GJ/t)				
Construction and repair	8.0	26.8	8.4	50.0
Farm operations	37.2	2.7	2.59	5.0
Embodied in fuel	36.5	5.3	34.4	–
Embodied in fertilizers	0.6	0.0	1.3	–
Embodied in disinfectants	0.0	0.0	2.4	–
Embodied in molasses	0.0	0.0	0.6	–
Embodied in other	0.0	0.0	0.2	–
Total	82.3	34.8	49.9	55.0
Embodied freshwater (m^3/t)				
Feed	1582	0.0	1932	–
Fuel	157	0.1	150	–
Fertilizers and minerals	1	0.0	134	–
Disinfectants	0.0	0.0	4	–
Molasses	0.0	0.0	68	–
Total	1740	0.1	2288	–

TABLE 1.3 Resources embodied in 1 t of shrimp feed and in 1 t of cane molasses. Credit: Modified from Chatvijitkul et al. (2017).

Resource	*L. vannamei feed*	*P. monodon feed*	*Cane molasses*
Land (ha/t)			
Mean	0.249	0.232	0.467
Range	0.115–0.352	0.187–0.314	–
Water (m^3/t)			
Mean	1116	1130	527
Range	819–1666	809–1543	–
Energy (GJ/t)			
Mean	9.07	10.64	5.81
Range	7.21–10.14	8.22–12.73	–
Wild fish (kg/t)			
Mean	818	1302	–
Range	270–1350	585–1935	–
Nitrogen fertilizer (kg N/t)			
Mean	8.63	7.00	26.7
Range	3.93–13.39	3.97–11.12	–
Phosphorus fertilizer (kg P/t)			
Mean	3.11	3.54	10.0
Range	1.29–3.54	2.13–5.01	–

Use of land

Building new shrimp farms or expanding existing ones often results in the conversion of natural areas into farm ponds and constitutes the largest environmental impact from shrimp farming because of the destruction of habitat and the resulting loss of native species. Most of this conversion occurs in coastal areas, but inland areas have also been converted to low-salinity shrimp farms in some countries. Apart from the area used for ponds, shrimp farms require additional land for canals, reservoirs, nurseries, settling basins, feed storage warehouses, living areas, roads, workshops, parking lots, offices, laboratories, and other facilities required for operations. A satellite imagery study of 100 aquaculture farms (2783 production ponds) in 26 countries revealed that the total farm area was on average 1.48 times greater than the pond water surface area (Jescovitch et al., 2016). In this study, the factor for converting the average production pond area of a farm to the total area was called the land-to-water ratio (LWR), which was calculated as $LWR = 1.5583X^{-0.102}$ where X is the average area (ha) of the production ponds on an aquaculture farm. The LWR decreases as the average area of ponds increases. Using the equation above, a farm with 0.1-ha ponds would require a total of 1.97 ha for each hectare of ponds. The corresponding LWR values for farms with larger average pond areas were 1.67 for 0.5-ha production ponds, 1.56 for 1-ha production ponds, and 1.23 for 10-ha production ponds. The LWR equation was based on data from aquaculture farms growing a variety of species and not specifically shrimp. The LWR values for small shrimp farms in four Asian countries where pond water surface areas were less than 1 ha and usually less than 0.5 ha were: India, 1.33; Indonesia, 1.75; Vietnam, 1.96; Thailand, 2.12 (Boyd et al., 2017, 2018). The low LWR in India resulted from farms typically using less area for reservoirs and settling basins. The much larger ponds used in Ecuador had an overall LWR of 1.32 (Boyd et al., 2021). The additional farm area necessary to support each hectare of production ranged from 0.31 in Ecuador to 1.85 in Thailand (Table 1.4). This variation resulted from differences in the proportion of farm area devoted to reservoirs and settling basins. The low LWR for Ecuador was the result of large ponds, averaging 6.5 ha. These numbers represent only land area and do not reflect the different ecological values of the land where ponds are built; for example, a case could be made for much higher unit area value for mangrove forests than for salt flats.

TABLE 1.4 Average use of land by shrimp farms in different countries and related land-to-water ratios (LWR).

Item	*India*	*Indonesia*	*Thailand*	*Vietnam*	*Ecuador*[1]	*Asian 'good example'*
Total area (ha)	6.84	4.90	14.76	3.46	1.31	8.36
Production pond area (ha)	5.15	2.80	6.96	1.76	1.00	5.00
Canals (ha)	0.20	0.30	0.34	0.16	0.10	0.30
Reservoirs (ha)	0.34	1.00	4.75	0.31	–	1.00
Settling basins (ha)	0.00	0.10	0.86	0.09	–	0.50
Support land (ha)	1.15	0.70	1.85	1.14	0.31	1.56
LWR	1.33	1.75	2.12	1.96	1.31	1.67

[1] Based on 1 ha of production pond surface area. Ponds average around 6.5 ha of water surface area each.

A significant amount of land is necessary to produce the legumes and grains required for shrimp feed (Chatvijitkul et al., 2017). Land is also used to produce raw sugar or molasses, which are sometimes used as pond amendments. The most conventional way to calculate land use for these plant ingredients is to base the estimate on global yields per hectare of processed feed ingredients (e.g., soybean meal, soybean oil, cornmeal, wheat flour, corn oil, sugar cane) and on the typical feed formulations. The yield of processed feed ingredients differs among countries and feed manufacturers, and most feed producers will not make their formulations public. Many manufacturers use least cost formulation, and the plant ingredient composition may change frequently (Chatvijitkul et al., 2017). The average amount of embodied land in feed for *L. vannamei* and *P. monodon* culture based on several feed formulations is provided in Table 1.3.

Total shrimp farmland use was 0.077–0.196 ha/t of shrimp (average = 0.136) in four Asian countries producing *L. vannamei* and 0.19 ha/t for semi-intensive shrimp farming in Ecuador, but only 0.02 ha/t in the hyper-intensive farm in Texas (Table 1.1). The inclusion of land used for feed and sugar or molasses production increased the total land use to an average of 0.510 ha/t shrimp in the four Asian countries, to 0.54 ha/t shrimp in Ecuador, and to 0.03 ha/t in the Texan farm. The land necessary at the shrimp farm level decreases exponentially with greater culture intensity, but the land use for growing feed per tonne of shrimp is fixed by the feed conversion ratio (FCR). At pond production above 10–15 t/ha, the land used for feed production represents nearly all the land use, and further intensification has little effect on total land use (Fig. 1.3).

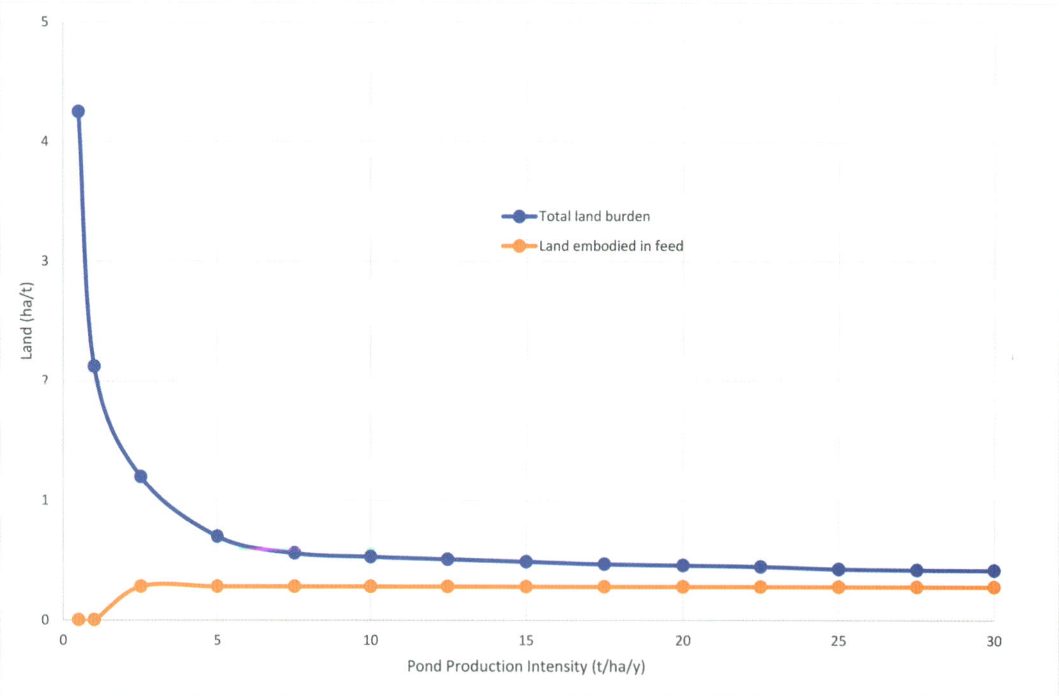

FIGURE 1.3 Total land burden and land embodied in feed as a function of shrimp farming culture intensity.

In the overall scheme of meat production and consumption, when shrimp are eaten, most likely a substitute meat is not consumed. While intensification requires 0.3–0.4 ha land/t of shrimp (Chatvijitkul et al., 2017), the agricultural land used to produce traditional sources of protein is not as ecologically sensitive or as important for ecological services as the coastal land where shrimp are farmed. If there is a fixed amount of meat consumed and shrimp are exchanged for another meat, perhaps there is no increase in agricultural land due to shrimp farming. This is because the production of other meats also requires feed that contains soybeans, grains, and other plant products.

For comparison with other sources of animal protein, land used for the production of three common meats (adjusted to the same live weight basis) was estimated for pig as 0.30 ha/t (Zu-Ermgassen et al., 2016), for chicken as 0.63 ha/t (Ibidhi et al., 2017), and for beef as 2.84 ha/t (Asem-Hiablie et al., 2019). Semi-intensive production of feed-based *L. vannamei* in Asia requires about 0.5 ha/t (Table 1.1), which is less land than for beef, more than for pig, and about the same as for chicken. Feed-based production of *P. monodon* requires total land use of 0.79 and 1.07 ha/t in India and Vietnam, respectively (Table 1.2). Because of the low yield per hectare, the total land required for extensive farming of *P. monodon* is higher, at 2.35 ha/t shrimp in Indonesia and 3.60 ha/t shrimp in Thailand (Table 1.2), even though no feed-embodied land is involved.

Mangroves and shrimp aquaculture

Shrimp farming has been implicated in the destruction or modification of coastal wetlands, particularly of mangroves (Naylor et al., 2000). Mangrove forests provide an extremely valuable range of ecosystem services and economic benefits, as reported in numerous studies (Table 1.5). These benefits include habitat for fisheries production, food for local communities, products for traditional medicine, and fuel and construction materials for local populations (Bandaranayake, 1998; Guo et al., 2017; Nagelkerken et al., 2008; Rönnbäck et al., 2007). In addition, mangroves sequester large amounts of carbon, protect the coast from storm surges and other natural disasters, prevent erosion, and aid in nutrient cycling (Donato et al., 2011; Guannel et al., 2016). Intact mangroves act as fish nurseries and foraging areas. Fish larvae naturally recruit into mangroves, where they can evade predation and mature in a structurally complex habitat (Nagelkerken et al., 2008) before migrating to adjacent deeper water habitats, including coral reefs (Jones et al., 2010; Nagelkerken et al., 2000). Tidal exchange allows predatory fish to enter and exit the mangroves to forage (Harborne et al., 2016) and juvenile fish to traverse through (Layman et al., 2004). The presence of intact mangroves has been shown to enhance adjacent reef fisheries by serving as an intermediate nursery habitat that increases survivorship of young fish (Mumby et al., 2004). In addition, carbon storage and regulation services provided by mangroves are important mitigators of climate change, and mangroves sequester carbon at disproportionately higher rates in comparison to other forest types. Numerous studies have found that intact mangroves provide wave attenuation and storm protection. Danielsen et al. (2005) found that the quality of a mangrove stand – its level of intactness – is an important factor in mitigating the effects of storm surges. Mangrove forests are an important part of the communities that they border. Intact mangrove forests can also serve as ecotourism destinations (Murtini et al., 2018) and can provide significant benefits to local communities and improve their livelihoods while remaining functional and protected.

TABLE 1.5 Estimates of economic benefit for mangrove forests from selected studies.

Study	*Location*	*Ecosystem service*	*Economic benefit*
Jerath *et al.* (2016)	Everglades, USA	Carbon sequestration	~56,000 US$/ha
Macreadie *et al.* (2017)	Australia	Carbon sequestration	~5100 US$/ha
Chong (2007)	Malaysia	Fisheries	846 US$/ha
Aburto-Oropeza (2008)	Mexico	Fisheries	37,500 US$/ha
Rönnbäck (1999)	Worldwide	Fisheries	750–167,500 US$/ha
Anneboina and Kumar (2017)	India	Fisheries	~2400 US$/ha
Das (2017)	Gujarat, India	Fisheries	~7000 US$/ha
Sathirathai and Barbier (2001)	Thailand	All services	~27,000–35,000 US$/ha
Costanza *et al.* (1997)	Worldwide	All services	9900 US$/ha
Malik *et al.* (2015)	South Sulawesi, Indonesia	All services	~4000–7600 US$/ha

Aquaculture has been associated with widespread destruction of mangroves in tropical and subtropical coastal areas, with 38% of the historic global loss being attributed to shrimp culture (Duke et al., 2007), resulting in decreased nutrient assimilation, increased erosion, exposure to storm surge, loss of fishing grounds, reduced abundance of spawning and juvenile fish, and other losses in ecosystem function. Rates of conversion of mangroves to shrimp aquaculture have varied across years and locations. For example, the Philippines lost over 66% of its mangrove cover in the second half of the 20th century following governmental promotion of aquaculture (Hishamunda et al., 2009). Recent surveys in India, Thailand, and Vietnam suggest that extensive aquaculture is more strongly associated with continued mangrove loss than its intensive counterpart (Boyd et al., 2017, 2018). Hamilton (2013) estimated that aquaculture accounted for 544,000 ha of mangrove forest conversion in eight tropical countries that account for 83% of the world's shrimp aquaculture production.

Because mangrove forests are ecologically important and because their soils are rich in organic content, generally acidic, and difficult to drain, shrimp farmers and regulatory bodies have long realized that building shrimp ponds in mangrove areas is environmentally destructive and not conducive to long-term productivity, and therefore should be avoided. Hence, it is likely that the future contribution of shrimp farming to the global destruction of mangroves will continue to decrease and will become insignificant relative to other human activities (Moss et al., 2001). In spite of the above, concerns remain about loss of mangrove habitat in areas where government agencies and marketers have promoted a type of aqua-farming known as integrated mangrove aquaculture (IMA) and of the resulting 'integrated mangrove-shrimp' as a high-end product that is friendly to mangrove forests (Ha et al., 2012, 2014). The majority of *P. monodon* from Vietnam are produced in such systems. This is a worrisome trend for a sector that spent decades attempting to distance itself from any form of mangrove destruction. Incredibly, there are cases of members of the environmental conservation community that are promoting this type of production. IMA is a form of low-density shrimp and fish aquaculture in which mangrove trees are incorporated into the farming system. IMA systems usually come in one of two basic designs, each of which has different environmental impacts. In one type, the mangroves are grown within the ponds, either in rows or in a central mass in the middle of the pond. In the second type, mangroves are grown directly adjacent to the ponds or on the pond embankments, usually with some connectivity to the tidal exchange (Bush et al.,

2010; Clough et al., 2002). One of the reasons given for the inclusion of mangroves in shrimp aquaculture is to partially restore the ecosystem functions that are lost when mangrove forests are cleared. However, the underlying assumption that integrated systems are beneficial to the environment is largely unfounded. Unfortunately, the biophysical conditions required for most of the services provided by mangroves are unlikely to be present in IMA systems, due to their fragmented nature. Balmford et al. (2002) found that intact mangrove forests have a total economic value 70% higher than that of mangroves converted to shrimp farms. IMA systems in which mangroves are located within the ponds do not allow fish and other fauna to enter and exit freely because of the lack of tidal exchange (Joffre et al., 2015). Mangrove fragmentation intrinsic to IMA systems has been found to adversely impact fish diversity. In Ca Mau Province, Vietnam, IMA areas had approximately half of the fish diversity found in intact mangrove stands (Tran and Fischer, 2017). Donato et al. (2011) noted that intact mangroves hold up to ten times as much carbon as similar areas of terrestrial forest. However, soils in cleared mangrove areas lose their carbon stocks, limiting their ability to function as sediment and carbon stores (Grellier et al., 2017; Otero et al., 2017; Perez et al., 2017). IMA systems are not likely to function in the same capacity as intact mangroves, although some authors have suggested that they act as carbon sinks (Ahmed et al., 2018). As the density of mangroves is lower in IMA areas than in intact forests, IMA-impacted areas have a reduced ability to buffer storm surges. One need only look at satellite pictures of the coastline of Ca Mau Province (Hauser et al., 2017) to consider the magnitude of mangrove areas deforested for shrimp farming and the significant habitat fragmentation that comes with this approach. It is because of these types of systems that 42% of the global shrimp pond area produces only 6% of farmed shrimp (Boyd and McNevin, 2018). If all shrimp farming were to be conducted through such an extensive and inefficient approach, the total land necessary to produce the global shrimp supply would be several orders of magnitude larger than it currently is. For these reasons, IMA is seen by many in the shrimp farming and environmental communities as a special interest scam to wrap an ecological tragedy in a deceptive environmental dressing and call it a good thing for the farmers who at the same time are being dreadfully exploited.

The conversion of highly valuable wilderness into extensive shrimp farms constitutes a waste of natural resources because of the high ecological value of the land and the low productivity of these farming systems. Mangrove preservation and restoration have emerged as key attributes denoting responsible aquaculture. If more incentives were created to intensify production and if proper spatial management were to be implemented, the remaining mangroves could be preserved, and some current production areas could be allowed to revert into mangrove habitats. While restoration can return some of the benefits of intact habitats over time, protecting the remaining mangroves from conversion remains a better option than allowing conversion to occur in the first place. Protecting the intact mangrove stands that remain, incentivizing restoration and increased connectivity, and shifting aquaculture production from coastal habitats to closed systems further inland could all serve to meet global shrimp demand while reducing the impacts of shrimp farming on mangrove habitats (Ahmed and Glaser, 2016; Bush et al., 2010).

Use of manpower

Shrimp farming has been promoted by agencies and organizations as a good way to encourage socio-economic development and alleviate poverty, but the balance between growth and the concentration of wealth remains a challenge. As shrimp becomes more commoditized, the share of the end price that is passed to farmers and intermediaries has decreased. The justification for shrimp farming as a source of income stems from the seemingly endless demand for shrimp in global markets and from economic returns that can be higher than those of alternate land uses. In 1996, global revenues to producers were estimated at over US$6 billion, with a retail value three times that amount (Flaherty et al., 1999); by 2017, farmer revenues had increased to about US$32 billion (Tacon, 2020). Thus, shrimp aquaculture has the potential to improve the livelihoods of people in tropical coastal areas where economic development has been weak, and where traditional fisheries are in decline because of overexploitation.

Shrimp farming and ancillary activities employ significant numbers of people. At the start of the shrimp boom in Ecuador, local sources estimated about 40,000 farm workers (Sutinen et al., 1989). A comprehensive survey of shrimp farming in 13 Asian countries (ADB/NACA, 1995) found that labour use on intensive farms ranged from 2.3 person-months/ha/y in Korea to 39 in Sri Lanka, with an average of 19. For semi-intensive farms the figures ranged from 2.4 in China to 117 in Vietnam, with an average of 31. Extensive farms required from 1.5 person-months/ha/y in the Philippines to 17.1 in Sri Lanka, with an average of 6.4. Boyd and Clay (2002) assessed a high-intensity farm in Belize that had a starting base salary approximately 40% higher than comparative rural employment. Daily wages for occasional labour were 50% higher at the shrimp farm. Huang (2015) studied shrimp farm wages in southeast China and found that in general they were on a par with the average for workers in the major cities of Guangdong Province. Boyd and McNevin (unpublished results, 2019) surveyed shrimp farms in Surat Thani, Thailand, in 2014, and found that on average one salaried labourer making US$2710/y is needed per hectare at both large and small farms. Overall, labour was estimated to account for 6.83% of the total farm operating costs, which is low by industry standards but does not account for the owners' wages or those of their immediate family (Table 1.6).

TABLE 1.6 Cost structure for farming shrimp (*L. vannamei*) in Thailand compared with the previous assessment of Clay (2004).

	Clay (2004)		*Boyd and McNevin (unpublished data)*	
Item	*($/kg of shrimp)*	*(% of total costs)*	*($/kg of shrimp)*	*(% of total costs)*
Depreciation of investment				
Land	–	–	0.385	8.27
Construction	–	–	0.030	0.65
Electrification	–	–	0.053	1.14
Equipment	–	–	0.519	11.15
Operating costs				
Postlarvae	0.40	9.35	0.320	6.88
Feed	2.39	55.84	2.193	47.12
Energy	0.50	11.68	0.754	16.20
Chemicals	0.01	0.23	0.082	1.76
Labour[1]	0.30	7.01	0.318	6.83
Overhead[2]	0.68	15.89	–	–
Total	4.28	100.00	4.65	100.00

[1] Labour costs exclude the owners' wages and those of their immediate family.

[2] Includes amortization of the capital costs of land, construction, electrification, and equipment.

Manpower used in shrimp aquaculture is distributed through the different links of the value chain: feed manufacturing, hatchery, farm, trading, processing, importing, and retailing. As these steps get closer to the consumer, the separation and attribution of labour become more difficult because of the different streams of products that converge into the same entities. Labour in shrimp farming has received considerable attention in recent years because of reports of forced and bonded labour in Thailand (Hodal et al., 2014; Kelly, 2016; Urbina, 2015). The reports focused on the capture of fish stocks used to manufacture feeds for the shrimp industry. There are also incidences of forced labour in the processing of shrimp (Mason et al., 2015). These situations reveal a lack of transparency in the Thai labour market. There are significant concerns about the effectiveness of the many eco-labels and certification programmes that were present there and yet had little knowledge of the growing labour issues. In order to avoid such problems in the future, the retail and food service components of the value chain will need better awareness of the supply dynamics of their shrimp, as it is ultimately their brands that are exposed to the greatest reputational risk.

Employment in the shrimp aquaculture sector has become increasingly volatile because of epizootics. In Madagascar, five of seven existing shrimp farms and processing facilities were forced to suspend operations in 2012 because of white spot syndrome virus (WSSV). In Mozambique, the outbreak of WSSV in 2011 led to the termination of several hundred employees' contracts (FAO, 2013). In Thailand, production volumes were cut in half in 2013 because of an outbreak of early mortality syndrome (EMS), with many farms having to suspend operations.

Living arrangements for workers at shrimp facilities vary by region and by proximity to population centres. Remote farms and processing plants typically provide living quarters for workers, whereas those that are closer to urban settings do not. When provided, living accommodation on farms is typically more rudimentary than that at processing plants, which usually require higher numbers of employees.

Use of energy

Shrimp farms require energy for pumping and aerating water, for routine operations such as feeding, harvesting, and removing pond sediments, and for powering miscellaneous equipment such as tractors and other vehicles, tools, lighting, and so on. Farms also use significant amounts of energy for construction, repairs, and renovation of infrastructure, which can be amortized over time and constitute only a minor fraction of the energy used per kg of shrimp produced. Also to be considered is the energy embodied in the physical infrastructure, such as concrete structures, pipes, plastic liners, and electrical grids. Moreover, fuels, feeds, and other pond inputs contain embodied energy resulting from their production. Embodied energy is a sizeable component of energy use in shrimp aquaculture.

The main sources of energy are grid electricity, diesel fuel, gasoline and, in some cases, liquefied propane gas. Some farms use fuel-powered generators to produce their own electricity on-site. There is a lot of interest in and some projects pioneering the use of alternative sources of energy such as wind and solar, and on the utilization of 'waste heat' from other industries or from geothermal sources in hyper-intensive farms and hatcheries.

The amounts of energy used in *L. vannamei* production in several important farming areas of the world are shown in Table 1.1. The direct energy used for shrimp production ranged from 24.2 to 62.1 GJ/t shrimp (average 40.4 GJ/t). In India, Indonesia, Thailand, and Vietnam, aeration accounted for most of the energy used, while in Ecuador the largest direct energy use was pumping for water exchange in production ponds. Embodied energy was substantial, ranging from 37.0 to 114.2 GJ/t shrimp (average 77.3 GJ/t). Total energy use ranged from 61.2 to 161.3 GJ/t shrimp (average = 117.6 GJ/t). The largest amount of embodied energy at farms in the four Asian countries was in fuel, with feed a distant second. In Ecuador, feed had the largest embodied energy component and fuel was a close second.

Table 1.2 shows the average energy required for producing *P. monodon* in four Asian countries. Energy use was higher in India and Vietnam, where feed-based systems are used (but lower than that used for *L. vannamei* as shown in Table 1.1). For Indonesia and Thailand, which mostly farm *P. monodon* in non-fed systems, energy use was largely the result of farm construction, but because of the low yields characteristic of extensive systems, total use was 34.8 GJ/t shrimp for Indonesia and 55.0 GJ/t shrimp for Thailand. As expected, energy use was lower for extensive production of *P. monodon* than for intensive production of either *L. vannamei* or *P. monodon*. However, much more land is required for extensive culture, and such land is usually in the intertidal zone, which is an ecologically sensitive area of high biodiversity and an important source of ecological services. For comparison, the production of beef, pig, and chicken requires on average 249, 100, and 35 GJ/kg (Cherry et al., 2021).

Carbon footprint is an important indicator of the environmental impacts of meat production. Carbon emissions tend to increase as the use of energy increases, but the relationship is complicated by the difference in emissions per unit of energy produced with different fuels. For example, CO_2 emissions are 136.7 kg/GJ for electricity, 71.2 kg/GJ for diesel fuel, and 70.9 kg/GJ for gasoline. In two recent assessments of resource use (Boyd et al., 2017, 2018) these differences were considered when estimating the carbon footprint of shrimp farmed in four Asian countries. In Indonesia and Thailand, pond aeration was powered almost exclusively by electricity, whereas in India and Vietnam there was a mix of diesel- and electricity-powered aeration. Assuming a 50:50 mix of electricity to diesel in India and Vietnam and an 80:20 mix in Thailand and Indonesia, CO_2 emissions per tonne of shrimp produced in the four countries were: India, 16.8 kg/kg shrimp; Vietnam, 13.1 kg/kg shrimp; Thailand and Indonesia, 12.8 kg/kg shrimp. From these estimations, the carbon footprint for farmed shrimp is lower than for beef at 29 kg CO_2/kg (Asem-Hiablie et al., 2019) but greater than for pig at 12.1 kg CO_2/kg and chicken at 6.9 kg CO_2/kg (Green Eatz, 2020).

Use of shrimp feeds

Most shrimp production is reliant on formulated feeds, which typically account for the largest share of operating costs, so feed manufacturers and farmers have strong incentives to manufacture and use feeds and feeding practices that result in efficient conversion of feeds into shrimp. Feeds also constitute the initial source of pollutants; hence, overfeeding or the use of poor-quality feeds can have negative impacts on water quality and shrimp production. The feeding strategies used in shrimp farming have a major effect on FCR and they vary widely with the species and age of the shrimp, and with the farming system and technical sophistication of the farmer (Tacon et al., 2005). The composition and physical properties of the feed also have a major effect on FCR. Feed management should therefore be aimed at optimizing utilization and nutrition, thus improving FCR and reducing water pollution.

Feeding behaviour begins with the detection of the feed by the shrimp, followed by the initiation of feeding and continuation to satiation. Lee and Meyers (1997) offered an extensive review on the attractability and palatability of feeds to shrimp. The review established that aquatic animals use water-borne chemical signals to identify and move towards food items. Low-molecular-weight compounds (<1000 Da) such as free amino acids, short-chain peptides, and nucleotides are the most effective attractants in food for marine crustaceans.

Feed efficiency is affected by the age, sex, and health of the shrimp, and by environmental factors such as water temperature and salinity, but the major factor is usually the efficiency of protein absorption. Feeds are often characterized in terms of their content of crude proteins, fats, and carbohydrates, but increasingly the emphasis is on the quality and digestibility of these nutrients. The amount and quality of protein vary significantly between feed ingredients and sources, but most feed formulations stay within tight ranges for each raw material, leaving little flexibility to adjust for variations in quality. Some tools have proven effective in the search for nutritious, cost-effective, and sustainable feeds. The lower use of wild fish in Ecuador shown in Table 1.1 resulted from reduced inclusion of fish meal and oil in feeds (Naylor et al., 2021). The arrival of rapid and consistent *in vitro* methods such as pH-stat protein digestibility with species-specific enzymes is a significant advance for the assessment of feed ingredients required by industrial processes (Ezquerra et al., 1997).

One of the important continuing discussions in aquaculture is the use of fishmeal and fish oil in feeds and the amount of wild fish it takes to produce farm-raised shrimp (Tacon and Metian, 2008). Dependence of aquaculture feeds on wild-origin fishmeal and fish oil has long been recognized as a major sustainability issue and possible limitation to the growth of farming of carnivorous species (Kristofferson and Anderson, 2006). This 'fishmeal trap' (finite supply of fishmeal versus growing demand as an ingredient in aquafeeds) has been dissected by Rust et al. (2014) into three separate issues: (1) responsible management of the fisheries used for industrial reduction to fishmeal and fish oil; (2) a classic case of resource allocation among competing uses; and (3) the need and search for novel ingredients to reduce or replace the use of fishmeal and fish oil in the medium to long term, to allow aquaculture production to grow unimpeded.

A recent report on the sustainability of reduction fisheries (Sustainable Fisheries Partnership, 2018) concluded that there has been a steady improvement in the overall management and sustainability of the 26 fish stocks analysed in the report, and that 91% of the fish used for fishmeal and fish oil from the Atlantic and eastern Pacific now come from fisheries that can be classified as 'reasonably well-managed or better'. Only 9% of the catch reported still comes from poorly managed fisheries, and this percentage has been decreasing and is currently at its lowest since the report began publication in 2012. Unfortunately, this report covers only 45.8% of the global catches of reduction fisheries, with the remaining 54.2% coming mostly from poorly managed Southeast Asian fisheries that still suffer from bad governance, have serious reporting and data deficiencies, and result in severe environmental impacts. The result is that about 58% of the global reduction catch still comes from poorly managed fisheries. Fig. 1.4 presents landings for the main fish species traditionally used for reduction, and average prices for fishmeal between 1998 and 2017. Fishmeal production has followed a fluctuating but overall declining trend during the period (FAO, 2018), while prices have increased.

Fishmeal and fish oil have been used in many industries, most frequently as components of animal feeds for pigs, chickens, and pets. In the past, fishmeal was also used as an agricultural fertilizer and fish oil was used in the manufacture of edible oils, paints, and varnishes, and more recently as a human nutritional supplement for omega-3 fatty acids. The increased scarcity and price of fishmeal and fish oil has resulted in a shift to its most economical uses, which include prominently aquaculture feeds. About 70% of the global production of fishmeal and 75% of fish oil are used for manufacturing aquafeeds

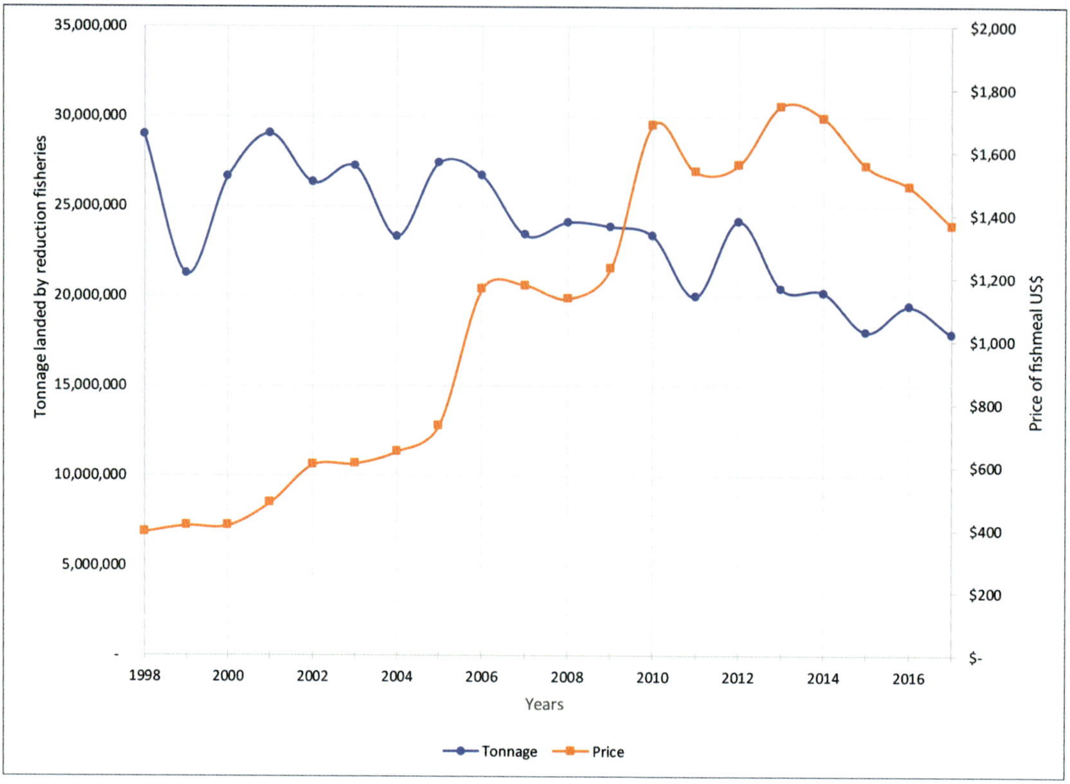

FIGURE 1.4 Fish landings by the principal reduction fisheries and average price of fishmeal (US$/t) between 1998 and 2017. Data on landings from FAO (2019b) and prices from IndexMundi (2019). Some of the catch is directed to non-reduction uses, i.e. canning for direct human consumption.

(OECD/FAO, 2019). Shrimp feeds accounted for about 15% of all fishmeal used in aquaculture feeds in 2017 (A. Tacon, Kailua, Hawaii, 2020, personal communication).

With the increased demand and cost of fishmeal and fish oil, and with caps on world supply, there are strong incentives and research efforts to reduce and replace their use in aquafeeds. Fishmeal is not an essential requirement in shrimp feeds, and formulations using no fishmeal or fish oil have been used experimentally, resulting in growth and survival similar to those obtained with traditional feeds (Amaya et al., 2007; Olmos et al., 2011; Sookying, 2010). Development of alternative feed ingredients continues to be one of the most active areas of aquaculture research. Alternatives for the substitution or reduction of fishmeal and fish oil include soybean meal, cottonseed meal, by-products from corn and wheat, legumes, algae, by-products of fermentation, and insect-based meals. It is estimated that by-products account for 25–35% of the total volume of fishmeal and fish oil produced (FAO, 2018). There is also a growing trend towards the use of fish trimmings and other fish-processing wastes that were historically discarded (Jackson, 2012; OECD/FAO, 2019). As fishmeal prices continue to increase and human uses for reduction fisheries' products continue to expand, it is expected that there will be increased incentives for the use of these and other ingredients to reduce or replace fishmeal and fish oil in shrimp feeds.

The 5.5 Mt of shrimp produced globally through aquaculture in 2017 required some 7.6 Mt of feed (Tacon, 2020). Because only an estimated 86% of the shrimp are fed (the rest depend on natural productivity), the resulting FCR is about 1.6, indicating that 1.6 kg of feed is required to produce 1 kg of shrimp. To manufacture these feeds, an estimated 753,300 t of fishmeal and 113,700 t of fish oil were used (Albert Tacon, Kailua, Hawaii, 2020, personal communication). The amount of wild fish used in feeds to produce farmed shrimp, also known as the [fish-in:fish-out ratio] (FIFO), is an important indicator of environmental performance for the aquafeeds industry. It varies between species, regions, feeds, and systems. Tables 1.1 and 1.2 show the figures for four Asian countries as related to the production of *L. vannamei and P. monodon*, respectively. For *L. vannamei*, FIFO was lowest in India and Thailand at 1.21 t of wild fish used to produce 1 t of shrimp, and highest in Indonesia at 1.48 t/t (Table 1.1). For *P. monodon*, FIFO was lowest for India at 1.77 t/t and highest for Vietnam at 2.23 t/t (Table 1.2). The Sustainable Fisheries Partnership (2018) gives a global average FIFO of 0.5 t/t for farmed crustaceans, which are assumed to be mostly shrimp and may include systems that are not fed, such as most of the *P. monodon* produced in Indonesia. Data from the International Fishmeal and Fish Oil Organization (IFFO, 2020) for 2000–2015 indicate that in general, aquaculture feeds have shown a reduction in their FIFO. However, the 2015 data showed that farmed shrimp FIFOs are similar to those of 2010. This has been explained by the impact of disease outbreaks in the sector and by the resulting drop in harvests, which, combined with higher FCRs, affected the calculations of fishmeal inclusion in the feeds (IFFO, 2020).

Use of by-products in shrimp feeds

Aquaculture has become increasingly dependent upon the use of compound aquafeeds, and improvements in technology and in feed ingredients have become important for the industry. The rapid growth of aquaculture and the concurrent increase in the proportion of fishmeal and fish oil used in aquafeeds have encouraged research and development on alternative sources. Shrimp are being selectively bred for maximum efficiency over shorter growth cycles and for better feed conversions; thus, it is crucial to provide them with the correct mix of nutrients in their diets. Pressures to reduce the use of fishmeal and fish oil and to reduce costs have intensified the search for suitable replacements. The formulation of low-fishmeal shrimp feeds requires the use of combinations of ingredients, as most replacements have significant nutritional and functional limitations that make them unsuitable at high levels in the diets of most species. Fishmeal and fish oil have traditionally been the preferred sources of proteins and lipids, but their increased price and market volatility have made it a priority for the industry to find more cost-effective replacements. By-products of animal and vegetable processing are a logical choice, as they provide additional advantages in sustainability.

ANIMAL BY-PRODUCTS

Animal by-products are well established as aquafeed ingredients; their protein content is higher than that of plant materials and their balance of essential amino acids is also superior (Cheng et al., 2002). Most animal by-products are significantly less expensive than fishmeal. In general, single by-products lack the balance of nutrients needed to function as complete replacements; therefore, by-product blends are most often used. While a feed manufacturer could source different by-products and use them directly in the feeds they produce, an alternative is to use premixed blends. These blends are high in protein and considerably less expensive than fishmeal when expressed as cost per unit of digestible protein. By-products from the rendering industry commonly used in aquafeeds include chicken meal, blood meal, meat and bone meal, and hydrolysed feathers. Rendered animal proteins are especially valuable in the context of fishmeal replacement. Among the alternatives, meat and feather meals have been more widely studied and their use as components of aquaculture feeds for shrimp has been well established (Amaya et al., 2007; Kureshy et al., 2000; Smith et al., 2000). Significant concerns remain regarding the relatively low digestibility of rendered meals (Table 1.7) and their effect in increasing the nitrogen and phosphorus loads in ponds and discharge waters. The nutritional quality of some by-product meals has been significantly enhanced by better sorting of the raw materials and optimization of their processing conditions (Bureau, 2006; Carvalho et al., 2016; Qiu et al., 2017; Yang et al., 2009).

Numerous studies have shown rendered animal protein ingredients to be effective sources of essential amino acids such as lysine, histidine, and arginine, and of sulphur-containing amino acids such as methionine, cysteine, and taurine. For this reason, they are good complements to plant protein meals, which are deficient in some of these nutrients. Amino acids in blood meals are highly digestible and appear to be utilized more efficiently than their synthetic counterparts; for example, the difference in bioavailability between lysine from blood meal and synthetic lysine is enough to be of value for least

TABLE 1.7 Protein content and digestibility of common ingredients used in the manufacture of aquaculture feeds. Credit: From Nates and Bureau (2007). Courtesy of Global Aquaculture Alliance.

Ingredient	*Crude protein (%)*	*Digestibility of protein (%)*
Fishmeal	65	89
Feather meal	80	75
Meat and bone meal	50	80
Poultry by-product meal	57	85
Soybean meal	48	89
Corn gluten meal	60	93

TABLE 1.8 Cholesterol content of rendered animal protein ingredients used in aquaculture feed manufacturing. Credit: From Nates and Bureau (2007). Courtesy of Global Aquaculture Alliance.

Ingredient	*Cholesterol (mg/100 g)*
Fishmeal, menhaden	237
Fishmeal, herring	302
Blood meal, avian, disc-dried	407
Blood meal, mammalian, flash-dried	255
Blood meal, bovine, ring-dried	241
Feather meal, steam-hydrolysed	90
Meat and bone meal, 43% crude protein	98
Meat and bone meal, 56% crude protein	103
Poultry by-product meal, 6% crude protein	168

cost formulation. In addition, animal proteins are generally more palatable to aquatic species than those of plant origin. They are also good sources of nutrients abundant in fishmeal, such as arachidonic acid, taurine, and hydroxyproline (Villareal-Cavazos et al., 2014). Rendered products can also be cost-effective sources of cholesterol for shrimp diets (Table 1.8). Furthermore, the addition of speciality animal by-products, including protein hydrolysates and spray-dried plasmas, has been shown to improve growth in shrimp (Suresh et al., 2011). Products derived from pig intestines have resulted in adequate and sanitary feeds with improved palatability (Suresh et al., 2011). Pig intestine peptides are a by-product of medical heparin production and incorporate high standards of quality; they have been shown to be beneficial in diets for rearing early stages of shrimp, an application for which fishmeal was considered indispensable (Nates and Swisher, 2013).

PLANT BY-PRODUCTS

The demand for improved feeds has spurred the development of new ingredients, especially those that serve specific purposes or provide cost advantages. Conspicuously, soybean meal and soy co-products are increasingly contributing to the sustainable production of aquafeeds. Microbial fermentation by fungal or bacterial strains has been used to reduce antinutritional factors and to increase the nutritional value, digestibility, and palatability of plant ingredients (Mukherjee et al., 2016; Shao et al., 2018; Shiu et al., 2015). Fermented soybean meals have been used successfully as protein sources in feeds for all life stages of shrimp. During fermentation, oligosaccharides are practically eliminated, and protein content and quality are improved. Ding et al. (2015) showed that 20–40% of the fishmeal could be replaced by fermented soybean meal in diets for *Macrobrachium nipponense*, without affecting the growth performance of the prawns. Sharawy et al. (2016) pointed to economic advantages when replacing up to 28% of the fishmeal with fermented soybean meal in diets for *Fenneropenaeus indicus*.

Dried distillers' grains with solubles (DDGS) are another class of plant by-product that can be used to replace fishmeal in aquafeeds while also reducing their cost. DDGS constitute an economical alternative to corn and soybean meal and have become a choice ingredient for shrimp feeds. The inclusion level of DDGS can be adjusted in relation to their nutritional composition and be accompanied by other plant proteins to balance the nutrient content of the feed. The quality and composition of DDGS depends on the grain source, the fermentation yeast, and the processing methods used (Lemos et al., 2009; Rhodes et al., 2015). Zhou et al. (2015) found that sorghum DDGS can be used in practical diets for *L. vannamei* in clear-water systems at inclusion levels of up to 40% without adverse effects on growth, survival, or FCR. Rhodes et al. (2015) found that lysine supplementation of the feed may be needed to meet the nutritional requirements of shrimp at higher inclusion levels of DDGS. Achupallas et al. (2015) showed that supplementing shrimp feeds with DDGS at 10 g/kg resulted in increased tolerance of the shrimp to bacterial infections.

Algal meals have also been considered for use in shrimp feeds. DeMicco (2015) replaced 60% of the fishmeal in shrimp diets with a combination of three marine macroalgae and showed that growth was affected only negligibly compared with controls. Using each one of the algal species separately resulted in significantly lower growth than with the full fishmeal control diet. Niu et al. (2019) reported that there is a species-specific response for shrimp to macroalgae in the diet and that low-level inclusion exerted a beneficial effect on shrimp. The relatively low nutritive value of macroalgae can be attributed to their low content of protein and carotenoids and to their high content of ash, polysaccharides, and fibre, which affect their digestibility.

Shrimp pond effluents

The main environmental concern over shrimp pond effluents is that they contribute nitrogen and phosphorus to the receiving waters, which encourages eutrophication. Effluents also contain dissolved and particulate organic matter that exerts an oxygen demand in the receiving waters. The amount of feed required to produce 1 t of live shrimp ranges from 1 to 2 t (FCR = 1.0–2.0), but most farms obtain FCRs lower than 1.5. Typically, 1 t of feed contains about 400 kg organic carbon, 56 kg nitrogen, and 10 kg phosphorus, while 1 t of live shrimp contains about 110 kg organic carbon, 27.5 kg nitrogen, and 2.5 kg phosphorus. The resulting waste loads are about 290–690 kg organic carbon, 28.5–84.5 kg nitrogen, and 7.5–17.5 kg phosphorus. At an FCR of 1.3 (easily achievable with quality feeds and good management), loads would be closer to the lower limits: 410 kg organic carbon, 45.3 kg nitrogen, and 10.5 kg phosphorus. These quantities represent the nutrient loads of the production system (pond), with most of the organic carbon not recovered in shrimp biomass being oxidized by shrimp during respiration and with the organic carbon in uneaten feed and faeces being oxidized by bacteria in the pond. By the same token, nitrogen not recovered in shrimp tissue is converted to ammonia by shrimp and microbial metabolism, and is lost to the atmosphere by ammonia volatilization (diffusion to air). Nitrification converts ammonia to nitrate, and denitrification transforms nitrate to gaseous nitrogen, which diffuses to the air. Much of the phosphorus not retained in shrimp tissue is sequestered by the pond bottom soil, and some organic matter is also stored in the soil. Naturally, some nitrogen and phosphorus are converted to algal biomass, and photosynthesis produces additional particulate organic matter that remains in effluents. Nevertheless, the environmental loads of organic carbon, nitrogen, and phosphorus contained in shrimp pond effluent are less than the load imposed upon the production system itself. In other words, the pond assimilates a large portion of the system load, and the environmental load is considerably less than the differences between the amounts of organic matter, nitrogen, and phosphorus in the feed and the amounts of these substances recovered in the harvested shrimp.

There is a large variation in the discharge loads of different farms per tonne of shrimp produced. This results from differences in FCR, water exchange, mechanical aeration, and other management practices and water quality characteristics. There is no generally accepted rule for the allowable amount of waste per unit of shrimp production, and governmental water pollution control agencies and aquaculture certification programmes simply impose water quality standards for shrimp farm effluents.

The most common practice for establishing water quality standards for shrimp farm discharges is to put limits on the concentrations of biochemical oxygen demand, dissolved oxygen, total suspended solids, nitrogen, and phosphorus in the effluents. Unfortunately, this is not the best method to regulate discharges from aquaculture facilities. If a body of water is available, it can be used to dilute the concentration of pollutants to an acceptable level, but the actual load of pollutants (concentration of pollutant per unit volume multiplied by volume discharged) may increase. The concentration limit only assures that the concentration of pollutants in the mixing zone does not exceed a selected value; it does not assure that the farm effluent does not result in increased oxygen demand, suspended solids, or nutrients in the receiving water body, which could still result in an acceleration of the eutrophication process.

An alternative to establishing concentration limits is to require farms to report loads of pollutants calculated from concentrations multiplied by effluent volumes. Load limits also require concentration limits, because a farm might discharge a small volume of a highly concentrated effluent. Such an effluent might represent a small load yet cause an unacceptable concentration of the pollutant or oxygen depletion in the mixing zone where the effluent enters the receiving water body. A related issue is that the daily load of a pollutant necessary to exceed the assimilation capacity of receiving water body is seldom known. Moreover, there is not usually a reliable way of determining the acceptable daily load for each pollutant.

Some governmental pollution control agencies determine the assimilation capacity of receiving water bodies for a list of pollutants from all sources and impose total maximum daily loads (TMDLs). This procedure requires an effort that accounts for all possible sources of pollution and is not applicable to shrimp aquaculture certification programmes concerned only with shrimp farms irrespective of other sources of pollution to a given water body.

While it is likely prudent to retain the traditional concentration limits for shrimp farm effluents as standards to protect the mixing zone, a requirement of a maximum acceptable FCR would be a good addition for restricting the amounts of organic matter, nitrogen, and phosphorus discharged per tonne of shrimp. This method, however, would not necessarily avoid the possibility of eutrophication, because the assimilation capacity of the receiving water body and the amount of pollution from other sources are usually unknown. Water quality monitoring to prevent a tipping point in nutrient concentrations can be implemented, but this practice will not reveal with certainty whether shrimp farms or other sources are causing eutrophication. The tipping point concept reveals that even careful water quality monitoring is not an adequate means of detecting when a water body has reached incipient conditions for progressing to a greater degree of eutrophication (Boyd, 2020).

Shrimp certification programmes should work in concert with governmental water pollution control agencies to address these issues in a unified manner. Where governments are using TMDLs to regulate effluents, the limits set based on TMDLs should be adequate for the certification standards. The idea that the certification standard based on concentration limits alone should be stricter than governmental standards is not correct in many situations. Unfortunately, it is problematic (and could be impossible) to formulate an effluent water quality regulation that will eliminate the possibility of eutrophication. The goal of certification programmes should be to minimize the quantities of pollutants discharged per tonne of shrimp to a level that is technologically and economically feasible. For some water bodies, it may be necessary to restrict the quantity of shrimp produced by the discharging farms, because there are limits to the reduction in pollutants that can be economically achieved with existing technology. We should strive to reduce pollution loads from shrimp farms and to prevent water quality deterioration in water bodies receiving effluents from large farms or multiple small farms. As technology advances, the emphasis should be on progressive improvements in the efficiency of converting feed to shrimp.

One of the most troublesome facets of aquaculture certification is the apparent necessity to have a standard review board that is reluctant to revise standards as our knowledge of shrimp farming and its effects on the environment advances. Certification programmes also strive to treat all farms in the same manner with respect to standards. This is a fatal error in environmental protection, because not all shrimp farms use the same protocols and they normally differ in their production capacity, hydraulic retention times, and the conditions in the receiving water bodies. It should be noted that in countries with good water pollution oversight by governmental agencies, each facility that discharges effluent is considered as a unique situation and a permit is drafted specifically for each facility. Therefore, it is not uncommon for facilities discharging the same type of effluent to have drastically different water quality standards imposed in their discharge permits.

Use of agrochemicals

Agrochemicals are used extensively in all forms of agriculture and animal production. The main categories of agrochemicals used in shrimp farming are liming agents, fertilizers, and therapeutants. There is also a recent trend for the use of prebiotics, probiotics, and vitamins. Each category will be discussed below.

Liming agents include different forms of agricultural limestone and lime, which is made by burning limestone in a kiln at high temperature to produce CaO, which may be used directly or hydrated to become CaOH. Liming materials are commonly used to neutralize the acidity of soils and waters in ponds. Lime also reacts with the CO_2 dissolved in water to increase the levels of bicarbonate and thus alkalinity. Liming is generally not harmful to the environment, as it does not increase the pH enough to harm aquatic life at the dosages used in aquaculture. The maximum recommended application of lime in shrimp ponds is 50 kg/ha and treated water should not be discharged for 3 or 4 days after application (Boyd and McNevin, 2015). Agricultural limestone application rates are usually between 5000 and 2000 kg/Ha (Boyd and McNevin, 2015; Boyd and Tucker, 1998). Lime is also used in shrimp farms to disinfect bottom soils between crops at application rates of 1000–3000 kg/ha (Boyd and Tucker, 1998). Any harmful environmental effects of liming can be prevented with proper dosing and retention time before discharging.

Fertilizers are used in low-intensity shrimp farming to increase nutrient levels and stimulate primary productivity above what is naturally present in ponds. Fertilizers can be organic wastes, agricultural by-products, or chemical compounds, and the primary plant nutrient elements released are nitrogen, phosphorus, and potassium. Boyd and McNevin (2015) calculated that fertilizer used by aquaculture represents only 0.20%, 0.27%, and 0.30% of the global annual use of nitrogen, potassium, and phosphates used in agriculture, respectively, and concluded that aquaculture is not a major contributor to the world's fertilizer discharges, but that excessive or concentrated discharges from aquaculture can lead to water body eutrophication. As in the case of liming agents, judicious use involves proper dosing (applying only the amounts that are needed) and allowing the time required for the nutrients to be fixed in plant and animal tissues before discharging the water.

Therapeutants, including antibiotics, are used in human and veterinary medicine to treat diseases, but the presence of antibiotic residues in the food supply is a serious human health concern as it can lead to the development and spread of antibiotic-resistant pathogens. In consequence, antibiotics approved for human medicine are banned in most countries for use in agriculture and aquaculture. Antibiotic use has been a serious challenge for the shrimp-farming industry. While the larger and better-organized producers have rigorous procedures to avoid antibiotic use, and secure a price premium for their 'chemical-free' shrimp, the issue is much harder to control among the myriad of small, disaggregated producers operating through elongated value chains, which often mask accountability. As the industry consolidates and matures, increasing numbers of producers are appreciating the value of preventing rather than treating diseases.

There are numerous other products used as pond amendments in shrimp farming, including prebiotics, probiotics, vitamins, and plant extracts. The effectiveness of some of these products is dubious, but in general they are not harmful to the environment. Engle et al. (2017) reported the use of 90 different pond amendments in Vietnam and Thailand that could account for up to 15% of farm operating costs.

Translocation of shrimp species

Non-native shrimp species used for aquaculture have the potential to escape into the wild and to have deleterious effects on native shrimp populations, on ecosystems, or on natural or cultured populations through contamination with pathogens. To prevent this, several international codes of practice exist to aid with the process of judicious transboundary transfer of species for aquaculture purposes. Among the most prominent are those stemming from the International Council for the Exploration of the Sea (ICES), the Cartagena Protocol of the Convention on Biological Diversity (CBD), the World Organization for Animal Health (OIE), the Sanitary and Phytosanitary Agreement of the World Trade Organization (WTO/SPS), the FAO Code of Conduct for Responsible Fisheries (CCRF), and the Asian Technical Guidelines on Health Management for the Responsible Movement of Live Aquatic Animals (Subasinghe and Reantaso, 2008; Wang, 2018). It is unfortunate that these procedures are not always followed in practice and that transboundary movements of shrimp for aquaculture purposes have been responsible for the translocation of species and/or the dissemination of shrimp pathogens into new geographical areas. Escapes from farms and hatcheries can occur accidentally during harvests or because of natural events such as storms or floods. The following sections deal with the main species involved.

L. VANNAMEI

Whiteleg shrimp, *L. vannamei*, is naturally distributed throughout the Pacific coast of the Americas, from northern Peru to Baja California, Mexico. It is by far the most widely farmed shrimp in the world and has been moved across regions more than any other shrimp species. There are anecdotal reports of escapes of *L. vannamei* from Thai farms into the Andaman Sea and the Gulf of Thailand, and of subsequent capture of limited numbers by fishermen (Briggs et al., 2004), but there are no reports of perceivable impacts on wild shrimp populations, or that the species has reproduced or become established in those areas. The same can be said for other regions of Asia where escapees are regularly caught but with no reports that breeding populations have been established. There are also reports of catches of *L. vannamei* from the wild in Brazil (Barbieri et al., 2016; Barbieri and Schmidt, 2006; Leão et al., 2011; Pereira and Netto, 2007). These reports are of tens of individuals caught in estuaries near farms and the authors conjectured they were escapees. The data give no indication that *L. vannamei* is reproducing in the wild or that it is causing problems. In that regard, CABI (2019) lists numerous areas of the world where *L. vannamei* has been introduced for culture and found in the wild, but it does not seem that the species has become established outside its native range (Briggs et al., 2004).

P. MONODON

Because of its biological characteristics and because it was a common aquaculture species during the 1970s, *P. monodon* is the most widely translocated shrimp species in the world. *P. monodon* prawns are native to the Indo-West Pacific Ocean, from the eastern coast of Africa to the Arabian Peninsula, Southeast Asia, Taiwan, China, the Sea of Japan, New Guinea, and northern Australia (Dore and Frimodt,

1987; Perez-Farfante and Kensley, 1997). The species has been introduced through aquaculture to many regions of the world and appears to have established breeding populations in some of them.

On the West coast of Africa *P. monodon* has become established from Senegal to South Africa, including Equatorial Guinea (G. Lutz, Baton Rouge, 2019, personal communication) and the whole of the Gulf of Guinea, including Nigeria, which now has a specialized fishery for the species (H. Lucien-Brun, Saint Amoult en Yvelines, 2019, personal communication), which supplies broodstock to shrimp-farming operations in Nigeria and Ghana. These shrimp populations may have originated in escapes from early farming trials in Ivory Coast, Gambia, Senegal, and/or Ghana in the 1980s (H. Lucien-Brun, Saint Amoult en Yvelines, 2019; A. Michel, Paris, 2019; E. De Muylder, Ternat, Belgium, 2019, personal communications).

P. monodon has also established breeding populations in the Atlantic coast of the Americas, from the Carolinas (USA), throughout the Caribbean, and south to Colombia, Venezuela, and Brazil (Fig. 1.5). One of the authors (Juárez) recalls visiting a Taiwanese-sponsored *P. monodon* farming project on the north coast of the Dominican Republic near the city of Monte Cristi around 1988. On the nearby beach, fishermen were using beach seines to capture the local shrimp, *Litopenaeus schmitti*, together with a few

FIGURE 1.5 Locations where breeding populations of invasive *P. monodon* have been reported in the eastern Atlantic, Gulf of Mexico, and Caribbean Sea between 2000 and 2019, according to the following sources: (01) and (02) Shrimp News International (2015) reports for 2000 and 2014, respectively, Knott et al. (2011); (03) CABI (2019) reports for the Caribbean basin as of 2007; (04) and (06) Wakida-Kusunoki et al. (2013) for catches off Tamaulipas and Campeche, Mexico, in 2012–2013; (05) Moran-Silva et al. (2014) for Veracruz, Mexico, in 2014; (07) Alfaro-Montoya et al. (2015) report of first record in Costa Rica; (08) Records from Brazil (Coelho et al., 2001; Santos and Coelho, 2002; Leão et al., 2011). Map template from National Geographic Mapmaker, http://mapmaker.nationalgeographic.org/.

juvenile *P. monodon*, which had most probably escaped from the farm. Recently, F. Richardson (Santo Domingo, 2019, personal communication) confirmed that the species is still present there, although the Taiwanese project is long gone. *P. monodon* were introduced into the Atlantic coast of the USA and were accidentally released by the Waddell Mariculture Center (South Carolina) in 1988. By 1990 they had spread as far south as Florida, and since 2006 they have been found in the Gulf of Mexico along the coastlines of Alabama, Florida, Georgia, Louisiana, Mississippi, and Texas (CABI, 2019; Duda and Palumbi, 1999; Knott et al., 2011; South Carolina Department of Natural Resources, 2008). On the Caribbean coast of Venezuela, it is common to find specimens as large as 600 g (C. Conroy, Guayaquil, 2019, personal communication). In Brazil, imported *P. monodon* were used for aquaculture during the 1970s. Escapes resulted in the species establishing reproducing populations on the coast of Pernambuco (Coelho et al., 2001; Leão et al., 2011; Santos and Coelho, 2002). Fig. 1.5 shows locations where breeding populations of invasive *P. monodon* have been reported in the eastern Atlantic, Gulf of Mexico, and Caribbean Sea between 2000 and 2019.

M. JAPONICUS

The kuruma prawn, *M. japonicus,* is a commercially important penaeid with a native range encompassing the Indo-West Pacific from the Red Sea, East and Southeast Africa, to Korea, Japan, and the Malay Archipelago, and it has also been reported from Fiji (Holthuis, 1980). It is thought to have entered the Mediterranean through the Suez Canal, with capture records as early as 1924 off the coast of Egypt (Galil et al., 2006), and later through aquaculture releases from farms in Italy, France, Spain, Portugal, and Greece. It is now well established in the eastern Mediterranean. D'Acoz (1999) reported that *M. japonicus* has almost excluded the native penaeid prawn, *Melicertus kerathurus*, from the easternmost part of the Mediterranean. Kuruma prawns can now be found as far west as the Atlantic coasts of Spain and North Africa, and up the Atlantic coast of Europe as far north as the Celtic Sea and the coast of northwest France (Quigley et al., 2013).

Disease translocation

Shrimp farming is one of the aquaculture enterprises with the highest potential economic return, but also one with high risks and failure rates. Losses from diseases are the number one problem facing the global shrimp-farming industry, although there are strategies to minimize disease transmission (Fegan and Clifford, 2001). Shrimp moved between regions for aquaculture purposes may carry pathogens that could affect both farmed and wild shrimp and other crustaceans. For example, it is suspected that diseases such as Taura syndrome, infectious myonecrosis, and necrotizing hepatopancreatitis were imported to Asia from Latin America with introductions of *L. vannamei* (CABI, 2019). Pathogens can also be transmitted through ocean currents, ship ballast water, processing plant discharges, birds and other animal carriers, and in refrigerated or frozen shrimp for market. Fig. 1.6 shows the major shrimp epizootics from 1986 to 2017 and their effects on global production of farmed shrimp.

The first significant global shrimp epizootics occurred in the 1990s and were those of yellow head virus (YHV), which started in Thailand (Boonyaratpalin et al., 1993), Taura syndrome virus (TSV), which started in in Ecuador (Jimenez, 1992), and WSSV, which started in China and Taiwan (Jiang, 2001; Zhan

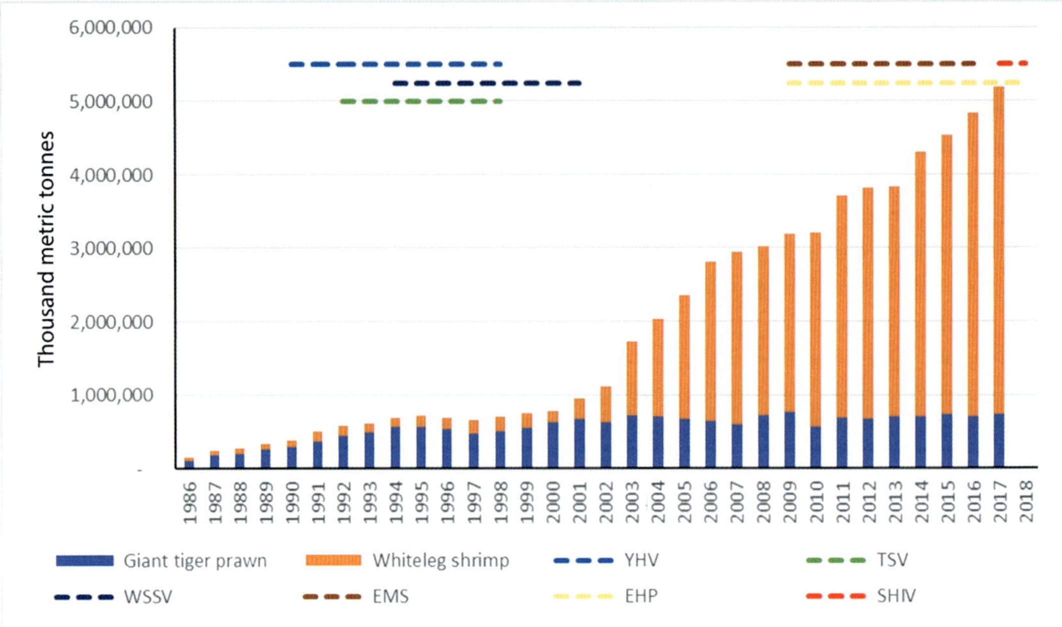

FIGURE 1.6 World aquaculture production of whiteleg shrimp, *L. vannamei*, and giant tiger prawn, *P. monodon* (FAO, 2019a), and major shrimp epizootics (Hasson et al., 1995; Inouye et al., 1994; Lightner, 2003; Limsuwan, 1991; Tourtip et al., 2009; Tran et al., 2013; Walker and Mohan, 2009). SHIV = shrimp hemocyte iridescent virus; EHP = *Enterocytozoon hepatopenaei.*

et al., 1998). WSSV spread rapidly, first throughout Asia and then through the Americas (Flegel, 2019), causing aquaculture production to level off during the late 1990s (Fig. 1.6). TSV first appeared in Ecuador in 1991 and by 1996 had spread to surrounding countries and then to Central and North America, causing losses estimated at between 1 and 2 billion dollars (Lightner, 2003). In 1999, TSV was introduced to Asia and by 2004 it had become endemic to most major shrimp-farming countries of East and Southeast Asia (Do et al., 2006; Nielsen et al., 2005). Acute hepatopancreatic necrosis disease (AHPND), also known as EMS, is another shrimp disease caused by a unique strain of *Vibrio parahaemolyticus*, which has resulted in significant losses in Southeast Asian and Latin American shrimp farms. EMS first appeared in southern China around 2009 (NACA-FAO, 2011). By 2010, the range of affected farms in China had expanded, and by 2011 EMS was confirmed in Vietnam and Malaysia (Lightner et al., 2012). EMS reached Thailand in 2012 (Flegel, 2012; Leaño and Mohan, 2012), Mexico in 2013 (Gomez-Gil et al., 2014; Nunan et al., 2014), and Ecuador in 2016 (Restrepo et al., 2016). EMS/AHPND was partially responsible for the decline in growth of farmed shrimp production between 2011 and 2013 (Fig. 1.6). Several other shrimp diseases have caused regional epizootics and continue to affect shrimp production from affected areas.

Other effects on biota

The major negative impacts of shrimp farming on the environment have been identified throughout this chapter, but it is important to acknowledge other impacts on marine biota. In this section we will briefly consider effects on shorebirds, the recently highlighted problem of plastics in the sea, the control of shrimp farm predators, and the possible impingement of marine mammals on water intake structures. The last two impacts are relatively minor compared with others previously discussed in this chapter, but they still should be mitigated to prevent further harm to marine biota.

Many shrimp farms are developed in or near natural salt flats, intertidal mudflats, mangroves, beaches, marshes, and/or seasonal wetlands, all habitats that are important to migratory and resident shorebirds. Morales et al. (2019) highlight examples from Central America where shrimp farmers, conservation organizations, and local communities have worked together towards the conservation of birds, including developing a manual of good management practices for the protection of shorebirds and standards for the production of 'shorebird-friendly' shrimp.

Pollution by plastics is a serious problem, to which aquaculture and shrimp farming contribute only a small fraction. Nonetheless, all contributing activities need to be addressed if the problem is to be solved. Although plastics have been observed suffocating birds, turtles, fish, and mammals, the greatest effect of plastics on the marine environment, and especially that of small plastic particles, is likely not yet fully realized. Fishmeal is beginning to show plastic residues and tests of small pelagic fish are demonstrating bioaccumulation (Collard et al., 2017; Hanachi et al., 2019). These plastics, ranging in size from the macro to the nanoscale, will further bioaccumulate through the food chain, but also through feeding with contaminated fishmeal. Systemic-wide solutions such as designing biodegradable plastics and country-wide regulations will have larger effects; for example, the use of plastic bags is beginning to be phased out in some nations. The better-organized shrimp farmers have instituted company policies regarding the use and disposal of plastics, including those used directly by the company and those used personally by employees.

While not a significant issue, the possible impingement of marine mammals in farm water intakes has been identified as an impact that should be minimized. Reduced populations of aquatic mammals have prompted greater awareness of this problem. The shrimp-farming industry continues to reduce the amount of water used per unit of product, but there are still regions and systems that rely on exchanging large volumes of water through ponds. Boyd (2008) suggested that screens with openings of 1 or 2 mm be installed across entrances to pump intakes to avoid impingement of larger aquatic organisms.

Predator control is an ongoing challenge at shrimp farms. In nature, shrimp are commonly preyed upon by a variety of fish, birds, and reptiles. Active predator control using lethal techniques has been largely phased out of aquaculture because of the futility of eliminating all the predators in a biologically dense and diverse environment such as an estuary. A more appropriate approach, which is growing in popularity, is to protect the farmed stocks through exclusionary devices such as walls, fences, or nets.

Conclusions and the future

Shrimp farming has sustained tremendous growth since its humble beginnings in the 1970s. With this growth also came the realization that environmental impacts need to be proactively minimized, and that environmental responsibility is critical to the survival and well-being of the industry.

Sustainability is a poorly defined concept but its applications to shrimp farming can be made useful by quantifying the different resources needed to produce a unit weight of shrimp, and by choosing production practices and procedures that minimize environmental impacts. In general, the more intensive production systems are highly productive and thus result in more efficient use of resources than their more extensive counterparts.

Freshwater used in shrimp farming is mainly embodied in the feed, fuel, and other farming inputs. Farmed shrimp requires considerably less freshwater per kg than beef, chicken, or pig production.

The land used to convert natural areas to shrimp farms constitutes the largest environmental impact of shrimp farming, especially when it occurs in wetlands or in other highly valuable ecosystems. Extensive farming systems require large areas and result in low shrimp yields. Semi-intensive production of shrimp requires less land than beef production, more than pig, and about the same as chicken. Hyper-intensive farms make by far the most efficient use of land.

The shrimp-farming industry has spent several decades distancing itself from the conversion of mangroves into shrimp ponds that occurred in the 1980s and 1990s, but the recent development of IMA, where large tracts of mangrove forests are modified to accommodate extensive, low-yielding shrimp farms, is particularly worrisome, as mangrove partition reduces many of the valuable ecosystem services and economic benefits that intact mangroves produce. Although dressed as environmentally sensitive, IMA systems are wasteful and destructive. Mangrove preservation and restoration are key attributes of responsible aquaculture.

Shrimp farming can be of value to foster economic development and poverty alleviation in underdeveloped regions, as it employs significant numbers of people and generates substantial revenues to farmers, about US$32 billion worldwide in 2017. However, balancing growth with the distribution of the wealth generated remains a challenge.

The use of energy in shrimp farming typically increases with culture intensity and ranges from 61.2 to 161.3 GJ/t of shrimp produced, with the lower values corresponding to semi-intensive farming systems, similar to those in Ecuador, and the higher values to less energy-efficient systems such as those in India, or to hyper-intensive, indoor farming operations. These amounts of energy are generally lower than that used to produce beef, about the same as for pig, and higher than for chicken. The carbon footprint of farmed shrimp was estimated to be between 12 and 17 kg of CO_2 per kg of shrimp produced, lower than for beef, but higher than for pig or chicken.

Shrimp feeds constitute the original source of pollutants in farm effluents and also account for the largest portion of farm operating costs, therefore FCR is closely watched by farmers, who have strong economic incentives to use feeds and practices that result in the efficient transformation of feed into shrimp. Fishmeal and fish oil are important ingredients in shrimp feeds, but finite supplies, declining availability, and increasing prices have promoted the development of replacements from multiple sources. Another way to make feeds more efficient is to use by-products of animal and plant processing as ingredients. By-product blends including meat and feather meals, fermented soybean meals, and DDGS are used in practical diets to supplement or replace fishmeal in shrimp diets while providing cost advantages.

Nitrogen and phosphorus from feeds can contribute to eutrophication of the bodies of water that receive discharges from shrimp farms. The use of quality feeds and good management practices maximize the conversion of feed into shrimp, and well-managed ponds typically assimilate most of the differences between the amounts of nitrogen, phosphorus, and organic matter in the feed and those recovered with the harvested shrimp. Water quality limitations for specific variables in farm discharges assure only that their concentration in the discharge mixing zone stays under certain values, and do not preclude the possibility of eutrophication. A better approach to regulation is to establish TMDLs, which consider the assimilation capacity of the receiving body and account for all possible sources of each pollutant. A good addition to the regulatory arsenal would be to establish a requirement for a maximum acceptable FCR, and to maintain water quality monitoring of the receiving bodies. For some water bodies it may be necessary to restrict the quantity of shrimp that can be produced in the area.

The use of agrochemicals such as lime, fertilizers, and therapeutants is another area of concern. Lime and fertilizers are normally not a problem if they are dosed properly and allowed the time prescribed to act before discharging. Conversely, the use of antibiotics in shrimp farming is a serious human health concern and has been banned in most producing countries.

Unchecked movements of live shrimp across regional or international boundaries have resulted in the translocation of species and dissemination of diseases. There are no reports of established breeding populations of translocated *P. vannamei*, the species most widely farmed around the globe, but there are numerous such reports for *P. monodon* and *M. japonicus*. Shrimp diseases have been widely translocated, particularly between the Americas and Southwest Asia, causing epizootics that have affected the global supply of farmed shrimp. Observance of established national regulations and international protocols for the movement of animals should prevent or minimize translocations of shrimp species and diseases.

Aquaculture is more closely related to agriculture than to fishing, with which it is more often associated. Shrimp farming competes for many of the same natural resources as traditional livestock production, transforming land, water, energy, and feed into shrimp. This dependency and competition between agriculture and other uses is shaping the future of aquaculture, as there is ever more pressure to maximize returns on scarce resources.

As shrimp become more commoditized, prices flatten and margins become smaller. Producers trying to remain competitive can opt for a place in a range of two opposing solutions: to accept lower margins but produce high volumes more efficiently, like Asian producers have, or to develop niche markets for differentiated products (by size, quality, or specific labelling), as some semi-intensive Latin American producers have done. These two approaches are not absolute or mutually exclusive. On the one hand, Asia is not uniformly intensified, and there remain many small-scale and low-intensity shrimp farmers. On the other hand, the traditional semi-intensive production systems of Latin America are evolving towards greater control; for example, in Ecuadorian shrimp farms water exchange has been reduced and some groups are moving towards more controlled production systems. In general, the larger, more organized, and better-financed groups will continue to innovate, further driving industry intensification and consolidation.

Several areas show promise for improving the sustainability of shrimp farming. For example, advances in electronics and artificial intelligence have the potential to greatly improve the monitoring and control of water quality, and to devise more precise methods for delivering feed in the required amounts and at the most appropriate times and locations within the culture systems. Recent pioneering examples include the use of demand feeders controlled by sound, sonar, heat sensing, image analysis, and other high-tech solutions. These techniques will contribute to maximizing growth and avoiding waste. The future will also bring better shrimp feeds that deliver improved nutrition and better physical properties with less waste. Fishmeal and fish oil will be progressively substituted with by-products and replaced with novel solutions based on biotechnology. Antibiotics and other agrochemicals are progressively being made unnecessary by the development of 'soft' solutions, such as probiotics.

Domestication of the farmed species is another technological approach that can lead to greater resource efficiency. During the past three decades, research institutions and breeding companies have implemented mass-selection and family-based genetic improvement programmes focused on improving growth rates, tolerance to diseases, and reproductive ability in both *L. vannamei* and *P. monodon*. By helping to produce more shrimp with less resources, these technologies make farming more sustainable. It is reasonable to assume that genetic gains will continue and will possibly accelerate as a result of the incorporation of advanced molecular biology tools such as marker-assisted selection, genomics, gene editing, and possibly transgenics. Growth rate and tolerance to diseases still have a lot of room to improve, but other commercially important traits such as FCR, performance on plant protein, tail-to-head ratio, and tolerance to high culture densities will increasingly enhance resource efficiency and sustainability. Other related technologies that show promise are the production of mono-sex shrimp and the long-term cryopreservation of gametes and embryos. There appears to be a growing trend towards the use of molecular biology tools with the potential to take shrimp aquaculture to new heights.

Culture systems continue to develop towards profitable shrimp production in hyper-intensive systems based on biofloc or recirculating aquaculture systems. About 20 or so start-up companies have sprung up in recent years in the USA, Canada, and Europe, prompted by high shrimp prices in local markets. These systems incorporate enhanced control of the variables that affect production, and they make the production environment more stable and predictable. While the cost of energy and the amortization of initial capital are still challenging, these systems may represent the future of sustainability in shrimp aquaculture.

Sustainability and traceability initiatives are increasingly being demanded by importers and consumers. Digital technologies allow importers to track shrimp from harvest to market and are helping to increase accountability. Other initiatives are driving the industry towards increased environmental sustainability, for example, the worthy new paradigm of no further conversion of intact natural habitat into shrimp farms. Companies that violate this rule may not be allowed to enter certain international markets. Such conversion-free commitments have been made by several major retailers and brands for other commodities such as soy and beef. It is only a matter of time until this permeates the shrimp aquaculture sector.

The overwhelming number of shrimp production certification programmes and standards will likely give way to refocusing on the key resources needed for production. In their attempt to address everything, these programmes have resulted in dilution of the most important impacts and have increasingly become perceived as a mere tax to access ecologically sensitive consumers and markets. Radical transparency will become the norm, with supply chain owners taking back control of their raw material suppliers from the external auditors, trade associations, and certification bodies. New approaches are beginning to disrupt the conventions around shrimp aquaculture, with new models for supply chain oversight that are less burdensome, less expensive, and more effective than before. Greater importing oversight and new forensic tools are being developed to verify country of origin. Shrimp farming needs to continue evolving in an increasingly sustainable manner so that it can contribute responsibly to the protein supply for the years to come.

Acknowledgements

The authors wish to thank Anthonie Schuur for critically reviewing the manuscript and offering useful suggestions.

References

Aburto-Oropeza, O., Ezcurra, E., Danemann, G., Valdez, V., Murray, J., and Sala, E. (2008) Mangroves in the Gulf of California increase fishery yields. *Proceedings of the National Academy of Sciences of the United States of America* 105, 10456–10459.

Achupallas, J.M., Zhou, Y., and Davis, D.A. (2015) Pond production of Pacific white shrimp, *Litopenaeus vannamei*, fed grain distillers dried yeast. *Aquaculture Nutrition* 22, 1222–1229.

ADB/NACA (1995) Regional shrimp and carp survey summary data. In: *Regional Study and Workshop on Aquaculture Sustainability and the Environment*. Asian Development Bank (ADB) and Network of Aquaculture Centres in Asia (NACA), Bangkok, Thailand.

Ahmed, N. and Glaser, M. (2016) Coastal aquaculture, mangrove deforestation and blue carbon emissions: Is REDD+ a solution? *Marine Policy* 66, 58–66.

Ahmed, N., Thompson, S., and Glaser, M. (2018) Integrated mangrove-shrimp cultivation: potential for blue carbon sequestration. *Ambio* 47, 441–452.

Alfaro-Montoya, J., Monge-Ortiz, A.M., Martínez-Fernández, D., and Herrera-Quesada, E. (2015) First record of the nonindigenous *Penaeus monodon* Fabricius, 1798 (Penaeidae) in the Caribbean Sea of Costa Rica, Central America, with observations on selected aspects of its reproductive biology. *BioInvasions Records* 4, 217–222.

Amaya, E., Davis, D.A., and Rouse, D.B. (2007) Replacement of fish meal in practical diets for the Pacific white shrimp (*Litopenaeus vannamei*) reared under pond conditions. *Aquaculture* 262, 393–401.

Anneboina, L.R. and Kumar, K.S. (2017) Economic analysis of mangrove and marine fishery linkages in India. *Ecosystem Services* 24**,** 114–123.

Asem-Hiablie, S., Battagliese, S., Stackhouse-Lawson, K.R., and Rotz, C.A. (2019) A life cycle assessment of the environmental impacts of a beef system in the USA. *The International Journal of Life Cycle Assessment* 24, 441–455.

Balmford, A., Bruner, A., Cooper, P., Costanza, R., Farber, S., Green, R.E., Jenkins, M., Jefferiss, P., Jessamy, V., Madden, J., Munro, K., Myers, N., Naeem, S., Paavola, J., Rayment, M., Rosendo, S., Roughgarden, J., Trumper, K., and Turner, K. (2002) Economic reasons for conserving wild nature. *Science* 297, 950–953.

Bandaranayake, W.M. (1998) Traditional and medicinal uses of mangroves. *Mangroves and Salt Marshes* 2, 133–148.

Barbieri, E., Coa, F., and Fernandes, K. (2016) Biodiversidade: ocorrência da espécie exótica *Litopenaeus vannamei* (Boone, 1931) no complexo estuarino-lagunar de Cananéia-Iguape-Ilha Comprida. *Boletim do Instituto de Pesca, São Paulo* 42, 479–485.

Barbieri, E. and Schmidt, G.A. (2006) Biodiversidade: ocorrência da espécie exótica *Litopenaeus Vannamei* (Boone, 1931) no complexo estuarino-lagunar de Cananéia-Iguape-Ilha Comprida. *O Mundo da Saúde São Paulo* 30, 654–659.

Boonyaratpalin, S., Supamattaya, K., Kasornchandra, J., Direkbusaracom, S., Aekpanithanpong, U., and Chantanachookin, C. (1993) Non-occluded baculo-like virus, the causative agent of yellow head disease in the black tiger shrimp (*Penaeus monodon*). *Fish Pathology* 28, 103–109.

Boyd, C.E. (2008) Better management practices for marine shrimp aquaculture. In: Tucker, C.S. and Hargreaves, J.A. (eds) *Environmental Best Management Practices for Aquaculture*. Blackwell, Ames, IA, pp. 227–259.

Boyd, C.E. (2020) *Water Quality; An Introduction*. 3rd edn. Springer, New York, NY.

Boyd, C.E. and Clay, J.W. (2002) *Evaluation of Belize Aquaculture Ltd: A Super Intensive Shrimp Aquaculture System*. Food and Agriculture Organization of the United Nations, Rome, Italy.

Boyd, C.E. and McNevin, A.A. (2015) *Aquaculture, Resource Use, and the Environment*. Wiley-Blackwell, Hoboken, NJ.

Boyd, C.E. and McNevin, A.A. (2018) Land use in shrimp aquaculture. *World Aquaculture* 49, 28–34.

Boyd, C.E. and Tucker, C.S. (1998) *Pond Aquaculture Water Quality Management*. Kluwer Academic Publishers, Boston, MA.

Boyd, C.E., McNevin, A.A., Davis, R., Godumala, R., and Mohan, A. (2018) Production methods and resource use at *Litopenaeus vannamei* and *Penaeus monodon* farms in India compared with previous findings from Thailand and Vietnam. *Journal of the World Aquaculture Society* 49, 551–569.

Boyd, C.E., McNevin, A.A., Racine, P. Tinh, H.Q., Minh, H.N., Viriyatum, R., Paungkaew, D., and Engle, C.R. (2017) Resource use of shrimp *Litopenaeus vannamei* and *Penaeus monodon* production in Thailand and Vietnam. *Journal of the World Aquaculture Society* 48, 201–226.

Boyd, C.E., Davis, R.P., Wilson, A.B., Mareillo, F., Brian, S. and McNevin, A.A. (2021) Resource use in Whiteleg shrimp *Litopenaeus vannamei* farming in Ecuador. *Journal of the World Aquaculture Society* 52, 772–888.

Briggs, M., Funge-Smith, S., Subasinghe, R., and Phillips, M (2004*) Introductions and movement of Penaeus vannamei and Penaeus stylirostris in Asia and the Pacific*. FAO Regional Office for Asia and the Pacific, Bangkok, Thailand.

Bureau, D.P. (2006) Rendered products in fish aquaculture feeds. In: Meeker, D.L. (ed.) *Essential Rendering: All About the Animal By-Products Industry.* National Renderers Association, Alexandria, VA, pp. 179–184.

Bush, S.R., Van Zwieten, P.A., Visser, L., Van Dijk, H., Bosma, R., De Boer, W.F., and Verdegem, M. (2010) Scenarios for resilient shrimp aquaculture in tropical coastal areas. *Ecology and Society* 15, 1–15.

CABI (2019) Invasive Species Compendium, *Litopenaeus vannamei* (whiteleg shrimp). Available at: https://www.cabi.org/isc/datasheet/71097 (accessed 6 June 2020).

Cao, L., Wang, W., Yang, Y., Yang, C., Yuan, Z., Xiong, S., and Diana, J. (2007) Environmental impact of aquaculture and countermeasures to aquaculture pollution in China. *Environmental Science and Pollution Research - International* 14, 452–462.

Carvalho, R.A., Haruo, O.R., Oliveira, K.V., Tacon, A.G., and Lemos, D. (2016) Apparent digestibility of protein, energy and amino acids of six protein sources included at three levels in diets for juvenile white shrimp *Litopenaues vannamei* reared in high performance conditions. *Aquaculture* 465, 223–234.

Chamberlain, G. (2010) History of shrimp farming. In: Alday-Sanz, V. (ed.) *The Shrimp Book*. Nottingham University Press, Nottingham, UK, pp. 1–34.

Chapin, F.S., Kofinas, G.P., and Folke, C. (2009) *Principles of Ecosystem Stewardship*. Springer, New York, NY.

Chatvijitkul, S., Boyd, C.E., Davis, D.A., and McNevin, A.A. (2017) Embodied resources in fish and shrimp feeds. *Journal of the World Aquaculture Society* 48, 7–19.

Cheng, Z.J., Behnke, K.C., and Dominy, W.G. (2002) Effects of poultry by-product meal as a substitute for fish meal in diets on growth and body composition of juvenile Pacific white shrimp, *Litopenaeus vannamei. Journal of Applied Aquaculture* 12, 71–83.

Cherry, B., Shannon, L., and Mary J. (2021) How much ENERGY (in BTUs) is expended or used to produce 1 pound of beef, 1 pound of chicken, and 1 pound of p.ork for retail consumption? Available at: https://askwonder.com/research/energy-in-btus-expended-used-produce-1-pound-beef-chicken-pork-retail-k5gwibwm3 (accessed 4 August 2021).

Chong, V.C. (2007) Mangroves-fisheries linkages – the Malaysian perspective. *Bulletin of Marine Science* 80, 755–772.

Clay, J.W. (2004) *World Agriculture and the Environment: A Commodity-by-Commodity Guide to Impacts and Practices*. Island Press, Washington, DC.

Clough, B., Johnston, D., Xuan, T.T., Phillips, M.J., Pednekar, S.S., Thien, N.H., Dan, T H., and Thong, L.P. (2002) *Silvofishery Farming Systems in Ca Mau Province, Vietnam*. World Bank, NACA, WWF and FAO Consortium Program on Shrimp Farming and the Environment, Rome, Italy.

Coelho, P.A., Santos, M.D., and Ramos-Porto, M. (2001) Ocorrência de *Penaeus monodon Fabricius*, 1798 no litoral dos estados de Pernambuco e Alagoas (Crustacea, Decapoda, Penaeidae). *Boletim Técnico Científico do CEPENE* 9, 149–153.

Collard, F., Gilbert, B., Compere, P., Eppe, G., Das, K., Jauniaux, T., and Parmentier, E. (2017) Microplastics in livers of European anchovies (*Engraulis encrasicolus, L.*). *Environmental Pollution* 229, 1000–1005.

Cook, H.L. and Murphy, A.M. (1966) Rearing penaeid shrimp from eggs to postlarvae. In: *Proceedings of the Southeastern Association of Game and Fish Commissioners.* Southeastern Association of Fish and Game Commissioners, Frankfort, KY, 283–288.

Costanza, R., D'Arge, R., de Groot, R., Farber, S., Grasso, M., Hannon, B., Limburg, K., Naeem, S., Oneill, R.V., Paruelo, J., Raskin, R.G., Sutton, P., and van den Belt, M. (1997) The value of the world's ecosystem services and natural capital. *Nature* 387**,** 253–260.

D'Acoz, U.C. (1999) Inventaire et distribution des crustacés décapodes de l'Atlantique oriental, de la Mediterranée et des eaux continentales au nord de 25° N. *Patrimoines Naturels* 40, 1–383.

Danielsen, F., Sørensen, M.K., Olwig, M.F., Selvam, V., Parish, F., Burgess, N.D., Hiraishi, T., Karunagaran, V.M., Rasmussen, M.S., Hansen, L.B., Quarto, A., and Suryadiputra, N. (2005) The Asian tsunami: a protective role for coastal vegetation. *Science* 310, 643–644.

Das, S. (2017) Ecological restoration and livelihood: contribution of planted mangroves as nursery and habitat for artisanal and commercial fishery. *World Development* 94, 492–502.

DeMicco, E.D. (2015) Feasibility of using biofuel by-products as a sustainable nutritional resource for aquaculture production of *Litopenaeus vannamei*. PhD dissertation. Nova Southeastern University, Davie, FL, USA. Available at: https://nsuworks.nova.edu/occ_stuetd/387 (accessed 5 June 2020).

Dierberg, F.E. and Kiattisimkul, W. (1996) Issues, impacts, and implications of shrimp aquaculture in Thailand. *Environmental Management* 20, 649–666.

Ding, Z., Zhang, Y., Ye, J., Du, Z., and Kong, Y. (2015) An evaluation of replacing fish meal with fermented soybean meal in the diet of *Macrobrachium nipponense*: growth, nonspecific immunity, and resistance to *Aeromonas hydrophila*. *Fish & Shellfish Immunology* 44, 295–301.

Do, J.W., Cha, S.J., Lee, N.S., Kim, Y.C., Kim, J.W., and Kim, J.D. (2006) Taura syndrome virus from *Penaeus vannamei* shrimp cultured in Korea. *Diseases of Aquatic Organisms* 12, 171–174.

Donato, D.C., Kauffman, J.B., Murdiyarso, D., Kurnianto, S., Stidham, M., and Kanninen, M. (2011) Mangroves among the most carbon-rich forests in the tropics. *Nature Geoscience* 4, 293–297.

Dore, I. and Frimodt, C. (1987) *An Illustrated Guide to Shrimp of the World*. Van Nostrand Reinhold, New York, NY.

Duda, T.F. and Palumbi, S.R. (1999) Population structure of the black tiger prawn, *Penaeus monodon*, among Western Indian Ocean and Western Pacific populations. *Marine Biology* 134, 705–710.

Duke, N.C., Meynecke, J.O., Dittmann, S., Ellison, A.M., Anger, K., Berger, U., Cannicci, S., Diele, K., Ewel, K.C., Field, C.D., Koedam, N., Lee, S.Y., Marchand, C., Nordhaus, I., and Dahdouh-Guebas, F. (2007) A world without mangroves? *Science* 317, 41–42.

Engle, C.R. (2019) Bringing aquaculture sustainability down to earth. *Journal of the World Aquaculture Society* 50, 246–248.

Engle, C.R. and D'Abramo, L.D. (2016) Showcasing research focusing on sustainability of aquaculture enterprises and global food security. *Journal of the World Aquaculture Society* 47, 311–313.

Engle, C.R., McNevin, A.A., Racine, P.E., Boyd, C.E., Paungkaew, D.N., Viriyatum, R., Tinh, H., and Mihn, H.N. (2017) Economics of sustainable intensification of aquaculture: evidence from shrimp farms in Vietnam and Thailand. *Journal of the World Aquaculture Society* 48, 227–239.

Ezquerra, J.M., García-Carreño, F.L., Civera, R., and Haard, N.F. (1997) pH-stat method to predict protein digestibility in white shrimp (*Penaeus vannamei*). *Aquaculture* 157, 251–262.

FAO (2013) *Development of a sub-regional strategy for improving biosecurity (aquatic animal health) in the sub-regional countries of the Mozambique Channel.* Food and Agriculture Organization of the United Nations, Rome, Italy.

FAO (2018) The state of world fisheries and aquaculture 2018 – meeting the sustainable development goals. Available at: http://www.fao.org/3/i9540en/i9540en.pdf (accessed 22 October 2019).

FAO (2019a) Global aquaculture production. Available at: http://www.fao.org/fishery/statistics/global-aquaculture-production/query/en (accessed 22 October 2019).

FAO (2019b) Global capture production 1950–2018. Available at: http://www.fao.org/fishery/statistics/global-capture-production/query/en (accessed 22 October 2019).

Fegan, D.F. and Clifford III, H.C. (2001) Health management for viral diseases in shrimp farms. In: Browdy, C.L. and Jory, D.E. (eds) *The New Wave: Proceedings of the Special Session on Sustainable Shrimp Culture.* The World Aquaculture Society, Baton Rouge, LA, USA. pp. 168–198.

Flaherty, M., Vandergeest, P., and Miller, P. (1999) Rice paddy or shrimp pond: tough decisions in rural Thailand. *World Development* 27, 2045–2060.

Flegel, T.W. (2012) Historic emergence, impact and current status of shrimp pathogens in Asia. *Journal of Invertebrate Pathology* 110, 166–173.

Flegel, T.W. (2019) A future vision for disease control in shrimp aquaculture. *Journal of the World Aquaculture Society* 50, 249–266.

Galil, B.S. (2006) *Marsupenaeus japonicus*. Delivering alien invasive species inventories for Europe (DAISE). Available at: https://web.archive.org/web/20111201023842/http://www.europe-aliens.org/pdf/Marsupenaeus_japonicus.pdf (accessed 21 July 2019).

Gomez-Gil, B., Soto-Rodriguez, S., Lozano, R., and Betancourt-Lozano, M. (2014) Draft genome sequence of *Vibrio parahaemolyticus* strain M0605, which causes severe mortalities of shrimps in Mexico. *Genome Announcements* 2, e00055-14.

Green Eatz (2020) Food's carbon footprint. Available at: https://www.greeneatz.com/foods-carbon-footprint.html (accessed 25 May 2020).

Grellier, S., Janeau, J.L., Nhon, D.H., Cuc, N.T., Quynh, L.T., Thao, P.T., Nhu-Trang, T.T., and Marchand, C. (2017) Changes in soil characteristics and C dynamics after mangrove clearing (Vietnam). *Science of the Total Environment* 593, 654–663.

Guannel, G., Arkema, K., Ruggiero, P., and Verutes, G. (2016) The power of three: coral reefs, seagrasses and mangroves protect coastal regions and increase their resilience. *PLoS One* 11, e0158094.

Guo, H., Weaver, C., Charles, S.P., Whitt, A., Dastidar, S., D'Odorico, P., Fuentes, J.D., Kominoski, J.S., Armitage, A.R., and Pennings, S.C. (2017) Coastal regime shifts: rapid responses of coastal wetlands to changes in mangrove cover. *Ecology* 98, 762–772.

Ha, T.T., Bush, S.R., Mol, A.P., and Van Dijk, H. (2012) Organic coasts? Regulatory challenges of certifying integrated shrimp-mangrove production systems in Vietnam. *Journal of Rural Studies* 28, 631–639.

Ha, T.T., Van Dijk, H., and Visser, L. (2014) Impacts of changes in mangrove forest management practices on forest accessibility and livelihood: a case study in mangrove-shrimp farming system in Ca Mau Province, Mekong Delta, Vietnam. *Land Use Policy* 36, 89–101.

Hall, S.J., Delaporte, A., Phillips, M.J., Beveridge, M., and O'Keefe M. (2011) *Blue Frontiers: Managing the Environmental Costs of Aquaculture*. The World Fish Center, Penang, Malaysia.

Hamilton, S. (2013) Assessing the role of commercial aquaculture in displacing mangrove forest. *Bulletin of Marine Science* 89, 585–601.

Hanachi, P., Karbalaei, S., Walker, T.R., Cole, M., and Hosseini, S.V. (2019) Abundance and properties of microplastics found in commercial fish meal and cultured common carp (*Cyprinus carpio*). *Environmental Science and Pollution Research* 26, 23777–23787.

Harborne, A.R., Talwar, B., and Brooks, E.J. (2016) The conservation implications of spatial and temporal variability in the diurnal use of Bahamian tidal mangrove creeks by transient predatory fishes. *Aquatic Conservation: Marine and Freshwater Ecosystems* 26, 202–211.

Hasson, K.W., Lightner, D.V., Poulos, B.T., Redman, R.M., White, B.M., Brock, J.A., and Bonami, J.R. (1995) Taura syndrome in *Penaeus vannamei*: demonstration of a viral etiology. *Diseases of Aquatic Organisms* 23, 115–126.

Hauser, L.T., Vu, G.N., Nguyen, B.A., Dade, E., Nguyen, H.M., Nguyen, T.T., Le, T.Q., Vu, L.H., Tong, A.T., and Pham, H.V. (2017) Uncovering the spatio-temporal dynamics of land cover change and fragmentation of mangroves in the Ca Mau peninsula, Vietnam using multi-temporal SPOT satellite imagery (2004–2013). *Applied Geography* 86, 197–207.

Hirono, Y. and Leslie, M. (1992) Shrimp culture industry in Ecuador. In: Fast, A.W. and Lester, L.J. (eds) *Marine Shrimp Culture: Principles and Practices*. Elsevier, Amsterdam, The Netherlands, pp. 783–815.

Hishamunda, N., Ridler, N.B., Bueno, P., and Yap, W.G. (2009) Commercial aquaculture in Southeast Asia: some policy lessons. *Food Policy* 34, 102–107.

Hodal, K., Kelly, C., and Lawrence, F. (2014) Revealed: Asian slave labour producing prawns for supermarkets in US, UK. *The Guardian* (June 10). Available at: https://www.theguardian.com/global-development/2014/jun/10/supermarket-prawns-thailand-produced-slave-labour (accessed 16 July 2021).

Holthuis, L.B. (1980) *Shrimps and prawns of the world. An annotated catalogue of species of interest to fisheries. FAO Fisheries Synopsis*. FAO, Rome, Italy.

Hove, H. (2004). Critiquing sustainable development: a meaningful way of mediating the development impasse? *Undercurrent* 1, 48–54.

Huang, Y. (2015) Can capitalist farms defeat family farms? The dynamics of capitalist accumulation in shrimp aquaculture in South China. *Journal of Agrarian Change* 15, 392–412.

Hudinaga, M. (1935) Studies on the development of *Penaeus japonicus* (Bate). *Hayatomo Fishery Institute Report* 1, 1–51.

Ibidhi, R., Hoekstra, A.Y., Gerbens-Leenes, P.W. and Chouchane, H. (2017) Water, land and carbon footprints of sheep and chicken meat produced in Tunisia under different farming systems. *Ecological Indicators* 77, 304–313.

IFFO (2020) Fish in: Fish out (FIFO) ratios for the conversion of wild feed to farmed fish, including salmon. Available at: https://www.iffo.net/fish-fish-out-fifo-ratios-conversion-wild-feed (accessed 6 May 2020).

IndexMundi (2019) Fishmeal monthly price. Available at: https://www.indexmundi.com/commodities/?commodity=fish-meal&months=240 (accessed 25 Jan 2020).

Inouye, K., Miwa, S., Oseko, N., Nakano, H., and Kimura, T. (1994) Mass mortalities of cultured kuruma shrimp, *Penaeus japonicus* in Japan in 1993: electron microscopic evidence of the causative virus. *Fish Pathology* 29, 149–158

Jackson, A. (2012) Fishmeal and fish oil and its role in sustainable aquaculture. *International Aquafeed* 15, 18–21.

Jerath, M., Bhat, M., Rivera-Monroy, V.H., Castaneda-Moya, E., Simard, M., and Twilley, R.R. (2016) The role of economic, policy, and ecological factors in estimating the value of carbon stocks in Everglades mangrove forests, South Florida, USA. *Environmental Science & Policy* 66, 160–169.

Jescovitch, L.N., Chaney, P.L., and Boyd, C.E. (2016) A preliminary assessment of land-to-water surface area ratios (LWR) for sustainable land use in aquaculture. *Papers in Applied Geography* 2, 178–188.

Jiang Y. (2001) *Thematic review on management strategies for major diseases in shrimp aquaculture*. In: Subasinghe, R.P., Arthur, J.R., Phillips, M.J., and Reantaso, M.B. (eds) *Proceedings of a workshop held in Cebu, Philippines*. The World Bank, Network of Aquaculture Centers Asia-Pacific, World Wildlife Fund and Food and Agricultural Organization of the United Nations Consortium Program on Shrimp Farming and the Environment, Rome, Italy, pp. 74–78.

Jimenez, R. (1992) Sindrome de Taura. *Revista Especializada de la Camara de Productores de Camaron de Guayaquil, Ecuador* 1, 1–16.

Joffre, O.M., Bosma, R.H., Bregt, A.K., Van Zwieten, P.A., Bush, S.R., and Verreth, J.A. (2015) What drives the adoption of integrated shrimp mangrove aquaculture in Vietnam? *Ocean & Coastal Management* 114, 53–63.

Jones, D.L., Walter, J.F., Brooks, E.N., and Serafy, J.E. (2010) Connectivity through ontogeny: fish population linkages among mangrove and coral reef habitats. *Marine Ecology Progress Series* 401, 245–258.

Kelly, A. (2016) Nestlé admits slavery in Thailand while fighting child labour lawsuit in Ivory Coast. *The Guardian* (February 1). Available at: https://www.theguardian.com/sustainable-business/2016/feb/01/nestle-slavery-thailand-fighting-child-labour-lawsuit-ivory-coast (accessed 16 July 2021).

Knott, D., Fuller, P., Benson, A., Neilson, M. (2011) United States Geological Service. Non-indigenous aquatic species. Available at: http://nas.er.usgs.gov/queries/FactSheet.aspx?speciesID=1209 (accessed 9 August 2019).

Kristofferson, D. and Anderson, J.L. (2006) Is there a relationship between fisheries and farming? Interdependence of fisheries, animal production and aquaculture. *Marine Policy* 30. 721–725.

Kureshy, N., Allen, D.A., and Arnold, C.R. (2000) Partial replacement of fish meal with meat-and-bone meal, flash-dried poultry by-product meal, and enzyme-digested poultry by-product meal in practical diets for juvenile red drum. *North American Journal of Aquaculture* 62, 266–272.

Layman, C.A., Arrington, D.A., Langerhans, R.B., and Silliman, B.R. (2004) Degree of fragmentation affects fish assemblage structure in Andros Island (Bahamas) estuaries. *Caribbean Journal of Science* 40, 232–244.

Leaño, E.M. and Mohan, C.V. (2012) Early mortality syndrome threatens Asia's shrimp farms. *Global Aquaculture Advocate* (July/August), 38–39.

Leão, T.C., Almeida, W.R., Dechoum, M.S., and Ziller, S.R. (2011) *Espécies exóticas invasoras no Nordeste do Brasil: contextualização, manejo e políticas públicas*. Centro de Pesquisas Ambientais do Nordeste e Instituto Hórus de desenvolvimento e conservação ambiental, Recife, Brazil.

Lee, P.G. and Meyers, S.P. (1997) Chemoattraction and feeding stimulation. In: D'Abramo, L.R., Conklin, D.E., and Akiyama, D.M. (eds) *Crustacean Nutrition Advances in World Aquaculture*. World Aquaculture Society, Baton Rouge, LA, pp. 292–352.

Lemos, D., Lawrence, A.L., and Siccardi III, A.J. (2009) Prediction of apparent protein digestibility of ingredients and diets by *in vitro* pH-stat degree of protein hydrolysis with species-specific enzymes for juvenile Pacific white shrimp *Litopenaeus vannamei*. *Aquaculture* 295, 89–98.

Lightner, D.V. (2003) The penaeid shrimp viral pandemics due to IHHNV, WSSV, TSV and YHV: history in the Americas and current status. In: *Proceedings of the 32nd Joint United States-Japan Natural Resources Panel on Aquaculture Symposium*. Davis and Santa Barbara, CA, pp. 17–20.

Lightner, D.V., Redman, R.M., Pantoja, C.R., Noble, B.L., and Tran, L. (2012) Early mortality syndrome affects shrimp in Asia. *Global Aquaculture Advocate* 15, 40.

Limsuwan, C. (1991) *Handbook for Cultivation of Black Tiger Prawns*. Tansetakit, Bangkok, Thailand (in Thai).

Lopez-Alvarado, J.L., Ruiz, W., and Moncayo, E. (2016) Offshore aquaculture development in Ecuador. *International Journal of Research and Education* 1, 1–6.

Macreadie, P.I., Ollivier, Q.R., Kelleway, J.J., Serrano, O., Carnell, P.E., Lewis, C.J., Atwood, T.B., Sanderman, J., Baldock, J., Connolly, R.M., Duarte, C.M., Lavery, P.S., Steven, A., and Lovelock, C.E. (2017) Carbon sequestration by Australian tidal marshes. *Scientific Reports* 7, 44071.

Malik, A., Fensholt, R., & Mertz, O. (2015) Economic valuation of mangroves for comparison with commercial aquaculture in South Sulawesi, Indonesia. *Forests* 6, 3028–3044.

Mason, M., McDowell, R., Mendoza, M., and Htusan, E. (2015) Global supermarkets selling shrimp peeled by slaves. *The Associated Press* (14 December). Available at: https://apnews.com/article/8f64fb25931242a985bc30e3f5a9a0b2 (accessed 16 July 2021).

Mekonnen, M.M. and Hoekstra, A.Y. (2011) The green, blue and grey water footprint of crops and derived crop products. *Hydrology and Earth System Sciences* 15, 1577–1600.

Morales, S., Jarquín, O., Reyes, E., and Navedo. J.G. (2019) Shorebirds and shrimp farming: assessment of shrimp farming activities on shorebirds in Central America. Available at: https://pacificflywayshorebirds.org/downloads/Shorebird-and-Shrimp-Farming-Assessment-2019.pdf (accessed 16 September 2019).

Morán-Silva, A., Jiménez-Badillo, M.L., Cházaro-Olvera, S., Meiners, C., Galindo-Cortes , G., and Oviedo, J. L. (2014) First record of tiger shrimp (*Penaeus monodon Fabricius*, 1798), in the south-central coast of Veracruz, Gulf of Mexico, *Biología, Ciencia y Tecnología* 7, 509–514.

Moss, S.M., Arce, S.M., Argue, B.J., Otoshi, C.A., Calderon, F.R., and Tacon, A.G. (2001) Greening of the blue revolution: efforts toward environmentally responsible shrimp culture. In: Browdy, C.L. and Jory, D.E. (eds) *The New Wave: Proceedings of the Special Session on Sustainable Shrimp Farming, Aquaculture 2001*. World Aquaculture Society, Baton Rouge, LA, pp. 1–19.

Motoh, H., Guartatanga, R., and Blacio, E. (1994) Shrimp fry collecting gear used in Ecuador. *Crustacean Research* 23, 61–68.

Mukherjee, R., Chakraborty, R., and Dutta, A. (2016) Role of fermentation in improving nutritional quality of soybean meal; a review. *Asian-Australasian Journal of Animal Sciences* 29, 1523–1529.

Mumby, P.J., Edwards, A.J., Arias-Gonzalez, J.E., Lindeman, K.C., Blackwell, P.G., Gall, A., Gorczynska, M.I., Harborne, A.R., Pescod, C.L., Renken, H., and Wabnitz, C.C. (2004) Mangroves enhance the biomass of coral reef fish communities in the Caribbean. *Nature* 427, 533–536.

Murtini, S., Kuspriyanto, and Kurniawati, A. (2018) Mangrove area development strategy wonorejo as ecotourism in Surabaya. Available at: https://iopscience.iop.org/article/10.1088/1742–6596/953/1/012174/pdf (accessed 6 February 2020).

NACA-FAO (2011) Quarterly aquatic animal disease report (Asia and Pacific Region). Available at: http://library.enaca.org/Health/QAAD/qaad-2011-2.pdf (accessed 3 March 2020).

Nagelkerken, I., Van der Velde, G., Gorissen, M.W., Meijer, G.J., Van't Hof, T., and Hartog, D. (2000) Importance of mangroves, seagrass beds, and the shallow coral reef as a nursery for important coral reef fishes, using a visual census technique. *Estuarine, Coastal and Shelf Science* 51, 31–44.

Nagelkerken, I.S., Blaber, S.J., Bouillon, S., Green, P., Haywood, M., Kirton, L.G., Meynecke, J.O., Pawlik, J., Penrose, H.M., Sasekumar, A., and Somerfield, P.J. (2008) The habitat function of mangroves for terrestrial and marine fauna: a review. *Aquatic Botany* 89, 155–185.

Nates, S.F. and Bureau, D.P. (2007) Opportunities for rendered animal proteins in aquaculture feeds. Available at:

https://www.aquaculturealliance.org/advocate/opportunities-for-rendered-animal-proteins-in-aquaculture-feeds/ (accessed 22 February 2020).

Nates, S. and Swisher, K., (2013) Animal byproduct concentrates useful tools in formulation. Available at: https://www.aquaculturealliance.org/advocate/animal-by-product-concentrates-useful-tools-in-aquafeed-formulation/ (accessed 22 February 2020).

Naylor, R.L., Golburg, R.J., Primavera, J.H., Kautsky, N., Beveridge, M.C., Clay, J., Folke, C., Lubchenco, J., Mooney, H., and Troell, M. (2000) Effect of aquaculture on world fish supplies. *Nature* 405, 1017–1024.

Naylor, R.L., Hardy, R.W., Buschmann, A.H., Bush, S.R., Cao, L., Klinger, D.H., Little, D.C., Lubchenco, J., Shumway, S.E., and Troell, M. (2021) A 20-year retrospective review of global aquaculture. *Nature* 591, 551–562.

Nielsen, L., Sang-oum, W., Cheevadhanarak, S., and Flegel, T.W. (2005) Taura syndrome virus (TSV) in Thailand and its relationship to TSV in China and the Americas. *Diseases of Aquatic Organisms* 63, 101–106.

Niu, J., Xie, J.J., Guo, T.Y., Fang, H.H., Zhang, Y.M., Liao, S.Y., Xie, S., Liu, Y.J., and Tian, L.X. (2019) Comparison and evaluation of four species of macro-algaes as dietary ingredients in *Litopenaeus vannamei* under normal rearing and WSSV challenge conditions: effect on growth, immune response, and intestinal microbiota. *Frontiers in Physiology* 9, 1880.

Nunan, L., Lightner, D., Pantoja, C., and Gomez-Jimenez, S. (2014) Detection of acute hepatopancreatic necrosis disease (AHPND) in Mexico. *Diseases of Aquatic Organisms* 111, 81–86.

OECD/FAO (2019) *OECD-FAO Agricultural Outlook 2019–2028*. OECD Publishing, Paris, France.

Olmos, J.L., Paniangua-Michel, J., and Contreras, R. (2011) Functional feed assessment on *Litopenaeus vannamei* using 100% fish meal replacement by soybean meal, high levels of complex carbohydrates and *Bacillus* probiotic strains. *Marine Drugs* 9, 1119–1132.

Otero, X.L., Méndez, A., Nóbrega, G.N., Ferreira, T.O., Santiso-Taboada, M.J., Meléndez, W., and Macías, F. (2017) High fragility of the soil organic C pools in mangrove forests. *Marine Pollution Bulletin* 119, 460–464.

Pereira, T.J. and Netto, S. (2007) Ocorrência do camarão exótico *Litopenaeus vannamei* Boone, 1931 (Crustacea, Decapoda, Penaeidae) no sistema estuarino de laguna, SC. *Biociências* 15, 157–159.

Perez, A., Machado, W., Gutierrez, D., Stokes, D., Sanders, L., Smoak, J.M., Santos, I., and Sanders, C.J. (2017) Changes in organic carbon accumulation driven by mangrove expansion and deforestation in a New Zealand estuary. *Estuarine Coastal and Shelf Science* 192, 108–116.

Perez-Farfante, I. and Kensley, B. (1997) Penaeid and Sergestoid shrimps and prawns of the world: keys and diagnoses for the families and genera. *Memoirs Du Museum National D'Historie Naturelle Paris* 175, 1–233.

Qiu, X., Nguyen, L., and Davis, D.A. (2017) Apparent digestibility of animal, plant and microbial ingredients for Pacific white shrimp *Litopenaeus vannamei*. *Aquaculture Nutrition* 24, 930–939.

Quigley, D.T., Herdson, D., and Flannery, K. (2013) Occurrence of the kuruma prawn *Marsupenaeus japonicus* (Spence Bate, 1888) in the Celtic Sea, English Channel, and North-West France. *BioInvasions Records* 2, 51–55.

Reddy, V.R., Reddy, P.P., and Kumar, U.H. (2004) Ecological and economic aspects of shrimp farming in Andhra Pradesh. *Indian Journal of Agricultural Economics* 20, 435.

Restrepo, L., Bayot, B., Betancourt, I., and Pinzón, A. (2016) Draft genome sequence of pathogenic bacteria *Vibrio parahaemolyticus* strain Ba94C2, associated with acute hepatopancreatic necrosis disease isolate from South America. *Genomics Data* 9, 143–144.

Rhodes, M.A., Yu, D., Zhou, Y., and Davis, D.A. (2015) Use of lipid extracted distilled dried grain with solubles (DDGS) in diets for Pacific white shrimp. *North American Journal of Aquaculture* 77, 539–546.

Rönnbäck, P. (1999) The ecological basis for economic value of seafood production supported by mangrove ecosystems. *Ecological Economics* 29, 235–252.

Rönnbäck, P., Crona, B., and Ingwall, L. (2007) The return of ecosystem goods and services in replanted mangrove forests: perspectives from local communities in Kenya. *Environmental Conservation* 34, 313–324.

Roy, L.A., Davis, D.A., Saoud, I.P., Boyd, C.A., Pine, H.J., and Boyd, C.E. (2010) Shrimp culture in inland low salinity waters. *Reviews in Aquaculture* 2, 191–208.

Rust, M.B., Amos, K.H., Bagwill, A.L., Dickhoff, W.W., Juarez, L.M., Price, C.S., Morris, J.A., and Rubino, M.C. (2014) Environmental performance of marine net-pen aquaculture in the United States. *Fisheries* 39, 508–524.

Santos, M.D. and Coelho, P.A. (2002) Espécies exóticas de camarões Peneídeos (*Penaeus monodon Fabricius*, 1798 e *Litopenaeus vannamei Boone*, 1931) nos ambientes estuarino e marinho do Nordeste do Brasil. *Boletim Técnico Científico do CEPENE* 10, 209–222.

Sathirathai, S. and Barbier, E.B. (2001) Valuing mangrove conservation in southern Thailand. *Contemporary Economic Policy* 19, 109–122.

Shao, J.W., Zhao, W., Han, S., Chen, Y., Wang, B., and Wang, L. (2018) Partial replacement of fishmeal by fermented soybean meal in diets for juvenile white shrimp (*Litopenaeus vannamei*). *Aquaculture Nutrition* 25, 145–153.

Sharawy, Z., Goda, A.M., and Hassaan, M.S. (2016) Partial or total replacement of fish meal by solid state fermented soybean meal with *Saccharomyces cerevisiae* in diets for Indian prawn shrimp, *Fenneropenaeus indicus*, postlarvae. *Animal Feed Science and Technology* 212, 90–99.

Shiu, Y.L., Hsieh, S.L., Guei, W.C., Tsai, Y.T., Chiu, C.H., and Liu, C.H. (2015) Using *Bacillus subtilis* E20-fermented soybean meal as replacement for fish meal in the diet of orange-spotted grouper (*Epinephelus coioides*, Hamilton). *Aquaculture Research* 46, 1403–1416.

Shrimp News International (2015) Fisherman catches twelve-foot-long giant tiger shrimp. Available at: https://www.shrimpnews.com/FreeReportsFolder/NewsReportsFolder/USAflFishermanCatchesTwelveFootLongTigerShrimp.html (accessed 6 May 2020).

Smith, D.M., Allan, G.L., Williams, K.C., and Barlow, C.G. (2000) Fishmeal replacement research for shrimp feed in Australia. In: Cruz Suárez, L.E., Ricque M., D., Nieto-López, M.G., Villarreal, D., Scholz, U., and González, M. (eds) *Memorias del V Simposium Internacional de Nutrición Acuícola*. Universidad Autónoma de Nuevo Leon, Monterrey, NL, Mexico, pp. 277–286.

Sookying, D. (2010) Development and application of soybean-based diets for Pacific white shrimp *Litopenaeus vannamei*. Doctoral dissertation. Auburn University, AL, USA.

South Carolina Department of Natural Resources (2008) South Carolina Aquatic Invasive Species Management Plan. Available at: http://www.dnr.sc.gov/invasiveweeds/aisfiles/SCAISplan.pdf. (accessed 9 February 2012).

Spies, R.B., Senner, S.E., and Robbins, C.S. (2016) An overview of the Northern Gulf of Mexico ecosystem. *Gulf of Mexico Science* 1, 98–121.

Stefanovich, I.L. (2000) *Safeguarding our Common Future: Rethinking Sustainable Development*. State University of New York Press, Albany, NY.

Stickney, R.R., Treece, G.D., and Tidwell, J.H. (2000) History of aquaculture. In: Stickney, R.R. (ed.) *Encyclopedia of Aquaculture*. John Wiley & Sons, New York, NY, 436–445.

Subasinghe, R.P. and Reantaso, M.G. (2008) The FAO/NACA Asia regional technical guidelines on health management for the responsible movement of live aquatic animals: lessons learned from their development and implementation. *Revue Scientifique et Technique Office International des Épizooties 27* 55–63.

Suresh, A.V., Kumaraguru Vasagam, K.P., and Nates, S. (2011) Attractability and palatability of protein ingredients of aquatic and terrestrial animal origin, and their practical value for blue shrimp, *Litopenaeus stylirostris* fed diets formulated with high levels of poultry byproduct meal. *Aquaculture* 319, 132–140.

Sustainable Fisheries Partnership (2018) Reduction fisheries: sustainable fisheries partnership fisheries sustainability overview 2018. Available at: https://www.sustainablefish.org/Publications?year=2018-01-01T00:00:00Z (accessed 4 June 2019).

Sutinen, J., Broadus, J., and Spurrier, W. (1989) An economic analysis of trends in the shrimp cultivation industry for Ecuador. In: Olsen, S. and Arriaga, L. (eds) *A Sustainable Shrimp Mariculture Industry for Ecuador*. Coastal Resources Center, University of Rhode Island, Narragansett, RI, pp. 19–44.

Tacon, A.G. (2020) Trends in global aquaculture and aquafeed production: 2000–2017. *Reviews in Fisheries Science & Aquaculture* 28, 43–56.

Tacon, A.G. and Metian, M. (2008) Global overview on the use of fish meal and fish oil in industrially compounded aquafeeds: trends and future prospects. *Aquaculture* 285, 146–158.

Tacon, A.G., Nates, S.F., and McNeil, R.J. (2005) Feeding strategies vary for shrimp producers. *Global Aquaculture Advocate* 8, 79–81.

Tang, C.P. and Tang, S.Y. (2006) Democratization and capacity building for environmental governance: managing land subsidence in Taiwan. *Environment and Planning A: Economy and Space* 38, 1131–1147.

Teng, S.K. (2008) Risk analysis of the soil salinisation due to low-salinity shrimp farming in central plain of Thailand. In: *Assessment and Communication of Environmental Risks in Coastal Aquaculture*. Joint Group of Experts on the Scientific Aspects of Marine Environmental Protection (UN/GESAMP). FAO, Rome, Italy, pp. 162–175.

Tourtip, S., Wongtripop, S., Stentiford, G.D., Bateman, K.S., Sriurairatana, S., Chavadej, J., Sritunyalucksana, K., and Withyachumnarnkul, B. (2009) *Enterocytozoon hepatopenaei* sp. nov. (Microsporida: Enterocytozoonidae), a parasite of the black tiger shrimp *Penaeus monodon* (Decapoda: Penaeidae): fine structure and phylogenetic relationships. *Journal of Invertebrate Pathology 102*, 21–29.

Tran, L., Nunan, L., Redman, R.M., Mohney, L.L., Pantoja, C.R., Fitzsimmons, K., and Lightner, D.V. (2013) Determination of the infectious nature of the agent of acute hepatopancreatic necrosis syndrome affecting penaeid shrimp. *Diseases of Aquatic Organisms 105*, 45–55.

Tran, L.X. and Fischer, A. (2017) Spatiotemporal changes and fragmentation of mangroves and its effects on fish diversity in Ca Mau Province (Vietnam). *Journal of Coastal Conservation* 21, 355–368.

Urbina, I. (2015) 'Sea Slaves': The human misery that feeds pets and livestock. *The New York Times* (27 July), p. 27. Available at: https://www.nytimes.com/2015/07/27/world/outlaw-ocean-thailand-fishing-sea-slaves-pets.html (accessed 16 July 2021).

Villareal-Cavazos, D.A., Ricque-Marie, D., Rodriguez, A.P., Lopez, M.N., Salazar, M.T., Lemme, A., Delgado, J.G., and Cruz-Suarez, L.E. (2014) Apparent digestibility of dry matter, crude protein, and amino acids of six rendered by-products in juvenile *Litopenaues vannamei*. *Ciencias Marinas* 40, 163–172.

Wakida-Kusunoki, A.T., Rojas-González, R.I., González-Cruz, A., Amador-del-Ángel, L.E., Sánchez-Cruz, J.L., and López Tellez, N. (2013) Presence of giant tiger shrimp *Penaeus monodon* Fabricius, 1798 on the Mexican coast of the Gulf of Mexico. *BioInvasions Records* 2, 325–328.

Walker, P.J. and Mohan, C.V. (2009) Viral disease emergence in shrimp aquaculture: origins, impact and the effectiveness of health management strategies. *Reviews in Aquaculture* 1, 125–154.

Wang, J. (2018) OIE international standards on aquatic animals. In: Tendencia, E.A., de la Peña, L.D., and de la Cruz, J.M. (eds) *Aquatic Emergency Preparedness and Response Systems for Effective Management of Transboundary Disease Outbreaks in Southeast Asia: Proceedings of ASEAN Regional Technical Consultation*. Aquaculture Department, Southeast Asian Fisheries Development Center, Tigbauan, Iloilo, Philippines, p. 80.

World Commission on Environment and Development (1987) *Our Common Future*. World Commission on Environment and Development and Oxford University Press, Oxford, UK.

Yang, Q., Zhou, X., Zhou, Q., Tan, B., Chi, S., and Dong, X. (2009) Apparent digestibility of selected feed ingredients for white shrimp *Litopenaeus vannamei*, Boone. *Aquaculture Research* 41, 78–86.

Zhan, W.B., Wang, Y.H., Fryer, J.L., Yu, K.K., Fukuda, H., and Meng, O.X. (1998) White spot syndrome infection of cultured shrimp in China. *Journal of Aquatic Animal Health* 10, 405–410.

Zhou, Y., Davis, D.A., and Buentello, A. (2015) Use of new soybean varieties in practical diets for the Pacific white shrimp, *Litopenaeus vannamei*. *Aquaculture Nutrition* 21, 635–643.

Zu-Ermgassen, E.K., Phalan, B., Green, R.E., and Balmford, A. (2016) Reducing the land use of EU pork production: where there's swill, there's a way. *Food Policy* 58, 35–48.

Chapter 2

AQUACULTURE INSURANCE

CÉSAR REAL R. LEGÍSIMA AND JASON SCOTT

Summary

Aquaculture insurance has been around since the 1970s but remains an unconsolidated sector of the insurance market. Within the sector, different species face different challenges, and as aquaculture is developing rapidly in comparison to other agri-industries, it is constantly employing new technologies, which create new risks and thus new challenges for insurers. Their approaches regularly have to change to reflect the risk levels inherent in each new or modified industry process. This chapter will take the reader on a short journey from the early origins of insurance to the present day, describing its function, constraints, current solutions, and proposals for the future – both for the aquaculture industry as a whole and for the shrimp industry in particular.

An overview of insurance

ORIGINS OF INSURANCE

We can link the origin of insurance with the origin of Lloyd's of London. Lloyd's evolved in the 17th century in the London coffee house of a Mr. Lloyd, which became a meeting place for merchants, traders, ships' captains, and business people. The very first mention of Lloyd's goes back to 1688, in an article that declared a reward for five stolen watches; it encouraged anyone with information to contact Lloyd at his coffee house. This was an early example of a nascent insurance industry.

Lloyd's Coffee House became a place to exchange information about shipping and the success and failure of the merchant ventures of that time. It became an essential location where risk-spreading activities were carried on, underpinning the same business ventures. "Underwriters" (as they became known) at Lloyd's Coffee House started to enjoy high profits from their risk-spreading activities, especially during the Seven Years' War (1756–1763). When the war came to an end, marine insurance premiums returned to lower levels, driving these underwriters to become more speculative and to widen the scope of their activities. The evolving market thus opened many other business lines, covering a much greater range of insurable interests. The following two centuries saw a continuous improvement in professionalism, regulation, and good practice, ultimately leading the insurance industry to become an indispensable part of modern society. The industry's overall contribution to society has become so wide that it is difficult to find an industry or sector of society that has not benefited from insurance.

PURPOSES OF INSURANCE

For a better understanding of the importance of insurance and its relevance, we highlight its two main purposes.

- **Economic purpose.** The objective of the insurance process is to compensate an insured individual (or business) who has suffered a loss, by returning them to the position they would have been in if the loss had not occurred. It is important to recognize that the purpose of insurance is *not* the enrichment of the insured! Insurance is designed to be an effective tool for enabling people and businesses to protect themselves against a wide variety of risks and hazards, putting them in a position of being no better or worse off, just equal. In this way insurance, if well designed, permits peace of mind and financial security.
- **Social purpose.** Insurance requires wide participation. It works by collecting numerous premiums into a fund from which losses can be compensated. As the number of insureds grows, and thus the pool of funds grows, its stability increases and its efficiency at safeguarding the wealth and fortune of insureds improves and becomes more secure.

Insurance has many side benefits, including making a major contribution to society through its risk-management approach. It has always been in insurers' interest to try to eliminate and reduce the effects of the risks they provide coverage against. As a result, throughout the history of insurance, there has been a tremendous risk-management feedback, to the overall betterment of society. Sprinkler systems in buildings and restrictions on pilots' flying hours are two examples, among very many, of where the lessons learned by paying insurance claims have caused underwriters to demand that risk-improving activities be implemented. This is a very important aspect of insurance, from which aquaculture has already benefited and stands to benefit greatly in the future.

REQUIREMENTS OF INSURANCE

To work properly and effectively, insurance has to be subject to some basic principles and fundamental conditions.

- **Unpredictability.** Insured risks must be unpredictable and accidental. If they are not, insureds could benefit from making claims. As underwriters' objective is to make a profit from the service they provide, they could tolerate a situation in which individuals/businesses seek to insure themselves against predictable and non-accidental risks only by increasing premiums.

- **Economic interest.** The insured must always have a direct economic interest in an insured risk. As mentioned earlier, the purpose of the insurance contract is to reinstate an insured's financial position, as if the loss has not occurred. If the insured has no interest in the insured item, by definition, no actual loss can be suffered by the insured.

CALCULATION OF INSURANCE PREMIUMS

The insurance premium is the amount the policyholder must pay according to the terms agreed in the policy. At a technical level, premiums comprise three main components.

- **Actuarial risk premium.** The calculation of risk premium can be simplified as: probability of an occurrence (P) multiplied by the average cost of the claim (A), where P = number of claims / number of similar risk insurance policies, and A = total cost of claims / number of claims.
- **Other insurance-related expenses** must be taken into account, including administration costs, costs of surveyors' and loss adjusters' fees, capital costs, and other business expenses of insurers.
- **Expected returns**. Insurance is like any other business, in that insurance underwriters and their shareholders expect the business to be profitable and to provide a return on their investments.

TYPES OF RISKS FACED BY OPERATORS

Insurance is about risk, and in principle there are three basic categories of risk.

- **Preventable risks.** These are internal risks arising from within the organization. They are controllable and should be eliminated or avoided. In other words, companies do not get any strategic benefits from paying the premium to insure these risks when they can be eliminated.
- **Business or strategy risks.** A company voluntarily accepts some risk in order to pursue its economic objectives.
- **External risks.** These are risks that arise from events outside a company that are beyond the company's influence or control. Natural catastrophe risks such as floods, storms, and earthquakes fall into this category – a category that these days is heavily affected by climate change.

Not all risks can be efficiently managed by insurance. Risks that are somehow predictable or avoidable would be tackled more efficiently with suitable management and business practices, and (particularly in the case of aquaculture) by adequate initial capitalization of the business. Taking the real estate industry as an example of management responsibility, building apartments in a hurricane-prone region, using materials or construction techniques that cannot resist strong winds, would not make business sense, and such structures would be uneconomically expensive to insure.

Aquaculture insurance

ORIGINS OF AQUACULTURE INSURANCE

Insuring fish farming began in the 1970s alongside the newly developing salmon cage farming industry. Many of the industry's insurable interests were quite familiar and normal to insurers. Fish farmers have buildings and equipment such as cages, pontoons, and boats, as well as the nets and processing equipment that all aquaculture businesses use. They also have liability and health and safety exposures, and many other insurable interests that, although individually handled very differently as far as aquaculture is concerned, are common to other industries and can be insured relatively easily in conventional insurance markets.

AQUACULTURE STOCK INSURANCE

One insurable interest that fish farmers have is quite unique to their business – their stocks of aquatic livestock and plants. Arguably, the stock is a fish farmer's most important asset. For this reason, the balance of this chapter focuses on insuring aquaculture stock, especially stocks of shrimp,

The first aquaculture stock policies offered insurance against "named perils". Subsequently, "all risks" policies were developed. As aquaculture was a new industry at the time, farming methods and technology were (and still are in some sectors) comparatively experimental. Thus, some of the risks to stock in these sectors fall into the category of "inevitabilities" and may be excluded by underwriters. For example, as new species are brought into husbandry – and in aquaculture there are many possible candidate species and plants for future cultivation – risks such as disease may well be difficult to insure. As the industry and the husbandry practices used within it evolve, production technology will stabilize and excluded perils may become insurable. Continuous investment and research will enable many sectors of aquaculture to gain better control of many of their risks, allowing insurers to consider bringing more sectors of aquaculture and their risks under the umbrella of insurance cover.

Aquaculture stock valuation

Most aquaculture stock policies are written on an "indemnity basis"; that is, every individual animal is accounted for and insured at an agreed value. The value is usually based on the sum of the production costs, plus an element of profit, the level of which is generally negotiable.

In the stock insurance process, information is provided by the individual/business seeking insurance (the proposer) completing one or more proposal forms for the production centers involved. The proposal forms contain questions about each operation and the risks it faces. The questions must be answered in detail, and the forms must be submitted with accompanying pictures, maps, diagrams, and charts. The proposer *must* disclose all relevant information, including any material information that may not be asked for in the proposal form itself. Proposal forms must be signed by applicants, and they form a part of any insurance policy issued. In addition, underwriters may require a survey of the site(s) to be carried out by a specialist surveyor.

A specialist underwriter will assess the information provided in the proposal form and accompanying information (and, if applicable, survey reports) and develop a risk profile for each growing site, each species, and for the overall management of the operation. At the end of this evaluation process, the underwriter will determine what rate and associated conditions (e.g. exclusions if the underwriter thinks that some risks are too high to be covered) are appropriate for the coverage offered. Generally, the rate is then applied to the projected average total stock value at risk, to calculate a deposit premium. The deposit must be paid to implement the insurance policy. During the term of the policy, the insured is required to provide monthly declarations of the actual stock value on each insured site. At the end of the policy period, the premium is adjusted to account for any difference between the projected (i.e. the deposit) and actual average stock value on site.

It is important to re-state that indemnity-based insurance is underpinned by the ability to accurately quantify the number and/or biomass of the insured stock at any point in time. This is crucial for the purposes of both premium calculation and loss settlement. It does not present a problem with some farming operations (e.g. salmon or seabass), where the stock can be quite accurately counted and whose size is relatively consistent and can be measured regularly and recorded. However, it does pose problems for farming operations that rear huge numbers of animals of individually very low value, into which category shrimp farming falls, along with a number of other aquaculture species.

Different stock insurance options

As mentioned earlier, aquaculture indemnity-based stock policies can be written on either an "all risks" or a "named perils" basis. There are important differences between these types of policies.

- **"All risks" policies** are generally more expensive, but are advantageous to the insured in the important area of "proof of loss". In the event of a loss under an all risks policy, the burden of proof to show that a loss is *not* covered falls on the insurer. In other words, a loss is covered unless the insurer can show that it is specifically excluded by the policy wording.
- **"Named perils" policies** offer cover against a restricted list of specific perils that are insured. Losses due to disease, storm, mechanical failure, drought, flood, earthquake, and fire (which has been known to cause the death of aquatic creatures in water!) are examples of "named perils". The insured can choose to purchase cover against any or all of these perils, and maybe others. But, very importantly, in the event of a loss, the burden of proof is on the insured to demonstrate that the loss was due to one of the specific insured perils listed in the policy. In aquaculture, where everything is in water and mostly out of sight, proving what has happened can be difficult.

An alternative to indemnity-based cover is parametric, or index-based, cover. This functions in a fundamentally different way to indemnity cover. Importantly, it does not require quantification of stock for either premium or loss calculation. Parametric policies also have the advantage that they do not require comprehensive proposal information or assessment of individual farms.

The basic principle of index-based cover is very simple. A predetermined payment is made to the insured if a predetermined event occurs. Applicable such events in the aquaculture class are generally quantifiable natural perils, such as flood, drought, extreme temperature, windstorm, and so on. For example, a wind speed of 175 km/h (approximately 110 miles per hour) could be set as the predetermined parameter, to be verified by a nearby reliable and independent weather station that is specified in the policy terms. Loss settlement can be very rapid. Continuing the above example, if the wind speed equals or exceeds the specified 175 km/h, the agreed payment is made.

Once such an arrangement has been set up, administrative demands are low, so it is feasible to offer cover to large numbers of farms without the need for quantification of stock. However, there are significant restrictions. Only perils that can be quantified and independently verified can be covered. This means that disease cannot be covered, and there must be an independent historical dataset available for each covered peril that is sufficiently long and complete for insurers to conduct their statistical analysis (e.g. from a nearby weather station, as in the above example). There must be an independent recording method that can be used to determine when a trigger is exceeded (e.g. again, a nearby weather station) and there must be a direct correlation between trigger events and actual losses suffered by the insured.

Sources of stock insurance cover

There are basically three categories of insurance providers: conventional stock companies, state schemes, and mutual insurance companies. We will discuss each in turn.

Conventional stock companies and Lloyd's of London

The stock companies are, generally speaking, either publicly quoted companies or part of publicly quoted conglomerates. Lloyd's of London is structured differently, but Lloyd's and the stock companies all exist to make underwriting profits for their owners and shareholders. Apart from Lloyd's, which is based in London, stock companies exist in almost every country in the world. They do not necessarily underwrite aquaculture stock policies, which is a highly specialized insurance class, but many of them will insure the more usual coverages aquaculture practitioners need, such as buildings, equipment, and vehicles.

State schemes

These are "pools" (i.e. groups) of operators and their farms, in either the agriculture or the aquaculture sector. They are normally established by a government, which implements, by statute, an insurance code facilitating the establishment of a management company, and the setting out of insurance provisions for transferring risk in a reasonable way. Governments bring in outside insurers and reinsurers whenever they can.

The management company is shared by conventional insurance companies, which participate in the "pool". The pool is operated on the basis that aquaculture insurance risk is first offered to a company that is participating in the pool, and the company presents it to the pool. The pool sets the premium, terms, and conditions of the insurance. The company then has the option of either passing 100% of the premium for the risk to the pool, or retaining a share of a risk from the pool, for its own account. For example, a company might keep 10% of the risk for its own account, and pass 90% to the pool.

One of the main objectives of a pool is to provide coverage for catastrophe risks (systemic risks) that cannot be handled by the insurance market because the risk is too high and the potential losses too great. Secondary objectives include encouraging the participation of the international reinsurance market, preventing unfair pricing competition, using local human, financial, and information resources to provide insurance, and encouraging the take-up of insurance in the aquaculture and agriculture sectors.

There are also some disadvantages of this approach. The pool is not solely designed to benefit policyholders (i.e. producers) so there is the potential for separation of policyholders' interests and needs from pool objectives. The pool is reliant on ongoing government support, both legislative and financial.

One weakness of a pool scheme is that there tends to be a disconnect between good risk practice, low claims, and the cost of cover. For example, if all the business goes into a pool, who decides the pricing for individual policyholders? Is the pricing a broad-brush approach with, for example, the 20% of claimants who produce 80% of the claims (based on the Pareto principle) being allowed to get the same terms and conditions as other, lower-risk, insureds, and these higher-risk, lower-quality operators being able to continue to make claims without penalty?

Mutual insurance companies

A mutual insurance company is an insurance company that is owned entirely by its policyholders (who are called "members"). Profits earned by a mutual insurance company belong to its members. Profits can be rebated to members in the form of dividend distributions or reduced premiums, or they can be retained and used to the benefit of members and their businesses, such as for risk management and other research that benefits members, or other uses that members agree to. This contrasts with the alternative to a mutual insurance company, a stock insurance company, which is owned by investors who have purchased stock in the company. Profits generated by a stock insurance company are distributed to the investors without necessarily benefiting the policyholders.

The attitude to claims of a mutual insurance company is very different from that of a conventional stock company. The mutual will try to make sure that members get a favorable settlement, whereas a stock company will always try to reduce claim settlements, as that will benefit its shareholders and investors. However, a mutual generally requires sufficient membership to generate the scale and spread of risk required to attain "critical mass" and sustainable financial outcomes. Cover beyond the appetite of the insurance and reinsurance marketplace (principally systemic risks) would require government support. Governments are generally supportive of mutual companies because they generally are more socially involved and active than traditional insurance companies.

Snapshot of the world's aquaculture market production

The total world farm-gate sale value of aquaculture production reached US$263.6 billion in 2018. Global aquaculture production of fish, crustaceans, mollusks, and other aquatic animals reached 82.1 Mt. Wild catch has been relatively stable since the 1980s, so nowadays more than 50% of the fish for human consumption comes from aquaculture. Continuous growth is predicted, as aquaculture production will need to double by the mid-21st century in order to meet the world's protein demand.

Asia accounts for nearly 90% of global aquaculture production, and China alone for 62%. Other Asian countries such as Vietnam (3.6 Mt) and Indonesia (16.6 Mt) currently produce more than Europe (3.08 Mt) or the Americas (3.79 Mt).

SNAPSHOT OF THE AQUACULTURE INSURANCE MARKET

It is generally acknowledged that only 5–7% of the world's fish farms are currently insured. A large proportion of farming operations are considered high risk and therefore almost uninsurable using the traditional insurance products.

Risk analysis and management is limited across the sector, and even where an organization conducts risk analysis and management, this has not protected insurance companies from losses, which has resulted in increased premiums. Insurers have experienced a number of years of poor results due to a reducing premium pool and a high frequency of large events. This has led to a number of insurers leaving the market or concentrating their capacity into specialist underwriting. Currently, almost no premium at all is coming from Southeast Asia into the direct global underwriting market; however, an increasing amount of aquaculture business is being written by domestic insurers.

The premium of the direct global aquaculture underwriting markets is highly concentrated in:

- **species:** salmon, seabass, seabream, tuna
- **regions:** Chile, Scandinavia, Canada, the Mediterranean.

Owing to this concentration, a single event, such as the 2016 algae bloom in Chile, can be enough to cancel out the entire annual premium income of the global aquaculture underwriters.

The shrimp aquaculture insurance market

Many people involved in insuring and risk managing aquaculture point to the fact that the main sequential losses of a biological nature – due to diseases, parasites, feed deficiencies, and so on – have occurred in the salmon industry, and more recently also the shrimp industry, on a regular basis. No sooner has one problem manifested itself and a solution been found than another one comes along and causes more problems – which are often expensive for insurers. This is certainly true of one of the biggest sectors of aquaculture, the salmon industry. For those involved in aquaculture of other species, it is hard to argue that their particular species is not going to follow the path of the salmon industry. This is particularly important as far as shrimp production is concerned.

The shrimp industry has experienced major biological problems. These include the disease early mortality syndrome (EMS), subsequently named acute hepatopancreatic necrosis syndrome (AHPNS), which devastated shrimp crops in most parts of Asia, and also in Mexico, causing massive financial losses. At the time of writing (2020) there are indications that another "situation" is developing in the Indian shrimp industry; a new problem has appeared, which is being called "continuous mortality syndrome". The danger this poses is unknown at this stage, but parallels with EMS are being drawn. The only thing to be said is that the shrimp industry is much wiser and better organized than it was when EMS started to be noted. The shrimp-producing nations are much more aware, first, of the national importance of their shrimp industries, and second, of the biosecurity steps that they must take to avoid the geographic spread of pathogens across wide sectors of the industry.

These threats and disasters pose real problems for insurers. They reinforce the perception that aquaculture is "very high risk". Consequently, the options for shrimp farmers seeking to obtain cover against loss or mortality of their stock are extremely limited. The global aquaculture insurance market has suffered over the past few years and at the time of writing no insurer is offering traditional indemnity-based cover to shrimp farmers, due to the perceived high-risk nature of the industry as well as the difficulties of accurately quantifying the stock at any point in the production cycle.

With specific reference to stock mortality cover for shrimp-farming operations, there have been various programs offered in various parts of the world (e.g. China, Vietnam, Mexico) at one time or another, but none of them have been successful as they have generally cost insurers heavily and have therefore been short-lived. There are various reasons why these programs have failed, but the one fundamental issue with indemnity-based programs covering shrimp is accurate quantification of each loss.

Indemnity-based cover requires the ability to accurately quantify the number and/or biomass of the insured stock at any point in time, for both premium calculation and loss settlement. This does not present a problem with farming operations that use relatively small stock-holding units containing tens of thousands or even a few hundred thousand individuals per unit. The number of stock first introduced is accurately known and can be reduced (and recorded) each time a normal trade mortality occurs. This is generally true in the case of most finned fish species, where carcasses can usually be easily recovered. Equally, if a substantial loss occurs there are either carcasses to count in the case of mortality, or the remaining number of fish to count in the event of escape or theft.

The necessity for accurate quantification of stock at any point in time severely restricts the application of indemnity-based cover to various sectors of the global aquaculture industry, including the majority of shrimp-farming operations, where there are millions of individuals and, in the event of any mortalities, carcasses are swiftly consumed by the survivors. In such operations, effectively, the only time the biomass is really accurately known is at harvest. There are, of course, various methods of sampling in order to estimate shrimp biomass; however, insurers are generally not confident of their accuracy, and as they require sampling to be conducted by an independent third party this would incur additional cost, which is likely to make the cover economically unviable.

Equally, it is theoretically possible to offer cover on a total-loss-only basis; however, that would not completely negate the issues related to quantification, and insurers feel it may introduce a moral hazard, as completely accurate measurement of biomass can still occur only at harvest. It is also the case that in even the most disastrous of events there tend to still be some stock left, which necessitates defining what a "total loss" is.

The future of aquaculture insurance

Before discussing changes to the current shrimp aquaculture insurance market, it is useful to recap the present situation.

- World aquaculture production is highly concentrated in Asia (90%) with China playing the leading role (60% of global production alone).
- The vast majority of global aquaculture production still uninsured. Although there is no exact data, it is estimated that the insured proportion is only around 5–7%.
- The salmon industry is the most widely insured sector, as it is a mature industry with proven technology and experience, the current insurance products were specifically designed for it, and third-party investment is common; hence it is attractive to insurers and insureds alike. However, as salmon can be harvested only in certain regions, the aquaculture insurance industry does not have wide geographic spread and is therefore exposed to concentrations of risks. There is a need for diversification, but it must be profitable for insurers.
- Aquaculture indemnity-based insurance is a very technical risk to underwrite, and it is done on

a case-by-case basis. This has led to a concentration of knowledge in the regions where sea cage aquaculture occurs, which has limited its application in other sectors.

- Aquaculture insurers have suffered large loss events, at times when there are insufficient premium reserves to absorb them. This has led to insurers withdrawing from the class and the remaining insurers raising premiums and introducing restrictions on cover. This limits the cover available to new farmers and hampers the insurance industry itself.
- For shrimp insurance, stock health and disease control, along with the quantification of losses, represent the major barriers.

How to move forward?

Hopefully the reader has started to realize what is in our opinion the only way to break through the current limitations of aquaculture insurance and, consequently, for the insurance industry to be able to provide meaningful, affordable (and profitable) insurance products for shrimp farmers. We propose two lines of action: (1) making the risk attractive and sustainable, and (2) taking advantage of developments in computer and communications technology.

THE "WHAT": APPROACHES TO MAKE THE RISK ATTRACTIVE AND SUSTAINABLE

Diversification

Insuring shrimp species would help diversify the geographic exposure of insurance carriers, expanding risks into Asia, Central America, and different parts of North and South America. It would dilute the focus on salmon and give insurers a broader base from which to maintain a consistent market presence.

Adaptation

Generally speaking, aquaculture insurance policies have developed around the needs of particular species, hence the current geographic distribution, and a particular type of farmer (large corporations). Such policies are not tailored toward the needs of other species, such as shrimp, which is a very large market.

Mass consolidation

It has been demonstrated that one of the key success factors of the industry in Norway is linked to its concentration and collaboration among peers. Size brings economies of scale that allow research and development, meaningful investments, and easiness of policy coordination. Although private aggregation requires time, it is also true that strong government or private clusters of bodies could play an accumulator role in these types of matters.

Global economic approach

In line with the previous point, subsidies should be focused not on isolated actions but on the whole local aquaculture industry, with a clear aim of making the "whole risk" safer and more secure. Examples could be promoting and actively guiding certification, setting up national health and biosecurity programs, granting access to the required technology (as discussed in the next section of this chapter), unified research and development programs, designing practical educational programs for the small farmers, and, of course, playing a representative role in front of aquaculture underwriters.

Pooling expertise

It is not just insurers that should be leading the way. Global brokers have been involved in the aquaculture market for many years, but most deal with clients on an individual basis and have done little to pool the various silos of knowledge held in offices around the world. Solutions toward the aggregation and exchange of information between the current clusters of local expertise might be something to reflect upon.

THE "HOW": WHY NEW TECHNOLOGY MIGHT BE THE ANSWER

Things that not so many years ago seemed impossible, or at least economically prohibitive, have radically changed. Technology, software, and other related issues, combined with the dramatic reduction in cost of access, have completely reshaped frontiers and barriers in almost all industries. We think that shrimp insurance is no different, and it could enormously benefit from this new era. Becoming an attractive risk may be more feasible than one would think, with the help of the Internet of Things (IoT), artificial intelligence (AI), mobile phones, or even blockchain if it is applied in a sensible and coordinated way.

The Internet of Things

The IoT is an array of combined connected devices, which collect reliable data in real time based on which immediate conclusions are obtained for decision making, automated action, among other things. Some possible applications of IoT technologies are:

- underwater drones that permit real-time visual underwater inspection
- flying drones equipped with advanced optical and thermal sensing devices that can survey and map remote locations almost immediately
- advanced and extremely cheap underwater sensors that can monitor almost any water quality parameter.

Blockchain

Structurally Blockchain is just a log or ledger recording events that happen over time. Its beauty is that it solves three key issues in a row: transparency + accountability + trustworthiness. Blockchain is perfect for situations where more than two players need to trade together but neither entirely trusts the other. It is also possible to include smart (auto-triggered) contracts linked to insurance and financing. For example, a farmer's data relating to feeding, water temperature, water components, and other parameters can be collected via IoT devices, introduced into a blockchain scheme, and be visible to insurers (anonymously). If that were to be achieved, the information gap that now dominates the insurance of shrimp farming would be cancelled out.

Artificial intelligence

Artificial intelligence (AI) can help to detect patterns, find hidden correlations, and even predict likely events. AI could be a powerful resource if fed correctly with the correct data (another reason why aggregation of data is important), leading to better understanding and control of the major risks that have been punishing the industry for so long.

Modern distribution (mobile phones and clusters)

As stated earlier, a great deal of aquaculture, and shrimp farming in particular, occurs in highly populated developing countries. In many of these countries we find the paradox that mobile communications technology is more widely spread than banking. In other words, having access to a mobile phone is easier than having a bank account.

This characteristic should be seriously considered when designing and distributing an effective insurance product. Enabling the purchase of aquaculture insurance through mobile phones could make a significant difference in reaching the desired target population.

Collaboration agreements between the top tech schools

It is difficult for a small, remote shrimp farmer to access the best human capital, but there is no doubt that a government or national association could easily achieve this. Size matters, but what matters most is consistency. It is not enough that one single farmer could have the best-in-class technical solutions; a minimum standard must be guaranteed right across the industry in order to make such solutions attractive. An initiative (public or private) of collaboration agreements between the best global technical schools aimed at solving the technological challenges and requirements described above, in an efficient and effective way, would be a clear winner.

SOME INSPIRING EXAMPLES

- **Worldcover:** This US-based company, which specializes in insuring farmers in developing countries, has developed an insurance product based on blockchain technology. Farmers in Uganda, Kenya, and Ghana now have access to an affordable crop insurance product that is paperless and where payouts are triggered automatically in the event of adverse weather conditions.
- **Zipline:** This company has developed the world's first automated, on-demand blood delivery service, using electrically powered autonomous drones (with a range of 110 km) in Rwanda. If something as delicate as human blood can be safely delivered in this way, almost any other component could be.

- **ecoRobotix:** Based in Switzerland, this company has developed a lightweight, eco-friendly robotic weeding machine. The machine is solar powered, and works up to 12 h a day injecting micro amounts of herbicides precisely and directly into weeds growing in agricultural fields. It is controlled by phone and is otherwise autonomous. A similar solution could be used to release the requested inputs on to shrimp ponds in exact measures and in specific zones.
- **IBM Food Trust:** Launched in 2018, the Food Trust is a blockchain-based food traceability platform for global use by retailers, wholesalers, and suppliers across the food supply chain. Following successful testing with major multinationals such as Nestle, Walmart, Carrefour, Tyson Foods, Kroger, and Unilever, IBM has decided to make the cloud solution available to everyone. Participants have insights into each item being tracked from the farm to the factory, to the shipper, and to the retail store.

Conclusions

Aquaculture insurance is like any other industry: it needs to be profitable in the long term to survive! In the current situation, a justifiable question can be raised about what the future might bring if no changes are introduced. A leap forward is needed and, as in previous times in history, answers to these issues might come not from inside the industry but from adjacent ones. Technology has been proving its potential added value to almost all production process, and right now we are immersed in a new wave of developments that could drastically change the state of the aquaculture insurance industry. Aquaculture cannot be ignored because there is a real increasing demand for animal protein for human consumption – a demand that is almost impossible to meet without it. However, for the aquaculture industry to develop efficiently and rapidly, it needs insurance. Problems that were until very recently unresolvable or economically unbearable may now be solvable. New technologies such as IoT, blockchain, AI, or mobile phones can help to widen access, help with data gathering, and make risks more attractive. But for that change to happen, all parties involved need to widen their view and start looking further afield for solutions to problems.

Further reading

Harvard Business Review (2020) How to lead in a time of crisis. HBR Special Issue (Summer 2020).

FAO (2020) The State of World FAO, Fisheries and Aquaculture (SOFIA), 2020. Available at: http://www.fao.org/publications/sofia/2020/en/ (accessed 23 July 2021).

Kaplan, R.S. and Mikes, S. (2012) Managing risks: a new framework. *Harvard Business Review* (June). Available at: https://hbr.org/2012/06/managing-risks-a-new-framework (accessed 23 July 2021).

Chapter 3

FUNCTIONAL MORPHOLOGY OF THE NEPHROCOMPLEX

GAËTAN M.A. DE GRYSE AND HANS J. NAUWYNCK

Summary

The shrimp antennal gland or nephrocomplex is a complex organ with extensions throughout the cephalothorax. The organ follows the basic blueprint of other crustacean excretory organs, but has many morphological alterations, including an extensive network of bladders and an expansion of the labyrinth. The gland performs many functions, such as the regulation of body volume, acid/base, and ion content, and plays a major role in the molting process and possibly in social functions.

Introduction

In contrast to mammals, crustaceans seem to have multiple organs responsible for excretion, osmoregulation, and ion metabolism, namely the antennal gland (also known as the green gland, renal gland, and antennary gland), the maxillary gland, and the gills. The gills are known not only for their respiratory function, as the site where oxygen and carbon dioxide are exchanged between the water and the hemolymph, but are also important for ammonium regulation, osmoregulation, calcium uptake, and acid/base regulation (Ahearn et al., 2004; Bauer, 1999; Foster and Howse, 1978; Freire et al., 2008; McMahon, 1995; Taylor and Taylor, 1992). Except for the respiratory function, the antennal gland and the maxillary gland perform more or less the same functions as the gills. However, the antennal gland and the maxillary gland do not appear together in the same species (Potts and Parry, 1964). They both have a similar blueprint, cytology, and function, but when the organ leaves the animal at the base of the antennal coxipodite it is called the antennal gland, whereas when it opens at the maxillae it is called the maxillary gland. Shrimp possess antennal glands (Fig. 3.1).

The crustacean excretory organ is bilaterally built and can generally speaking be subdivided into three main parts: (i) a coelomosac, (ii) an efferent duct, and (iii) a terminal duct, leading to a nephropore (Freire et al., 2008; Icely and Nott, 1979; Young, 1959). The coelomosac is of mesodermal origin and is responsible

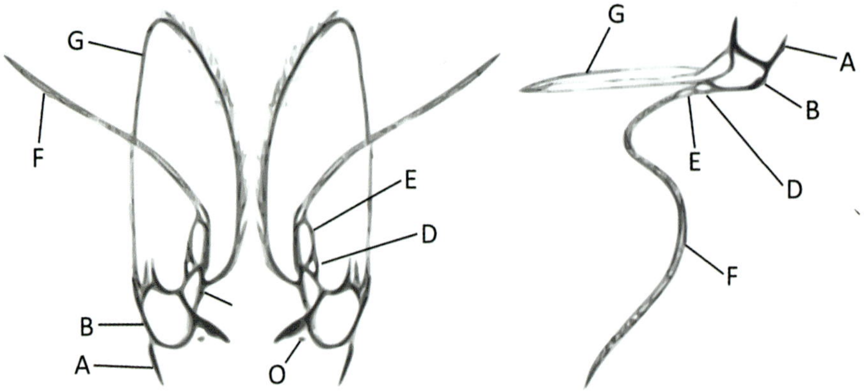

FIGURE 3.1 Drawing of the antennal appendix of the shrimp *Penaeus vannamei*. A: coxipodite, B: basipodite, C: ischiocerite, D: merocerite, E: carpocerite, F: flagellum, G: scaphocerite, O: nephropore, A+B+C+D+E: pediculum of the antenna, A+B: protopodite of the second antenna, C+D+E, endopodite of the second antenna. Credit: De Gryse et al. (2020).

for primary ultrafiltration of the hemolymph; it is thought to be a remnant of the coelom. Next to the coelomosac is the efferent duct. In decapods, this has evolved into an anastomosing network of tubuli, called the labyrinth (Peterson and Loizzi, 1973). The terminal duct sometimes has a urinary bladder, but always leads to the exit pore. Although the excretory organ in shrimp opens at the base of the antennal coxipodite, it has nothing to do with the antennae itself. Moreover, the excretory organ is much more complex in penaeid shrimp compared with what is described in other crustaceans. Also, the description "gland" does not fit the majority of the organ's functions. Therefore, a new name, the nephrocomplex, was proposed for penaeids (De Gryse et al., 2020). Many hiatuses still exist concerning the different structures of the nephrocomplex. Much of today's knowledge is extrapolated from other crustaceans, mainly crayfish and lobster. The aim of this chapter is to describe the full anatomy, morphology, and histology of the nephrocomplex in shrimp, while providing information on (possible) functions of the compartments of the organ. We hope that this chapter might provide a strong encouragement for other researchers to dive into the functionality of this organ, as there is much left to discover.

General anatomy

The morphology of the penaeid shrimp nephrocomplex is quite complex and consists of multiple sections (Fig. 3.2). The three main divisions found in other crustaceans (hemocele, labyrinth, and terminal duct with urinary bladder) can also be found in penaeid shrimp, but several modifications have been found reaching throughout the entire cephalothorax. The coelomosac and labyrinth are gathered together in a compact glandular compartment (CGC), except for the rostral compartment (RC), which is a continuation of the labyrinth's tubuli reaching as far as the supra-esophageal ganglion. These tubuli eventually return to the CGC, before flowing into a ventral urinary bladder (VB), the first part of the terminal ductus. A small duct leads to the outside world via the nephropore. Apart from some ramifications of minor importance, the VB continues also caudally via an extension. This extension of the VB forms a pre-esophageal connection (PEC) with the contralateral VB. From this PEC, a square duct gives rise to the median compartment (MC). The MC is a voluminous structure, which is located rostrodorsally to the stomach and can be subdivided into two lobes, a ventral lobe (VL) and a dorsal lobe (DL). To the lateral side of the esophagus, the MC continues as the lateral compartment (LC) laterally flanking the gut, and in turn it is flanked by the large protocephalon adductor. The extension of the VB is also connected to the LC at the lateral side of the esophagus. Caudally, the LC diminishes in height to form the caudal extensions (CE), which end blindly at the height of the hepatopancreas (De Gryse et al., 2020).

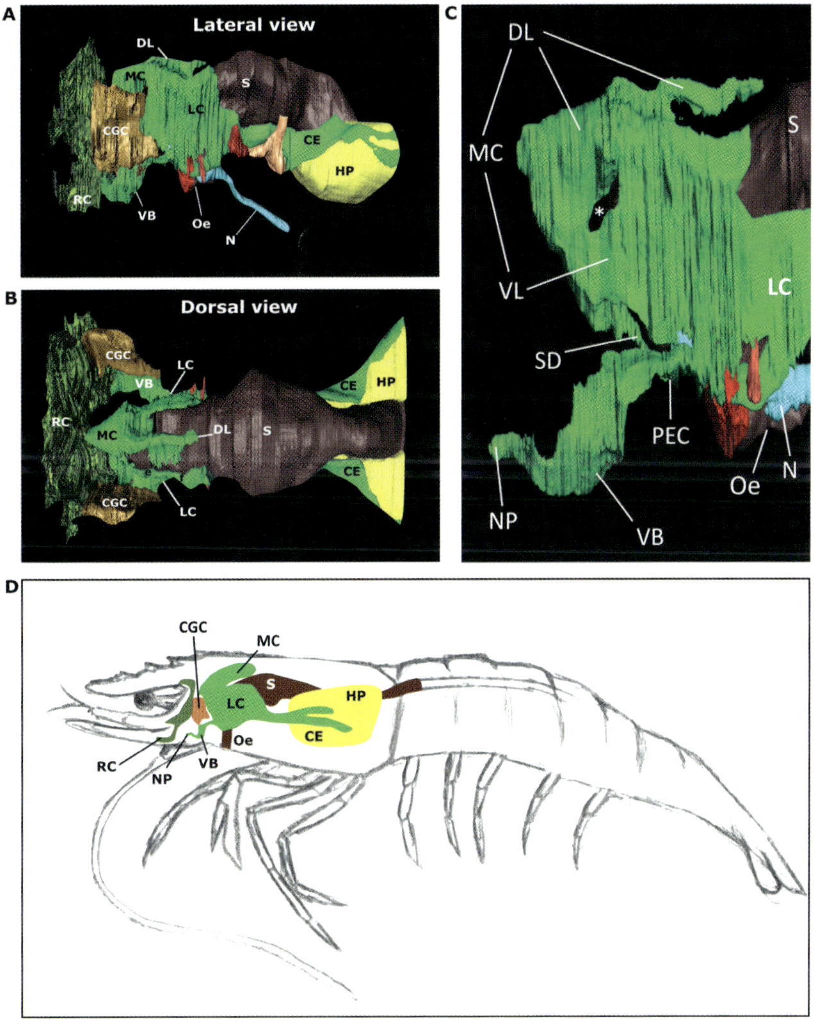

FIGURE 3.2 The general morphology of the nephrocomplex. (A) Lateral view. (B) Dorsal view. (C) Lateral view without compact glandular compartment and rostral compartment. (D) localization of the organ in the shrimp. RC: rostral compartment, CGC: compact glandular compartment, VB: ventral bladder, MC: median compartment, DL: dorsal lobe of the MC, VL: ventral lobe of the MC, LC: lateral compartment, CE: caudal extensions, PEC: pre-esophageal connection, SD: square duct, NP: nephropore, Oe: esophagus, S: stomach, N: nerve, HP: hepatopancreas. Credit: De Gryse et al. (2020).

Nephropore

The nephropore is the only part of the nephrocomplex covered by a cuticle. The nephropore controls the access to the nephropore and is a first-line defense mechanism, protecting the nephrocomplex from invading pathogens. The opening itself is slit-shaped and lies on the bottom of a caudal-facing, concave protuberance of the medial side of the coxipodite. The slit is formed by two valves, of which the rostral valve is superimposed by the caudal valve (Fig. 3.3). Pressure from the outside on the valve system does not result in an influx of water, whereas pressure from the inside of the valve passively opens the nephropore, allowing efflux. This demonstrates that the structure acts like a non-return valve, allowing only a one-way stream of fluid (De Gryse et al., 2020).

The integrity of this valve system is very robust, compromised only when the shrimp urinates. During this mictio process, the pressure is first highest inside the VB, passively opening the nephropore and expelling urine to the outside world. The more urine leaves the VB, the lower this pressure becomes, until the VB contents are so low that a pressure equilibrium exists between the inside and outside of the shrimp. When the muscles surrounding the VB inside the shrimp relax, a brief vacuum is created, allowing for a little water influx. Sometimes this small volume contains enough pathogens to initiate an infection (De Gryse et al., 2020).

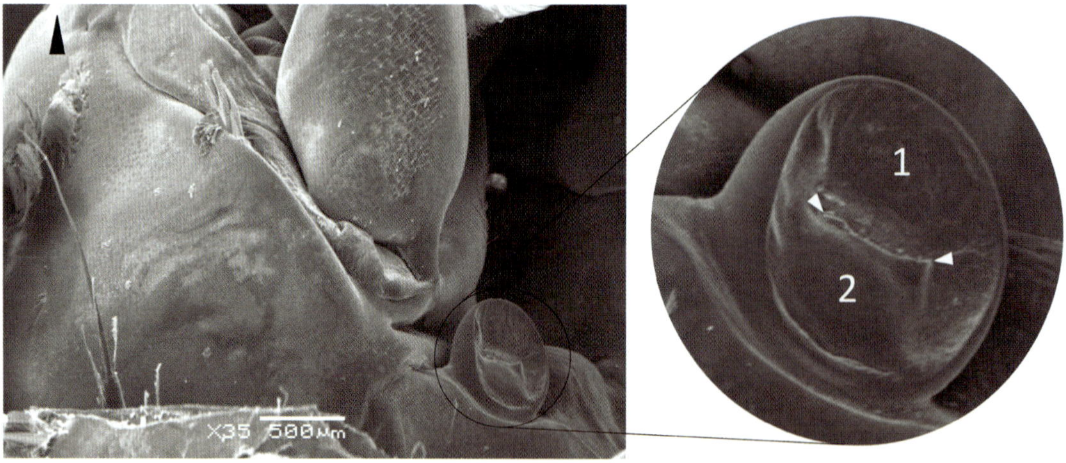

FIGURE 3.3 Scanning electron microscopy of the antennal coxipodite with the nephropore. 1: rostral valve, 2: caudal valve. The white arrowheads point to the nephropore split and the black arrowhead points to the rostrum. Credit: De Gryse et al. (2020).

Compact glandular compartment and rostral compartment

Situated at the transition from cephalothorax to the antennal coxipodite lies the bean-shaped compact glandular part (CGC) (Fig. 3.4(A)). The coelomosac can be found at the center of the compartment, surrounded by the tubuli of the labyrinth. One cell layer of podocytes lines the coelomosac. They have large, round, irregular-shaped nuclei (Buranajitpirom et al., 2010; Xugang et al., 2013). The lumen itself consists of large collection spaces, surrounded by coelomosac domes. These domes are in turn

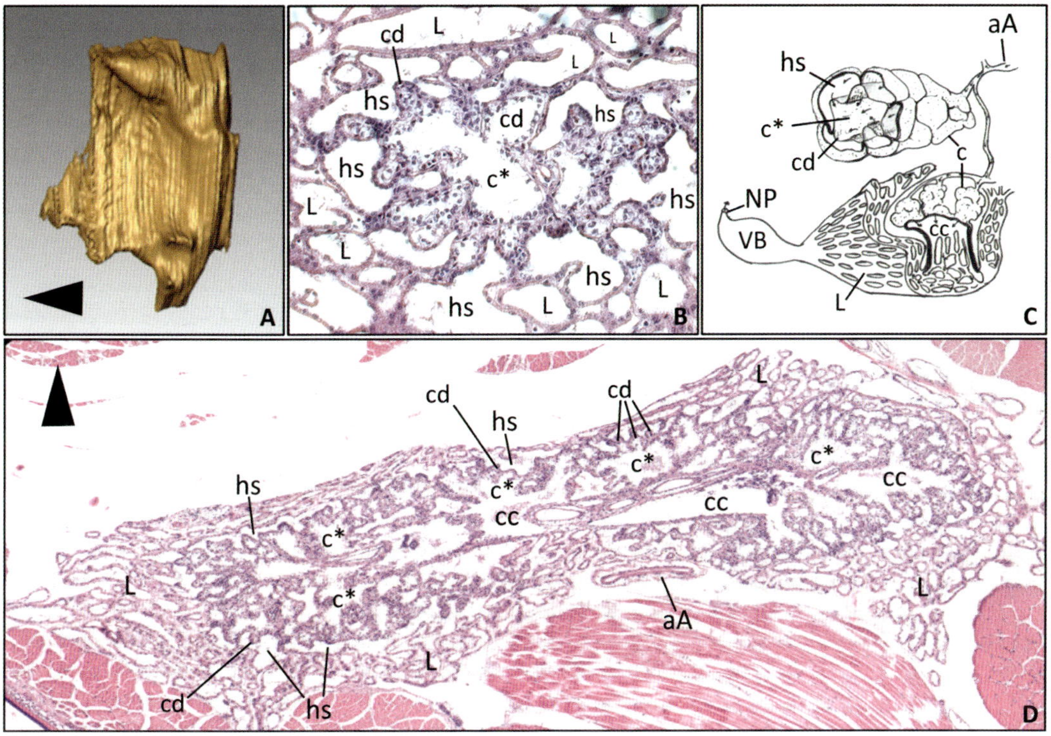

FIGURE 3.4 The coelomosac (c) is surrounded by the labyrinth (L) and together they make up the CGC. The black arrowhead points to the rostrum. (A) Three-dimensional model of the CGC. (B) Hematoxylin and eosin (HE)-stained section of the coelomosac lumen (c*), surrounded by coelomosac domes (cd) and hemolymph sinuses (hs). (C) Schematic overview of the filtrate flow. aA: antennal artery, cc: central coelomosac lumen, NP: nephropore, VB: ventral bladder. (D) HE-stained section of the CGC. Credit: adapted from De Gryse (2015).

surrounded by large hemolymph sinuses with flattened epithelia (Fig. 3.4(B)). Hemolymph is supplied to the coelomosac via the antennal artery, which penetrates the CGC and branches out throughout the structure. These branches debouch into the hemolymph sinuses enclosing the coelomosac domes, through which the ultrafiltrate is produced. Next, the fluid is collected by several large coelomosac lumens, which eventually merge into one larger central coelomosac lumen (Fig. 3.4(C, D)) (De Gryse, 2015).

The coelomosac is the site where ultrafiltration takes place (Behnke et al., 1990; Felgenhaur, 1992; Nakamura and Nishigaki, 1991). Its functional cells (podocytes) highly resemble the podocytes found in the Bowman capsule of the kidney of mammals. Intracellular channels between the podocytes can be found and the adjacent cells are connected by desmosome-like junctions. Cytoplasmatic cytopodia connect the podocytes with the basement membrane (Al-Mohsen, 2009; Khodabandeh et al., 2005; Ueno et al., 1997; Xugang et al., 2013). On top of the basement membrane, a slit diaphragm has been found in crayfish (Fig. 3.5). This diaphragm acts as a secondary filtration barrier, blocking macromolecules that managed to pass the basement membrane. These blocked macromolecules form a lamina densa between the slit diaphragm and the basement membrane (Schaffner and Rodewald, 1978; Vogt, 2002). The podocytes contain organelles that indicate active producing cells (Golgi apparatus, mitochondria, rough endoplasmic reticulum, vesicles, vacuoles, ribosomes, and a large nucleus with clear nucleoli; Fig. 3.5).

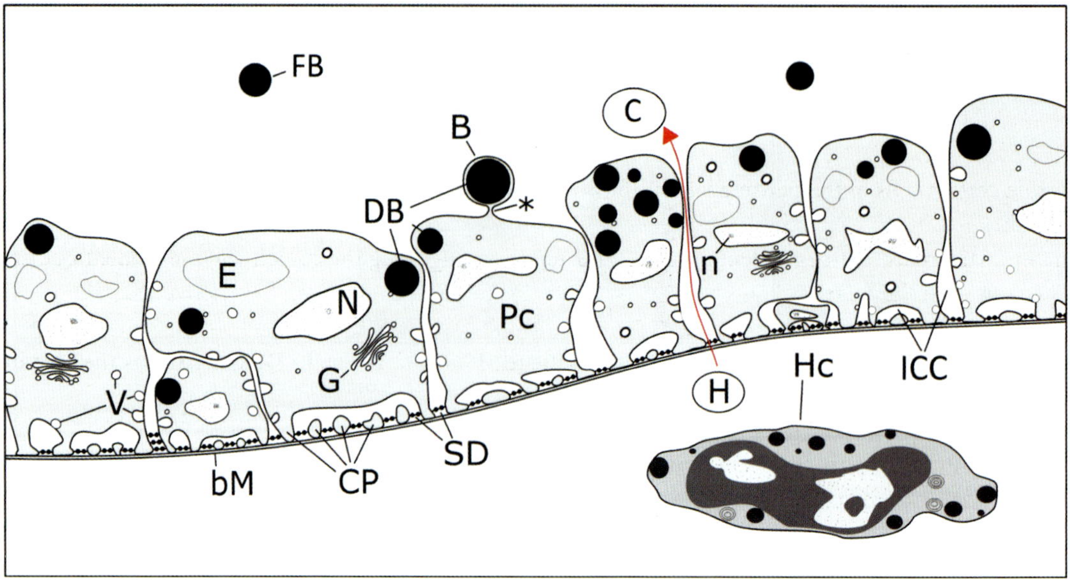

FIGURE 3.5 Ultrastructure of the podocytes. Pc: podocyte, bM: basement membrane, H: hemolymph, C: coelomosac lumen, CP: cytopodia, ICC: intercellular channel, SD: slit diaphragm, V: pinocytotic vacuoles, G: golgi apparatus, E: endosome, N: nucleus, n: nucleolus, DB: dense body, B: blebbing, FB: formed body, Hc: hemocyte. The red arrow represents the stream of fluid from the hemolymph sinus to the coelomosac dome and lumen. The DBs are excreted by a process called blebbing. The DB is pinched off (*) and is now part of the filtrate. Credit: De Gryse (2015).

When the ultrafiltrate passes the intercellular channels, the podocytes modify the fluid by pinocytosis (Peterson and Loizzi, 1974; Ueno and Inoue, 1996). Likewise, substances from the coelomosac lumen are endocytosed (Kirschner and Wagner, 1965). In doing this, podocytes are able to capture and sometimes metabolize ions (Holliday and Miller, 1984; Schmidt-Nielsen et al., 1968), metallic particles (detoxification; Doughtie and Rao, 1984; Roldan and Shivers, 1987), proteins, and (macro)molecules (Ueno and Inoue, 1996; Ueno et al., 1997). Both endocytotic and pinocytotic vesicles will fuse together to form dense bodies, which are excreted by a process called blebbing (Fig. 3.5) (Ueno and Inoue, 1996). Once in the coelomosac lumen, the dense bodies are called formed bodies (Khodabandeh et al., 2005; Riegel, 1966a). These formed bodies have been studied best in crayfish. It is believed that when they are in the coelomosac lumen, they still contain large, undigested macromolecules (Riegel, 1966a, b). Finally, at the basal side of the cytopodia, Na^+/K^+-ATPases are present in crabs, as well as H^+-ATPases, $Na^+/K^+/2Cl^-$-cotransporters, and Na^+/H^+ exchangers in the apical side of the podocytes (Tsai and Lin, 2014). Khodabandeh et al. (2005), however, could not find Na^+/K^+-ATPases in the lobster coelomosac. In *Penaeus monodon*, Buranajitpirom et al. (2010) found Na^+/K^+-ATPase immunoreactivity in the coelomosac. When shrimp were challenged with different salinities, the enzyme activity inside the nephrocomplex was altered in adaptation to these conditions.

Eventually, the collected filtrate inside the coelomosac will flow through to the labyrinth. The labyrinth is made up of anastomosing tubuli, which are bathed in the hemolymph (Nakamura and Nishigaki, 1991). In contrast to crayfish, lobster, and *Macrobrachium* species, in shrimp only one type of labyrinth cell exists, namely a single-cell-layer thick cuboidal to high columnar epithelium, with a microvilli brush border and a basally located nucleus (De Gryse et al., 2020; Xiaoyun et al., 2003; Xugang et al., 2013). This cytomorphology strongly suggests a function in absorption, reabsorption, and/or other modification of the filtrate. Indeed, the labyrinth cells are involved in the reabsorption of large molecules, fluid, and ions. Waste elimination and the regulation of osmotic pressure are also attributed to the labyrinth tubuli (Xiaoyun et al., 2003). Like in the lobster (Khodabandeh et al., 2005), Na^+/K^+ channels were found inside certain cells of the penaeid labyrinth (Buranajitpirom et al., 2010). Formed bodies coming from the coelomosac start to undergo some changes. Once they arrive in the lumen of the labyrinth, where a lower pH exists, pH-dependent proteases are activated, digesting the contents of the formed bodies. Subsequently, they attract water and burst. It is believed that in crayfish, this process, together with hydrolysis processes, creates a concentration gradient between the hemolymph and the lumen of the labyrinth, effectively aiding in osmoregulation (Riegel, 1966a, b, 1970a, b) Whether this is the case in shrimp remains to be investigated.

When the tubuli of the labyrinth are traced to the front of the animal, they can be seen leaving the CGC and gradually filling the entire hemolymph space in the cephalon, surrounding the supra-esophageal ganglion (Fig. 3.6(A)). Here, they connect to the tubuli of the contralateral side, forming the RC, before returning to the CGC to end in the VB (Fig. 3.6(B)). The cytology of the RC tubuli is mostly cubical with large round nuclei (De Gryse, 2015; De Gryse et al., 2020) The function of the RC remains obscure, but the conformation strongly resembles the loops of Henle found in reptile, avian, and mammalian kidneys. Only briefly, at both sides of the cephalon, the tubuli of the RC are closely associated with the continuation of the antennal artery coming from the CGC (Fig. 3.6(C)).

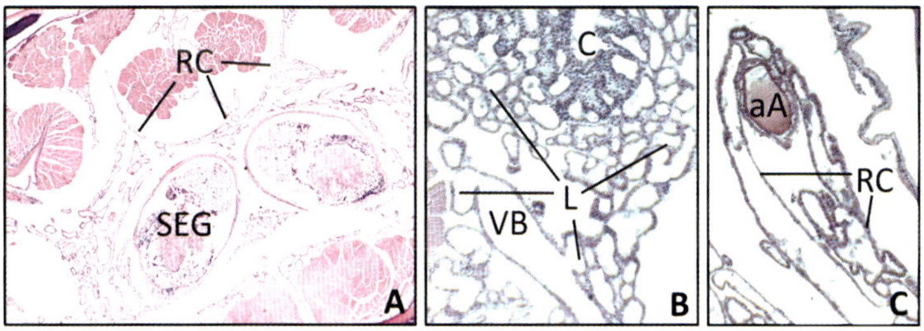

FIGURE 3.6 (A) The rostral compartment (RC) engulfs the supra-esophageal ganglion (SEG). (B) The tubuli of the RC return to the CGC labyrinth (L) and end in the ventral bladder (VB). C: coelomosac. (C) Association of the RC with the antennal artery (aA). Credit: Adapted from De Gryse (2015) and De Gryse et al. (2020).

Ventral bladder

The bilaterally present VB receives its contents from the tubuli of the CGC. Connections with the CGC can occur at a variable number of places. The bladder is a widened modification of the terminal ductus and acts mainly as a storage structure (Fig. 3.7(A)). It is one of the few sections of the shrimp nephrocomplex in possession of multiple cell layers. It is topped by a single squamous epithelium with bulging nuclei and a microvilli brush border. Underneath, a lamina muscularis above loose connective tissue can be found. Finally, a lamina serosa defines the outer surface of the VB. Sometimes, the epithelium can be observed to be two cell layers thick. To the caudal side, the VB has an extension, which has a narrow lumen and splits into three structures (Fig. 3.7(B)). Medially, a connection with the contralateral VB is made in the form of PEC. To the caudal side, the extension of the VB ends in the LC (De Gryse, 2015; De Gryse et al., 2020; Nakamura and Nakashima, 1992; Xiaoyun et al., 2003). On the basal side and around the extension

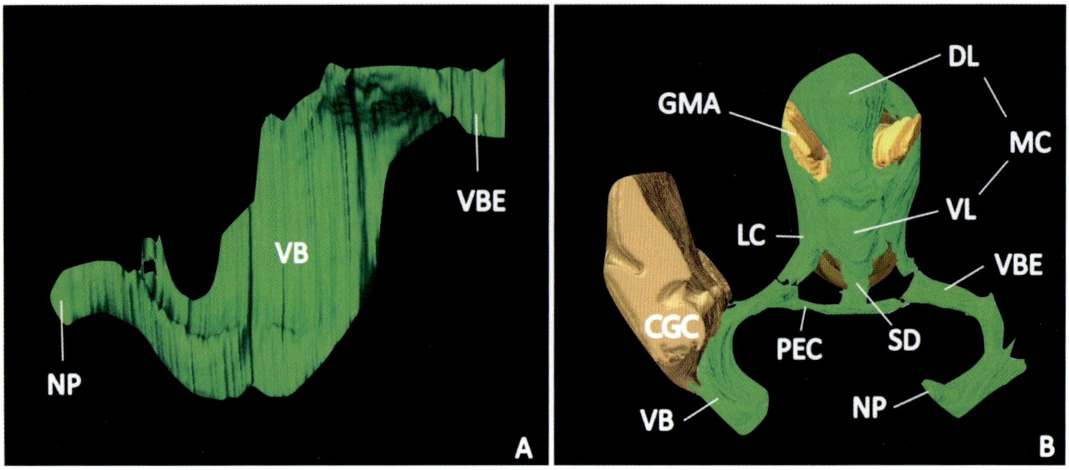

FIGURE 3.7 (A) Medial view of the right ventral bladder (VB). NP: nephropore, VBE: ventral bladder extension. (B) Rostral view of the VB, compact glandular compartment (CGC), lateral compartment (LC), and median compartment (MC) with ventral lobe (VL) and dorsal lobe (DL). PEC: pre-esophageal connection, SD: square duct, GMA: gastric mill attractor muscle. Credit: Adapted from De Gryse (2015).

of the VB, ramifications of unknown significance can be found (Fig. 3.8). Lastly, the VB has a short duct via which the evacuation of urine takes place. Departing from the VB, this duct starts as triangle-shaped, changing into a square, and ending as a moon-shaped duct before connecting to the outside world via the nephropore (De Gryse, 2015; Nakamura and Nakashima, 1992).

Around the VB, three important muscles can be found. These are the coxipodite adductor (CA), the small scaphocerite abductor (SSA), and the large scaphocerite abductor (LSA) (Fig. 3.8). Both the CA and the SSA originate from the LSA. The LSA muscle attaches in the scaphocerite, to run through the coxipodite. Just rostrally from the VB, the SSA and the CA branches split from the main muscle. The SSA runs lateral from the VB and the CA runs medially, while the LSA lies ventrally from the VB. All three muscles end on the inside of the exoskeleton, caudally to the VB. This conformation effectively encloses the VB. A coordinated engagement of these three muscles, potentially aided by the smooth muscle layer in the VB wall, would result in a higher pressure inside the VB, causing the passive opening of the nephropore valves and the evacuation of urine (mictio). According to Young (1959), the function of the SSA and LSA is to abduct the scaphocerite. Lateral movement of the

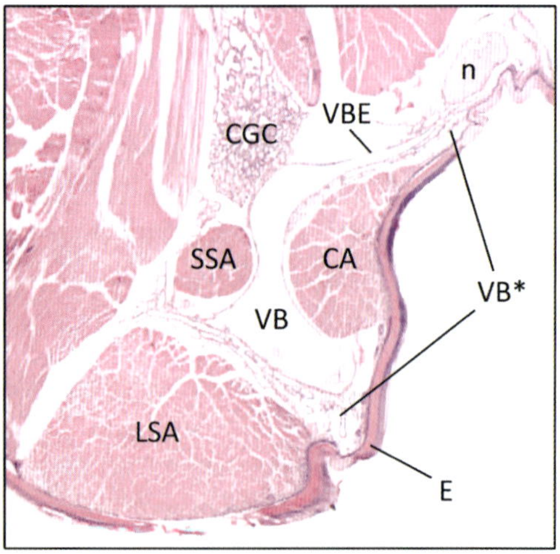

FIGURE 3.8 Muscles associated with the ventral bladder (VB). LSA: large scaphocerite abductor, SSA: small scaphocerite abductor, CA: coxipodite adductor, CGC: compact glandular compartment, VBE: ventral bladder extension, VB*: ramifications of the VB, n: nerve, E: exoskeleton. Credit: Adapted from De Gryse (2015).

scaphocerite (Fig. 3.1) is associated with grooming of the cephalon appendices and a sudden flush from the gill chamber. Given the orientation of the nephropore in the direction of the gills, the function of the SSA and LSA, and the position of these muscles around the VB, it is plausible that a forceful scaphocerite abduction results in urine flow (De Gryse, 2015). It is described that the gill chamber protects against pathogens but that the conformation of the chamber allows sediments to be trapped (Bauer, 1999). A powerfully aimed flow of urine could help in the process of mechanically cleaning the gill chamber.

Median compartment

As mentioned before, the extension of the VB has a PEC with the contralateral VB. From the PEC, a square duct runs dorsally, giving rise to the MC (Fig. 3.2(C), Fig. 3.7(B)). The MC is a large bladder structure, covering the stomach rostrodorsally. It can be divided into two parts: a VL and a DL. These lobes are separated by the anterior gastric mill attractor muscles. Caudally, the structure will gradually turn into the LC (Fig. 3.2). The function of the MC is unknown; however, like the VB, several muscles are associated with the MC. Besides the anterior gastric mill attractor dividing the two lobes, the large protocephalon attractor runs laterally from the MC and the epistomal stator muscle runs lateral from the DL, which is also dorsally covered by two posterior protocephalon levator muscles. Again, a coordinated action of these muscles could modulate the contents of the MC. Histologically, the MC is similar to the VB, but with a slightly thicker smooth muscle layer, hinting at a dynamic function (De Gryse 2015; De Gryse et al., 2020).

Lateral compartment and caudal extensions

Bilaterally from the esophagus, a large longitudinal bladder structure arises: the LC (Fig. 3.2, Fig. 3.7(B)). It is connected with the extension of the VB and to both the VL and the DL of the MC. The two halves connect via a post-esophageal connection, similar to the PEC. Laterally flanking the esophagus, stomach, and gut, the LC runs caudally, gradually decreasing in height. It is, in turn, flanked by the two large protocephalon attractor muscles (Fig. 3.9). The epithelium of the LC contains only one layer of flattened epithelial cells. The LC is attached to both the stomach and the protocephalon attractor muscle. Occasionally, a small muscle layer can be observed. The LC is attached to the large protocephalon attractor muscle on the lateral side (Fig. 3.9(A)). To the caudal side, the LC continues as the CE, limited dorsally by the pylorus stomach and ventrally by muscle layers and/or exoskeleton. Just rostrally of the hepatopancreas, a last connection to the contralateral side can be found. Next, the CE diverges around the hepatopancreas, ending blindly (De Gryse 2015; De Gryse et al., 2020).

Other muscles associated with the LC are the esophageal dilators, the maxillary tensor muscles, the mandibular adductor, and two oblique muscles. All these muscles have the potential to isolate (parts) of the LC, meaning that the shrimp can regulate the volume of the LC and its CE (De Gryse, 2015). This was observed in micro-magnetic resonance imaging studies, in which the contents of the CE of the gland were compared to the hepatopancreas. The relative size of the EC contents is higher in the post-molting stages, with a maximum around ecdysis (De Gryse et al., 2020).

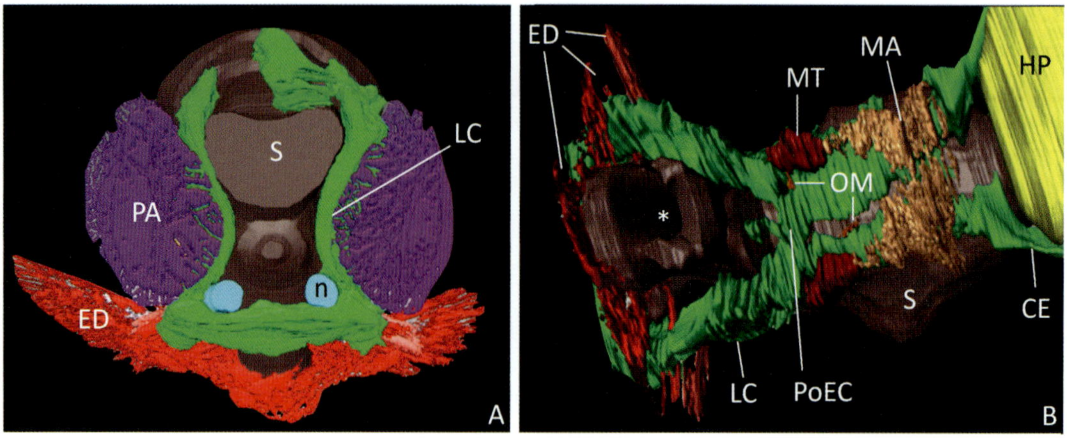

FIGURE 3.9 Muscles associated with the lateral compartment (LC) and caudal extensions (CE). (A) Frontal view of a cross section of the LC. PA: protocephalon attractor, ED: esophageal dilator, S: stomach, n: nerve. (B) Ventral view of the LC and EC. PoEC: post-esophageal connection, MT: maxillary tensor, MA: mandibular adductor, OM: oblique muscle, HP: hepatopancreas. Credit: Adapted from De Gryse (2015).

The role of the nephrocomplex in molting

Molting is a very impactful event on shrimp: drastic changes occur in the shrimp body when the metabolism changes from homeostasis to allostasis while the animal prepares to shed its old cuticle and grows new tissue. Given its complexity and an almost unrivaled extensiveness, it is not surprising that the nephrocomplex plays a significant role in this process. A first part is played by metabolizing the molting hormone ecdysone from the Y-organ and excretion of its metabolites. In this way, the organ can regulate the balance between the molt-inhibiting hormone and ecdysone, which determines the progression through the subsequent molting stages (Chang, 1985; Mykles, 2011). The molting hormone ecdysone reaches a peak in stage D1 (early pre-molt), signaling to the cells to prepare for ecdysis. In late pre-molt, these hormone levels drop significantly (Corteel and Nauwynck, 2010). Immediately before ecdysis, coordinated actions of muscles associated with the nephrocomplex pump extra fluids to the CE of the LC. The contents of the CE are now at a maximum (Fig. 3.10). This higher volume aids in raising the inside pressure on the carapax, putting ample pressure on the ligament between the carapace and the first abdominal segment. When the shrimp contracts its other muscles, the added pressure from the CE helps in shedding of the exuvium. Immediately after the shedding of the old cuticula, the shell is still soft. This is the only time during the molting cycle when the absolute volume of the shrimp can be increased. Shortly after the ecdysis, the volume inside the CE remains high, helping the shrimp to increase its volume as much as possible while the exoskeleton hardens (De Gryse et al., 2020). The hardening of the new, still soft shell happens through the process of mineralization. Calcium plays a pivotal role in the mineralization of the new exoskeleton by forming calcite ($CaCO_3$) and also acts as an intracellular messenger in the hormonal signaling (Ahearn et al., 2004; Wheatly et al., 2004). Without doubt, the nephrocomplex fulfills an important part in regulating the calcium concentrations inside the hemolymph, most probably by the cells of the labyrinth and by the coelomosac, and quite possibly by other parts of the organ as well (e.g.

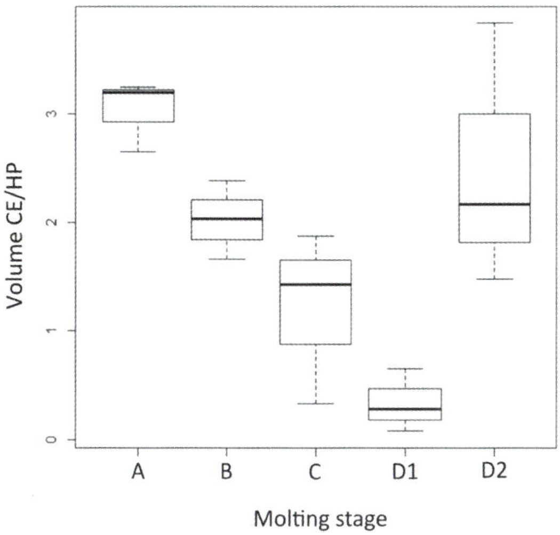

FIGURE 3.10 Dynamic change of caudal extension (CE) volume, relative to hepatopancreas (HP) volume, over different molting stages. Credit: De Gryse et al. (2020).

the RC). However, this remains to be investigated. Next, when the shrimp progresses through the molting stages (A and B), the volume inside the CE will gradually drop, making room for newly grown tissue. This diminution in volume is most likely attained by frequently expelling small volumes of urine, thus compromising the defensive integrity of the nephropore during the postmolting stages (De Gryse et al., 2020). During intermolt (C) and early pre-molt (D1), the volume inside the CE will be at its lowest and the molt-inhibiting hormone will be in balance with the molting hormone ecdysone (Corteel and Nauwynck, 2010; De Gryse et al., 2020).

Other functions associated with the nephrocomplex

The shrimp nephrocomplex is vast and complex; it spreads throughout the entire cephalothorax and is in close contact with many other organs. Besides its pivotal role in the molting and growing process, the organ also performs other functions. One of the most evident of these is maintenance of body volume and hemolymph osmolarity. When shrimp swim in low-salinity water, they produce more urine to keep the shrimp's body and its cells from swelling and bursting. Conversely, when shrimp are in a high-salinity environment, the production of urine is halted to prevent them from dehydrating (Lin et al., 2000). As a reaction to a change in salinity, but also in other circumstances, the shrimp's nephrocomplex is known to regulate body levels of ions such as magnesium, calcium, potassium, sulfate, copper, iron, and sodium (Cheng and Liao, 1986; Dall and Smith, 1981; Lin et al., 2000; Vogt and Quinitio, 1994). Na^+/K^+-ATPase activity in the nephrocomplex is related to the salinity of the water in which shrimp are reared (Buranajitpirom et al., 2010). Another saline-sensitive activity in the nephrocomplex is that of the carbonic anhydrase metalloenzyme. Carbonic anhydrase catalyzes the interconversion between carbon dioxide and bicarbonate in the presence of water. It is believed to rapidly supply counterions for transport processes involved in ion regulation, for example, NaCl uptake against a chemical gradient (Henry et al., 2006; Pongsomboon et al., 2009). Carbonic anhydrase has been reported in many invertebrates and is considered to play an active role in the acid/base regulation of *Penaeus vannamei* as well (Liu et al., 2015).

When shrimp are exposed to heavy metals, the nephrocomplex, along with several other organs, acts as a chelator and excretes these toxic components. Mercury, lead, chromium, nickel, zinc, cadmium, and manganese have been described to be sequestered and secreted via the urine when animals were challenged with these toxic elements. Thus, the nephrocomplex plays an important role in detoxification in shrimp (Doughtie and Rao, 1984; Tikær Andersen and Baatrup, 1988; Páez-Osuna and Tron-Mayen, 1996; Vogt and Quinitio, 1994; Xugang et al., 2013).

Like many other animals, shrimp fight for feed and to establish social relations and/or dominance. In *Macrobrachium* species and lobsters, social dominance involves the excretion of hormones via the urine (Al-Mohsen, 2009; Breithaupt and Atema, 2000; Katoh et al., 2008). When shrimp were getting ready to mate, the "egg-laying hormone" was found in the shrimp nephrocomplex, potentially regulating the hemolymph hormone level, or perhaps acting as a peptide signal to other shrimp (Liu et al., 2006). In *Macrobrachium rosenbergii*, the male-to-female attraction-stimulating hormone *N*-acetylglucosamine-1,5-lactone was found to be expressed in its excretory organ (Bose et al., 2017), and in *Sagmariasus verreauxi*, the expression of genes with pheromone functions in the organ was identified (Chandler et al., 2016). Together with the discovery of the receptor "trapped in endoderm 1" (suggested to be crucial in courtship and mating behavior in *Drosophila melanogaster*) in the nephrocomplex of *P. monodon*, these studies strongly hint at the antennal gland as a pheromone-releasing organ in shrimp (Viet Nguyen et al., 2020).

Future directions

The morphology of the nephrocomplex is now fairly completely described. However, the functions of many parts of the organ remain unclear; for example, does the RC resemble the loop of Henle only morphologically or also functionally? Many of the functions of the CGC remain vague and are molecularly poorly described. A lot of the functional knowledge is still extrapolated from other crustaceans and sometimes even insects. Therefore, functional and molecular characterization studies of the coelomosac, labyrinth, and RC cells is warranted. For example, it would be interesting to know where Na^+/k^+-ATPases are present besides in the coelomosac and labyrinth. Information on the embryological development of the nephrocomplex would also be welcome.

Another path wide open to scientific research is to further unravel the complete role of the nephrocomplex in the molting process. For instance, the MC is most likely involved in regulating absolute body volume after ecdysis, like the CE are, but evidence has not yet been provided.

Since the nephrocomplex has been identified as a major portal for pathogen entry, the pathogenesis of bacteria, viruses, parasites, and unknown syndromes can be studied. It would also be beneficial to examine the shrimp's urine and nephrocomplex for immunological parameters such as antibacterial proteins.

References

Ahearn, G.A., Mandal, P.K., and Mandal, A. (2004) Calcium regulation in crustaceans during the molt cycle: a review and update. *Comparative Biochemistry and Physiology Part A: Molecular & Integrative Physiology* 137, 247–257.

Al-Mohsen, I. (2009) *Macrobrachium rosenbergii* (de Man 1879): the antennal gland and the role of pheromones in mating behaviour. PhD thesis. Institute of Aquaculture, University of Stirling, Stirling, UK, pp. 60–120.

Bauer, R.T. (1999) Gill-cleaning mechanisms of a dendrobranchiate shrimp, *Rimapenaeus similis* (Decapoda, Penaeidae): description and experimental testing of function. *Journal of Morphology* 242, 125–139.

Behnke, R.D., Wong, R.K., Huse, S.M., Reshkin, S.J., and Ahearn, G.A. (1990) Proline transport by brush-border membrane-vesicles of lobster antennal glands. *American Journal of Physiology* 258, F311–F320.

Bose, U., Kruangkum, T., Wang, T., Zhao, M., Ventura, T., Mitu, S.A., Hodson, M.P., Shaw, P.N., Sobhon, P., and Cummins, S.F. (2017) Biomolecular changes that occur in the antennal gland of the giant freshwater prawn (*Machrobrachium rosenbergii*). *PLoS One* 12, e0177064.

Breithaupt, T. and Atema, J. (2000) The timing of chemical signaling with urine in dominance fights of male lobsters (*Homarus americanus*). *Behavioral Ecology and Sociobiology* 49, 67–78.

Buranajitpirom, D., Asuvapongpatana, S., Weerachatyanukul, W., Wongprasert, K., Namwong, W., Poltana, P., and Withyachumnarnkul, B. (2010) Adaptation of the black tiger shrimp, *Penaeus monodon*, to different salinities through an excretory function of the antennal gland. *Cell and Tissue Research* 340, 481–489.

Chandler, J.C., Aizen, J., Fitzgibbon, Q.P., Elizur, A., and Ventura, T. (2016) Applying the power of transcriptomics: understanding male sexual development in decapod Crustacea. *Integrative and Comparative Biology* 56, 1144–1156.

Chang, E.S. (1985) Hormonal control of molting in decapod crustacea. *American Zoologist* 25, 179–185.

Cheng, J.-H. and Liao, I.-C. (1986) Effect of salinity on the osmotic and ionic concentrations in the hemolymph of *Penaeus monodon* and *P. penicillatus*. In: Maclean, J.L., Dizon, L.B., Hosillos, L.V. (eds) *First Asian Fisheries Forum*. Manila, Philippines, 26–31 May 1986.

Corteel, M. and Nauwynck, H.J. (2010) The integument of shrimp. In: Alday-Sanz, V. (ed.), *The Shrimp Book*. Nottingham University Press, Nottingham, UK, pp. 73–88.

Dall, W. and Smith, D. (1981) Ionic regulation of four species of penaeid prawn. *Journal of Experimental Marine Biology and Ecology* 55, 219–232.

De Gryse, G.M. (2015) Driedimensionale morfologie van de antennale klier bij de *Penaeus Vannamei* en de ontwikkeling van een kwantificatiemethode voor de studie van het White Spot Syndrome Virus. MSc thesis, Department of Virology, Immunology and Parasitology. Ghent University, Ghent, Belgium, p. 84.

De Gryse, G.M.A., Khuong, T.V., Descamps, B., Van Den Broeck, W., Vanhove, C., Cornillie, P., Sorgeloos, P., Bossier, P., and Nauwynck, H.J. (2020) The shrimp nephrocomplex serves as a major portal of pathogen entry and is involved in the molting process. *Proceedings of the National Academy of Sciences of the United States of America* 117, 28374–28383.

Doughtie, D.G. and Rao, K.R. (1984) Histopathological and ultrastructural changes in the antennal gland, midgut, hepatopancreas, and gill of grass shrimp following exposure to hexavalent chromium. *Journal of Invertebrate Pathology* 43, 89–108.

Felgenhaur, B., (1992) Decapod internal anatomy. In: Harrison, F.W., Humes, A.G. (eds), *Microscopic Anatomy of Invertebrates: Decapod Crustacea*. Wiley-Liss, New York, NY, pp. 7–43.

Foster, C.A. and Howse, H.D. (1978) A morphological study on gills of the brown shrimp, *Penaeus aztecus*. *Tissue and Cell* 10, 77–92.

Freire, C.A., Onken, H., and McNamara, J.C. (2008) A structure–function analysis of ion transport in crustacean gills and excretory organs. *Comparative Biochemistry and Physiology Part A: Molecular & Integrative Physiology* 151, 272–304.

Henry, R.P., Thomason, K.L., and Towle, D.W. (2006) Quantitative changes in branchial carbonic anhydrase activity and expression in the euryhaline green crab, *Carcinus maenas*, in response to low salinity exposure. *Journal of Experimental Zoology Part A: Comparative Experimental Biology* 305, 842–850.

Holliday, C.W. and Miller, D.S. (1984) Cellular mechanisms of organic anion transport in crustacean renal tissue. *American Zoologist* 24, 275–284.

Icely, J.D. and Nott, J.A. (1979) The general morphology and fine structure of the antennary gland of *Corophium volutator* (Amphipoda: Crustacea). *Journal of the Marine Biological Association of the United Kingdom* 59, 745–755.

Katoh, E., Johnson, M., and Breithaupt, T. (2008) Fighting behaviour and the role of urinary signals in dominance assessment of Norway lobsters, *Nephrops norvegicus*. *Behaviour* 145, 1447–1464.

Khodabandeh, S., Charmantier, G., and Charmantier-Daures, M. (2005) Ultrastructural studies and Na^+,K^+-ATPase immunolocalization in the antennal urinary glands of the lobster *Homarus gammarus* (Crustacea, Decapoda). *Journal of Histochemistry and Cytochemistry* 53, 1203–1214.

Kirschner, L.B. and Wagner, S. (1965) The site and permeability of the filtration locus in the crayfish antennal gland. *Journal of Experimental Biology* 43, 385–395.

Lin, S.-C., Liou, C.-H., and Cheng, J.-H. (2000) The role of the antennal glands in ion and body volume regulation of cannulated *Penaeus monodon* reared in various salinity conditions. *Comparative Biochemistry and Physiology Part A: Molecular & Integrative Physiology* 127, 121–129.

Liu, M., Liu, S., Hu, Y., and Pan, L. (2015) Cloning and expression analysis of two carbonic anhydrase genes in white shrimp *Litopenaeus vannamei*, induced by pH and salinity stresses. *Aquaculture* 448, 391–400.

Liu, Z., Sobhon, P., Withyachumnarnkul, B., and Hanna, P. (2006) Identification of a putative egg-laying hormone in neural and ovarian tissues of the black tiger shrimp, *Penaeus monodon*, using immunocytochemistry. *Invertebrate Neuroscience* 6, 41–46.

McMahon, B.R. (1995) The physiology of gas exchange, circulation, ion regulation, and nitrogenous excretion: an integrative approach. In: Factor, J.R. (ed.), *Biology of the Lobster*. Academic Press, San Diego, CA, pp. 497–517.

Mykles, D.L. (2011) Ecdysteroid metabolism in crustaceans. *Journal of Steroid Biochemistry and Molecular Biology* 127, 196–203.

Nakamura, K. and Nakashima, N. (1992) Structure of antennal gland coelomasac in the kuruma prawn. *Nippon Suisan Gakkaishi* 58, 1551.

Nakamura, K. and Nishigaki, Y. (1991) Structure of antennal gland in kuruma prawn *Penaeus japonicus*. *Nippon Suisan Gakkaishi* 57, 1859–1863.

Páez-Osuna, F. and Tron-Mayen, L. (1996) Concentration and distribution of heavy metals in tissues of wild and farmed shrimp *Penaeus vannamei* from the northwest coast of Mexico. *Environment International* 22, 443–450.

Peterson, D.R. and Loizzi, R.F. (1973) Regional cytology and cytochemistry of the crayfish kidney tubule. *Journal of Morphology* 141, 133–145.

Peterson, D.R. and Loizzi, R.F. (1974) Ultrastructure of the crayfish kidney—coelomosac, labyrinth, nephridial canal. *Journal of Morphology* 142, 241–263.

Pongsomboon, S., Udomlertpreecha, S., Amparyup, P., Wuthisuthimethavee, S., and Tassanakajon, A. (2009) Gene expression and activity of carbonic anhydrase in salinity stressed *Penaeus monodon*. *Comparative Biochemistry and Physiology Part A: Molecular & Integrative Physiology* 152, 225–233.

Potts, W.T.W. and Parry, G. (1964) *Osmotic and Ionic Regulation in Animals*. International Series of Monographs on Pure and Applied Biology, Zoology Division, Vol. 19. Pergamon Press, Oxford, UK.

Riegel, J.A. (1966a) Analysis of formed bodies in urine removed from crayfish antennal gland by micropuncture. *Journal of Experimental Biology* 44, 387–395.

Riegel, J.A. (1966b) Micropuncture studies of formed-body secretion by excretory organs of crayfish frog and stick insect. *Journal of Experimental Biology* 44, 379–385.

Riegel, J.A. (1970a) In vitro studies of fluid and ion movements due to the swelling of formed bodies. *Comparative Biochemistry and Physiology* 35, 843–856.

Riegel, J.A. (1970b) A new model of transepithelial fluid movement with detailed application to fluid movement in crayfish antennal gland. *Comparative Biochemistry and Physiology* 36, 403–408.

Roldan, B.M. and Shivers, R.R. (1987) The uptake and storage of iron and lead in cells of the crayfish (*Orconectes propinquus*) hepatopancreas and antennal gland. *Comparative Biochemistry and Physiology Part C: Comparative Pharmacology* 86, 201–214.

Schaffner, A. and Rodewald, R. (1978) Filtration barriers in the coelomic sac of the crayfish, *Procambarus clarkii*. *Journal of Ultrastructure Research* 65, 36–47.

Schmidt-Nielsen, B., Gertz, K.H., and Davis, L.E. (1968) Excretion and ultrastructure of the antennal gland of the fiddler crab *Uca mordax*. *Journal of Morphology* 125, 473–496.

Taylor, H.H. and Taylor, E.W. (1992) Gills and lungs: the exchange of gases and ions. In: Harrison, F.W., Humes, A.G. (eds.), *Microscopic Anatomy of Invertebrates, Decapod Crustacea*. Wiley-Liss, New York, NY, pp. 203–293.

Tikær Andersen, J., and Baatrup, E. (1988) Ultrastructural localization of mercury accumulations in the gills, hepatopancreas, midgut, and antennal glands of the brown shrimp, *Crangon crangon*. *Aquatic Toxicology* 13, 309–324.

Tsai, J.R. and Lin, H.C. (2014) Functional anatomy and ion regulatory mechanisms of the antennal gland in a semi-terrestrial crab, *Ocypode stimpsoni*. *Biology Open* 3, 409–417.

Ueno, M. and Inoue, Y. (1996) The fine structure of podocytes in crayfish antennal glands. *Journal of Electron Microscopy* 45, 395–400.

Ueno, M., Inoue, Y., and Niwa, N. (1997) Podocytes of the freshwater shrimp—fine structure and effect of injected trypan blue. *Journal of Electron Microscopy* 46, 485–490.

Viet Nguyen, T., Ryan, L.W., Nocillado, J., Le Groumellec, M., Elizur, A., and Ventura, T. (2020) Transcriptomic changes across vitellogenesis in the black tiger prawn (*Penaeus monodon*), neuropeptides and G protein-coupled receptors repertoire curation. *General and Comparative Endocrinology* 298, 113585.

Vogt, G. (2002) Functional anatomy. In: Holdich, D.M. (ed.), *Biology of Freshwater Crayfish*. Blackwell Science, Oxford, UK, pp. 53–151.

Vogt, G. and Quinitio, E.T. (1994) Accumulation and excretion of metal granules in the prawn, *Penaeus monodon*, exposed to water-borne copper, lead, iron and calcium. *Aquatic Toxicology* 28, 223–241.

Wheatly, M.G., Gao, Y.P., and Nade, M. (2004) Integrative aspects of renal epithelial calcium transport in crayfish: temporal and spatial regulation of PMCA. *International Congress Series* 1275, 96–103.

Xiaoyun, L., Wei, X., and Zhenmin, B. (2003) Histology and functions study of the antennal gland of *Penaeus chinensis*. *Journal of Ocean University of Qingdao* 33, 854–860.

Xugang, H., Guangfu, H., and Guangtao, L. (2013) The effect of lower salinity on microstructure of antennary gland of *Litopenaeus vannamei*. *Journal of Food, Agriculture & Environment* 11, 782–785.

Young, J. (1959) Morphology of the white shrimp *Penaeus setiferus* (Linnaeus 175). *Fishery Bulletin 145*. The Fish and Wildlife Service, Washington, DC, p. 173.

Chapter 4

SHRIMP GUT MICROBIOTA AND ITS POTENTIAL IN SUSTAINABLE AQUACULTURE PRODUCTION

SOPACHA ARAYAMETHAKORN, SAGE CHAIYAPECHARA AND WANILADA RUNGRASSAMEE

Summary

Fisheries and aquaculture are important food-producing industries helping to support the world's growing population. However, intensive cultivation of aquatic species such as shrimp can result in frequent outbreaks of disease, causing major production losses. Maintenance of a beneficial microbial balance in an animal host and its rearing environment might be a promising option to mitigate the risk of disease outbreaks in aquaculture. In particular, intestinal bacteria play important roles in maintaining immune and nutrient homeostasis and protecting against pathogens in many animal hosts, including shrimp. Several factors have been reported to influence the intestinal microbial establishment, such as diet, life history, genetics, and environment. Here, we summarize our current understanding of the characterization of the intestinal microbiota in shrimp from a historical perspective, and describe its impact on host health and its potential applications for sustainable shrimp production.

Introduction

Fisheries and aquaculture are among the important industries expected to serve as a food source to meet the demands of the world's growing population. Currently, aquaculture production provides over 50% of fish products to consumers, making it one of the most important global food systems. Among the ~600 aquatic species that are cultured, penaeid shrimps are cultivated in over 50 countries (Whetstone et al., 2002). Despite the increasing growth in the global production of cultured shrimp, shrimp aquaculture has been affected by periodic losses due to disease outbreaks. A global survey conducted by the Global Aquaculture Alliance in 2018 revealed that disease was one of the top challenges for the shrimp industry (Anderson et al., 2019). In addition, the prophylactic use of antibiotics to minimize the risks of disease outbreaks has been banned in aquaculture due to growing concerns regarding food safety for consumers.

While aquatic diseases are emerging, treatments remain limited. Thus, it is crucial to explore alternative means for controlling diseases in shrimp farming.

To tackle disease outbreaks, understanding the contributing factors that promote shrimp health is essential. Although most relevant research focuses on the treatment of shrimp diseases after infection has occurred, disease prevention by good farming practice should be seriously considered. Because shrimp are always surrounded by a dynamic environment containing a variety of microorganisms, they are likely to be exposed to potential disease-causing agents on a regular basis. One of the factors to consider for preventing diseases is to monitor the normal gut microflora in shrimp and their rearing environments, and to maintain a beneficial balance of intestinal bacteria as a preventive measure against pathogen infection. This review aims to provide perspectives on shrimp gut microbiota and their potential applications for sustainable shrimp production.

Historical perspective on microbiome study and techniques to study gut microbiota

Historically, most of the microbial community studies in aquatic animals were primarily done based on culture-dependent methods (Shakibazadeh et al., 2009), which were applied to identify features such as the morphological characteristics of colonies, microbial growth on different artificial media, and their metabolic capabilities. Although culture-dependent approaches were able to distinguish many microorganisms, such techniques might not be able to differentiate specific bacterial types at lower taxonomic levels (Morgan and Huttenhower, 2012). Furthermore, bacterial communities associated with the gastrointestinal tract usually have complex structures, and it might be difficult to isolate many types of bacteria by using selective media (Greetham et al., 2002; Kim et al., 2007). Only ~5–20% of the microbial species are estimated to be culturable under laboratory conditions and taxonomically identified (Eckburg et al., 2005). Therefore, the culture-dependent approach has limitations in terms of the range of microorganisms that can be identified, the selectivity of growth media, and the culture conditions (Li et al., 2015). Despite these limitations, most studies of the microbiome began from culture-based assays, and were able to provide important knowledge. For example, (Huang et al., 2016a) reported that multiple bacterial species involved in hepatopancreas necrosis syndrome of *Penaeus vannamei* can be determined using enriched media; in this study, 111 bacterial species were identified from 13 taxonomic groups.

Advances in microbiological and molecular techniques have led to the development of culture-independent methods, whose main advantage is the elimination of culture selection biases. These approaches are DNA-based techniques for identifying bacteria based on the 16S ribosomal RNA subunit gene (16S rRNA) for better taxonomic identification. The 16S rRNA gene is ubiquitous and highly conserved among bacterial species. It is useful as a molecular marker to investigate several features of microbial diversity analyses (Tringe and Hugenholtz, 2008), in terms of qualitative and quantitative information on different species. The construction of clone libraries was the earliest DNA-based method used to assess bacterial populations and diversity. Sequencing analysis of clone libraries, based on phylogenetic marker genes, can be performed by Sanger sequencing. However, the Sanger sequencing platform allows determination of the sequence of only a single DNA fragment at a time. In order to provide enough coverage of different bacteria, examination of a sufficiently large number of 16S rRNA genes in a clone library would be necessary, a process that is costly and time-consuming (Chaiyapechara et al., 2012; Johnson et al., 2008). To provide comprehensive data on microbial communities, several conventional molecular methods can be employed in parallel to generate better bacterial community profiles (Zoetendal et al., 1998). Examples of these DNA-based methods are denaturing gradient gel electrophoresis (DGGE), terminal restriction fragment length polymorphism (T-RFLP), fluorescence *in situ* hybridization (FISH), and quantitative real-time PCR (qPCR).

DGGE AND T-RFLP

DGGE and T-RFLP are DNA fingerprinting techniques allowing the separation of 16S rRNA gene amplicons according to their physical or chemical properties (e.g. their G-C content or size) (Costa and Weese, 2019). Patterns of DNA bands can be qualitatively compared between communities. These methods have been used to characterize various microbial communities, such as those in freshwater aquaculture system farming (Tang et al., 2014), pond sediment (Cornejo-Granados et al., 2017), and the digestive tract of shrimp (Rungrassamee et al., 2016). The T-RFLP approach is applied for the estimation of phenotype richness and community composition. It has greater sensitivity than DGGE, especially in terms of its ability to identify the different taxa within bacterial assemblages (Hewson and Fuhrman, 2004). The T-RFLP technique was used to investigate bacterial diversity in different aquatic environments, including sediments, coastal seawater, and groundwater (Danovaro et al., 2006). The techniques are semi-quantitative, rapid, and cheap. However, the drawbacks of these methods are that they are primarily low-throughput, and their lack of phylogenetic identification and underestimation of the overall diversity due to band overlapping.

FISH

FISH is a probe hybridization technique that uses fluorescently labelled oligonucleotides to target complementary DNA sequences in the genome of a specific microbial community (Amann et al., 1995). The method can be applied to detect bacterial abundance at low levels of taxonomy, from species to phylum, and can be used for bacterial quantification. However, FISH has limited application in the detection of unknown bacterial species because probes need to be designed to target specific groups of bacteria. FISH has been used to study the composition of bacterial communities in the gut of abalone, revealing that the *Vibrio* group can be a dominant component (Tanaka et al., 2004), and to track specific probiotic bacteria in the gut of fish (Del'Duca et al., 2013).

QPCR

qPCR can be used for the quantification of DNA present in a sample to detect specific groups of organisms. This method requires specific primers to target each bacterial group. The main advantages of qPCR are its quantitative determination and discrimination of bacterial groups. The disadvantages of qPCR are that a set of primers must be designed to target each bacterial taxon to increase resolution, which can be labour intensive, time consuming, and technically challenging. Additionally, qPCR might not be a useful tool to assess the broad microbiome; however, it can be used in combination with other techniques to provide more information on the abundance of gut microbiota. For example, Mongkol et al. (2018) compared the bacterial distribution in different segments of the shrimp gastrointestinal tract using automated ribosomal intergenic spacer analysis, 16S rRNA clone libraries, and qPCR. They demonstrated distinct bacterial community profiles in different segments of gastrointestinal tract.

To provide better and more comprehensive coverage of the intestinal microbiota, high-throughput sequencing technologies have been developed, which obtain a higher output of sequence reads that can be used for bacterial diversity analysis (Sun et al., 2014). The recent next-generation sequencing technologies include 454 Pyrosequencing® (Roche), Illumina® (Solexa and Lynx Therapeutics), Ion Torrent™ (Thermo Fisher Scientific), and Single Molecule Real Time (SMRT®) sequencing (Pacific Biosciences). These next-generation sequencing technologies allow high-throughput sequencing analysis, and massive numbers of DNA fragments are sequenced simultaneously. These platforms are powerful and can provide efficient methods for characterization of the microbial community in various environments. However, their major limitation is the sequence length of DNA fragments (~300–500 nucleotides), which limits their capability and accuracy for classifying organisms to lower taxonomic levels. Many studies have applied high-throughput sequencing techniques to characterize the gut microbiota of aquatic animals such as gilt-head seabream (Califano et al., 2017), zebrafish (Siriyappagouder et al., 2018), and shrimp (Rungrassamee et al., 2016; Zheng et al., 2017).

Factors influencing gut microbial diversity in shrimp

The shrimp gut microbiota is a complex community of bacteria that has integral roles in the health of its animal host, which can include protecting the host from harmful microbes, regulating metabolic process, and developing and maintaining the host immune response (Cornejo-Granados et al., 2017; Ye et al., 2014). Several factors can influence the diversity of the gut microbiota, including the host genetics, developmental stage, environment, and diet, and the use of antibiotics in disease management.

HOST GENETICS

Naturally, aquatic animals are directly exposed to the aquatic environment, and thus are both evolutionarily adapted to this environment and in contact with the microorganisms within it. Bacterial dynamics can be influenced by the host genetic background, ecology, and evolution, which can affect their health and development (Pérez et al., 2010). Previous studies on shrimp have documented that host genetic divergence at a species level is correlated with the taxonomic composition of the gut microbiota. For example, the effect of species (e.g. *Macrobrachium nipponense* versus *Macrobrachium asperulum*) had a greater impact on the divergence of the gut microbial compositions than the host habitats (Tzeng et al., 2015). The development of shrimp genetic lines has focused on growth performance and disease resistance (Andriantahina et al., 2013). The host genetic background was found to affect the composition of the gut bacterial communities of Pacific white shrimp in an indoor aquaculture facility (Landsman et al., 2019). These studies indicated that the interaction between host genetics and gut microbial divergence has an effect on the evolution of the microbiome, which can be used to assess the shape and structure of gut microbiota.

DEVELOPMENTAL STAGE

Microbial intestinal colonization is an important process in the early life stages of animal hosts that could potentially help to establish the effects on health at the later life stages. Recently, several studies have shown that the intestinal bacterial microbiota of the Pacific white shrimp is tightly correlated with host age (Cornejo-Granados et al., 2017; Xiong et al., 2015). The differences in the bacterial composition of the microbiota have been found to depend on host physiology and dietary requirements at each stage (Ronquillo et al., 2006). For example, in one study, shrimp larvae were fed with *Artemia* nauplii (brine shrimp) during their primary development, and then on commercial pellets that have a high protein content (Li et al., 2018). At each developmental stage, *Bacteroidetes*, *Proteobacteria*, *Actinobacteria*, and *Planctomycetes* dominated as core bacterial phyla in the gut (Angthong et al., 2020; Rungrassamee et al., 2013). In addition, there are reports that some *Vibrio* species, which are established in the microbiota since the nauplius stage, might provide benefits in increasing disease resistance in the host (Balcázar et al., 2007; Oxley et al., 2002). Therefore, these findings revealed that developmental stages were a major factor that affected the dynamics of bacterial communities and could help in disease management in aquaculture.

ENVIRONMENT

The digestive tract of aquatic animals, including shrimp, is known to harbour a complex community of bacteria. It can be constantly influenced by environmental factors such as salinity, pressure, and temperature, which may stress the host. Many studies have reported that Pacific white shrimp show significantly lower survival rates and growth performance at low salinity, and the salinity can influence gut microbial diversity (Li et al., 2008; Zhang et al., 2016b). Water quality parameters such as temperature, pH, and nitrite and ammonia concentrations induced different changes of the microbial communities in water, sediment, and intestine (Engering et al., 2013; Fan et al., 2017; Zhang et al., 2016a). Certain environmental conditions can have positive impacts on the ecological strategy for certain bacteria to dominate the shrimp gut microbiota and prevent pathogenic invasion (De Schryver and Vadstein, 2014; Xiong et al., 2016).

DIET

The digestive system is able to break down macronutrients and micronutrients, which may positively affect shrimp growth, gut microbiota composition, immune response, and survival rates (Huang et al., 2016b). Hence, diet is one of the critical factors that can affect the intestinal microbiota in shrimp. Several studies have examined the effects of dietary components such as lipid, protein, carbohydrate, and feed additives on shrimp gut microbiota (Li et al., 2018; Niu et al., 2018; Shao et al., 2019). For example, many studies have provided evidence linking intestinal bacteria and nutrient absorption, as well as the fatty acid metabolism of the host animal (Cheng and Hardy, 2004; Yang et al., 2011). Additionally, different lipid sources, containing different combinations of fatty acids, could also alter the composition of the gut microbiota (Zhang et al., 2014). The use of plant-based protein sources, which contain higher carbohydrate levels than animal-based protein sources, also results in changes in the intestinal bacterial population in shrimp (Glencross et al., 2013; Simon et al., 2020). Thus, the gut bacterial community can be manipulated by varying the ingredients of the diet.

ANTIBIOTIC USE

Although this usage is now being regulated, antibiotics have previously been applied prophylactically for short periods of time to prevent disease outbreaks in aquaculture (Okocha et al., 2018). Studies have suggested that antibiotic treatment has a negative effect on the commensal gut microbiota, causing an imbalance in the stability of the composition of gut microbial populations (Francino, 2015; Jernberg et al., 2010; Kim et al., 2017). For example, florfenicol and oxolinic acid could impact the intestinal bacterial community composition in Atlantic salmon (Gupta et al., 2019). In seabass, oxytetracycline treatment caused a reduction in the abundance of *Photobacterium* in the gills (Rosado et al., 2019). Moreover, oxytetracycline significantly affected the microbial community in the intestine of grass shrimp (Uyaguari et al., 2009). Importantly, these studies showed that the use of antibiotics can also lead to a decrease in microbial core diversity, indicating that host immunity is reduced (Carlson et al., 2017; Legrand et al., 2017).

Applications of probiotics for sustainable shrimp production and their mode of action

One of the key factors for sustainable health management in aquaculture is the development of an alternative, non-antibiotic treatment, such as probiotics. The use of probiotics in aquaculture can provide benefits in many areas, including growth performance, nutritional status, and immune responses (Lazado et al., 2015). Several probiotic species have been applied in aquaculture, including *Lactobacillus*, *Bacillus*, *Streptomyces*, *Enterococcus*, *Aeromonas*, *Microbacterium*, and *Pseudomonas*. For example, *Lactobacillus* sp., isolated from the intestine of Pacific white shrimp, can increase resistance against white spot syndrome virus, growth performance, and digestive enzyme activities (Zuo et al., 2019). Application of *Lactobacillus brevis* results in higher growth performance and increased digestive enzyme levels, and also modulates the gut microbiota of *Macrobrachium rosenbergii* (Madhayan and Bhavan, 2018). Application of the four bacterial strains (*Bacillus subtilis* UTM126, *Vibrio alginolyticus* UM102, *Roseobacter gallaeciensis* SLV03, and *Pseudomonas aestumarina* SLV22) isolated from the intestinal tract of *Penaeus vannamei* helped to increase survival against the pathogen *Vibrio parahaemolyticus* PS-017 in shrimp (Balcázar et al., 2007). *Vibrio fluvialis* influenced the survival of *Penaeus japonicus* larvae (El-Sersy et al., 2006), and strain PM17 was used as a probiotic for black tiger shrimp, *Penaeus monodon* (Alavandi et al., 2004). *Vibrio gazogenes* NCIMB 2250 with chitin-supplemented feed was reported to decrease levels of vibrios in the fore- and hindgut of Pacific white shrimp (Thompson et al., 2010). Applications of beneficial bacteria have also been reported to enhance cellular and humoral immune responses in shrimp (Liu et al., 2010). *Enterococcus faecium* isolated from the intestine of giant tiger prawn has been reported to help increase disease resistance and bodyweight when applied as a feed additive (Sha et al., 2016; Shefat, 2018).

Probiotics provide inhibitory effects against pathogens through different mechanisms. During the past decades, several studies have reported on the modes of action of probiotics, which include modulation of the host digestive enzymes and nutrient sources, improvement of water quality, and modulation of the host immune response.

MODULATION OF HOST DIGESTIVE ENZYMES AND NUTRIENT SOURCES

Probiotics have beneficial effects on the digestive process of cultured aquatic animals. Several studies concluded that probiotic organisms can improve host nutrition by producing extracellular enzymes, including proteases, amylases, and lipases, which detoxify compounds in the diet, as well as improving host digestion of nutrients and growth (Giraffa et al., 2010; Zorriehzahra et al., 2016).

IMPROVEMENT OF WATER QUALITY

Bacteria in genera such as *Bacillus*, *Pseudomonas*, *Acinetobacter*, *Cellulomonas*, *Rhodopseudomonas*, and *Nitrobacter* can regulate the microbiota of water and control diseases to improve water quality for sustainable aquaculture (Kuebutornye et al., 2019). It has been reported that *Bacillus* sp. can remediate toxic wastes to improve water quality, resulting in improvements in growth and survival rates in shrimp (Devaraja et al., 2013). The bacteria can decompose organic matter in the water to CO_2 (Dalmin et al., 2001).

MODULATION OF HOST IMMUNE RESPONSE

Many studies have reported that probiotic strains of bacteria can act as immunostimulants in fish and shrimp. Probiotics can generate signalling molecules that enhance host immunity against pathogenic bacteria (Merrifield et al., 2010). Various studies revealed that lactic acid bacteria and *Bacillus* sp. have inhibitory effects on the growth of pathogenic organisms and stimulate the innate immune response, which improve the health of the host animal (Hoseinifar et al., 2019; Laranja et al., 2017; Ringø et al., 2018). In Nile tilapia (*Oreochromis niloticus*) and other fish, it has been shown that the supplementation of probiotics could increase the level of intraepithelial lymphocytes and acidophilic granulocytes, lysozyme activity, and bactericidal activity (Magda et al., 2011).

Conclusion: unanswered questions and future directions for research on shrimp gut microbiota

Our understanding of the shrimp gut microbiota and factors affecting it have expanded greatly in the past several years. Many researchers in different parts of the world are working towards better understanding of the roles of the gut microbiota in shrimp health. To fully take advantage of the potential of the gut microbiota in disease management in the future, the gap in our knowledge of the interaction between the gut microbiota and the host animal needs to be addressed. Here, we list a few important areas that we think should receive priority.

Gaining fundamental knowledge on the dynamics of the microbiota and its homeostasis in the intestines should be one of the future goals. The changes in the gut microbial community between the normal state and a disrupted state caused by either disease, stress, or environmental changes should be examined. So far, there have been a few reports in the literature on the disruption of the shrimp gut microbiota associated with disease (Xiong, 2018; Zheng et al., 2017). The mechanisms of how the gut microbiota is affected, and the interactions between the pathogen, the microbiota, and the host's immunity during disease outbreaks, remain to be elucidated at the time of writing. The use of modern -omics techniques such as shotgun metagenomics, transcriptomics, meta-transcriptomics, and metabolomics alongside other relevant traditional techniques could help to explain the underlying mechanisms. Owing to the wide-ranging influences of the gut microbiota, a systemic approach could be employed to study the interactions. Understanding these interactions could be useful for both research and applications to utilize the gut microbiota to help prevent disease outbreaks, minimize loss from disease, or stimulate host immunity.

Understanding the mechanisms and mode of actions of probiotic bacteria is another research priority. While probiotics are in widespread use among shrimp farmers, many probiotic products are applied without much background knowledge. Given that many types of probiotics are being used in aquaculture, there may be different ways in which these products can affect disease resistance or host immunity, either directly in the shrimp itself or indirectly via improvement of the rearing environment. Progress in this research area will aid our understanding of how certain probiotics function, or how different probiotics interact among one another. With that knowledge, it would then be possible to measure the exact effects of probiotics, which could lead to better targeting of specific probiotics or meaningful combinations of probiotics with desirable functions. Improvements could also be made in either selecting or screening for high-functioning new strains and in optimizing their usage for maximum effect.

The shrimp gut microbiota can play important roles in preventing disease, minimizing the use of antibiotics, and maintaining healthy animals. The use of probiotics (and, similarly, prebiotics) will most likely remain a popular means of manipulating the microbiota in aquaculture, given its long history of use. Other practical methods to shape the gut microbiota could involve manipulating factors that influence the gut microbiota and host immunity, such as the rearing environment (Pilotto et al., 2020; Vadstein et al., 2018). In the further future, for the shrimp gut microbiota to function as a potential monitoring tool of reliable production, some form of equipment or test kit for on-site analysis of the microbiota that can provide both qualitative and quantitative information would be needed in place of the qPCR or next-generation sequencing techniques currently used in research.

References

Alavandi, S.V., Vijayan, K.K., Santiago, T.C., Poornima, M., Jithendran, K.P., Ali, S.A., and Rajan, J.J. (2004) Evaluation of *Pseudomonas* sp. PM 11 and *Vibrio fluvialis* PM 17 on immune indices of tiger shrimp, *Penaeus monodon*. *Fish & Shellfish Immunology* 17, 115–120.

Amann, R.I., Ludwig, W., and Schleifer, K.H. (1995) Phylogenetic identification and in situ detection of individual microbial cells without cultivation. *Microbiology Reviews* 59, 143–169.

Anderson, J.L., Valderrama, D., and Jory, D.E. (2019) Goal 2019: Global Shrimp Production Review; GOAL: Portsmouth, NH. Global Aquaculture.

Andriantahina, F., Liu, X., Feng, T., and Xiang, J. (2013) Current status of genetics and genomics of reared penaeid shrimp: information relevant to access and benefit sharing. *Marine Biotechnology* 15, 399–412.

Angthong, P., Uengwetwanit, T., Arayamethakorn, S., Chaitongsakul, P., Karoonuthaisiri, N., and Rungrassamee, W. (2020) Bacterial analysis in the early developmental stages of the black tiger shrimp (*Penaeus monodon*). *Scientific Reports* 10, 4896.

Balcázar, J.L., Rojas-Luna, T., and Cunningham, D.P. (2007) Effect of the addition of four potential probiotic strains on the survival of pacific white shrimp (*Litopenaeus vannamei*) following immersion challenge with *Vibrio parahaemolyticus*. *Journal of Invertebrate Pathology* 96, 147–150.

Califano, G., Castanho, S., Soares, F., Ribeiro, L., Cox, C.J., Mata, L., and Costa, R. (2017) Molecular taxonomic profiling of bacterial communities in a gilthead seabream (*Sparus aurata*) hatchery. *Frontiers in Microbiology* 8, 204.

Carlson, J.M., Leonard, A.B., Hyde, E.R., Petrosino, J.F., and Primm, T.P. (2017) Microbiome disruption and recovery in the fish *Gambusia affinis* following exposure to broad-spectrum antibiotic. *Infection and Drug Resistance* 10, 143–154.

Chaiyapechara, S., Rungrassamee, W., Suriyachay, I., Kuncharin, Y., Klanchui, A., Karoonuthaisiri, N., and Jiravanichpaisal, P. (2012) Bacterial community associated with the intestinal tract of *P. monodon* in commercial farms. *Microbial Ecology* 63, 938–953.

Cheng, Z.J. and Hardy, R.W. (2004) Protein and lipid sources affect cholesterol concentrations of juvenile Pacific white shrimp, *Litopenaeus vannamei* (Boone). *Journal of Animal Science* 82, 1136–1145.

Cornejo-Granados, F., Lopez-Zavala, A.A., Gallardo-Becerra, L., Mendoza-Vargas, A., Sánchez, F., Vichido, R., Brieba, L.G., Viana, M.T., Sotelo-Mundo, R.R., and Ochoa-Leyva, A. (2017) Microbiome of Pacific Whiteleg shrimp reveals differential bacterial community composition between Wild, Aquacultured and AHPND/EMS outbreak conditions. *Scientific Reports* 7, 11783.

Costa, M. and Weese, J.S. (2019) Methods and basic concepts for microbiota assessment. *The Veterinary Journal* 249, 10–15.

Dalmin, G., Kathiresan, K., and Purushothaman, A. (2001) Effect of probiotics on bacterial population and health status of shrimp in culture pond ecosystem. *Indian Journal of Experimental Biology* 39, 939–942.

Danovaro, R., Luna, G.M., Dell'Anno, A., and Pietrangeli, B. (2006) Comparison of two fingerprinting techniques, terminal restriction fragment length polymorphism and automated ribosomal intergenic spacer analysis, for determination of bacterial diversity in aquatic environments. *Applied and Environmental Microbiology* 72, 5982–5989.

De Schryver, P. and Vadstein, O. (2014) Ecological theory as a foundation to control pathogenic invasion in aquaculture. *The ISME Journal* 8, 2360–2368.

Del'Duca, A., Cláudio Galuppo, D., Dionéia Evangelista, C., Paulo César, A. (2013) Evaluation of the presence and efficiency of potential probiotic bacteria in the gut of tilapia (*Oreochromis niloticus*) using the fluorescent in situ hybridization technique. *Aquaculture* 388, 115–121.

Devaraja, T., Banerjee, S., Yusoff, F., Shariff, M., Khatoon, H. (2013) A holistic approach for selection of *Bacillus* spp. as a bioremediator for shrimp postlarvae culture. *Turkish Journal of Biology* 37, 92-100.

Eckburg, P.B., Bik, E.M., Bernstein, C.N., Purdom, E., Dethlefsen, L., Sargent, M., Gill, S.R., Nelson, K.E., Relman, D.A. (2005) Diversity of the human intestinal microbial flora. *Science* 308, 1635–1638.

El-Sersy, N., AbdelRazek, F.A., Taha, S.M. (2006) Evaluation of various probiotic bacteria for the survival of *Penaeus japonicus* larvae. *Fresenius Environmental Bulletin* 15, 1506-1511.

Engering, A., Hogerwerf, L., Slingenbergh, J. (2013) Pathogen-host-environment interplay and disease emergence. *Emerging Microbes & Infections* 2, 1–7.

Fan, L., Barry, K., Hu, G., Meng, S., Song, C., Qiu, L., Zheng, Y., Wu, W., Qu, J., Chen, J., and Xu, P. (2017) Characterizing bacterial communities in tilapia pond surface sediment and their responses to pond differences and temporal variations. *World Journal of Microbiology and Biotechnology* 33, 1.

Francino, M.P. (2015) Antibiotics and the human gut microbiome: dysbioses and accumulation of Resistances. *Frontiers in Microbiology* 6, 1543.

Giraffa, G., Chanishvili, N., and Widyastuti, Y. (2010) Importance of lactobacilli in food and feed biotechnology. *Research in Microbiology* 161, 480–487.

Glencross, B., Tabrett, S., Irvin, S., Wade, N., Anderson, M., Blyth, D., Smith, D., Coman, G., and Preston, N. (2013) An analysis of the effect of diet and genotype on protein and energy utilization by the black tiger shrimp, *Penaeus monodon* – why do genetically selected shrimp grow faster? *Aquaculture Nutrition* 19, 128–138.

Greetham, H.L., Giffard, C., Hutson, R.A., Collins, M.D., and Gibson, G.R. (2002) Bacteriology of the Labrador dog gut: a cultural and genotypic approach. Journal of Applied Microbiology 93, 640–646.
Gupta, S., Fernandes, J., and Kiron, V. (2019) Antibiotic-induced perturbations are manifested in the dominant intestinal bacterial phyla of Atlantic salmon. *Microorganisms* 7, 233.
Hewson, I. and Fuhrman, J.A. (2004) Richness and diversity of bacterioplankton species along an estuarine gradient in Moreton Bay, Australia. *Applied and Environmental Microbiology* 70, 3425–3433.
Hoseinifar, S.H., Roy, S., Kumar, V., Lymbery, A., and Ringoe, E. (2019) Genus *Bacillus*, Promising probiotics in aquaculture: aquatic animal origin, bio-active components, bioremediation and efficacy in fish and shellfish. *Reviews in Fisheries Science & Aquaculture* 27, 331–379.
Huang, Z., Chen, Y., Weng, S., Lu, X., Zhong, L., Fan, W., Chen, X., Zhang, H., and He, J. (2016a) Multiple bacteria species were involved in hepatopancreas necrosis syndrome (HPNS) of *Litopenaeus vannamei*. *Journal of Sun Yat-sen University (Natural Science Edition)* 55, 1–11.
Huang, Z., Li, X., Wang, L., and Shao, Z. (2016b) Changes in the intestinal bacterial community during the growth of white shrimp, *Litopenaeus vannamei*. *Aquaculture Research* 47, 1737–1746.
Jernberg, C., Löfmark, S., Edlund, C., and Jansson, J.K. (2010) Long-term impacts of antibiotic exposure on the human intestinal microbiota. *Microbiology* 156, 3216–3223.
Johnson, C.N., Barnes, S., Ogle, J., Grimes, D.J., Chang, Y.-J., Peacock, A.D., and Kline, L. (2008) Microbial community analysis of water, foregut, and hindgut during growth of Pacific white shrimp, *Litopenaeus vannamei*, in closed-system aquaculture. *Journal of the World Aquaculture Society* 39, 251–258.
Kim, D.-H., Brunt, J., and Austin, B. (2007) Microbial diversity of intestinal contents and mucus in rainbow trout (*Oncorhynchus mykiss*). *Journal of Applied Microbiology* 102, 1654–1664.
Kim, D., Zeng, M.Y., and Núñez, G. (2017) The interplay between host immune cells and gut microbiota in chronic inflammatory diseases. *Experimental & Molecular Medicine* 49, e339.
Kuebutornye, F.K.A., Abarike, E.D., and Lu, Y. (2019) A review on the application of Bacillus as probiotics in aquaculture. *Fish & Shellfish Immunology* 87, 820–828.
Landsman, A., St-Pierre, B., Rosales-Leija, M., Brown, M., and Gibbons, W. (2019) Investigation of the potential effects of host genetics and probiotic treatment on the gut bacterial community composition of aquaculture-raised Pacific whiteleg shrimp, *Litopenaeus vannamei*. *Microorganisms* 7, 217.
Laranja, J.L., Amar, E., Pascual, G., Niu, Y., Geaga, M., De Schryver, P., and Bossier, P. (2017) A probiotic *Bacillus* strain containing amorphous poly-beta-hydroxybutyrate (PHB) stimulates the innate immune response of *Penaeus monodon* postlarvae. *Fish & Shellfish Immunology* 68, 202–210.
Lazado, C.C., Caipang, C.M.A., and Estante, E.G. (2015) Prospects of host-associated microorganisms in fish and penaeids as probiotics with immunomodulatory functions. *Fish & Shellfish Immunology* 45, 2–12.
Legrand, T., Catalano, S.R., Wos-Oxley, M.L., Stephens, F., Landos, M., Bansemer, M.S., Stone, D.A.J., Qin, J.G., and Oxley, A.P.A. (2017) The inner workings of the outer surface: skin and gill microbiota as indicators of changing gut health in yellowtail kingfish. *Frontiers in Microbiology* 8, 2664.
Li, E., Xiong, Z., Chen, L., Zeng, C., and Li, K. (2008) Acute toxicity of boron to juvenile white shrimp, *Litopenaeus vannamei*, at two salinities. *Aquaculture* 278, 175–178.
Li, E., Xu, C., Wang, X., Wang, S., Zhao, Q., Zhang, M., Qin, J.G., Chen, L. (2018) Gut microbiota and its modulation for healthy farming of Pacific white shrimp *Litopenaeus vannamei*. *Reviews in Fisheries Science & Aquaculture* 26, 381–399.
Li, T., Long, M., Gatesoupe, F.-J., Zhang, Q., Li, A., and Gong, X. (2015) Comparative analysis of the intestinal bacterial communities in different species of carp by pyrosequencing. *Microbial Ecology* 69, 25–36.
Liu, K.F., Chiu, C.H., Shiu, Y.L., Cheng, W., and Liu, C.H. (2010) Effects of the probiotic, *Bacillus subtilis* E20, on the survival, development, stress tolerance, and immune status of white shrimp, *Litopenaeus vannamei* larvae. *Fish & Shellfish Immunology* 28, 837–844.
Madhayan, K. and Bhavan, P. (2018) Supplementation of *Lactobacillus brevis* for growth promotion of the freshwater prawn *Macrobrachium rosenbergii* post larvae and identification of gut microflora through 16s rDNA. *Research Journal of Biotechnology* 13, 34–50.
Merrifield, D.L., Dimitroglou, A., Foey, A., Davies, S.J., Baker, R.T.M., Bøgwald, J., Castex, M., and Ringø, E. (2010)

The current status and future focus of probiotic and prebiotic applications for salmonids. *Aquaculture* 302, 1–18.

Mongkol, P., Bunphimpapha, P., Rungrassamee, W., Arayamethakorn, S., Klinbunga, S., Menasveta, P., and Chaiyapechara, S. (2018) Bacterial community composition and distribution in different segments of the gastrointestinal tract of wild-caught adult *Penaeus monodon*. *Aquaculture Research* 49, 378–392.

Morgan, X.C. and Huttenhower, C. (2012) Chapter 12: Human microbiome analysis. *PLoS Computational Biology* 8, e1002808.

Magda M.E., Sabry S.E., Mohamed A.E., Santoch L., Said M.D., and Neven A.E. (2011) The viability of probiotics as a factor influencing the immune response in the Nile tilapia (*Oreochromis niloticus*). *Egyptian Journal of Aquatic Biology and Fisheries* 15, 105–124.

Niu, J., Xie, J.J., Guo, T.Y., Fang, H.H., Zhang, Y.M., Liao, S.Y., Xie, S.W., Liu, Y.J., and Tian, L.X. (2018) Comparison and evaluation of four species of macro-algaes as dietary ingredients in *Litopenaeus vannamei* under normal rearing and WSSV challenge conditions: effect on growth, immune response, and intestinal microbiota. *Frontiers in Physiology* 9, 1880.

Okocha, R.C., Olatoye, I.O., and Adedeji, O.B. (2018) Food safety impacts of antimicrobial use and their residues in aquaculture. *Public Health Reviews* 39, 21.

Oxley, A.P., Shipton, W., Owens, L., and McKay, D. (2002) Bacterial flora from the gut of the wild and cultured banana prawn, *Penaeus merguiensis*. *Journal of Applied Microbiology* 93, 214–223.

Pérez, T., Balcázar, J.L., Ruiz-Zarzuela, I., Halaihel, N., Vendrell, D., de Blas, I., and Múzquiz, J.L. (2010) Host–microbiota interactions within the fish intestinal ecosystem. *Mucosal Immunology* 3, 355–360.

Pilotto, M.R., Argenta, N., Forte, J.M., Hostins, B., Menezes, F.G.R., Maggioni, R., de Sousa, O.V., Wasielesky, W., Jr., Rosa, R.D., and Perazzolo, L.M. (2020) Environmental rearing conditions are key determinants of changes in immune gene expression patterns in shrimp midgut. *Developmental & Comparative Immunology* 106, 103618.

Ringø, E., Hoseinifar, S.H., Ghosh, K., Doan, H.V., Beck, B.R., and Song, S.K. (2018) Lactic acid bacteria in finfish—an update. *Frontiers in Microbiology* 9, 1818.

Ronquillo, J.D., Saisho, T., and McKinley, R.S. (2006) Early developmental stages of the green tiger prawn, *Penaeus semisulcatus* de Haan (Crustacea, Decapoda, Penaeidae). *Hydrobiologia* 560, 175–196.

Rosado, D., Xavier, R., Severino, R., Tavares, F., Cable, J., and Pérez-Losada, M. (2019) Effects of disease, antibiotic treatment and recovery trajectory on the microbiome of farmed seabass (*Dicentrarchus labrax*). *Scientific Reports* 9, 18946.

Rungrassamee, W., Klanchui, A., Chaiyapechara, S., Maibunkaew, S., Tangphatsornruang, S., Jiravanichpaisal, P., and Karoonuthaisiri, N. (2013) Bacterial population in intestines of the black tiger shrimp (*Penaeus monodon*) under different growth stages. *PLoS One* 8, e60802.

Rungrassamee, W., Klanchui, A., Maibunkaew, S., and Karoonuthaisiri, N. (2016) Bacterial dynamics in intestines of the black tiger shrimp and the Pacific white shrimp during *Vibrio harveyi* exposure. *Journal of Invertebrate Pathology* 133, 12–19.

Sha, Y., Wang, L., Liu, M., Jiang, K., Xin, F., and Wang, B. (2016) Effects of lactic acid bacteria and the corresponding supernatant on the survival, growth performance, immune response and disease resistance of *Litopenaeus vannamei*. *Aquaculture* 452, 28–36.

Shakibazadeh, S., Saad, C.R., Christianus, A., Kamarudin, M.S., Sijam, K., Nor Shamsudin, M., and Neela, V.K. (2009) Bacteria flora associated with different body parts of hatchery reared juvenile *Penaeus monodon*, tanks water and sediment. *Annals of Microbiology* 59, 425–430.

Shao, J., Wang, B., Liu, M., Jiang, K., Wang, L., and Wang, M. (2019) Replacement of fishmeal by fermented soybean meal could enhance the growth performance but not significantly influence the intestinal microbiota of white shrimp *Litopenaeus vannamei*. *Aquaculture* 504, 354–360.

Shefat, S.H. (2018) Use of probiotics in shrimp aquaculture in Bangladesh. *Acta Scientific Microbiology* 1, 20–27.

Simon, C.J., Truong, H.H., Noble, T.H., Osborne, S.A., Wynne, J.W., and Wade, N.M. (2020) Microbial biomass, marine invertebrate meals and feed restriction influence the biological and gut microbiota response of shrimp *Penaeus monodon*. *Aquaculture* 520, 734679.

Siriyappagouder, P., Galindo-Villegas, J., Lokesh, J., Mulero, V., Fernandes, J.M.O., and Kiron, V. (2018) Exposure to yeast shapes the intestinal bacterial community assembly in zebrafish larvae. *Frontiers in Microbiology* 9, 1868.

Sun, Z., Li, G., Wang, C., Jing, Y., Zhu, Y., Zhang, S., and Liu, Y. (2014) Community dynamics of prokaryotic and eukaryotic microbes in an estuary reservoir. *Scientific Reports* 4, 6966.

Tanaka, R., Ootsubo, M., Sawabe, T., Ezura, Y., and Tajima, K. (2004) Biodiversity and in situ abundance of gut microflora of abalone (*Haliotis discus hannai*) determined by culture-independent techniques. *Aquaculture* 241, 453–463.

Tang, Y., Tao, P., Tan, J., Mu, H., Peng, L., Yang, D., Tong, S., and Chen, L. (2014) Identification of bacterial community composition in freshwater aquaculture system farming of *Litopenaeus vannamei* reveals distinct temperature-driven patterns. *International Journal of Molecular Sciences* 15, 13663–13680.

Thompson, J., Gregory, S., Plummer, S., Shields, R.J., and Rowley, A.F. (2010) An *in vitro* and *in vivo* assessment of the potential of *Vibrio* spp. as probiotics for the Pacific white shrimp, *Litopenaeus vannamei*. *Journal of Applied Microbiology* 109, 1177–1187.

Tringe, S.G. and Hugenholtz, P. (2008) A renaissance for the pioneering 16S rRNA gene. *Current Opinion in Microbiology* 11, 442–446.

Tzeng, T.D., Pao, Y.Y., Chen, P.C., Weng, F.C., Jean, W.D., and Wang, D. (2015) Effects of host phylogeny and habitats on gut microbiomes of oriental river prawn (*Macrobrachium nipponense*). *PLoS One* 10, e0132860.

Uyaguari, M., Key, P., Moore, J., Jackson, K., and Scott, G. (2009) Acute effects of the antibiotic oxytetracycline on the bacterial community of the grass shrimp, *Palaemonetes pugio*. *Environmental Toxicology and Chemistry* 28, 2715–2724.

Vadstein, O., Attramadal, K.J.K., Bakke, I., and Olsen, Y. (2018) K-selection as microbial community management strategy: a method for improved viability of larvae in aquaculture. *Frontiers in Microbiology* 9, 2730.

Whetstone, J.M., Treece, G.D., Browdy, C.L., and Stokes, A.D. (2002) Opportunities and constraints in marine shrimp farming. Southern Regional Aquaculture Center Publishcation No. 2600.

Xiong, J. (2018) Progress in the gut microbiota in exploring shrimp disease pathogenesis and incidence. *Applied Microbiology and Biotechnology* 102, 7343–7350.

Xiong, J., Dai, W., and Li, C. (2016) Advances, challenges, and directions in shrimp disease control: the guidelines from an ecological perspective. *Applied Microbiology and Biotechnology* 100, 6947–6954.

Xiong, J., Wang, K., Wu, J., Qiuqian, L., Yang, K., Qian, Y., and Zhang, D. (2015) Changes in intestinal bacterial communities are closely associated with shrimp disease severity. *Applied Microbiology and Biotechnology* 99, 6911–6919.

Yang, L., Liu, J., Liu, M., Qian, M., Zhang, M., and Hu, H. (2011) Identification of fatty acid synthase from the Pacific white shrimp, *Litopenaeus vannamei* and its specific expression profiles during white spot syndrome virus infection. *Fish & Shellfish Immunology* 30, 744–749.

Ye, L., Amberg, J., Chapman, D., Gaikowski, M., and Liu, W.T. (2014) Fish gut microbiota analysis differentiates physiology and behavior of invasive Asian carp and indigenous American fish. *The ISME Journal* 8, 541–551.

Zhang, H., Sun, Z., Liu, B., Xuan, Y., Jiang, M., Pan, Y., Zhang, Y., Gong, Y., Lu, X., Yu, D., Kumar, D., Hu, X., Cao, G., Xue, R., and Gong, C. (2016a) Dynamic changes of microbial communities in *Litopenaeus vannamei* cultures and the effects of environmental factors. *Aquaculture* 455, 97–108.

Zhang, M., Sun, Y., Chen, K., Yu, N., Zhou, Z., Chen, L., Du, Z., and Li, E. (2014) Characterization of the intestinal microbiota in Pacific white shrimp, *Litopenaeus vannamei*, fed diets with different lipid sources. *Aquaculture* 434, 449–455.

Zhang, M., Sun, Y., Liu, Y., Qiao, F., Chen, L., Liu, W.-T., Du, Z., and Li, E. (2016b) Response of gut microbiota to salinity change in two euryhaline aquatic animals with reverse salinity preference. *Aquaculture* 454, 72–80.

Zheng, Y., Yu, M., Liu, J., Qiao, Y., Wang, L., Li, Z., Zhang, X.H., and Yu, M. (2017) Bacterial community associated with healthy and diseased Pacific white shrimp (*Litopenaeus vannamei*) larvae and rearing water across different growth stages. *Frontiers in Microbiology* 8, 1362.

Zoetendal, E.G., Akkermans, A.D., and De Vos, W.M. (1998) Temperature gradient gel electrophoresis analysis of 16S rRNA from human fecal samples reveals stable and host-specific communities of active bacteria. *Applied and Environmental Microbiology* 64, 3854–3859.

Zorriehzahra, J., Delshad, S., Adel, M., Tiwari, R., Karthik, K., Dhama, K., and Lazado, C. (2016) Probiotics as beneficial microbes in aquaculture: an update on their multiple modes of action: a review. *The Veterinary Quarterly* 36, 228–241.

Zuo, Z.H., Shang, B.J., Shao, Y.C., Li, W.Y., and Sun, J.S. (2019) Screening of intestinal probiotics and the effects of feeding probiotics on the growth, immune, digestive enzyme activity and intestinal flora of *Litopenaeus vannamei*. *Fish & Shellfish Immunology* 86, 160–168.

Chapter 5

GENOME OF SHRIMP AND THEIR POTENTIAL APPLICATIONS

XIAOJUN ZHANG, JIANBO YUAN, YANG YU, FUHUA LI AND JIANHAI XIANG

Summary

Over the past decades, the genome-based "omics" techniques have made enormous impacts on our understanding of shrimp biological function, trait physiology, and inheritance, which is very important for genetic improvement of the aquaculture species that are of high value. In this chapter, we summarize the progress in genomic research into penaeid shrimp, introduce the process of whole-genome sequencing and the genetic underpinnings of specific biological processes of the Pacific white shrimp *Litopenaeus vannamei*, and describe the prospects for the potential application of these techniques in genetic breeding of shrimp.

Introduction

Penaeid shrimp (family Penaeidae), comprising more than 200 species, represent the most commercially important crustaceans in marine aquaculture (Hurzaid et al., 2020). However, the production of shrimp aquaculture is unstable due to outbreaks of various diseases, the degradation of germplasm, and poor understanding on the biological and genetic basis of production traits. Genetic improvement of aquaculture species offers an opportunity for increased production efficiency, better health management, and ultimately, increases profitability of aquaculture enterprises (Davis and Hetzel, 2000). From the 1990s, research on the genetics and genomics of penaeid shrimp has developed rapidly, and many efforts have been made to utilize genetic resources to investigate the genomic characterization, mechanisms of growth development, and selective breeding in shrimp. However, these studies cannot yet match the pressing needs of the shrimp aquaculture industry.

In the past ten years, "omics" approaches, including high-throughput next-generation sequencing (NGS), single nucleotide polymorphism (SNP) genotyping, high-density genetic map, transcriptomics, proteomics, and so on, have been developed rapidly, offering unprecedented power to uncover functionally important genetic variation at the level of whole genomes, bringing many important scientific and technological achievements. A series of omics studies based on genome sequencing were developed in shrimp, and a vast amount of omics data of high quality were obtained. These data have played a leading role in crustacean genetic studies, and in particular have facilitated the understanding of the basic molecular biology, conservation of germplasm resources, and genetic breeding in shrimp.

As early as 1997, an international workshop on genome mapping of aquaculture animals was founded. The penaeid shrimp was assigned as one of five target aquaculture organisms for genome sequencing, together with salmon, catfish, tilapia, and oyster (Alcivar-Warren et al., 1997). The genomes of the other four species were published earlier (Berthelot et al., 2014; Lien et al., 2016; Liu et al., 2016; Zhang et al., 2012). However, the sequencing of the shrimp genome has encountered many obstacles because of its comparatively large size (~2.6 Gb for the Pacific whiteleg shrimp, *Litopenaeus vannamei*), the difficulty in isolating large fragments of DNA, and the high proportion (~80%) of repetitive sequences (Abdelrahman et al., 2017; Yu et al., 2015), and as a consequence the shrimp genome is considered to be the most difficult for sequencing and assembly among all the genomes of aquaculture animals. At the beginning of 2019, the first shrimp *de novo* reference genome had finally been completed (Zhang et al., 2019d). After that, the genomes of three other economically important penaeid shrimp, *Fenneropenaeus chinensis*, *Penaeus monodon*, and *Marsupenaeus japonicus*, have also been successfully sequenced consecutively (Kawato et al., 2021; Uengwetwanit et al., 2021; Wang et al., 2021; Yuan et al., 2021a). The accomplishment of these reference genomes provided a crucial basis for understanding the molecular mechanisms controlling traits of biological and commercial interest (e.g. growth, reproduction, disease resistance, nutrition, and pigmentation) in shrimp.

Genome research in penaeid shrimp

GENOME SIZE, CHROMOSOME NUMBER, AND POLYPLOIDY

In general, penaeid shrimp have a large genome. The C-value (haploid DNA contents in pictograms (pg), 1 pg = 978 Mb) of Penaeidae species range from 2.14 pg to 3.00 pg (Swathi et al., 2018) (http://www.genomesize.com), suggesting that their genome sizes are generally larger than 2 Gb.

There has been relatively little research undertaken on the chromosome number, structure, and composition in the Penaeidae. In the 1990s, Xiang and other researchers reported techniques for obtaining high-quality somatic and meiotic chromosomes from marine shrimp at developmental stages from embryos to adults (Chow et al., 1990; Xiang, 1988; Xiang et al., 1993, 1996). The chromosome numbers of six species of Penaeidae ranged from 68 to 92 (Xiang et al., 1994). Most of the chromosomes of penaeid shrimp are metacentric and sub-telocentric or telocentric, but it is difficult to define the karyotypes due to their quite small size, similar morphology, and the changes in different tissues or with different methods of preparation.

There is no confirmed evidence of genome duplication in the Penaeidae, even though they have large genome sizes and numerous chromosomes. Artificial triploids have been successfully induced in several shrimp species, including *F. chinensis*, *Fenneropenaeus indicus*, *Fenneropenaeus merguiensis, M. japonicus*, *Metapenaeus ensis*, *L. vannamei*, and *P. monodon* (Coman et al., 2008; Li et al., 2003, 2006a; Sellars et al., 2010, 2012; Xiang et al., 2006), and the optimized triploid induction rate has reached 100%. In triploid shrimp, females are predominant with sterile ovaries, so the triploid ovary provides an ideal model to study the mechanism of molecular regulation of ovarian development (Li et al., 2003; Xie et al., 2008).

MOLECULAR MARKERS AND GENETIC LINKAGE MAPPING

The main purpose of developing molecular markers in shrimp is for locating traits of interest and for molecular breeding. The first molecular marker in shrimp is isozyme, which is used to determine the genetic distance between different populations (Lester, 1979). Isozyme markers were limited in a phenotype-based analysis, and relatively few biochemical assays are available for detection. Thus, they have been replaced by DNA-based molecular markers, such as restriction fragment length polymorphism (RFLP), random amplified polymorphic DNA (RAPD), amplified fragment length polymorphism (AFLP), simple sequence repeat (SSR), and SNP. Especially in recent years, thousands of SNPs and insertion–deletion (indel) markers have been identified in the transcriptome or genome, based on NGS technology. The traditional marker sets described above have been replaced by these powerful and low-cost modern markers.

Genetic linkage maps are highly versatile genomic resources that provide abundant genomic information and facilitate understanding of the genetic architecture of commercially and biologically important traits (Guppy et al., 2018). Markers on the maps can represent DNA regions (genes) or DNA segments with unknown function but whose inheritance pattern can be followed and related to biological characteristics. So, linkage mapping is essential for quantitative trait locus (QTL) mapping of economically important traits. In penaeid shrimp, research on genetic linkage mapping has been conducted for nearly 20 years. Early versions of linkage maps contained a limited number of genetic markers. However, in the past decade, the quality of the linkage map for shrimp has improved dramatically, as thousands of polymorphic SNPs based on NGS technology have made the maps more accurate (Table 5.1). For example, Four high-density linkage maps were constructed in *L. vannamei*, *M. japonicus*, *P. monodon*, and *F. chinensis* (Guo et al., 2019; Lu et al., 2016b; Meng et al., 2021; Yu et al., 2015), with 6146, 9298, 6524 and 12,884 SNP markers, respectively. The average inter-marker distances of 0.7, 0.39, 0.9 and 0.41 cM, respectively, equate to an estimated physical genome distance of ~598.89 kb/cM for *L. vannamei*, ~657.89 kb/cM for *M. japonicus*, ~685 kb/cM for *P. monodon*, and ~503.05 kb/cM for *F. chinensis*. The increase in the density of polymorphic markers achieved has dramatically improved the applicability of these resources, including assisting genomic assembly, mapping QTLs, and undertaking comparative mapping.

EXPRESSED SEQUENCE TAG AND TRANSCRIPTOME SEQUENCING

Expressed sequence tag (EST) sequencing is the earliest method for providing transcriptomic data of organisms with an uncharacterized genome. Since the first shrimp EST analysis (Lehnert et al., 1999), 20,318 EST tags have been submitted to the NCBI database (https://www.ncbi.nlm.nih.gov/). Several laboratories have established their own database websites, such as Marine Genomics (https://www.marinegenomics.org), ShEST Shrimp Genome Project (http://shrimp.ufscar.br/en/introduction/), and Penaeus Genome Database (PAGE) (http://sysbio.iis.sinica.edu.tw/page/). These genome resources have provided vast numbers of gene sequences and annotations, and have given important clues for gene

TABLE 5.1 Genetic linkage maps available for penaeid shrimp.

Species	*Marker type*	*Individuals*	*Linkage groups*	*Mapped markers*	*Map length (cM)*	*Average map distance (cM)*	*Genome coverage (%)*	*Reference*
L. vannamei	SNPs	631	44	4817	4552.5	0.97	98.12	Jones et al. (2017a)
	SNPs	205	44	6146 (4201♂, 4396♀)	4271.43	0.7	98.39	Yu et al. (2015)
	AFLP	192	18	157	NA	NA	NA	Gonçalves et al. (2014)
	AFLP, SSR	43	49	451	313.9	7.6	72.63	Andriantahina et al. (2013)
	SNPs	144	45	418 (413♂, 418♀)	2262.30	NA	37.25♂, 48.27♀	Du et al. (2010)
	AFLP	42	51♂, 47♀	394 (182♂, 212♀)	2116♂, 2771♀	15.6♂, 17.1♀	59♂, 62♀	Pérez et al. (2004)
	AFLP, SSR	94	45	267♂, 319♀	3220.9♂, 4134.4♀	14.5♂, 15.1♀	69♂,75♀	Zhang et al. (2007)
	SSR	91	14	67	689.1	22.2	26	Alcivar-Warren et al. (2007)
P. monodon	SNPs	100	44	6524	3275.4	1.1	96.1	Guo et al. (2019)
	SNPs	1024	44	3959 (757♂, 494)♀	2917♂, 4060♀	0.9♂, 1.2♀	91♂, 82♀	Baranski et al. (2014)
	SSR, EST	42	47♂, 36♀	157♂, 111♀	1101♂, 891.4♀	7.8	NA	Maneeruttanarungroj et al. (2006)
	SSR, EST, SCAR	76	9	27	103.6	3.8	NA	Wuthisuthimethavee et al. (2005)
	AFLP	344	21 (44♂, 43♀)	70 (757♂, 494♀)	2378♂, 2362♀	2.1♂, 2.8♀	NA	Staelens et al. (2008)
	AFLP	126	20	116	1412♂	NA	NA	Wilson et al. (2002)
M. japonicus	SNPs	150	41	9298 (3927♂, 5849♀)	3610.90	0.388	99.06	Lu et al. (2016a)
	AFLP	102	43♂, 31♀	217♂, 125♀	1780♂, 1026♀	NA	NA	Li et al. (2003)
F. chinensis	SNPs	275	44	12,884 (6509♂, 7138♀)	3711♂, 5070♀	0.58♂, 0.72♀	99.43	Meng et al. (2021)
	SNPs	100	21	180 (115♂, 119♀)	899.3	5.4	51.94♂, 53.77♀	Zhang et al. (2013)
	AFLP, SSR	100	47	354	4580.50	11.3	75.80	Liu et al. (2010)
	AFLP	110	44 (35♂, 28♀)	216 (144♂, 103♀)	1772.1	10.42	63.17	Tian et al. (2008)
	SSR, RAPD	82	10♂, 8♀	46♂, 49♀	1144.6♂, 1173♀	12.5♂, 11.28♀	62.01♂, 59.36♀	Sun et al. (2008)
	AFLP	100	36♂, 35♀	194♂, 197♀	1737.3♂, 2191.1♀	11♂, 13.5♀	68.10♂, 69.60♀	Li et al. (2006b)

Notes: AFLP, amplified fragment length polymorphism; EST, expressed sequence tag; RAPD, random amplified polymorphic DNA; SCAR, sequence characterized amplified region; SNPs, single nucleotide polymorphisms; SSR, microsatellite or simple sequence repeats; ♂, male map; ♀, female map; NA, not available.

function studies in shrimp, including studies related to viral resistance, stress tolerance, growth, and reproduction (Aoki et al., 2011; Wang et al., 2006). However, high-throughput whole-transcriptome sequencing approaches are now becoming the preferred method to study the RNA profile of shrimp, due to the ability of these approaches to simultaneously provide more and longer sequence data and differential gene expression information.

Transcriptome sequencing, or RNA-Seq, is based on NGS technology and can potentially cover the entire transcripts in a given tissue. Therefore, it provides a general view of gene expression, which is especially important for the species without a fully sequenced and annotated genome. The ability to produce complete transcriptome profiles has increased dramatically in penaeid shrimp species, and dozens of shrimp transcriptomes are now published every year. At the time of writing, transcriptome sequencing has already contributed to studies on shrimp resistance to pathogens (Chen et al., 2013; Li et al., 2013; Maralit et al., 2018; Ren et al., 2019; Shi et al., 2016; Sookruksawong et al., 2013; Xue et al., 2013; Zeng et al., 2013), abiotic responses (Chen et al., 2015; Guo et al., 2013; Huang et al., 2018; Liu et al., 2020; Lu et al., 2016a; Meng et al., 2019; Zheng et al., 2019; Zhuo et al., 2020), reproductive biology (Peng et al., 2015; Uengwetwanit et al., 2018; Wang et al., 2019a), embryonic development (Li et al., 2012; Sellars et al., 2015; Wei et al., 2014), sexual differentiation (Wang et al., 2019a; Wang et al., 2020a), growth (Santos et al., 2021; Zhao et al., 2021) and molting (Gao et al., 2015) (Table 5.2). These transcriptome data provide a huge quantity of information for identifying novel genes and clarifying the regulatory mechanisms for growth, development, immunity, resistance, and so on, in shrimp.

TABLE 5.2 Summary of currently transcriptome for *penaeid* shrimp.

Species	*Treatment/trait*	*Tissues/organs*	*Number of unigenes*	*Annotation rate*	*Reference*
L. vannamei	Growth	Muscle	33,274	NA	Santos et al. (2021)
	Cold stress	Hepatopancreas	143,029	46.6%	Zhou et al. (2021)
	Sex determination and differentiation	Zoea I larvae	41,270	40.69%	Wang et al. (2020a)
	Spiroplasma infection	Hemolymph	8077	90.97%	Ren et al. (2020)
	WSSV and neuroendocrine-immune system	Hemolymph	44,793	32.91%	Wang et al. (2019)
	Ammonia stress	Hemocytes	61,283	20.28%	Liu et al. (2020)
	High-pH stress	Gill	36,589	29.59%	Huang et al. (2018)
	Ovarian maturation	Eyestalk	127,031	21.71%	Wang et al. (2019)
	VP_{AHPND} and VP_{AHPND} toxin	Hemocytes	127,205	78%	Maralit et al. (2018)
	WSSV and growth	Hepatopancreas and muscle	14,124	33%	Santos et al. (2018)
	Feed efficiency	Muscle	72,120	28.63%	Dai et al. (2017)
	Salinity	Gills	349,012	7.03%	Zhang et al. (2016)
	Acute ammonia stress	Hepatopancreas	78,636	28.84%	Lu et al. (2016b)
	Gonadal development	Testis and ovary	30,304	46.50%	Peng et al. (2015)
	Osmoregulatory stress	Hepatopancreas	38,237	25.20%	Chen et al. (2015)

TABLE 5.2 *continued*

Species	*Treatment/trait*	*Tissues/organs*	*Number of unigenes*	*Annotation rate*	*Reference*
	Molting	Whole shrimp	93,756	16.60%	Gao et al. (2015)
	Larval development	Embryo, nauplius, zoea, mysis, postlarva	66,815	48.40%	Wei et al. (2014)
	WSSV	Hemocytes	52,073	45.30%	Xue et al. (2013)
	WSSV	Hepatopancreas	14,583	73.24%	Chen et al. (2013)
	TSV	Hemolymph, and hemocytes	61,937	20.00%	Sookruksawong et al. (2013)
	TSV	Hepatopancreas	15,004	69.50%	Zeng et al. (2013)
	Nitrite	Hepatopancreas and hemocytes	42,336	55.60%	Guo et al. (2013)
	Embryo development	Larvae (20 days post spawning)	73,505	44.10%	Li et al. (2012)
P. monodon	Reproduction	Ovary and testis	41,136	48.39%	Uengwetwanit et al. (2019)
	Ammonia stress	Hepatopancreas	64,475	34.12%	Li et al. (2018)
	Disease resistance, reproduction, and growth	Eyestalk, stomach, female gonad, male gonad, gill, haemolymph, hepatopancreas, lymphoid organ and tail muscle; embryo, nauplii, zoea, and mysis; post-larval stages ranging from days 1, 4, 10 and 15	236,388	~30%	Huerlimann et al. (2018)
	Growth	Muscle, eyestalk, hepatopancreas, and heart	69,089	17.83%	Nguyen et al. (2016)
	Reproduction, lipid metabolism, and reproduction	Hepatopancreas and ovary	15,867	NA	Rotllant et al. (2015)
M. japonicus	Growth	Muscle	10,194	82.79%	Zhao et al. (2021)
	Innate immune response	Hepatopancreas, gill, hemocytes, stomach and intestine	11,222	94.84%	Zheng et al. (2020)
	Heat stress	Hepatopancreas and gills	95,763	NA	Zheng et al. (2019)
	WSSV	Hepatopancreas	29,180	52%	Ren et al. 2019
	Embryo development	Fertilized egg, embryo, vegetal halves	41,567	7.4–10%	Sellars et al. (2015)
	YHV	Hemocyte	759	47% (DEG)	Pongsomboon et al. (2008)
F. chinensis	Sexual dimorphism	Muscle and gonad	10,795	85.01%	Wang et al. (2019a)
	Low-temperature stress	Juvenile	127,939	46%	Meng et al. (2019)
	WSSV	Hepatopancreas	59,137	75 genes	Shi et al. (2016)
	WSSV	Cephalothorax	46,676	46.28%	Li et al. (2013)

TABLE 5.2 *continued*

Species	*Treatment/trait*	*Tissues/organs*	*Number of unigenes*	*Annotation rate*	*Reference*
F. merguiensis	Reproduction and development	Hepatopancreas, stomach, eyestalk, nerve cord, male gonad, female gonad, androgenic gland region, muscle/cuticle	59,179 (proteins)	34.52%	Powell et al. (2015)
	Body colouration	Cuticle, muscle, androgenic gland, hepatopancreas, stomach, nervous system, eyestalk, male gonad, female gonad	5990	59.80%	Ertl et al. (2013)

Notes: AHPND, Acute hepatopancreatic necrosis disease; TSV, Taura syndrome virus; VP, *Vibrio parahaemolyticus*; WSSV, White spot syndrome virus; YHV, Yellow head virus; DEG. Differentially expressed gene; NA, not available.

Meanwhile, new transcriptome sequencing and analysis technologies are still developing. Long-read RNA sequencing methods, including PacBio Iso-Seq and Nanopore direct RNA sequencing, can sequence full-length RNA fragments without assembly, which provides increased insights into isoform diversity in shrimp (Zeng et al., 2018; Zhang et al., 2019c; Ren et al., 2020; Zheng et al., 2020). Besides, single-cell RNA-Seq has been demonstrated as a powerful technique for the classification of tissue-specific cells and studying cell differentiation in the development of an organ, or even in an individual organism (embryo). These efficient methods have been applied to shrimp (Koiwai, et al., 2021), and are becoming key tools to explore cell differentiation process and figure out the complex gene networks in shrimp species.

PROTEOME ANALYSIS

In penaeid shrimp, variations in the protein expression profiles have been analyzed, with a focus on understanding the effects of pathogen infection (Chai et al., 2010; Chaikeeratisak et al., 2012; Chen et al., 2016; Chongsatja et al., 2007; Kulkarni et al., 2014; Li et al., 2014; Rattanarojpong et al., 2007; Ren et al., 2019; Robalino et al., 2009; Somboonwiwat et al., 2010; Wang et al., 2007; Wu et al., 2007; Zhang et al., 2010a), immunity (Robalino et al., 2009; Wang et al., 2020b), environmental stress (Guo et al., 2020; Fan et al., 2013, 2016, 2019; Jiang et al., 2009; Li et al., 2021; Lin et al., 2021; Xu et al., 2017), and nutrition (Qiao et al., 2011; Silvestre et al., 2010) (Table 5.3). Comparative proteomic profiling during ovarian development and cold stress was also used to identify proteins in shrimp species such as *M. ensis* and *L. vannamei* (Cui et al., 2014; Fan et al., 2019). These results demonstrate the application of proteomic analysis for the identification of proteins involved in immunity, reproduction and stress tolerance, and they provide new insights into the physiological process and metabolism in shrimp.

A publicly available peptide identification database has critical importance for proteomic research. However, these resources rely on existing homologous protein-coding sequence databases, which remain very insufficient for many non-model organisms like shrimp. High-quality transcriptome data can greatly

TABLE 5.3 Summary of currently penaeid shrimp proteomic studies.

Species	*Treatment/trait*	*Tissues/organs*	*Proteins identified*	*Differentially expressed protein*
L. vannamei	Hydroxyl radical	Muscle	NA	116
	Copper stress	Gill	5034	385
	Temperature fluctuation	Intestine	5796	27
	Cold stress	Hepatopancreas	4062	137
	Microsporidia	Hepatopancreas	3417	266
	Low-salinity stress	Hepatopancreas	533	84
	WSSV	Hemocytes	1528	123
	WSSV	Hemolymph	486	18
	T-2 toxin	Muscle	11	30
	Cold stress	Hepatopancreas	3349	1149 (up-regulated), 815 (down-regulated)
	Ammonia stress	Hepatopancreas	738	202
	WSSV	Gills	~500 spots, 40 proteins	20
	Cold stress	Hepatopancreas	913 spots, 37 proteins	37
	Cold stress	Hemocytes	755 spots, 30 proteins	30
	Vitamin C and Chinese herbs	Hemocytes, Hepatopancreas	220–400 spots, 10 proteins	29, 28
	WSSV	Gills	ND spots, 37 proteins	37
	TSV	Hemocytes	320–400 spots, 20 proteins	32
	YHV	Gills	>200 spots, 20 proteins	18
	WSSV	Stomach	500 spots, 53 proteins	27
P. monodon	Antibiotics	Hemolymph	2394 spots, 2 proteins	9
	Vibrio harveyi	Lymphoid organ	~1000 spots, 20 proteins	27
	Vibrio harveyi	Hemocytes	29 proteins	27
	WSSV	Stomach	>200 spots, 17 proteins	20
	YHV	Lymphoid organ	370–420 spots, 26 proteins	33
	WSSV	Epithelium	429 proteins	12
F. chinensis	WSSV	Hemocytes	24 proteins in granulocytes and 23 proteins in hyalinocytes	26, 24
	Hypoxic stress	Hepatopancreas	620–640 spots, 33 proteins	67
	WSSV	Hepatopancreas	580–600 spots, 81 proteins	81
	WSSV	Hemocytes	580 spots, 33 proteins	45
	Vibrio anguillarum, WSSV	Hemocytes	680–710 spots	48 (Vibrio-challenged), 49 (WSSV-challenged)
	Vibrio anguillarum	Lymphoid organ	700 spots, 95 proteins	17

Detailed technique	Reference
HPLC-MS/MS	Lin et al. (2021)
iTRAQ, PRM, HPLC	Guo et al. (2021)
TMT, LC-MS/MS	
NanoLC-ESI-MS/MS	Fan et al. (2019)
HPLC, LC-MS/MS, GC/MS	Ning et al. (2019)
iTRAQ, LC-MS/MS	Xu et al. (2019)
iTRAQ, LC-MS/MS	Hernández-Pérez et al. (2019)
iTRAQ, nanoLC-MS/MS	Tao et al. (2019)
MALDI-TOF-TOF MS	Huang et al. (2020)
2D gels, LC-MS/MS	Peng et al. (2018)
iTRAQ, LC-MS/MS	Lu et al. (2018)
nanoLC-ESI-MS/MS	Chen et al. (2016)
2D gels, MALDI-TOF/TOF MS	Fan et al. (2016)
2D gels, MALDI-TOF/TOF MS	Fan et al. (2013)
2D gels, MALDI-TOF/MS	Qiao et al. (2011)
Not described	Robalino et al. (2009)
2D gels, nanoLC-ESI-MS/MS	Chongsatja et al. (2007)
2D gels, nanoLC-ESI-MS/MS	Rattanarojpong et al. (2007)
2D gels, nanoLC-ESI-MS/MS	Wang et al. (2007)
2D DIGE, nanoLC-ESI-MS/MS	Silvestre et al. (2010)
2D gels, nanoLC-ESI-MS/MS	Chaikeeratisak et al. (2012)
2D gels, nanoLC-ESI-MS/MS	Somboonwiwat et al. (2010)
2D gels nanoLC-ESI-MS/MS	Kulkarni et al. (2014)
MALDI-TOF MS, nanoLC-ESI-MS/MS	Bourchookarn et al. (2008)
iCAT or SCX, MALDI-TOF/TOF-MS/MS	Wu et al. (2007)
2D gels, MALDI-TOF/TOF MS	Zhu et al. (2018)
2D gels LC-ESI-MS/MS	Jiang et al. (2009)
2D gels, MALDI-TOF/TOF MS	Chai et al. (2010)
2D gels, MALDI-TOF MS/MS	Li et al. (2014)
2D gels, nanoLC-ESI-MS/MS	Jiang et al. (2014)
2D gels, nanoLC-ESI-MS/MS	Zhang et al. (2010a)

Notes: HPLC, High Performance Liquid Chromatography; LC, liquid chromatography; MS, mass spectrometry; MS/MS, tandem mass spectrometry; iTRAQ, Multiplexed isobaric tagging technology for relative quantitation; PRM, Parallel reaction monitoring; TMT, Tandem mass tags; ESI, electrospray ionization; GC/MS, Gas Chromatography Mass Spectrometry; MALDI, matrix-assisted laser desorption/ionization; TOF, time of flight; 2D gel, Two dimensional gel electrophoresis; 2D DIGE, Two-dimensional difference gel electrophoresis; iCAT, isotope-coded affinity tags; SCX, strong cation exchange; NA, not available.

Source: This table is modified and supplemented from Guppy et al. (2018).

improve the identification of peptide fragments and proteins within proteomics workflows, even if the proteins have no known functions. So far, in the proteomic analysis of shrimp, most of the tandem mass spectra were searched against a protein database that was constructed with the deduced peptide sequences from the transcriptomes, optimally from the research organism under investigation (Guppy et al., 2018). Since the transcriptome data of shrimp are abundant enough now, the penaeid shrimp research community should organize and construct a more comprehensive protein database to improve data utilization and sharing.

WHOLE-GENOME SEQUENCING

Comprehensive and complete genomic assemblies are most crucial omics resource. However, the shrimp genome is considered to be the most difficult for sequencing and assembly among all the genomes of aquaculture animals, due to the difficulties in isolating the large genomic DNA fragments, the large number of chromosomes, and the high level of repetitive sequences (Abdelrahman et al., 2017). Therefore, the progress of genome sequencing in shrimp, even relative crustaceans, has been slow and tortuous (Table 5.4). Previously, several genome surveys were carried out in *L. vannamei*, *P. monodon*, and *M. japonicus* (Yu et al., 2015; Yuan et al., 2018); some genome data for *F. chinensis* were also submitted to the database (Abdelrahman et al., 2017). However, these low-coverage data cannot be effectively applied to the analysis of genetic structure and used as a reference for genetic breeding. So far, the only officially published shrimp genomes are that of *L. vannamei* (Zhang et al., 2019d) and, very recently published, that of *P. monodon* (Uengwetwanit et al., 2021), *F. chinensis* (Wang et al., 2021; Yuan et al., 2021a), and *M. japonicus* (Kawato et al., 2021). These four species are the most predominant farmed penaeid shrimp. Among them, *L. vannamei* accounts for ~80% of the world's production of aquaculture shrimp (Zhang et al., 2019d). These genomic data provide the most important sequence data basis for shrimp biology and genetic breeding. The sequencing process and structural characteristics of the *L. vannamei* genome will be discussed in the following sections.

Genome sequencing and genome structure of *L. vannamei*

As *L. vannamei* is the most economically important aquaculture shrimp, the sequencing of its genome has been paid more attention than that of other species. A detailed workflow of the Shrimp Genome Project was designed in 2009 (Fig 5.1), when NGS technology was just beginning to be applied in aquaculture organisms. After long and arduous efforts, the core work of the Shrimp Genome Project, a high-quality reference genome map of *L. vannamei*, was published in 2019 (Zhang et al., 2019d).

LARGE-INSERT GENOMIC LIBRARY CONSTRUCTION

A large-insert genomic DNA library was an essential platform for genome analysis, positional cloning, physical mapping, and genome sequencing. A bacterial artificial chromosome (BAC) library is a typical large insert library, which contains insert sizes of 100–300 kb, and has been a crucial tool in the sequencing

TABLE 5.4 Status of whole-genome sequencing and assembly of penaeid shrimp and other crustaceans.

Order	*Species*	*Genome size (Gb)*	*Contig N50 (kb)*	*Scaffold N50 (kb)*	*Percentage on chromosome*	*Number of scaffolds*	*Assembly size (Gb)*	*Sequencing platform*	*Reference*
Class Branchiopoda									
Diplostraca	*Daphnia pulex*	0.2	–	318.5	–	5191	0.26	Sanger	Colbourne et al. (2011)
Class Copepoda									
Calanoida	*Eurytemora affinis*	0.587–0.685	5.7	862.6	–	6899	0.495	Illumina	Eyun et al. (2017)
Class Malacostraca									
Amphipoda	*Parhyale hawaiensis*	3.6	0.627	161.89	–	133,035	4.02	Illumina	Kao et al. (2016)
	Hyalella azteca	1.05	5.445	987.997	–	10,380	1.18	Illumina	Poynton et al. (2018)
Decapoda	*Penaeus monodon*	2.59	79	79	82.9	26,887	2.39	PacBio, Chicago, Hi-C	Uengwetwanit et al. (2021)
	Litopenaeus vannamei	2.6	57.65	605.56	87.34	4683	1.66	Illumina; PacBio	Zhang et al. (2019d)
	Marsupenaeus japonicus	1.93	113.129	234.949	88	18,210	1.70	Illumina, Nanopore	Kawato et al. (2021)
	Fenneropenaeus chinensis	1.88	59	28,917	–	8768	1.58	Illumina, PacBio, HiC	Yuan et al. (2021)
		1.38	472.84	–	99	9,015	1.47	Illumina, PacBio, HiC	Wang et al. (2021)
	Procambarus virginalis	3.5	–	39.3	–	–	3.5	Illumina	Gutekunst et al. (2018)
	Cherax quadricarinatus	~5	–	33.2	–	508,682	3.24	Illumina, Nanopore	Tan et al. (2020)
	Homarus americanus	3.2	133.3	759.6	90	47,245	2.29	Illumina, Nanopore, Chicago, Hi-C	Polinski et al. (2021)
	Eriocheir sinensis	1.66	0.224	111.8	–	17,553	1.12	Illumina	Song et al. (2016)
		1.19	–	3,186	–	94	1.27	Illumina, Nanopore	Tang et al. (2020a)
		1.45	26,045	17,128	80.82	10,791	1.57	Illumina, PacBio, 10×Genomics	Cui et al. (2021)
	Portunus trituberculatus	~0.919	4,121	21,794	–	17	1.01	Illumina, Nanopore, Hi-C	Tang et al. (2020b)

Notes: –, not available; Chicago, in vitro proximity ligation, an in vitro method for generating long-range linkage data that improves the scaffolding of de novo assembled genomes; Hi-C, High-through chromosome conformation capture; 10×Genomics, 10× Genomics Chromium sequencing technology.

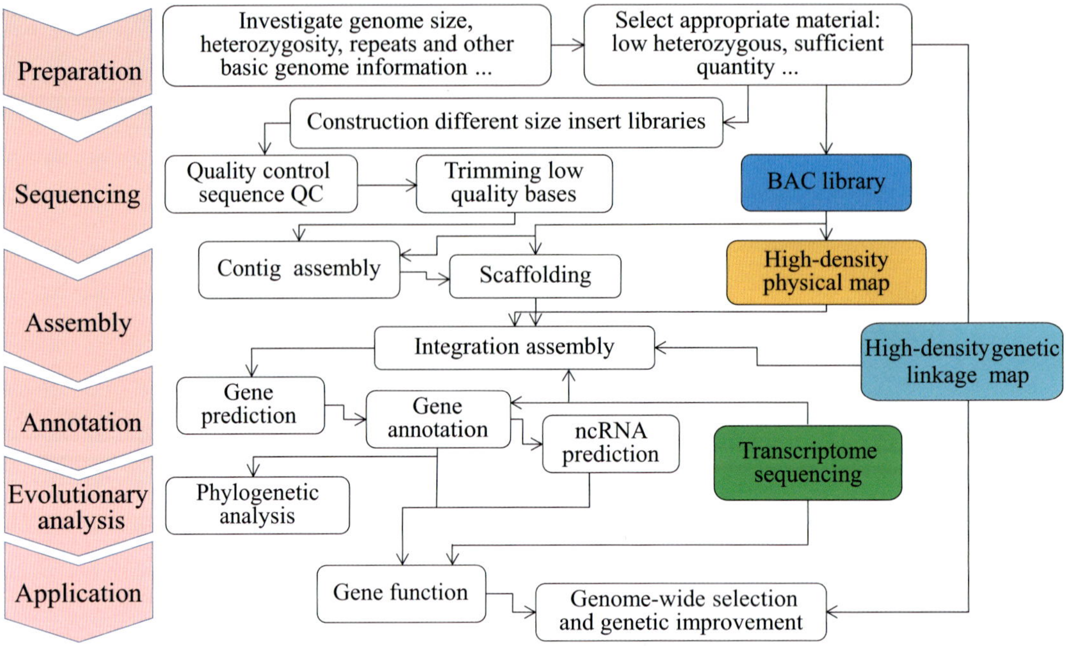

FIGURE 5.1 The workflow of the Shrimp Genome Project.

of the human genome. However, for shrimp, it is difficult to generate a stable large insert BAC library due to difficulties in isolating sufficient unique BAC fragments and clones, only two BAC libraries are available in *L. vannamei* and *M. japonicus* so far. The *L. vannamei* BAC library comprises 101,760 clones, with an average insert size of 101 kb and 5× genome coverage (Zhang et al., 2010b); the *M. japonicus* BAC library consists of 49,152 BAC clones, with an average insert size of ~135 kb (Koyama et al., 2010). In consideration of the difficulty in constructing a large insert library, several mid-sized (~40 kb) shrimp fosmid libraries (for *P. monodon* and *L. vannamei*) had been constructed (Huang et al., 2011; Saski et al., 2009). Approximately 20,926 fosmid end sequences and 28,000 BAC end sequences were sequenced in *P. monodon* and *L. vannamei*, respectively (Huang et al., 2011; Zhao et al., 2012).

GENOME SEQUENCING, ASSEMBLY, AND ANNOTATION

The genome size of the *L. vannamei* was measured to be 2.45 Gb by flow cytometry (Chow et al., 1990), which was similar to the size (2.60 Gb) estimated by k-mer analysis with Illumina short reads (Zhang et al., 2019d). In earlier attempts, the genome survey analysis highlighted the complexity of the *L. vannamei* genome. Approximately 80% of the genome was occupied by repetitive sequences (Yu et al., 2015). In the genome survey of two other economically important shrimp species, *M. japonicus* and *P. monodon*, at least 626 Mb (34.96%) and 833 Mb (46.68%) of repeats were identified, respectively (Yuan et al., 2018). Therefore, the shrimp genomes have a high proportion of repetitive sequences. When short-read sequencing methods (Illumina) are applied to highly repetitive regions, the short reads (150–350 bp fragments) are of insufficient length to span most repeat regions, resulting in a breakdown in the genome assembly. Moreover, high levels of heterozygosity exacerbate this difficulty, as they are erroneously joined adjacently to each other (Henson et al., 2012; Pryszcz and Gabaldon, 2016). In the early short-read assembly (41.86-fold Illumina data), a total of 6,908,022 contigs with an N50 (a metric that measures the quality of assembled genomes) size of 409 bp were produced (Yu et al., 2015). Likewise, early *P. monodon* assemblies have resulted in a contig average length of 1115bp (Montenegro et al., 2018). In addition, the 10× Genomics platform cannot reconstruct such complex genomes, and the other approach, chromosome conformation capture (Hi-C), requires access to a draft genome which already has a relatively large average size (>100 kb) for effective short-read mapping and further improvement of genome scaffolding for shrimp (Guppy et al., 2018). The traditional genome-assembly methods therefore need to be modified to overcome these difficulties in sequencing shrimp genomes. In particular, the third-generation sequencing technologies (e.g. Pacific Biosciences SMRT) are needed to produce longer sequencing reads (averaging >10 kb) to improve the accuracy and integrity of shrimp genome assembly (Yuan et al., 2021b).

A variety of sequencing technologies were applied in the Shrimp Genome Project for *L. vannamei*, which generated 828 Gb of Illumina clean sequences (338×), 133 Gb of PacBio long reads (54×), and 34,266 BAC end sequences (0.46×). Furthermore, with regard to assembling software, numerous approaches were conducted (Yuan et al., 2021b). Finally, a WTDBG approach (Ruan and Li, 2020) was developed to obtain the best assembly for the shrimp genome. The final assembly, with a size of 1.66 Gb, displayed a higher continuity and accuracy than the earlier short-read assembly, with a contig N50 of 57.65 kb and scaffold N50 length of 605 kb (Zhang et al., 2019d). Over 93% of reads from the Illumina data could be successfully mapped to the genome, and over 94% of the unigenes assembled from RNA-Seq data were covered by the genome, as well as 94.76% of conserved core eukaryotic genes, suggesting high integrity and quality. Further, the genome was aligned to 14 sequenced BAC clones, which were completely covered by corresponding scaffolds with high synteny (Zhang et al., 2019d). To obtain chromosome-level assemble, a high-density linkage map was used to anchor the scaffolds (Yu et al., 2015). A total of 3275 scaffolds were anchored to 44 linkage groups, representing 87.34% of the total genome (Fig 5.2(A)). In addition, a total of 25,596 protein-coding gene models were annotated in the *L. vannamei* genome. Compared with other crustaceans, *L. vannamei* has a longer average exon size (259 bp) and a larger number of exons per gene (5.94).

REPETITIVE SEQUENCES

The repetitive sequences account for 49.38% of the final genome assembly of *L. vannamei*, which is lower than in the previous genome survey (~78%) (Yu et al., 2015), indicating that some repetitive sequences are still missing from the genome assembly. Transposable elements account for 16.17% of the genome, with DNA transposons (En-Spm) and long interspersed elements (LINEs, RTE-BovB, and Penelope) comprising the two major classes. SSRs make up 23.93% of the *L. vannamei* genome, which is a much higher proportion than that found in any other sequenced animal genome (Fig 5.2(B)). The longest SSR, of 13,769 bp, belongs to $(AACCT)_n$, a telomere component that was identified by fluorescence *in situ* hybridization (Fig 5.2(D)) (Zhang et al., 2019d). Dinucleotide repeats are the dominant type of SSRs, with $(AT)_n$, $(AC)_n$, and $(AG)_n$ accounting for 81.40% of total SSRs (Fig 5.2(C)). Most SSRs were located in intergenic regions and introns, with much fewer found in exon regions. In other penaeid

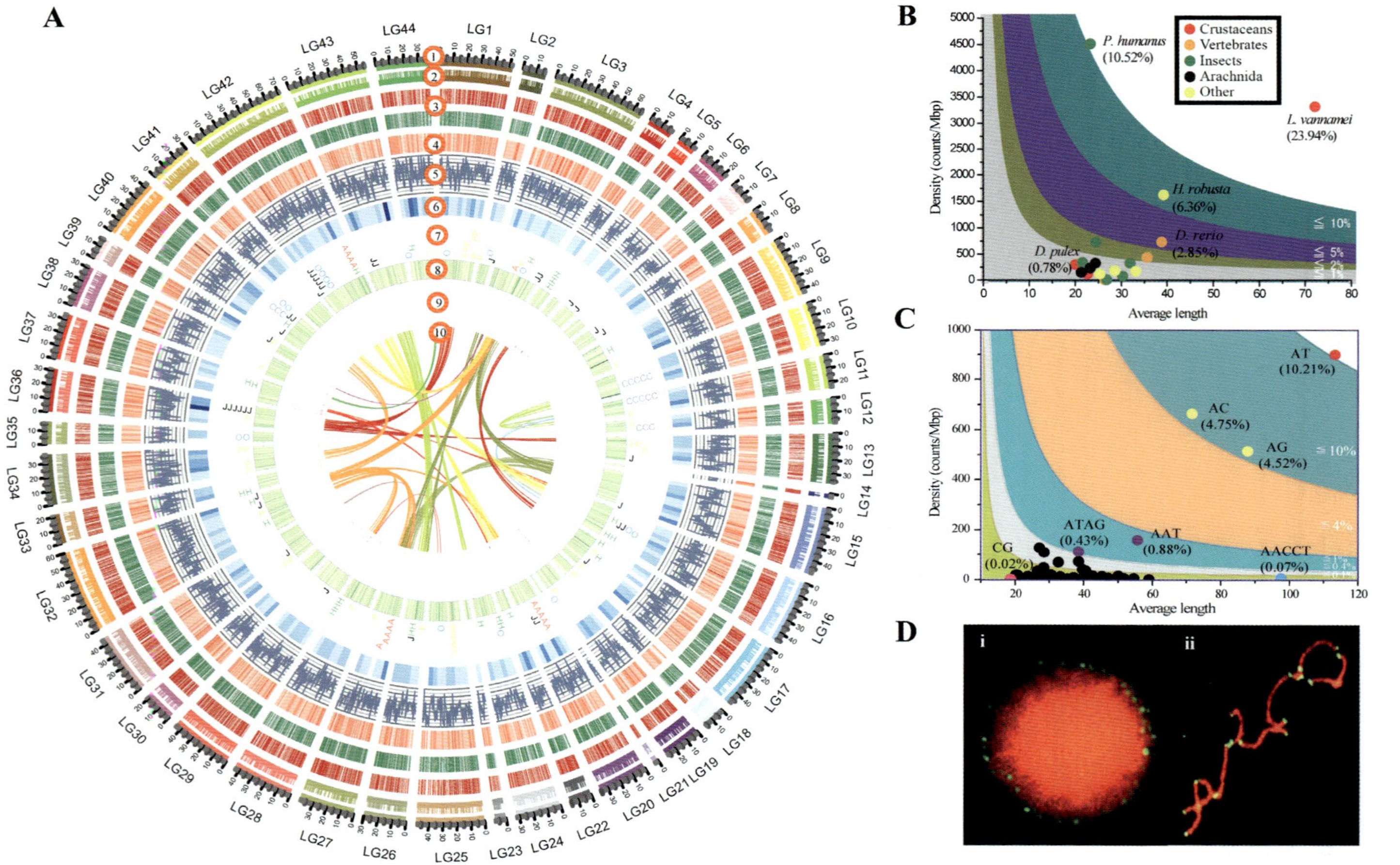

FIGURE 5.2 The genomic characteristics of *L. vannamei.* (A) Schematic representation of the genomic characteristics of *L. vannamei.* (B) SSR content among different animal genomes. (C) Distribution of different types of SSRs in the *L. vannamei* genome. (D) Fluorescence *in situ* hybridization of $(AACCT)_n$-type SSRs in the nucleus (i) and the chromosomes (ii) of *L. vannamei.* Credit: Zhang et al. (2019d).

shrimp, whole genome sequencings also showed that SSR accounted for a large proportion (19.12% for *F. chinensis*; 23.93% for *P. monodon*, and 27.44% for *M. japonicus*) in the assemblies (Kawato et al., 2021; Uengwetwanit et al., 2021; Yuan et al., 2021a). By comparison the genomes of the marble crayfish *Procambarus virginalis* (Gutekunst et al., 2018) and the amphipod *Parhyale hawaiensis* (Kao et al., 2016) contain only 0.99% and 1.27% of SSRs in their genome, respectively. Therefore, the expansion of SSRs was indicated as stems from the common ancestor of penaeid shrimp.

EXPANDED GENE FAMILIES AND TANDEM DUPLICATED GENES

The *L. vannamei* genome has 762 expanded gene families and 16,291 species-specific genes (>57% of the entire gene repertoire), including genes related to the myosin complex, chitin binding, metabolic processes, and signal transduction (Zhang et al., 2019d). Gene tandem duplication was one of the specific genomic characters identified in the *L. vannamei* genome. A total of 4662 genes (18.21%) were duplicated in tandem along the genome with high sequence similarities (identity >98%), suggesting that they were recent duplications. These genes mainly consist of genes coding for opsins, crustacyanins, chitinases, cuticle proteins, myosins, actins, and heat shock proteins, which suggests that they are important for the adaptive evolution of *L. vannamei* (Zhang et al., 2019d). Lineage-specific tandem gene duplication events are a particularly interesting category; they are often related to the unique biology of a taxon (Shingate et al., 2020), which may provide important information on genome evolution related to adaptations.

EVOLUTIONARY ANALYSIS

The clade Pancrustacea comprises all crustaceans and hexapods (Regier et al., 2005). However, the relationships among the constituent lineages of Pancrustacea remain controversial. Many studies supported Branchiopoda as the closest group to Hexapoda, while Malacostraca (the largest class of Crustacea) occupied a sister position to the Hexapoda + Branchiopoda group (Aleshin et al., 2009; Baurain et al., 2007; Regier et al., 2005). However, the results of homologous gene analysis of the shrimp genome did not support Branchiopoda as being close to Hexapoda but, rather, as being close to Malacostraca.

As members of Malacostraca, *L. vannamei* (Decapoda) and *P. hawaiensis* (Amphipoda) diverged at about 240 million years ago, that is, after the Permian-Triassic mass extinction, during which about 96% of marine species became extinct, especially marine invertebrates with calcium carbonate exoskeletons (Shen et al., 2011), followed by the radiation of many new life forms. Furthermore, this divergence period is in accordance with the radiation of shrimp-like decapods at the early Middle Triassic (Feldmann et al., 2017; Fransen and Grave, 2009). The recovery from the mass extinction led to a quick and dynamic rebound in the marine ecosystem, which might have prompted the rapid evolution of decapod shrimp species.

HORIZONTAL GENE TRANSFER

Through sequence homology comparison, phylogenetic analysis, and experimental verification, 14 genes originating from horizontal gene transfer (HGT) were detected in the *L. vannamei* genome, 12 of them transferred from bacteria and two from fungi. Spliceosomal-type introns were found to be inserted in four HGT genes, while non-coding regions of two large horizontally transferred segments were lacking in the shrimp genome. These structural alterations provide evidence that mature RNA may be the substrate for HGT events. Among the 14 HGT genes, most were detected to be exposed to negative selection pressure, and they appeared to be functional. Prediction of the function of these genes annotated them to be associated with energy metabolism and defense mechanisms (Yuan et al., 2013).

Genome understanding for the different aspects of shrimp biology

VISUAL SENSITIVITY AND BENTHIC ADAPTATION

L. vannamei live as benthic organisms in shallow and turbid waters; therefore, they are expected to have developed photoreceptors that are able to capture light in this environment (Fig. 5.3(A)). A total of 42 structurally complete opsin genes were identified in the *L. vannamei* genome. All of these opsins belong to the visual type related to color perception, which is the largest number of opsins reported among all animals so far – even more than that of animals with excellent vision, such as mantis shrimp (6–33 opsins) and dragonfly (20 opsins). The 42 visual-opsin genes were divided into three subfamilies: long-wavelength type (LW; 33 members), middle-wavelength type (MW; seven members), and short-wavelength type (SW; two members) (Fig. 5.3(B)). Most of the LW subfamily genes belong to two clusters in the genome (Fig. 5.3(C)), suggesting a major expansion via tandem duplication (Zhang et al., 2019d).

The expansion of LW opsins sensitive to light with a wavelength of 560 nm is highly correlated with the adaptation of *L. vannamei* to benthic environments, since yellow and red (long-wavelength) substrates could enhance the visual detection of food, with a consequent effect on the feeding efficiency and growth rate of the shrimp (Luchiari et al., 2012). Opsin genes were also found to be expanded in the genome of the water flea *Daphnia pulex* (36 opsins); however, this expansion is restricted to the MW subfamily, which is sensitive to light at wavelengths of about 530 nm that are associated with their planktonic habitat. Considering the shift from the planktonic to the benthic habitat during the shrimp's life cycle, the MW opsins showed higher expression in the planktonic larval stages (nauplius, zoea, and mysis), whereas the LW-type opsins showed significantly higher expression levels in the benthic postlarvae (Fig. 5.3(D)). Thus, the differential expression among the types of visual opsin genes was highly correlated with the planktonic–benthic habitats of the shrimp, suggesting that *L. vannamei* has a highly effective visual system adapted to its planktonic–benthic transition of habitats.

RAPID NERVE SIGNAL CONDUCTION AND LOCOMOTION

The giant nerve fibers of penaeid shrimp have a conduction speed of about 200 m/s, which is regarded as the fastest conduction speed in animals (Fan et al., 1961). In the *L. vannamei* genome, the gene categories for signal recognition and neural development are significantly enriched. Several notable gene families are expanded, including 457 G Protein-Coupled Receptors (GPCRs), 53 transient receptor potential (TRP) channels, 21 innexins, and 47 protocadherins (Zhang et al., 2019d). These gene families are related to the core neural development across bilaterians (Albertin et al., 2015). The rapid nerve conduction in

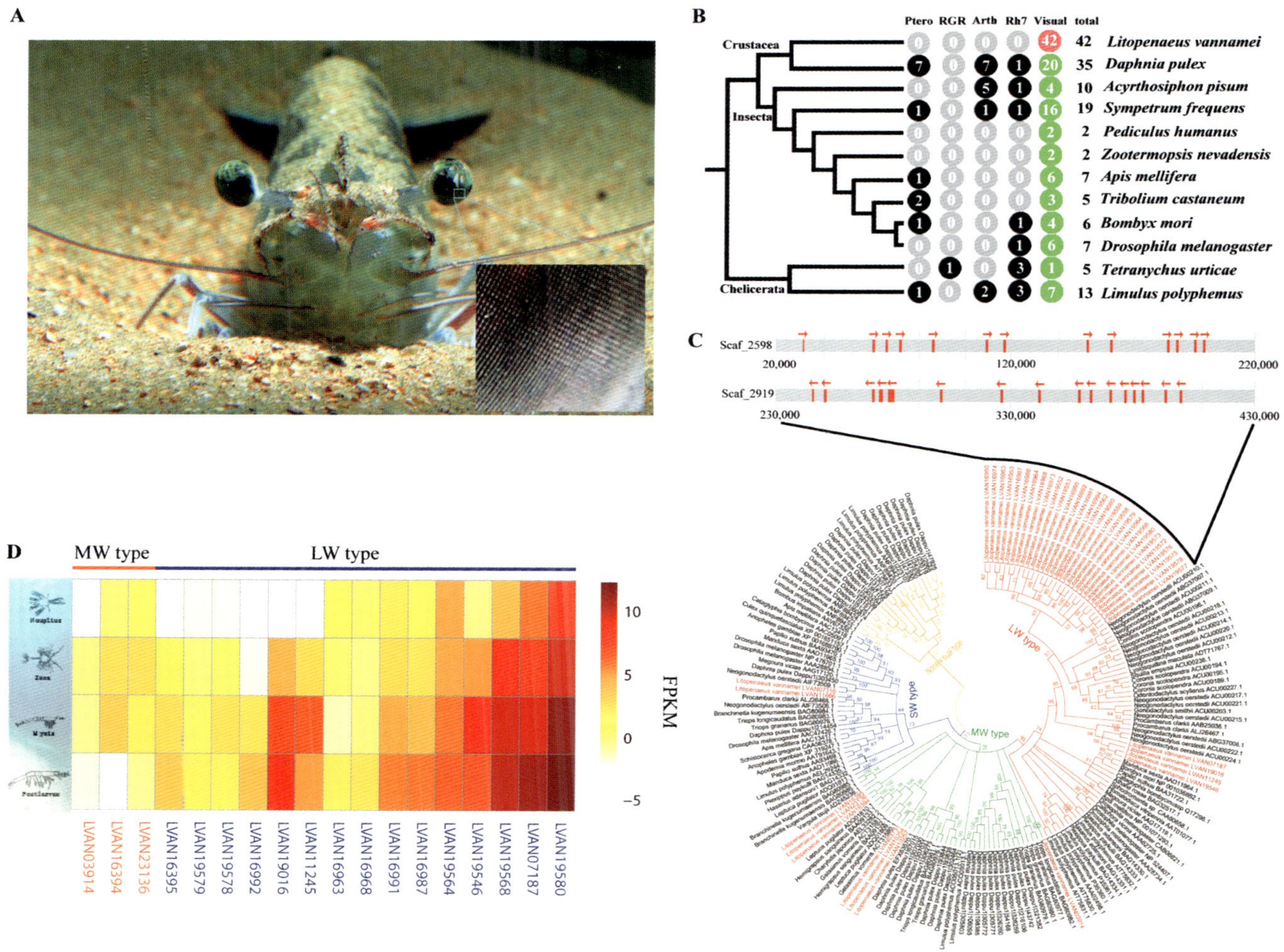

FIGURE 5.3 The opsin gene family in the shrimp genome. (A) The *L. vannamei* eyes and ommatidia. (B) Opsin genes of *L. vannamei* in comparison with those in the genomes of various other arthropods. (C) Phylogenetic tree of the opsin gene family in arthropods. Six clades of opsins in the *L. vannamei* genome were observed (red). (D) Expression of opsin genes (FPKM >1) during different larval stages of *L. vannamei*. Credit: Zhang et al. (2019d).

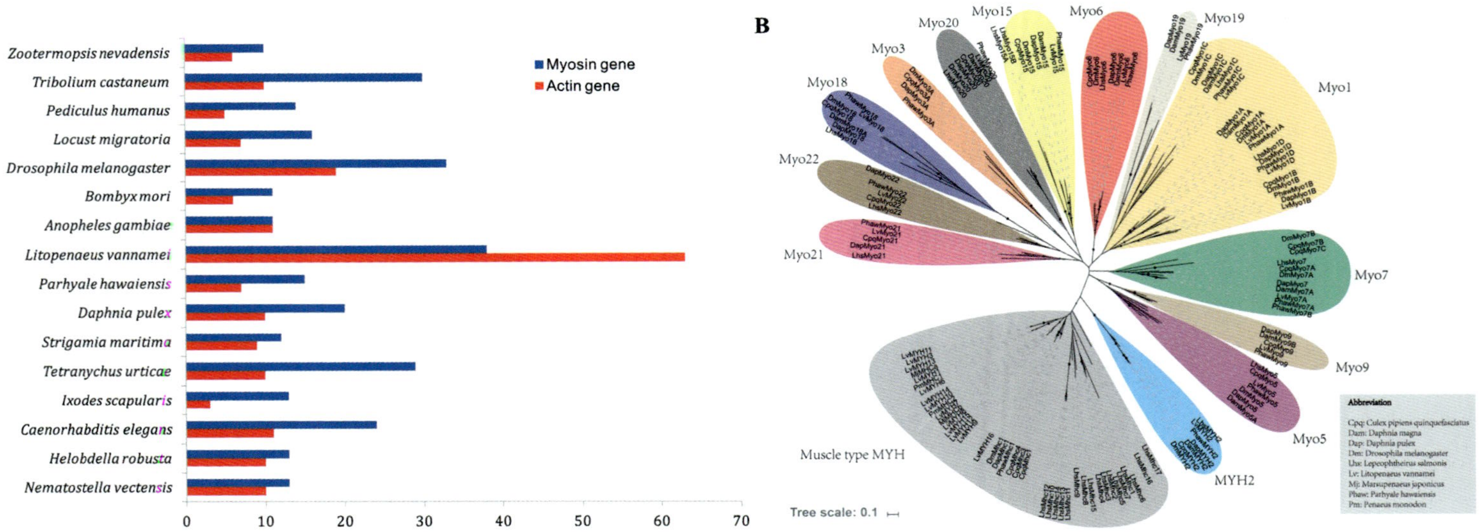

FIGURE 5.4 The actin and myosin genes in the shrimp genome. (A) Copy number of actin and myosin genes in 16 arthropod species. (B) Phylogenetic tree of myosin head domains. Credit: Zhang et al. (2019b).

shrimp is based on the heavy myelin sheath of their axons (Xu and Terakawa, 1999). Genes encoding the crustacyanins (apolipoprotein family genes) are also significantly expanded and duplicated in tandem in the *L. vannamei* genome, and most of them are highly expressed in neural tissues (Zhang et al., 2019d). In addition, 169 genes coding for ionotropic glutamate receptors (iGluRs) and 148 genes coding for glycine receptors (GlyRs) are present in the genome, and they are mainly expressed in the nervous system as well as in muscle. These expanded gene families are suggested to be the molecular basis of the observed rapid nerve signal conduction.

Moreover, the genes involved in muscle contraction, such as those encoding actins and myosins, are significantly expanded as well (Fig 5.4) (Zhang et al., 2018, 2019b). These features might equip the shrimp to perform neural signal transduction rapidly, contributing to responsiveness and strong locomotory activities, including escape reactions, in adaptation to its benthic life swimming in the shallow seas.

SIGNALING PATHWAY AND ENVIRONMENTAL FACTORS IN THE REGULATION OF MOLTING

Molting in crustaceans differs from the process in insects. In insects, molting mainly occurs at the larval stages for growth and metamorphosis, whereas molting occurs throughout the life of crustaceans. *L. vannamei* experiences about 50 molts during a lifetime, which is much more frequent than in other arthropods (Fig 5.5(A)). The global gene expression changes in the molting cycle of *L. vannamei* were investigated by transcriptomic sequencing (Gao et al., 2015). A total of 5117 differentially expressed genes were identified in adjacent molt stages; their expression patterns involved several molecular events critical for molting, such as hormone regulation, phase triggering, cuticle apolysis, exoskeleton sclerotization, and immune responses.

The immediate function of the hormone ecdysone in promoting molting is amplified by the expansion of downstream genes in the ecdysone signaling pathway (Fig 5.5b). The major ecdysone response gene, *Broad-Complex* (*Br-C*) (Chaharbakhshi and Jemc, 2016), is apparently expanded in the *L. vannamei* genome (Fig 5.5(C)). *Br-C* genes are highly expressed at stages D1 and P1 (Fig 5.5(D)), which is synergistic with the expression pattern of their downstream effectors (such as chitinase and cuticular proteins in molting. The genes encoding these downstream effectors are also significantly expanded (Fig 5.5(C)). The expansion of downstream genes in the ecdysone signaling pathway might provide an enhanced positive regulation on ecdysone function in shrimp molting. Upstream of the ecdysone signaling pathway, seven molt-inhibiting hormone (MIH) genes and 19 type I crustacean hyperglycemic hormone (CHH) genes are all duplicated in tandem. Furthermore, genes involved in the "triggering" phase of MIH regulation (Chang and Mykles, 2011), including cAMP-dependent protein kinase (PKA), phosphodiesterase 1 (PDE1), calmodulin (CAM) and Ca^{2+}/calmodulin-dependent protein kinase (CAMK) genes, are all under positive selection in the *L. vannamei* genome (Zhang et al., 2019d). Thus, the expansion and tandem duplication of MIH and CHH genes and the positive selection of MIH "triggering"-phase genes might strengthen the negative control of ecdysone synthesis.

The molting of crustaceans is influenced by many environmental factors, such as nutrition and photoperiod. In the eyestalk of *L. vannamei*, all detected transcripts of *MIH* and its transcription factor SREBP had similar expression patterns, with high expression levels, showing opposite trends to those of the ecdysone level and the *in vivo* cholesterol level during molting. Knockdown of the SREBP gene in

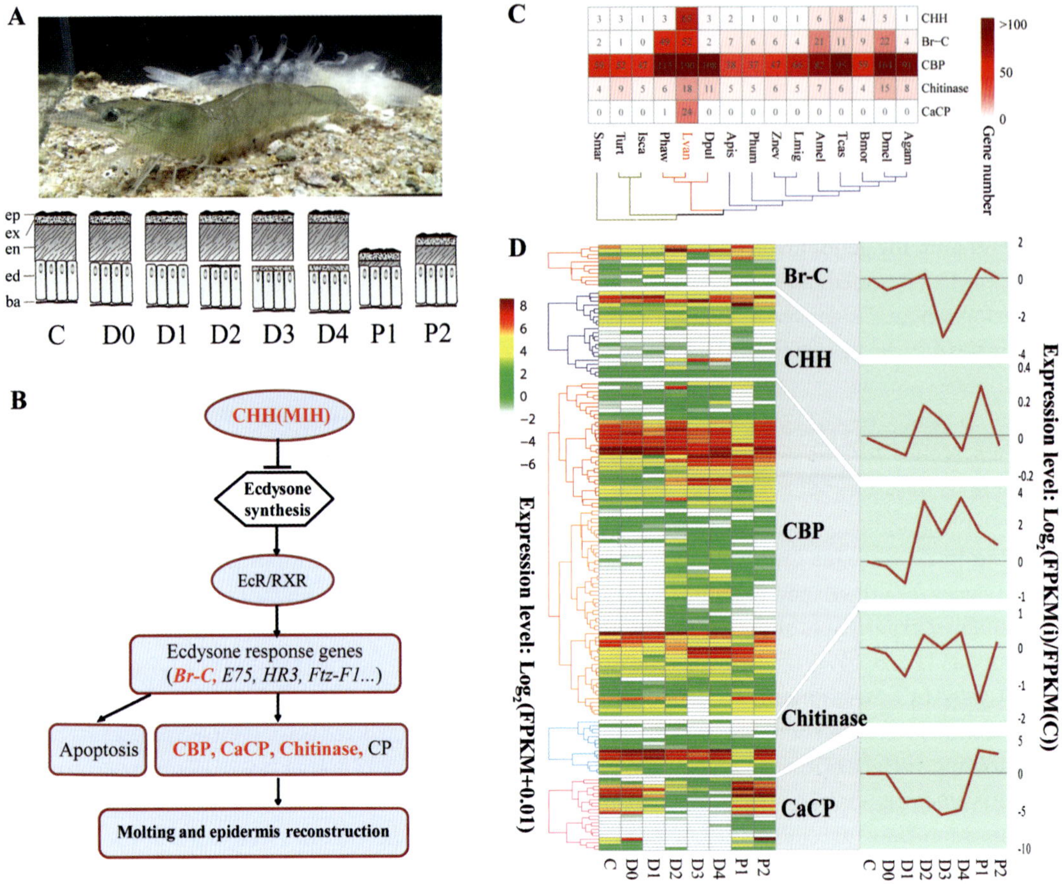

FIGURE 5.5 Key gene families related to the ecdysone signaling pathway in the *L. vannamei* genome. (A) *L. vannamei* and its shed exoskeleton. Schematic diagrams below the photograph show the changes in the epidermis during the molting cycle. (B) The shrimp ecdysone signaling pathway. Expanded gene families are highlighted in red. (C) The expanded gene families in the *L. vannamei* (Lvan) ecdysone signaling pathway and comparison with other arthropods. (D) Expression patterns of the expanded gene families and overall expression trends at the different molting stages of *L. vannamei*. Credit: Zhang et al. (2019d).

the eyestalk significantly decreased the expression levels of all detected MIH genes by about 95%. The injection of cholesterol into shrimp also reduced the expression levels of MIH genes by more than 90%. By contrast, the injection of Lipitor, a lipid-lowering agent, apparently retarded molting at the D3 stage, as shrimp in the control group progressed into the postmolt stage (Zhang et al., 2019d). These data indicated that SREBP mediates MIH-regulated ecdysone synthesis.

Photoperiod has obvious effects on molting in crustaceans (Guo et al., 2011; Rice and Armitage, 1974). Opsin is the key in visual transduction. Knockdown of several LW-type opsins up-regulated the expression levels of MIH genes significantly, indicating an inhibitory role of the opsins on MIH expression. After ten days of opsin knockdown, shrimp showed a much lower molting rate than that in the control group. These data suggest that opsin might be the mediator between photoperiod and

MIH-regulated molting. Taken together, these findings indicate that SREBP and opsin are key regulators that mediate cholesterol supply and photoperiod, respectively, in regulating molting.

ALTERNATIVE SPLICING IN THE STRESS RESPONSE

Alternative splicing (AS) can increase transcript diversity dramatically and plays an important role in stress adaptation. A total of 38,781 AS events were identified in the *L. vannamei* genome; these AS events covered 9209 genes, accounting for 36% of protein-coding genes in the species (Zhang et al., 2019e). The number of AS events increased after viral or bacterial infection or low-salinity stress. Type I AS genes (those for which AS was initially activated) were mainly enriched in substance and energy metabolism, such as carbon metabolism and amino acid metabolism. However, type II AS genes (those for which AS events changed) displayed specific enrichment in response to different stress challenges. Specifically, under biotic stresses, type II AS genes were mainly enriched in the pathogenic pathway and immune network. In addition, under low-salinity stress, the AS genes were significantly enriched for betalain biosynthesis. In summary, AS events are complex in shrimp and may be closely related to stress adaptation.

Application of the shrimp genome

The availability of whole-genome sequences of penaeid shrimp has provided an enormous opportunity for studies of gene numbers, corresponding functions, and conservation and diversification compared with other crustaceans. On this basis, comparative genomics can be developed, and targeted gene knockdown (RNAi and CRISPR-Cas9) and gene regulation (Methyl-Seq, ChIP-Seq, and ATAC-Seq) techniques can be carried out to reveal the molecular mechanisms of physiological processes in shrimp. In particular, as important aquaculture species, high-quality reference shrimp genomes will have a massive impact on genetic breeding as a result of the deep understanding of the genetic architecture of important traits, and will aid in marker selection and candidate gene detection.

RESEQUENCING

L. vannamei was introduced into China in 1988 and now contributes to about 80% of total cultured penaeid shrimp production in China and around the world. To examine the genomic effects of such intensive selective breeding, 22 individuals were selected for genome resequencing, including eight individuals from the wild population (Mexico) and 14 individuals from five aquaculture broodstocks from different countries. A total of 31,993,474 SNPs were identified, most SNPs (86.68%) were located in the intergenic regions. Significant differences in genetic diversity were observed between the wild shrimp and various breeding lines. Phylogenetic analyses based on the whole-genome SNPs placed the wild individuals captured in Mexico into a group. The breeding individuals were grouped into two clades: the individuals from Ecuador were clustered into a separate clade because they had undergone only four generations of selection, while all the individuals from commercial breeding lines were clustered in the other clade (Zhang et al., 2019d) (Fig. 5.6(A)). These genomic analyses revealed that almost 30 years of breeding has exerted a significant impact on the genome of *L. vannamei* broodstocks.

The divergence index (F_{ST}) value and the $\theta\pi$ ratios ($\pi_{wild/}\pi_{breeding}$) were generally used to identify the chromosome regions and genes under selective pressure. In the *L. vannamei* genome, 14 regions (1.66 Mb in size) were under significant selection; these regions contained 28 predicted genes (Fig. 5.6(B, C)). These genes were majorly enriched in the Rho protein signal transduction pathway and the glycosphingolipid

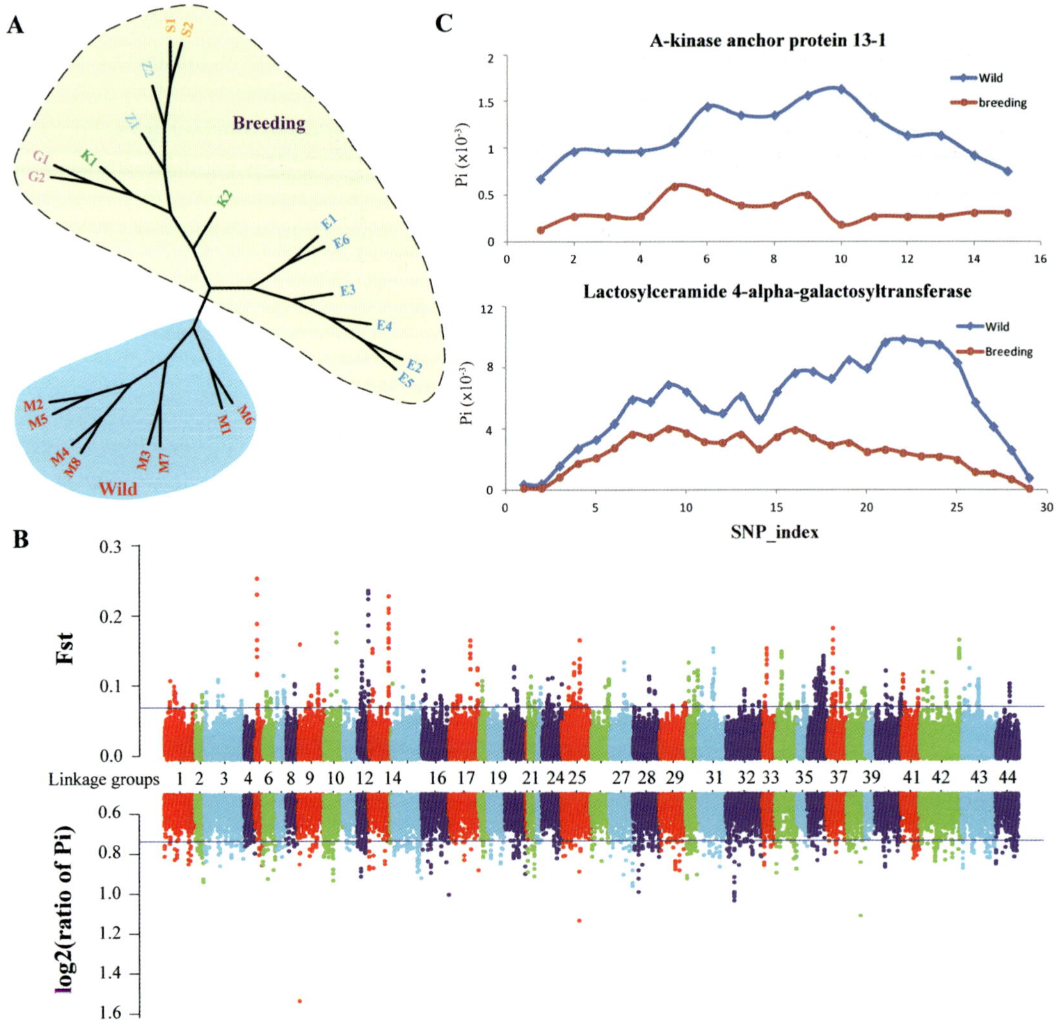

FIGURE 5.6 Genetic diversity and selective sweep analyses in *L. vannamei*. (A) The phylogenetic tree of wild (blue) and breeding (yellow) shrimp, constructed using whole-genome SNPs. (B) Genome-wide view of differentiation (F_{ST}) and reduction in diversity ($\pi_{wild}/\pi_{breeding}$) statistics associated with the wild and breeding shrimp. (C) The genetic diversity value $\theta\pi$ in the wild and breeding populations for the genes encoding A-kinase anchor protein 13 and lactosylceramide 4-alpha-galactosyltransferase. Credit: Zhang et al. (2019d).

biosynthesis pathway. Two A-kinase anchor protein 13 (AKAP13) genes, a lactosylceramide 4-alpha-galactosyltransferase (A4GALT) gene and an anti-lipopolysaccharide factor like protein (ALF) gene (Fig. 5.6(C)), were found to be under strong selection (Zhang et al., 2019d). All the genes are involved in innate immunity; this result illustrates that artificial selective breeding under culture conditions influences the disease resistance of penaeid shrimp.

HIGH-THROUGHPUT MICROARRAYS

With the completion of the shrimp genomes, the application of high-density microarrays is attracting more attention for marker assisted-selection and genomic selection (GS) breeding of shrimp. SNPs are the most abundant polymorphic markers that are fundamental to modern genomic studies. Although the experimental workflows and bioinformatics analyses are more complex, the repeatability and reproducibility are higher for SNP arrays than for DNA sequencing (Robledo et al., 2018). Based on SNP discovery, commercial genotyping arrays containing >50,000 SNPs have been developed as tools for the study and management of many aquaculture species, such as catfish, *Ictalurus punctatus* (250K, 690K) (Liu et al., 2014; Zeng et al., 2017), Atlantic salmon, *Salmo salar* (50K) (Wellband et al., 2019), common carp, *Cyprinus carpio* (250K) (Xu et al., 2014), and Pacific oyster, *Crassostrea gigas* (190K, 23K) (Gutierrez et al., 2018; Qi et al., 2017). Before the genome was completed, only two studies produced validated SNP genotyping arrays for the shrimp species, *P. monodon* (Baranski et al., 2014), and *L. vannamei* (Jones et al., 2017b). These arrays contain 6000 and 6400 genome-wide SNPs, respectively, and present a significant advancement in commercially available breeding chips for these two penaeid shrimp industries. These microarrays were also used to obtain a genomic gene-based resource for commercially important traits through linkage mapping, providing the foundation for further trait-mapping studies (Jones et al., 2017a; Robinson et al., 2014). Based on the assembled reference shrimp genomes and the resequencing data, a high throughput SNP genotyping method based on AmpliSeq target sequencing were developed, 581 amplicons and 2130 high quality SNPs were used for genotyping (Yu et al., 2020), further, a genome-wide 600K SNP array will be developed for shrimp breeding. In black tiger shrimp, 4194 SNPs were screened and constructed a finalized target-capture genotype-by-sequence assay (DArTcap) for cost effective commercial breeding programs (< $15 AUD per sample) (Guppy et al., 2020). The application of this high-density array has been proved to be powerful in gene mapping and GS breeding. With these genomic resources and modern genotyping tools, a series of low-density arrays may provide a cost-effective approach for guided farm management and the improvement of shrimp production in the near future.

QUANTITATIVE TRAIT LOCUS MAPPING

Identifying genomic loci that control traits is the first step toward revealing the genetic architecture of traits. Early studies on the mapping of QTLs were limited in shrimp (Alcivar-Warren et al., 2007; Du et al., 2010). Based on the linkage map of *M. japonicus*, the QTLs that regulate growth traits have been detected; one suggestive QTL marker mapped in the linkage group LG1 of the male map (Li et al., 2006b). In the Chinese shrimp *F. chinensis*, five potential QTLs associated with body length and anti-WSSV defense were detected on female LG1 and male LG44, respectively (Wang et al., 2012). In *L. vannamei*, 14 QTLs were identified for growth-related traits, which explained 2.62–61.42% of phenotypic variation (Andriantahina et al., 2013). Several factors have contributed to the slow progress of QTL mapping in shrimp. The lack of a high-quality reference genome in shrimp makes it difficult to investigate sequence variation (SNPs, indels, and presence/absence variation) in the QTL regions, and the limited marker number and small population size also severely limited the discovery and identification of the precise locations of shrimp QTLs. On the other hand, the short domestication history results in the lack of characteristic broodstock in shrimp.

The situation has improved with the advancement of NGS technology, high-throughput genotyping, and the completion of the whole genomes. Many recent QTL studies have greatly increased the marker density (4000–5000 SNPs) and improved the ability to analyze the potential genetic variation of traits. Yu et al. (2015) and Lu et al. (2016a) identified QTLs associated with growth in *L. vannamei*, and with growth and high-temperature tolerance in *M. japonicus*. The integration of linkage maps with the genome sequence data has been completed, and the fine mapping for various economically relevant traits is becoming available. Yu et al. (2017) reported a small region at LG18 to be highly associated to the phenotypic sex, which was used to design a DNA probe for sex identification in *L. vannamei* (Xiang et al. 2019). This confirms what previous studies suggested, that is, that the sex-determination system in penaeids follows a ZW/ZZ model in which females are the heterogametic sex (Robinson et al., 2014). However, although these studies provide a basic understanding of the genetic structure of these traits, it is still difficult to provide an opportunity to use these findings to improve the selection of broodstock. The integration of high-density linkage maps, genome assemblies, and transcriptome data will contribute to QTL characterization and will further expand the practicability of research results, which should be the main goal of future QTL studies in penaeid shrimp species (Guppy et al., 2018).

GENOME-WIDE ASSOCIATION STUDIES

Quantitative traits such as growth or disease resistance are likely underpinned by tens to hundreds of genes, or regions of small effect, often dispersed across multiple chromosomes of the genome. Genome-wide association studies (GWAS) have been successfully performed to identify genes participating in the regulation of complex traits, and some candidate genes associated with growth traits or disease resistance have been reported in a number of aquatic animals, including Atlantic salmon (Jin et al., 2017, Sodeland et al., 2013), rainbow trout (Gonzalez-Pena et al., 2016), and catfish (Geng et al., 2015). To date, only a few GWAS works have been published for penaeid shrimp species; most of these studies have successfully identified a number of loci significantly associated with sex QTL (Khatkar et al., 2017; Jones et al., 2020; Robinson et al., 2014; Staelens et al., 2008), concluding that sex was likely a simple genetically determined trait in shrimp (Guppy et al., 2018). Recently, the gene *LvSRC*, encoding a class C scavenger receptor, was identified by GWAS as a growth-related candidate gene in *L. vannamei*. Several non-synonymous mutations within *LvSRC* were significantly associated with bodyweight ($P<0.01$), and the most significant marker could explain 13% of phenotypic variance (Wang et al., 2019b). Likewise, another two genes, *PKC-delta* and *Rap-2a*, were also significantly associated with the growth trait of *L. vannamei* (Yu et al., 2019). In a screen for disease-resistance-related SNPs, 30 SNPs or genes were identified to be related to resistance of the shrimp against the pathogen *Vibrio parahaemolyticus* (Zhang et al., 2019a). These studies illustrated that GWAS are effective in identifying markers associated with economic traits, provide a convincing result for revealing the molecular mechanism of growth and disease-resistant traits, and could be used as molecular markers for breeding in shrimp.

GENOMIC SELECTION

The application of GS offers significant advantages over traditional aquatic breeding programs through being able to accurately predict complex polygenic traits, increase genetic gain, and minimize inbreeding (Zenger et al., 2018). Compared with the long breeding history in plants, terrestrial livestock, and poultry (Hake and Ross-Ibarra, 2015; Hickey et al., 2017), the breeding history of shrimp is relatively short. However, the selection pressure in shrimp is very high, as the shrimp could be selected at a rate of one generation per year and a single spawning female can produce more than 300,000 offspring. The prediction accuracy of different models of GS has been investigated in *L. vannamei*. Three GS models (RR-BLUP, BayesA, and Bayesian LASSO) were used to predict the genomic estimated breeding value (GEBV), and the three models performed very similarly to each other in terms of prediction ability. The marker density is also a factor influencing the prediction accuracy; however, after reaching 3000–5000 SNPs, the marker density has little influence on the prediction accuracy. Differences in population structure between the reference population and the prediction population also significantly affect the prediction accuracy, which provides an important suggestion for the application GS to shrimp breeding (Fig 5.7) (Wang et al., 2017).

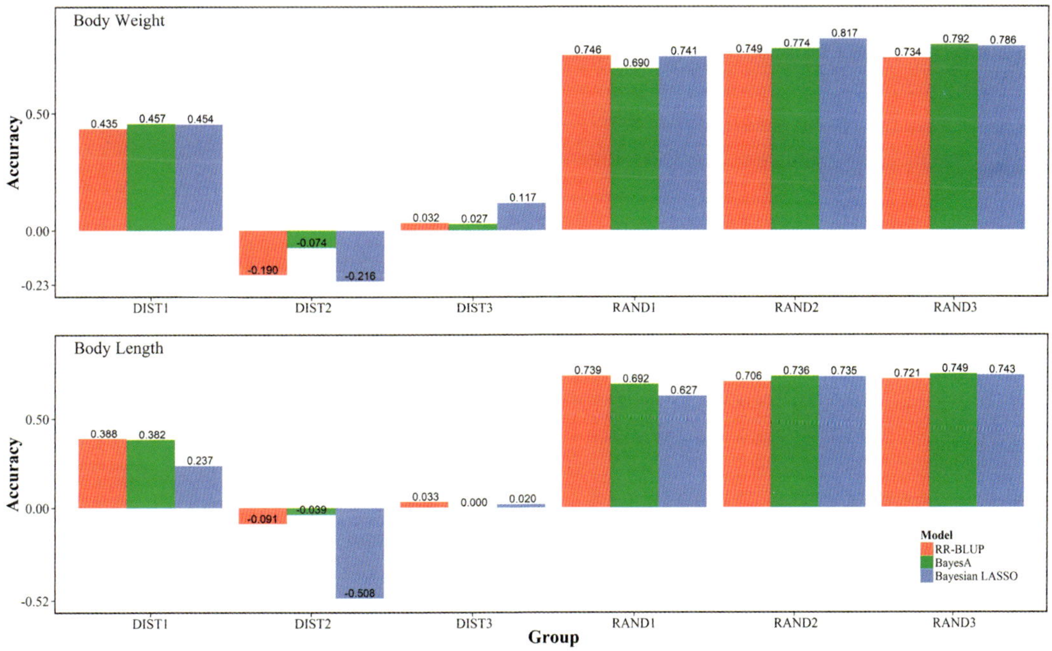

FIGURE 5.7 Accuracy of cross-validation for the genomic prediction of body weight and body length in *L. vannamei*. DIST, Distant reference population; RAND, random reference population. Credit: Wang et al. (2017).

Compared with the traditional selective method, the GS method has advantages in terms of selecting for disease resistance traits and growth traits. A trial of the application of GS in increasing resistance against *V. parahaemolyticus* has been performed in *L. vannamei*. A 6.8% increase in the prediction accuracy for survival time was obtained by using the genomic best linear unbiased prediction method (Wang et al., 2019b). This study suggested that GS could be an alternative approach to improve the genetic gains for disease resistance. Meanwhile, another study also found that the GWAS-GS strategy is accurate for predicting the disease resistance traits with a low cost (Luo et al., 2021).

After the several economically important shrimp genomes were completed, the application of GS in breeding programs has become more noticeable. Efforts have been made on the prediction model, high-throughput and low-cost SNP genotyping, and optimization of the breeding program to improve the efficiency of GS. Several reports have illustrated that using the low-density markers can obtain similar prediction results to those from the high-density markers (Wang et al., 2017; Kriaridou et al., 2020). This could decrease the cost of GS dramatically. However, improving the prediction accuracy and practicability of GEBV is still the main problem of GS research. Further exploration of GS breeding program designs, the density of genotyping markers, and multiple parameters such as genotype-by-environment interactions, the application of heterosis in multi-trait selection, realistic linkage disequilibrium, bio-economical modeling, and index selection should be considered in future research to ensure the optimization of genetic gain and to improve breeding efficiency (Lillehammer et al., 2013; Zenger et al., 2018).

Conclusions

The shrimp genome had been considered to be the most difficult genome to sequence among all aquaculture species. At the time of writing, it has been sequenced and assembled through multiple approaches. The high-quality reference genome of *L. vannamei* provides an opportunity to understand various biological processes of shrimp at the genome level. The most prominent characteristic of the coding region of the shrimp genome is the expansion of a series of genes related to the visual system, nerve signal conduction, and molting regulation. These gene expansions might be associated with the excellent eyesight, rapid nerve signal conduction, and frequency of molting of shrimp. Combined with multi-omics studies, the shrimp genome assembly also sheds light on the mechanism for benthic adaptation, fast growth, high fertility, and high disease-and abiotic resistance, which will provide useful clues for genetic analysis and improvement of these important processes.

Shrimp are among the most highly traded seafood species in the world, and have played an important role in providing high-quality animal proteins for humans. Although the relatively young nature of shrimp aquaculture and the insufficient basic research have constrained omics resource development, this also means the latent natural variation of many species and the genetic potential for increased productivity remains unharnessed. The study of genomics has had a revolutionary impact on the production of plants and animal husbandry. Therefore, similar efforts to improve aquaculture species by using genome-based omics technologies are expected to greatly improve aquaculture production. The shrimp genome, and in particular the large number of SNP markers and genes, will serve as a useful resource for the application of GWAS, GS, molecular design breeding, and thus accelerate genetic improvements in the shrimp industry.

References

Abdelrahman, H., Elhady, M., Alcivar-Warren, A., Allen, S., Al-Tobasei, R., Bao, L.S., Beck, B., Blackburn, H., Bosworth, B., Buchanan, J., Chappell, J., Daniels, W., Dong, S., Dunham, R., Durland, E., Elaswad, A., Gomez-Chiarri, M., Gosh, K., Guo, X.M., Hackett, P., Hanson, T., Hedgecock, D., Howard, T., Holland, L., Jackson, M., Jin, Y.L., Kahlil, K., Kocher, T., Leeds, T., Li, N., Lindsey, L., Liu, S.K., Liu, Z.J., Martin, K., Novriadi, R., Odin, R., Palti, Y., Peatman, E., Proestou, D., Qin, G.Y., Reading, B., Rexroad, C., Roberts, S., Salem, M., Severin, A., Shi, H.T., Shoemaker, C., Stiles, S., Tan, S.X., Tang, K.F.J., Thongda, W., Tiersch, T., Tomasso, J., Prabowo, W.T., Vallejo, R., Van Der Steen, H., Vo, K., Waldbieser, G., Wang, H.P., Wang, X.Z., Xiang, J.H., Yang, Y.J., Yant, R., Yuan, Z.H., Zeng, Q.F., and Zhou, T. (2017) Aquaculture genomics, genetics and breeding in the United States: current status, challenges, and priorities for future research. *BMC Genomics* 18, 191.

Albertin, C.B., Simakov, O., Mitros, T., Wang, Z.Y., Pungor, J.R., Edsinger-Gonzales, E., Brenner, S., Ragsdale, C.W., and Rokhsar, D.S. (2015) The octopus genome and the evolution of cephalopod neural and morphological novelties. *Nature* 524, 220–224.

Alcivar-Warren, A., Dunham, R., Gaffney, P., Kocher, T., and Thorgaard, G. (1997) First Aquaculture Species Genome Mapping Workshop. *Animal Genetics* 28, 451–452.

Alcivar-Warren, A., Meehan-Meola, D., Park, S.W., Xu, Z., Delaney, M., and Zuniga, G. (2007) A low-density, microsatellite-based linkage map of the pacific whiteleg shrimp, *Litopenaeus vannamei*: identification of sex-linked markers in linkage group 4. *Journal of Shellfish Research* 26, 1259–1277.

Aleshin, V.V., Mikhailov, K.V., Konstantinova, A.V., Nikitin, M.A., Rusin, L.Y., Buinova, D.A., Kedrova, O.S., and Petrov, N.B. (2009) On the phylogenetic position of insects in the Pancrustacea clade. *Molecular Biology* 43, 804–818.

Andriantahina, F., Liu, X., and Huang, H. (2013) Genetic map construction and quantitative trait locus (QTL) detection of growth-related traits in *Litopenaeus vannamei* for selective breeding applications. *PLoS One* 8, e75206.

Aoki, T., Wang, H.C., Unajak, S., Santos, M.D., Kondo, H., and Hirono, I. (2011) Microarray analyses of shrimp immune responses. *Marine Biotechnology* 13, 629–638.

Baranski, M., Gopikrishna, G., Robinson, N.A., Katneni, V.K., Shekhar, M.S., Shanmugakarthik, J., Jothivel, S., Gopal, C., Ravichandran, P., Kent, M., Arnyasi, M., and Ponniah, A.G. (2014) The development of a high density linkage map for black tiger shrimp (*Penaeus monodon*) based on cSNPs. *PLoS One* 9, e85413.

Baurain, D., Brinkmann, H., and Philippe, H. (2007) Lack of resolution in the animal phylogeny: closely spaced cladogeneses or undetected systematic errors? *Molecular Biology and Evolution* 24, 6–9.

Berthelot, C., Brunet, F., Chalopin, D., Juanchich, A., Bernard, M., Noel, B., Bento, P., Da Silva, C., Labadie, K., Alberti, A., Aury, J.M., Louis, A., Dehais, P., Bardou, P., Montfort, J., Klopp, C., Cabau, C., Gaspin, C., Thorgaard, G.H., Boussaha, M., Quillet, E., Guyomard, R., Galiana, D., Bobe, J., Volff, J.N., Genet, C., Wincker, P., Jaillon, O., Roest Crollius, H., and Guiguen, Y. (2014) The rainbow trout genome provides novel insights into evolution after whole-genome duplication in vertebrates. *Nature Communications* 5, 3657.

Bourchookarn, A., Havanapan, P.O., Thongboonkerd, V., and Krittanai, C. (2008) Proteomic analysis of altered proteins in lymphoid organ of yellow head virus infected *Penaeus monodon*. *Biochimica et biophysica acta* 1784, 504–511.

Chaharbakhshi, E. and Jemc, J.C. (2016) Broad-complex, tramtrack, and bric-a-brac (BTB) proteins: Critical regulators of development. *Genesis* 54, 505–518.

Chai, Y.M., Yu, S.S., Zhao, X.F., Zhu, Q.A., and Wang, J.X. (2010) Comparative proteomic profiles of the hepatopancreas in *Fenneropenaeus chinensis* response to white spot syndrome virus. *Fish & Shellfish Immunology* 29, 480–486.

Chaikeeratisak, V., Somboonwiwat, K., Wang, H.C., Lo, C.F., and Tassanakajon, A. (2012) Proteomic analysis of differentially expressed proteins in the lymphoid organ of *Vibrio harveyi*-infected *Penaeus monodon*. *Molecular Biology Reports* 39, 6367–6377.

Chang, E.S., and Mykles, D.L. (2011) Regulation of crustacean molting: a review and our perspectives. *General and Comparative Endocrinology* 172, 323–330.

Chen, K., Li, E.C., Xu, Z.X., Li, T.Y., Xu, C., Qin, J.G., and Chen, L.Q. (2015) Comparative transcriptome analysis in the hepatopancreas tissue of Pacific white shrimp *Litopenaeus vannamei* fed different lipid sources at low salinity. *PLoS One* 10, e0144889.

Chen, L.H., Lin, S.W., Liu, K.F., Chang, C.I., Hseu, J.R., and Tsai, J.M. (2016) Comparative proteomic analysis of *Litopenaeus vannamei* gills after vaccination with two WSSV structural proteins. *Fish & Shellfish Immunology* 49, 306–314.

Chen, X., Zeng, D., Xie, D., Zhao, Y., Yang, C., Li, Y., Ma, N., Li, M., Yang, Q., Liao, Z., and Wang, H. (2013) Transcriptome analysis of *Litopenaeus vannamei* in response to white spot syndrome virus infection. *PLoS One* 8, e73218.

Chongsatja, P.O., Bourchookarn, A., Lo, C.F., Thongboonkerd, V., and Krittanai, C. (2007) Proteomic analysis of differentially expressed proteins in *Penaeus vannamei* hemocytes upon Taura syndrome virus infection. *Proteomics* 7, 3592–3601.

Chow, S., Dougherty, W.J., and Sandifer, P.A. (1990) Meiotic chromosome complements and nuclear-DNA contents of four species of shrimps of the genus *Penaeus*. *Journal of Crustacean Biology* 10, 29–36.

Colbourne, J.K., Pfrender, M.E., Gilbert, D., Thomas, W.K., Tucker, A., Oakley, T.H., Tokishita, S., Aerts, A., Arnold, G.J., Basu, M.K., Bauer, D.J., Cáceres, C.E., Carmel, L., Casola, C., Choi, J.H., Detter, J.C., Dong, Q., Dusheyko, S., Eads, B.D., Fröhlich, T.,Geiler-Samerotte, K.A., Gerlach, D., Hatcher, P., Jogdeo, S., Krijgsveld, J., Kriventseva, E.V., Kültz, D., Laforsch, C., Lindquist, E., Lopez, J., Manak, J.R., Muller, J., Pangilinan, J., Patwardhan, R.P., Pitluck, S., Pritham, E.J., Rechtsteiner, A., Rho, M., Rogozin, I.B., Sakarya, O., Salamov, A., Schaack, S., Shapiro, H., Shiga, Y., Skalitzky, C., Smith, Z., Souvorov, A., Sung, W., Tang, Z., Tsuchiya, D., Tu, H., Vos, H., Wang, M., Wolf, Y.I., Yamagata, H., Yamada, T., Ye, Y., Shaw, J.R., Andrews, J., Crease, T.J., Tang, H., Lucas, S.M., Robertson, H.M., Bork, P., Koonin, E.V., Zdobnov, E.M., Grigoriev, I.V., Lynch, M., Boore, J. L. (2011) The ecoresponsive genome of *Daphnia pulex*. *Science* 331, 555–561.

Coman, F.E., Sellars, M.J., Norris, B.J., Coman, G.J., and Preston, N.P. (2008) The effects of triploidy on *Penaeus* (*Marsupenaeus*) *japonicus* (Bate) survival, growth and gender when compared to diploid siblings. *Aquaculture* 276, 50–59.

Cui, J., Wu, L.T., and Chu, K.H. (2014) Comparative proteomic profiling during ovarian development of the shrimp *Metapenaeus ensis*. *Molecular Biology Reports* 41, 519–528.

Cui, Z., Liu, Y., Yuan, J., Zhang, X., Ventura, T., Ma, K.Y., Sun, S., Song, C., Zhan, D., Yang, Y., Liu, H., Fan, G., Cai, Q., Du, J., Qin, J., Shi, C., Hao, S., Fitzgibbon, Q.P., Smith, G.G., Xiang, J., Chan, T.Y., Hui, M., Bao, C., Li F, and Chu, K.H. (2021) The Chinese mitten crab genome provides insights into adaptive plasticity and developmental regulation. *Nature communications* 12, 2395.

Dai, P., Luan, S., Lu, X., Luo, K., and Kong, J. (2017) Comparative transcriptome analysis of the Pacific White Shrimp (*Litopenaeus vannamei*) muscle reveals the molecular basis of residual feed intake. *Scientific reports* 7, 10483.

Davis, G.P. and Hetzel, D.J.S. (2000) Integrating molecular genetic technology with traditional approaches for genetic improvement in aquaculture species. *Aquaculture Research* 31, 3–10.

Du, Z.Q., Ciobanu, D.C., Onteru, S.K., Gorbach, D., Mileham, A.J., Jaramillo, G., and Rothschild, M.F. (2010) A gene-based SNP linkage map for pacific white shrimp, *Litopenaeus vannamei*. *Animal Genetics* 41, 286–294.

Ertl, N.G., Elizur, A., Brooks, P., Kuballa, A.V., Anderson, T.A., and Knibb, W.R. (2013). Molecular characterisation of colour formation in the prawn *Fenneropenaeus merguiensis*. *PLoS ONE* 8, e56920.

Eyun, S.I., Soh, H.Y., Posavi, M., Munro, J.B., Hughes, D., Murali, S.C., Qu, J., Dugan, S., Lee, S.L., Chao, H., Dinh, H., Han, Y., Doddapaneni, H., Worley, K.C., Muzny, D.M., Park, E.O., Silva, J.C., Gibbs, R.A., Richards, S., and Lee, C.E. (2017) Evolutionary history of chemosensory-related gene families across the Arthropoda. *Molecular biology and evolution* 34, 1838–1862.

Fan, L.F., Wang, A.L., and Wu, Y.X. (2013) Comparative proteomic identification of the hemocyte response to cold stress in white shrimp, *Litopenaeus vannamei*. *Journal of Proteomics* 80, 196–206.

Fan, L.F., Wang, A.L., Miao, Y.T., Liao, S.A., Ye, C.X., and Lin, Q.C. (2016) Comparative proteomic identification of the hepatopancreas response to cold stress in white shrimp, *Litopenaeus vannamei*. *Aquaculture* 454, 27–34.

Fan, L.F., Wang, L., and Wang, Z.L. (2019) Proteomic characterization of the hepatopancreas in the Pacific white shrimp *Litopenaeus vannamei* under cold stress: Revealing the organism homeostasis mechanism. *Fish & Shellfish Immunology* 92, 438–449.

Fan, S., Hsu, K., Chen, F., and Ho, B. (1961) On the high conduction velocity of the giant nerve fiber of shrimp *Penaeus orientalis*. *Chinese Science Bulletin* 12, 51–52.

Feldmann, R.M., Schweitzer, C.E., Hu, S.X., Huang, J.Y., Zhang, Q.Y., Zhou, C.Y., Wen, W., Xie, T., and Maguire, E.P. (2017) A new Middle Triassic (Anisian) cyclidan crustacean from the Luoping Biota, Yunnan Province, China: morphologic and phylogenetic insights. *Journal of Crustacean Biology* 37, 406–412.

Fransen, C.H.J.M. and Grave, S.D. (2009) Evolution and radiation of shrimp-like decapods: an overview. In: Martin, J.W., Crandall, K.A., and Felder, D.L. (eds) *Decapod Crustacean Phylogenetics*. Taylor and Francis/CRC Press, Boca Raton, FL, pp. 245–259.

Gao, Y., Zhang, X.J., Wei, J.K., Sun, X.Q., Yuan, J.B., Li, F.H., and Xiang, J.H. (2015) Whole transcriptome analysis provides insights into molecular mechanisms for molting in *Litopenaeus vannamei*. *PLoS One* 10, e0144350.

Geng, X., Sha, J., Liu, S., Bao, L., Zhang, J., Wang, R., Yao, J., Li, C., Feng, J., Sun, F., Sun, L., Jiang, C., Zhang, Y., Chen, A., Dunham, R., Zhi, D., and Liu, Z. (2015) A genome-wide association study in catfish reveals the presence of functional hubs of related genes within QTLs for columnaris disease resistance. *BMC Genomics* 16, 196.

Gonçalves, M.M., de Almeida Regitano, L.C., Dami, C., Salgado, C.C., de Freitas, P.D., Teixeira, A.K.G., and Galetti-Junior, P.M. (2014) Inheritance of AFLP markers and genetic linkage analysis in two full-sib families of the marine shrimp *Litopenaeus vannamei* (Crustacea, Decapoda). *Advances in Bioscience and Biotechnology* 5, 273–281.

Gonzalez-Pena, D., Gao, G., Baranski, M., Moen, T., Cleveland, B.M., and Kenney, P.B. (2016) Genome-wide association study for identifying loci that affect fillet yield. carcass, and body weight traits in rainbow trout (*Oncorhynchus mykiss*). *Frontiers in Genetics* 7, 203.

Guo, B., Wang, F., Dong, S.L., and Gao, Q.F. (2011) The effect of rhythmic light color fluctuation on the molting and growth of *Litopenaeus vannamei*. *Aquaculture* 314, 210–214.

Guo, H., Ye, C.X., Wang, A.L., Xian, J.A., Liao, S.A., Miao, Y.T., and Zhang, S.P. (2013) Trascriptome analysis of the Pacific white shrimp *Litopenaeus vannamei* exposed to nitrite by RNA-seq. *Fish & Shellfish Immunology* 35, 2008–2016.

Guo, L., Xu, Y.H., Zhang, N., Zhou, F.L., Huang, J.H., Liu, B.S., Jiang, S.G., and Zhang, D.C. (2019) A high-density genetic linkage map and QTL mapping for sex in black tiger shrimp (*Penaeus monodon*). *Frontiers in Genetics* 10, 326.

Guo, H., Chen, T., Liang, Z., Fan, L., Shen, Y., and Zhou, D. (2021) iTRAQ and PRM-based comparative proteomic profiling in gills of white shrimp *Litopenaeus vannamei* under copper stress. *Chemosphere* 263, 128270.

Guppy, J.L., Jones, D.B., Jerry, D.R., Wade, N.M., Raadsma, H.W., Huerlimann, R., and Zenger, K.R. (2018) The state of "omics" research for farmed penaeids: advances in research and impediments to industry utilization. *Frontiers in Genetics* 9, 282.

Guppy, J.L., Jones, D.B., Kjeldsen, S.R., Le Port, A., Khatkar, M.S., Wade, N.M., Sellars, M.J., Steinig, E.J., Raadsma, H.W., Jerry, D.R., and Zenger, K.R. (2020) Development and validation of a RAD-Seq target-capture based genotyping assay for routine application in advanced black tiger shrimp (*Penaeus monodon*) breeding programs. *BMC genomics* 21, 541.

Gutekunst, J., Andriantsoa, R., Falckenhayn, C., Hanna, K., Stein, W., Rasamy, J., and Lyko, F. (2018) Clonal genome evolution and rapid invasive spread of the marbled crayfish. *Nature Ecology & Evolution* 2, 567–573.

Gutierrez, A.P., Matika, O., Bean, T.P., and Houston, R.D. (2018) Genomic selection for growth traits in pacific oyster (*Crassostrea gigas*): potential of low-density marker panels for breeding value prediction. *Frontiers in Genetics* 9, 391.

Hake, S., and Ross-Ibarra, J. (2015) Genetic, evolutionary and plant breeding insights from the domestication of maize. *eLife* 4, e05861.

Henson, J., Tischler, G., and Ning, Z.M. (2012) Next-generation sequencing and large genome assemblies. *Pharmacogenomics* 13, 901–915.

Hernández-Pérez, A., Zamora-Briseño, J.A., Ruiz-May, E., Pereira-Santana, A., Elizalde-Contreras, J.M., Pozos-González, S., Torres-Irineo, E., Hernández-López, J., Gaxiola-Cortés, M.G., and Rodríguez-Canul, R. (2019) Proteomic profiling of the white shrimp *Litopenaeus vannamei* (Boone, 1931) hemocytes infected with white spot syndrome virus reveals the induction of allergy-related proteins. *Developmental and comparative immunology* 91, 37–49.

Hickey, J.M., Chiurugwi, T., Mackay, I., Powell, W., and Implementing Genomic Selection in CGIAR Breeding Programs Workshop Participants (2017) Genomic prediction unifies animal and plant breeding programs to form platforms for biological discovery. *Nature Genetics* 49, 1297–1303.

Huang, S.W., Lin, Y.Y., You, E.M., Liu, T.T., Shu, H.Y., Wu, K.M., Tsai, S.F., Lo, C.F., Kou, G.H., Ma, G.C., Chen, M., Wu, D., Aoki, T., Hirono, I., and Yu, H.T. (2011) Fosmid library end sequencing reveals a rarely known genome structure of marine shrimp *Penaeus monodon. BMC Genomics* 12, 242.

Huang, W., Li, H.M., Cheng, C.H., Ren, C.H., Chen, T., Jiang, X., Cheng, K.M., Luo, P., and Hu, C.Q. (2018) Analysis of the transcriptome data in *Litopenaeus vannamei* reveals the immune basis and predicts the hub regulation-genes in response to high-pH stress. *PLoS One* 13, e0207771.

Huang, Z., Wang, Y., Sun, L., Wang, X., Lu, P., Liang, G., Pang, H., Wu, Q., Gooneratne, R., and Zhao, J. (2020) Effects of T-2 toxin on the muscle proteins of shrimp (*Litopenaeus vannamei*) - a proteomics study. *Journal of the science of food and agriculture* 100, 119–128.

Huerlimann, R., Wade, N.M., Gordon, L., Montenegro, J.D., Goodall, J., McWilliam, S., Tinning, M., Siemering, K., Giardina, E., Donovan, D., Sellars, M. J., Cowley, J. A., Condon, K., Coman, G.J., Khatkar, M.S., Raadsma, H.W., Maes, G.E., Zenger, K.R., and Jerry, D.R. (2018) De novo assembly, characterization, functional annotation and expression patterns of the black tiger shrimp (*Penaeus monodon*) transcriptome. *Scientific reports* 8, 13553.

Hurzaid, A., Chan, T.Y., Mohd Nor, S.A., Muchlisin, Z.A., Chen, W.J. (2020) Molecular phylogeny and diversity of penaeid shrimps (Crustacea: Decapoda) from South-East Asian waters. *Zoologica scripta* 49, 596– 613.

Jiang, H., Li, F.H., Xie, Y.S., Huang, B.X., Zhang, J.K., Zhang, J.Q., Zhang, C.S., Li, S.H., and Xiang, J.H. (2009) Comparative proteomic profiles of the hepatopancreas in *Fenneropenaeus chinensis* response to hypoxic stress. *Proteomics* 9, 3353–3367.

Jiang, H., Li, F., Zhang, J., Zhang, J., Huang, B., Yu, Y., and Xiang, J. (2014) Comparison of protein expression profiles of the hepatopancreas in *Fenneropenaeus chinensis* challenged with heat-inactivated *Vibrio anguillarum* and white spot syndrome virus. *Marine biotechnology* 16, 111–123.

Jin, Y., Zhou, T., Geng, X., Liu, S., Chen, A., Yao, J., Jiang, C., Tan, S., Su, B., and Liu, Z. (2017) A genome-wide association study of heat stress-associated SNPs in catfish. *Animal Genetics* 48, 233–236.

Jones, D.B., Jerry, D.R., Khatkar, M.S., Raadsma, H.W., Steen, H.V., Prochaska, J., Foret, S., and Zenger, K.R. (2017a) A comparative integrated gene-based linkage and locus ordering by linkage disequilibrium map for the Pacific white shrimp, *Litopenaeus vannamei. Scientific Reports* 7, 10360.

Jones, D.B., Zenger, K.R., Khatkar, M.S., Raadsma, H.W., Steen, H.a.M.V.D., and Prochaska, J. (2017b) Development of a low-density commercial genotyping array for the white legged shrimp, *Litopenaeus vannamei*. In: *Proceedings of the 22nd Conference of the Association for the Advancement of Animal Breeding and Genetics*. 2–5 July 2017, Townsville, Queensland, Australia, pp. 117–120.

Jones, D.B., Nguyen, H.T., Khatkar, M.S., Simma, D.B., Jerry, D.R., Raadsma, H.W., van der Steen, H., Prochaska, J., and Zenger, K.R. (2020) The identification of a major sex QTL in the white-leg shrimp, *Litopenaeus vannamei. Aquaculture* 529, 735673.

Kao, D.M., Lai, A.G., Stamataki, E., Rosic, S., Konstantinides, N., Jarvis, E., Di Donfrancesco, A., Pouchkina-Stancheva, N., Semon, M., Grillo, M., Bruce, H., Kumar, S., Siwanowicz, I., Le, A., Lemire, A., Eisen, M.B., Extavour, C., Browne, W.E., Wolff, C., Averof, M., Patel, N.H., Sarkies, P., Pavlopoulos, A., and Aboobaker, A. (2016) The genome of the crustacean *Parhyale hawaiensis*, a model for animal development, regeneration, immunity and lignocellulose digestion. *eLife* 5, e20062.

Kawato, S., Nishitsuji, K., Arimoto, A., Hisata, K., Kawamitsu, M., Nozaki, R., Kondo, H., Shinzato, C., Ohira, T., Satoh, N., Shoguchi, E., and Hirono, I. (2021) Genome and transcriptome assemblies of the kuruma shrimp, *Marsupenaeus japonicus. G3 Genes Genomes Genetics* jkab268.

Khatkar, M., Zenger, K., Jones, D., Prochaska, J., van der Steen, H., Jerry, D., and Raadsma, H. (2017) Quantitative genomic analyses in the Pacific whiteleg shrimp *Litopenaeus vannamei. Proceedings of the Association of Advancement in Animal Breeding and Genetics* 22, 121-124.

Koiwai, K., Koyama, T., Tsuda, S., Toyoda, A., Kikuchi, K., Suzuki, H., and Kawano, R. (2021) Single-cell RNA-seq analysis reveals penaeid shrimp hemocyte subpopulations and cell differentiation process. *Elife* 10:e66954.

Koyama, T., Asakawa, S., Katagiri, T., Shimizu, A., Fagutao, F.F., Mavichak, R., Santos, M.D., Fuji, K., Sakamoto, T., Kitakado, T., Kondo, H., Shimizu, N., Aoki, T., and Hirono, I. (2010) Hyper-expansion of large DNA segments in the genome of kuruma shrimp, *Marsupenaeus japonicus. BMC Genomics* 11, 141.

Kriaridou, C., Tsairidou, S., Houston, R. D., and Robledo, D. (2020) Genomic prediction using low density marker panels in aquaculture: performance across species, traits, and genotyping platforms. *Frontiers in genetics* 11, 124.

Kulkarni, A.D., Kiron, V., Rombout, J.H.W.M., Brinchmann, M.F., Fernandes, J.M.O., Sudheer, N.S., and Singh, B.I.S. (2014) Protein profiling in the gut of *Penaeus monodon* gavaged with oral WSSV-vaccines and live white spot syndrome virus. *Proteomics* 14, 1660–1673.

Lehnert, S.A., Wilson, K.J., Byrne, K., and Moore, S.S. (1999) Tissue-specific expressed sequence tags from the black tiger shrimp *Penaeus monodon*. *Marine Biotechnology* 1, 465–476.

Lester, L.J. (1979) Population genetics of penaeid shrimp from the Gulf of Mexico. *Journal of Heredity* 70, 175–180.

Li, F.H., Xiang, J.H., Zhang, X.J., Zhou, L.H., Zhang, C.S., and Wu, C.G. (2003) Gonad development characteristics and sex ratio in triploid chinese shrimp (*Fenneropenaeus chinensis*). *Marine Biotechnology* 5, 528–535.

Li, F.H., Zhang, C.S., Yu, K.J., Liu, X.L., Zhang, X.J., Zhou, L.H., and Xiang, J.H. (2006a) Larval metamorphosis and morphological characteristic analysis of triploid shrimp *Fenneropenaeus chinensis* (Osbeck, 1765). *Aquaculture Research* 37, 1180–1186.

Li, Y., Dierens, L., Byrne, K., Miggiano, E., Lehnert, S., and Preston, N.P. (2006b) QTL detection of production traits for the Kuruma prawn *Penaeus japonicus* (Bate) using AFLP markers. *Aquaculture* 258, 198–210.

Li, C.Z., Weng, S.P., Chen, Y.G., Yu, X.Q., Lu, L., Zhang, H.Q., He, J.G., and Xu, X.P. (2012) Analysis of *Litopenaeus vannamei* transcriptome using the next-generation DNA sequencing technique. *PLoS One* 7, e47442.

Li, S.H., Zhang, X.J., Sun, Z., Li, F.H., and Xiang, J.H. (2013) Transcriptome analysis on Chinese shrimp *Fenneropenaeus chinensis* during WSSV acute infection. *PLoS One* 8, e58627.

Li, W., Tang, X.Q., Xing, J., Sheng, X.Z., and Zhan, W.B. (2014) Proteomic analysis of differentially expressed proteins in *Fenneropenaeus chinensis* hemocytes upon white spot syndrome virus infection. *PLoS One* 9, e89962.

Li, Y., Zhou, F., Huang, J., Yang, L., Jiang, S., Yang, Q., He, J., and Jiang, S. (2018) Transcriptome reveals involvement of immune defense, oxidative imbalance, and apoptosis in ammonia-stress response of the black tiger shrimp (*Penaeus monodon*). *Fish & shellfish immunology* 83, 162–170.

Li, Z., Tang, X., Li, J., and He, Y. (2021) Comparative proteomic and transcriptomic analysis reveals high pH-induced expression signatures of Chinese shrimp *Fenneropenaeus chinensis*. *Functional & Integrative Genomics* 21, 299-311.

Lien, S., Koop, B.F., Sandve, S.R., Miller, J.R., Kent, M.P., Nome, T., Hvidsten, T.R., Leong, J.S., Minkley, D.R., Zimin, A., Grammes, F., Grove, H., Gjuvsland, A., Walenz, B., Hermansen, R.A., Von Schalburg, K., Rondeau, E.B., Di Genova, A., Samy, J.K.A., Vik, J.O., Vigeland, M.D., Caler, L., Grimholt, U., Jentoft, S., Vage, D.I., De Jong, P., Moen, T., Baranski, M., Palti, Y., Smith, D.R., Yorke, J.A., Nederbragt, A.J., Tooming-Klunderud, A., Jakobsen, K.S., Jiang, X.T., Fan, D.D., Liberles, D.A., Vidal, R., Iturra, P., Jones, S.J.M., Jonassen, I., Maass, A., Omholt, S.W., and Davidson, W.S. (2016) The Atlantic salmon genome provides insights into rediploidization. *Nature* 533, 200–205.

Lillehammer, M., Meuwissen, T.H.E., and Sonesson, A.K. (2013) A low-marker density implementation of genomic selection in aquaculture using within-family genomic breeding values. *Genetics Selection Evolution* 45, 39.

Lin, H.M., Qi, X.E., Shui, S.S., Benjakul, S., Aubourg, S.P., and Zhang, B. (2021) Label-free proteomic analysis revealed the mechanisms of protein oxidation induced by hydroxyl radicals in whiteleg shrimp (*Litopenaeus vannamei*) muscle. *Food & function* 12, 4337–4348.

Liu, B., Wang, Q., Li, J., Liu, P., and He, Y. (2010) A genetic linkage map of marine shrimp *Penaeus* (*Fenneropenaeus*) *chinensis* based on AFLP, SSR, and RAPD markers. *Chinese Journal of Oceanology and Limnology* 28, 815–825.

Liu, S., Sun, L., Li, Y., Sun, F., Jiang, Y., Zhang, Y., Zhang, J., Feng, J., Kaltenboeck, L., Kucuktas, H., and Liu, Z. (2014) Development of the catfish 250K SNP array for genome-wide association studies. *BMC Research Notes* 7, 135.

Liu, Z.J., Liu, S.K., Yao, J., Bao, L.S., Zhang, J.R., Li, Y., Jiang, C., Sun, L.Y., Wang, R.J., Zhang, Y., Zhou, T., Zeng, Q.F., Fu, Q., Gao, S., Li, N., Koren, S., Jiang, Y.L., Zimin, A., Xu, P., Phillippy, A.M., Geng, X., Song, L., Sun, F.Y., Li, C., Wang, X.Z., Chen, A.L., Jin, Y.L., Yuan, Z.H., Yang, Y.J., Tan, S.X., Peatman, E., Lu, J.G., Qin, Z.K., Dunham, R., Li, Z.X., Sonstegard, T., Feng, J.B., Danzmann, R.G., Schroeder, S., Scheffler, B., Duke, M.V., Ballard, L., Kucuktas, H., Kaltenboeck, L., Liu, H.X., Armbruster, J., Xie, Y.J., Kirby, M.L., Tian, Y., Flanagan, M.E., Mu, W.J., and Waldbieser, G.C. (2016) The channel catfish genome sequence provides insights into the evolution of scale formation in teleosts. *Nature Communications* 7, 11757.

Liu, F., Li, S.H., Yu, Y., Sun, M.Z., Xiang, J.H., and Li, F.H. (2020) Effects of ammonia stress on the hemocytes of the Pacific white shrimp *Litopenaeus vannamei*. *Chemosphere* 239, 124759.

Lu, X., Luan, S., Hu, L.Y., Mao, Y., Tao, Y., Zhong, S.P., and Kong, J. (2016b) High-resolution genetic linkage mapping, high-temperature tolerance and growth-related quantitative trait locus (QTL) identification in *Marsupenaeus japonicus*. *Molecular Genetics and Genomics* 291, 1391–1405.

Lu, X., Kong, J., Luan, S., Dai, P., Meng, X.H., Cao, B.X., and Luo, K. (2016a) Transcriptome analysis of the hepatopancreas in the Pacific white shrimp (*Litopenaeus vannamei*) under acute ammonia stress. *PLoS One* 11, e0164396.

Lu, X., Luan, S., Dai, P., Meng, X., Cao, B., Luo, K., and Kong, J. (2018) iTRAQ-based comparative proteome analysis for molecular mechanism of defense against acute ammonia toxicity in Pacific White shrimp *Litopenaeus vannamei*. *Fish & shellfish immunology* 74, 52–61.

Luchiari, A.C., Marques, A.O., and Freire, F.A.M. (2012) Effects of substrate solour preference on growth of the shrimp *Litopenaeus vannamei* (Boone, 1931) (Decapoda, Penaeoidea). *Crustaceana* 85, 789–800.

Luo, Z., Yu, Y., Xiang, J., and Li, F. (2021) Genomic selection using a subset of SNPs identified by genome-wide association analysis for disease resistance traits in aquaculture species. *Aquaculture* 539, 736620.

Maneeruttanarungroj, C., Pongsomboon, S., Wuthisuthimethavee, S., Klinbunga, S., Wilson, K.J., Swan, J., Li, Y., Whan, V., Chu, K.H., Li, C.P., Tong, J., Glenn, K., Rothschild, M., Jerry, D., and Tassanakajon, A. (2006) Development of polymorphic expressed sequence tag-derived microsatellites for the extension of the genetic linkage map of the black tiger shrimp (*Penaeus monodon*). *Animal genetics* 37, 363–368.

Maralit, B.A., Jaree, P., Boonchuen, P., Tassanakajon, A., and Somboonwiwat, K. (2018) Differentially expressed genes in hemocytes of *Litopenaeus vannamei* challenged with *Vibrio parahaemolyticus* AHPND (VPAHPND) and VPAHPND toxin. *Fish & Shellfish Immunology* 81, 284–296.

Meng, X.H., Dong, L.J., Shi, X.L., Li, X.P., Sui, J., Luo, K., Luan, S., Chen, B.L., Cao, B.X., Cao, J.W., and Kong, J. (2019) Screening of the candidate genes related to low-temperature tolerance of *Fenneropenaeus chinensis* based on high-throughput transcriptome sequencing. *PLoS One* 14, e0211182.

Meng, X., Fu, Q., Luan, S., Luo, K., Sui, J., and Kong, J. (2021) Genome survey and high-resolution genetic map provide valuable genetic resources for *Fenneropenaeus chinensis*. *Scientific reports* 11, 7533.

Montenegro, J.D., Huerlimann, R., Simmering, K., Gordon, L., Tinning, M., and Jabbari, J.S. (2018) Hybrid assembly of the black tiger shrimp genome (*Penaeus mondon*). *International Symposium on Genetics in Aquaculture*. 15 July 2018, Cairns, Queensland, Australia.

Nguyen, C., Nguyen, T.G., Van Nguyen, L., Pham, H.Q., Nguyen, T.H., Pham, H.T., Nguyen, H.T., Ha, T.T., Dau, T.H., Vu, H.T., Nguyen, D.D., Nguyen, N.T.T., Nguyen, N.H., Van Quyen, D., Chu, H,H., and Dinh, K.D. (2016) De novo assembly and transcriptome characterization of major growth-related genes in various tissues of *Penaeus monodon*. *Aquaculture* 464, 545–553.

Ning, M., Wei, P., Shen, H., Wan, X., Jin, M., Li, X., Shi, H., Qiao, Y., Jiang, G., Gu, W., Wang, W., Wang, L., and Meng, Q. (2019) Proteomic and metabolomic responses in hepatopancreas of whiteleg shrimp *Litopenaeus vannamei* infected by microsporidian Enterocytozoon hepatopenaei. *Fish & shellfish immunology* 87, 534–545.

Peng, J.X., Wei, P.Y., Zhang, B., Zhao, Y.Z., Zeng, D.G., Chen, X.L., Li, M., and Chen, X.H. (2015) Gonadal transcriptomic analysis and differentially expressed genes in the testis and ovary of the Pacific white shrimp (*Litopenaeus vannamei*). *BMC Genomics* 16, 1006.

Peng, J.X., He, P.P., Wei, P.Y., Zhang, B., Zhao, Y.Z., Li, Q.Y., Chen, X.L., Peng, M., Zeng, D.G., Yang, C.L., and Chen, X. (2018) Proteomic responses under cold stress reveal unique cold tolerance mechanisms in the Pacific White Shrimp (*Litopenaeus vannamei*). *Frontiers in physiology* 9, 1399.

Pérez, F., Erazo, C., Zhinaula, M., Volckaert, F., and Calderón, J. (2004) A sex-specific linkage map of the white shrimp *Penaeus* (*Litopenaeus*) *vannamei* based on AFLP markers. *Aquaculture* 242, 105–118.

Polinski, J.M., Zimin, A.V., Clark, K.F., Kohn, A.B., Sadowski, N., Timp, W., Ptitsyn, A., Khanna, P., Romanova, D.Y., Williams, P., Greenwood, S.J., Moroz, L.L., Walt, D.R., and Bodnar, A.G. (2021) The American lobster genome reveals insights on longevity, neural, and immune adaptations. *Science advances* 7, eabe8290.

Pongsomboon, S., Tang, S., Boonda, S., Aoki, T., Hirono, I., Yasuike, M., and Tassanakajon, A. (2008) Differentially expressed genes in *Penaeus monodon* hemocytes following infection with yellow head virus. *BMB reports* 41, 670–677.

Powell, D., Knibb, W., Remilton, C., and Elizur, A. (2015) De-novo transcriptome analysis of the banana shrimp (*Fenneropenaeus merguiensis*) and identification of genes associated with reproduction and development. *Marine genomics* 22, 71–78.

Poynton, H.C., Hasenbein, S., Benoit, J.B., Sepulveda, M.S., Poelchau, M.F., Hughes, D., Murali, S.C., Chen, S., Glastad, K.M., Goodisman, M., Werren, J.H., Vineis, J.H., Bowen, J.L., Friedrich, M., Jones, J., Robertson, H.M., Feyereisen, R., Mechler-Hickson, A., Mathers, N., Lee, C.E., Colbourne, J.K., Biales, A., Johnston, J.S., Wellborn, G.A., Rosendale, A.J., Cridge, A.G., Munoz-Torres, M.C., Bain, P.A., Manny, A.R., Major, K.M., Lambert, F.N., Vulpe, C.D., Tuck, P., Blalock, B.J., Lin, Y.Y., Smith, M.E., Ochoa-Acuña, H., Chen, M.M., Childers, C.P., Qu J., Dugan, S., Lee, S.L., Chao, H., Dinh, H., Han, Y., Doddapaneni, H., Worley, K.C., Muzny, D.M., Gibbs, R.A., and Richards, S. (2018) The toxicogenome of *Hyalella azteca*: a model for sediment ecotoxicology and evolutionary toxicology. *Environmental science & technology* 52, 6009–6022.

Pryszcz, L.P., and Gabaldon, T. (2016) Redundans: an assembly pipeline for highly heterozygous genomes. *Nucleic Acids Research* 44, e113.

Qi, H., Song, K., Li, C., Wang, W., Li, B., Li, L., and Zhang, G. (2017) Construction and evaluation of a high-density SNP array for the Pacific oyster (*Crassostrea gigas*). *PLoS One* 12, e0174007.

Qiao, J., Du, Z.H., Zhang, Y.L., Du, H., Guo, L.L., Zhong, M.Q., Cao, J.S., and Wang, X.Y. (2011) Proteomic identification of the related immune-enhancing proteins in shrimp *Litopenaeus vannamei* stimulated with vitamin C and Chinese herbs. *Fish & Shellfish Immunology* 31, 736–745.

Rattanarojpong, T., Wang, H.C., Lo, C.F., and Flegel, T.W. (2007) Analysis of differently expressed proteins and transcripts in gills of *Penaeus vannamei* after yellow head virus infection. *Proteomics* 7, 3809–3814.

Regier, J.C., Shultz, J.W., and Kambic, R.E. (2005) Pancrustacean phylogeny: hexapods are terrestrial crustaceans and maxillopods are not monophyletic. *Proceedings of the Royal Society B: Biological Sciences* 272, 395–401.

Ren, X.Y., Liu, P., and Li, J. (2019) Comparative transcriptomic analysis of *Marsupenaeus japonicus* hepatopancreas in response to *Vibrio parahaemolyticus* and white spot syndrome virus. *Fish & Shellfish Immunology* 87, 755–764.

Ren, Y., Li, J., Guo, L., Liu, J. N., Wan, H., Meng, Q., Wang, H., Wang, Z., Lv, L., Dong, X., Zhao, W., Zeng, Q., and Ou, J. (2020) Full-length transcriptome and long non-coding RNA profiling of whiteleg shrimp *Penaeus vannamei* hemocytes in response to *Spiroplasma eriocheiris* infection. *Fish & shellfish immunology* 106, 876–886.

Rice, P.R. and Armitage, K.B. (1974) The effect of photoperiod on oxygen consumption of the crayfish *Orconectes nais* (Faxon). *Comparative Biochemistry and Physiology Part A: Molecular & Integrative Physiology* 47, 261–270.

Robalino, J., Carnegie, R.B., O'leary, N., Ouvry-Patat, S.A., De La Vega, E., Prior, S., Gross, P.S., Browdy, C.L., Chapman, R.W., Schey, K.L., and Warr, G. (2009) Contributions of functional genomics and proteomics to the study of immune responses in the Pacific white leg shrimp *Litopenaeus vannamei*. *Veterinary Immunology and Immunopathology* 128, 110–118.

Robinson, N.A., Gopikrishna, G., Baranski, M., Katneni, V.K., Shekhar, M.S., Shanmugakarthik, J., Jothivel, S., Gopal, C., Ravichandran, P., Gitterle, T., and Ponniah, A.G. (2014) QTL for white spot syndrome virus resistance and the sex-determining locus in the Indian black tiger shrimp (*Penaeus monodon*). *BMC Genomics* 15, 731.

Robledo, D., Palaiokostas, C., Bargelloni, L., Martinez, P., and Houston, R. (2018) Applications of genotyping by sequencing in aquaculture breeding and genetics. *Reviews in Aquaculture* 10, 670–682.

Rotllant, G., Wade, N.M., Arnold, S.J., Coman, G.J., Preston, N.P., and Glencross, B.D. (2015) Identification of genes involved in reproduction and lipid pathway metabolism in wild and domesticated shrimps. *Marine genomics* 22, 55–61.

Ruan, J., and Li, H. (2020) Fast and accurate long-read assembly with wtdbg2. *Nature Methods* 17, 155–158.

Santos, C.A., Andrade, S., Teixeira, A.K., Farias, F., Kurkjian, K., Guerrelhas, A.C., Rocha, J.L., Galetti, P.M., Jr, and Freitas, P.D. (2018) *Litopenaeus vannamei* Transcriptome Profile of Populations Evaluated for Growth Performance and Exposed to White Spot Syndrome Virus (WSSV). *Frontiers in genetics* 9, 120.

Santos, C.A., Andrade, S.C.S., Teixeira, A.K., Farias, F., Guerrelhas, A.C., Rocha, J.L., and Freitas, P.D. (2021) Transcriptome differential expression analysis reveals the activated genes in *Litopenaeus vannamei* shrimp families of superior growth performance. *Aquaculture* 531, 735871.

Saski, C.A., Hlederman, R., Chapman, R.W., and Benzie, J. (2009) Advancing shrimp genomics. *Plant and Animal Genome XVII Conference*. January 2009, San Diego, CA, USA.

Sellars, M.J., Li, F., Preston, N.P., and Xiang, J. (2010) Penaeid shrimp polyploidy: global status and future direction. *Aquaculture* 310, 1–7.

Sellars, M.J., Arce, S.M., and Hertzler, P.L. (2012) Triploidy induction in the Pacific white shrimp *Litopenaeus vannamei*: an assessment of induction agents and parameters, embryo viability, and early larval survival. *Marine Biotechnology* 14, 740–751.

Sellars, M.J., Trewin, C., Mcwilliam, S.M., Glaves, R.S.E., and Hertzler, P.L. (2015) Transcriptome profiles of *Penaeus (Marsupenaeus) japonicus* animal and vegetal half-embryos: identification of sex determination, germ line, mesoderm, and other developmental genes. *Marine Biotechnology* 17, 252–265.

Shen, S.Z., Crowley, J.L., Wang, Y., Bowring, S.A., Erwin, D.H., Sadler, P.M., Cao, C.Q., Rothman, D.H., Henderson, C.M., Ramezani, J., Zhang, H., Shen, Y., Wang, X.D., Wang, W., Mu, L., Li, W.Z., Tang, Y.G., Liu, X.L., Liu, L.J., Zeng, Y., Jiang, Y.F., and Jin, Y.G. (2011) Calibrating the end-Permian mass extinction. *Science* 334, 1367–1372.

Shi, X.L., Kong, J., Meng, X.H., Luan, S., Luo, K., Cao, B.X., Liu, N., Lu, X., Deng, K.Y., Cao, J.W., Zhang, Y.X., Zhang, H.H., and Li, X.P. (2016) Comparative microarray profile of the hepatopancreas in the response of "Huanghai No. 2" *Fenneropenaeus chinensis* to white spot syndrome virus. *Fish & Shellfish Immunology* 58, 210–219.

Shingate, P., Ravi, V., Prasad, A., Tay, B.H., Garg, K.M., Chattopadhyay, B., Yap, L.M., Rheindt, F.E., and Venkatesh, B. (2020) Chromosome-level assembly of the horseshoe crab genome provides insights into its genome evolution. *Nature Communications* 11, 2322.

Silvestre, F., Huynh, T.T., Bernard, A., Dorts, J., Dieu, M., Raes, M., Nguyen, T.P., and Kestemont, P. (2010) A differential proteomic approach to assess the effects of chemotherapeutics and production management strategy on giant tiger shrimp *Penaeus monodon. Comparative Biochemistry and Physiology Part D: Genomics & Proteomics* 5, 227–233.

Sodeland, M., Gaarder, M., Moen, T., Thomassen, M., Kjøglum, S., and Kent, M. (2013) Genome-wide association testing reveals quantitative trait loci for fillet texture and fat content in Atlantic salmon. *Aquaculture* 408, 169–174.

Somboonwiwat, K., Chaikeeratisak, V., Wang, H.C., Lo, C.F., and Tassanakajon, A. (2010) Proteomic analysis of differentially expressed proteins in *Penaeus monodon* hemocytes after *Vibrio harveyi* infection. *Proteome Science* 8, 39.

Song, L., Bian, C., Luo, Y., Wang, L., You, X., Li, J., Qiu, Y., Ma, X., Zhu, Z., Ma, L., Wang, Z., Lei, Y., Qiang, J., Li, H., Yu, J., Wong, A., Xu, J., Shi, Q., and Xu, P. (2016) Draft genome of the Chinese mitten crab, *Eriocheir sinensis. GigaScience* 5, 5.

Sookruksawong, S., Sun, F.Y., Liu, Z.J., and Tassanakajon, A. (2013) RNA-Seq analysis reveals genes associated with resistance to Taura syndrome virus (TSV) in the Pacific white shrimp *Litopenaeus vannamei. Developmental and Comparative Immunology* 41, 523–533.

Staelens, J., Rombaut, D., Vercauteren, I., Argue, B., Benzie, J., and Vuylsteke, M. (2008) High-density linkage maps and sex-linked markers for the black tiger shrimp (*Penaeus monodon*). *Genetics* 179, 917–925.

Sun, Z., Liu, P., Li, J., Meng, X., and Zhang, X. (2008) Construction of a genetic linkage map in *Fenneropenaeus chinensis* (Osbeck) using RAPD and SSR markers. *Hydrobiologia* 596, 133–141.

Swathi, A., Shekhar, M.S., Katneni, V.K., and Vijayan, K.K. (2018) Genome size estimation of brackishwater fishes and penaeid shrimps by flow cytometry. *Molecular Biology Reports* 45, 951–960.

Tan, M.H., Gan, H.M., Lee, Y.P., Grandjean, F., Croft, L.J., and Austin, C.M. (2020) A giant genome for a giant crayfish (*Cherax quadricarinatus*) with insights into cox1 pseudogenes in decapod genomes. *Frontiers in genetics* 11, 201.

Tang, B., Wang, Z., Liu, Q., Zhang, H., Jiang, S., Li, X., Wang, Z., Sun, Y., Sha, Z., Jiang, H., Wu, X., Ren, Y., Li, H., Xuan, F., Ge, B., Jiang, W., She, S., Sun, H., Qiu, Q., Wang, W., Qiu, G., Zhang, D., and Li, Y. (2020a) High-quality genome assembly of *Eriocheir japonica sinensis* reveals its unique genome evolution. *Frontiers in genetics* 10, 1340.

Tang, B., Zhang, D., Li, H., Jiang, S., Zhang, H., Xuan, F., Ge, B., Wang, Z., Liu, Y., Sha, Z., Cheng, Y., Jiang, W., Jiang, H., Wang, Z., Wang, K., Li, C., Sun, Y., She, S., Qiu, Q., Wang, W., Li, X., Li, Y., Liu, Q., and Ren, Y. (2020b) Chromosome-level genome assembly reveals the unique genome evolution of the swimming crab (*Portunus trituberculatus*). *GigaScience* 9, giz161.

Tao, M., Zhou, H., Luo, K., Lu, J., Zhang, Y., and Wang, F. (2019) Quantitative serum proteomics analyses reveal shrimp responses against WSSV infection. *Developmental and comparative immunology*, 93, 89–92.

Tian, Y., Kong, J., and Wang, W. (2008) Construction of AFLP-based genetic linkage maps for the Chinese shrimp *Fenneropaeneus chinensis. Chinese Science Bulletin* 53, 1205–1216.

Uengwetwanit, T., Ponza, P., Sangsrakru, D., Wichadakul, D., Ingsriswang, S., Leelatanawit, R., Klinbunga, S., Tangphatsornruang, S., and Karoonuthaisiri, N. (2018) Transcriptome-based discovery of pathways and genes related to reproduction of the black tiger shrimp (*Penaeus monodon*). *Marine Genomics* 37, 69–73.

Uengwetwanit, T., Pootakham, W., Nookaew, I., Sonthirod, C., Angthong, P., Sittikankaew, K., Rungrassamee, W., Arayamethakorn, S., Wongsurawat, T., Jenjaroenpun, P., Sangsrakru, D., Leelatanawit, R., Khudet, J., Koehorst, J.J., Schaap, P.J., Martins Dos Santos, V., Tangy, F., and Karoonuthaisiri, N. (2021) A chromosome-level assembly of the black tiger shrimp (*Penaeus monodon*) genome facilitates the identification of novel growth-associated genes. *Molecular Ecology Resources* 21, 1620–1640.

Wang, B., Li, F., Dong, B., Zhang, X., Zhang, C., and Xiang, J. (2006) Discovery of the genes in response to white spot syndrome virus (WSSV) infection in *Fenneropenaeus chinensis* through cDNA microarray. *Marine Biotechnology* 8, 491–500.

Wang, H.C., Wang, H.C., Leu, J.H., Kou, G.H., Wang, A.H.J., and Lo, C.F. (2007) Protein expression profiling of the shrimp cellular response to white spot syndrome virus infection. *Developmental and Comparative Immunology* 31, 672–686.

Wang, W., Tian, Y., Kong, J., Li, X., Liu, X., and Yang, C. (2012) Integration genetic linkage map construction and several potential QTLs mapping of Chinese shrimp (*Fenneropenaeus chinensis*) based on three types of molecular markers. *Genetika* 48, 508–521.

Wang, Q., Yu, Y., Yuan, J., Zhang, X., Huang, H., Li, F., and Xiang, J. (2017) Effects of marker density and population structure on the genomic prediction accuracy for growth trait in Pacific white shrimp *Litopenaeus vannamei*. *BMC Genetics* 18, 45.

Wang, Q., He, Y., and Li, J. (2019a) Conjoint analysis of SMRT- and Illumina-based RNA-sequencing data of *Fenneropenaeus chinensis* provides insight into sex-biased expression genes involved in sexual dimorphism. *Frontiers in genetics* 10, 1175.

Wang, Q.C., Yu, Y., Zhang, Q., Zhang, X.J., Huang, H., Xiang, J.H., and Li, F.H. (2019b) Evaluation on the genomic selection in *Litopenaeus vannamei* for the resistance against Vibrio parahaemolyticus. *Aquaculture* 505, 212–216.

Wang, Y., Yu, Y., Li, S., Zhang, X., Xiang, J., and Li, F. (2020a) Sex-specific transcriptome sequencing of zoea i larvae and identification of sex-linked genes using bulked segregant analysis in Pacific white shrimp *Litopenaeus vannamei*. *Marine biotechnology* 22, 423–432.

Wang, Z., Wang, L., Zhou, J., Zou, J., and Fan L. (2020b) New insights into the immune regulation and tissue repair of *Litopenaeus vannamei* during temperature fluctuation using TMT-based proteomics. *Fish & Shellfish Immunology* 106, 975-981.

Wang, Q., Ren, X., Liu, P., Li, J., Lv, J., Wang, J., Zhang, H., Wei, W., Zhou, Y., He, Y., and Li, J. (2021) Improved genome assembly of Chinese shrimp (*Fenneropenaeus chinensis*) suggests adaptation to the environment during evolution and domestication. *Molecular ecology resources* 10.1111/1755-0998.13463.

Wei, J.K., Zhang, X.J., Yu, Y., Huang, H., Li, F.H., and Xiang, J.H. (2014) Comparative transcriptomic characterization of the early development in Pacific white shrimp *Litopenaeus vannamei*. *PLoS One* 9, e106201.

Wellband, K., Merot, C., Linnansaari, T., Elliott, J.A.K., Curry, R.A., and Bernatchez, L. (2019) Chromosomal fusion and life history-associated genomic variation contribute to within-river local adaptation of Atlantic salmon. *Molecular Ecology* 28, 1439–1459.

Wilson, K., Li, Y.T., Whan, V., Lehnert, S., Byrne, K., Moore, S., Pongsomboon, S., Tassanakajon, A., Rosenberg, G., Ballment, E., Fayazi, Z., Swan, J., Kenway, M., Benzie, J. (2002) Genetic mapping of the black tiger shrimp *Penaeus monodon* with amplified fragment length polymorphism. *Aquaculture* 204, 297–309.

Wu, J., Lin, Q., Lim, T.K., Liu, T., and Hew, C.L. (2007) White spot syndrome virus proteins and differentially expressed host proteins identified in shrimp epithelium by shotgun proteomics and cleavable isotope-coded affinity tag. *Journal of Virology* 81, 11681–11689.

Wuthisuthimethavee, S., Lumubol, P., Toojinda, T., Tragoonrung, S., and Vanavichita, A. (2005) SSLP-based linkage map construction in black tiger shrimp (*Penaeus monodon* Fabricius). *ScienceAsia* 31, 91–97.

Xiang, J. (1988) The chromosome study on Chinese shrimp, *Penaeus orientalis* Kishinouye. *Oceanologia et Limnologia Sinica* 19, 205–209.

Xiang, J.H., Zhou, L.H., Liu, R.Y., Zhu, J.Z., Li, F.H., Li, X.H., and Liu, X.D. (1993) Induction of the tetraploids of the Chinese shrimp, *Penaeus chinensis. Biotechnology in Agriculture* 15, 496–501.

Xiang, J., Liu, R., and Zhou, L. (1994) Studies on the chromosomes of marine shrimps with special reference to different techniques. *Journal of Aquaculture* 111, 321.

Xiang, J., Courtney, A.J., and Zhou, L. (1996) Chromosome complements in the spermatogenesis of two penaeid prawns, *Penaeus merguiensis* and *P. esculentus. Cytologia* 61, 317–320.

Xiang, J.H., Li, F.H., Zhang, C.S., Zhang, X.J., Yu, K.J., Zhou, L.H., and Wu, C.G. (2006) Evaluation of induced triploid shrimp *Penaeus* (*Fenneropenaeus*) *chinensis* cultured under laboratory conditions. *Aquaculture* 259, 108–115.

Xiang, J., Yu Y., Li F., and Zhang X. (2019) DNA probe sequence for genetic sex identification of Litopenaeus vannamei and acquisition method. United States Patent, US10280460B2.

Xie, Y.S., Li, F.H., Zhang, C.S., Yu, K.J., and Xiang, J.H. (2008) Synaptonemal complex analysis in spermatocytes of diploid and triploid Chinese shrimp *Fenneropenaeus chinensis. Tissue & Cell* 40, 343–350.

Xu, J., Zhao, Z., Zhang, X., Zheng, X., Li, J., Jiang, Y., Kuang, Y., Zhang, Y., Feng, J., Li, C., Yu, J., Li, Q., Zhu, Y., Liu, Y., Xu, P., and Sun, X. (2014) Development and evaluation of the first high-throughput SNP array for common carp (*Cyprinus carpio*). *BMC Genomics* 15, 307.

Xu, C., Li, E., Liu, Y., Wang, X., Qin, J., and Chen, L. (2017) Comparative proteome analysis of the hepatopancreas from the Pacific white shrimp *Litopenaeus vannamei* under long-term low salinity stress. *Journal of Proteomics* 162, 1–10.

Xu, K., and Terakawa, S. (1999) Fenestration nodes and the wide submyelinic space form the basis for the unusually fast impulse conduction of shrimp myelinated axons. *Journal of Experimental Biology* 202, 1979–1989.

Xue, S., Liu, Y., Zhang, Y., Sun, Y., Geng, X., and Sun, J. (2013) Sequencing and de novo analysis of the hemocytes transcriptome in *Litopenaeus vannamei* response to white spot syndrome virus infection. *PLoS One* 8, e76718.

Yu, Y., Zhang, X.J., Yuan, J.B., Li, F.H., Chen, X.H., Zhao, Y.Z., Huang, L., Zheng, H.K., and Xiang, J.H. (2015) Genome survey and high-density genetic map construction provide genomic and genetic resources for the Pacific White Shrimp *Litopenaeus vannamei. Scientific Reports* 5, 15612.

Yu, Y., Zhang, X., Yuan, J., Wang, Q., Li, S., Huang, H., Li, F., and Xiang, J. (2017) Identification of sex-determining loci in Pacific white shrimp *Litopeneaus vannamei* using linkage and association analysis. *Marine biotechnology* 19, 277–286.

Yu, Y., Wang, Q., Zhang, Q., Luo, Z., Wang, Y., Zhang, X., Huang, H., Xiang, J., and Li, F. (2019) Genome scan for genomic regions and genes associated with growth trait in Pacific white shrimp *Litopeneaus vannamei. Marine Biotechnology* 21, 374–383.

Yuan, J.B., Zhang, X.J., Liu, C.Z., Wei, J.K., Li, F.H., and Xiang, J.H. (2013) Horizontally transferred genes in the genome of Pacific white shrimp, *Litopenaeus vannamei. BMC Evolutionary Biology* 13, 165.

Yuan, J., Zhang, X., Liu, C., Yu, Y., Wei, J., Lia, F., and Xiang, J. (2018) Genomic resources and comparative analyses of two economical penaeid shrimp species, *Marsupenaeus japonicus* and *Penaeus monodon. Marine Genomics* 39, 22–25.

Yuan, J., Zhang, X., Wang, M., Sun, Y., Liu, C., Li, S., Yu, Y., Gao, Y., Liu, F., Zhang, X., Kong, J., Fan, G., Zhang, C., Feng, L., Xiang, J., and Li, F. (2021a) Simple sequence repeats drive genome plasticity and promote adaptive evolution in penaeid shrimp. *Communications biology* 4, 186.

Yuan, J., Zhang, X., Li, F., and Xiang, J. (2021b) Genome sequencing and assembly strategies and a comparative analysis of the genomic characteristics in penaeid shrimp species. *Frontiers in genetics* 12, 658619.

Zeng, D.G., Chen, X.L., Xie, D.X., Zhao, Y.Z., Yang, C.L., Li, Y.M., Ma, N., Peng, M., Yang, Q., Liao, Z.P., Wang, H., and Chen, X.H. (2013) Transcriptome analysis of Pacific white shrimp (*Litopenaeus vannamei*) hepatopancreas in response to Taura syndrome virus (TSV) experimental infection. *PLoS One* 8, e57515.

Zeng, Q., Fu, Q., Li, Y., Waldbieser, G., Bosworth, B., Liu, S., Yang, Y., Bao, L., Yuan, Z., Li, N., and Liu, Z. (2017) Development of a 690 K SNP array in catfish and its application for genetic mapping and validation of the reference genome sequence. *Scientific Reports* 7, 40347.

Zeng, D.G., Chen, X.L., Peng, J.X., Yang, C.L., Peng, M., Zhu, W.L., Xie, D.X., He, P.P., Wei, P.Y., Lin, Y., Zhao, Y.Z., and Chen, X.H. (2018) Single-molecule long-read sequencing facilitates shrimp transcriptome research. *Scientific Reports* 8, 16920.

Zenger, K.R., Khatkar, M.S., Jones, D.B., Khalilisamani, N., Jerry, D.R., and Raadsma, H.W. (2018) Genomic selection in aquaculture: application, limitations and opportunities with special reference to marine shrimp and pearl oysters. *Frontiers in Genetics* 9, 693.

Zhang, L., Yang, C., Zhang, Y., Li, L., Zhang, X., Zhang, Q., and Xiang, J. (2007) A genetic linkage map of Pacific white shrimp (*Litopenaeus vannamei*): sex-linked microsatellite markers and high recombination rates. *Genetica* 131, 37–49.

Zhang, J.K., Li, F.H., Jiang, H., Yu, Y., Liu, C.Z., Li, S.H., Wang, B., and Xiang, J.H. (2010a) Proteomic analysis of differentially expressed proteins in lymphoid organ of *Fenneropenaeus chinensis* response to *Vibrio anguillarum* stimulation. *Fish & Shellfish Immunology* 29, 186–194.

Zhang, X., Zhang, Y., Scheuring, C., Zhang, H.B., Huan, P., Wang, B., Liu, C., Li, F., Liu, B., and Xiang, J. (2010b) Construction and characterization of a bacterial artificial chromosome (BAC) library of Pacific white shrimp, *Litopenaeus vannamei*. *Marine Biotechnology* 12, 141–149.

Zhang, G.F., Fang, X.D., Guo, X.M., Li, L., Luo, R.B., Xu, F., Yang, P.C., Zhang, L.L., Wang, X.T., Qi, H.G., Xiong, Z.Q., Que, H.Y., Xie, Y.L., Holland, P.W.H., Paps, J., Zhu, Y.B., Wu, F.C., Chen, Y.X., Wang, J.F., Peng, C.F., Meng, J., Yang, L., Liu, J., Wen, B., Zhang, N., Huang, Z.Y., Zhu, Q.H., Feng, Y., Mount, A., Hedgecock, D., Xu, Z., Liu, Y.J., Domazet-Loso, T., Du, Y.S., Sun, X.Q., Zhang, S.D., Liu, B.H., Cheng, P.Z., Jiang, X.T., Li, J., Fan, D.D., Wang, W., Fu, W.J., Wang, T., Wang, B., Zhang, J.B., Peng, Z.Y., Li, Y.X., Li, N., Wang, J.P., Chen, M.S., He, Y., Tan, F.J., Song, X.R., Zheng, Q.M., Huang, R.L., Yang, H.L., Du, X.D., Chen, L., Yang, M., Gaffney, P.M., Wang, S., Luo, L.H., She, Z.C., Ming, Y., Huang, W., Zhang, S., Huang, B.Y., Zhang, Y., Qu, T., Ni, P.X., Miao, G.Y., Wang, J.Y., Wang, Q., Steinberg, C.E.W., Wang, H.Y., Li, N., Qian, L.M., Zhang, G.J., Li, Y.R., Yang, H.M., Liu, X., Wang, J., Yin, Y., and Wang, J. (2012) The oyster genome reveals stress adaptation and complexity of shell formation. *Nature* 490, 49–54.

Zhang, J.Y., Wang, W.J., Kong, J., and Wang, Q.Y. (2013) Construction of a genetic linkage map in *Fenneropenaeus chinensis* using SNP markers. *Russian Journal of Marine Biology* 39, 136–142.

Zhang, D., Wang, F., Dong, S., and Lu, Y. (2016) De novo assembly and transcriptome analysis of osmoregulation in *Litopenaeus vannamei* under three cultivated conditions with different salinities. *Gene* 578, 185–193.

Zhang, X.X., Zhang, X.J., Yuan, J.B., Du, J.L., Li, F.H., and Xiang, J.H. (2018) Actin genes and their expression in pacific white shrimp, *Litopenaeus vannamei*. *Molecular Genetics and Genomics* 293, 479–493.

Zhang, Q., Yu, Y., Wang, Q., Liu, F., Luo, Z., Zhang, C., Zhang, X., Huang, H., Xiang, J., and Li, F. (2019a) Identifcation of single nucleotide polymorphisms related to the resistance against acute hepatopancreatic necrosis disease in the Pacifc white shrimp *Litopenaeus vannamei* by target sequencing approach. *Frontiers in Genetics* 10, 700.

Zhang, X., Yuan, J., Zhang, X., Liu, C., Li, F., and Xiang, J. (2019b) Genome-wide identification and expression profiles of myosin genes in the pacific white shrimp, *Litopenaeus vannamei*. *Frontiers in Physiology* 10, 610.

Zhang, X.J., Li, G.Y., Jiang, H.Y., Li, L.M., Ma, J.G., Li, H.M., and Chen, J.P. (2019c) Full-length transcriptome analysis of *Litopenaeus vannamei* reveals transcript variants involved in the innate immune system. *Fish & Shellfish Immunology* 87, 346–359.

Zhang, X.J., Yuan, J.B., Sun, Y.M., Li, S.H., Gao, Y., Yu, Y., Liu, C.Z., Wang, Q.C., Lv, X.J., Zhang, X.X., Ma, K.Y., Wang, X.B., Lin, W.C., Wang, L., Zhu, X.L., Zhang, C.S., Zhang, J.Q., Jin, S.J., Yu, K.J., Kong, J., Xu, P., Chen, J., Zhang, H.B., Sorgeloos, P., Sagi, A., Alcivar-Warren, A., Liu, Z.J., Wang, L., Ruan, J., Chu, K.H., Liu, B., Li, F.H., and Xiang, J.H. (2019d) Penaeid shrimp genome provides insights into benthic adaptation and frequent molting. *Nature Communications* 10, 356.

Zhang, X.X., Yuan, J.B., Zhang, X.J., Liu, C.Z., Xiang, J.H., and Li, F.H. (2019e) Genome-wide analysis of alternative splicing provides insights into stress response of the Pacific white shrimp *Litopenaeus vannamei*. *Frontiers in Genetics* 10, 845.

Zhao, C., Zhang, X.J., Liu, C.Z., Huan, P., Li, F.H., Xiang, J.H., and Huang, C. (2012) BAC end sequencing of Pacific white shrimp *Litopenaeus vannamei*: a glimpse into the genome of Penaeid shrimp. *Chinese Journal of Oceanology and Limnology* 30, 456–470.

Zhao, J., He, Z., Chen, X., Huang, Y., Xie, J., Qin, X., Ni, Z., and Sun, C. (2021) Growth trait gene analysis of kuruma shrimp (*Marsupenaeus japonicus*) by transcriptome study. Comparative biochemistry and physiology. *Part D, Genomics & proteomics* 40, 100874.

Zheng, J., Wang, P., Mao, Y., Su, Y., and Wang, J. (2020) Full-length transcriptome analysis provides new insights into the innate immune system of *Marsupenaeus japonicus*. *Fish & shellfish immunology* 106, 283–295.

Zheng, J.B., Cao, J.W., Mao, Y., Su, Y.Q., and Wang, J. (2019) Comparative transcriptome analysis provides comprehensive insights into the heat stress response of *Marsupenaeus japonicus*. *Aquaculture* 502, 338–346.

Zhu, L., Tang, X., Xing, J., Sheng, X., and Zhan, W. (2018) Differential proteome of haemocyte subpopulations responded to white spot syndrome virus infection in Chinese shrimp *Fenneropenaeus chinensis*. *Developmental and comparative immunology*, 84, 82–93.

Zhuo, X., Qin, Y., He, P., Wei, P., Zhang, B., Chen, X., and Peng, J. (2021) Transcriptomic analysis of *Litopenaeus vannamei* hepatopancreas under cold stress in cold-tolerant and cold-sensitive cultivars. *Gene* 764, 145090.

Chapter 6

SHRIMP GENETIC IMPROVEMENT PROGRAMS

HUGO H. MONTALDO AND JUAN C. QUINTANA-CASARES

Summary

This chapter provides an overview of the principles and most recent research results for designing shrimp breeding programs, emphasizing the elements needed for decision-making processes and the conceptual and theoretical bases underpinning such decisions. The first part discusses the principles of shrimp breeding and mathematical modeling of quantitative traits, and explains the primary issues related to this process. Specifically, this part examines the principal objectives involved in the use of mixed linear models in animal breeding, including estimation of genetic parameters; prediction of traditional breeding value; prediction of genomic breeding values; methods for the calculation of selection response and single and multiple trait selection, and the bases in the design of shrimp breeding programs; the principal issues related to genomic evaluation and selection; genotype x environment interactions; and those related to the maintenance of genetic variability in breeding populations, including effective population size, drift, and inbreeding in the context of shrimp breeding. Finally, this part analyzes issues related to the basic design of selection programs and genetic aspects related to migration and crossbreeding. The second part of this chapter, which focuses on practical shrimp breeding, presents data from current research with an impact on breeding, including traits considered in shrimp selection programs, with a particular emphasis on selection for disease resistance. Research on the use of marker-assisted selection, genomic selection, genome editing, and the effects of inbreeding and optimization of breeding programs for shrimp are also described. The chapter is intended to create a bridge between the scientific principles and current research on the one hand and practical shrimp breeding on the other, stimulating multidisciplinary dialogues between science and industry.

Introduction

In this chapter we will discuss some fundamental quantitative genetic principles as well as research and developments in shrimp breeding since the previous edition of *The Shrimp Book* was published. We refer the reader to the most relevant chapter in that book (Cock et al., 2010) and to other works (Gjedrem and Baranski, 2010; Gjedrem and Rye, 2018; Gjedrem and Thodesen, 2005; Moss et al., 2019, 2012) for historical aspects of aquaculture breeding and a general overview of shrimp breeding using experiences from shrimp and other aquaculture species.

The practical application of genetics and statistical methods is guided by basic breeding principles. Therefore, knowledge of a number of these fundamental tenets is necessary for the design of breeding programs for any species, and to promote a productive conversation among breeders, company managers, and other specialists involved in such programs. Unlike many plants with reproductive mechanisms such as self-fertilization and vegetative propagation and that are autopolyploid, shrimp share with cross-fertilizing plants, mammals, amphibians, reptilians, and other arthropod species the characteristics of diploidy, sexual reproduction, and outbred, naturally polymorphic populations (cf. Vela-Avitúa et al., 2013). Rather than engaging in a strictly technical discussion, here we will adopt a conceptually oriented approach emphasizing what should be discussed with breeders or the administrator overseeing breeding projects and providing a general overview of the fundamental principles and methods in order to facilitate a multidisciplinary dialogue. The primary focus will be on the development of breeding programs as a practical activity related to principles and planning, usually to fulfill business goals such as selling genetically improved broodstock for breeding and larvae for the shrimp growing industry.

Principles of shrimp breeding

In this first section we will provide a general overview of the main principles driving genetic improvement in shrimp and their application in actual breeding programs, discussing both well-established “traditional” quantitative and population genetics (Caballero, 2020; Falconer and Mackay, 1996; Hill, 2010, Lynch and Walsh, 1998; Walsh and Lynch, 2018), and newer aspects that have developed rapidly since the last *Shrimp Book* was published, most notably, methodologies based on the use of single nucleotide polymorphism (SNP) DNA markers. These new approaches are related to a great extent to genomic selection (GS) (Houston et al., 2020; Meuwissen et al., 2013, 2016; Misztal et al., 2020; Zenger et al., 2019) and the detection of individual loci associated with quantitative traits, called quantitative trait loci (QTLs) (Houston et al., 2020; Weller, 2016).

PHENOTYPE MODEL FOR A SINGLE POPULATION

The interest in genetics in aquaculture production has its origin in the fact that the observed performance of any population, that is, the phenotype (P) for the traits linked directly or indirectly to the marketing of breeding animals and shrimp products, is the result of two primary factors, genetic (G) and environmental (E), as well as the general mean (μ). The phenotype may be represented in the following equation:

$$P = \mu + G + E \qquad [1]$$

This model exemplifies both a methodology based on approximations and a good example of human ingenuity in using mathematical tools for practical purposes. Developed largely by R.A. Fisher in the early 20th century, it involves many assumptions and simplifications. Nonetheless, testing the model has demonstrated that for many uses it is a useful guide to optimize selection and other breeding tools for practical genetic improvement in animal populations (Caballero, 2020; Hill, 2010).

The basic idea behind this approach is that many loci (perhaps ranging from tens to thousands in number) accumulated in G, each with small additive effects (Caballero, 2020; Hill. 2010), that is, these traits are polygenic, as well as many intangible random individual environmental effects, are involved in E (after correction of measurements of P for any known identifiable environmental factor, e.g. tank, year, location, cages, date of measurement and age) (Isik et al., 2017). In such circumstances, both G and E will be independent random variables, normally distributed, as is P. In this context and for a single population, most G will be additive (A). Additive genetic effects are those that cause the most genetic progress by selection and the relationship between parents and offspring; moreover, other genetic effects involved in G, such as dominance (D) and epistatic interaction (I), are normally small compared with the size of A and are hard to track within or across generations (Hill et al., 2008). Therefore, the model used for analyzing quantitative traits with a view toward selection is:

$$P = \mu + A + E \qquad [2]$$

E in this model may contain fractions of genetic effects from D and I. The A effect is the additive genetic effect or the breeding value.

Despite the impressive number of academic publications searching for individual large-effect genes, this conceptualization continues to be the primary tool in understanding variation and inheritance in quantitative traits (Hill, 2010).

Variances, covariances, and genetic parameters

Just as the phenotypic value is decomposed in effects in equation [2], the phenotypic variation (assuming they are independent) can be defined in terms of the variances for these effects:

$$V_P = V_A + V_E,$$

where V_P is the phenotypic variance, V_A is the additive genetic variance, and V_E is the environmental variance. The means of A and E will be 0.

Phenotypic variance is related to genetic and environmental variances as: $V_P = V_P h^2 + V_P(1 - h^2)$. Here, h^2 is the narrow-sense heritability of the trait considered, which is defined as the proportion of the phenotypic variance represented by the additive genetic effects (V_A / V_P), and the proportion $(1 - h^2)$ is the proportion of V_P associated with environmental effects (and possibly some other genetic effects as well). This may indicate that heritability is also the linear regression slope of the additive genetic value on the phenotypic value ($b_{A,P}$). This property of h^2 makes it an important genetic parameter for the prediction of the genetic values of the candidates in selection and also to predict the advance by selection (Van Vleck, 1993). The variance values V_P, V_A, V_E, and h^2 are population genetic parameters and therefore change with the trait and the population studied within a species. In principle, the aim of selection is to change the mean value of the genetic effects, so that the mean value of the phenotype changes as well.

The description above is valid for a single quantitative trait that may be described by the variances defined and the associated h^2 value. When considering several traits at the same time, we also need to consider genetic and environmental covariances (and correlations) (Caballero, 2020; Falconer and Mackay, 1996; Mrode, 2014).

We can write model [2] in term of averages after selection is performed, as the following:

$$\bar{P}' = \mu + \bar{A}' + \bar{E}'$$

In order to change the original phenotypic average of a population ($\bar{P}$) to a new value $\bar{P}'$ in a breeding population, we need to change the original average of $\bar{A} = 0$, to the new value after selection ($\bar{A}'$). Changes in $\bar{P}$, $\bar{A}$ and $\bar{E}$ are denoted as ΔP, ΔG, and ΔE and are the phenotypic, genetic, and environmental changes, respectively. ΔG is called genetic progress or selection response.

ANIMAL MODEL

In the analysis of real data, the basic additive genetic model [2] corresponds to a statistical (animal) model for a quantitative trait.

In the animal model, phenotypic records are defined as:

$$y_{ijk} = f_i + a_j + e_{ijk},$$

where y_{ijk} is an observation of a trait corresponding to the i-th level of fixed/environmental effects (f_i), the j-th level of random animal effects/additive genetic effects (a_j), and the ijk-th level of the random residual environmental effects (e_{ijk}). This is a mixed model as it includes both fixed and random effects (Isik et al., 2017). One specific feature of this model is that the genetic relationships between the animal effects are considered through an additive numerator relationship matrix ($\mathbf{A}_M$) that estimates the expected proportion of genes identical by descent (IBD) that two animals share. The matrix $\mathbf{A}_M$ is obtained from the pedigree (Van Vleck, 1993).

The practical importance of the solutions obtained for genetic effects such as the animal effect using mixed models involves their statistical properties when the variances are known. These properties are: (1) Best (2) Linear (3) Unbiased (4) Prediction, better known by the acronym BLUP. *Best* refers to the fact that the prediction of estimated breeding values (EBV) of the true breeding value A will have the minimum prediction error variance (Var(EBV-A)) and maximum accuracy (correlation between EBV and A), and the probability of obtaining the correct ranking under multivariate normal distribution will be the maximum. *Linear* refers to the fact that the prediction is a linear function of data. *Unbiased* indicates that the "environmental" fixed effects are corrected in the model equations to force solutions to be unbiased. *Prediction* refers to obtaining the realized value of a level for a random variable (Henderson, 1984; Mrode, 2014; Van Vleck, 1993). A variant of BLUP with similar properties for genomic evaluations is Genomic BLUP (GBLUP), upon which a similar mixed model is fixed, but here a genomic relationship matrix ($\mathbf{G}_M$) built from many genetic markers (usually thousands) is used rather than an $\mathbf{A}_M$ matrix obtained from the pedigree, or a combined matrix ($\mathbf{H}_M$) formed from $\mathbf{A}_M$ and $\mathbf{G}_M$ is used in single-step GBLUP (ssGBLUP) to better estimate the breeding values of the animals using all available information on phenotypes and genotypes (Lourenco et al., 2020).

Other models in addition to the basic animal model

There are many possible changes and additions to this basic additive genetic model, with corresponding changes in the statistical models used to analyze data in animal breeding. Key changes include the following.

1. The effect of G on equation [1] can be decomposed into additive, dominant, and epistatic effects. In most applications that involve selection within a single population, G will contain only the additive genetic effects (A). Special family data structures are required to estimate dominant and epistatic variances, but in most cases, these estimates will be difficult to calculate and will have a relatively limited practical value, as most genetic variance is expected to be additive; therefore, most of the genetic response will be obtained from additive genetic differences (Hill et al., 2008). Non-additive genetic effects may be important in phenomena related to mating systems, such as in the presence of mating between related animals (inbreeding) in a population, and where different populations are mated (crossbreeding). In a single breeding line and in real life we usually limit ourselves to the additive effect, reserving non-additive effects mainly for the purposes of research.
2. In some cases, models such as nested sire/dam (Falconer and Mackay, 1996; Hernández-Ruíz et al., 2020) and sire–dam (Salte et al., 2010) can be used as an alternative to the classical animal model, depending on the characteristics of the information available. Moreover, alternative methods related to generalized linear models can be used in the analysis of traits with distributions that depart from normality (Gilmour et al., 1985).
3. The environmental effect can be subdivided into several components that correspond to effects/variances common to particular groups of records. These additional variances correspond to extra covariances between phenotypes of pairs of members of a given group, but not between members of different groups. An example of environmental effects that affect groups, besides the usual random individual environmental effects, are those common environmental effects associated with full-sib families (c-effects), modeled as additional independent dam effects (V_C), and permanent environmental effects (V_{PE}), modeled as an additional independent animal effect. Such random effects are included in the model in addition to the animal effects and can have an important impact on the estimates of genetic parameters in shrimp (Luan et al., 2020; Montaldo 2020; Montaldo et al., 2013b).
4. Many variants of models based on genetic markers to replace or complement those obtained from the pedigree are based on the genomic relationship matrix ($\mathbf{G}_M$), which comprises thousands of SNP genetic markers. Genomic relationship matrices are estimated in different ways (Meuwissen et al., 2013) but they always represent the genomic similarity between pairs of individuals from a population based on actual allelic similarity for a sample of genetic markers. Therefore, unlike relationships based on pedigree ($\mathbf{A}_M$) and IBD estimates, the values in G_M are not just expected but observed values, subject to sampling variation and dependent on the number of markers used (Lourenco et al., 2015; Meuwissen et al., 2013, 2016; Misztal et al., 2020; Weller, 2016). Genomic relationships estimate the proportion that two individuals share based on identical by state (IBS) alleles (Lourenco et al., 2015).

Uses of mixed models in animal breeding

The analysis of data with linear mixed models (Lynch and Walsh, 1998; Mrode, 2014) can have a number of objectives.

1. To estimate genetic parameters, that is, estimating variances associated with different genetic and environmental effects (and also covariances in the case of models for multiple-trait analyses) required for (a) the prediction of breeding values with BLUP or GBLUP methods and (b) predicting and optimizing the selection response from breeding programs. Methodologies for estimating genetic parameters can be based on different statistical approaches. In practice, currently only two methods are used: (a) Restricted Maximum Likelihood and (b) Bayesian. Both methods yield accurate and acceptable estimates provided that the model and the data are adequate. Basic categories of genetic parameters obtained directly from the data and the model equations are variances (additive genetics, environmental, phenotypic), covariances that measure associations between traits (additive genetics, environmental, phenotypic) and derived parameters as h^2 (additive genetics, environmental and phenotypic correlations) (Gjedrem and Thodesen, 2005). If the additive relationship coefficient between individuals A and B is a_{AB} and the variance of the trait is V_A, the additive covariance between them (COV_{AB}) is $a_{AB}V_A$. This covariance is the basic principle used in the estimation of genetic parameters and for the estimation of breeding values under additive genetic effects models [2] (Lynch and Walsh, 1998; Caballero, 2020).
2. To predict EBVs, the animal effect is also the additive genetic effect of an individual EBV with the so-called "traditional BLUP" estimate, composed of information from several sources, such as the parents' phenotypes, collateral relative phenotypes, individual phenotypic information, and phenotypic data from progeny and genotypes when available. In the latter case, the value of the genotypes for selection will be proportional to the availability of a large, genetically related population with both phenotypes and genotypes for the genetic markers (Meuwissen et al., 2013, 2016; Solberg et al., 2008).
3. In genomic evaluations, EBV is based on the genomic information of selection candidates (testing populations) and knowledge of the relationships between SNP and phenotypes through what is known as the training population. However, insofar as frequently some selection candidates are not genotyped, combined genotypes and phenotypes can be used jointly for genetic evaluation. Traditional BLUP evaluations are solely based on phenotypes connected through the pedigree. For traditional BLUP, the EBV for an animal i can be expressed as follows (VanRaden and Wiggans, 1991):

$$EBV_i = w_1 PA_i + w_2 YD_i + w_3 PC_i,$$

where PA_i is the parent average EBV for animal i, YD_i is the yield deviation (phenotype adjusted for the model effect's solutions other than additive genetic effects and errors) for animal i, and PC_i is the progeny contribution for animal i.

The EBV for an animal i when genomic information is available, that is, the genomic estimated breeding value (GEBV), is as follows (VanRaden and Wright, 2013):

$$EBV_i = w_1 PA_i + w_2 YD_i + w_3 PC_i + w_4 GI_i,$$

where GI_i contains information from the genotypes of animal i and all weights add up to 1.

4. The estimation of either EBV or GEBV can be used to predict genetic progress by averaging them by year of birth of animals. Estimations of genetic gains are important for evaluating the results of the selection programs (Gjedrem and Thodesen, 2005).

SELECTION

When developing a breeding program, there are generally two primary and not necessarily mutually exclusive options to consider: (1) selection within a single population and (2) the use of several populations in a selection across populations and crossbreeding.

Selection, defined as the differential reproduction of individuals/genotypes (Caballero, 2020), is one of the most important forces of change effecting the genetic structure of a population. Selection, whether without (natural selection) or – primarily – with human intervention (artificial selection), will generate a change in the allele frequencies for loci involved in the phenotypic control of quantitative traits, increasing the frequency of favorable alleles and decreasing the frequency of unfavorable ones. For a simple inheritance trait controlled by only one or two loci, it is theoretically possible to follow these changes in the causal loci, but to make these calculations possible, either these loci must be known or a genetic marker must exist for each causal gene. For complex quantitative traits controlled by tens to thousands of loci, each with small effects, it is possible in principle to follow these changes, but substantial technical and statistical difficulties remain because the location for most of the genes controlling the trait in the genome is unknown.

Therefore, as with the procedures described above to estimate genetic variances and breeding values, here we rely on predictions made with the help of values obtained at the population level, and such predictions will refer above all to phenotypic averages. Changes in variances can also occur with selection, but this is difficult to predict (Walsh and Lynch, 2018). Frequently, changes in the means are those considered in a few generations, such as in optimizing commercial breeding programs.

Single-trait selection response

A general equation for the calculation of selection response per year for a single quantitative trait in a selected population is:

$$\Delta G_y = (D_m \sigma_{EBVm} + D_f \sigma_{EBVf}) / (L_m + L_f), \quad [3]$$

where ΔG_y is the selection response for the selected trait in the units of measurement per year; D_m and D_f are the selection intensities in males and females, respectively; σ_{EBVm} and σ_{EBVf} are the standard deviations for the EBV in males and females; and L_m and L_f are the generation intervals in males and females. Nicholas (1980) reviews the calculations needed in many types of populations in detail, providing derivations to calculate the expected variability of response accounting for population size.

A detailed account of each component is outside the scope of this chapter, but a general description would be as follows. The selection intensity (D) on each sex is the superiority expressed in standard deviation units of the selection criterion for a given selected proportion of animals, ranked according to the EBV. A normal distribution is usually assumed for the EBV, and selected animals are assumed to be all those with EBVs above a given fixed level, called the truncation point (truncation selection).

Selection intensity values for some selection proportions are shown in Table 6.1. A low selected proportion (greater difference between the average of selected individuals with respect to the population mean) implies a high D value: selecting the best 0.01 fraction for breeding, D = 2.660. Conversely, if one were to keep all animals for breeding, or choose breeding animals at random, that would be equivalent to selecting average animals; thus, the selection proportion will be 1 and therefore D = 0. For complete tables and further discussion of the concept of D, see Falconer and Mackay (1996) and Van Vleck (1993). As the D value is a standardized selection "pressure" it can be compared across traits and populations.

The standard deviation of the EBVs is the product of the selection accuracy multiplied by

TABLE 6.1 Selection intensity according to the selected proportion for a normally distributed trait in a large population.

Selected proportion	*Selection intensity (D)*
1.00	0.000
0.80	0.350
0.50	0.798
0.20	1.400
0.10	1.755
0.05	2.064
0.01	2.660

TABLE 6.2 Some examples of selection accuracy ($r_{A,EBV}$) in animal evaluations with different selection indices using individual, family, and genomic data for a quantitative trait based on the accuracy of parent EBV/data available, h^2, and c^2.

	h^2			
Information available (EBV accuracy)	*0.02*	*0.10*	*0.20*	*0.40*
Sire EBV (0.30) + dam EBV (0.30)	0.21	0.21	0.21	0.21
Sire EBV (0.50) + dam EBV (0.50)	0.35	0.35	0.35	0.35
Sire EBV (0.90) + dam EBV (0.90)	0.64	0.64	0.64	0.64
Own phenotype	0.14	0.32	0.45	0.63
10 progeny	0.22	0.45	0.59	0.73
50 progeny	0.45	0.75	0.85	0.92
10 full-sibs; $c^2 = 0$	0.21	0.42	0.51	0.60
50 full-sibs; $c^2 = 0$	0.41	0.60	0.65	0.68
10 full-sibs; $c^2 = h^2$	0.20	0.33	0.37	0.10
50 full-sibs; $c^2 = h^2$	0.32	0.39	0.40	0.41
Own phenotype + 10 full-sibs; $c^2 = 0$	0.25	0.49	0.62	0.74
Own phenotype + 50 full-sibs; $c^2 = 0$	0.43	0.64	0.70	0.78
Own phenotype + 10 full-sibs; $c^2 = h^2$	0.24	0.40	0.48	0.64
Own phenotype + 50 full-sibs; $c^2 = h^2$	0.33	0.43	0.49	0.65
BLUP[1]; 50 full-sibs; $c^2 = 0$	0.48	0.62	0.66	0.68
BLUP[1]; 50 full-sibs; $c^2 = h^2$	0.41	0.47	0.50	0.57
Genomic evaluation[2]	0.47	0.76	0.85	0.90

Notes: c^2, Common environmental full-sib variance (c-effects) as a fraction of phenotypic variance = Vc / Vp. [1] PseudoBLUP with no individual information and 1.4 females per male. [2] Genomic evaluation accuracy estimated according to method described by Lee et al. (2017) with a training population of 4000 individuals, an effective population size of 100, a genome of 30 Morgans, and 5000 SNP markers.

the additive genetic standard deviation of the trait () within each sex. Selection accuracy ($r_{A,EBV}$) is the correlation between the true additive genetic value (A) and the EBV.

The $r_{A,EBV}$ values will be estimated depending on the information used to obtain the EBVs and h^2. An approximate guide to the possible values of $r_{A,EBV}$, obtained with the STEBV accuracy Excel spreadsheet developed by Van der Werf (2017) is provided in Table 6.2.

Selection accuracy ranges from 0 to 1. Values depend on the type and amount of information used in the genetic evaluation of animals. BLUP procedures are based on combined information with no individual information considered, but the value of additional information depends on the type and amount of existing information and genetic parameter values. In Table 6.2 we can observe that family information (parent EBV, progeny, full-sibs) provides greater accuracy in relative terms with small values, compared with larger values for h^2. Sib selection, which is widely implemented in aquaculture and in shrimp breeding in particular for traits that cannot be measured in individual animals, or when selecting families for disease resistance from results in a group of exposed sibs, yields relatively high accuracies, but these are negatively affected by the value of c^2, the common environmental full-sib variance as fraction of phenotypic variance. As with combined selection, BLUP-based EBVs are useful in obtaining high accuracy even in the presence of c^2 effects.

GS under a model that makes use of closely related individuals in the training and testing populations (Lee et al., 2017) would yield an accuracy very similar to that obtained from the mean of 50 progeny, because in both cases within-family variances would be considered. The use of progeny testing is generally discouraged in shrimp to maintain short generation intervals that are biologically feasible, and therefore most advanced breeding strategies considered will be sib and genomic evaluations (Castillo-Juárez et al., 2015; Yáñez et al., 2020), while simpler programs could be based on single-pair mating individual selection (Bentsen and Olesen, 2002), within-family (Montaldo and Castillo-Juárez, 2017), and simple mass selection programs (Gjedrem and Baranski, 2010).

The generation interval (L) is the average age of the parents when the replacement animals are born. L values depend on the biology of the species, the management of the age structure of the population, and the time required to obtain EBV from selection candidates (Nicholas, 1980).

Equation [3] is a key tool to optimize breeding programs with the addition of methods to estimate effective population size and associated inbreeding increase and loss of genetic variation.

Multi-trait breeding goals and selection criteria

In most cases, several traits are considered in genetic improvement. Breeding goal (BG) is a key factor determining both the direction and the speed of changes occurring through breeding in any selected population. Perhaps the first issue to be defined, the BG is an overall assessment of the value of each trait contributing to profit. Hence, for example, the BG considering three traits is expressed as:

$$BG = g_1v_1 + g_2v_2 + g_3v_3,$$

where g_i = breeding value of the i-th trait and v_i = economic value of the i-th trait.

The BG is always an abstraction from reality and involves making several decisions. It is likely that many traits would contribute to profit in any economic and biological circumstance, but often it is not possible to include all of them, as this would require knowledge of at least an approximate economic value and the genetic parameters, not only of the trait considered, but also of parameters linking the trait with the others in the BG and those considered as selection criteria. The traits in the BG are usually different from those considered in the selection criteria or selection index (SI). The SI is the value used to rank and select selection candidates and has the following form, here using an example considering five measured traits:

$$SI = b_1y_1 + b_2y_2 + b_3y_3 + b_4y_4 + b_5y_5,$$

where b_i = index weight of the i-th trait and y_i = measurement of the i-th trait. In this equation, the y's are considered corrected for any environmental and genetic factors, so they are perfectly (unbiasedly) comparable.

When using BLUP-EBV estimates rather than SI-EBV estimates for selection, the optimal way to proceed is to multiply the EBV for each trait obtained in a multi-trait analysis (Mrode, 2014) by its relative economic weight to estimate what we call here an economic index (EI). Each EBV will be a synthesis of all the information on each trait. The use of this method over single-trait-analysis EBVs would yield very good approximations in many applications.

$$EI = v_1EBV_1 + v_2EBV_2 + v_3EBV_3$$

The development of relative economic weights is an important step in the implementation of any selection program. There are several techniques that may be used to this end (Weller, 1994). The technique used by Ponzoni et al. (2007) in fish is straightforward and may be applied to different profit equations on an individual-animal basis to consider multiple traits (Campos-Montes et al., 2017; Janssen et al., 2018). In cases where no economic information is available, weights can sometimes be obtained from desired values for relative genetic gains of the traits in the selection objective (Weller, 1994). Many caveats exist for the calculation of the exact economic value for each trait, with several possible options regarding the sector of the production chain that obtains the profit and whether the optimization should be made per animal or per unit of product (Weller, 1994). In any case, examining the problem and obtaining at least approximations is of great value, in particular for breeding companies.

Responses to selection for multi-trait selection are based on a generalization of single-trait equations [3] and require the use of SI methodology (Van Vleck, 1993) and estimates of variance–covariance matrices of genetic parameters (Gjedrem and Thodesen, 2005; Lynch and Walsh, 1998).

Marker-assisted selection and genomic selection

Ideas on the use of DNA genetic markers in animal breeding initially centered around the concept of marker-assisted selection (MAS). MAS was conceptualized as adding information about genotypes to the usual polygenic-based EBV for a small number of genes of large effect identified in the genome (Weller, 2016). Information from a few molecular markers linked to QTLs (i.e., genomic regions harboring genes with a significant effect on the trait) could be implemented in breeding schemes through MAS, if they explain a high proportion of genetic variation in the trait (Houston et al., 2020; Weller, 2016; Yáñez et al., 2020; Zenger et al., 2019). While in principle there is a possibility that any quantitative trait might be controlled by a few discernible loci with large effects (major-effects QTLs) in MAS, in general the practical effectiveness of this procedure has been limited because the number of large QTLs identified is usually small and the proportion of the additive genetic variability explained by individual QTLs is low for most quantitative traits (Hill, 2010; Weller, 2016; Weller and Ron, 2011). An example of a remarkable exception to the above in aquaculture is infectious pancreatic necrosis, an important disease in Atlantic salmon. In this disease, a substantial proportion of the total variance for resistance was explained by a single QTL, which has allowed the incorporation of DNA testing for the genetic improvement of this trait in Norway and Scotland (Yáñez et al., 2014). Nevertheless, in *Litopenaeus vannamei*, results from a genome scan study did not show any evidence of major genes for growth (Khatkar et al., 2017).

Additionally, the information of thousands of markers could be simultaneously incorporated into genetic evaluation to estimate genomic breeding values (GEBVs) in what is known as genomic selection (GS) (Meuwissen et al., 2013, 2016). The basis of GS is to track the effects of QTLs on quantitative traits based on DNA marker information. Markers could help estimate the QTL effects due to the linkage between them. This creates a statistical association at the population level known as linkage disequilibrium. In family-based programs such those most common in shrimp, the increase in response using GS is likely moderate for weight, but can be substantial in disease resistance if a large population is genotyped and phenotyped in each generation (Dai et al., 2020; Houston, 2017; Houston et al., 2020; Nielsen et al., 2009; Sonneson and Meuwissen, 2009; Yáñez et al., 2014, 2020). GS is a promising method for the genetic improvement of any animal or plant species, but is particularly well suited to application for disease resistance selection in aquaculture species because there is no need to expose selection candidates (Castillo-Juárez et al., 2015; Houston et al., 2020; Yáñez et al., 2014). Potential advantages of GS over other selection strategies in shrimp breeding would depend on the selection accuracy, generation interval, and selection intensity. Its practical use would also depend on genotyping costs. Although the expected advantages of GS in shrimp are not as high as in cattle (Weller, 2016), they are interesting nonetheless (Zenger et al., 2019).

Genotype x environment interactions

In cases where environmental effects impact different genotypes to varying degrees, the independence of A and E is lost, and we may speak of genotype × environment interaction (G×E) effects. In this case the phenotype model contains an additional interaction term, and the general model [1] is modified to:

$$P = \mu + G + E + G \times E$$

The presence of G×E may reduce the expected results from the use of populations or animals selected in other environmental conditions or may reduce the selection responses, acting as an additional source of variation and therefore reducing the h^2 across testing environments (Isik et al., 2017; Montaldo, 2001; Sae-Lim et al., 2016). A general remedy for the presence of G×E is to perform the selection of animals under conditions similar to those of the commercial population, or at least to try to keep the basic environmental characteristics as similar as possible (Ponzoni et al., 2008). A notable example of high G×E effects was found for biomass production in comparing two *L. vannamei* lines, Resistance and Susceptible (the genotypes), in ponds with (+) or without (–) acute hepatopancreatic necrosis disease (AHPND)/ white spot disease (WSD) outbreaks (the environments) in northwestern Mexico. The average ratio of biomass production in Resistance:Susceptible lines in + ponds was 16.3, whereas that in – ponds was

0.72 (Gallaga-Maldonado et al., 2020). Here the G×E effects are between cross/pure-line populations and environmental levels in the context of selecting the best population/cross.

G×E effects for genetic relationships between harvest size weight and survival for the grow-out period (65–130 days) within the WSD Resistance line described above were also found using unpublished results from analyses of field data (2017) from northwestern Mexico involving ponds with (+) or without (–) AHPND/WSD outbreaks obtained in 2015-2017. The genetic correlation for survival between + and – ponds was lower (0.19) than the correlation between – and – ponds (0.64), indicating a strong G×E effect. This implies that selecting for increasing survival in – ponds should be made based on data obtained in these ponds, because information about survival in – ponds depends on other genes. For weight at harvest, genetic correlations between + and – ponds (0.74) and between – and – ponds (0.94) were approximately similar, indicating comparatively smaller G×E effects, which is consistent with previous estimates with no infections (Campos-Montes et al., 2009; Castillo-Juárez et al., 2007). These large and small G×E effects for survival and weight, respectively, affect the optimal selection strategy and use of data from different sources. Here the G×E effects are between genetic additive values and environmental levels in the context of selection within a single population.

MAINTENANCE OF GENETIC DIVERSITY AND INBREEDING CONTROL

Effective population size

Size is one of the most important genetic features of any population. Population size, associated with the number of parents, sex ratio, mating system, and selection procedures, will influence the evolution of the loss of genetic variation, accumulation of inbreeding, and variability of selection response (de Rochambeau et al., 2000; Nicholas, 1980). In practical terms, a simple approximation to the "true" genetic population size is effective population size (N_e), defined as the size of an idealized population undergoing the same rate of genetic drift as the population under study (Caballero, 2020; de Rochambeau et al., 2000; Falconer and Mackay, 1996). This is estimated as follows:

$$N_e = 4N_mN_f / (N_m + N_f), \quad [4]$$

where N_m is the number of males and N_f is the number of females used per generation. This N_e formula tends to overestimate the real value in selected populations by a factor of two or more due to a number of assumptions that are not fulfilled in most real populations (e.g. random mating, equal number of progeny per parent, no selection). However, the basic estimation of N_e provides an approximate idea in practice of the possible evolution of the population diversity. Several methods have been proposed to estimate N_e from demographic, pedigree, or molecular data (Caballero, 2020; Howard et al., 2017). Moreover, given that N_e is related to the variation in the ΔG_y another reason to maintain a minimum value is to ensure a high probability of obtaining a proportion of the expected selection response (Nicholas, 1980).

Selection and loss of genetic variability in closed populations

As any selected population has a finite size (Caballero, 2020), additive genetic variance will be lost by random sampling (random genetic drift) at a rate of $1 / (2N_e)$ per generation (Hill, 2016). In this case, additive genetic variance is also expected to be progressively reduced by selection as the allele frequencies will tend toward extreme values. Moreover, selection will reduce additive genetic variance due to the so-called Bulmer effect that creates negative correlations between favorable and unfavorable alleles in the first generations of a selected population until an equilibrium is reached (Caballero, 2020; de Rochambeau et al., 2000). Although all these mechanisms are known, the methods to predict such changes are complex, and the best we can do is to provide approximations to real values. In many cases the base population is not clearly defined and the parameters required for calculation of loss in variability are not available. As randomness is involved in the result of each actual finite-sized population, no precise deterministic prediction is possible. Therefore, simulation is seldom needed to measure all these factors acting in conjunction. In addition to these factors, which tend to reduce variance, mutation has the tendency to replenish genetic variation, even over the course of a relatively small number of generations and for a small N_e, as shown with selection response and molecular variation in the Illinois maize selection experiment and selected poultry lines for more than 50 generations (Hill, 2016).

Approximate minimum values for N_e in breeding programs to prevent serious declines in genetic variation for selection and inbreeding depression are between 50 and 100, depending on the aim and time horizon considered. For the conservation of most genetic variability, figures are higher, between 500 and 1000 (Hill, 2016; Lynch and Lande, 1998). Montaldo et al. (2011) estimated relatively stable values for N_e (81–132) from an analysis of six generations in a selected Mexican population of *L. vannamei* using equation [4]. The values of N_e estimated on the basis of the genetic relationships decreased from 193 to 58.

Inbreeding

Inbreeding is the result of mating between related animals. The main consequences of inbreeding are an increased proportion of homozygous individuals in the population and an increased proportion of homozygous loci at the individual level. Inbreeding is due to the additional probability (compared with non-inbred base animals in a pedigree) that IBD alleles duplicated from common ancestors are joined in individuals through their parents (Pirchner, 1985; Wright, 1922). A concept related to many important implications in animal breeding is the additive genetic relationship (de Rochambeau et al., 2000). The additive genetic relationship (a_{AB}) between individuals A and B is defined as the proportion of shared IBD alleles and can be determined from the genealogy (pedigree). This (numerator) additive genetic relationship is the basis for estimating the inbreeding coefficient. The inbreeding coefficient for individual X, which is a progeny of individuals A and B, is half of a_{AB}: $(F_X) = 1/2a_{AB}$, as defined by Wright (1922).

In a closed finite population, inbreeding would be the product of the increasing average relationship between individuals across generations as progressively more individuals share alleles by descent over time (Caballero, 2020). Random changes in allele frequencies (genetic drift) also help increase the proportion of homozygous loci. Genetic drift and inbreeding are frequently treated as similar manifestations of the same basic issue because both are linked to N_e; however, they are different phenomena. Many conceptual and practical complexities emerge in real populations in relation to this topic, and have been discussed elsewhere (Lynch and Walsh, 1998; Walsh and Lynch, 2018).

At the population level, inbreeding change per generation (Falconer and Mackay, 1996) is defined as:

$$\Delta F = 1 / (2N_e)$$

The level of inbreeding at time t, expressed in years (Pirchner, 1985), would thus be:

$$F_t = 1 - (1 - [1 / N_e])^{t/L},$$

where L is the average generation interval for sires and dams in years.

In outbred animal populations, inbreeding is mainly considered an unintended and undesirable consequence of having small breeding populations. Typically, aquaculture breeding lines have relatively small effective population sizes, and therefore an increase in inbreeding is considered something to be controlled. The primary reason for this is inbreeding depression, a decrease in the productivity and fitness traits related to the increase in the average inbreeding coefficient in the population.

Most inbreeding depression effects are associated with increasing proportions of harmful recessive genes; therefore, in broad terms, dominance genetic effects are considered to be associated with these effects (Caballero, 2020; Dickerson, 1973; Pirchner, 1985). Theoretically, inbreeding depression would affect most fitness traits; however, these traits are subject to strong natural selection, and both theoretical and experimental results have shown that inbreeding depression effects on survival traits in shrimp are smaller than on growth after a few generations (De los Ríos-Pérez et al., 2015, 2017; Moss et al., 2007). A decrease in the depressing effects of inbreeding for reproduction traits in females has been observed across generations in highly inbred shrimp in what is interpreted as the effect of natural selection against recessive homozygous harmful alleles, which counteracts inbreeding depression by reducing the frequency of such alleles in what is known as genetic purge (De los Ríos-Pérez et al., 2017).

Controlling losses in genetic diversity

Knowledge of the genetic relationships between all the animals in a population (relationship matrix), estimated through either pedigree information or DNA genetic markers, allows breeders to implement selection-mating strategies to maintain genetic diversity in the population and to reduce any increase in inbreeding rates by using either methods or rules to reduce the average relationships between mates and breeding animals of each sex (Meuwissen, 2009). Simple rules can be implemented, such as trying to maintain uniform family sizes, setting a maximum number of replacements from each sire and dam, and forming circular mating groups (de Rochambeau et al., 2000). Additional methods have been developed that use algorithms to restrict mating between related animals at a similar level of selection response, or to maximize selection response at a maximum admissible level of increase in inbreeding with appropriate algorithms (D'Agaro et al., 2007; de Rochambeau et al., 2000; Henryon et al., 2019; Howard et al., 2017; Meuwissen, 2009). Few estimates are available on the possible reductions of genetic variability across time for breeding programs. Cala et al. (2014) estimated changes in variances and heritability for harvest size weight in a selected Mexican *L. vannamei* population. Based on data from eight generations, this study found no negative changes in the variances or heritabilities for this population, which was compatible with the N_e value, which had an average of 74. Andriantahina et al. (2013) found moderate reductions in heritabilities in a *L. vannamei* population selected for four generations; however, the trend is probably not significant from a statistical point of view.

Most improvements to better balance selection response and inbreeding/loss of diversity in a closed population are temporary. Opening the population to other genetic material from outside, which in the case of shrimp can incorporate broodstock from other genetic nuclei, unrelated animals from commercial ponds in the same or other regions, and even wild animals, represents a dramatic but sometimes necessary solution to rapidly reduce inbreeding levels and introduce genetic variation to closed populations for specific purposes, such as the rapid incorporation of disease resistance genes. In the most extreme cases, the total replacement of a population may be indicated, at least temporarily, to obtain survival rates compatible with profitable production (cf. Gallaga-Maldonado et al., 2020). Decisions regarding which approach is most appropriate depend on the specific circumstances of the population.

One option for preserving genetic variability is to maintain replicated lines with a relatively high N_e. Selection may be limited to within-family evaluations (Montaldo and Castillo-Juárez, 2017). In that case the effective population size is twice the N_e for the same population under random mating (Falconer and Mackay, 1996). These lines may have other functions, such as acting as a back-up against catastrophic events, selecting for additional traits (Hill, 2016), and testing novel breeding strategies.

Creating base populations and bottlenecks

The existence of genetic variability for the traits to be selected in a base population is a prerequisite for genetic improvement. Reduced genetic variability will decrease the rates of genetic response over a given time horizon. Thus, in shrimp breeding programs, the creation of base populations with high genetic variation, and maintaining that variation, is of great importance for continuous improvement and therefore highly relevant to practical breeding programs.

Most loss in genetic variability occurs during a preliminary sharp reduction from a natural shrimp population (genetic bottleneck) when creating the initial population to begin a breeding line (cf. Vela-Avitúa et al., 2013). Some attempts to increase initial variability can be made by using large numbers of sires and dams from several populations to create a base line for breeding. One option is to use a mixed population composed of several sources followed by random mating to create synthetic or composite populations (Dickerson, 1973; Holtsmark et al., 2006). Research cannot indicate the exact ideal procedure for creating new populations, because there is a trade-off between the genetic level of the original population, the number of populations used, and the size of the total population and each of the contributing populations. Managing a large number of populations and implementing strategies to control loss of genetic diversity by using pedigree or genomic relationships would be complicated from an operational point of view (Fernández et al., 2014). It is likely that creating a mixed population may be a good option in principle to combine attributes from several populations into one, making it possible to concentrate on selection. However, this is clearly a temporary solution, because the need to use a small number of selected superior animals decreases N_e across the breeding generations.

A common cause of severe genetic bottlenecks in shrimp is the use of survivors of catastrophic disease outbreaks for breeding. In the case of *L. vannamei*, it is likely that many populations passed through this situation during the selection against Taura syndrome virus (TSV), where survivors representing less than 2% of the total population were used as breeders, apparently without displaying any important inbreeding depression, although this was not measured (Cock et al., 2009).

Use of inbreeding in breeding programs

Inbreeding can be used as a tool to increase genetic variation in populations of any size by deliberate mating between relatives. Interestingly, the additive genetic variance in an inbred population consisting of a collection of inbred lines is predicted to *increase* compared with that in a random mating population with no inbreeding. Therefore, not only can inbreeding redistribute additive genetic variance in a population, but it can also create some of it. This is the main reason why inbred lines are used in plant breeding (Dickerson, 1973; Hickey et al., 2017). Such a method would be particularly useful in self-fertilizing species, where the cost of losing a line is low. Developing improvement programs based on inbred lines within breeds has been considered infeasible in large terrestrial animals since the 1930s, because of economic and practical issues related to the high individual cost of animals, the expense of maintaining large populations, long generation intervals, low reproductive rates, and difficulties in obtaining highly inbred lines in a few generations (Dickerson, 1973). In terrestrial animals, the advantages related to crossbreeding (as discussed in the section later in this chapter) are generally greater when using crosses between breeds that are mildly inbred populations, without having the problems associated with the creation of inbred lines. As some of these restrictions are not present in shrimp, and because well-defined lines or strains are not yet known in most shrimp species, inbreeding could perhaps be used in creative ways to enhance selection responses (Dickerson, 1973; Pirchner, 1985). However, research based on model planning of breeding programs, including economic considerations, is warranted before implementation.

DESIGN AND OPTIMIZATION OF SELECTION PROGRAMS

The evaluation of breeding programs makes use of biological and economic considerations related to the selection responses, the associated costs for a particular testing and selection strategy, and the costs associated with the increase in inbreeding rate/loss of genetic variation for a given time horizon (D'Agaro et al., 2007; Dupont-Nivet et al., 2006; Gjedrem and Thodesen, 2005; Janssen et al., 2018; Lind et al., 2012; Nguyen, 2016; Ponzoni et al., 2007). In practice, decisions on breeding operations should be based on considerations of economic efficiency and feasibility (Montaldo, 2020).

Several methods may be implemented when using selection for the genetic improvement of shrimp, including individual selection, family selection, within-family selection, combined selection, and GS (Bentsen and Olesen, 2002; Castillo-Juárez et al., 2015; Gjedrem and Thodesen, 2005; Zenger et al., 2019). Not all methods are suitable for the improvement of all traits, because in many instances combinations of methods can be used in programs that may use multi-stage selection (Montaldo, 2020).

An important part of designing breeding programs is determining the best mating structure design. Most family-based selection programs in shrimp are currently based on nested sire/dam structures obtained with artificial insemination with modest male:female ratios of between 1:1 and 1:1.5, or single pair-mating schemes where sires and dams (with male:female ratios of 1:1) are used to create a number of full-sib families (Bentsen and Olesen, 2002; Gjedrem and Thodesen, 2005; Nguyen, 2016). Factorial design, where each sire and dam are mated to a number of partners, can be useful in separating the effects of additive genetics, dominance genetics, and maternal effects; therefore, it is another option to increase selection accuracy, but this depends on further development of assisted reproduction techniques in shrimp (Montaldo, 2020). Nested designs with full- and half-sib relationships could help improve the separation of c-effects from additive genetic effects compared with single-pair mating (Montaldo et al., 2013b). Factorial mating designs would allow for increased rates of genetic response and a reduced loss of genetic variability along the same line.

A type of selection program frequently used in selecting shrimp in some countries, such as Ecuador, is one that uses genetic markers to recreate the pedigree of each generation (accepting some degree of error) to try to minimize genetic relationships between untagged breeding candidates recovered from grow-out field ponds, and therefore to avoid the use of highly related animals as breeders to prevent a rapid loss of genetic diversity in the population. These programs are in many respects different from family-based systems such as the traditional sib shrimp programs in which all candidates are measured and tagged (Campos-Montes et al., 2013; Gjedrem and Thodesen, 2005). Such programs are based on retaining the higher-weight animals recovered from ponds for reproduction, as they have survived the environmental conditions/challenges present in the pond, making them suitable candidates for breeding to increase yield (biomass production) under commercial conditions. Survival rate is also selected individually, because the recovered shrimp are survivors of any environmental challenge, including disease. A similar procedure, developed in Colombia to create a population with higher genetic resistance levels to TSV based on surviving animals, may be described as successful, even if some of the populations were very small (Cock et al., 2009). The main issue with this type of program is that it is very difficult to know the actual values of the genetic parameters or to predict selection responses.

Because programs based on the recovery of animals from grow-out ponds avoid some of the G×E issues found in nucleus-population breeding systems, such programs should, in principle, "work" by creating genetic progress for the required production conditions (Ponzoni et al., 2008). However, unlike family-based programs (cf. Campos-Montes et al., 2013, Gjedrem and Thodesen, 2005), there are no publications on predicted or estimated results regarding genetic progress or loss in genetic variability.

Pedigree reconstruction can be used in many breeding contexts for shrimp genetic improvement programs. It is frequently argued that programs where shrimp (or other aquaculture species) are communally reared from an early age would totally resolve the problem related to common environmental variance as a factor that reduces selection accuracy. However, that is only true depending on the time the shrimp are kept separate and the contribution to these effects from maternal effects, which are largely unknown and should vary with the traits considered. Thus, perhaps even in a communally-raised population the optimal design to avoid c-effects may be one implementing a nested or factorial mating structure.

In shrimp, the diffusion of genetic superiority from a breeding nucleus population to the commercial population is easy because of the high fecundity of females in the absence of any artificial reproduction technique. Therefore, there is a possibility, at least theoretically, of eliminating any lag created by the presence of a multiplier stratum by directly producing all commercial larvae directly from the nucleus. However, the presence of such strata may be necessary in most cases for logistic reasons and to reach the number of postlarvae that the company needs to supply to the grow-out sector.

MIGRATION AND CROSSBREEDING

Whether used to replace the current population, to create a new synthetic population based on a stable mixture of genes from two or more populations, or to create a crossed commercial population by mating pure lines, migration is an important tool in changing populations (Benzie, 2009; Gjedrem and Thodesen, 2005). This is because as more populations exist within a species, more opportunities are created to use that diversity to increase productivity in commercial animals (Caballero, 2020; Dickerson, 1973).

The potential advantages to be obtained from crossbreeding systems include heterosis, which is the advantage of crossed animals over the average of contributing purebred animals, as well as the possibility of combining traits from several populations, for example, by combining disease resistance with a high growth rate in a synthetic population, which will also retain some proportion of the heterosis of the first two-way cross, or F1. Crossbreeding systems also allows breeders to produce commercial animals with no inbreeding.

Unlike poultry and other terrestrial animals, in shrimp and other aquaculture species only closed breeding lines have been used thus far, and systematic crossbreeding systems have not been used to produce commercial shrimp for aquaculture. Several factors are associated with this difference, but the main one is that as shrimp are recently domesticated species, only relatively limited diversity exists between selected lines available for crossing (Dickerson, 1973; Pirchner, 1985). Therefore, heterosis levels have been found for only some traits, making the economic value of the crosses inferior to that of pure lines (cf. Gallaga-Maldonado, 2020). In the future, it is likely that different inbred lines will exist in shrimp, so that crossbreeding may become a more common system in the production of commercial animals. Two advantages in shrimp to exploit heterosis are the high female fecundity and short generation interval. The creation of many inbred lines to be systematically crossed to look for heterosis is another possibility (Dickerson, 1973; Pirchner, 1985); however, that strategy would entail a considerable amount of time, investment, and even luck.

Practical shrimp breeding

In this second section we review research results concerning practical aspects of shrimp breeding based on published material, and present unpublished information generated as part of the research undertaken with the Mexican hatchery Maricultura del Pacífico (MARPAC), located in the state of Sinaloa, in northwestern Mexico.

TRAITS TO CONSIDER IN SHRIMP SELECTION PROGRAMS

Traits to be included in selection programs should have a value for the industry or society. Shrimp breeding is unusual insofar as selection to this point has been based on only a few selection criteria, mostly growth/size-related traits, specific disease-resistance-related, traits and general survival. Unlike other aquaculture species such as fish, carcass quality or yield traits have generally been considered less important in shrimp breeding programs (Gjedrem and Baranski, 2010).

Weight and feed efficiency ratio

In all aquaculture animals, including shrimp, the first traits to be considered in selection are those related to size (Gjedrem and Baranski, 2010). These traits, measured as weight, length, or other body dimensions, are closely related to the economic value of the harvest. In addition, a shorter growing period would imply less feed, management, and other costs, besides a reduction in risk. Harvest size weight is therefore a logical priority in shrimp.

Studies on the genetic relationships between size traits of shrimp harvested at different ages – close to 130 days as the chosen harvest age in northwestern Mexico, solely as a point of reference, and 150 days looking at heavier shrimp for specific markets – have shown that the values are high and positive (>0.90); therefore, selection for weight at 130 days would be adequate to increase harvest size at higher ages. The genetic relationships between early weights and sizes with harvest weights are positive and moderate, indicating that programs involving early selection may have some merit (Andriantahina et al., 2013; Campos-Montes et al., 2013), particularly to take advantage of the reductions in numbers of animals made in individual family cages when performing within-family selection (Montaldo and Castillo-Juárez, 2017).

The heritabilities of growth traits at harvest size, the most studied size selection criteria, are generally between approximately 0.20 and 0.40 in *L. vannamei* (Andriantahina et al., 2013; Campos-Montes et al., 2013; Castillo-Juárez et al., 2007; Garcia et al., 2021). As is typical for any trait, it is very difficult to identify the factors associated with these changes in heritability values because the original genetic variation, previous population size management, mating design, statistical model, estimation methodology, and rearing conditions, among others, are all possible factors. Moreover, these values depend on the size of the measured population and the quality of the pedigree or genomic matrix used to measure the genetic relationships. A common problem that inflates heritability estimates in shrimp is the analysis of data from full-sib families only. Although a high c-effect will reduce selection accuracies only moderately (Table 6.2), estimates of heritability can be severely inflated (Montaldo, 2020; Montaldo et al., 2013b).

The use of size-related traits as selection criteria or objectives in aquaculture breeding programs has been criticized, with the argument that increased size would not necessarily increase the feed efficiency ratio. However, in practical terms we can perhaps consider these traits as acceptable surrogates for efficiency, based on results for terrestrial species (Ponzoni et al., 2007), until a practical technology is developed in aquaculture to estimate feed efficiency ratio at the individual-animal level with reasonable costs. The only result for *L. vannamei* regarding feed efficiency ratio and weight (Dai et al., 2017) indicates favorable genetic correlations of between 0.62 and 0.85. Another possibility for measuring feed efficiency in shrimp with reasonable costs and accuracy is by evaluating pooled family data (cf. Hernández-Ruiz et al., 2020).

Selection responses in shrimp in programs designed to increase growth have always been positive, with an average close to 8% per year (albeit very variable, ranging between 2% and 21%) (Gjedrem and Rye, 2018). These results confirm a general presence of additive genetic variance in every population, but the heterogeneity of estimates also reflects differences between traits measured, genetic parameters, emphasis on different selected traits, and selection evaluation and procedures.

Survival

Survival is a binary trait (0 = dead; 1 = alive) of immense economic importance in any shrimp-growing enterprise (Ponzoni et al., 2007). Income is nearly linearly related to the product of average harvest weight × survival rate (biomass production). Nonetheless, survival is subject to considerable environmental variation, and therefore its heritability is likely to be small, generally less than 0.10 (Campos-Montes et al., 2013; Fjalestad et al., 1993; Garcia et al., 2021). In addition, survival has been questioned as a selection criterion because just as environmental conditions may change dramatically each growing season, factors affecting survival may also change, and therefore the genes that allow animals to survive a given year could confer less survival advantage in another year. However, a positive genetic trend for survival has been obtained in a shrimp nucleus population in northwestern Mexico from 2004 to 2010 using linear mixed models (0.26% per year) in the absence of any catastrophic disease outbreaks, indicating a favorable correlated response from selection to increase growth, which is expected from the positive genetic correlation between growth and survival (Campos-Montes et al., 2013). This experience indicates a cumulative increase in breeding value for "general" survival, contradicting the skepticism regarding the choice of survival as a selection criterion. However, such general survival is a different trait from survival of outbreaks caused by specific diseases (Cock et al., 2009), and evidence suggests that selection for survival in the absence of disease would probably not increase disease resistance against new diseases. Survival associated with specific diseases is discussed in the next section.

Disease resistance

Disease resistance is probably the most important single aspect of shrimp genetic populations apart from growth in the current open grow-out ponds, the most prevalent system worldwide (Cock et al., 2017). Totally biosecure shrimp production systems are conceptually possible but remain in a preliminary phase of development or are demonstrably profitable in only a few situations. Moreover, based on the experience with terrestrial animals, we know that new diseases can develop in any system, and closed systems tend to have traits, such as high animal density, that put them at risk for diseases that can bypass safety measures (Cock et al., 2017). These conditions create a need for the development of disease-resistant animals (Houston, 2017).

Experiences with major shrimp diseases and genetic resistance involve at least two, and probably three, diseases. These are Taura syndrome, caused by TSV; WSD, caused by the white spot syndrome virus (WSSV); and possibly AHPND. Each of these diseases has its own history. When TSV decimated shrimp populations in Latin America in the early 1990s, new resistant populations were created and disseminated across Latin American countries. These populations were created by selecting small numbers of survivors, and it is believed that they may contain mutant resistant alleles having a dominant genetic effect (Cock et al., 2009). Additionally, animals with a higher degree of resistance for TSV were obtained in family-based programs using challenge tests for this virus (Moss et al, 2009, 2012). Relatively high heritability values of 0.20–0.40 allowed for high selection responses of up to 18.4% per generation (Noble et al., 2017).

In the case of WSD, despite early setbacks and the very low genetic variation initially found (Cock et al., 2009), some evidence for resistance has begun to emerge since 2010, and finally, around 2012, evidence of some families with high resistance to WSD was found (Cuéllar-Anjel et al., 2012). This discovery was followed by the development of commercial resistant lines. Animals from Ecuador were recognized as being much more resistant to field infection in Mexico than local Mexican high-growth lines that were selected in the absence of disease in crossbreeding studies (Gallaga-Maldonado et al., 2020), and there is also anecdotal evidence to this effect. Since then, the use of WSD-resistant lines has been widespread. As a dramatic increase in WSD mortality in Mexico in 2013 was associated with the introduction of AHPND as a major cause of loss in the shrimp industry, it is highly likely that these diseases act together and that shrimp that are resistant to both diseases are needed to increase productivity in the field. In fact, evidence from the field and challenge testing in Mexico indicate that WSD-resistant lines are also more resistant to AHPND (Castillo-Juárez et al., 2018; Gallaga-Maldonado, 2020). Moreover, genetic correlations within lines also indicate a positive genetic correlation (0.18–

0.56) between survival time for both diseases in the resistant line, but interestingly, these estimates were slightly negative and close to 0 in a growth line, which suggests a different genetic architecture of the traits within each line (Campos-Montes et al., 2020). However, genetic correlations between resistance for the two diseases using challenge survival were consistently high and positive in both lines (0.25–0.61).

Heritability estimates for survival time in challenge tests for WSD ranged from 0.08 to 0.26 for a resistant line and from 0.06 to 0.10 for a growth line; similar estimates for survival for the full experimental period were 0.06–0.09 for a resistant line and 0.02–0.06 for a growth line (Campos-Montes et al., 2020). Heritability of up to 0.38 was obtained in *L. vannamei* by Trang et al. (2019a) for survival in challenges with WSD when information from days 1–3 of the challenge was used. In that study, the heritability dropped to near 0 when the full experimental survival period, days 1–15, was considered. This difference may be due to a decrease in the differences between families. In another study of WSSV resistance (survival in a challenge test), heritabilities were between 0.19 and 0.27 depending on the statistical model used (Trang et al., 2019b). Heritability for survival in ponds naturally affected by WSD during the growth period (75–130 days) in a susceptible population ranged from 0.09 to 0.11 (Caballero-Zamora et al., 2015b). In short, heritability levels for WSD resistance traits allow selection responses to be obtained, but the time and variables chosen to measure survival in challenge tests must be carefully selected. Heritabilities for survival rates are always dependent on average survival (Fjalestad et al., 1993; Ødegård et al., 2011).

Estimates for survival time in challenge tests with AHPND ranged from 0.09 to 0.15 for a resistant line and from 0.09 to 0.18 for a growth line for a nested model (Castillo-Juárez et al., 2018). Estimates for another analysis for the same variable in the same population using an animal model ranged from 0.10 to 0.13 for a resistant line and from 0.06 to 0.13 for a growth line (Campos-Montes et al., 2020). Similar estimates for survival for the full experimental period were 0.09 for a resistant line and 0.11–0.14 for a growth line (Campos-Montes et al., 2020). Heritabilities from 0.20 to 0.23 were obtained in *L. vannamei* by Lyu et al. (2020) for survival time in challenges with AHPND. In another study with AHPND challenge, the heritability was estimated to be 0.15 for survival and 0.24 for survival time (Wang et al., 2019).

Analyses of data for 2015–2017 from northwestern Mexico yielded an average genetic correlation of 0.33 between survival time in WSD challenges and survival in infested ponds, and an average genetic correlation of 0.03 between survival time in WSD challenges and survival in non-infested ponds. Using data from the same origin, an average genetic correlation of 0.20 between survival time in AHPND challenges and survival in infested ponds, and an average genetic correlation of 0.06 between survival time in AHPND challenges and survival in non-infested ponds, were found (MARPAC, unpublished results, 2017). On the whole, these findings illustrate that both WSD and AHPND may be involved in mortality in infested ponds in Mexico during that period. The average heritability for survival in infested ponds was 0.17, indicating that direct selection for survival with data from ponds can be an effective tool to increase resistance to both WSD and AHPND.

Indirect selection for disease resistance

Fjalestad et al. (1993) analyzed the conditions in which indirect selection criteria can be useful in the genetic improvement of disease resistance traits in aquaculture. Many considerations make disease resistance traits good prospects for indirect selection. Most of them are practical, related to contributing to a decrease in the cost and time required to score animals. Despite a longer history of research in salmonids, little evidence has been found about the existence of useful indirect traits with the required high levels of heritability and favorable genetic correlations with the disease resistance traits (Yáñez et al., 2014). Examples of these types of traits from studies in shrimp involve pathogen loads, which are considered to be possible indicator traits for disease resistance.

The most obvious trait to be considered when selecting for increased disease resistance in shrimp is harvest size weight. When survival and harvest size weight were measured in WSD/AHPND-infected ponds in northwestern Mexico, the genetic correlation was favorable (0.21); however, the genetic correlation between harvest size weight measured in uninfected ponds and survival in infected ponds was negative (–0.12), indicating that pond disease status determines the usefulness of harvest size weight to increase resistance. Modeling of genetic responses indicates that the incorporation of survival in the indices is needed and that the greatest economic improvement rates are obtained when the selection criteria used to increase survival in infested ponds are obtained solely from these ponds, even if the numbers of candidates for body weight – and, in turn, selection intensity – decrease (MARPAC, unpublished results, 2017). This suggests the presence of highly significant G×E for these traits when comparing infested versus non-infested environments.

Viral loads (VL) were used in several studies to estimate genetic relationships with disease resistance traits in shrimp. Phuthaworn et al. (2016) estimated the heritability of the log-transformed hepatopancreatic parvo-like virus (HPV) VL measured by quantitative polymerase chain reaction in banana shrimp to be 0.41. These authors observed a favorable genetic correlation of HPV VL with body weight (–0.33) and body length (–0.38). Hernández-Ruíz et al. (2020) estimated the heritability of VL for infectious hypodermal and hematopoietic necrosis virus (IHHNV) in *L. vannamei* at 0.08, and observed favorable genetic correlations with survival (–0.57) and harvest size weight (–0.04). However, it is unlikely that even heritability for the family average of VL based on a pool of six animals (0.42) would be high enough to replace survival measurement.

Huang et al. (2011) reported a relationship between VL and disease resistance in a selection experiment for WSSV in *L. vannamei*; WSSV VL in surviving shrimp from high-resistance families were found to be significantly lower than those from unselected families. Studying the same disease, Trang et al. (2019b) found a negative (favorable) genetic correlation between VL and survival of WSSV challenges (–0.55), indicating the possibility of using VL as a trait to select for increasing resistance to WSD.

In conclusion, survival, in particular with disease exposure either to natural infestation or in challenge tests, or survival time, continues to be the basis for the selection of more resistant populations. Whenever possible, data from field evaluations from natural infestations should be used in selection, to avoid the potential G×E problems created by challenge testing.

OTHER TRAITS

Several other traits have been considered as part of a selection objective in shrimp breeding programs, including quality-related traits and fertility traits. Campos-Montes et al. (2017) evaluated the contribution of the incorporation of tail weight and tail yield at harvest size in selection programs for *L. vannamei*, estimating relevant genetic parameters and including economic aspects. They concluded that the incorporation of tail yield at harvest size (130 days) to a selection objective to increase economic response is not necessary, as selection for increased weight would yield similar economic results.

Tail or body color, fresh or cooked, is a trait that has sometimes been presented as a possible selection objective in shrimp, and heritability has been determined for some color-related traits in banana shrimp and *L. vannamei* (Giang et al., 2019; Nguyen et al., 2014, 2020). However, before selecting to genetically modify color, an economic advantage should be demonstrated, as well as determining which color is more desirable for the industry. Because color is a trait controlled by the presence of many components in food and water (Bernal Rodríguez et al., 2017), genetic effects may have strong interactions with the environmental effects, so that selection success may be restricted to specific environments.

It has been argued that some highly unsaturated fatty acid concentrations may play a role in the stress response and immune system of shrimp; therefore, genetic parameters have been estimated for fatty acid content in the hemolymph of healthy *L. vannamei* not exposed to disease (Nolasco-Alzaga et al., 2018). The estimate of heritability for the proportions of fatty acids were small (the heritabilities of actual concentrations were not provided). Insofar as the consequences of changing these proportions by selection are unknown and selection responses would likely be small, no attempt is advised for the incorporation of these measurements as criteria for selecting more profitable *L. vannamei* populations.

Fertility traits are of economic interest for breeding companies, and the multipliers of nucleus populations and heritability range from small to moderate in several studies (Caballero-Zamora et al., 2015a; Tan et al., 2017, 2019). However, because average female fecundity in *L. vannamei* is high, the incorporation of these traits as selection criteria would be justified only if genetic correlations were found to be unfavorable between fertility- or fecundity-related traits or if these fertility traits decreased below a practical threshold as a product of the selection program for increasing growth rate and other traits. Another way of monitoring these traits is by estimating their genetic or phenotypic trends. Available evidence indicates some small negative values for genetic relationships between fertility traits and harvest size weight (Caballero-Zamora et al., 2015a), but these were not significant and were in contradiction with the relationship between female fertility traits and female weight at insemination, which were positive and significant (0.49–0.54). Harvest size weight and female weight at insemination had a positive genetic correlation (0.30). Similar positive genetic correlations between mature female body weight and egg number were found by Tan et al. (2017). Bearing in mind these correlation estimates and the relatively low heritability estimates for the female reproductive traits, there is little reason for concern about rapid deterioration in fertility traits as a result of selection for harvest size weight.

Some studies have examined selection to increase uniformity in shrimp. Uniformity is a desirable trait in any meat-producing species, but results from research show that selection to increase the mean of this growth trait in several species is usually related to an increase in both the environmental variance and the coefficient of variation (Garcia et al., 2021). An initial study in shrimp analyzing the log of squared residuals of weight as a measure of environmental variation from a mixed linear model (Castillo-Juárez et al., 2012) found that residual variance for body weight in *L. vannamei* had a heritability value above zero (0.04) and a genetic correlation of 0.18 with body weight. Garcia et al. (2021) estimated multi-trait genetic parameters for residual variance for body weight and survival using a sire–dam double hierarchical generalized linear model in two environments. The results of this study indicate that environmental variance is heritable and that there is a negative genetic association between the residual variance for body weight and survival in the more restricted environment, suggesting that selection for increased uniformity could help increase survival in extensive grow-out conditions.

USE OF DNA MARKERS FOR IMPROVEMENT OF DISEASE RESISTANCE IN SHRIMP

Marker-assisted selection

In addition to selection methods based on quantitative genetics, new genomic methodologies are being studied for the genetic improvement of disease resistance in shrimp populations (Houston et al., 2020). The first approach is MAS (Guppy et al., 2018; Houston, 2017), which requires a precise location of loci with a high phenotypic effect on specific traits. The second is the very promising genomic evaluation methodology (Zenger et al., 2019). Both technologies are applicable in principle to any trait, but as it is more difficult to improve disease resistance with traditional methods, it is expected that these technologies may be used in the future to increase the rates of genetic progress for these traits in shrimp breeding (Ødegård et al., 2011).

MAS is a promise that is unlikely to meet the expectations that were once optimistically imagined. The reasons are, first of all, the lack of any large-effect QTL detected for disease resistance traits in shrimp, which appears to reflect the basic biological reality that quantitative traits are mostly polygenic in nature (Hill, 2010). This situation is illustrated by the lack of available genetic markers from individual loci to identify WSSV-sensitive and resistant animals, despite efforts in this direction (Cock et al., 2017). In other aquaculture species, the number of major genes detected that explain a relatively large proportion of the genetic variance of the traits is small, indicating a need to use DNA markers in a different way (Guppy et al., 2018; Houston et al., 2020; Ødegård et al., 2011; Zenger et al., 2019). According to Misztal et al. (2020), there is a dilemma as to whether causative large-effect SNPs, if found, should be used in selection programs for strongly selected traits. With long-term selection, alleles with a positive effect for most traits are likely to be fixed or close to fixation, and genes that have a large effect but are not fixed are likely to show undesirable effects on other important traits. In the future, technological advances in genomics, decreasing prices of genomic information, and more sophisticated statistical methods may allow some major genes to be incorporated into shrimp selection programs (Ali et al., 2020; Houston et al., 2020).

Genomic selection

GS based on the tracking of possibly causative genes indirectly with thousands of linked DNA genetic markers in the genome rather than searching for individual large-effect genes is emerging as the most useful option for using the new genomic knowledge to accelerate breeding advances and better manage genetic diversity in aquaculture (Houston et al., 2020; Zenger et al., 2019). In doing so, SNP typing information obtained through SNP panels with a lower number of markers typed through targeted genotyping-by-sequencing techniques, is expected to facilitate GS by lowering costs. Imputation techniques may allow the accurate assignment of genotypes not included in the SNP array actually measured, thus decreasing genotyping costs (Houston et al., 2020; Zenger et al., 2019). In addition, the same type of marker information used to perform GS can be used to better manage genetic diversity, as discussed above, by performing parent assignment in the absence of control of mating and tagging. GS may allow for the genetic evaluation of a set of selection candidates without the need for measuring phenotypes or infesting potential breeders, which makes improvement of the genetic resistance of the population with animals totally free of infection possible. This may also be obtained by dividing families into tested and non-tested animals; however, GS allows within-family variance to be included in selection, thereby improving accuracy (Castillo-Juárez et al., 2015; Nielsen et al, 2009; Sonneson and Meuwissen, 2009; Zenger et al., 2019). Studies involving actual estimates in terms of the accuracy of genomic evaluations versus traditional pedigree-based evaluations for disease resistance in shrimp are scarce, and those that have been published were obtained using small samples (Houston et al., 2020; Lyu et al., 2020; Nguyen et al., 2020; Wang et al., 2017a, b, 2019). Nonetheless, based on these results, the future of this type of selection is encouraging, because the small advantages in selection accuracy of AHPND resistance estimated for genomic evaluations over traditional pedigree-based evaluations are probably the product of the small training population size and low marker density, and therefore larger populations and the use of a larger number of markers should provide greater advantages (Castillo-Juárez et al., 2015; Sonesson and Meuwissen, 2009; Wang et al., 2017a, b; Zenger et al., 2019). The prospects for GS selection for growth in *L. vannamei* and other shrimp species are therefore good without any indication of the presence of any major gene (Khatkar et al., 2017).

In conclusion, as has been demonstrated in other animal species, GS is probably the best way yet to take advantage of the advances in the knowledge of shrimp genomics in shrimp breeding. The implementation of GS programs in shrimp will depend on the cost of genotyping technology and development of optimized programs.

Within-family estimation of heritability using markers

A novel option to estimate the heritability for any quantitative trait in an accurate and unbiased manner using a sample of a small number of large families (or even a single family) and estimating the IBD relationships between individuals within the family using DNA markers has been described by Ødegård and Meuwissen (2012). This method may be useful to estimate heritability in shrimp, where large families are easily produced.

GENOME-EDITING TOOLS

The use of efficient genome-editing tools, such as CRISPR-Cas9, to increase disease resistance could have conceptually important applications in shrimp breeding (Houston et al., 2020). The main requirement for the use of this technology is to locate the regions that need to be modified in the genome (Zenger et al., 2019); therefore, the primary limitation is related not to the technique per se, but, like MAS, to the polygenic nature of most quantitative traits. Unique gene modification via CRISPR-Cas9 may be used to enhance disease resistance and for other important applications such as changing adaptation traits of animals to better use resources. This can be obtained by enabling the fixation of favorable alleles from other populations/species or creating *de novo* alleles on the basis of the biology of the traits (Houston et al., 2020). Although specific technical difficulties still need to be overcome in shrimp and the results thus far are only experimental, this is an area of intense research (Gratacap et al., 2019).

INBREEDING DEPRESSION EFFECTS

Inbreeding depression effects for growth and survival traits have been examined in *L. vannamei* and other shrimp species. Effects on weight at harvest were found to be approximately similar to those found in the experimental studies of Moss et al. (2007) in a *L. vannamei* selected population from Hawaii, and from another experimental study by De los Ríos-Pérez et al. (2015) in a "conservation" line in Mexico, where inbreeding level (F) depression effects were estimated as the percentage change in the non-inbred trait mean per 10% increase in F (ID10). The results of these two studies did not show significant effects for survival. A retrospective study on another selected commercial growth line from the same hatchery (Montaldo et al., 2013a) gave approximately comparable results for ID10 for weight, and unreliable (not different from 0) results for survival, again coinciding with the results of Moss et al. (2007) (Table 6.3). However, no differentiation of ID10 between different generations was made in these studies, and it is not clear whether the lack of significance for survival is related to selection effects or statistical issues resulting from insufficient sample sizes.

Luo et al. (2014) estimated ID10 for inbreeding levels of 0.25, 0.375, and 0.50 (F1, F2, and F3 full-sib mating progeny, respectively) for growth traits, survival, and resistance to WSSV challenges in Chinese shrimp measured at 80, 100, and 140 days of age. Values of ID10 ranged from −3.06 to −4.55 for growth traits and −0.39 to −2.52 (not significant) for survival in the absence of disease challenge for each growing period. Similar values were observed across generations; therefore, linear effects are to be expected, leading to results similar to those obtained previously. An analysis of results data with linear regression models indicates that the predicted means for totally inbred animals as a percentage of the means of non-inbred animals (100) will be 53.1 for weight at 140 days and 51.9 for cumulative survival of 0–140 days, which implies considerable loss. No purging (selection) effects were found for the studied traits at those levels of inbreeding. In an experiment using three generations of inbred individuals for female reproduction traits (fertilization rate, hatching rate, and number of nauplii) within a *L. vannamei* population, De los Ríos-Pérez et al. (2017) found ID10 using the F of the female of approximately −25% in the first generation, approximately −13% to −16% in the second generation, and approximately −1% to −3% in the third generation. These increases in the ID10 values across generations were interpreted as being related to purging effects. Unlike growth, some reproduction traits are likely subject to large natural selection pressure. Another trait studied, the total number of eggs, did not display any inbreeding depression effects. These results demonstrate that inbreeding depression in first generations can be very high for some traits and that the purging effect could recover depression in a few generations if those traits are related to fitness.

TABLE 6.3 Inbreeding effects for harvest size weight and general survival rate in *L. vannamei*.

Trait		*Change in the mean (%) per 10% increase in inbreeding*		
		Experimental results		*Retrospective analysis (C)*
	Mean	*A*	*B*	
Harvest size body weight	19.6 g	−2.2±0.4**	−3.6	−3.4±0.3*
Survival	81.7%	−0.01±0.01	−0.3	−1.9±5.9

Source: A: De los Ríos-Pérez et al. (2015); B: Moss et al. (2007); C: Montaldo et al. (2013).

Doyle (2016) suggested that an incorrect use of commercial genetically improved postlarvae obtained from breeding companies as breeders by shrimp growers may have caused a large number of inbred progeny in the grow-out ponds, thereby increasing disease risk. In that theory, most of the animals would be highly inbred (F close to 25%) and progeny of full-sib parents, because companies would produce all progeny to sell from one pair of mated shrimp, and shrimp growers would not only always use sires and dams from a single source, but also inbreeding would cause a notable increase in disease susceptibility. This chain of events is difficult to demonstrate or disprove, with the exception of the last step. Therefore, it is interesting to examine the evidence regarding the possible increased sensitivity to disease in inbred animals, in particular for WSD and AHPND.

As noted earlier in this chapter, the first challenge testing for diseases in shrimp was performed successfully for TSV, thereby generating some information on inbreeding effects. Moss et al. (2007) reported some results from inbreeding depression for TSV resistance. Values for ID10 were in a wide range (−8.3 to −38.7), but regression slopes with the higher values are clearly imprecise and not significantly different from zero due to their large standard error values. Under the currently accepted linear inbreeding effects based on a dominance model (Dickerson, 1973; Hill, 2016; Pirchner, 1985), the minimum possible value for ID10 is −10, so an animal with an inbreeding coefficient of 100% will have a mean of 0% compared with that in non-inbred animals.

Several estimates of inbreeding effects for WSSV and AHPND tests have been obtained from experimental challenges. Luo et al. (2014) estimated small and non-significant ID10 values ranging from −0.27 to −0.44. With these values, a 100% inbred population would have a mean survival time for WSSV only 4.4% lower than that of the non-inbred population, and any purge effect would decrease the depression effect. Results on inbreeding from data obtained from full-sib mating in one generation for a WSD-resistant line composed of Ecuadorean shrimp yielded estimates for ID10 of between −1.3 and −9.4 for survival time for WSSV (Gallaga-Maldonado, 2020). In the same study, ID10 results for survival times in AHPND challenges were smaller but also negative, between −0.5 and −5.1. ID10 estimates for biomass production in disease-infested ponds were all negative but also with a wide range, from −1.0 to −14.2. Regarding the possible contribution of inbreeding to catastrophic mortalities, levels of 25% of inbreeding alone are unlikely to explain the high mortalities observed with WDS and AHPND, but inbreeding can be a contributing factor.

The lack of evidence for inbreeding depression following the use of populations of small effective size created from a few survivor animals recovered from the devastating generalized outbreaks of TSV in Colombia and tropical Latin American countries in the 1990s may be the product of natural and artificial selection counteracting inbreeding depression (Cock et al., 2009, 2017). It is likely that purges of inbreeding effects for survival are possible not only across generations, but also within generations in early stages of life, particularly in the case of large shrimp families (Moss et al., 2007).

As mentioned earlier, the other undesirable effect of small effective population size is a loss of genetic variability. Variability, however, can be at least partially restored by the incorporation of mutations in a natural process across many generations, or rapidly by crossbreeding. In the case of shrimp, both domesticated and wild animals could be strategically incorporated to increase genetic variability and possible disease resistance alleles (Cock et al., 2009, 2017), and new genomic technology could help expedite the introgression of some genes from one population to another (Houston et al., 2020; Yáñez et al., 2020). It is important to distinguish between inbreeding and loss of genetic variability, because in many real-life shrimp populations they will be different (Caballero, 2020); therefore, decisions about the management of each should be taken based on the specific population history regarding their origin and management. It is also very important to remember that both inbreeding and genetic drift and mutation are random processes acting on populations that are originally diverse. Therefore, making precise predictions of inbreeding and drift effects on a particular population is impossible.

MEASUREMENT OF REALIZED SELECTION RESPONSE

One important issue in breeding programs is the measurement of genetic progress. There are basically two strategies for making such estimations in shrimp. The first is to use control populations. A control population can be managed without selection (breeders can be selected at random) and maintained under the same environmental conditions as the selected populations. One way to deal with this type of comparison is to tag the animals from the control populations as an additional family to obtain the measurements under similar environmental conditions (Andriantahina et al., 2013; Gjedrem and Baranski, 2010; Gjedrem and Thodesen, 2005). One option to avoid the need to maintain a small control population that likely will be affected by genetic drift and inbreeding is to create a number of families each year derived from mating with randomly chosen animals or animals with average EBVs (Gjedrem and Thodesen, 2005). The differences in the progeny averages will indicate the result of the selection performed on the previous generation. As mentioned earlier in this chapter, another option to estimate genetic progress is using a mixed linear model, comparing the average EBV or GEBV of animals born in different years (Gjedrem and Thodesen, 2005; Mrode, 2014). Studies comparing the use of a control population and mixed linear models to estimate selection response are scarce in shrimp (Sui et al., 2016). It is likely that methods based on the use of control populations are more precise than those based on comparing the average EBV of different generations, especially if no repeated mating of the same animals is used across generations as occurs in shrimp breeding programs (Montaldo, 2020); therefore, further research is needed on this topic.

OPTIMIZATION OF BREEDING PROGRAMS FOR SHRIMP

There are a number of publications containing general principles for considering economic aspects in the design of breeding programs and program design optimization that can be used as a basis to do this type of applied work (Rutten et al., 2002; Weller, 1994). A good deal of information on the optimization of breeding programs in fish and other species is applicable to shrimp (Gjedrem and Rye, 2018; Gjedrem and Thodesen, 2005; Janssen et al., 2018; Nguyen and Ponzoni, 2006; Ponzoni et al., 2007, 2008), but their biological differences must be considered through research specifically directed to shrimp breeding (Montaldo, 2020). In most shrimp programs based on sib evaluation, total operation costs depend primarily on the number of animals tagged, seeded, and measured; hence it is important to make calculations providing a general guide that can help companies make rational decisions regarding costs and profit. The use of modeling tools such as estimating selection responses with index methodology can help in the design of programs that make better use of the resources invested.

Based on the analysis conducted for optimizing programs for MARPAC, we made a calculation to show how factors affecting a breeding program can be considered. The objective is to choose the number

TABLE 6.4 Optimal parameters for different size sib programs with restricted inbreeding.

Number of families	*Progeny per female*	*Total tagged/ scored*[a]	*Optimal selected proportion*	*Genetic gain/ gen.*[b]	*ΔF/gen. (%)*
100	40	4000	0.5300	0.168	0.995
200	40	8000	0.3700	0.224	0.992
400	40	16000	0.2600	0.270	0.977
800	40	32000	0.1825	0.311	0.997
100	10	1000	0.5000	0.152	0.997
200	10	2000	0.3600	0.194	0.980
200	10	2000	0.5100	0.151	0.491
400	10	4000	0.2525	0.233	0.995
800	10	8000	0.1825	0.268	0.997

Notes: [a] Number of animals seeded and tagged would be proportional to 1 / survival rate. [b] In phenotypic standard deviations. ΔF, inbreeding increase; gen., generation.

of families tested and the selected proportion of parents in each breeding cycle by balancing selection response and expected increase of inbreeding (Table 6.4). In this case we used a range of families common in actual aquaculture sib breeding programs along the lines of that described by Ponzoni et al. (2007).

Calculations were for a trait with a heritability of 0.20, which approximates a case for increasing a selection objective that includes growth and survival rate in shrimp programs (Garcia et al., 2021). Results were obtained using SelAction software (Rutten et al., 2002). For simplicity, only full-sib families were considered with female:male ratios = 1, and c-effects were assumed to be 0. Selection was based on BLUP estimates of breeding values based on parent and full-sib information, therefore the animals are only family tagged with elastomers. Results are shown restricting inbreeding increase (ΔF) to a maximum of 1% per generation, thus having N_e close to 50. In one case we examined the effect of restricting ΔF to a maximum of 0.5% per generation. Genetic gains are expressed in phenotypic standard deviations of the selection objective.

These findings suggest that even programs based on a relatively small number of families can be efficient; programs with the same maximum expected inbreeding levels with a larger number of families would be expected to have higher genetic responses because admissible selection proportions are lower with a larger number of candidate families. Increases in the expected response for 200, 400, and 800 families over 100 families were 33%, 61%, and 85%, with 40 progeny per female, and 28%, 53%, and 76% with 10 progeny measured per female, respectively (Table 6.4). Selection responses using 40 versus 10 progeny per female increased 11%, 15%, 16%, and 16% for programs using 100, 200, 400, and 800 families, respectively. Decreasing the maximum ΔF to 0.5% in a population of 200 families would decrease genetic progress by 22% (Table 6.4).

In theory, a sib-based selection program that uses just 100 families and approximately 1250 tagged and seeded animals (assuming a survival rate of 0.80) could have a genetic gain per generation of approximately 0.152 phenotypic standard deviations (Table 6.4). For harvest size weight (130 days) in northwestern Mexico, the phenotypic standard deviation for this trait was close to 3.0 g, yielding a genetic gain of 0.46 g per generation. The effects of the progeny number per female were modest, but adequate estimation of breeding values for survival demands larger numbers, in particular with low survival rates. The responses in Table 6.4 are the result of between-family selection. The incorporation of within-family selection when possible can increase these figures if large families can be produced and measured (Montaldo and Castillo-Juárez, 2017). In addition, the incorporation of individual tagging can also increase the response by making within-family variation available to selection. Another way to incorporate within-family selection is using GS, but the size of training populations, as well as proper genotyping technology and marker numbers, are key issues in that respect (Castillo-Juárez et al., 2015; Houston et al., 2020; Zenger et al., 2019).

More elaborate and specific calculations involving real parameters, multi-trait selection, and relative economic values, as well as the incorporation of GS, are needed to optimize breeding programs depending on the perspectives and needs of companies and priorities for economic considerations. Risk considerations and accounting for the possible effects of G×E are also key issues in shrimp breeding programs.

Conclusions

Breeding offers many opportunities to increase the profitability of shrimp culture. In addition to increasing traits usually considered in selection, such as growth and size, there is a great deal of evidence on important results for increasing disease resistance. Such evidence points to multi-loci (polygenic) genetic control of most traits of economic importance in shrimp production. Most important concerns regarding the consequences of selection are related to decreased genetic variability and concomitant increases in inbreeding rates as the result of the increased cumulative relationships between candidate breeders that occur in breeding populations. Therefore, there is a need to seek an appropriate balance between increasing selection responses and preserving the genetic diversity of selected populations.

The optimization of breeding programs should be conducted on the basis of well-established principles of quantitative genetics and economics. New genomic technologies can be used in shrimp genetic improvement and will probably be increasingly incorporated to obtain more profitable lines for specific environmental conditions. GS based on the tracking of possibly thousands of causative genes linked to DNA genetic markers in the genome rather than searching for individual large-effect loci is emerging as the most useful option for using the new genomic knowledge in accelerating genetic improvement and better managing genetic diversity. For disease resistance and other traits as well, the presence of important G×E effects indicates a need for caution in choosing appropriate selection criteria for well-defined environments. Marker and phenotype information need to be used jointly in the development of more effective genetic improvement programs in shrimp using genetic variability both within and between populations. New technologies for genetic modification, such as CRISPR-Cas9 systems, may be used to develop more efficient breeding programs in shrimp, using all populations available for breeding, but will require a thorough knowledge of how to locate and modify specific genes associated with important traits.

Acknowledgments

We are grateful to Cesáreo Cabrera-Villela, Director of Maricultura del Pacífico, for his generous sharing of knowledge on shrimp breeding, and to CONACyT Mexico and UNAM for funding support. We would also like to thank Kristine Ibsen for her assistance in revising the English in this chapter.

References

Ali, A., Al-Tobasei, R., Lourenco, D., Leeds, T., Kenney, B., and Salem, M. (2020) Genome-wide identification of loci associated with growth in rainbow trout. *BMC Genomics* 21, 209.

Andriantahina, F., Liu, X., Huang, H., and Xiang, J. (2013) Selection for growth performance of tank-reared Pacific white shrimp, *Litopenaeus vannamei*. *Chinese Journal of Oceanology and Limnology* 31, 534–541.

Bentsen, H.B. and Olesen, I. (2002) Designing aquaculture mass selection programs to avoid high inbreeding rates. *Aquaculture* 204, 349–359.

Benzie, J.A. (2009) Use and exchange of genetic resources of penaeid shrimps for food and aquaculture. *Reviews in Aquaculture* 1, 232–250.

Bernal Rodríguez, C.E., García, A.C., Ponce-Palafox, J.T., Spanopoulos-Hernández, M., Puga-López, D., Arredondo-Figueroa, J.L., and Martínez-Cárdenas, L.M. (2017) The color of marine shrimps and its role in the aquaculture. *International Journal of Aquaculture and Fishery Sciences* 3, 062-065.

Caballero, A. (2020) *Quantitative Genetics*. Cambridge University Press, Cambridge, UK.

Caballero-Zamora, A., Cienfuegos-Rivas, E.G., Montaldo, H.H., Campos-Montes, G.R., Martínez-Ortega, A., and Castillo-Juárez, H. (2015a) Genetic parameters for spawning and growth traits in the Pacific white shrimp (*Penaeus (Litopenaeus) vannamei*). *Aquaculture Research* 46, 833–839.

Caballero-Zamora, A., Montaldo, H.H., Campos-Montes, G.R., Cienfuegos-Rivas, E.G., Martínez-Ortega, A., and Castillo-Juárez, H. (2015b) Genetic parameters for body weight and survival in the Pacific White Shrimp *Penaeus (Litopenaeus) vannamei* affected by a White Spot Syndrome Virus (WSSV) natural outbreak. *Aquaculture* 447, 102–107.

Cala, N., Montaldo, H.H., Campos-Montes, G.R., and Castillo-Juárez, H. (2014) Evolution of genetic variability for growth-out survival rate in a selected population of Pacific white shrimp *Penaeus (Litopenaeus) vannamei*. In: *Proceedings of the 10th World Congress of Genetics Applied to Livestock Production*. Vancouver, Canada, 274–275.

Campos-Montes, G.R., Montaldo, H.H., Martínez-Ortega, A., and Castillo-Juárez, H. (2009) Genotype by environment interaction effects for body weight at 130 days of age in the Pacific white shrimp [*Penaeus (Litopenaeus) vannamei*]. *Veterinaria México* 40, 255–268.

Campos-Montes, G.R., Montaldo, H.H., Martínez-Ortega, A., Jiménez, A.M., and Castillo-Juárez, H. (2013) Genetic parameters for growth and survival traits in Pacific white shrimp *Penaeus (Litopenaeus) vannamei* from a nucleus population undergoing a two-stage selection program. *Aquaculture International* 21, 299–310.

Campos-Montes, G.R., Montaldo, H.H., Armenta-Córdova, M., Martínez-Ortega, A., Caballero-Zamora, A., and Castillo-Juárez, H. (2017) Incorporation of tail weight and tail percentage at harvest size in selection programs for the Pacific white shrimp *Penaeus (Litopenaeus) vannamei*. *Aquaculture* 468, 293–296.

Campos-Montes, G.R., Caballero-Zamora, A., Montaldo, H.H., Montoya-Rodríguez, L., Rodríguez-Sala, B.G.G., Rodríguez, S.A.S., Martínez-Ortega, A., Quintana-Casares, J.C., and Castillo-Juárez, H. (2020) Genetic (co) variation in resistance of Pacific white shrimp *Litopenaeus vannamei* to acute hepatopancreatic necrosis disease (AHPND) and white spot syndrome virus (WSSV) in challenge tests. *Aquaculture* 520, 734994.

Castillo-Juárez, H., Campos-Montes, G.R., Caballero-Zamora, A., and Montaldo, H.H. (2015) Genetic improvement of Pacific white shrimp [*Penaeus (Litopenaeus) vannamei*]: perspectives for genomic selection. *Frontiers in Genetics* 6, 93.

Castillo-Juárez, H., Quintana-Casares, J.C., Campos-Montes, G., Cabrera-Villela, C., Martínez-Ortega, A., and Montaldo, H.H. (2007) Heritability for body weight at harvest size in the Pacific white shrimp, *Penaeus (Litopenaeus) vannamei*, from a multi-environment experiment using univariate and multivariate animal models. *Aquaculture* 273, 42–49.

Castillo-Juárez, H., Montaldo, H.H., and Campos-Montes, G.R. (2012) Genetic parameter estimates of the environmental variation for body weight at harvest size in a Pacific white shrimp breeding population. In: *4th International Conference on Quantitative Genetics: Understanding Variation in Complex Traits*. Edinburgh, UK, p. 167.

Castillo-Juárez, H., Montaldo, H.H., Campos-Montes, G.R., Quintana-Casares, J.C., Soto-Rodríguez, S.A., Betancourt-Lozano, M., Martínez-Ortega, A., Lozano-Olvera, R., Gómez-Gil, B., Caballero-Zamora, A., and

Gallaga-Maldonado, E.P. (2018) Heritability, genetic line and inbreeding effects on resistance of whiteleg shrimp *Penaeus vannamei* Boone 1931 to Acute Hepatopancreatic Necrosis Disease (AHPND) in Mexico. *Asian Fisheries Science* 31, 88–101.

Cock, J., Gitterle, T., Salazar, M., Hayes, B., and Rye, M. (2010) Selective breeding of shrimp. In: Alday-Sanz, V. (ed.) *The Shrimp Book*. Nottingham University Press, Nottingham, UK, pp. 377–427.

Cock, J., Gitterle, T., Salazar, M., and Rye, M. (2009) Breeding for disease resistance of Penaeid shrimps. *Aquaculture* 286, 1–11.

Cock, J., Salazar, M., and Rye, M. (2017) Strategies for managing diseases in non-native shrimp populations. *Reviews in Aquaculture* 9, 211–226.

Cuéllar-Anjel, J., White-Noble, B., Schofield, P., Chamorro, R., and Lightner, D.V. (2012) Report of significant WSSV-resistance in the Pacific white shrimp, *Litopenaeus vannamei*, from a Panamanian breeding program. *Aquaculture* 368, 36–39.

D'Agaro, E, Woolliams, J.A., Haley, C.S., and Lanari, D. (2007) Optimizing mating schemes in fish breeding. *Italian Journal of Animal Science* 6, 795–796.

Dai, P., Kong, J., Liu, J., Lu, X., Sui, J., Meng, X., and Luan, S. (2020) Evaluation of the utility of genomic information to improve genetic evaluation of feed efficiency traits of the Pacific white shrimp *Litopenaeus vannamei*. *Aquaculture* 527, 735421.

Dai, P., Luan, S., Lu, X., Luo, K., Meng, X., Cao, B., and Kong, J. (2017) Genetic assessment of residual feed intake as a feed efficiency trait in the Pacific white shrimp *Litopenaeus vannamei*. *Genetics Selection Evolution* 49, 61.

De los Ríos-Pérez, L., Campos-Montes, G.R., Martínez-Ortega, A., Castillo-Juárez, H., and Montaldo, H.H. (2015) Inbreeding effects on body weight at harvest size and grow-out survival rate in a genetic selected population of Pacific white shrimp *Penaeus (Litopenaeus) vannamei*. *Journal of the World Aquaculture Society* 46, 53–60.

De los Ríos-Pérez, L., Campos-Montes, G.R., Martínez-Ortega, A., Castillo-Juárez, H., and Montaldo, H.H. (2017) Inbreeding effects on reproductive traits in a breeding population of Pacific white shrimp *Penaeus (Litopenaeus) vannamei*. *Aquaculture* 479, 442–446.

De Rochambeau, H., Fournet-Hanocq, F., and Khang, J.V.T. (2000) Measuring and managing genetic variability in small populations. *Annales de Zootechnie* 49, 77–93.

Dickerson, G.E. (1973) Inbreeding and heterosis in animals. *Journal of Animal Science* 1973 (Issue Symposium), 54–77.

Doyle, R.W. (2016) Inbreeding and disease in tropical shrimp aquaculture: a reappraisal and caution. *Aquaculture Research* 47, 21–35.

Dupont-Nivet, M., Vandeputte, M., Haffray, P., and Chevassus, B. (2006) Effect of different mating designs on inbreeding, genetic variance and response to selection when applying individual selection in fish breeding programs. *Aquaculture* 252, 161–170.

Falconer, D.S. and Mackay, T.F.C. (1996) *Introduction to Quantitative Genetics*, 4th ed. Longmans Green, Harlow, UK.

Fernández, J., Toro, M.Á., Sonesson, A.K., and Villanueva, B. (2014) Optimizing the creation of base populations for aquaculture breeding programs using phenotypic and genomic data and its consequences on genetic progress. *Frontiers in Genetics* 5, 414.

Fjalestad, K.T., Gjedrem, T., and Gjerde, B. (1993) Genetic improvement of disease resistance in fish: an overview. *Aquaculture* 111, 65–74.

Gallaga-Maldonado, E.P. (2020) Efectos de cruzamiento y consanguinidad sobre la resistencia a la necrosis aguda del hepatopáncreas, la enfermedad de la mancha blanca y el crecimiento en camarón blanco del pacífico *Litopenaeus vannamei* [Crossbreeding and inbreeding effects on disease resistance to acute hepatopancreas necrosis, white spot and growth in the Pacific white shrimp *Litopenaeus vannamei*]. PhD thesis. Universidad Nacional Autónoma de México, Mexico.

Gallaga-Maldonado, E.P., Montaldo, H.H., Castillo-Juárez, H., Campos-Montes, G.R., Martínez-Ortega, A., Quintana-Casares, J.C., Montoya-Rodríguez, L., Betancourt-Lozano, M., Lozano-Olvera, R., and Vázquez-Peláez, C. (2020) Crossbreeding effects for White Spot Disease resistance in challenge tests and field pond performance in Pacific white shrimp *Litopenaeus vannamei* involving susceptible and resistance lines. *Aquaculture* 516, 734527.

Garcia, B.F., Montaldo, H.H., Iung, L.H.S., and Carvalheiro, R. (2021) Effect of harvest weight and its uniformity on survival in *Litopenaeus vannamei* reared in different systems. *Aquaculture* 531, 735891.

Giang, C.T., Knibb, W., Muu, T.T., Ninh, N.H., Nguyen, H.N., and Nguyen, H.N. (2019) Prospects for genetic improvement in objective measurements of body color in Pacific Whiteleg Shrimp (*Litopenaeus vannamei*). *Journal of Marine Science and Engineering* 7, 460.

Gilmour, A.R., Anderson, R.D., and Rae, A.L. (1985) The analysis of binomial data by a generalized linear mixed model. *Biometrika* 72, 593–599.

Gjedrem, T. and Baranski, M. (2010) *Selective Breeding in Aquaculture: An Introduction*. Springer Science & Business Media, Dordrecht, The Netherlands.

Gjedrem, T. and Rye, M. (2018) Selection response in fish and shellfish: a review. *Reviews in Aquaculture* 10, 168–179.

Gjedrem, T. and Thodesen, J. (2005) *Selection and Breeding Programs in Aquaculture*. Akvaforsk/Springer, Dordrecht, The Netherlands.

Gratacap, R.L., Wargelius, A., Edvardsen, R.B., and Houston, R.D. (2019) Potential of genome editing to improve aquaculture breeding and production. *Trends in Genetics* 35, 672–684.

Guppy, J.L., Jones, D.B., Jerry, D.R., Wade, N.M., Raadsma, H.W., Huerlimann, R., and Zenger, K.R. (2018) The state of "omics" research for farmed penaeids: advances in research and impediments to industry utilization. *Frontiers in Genetics* 9, 282.

Henderson, C.R. (1984) *Applications of Linear Models in Animal Breeding*. University of Guelph, Guelph, Canada.

Henryon, M., Liu, H., Berg, P., Su, G., Nielsen, H.M., Gebregiwergis, G.T., and Sørensen, A.C. (2019) Pedigree relationships to control inbreeding in optimum-contribution selection realise more genetic gain than genomic relationships. *Genetics Selection Evolution* 51, 39.

Hernández-Ruíz, H., Montaldo, H.H., Bustos-Martínez, J., Campos-Montes, G.R., and Castillo-Juárez, H. (2020) Heritability and genetic correlations for infectious hypodermal and hematopoietic necrosis virus load, body weight at harvest, and survival rate in Pacific white shrimp (*Litopenaeus vannamei*). *Journal of the World Aquaculture Society* 51, 312–323.

Hickey, J.M., Chiurugwi, T., Mackay, I., Powell, W., Eggen, A., Kilian, A., Jones, C., Canales, C., Grattapaglia, D., Bassi, F., and Atlin, G. (2017) Genomic prediction unifies animal and plant breeding programs to form platforms for biological discovery. *Nature Genetics* 49, 1297.

Hill, W.G. (2010) Understanding and using quantitative genetic variation. *Philosophical Transactions of the Royal Society B: Biological Sciences* 365, 73–85.

Hill, W.G. (2016) Is continued genetic improvement of livestock sustainable? *Genetics* 202, 877–881.

Hill, W.G., Goddard, M.E., and Visscher, P.M. (2008) Data and theory point to mainly additive genetic variance for complex traits. *PLoS Genetics* 4, e1000008.

Holtsmark, M., Sonesson, A.K., Gjerde, B., and Klemetsdal, G. (2006) Number of contributing subpopulations and mating design in the base population when establishing a selective breeding program for fish. *Aquaculture* 258, 241–249.

Houston, R.D. (2017) Future directions in breeding for disease resistance in aquaculture species. *Revista Brasileira de Zootecnia* 46, 545–551.

Houston, R.D., Bean, T.P., Macqueen, D.J., Gundappa, M.K., Jin, Y.H., Jenkins, T.L., Selly, S.L.C., Martin, S.A.M., Stevens, J.R., Santos, E.M. Davie, A., and Robledo, D. (2020) Harnessing genomics to fast-track genetic improvement in aquaculture. *Nature Reviews Genetics* 21, 389–409.

Howard, J.T., Pryce, J.E., Baes, C., and Maltecca, C. (2017) Invited review: Inbreeding in the genomics era: inbreeding, inbreeding depression, and management of genomic variability. *Journal of Dairy Science* 100, 6009–6024.

Huang, Y.C., Yin, Z.X., Ai, H. S., Huang, X.D., Li, S.D., Weng, S.P., and He, J.G. (2011) Characterization of WSSV resistance in selected families of *Litopenaeus vannamei*. *Aquaculture* 311, 54–60.

Isik, F., Holland, J., and Maltecca, C. (2017) *Genetic Data Analysis for Plant and Animal Breeding*. Springer International Publishing, New York, NY.

Janssen, K., Saatkamp, H., and Komen, H. (2018) Cost-benefit analysis of aquaculture breeding programs. *Genetics Selection Evolution* 50, 1–16.

Khatkar, M., Zenger, K., Jones, D., Prochaska, J., van der Steen, H., Jerry, D., and Raadsma, H.W. (2017) Quantitative genomic analyses in the Pacific whiteleg shrimp *Litopenaeus vannamei*. *Proceedings of the 22nd Conference of the Association for the Advancement of Animal Breeding and Genetics.* 2–5 July 2017, Townsville, Queensland, Australia, pp. 121–124.

Lee, S.H., Clark, S., and Van der Werf, J.H. (2017) Estimation of genomic prediction accuracy from reference populations with varying degrees of relationship. *PLoS One* 12, e0189775.

Lind, C.E., Ponzoni, R.W., Nguyen, N.H., and Khaw, H.L (2012) Selective breeding in fish and conservation of genetic resources for aquaculture. *Reproduction in Domestic Animals* 47, 255–263.

Lourenco, D., Legarra, A., Tsuruta, S., Masuda, Y., Aguilar, I., and Misztal, I. (2020) Single-step genomic evaluations from theory to practice: using SNP chips and sequence data in BLUPF90. *Genes* 11, 790.

Lourenco, D.A.L., Fragomeni, B.O., Tsuruta, S., Aguilar, I., Zumbach, B., Hawken, R.J., Legarra, A., and Misztal, I. (2015) Accuracy of estimated breeding values with genomic information on males, females, or both: an example on broiler chicken. *Genetics Selection Evolution* 47, 56.

Luan, S., Qiang, G., Cao, B., Luo, K., Meng, X., Chen, B., and Kong, J. (2020) Feed competition reduces heritable variation for body weight in *Litopenaeus vannamei*. *Genetics Selection Evolution* 52, 45.

Luo, K., Kong, J., Luan, S., Meng, X.H., Zhang, T.S., and Wang, Q.Y. (2014) Effect of inbreeding on survival, WSSV tolerance and growth at the postlarval stage of experimental full-sibling inbred populations of the Chinese shrimp *Fenneropenaeus chinensis*. *Aquaculture*, 32–37.

Lynch, M. and Lande, R. (1998) The critical effective size for a genetically secure population. *Animal Conservation* 1, 70–72.

Lynch, M. and Walsh, B. (1998) *Genetics and Analysis of Quantitative Traits* Vol. 1. Sinauer Associates, Sunderland, MA, USA.

Lyu, D., Yu, Y., Zhang, Q., Luo, Z., Wang, Q., Xiang, J., and Li, F. (2020) Estimating genetic parameters for resistance to *Vibrio parahaemolyticus* with molecular markers in Pacific white shrimp. *Aquaculture* 527, 735439.

Meuwissen, T. (2009) Genetic management of small populations: a review. *Acta Agriculturae Scandinavica, Section A–Animal Science* 59, 71–79.

Meuwissen, T., Hayes, B., and Goddard, M. (2013) Accelerating improvement of livestock with genomic selection. *Annual Review of Animal Biosciences* 1, 221–237.

Meuwissen, T., Hayes, B., and Goddard, M. (2016) Genomic selection: a paradigm shift in animal breeding. *Animal Frontiers* 6, 6–14.

Misztal, I., Lourenco, D., and Legarra, A. (2020) Current status of genomic evaluation. *Journal of Animal Science* 98, 1–14.

Montaldo, H.H. (2001) Genotype by environment interactions in livestock breeding programs: a review. *Interciencia* 26, 229–235.

Montaldo, H.H. (2020) Gamete manipulation and conservation for genetic improvement in penaeid shrimp. In: *Cryopreservation of Fish Gametes.* Springer, Singapore, pp. 345–352.

Montaldo, H.H. and Castillo-Juárez, H. (2017) Response to strict within-family selection with special reference to aquaculture. *Aquaculture Research* 48, 5175–5178.

Montaldo, H.H., Castillo-Juárez, H., Campos-Montes, G.R., and De Los Ríos Pérez, L. (2013a) Passive inbreeding effects on harvesting body weight and survival rate in the Pacific white shrimp (*Litopenaeus (Penaeus) vannamei*). Aquaculture 2013, 3–7 November, 2013, Gran Canaria, Spain.

Montaldo, H.H., Castillo-Juárez, H., Campos-Montes, G.R., Peralta-Aban, A., and Martínez-Ortega, A. (2011) Changes in inbreeding and effective population size in a breeding nucleus population of Pacific white shrimp Penaeus (Litopenaeus) vannamei. *World Aquaculture Society 2011*, 6–10 June, 2011, Natal, Brazil, 760.

Montaldo, H.H., Castillo-Juárez, H., Campos-Montes, G., and Pérez-Enciso, M. (2013b) Effect of the data family structure, tank replication and the statistical model, on the estimation of genetic parameters for body weight at 28 days of age in the Pacific white shrimp (*Penaeus (Litopenaeus) vannamei* Boone, 1931). *Aquaculture Research* 44, 1715–1723.

Moss, D.R., Arce, S.M., Otoshi, C.A., Doyle, R.W., and Moss, S.M. (2007) Effects of inbreeding on survival and growth of Pacific white shrimp *Penaeus (Litopenaeus) vannamei*. *Aquaculture* 272, S30–S37.

Moss, S.M., Moss, D.R., Arce, S.M., Lightner, D.V. and Lotz, J.M. (2012) The role of selective breeding and biosecurity in the prevention of disease in penaeid shrimp aquaculture. *Journal of Invertebrate Pathology* 110, 247–250.

Moss, S.M., Moss, D.R., Shumway, S.E., and Rodrick, G.E. (2009) Selective breeding of penaeid shrimp. In: Shumway, S.E. and Rodrick, G.E. (eds) *Shellfish Safety and Quality.* Woodhead Publishing, Cambridge, UK, pp. 425–452.

Mrode, R.A. (2014) *Linear Models for the Prediction of Animal Breeding Values*, 3rd ed. CABI, Wallingford, UK.

Nguyen, N.H. (2016) Genetic improvement for important farmed aquaculture species with a reference to carp, tilapia and prawns in Asia: achievements, lessons and challenges. *Fish* 17, 483–506.

Nguyen, N.H., Phuthaworn, C., and Knibb, W. (2020) Genomic prediction for disease resistance to *Hepatopancreatic parvovirus* and growth, carcass and quality traits in Banana shrimp *Fenneropenaeus merguiensis. Genomics* 112, 2021–2027.

Nguyen, N.H. and Ponzoni, R.W. (2006) Perspectives from agriculture: advances in livestock breeding-implications for aquaculture genetics. *NAGA, WorldFish Center Quarterly* 29, 39–45.

Nguyen, N.H., Quinn, J., Powell, D., Elizur, A., Thoa, N.P., Nocillado, J., Lamont, R., Remilton, C., and Knibb, W. (2014) Heritability for body color and its genetic association with morphometric traits in Banana shrimp (*Fenneropenaeus merguiensis*). *BMC Genetics* 15, 132.

Nicholas, F.W. (1980) Size of population required for artificial selection. *Genetics Research* 35, 85–105.

Nielsen, H.M., Sonesson, A.K., Yazdi, H., and Meuwissen, T.H. (2009) Comparison of accuracy of genome-wide and BLUP breeding value estimates in sib-based aquaculture breeding schemes. *Aquaculture* 289, 259–264.

Noble, T.H., Wade, N., Coman, G.J., and Jerry, D.R. (2017) Breeding for disease resistance in Australian shrimp: how do we get there? In: *Proceedings of the 22nd Conference of the Association for the Advancement of Animal Breeding and Genetics*. 2–5 July 2017, Townsville, Queensland, Australia, pp. 129–132.

Nolasco-Alzaga, H.R., Perez-Enriquez, R., Enez, F., Bestin, A., Palacios-Mechetnov, E., and Haffray, P. (2018) Quantitative genetic parameters of growth and fatty acid content in the hemolymph of the Whiteleg shrimp *Litopenaeus vannamei. Aquaculture* 482, 17–23.

Ødegård, J., Baranski, M., Gjerde, B., and Gjedrem, T. (2011) Methodology for genetic evaluation of disease resistance in aquaculture species: challenges and future prospects. *Aquaculture Research* 42, 103–114.

Ødegård, J. and Meuwissen, T.H. (2012) Estimation of heritability from limited family data using genome-wide identity-by-descent sharing. *Genetics Selection Evolution* 44, 16.

Phuthaworn, C., Nguyen, N.H., Quinn, J., and Knibb, W. (2016) Moderate heritability of hepatopancreatic parvovirus titre suggests a new option for selection against viral diseases in banana shrimp (*Fenneropenaeus merguiensis*) and other aquaculture species. *Genetics Selection Evolution* 48, 64.

Pirchner, F. (1985) Genetic structure of populations. 1. Closed populations or matings among related individuals. In: Chapman, A.B. (ed.) *General and Quantitative Genetics*. Elsevier Science, Amsterdam, The Netherlands, pp. 227–250.

Ponzoni, R.W., Nguyen, N.H., and Khaw, H.L. (2007) Investment appraisal of genetic improvement programs in Nile tilapia (*Oreochromis niloticus*). *Aquaculture* 269, 187–199.

Ponzoni, R.W., Nguyen, N.H., Khaw, H.L., and Ninh, N.H. (2008) Accounting for genotype by environment interaction in economic appraisal of genetic improvement programs in common carp *Cyprinus carpio*. *Aquaculture* 285, 47–55.

Rutten, M.J.M., Bijma, P., Woolliams, J.A., and Van Arendonk, J.A.M. (2002) SelAction: Software to predict selection response and rate of inbreeding in livestock breeding programs. *Journal of Heredity* 93, 456–458.

Salte, R., Bentsen, H.B., Moen, T., Tripathy, S., Bakke, T.A., Ødegård, J., Omholt, S., and Hansen, L.P. (2010) Prospects for a genetic management strategy to control *Gyrodactylus salaris* infection in wild Atlantic salmon (*Salmo salar*) stocks. *Canadian Journal of Fisheries and Aquatic Sciences* 67, 121–129.

Sae-Lim, P., Gjerde, B., Nielsen, H.M., Mulder, H., and Kause, A. (2016) A review of genotype-by-environment interaction and micro-environmental sensitivity in aquaculture species. *Reviews in Aquaculture* 8, 369–393.

Solberg, T.R., Sonesson, A.K., Woolliams, J.A., and Meuwissen, T.H. (2008) Genomic selection using different marker types and densities. *Journal of Animal Science* 86, 2447–2454.

Sonesson, A.K. and Meuwissen, T.H. (2009) Testing strategies for genomic selection in aquaculture breeding programs. *Genetics Selection Evolution* 41, 37.

Sui, J., Luan, S., Luo, K., Meng, X., Lu, X., Cao, B., Li, W., Chai, Z., Liu, N., Xu, S., and Kong, J. (2016) Genetic parameters and response to selection for harvest body weight of Pacific white shrimp, *Litopenaeus vannamei. Aquaculture Research* 47, 2795-2803.

Tan, J., Kong, J., Cao, B., Luo, K., Liu, N., Meng, X., Xu, S., Guo, Z., Chen, G., and Luan, S. (2017) Genetic parameter estimation of reproductive traits of *Litopenaeus vannamei*. *Journal of Ocean University of China* 16, 161–167.

Tan, J., Luan, S., Cao, B., Luo, K., Meng, X., and Kong, J. (2019) Evaluation of genetic parameters for reproductive traits and growth rate in the Pacific white shrimp *Litopenaeus vannamei* reared in brackish water. *Aquaculture* 511, 734244.

Trang, T.T., Hung, N.H., Ninh, N.H., Knibb, W., and Nguyen, N.H. (2019a) Genetic variation in disease resistance against White Spot Syndrome Virus (WSSV) in *Litopenaeus vannamei*. *Frontiers in Genetics* 10, 264.

Trang, T.T., Hung, N.H., Ninh, N.H., and Nguyen, N.H. (2019b) Selection for improved white spot syndrome virus resistance increased larval survival and growth rate of Pacific Whiteleg shrimp, *Litopenaeus vannamei*. *Journal of Invertebrate Pathology* 166, 107219.

Van der Werf, J.H.J. (2017) Single trait selection index. Excel spreadsheet. Available at: https://jvanderw.une.edu.au/software.htm (accessed July 2020).

VanRaden, P.M. and Wiggans, G.R. (1991) Deviation, calculation, and use of national animal model information. *Journal of Dairy Science* 74, 2737–2346.

VanRaden, P.M. and Wright, J.R. (2013) Measuring genomic pre-selection in theory and in practice. *Interbull Bulletin* 47, 147–150.

Van Vleck, L.D. (1993) *Selection Index and Introduction to Mixed Model Methods for Genetic Improvement of Animals: The Green Book*. CRC Press, Boca Raton, FL.

Vela-Avitúa, S., Montaldo, H.H., Márquez Valdelamar, L., Campos Montes, G.R., and Castillo Juárez, H. (2013) Decline of genetic variability in a captive population of Pacific white shrimp *Penaeus (Litopenaeus) vannamei* using microsatellite and pedigree information. *Electronic Journal of Biotechnology* 16, 9.

Walsh, B. and Lynch, M. (2018) *Evolution and Selection of Quantitative Traits*. Oxford University Press, Oxford, UK.

Wang, Q., Yu, Y., Li, F., Zhang, X., and Xiang, J. (2017a) Predictive ability of genomic selection models for breeding value estimation on growth traits of Pacific white shrimp *Litopenaeus vannamei*. *Chinese Journal of Oceanology and Limnology* 35, 1221–1229.

Wang, Q., Yu, Y., Yuan, J., Zhang, X., Huang, H., Li, F., and Xiang, J. (2017b) Effects of marker density and population structure on the genomic prediction accuracy for growth trait in Pacific white shrimp *Litopenaeus vannamei*. *BMC Genetics* 18, 45.

Wang, Q., Yu, Y., Zhang, Q., Zhang, X., Huang, H., Xiang, J., and Li, F. (2019) Evaluation on the genomic selection in *Litopenaeus vannamei* for the resistance against *Vibrio parahaemolyticus*. *Aquaculture* 505, 212–216.

Weller, J.I. (1994) *Economic Aspects of Animal Breeding*. Chapman & Hall, London, UK.

Weller, J.I. (2016) *Genomic Selection in Animals*. John Wiley & Sons Inc., Hoboken, NJ.

Weller, J.I. and Ron, M. (2011) Invited review: Quantitative trait nucleotide determination in the era of genomic selection. *Journal of Dairy Science* 94, 1082–1090.

Wright, S. (1922) Coefficients of inbreeding and relationship. *The American Naturalist* 56, 330–338.

Yáñez, J.M., Houston, R.D., and Newman, S. (2014) Genetics and genomics of disease resistance in salmonid species. *Frontiers in Genetics* 5, 415.

Yáñez, J.M., Joshi, R., and Yoshida, G.M. (2020) Genomics to accelerate genetic improvement in tilapia. *Animal Genetics* 51, 658–674.

Zenger, K.R., Khatkar, M.S., Jones, D.B., Khalilisamani, N., Jerry, D.R., and Raadsma, H.W. (2019) Genomic selection in aquaculture: application, limitations and opportunities with special reference to marine shrimp and pearl oysters. *Frontiers in Genetics* 9, 693.

Chapter 7

AUTOTROPHIC AND HETEROTROPHIC WATER TREATMENT IN SEMI-INTENSIVE, INTENSIVE, AND SUPER-INTENSIVE FISH AND SHRIMP CULTURE

D. E. BRUNE

Summary

Aquaculture productivity has progressively increased over the past 50 years. Unaerated, still-water aquaculture, referred to as extensive systems, yields 1121–2242 kg/ha (1000–2000 lb/ac) of fish or shrimp in freshwater or marine systems. Feed rates average 11–17 kg/ha-d (10–15 lb/ac-d), the limiting factor being wind-driven surface reoxygenation rates. Algal photosynthesis controls ammonia levels at algal productivity of 0.5–1.0 g carbon (C)/m^2-d with water column volatile suspended solids levels of 10–20 mg/l. In the 1980s, farmers added 1.8–3.7 kW/ha (1–2 hp/ac) supplemental aeration to marine shrimp and freshwater fish ponds, pushing yields to 4483–7846 kg/ha (4000–7000 lb/ac) at average feed rates of 56 kg/ha-d (50 lb/ac-d) and maximum feed rates of 112 kg/ha-d (100 lb/ac-d). Algal photosynthesis of 2–3 g C/m^2-d matches the ammonia generation rate in such systems. Exceeding this level of production requires enhanced algal production or pond flushing with discharge of ammonia to the environment. In marine systems in Asia, flushing of ammonia-laden water and increasing pond aeration to 19–37 kW/ha (10–20 hp/ac) allowed shrimp production in excess of 11,209 kg/ha per cycle (10,000 lb/ac per cycle). In the USA, discharge of water was increasingly prohibited, and consequently US aquaculturists turned to enhanced photosynthetic systems such as partitioned aquaculture systems, split-ponds, in-pond raceways, and intensively aerated ponds to increase carrying capacity. Fish yields of 11,209–21,296 kg/ha (10,000–19,000 lb/ac) were achievable with feed loadings of 112–280 kg/ha-d (100–250 lb/ac-d), with algal photosynthesis peaking at 6–12 g C/m^2-d. To further increase production to "super-intensive" levels, farmers turned to alternative biological treatment techniques, fixed-film nitrifying reactors, or suspended-culture microbial reactors. Biofloc treatment using nitrifying systems (NS) or heterotrophic systems (HS) is economically favored in shrimp culture. Optimal microbial levels in the range 200–400 mg/l keep pace with the required ammonia removal without imposing an excessive oxygen demand. Shrimp production in excess of 44,834 kg/ha per cycle (40,000 lb/ac per cycle) is achievable. The major advantage of HS culture is the rapid growth rate of heterotrophic microbes allowing nearly instantaneous control of culture ammonia levels. HS culture requires organic supplementation of 40–60% of feed and removal of as much as 10% of microbial mass per day. Solids production from NS is typically one-tenth that of HS, and no additional organic supplementation is required. Enterprise budgets suggest that shrimp production costs in NS and HS at 44,834 kg/ha per 120-day cycle (40,000 lb/ac 120-day cycle) are similar at US$8.00/kg shrimp. An added advantage of super-intensive aquaculture is the potential to provide zero discharge systems, eliminating the need for water and waste sludge discharge by converting microbial biomass production to useful by-products.

Introduction

Aquaculture productivity has progressively increased over the past 50 years (Table 7.1). The primary requirement to increase the carrying capacity of aquaculture production systems is to maintain water oxygen, carbon dioxide, and ammonia concentrations within acceptable levels. Natural or enhanced aeration is used to maintain oxygen and carbon dioxide levels, while photoautotrophic (algae), chemoautotrophic (nitrification), or heterotrophic microbial growth may be used to maintain control of ammonia nitrogen levels. In the 1960s, unaerated aquaculture ponds, referred to as extensive systems, were observed to yield 1121–2242 kg/ha (1000–2000 lb/ac) of fish or shrimp in freshwater or marine systems (Brune and Drapcho, 1991). At these yields, feed application rates average 11–17 kg/ha-d (10–15 lb/ac-d), the limiting factor being wind-driven re-aeration rates. In such ponds, algal photosynthesis suppresses ammonia levels at algal productivities of 0.5–1.0 g C/m^2-d at water column volatile suspended solids (VSS) levels of 10–20 mg/l.

Beginning in the 1980s, farmers added 1.8–3.7 kW/ha (1–2 hp/ac) supplemental aeration to marine shrimp and freshwater fish ponds, pushing yields to 4483–7846 kg/ha (4000–7000 lb/ac) at average feed rates of 56 kg/ha-d (50 lb/ac-d) with a maximum of 112 kg/ha-d (100 lb/ac-d) (Brune and Drapcho, 1991). At these feed-loading rates, algal photosynthesis of 2–3 g C/m^2-d matches the rate of ammonia generation. Exceeding this level of production requires enhanced algal productivity and/or pond flushing with discharge of ammonia and VSS to the environment (Fig. 7.1).

In early marine systems in the USA and Asia, flushing of ammonia-laden water and increased pond aeration to 19–37 kW/ha (10–20 hp/ac) allowed shrimp production in excess of 11,209 kg/ha per cycle

TABLE 7.1 Yield and productivity parameters of different aquaculture systems and their approximate date of inception. PAS: partitioned aquaculture system, SP: split-ponds, VSS: volatile suspended solids.

System	*Yield (kg/ha)*	*Feed (kg/ha-d)*	*Aeration (kW/ha)*	*Type (g C/m^2-d)*	*VSS (mg/l)*	*Timeline*
Extensive	1124–2247	11–34	Wind	Algal (0.5–1)	10–20	1960
Semi-intensive	4494–6742	56–112	1.8–3.7	Algal (2–3)	50–100	1980
Intensive pond	11,235–13,483	112–168	11–22	Mixed (3–4)	100+	1990
PAS/SP	16,854–21,348	225–280	13–18	Algal (6–12)	50–100	2000
Super-intensive nitrifying	44,943+	1123	92–111	Nitrification	300–400	2006
Super-intensive heterotrophic	44,943+	1123/674	111–148	Heterotrophic	300–400	2006

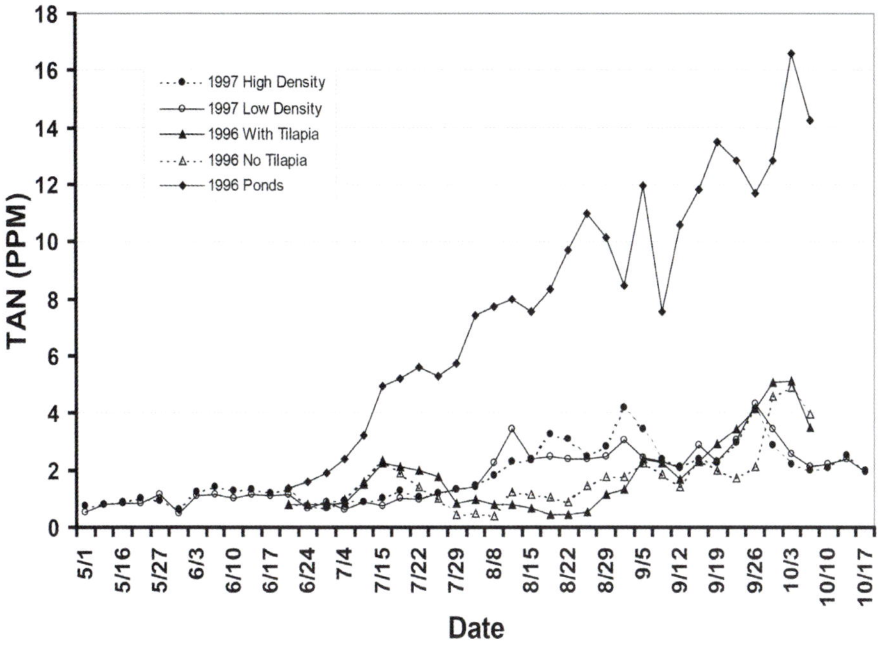

FIGURE 7.1 Total ammonia nitrogen (TAN) levels in conventional ponds versus partitioned aquaculture systems at feed levels of 56–224 kg/ha-d (50–200 lb/ac-d) (from Brune et al., 2012).

(10,000 lb/ac per cycle) (Avnimelech, 2010). In the USA, discharge of water was increasingly prohibited, and beginning in the 1990s US aquaculturists turned to enhanced photosynthetic systems such as partitioned aquaculture systems (PAS), split-ponds, in-pond raceways, and intensively aerated ponds, further increasing carrying capacity (Brune, 2014; Brune et al., 2012). Fish yields of 11,209–21,296 kg/ha (10,000–19,000 lb/ac) were achievable with feed loadings of 112–280 kg/ha-d (100–250 lb/ac-d). Under these conditions, algal photosynthesis peaks at levels of 6–12 g C/m^2-d.

To further increase production to "super-intensive" levels, farmers turned to alternative biological treatment techniques, such as fixed-film nitrifying reactors (trickling filters) or suspended-culture (biofloc) microbial reactors (Avnimelech, 2010, 2015). Trickling filters are economically favored in most finfish culture because of the potential for higher volumetric density in fish as opposed to shrimp culture. In general, biofloc treatment is economically favored in shrimp culture, using nitrifying systems (NS) or heterotrophic systems (HS). HS require external organic supplementation in addition to feed application. Optimal microbial levels in the range of 200–400 mg/l keep pace with the required ammonia removal while not imposing excessive oxygen demand. Shrimp production in excess of 44,834 kg/ha per cycle (40,000 lb/ac per cycle) is achievable. NS require aeration power of 93–112 kW/ha (50–60 hp/ac), whereas aeration requirements in HS can exceed 112–149 kW/ha (60–80 hp/ac). The major advantage of heterotrophic culture is the rapid growth rate of heterotrophic microbes, providing nearly instantaneous control of ammonia levels. Heterotrophic culture requires organic supplementation of 40–60% of feed and removal of as much as 10% of the microbial mass per day. Production of microbial solids from NS is typically 10% that of HS and no additional organic supplementation is required. Enterprise budgets suggest that shrimp production costs in NS and HS at 44,834 kg/ha per 120-day cycle (40,000 lb/ac per 120-day cycle) are similar at US$8.00/kg shrimp. An added advantage of super-intensive aquaculture is the potential to provide zero discharge operation, eliminating water and sludge discharge by converting microbial biomass production to useful by-products.

Extensive pond aeration

Observed yields in unaerated ponds range from 1121 to 2242 kg/ha (1000–2000 lb/ac) (Brune and Drapcho, 1991). Boyd and Teichert-Coddington (1992) give the surface reaeration rate of small fishponds as:

$$K_La\ (/h^{-1}) = 0.017\ (\text{Wind (m/s)}) - 0.014, \qquad [1]$$

where K_La is the standard oxygen transfer coefficient, yielding a value of 0.071/h at an average oxygen concentration of ~5.0 mg/l, an average wind speed of 3.5 m/s, and an effective mixed depth of 1 m, suggesting a surface re-aeration rate of 30–50 kg O_2/ha-d (27–45 lb O_2/ac-d). The typical average seasonal oxygen demand of fish and shrimp ranges from 1.5% to 2.0% of body weight per day (Brune et al., 2003), suggesting a surface re-aeration oxygen-carrying capacity of 1513–3363 kg/ha (1350–3000 lb/ac), in agreement with observed catfish pond harvests of 1121–2242 kg/ha (1000–2000 lb/ac). On the other hand, observed shrimp yields in unaerated ponds are typically less than 1121 kg/ha (1000 lb/ac). Garcia and Brune (1991) suggested that enhanced mixing is needed to supply oxygen to benthic shrimp ponds, leading to re-aeration energy requirements for large shrimp ponds (>4.1 ha (10 ac)) two to three times higher than for freshwater catfish ponds, and 1.1 to 1.5 times higher for smaller shrimp ponds (<0.41 ha (1 ac)).

Mechanically aerated conventional ponds

Beginning in the 1980s, shrimp and catfish farmers were supplementing pond surface re-aeration with electrically driven paddlewheel or fountain aerators of 1.8–3.7 kW/ha (1–2 hp/ac) in ponds 2–4 ha (5–10 ac) in size (Brune and Drapcho, 1991). Under these conditions, typical observed catfish yields are 4483–7846 kg/ha (4000–7000 lb/ac). Mechanical aerators are capable of supplying 0.9–1.2 kg O_2/kW-h (1.5–2 lb O_2/hp-h). These so-called "conventional ponds" are operated with an aeration energy of 1.8–3.7 kW/ha (1–2 hp/ac), operating 12–24 h per day and transferring typically 1.2 kg O_2/kW-h (2.0 lb O_2/hp-h). Under these conditions, aerators supply between 27 and 108 kg/ha-d (24–96 lb/ac-d) of oxygen, with surface re-aeration supplying an additional 30–50 kg/ha-d (27–45 lb/ac-d), providing a net oxygen flux of 57–158 kg/ha-d (51–141 lb/ac-d). At a catfish oxygen demand of 2.0% of body weight per day, this oxygen supply is predicated to support a catfish-carrying capacity of 3194–7902 kg/ha (2850–7050 lb/ac), in agreement with field-observed yields. In contrast, shrimp pond yields at 1.8–3.7 kW/ha (1–2 hp/ac) aeration typically range from 1121 to 2914 kg/ha (1000–2600 lb/ac), one-third of the yield of freshwater catfish ponds at similar aeration energy levels (Avnimelech, 2010).

Semi-intensive ponds

By the 1990s, catfish and shrimp farmers were seen to increase pond carrying capacity by increasing aeration power levels to 11–19 kW/ha (6–10 hp/ac) in catfish ponds and 19–37 kW/ha (10–20 hp/ac) in shrimp ponds (Tucker et al., 2014). Under these conditions catfish yields of 11,208–13,450 kg/ha (10,000–12,000 lb/ac) are possible, with similar shrimp yields at twice the aeration energy levels. At 11 kW/ha (6 hp/ac) with aerators running 12–24 h per day, oxygen transfer rates of 161–323 kg/ha-d (144–288 lb/ac-d) suggest a fish-carrying capacity of 8070–16,140 kg/ha (7200–14,400 lb/ac). In shrimp ponds, typical carrying capacity is given at 660–887 kg/kW (1085–1459 lb/hp) aeration energy (Avnimelech, 2010; FAO, 2006; Mohanty, 2001), predicting shrimp yields ranging from 12,329 kg/ha (10,085 lb/ac) at 19 kW/ha (10 hp/ac) to 32,706 kg/ha (29,180 lb/ac) at 37 kW/ha (20 hp/ac). However, at fish- or shrimp-carrying capacities in excess of 13,450 kg/ha (12,000 lb/ac), field observations suggest that treatment capacity in conventional ponds or intensively aerated ponds is not capable of keeping up with ammonia generation rates (as seen in Fig. 7.1).

Intensive photoautotrophic ponds

Feed levels applied to unaerated extensive ponds range from 11 to 22 kg/ha-d (10–20 lb/ac-d). At feed protein levels of 32–35% and 75% of feed nitrogen (N) released to the water column, this level amounts to 0.43–0.94 kg N/ha-d (0.38–0.84 lb N/ac-d), or 0.043–0.094 g N/m^2-d, or 0.43–0.094 mg N/l-d at an effective depth of 1 m. At an average algal C/N ratio of 5.6:1, this suggests an algal fixation rate of 0.24–0.53 g C/m^2-d, or as much as 1.0 g C/m^2-d with an internal nitrogen recycling rate of 100%. Significant nitrogen recycling has been documented to occur in intensive aquaculture systems (Brune, 2018a, 2018b; Brune et al. 2003, 2012). At a typical algal cell age in unmixed ponds of 10–20 days, this suggests an algal VSS concentration of 5–20 mg/l. In conventional aerated ponds, feed levels are ramped to 56–112 kg/ha-d (50–100 lb/ac-d), increasing algal fixation to 2–3 g C/m^2-d at VSS of 50–100 mg/l. As feed levels are increased further to 112–202 kg/ha-d (100–180 lb/ac-d), carbon fixation levels of 3–4 g C/m^2-d are observed at algal VSS exceeding 100 mg/l depending on algal cell age. Attempts to apply feed nitrogen at levels in excess of 202 kg/ha-d (180 lb/ac-d) to aerated ponds (without flushing) are typically not successful, as such pond configurations are not capable of providing algal nitrogen assimilation in excess of 4–6 g C/m^2-d.

The PAS developed at Clemson University, South Carolina (Fig. 7.2) builds upon the increased algal productivity in "high-rate" ponds, originally developed for wastewater treatment in the 1960s (Oswald, 1988). Paddlewheel water mixing increases algal exposure to sunlight while simultaneously imposing uniform water velocity throughout the pond (Schwartz, 1998) (Fig. 7.3). These effects increase algal carbon fixation rates to 6–12 g C/m^2-d, providing a demonstrated fish-carrying capacity of as much as 21,296 kg/ha (19,000 lb/ac) for catfish production, with an additional 5,044 kg/ha (4,500 lb/ac) for tilapia (Brune et al., 2012). Under such conditions, maximum feed levels may be sustained at 280 kg/ha-d (250 lb/ac-d). Tilapia co-culture is used to continually crop algal population to levels of 50–100 mg/l, at an average algal cell age of 5–6 days, while also eliminating cyanobacterial dominance and minimizing zooplankton populations, thereby stabilizing the pond algal population (Brune, 2017a, 2017b).

In addition to the PAS, other enhanced photosynthetic systems include split-ponds and semi-intensive ponds. Split-ponds represent an attempt by farmers and researchers at the Warmwater Aquaculture

FIGURE 7.2 A 2 acre (0.8 ha) PAS at Clemson University producing catfish (from Brune et al., 2012).

Center (WAC) in Stoneville, Mississippi, to duplicate the increased carrying capacity of the PAS in a lower-cost system design (Tucker et al., 2014). Split-ponds typically consist of higher-density earthen fish-culture ponds of 0.8–2.0 ha (2–5 ac) coupled to a large waste-treatment pond of 2.4–8.1 ha (6–20 ac) using slowly rotating paddlewheels (1–3 rpm) or axial screw pumps to exchange water between the two.

At the WAC, and at Clemson University, comparisons between PAS, split-ponds, semi-intensive ponds, and conventional ponds of 0.8, 2.8, 1.6, and 0.8 ha (2.0, 7.0, 4.0, and 2.0 ac) in size indicate fish culture densities of 96, 4, 0.96, and 0.64 kg/m^3 (6.0, 0.25, 0.06, and 0.04 lb/ft^3) at average photosynthetic rates of 7.5, 5.5, 6.6 and 1.4 g C/m^2-d, at feed rates of 179–280, 123–280, 94–202, and 67–112 kg/ha-d (160–250, 110–250, 84–180, and 60–100 lb/ac-d). Algal VSS levels were, respectively, 80 mg/l dominated by green algae, 115 mg/l dominated by cyanobacteria, 110 mg/l of cyanobacteria, and 110 mg/l of slowly growing cyanobacteria. Fish production in the four systems averaged 20,175, 15,132, 12,890, and 8406 kg/ha (18,000, 13,500, 11,500, and 7500 lb/ac), respectively. In the semi-intensive ponds, as much as 20% of added nitrogen was removed via nitrification. The major advantage of the split-pond is in providing 75% of the production capacity of the PAS at 50% of the cost. The increase in nitrification in semi-intensive ponds indicates the beginning of a transition from algal-dominated culture to nitrifying culture, with required algal productivity exceeding the pond limit of 5–6 g C/m^2-day. The semi-intensive ponds lack

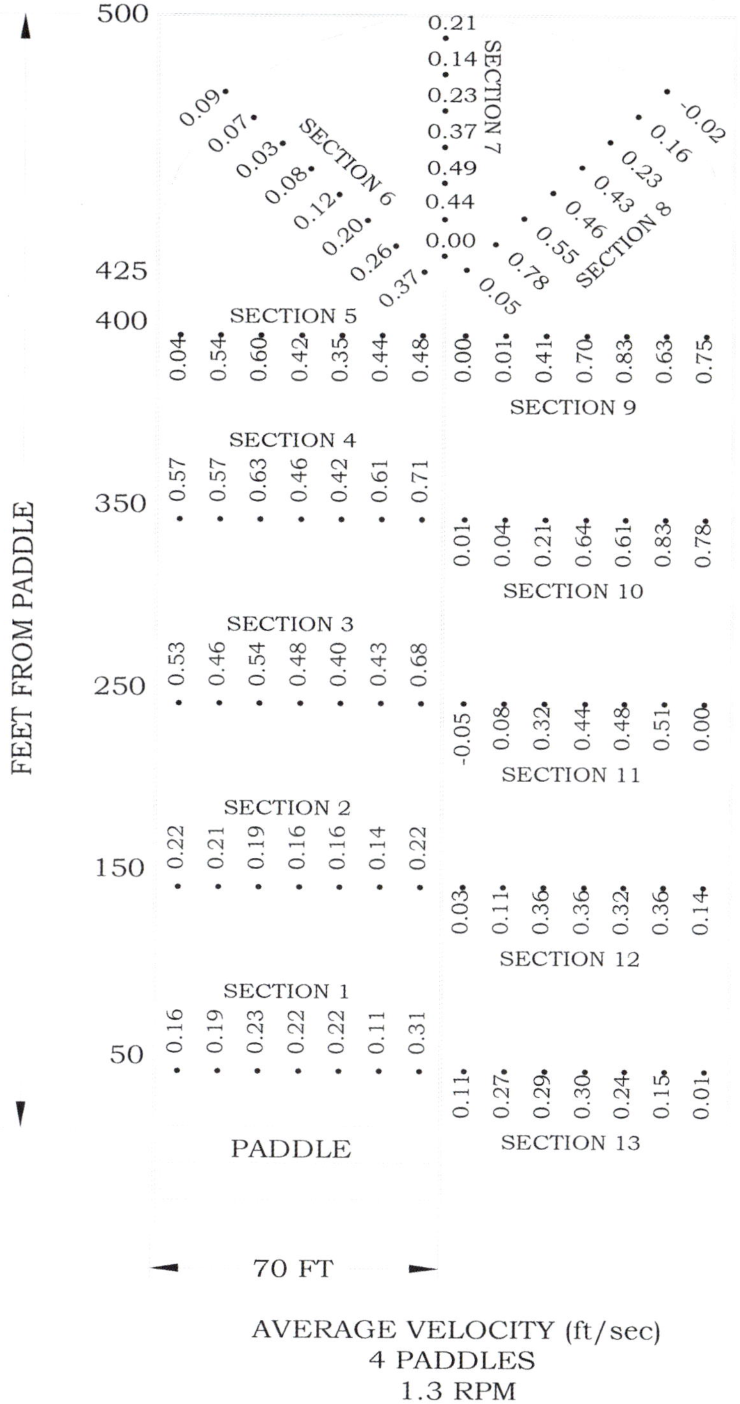

FIGURE 7.3 Velocity profile in a 2 acre (0.8 ha) PAS at a paddlewheel speed of 1.3 rpm (from Schwartz, 1998).

the uniform water velocity of the PAS and therefore algal productivity at ~6 g C/m^2-d is failing to keep pace with nitrogen loading; consequently, nitrifying microbial biomass production gains influence.

Super-intensive aquaculture

Within the past 20 years, researchers, farmers, and investors have installed greenhouse-enclosed raceways and lined ponds capable of producing yields of 39,230–50,438 kg/ha (35,000–45,000 lb/ac). In such systems, feed rates may exceed 897 kg/ha-d (800 lb/ac-d), with additional organic carbon loading surpassing 40 g C/m^2-d, a level far beyond practical sustainable algal productivity limits of 12 g C/m^2-d. In such cases it becomes necessary to employ alternative biological treatment techniques, such as fixed-film nitrifying reactors (trickling filters) or suspended-culture microbial reactors, referred to as biofloc systems. Trickling filters are economically favored in most finfish culture because of the potential for higher volumetric densities in fish, as opposed to shrimp culture systems. In fish culture systems, fish densities of 96 kg/m^3 (6 lb/ft^3) 1 m deep result in nitrogen loadings of 70–80 mg N/l-d. This level of nitrogen loading may be removed in a trickling filter of 10–50% of fish culture volume. On the other hand, shrimp culture is typically two-dimensional, at super-intensive shrimp culture densities of 3.6–4.5 kg/m^2 (8–10 lb/m^2), at feed nitrogen loading of 3–4 mg N/l. This level of nitrogen loading can be removed with 200–400 g of microbial biomass per m^3, whereas a trickling filter would require ten times the water pumping and loading to treat ammonia at a rate similar to a shrimp biofloc system. Consequently, biofloc treatment is economically favored in shrimp culture, using nitrifying systems (NS) or heterotrophic systems (HS). In biofloc systems, aquatic animal rearing and water treatment occur predominantly in the same water footprint, as opposed to separate water treatment operations such as fixed-film biofilters or gas exchange columns. Water treatment in these systems may include a combination of suspended-cell microbial growth and/or other microbially mediated reactions, including photosynthesis, nitrification, denitrification, and/or heterotrophic bacterial growth.

Super-intensive nitrifying biofloc systems

Nitrifying biofloc systems have been demonstrated to yield shrimp at 39,230–44,834+ kg/ha per cycle (35,000–40,000+ lb/ac per cycle). Nitrifying cycles typically range from 120 to 150 days producing jumbo-sized (20–25 g) shrimp (Brune, 2018a, 2018b). Feed-loading rates range from 785 to 1121 kg/ha-d (700–1000 lb/ac-d). At 44,834 kg/ha (40,000 lb/ac) of carrying capacity, system oxygen demand ranges from 673 to 897 kg O_2/ha-d (600–800 lb O_2/ac-d). With continuous aeration and 1.2 kg O_2 transferred per kW-h (2 lb O_2/hp-h), predicted aeration demand is 31 kW/ha (16.6 hp/ac); however, practical experience has indicated field energy requirements of 93–112 kW/ha (50–60 hp/ac) to support the increasing shrimp biomass (Brune, 2017a, 2017b). This three-fold requirement of energy input is likely due to the need to maintain complete water mixing, ensuring adequate VSS suspension and water treatment. Metcalf & Eddy, Inc. (2003) suggest 131–262 kW/ha (70–140 hp/ac) to maintain completely mixed conditions in aerobic lagoons, three-fold over the energy needed for re-aeration. Garcia and Brune (1991) suggested that this two- to three-fold mixing energy over aeration requirement is required to support large shrimp ponds (2–4 ha (5–10 ac)), and 1.5–2.0-fold for small ponds (0.4–0.8 ha (1–2 ac)), to maintain oxygen transport to benthic shrimp populations and maintain VSS suspension. In such systems, mixing of VSS and uniform distribution of oxygen is more limiting than the actual oxygen transfer rate.

Typical VSS levels in super-intensive nitrifying systems range from 200 to 400 mg/l, with 300 mg/l VSS considered optimum, keeping pace with ammonia removal while not imposing excessive oxygen demand. However, ammonia removal rates may be maintained at reduced solids levels of 70–80 mg/l when aggressive removal of solids using air jets in tanks combined with discharge to bead filters is used (P.R. McIntosh, Charoen Pokphand, Bangkok, Thailand, 2019, personal communication). Field-observed ammonia removal rates range between 0.02 and 0.05 mg N/g VSS-d (Brune et al., 2007). In contrast, observed ammonia removal rates in wastewater nitrifying systems typically range from 0.07 to 0.09 mg N/g VSS-d (Metcalf & Eddy Inc., 2003; Scheible et al., 1993), suggesting that nitrifying biomass is 22–55% of total microbial biomass in super-intensive nitrifying aquaculture systems. In contrast, in the McIntosh prototype system (at 70–80 mg/l), nitrifying biomass likely approaches 80–100% of the total microbial biomass. At 300 mg/l VSS, the potential ammonia uptake rate would be 6–15 mg N/l-d. At a 35% protein feed application rate of 897 kg/ha-day (800 lb/ac-d), nitrogen loading would be expected to be 5 mg/l-d, well within the projected microbial capacity. Solids production from nitrification ranges from 0.1 to 0.2 mg VSS/mg N removed. Therefore, the expected accumulation of nitrifying biomass is ~1.0 mg/l-d. At a solids level of 300 mg/l, the average solids removal rate is predicted at 1/300 of culture volume per day. In contrast, the solids production rate from shrimp fecal production is around 20% of feed, or 27 mg/l-d, suggesting that nitrifying biomass removal is not an essential requirement.

Super-intensive heterotrophic biofloc systems

Heterotrophic biofloc systems require the input of external organic supplementation in addition to feed application. In this case, ammonia removal is accomplished by the uptake of nitrogen to build microbial biomass using added carbohydrate energy. As in the case of NS, shrimp production in excess of 44,834 kg/ha per cycle (40,000 lb/ac per cycle) is achievable. Browdy et al. (2012) provide a predictor of the organic supplementation required as

$$\text{Organic supplementation (g)} = (\text{Feed}) \times (\%\text{N}) \times 20 \times (\%\text{N excreted}) \qquad [2]$$

At 897 kg feed/ha-d (800 lb feed/ac-d) and 35% protein, with 75% nitrogen excreted, this amounts to 84% of 897 kg (800 lb) or 753 kg organic/ha-d (672 lb organic/ac-d), for a total feed + organic application of 1650 kg/ha-d (1472 lb/ac-d). The combined oxygen demand of fish and shrimp feed and organic supplementation ranges from 0.8 to 1.25 kg O_2/kg feed (Brune and Drapcho, 1991; Chatvijitkul et al., 2017). Therefore, at 1.0 kg O_2/kg organic added, and at an oxygen addition rate of 1.2 kg O_2/kW-h (2.0 lb-O_2/hp-h), this suggests an aeration requirement of 56 kW/ha (30 hp/ac). However, as with previous cases of observed shrimp aeration requirements, the actual energy required is two to three times the oxygenation requirement, or 112–168 kW/ha (60–90 hp/ac). Metcalf & Eddy, Inc. (2003) recommend 0.004–0.008 kW/m^3 (0.5–1.0 hp/1000 ft^3) to prevent solids settling in activated sludge reactors, which translates to 133–267 kW/ha (71–143 hp/ac) in 3 m deep wastewater treatment lagoons. The major advantage of heterotrophic culture is the rapid growth rate of heterotrophic microbes, providing nearly instantaneous control of ammonia levels. Heterotrophic growth adds additional microbial solids loading of as much as 50% of the mass of carbohydrate addition. In the example given above, this would amount to as much as 38 mg/l-d of VSS in addition to 27 mg/l VSS from shrimp fecal waste, adding 65 mg/l of solids that must be removed each day, or 20% of standing VSS levels – very different from nitrifying system solids removal and treatment needs.

Transitional and mixed microbial systems

As stated earlier, measurements of photosynthesis and nitrification in semi-intensive catfish production systems demonstrated that as much as 20% of nitrogen loading was treated via nitrification (Brune et al., 2015). In these ponds, 11–19 kW/ha (6–10 hp/ac) is insufficient to insure complete mixing of the pond volume. Consequently, maximum sustained photosynthesis does not typically exceed 6 g C/m^2-d. Under such conditions, elevated ammonia concentrations favor the development of nitrifying populations. McIntosh (2001) presented data showing the success of early super-intensive ponds, dependent upon both heterotrophic and nitrifying microbial populations to maintain control of accumulating ammonia levels. The green visual appearance of these ponds suggests that algal populations were also present, but likely were not contributing a major portion of the ammonia treatment. These ponds, as with most intensive and super-intensive culture systems, contain a mixture of microbial types.

In shrimp production trials at Clemson University, Kirk (2010) and Kirk and Brune (2011) quantified water column VSS concentrations, gross and net photosynthesis, as well as the apparent color of the culture system as a function of feed loading (Figs 7.4, 7.5, and 7.6). Net photosynthesis was seen to peak at 13.8 g C/m^2-d at feed loading of 179–247 kg/ha-d (160–220 lb/ac-d). During this transitional period, water column VSS levels increased from 80 mg/l to nearly 400 mg/l. As feed levels increased to 673 kg/ha-d (600 lb/ac-d), a green color from algal biomass was no longer observable by the naked eye, in spite of a persistent low level of photosynthesis in the order of 1–2 g C/m^2-d. These observations illustrate the origins and nature of the "invention" of non-photosynthetic super-intensive aquaculture. At feed levels exceeding carbon and nitrogen fixation within enhanced photosynthesis systems (6–12 g C/m^2-d),

FIGURE 7.4 Photosynthetic and nitrifying biofloc in a super-intensive shrimp production system at Clemson University. (A) Shrimp PAS at maximum photosynthesis rate of 20 g C/m^2-d. (B) Biofloc: dilute activated sludge (from Brune, 2018a).

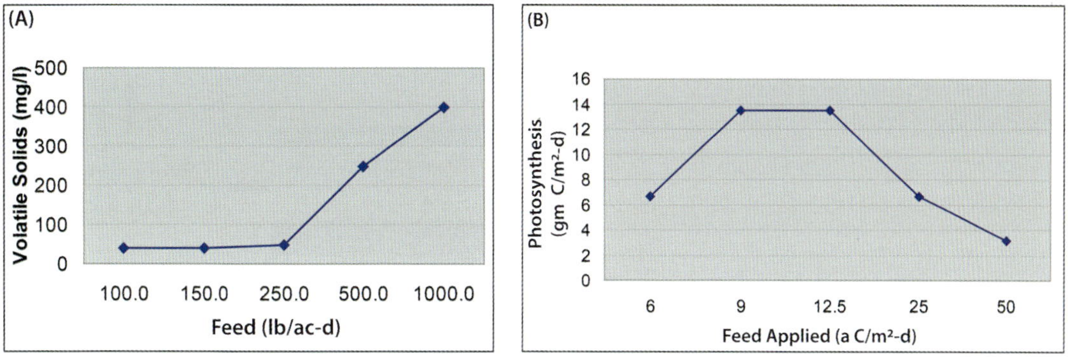

FIGURE 7.5 (A) Volatile solids concentration versus feed application rate and (B) photosynthetic carbon fixation versus feed carbon applied in intensive and super-intensive shrimp production systems (from Brune et al., 2010).

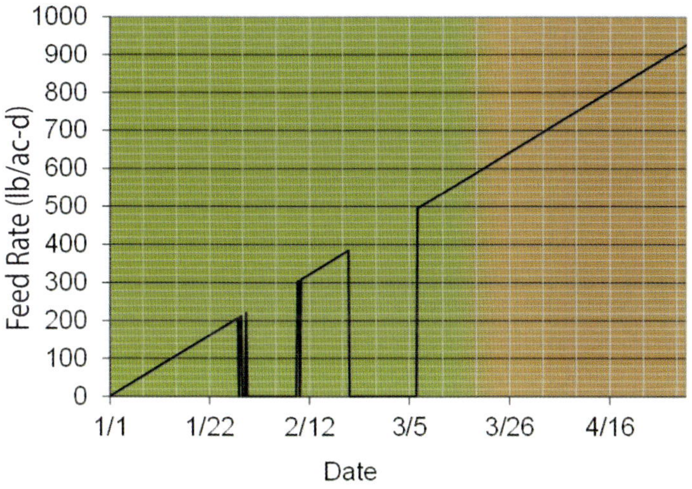

FIGURE 7.6 Water color in a shrimp culture system at feed rates ranging from 112 to 1009 kg/ha-d (100–900 lb/ac-d) (from Kirk and Brune, 2011).

rising system ammonia levels provide the driving force for increased nitrifying microbial growth and accumulation. If sufficient organic matter is available in the water column from either uneaten suspended feed, fecal matter, or intentionally supplemented organics, microbial populations will shift to dominance by faster-growing heterotrophic populations.

Nitrogen fate and impact on system alkalinity

In most pond culture systems, the ultimate fate of nitrogen beyond that harvested in the fish or shrimp is through denitrification and loss of nitrogen gas from the anoxic zone at the interface of the aerobic pond water and the anaerobic sediment (Fig. 7.7).

In super-intensive HS, the bulk of nitrogen will be transformed into microbial sludge that must be either treated and disposed of as a potential crop fertilizer or captured via additional aquatic animal

FIGURE 7.7 Loading rate and fate of nitrogen in a nitrifying/denitrifying shrimp production system at 897 kg/ha-d (800 lb/ac-d) of feed application (from Brune et al., 2010).

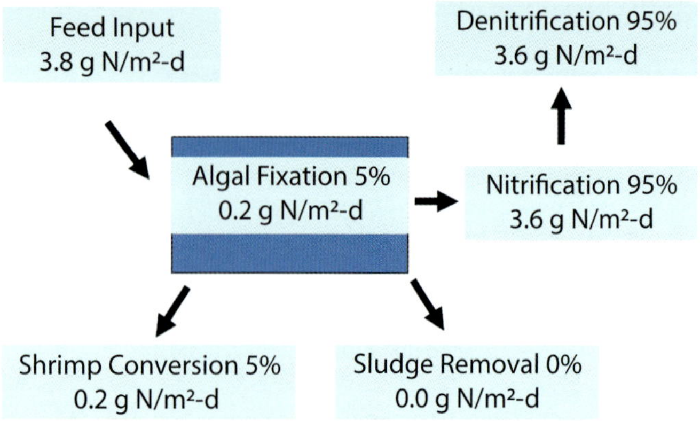

culture, providing valuable by-products. In NS, the bulk of the nitrogen is converted to nitrate, which accumulates in the culture water and must be either removed for land application, which is difficult in the case of saltwater systems, or treated in a denitrifying reactor. At Clemson University, and at the University of Missouri, sludge-settling tanks were demonstrated to be capable of serving the double purpose of removal of excess microbial solids and also serving as low-cost denitrifying reactors (Brune 2017a, 2018b). An added advantage of denitrification is that the process replaces alkalinity lost in the nitrification process. An integrated nitrifying/denitrifying culture system does not require replacement of alkalinity. In contrast, a system producing 44,834 kg/ha (40,000 lb/ac) of shrimp requiring 89,668 kg/ha (80,000 lb/ac) of 35% protein feed would require 22,417 kg/ha (20,000 lb/ac) of $NaHCO_3$ to neutralize acid production from the nitrification reactions (Brune, 2018b).

Cost of production

Kumar and Engle (2017) recently showed that catfish production costs in semi-intensive ponds at aeration levels of 11–19 kW/ha (6.0–10.0 hp/ac) ranged from US$1.85 to US$2.16/kg (US$0.84–0.98/lb), while costs of a variety of catfish production strategies range from US$1.96 to US$2.84/kg (US$0.89–1.29/lb) (Johnson et al., 2014). Kumar and Engle (2017) suggested costs of US$3.75–8.40/kg (US$1.70–3.81/lb) for shrimp system production yields ranging from 2018 to 10,088 kg/ha per cycle (1800–9000 lb/ac per cycle). Hanson et al. (2014) reported US$7.21/kg (US$3.27/lb) production costs of for super-intensive biofloc shrimp production at 39,230 kg/ha per cycle (35,000 lb/ac per cycle).

A recently presented enterprise budget (Fig. 7.8) compares the costs of shrimp production in super-intensive photoautotrophic, nitrifying, and heterotrophic systems. Twelve system configurations are compared, at productivities ranging from 10,984 to 50,438 kg/ha per cycle (9800–45,000 lb/ac per cycle) with cycle times of 80–120 days. The predicted costs ranged between US$8.00 and US$14.66/kg (US$3.63–6.65/lb), with costs of NS and HS at 50,438 kg/ha per 120-day cycle (45,000 lb/ac per 120-day cycle) being similar at US$8.00/kg shrimp (US$3.63/lb shrimp) (Fig. 7.8), comparable to Asian semi-intensive shrimp production costs of US$6.06/kg (US$2.75/lb) as opposed to retail prices of US$7.71–20.94/kg (US$3.50–9.50/lb).

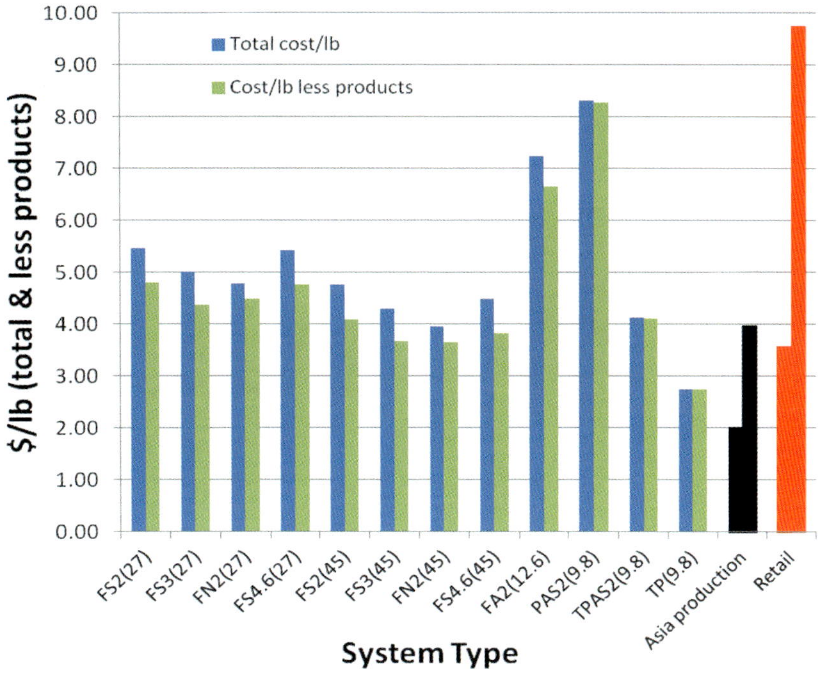

FIGURE 7.8 Cost of shrimp produced in 12 different heterotrophic, nitrifying, and photoautotrophic super-intensive shrimp culture systems (from Brune, 2018a).

System sustainability

An added advantage of super-intensive aquaculture is the potential to provide zero discharge systems, eliminating water and sludge discharge by converting microbial biomass production to useful by-products (Figs 7.9 and 7.10).

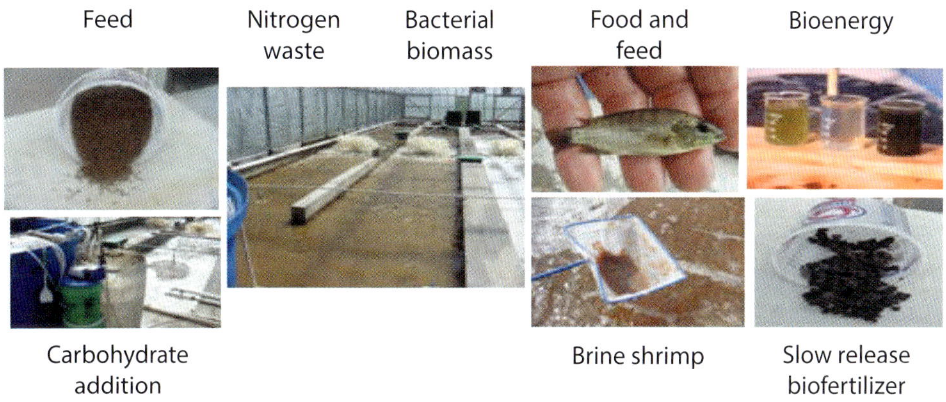

FIGURE 7.9 Conversion of excess heterotrophic microbial solids treating shrimp culture to tilapia and brine shrimp co-products. (from Brune, 2018b).

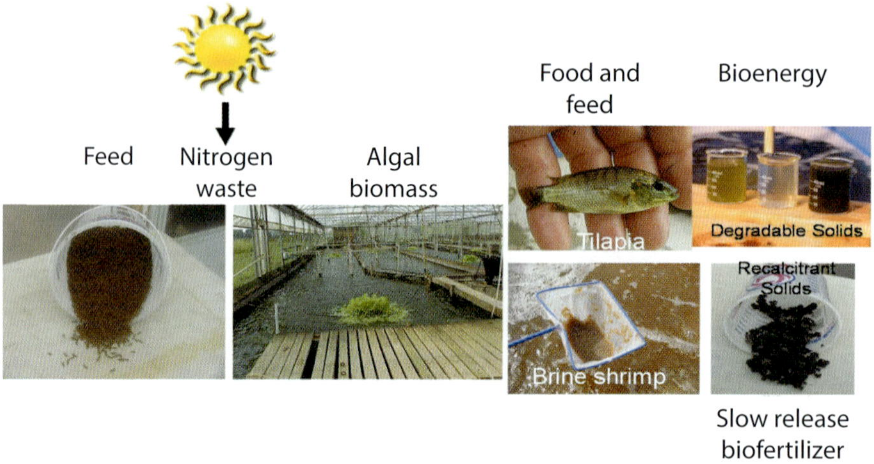

FIGURE 7.10 Conversion of excess algal solids treating shrimp cultures to tilapia and brine shrimp co-products (from Brune, 2018b).

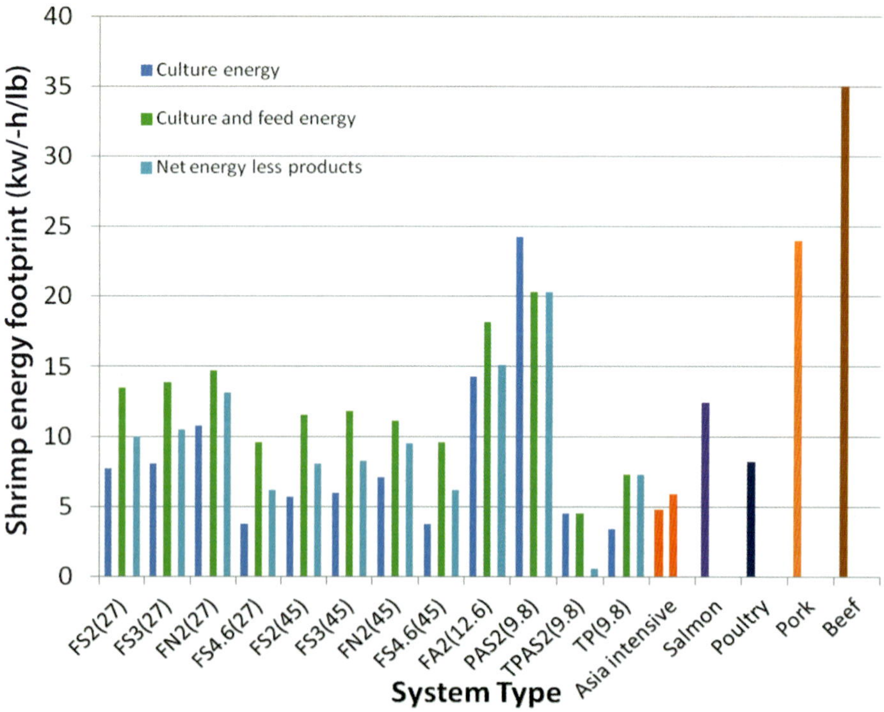

FIGURE 7.11 Energy investment in shrimp, projected for 12 different super-intensive systems (from Brune, 2018b).

Solids produced from pond culture typically settle to the bottom of the pond and degrade anaerobically, releasing ammonia and organic compounds back into the water column, adding to the water treatment and aeration needs. However, VSS production from super-intensive systems can be diverted to filter-feeding cultures, such as tilapia or brine shrimp, which eliminate the need for water or sludge discharge and disposal, and can also provide additional income as fishmeal replacement or additional human seafood products. In spite of the high energy requirement for mixing and aeration in super-intensive systems, because of the high productivity per unit area, the energy requirement per kg of product in these systems can be less than the 3.6 kW-h/kg (8 kW-h/lb) required for chicken production (Fig. 7.11).

Conclusions

Over the past 50 years, aquaculture productivity has been increased from 1121–2242 kg/ha (1000–2000 lb/ac) to over 50,438 kg/ha (45,000 lb/ac) by adding increasing levels of aeration from 1.9–3.7 kW/ha (1–2 hp/ac) to as much as 168 kW/ha (90 hp/ac) or even more. Natural levels of algal productivity typically control ammonia levels in extensive or conventional fish culture ponds, but as feed loading exceeds 112 kg/ha-d (100 lb/ac-d) pond algal productivity cannot keep pace with ammonia removal and carbon fixation requirements. Enhanced photosynthetic techniques such as PAS and split-ponds allow feed loading in excess of 224 kg/ha-d (200 lb/ac-d), supporting fish production in excess of 11,209–15,692 kg/ha (10,000–14,000 lb/ac) at algal fixation rates of 6–12 g C/m^2-d. Productivity exceeding sustainable algal growth capacity requires culture systems utilizing nitrifying and/or heterotrophic microbial growth to keep pace with nitrogen loading at feeding rates exceeding 673–897 kg/ha-d (600–800 lb/ac-d). In spite of its intense energy usage and elevated feeding rates, super-intensive aquaculture offers the potential to provide sustainable seafood production in terms of energy and water requirements per unit of aquatic biomass produced. Integrated super-intensive systems, harvesting and converting microbial biomass used to maintain water quality, can provide valuable by-products while simultaneously reducing, or even eliminating, water or waste discharge to the environment.

References

Avnimelech, Y. (2010) Intensive production of shrimp. In: Alday-Sanz, V. (ed.) *The Shrimp Book*. Nottingham University Press, Nottingham, UK, pp. 233–246.

Avnimelech, Y. (2015) *Biofloc Technology: A Practical Guidebook*. World Aquaculture Society, Baton Rouge, LA.

Boyd, C.E. and Teichert-Coddington, D. (1992) Relationship between wind speed and reaeration in small aquaculture ponds. *Aquaculture Engineering*, 11, 121–131.

Browdy, C.L., Ray, A., Leffler, J.W., and. Avnimelech, Y. (2012) Biofloc-based aquaculture systems. In: Tidwell, J.H. (ed.) *Aquaculture Production Systems*. Wiley-Blackwell, Ames, IA, pp. 278–307.

Brune, D.E. (2014) Bioprocessing of microbial biomass enabling sustainable marine shrimp production. *Aquaculture America*, Seattle, WA.

Brune, D.E. (2017a) Tilapia in high-rate aquaculture processes, In Perschbacher, P.W. and Stickney, R.R. (eds) *Tilapia in Intensive Co-Culture*. John Wiley & Sons, Oxford, UK, pp. 186–210.

Brune, D.E. (2017b) Zero discharge biofloc aquaculture. *Presentation to the North Carolina Aquaculture Workshop*, New Bern, NC.

Brune, D.E. (2018a) Zero-discharge marine shrimp production. *North Central Regional Aquaculture Meeting*, Kansas City, MO.

Brune, D.E. (2018b) Impact of nitrification and denitrification on alkalinity in recirculating. aquaculture systems. *World Aquaculture*, 49(4), 58–62.

Brune, D.E., Brown, T.W., and Tucker, C.S. (2015) Oxygen and nitrogen dynamics in spilt ponds vs. conventional catfish production ponds. *Catfish Farmers of America*, Natchez, MS.

Brune, D.E. and Drapcho, C.M. (1991) Fed pond aquaculture, aquaculture systems engineering. In: P. Giovannini (ed.), *Aquacultural Systems Engineering: Proceedings of the World Aquaculture Society*, St. Joseph, MI, pp. 15–33.

Brune, D.E., Henrich, C., Kirk, K., and Eversole, A.G. (2010) Suspended cell co-culture for limited discharge aquaculture. *Eighth International Conference on Recirculating Aquaculture*, Roanoke VA.

Brune, D.E., Kirk, K., and Eversole, A.G. (2007) Nitrogen transformation rates in a zero-discharge suspended-culture shrimp production system. *World Aquaculture Society*, San Antonio, TX.

Brune, D.E., Schwartz, G., Collier, J.A., Eversole, A.G., and Schwedler, T.E. (2004) Partitioned aquaculture systems. In: Tuckers, C.S. and Hargreaves, J.A. (eds) *Biology and Culture of Channel Catfish*, Elsevier, Amsterdam, The Netherlands, pp. 561–584.

Brune, D.E., Schwartz, G., Eversole, A.G., Collier, J.A., and Schwedler, T.E. (2003) Intensification of pond aquaculture and high rate photosynthetic systems. *Aquacultural Engineering* 28, 65–86.

Brune, D.E., Tucker, C., Massingill, M., and Chappell, J. (2012) Partitioned aquaculture systems. In: Tidwell, J.H. (ed.) *Aquaculture Production Systems*, Wiley-Blackwell, Ames, IA, pp. 308–342.

Chatvijitkul, S., Boyd, E.E., Davis, D.A., and McNevin, A.A. (2017) Pollution potential indicators for feed based fish and shrimp culture, *Aquaculture* 477, 43–49.

FAO (2006) Cultured Aquatic Species Information Programme. *Penaeus vannamei* (Boone, 1931). FAO Fisheries Division. Available at: http://www.fao.org/fishery/culturedspecies/Penaeus_vannamei/en (accessed 29 July 2021).

Garcia, A. and Brune, D.E. (1991) Transport limitation of oxygen in shrimp culture ponds. *Aquacultural Engineering* 10, 269–279.

Hanson, T.R., Castro, L.R., Zeigler, T.R., Markey, T.A., and Samocha, T.M. (2014) Economic analysis of feeds for super intensive white shrimp grow-out. Global Aquaculture Advocate. Available at: https://www.globalseafood.org/advocate/economic-analysis-of-feeds-for-super-intensive-white-shrimp-growout/ (accessed 28 Spetember 2021).

Johnson, K., Engle, C., and Wagner, B. (2014) Comparative economics of US catfish production strategies. *Journal of the World Aquaculture Society*, 45, 279–289.

Kirk, K. (2010) Modeling microbial and nutrient dynamics in zero discharge aquaculture systems. PhD thesis. Clemson University, Clemson, SC, USA.

Kirk, K.R. and Brune, D.E. (2011) Modeling microbial and nutrient dynamics in zero-discharge aquaculture systems. *Institute of Biological Engineering Annual Conference*, Atlanta, GA.

Kumar, G. and Engle, C. (2017) Economics of intensively aerated catfish ponds. *Journal of World Aquaculture*, 48, 320–332.

McIntosh, P.R. (2001) Changing paradigms in shrimp farming; establishment of heterotrophic bacterial communities. *Global Aquaculture Advocate* 4, 53–58.

Metcalf & Eddy, Inc. (2003) *Wastewater Engineering: Treatment and Reuse*, 4th ed. McGraw-Hill, New York, NY.

Mohanty, R.K. (2001) Effect of pond aeration on growth and survival of *Penaeus monodon* Fab. *Bangladesh Journal of Fisheries Research* 5, 59–65.

Oswald, W.J. (1988) Micro-algae and waste-water treatment. In: Borowitzka, M.A. and Borowitzka, L.J. (eds) *Micro-Algal Biotechnology*. Cambridge University Press, Cambridge, UK, pp. 305–328.

Scheible, O., Mulbarger, M., Sutton, P., Simpkin, T., Daigger, G., Heidman, J., Yoder, M., Schwinn, D., and Storrier, D. (1993) *Process Design Manual: Nitrogen Control*. EPA/625/R-93/010 (NTIS PB94159142). United States Environmental Protection Agency, Washington, DC.

Schwartz, G. (1998) Photosynthesis and oxygen dynamics in the partitioned aquaculture system. MS Thesis, Clemson University, Clemson, SC, USA.

Tucker, C.S., Brune, D.E., and Torrans, E. L. (2014) Partitioned pond aquaculture systems. *World Aquaculture* 45(2), 9–14.

Chapter 8

MICROBIAL MANAGEMENT: A KEY CONCEPT FOR SUSTAINABLE SHRIMP CULTURE

PETER DE SCHRYVER, OLAV VADSTEIN, TOM DEFOIRDT AND PATRICK SORGELOOS

Summary

At the FAO Global Conference on Aquaculture in Phuket (Thailand) in 2010 we postulated that modern aquaculture requires a turning point (Sorgeloos, 2013). Although the farming of aquatic plants and animals has evolved into a mature industry, too much of the technology is still based on empirical approaches. Knowledge-based development is needed to make aquaculture a blue biotechnology for the future. 'More microbial management for more sustainable production' and 'More integrated production systems for plant and animal farming' were identified as top priorities at this conference.

Ten years later, we can report significant progress, mainly based on novel insights regarding the role of the microbiota in aquaculture. The traditional approaches of plate count analysis and microscopy have been instrumental in making us aware of the huge numbers of bacteria that may negatively interfere in aquaculture systems. However, it has been shown that not the numbers but the diversity and functionality of bacteria in aquaculture systems should be used as indicators for microbial dysbiosis (Vadstein et al., 2018b). This insight was generated by the application of innovative microbial research platforms such as next-generation sequencing and gnotobiotic animal models. An example of a crucial finding was that certain bacterial pathogens in aquaculture systems can regulate virulence as a function of population density by means of quorum sensing mechanisms, and in this way collectively organize infection in fish or shrimp (Defoirdt et al., 2011a). This explained why the detection of pathogens should not be used as the sole predictor of compromised animal performance and why we have to take the overall status of the bacterial community into consideration.

The lack of reproducibility and predictability observed during culture despite 'full control' over the system remains problematic. This has been particularly accepted in marine fish larviculture systems. When referring to 'full control' this does not include microbial control. Indeed, to date, we have operated our production systems as microbial 'black boxes', and we have tried to control pathogenic diseases merely by repressive methods under the general assumption that 'the only good bug is a dead bug'. However, based on recent microbial insights, we believe that repression alone is not the best approach to minimize pathogenic disease risk (De Schryver and Vadstein, 2014; Vadstein et al., 2018a,b). A non-selective reduction of unwanted bacteria by disinfection to maximize biosecurity should always be followed by a selective enhancement of desirable microbes (Vadstein et al., 2018a). Up to now, this has almost exclusively been done by the use of probiotics. However, by manipulating resource competition among the microbes, that is, deliberately changing resource supply relative to the number of bacteria, K-strategic bacterial populations can be selected for and will outcompete r-strategic bacteria; the latter is the ecological strategy of most opportunistic pathogens. This approach demonstrated 50–100% improvement in several performance indicators (Vadstein et al., 2018b). The use of probiotics is complementary to this approach, but the success of probiotics can be substantially increased when they are applied in system settings that support their ecology. The result is a substantially lower disease risk compared with a repression/decimation approach used alone.

Both experimental and field observations that have accumulated during the past 10 years provide support for the importance of the approach explained above. An example is the improved performance seen in recirculation systems with biofilters for hatchery and nursery production of marine fish and shrimp. Support is also coming from intensive shrimp farming in zero water exchange ponds with shrimp toilets for increased waste removal, and from extractive ponds stocked with tilapia or seaweed in recirculation mode. In such systems, there seems to be an intrinsic microbial selection mechanism that counteracts the (selection for and) dominance of opportunistic pathogens such as vibrios. This results in a lower risk of unpredictable interference by opportunistic pathogens by maintaining their densities below the critical quorum sensing density.

The currently available experimental and empirical observations can form the foundation for a science-driven and knowledge-based aquaculture. The use of novel methods, such as flow cytometry coupled to cell sorting and next-generation sequencing, and appropriate scientific theory will substantially increase our level of understanding and contribute to a more predictive and sustainable aquaculture.

Negative interference of microbes in aquaculture

INDUSTRIALIZATION IN AQUACULTURE LEADING TO DYSBIOSIS AND DISEASE

The way aquaculture is practised nowadays is not in any way comparable to the insignificant production 70 years ago. In the 1950s and 1960s, aquaculture production grew at a rate of about 5% per year, increasing to about 8% per year in the 1970s and 1980s, and as high as 10% since the 1990s (FAO, 2000). These numbers indicate a transition towards an industrial approach in aquatic food production. The industrialization of the aquaculture industry has been the driver for significant advances in biological knowledge, improvements in operational conditions, perfections in feed technology, and increased application of biotechnology. However, industrialization also comes with the challenges typical of industrial animal production. It may not only create substantial impact on the environment, but also compromise its own performance. With regard to the latter, the increased incidence of infectious disease outbreaks is considered one of the most confounding factors in modern aquaculture production. During the GOAL conference in 2019 (Chennai, India), diseases were identified as the number one global constraint to growth of the shrimp aquaculture industry (Fig. 8.1), the same as it was for the years 2017 and 2018 (at the GOAL conferences in Dublin, Ireland, and Guayaquil, Ecuador, respectively).

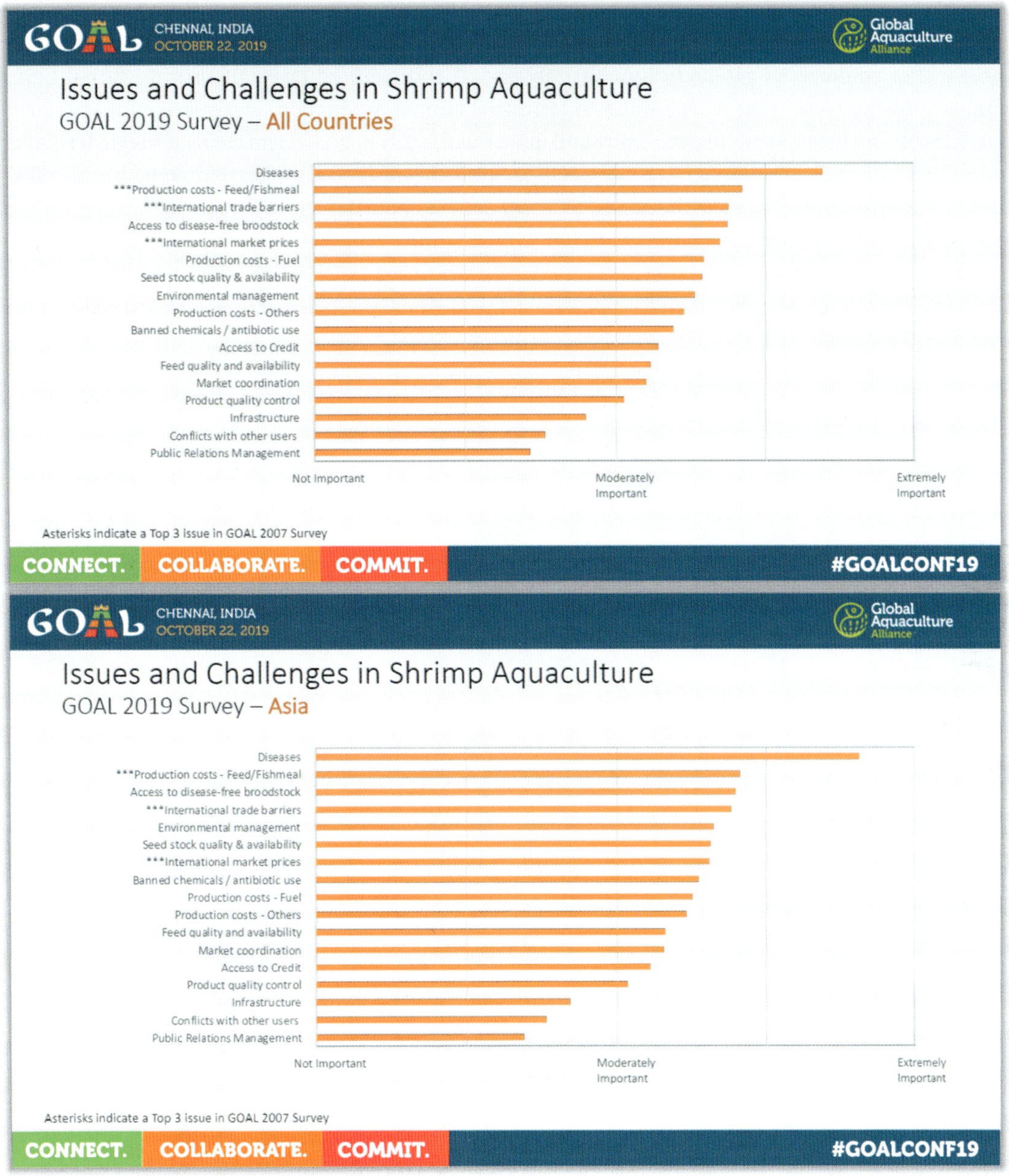

FIGURE 8.1 Worldwide top issues and challenges in shrimp aquaculture as identified from a global survey for the GOAL 2019 conference. Credit: Courtesy of James L. Anderson (from Anderson et al., 2019).

When considering the episodes of the infectious diseases the shrimp aquaculture sector has known over the years, it is remarkable that the frequency with which new diseases started appearing increased shortly after commercial production of shrimp started and elevated levels of shrimp started being produced by the sector (Fig. 8.2). Similar to terrestrial animal production, increased output has been possible only because systems became more and more industrially operated and characterized by a higher output per unit area or volume (Ackefors, 2009). In other words, shrimp started being produced in confined areas at higher stocking densities, by using (inter)national distribution of performing shrimp strains, an increase in the genetic uniformity of shrimp strains being produced, more inputs (feed and products) into the systems, and with higher levels of stress being experienced by the animals. All these factors are known to contribute to the proliferation of pathogens and increased risk for disease outbreaks (Kautsky et al., 2000).

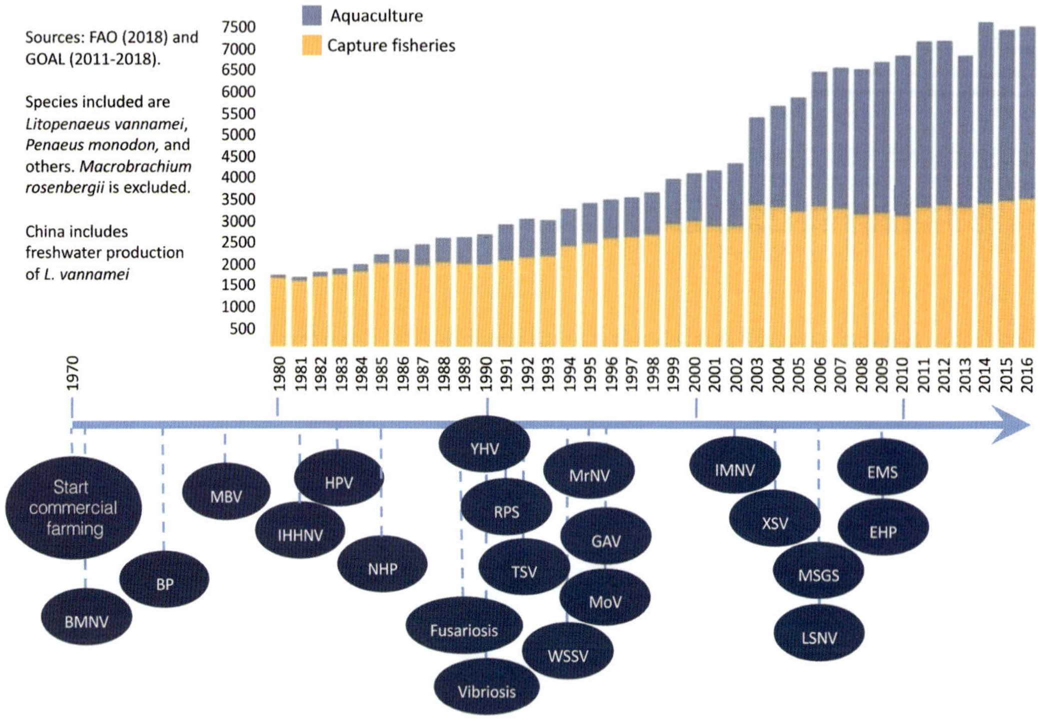

FIGURE 8.2 Overview of world production of shrimp in correspondence to episodes of new disease outbreaks. BMNV: *Baculovirus* mid-gut gland necrosis virus, BP: *Baculovirus penaei*, EHP: *Enterocytozoon Hepatopenaei*, EMS: Early mortality syndrome, GAV: Gill-associated virus, HPV: Hepatopancreatic parvovirus, IHHNV: Infectious hypodermal and haematopoietic necrosis virus, IMNV: Infectious myonecrosis Virus, LSNV: Laem-Singh virus, MBV: monodon-type baculovirus, MoV: Mourilyan virus, MrNV: *Macrobrachium rosenbergii* nodavirus, MSGS: Monodon slow growth syndrome, NHP: Necrotising hepatopancreatitis, RPS: Rhabdovirus of panaeid shrimp, TSV: Taura syndrome virus, WSSV: white spot syndrome virus, XSV: Extra small virus, YHV: Yellow head virus. The figure is redrawn from illustrations by Mayes (2018), Walker and Mohan (2009), and Dhar (2019).

THE NEED FOR A HOLISTIC UNDERSTANDING OF PATHOGENIC DISEASE IN AQUACULTURE SYSTEMS

There are several types of microorganisms that can cause devastating disease during shrimp culture, at both the larval and ongrowing stages. Such pathogens include viruses, fungi, parasites, and bacteria (Karunasagar et al., 2014). The history of shrimp farming contains a high number of disease episodes, as illustrated in Fig. 8.2. At each episode, attempts were made to identify the causative agent as soon as possible as the starting point for the development of specific preventive or therapeutic strategies. Such strategies then preferentially consisted of preventive approaches to avoid the occurrence of the disease. Examples of such strategies are to include the pathogen in the list to make up specific pathogen-free (SPF) shrimp, the development of selective bacteriophages (Choudhury et al., 2017), or more recent technologies under development such as treating the shrimp with highly specific interfering RNA (RNAi). Curative approaches against viral and parasitic diseases are almost non-existent (Seibert and Pinto, 2012), while the occurrence of bacterial or fungal diseases has most often been treated by the use of chemicals, with antibiotics being the most notorious example. The latter is an approach that should be abandoned as soon as possible, as is becoming more and more accepted within the industry and academia (Santos and Ramos, 2018).

In cases where it is not possible to find a specific (preventive or curative) method to control a disease, the best approach is a holistic one. This means an attempt to maximize the strength of the cultured shrimp by providing high-quality diets, immunostimulation, and genetic selection, while at the same time trying to minimize stress and disease pressure (Fig. 8.3).

For years, obligate pathogens have been tormenting the industry at a high frequency and severity. Therefore, these have drawn full attention in the development of disease control strategies. This focus on obligate pathogens is illustrated in the reviews that have been published on pathogenic disease in shrimp

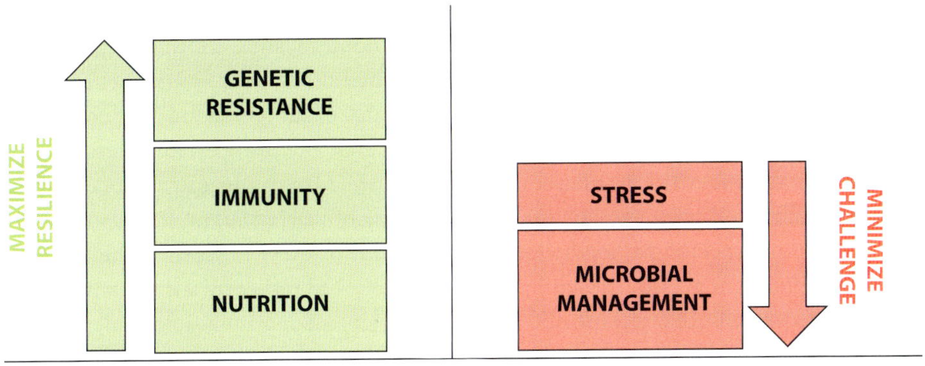

FIGURE 8.3 A holistic approach to minimize the impact of pathogenic diseases in shrimp farming. The concept consists of supporting the animals' resistance to pathogenic interference by providing optimal nutrition, stimulation of the immune system, and rendering them intrinsically more resistant. At the same time, it is important to minimize the level of stress of the animals and the level of pathogenic pressure they experience within the system. As the current approach of using chemicals to combat pathogens has been shown to be unsustainable and often ineffective, the latter is where new approaches for microbial management should come in. This requires a thorough understanding of the dynamics and functionality of the microbiota that develop in shrimp aquaculture systems.

culture (e.g. Thitamadee et al., 2016). The finding of practical solutions in essence implied destructive elimination of the obligate pathogen from the system. This eradication approach is still absolutely essential in modern shrimp culture practices when applied sustainably, for example, in the form of disinfection to clean the environment and the water before shrimp are introduced. Indeed, biosecurity lowers the risk of disease outbreaks caused by a large number of obligate pathogens. Unfortunately, it does not provide full protection against all types of pathogens, even when adequately applied. In the case of some pathogens, mainly the opportunistic ones, the application of biosecurity measures alone makes it even more challenging to minimize disease risk. This is related to their ecology being different from that of obligate pathogens. In the section 'Microbial stability as central concept in sustainable microbial management' later in this chapter, we explain the reasons for this in more detail. An example is the acute hepatopancreatic necrosis disease (AHPND) caused by bacterial agents of the genus *Vibrio* (De Schryver et al., 2014). Despite applying adequate biosecurity measures, farmers are still confronted with this disease. Now, we know that AHPND is caused by a variety of vibrios (Xiao et al., 2017), which are bacteria that can survive independent of the host in the environment and are typical examples of opportunistic pathogens. So, despite also causing a disease with very specific clinical signs, the ecology of these pathogens is completely different from that of obligate pathogens, which need a host to survive. Therefore, additional considerations are needed for the control of opportunistic pathogens.

The causative agents of AHPND belong to the same group of bacteria as those responsible for the outbreaks of luminescent vibriosis in the early 1990s. This disease was caused by *Vibrio harveyi* and closely related species, all belonging to the Harveyi clade of vibrios (Defoirdt et al., 2007). Similar to AHPND, luminescent vibriosis occurred during the first 10–45 days after stocking of shrimp postlarvae in the grow-out ponds. The outbreak of the disease was preceded by a substantial increase in the number of opportunistic vibrios in the pond water (Lavilla-Pitogo et al., 1998). This increase followed pond disinfection (Bratvold et al., 1999; Lavilla-Pitogo et al., 1998), but at the time no consequential link was made.

In the same category of pathogens can also be included the ones most likely to be responsible for negatively affecting the earliest life stages of fish and shrimp. Vadstein et al. (2018b) described that for marine fish – depending on the species and the rearing technology for the given species – typically the survival is 3–10% at metamorphosis at an early stage of domestication. However, there is no reason why most of the eggs (>80%) produced from a well-nourished broodstock would die, and therefore it was claimed that survival of around 50% or lower is an indicator of suboptimal rearing conditions. From the variety of studies performed by Vadstein and his colleagues on the youngest life stages of marine fish, it became clear that detrimental fish–microbe interactions are a key factor causing poor performance and low reproducibility and predictability during first feeding of marine larvae, and also that it is rarely possible to identify the specific pathogens that are the cause of these problems. The main reason is that the observed problems are not caused by true pathogens, but rather by a dysbiosis of the normal microbiota. This may consist of opportunistic bacteria (over)colonizing a physiologically immature and stressed animal, directly resulting in disease, or acting through a secondary infection. The same group of researchers found that especially when non-selective eradication methods are applied without any control of subsequent recolonization there seems to be a selection for opportunistic bacterial species. A practical example of the impact of such seemingly random microbial interference leading to unpredictability in survival during marine fish culture is given in Fig. 8.4.

For the earliest life stages of shrimp, similar claims of seemingly random loss events associated with microbial dysbiosis have been made (Zheng et al., 2017). A widely known example is the zoea-2

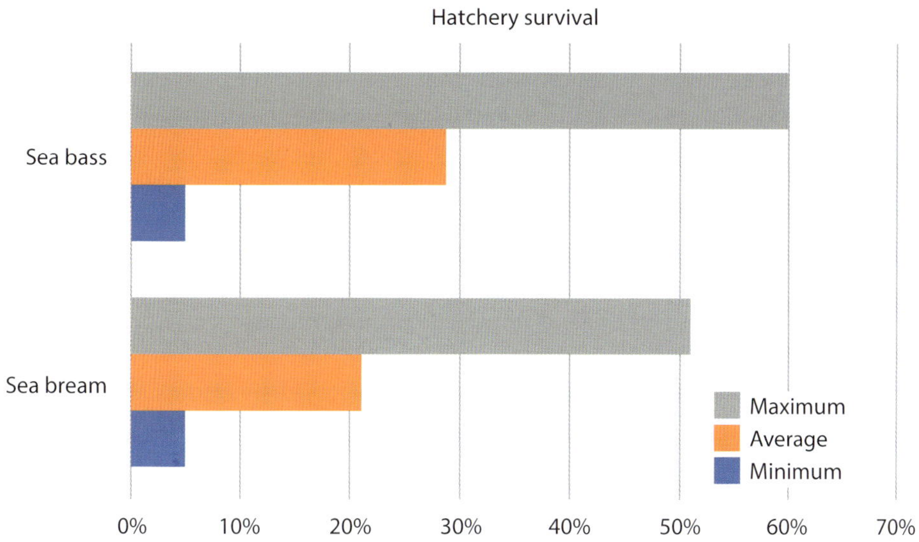

FIGURE 8.4 Data from 2019 illustrating the variation in larval performance that is observed in the hatchery phase of marine fish culture, in this case hatchery survival of sea bass and sea bream in the Mediterranean area. Scientific information indicates that random negative microbial interference by opportunistic pathogens is the likely cause for a large part of this variation.

syndrome, in which larvae around the zoea 2 stage mysteriously start to show heavy mortality without any clear reason. To date no specific infectious agent has been identified, but it has been suggested that negative microbial interference, and most likely opportunistic pathogens of the *Vibrio* type, may be the cause (Kumar et al., 2017). A similar situation exists for the bolitas syndrome observed in shrimp larvae in Latin America, which manifests as detachment of the epithelial cells in the intestine and hepatopancreas followed by the formation of small spheres within the intestine. A bacterial cause has been demonstrated for this syndrome, and the research of Vandenberghe et al. (1999) suggests a role of different *Vibrio* species. In addition, the problem of luminescent vibriosis in shrimp hatcheries still exists today (Sotomayor et al., 2019). Despite it being known to be caused by opportunistic pathogens of the *Vibrio* genus it remains a problem with low predictability and high impact.

THE CURRENT TUNNEL VIEW IN MICROBIAL MANAGEMENT FOR AQUACULTURE

Despite the fact that farmers are aware of the presence of detrimental opportunistic microorganisms in their culture systems, we are still having difficulties with controlling and combatting these microbes. For a long time, the biased focus on physical and chemical parameters (temperature, pH, dissolved oxygen, ammonia, nitrite, nitrate, alkalinity, etc.) created the feeling that all possible variables during culture were under control. Practices such as the clearly separated use of tanks or ponds, the application of water and tank/pond disinfection, the use of equipment that cannot be commonly used between different culture units, and the disinfection of workers' hands and boots were then introduced when it was realized that we were not controlling microbial interference. Although these 'microbial control methods' are very important to maintain and apply, they are only partly effective, as evidenced by the fact that infectious disease caused by opportunistic pathogens still is a big problem. This is because still today there is inadequate awareness of the importance of the microbial composition, dynamics, and functionality during the culture of shrimp. Consequently, we are still today culturing shrimp in bacterial 'black boxes'. It is therefore safe to put forward as a hypothesis that the unpredictable performance is due to uncontrolled microbial community composition and activity in the culture systems.

From the expert panel discussions on 'Improving biosecurity: a necessity for aquaculture sustainability' during the Global Conference on Aquaculture 2010 (FAO/NACA, 2012) a few statements can be cited that reflect the status of knowledge on the bacterial microbiota and its management in shrimp culture practices:

> By comparison to viral pathogens, work on control of bacterial pathogens of shrimp has been less intensive and has focused mainly on farm management practices related to control of the environment in hatchery tanks and rearing ponds. Much of this has been focused on the use of probiotics and immunostimulants.
>
> Turning to the application of new technologies such as probiotics, immunostimulants, and vaccines, there has been little change in the situation since 2005. Despite the widespread use of probiotics and to a lesser extent immunostimulants in shrimp farming, there have been no published results from large-scale field trials to prove by statistical analysis that they are really effective. Field trials and more research are also needed on quorum sensing control of bacterial pathogens.

Despite the lack of in-depth knowledge on microbial composition and dynamics, farmers have been applying the available microbial intervention tools in an attempt to gain control over the interfering microbiota during shrimp culture. As is illustrated by the many reviews on this topic (e.g. Dash et al., 2017; Hoseinifar et al., 2018; Jamal et al., 2019; Ninawe and Selvin, 2009; Ramachandran, 2017; Soltani et al., 2019; Van Hai and Fotedar, 2010), there currently is a tunnel view that considers disinfection and probiotics to be *the* microbial management tools for the industry. Probiotics were introduced into aquaculture practices by Gatesoupe (1991) and since then have gained a lot of popularity. The beneficial effects that have been attributed to the use of probiotics are (1) the control of pathogens through production of antimicrobial compounds and competition for nutrients and adhesion sites, (2) the promotion of animal health by induction of immune responses and contribution to digestion and supplementation of macro- and micronutrients, and (3) improvement of the quality of the environment (Balcázar et al., 2006; Gatesoupe, 1999; Irianto and Austin, 2002; Verschuere et al., 2000). There are numerous papers on the application of probiotics in shrimp farming (e.g. Dash et al., 2017; Hoseinifar et al., 2018; Jamal et al., 2019; Ninawe and Selvin, 2009; Ramachandran, 2017; Soltani et al., 2019; Van Hai and Fotedar, 2010). It is worth mentioning that almost all studies report on how the application of probiotics can contribute to an overall improvement in the performance of the animals in terms of growth and/or increased survival, whether or not they are under conditions of a pathogenic challenge. Moreover, from the field, there is clear feedback that the application of probiotics can result in clear benefits in terms of environmental culture conditions or shrimp performance. All of this illustrates that the application of probiotics may indeed contribute to lowering the risk of pathogenic diseases. Nonetheless, it would be wrong to state that the application of probiotics is *the* answer to the question of how to control pathogenic disease in shrimp farming (whether caused by specific pathogens or opportunistic ones). There are equally examples – both from academia and from industry – in which the application of probiotics did not prevent disease. So, while there is intrinsic potential in their use, there is still a lot of progress to be made. The generation of knowledge on how probiotics actually work *in vivo* and how they interact with the microbiota in aquaculture systems is essential to have realistic expectations of their activity, as well as understanding that microbial management requires more than only the application of probiotics.

The majority of the studies that have been conducted over the years on the use of probiotics in shrimp culture report on the zootechnical performance of probiotic-treated shrimp (e.g. growth, survival, feed conversion ratio, stress tolerance). Some studies also include analyses such as immune variables (e.g. total haemocyte count, respiratory burst, phagocytic activity, gene expression), activity of enzymes (e.g. superoxide dismutase, catalase, phenoloxidase), or bacterial counts (e.g. total heterotrophic count, presumed *Vibrio* count). However, almost all studies have failed to show the actual *in vivo* mode of action (or, in other words, causality) of the probiotics, and thus how and why they worked, or did not work, in that specific case. Therefore, despite being introduced into the industry about 30 years ago, probiotics are still being applied mainly in a generic way. The practical experience, however, shows that this is not enough to take on the pathogenic disease challenges the industry faces nowadays. Grover et al. (2012) stated that even for application in humans, which is a much more scientifically advanced field than aquaculture, proving specific health claims associated with probiotics is still required. This will be possible only when we can start from a better understanding of the composition, dynamics, and functionality of the bacterial microorganisms in the system under consideration. This basic knowledge on both the 'good' and the 'bad' bacteria would show what are the modifications we need to introduce (e.g. by the use of probiotics) in order to improve the activity of the microbes and maximize their contribution to the system's performance. Taking niche differentiation throughout the developmental stages of shrimp into consideration, the use of probiotics can become more successful (Xiong et al., 2016).

Innovative research platforms generate insights into microbial community composition and functionality

GNOTOBIOTIC RESEARCH MODELS

Another conclusion from the Global Conference on Aquaculture 2010 (FAO/NACA, 2012) was that

> Work on molecular ecology (i.e. metagenomics) and biochemical engineering to control the microbial dynamics in shrimp ponds and hatchery tanks has been relatively neglected.

For a long period, we relied on culture-based methodologies to provide insight into the presence of interfering microorganisms in shrimp culture systems (e.g. Zheng et al., 2016). Due to the culture bias, such studies reveal only part of the information required to assess how microbes interfere with shrimp performance. Fortunately, over the past 20 years the methodologies available have advanced substantially, which have to date mainly been applied to increase the possibilities for pathogen detection (Fig. 8.5). But, as microbial management requires more than pathogen detection alone, more and more research platforms are being applied to determine the functionality of the bacterial microorganisms – both pathogens and commensals – that live in and around the animals that we culture.

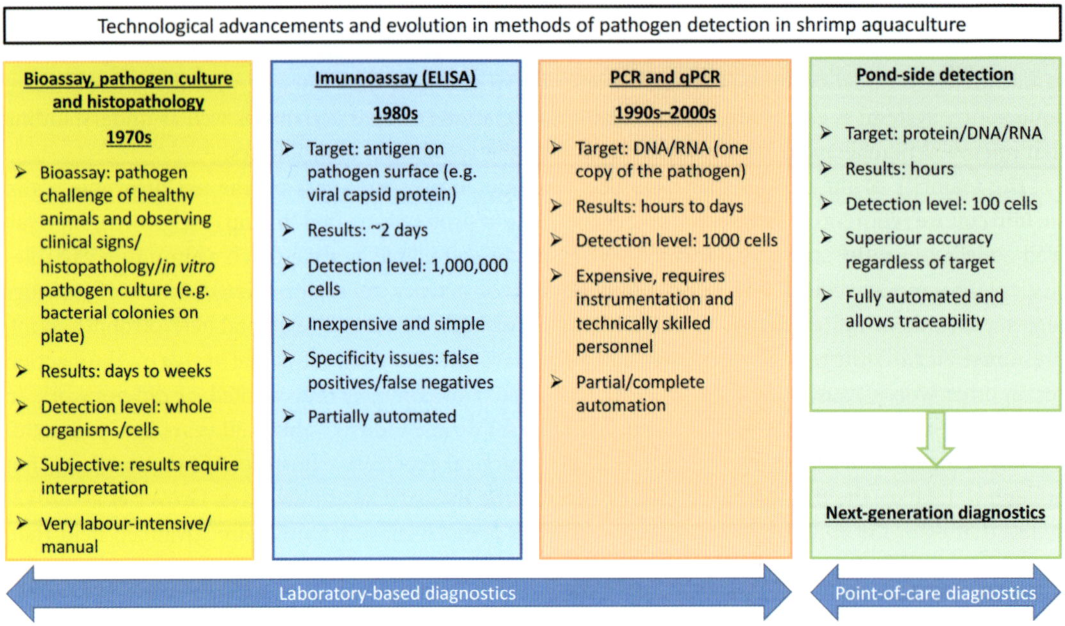

FIGURE 8.5 The evolution in technological advancements for shrimp pathogen detection as presented by Prof. Arun Dhar at the Asian-Pacific Aquaculture 2019 conference in Chennai (India). The detection of shrimp pathogens during culture is today considered as an important variable to evaluate disease risk. A question to ask, however, is whether detection of a single pathogen is the most trustworthy predictor for a disease outbreak and to use it as a decision tool to take action accordingly. As the study of Xiong et al. (2017) indicated, changes in the shrimp gut microbial signature occur earlier than the emergence of apparent disease signs, and may therefore be used as a next-generation early warning of disease during shrimp culture.

The use of, for example, gnotobiotic animal models as a research model is not new (Marques et al., 2005), but they have been increasingly combined with innovative approaches such as genomics, transcriptomics, epigenomics, or genetic manipulation of the host or bacteria to generate new and better insights into host–microbe interactions. This is illustrated in the advances that have been made with regard to how quorum sensing regulates virulence in opportunistic pathogens and what opportunities this presents for their control (see Box 8.1).

BOX 8.1. Quorum sensing, bacterial cell-to-cell communication, as a target for the development of novel therapeutics

Infection by bacterial pathogens is caused by the production of different virulence factors, that is, molecules produced by pathogens that enable them to colonize or harm the host (Dickey et al., 2017). Quorum sensing systems are mechanisms used by major shrimp pathogens such as *Vibrio* sp. belonging to the Harveyi clade (i.e. *V. harveyi* and closely related species such as *V. campbellii* and *V. parahaemolyticus*) to coordinate the production of such virulence factors (Defoirdt, 2018). Virulence factor genes are switched on by pathogens when their population density reaches an adequate level to cause infection in a host. To 'measure' population density, these bacteria use three different signal molecules, called Harveyi Autoinducer 1 (HAI-1), Autoinducer 2 (AI-2), and Cholerae Autoinducer 1 (CAI-1) (Fig. 8.6(A–C)) to activate phenotypes such as bioluminescence, the production of metalloprotease, siderophore, and chitinase, and flagellar motility (Defoirdt et al., 2010; Henke and Bassler, 2004; Mok et al., 2003; Natrah et al., 2011; Yang and Defoirdt, 2015). The three-channel quorum sensing system is also required for full virulence towards different aquatic organisms, including shrimp (Defoirdt and Sorgeloos, 2012; Noor et al., 2019; Pande et al., 2013b). More recently, other signal molecules have been added to the quorum sensing arsenal of vibrios, of which indole has been the most intensively studied. Indole signalling controls the virulence of *V. campbellii* towards brine shrimp and giant river prawn larvae (Yang et al., 2017).

One option for controlling pathogenicity regulated by quorum sensing is to identify agents that are able to interfere with signalling mechanisms in pathogenic bacteria. Halogenated furanones constitute one of the most intensively studied classes of quorum sensing inhibitory compounds, including both natural compounds produced by marine algae and synthetic derivatives (Fig. 8.6(D, E)) (Janssens et al., 2008), and have shown potential for protecting shrimp against vibriosis (Defoirdt et al., 2006; Pande et al., 2013a). Unfortunately, these compounds are too toxic to higher organisms to be applied in practice, with toxic concentrations being only slightly higher than quorum sensing-disrupting concentrations. Brominated thiophenones (Fig. 8.6(F)) have shown an interesting therapeutic potential to treat vibriosis in shrimp, with a therapeutic index of approximately 100 (Defoirdt et al., 2012). Another compound with quorum sensing-disrupting activity is the non-toxic flavour compound cinnamaldehyde (Fig. 8.6G). This compound has been shown to be effective in different aquatic host–microbe systems (Natrah et al., 2012; Pande et al., 2013a). A major advantage of this compound with respect to practical applications is that it is Generally Recognized As Safe (GRAS). Besides specific compounds, many marine organisms (microalgae, invertebrates, fungi, and bacteria) have been reported to produce quorum sensing inhibitors (Saurav et al., 2017; Torres et al., 2019). In most cases, the chemical structure of the inhibitors has not been elucidated. Some exceptions are malyngolide produced by the cyanobacterium *Lyngbya majuscula* (Dobretsov et al., 2010), phenetylamide metabolites produced by *Halobacillus salinus* (Teasdale et al., 2009), and 2,6-di-tert-butyl-4-methylphenol produced by the cyanobacterium *Chroococcus turgidus* (Santhakumari et al., 2018).

In addition to quorum sensing inhibitors, a second quorum sensing-disrupting mechanism is possible: enzymatic inactivation and biodegradation of quorum sensing molecules. The ability to degrade acylhomoserine lactones (AHLs) is widely distributed in the bacterial kingdom (Dong et al., 2007). The inactivation of the signal compound can be mediated by two major types of enzymes: AHL lactonases and AHL acylases. Genes encoding AHL-degrading lactonases are widespread in *Bacillus* sp. (Dong et al., 2002). AHL-degrading enzymes have been described in many marine bacteria, with acylases being more prevalent than lactonases (Torres et al., 2019), and AHL-degrading bacteria have been isolated from microalgal cultures and from the digestive tract of healthy shrimp (Defoirdt et al., 2011b; Pande et al., 2015).

Besides approaches that interfere in the activity of the quorum sensing system itself, an alternative option is to prevent the pathogen population density reaching the level at which virulence can be activated by means of quorum sensing. This is the basic idea of the approach described in the section 'Microbial stability as central concept in sustainable microbial management' later in this chapter.

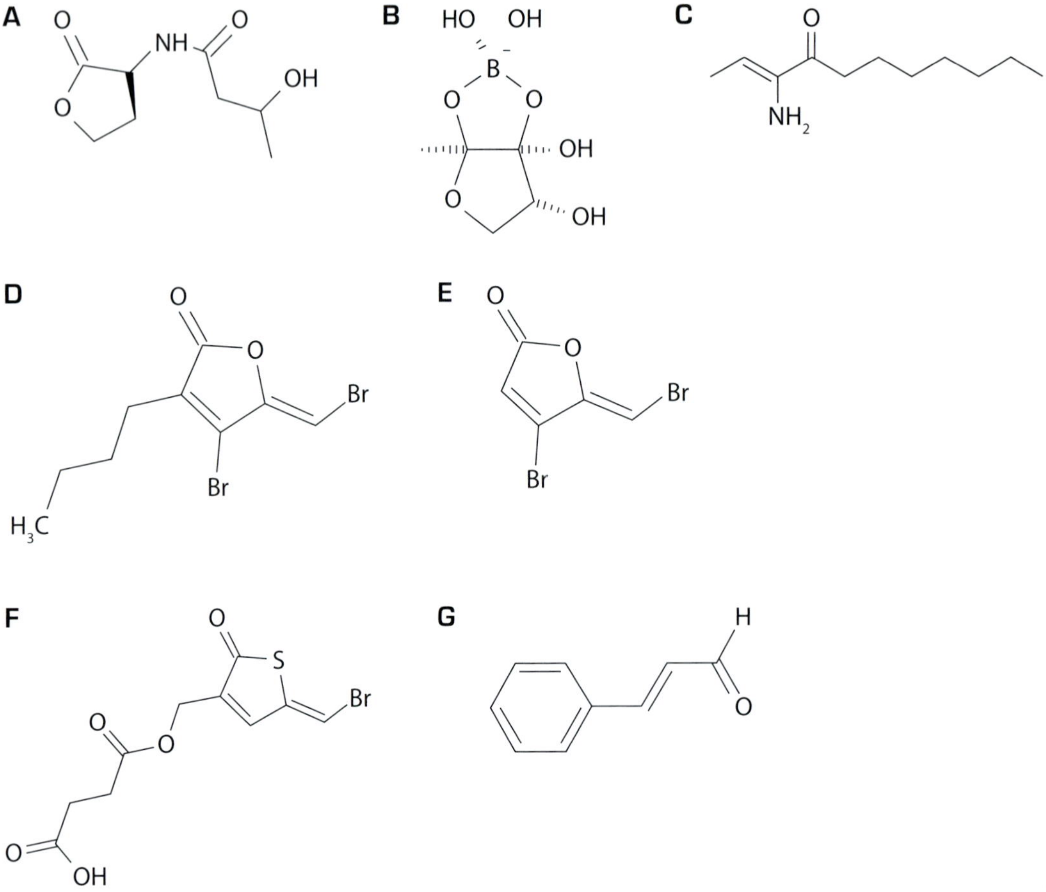

FIGURE 8.6 Chemical structure of quorum sensing signal molecules produced by *Harveyi* clade vibrios and of quorum sensing inhibitors. (A) Harveyi Autoinducer-1 (HAI-1), N-(3-hydroxybutanoyl)-L-homoserine lactone. (B) Autoinducer-2 (AI-2), the furanosyl borate diester 3A-methyl-5,6-dihydro-furo [2,3-D][1,3,2] dioxaborole-2,2,6,6 A tetraol. (C) Cholerae Autoinducer-1 (CAI-1), (S)-3-hydroxytridecan-4-one. (D) The natural furanone (5Z)-4-bromo-5-(bromomethylene)-3-butyl-2(5H)-furanone, produced by the red marine alga *Delisea pulchra*. (E) The synthetic derivative (5Z)-4-bromo-5-(bromomethylene)-2(5H)-furanone. (F) The brominated thiophenone (Z)-4-((5-(bromomethylene)-2-oxo-2,5-dihydrothiophen-3-yl)methoxy)-4-oxo-butanoic acid. (G) Cinnamaldehyde.

THE CONTRIBUTION OF OMICS TECHNOLOGIES TO EXPLAIN MICROBIOME FUNCTIONALITY

The increased application of metagenomics under field conditions shows partly promising results, and an increasing number of studies have focused on the characterization and functionality of shrimp microbiomes, as illustrated in Cornejo-Granados et al. (2018) and Li et al. (2018). Such studies are important as they generate information on what controls the development of the bacterial microbiota during the different phases of a shrimp culture cycle, but even more because they provide insight into how the overall bacterial community relates to the health status and performance of the shrimp. For example, Fan et al. (2019) tried to relate gut microbiota dynamics to shrimp growth performance; Cornejo-Granados et al. (2017) and Chen et al. (2017) tried to relate the gut microbiota composition to the outbreak of APHND; Zhang et al. (2019) tried to relate the microbiota of water and sediment to environmental quality; Yao et al. (2018) and Zhu et al. (2016) tried to relate gut microbiota composition to disease outbreaks; and Landsman et al. (2019) investigated the effects of shrimp culture practices on the intestinal gut bacterial community. It is worth noting that all these studies were published in a period of 5 years before the writing of this chapter. They contribute to the pool of knowledge on the microbiome during shrimp culture, but unfortunately generate little concrete information for practical use, because it is not known if their conclusions are system dependent or universally valid. In the context of disease, there is thus a need for coordinated research initiatives on a global level to find answers to the following initial questions.

- What does a 'normal' bacterial community composition in the intestinal tract or culture environment of healthy shrimp at different life stages look like?
- Are there patterns that are universally conserved, or should such patterns be determined for each location/species/culture cycle separately?
- What is the relationship between the intestinal bacterial community and the bacterial community in the culture environment of shrimp at different life stages?

To determine whether microbial management should target the establishment of a microbiota composed of specific bacterial phylotypes, we need to find out whether a specific baseline microbiome exists for healthy shrimp. Up to now, most studies have described only the dominating bacterial groups associated with shrimp at selected stages of development. Some of these studies were done under laboratory conditions (e.g. Huang et al., 2014), but most were field studies (e.g. Hou et al., 2016; Huang et al., 2014; Zoqratt et al., 2018). Although the number of descriptive studies is increasing, it is safe to state that the number is still far too limited to make conclusions with regard to universally conserved patterns. In particular, because the set-up of the studies (including factors such as sample location, sample choice, sample preservation, and analysis protocols) is very diverse, there is substantial bias when meta-analyses are attempted (Cornejo-Granados et al., 2018; Yu et al., 2018).

Besides the collection of compositional data on the intestinal and environmental microbiome of shrimp at different life stages, the relationship between the two is of equal importance. For the youngest life stages of cultured animals, it has been stated that the initial bacterial colonization could have long-lasting consequences (De Schryver and Vadstein, 2014), although the exact impact remains unknown. In addition, it currently remains unclear how initial colonization occurs – for example, what is the relative contribution of water microbiota versus feed microbiota, and how important is selection versus stochastic processes. Finding out how strong the compositional link is between the environmental and intestinal microbiota is important to determine whether initial colonization can be controlled and whether targeting a specific microbiota in the early developmental stages of shrimp is realistic. Furthermore, if a core microbiota exists throughout the life of shrimp, it would indicate that manipulation of the gut microbiota at the earliest life stages would have an impact on the microbiota later in life. Determining the existence of a core microbiota would also provide information on whether manipulation at later developmental stages would still be feasible and on the degree of invasion resistance by an established microbiota. The latter two aspects would be extremely relevant for the concept of applying probiotics as a microbial management strategy.

A limited number of studies have focused on answering these questions. Xiong et al. (2017) described that for shrimp larvae there was a similar importance of deterministic (i.e. niche/selection-based) and stochastic (i.e. random) processes in controlling the gut microbiota. This indicates that there may be opportunities to establish a 'wanted' microbial community in the gut of shrimp larvae. Xiong et al. (2015) analysed the bacterial communities in water and gut samples from shrimp ponds. They found that the abundance of the dominant classes differed substantially between the two sample types. This confirmed that deterministic selective forces are important for the composition of the microbiota in the gut of shrimp, and that the gut microbiota does not simply reflect the microbiota in the water. Hou et al. (2018) came to a similar conclusion. They observed that, despite sharing a substantial number of operational taxonomic units, the most abundant taxa were significantly different between the gut and environmental bacterial communities. This indicated that environmental factors in the intestine shape its bacterial community. In addition, Xiong et al. (2019b) found that the similarities in the gut microbiota between shrimp reared in distant locations was higher than the similarity between shrimp and their respective rearing water. All this information indicates that targeted management of the gut microbiota through manipulating the composition in the water could be possible but is not straightforward. A possible strategy could be to increase the number of bacteria in the water in a composition that is more similar to the microbiota of healthy shrimp. As shrimp are more likely to select bacteria from higher-abundance pools, one has a higher chance to steer the microbiota towards a healthy one as the probability for the establishment of a healthy microbiota is increased. For fish larvae it has been shown that one can control viability, for example, a 70% increase in larvae survival, through manipulation of the water microbiota (Vadstein et al., 2018b). It should be mentioned that the cited studies on shrimp did not consider the moments of initial colonization (i.e. at mouth opening) in their analysis. This moment of microbial colonization of the gut environment in shrimp requires some special attention in further research, as it is more likely to control this initial colonization than to affect an already established microbiota (De Schryver and Vadstein, 2014). Further, it is known that the intestinal bacterial composition changes substantially from one life stage to another, probably because the gut environment is under continuous development. Due to a lower selective pressure in the intestine compared with adult stages, the larval stages of shrimp are likely to be the best period for managing the gut microbiota (Xiong et al., 2017, 2019a).

When evaluating the possibilities for manipulating the gut bacterial community at later developmental stages, Xiong et al. (2017) found that from postlarvae (at 28 days) until adults (at 77 days) the assembly was mainly determined by deterministic processes. This indicates that management of the gut bacterial community at these stages should be based on knowledge on available niches in the gut environment (e.g. when using probiotics). The results from the study of Miranda-Baeza et al. (2020) further suggest that bacterial conglomerates in water during mature shrimp stages contain well-established bacterial communities that are difficult to alter in a directed way by the addition of probiotics. The results of these studies do not mean, however, that probiotics cannot have any positive effect. Even transient microbes that do not become members of established conglomerates in the gut are active and as such influence the host (e.g. by the production of metabolites). Especially in shrimp, due to defaecation and coprophagic behaviour, there is continuous dispersal of microbes between individual animals and the environment. This suggests that actual establishment of the provided probiotics in the gut microbiota of shrimp should not be considered a necessity to conclude that the application of probiotics is successful. For fish, it has even been shown that dispersal among hosts selects for bacterial taxa with life history strategies favouring dispersal and repeated colonization of multiple host types (Burns et al., 2017).

Even more important than the compositional status of the bacterial community in the shrimp intestine and the environment is the functionality of different bacterial communities for viability of the host. This raises the following questions.

- Can bacterial phylotypes be identified in the gut or the culture environment of shrimp to serve as indicators of a healthy status of shrimp at different life stages, or as indicators or 'forecasters' of a diseased status?
- Can the dynamics in the bacterial community composition in the intestinal tract or culture environment of shrimp at different life stages be used as an indicator of disease risk?

During his talk 'Diseases in shrimp aquaculture – possible remedial pathways' in the Society of Aquaculture Professionals shrimp session at the Asian-Pacific Aquaculture 2019 conference in Chennai (India), Prof. Arun Dhar emphasized the need for diagnostic methods that allow the detection of pathogens well before the point at which they actually cause a disease outbreak. We agree, but add that detection of pathogens alone is not enough. Prof. Jinbo Xiong and his colleagues (Xiong et al., 2016) stated that water quality indices and abundance of specific pathogens, which currently are the main indicators of shrimp disease, provide inaccurate predictions of shrimp disease. For example, the use of *Vibrio* sp. counts alone to predict disease outbreaks is not optimal, as vibrios are also frequently observed in and around healthy shrimp. There is thus a need for alternative approaches to predict disease at an early stage with high accuracy (Fig. 8.7). The use of a group of indicator bacterial types is an interesting concept, as it goes beyond mere detection of known causative agents to evaluate disease risks. The concept relies on the detection of a specific range of bacteria, not necessarily the causative agents of a specific disease, which most often will represent a minor fraction of the total bacterial community but nonetheless are indicative of the health status of the cultured shrimp. In other words, they have the potential to be early indicators of dysbiosis. Dai et al. (2018) referred to this as the microbial gatekeeper concept, and changes in the relative abundances of the gatekeeper bacteria would disproportionally endanger the whole community and affect its stability and functionality.

With regard to the concept of indicator bacteria, Zhang et al. (2014) found a group of six bacterial orders serving as an indicator for a healthy status and a group of five bacterial orders indicating diseased shrimp. In addition, Xiong et al. (2014b) found that disease indicator bacterial phylotypes could be identified in cultured shrimp. The changes in the relative abundances of 13 indicator phylotypes at family level (seven indicative of a healthy status and six indicative of a diseased status) could predict a diseased status with an accuracy of more than 79%. Further, the study of Xiong et al. (2015) identified a range of indicator taxa for a healthy or diseased status of shrimp. High abundances of Bacilli, Flavobacteriales, Acidimicrobiales, and Alteromonadales were indicative of healthy shrimp, whereas high abundances of Actinomycetales, Sphingobacteriales, and Vibrionales were indicative of diseased shrimp. These observations did not imply any conclusions regarding causality, and assumptions regarding actual functionality were made based only on known ecology of the indicator phylotypes. In further studies by Xiong et al. (2017, 2019b) it became clear that the age of the cultured shrimp should be taken into consideration when applying the concept of relative abundances of disease-discriminatory taxa to predict disease. Indeed, bacterial indicator phylotypes for shrimp age were also identified, and when taking these age-discriminatory bacterial taxa into account the accuracy of predicting a diseased status of the cultured shrimp could be increased to 91.5–95.5%.

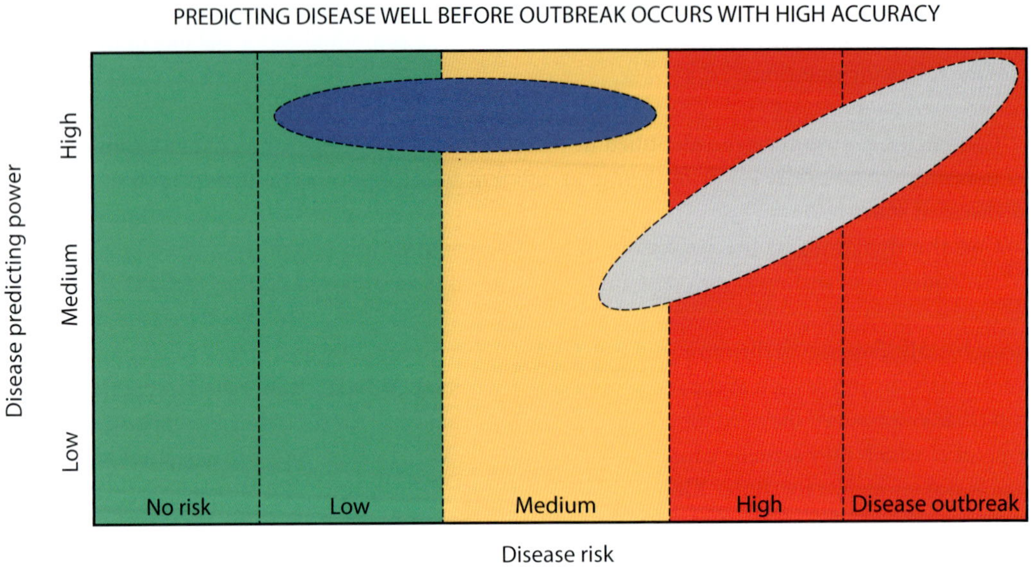

FIGURE 8.7 Disease diagnostic methods are applied in cases when shrimp farmers suspect disease. In practice, these often provide an effective diagnosis when it is too late to remedy the situation (grey zone) (Loomis, 2018). This applies even for the pond-side diagnostic tools that are currently finding their way to the market. The goal of future disease-prediction tools should be to accurately predict the risk of disease at a very early stage, giving the farmer ample time to implement a remedy (blue zone).

Besides using a specific group of bacteria as an indicator of health or disease, the overall temporal changes in the bacterial community can be a more accurate predictor of upcoming disease. Dysbiosis (i.e. microbial instability) is often related to disease caused by opportunistic pathogens, a situation often amplified under adverse environmental conditions leading to stress and higher disease susceptibility of the cultured animals (Fig. 8.8). As microorganisms in general are the first responders to perturbations, the detection of dysbiosis in shrimp is expected to predate the physical symptoms of shrimp disease (Xiong, 2018). The early detection of dysbiosis may thus be used as an alternative approach for disease prediction. It would represent a significant advance to the current way in which disease in cultured shrimp is diagnosed. For such an approach, there is a need to establish a baseline for the dynamics in intestinal microbial assemblages that are indicative of a normal healthy situation. Xiong et al. (2014a) found that variations in the bacterial communities in different ponds showed similar succession patterns, indicating that temporal dynamics in the bacterial community were predictable to some extent. Xiong et al. (2017, 2018, 2019b) further described that the change in the gut bacterial community composition progresses consistently during shrimp ageing in a predictable way. Interestingly, shrimp disease disturbs the expected patterns. A deviation from this baseline could thus be indicative, or even predictive, of upcoming shrimp disease. The work of Xiong et al. (2017) relates closely to this concept, as it builds upon the use of age-discriminatory bacterial taxa to determine whether temporal dynamics are driven by age or dysbiosis (Fig. 8.9). A final remark made by Xiong and colleagues is that it remains to be determined whether the concept of age- and disease-discriminatory taxa holds true over crops. Moreover, the validity of this promising concept should be evaluated for a range of different culture methods, conditions, and locations to check its universality.

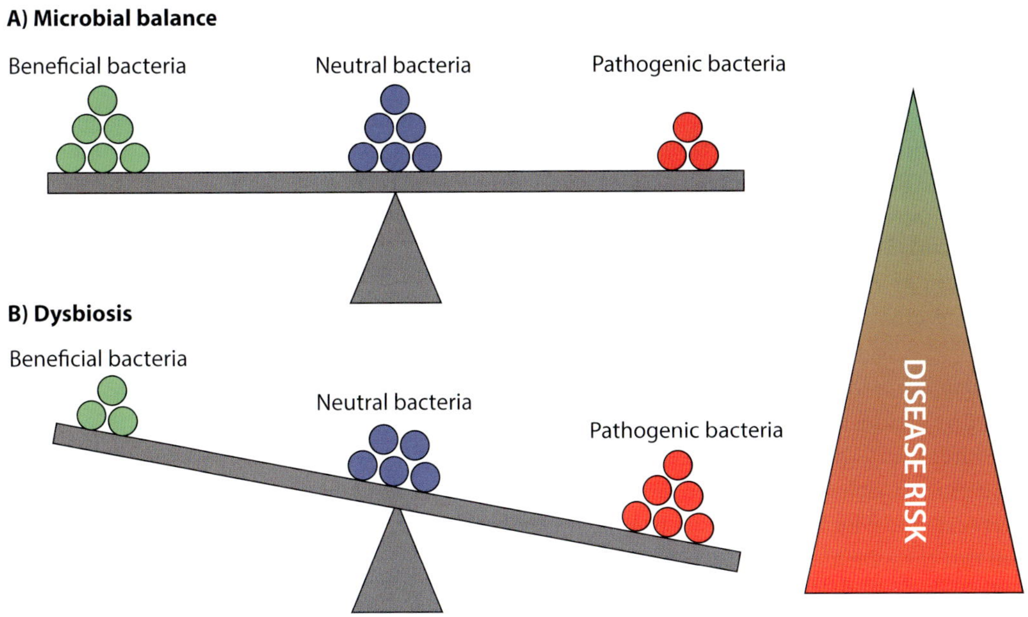

FIGURE 8.8 A conceptual model for the relationship between microbial community composition and shrimp disease risk, as presented by Xiong et al. (2016). In a situation of microbial balance, the numbers and the activities of beneficial and neutral bacteria keep the activity of the pathogenic bacteria under control (A). This results in low pathogenic disease risk for the cultured shrimp. In a situation of dysbiosis, the microbial balance is disrupted, creating possibilities for the pathogenic bacteria to increase in number or activity (B). The detection of dysbiosis at an early stage may be one of the earliest indicators of upcoming disease. Credit: Courtesy of Prof. Jinbo Xiong.

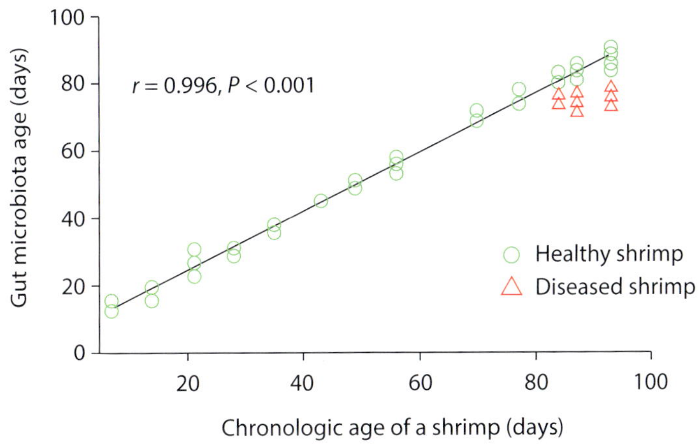

FIGURE 8.9 Xiong et al. (2017) found that the relative abundance of specific bacterial taxa in the gut microbiome of shrimp is predictive of shrimp age under microbial balanced conditions (green data points), and that deviations from normal conditions could be an indicator of bacterial dysbiosis and thus increased disease risk (red data points). Credit: Courtesy of Prof. Jinbo Xiong.

The above concepts refer to specific assemblages of bacteria, or the dynamics between them, being used as predictors of disease outbreaks. However, disease is context dependent, which means that it is influenced by host and environmental conditions. For example, the continuous feeding of shrimp during a culture cycle will have an impact on the physicochemical variables of the water. Deterioration of the water quality will impose stress on the shrimp, making them more susceptible to the negative interference of opportunistic pathogens. Consequently, both the development of potentially harmful bacterial populations and the susceptibility of the animals to these populations depend on the physicochemical variables in the water and the gut. The combination eventually leads to a diseased status of the shrimp. Conversely, a diseased status of the shrimp may affect the composition of the microbiota associated with them, making them more prone to be further negatively affected by bacteria. Bass et al. (2019) stated that there is a shift from the 'one pathogen–one disease' paradigm towards the 'pathobiome' concept, integrating the interaction of multiple symbionts, host, and environment in a new understanding of disease aetiology. This means that besides the microbiome, other biological variables (the host) and physicochemical variables (the environment) should be included when considering disease risk. This results in a huge complexity of interrelating variables that are impossible to interpret without adequate tools. Xiong (2018) presented a conceptual model for the interaction between the microbiome, external factors, and shrimp disease. He stated that invasion and colonization by shrimp pathogens are antagonized by environmental pressure, host selection, and resistance to colonization imposed by commensals. In diseased shrimp, the balance in the protective commensals and host immunity is thus compromised by external disturbances. For example, deteriorating water and sediment qualities could stress shrimp and alter their gut microbiota, promoting opportunistic pathogen growth and/or inducing the expression of virulence factors. These changes then cumulatively open the ecological niches for potentially more invasion by pathogens.

The pathobiome concept is a very recent introduction in aquaculture production, and to our knowledge no studies have focused on disentangling the complex interactions that may occur between the host, the microbiome, and the environment in the context of disease development. Multi-omics approaches (including metagenomics, metatranscriptomics, and metaproteomics) will be needed to generate the necessary data for such an analysis, and machine learning technology can be used to discover the factors with predictive power. Based on the pathobiome concept, an altered status of the symbiome may provide a diagnosis at an early stage of disease, well before specific symptoms and signs of disease become apparent at the level of the host.

Better understanding of microbial community functionality in shrimp culture systems: requirement for a shrimp microbiome project

The studies mentioned in the previous section that deal with quantitatively predicting disease based on indicator taxa, dynamics in gut microbiota composition, or the development of a pathobiome were done with a limited variety of experimental conditions, a limited number of samples, or limited age and temporal variability. For example, Xiong et al. (2017) sampled from six ponds, split up into healthy and diseased ponds during a single crop, while Xiong et al. (2019b) used samples from 11 ponds at two locations during a single crop. All 18 studies that were included in the meta-analysis performed by Cornejo-Granados et al. (2018) together represented a total of 199 samples, whereas the five studies included in the meta-analysis of Yu et al. (2018) represented a total of 99 samples. These meta-analyses were performed to describe the shrimp microbiome and determine the host and environmental factors that shape it (Cornejo-Granadas et al., 2018), or to determine universal signatures that can be used to diagnose incidence of disease (Yu et al., 2018). Unfortunately, these efforts clearly suffered from different experimental and bioinformatics protocols resulting in a low number of studies that could be included in the meta-analyses. Consequently, it is difficult to reach general conclusions. For this reason, there is a need for a global research initiative on the shrimp microbiome, including both academia and industry. Aligned metagenomics and bioinformatics protocols should be applied in a series of experiments on a long-term basis and in a large variety of culture systems and conditions. Only then can the necessary volume of descriptive information on the microbiome in the gut and the culture environment of shrimp at different life stages be generated. Such information is essential to progress to the next generation of microbial management approaches, as summarized in Box 8.2.

BOX 8.2. Shrimp Microbiome Project

The goal of a 'shrimp microbiome project' would be to establish the baseline statuses in shrimp microbiome composition, and to develop a predictive model for microbial management serving as a tool to quantitatively predict the incidence and severity of disease outbreaks. Realistic evaluation variables based on temporal changes in the microbiota should then be identified, to predict the upcoming incidence and severity of disease. Such an approach may be independent from the detection of specific causal pathogens, although the latter will remain valuable for obligate pathogens that are known to cause disease upon detection. The quantitative presence of globally common bacterial indicator taxa for disease could also be used to evaluate disease risk and potentially predict disease severity.

Most of the research so far has focused on analysing the microbiota in the gut of cultured shrimp, as it is logically assumed that these microorganisms are of highest relevance for the health status of the animals. Xiong et al. (2014b), however, used the microbial signatures in the water to show that indicator bacterial assemblages in the water can also be used to predict the incidence of shrimp disease, and they obtained a prediction accuracy of more than 79%. The use of water samples to predict disease would be easier than gut samples, as they can easily be taken at any time point during culture, and do not require the dissection of sampled shrimp. Such an approach would substantially increase the ease of sampling and lower the risk of protocol-related biases. The predictive power of gut samples versus water samples should thus be compared, and this would require both sample types to be taken at the same time.

Another aspect of omics-based analyses to predict disease during shrimp culture is that the technology currently does not allow for on-site and in-line monitoring. After sampling, the samples need to be processed and analysed in specialized laboratories with specific sequencing equipment, followed by data processing using bioinformatics tools. This procedure still requires substantial time, which impacts the concept of using the bacterial communities as predictors of upcoming disease. This problem is emphasized by the fact that bacterial dynamics can change within a matter of minutes to hours. It was stated at the Global Conference on Aquaculture 2010 (FAO/NACA, 2012), during the expert panel discussions on 'Improving biosecurity: a necessity for aquaculture sustainability', that

> Rapid disease diagnosis is crucial to the sustainability of aquaculture, and rapid progress in biotechnology over the last decade has enabled the development and improvement of a wide range of immunodiagnostic and molecular techniques, and reagents and kits have become more widely available. In recent years, methods developed for clinical and veterinary medicine have been adapted and optimized for use in aquaculture. Despite this, identification of certain pathogens is difficult to achieve, and some of the methods developed are too complicated to implement and interpret. Traditional methods of pathogen isolation and characterization tend to be costly, labour intensive, slow and may not give a definitive diagnosis. For many rapid methods, live and dead pathogens cannot be distinguished; therefore, enrichment methods and the use of live/dead kits are useful supplementary methods. Interpretation of results using rapid methods should be considered with other clinical evidence.

The above indicates that besides the generation of knowledge regarding microbiome functionality in the context of disease, the development of new technologies that allow high-throughput analysis (within hours or even minutes) of microbial community status within shrimp culture systems would be extremely relevant. An example with potential is the flow cytometry approach, by which water samples can be screened within a matter of minutes. Although this approach has been mainly investigated for use in other domains, it seems that flow cytometry may present an alternative, but much faster, method than Illumina sequencing to generate information on the dynamics of the microbiome in aquaculture systems (Props et al., 2016, 2018).

Microbial stability as central concept in sustainable microbial management

Despite the advances that have been made with regard to the functionality of microbes in aquaculture systems, there is still a lot of knowledge missing that would be required to achieve optimal microbial management. At the same time, however, the industry keeps on developing and thus tries to find its own approaches to deal with the disease problems it is facing in the field. Often, improvements are made on a trial-and-error basis without a true understanding of how the modifications impact the microbial compartment within the culture system. From these ad hoc cases, a lot of valuable information can be deduced that may contribute to the understanding of how microbes in shrimp culture systems can be managed.

An example is the new concept of intensive zero water exchange shrimp farming integrated with shrimp toilets for increased waste removal. Recirculating aquaculture systems in hatcheries and nurseries with over-dimensioned biofilters yield improved results in terms of the survival, growth, and disease resistance of shrimp. The same goes for shrimp pond farming with reconfiguration of culture ponds into a recirculation modus, including extractive ponds pre-stocked with tilapia or with *Caulerpa* or *Gracilaria* sp. seaweed (Fig. 8.10). Biofloc technology systems can also be considered in the same category. It is said that the opportunistic pathogen *V. parahaemolyticus* – a known causative agent of AHPND – can still be detected in such systems but it does not easily reach critical densities for the expression of virulence. It can thus be hypothesized that such integrated systems provide a microbially stable environment, resulting in a lower probability of dysbiosis.

Although the adjustments that have been made to shrimp culture configurations over recent years were not inspired by in-depth knowledge of microbial ecology, they seem to have as a common feature

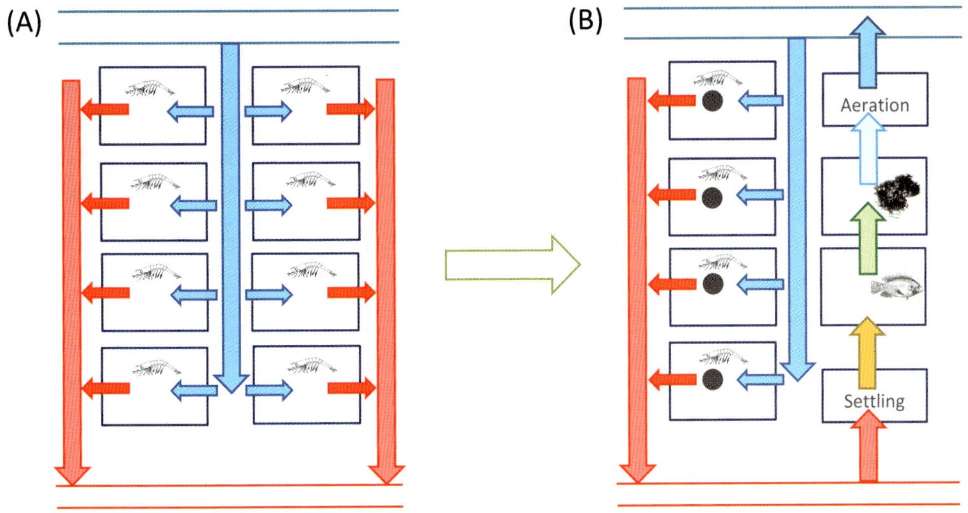

FIGURE 8.10 Over the years, farmers in Asia started to adjust the configuration of their shrimp ponds to cope with disease and minimize environmental impact. Initially, all ponds were used to culture shrimp and each pond was provided with water in a flow-through mode (A). Increasingly, only some of the ponds are used to culture shrimp, and these ponds are provided with a shrimp toilet to remove excess waste during culture. The remainder of the ponds are used as treatment ponds through which the water is recirculated before reuse or discharge (B). The result is also that at pond level, farming is done in a recirculation system. Redrawn after Taw (2017).

that they contribute to creating microbial stability in the system in accordance with ecological theory. The observations made in these systems therefore allow us to formulate a theory of microbial stability leading to lower disease risk and, combined with the scientific knowledge available, suggest the elements for best shrimp culture practices as described in the following sections. As no approach can guarantee that disease will not break out in the culture system, the target should be to create a situation that minimizes disease risk.

BIOSECURITY AS ESSENTIAL STARTING POINT OF ADEQUATE MICROBIAL MANAGEMENT

We are still working in microbial 'black boxes' when culturing shrimp. Except for the sporadic plating of (mostly) presumptive vibrios and total heterotrophic bacteria, we have no knowledge of which bacteria are present, and in what quantities. In fact, it is not required to know in detail which taxonomic groups (i.e. names) are present; rather, we need to know (and try to control) functional groups. In the context of disease, the bacteria present in the system can therefore broadly be categorized into three groups (Fig. 8.11), to which Dr Loc Tran referred as 'the good, the bad, and the ugly' during his talk at the Asian-Pacific Aquaculture 2018 conference (Taipei, Taiwan).

1. The bad would be obligate pathogens that need a host to survive and thus pose a high risk for disease when they are in the system.

2. The ugly are opportunistic pathogens that can survive independently of the host and are normal members of the microbial community in every aquaculture system. They become problematic depending on prevailing conditions, e.g. the quorum in which they are present, and are thus difficult to quantitatively correlate with disease risk. They are the causative agents of the typically dysbiosis-associated diseases.
3. The good are the bacteria with a neutral or beneficial activity. Most often, probiotic bacteria are placed into this category, but naturally prevailing bacteria also belong to this group.

Based on the above, it is easy to understand that a desired situation is one in which the presence of obligate pathogens is avoided at all times, while the levels of opportunistic pathogens during shrimp culture are kept as low as possible to minimize the risk for disease breaking out. As a first line of biosecurity, it is therefore essential to start with disinfection of intake water and the culture environment (pond and tank surfaces) to eliminate as far as possible obligate pathogens and opportunistic pathogens from the system in which the shrimp will be stocked. In combination with other strict biosecurity measures, such as using SPF shrimp, using products and feeds that are certified to be free of pathogens, using live feeds with specific pathogen control technology (e.g. anti-*Vibrio* sp. *Artemia* cysts), regular disinfection of equipment used in the culture systems, and disinfection of workers' hands and boots, this is the only approach to minimize the chance of the introduction and proliferation of (in particular obligate) pathogens.

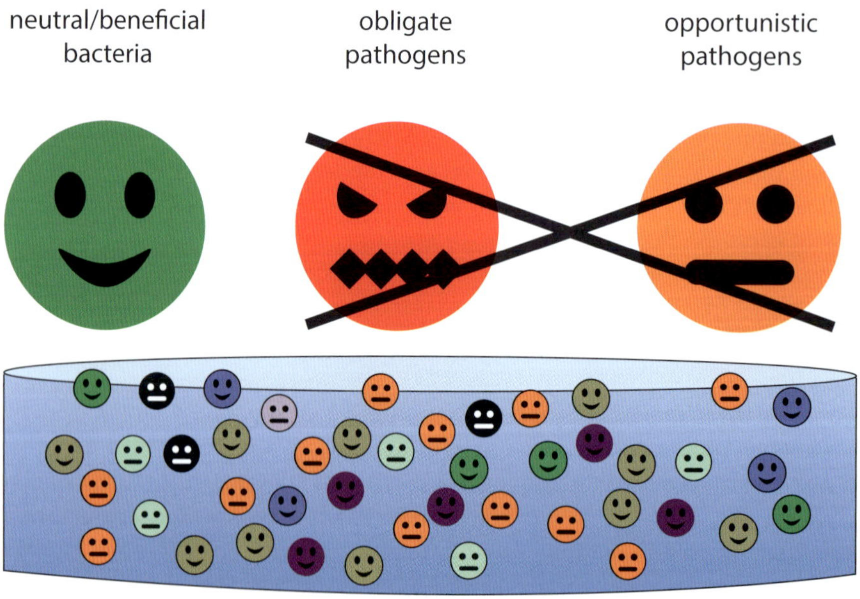

FIGURE 8.11 The bacteria in aquaculture systems consist of a randomly mixed microbial community, which – in the context of disease – can be categorized into three groups: neutral/beneficial bacteria, obligate pathogens, and opportunistic pathogens. Without the need to know actual taxonomic classifications, the final goal of microbial management consists of decreasing the presence of the pathogenic groups while increasing the presence of the beneficial group in order to minimize disease risk.

R/K SELECTION THEORY TO MINIMIZE DYSBIOSIS AND DISEASE

Thorough disinfection can eliminate unwanted microbes at the beginning of culture but will also eliminate the neutral/beneficial ones. However, disinfection is never absolute, nor is it possible to operate a shrimp culture unit in a sterile way. The introduction of shrimp with an existing gut microbiome, the open nature of the culture system, and the presence of bacteria in the (live) feed automatically imply the (re) introduction of microbes during culture. Nonetheless, if disinfection and other biosecurity measures are done adequately, problems with obligate pathogens (i.e. pathogens that need a host to multiply and survive) can be minimized. The non-sterile nature of shrimp culture systems further ensures that both neutral/beneficial bacteria and opportunistic bacteria can re-establish in the system. How they develop relative to each other, however, depends on the selection regime in the system and the difference in ecology between the two groups. A detailed explanation of this can be found in Vadstein et al. (2018a). In summary, opportunistic pathogens very often belong to the ecological group of the r-strategists (*Vibrio* sp. are typical examples), which means they show high maximum growth rates (r) but have a low competitive ability and specialization. As a consequence, they can rapidly colonize environments that have a high availability of nutrients relative to the number of bacteria present. By contrast, beneficial/neutral bacteria often belong to the group of the K-strategists. They show a lower maximum growth rate and are competition specialists, that is, they have traits that make them successful in environments with bacterial densities close to the carrying capacity (K) (i.e. low nutrient availability relative to the number of bacteria). r-strategic bacteria typically dominate so-called pioneer communities, which are the unstable bacterial communities that establish first in unexploited ‘bacteria-free’ environments. K-strategic bacteria dominate in so-called mature communities, which are much more resilient communities that take over the environment while the water matures. The properties of r- and K-strategic microbial species and of pioneer and mature communities are shown in Table 8.1.

In a desired situation, r/K selection is implemented as such in aquaculture systems so that at the species level it results in a minimal presence of r-strategic bacteria because of their potentially harmful

TABLE 8.1 Characteristics of r/K-strategic bacterial species and of pioneer/mature communities (Vadstein et al. 2018a).

Species level	*r-strategists*	*K-strategists*
Maximum growth rate	High	Low
Biomass at carrying capacity	Unstable	Stable
Effect of enrichment/instability	Rapid growth	Slow growth
Competitive ability in low substrate supply per capita	Poor	Good
Affinity to substrate	Low	High
Community level	*Pioneer community*	*Mature community*
Biological control	Low	High
Stability to perturbations	Poor	Good
Diversity (species, biochemical)	Low	High
Niche width	Wide	Narrow
Specialization	Low	High

nature. At the community level, it is targeting a mature community because this offers stability in terms of microbial composition and functionality. Shortly after disinfection, when shrimp are stocked in the systems, there is a low bacterial load in combination with a high availability of nutrients originating from uneaten feed, excretion of faeces, dead shrimp, and so on. Under these conditions, r-strategist bacteria will grow very fast and their numbers will explode. At the same time, K-strategic, slow-growing specialists also develop in the environment, but at a much slower rate. As a result, there is an initial dominance of r-strategic bacteria in the environment (Fig. 8.12(A)), and since opportunistic pathogens such as vibrios belong to this group there is a higher risk for disease in the cultured shrimp. The exact relative levels of the two groups of bacteria in the period following disinfection are impossible to predict, as these will depend on the conditions in the system that determine the bacterial carrying capacity and growth characteristics (i.e. nutrient levels, water exchange rate, temperature, salinity, etc.). However, with successful disinfection and normal supply of organic matter, r-strategists may easily constitute more than 90% of the community.

What happens with the two groups in the long run depends on how the system is operated. Some systems support the long-term establishment of high numbers of r-strategic bacteria relative to the K-strategists. Examples of such systems are flow-through systems with short water retention times in which K-strategists are consistently flushed out, or systems without adequate removal of waste resulting in a constantly high substrate:bacteria ratio (Fig. 8.12(B)). It is typically such systems that intrinsically suffer from a higher disease risk because the microbial community is dominated by unstable r-strategic bacteria. Such systems contain microbially immature pioneer communities. Others systems favour the dominance of K-strategic bacteria. These include, for example, minimal water exchange systems and the use of measures to minimize waste load, ensuring a low substrate:bacteria ratio (Fig. 8.12(C)). In such systems, the r-strategic populations are outcompeted from the system by the specialist K-strategic bacteria. This creates a microbially stable mature community that is less susceptible to changing environmental conditions and thus is also more resistant to dysbiosis.

Most often without realizing the consequences at the microbial level, the new approaches of integrated shrimp culture or shrimp culture in recirculation systems promote microbial stability by K-selection and thus provide greater resistance to dysbiosis. However, not all farmers are able to adjust their system design to minimize disease risk by microbial K-selection. As mentioned earlier in this chapter, probiotics are applied most often to control the domination of opportunistic pathogens, although it is not always clear how exactly they interfere in the microbial environment. One can assume that probiotics are beneficial – or at least neutral – by nature for the cultured animals. In that sense, they can be considered to be part of the 'wanted' microbes. Depending on their ecology (growth rate, competitive ability, and specialization), probiotics can, however, belong to the r-strategic or the K-strategic group. When mainly showing

FIGURE 8.12 (A) Shortly following disinfection and the introduction of shrimp into the system, there is an ecological selection principle active that results in an increased risk of dominance of r-strategic bacteria (orange symbols), which are often opportunistic pathogens, over K-strategic bacteria (green symbols). The microbial community can be described as an unstable pioneer community. At this stage, there is an increased risk of disease. (B) Systems that allow a high substrate:bacteria ratio or provide unstable environmental conditions (flow-through set-up, minimal waste removal) allow for the continued dominance, or at least high presence, of potentially opportunistic pathogens in the system. Despite beneficial/neutral bacteria also being able to grow, these are typically systems with a remaining pioneer community, which are under continuous high risk of dysbiosis and disease outbreaks. (C) Other systems that ensure a minimal substrate:bacteria ratio and stable environmental conditions (zero water exchange, waste removal) lead to a dominance of beneficial/neutral bacteria, while bacteria with the ecology of opportunistic pathogens are outcompeted. In these systems a mature microbial community can establish, resulting in an intrinsically much lower risk of dysbiosis and disease.

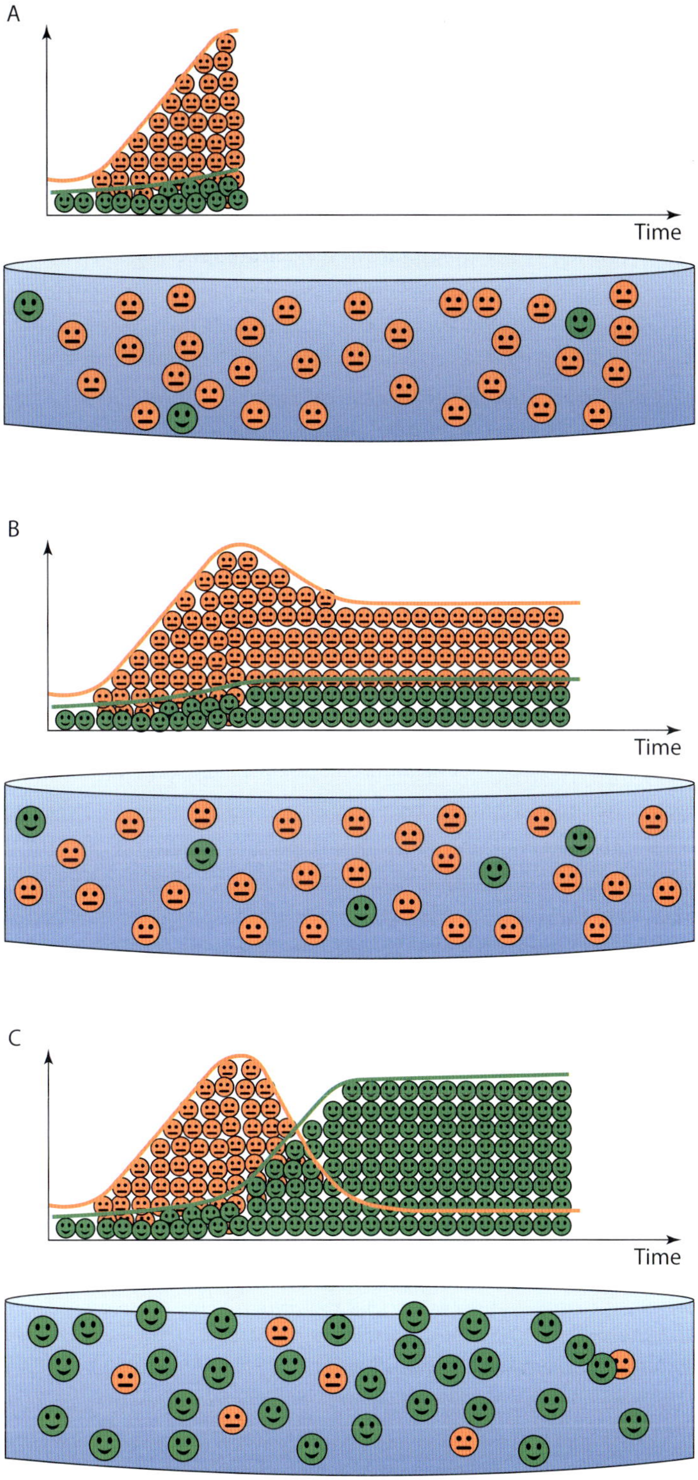
A
Time
B
Time
C
Time

K-strategic characteristics, the addition of probiotics can strengthen this group from the beginning and contribute to an improved (i.e. lower) substrate:bacteria ratio. From an ecological point of view this should result in a faster turnover to a mature microbial community. The success of establishment – and thus the frequency of application – still depends, however, on the system design. However, probiotics are often selected for their high growth rates, meaning that they often also show characteristics of r-strategists. In this case their use is also beneficial because they contribute to the initial peak of r-strategists in the system, thereby taking away substrate from the potentially dangerous opportunistic pathogens that now cannot grow as much in the system. Or, in other words, the peak of r-strategists still exists but it consists at least partly of probiotic bacteria with known beneficial or neutral characteristics for the animals, while it prepares for the succession of a beneficial mature community (if the system allows).

Conclusion

In this chapter, we have sought to document that new knowledge on the presence and activity of bacterial groups in the water and the gut of the cultured shrimp makes it possible to better frame the empirical observations of unexplained mortalities in shrimp hatchery and grow-out systems. The present approach of focusing on repressive methods alone to control pathogens in our aquaculture systems is not the best way to minimize pathogenic disease risks. Often, a non-selective reduction of unwanted bacteria by disinfection leads to the promotion of r-strategist bacteria, an ecological group that includes many opportunistic pathogens. A typical example are vibrios whose virulence is known to be switched on as a function of population density due to the quorum sensing phenomenon. We advise the industry to set up microbial selection regimes leading to balanced r/K microbiomes in culture systems to minimize the chances of disease-associated bacteria becoming established. In addition, both academia and industry should perform collaborative research to further fine-tune and develop more cost-effective microbial management protocols. In any case, it is clear that a more ecological approach to disease control will be increasingly implemented in future shrimp farming.

> Microbes and their activities have pervasive, remarkably profound and generally positive effects on the functioning, and thus health and well-being, of human beings, the whole of the biological world, and indeed the entire surface of the planet and its atmosphere (Timmis et al., 2019).

References

Ackefors, H. (2009) The evolution of a worldwide shrimp industry. *World Aquaculture* 40(3), 46–55.

Anderson J.L., Valderrama, D., and Jory, D. (2019) GOAL 2019 Shrimp Review. *GOAL Conference*, Chennai, India.

Balcázar, J.L., de Blas, I., Ruiz-Zarzuela, I., Cunningham, D. Vendrell, D., and Múzquiz, J.L. (2006) The role of probiotics in aquaculture. *Veterinary Microbiology* 114, 173–186.

Bass, D., Stentiford, G.D., Wang, H.-C., Koskella, B., and Tyler, C.R. (2019) The pathobiome in animal and plant diseases. *Trends in Ecology and Evolution* 34, 996–1008.

Bratvold, D., Lu, J., and Browdy, C.L. (1999) Disinfection, microbial community establishment and shrimp production in a prototype biosecure pond. *Journal of the World Aquaculture Society* 30, 422–432.

Burns, A.R., Miller, E., Agarwal, M., Rolig, A.S., Milligan-Myhre, K., Seredick, S., Guillemin, K., and Bohannan,

B.J.M. (2017) Interhost dispersal alters microbiome assembly and can overwhelm host innate immunity in an experimental zebrafish model. *Proceedings of the National Academy of Sciences, USA* 114, 11181–11186.

Chen, W.-Y., Ng, T.H., Wu, J.-H., Chen, J.-W., and Wang, H.-C. (2017) Microbiome dynamics in a shrimp grow-out pond with possible outbreak of acute hepatopancreatic necrosis disease. *Scientific Reports* 7, 9395.

Choudhury, T.G., Nagaraju, V.T., Gita, S., Paria, A., and Parhi, J. (2017) Advances in bacteriophage research for bacterial disease control in aquaculture. *Reviews in Fisheries Science & Aquaculture* 25, 113–125.

Cornejo-Granados, F., Gallardo-Becerra, L., Leonardo-Reza, M., Ochoa-Romo, J.P., and Ochoa-Leyva, A. (2018) A meta-analysis reveals the environmental and host factors shaping the structure and function of the shrimp microbiota. *PeerJ* 6, e5382.

Cornejo-Granados, F., Lopez-Zavala, A.A., Gallardo-Becerra, L., Mendoza-Vargas, A., Sánchez, F., Vichido, R., Brieba, L.G., Viana, M.T., Sotelo-Mundo, R.R., and Ochoa-Leyva, A. (2017) Microbiome of Pacific Whiteleg shrimp reveals differential bacterial community composition between Wild, aquacultured and AHPND/EMS outbreak conditions. *Scientific Reports* 7, 11783.

Dai, W., Chen, J., and Xiong, J. (2018) Concept of microbial gatekeepers: positive guys? *Applied Microbiology and Biotechnology* 103, 633–641.

Dash, P., Avunje, S., Tandel, R.S., Sandeep, K.P., and Panigrahi, A. (2017) Biocontrol of luminous vibriosis in shrimp aquaculture: a review of current approaches and future prospects. *Reviews in Fisheries Science and Aquaculture* 25, 245–255.

Defoirdt, T. (2018) Quorum-sensing systems as targets for antivirulence therapy. *Trends in Microbiology* 26, 313–328.

Defoirdt, T., Benneche, T., Brackman, G., Coenye, T., Sorgeloos, P., and Scheie, A.A. (2012) A quorum sensing-disrupting brominated thiophenone with a promising therapeutic potential to treat luminescent vibriosis. *PLoS ONE* 7, e41788.

Defoirdt, T., Boon, N., Sorgeloos, P., Verstraete, W., and Bossier, P. (2007) Alternatives to antibiotics to control bacterial infections: luminescent vibriosis in aquaculture as an example. *Trends in Biotechnology* 25, 472–479.

Defoirdt, T., Crab, R., Wood, T.K., Sorgeloos, P., Verstraete, W., and Bossier, P. (2006) Quorum sensing-disrupting brominated furanones protect the gnotobiotic brine shrimp *Artemia franciscana* from pathogenic *Vibrio harveyi*, *Vibrio campbellii* and *Vibrio parahaemolyticus* isolates. *Applied and Environmental Microbiology* 72, 6419–6423.

Defoirdt, T., Ruwandeepika, H.A.D., Karunasagar, I., Boon, N., and Bossier, P. (2010) Quorum sensing negatively regulates chitinase in *Vibrio harveyi*. *Environmental Microbiology Reports* 2, 44–49.

Defoirdt, T. and Sorgeloos, P. (2012) Monitoring of *Vibrio harveyi* quorum sensing activity in real time during infection of brine shrimp larvae. *ISME Journal* 6, 2314–2319.

Defoirdt, T., Sorgeloos, P., and Bossier, P. (2011a) Alternatives to antibiotics for the control of bacterial disease in aquaculture. *Current Opinion in Microbiology* 14, 251–258.

Defoirdt, T., Thanh, L.D., Van Delsen, B., De Schryver, P., Sorgeloos, P., Boon, N. and Bossier, P. (2011b). N-acylhomoserine lactone-degrading *Bacillus* strains isolated from aquaculture animals. *Aquaculture* 311, 258–260.

De Schryver, P., Defoirdt, T., and Sorgeloos, P. (2014) Early mortality syndrome outbreaks: a microbial management issue in shrimp farming? *PLoS Pathogens* 10, e1003919.

De Schryver, P. and Vadstein, O. (2014) Ecological theory as a foundation to control pathogenic invasion in aquaculture. *ISME Journal* 8, 2360–2368.

Dhar, A. (2019) Diseases in shrimp aquaculture – Possible remedial pathways. *Asia Pacific Aquaculture 2019 Conference*, Chennai, India.

Dickey, S.W., Cheung, G.Y.C., and Otto, M. (2017) Different drugs for bad bugs: antivirulence strategies in the age of antibiotic resistance. *Nature Reviews Drug Discovery* 16, 457–471.

Dobretsov, S., Teplitski, M., Alagely, A., Gunasekera, S.P., and Paul, V.J. (2010) Malyngolide from the cyanobacterium *Lyngbya majuscula* interferes with quorum sensing circuitry. *Environmental Microbiology Reports* 2, 739–744.

Dong, Y.H., Gusti, A.R., Zhang, Q., Xu, J.L., and Zhang, L.H. (2002) Identification of quorum-quenching *N*-acyl homoserine lactonases from *Bacillus* species. *Applied and Environmental Microbiology* 68, 1754–1759.

Dong, Y.H., Wang, L.H., and Zhang, L.H. (2007) Quorum-quenching microbial infections: mechanisms and implications. *Philosophical Transactions of the Royal Society B* 36, 1201–1211.

Fan, J., Chen, L., Mai, G., Zhang, H., Yang, J., Deng, D., and Ma, Y. (2019) Dynamics of the gut microbiota in developmental stages of *Litopenaeus vannamei* reveal its association with body weight. *Scientific Reports* 9, 734.

FAO (2000) *The state of world aquaculture and fisheries 2000. Part I: World review of fisheries and aquaculture.* Food and Agriculture Organization of the United Nations, Rome, Italy.

FAO/NACA (2012) R.P. Subasinghe, J.R. Arthur, D.M. Bartley, S.S. De Silva, M. Halwart, N. Hishamunda, C.V. Mohan & P. Sorgeloos (eds) *Farming the Waters for People and Food. Proceedings of the Global Conference on Aquaculture 2010*, Phuket, Thailand, 22–25 September 2010. FAO, Rome, Italy and NACA, Bangkok, Thailand.

Gatesoupe, F. J. (1991) *The Use of Probiotics in Fish Hatcheries: Results and Prospect.* ICES Mariculture Committee report F.4–7. ICES, Copenhagen, Denmark.

Gatesoupe, F. J. (1999) The use of probiotics in aquaculture. *Aquaculture* 180, 147–165.

Grover, S., Rashmi, H.M., Srivastava, A.K., and Kumar, V. (2012) Probiotics for human health – new innovations and emerging trends. *Gut Pathogens* 4, 15.

Henke, J.M. and Bassler, B.L. (2004) Quorum sensing regulates type III secretion in *Vibrio harveyi* and *Vibrio parahaemolyticus*. *Journal of Bacteriology* 186, 3794–3805.

Hoseinifar, S.H., Sun, Y.-Z., Wang A., and Zhou, Z. (2018) Probiotics as means of diseases control in aquaculture, a review of current knowledge and future perspectives. *Frontiers in Microbiology* 9, 2429.

Hou, D., Huang, Z., Zeng, S., Liu, J., Weng, S., and He, J. (2018) Comparative analysis of the bacterial community compositions of the shrimp intestine, surrounding water and sediment. *Journal of Applied Microbiology* 125, 792–799.

Hou, D., Zeng, S., Liu, J., Yan, M., Weng, S., He, J., and Huang, Z. (2016) Characterization of prokaryotic and eukaryotic microbial community in Pacific white shrimp ponds. *Journal of Aquaculture Research and Development* 7, 12.

Huang, Z., Li, X., Wang, L., and Shao, Z. (2014) Changes in the intestinal bacterial community during the growth of white shrimp, *Litopenaeus vannamei*. *Aquaculture Research* 47, 1737–1746.

Irianto, A. and Austin, B. (2002) Use of probiotics to control furunculosis in rainbow trout, *Oncorhynchus mykiss* (Walbaum). *Journal of Fish Diseases* 25, 333–342.

Janssens, J.C.A., De Keersmaecker, S.C.K., De Vos, D.E., and Vanderleyden, J. (2008) Small molecules for interference with cell-cell communication systems in Gram-negative bacteria. *Current Medicinal Chemistry* 15, 2144–2156.

Jamal, M.T., Abulrahman, I.A., Harbi, M.A., and Chithambaran, S. (2019) Probiotics as alternative control measures in shrimp aquaculture: a review. *Journal of Applied Biology and Biotechnology* 7, 69–77.

Karunasagar, I., Karunasagar, I., and Umesha, R.K. (2014) Microbial diseases in shrimp aquaculture. In: Ramaiah, N. (ed.) *Marine Microbiology: Facets & Opportunities*. National Institute of Oceanography, Goa, India, pp. 121–134.

Kautsky, N., Rönnbäcka, P., Tedengren, M., and Troell, M. (2000) Ecosystem perspectives on management of disease in shrimp pond farming. *Aquaculture* 191, 145–161.

Kumar, T.S., Vidya, R., Kumar, S., Alavandi, S.V., and Vijayan, K.K. (2017) Zoea-2 syndrome of *Penaeus vannamei* in shrimp hatcheries. *Aquaculture* 479, 759–767.

Landsman, A., St-Pierre, B., Rosales-Leija, M., Brown, M., and Gibbons, W. (2019) Impact of aquaculture practices on intestinal bacterial profiles of Pacific whiteleg shrimp *Litopenaeus vannamei*. *Microorganism* 7, 93.

Lavilla-Pitogo, C.R., Leaño, E.M., and Paner, M.G. (1998) Mortalities of pond-cultured juvenile shrimp, *Penaeus monodon*, associated with dominance of luminescent vibrios in the rearing environment. *Aquaculture* 164, 337–349.

Li, E., Xu, C., Wang, X., Wang, S., Zhao, Q., Zhang, M., Qin, J.G., and Chen, L. (2018) Gut microbiota and its modulation for healthy farming of Pacific white shrimp *Litopenaeus vannamei*. *Reviews in Fisheries Science & Aquaculture* 26, 381–399.

Loomis, I. (2018) The unmet promise of pondside PCR. *Global Aquaculture Advocate*, 20 August 2018. Available at: https://www.aquaculturealliance.org/advocate/unmet-promise-pondside-pcr (accessed 30 July 2021).

Marques, A., Ollevier, F., Verstraete, W., Sorgeloos, P., and Bossier, P. (2006) Gnotobiotically grown aquatic animals: opportunities to investigate host–microbe interactions. *Journal of Applied Microbiology* 100, 903–918.

Mayes, T. (2018) 18% increase in shrimp production predicted. *The Fish Site*, 28 September 2018. Available at: https://thefishsite.com/articles/18-increase-in-shrimp-production-predicted (accessed 30 July 2021).

Miranda-Baeza, A., Nolasco-López, M., Rivas-Vega, M.E., Huerta-Rábago, J.A., Martínez-Córdova, L.R., and Martínez-Porchas, M. (2020) Short-term effect of the inoculation of probiotics in mature bioflocs: Water quality parameters and abundance of heterotrophic and ammonia-oxidizing bacteria. *Aquaculture Research* 51, 255–264.

Mok, K.C., Wingreen, N.S., and Bassler, B.L. (2003) *Vibrio harveyi* quorum sensing: a coincidence detector for two autoinducers controls gene expression. *EMBO Journal* 22, 870–881.

Natrah, F.M.I., Alam, M.I., Pawar, S., Harzevili, A.S., Nevejan, N., Boon, N., Sorgeloos, P., Bossier, P., and Defoirdt, T. (2012) The impact of quorum sensing on the virulence of *Aeromonas hydrophila* and *Aeromonas salmonicida* towards burbot (*Lota lota* L.) larvae. *Veterinary Microbiology* 159, 77–82.

Natrah, F.M.I., Ruwandeepika, H.A.D., Pawar, S., Karunasagar, I., Sorgeloos, P., Bossier, P., and Defoirdt, T. (2011) Regulation of virulence factors by quorum sensing in *Vibrio harveyi. Veterinary Microbiology* 154, 124–129.

Ninawe, A.S. and Selvin, J. (2009) Probiotics in shrimp aquaculture: avenues and challenges. *Critical Reviews in Microbiology* 35, 43–66.

Noor, N.M., Defoirdt, T., Alipiah, N., Karim, M., Daud, H., and Natrah, I. (2019) Quorum sensing is required for full virulence of *Vibrio campbellii* towards tiger grouper (*Epinephelus fuscoguttatus*) larvae. *Journal of Fish Diseases* 42, 489–495.

Pande, G.S.J., Aamdal Scheie, A., Benneche, T., Wille M., Sorgeloos P., Bossier P., and Defoirdt T. (2013a) Quorum sensing-disrupting compounds protect larvae of the giant freshwater prawn *Macrobrachium rosenbergii* from *Vibrio harveyi* infection. *Aquaculture* 406–407, 121–124.

Pande, G.S.J., Natrah, F.M., Flandez, A.V., Kumar, U., Niu, Y., Bossier, P., and Defoirdt, T. (2015) Isolation of AHL-degrading bacteria from micro-algal cultures and their impact on algal growth and on virulence of *Vibrio campbellii* to prawn larvae. *Applied Microbiology and Biotechnology* 99, 10805–10813.

Pande, G.S.J., Natrah, F.M.I., Sorgeloos, P., Bossier, P., and Defoirdt, T. (2013b) The *Vibrio harveyi* quorum sensing signals have a different impact on virulence of the bacterium towards different crustacean hosts. *Veterinary Microbiology* 167, 540–545.

Props, R., Monsieurs, P., Mysara, M., Clement, L., and Boon, N. (2016) Measuring the biodiversity of microbial communities by flow cytometry. *Methods in Ecology and Evolution* 7, 1376–1385.

Props, R., Rubbens, P., Besmer, M., Buysschaert, B., Sigrist, J., Weilenmann, H., Waegeman, W., Boon, N., and Hammes, F. (2018) Detection of microbial disturbances in a drinking water microbial community through continuous acquisition and advanced analysis of flow cytometry data. *Water Research* 145, 73–82.

Ramanchandran, K. (2017) Present status and prospect trends of probiotics in shrimp aquaculture. *Journal of Fisheries Sciences* 11, 71–73.

Santhakumari, S., Jayakumar R., Logalakshmi, R., Prabhu, N.M., Abdul Nazar, A.K., Karutha Pandian, S., and Veera Ravi, A. (2018) *In vitro* and *in vivo* effect of 2,6-di-*tert*-butyl-4-methylphenol as an antibiofilm agent against quorum sensing mediated biofilm formation in *Vibrio* spp. *International Journal of Food Microbiology* 281, 60–71.

Santos, L. and Ramos, F. (2018) Antimicrobial resistance in aquaculture: current knowledge and alternatives to tackle the problem. *International Journal of Antimicrobial Agents* 52, 135–143.

Saurav, K., Costantino, V., Venturi, V., and Steindler, L. (2017) Quorum sensing inhibitors from the sea discovered using bacterial *N*-acyl-homoserine lactone-based biosensors. *Marine Drugs* 15, 53.

Seibert, C.H. and Pinto, A.R. (2012) Challenges in shrimp aquaculture due to viral diseases: distribution and biology of the five major penaeid viruses and interventions to avoid viral incidence and dispersion. *Brazilian Journal of Aquaculture* 43, 857–864.

Soltani, M., Ghosh, K., Hoseinifar, S.H., Kumar, V., Lymbery, A.J., Roy, S., and Ringø, E. (2019) Genus *Bacillus*, promising probiotics in aquaculture: aquatic animal origin, bio-active components, bioremediation and efficacy in fish and shellfish. *Reviews in Fisheries Science and Aquaculture* 27, 1–49.

Sorgeloos, P. (2013) Aquaculture: the blue biotechnology of the future. *World Aquaculture* 44(3), 18–27.

Sotomayor, M.A., Reyes, J.K., Restrepo, L., Dominguez-Borbor, C., Maldonado, M., and Bayot, B. (2019) Efficacy assessment of commercially available natural products and antibiotics, commonly used for mitigation of pathogenic *Vibrio* outbreaks in Ecuadorian *Penaeus* (*Litopenaeus*) *vannamei* hatcheries. *PLoS One* 14, e0210478.

Taw, N. (2017) A look at various intensive shrimp farming systems in Asia. *Global Aquaculture Advocate*, 24 July 2017. Available at: https://www.aquaculturealliance.org/advocate/intensive-shrimp-farming-asia/ (accessed 30 July 2021)

Teasdale, M.E., Liu, J., Wallace, J., Akhlaghi, F., and Rowley, D.C. (2009) Secondary metabolites produced by the marine bacterium *Halobacillus salinus* that inhibit quorum sensing-controlled phenotypes in Gram-negative bacteria. *Applied and Environmental Microbiology* 75, 567–572.

Thitamadee, S., Prachumwat, A., Srisala, J., Jaroenlak, P., Salachan, P., Sritunyalucksana, K., Flegel, T.W., and Itsathitphaisarn, O. (2016) Review of current disease threats for cultivated penaeid shrimp in Asia. *Aquaculture* 452, 69–87.

Timmis, K., Cavicchioli, R., Garcia, J.L., Nogales, B., Chavarría, M., Stein, L., McGenity, T.J., Webster, N., Singh, B.K., Handelsman, J., de Lorenzo, V., Pruzzo, C., Timmis, J., Martín, J.L.M., Verstraete, W., Jetten, M., Danchin, A., Huang, W., Gilbert, J., Lal, R., Santos, H., Lee, S.Y., Sessitsch, A., Bonfante, P., Gram, L., Lin, R.T.P., Ron, E., Karahan, Z.C., van der Meer, J.R., Artunkal, S., Jahn, D., and Harper, L. (2019) The urgent need for microbiology literacy in society. *Environmental Microbiology* 21, 1513–1528.

Torres, M., Dessaux, Y., and Llamas, I. (2019) Saline environments as a source of potential quorum sensing disruptors to control bacterial infections: a review. *Marine Drugs* 17, 191.

Vadstein, O., Attramadal, K.J.K., Bakke, I., Forberg, T., Olsen, Y., Verdegem, M, Giatsis, C., Skjermo, J., Aasen, I.M., Gatesoupe, F.-J., Dierckens, K., Sorgeloos, P., and Bossier, P. (2018a) Managing the microbial community of marine fish larvae: a holistic perspective for larviculture. *Frontiers in Microbiology* 9, 1820.

Vadstein, O., Attramadal, K.J.K., Bakke, I., and Olsen, Y. (2018b) K-selection as microbial community management strategy: a method for improved viability of larvae in aquaculture. *Frontiers in Microbiology* 9, 2730.

Vandenberghe, J., Verdonck, L., Robles-Arozarena, R., Rivera, G., Bolland, A., Balladares, M., Gomez-Gil, B., Calderon, J., Sorgeloos, P., and Swings, J. (1999) Vibrios associated with *Litopenaeus vannamei* larvae, postlarvae, broodstock, and hatchery probionts. *Applied and Environmental Microbiology* 65, 2592–2597.

Van Hai, N. and Fotedar, R. (2010) A review of probiotics in shrimp aquaculture. *Journal of Applied Aquaculture* 22, 251–266.

Verschuere, L., Rombaut, G., Sorgeloos, P., and Verstraete, W. (2000) Probiotic bacteria as biological control agents in aquaculture. *Microbiology and Molecular Biology Reviews* 64, 655–671.

Walker, P.J. and Mohan, C.V. (2009) Viral disease emergence in shrimp aquaculture: origins, impact and the effectiveness of health management strategies. *Reviews in Aquaculture* 1, 125–154.

Xiao, J., Liu, L., Ke, Y., Li, X., Liu, Y., Pan, Y., Yan, S., and Wang, Y. (2017) Shrimp AHPND-causing plasmids encoding the PirAB toxins as mediated by *pirAB*-Tn903 are prevalent in various *Vibrio* species. *Scientific Reports* 7, 42177.

Xiong, J. (2018) Progress in the gut microbiota in exploring shrimp disease pathogenesis and incidence. *Applied Microbiology and Biotechnology* 102, 7343–7350.

Xiong, J., Dai, W., and Li, C. (2016) Advances, challenges, and directions in shrimp disease control: the guidelines from an ecological perspective. *Applied Microbiology and Biotechnology* 100, 6947–6954.

Xiong, J., Dai, W., Qiu, Q., Zhu, J., Yang, W., and Li, C. (2018) Response of host–bacterial colonization in shrimp to developmental stage, environment and disease. *Molecular Ecology* 27, 3686–3699.

Xiong, J., Nie, L., and Chen, J. (2019a) Current understanding on the roles of gut microbiota in fish disease and immunity. *Zoological Research* 18, 70–76.

Xiong, J., Wang, K., Wu, J., Qiuqian, L., Yang, K, Qian, Y., and Zhang, D. (2015) Changes in intestinal bacterial communities are closely associated with shrimp disease severity. *Applied Microbiology and Biotechnology* 99, 6911–6919.

Xiong, J., Xuan, L., Yu, W., Zhu, J., Qiu, Q., and Chen, J. (2019b) Spatiotemporal successions of shrimp gut microbial colonization: high consistency despite distinct species pool. *Environmental Microbiology* 21, 1383–1394.

Xiong, J., Zhu, J., Dai, W., Dong, C., Qiu, Q., and Li, C. (2017) Integrating gut microbiota immaturity and disease-discriminatory taxa to diagnose the initiation and severity of shrimp disease. *Environmental Microbiology* 19, 1490–1501.

Xiong, J., Zhu, J., Wang, K., Wang, X., Ye, X., Liu, L., Zhao, Q., Hou, M., Qiuqian, L., and Zhang, D. (2014a) The temporal scaling of bacterioplankton composition: high turnover and predictability during shrimp cultivation. *Microbial Ecology* 67, 256–264.

Xiong, J., Zhu, J., and Zhang, D. (2014b) The application of bacterial indicator phylotypes to predict shrimp health status. *Applied Microbiology and Biotechnology* 98, 8291–8299.

Yang, Q. and Defoirdt, T. (2015) Quorum sensing positively regulates flagellar motility in pathogenic *Vibrio harveyi*. *Environmental Microbiology* 17, 960–968.

Yang, Q., Pande, G.S.J., Wang, Z., Lin, B., Rubin, R.A., Vora, G.J., and Defoirdt, T. (2017) Indole signalling and (micro) algal auxins decrease the virulence of *Vibrio campbellii*, a major pathogen of aquatic organisms. *Environmental Microbiology* 19, 1987–2004.

Yao, Z., Yang, K., Huang, L., Huang, X., Qiuqian, L., Wang, K., and Zhang, D. (2018) Disease outbreak accompanies the dispersive structure of shrimp gut bacterial community with a simple core microbiota. *AMB Express* 8, 120.

Yu, W., Wu, J.-H., Zhang, J., Yang, W., Chen, J., and Xiong, J. (2018) A meta-analysis reveals universal gut bacterial signatures for diagnosing the incidence of shrimp disease. *FEMS Microbiology Ecology* 94, fiy147.

Zhang, D., Wang, X., Xiong, J., Zhu, J., Wang, Y. Zhao, Q., Chen, H., Guo, A., Wu, J., and Dai, H. (2014) Bacterioplankton assemblages as biological indicators of shrimp health status. *Ecological Indicators* 38, 218–224.

Zhang, M., Pan, L., Huang, F., Gao, S., Su, C., Zhang, M., and He, Z. (2019) Metagenomic analysis of composition, function and cycling processes of microbial community in water, sediment and effluent of *Litopenaeus vannamei* farming environments under different culture modes. *Aquaculture* 506, 280–293.

Zheng, Y., Yu, M., Liu, J., Qiao, Y., Wang, L., Li, Z., Zhang, X.-H., and Yu, M. (2017) Bacterial community associated with healthy and diseased Pacific white shrimp (*Litopenaeus vannamei*) larvae and rearing water across different growth stages. *Frontiers in Microbiology* 8, 1362.

Zheng, Y. Yu, M., Liu, Y., Su, Y., Xu, T., Yu, M., and Zhang, X.-H. (2016) Comparison of cultivable bacterial communities associated with Pacific white shrimp (*Litopenaeus vannamei*) larvae at different health statuses and growth stages. *Aquaculture* 451, 163–169.

Zhu, J., Dai, W., Qiu, Q., Dong, C., Zhang, J., and Xiong, J. (2016) Contrasting ecological processes and functional compositions between intestinal bacterial community in healthy and diseased shrimp. *Microbial Ecology* 72, 975–985.

Zoqratt, M.Z.H.M., Eng, W.W.H., Thai, B.H., Austin, C.M., and Gan, H.M. (2018) Microbiome analysis of Pacific white shrimp gut and rearing water from Malaysia and Vietnam: implications for aquaculture research and management. *PeerJ* 6, e5826.

Chapter 9

DISSOLVED OXYGEN AND AERATION IN PENAEID SHRIMP AQUACULTURE PONDS

JOHN A. HARGREAVES AND CLAUDE E. BOYD

Summary

Nearly all penaeid shrimp are farmed in ponds in coastal areas of tropical and subtropical climate zones. Dissolved oxygen supply is a critical ecosystem service provided by ponds to support shrimp production and is a fundamental limiting factor affecting production capacity. The relative importance of processes that produce and consume oxygen in shrimp ponds varies by production intensity, with photosynthesis and respiration by phytoplankton being the main processes in semi-intensive ponds. Penaeid shrimp have evolved numerous physiological and behavioral adaptations to cope with limited oxygen supply. Shrimp growth potential is maximized at a dissolved oxygen concentration of 6 mg/l. Shrimp can tolerate repeated short-term exposure to hypoxia, but exposure to hypoxia reduces appetite, affects energy allocation to growth, and diminishes immune system function. Dissolved oxygen concentration is managed with tidal water exchange, pumped water exchange, and mechanical aeration as production intensity increases. Estimation of aeration requirements based on total respiration rate and aerator performance is unreliable. An empirical relationship between shrimp biomass and installed aeration capacity (400–500 kg/hp) is widely applied in shrimp farming. Energy used in aeration can vary with the efficiency of the aeration device selected, type of power source, fuel type, and drive-train configuration. Placement of aerators affects the size of the pond bottom area that can be inhabited by shrimp.

Introduction

Penaeid shrimp are mainly farmed in earthen ponds in coastal areas of tropical and subtropical climate zones around the world. Shrimp are cultured in waters with a broad salinity range, from fresh to hypersaline. Ponds located in the intertidal zone are filled and drained by tidal action, and those above the intertidal zone use saline water pumped from estuaries or the sea. Inland shrimp ponds can use groundwater from deep wells with a suitable mineral profile or surface waters that are subsequently supplemented with mineral additives. Globally, there are about 2.1 million ha of shrimp ponds, and

Vietnam, China, Bangladesh, Ecuador, and Indonesia are the countries with the greatest area of shrimp ponds (Boyd and McNevin, 2018).

Dissolved oxygen supply is a fundamental ecosystem service of aquaculture ponds and is an essential factor that limits the production capacity of fed aquaculture systems. Dissolved oxygen is needed to support respiration of the farmed shrimp crop, decomposition of organic matter, and respiration of pond biota other than the shrimp crop, and various important chemical oxidation reactions associated with nutrient cycling. At lower intensity, oxygen is provided by phytoplankton photosynthesis. Beyond the natural capacity of ponds to assimilate wastes and support the shrimp crop, oxygen must be supplemented by mechanical aeration. Around 70–80% of the roughly 5 Mt of farm-reared penaeid shrimp in 2016 was produced in ponds with mechanical aeration (Boyd and McNevin, 2018).

A comprehensive review paper on dissolved oxygen and aeration in channel catfish aquaculture ponds (Boyd et al., 2018) is the basis for the organization and content of this chapter. In large measure, the underlying ecological mechanisms and dynamics of channel catfish ponds and shrimp ponds are similar. This chapter will summarize the sections of that review as they apply and are relevant to shrimp farming. Similar to that review, this chapter will summarize the dynamics of dissolved oxygen in shrimp ponds, review the physiology of dissolved oxygen use by shrimp, and describe aeration of shrimp ponds as a function of production intensity.

Pacific white shrimp *Litopenaeus vannamei* has become the most important farmed penaeid species, responsible for about 80% of global production, and so this chapter will focus on aspects of oxygen and aeration related to that species, although all concepts are broadly applicable to the aquaculture of other penaeid shrimp species. The scope of this chapter is also limited to ponds used for grow-out of penaeid shrimp, although dissolved oxygen supply is critical for broodstock holding systems and larval rearing tanks in hatcheries. Penaeid shrimp are increasingly being grown in two- or three-phase production systems, and dissolved oxygen is also important in early nursery tanks and ponds.

Effect of salinity on solubility of oxygen in water

Atmospheric air is soluble in water, but each gas in air dissolves independently. Dry air contains 78.08% nitrogen (N_2), 20.95% oxygen (O_2), 0.93% argon (Ar), 0.04% carbon dioxide (CO_2), and smaller percentages of a few other gases. The atmosphere exerts pressure on Earth's land and water surfaces, and atmospheric pressure is equal to the weight of air pressing down on the surface. At sea level and 0°C this weight is equal to that of a column of mercury 760 mm in height. Standard atmospheric pressure is taken as 760 mm Hg (101.325 kPa) at sea level and 0°C.

Because gas molecules are in continuous movement, they exert pressure. In accordance with Dalton's law, the pressure of a gas in the atmosphere (partial pressure of that gas) is equal to its decimal fraction in the atmosphere multiplied by atmospheric pressure. Hence, the partial pressure of oxygen in air is 159.2 mm Hg (760 mm Hg × 0.2095). The partial pressure also may be given in atmospheres (1 atm = 760 mm Hg). The same principle can be applied to oxygen dissolved in water, and at standard temperature and pressure, water saturated with air would have an oxygen pressure (often referred to as oxygen tension) of 159.2 mm Hg and be at 100% saturation with dissolved oxygen.

When the atmosphere contacts a water surface, there will be an exchange of oxygen molecules across the surface in which the direction and rate of exchange depends upon the pressure of oxygen in the atmosphere and in the water. When water is below saturation with dissolved oxygen, there is net movement of oxygen from the atmosphere to the water. When water is supersaturated with dissolved oxygen, movement in the opposite direction will occur. When water is saturated with dissolved oxygen there is no net movement of oxygen between the atmosphere and the water. Of course, the same principle applies equally to the other atmospheric gases, but each gas has a characteristic solubility determined largely by the size of its molecules and its molecular weight.

The laws of thermodynamics govern the exchange of gases between the atmosphere and water. A chemical system will continuously move toward thermodynamic equilibrium in which Gibbs free energy is at a minimum (Silbey et al., 2005). Although oxygen molecules do not react chemically with water molecules, energy relationships in the process of oxygen exchange between air and water reaches an equilibrium in a manner analogous to the principle of Le Chatelier (law of mass action) that governs the direction and rate of chemical reactions. Equilibrium in chemical reactions also is reached at minimum Gibbs free energy. Thus, Henry's law states that the partial pressure of a gas in water is directly proportional to the partial pressure of that gas in the atmosphere above the water. The Henry's law constant can be thought of as the equilibrium constant for solubility:

$$K_H = \frac{[O_{2(aq)}]}{P_{O_{2(g)}}}$$

where K_H = Henry's law constant (M/atm), $O_{2(aq)}$ = dissolved oxygen concentration in water (M), and $P_{O_{2(g)}}$ = partial pressure of oxygen in the atmosphere or air above the water (atm). The value of Henry's law constant differs with temperature and salinity.

In recent years, it has become more common to use the Bunsen coefficient, which is the volume of a gas present in water at equilibrium with 1 atm pressure at a given temperature and salinity (Broecker and Peng, 1974; Forstner and Gnaiger, 1983). Colt (2012) provides tables of Bunsen gas solubility coefficients and explains how to use them in calculating the equilibrium concentrations of gases in water.

A dissolved oxygen solubility table is provided in Table 9.1. The concentrations in this table are for moist air because the air above a water surface is typically saturated with water vapor (which is gaseous). Water vapor has an increasingly large partial pressure (called vapor pressure for water) with higher temperature, for example, 4.570 mm Hg at 0°C, 9.209 mm Hg at 10°C, 17.535 mm Hg at 20°C, and 31.824 mm Hg at 30°C. The vapor pressure dilutes the other atmospheric gases slightly, and the moist air solubility values in Table 9.1 were adjusted for the dilution of dry atmospheric air upon which Henry's law constants and Bunsen coefficients were derived.

In Table 9.1, the solubility of dissolved oxygen at any particular salinity decreases as the temperature increases. Boyd (2020) gives a simple explanation of the cause of this phenomenon: decreasing water temperature results in less molecular energy in both water molecules and dissolved oxygen molecules. The water molecules fit closer together and all molecules move slower, resulting in the water being able to hold more oxygen at lower temperatures than at higher temperatures. Fernández-Prini (1982) gave a more precise explanation based on the fact that when oxygen or other gases dissolve in water, a small quantity of heat is released:

$$O_{2(g)} \leftrightarrow O_{2(aq)} + \text{heat}$$

TABLE 9.1 Standard air saturation concentration of oxygen (mg/l) as a function of temperature and salinity at 1 atm moist air (Colt, 2012).

Temperature (°C)	*Salinity (g/kg)*								
	0	*5*	*10*	*15*	*20*	*25*	*30*	*35*	*40*
15	10.08	9.78	9.49	9.20	8.92	8.65	8.39	8.14	7.89
16	9.87	9.58	9.29	9.01	8.74	8.48	8.22	7.98	7.74
17	9.67	9.38	9.10	8.83	8.57	8.31	8.06	7.82	7.59
18	9.47	9.19	8.92	8.65	8.40	8.15	7.91	7.68	7.45
19	9.28	9.01	8.74	8.49	8.24	8.00	7.76	7.53	7.31
20	9.09	8.83	8.57	8.32	8.08	7.85	7.62	7.40	7.18
21	8.91	8.66	8.41	8.17	7.93	7.70	7.48	7.26	7.05
22	8.74	8.49	8.25	8.01	7.79	7.56	7.34	7.13	6.93
23	8.58	8.33	8.10	7.87	7.64	7.43	7.22	7.01	6.81
24	8.42	8.18	7.95	7.73	7.10	7.30	7.09	6.89	6.69
25	8.26	8.03	7.81	7.59	7.38	7.17	6.97	6.77	6.58
26	8.11	7.89	7.67	7.46	7.25	7.05	6.85	6.66	6.47
27	7.97	7.75	7.53	7.33	7.12	6.93	6.73	6.55	6.37
28	7.83	7.61	7.40	7.20	7.00	6.81	6.62	6.44	6.26
29	7.69	7.48	7.28	7.08	6.89	6.70	6.52	6.34	6.16
30	7.56	7.35	7.16	6.96	6.77	6.59	6.41	6.24	6.07
31	7.43	7.23	7.04	6.85	6.66	6.48	6.31	6.14	5.97
32	7.31	7.11	6.92	6.74	6.56	6.38	6.21	6.04	5.88
33	7.18	6.99	6.81	6.63	6.45	6.28	6.11	5.95	5.79
34	7.07	6.88	6.70	6.52	6.35	6.18	6.02	5.86	5.70
35	6.95	6.77	6.59	6.42	6.25	6.09	5.93	5.77	5.62
36	6.84	6.66	6.49	6.32	6.15	5.99	5.83	5.68	5.53
37	6.73	6.55	6.38	6.22	6.06	5.90	5.75	5.60	5.45
38	6.62	6.45	6.28	6.12	5.96	5.81	5.66	5.51	5.37
39	6.52	6.35	6.19	6.03	5.87	5.72	5.58	5.43	5.29
40	6.41	6.25	6.09	5.94	5.78	5.64	5.49	5.35	5.22

Thus, in accordance with Gibbs free energy and Le Chatelier's principle, a rise in temperature (greater heat) drives the reaction to the left, lessening solubility of oxygen. A plot of oxygen solubility versus water temperature reveals a linear decrease in oxygen solubility with increasing temperature.

The solubility of oxygen also decreases as salinity increases (Table 9.1). This phenomenon results because water hydrates ions dissolved in it. The water molecule is polar, with a weak positive charge on the side where the hydrogen atoms are attached to the oxygen atom by covalent bonds and a weak negative charge on the opposite side (Boyd, 2020). These weak charges are attracted to the stronger negative charges on the anions (Cl^-, SO_4^-, and HCO_3^-) and cations (Na^+, K^+, Ca^{2+}, and Mg^{2+}) in natural water. The ions bind the water tightly to them, a process called hydration. Water so bound is not free for oxygen or other gases to dissolve in it. The salinity is the concentration of dissolved ions in water, and as salinity increases, the solubility of oxygen in water decreases in a linear relationship. Full-strength seawater (35 ppt salinity) holds about 20% less dissolved oxygen than freshwater at the same temperature.

The solubilities of oxygen given in Table 9.1 are for 760 mm Hg atmospheric pressure. The solubility of oxygen will be greater or less depending upon the actual atmospheric pressure. The values in Table 9.1 can be adjusted for atmospheric pressure as follows:

$$C_s = C_{ts}\left(\frac{BP}{760}\right)$$

where C_s = dissolved oxygen concentration at saturation (mg/l), C_{ts} = tabular dissolved oxygen concentration at saturation (mg/l), and BP = barometric pressure at the water surface (mm Hg).

Shrimp are usually farmed at locations near sea level (the base for atmospheric pressure measurement), but atmospheric pressure changes almost hourly, causing rather small changes in oxygen solubility, but these changes can usually be ignored for aquaculture purposes.

Atmospheric pressure acts on the surface of a body of water, but the pressure on a point at some depth beneath the water surface increases as a result of the weight of the water column above that point. The pressure holding dissolved oxygen in water increases with greater depth, and the weight of water also increases slightly with greater salinity. The weight of water (regardless of salinity) also decreases slightly with increasing temperature above 3.98°C. The result is that the solubility of oxygen in a body of water increases with depth in a salinity-dependent relationship. The hydrostatic pressures resulting per meter of water depth for different salinities and temperatures are provided in Table 9.2. The following equation allows calculation of dissolved oxygen saturation at any particular depth:

$$C_s = C_{ts}\frac{(BP + HP)}{760}$$

where HP = hydrostatic pressure from Table 9.2 (mm Hg/m).

Some dissolved oxygen meters will display milligrams per liter or percentage saturation. The percentage saturation is the percentage by which water is undersaturated or supersaturated with dissolved oxygen. The percentage saturation can be calculated from the measured dissolved oxygen concentration as follows:

TABLE 9.2 Specific weight of water (mm Hg/m depth) as a function of temperature and salinity (Colt, 2012).

Temperature (°C)	*Salinity (g/kg)*								
	0	*5*	*10*	*15*	*20*	*25*	*30*	*35*	*40*
0	73.54	73.84	74.14	74.44	74.73	75.03	75.33	75.62	75.92
5	73.55	73.85	74.14	74.43	74.72	75.01	75.30	75.59	75.88
10	73.53	73.82	74.11	74.39	74.68	74.97	75.25	75.54	75.83
15	73.49	73.77	74.05	74.34	74.62	74.90	75.18	75.47	75.75
20	73.42	73.70	73.98	74.26	74.54	74.82	75.10	75.38	75.66
25	73.34	73.62	73.89	74.17	74.44	74.72	75.00	75.27	75.55
30	73.24	73.51	73.78	74.06	74.33	74.60	74.88	75.15	75.43
35	73.12	73.39	73.66	73.93	74.20	74.48	74.75	75.02	75.30
40	72.98	73.25	73.52	73.79	74.06	74.34	74.61	74.88	75.15

$$PS = \frac{C_m}{C_s} \times 100$$

where PS = percentage saturation (%) and C_m = measured dissolved oxygen concentration (mg/l).

At a given temperature, the percentage saturation will increase as water temperature rises. For example, full-strength seawater at 25°C that contains 6.75 mg/l dissolved oxygen is at 100% saturation. At 30°C, 6.75 mg/l dissolved oxygen would be only 108.5% saturation, because at 30°C the saturation concentration is 6.22 mg/l.

Shrimp actually respond to the partial pressure of oxygen in water rather than the weight/volume concentration in milligrams per liter. In reality, percentage saturation would be the best way of assessing dissolved oxygen concentration in aquaculture. However, farmers are more familiar with concentration (mg/l), and great confusion would result from use of percentage saturation.

In terms of aeration, it is important to realize that aerators transfer oxygen to the water more effectively at low dissolved oxygen concentrations with maximum effectiveness at 0 mg/l. There is an exponential decrease in the rate of oxygen transfer as the dissolved oxygen concentration increases toward saturation. At saturation, net transfer of oxygen as the result of aerator action ceases. Above saturation, the aerator hastens the transfer of oxygen from water to the air (Boyd et al., 2018). At water temperatures and salinities typical of shrimp ponds, aerators are only about 50% as efficient in transferring oxygen as they are at 0 mg/l dissolved oxygen (Boyd et al., 2018).

Processes affecting dissolved oxygen concentration in shrimp ponds

In shrimp ponds, dissolved oxygen concentration measured at a particular time and place in the water column depends on the net result of simultaneously occurring processes that produce and consume oxygen at different rates. These processes are: (1) photosynthesis by phytoplankton, (2) respiration by phytoplankton and bacteria suspended in the water column, (3) sediment oxygen demand (SOD), (4) shrimp respiration, and (5) air–water gas transfer. These are discussed in turn.

PHYTOPLANKTON PHOTOSYNTHESIS

Phytoplankton photosynthesis is the main source of dissolved oxygen in extensive and most semi-intensive shrimp ponds. Oxygen is produced as a by-product of carbohydrate synthesis by the photosynthetic apparatus of phytoplankton. The rate of photosynthesis is dependent primarily on the availability of solar light energy in the water column. Solar light intensity diminishes as an exponential function of depth from the water surface. Suspended matter in the water column reduces the penetration of light relative to clear water. The compensation point is the depth at which underwater light intensity is 1% of the value at the surface. At this point, it is assumed that phytoplankton photosynthesis is not possible and net photosynthesis (gross primary productivity minus respiration) is 0. The compensation depth decreases as phytoplankton density and non-algal turbidity increases. In ponds with intermediate phytoplankton density and with a water transparency of 30–50 cm, the compensation depth may be around 1 m. In hypereutrophic ponds with high phytoplankton density, the compensation depth may be around 10 cm.

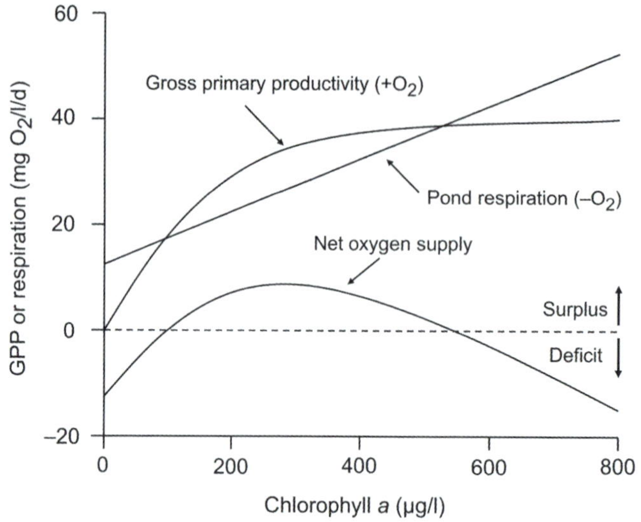

FIGURE 9.1 The relationship between gross primary productivity (GPP) and pond respiration as a function of phytoplankton density, as indicated by the concentration of chlorophyll *a*. The rate of oxygen production by phytoplankton reaches saturation once light in the water column becomes a limiting factor. Net oxygen production is maximum at intermediate levels of phytoplankton density.

The implication of this effect is that the proportion of the water column involved in photosynthetic oxygen production is restricted to a shallow surface layer when ponds are hypereutrophic.

The best net oxygen production occurs at intermediate phytoplankton density (Fig. 9.1). As phytoplankton density increases, gross primary productivity increases with density but then reaches a saturation level because further potential increases in photosynthesis from greater phytoplankton biomass are offset by the reduced penetration and availability of light in the water column. As phytoplankton density increases, respiration increases as a linear function of phytoplankton density. Net photosynthesis is maximum at intermediate levels of phytoplankton density, where the trade-offs between algal biomass and light availability in the water column are minimized.

Phytoplankton photosynthesis is also affected by temperature, with a Q10 temperature coefficient of about 2. Different species in the algal community have different temperature optima or ranges.

PLANKTON RESPIRATION

The main process that consumes oxygen in shrimp ponds is the respiration of all suspended planktonic organisms in the water column, including bacteria, phytoplankton, zooplankton, and biofloc. Respiration by phytoplankton is often the primary component of water column respiration.

In semi-intensive shrimp ponds, water column respiration is around 0.5 mg/l·h, depending on temperature and phytoplankton density (Madenjian, 1990; Santa and Vinatea, 2007). In intensive lined ponds with biofloc stocked at 125–140/m^2, water respiration ranged from 0.30 to 0.62 mg/l·h (Burford et al., 2003), not appreciably different from that of semi-intensive ponds. However, biofloc ponds often transition from a greenwater phase dominated by phytoplankton dynamics to a brownwater phase dominated by bacteria. After such a transition occurs, when biomass is around 13.5 t/ha and feeding rates are >300 kg/ha·d, water respiration can be >2.0 mg/l·h (McIntosh, 2000a). In super-intensive shrimp raceways with biofloc, water column respiration can reach 2.0–2.5 mg/l·/h (Browdy et al., 2012; Kirk, 2010; Vinatea et al., 2010) and remain remarkably stable up to a shrimp biomass density of 7 kg/m^3 (70 t/ha).

Many farmers using intensive and super-intensive systems apply various pond amendments that may affect water respiration. A supplemental carbohydrate source such as molasses or grain pellets is added to encourage growth of heterotrophic bacteria in biofloc ponds. Fermented rice bran is added as a pond supplement in so-called synbiotic systems. Bacterial amendments that are used as bioremediators to improve water quality or aid decomposition of organic matter on the pond bottom are also used by many shrimp producers. Following application of these materials, in all cases, dissolved oxygen concentration may decline rapidly within hours of application as bacteria proliferate and consume oxygen. Water respiration rates may peak at levels that may result in hypoxia, so sufficient aeration capacity must be in place to compensate for amendment loading.

SEDIMENT OXYGEN DEMAND

There are three main sources of organic matter that flow from water column to sediment in shrimp ponds: fecal solids derived from feeding, a daily fraction of the plankton biomass that dies and settles, and various forms added to ponds in source water. The sediment–water interface is the location of organic matter decomposition, coupled nitrification–denitrification, and the chemical oxidation of reduced substances that diffuse from deeper layers to the interface – all processes that consume oxygen.

There have been few direct measurements of SOD in shrimp ponds and measurement techniques have not been standardized. In shrimp ponds stocked at 5/m^2 without aeration, the maximum SOD was 40 mg/m^2·h of sediment surface. In ponds stocked at between 20 and 200/m^2, with aeration, SOD was about 60 mg/m^2·h (Garcia and Brune, 1991). SOD in shrimp ponds typically increases during the culture period. SOD increased from about 70 mg/m^2·h to about 160 mg/m^2·h after 9 weeks in ponds stocked at 21/m^2 (Santa and Vinatea, 2007). Suplee and Cotner (1996) measured a similar effect, with SOD ranging from 60 to 140 mg/m^2·h. In a survey of commercial shrimp farms stocking shrimp at 25–35/m^2, the average SOD was 120 mg/m^2·h (Burford and Longmore, 2001). In marine shrimp ponds, a mean SOD of about 290 mg/m^2·h was measured by Madenjian (1990). The range of SOD measured in channel catfish ponds (63–1038 mg/m^2·h of sediment surface; Steeby et al., 2004) is likely similar to that in shrimp ponds.

The SOD of shrimp ponds will vary depending on production system intensity and sediment characteristics. The high feeding rates characteristic of intensive systems drive feed-related solid waste loading to the pond bottom. These wastes are generally decomposed easily and thus the increase in SOD in response to increased feed-derived waste loading is rapid. However, overloading of the sediment, resulting in organic matter accumulation, can occur if the supply of oxygen diffusing from the water column to the sediment is insufficient to support the decomposition of continuously settling organic matter. As production intensity increases, the proportion of settling organic matter that is derived from feeding increases.

Many modern intensive shrimp ponds are lined with plastic. Although liners do not eliminate SOD, organic matter decomposition may occur more rapidly in lined ponds because oxygen availability at the pond bottom is improved because of greater aeration and mixing in these ponds. Furthermore, chemical oxidation of reduced substances diffusing from deeper sediment layers is eliminated by liners.

SHRIMP RESPIRATION

The oxygen consumption rate – sometimes called the routine metabolic rate (Walker et al., 2009, 2011) – of shrimp depends on their body size, stage of molt cycle, time since consuming feed, and activity level, as well as the water temperature, salinity, and dissolved oxygen concentration. Respirometry has been used to evaluate the combined effects of body weight, temperature, and salinity on oxygen consumption rate. Similar to other aquatic animals grown in aquaculture, in shrimp the respiration rate decreases as an inverse logarithmic function of body weight. Respiration per unit weight is greater for smaller shrimp, although the absolute respiration rate per individual is greater for large shrimp. At 30°C, the respiration rate of Pacific white shrimp weighing 1 g is about 450 mg O_2/kg·h and that of 10 g shrimp is about 200–250 mg O_2/kg·h (Ponce-Palafox et al., 2013). The respiration rate of Pacific white shrimp <5 g is around 650 mg O_2/kg·h and that of shrimp >15 g is 300–400 mg O_2/kg·h (Walker et al., 2009). At 30°C and a salinity of 25 ppt, a 5 g Pacific white shrimp will consume 1.6 mg O_2/h and a 20 g shrimp will consume 6.4 mg O_2/h (Bett and Vinatea, 2009).

As in other ectotherms, the routine metabolic rate of shrimp is directly related to temperature, with a linear slope of the line relating the natural logarithm of metabolic rate as a function of the natural logarithm of temperature. The thermal coefficient, or the relative metabolic rate, of Pacific white shrimp between 20 and 30°C is 1.81 (Ponce-Palafox et al., 2013), and shrimp respiration increases accordingly. For 6 g shrimp, the oxygen consumption rate is about 200 mg O_2/kg·h at 20°C, 300 mg O_2/kg·h at 24°C, 400 mg O_2/kg·h at 28°C, and around 500 mg O_2/kg·h at 32°C (Walker et al., 2011).

Salinity has only a minor effect on the respiration rate of many penaeid shrimp species. Over the salinity range of 15–45 ppt, there is little to no effect of salinity on oxygen consumption rate (Bett and Vinatea, 2009; Ponce-Palafox et al., 2013; Walker et al., 2009). The isosmotic salinity for most penaeid shrimp species ranges between about 20 and 25 ppt (680–768 mOs/kg; Castille and Lawrence, 1981). For Pacific white shrimp, the respiration rate is minimized at the isosmotic salinity of 24.7 ppt (718 mOs/kg) (Ponce-Palafox et al., 2013). The combined effect of high temperature and high salinity causes a large increase in shrimp respiration. At temperatures of 25–30°C and salinities of 14–40 ppt, specific respiration rate was around 250–350 mg O_2/kg·h. At a temperature of 33°C and salinity of 40 ppt, respiration rate increases to >750 mg O_2/kg h (Ponce-Palafox et al., 2013). Low salinity does not affect respiration rate of shrimp, despite the energetic requirements for osmoregulation (Walker et al., 2009).

Above a species-dependent threshold value, shrimp are oxygen regulators, and oxygen consumption is independent of dissolved oxygen concentration. Below that critical level, shrimp are oxygen conformers, and oxygen consumption depends on dissolved oxygen concentration. Maximum oxygen consumption occurs at 4 mg/l for *Penaeus monodon* (Liao and Murai, 1986), 5.5 mg/l for blue shrimp *Penaeus stylirostris* (Re and Díaz, 2011), and 6 mg/l for *L. vannamei* (Ponce-Palafox et al., 2013). Respiratory rate decreases with decreasing dissolved oxygen concentration from those critical levels. Oxygen consumption below those thresholds is directly related to oxygen concentration. The respiratory rate at dissolved oxygen concentrations between 3 and 4 mg/l is about 50% of normal maximum, indicating a significant reduction in metabolic capacity at those oxygen levels.

Shrimp respiration rate can be affected by the concentration of carbon dioxide, which may accumulate in intensive systems with elevated water respiration rates, such as those using biofloc technology. Maintaining carbon dioxide concentrations <30 mg/l results in a normal respiration rate of 233–330 mg O_2/kg·h. At 60 mg CO_2/l, oxygen consumption rate increases to 521 mg O_2/kg·h. At greater concentrations, oxygen consumption decreases and shrimp lose equilibrium (Furtado et al., 2016a).

Like all crustaceans, penaeid shrimp go through periodic molts of the exoskeleton (ecdysis) to allow further somatic growth. The molt cycle consists of phases (post-molt, intermolt, pre-molt, and molt), each with multiple stages. Oxygen consumption by *Penaeus esculentus* increases by 55% during the late pre-molt stage, 3 days before ecdysis, and then declines to normal levels 1 day after (Dall, 1986). The greatest rate of oxygen consumption of shrimp occurs during the post-molt stage (Ding et al., 2014). The post-molt stage is characterized by hardening of the new cuticle and absorption of a relatively large volume of water, leading to osmotic shock. Increased respiration is necessary to support osmoregulation to restore homeostasis. Although the duration of the post-molt stage is only about 0.5–1 d of a 10–12 d molt cycle for larger (15 g) shrimp, it represents a significant increase in the respiratory demand for oxygen by shrimp during that time. Furthermore, the concentration of the oxygen-carrying molecule hemocyanin is lowest in post-molt shrimp relative to other molt stages (Cheng et al., 2002).

Other factors may affect shrimp respiration, most notably their activity level. Shrimp prefer relatively quiescent water, with water current speeds <5 cm/s. In intensive ponds with substantial aeration, aerator currents may force shrimp to swim at greater than critical swimming speeds. At rest, shrimp (*P. esculentus*) respiration was oxygen-independent until about 25% saturation (~2 mg/l), but while actively swimming, respiration was oxygen-dependent up to full saturation (Dall, 1986).

As with other species in aquaculture, shrimp respiration rates increase after food ingestion, as oxygen is used for protein synthesis and other metabolic processes related to digestion. The magnitude of this increase, called apparent heat increment or specific dynamic action by nutritionists, has not been evaluated experimentally but is typically 1.5–2× the basal rate, with a relatively greater increase in shrimp fed higher-protein diets.

AIR–WATER TRANSFER (DIFFUSION)

Unlike the previously described processes that result in gains or losses of oxygen in shrimp ponds, air–water transfer can result in either, depending on the direction of oxygen movement. If the pond water is undersaturated with oxygen, then diffusion of oxygen into water can result in an oxygen gain, especially at times when dissolved oxygen concentrations are critically low. Diffusion can be enhanced by surface turbulence and pond mixing caused by winds interacting with the pond surface, especially in large ponds with a long fetch.

When dissolved oxygen concentration is supersaturated, then diffusion of oxygen from pond water to the atmosphere will result in a loss of oxygen from the pond. Mixing of ponds at this time can disrupt stratification, enhance storage of oxygen produced in surface waters throughout the water column, and minimize diffusive losses from pond water to the atmosphere.

Oxygen diffusion into ponds is slow, and the contribution of oxygen diffusing from the atmosphere to overall pond oxygen budgets is generally low, although it can be important at times such as during depletion of dissolved oxygen caused by sudden phytoplankton die-offs, when photosynthesis supplies negligible dissolved oxygen. Although the magnitude of natural diffusion is relatively minor in pond oxygen budgets, it has a profound impact on the performance of aerators because the difference between the measured concentration and the saturation concentration represents the driving force for air–water oxygen transfer by aerators.

Patterns of dissolved oxygen concentration in shrimp ponds

Shrimp pond oxygen budgets estimate the magnitude of gains and losses of oxygen, typically hourly over a 24 h period. In most extensive and semi-intensive shrimp ponds, phytoplankton photosynthesis provides the largest gain of oxygen and phytoplankton respiration is responsible for the largest loss of oxygen. Respiration in the water column by phytoplankton and bacteria, in the sediment by bacteria and by chemical oxidation of reduced substances, and by the shrimp crop is approximately constant over 24 h. During the day, phytoplankton photosynthesis produces dissolved oxygen faster than it is consumed by respiration, and dissolved oxygen concentration increases. At night, in the absence of photosynthesis, respiration results in a decrease in dissolved oxygen concentration.

The net result of simultaneously occurring gain and loss processes is a characteristic 24 h pattern of increasing dissolved oxygen concentration during the day and decreasing dissolved oxygen concentration at night. The magnitude of the fluctuation in dissolved oxygen concentration increases with phytoplankton biomass (density integrated through the water column). When shrimp ponds become highly eutrophic (Secchi disk visibility <30 cm) in response to daily feeding rates >30 kg/ha, the magnitude of fluctuation may be such that dissolved oxygen concentration reaches levels that affect shrimp growth, feed conversion efficiency, or even survival.

In quiescent, stratified ponds, the spatial pattern is one of greater oxygen concentration near the surface and decreasing concentration through the water column. The transport and diffusion of dissolved oxygen from surface waters to the pond bottom is a major limiting factor and engineering challenge in shrimp production (Garcia and Brune, 1991). When not swimming, shrimp are typically resting or walking on the pond bottom, where dissolved oxygen concentration may be chronically undersaturated. Turbulent mixing by wind- or aerator-driven currents can break down stratification and distribute oxygen from surface waters to the pond bottom.

Average daily dissolved oxygen concentration in shrimp ponds can vary through a crop cycle. Unlike channel catfish ponds that may not be drained for many years, shrimp production cycles are short, typically <120 days. In semi-intensive and intensive shrimp ponds, phytoplankton density gradually increases from a low level and phytoplankton communities go through ecological succession based on the availability of light, nutrients, and the initial inoculum. In intensive and super-intensive ponds, the succession proceeds through an algal phase to a predominantly bacterial phase. In general, dissolved oxygen concentrations are favorable in the early part of the crop cycle, when shrimp are small and feeding rates are low. As ponds become more eutrophic, the magnitude of fluctuation in dissolved oxygen concentration increases and may reach critical levels late at night during the last 1–2 months of the production cycle.

During the crop cycle, episodically and especially at night, the impact of many shrimp molting simultaneously can substantially increase the demand for oxygen associated with increased shrimp respiration just prior to and after molting. During this time, shrimp do not feed and may be overfed, resulting in increased oxygen demand associated with the decomposition of uneaten feed. An additional oxygen demand is associated with the decomposition of molted exoskeletons that are not consumed by shrimp.

Shrimp pond oxygen budgets are useful to understand the relative magnitude of gain and loss processes at different levels of intensity, pond configurations, or water management practices. The models used for pond dissolved oxygen do not have much predictive value for farmers. For extensive and many semi-intensive shrimp farms, dissolved oxygen is not measured and is difficult to control because pond size may be quite large (>10 ha). For those extensive and semi-intensive shrimp farmers who do measure dissolved oxygen concentration, the measurements are used to guide decisions about water exchange. Intensive systems (60–200/m^2 or more) are often aerated continuously and the dissolved oxygen concentration is usually not a problem if aeration level is sufficient for a given stocking density. However, the response time in intensive systems, especially those using biofloc technology, is very short, on the order of 0.5–1 h.

One of the emerging trends in pond aquaculture is continuous monitoring of pond dissolved oxygen concentration. In some cases, monitoring system software can activate aeration devices or pumps. Data collected from dissolved oxygen sensors has been used to regulate the activation of sound-based automatic feeders in shrimp ponds. However, the high cost and low reliability of dissolved oxygen sensors has limited their application.

Coping with limited oxygen supply

In nature, juvenile penaeid shrimp inhabit estuarine channels and nearshore bays where water quality can vary over short time scales and distances. Parts of estuarine channels not subject to tidal exchange or flushing from land runoff may be chronically hypoxic, particularly those impacted by anthropogenic inputs of nutrients and organic matter. It is difficult for shrimp to obtain oxygen, in part because dissolved oxygen is only sparingly soluble in water. In general, crustaceans are more sensitive to hypoxia than finfish, with a higher lethal concentration (LC_{50}) and a shorter time to mortality (LT_{50}). Penaeid shrimp have evolved numerous adaptations to cope with limited oxygen supply.

Gills are the main interface between a shrimp and its environment. Fundamentally, for respiration, gills provide surface area for gas exchange. Oxygen diffuses through the gills in response to a concentration gradient of oxygen between water moving through the branchial chamber and the hemolymph circulating within the shrimp. The diffusion rate is a function of the surface area of the gills, the concentration gradient of oxygen, and diffusive properties of oxygen gas.

Gills of penaeid shrimp are dendrobranchiate (highly branching), with a central branch, paired lateral branches, and secondary, often dendritic, branches arising from lateral branches. Shrimp gills are the primary sites of respiration, osmoregulation, ion regulation, and acid–base balance. Anterior gills have a relatively thinner epithelium and are used mostly in respiration. Posterior gills have a relatively thicker and highly folded epithelium and are used primarily in osmoregulation. During salinity acclimation, oxygen is preferentially used by posterior gills to support osmoregulation (Li et al., 2019a), and this oxygen is thus not available to support routine metabolism, reducing the scope for growth.

Hemocyanin is the oxygen-transport protein in penaeid shrimp. Unlike vertebrates, in which the oxygen-transport protein hemoglobin is contained within erythrocytes, hemocyanin is dissolved and circulates in the hemolymph. About 80% of the protein in the hemolymph is hemocyanin. The oxygen-binding sites of hemocyanin are copper-based. When oxygen binds to copper, the valence changes from

Cu(I) to Cu(II), forming oxyhemocyanin with a characteristic blue color. (Deoxyhemocyanin is colorless.) Although the primary role of hemocyanin is oxygen transport to sites of respiration, it has other functions in immune response, hardening of the cuticle after molting, and wound healing.

Hemocyanin concentrations can vary depending on shrimp rearing conditions, diet, and inherent species differences. Pacific white shrimp grown in aquaculture had greater concentrations of hemocyanin than wild-caught *L. vannamei*, *Penaeus aztecus*, and *Penaeus setiferus* (Tommerdahl et al., 2015). Exposure of shrimp to sustained hypoxia (2–2.6 mg/l) for 2–4 weeks resulted in an increase in hemocyanin, thereby increasing the oxygen-carrying capacity of the hemolymph (Racotta et al., 2002; Tommerdahl et al., 2015).

Shrimp have some behavioral actions that increase the partial pressure of oxygen in the hemolymph. Shrimp can increase their cardiac output by increasing stroke volume or can increase ventilation rate of the gills. Increased ventilation under hypoxic conditions facilitates carbon dioxide excretion, increasing the hemolymph pH, which increases the affinity of hemocyanin for oxygen. Shrimp have scaphognathites, part of the second maxillae, that function as pumps to move water through the branchial chamber. Shrimp also have positive rheotaxis to water currents and can purposefully move to the surface, where dissolved oxygen concentrations may be greater.

Dissolved oxygen requirements and the effects of hypoxia

As described earlier, the typical pattern of dissolved oxygen concentration in most commercial shrimp ponds is one of increasing dissolved oxygen concentration during day and decreasing concentration during night, alternating between a supersaturated condition during the day and early evening, and undersaturated conditions during the night and early morning. Furthermore, penaeid shrimp typically forage on the pond bottom, where dissolved oxygen concentrations may be chronically undersaturated. Thus, shrimp in aquaculture ponds are often exposed to hypoxic conditions, resulting in a deficiency of oxygen reaching tissues where it is required for various aerobic metabolic processes. The physiological responses of penaeid shrimp to various dissolved oxygen concentrations are summarized in Table 9.3.

TABLE 9.3 Summary of effects of dissolved oxygen concentration on shrimp.

Concentration (mg/l)	*Effect*
0.2	Lethal after 1 h of exposure
0.5	Lethal for 8–10 g shrimp
0.8	Lethal for 3 g shrimp
<1	Reduced heart rate
2 (1.5–2.5)	Limiting oxygen concentration (LOCr)
<2	Reduced survival
<2–3	Shift toward greater use of protein for energy
<3	Decrease in osmoregulatory capacity at low salinity
<3.1	Reduced appetite
<3.5	Reduced innate immune system function
1–4	No effect on apparent feed digestibility
3–4	Respiration rate is 50% of maximum
≥4–6	Maximum respiration rate; greatest scope for growth

Penaeid shrimp are able to detect low dissolved oxygen concentrations and actively avoid hypoxic areas by moving to areas with oxygenated water (Wu et al., 2002). In response to decreasing environmental dissolved oxygen concentrations, shrimp will increase their cardiac stroke volume and heart rate to increase perfusion of hemolymph through the gill lamellae and increase ventilation of water across the gills (Zhang et al., 2006a), all of which serve to maximize the partial pressure of oxygen in the hemolymph. As water becomes progressively more hypoxic, shrimp swimming activity increases and they may make short excursions to the surface. As the dissolved oxygen concentration declines further, activity decreases and swimming speed slows. Once the dissolved oxygen concentration declines to 1 mg/l, shrimp activity ceases (Allan and Maguire, 1991; Zhang et al., 2006a).

For penaeid shrimp species, lethal dissolved oxygen concentrations range from 0.2 to 1 mg/l (Zhang et al., 2006a). Acute exposure to 0.2 mg/l dissolved oxygen for 1 h is lethal to *L. vannamei* (Pérez-Rostro et al., 2004). There is a significant effect of body weight on lethal dissolved oxygen concentration, with smaller shrimp being more sensitive to lower dissolved oxygen concentrations. For *L. vannamei*, the lethal concentration is around 0.8 mg/l for 3 g shrimp and around 0.5 mg/l for 8–10 g shrimp (Zhang et al., 2006a). For *Metapenaeus ensis*, heart rate slows substantially when dissolved oxygen concentration declines to <1 mg/l and the 8 h LC_{50} for dissolved oxygen concentration is 0.77 mg/l (Wu et al., 2002). For *M. ensis* exposed to a dissolved oxygen concentration of 0.5 mg/l, the time to death is 6.65 h. For juvenile *P. monodon*, the 16 h LC_{50} is around 0.5 mg/l and the 96 h LC_{50} is around 0.9 mg/l (Allan and Maguire, 1991). Similarly, in ponds, the lethal dissolved oxygen concentration for *L. vannamei* is around 1 mg/l (Hopkins et al., 1991).

In shrimp aquaculture ponds, it is obviously undesirable to have dissolved oxygen concentrations near levels that cause asphyxiation. Every effort is made to avoid acute exposure to low dissolved oxygen concentrations sufficient to kill shrimp. It is much more likely that shrimp are exposed to sublethal but stressful dissolved oxygen concentrations that affect their growth, feed conversion efficiency, and immune response.

The standard metabolism and routine activity of shrimp under a specific set of environmental conditions exert an oxygen demand that must be supplied by pond water. Respiratory and circulatory systems must be able to function to supply oxygen at some minimum rate necessary to support basic metabolic requirements. The minimum dissolved oxygen concentration required to accomplish this is the limiting oxygen concentration (LOCr) (Walker et al., 2009, 2011). LOCr at temperatures ranging from 20 to 32°C is about 2 mg/l, with most values between 1.5 and 2.5 mg/l, whether shrimp are starved or fed (Walker et al., 2011). LOCr does not vary with temperature but decreases as shrimp body weight increases. At oxygen concentrations less than the LOCr, shrimp will experience acute hypoxia and will not grow well. Molting of shrimp (*Penaeus semisulcatus*) was inhibited by long-term (17 days) exposure to a dissolved oxygen concentration of 2 mg/l (Clark, 1986). Molted shrimp (*P. monodon*) are more susceptible to mortality from low dissolved oxygen concentration (Allan and Maguire, 1991). Low dissolved oxygen concentration (2.1 mg/l) also increases the acute toxicity of ammonia to *P. monodon* (Allan et al., 1990).

In an effort to simulate the stress of daily short-term exposure to hypoxia that typifies semi-intensive shrimp ponds, Allan and Maguire (1991) subjected *P. monodon* to repeated hypoxia (0.5–0.6 mg/l or 1.0–1.1 mg/l dissolved oxygen) for 4, 8, or 12 h followed by exposure to normoxia (near saturation) for 21 days. There were no differences in the growth or feed conversion efficiency of shrimp subjected to any of the dissolved oxygen regimes. Similarly, growth of *L. vannamei* and *P. monodon* at 28°C and dissolved oxygen concentrations of 1, 2, 3, or 4 mg/l was reduced only at the lowest concentration (Seidman and Lawrence, 1985). In shrimp ponds stocked at 33 ppt salinity, where aeration was activated to maintain nominal minimum dissolved oxygen concentrations of 1, 3, or 5 mg/l, shrimp growth was not affected (McGraw et al., 2001). Exposure of *L. vannamei* to sublethal hypoxia (<3 mg/l) caused a decrease in osmoregulatory capacity at low salinity and in seawater, but the effect was temporary and shrimp recovered after 24 h in oxygen-saturated water (Charmantier et al., 1994).

Juvenile Chinese white shrimp *Fenneropenaeus chinensis* were exposed to hypoxia (2–3 mg/l dissolved oxygen) for 10 days and then normoxia (5.6 mg/l) for 30 days (Wei et al., 2008). Growth was depressed by exposure to hypoxia but shrimp compensated for growth depression in the 30-day period after switching to normoxic water, mainly by increased feed consumption and improved feed conversion efficiency. These observations suggest that shrimp can tolerate and adapt to short-term exposure to sublethal hypoxia by compensatory growth. In a follow-up study, Chinese white shrimp were exposed to 2.1, 3.1, 4.1, 4.7, and 5.5 mg/l dissolved oxygen for 30 days (Wei et al., 2009). Dissolved oxygen concentrations <4.1 mg/l resulted in reduced growth.

Shrimp appetite is reduced by exposure to hypoxia (<3.1 mg/l dissolved oxygen) and the ingestion of less feed directly reduces the metabolic energy available for growth (Seidman and Lawrence, 1985; Wei et al., 2008, 2009). Of the feed ingested, apparent feed digestibility is not affected by dissolved oxygen concentrations between 1 and 4 mg/l (Seidman and Lawrence, 1985).

Exposure to hypoxia affects the allocation of energy to growth and other metabolic processes. The proportion of ingested energy available for growth decreases and the proportion of energy allocated to energetic losses from feces, ammonia excretion, oxygen consumption, and molting increases as the dissolved oxygen concentration to which shrimp are exposed decreases, with a significant threshold at 4 mg/l (Wei et al., 2009, Re and Díaz, 2011). Energy partitioning that preserves essential metabolic functions at the expense of the scope for growth of shrimp exposed to hypoxia contributes to reduced feed conversion efficiency. The greatest proportion of energy allocated to scope for growth occurs in shrimp at 6 mg/l (Re and Díaz, 2011).

Exposure to hypoxia affects the catabolic pathways used to obtain energy from ingested feed. Under normoxic conditions, shrimp use a mixture of lipids, carbohydrate, and protein for energy. Under hypoxic conditions, shrimp shift to use predominantly protein for energy (Re and Díaz, 2011; Rosas et al., 1999). This shift is indicated by a reduction in the O/N ratio (oxygen consumption/ammonia excretion), where a low O/N ratio indicates proportionally greater ammonia excretion associated with the use of more protein for energy under hypoxic conditions. An abrupt reduction in the O/N ratio occurs at a dissolved oxygen concentration of around 2–3 mg/l (Re and Díaz, 2011; Rosas et al., 1999).

Chronic exposure to hypoxia leads to a shift from aerobic to anaerobic pathways to obtain energy. In *L. vannamei*, activation of anaerobic glycolysis results in the accumulation of lactate in the hemolymph as an end-product (Perez-Rostro et al., 2004; Racotta et al., 2002). After exposure to 14–16 days of chronic hypoxia, lactate and protein also increased in the hepatopancreas and muscle of *P. monodon* and *L. vannamei* (Perez-Rostro et al., 2004; Seidman and Lawrence, 1985). The mobilization of protein for energy to fuel routine metabolism under hypoxic conditions indicates that this protein is not available for somatic growth.

Hypoxia-inducible factor-1 (HIF-1) is a transcription factor that is induced during hypoxia to regulate the expression of genes that mediate the effects of hypoxia (Ulaje et al., 2019). Among the molecular responses, genes involved in producing glycolytic enzymes are upregulated by HIF-1. The HIF-1 in

L. vannamei has been characterized as a heterodimer with α and β subunits (Soñanez-Organis et al., 2009). The α subunit is more abundant in gills and the β subunit is more abundant in the hepatopancreas. In response to hypoxia, there is a large increase in the α subunit in gills but little change in the β subunit. HIF-1 is a key regulator of oxygen homeostasis in shrimp, as it is in other metazoans.

Acute hypoxia can induce oxidative stress, DNA damage, and lipid peroxidation in the gills, hepatopancreas, and hemolymph of *L. vannamei* (Li et al., 2016). Oxidative stress results from the accumulation of reactive oxygen species (ROS) such as superoxide, hydrogen peroxide, hydroxyl ions, and singlet oxygen. Overproduction of ROS during recovery from hypoxia can cause DNA damage in the gill, hepatopancreas, and hemolymph tissues (Li et al., 2016). Increased ROS induces production of HIF-1, which leads to upregulation of genes that produce antioxidant enzymes (Kniffin et al., 2014). Metallothionein proteins can reduce oxidative stress in shrimp by detoxifying ROS (Felix-Portillo et al., 2014). In *L. vannamei*, metallothionein genes are expressed at maximum levels in the hepatopancreas during hypoxia. Metallothionein genes are also activated by HIF-1. In general, genes that are activated during exposure to hypoxia contribute to reducing metabolic energy demand and increasing protection from oxidative stress (Rathburn et al., 2013).

Although exposure to hypoxia results in reduced growth associated with reduced feed consumption and metabolic changes in energy allocation, the effects are modest, occur only under severe hypoxia, or are reversible. Perhaps the greatest impact of exposure to hypoxia on shrimp aquaculture is its effect on shrimp survival. Poor survival will reduce crop yield and be the main factor affecting feed conversion efficiency. Survival is significantly lower when minimum dissolved oxygen concentration is <2 mg/l (McGraw et al., 2001; Wei et al., 2009). The time required for carapace hardening increases in post-molt shrimp exposed to hypoxia. During such an extended post-molt phase, shrimp with a soft carapace are subject to cannibalism. Shrimp (*P. semisulcatus*) exposed to chronic hypoxia (2 mg/l for 17 days) did not molt (Clark, 1986).

Shrimp survival is also affected by immune status, which is affected by hypoxia. Exposure to hypoxic stress decreases the total count of hemocytes, which are important in the encapsulation and phagocytosis of invading pathogens, increases prophenoloxidase activity, which is important in the part of the innate immune system of shrimp that controls melanization, and reduces the bactericidal activity of the hemolymph (Hu et al., 2009; Jiang et al., 2005; Le Moullac et al., 1998; Nonwachai et al., 2011). These effects are pronounced when dissolved oxygen concentration is <3.5 mg/l (Jiang et al., 2005). A lower total hemocyte count can increase susceptibility to pathogens such as *Vibrio alginolyticus* (Le Moullac et al., 1998). Exposure to hypoxia (<3 mg/l) can also increase the concentration of dopamine in the hemolymph (Hu et al., 2009).

The dissolved oxygen requirements of shrimp and the effects of hypoxia have implications for the factors that affect productivity and potential profitability of shrimp aquaculture. Lethal dissolved oxygen concentrations range from 0.2 to 1 mg/l, and the dissolved oxygen concentration needed to support basic metabolism ranges from 1.5 to 2.5 mg/l. Exposure to short-term sublethal hypoxia depresses shrimp growth but compensatory growth can occur once normoxia is restored. Hypoxia can result in reduced appetite, shift the allocation of dietary energy to reduce the quantity available for growth, shift to greater use of protein for energy, and shift from aerobic to anaerobic pathways to obtain energy. Exposure to hypoxia also compromises the immune status of shrimp, increasing susceptibility to disease and mortality. In shrimp production systems, dissolved oxygen concentration should always be >4 mg/l and ideally between 6 and 8 mg/l.

Factors affecting the severity of hypoxia

The discussion in the previous section on the effects of hypoxia emphasized the role of low dissolved oxygen concentrations on shrimp growth, feed conversion, survival, and physiological responses. Those affects are exacerbated when episodes of low dissolved oxygen concentration occur in conjunction with other sets of conditions that (1) reduce the rate of diffusion of oxygen from water to hemolymph, (2) decrease oxygen carried by hemolymph, (3) reduce the capacity of tissues to utilize oxygen, or (4) increase metabolic oxygen demand. Under these conditions, shrimp may experience hypoxia when dissolved oxygen concentrations are otherwise adequate.

Shrimp gills represent a thin barrier to the diffusion of oxygen from water to the hemolymph. Water quality degradation can lead to alterations to gill structure and function that slow the rate of diffusion, from increased thickness of the diffusion barrier, or reduce gill function and respiratory efficiency, as a result of lesions and other histopathological alterations to gill structure. In lined ponds operated with low rates of water exchange, Hopkins et al. (1993) observed moribund shrimp with gills that had trapped debris with associated epicommensal bacteria in ponds with adequate levels of dissolved oxygen. Similarly, histological alterations were observed in gills of Pacific white shrimp in high-density biofloc systems with high concentrations of total ammonia-N (13.1 mg/l), nitrite-N (1.4 mg/l), and total suspended solids (>100 mg/l) (Fregoso-López et al., 2017). Once water quality deteriorated, bioflocs and phytoplankton attached to or were trapped by branchial lamellae and obstructed gill chambers, followed by invasion of the parasites *Leucothrix mucor* and *Zoothanmnium* sp. Histological alterations of gills with increasing ammonia and nitrite concentrations occurred in three phases: (1) hemolymph infiltration in branchial lamellae, (2) branchial inflammation (epithelial edema and hemocyte infiltration) and melanization, and (3) necrosis.

The respiratory function of shrimp gills can be impaired during normal operation of the immune system (Burnett and Burnett, 2015). In response to introduced pathogens, circulating hemocytes aggregate with the pathogens to form nodules. As the hemolymph containing these aggregates circulates through the gills, the nodules become trapped in the narrow circulatory channels of the gill lamellae and are subsequently melanized. These trapped aggregates can impair gill perfusion by hemolymph and create a barrier to the diffusion of oxygen from water to hemolymph. This restriction on diffusion of oxygen incurs metabolic costs. Relief is obtained when melanized aggregates are eliminated during molting.

Several substances affect the amount of oxygen transported by hemolymph, primarily by affecting the binding of oxygen to hemocyanin. In estuarine environments and eutrophic shrimp ponds, photosynthesis and respiration result in a diel pattern where dissolved oxygen concentration is least and dissolved carbon dioxide concentration is greatest in the early morning. dissolved carbon dioxide concentration can also be elevated in super-intensive systems, especially those such as biofloc systems with high water column respiration rates. Carbon dioxide acts as an acid in water, increasing the hydrogen ion concentration. Thus, low dissolved oxygen concentration (hypoxia) occurs at the same time as elevated carbon dioxide concentration (hypercapnia) and hydrogen ion concentration. Penaeid shrimp have a strong positive Bohr effect, where a greater hydrogen ion concentration and, independently, hypercapnia decreases the oxygen-binding affinity of hemocyanin. The oxygen-binding affinity of hemocyanin can be increased by calcium and chloride ions (Brouwer et al., 1978).

Nitrite is an intermediate product of nitrification and may accumulate to toxic levels during the initial start-up period (first 3–6 weeks) of intensive and super-intensive ponds. Exposure to nitrite reduces the oxygen-binding affinity of hemocyanin, even though the partial pressure of oxygen in the hemolymph increases in response to nitrite exposure (Chen and Cheng, 1995; Cheng and Chen, 1999; Cheng et al., 2013). Nitrite binds with deoxyhemocyanin to form methemocyanin, a form that is incapable of binding oxygen. Nitrite exposure also reduces the oxyhemocyanin content of the hemolymph and inhibits hemocyanin synthesis (Li et al., 2019b). Salinity is a mitigating factor, and nitrite toxicity decreases as salinity increases (Barbieri et al., 2014; Lin and Chen, 2003). Chloride and nitrite ions have a similar molecular diameter and the same charge, and are transported across gill epithelia by similar mechanisms. The recommended safe concentration of nitrite-N in low-salinity (2 g/l) shrimp ponds is <0.5 mg/l (Gross et al., 2004). Shrimp that are molting are particularly susceptible to nitrite toxicity (Furtado et al., 2016b).

Hydrogen sulfide is an example of a substance that reduces the capacity of tissues to utilize oxygen. Hydrogen sulfide is produced in anaerobic sediments by the reduction of sulfate, an abundant anion in brackish water and seawater. In undisturbed sediment with an aerobic sediment–water interface, hydrogen sulfide is prevented from diffusing to the overlying water. However, periodic disturbance of sediment by water currents or some farm operations, or low dissolved oxygen concentrations at the pond bottom that remove the barrier to diffusion from sediment to water, can result in hydrogen sulfide entering the water column. Sulfide binds to and inhibits the activity of cytochrome C oxidase, the enzyme that catalyzes the final step in electron transfer in mitochondria, thereby blocking oxygen transport and energy production during oxidative phosphorylation. Although the partial pressure of oxygen in hemolymph may be otherwise sufficient, sulfide prevents oxygen from being utilized by tissues. Hypoxia and low pH can intensify sulfide toxicity (Thuluasi et al., 2020).

The final set of conditions that may cause tissue hypoxia are those that increase metabolic oxygen demand. In teleosts with a true stomach, a large increase in metabolic oxygen demand may occur following feeding. Post-prandial peaks in oxygen demand by fish may result in hypoxia if the oxygen supply is insufficient and dissolved oxygen concentration is low. The gut clearance time of penaeid shrimp is rapid (<1 h) and continuous consumption of small meals is preferred. Therefore, the magnitude of post-prandial oxygen demand relative to basal metabolism is small. A more important factor affecting shrimp metabolic oxygen demand is molting status. As discussed earlier in the section on factors affecting shrimp respiration, respiration rate is maximum during the post-molt stage (Ding et al., 2014), largely in response to increased energetic requirements for osmoregulation when shrimp absorb a relatively large volume of water following molting. Shrimp are particularly susceptible to the effects of hypoxia during the post-molt stage.

Production intensity and dissolved oxygen management

Production system intensity can be defined in terms of stocking density, crop yield, or sustainable feed loading rate, among other indicators. Commonly, intensity levels are distinguished relative to each other using the qualitative terms extensive, semi-intensive, intensive, and super-intensive (Table 9.4). Conceptually, production intensity increases on a continuum but, in practice, production intensity increases as a discrete step function, with different levels of intensity being associated with different packages of practices and technology that have evolved to overcome limiting factors and improve system productivity. The main components of those packages are nutrition and feeding, species and breeds, health and disease management, and water quality management. The elements of each package

TABLE 9.4 Characteristics and indicators of the range of production intensity of penaeid shrimp and the associated principal mechanism for dissolved oxygen (DO) management.

Intensity descriptor	*Pond size (ha)*	*Stocking density (no./m²)*	*Yield per crop (t/ha)*	*Primary mechanism of DO management*
Extensive	2–3	0.5–3	0.2–0.5	Tidal water exchange
Semi-intensive	<0.5–>20	5–15	1–2	Pumped water exchange
		15–30	2–4	Pumped water exchange, some aeration
		25–45	2.5–5	≤10 hp/ha aeration
Intensive	0.2–1	60–100	5–10	15–25 hp/ha aeration
Super-intensive	0.15–0.25	150–250	15–35	>50 hp/ha aeration
	0.08–0.2	300–400	40–80	>80 hp/ha aeration

are complementary because technology development of one component may be limited by technology development in another area. For example, full realization of the production potential of genetically improved shrimp breeds requires high-health seed, feeds that meet nutritional needs at high densities, biosecurity and probiotics that maintain health, and management practices to maintain stable water quality in the culture environment at optimum levels.

Shrimp production intensity varies by at least two orders of magnitude, from 0.5 t/ha per crop for extensive systems to more than 50 t/ha for super-intensive systems. In general, as in all fed aquaculture systems, there is a direct relationship between stocking density and yield, and an inverse relationship between pond size and yield. Obtaining high yields in large (>10 ha) ponds is not possible because it is difficult to control water quality, especially dissolved oxygen concentration, at levels that support high feeding rates.

In part, the history of shrimp farming has been one of progressive intensification. During the early years of commercial shrimp farming, and remaining important to the present day, most shrimp ponds were extensive. Through the 1980s, most extensive shrimp ponds were large (20–100 ha), with stocking densities of 2–8/m² and production levels of 200–500 kg/ha per crop. Extensive ponds remain important: although they represent only 13% of global shrimp production volume, extensive ponds represent about 46% of the total land area used for shrimp farming (Boyd and McNevin, 2018). The largest areas of extensive ponds are in Vietnam, Bangladesh, Myanmar, Indonesia, and China.

In the mid- to late 1990s, farmers began to shift from culture of black tiger shrimp (*P. monodon*) to culture of domesticated Pacific white shrimp and feeds became more widely available. Semi-intensive farming, with stocking densities up to 30/m², resulted in production levels of 1.5–3 t/ha in 10 ha ponds.

By the late 2000s, the availability of reliable electric power was more widespread, allowing more aeration to support the feed loading necessary to grow shrimp at higher densities. Pond sizes were reduced to 1–2 ha, with stocking densities of 60–90/m² and production levels of 5–10 t/ha. In the 2010s, super-intensive systems in 0.25–0.5 ha plastic-lined ponds were stocked at 100–200/m², with production levels of 15–25 t/ha. In the late 2010s, even smaller (0.1–0.3 ha) super-intensive ponds are stocked at >250/m², with production levels of 30–60 t/ha.

Although intensification of shrimp farming has been the trend over the past few decades, the availability, reliability, and cost of electric power are major factors limiting the expansion of intensive forms of shrimp production. In systems dependent on electric power, the response times to initiate remedial action may be quite short (<1 h) before dissolved oxygen depletion occurs and shrimp suffocate from lack of dissolved oxygen. Power requirements and energy use for aeration are addressed in later sections of this chapter.

In addition to containment, shrimp farmers must provide nutrition, oxygen, and waste treatment as essential ecosystem services to support production. The supply of each service has a limit, with the most limiting service defining carrying capacity. The supply of services can be augmented by technology, such as aeration, that can increase carrying capacity. Essential ecosystem services are interrelated, with changes in the availability of one service having knock-on effects on the supply of other services. For example, increasing the feeding rate to support higher stocking densities creates greater demands on oxygen supply and waste treatment services. The approaches, practices, and technologies for managing oxygen supply vary across the production intensity gradient. As production intensity increases, the approach to managing each service changes, with progressively greater dependence on inputs from outside the pond.

EXTENSIVE SYSTEMS

Extensive ponds closely reflect conditions in nearby intertidal estuarine areas, where such systems are commonly located. Oxygen supply is seldom a problem because stocking densities and pond inputs to support shrimp nutrition are low. Shrimp obtain nutrition from natural food webs, with some supplementation with agricultural by-products or commercial feed, but the overall input rarely impacts oxygen supply. In such a low-management natural system, the risk of dissolved oxygen depletion is generally low, although episodes may occur. When those events occur, air-to-water diffusion of oxygen can be of critical importance in supplying sufficient oxygen to support the shrimp crop.

Extensive ponds have traditionally used tidal water exchange as the principal management tool to avoid pond stagnation and episodes of hypoxia. Depending on farm location with respect to the intertidal zone, water exchange occurs with variable frequency and intensity, and is thus inconsistent from day to day and does not necessarily meet the needs for oxygen as shrimp biomass increases. However, for the most part, tidal water exchange creates conditions favorable for algal photosynthesis to provide sufficient oxygen to support the low-density shrimp crop.

SEMI-INTENSIVE SYSTEMS

Semi-intensive systems are exemplified by commercial shrimp farms in Ecuador, where shrimp are stocked at 10–12/m^2. The typical pond size (~10 ha) is too large to allow a high degree of control over water quality, including dissolved oxygen concentration. Shrimp are provided with high-quality feeds, supplemented by natural food. Maximum sustained feeding rates at the end of the production cycle are near 30–35 kg/ha·d, which is the recommended feeding rate limit to avoid depletion of dissolved oxygen in channel catfish ponds. Yields range from 1–2 t/ha per crop, with two crops per year.

With the greater stocking densities of semi-intensive ponds and siting of ponds above the intertidal zone, shrimp farmers replaced inconsistent tidal water exchange with regular pumped water exchange. Water exchange of semi-intensive ponds is not practiced during the first month of the crop cycle because dissolved oxygen concentration is rarely a problem when shrimp are small and biomass and feeding rate are low. Water exchange can be contingent on specific water quality thresholds of dissolved oxygen concentration or transparency. After the first month, the water exchange rate of semi-intensive shrimp ponds ranges from 5–20%/d and is typically 10–15%/d.

Water exchange is traditionally practiced for semi-intensive shrimp culture, although its value is questionable. Some farmers believe that water exchange increases oxygen supply by providing oxygen with incoming pumped water. However, at water exchange rates that are typical in semi-intensive shrimp culture, the addition of new dissolved oxygen with incoming water is modest. Assume a 1 ha, 1 m deep (for simplicity) shrimp pond with a dissolved oxygen concentration of 3.0 mg/l. Then, at a 10%/d water exchange rate, 10 cm of water is discharged before refilling with 10 cm of new water with a dissolved oxygen concentration of 6.0 mg/l. The quantity of dissolved oxygen in the 1 ha pond after draining is 90 cm × 3.0 mg/l × 1 ha = 27 kg and the quantity of dissolved oxygen in the 10 cm of new water is 10 cm × 6.0 mg/l × 1 ha = 6 kg. Thus, the new total quantity of dissolved oxygen is 33 kg, resulting in a concentration of 3.3 mg/l, or a quite modest 10% increase, providing limited relief from hypoxia. The dilution of nutrients leading to pond eutrophication is similarly limited by water exchange at this level.

Research conducted at the Waddell Mariculture Institute in the 1990s (Hopkins et al., 1991, 1993, 1995, 1996) established the limited value of water exchange as a dissolved oxygen management tool. In small (0.1 ha) lined ponds stocked at 22–66/m^2, there was no difference in shrimp production from ponds managed with a daily water exchange between 2.5% and 25%. Production levels of 6–8 t/ha per crop were achieved with no or limited (2.5%) water exchange. There was no relationship between water exchange rate and early morning dissolved oxygen concentration. As long as the early morning dissolved oxygen concentration was adequate, water exchange was not necessary to obtain good shrimp growth and survival. Based on this research, water exchange may be effective only when the daily exchange rate is more than 25%.

Water exchange can be an effective practice to manage dissolved oxygen concentration if the source water has a low concentration of suspended organic matter (<25 mg/l total suspended solids) and the practice is implemented in a timely way and with an understanding of underlying mechanisms. In brief, water exchange can improve oxygen supply by maintaining an intermediate water transparency that increases the availability of underwater solar radiation to drive phytoplankton photosynthesis. This approach to water exchange considers the fundamental relationship between phytoplankton biomass density and net oxygen production (Fig. 9.1). As phytoplankton density increases, gross primary productivity increases but reaches a point of saturation where further increases in density do not result in an increase in gross primary productivity, an effect described as self-shading. In contrast, phytoplankton respiration increases linearly with phytoplankton density. Net photosynthesis is gross primary productivity minus respiration. Net primary productivity is maximum at intermediate levels of phytoplankton density. At low phytoplankton density, gross primary productivity and water respiration are low and net primary productivity is often negative (oxygen consumption > oxygen production). At high phytoplankton density, respiration exceeds gross primary productivity and net primary productivity is likewise negative. It is only at intermediate density levels that gross primary productivity exceeds phytoplankton respiration, resulting in positive net primary productivity. This relationship, then, is the underlying basis for the recommendation to maintain water transparency (as measured by Secchi disk) at intermediate levels between 30 and 50 cm to avoid episodes of low early morning dissolved oxygen concentration.

If the transparency of the water source for water exchange is good, such as oceanic water, then water exchange can be effective at flushing away a fraction of the phytoplankton biomass and diluting nutrients, increasing water transparency and the availability of light in the water column to drive photosynthesis. Estuarine water quality can be quite variable, ranging from clear to extremely turbid, and with a dissolved oxygen concentration that may be less than that of pond water. Pumping turbid, low dissolved oxygen concentration source water for water exchange is self-defeating and provides no enhancement of the underlying processes that improve dissolved oxygen concentration and pond carrying capacity in non-aerated ponds. Managing pond water transparency is the best rationale for the use of water exchange to manage dissolved oxygen concentration in semi-intensive shrimp ponds.

Although its value as a tool for the management of dissolved oxygen concentration is limited, water exchange can improve water quality in other beneficial ways. Water exchange continues to be practiced widely in semi-intensive shrimp farming, despite the elevated biosecurity risks associated with operating an open system. Some shrimp farmers use water exchange as part of internal, farm-level water recirculation. Water exchange and internal recirculation support and enhance the waste treatment service, especially in drainage canals. Water exchange can aid pond mixing and avoidance of pond stratification. The area of water movement adjacent to pond inlets may be preferred and sought out by shrimp.

The carrying capacity of large (~10 ha) semi-intensive shrimp ponds operated with water exchange is about 2.0–2.5 t/ha. Given the relative ineffectiveness of water exchange as shrimp production is progressively intensified, the most effective way to increase pond carrying capacity is to augment the oxygen supply through mechanical aeration.

Effective aeration of 10 ha ponds is difficult. Aerators are often scattered along perimeter embankments or deployed at intervals or in a grid throughout the pond area. Each aerator creates a zone of influence that can support the respiration of a certain biomass of shrimp. The effect on overall oxygen supply is modest and the main objective is to address the risk of occasional episodes of dissolved oxygen depletion derived from calm weather, high temperature, and high phytoplankton biomass.

With increasingly widespread use of automatic feeding equipment, aeration of feeding zones to maintain favorable dissolved oxygen concentrations is a prudent approach that focuses placement of a limited number of aeration devices where they may be most beneficial. During feeding, shrimp density may increase by an order of magnitude in the feeding area surrounding automatic feeders as many shrimp gather to consume feed. The carrying capacity of large (10 ha), intensively aerated (5–6 hp/ha) semi-intensive ponds is around 3.5–4 t/ha, although production of 6 t/ha is possible with partial harvests.

INTENSIVE AND SUPER-INTENSIVE SYSTEMS

Further intensification of production has been accompanied by a reduction in pond size. Productivity is inversely related to pond size, with production levels of up to 10 t/ha in 1–2 ha ponds, 20–30 t/ha in 0.5 ha ponds, 30–40 t/ha in 0.1–0.2 ha ponds, and 50–60 t/ha in ponds <0.1 ha. In intensive shrimp ponds (>60/m^2), oxygen supplied by mechanical aeration is essential to support respiration of large shrimp standing crops and elevated oxygen demand for respiration by phytoplankton and bacteria in the water column and sediment. Although a production sector developed around intensive farming of black tiger shrimp in Taiwan from the late 1970s until its collapse in the late 1980s, most intensive farming of penaeid shrimp currently uses Pacific white shrimp.

Intensive systems are operated in hydraulic modes from open, flow-through to closed, recirculating. One of the drivers toward intensification has been concern about viral biosecurity and thus many ponds are now operated with low rates of water exchange, including so-called zero exchange ponds and biofloc ponds. The system operated by the commercial shrimp farm Belize Aquaculture Limited, starting in the late 1990s, was an early successful example of operating ponds with low water exchange rates. However, intensive ponds have been successfully operated with high water exchange rates (>1000%/cycle) and moderate water exchange rates (5–10%/d) but with all water recirculated internally at the farm level.

Oxygen dynamics in many intensive ponds are in part related to a pattern of temporal succession from an algal-dominated system to a mixture of algae and bacteria to a bacteria-dominated system as feeding rate increases. At the transition point, at a feeding rate of about 300 kg/ha·d, occurring at 8–10 weeks at Belize Aquaculture (McIntosh, 2000b), there is a rather abrupt shift that is marked by a decline in phytoplankton photosynthesis and a large increase in water respiration rate as the system becomes bacteria-dominated. Beyond this transition, further increases in production are supported by proportionally greater aeration.

Another characteristic of the transition from an algal-dominated to a bacteria-dominated system is the reduction in the magnitude of diel fluctuation of dissolved oxygen concentration (McIntosh, 2000b). Most intensive ponds are operated with a mixture of algae and bacteria, with the relative importance of each process dependent on the feeding rate. In super-intensive ponds <0.1 ha, many producers partially cover ponds with shade cloth to reduce the role of phytoplankton and thereby obtain greater stability in water quality with respect to fluctuations in dissolved oxygen and pH.

Ponds <0.5 ha stocked at densities >60/m^2 are typically lined with high-density polyethylene plastic. Among other advantages, lining ponds will avoid the scouring and deposition of soil caused by high aeration rates required to support the respiratory demand associated with high feeding rates. With respect to oxygen budgets, lining ponds with plastic can reduce SOD relative to earthen ponds by removing the requirement for oxygen to oxidize reduced substances diffusing to the sediment surface and to oxidize the reservoir of organic matter normally contained in pond soils. Of course, organic matter settling to the bottom of lined ponds will exert an oxygen demand as it is decomposed.

Smaller (<0.5 ha) intensive ponds often include a central cone for collection of fecal solids, molted exoskeletons, dead phytoplankton and bacteria, and heavier bioflocs. Removal of these solids reduces the feed oxygen demand, with the net effect that more of the oxygen produced by mechanical aeration is available to support the shrimp crop and proportionally less is used for water column and sediment respiration. In comparison to extensive and semi-intensive ponds, a greater proportion of the overall respiratory demand for oxygen is required by the shrimp crop in intensive ponds.

Growing shrimp in biofloc or semi-biofloc ponds has been successful at commercial scale (Taw et al., 2013). Bioflocs are aggregates of detritus, algae, bacteria, and grazing organisms that are maintained in suspension by vigorous continuous aeration. Heterotrophic and nitrifying bacteria that are part of suspended bioflocs are responsible for controlling ammonia concentration. Paddlewheel aerators provide oxygen, circulate water, and force waste solids to concentrate at the pond center, where they are periodically removed by draining or siphoning to limit oxygen demand. Aspirator aerators can be positioned to aid the process of concentrating solids near the center of the pond bottom. Shrimp are stocked at 130–150/m^2 and ponds are aerated at 23–32 hp/ha and operated with low water exchange (<1%/d), resulting in shrimp yields of 20 t/ha or more (Taw et al., 2013). Without the turbulence caused by aeration, flocs would settle quickly and dissolved oxygen depletion to critical levels would occur with 0.5–1 h.

Increasingly, super-intensive shrimp-farming systems are operated as farm-level recirculating systems where relatively small, lined tanks (500–1000 m^3) are stocked at high density (>200/m^2) and integrated with a waste treatment system consisting of combinations of settling ponds, facultative ponds, and oxidation ponds that often include sanitary species such as tilapia. If properly designed, these systems can be operated with reuse of water for many production cycles over several years. Aeration is concentrated in shrimp production tanks, with long-arm or conventional paddlewheel aerators, mainly to circulate water, and either an array of air diffuser tubing outlets across the pond bottom or a Venturi pump and aspirator to produce microbubbles. In the waste treatment components, dissolved oxygen concentration is not controlled in sections operated in an anaerobic or facultative mode. In oxidation ponds, oxygen from phytoplankton photosynthesis, supplemented by mechanical aeration, especially if fish are included, is used to supply requirements. The concentration of aeration placement in high-density shrimp tanks increases the proportion of oxygen used by the shrimp crop because oxygen demand in the form of waste solids and ammonia is displaced to treatment ponds with recirculating water exchange.

Shrimp production in super-intensive biofloc systems in shallow raceway tanks covered by greenhouses has been evaluated by members of the US Marine Shrimp Farming Program. Plastic-lined raceways (40–100 m^3) have been stocked at 300–500/m^3. Biofloc concentration was controlled by settling tanks and foam fractionators. Diffused air and Venturi injectors provided sufficient dissolved oxygen to produce yields of 5–6 kg/m^3 up to 7.5 kg/m^3 (Browdy et al., 2012; Samocha, 2019). Beyond this level, dissolved oxygen supply limits production. With pure oxygen supplementation, yields of around 10 kg/m^3 are possible (Samocha, 2019).

Aerators for shrimp farming

The capacity of an aerator to transfer oxygen to water is reported as the standard aeration efficiency (SAE), in kg/hp·h or kg/kW·h. The SAE is determined under a standard set of conditions (clean water, 20°C, 0 mg/l dissolved oxygen) and is a useful value to compare aeration devices when selecting aerators. However, under the practical conditions that prevail during operation in shrimp ponds, aerators do not transfer as much oxygen as indicated by the SAE. The main factors that affect oxygen transfer rate are dissolved oxygen concentration during aerator operation, water temperature, total dissolved solids and surfactants in water, type and power of aeration device, water turbulence, and pond geometry.

The main factor that affects the oxygen transfer rate is dissolved oxygen concentration during aerator operation (Fig. 9.2). The difference between the saturation dissolved oxygen concentration and the concentration at which aerators are operated provides the driving force for gas transfer. At a pond dissolved oxygen concentration of 3.0 mg/l, actual oxygen transfer efficiency will be about half of measured aerator SAE (Boyd, 1998). At a dissolved oxygen concentration near saturation, aerators transfer very little oxygen, and when pond waters are supersaturated with respect to dissolved oxygen, aerators transfer oxygen from water to air.

Penaeid shrimp are farmed commercially in water with a very broad range of salinities and therefore aerators may be operated in fresh to hyper-saline water. The oxygen transfer efficiency of paddlewheel, Venturi, diffused, and Venturi injector aerators increases with salinity (Fast et al., 1999; Ruttanagosrigit et al., 1991; Vinatea and Carvalho, 2007; Yang and Park, 2011). For paddlewheel and Venturi aerators, there is a large increase in standard oxygen transfer rate (SOTR) and SAE between 0 ppt and 10–15 ppt (Fast and Boyd, 1992; Vinatea and Carvalho, 2007). The SOTR of a paddlewheel aerator at 15 ppt is 68% greater than at 0 ppt, and that of a Venturi aerator was 151% greater (Vinatea and Carvalho, 2007). Across

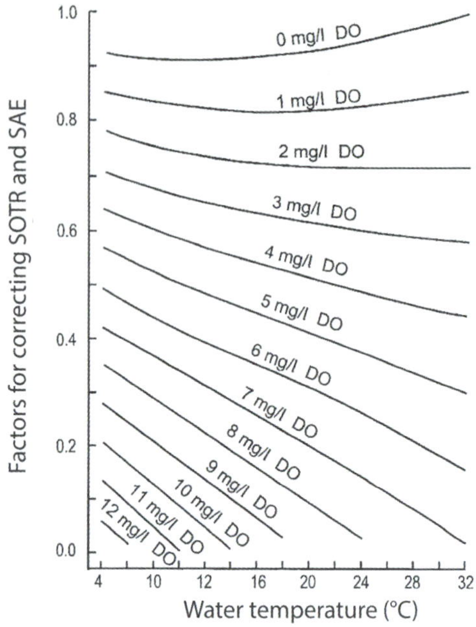

FIGURE 9.2 Factors that can be applied to correct the standard oxygen transfer rate (SOTR) or SAE as a function of water temperature for various pond dissolved oxygen (DO) concentrations. At typical water temperatures for shrimp farming and at the dissolved oxygen concentrations at which aerators are operated (4–5 mg/l), the actual oxygen transfer rate and aeration efficiency is about 50% of standard values determined by aerator tests.

the salinity range from 0 to 60 ppt, the best SAE occurs at 30 ppt, with slightly lower SAE at even greater salinity (Vinatea and Carvalho, 2007). Bubbles created by aeration of saline water are smaller and more numerous than those created in freshwater (Fast et al., 1999; Pipoppinyo and Boyd, 1993; Ruttanagosrigit et al., 1991), presumably because surface tension increases with salinity. Smaller bubbles increase the surface area for gas transfer.

Broadly, aerators in shrimp aquaculture can be distinguished between surface splasher types that throw water into the air and submerged bubbler types that inject air into water. Paddlewheel aerators and Venturi aerators are the most commonly used types in commercial shrimp farming, although diffused aeration is widely used in hatchery and nursery applications and increasingly in super-intensive culture.

PADDLEWHEEL AERATOR

Paddlewheel aerators have been in widespread use in shrimp farming since the 1980s and are by far the most common aerator type in shrimp farming. At its core, a paddlewheel aerator consists of a rotor with paddles arranged around it. Oxygen transfer is accomplished in at least two ways. First, as a paddle enters the water surface, an air cavity is created behind the paddle. As that paddle rotates through the water, the air cavity collapses, creating a swarm of bubbles behind the paddle that are pushed forward when the next paddle rotates through the water. Second, as paddles exit the water surface, water is thrown into the air as droplets to a height and distance from the aerator depending on the rotational speed, paddle type, paddle arrangement on the hub, and paddle tip immersion. As droplets travel through the air, breaking up as they collide with other droplets, a large surface area for gas transfer is created. Turbulence created by rotating paddles creates water currents that move water aerated by both mechanisms away from the aerator and replaces it with low-oxygen water.

FIGURE 9.3 The most common aerators in shrimp farming are Asian-style, electric paddlewheel aerators, like these 2 hp models. There are typically two impellers per horsepower.

Since its introduction in the early 1980s, the most popular aerator in shrimp farming has been the 1–3 hp Taiwan-style paddlewheel aerator (hereafter called "Asian-style" because there are now manufacturers in several Asian countries) (Fig. 9.3). The aerator includes two plastic impellers per horsepower, each impeller being 65 cm in diameter and having eight paddles that are 20 cm wide and 18 cm long. The typical depth of submergence of each paddle is about 5–6 cm. Each paddle has a grid of round 20 mm diameter holes intended to create additional turbulence and effect oxygen transfer. Impellers rotate on the drive shaft at 80–100 rpm.

Often, impellers are aligned in such a way that paddles on each impeller enter the water at the same time. This creates pulsed loading and unloading of the drive motor. However, aerator research has indicated that paddlewheel aerators with solid paddles staggered in a spiral arrangement around the hub for more continuous motor loading are more efficient. The SAE of Asian-style paddlewheel aerators is about 0.9–1.2 kg/kW·h (Boyd, 1998; Peterson and Walker, 2002), on the low end of potential paddlewheel performance (Table 9.5). More efficient designs of paddlewheel aerators have an SAE of about 2.0–2.5 kg/kW·h to as high as 3.0 kg/kW·h. Efficiency depends on impeller (hub) diameter, paddle design, depth of paddle tip submergence, and rotational velocity.

Although the Asian-style paddlewheel aerators are not very efficient in comparison to other designs, they are relatively inexpensive (<\$500 for a 2 hp unit in 2020) and have a relatively short useful life of

TABLE 9.5 Typical ranges in SAE of common aquaculture aerators (Boyd and McNevin, 2021).

Type	*SAE (kg O_2/hp·h)*
Paddlewheel	
1–2 hp floating, electric (Asian style)	0.9–1.2
10 hp floating, electric (US catfish pond style)	1.4–1.8
Venturi	0.9–1.2
Vertical pump	0.8–1.1
Diffuser	0.4–0.8

3–5 years. They are the workhorse of shrimp pond aeration in semi-intensive, intensive, and some super-intensive ponds.

In the past ten years, long-arm paddlewheel (LAPW) aerators (Fig. 9.4) have become popular with intensive and super-intensive shrimp farmers because of their local availability, low cost, simple construction and installation, and ease of operation. Although commercial models are available, most LAPWs are fabricated on-farm or by local machine shops. They are made by extending the shaft (hub) to a custom length, up to 20 m, often using inexpensive galvanized iron water pipe. Impellers from 1–3 hp Asian-style paddlewheel aerators or custom-made paddles are staggered on the hub, 30–50 cm apart, and floats to support bearing blocks are likewise staggered along the hub. Paddlewheel assemblies are connected through a drive shaft to a diesel or electric motor on the pond embankment.

Although the primary purpose of LAPW aerators is aeration, they are also valued for their water pumping capacity, which provides mixing and circular water currents in square or round ponds. Establishing these currents is necessary to concentrate waste solids near the pond center, where they can be removed by siphoning or via central drains, thereby keeping the pond bottom clean. Increasing the depth of submergence of the paddle tip creates strong water currents that circulate water but reduces oxygen transfer compared with conventional paddlewheel aerators. Increasing the depth of paddle tip submergence also increases the wetted surface area, thereby increasing the torque (power) requirement. In general, deeper paddle submergence and slower rotational speed favor water pumping and movement, and shallower paddle submergence and faster rotational speed favor oxygen transfer. In some super-intensive ponds, some LAPW aerators are dedicated to water movement and mixing and others are dedicated to oxygenation.

There has been no systematic evaluation of the SAE of LAPW aerators, nor has there been standardization of their design. However, there are many reasons to believe that many LAPW aerators are inefficient with respect to oxygen transfer. Custom-made LAPW aerators are often designed and operated with poor appreciation of best practices. Specifically, the depth of paddle tip submergence and rotational speed are not well controlled. The rotational speed of LAPW aerators powered by diesel engines can be adjusted using the fuel throttle. At low speeds, oxygen transfer is low but the aerator is efficient as a water pump, especially if operated with a paddle tip submergence >10 cm.

FIGURE 9.4 A LAPW aerator powered by a motor placed on the pond embankment. Such aerators may extend to a length of 20 m.

FIGURE 9.5 A 3 hp spiral aerator with plastic components that resist corrosion. The blades of the aerator consist of 3.8 cm hollow plastic tubes with tips cut at a diagonal and 70 staggered blades per spiral. This 3 hp model has three spirals, each of which has an SOTR of 2 kg/h. The shaft rotates at 120 rpm. Photograph: David Kawahigashi.

A variation of the paddlewheel aerator is the spiral aerator. The aerator pictured in Fig. 9.5 has three hubs that are 150 cm long each, and has blades consisting of 70 cm lengths of 3.8 cm HDPE pipe that extend through the hub. Blades are staggered around the hub in a spiral pattern, approximately 2 cm apart, with 70 blades per hub. The tip of each blade is cut on a diagonal. The mechanism of oxygen transfer of a spiral aerator is similar to that of a paddlewheel aerator. One advantage of arranging blades in a spiral pattern is more continuous loading of the drive motor, thus increasing its longevity. The plastic components of the spiral aerator are also resistant to corrosion caused by operation in saline waters.

VENTURI AERATOR

Venturi aerators (Fig. 9.6), also known as propeller aspirator pump aerators or simply aspirator aerators, have also found use in semi-intensive shrimp farming, although not nearly to the same extent as paddlewheel aerators. Venturi aerators are often deployed to aerate and destratify deeper areas of shrimp ponds. This aerator type is also known by the popular brand name Aire-O_2®, manufactured by Aeration Industries, although other companies manufacture aerators of this type. The aerator consists of a direct-drive, rotating, hollow shaft with a propeller at the end to jet water. As the propeller rotates, the movement of water away from the propeller creates a partial vacuum (Venturi effect) in the shaft. Air is drawn through holes in the part of the hollow shaft that is above water and ejected below water near the propeller in a stream of fine bubbles. The angle of the shaft of the aerator can be adjusted from 30–60° to force

FIGURE 9.6 Venturi aerators are often used in shrimp ponds for destratification or for aeration of deeper pond zones, or are positioned near the center of intensive ponds to concentrate wastes at the pond center. The angle of the hollow rotating shaft is adjustable.

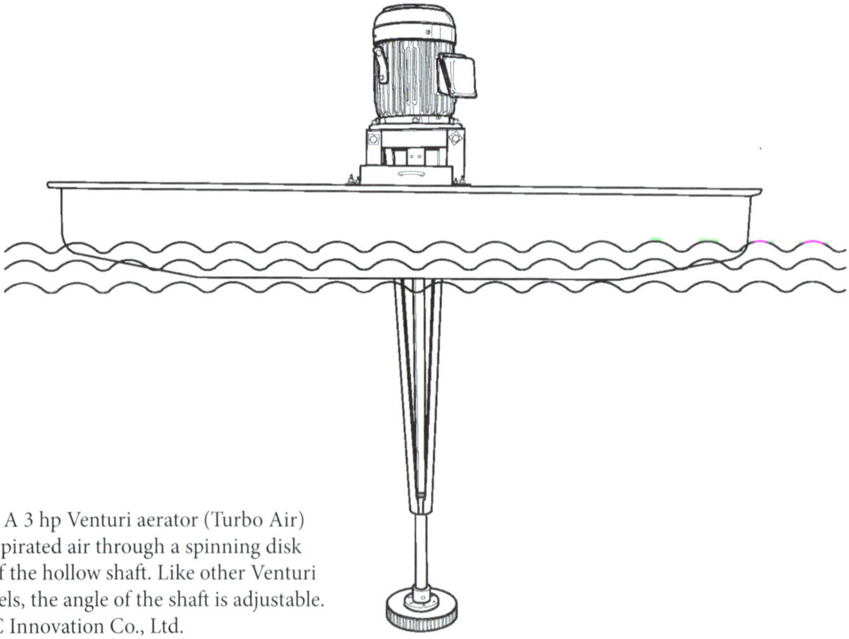

FIGURE 9.7 A 3 hp Venturi aerator (Turbo Air) that ejects aspirated air through a spinning disk at the base of the hollow shaft. Like other Venturi aerator models, the angle of the shaft is adjustable. Source: TNC Innovation Co., Ltd.

aerated water into greater depths than surface paddlewheel aerators. A more vertical angle increases the retention time of bubbles in the water column, increasing oxygen transfer.

A variation of the Venturi aerator consists of a motor connected to a narrow hollow shaft that is attached to a bilayer disk with numerous channels that extend outward from the center around the circumference of the disk (Fig. 9.7). When the disk rotates, water is pushed outward from the center of rotation through the channels, drawing air through the hollow shaft and ejecting it at the center of the spinning disk. Operation creates a swarm of medium-sized macrobubbles that flow away from the disk. The shaft can be oriented at angles ranging from about 15° to 90° (vertical) relative to the water surface.

VERTICAL PUMP AERATOR

Vertical pump or fountain aerators are not used much in shrimp grow-out ponds, although they may be installed in smaller nursery ponds or tanks. The aerator consists of an electric motor inside a waterproof housing that is attached to a propeller. The motor assembly is submersed and attached to a flotation collar or ring. In operation, the pump propels water into the air, where it breaks into droplets and oxygen transfer takes place. Some vertical pump aerators have deflectors that increase the diameter of the spray pattern. Propeller pitch is a key design variable because it affects height of the spray. Although fountain aerators that spray water high into the air appear spectacular, they are not as efficient as aerators that spray a greater volume of water to a moderate height (1–2 m) with the same energy. In grow-out ponds, vertical pump aerators are not good mixers because water circulates within a restricted zone near the aerator.

DIFFUSED AERATOR

Diffused aerators are not used in semi-intensive and many intensive grow-out ponds but have found use in super-intensive ponds and in nursery tanks and ponds. A diffused aeration system consists of a high-volume, low-pressure regenerative air blower connected to an air-distribution manifold that supplies air to diffusers. Diffusers in shrimp grow-out ponds most often consist of lengths of porous tubing arranged in lines or in spiral or coiled arrays. The tubing must be weighted or attached to the pond bottom. Spiral or coiled lengths of porous tubing are often attached to metal frames that position the diffuser about 10 cm above the pond bottom.

Air bubbles (3–5 mm) are emitted from the diffuser and rise through the water column. Gas transfer takes place across the two-film air–water interface of each bubble. Oxygen transfer rate is a function of gas flow rate and water depth, which affects the time that bubbles remain in the water column before bursting at the surface. The rate at which bubbles rise in the water column is a function of bubble size and follows Stokes' law. Bubbles that are 1.5–4 mm in diameter have a rising velocity of 25–40 cm/s (Jamnongwong et al., 2016). Most shrimp ponds are shallow (1.2–1.5 m depth) and thus the duration that bubbles remain in the water column is quite short (<10 s) and so there is a limited opportunity for gas transfer to take place. For this reason, the SAE of diffused aerators in aquaculture ponds is low (0.5–1 kg/kW·h). Some super-intensive ponds are deeper (1.5–1.8 m depth) than semi-intensive ponds and thus the SAE is improved. However, there is an inverse relationship between the volume of air that can be delivered by a blower of a particular power rating and water depth that affects the air pressure required to exceed hydrostatic head pressure.

Blowers produce upwelling water currents that extend from the depth of release, often near the pond bottom, to the pond surface. In super-intensive ponds, a grid of diffuser points is intended to avoid areas of low dissolved oxygen where solids accumulate. Solids suspended by the upwelling action of diffusers can then be progressively moved to the pond center by circular water currents created by LAPW aerators. Diffusers can also be used in directional aerators or airlift pumps to effect horizontal water movement, although these have not yet been applied in commercial systems.

MICROBUBBLE GENERATOR

In recent years, microbubble generators have been used in some super-intensive grow-out ponds. These generators produce much smaller bubbles than conventional diffusers and provide an enormous surface area relative to volume for gas transfer and extended retention time (weeks to months) in the water column. Despite their considerable potential, there has been no systematic evaluation of the SAE of microbubble generators in aquaculture.

Bubbles can be divided broadly by size into macrobubbles, microbubbles, and ultrafine bubbles. Macrobubbles are greater than 0.1 mm in diameter, individually visible, and increase in size as they rise through the water column and burst at the surface. Most bubbles produced by conventional diffuser systems in aquaculture are macrobubbles. The diameter of macrobubbles produced by different diffuser types are 5–10 mm for coarse, 1–3 mm for medium, and 0.5–2 mm for fine pore (Rogers, 2009).

Microbubbles are between 10 and 100 μm in diameter and appear as a milky suspension in water. (Bubbles as small as 50–60 μm are visible to the human eye.) Microbubbles slowly shrink as they rise, providing an extended period for gas transfer.

Ultrafine bubbles or nanobubbles are between 1 nm and 1000 nm, with most nanobubbles produced by generators being around 100–300 nm. Nanobubbles are invisible and are stable in the water column for weeks to months. Ultrafine bubbles can be generated in various ways but typically require a water pump connected to the bubble generator.

VENTURI INJECTOR

Venturi injectors have been used in some super-intensive systems in small (<300 m^3) greenhouse tanks but are not widely used in commercial shrimp farming. Venturi injectors consist of a section of pipe or tubing that includes a constriction. As water is compressed in the constriction, water velocity increases and water pressure decreases. This pressure drop creates a partial vacuum and draws in air through a suction port in the constriction that is connected to tubing that extends above the water surface. As the air–water mixture passes into a wider section of pipe on the downstream side of the constriction, higher pressure and turbulence create abundant fine bubbles that are then discharged from the injector. Most Venturi injectors operate with a gas-to-liquid ratio of around one, but some other injector designs allow operation at a ratio that may reach three. Venturi injectors are now more widely used in intensive shrimp farming in Vietnam. In 1500 m^2 ponds, a combination of four LAPW aerators and a 2–3 hp Venturi injector can maintain dissolved oxygen concentrations >7 mg/l or support a standing biomass up to 6–7 kg/m^3 (T. Hoang, personal communication, Brisbane, 2020).

EMERGENCY OXYGEN SUPPLY

Emergency aeration with tractor-powered paddlewheel aerators is common practice on many commercial channel catfish farms to address dissolved oxygen depletion. However, on commercial shrimp farms, water exchange at greater than normal rates is most commonly used when dissolved oxygen depletion occurs. The value of water exchange in ameliorating the effects of hypoxia depends on available water-pumping capacity, whole-pond respiration rate, source water respiration rate, and dissolved oxygen concentrations in pond water and source water. Despite the focus on water exchange as a tool to manage dissolved oxygen emergencies, it is clear that tractor-powered paddlewheel aeration is much more effective. The recommendation of one emergency aerator powered by a tractor (60–100 hp) for every three channel catfish ponds likely applies equally to semi-intensive shrimp ponds, where the objective is to have a mechanism in place to address occasional episodes of dissolved oxygen depletion.

Although it is primarily used as a disinfectant, some shrimp farmers apply hydrogen peroxide to ponds during dissolved oxygen emergencies. In water, hydrogen peroxide spontaneously decomposes to water and molecular oxygen according to the reaction

$$2H_2O_2 \leftrightarrow 2H_2O + O_2$$

The stoichiometry of this equation indicates that 2.1 mg/l of hydrogen peroxide will produce 1 mg/l of dissolved oxygen. About 0.5 ml of 6% hydrogen peroxide in 1 l of water will produce 1.5 mg/l of dissolved oxygen.

In a field study, hydrogen peroxide was added to tanks containing pond water and shrimp at 5 mg/l. After 1 h, the dissolved oxygen concentration increased by 1.2–2.2 mg/l, although the duration of the effect was quite short (2–3 h) (de Lima et al., 2012). Hydrogen peroxide at the concentrations used to add oxygen is safe for shrimp. The 96 h LC_{50} for hydrogen peroxide to juvenile Pacific white shrimp is 493 mg/l (Furtado et al., 2014). Multiplying the 96 h LC_{50} by a safety factor of 0.1 to obtain the no-effect concentration indicates that 50 mg/l hydrogen peroxide can be used safely for shrimp.

Pure oxygen or oxygen-enriched air can be used during emergency episodes of hypoxia. Given the cost and practicalities of installing and operating supplemental pure oxygen systems, they are most appropriate for super-intensive systems where the value of the shrimp crop justifies their use. Pure oxygen can be purchased as compressed gas cylinders or in liquid form that is held in on-site storage tanks. Pressure swing adsorption generators can produce oxygen-enriched air (up to 90% oxygen). Oxygen from these sources is then added to gas–water contact chambers such as Speece cones (downflow bubble contactors), U-tubes, or pressurized packed columns. Oxygen can also be added to the gas stream that supplies Venturi injectors or microbubble generators.

Aeration rate requirement

Given the critical need to provide a supply of oxygen sufficient to support the respiratory demands of the shrimp crop and the waste treatment ecosystem service of the pond, establishing a mechanism to determine the aeration requirements of shrimp ponds at various levels of production intensity is potentially valuable for system design, planning, and operation. Dissolved oxygen dynamics in shrimp ponds are complex, rendering efforts to predict aeration requirements with any certainty difficult.

Methods that address the underlying mechanisms related to oxygen demand from various pond components and others that take a more empirical approach have been reported. Results are usually given as installed aeration capacity per unit area, but aeration rate also includes consideration of operational scheduling through the crop cycle and at specific times during the day. Although the appropriate metric system unit for aeration power divided by pond area is kW/ha, here we use the mixed measurement system unit of hp/ha because it is more widely used by commercial shrimp farmers.

Hopkins et al. (1991) established an empirical relationship among daily feeding rate, aeration rate, and dissolved oxygen concentration at dawn using 6 years of data from 0.1–0.5 ha intensive lined ponds at the Waddell Mariculture Center. Water exchange of 15–20%/d did not appear to affect the relationship, which was described as

$$\frac{F}{A} = 28.83 - (4.31 \times DO)$$

where F = feeding rate (kg/ha·d), A = aeration rate (hp/ha), and DO = dissolved oxygen concentration at dawn (mg/l). For a dissolved oxygen concentration at dawn of 4 mg/l, F/A is 11.6 kg feed/d per hp and for 3 mg/l it is 15.9 kg feed/d per hp. At a feeding rate of 2.5% body weight daily near the end of the production cycle, these aeration rates are equivalent to levels that support shrimp crops of 464–636 kg/hp.

Another common method is to determine the number or power of aerators needed for a pond based on measurements or estimates of total hourly oxygen demand, the hourly oxygen transfer rate of the aerator type and size selected, and minimum dissolved oxygen concentration required by shrimp (Fast and Boyd, 1992; Santa and Vinatea, 2007; Vinatea and Carvalho, 2007; Vinatea et al., 2011). The general procedure for estimating aeration requirements based on total oxygen demand is as follows:

1. Obtain estimates for respiration of the shrimp crop, water column, and sediment (mg/l·h)
2. Sum these estimates to obtain total oxygen demand (mg/l·h)
3. Consider pond volume to express total oxygen demand as kg/h
4. Select aerator type and power
5. Using the SAE for selected aerator at salinity of operation, convert to SOTR under standard conditions
6. Correct SOTR for temperature of operation
7. Establish a minimum dissolved oxygen threshold (3–4 mg/l)
8. Calculate oxygen transfer rate at dissolved oxygen threshold
9. Divide total oxygen demand by transfer rate per aerator to determine number of aerators.

Based on worst-case estimates of respiration by the water column (0.86 mg/l·h), sediment (0.78 mg/l·h), and shrimp crop (0.16 mg/l·h) in semi-intensive ponds, the calculated aeration rate was 21 hp/ha (Fast and Boyd, 1992), about twice the rate used in practice. Whole-pond respiration in semi-intensive shrimp ponds ranges from about 0.5 to 2 mg/l·h and is usually no more than 1 mg/l·h. Given variation in hourly respiration from day to day and through the crop cycle, this method of estimating aeration rate has limited utility. In addition, using worst-case estimates for whole-pond respiration to determine aeration rate means that installed aeration capacity will not be fully utilized for most of the crop cycle.

An alternative method to estimate aeration requirement is to consider the oxygen demand associated with feed inputs to shrimp ponds (Boyd, 2009; Boyd et al., 2018). The total feed oxygen demand (FOD_T) is the amount of oxygen required to oxidize the organic carbon (FOD_C) and nitrogen (FOD_N) in feed that is not converted to harvested shrimp. Part of that organic carbon represents the energy content of feed that is converted to carbon dioxide by shrimp respiration during routine metabolism. The remainder consists of organic carbon from uneaten feed and excreted as fecal wastes. Ammonia is produced by shrimp excretion through gills and by decomposition of organic matter at the pond bottom. Although nitrogen can cycle multiple times among inorganic and various organic forms (phytoplankton, heterotrophic bacteria, and detritus), the ultimate fate of nitrogen in ponds is nitrification of ammonia, a process that requires oxygen.

The amount of oxygen needed to oxidize waste from 1 kg of feed may be expressed as follows:

$$FOD_C = \left(\left[\frac{C_f}{100} \right] - \left[\left(\frac{C_s}{100} \right) (FCE) \right] \right) 2.67$$

$$FOD_N = \left(\left[\frac{N_f}{100} \right] - \left[\left(\frac{N_s}{100} \right) (FCE) \right] \right) 4.57$$

where FOD_C is expressed as kg O_2/kg feed and C_f, C_s, N_f, and N_s are the carbon and nitrogen in feed and shrimp, respectively (%). Feed conversion efficiency (FCE) is the inverse of feed conversion ratio, or the amount of shrimp produced per kg of feed. The FOD_T is the sum of FOD_C and FOD_N. Using values for C_f, C_s, N_f, and N_s for Pacific white shrimp (Chatvijitkul et al., 2017, 2018) and a feed conversion ratio of 1.5, the FOD_C is 0.85 kg O_2/kg feed and the FOD_N is 0.16 kg O_2/kg feed, resulting in a FOD_T of 1.01 kg O_2/kg feed. Although reducing the protein content of the feed can improve FOD, improving feed conversion efficiency has a greater effect.

Boyd et al. (2018) suggest that aeration requirements should be based on the maximum rate of oxygen loss during the night, calculated as the FOD plus the water column respiration rate. The FOD would be expressed as an hourly rate at the expected maximum feeding rate, and respiration by phytoplankton and bacteria in the water column can be estimated using short-term dark-bottle incubation of a pond water sample. Then, similar to the procedure outlined earlier for using total oxygen demand to estimate aeration rate, the oxygen supply rate from selected aeration equipment can be used to determine the aeration requirement to maintain dissolved oxygen concentration at the desired threshold.

Estimation of the potential oxygen demand of feed is complicated by several factors when applied to calculation of aeration requirements. The expression of FOD is not instantaneous; there is a lag as the various components respond. The increase in shrimp respiration occurs within minutes of feeding, organic waste decomposition requires days to weeks, and nitrification requires weeks of feed application. Waste nutrients from feeding stimulate algal growth. Over the time frame of a cropping cycle, the amount of oxygen produced in photosynthesis is about the same as the amount of oxygen used in phytoplankton respiration plus that used in eventual decomposition of dead algae. Thus, the difference between oxygen supply and oxygen demand is small and can be neglected. Therefore, the long-term average oxygen demand can be calculated from feed inputs, without consideration of phytoplankton metabolism.

Over the short term (hours to days), using FOD to estimate aeration requirements is difficult because dissolved oxygen dynamics in fed aquaculture ponds are complex. Water temperature and dissolved oxygen concentration are in constant flux over a diel cycle and from the surface to the pond bottom through the water column. Algal biomass can change rapidly within days, with occasional sudden algal die-offs that dramatically increase oxygen demand. The supply of oxygen is shared among the shrimp crop and the algae and bacteria that are necessary for the pond to function properly to process wastes from feeding, and often, at least in the context of semi-intensive shrimp production, there are limited opportunities to direct oxygen supplied from aeration to the shrimp crop to the exclusion of other processes that place a demand. In short, the large daily variability in pond oxygen budgets and the inability to calculate the hourly oxygen demand accurately renders use of FOD to estimate aeration rate to be of limited value. The calculated aeration rate tends to overestimate requirements for most of the production cycle but underestimates requirements during unexpected events such as phytoplankton die-offs.

The complexities of using FOD to calculate aeration rate are most pronounced in semi-intensive shrimp ponds where the role of phytoplankton in oxygen dynamics is paramount. In such ponds, even those with daily water exchange rates of 20%, essentially all of the waste organic matter produced as a result of feed inputs remains in the pond, where decomposition exerts an oxygen demand. This is also the case in so-called zero exchange ponds where waste solids are suspended in the water column rather than allowed to settle on the pond bottom. As production intensity increases, respiration by the shrimp crop can dominate the dissolved oxygen budget and the use of FOD to predict aeration requirement is more accurate.

Some modern super-intensive shrimp ponds include a central cone that serves as a sump to collect waste solids. Removal of these solids can reduce the FOD_C and, to a limited extent, FOD_N. In the example given above, the FOD_C is 0.85 kg O_2/kg feed. Here it is useful to partition FOD_C into oxygen demand by shrimp respiration and that by heterotrophic bacterial respiration. At the end of a production cycle, shrimp are fed at about 2.5% body weight daily. Thus, 1 kg of feed will support 40 kg of shrimp (1 kg ÷ 0.025). The respiration rate of those shrimp is 40 kg × 250 mg O_2/kg·h × 24 h/d = 240 g O_2, or 0.24 kg O_2/kg feed. Subtracting shrimp respiration from FOD_C gives the amount of oxygen required to oxidize waste solids from feeding: 0.85 – 0.24 = 0.61 g O_2/kg feed. Removing 50% of the waste organic solids can reduce FOD_T by 30%. If waste solids are removed before they can be decomposed, the aeration requirement can be calculated to meet the demand by shrimp respiration and FOD_N only.

To summarize this section, a method for estimating the amount of aeration capacity to install in an individual shrimp pond is desirable. The aeration requirement for a pond depends on the minimum acceptable dissolved oxygen concentration, the respiration rate of the shrimp crop, the water column photosynthesis and respiration rates, sediment oxygen demand, the exchange of oxygen between air and water (diffusion), and dissolved oxygen in inflow and outflow water. Other than minimum acceptable dissolved oxygen concentration, other factors are variable, external conditions controlled mainly by wind velocity, solar radiation, water temperature, nutrient concentrations, sediment organic matter concentration, and water exchange rate. Meeting this oxygen demand in an aerated pond requires that a specific quantity of dissolved oxygen be transferred to the water per hour. The amount of aeration necessary to accomplish this depends on the SAE of the aerators selected and the dissolved oxygen concentration at which aerators are operated. The complexity of aquaculture pond oxygen budgets makes application of equations for estimating aeration capacity unreliable in ponds, especially accounting for rare and extreme events such as phytoplankton die-offs. Practical experience and monitoring of dissolved oxygen concentration remain the best ways of deciding how much aerator power to install in ponds.

SHRIMP YIELD TO INSTALLED AERATION CAPACITY RATIO

The carrying capacity of semi-intensive shrimp ponds operated with water exchange and no aeration is about 1.5–2 t/ha. In these ponds, the shrimp crop and waste treatment ecosystem service are supported by "free" oxygen provided by phytoplankton photosynthesis and air–water diffusion. To achieve greater yields, mechanical aeration must be provided to support the increased demands for oxygen. As the previous section concluded, the quantity of aeration required is a complex question that cannot be answered by solving an equation based on rates of pond processes or feed loading.

An empirical relationship between shrimp biomass and installed aeration capacity is widely applied in intensive and super-intensive systems. Based on experience with intensive round shrimp ponds in Taiwan in the 1980s, I.-C. Liao reported that 650 kg of shrimp can be supported by the oxygen provided by 1 hp of Asian-style paddlewheel aeration (Hopkins et al., 1991). Boyd (1998) indicated that 1 kW of aeration should support about 500 kg of shrimp (373 kg/hp). In zero exchange intensive ponds stocked at 100–150/m^2, with a carrying capacity of 11–15 t/ha, the relationship was 450–550 kg/hp (McIntosh, 2000a). Most published estimates indicate a range of 400–500 kg of shrimp for every 1 hp of installed aeration capacity using Asian-style paddlewheel and Venturi aerators (Boyd et al., 2019).

It is instructive to understand how the oxygen added to ponds by aeration is partitioned between the shrimp crop and other processes that exert an oxygen demand. Assume an Asian-style paddlewheel aerator has a SAE of 1.0 kg O_2/kW·h. Operating that aerator at a minimum dissolved oxygen concentration threshold of around 3.5–4 mg/l and pond temperatures of 25–30°C reduces the actual oxygen transfer rate to about 50% of the SAE (Fig. 9.2), or 0.5 kg O_2/kW·h. Assuming a shrimp respiration rate of 250 mg O_2/kg·h, 1 hp of Asian-style paddlewheel aeration can support 1492.5 kg of shrimp [(0.5 kg O_2/kW·h × 10^6 mg/kg) ÷ (250 mg O_2/kg·h × 1.34 hp/kW)]. That value assumes that all the oxygen added by aeration is used for shrimp respiration. Based on empirical results from commercial farms with 400–500 kg/hp, of the available oxygen supply provided by aeration, shrimp use 27–34% and the remainder is used to support other processes that consume oxygen in the pond. This range is consistent with studies of shrimp pond oxygen budgets indicating that shrimp typically use about 20% of the oxygen supply (Boyd, 1998), including a small safety factor.

Using the 400–500 kg/hp range as a benchmark, producers of shrimp in intensive and super-intensive ponds can use yield to installed aeration capacity ratio as a performance indicator. Values less than the benchmark indicate excessive aeration and values greater than the benchmark indicate elevated efficiency of aeration.

In general, the yield to installed aeration capacity ratio increases and the proportion of total variable costs represented by energy decreases as production intensity increases. Yield to installed aeration capacity ratio can be increased by using aerators with a greater SAE than the Asian-style paddlewheel aerators used to develop the general benchmark, increasing water exchange rates to improve water

TABLE 9.6 Crop yield, aeration rate, aeration efficiency, and proportion of total variable costs (TVC) represented by energy use from a survey of shrimp farms in Vietnam and Thailand (from Engle et al., 2017).

Country/ species[1]	*Production intensity*	*Stocking density (no./m²)*	*Yield (kg/ha per crop)*	*Aeration rate (hp/ha)*	*Aeration efficiency (kg/hp)*	*E/TVC (%)*[2]
V - PM	Low	12	895	8	112	7.5
	Medium	23	3710	10	371	16.3
	High	58	7962	21	379	6.2
	Very high	59	11,324	21	539	3.0
V - LV	Low	26	265	4	66	6.3
	Medium	31	3469	12	289	3.9
	High	66	6974	28	249	3.2
	Very high	73	11,702	50	234	1.9
T - LV	Low	62.5	1301	9	145	16.9
	Medium	62.5	3483	33	106	20.8
	High	66	6982	35	199	12.8
	Very high	73	13,560	49	277	7.7

Notes: [1] Country code: V = Vietnam, T = Thailand; species code: PM = *P. monodon*, LV = *L. vannamei*. [2] Energy costs are not partitioned among aeration, water pumping, and other energy uses.

quality, and conducting partial harvests when ponds reach carrying capacity to increase overall yield. For example, Taw (2014) reported yield to installed aeration capacity ratios ranging from 650 to 1100 kg/hp in biofloc ponds operated with two partial harvests and one final harvest and 100–200 hp/ha of paddlewheel aeration.

Yield to installed aeration capacity ratios reported for shrimp farms in Vietnam and Thailand suggest that installed aeration capacity is excessive (Table 9.6). In a study of six intensive shrimp farms in Thailand using combinations of LAPW, Venturi, and diffused aerators, Boyd and McNevin (2021) measured yield to aeration capacity ratios ranging from 153 to 597 kg/hp with an average of 295±132 kg/hp. The ratio obtained by two farms using farm-made LAPW aerators was 300 and 333 kg/hp. Excessive amounts of installed aeration capacity in Asian shrimp ponds may be a risk management strategy by farmers attempting to avoid unexpected low dissolved oxygen concentrations in the absence of dissolved oxygen monitoring to verify aeration sufficiency.

OPERATIONAL SCHEDULING OF AERATION

The previous discussion referred to aeration rate in terms of installed aeration capacity. That rate is typically based on the maximum expected requirements for oxygen supply, which often occur late in the cropping cycle when shrimp are near market size and standing crop and feeding rates are near pond carrying capacity. It is also concurrent with high demands for oxygen by phytoplankton and bacteria in the water column and at the sediment surface. In this section, aeration rate refers to operational scheduling as a function of time in the production cycle and time of day. The objective is to optimize aeration to provide sufficient oxygen at the time and in the amount needed to maintain favorable conditions for growth while restricting use to control variable costs associated with aeration.

TABLE 9.7 Examples of aerator scheduling as a function of culture day for penaeid shrimp (based on Mohanty 2001; Mohanty et al., 2018).

Species	*Days of culture*	*hp/ha*	*h/d*
P. monodon ($15–35/m^2$)	1–45	8	8–14
	46–60	10	14–16
	61–90	12	15–18
	>90	16	17–22
L. vannamei ($40–60/m^2$)	1–50	16	10
	51–90	16	14[1]
	>90	16	18[1]

Note: [1] Does not include an additional five nighttime hours from 00:30 to 05:30.

During the first 30 days of a production cycle, shrimp biomass and corresponding feeding rates are quite low, and therefore oxygen supply from photosynthesis and air–water diffusion is normally sufficient. In intensive and super-intensive ponds <0.5 ha with LAPW aerators, at least one aerator should be operated continuously to maintain gentle water circulation and to aid the diffusion of photosynthetic oxygen to the pond bottom. If a pond has four LAPW aerators, these can be operated in rotation. The number and duration of aerators in operation should be increased as shrimp biomass and feeding rate increase. Examples of aeration scheduling are provided in Table 9.7.

The normal pattern of dissolved oxygen concentration in semi-intensive and many intensive shrimp ponds is supersaturation during the afternoon and early evening, and undersaturation during the late night and early morning. Operating aerators when waters are supersaturated with dissolved oxygen will accelerate the loss of oxygen from water to air, although it can also redistribute photosynthetic oxygen from surface waters to the pond bottom and reduce the intensity of thermal stratification. In many ponds, depending on density, it is not necessary to operate all aerators during the day to take advantage of the free oxygen from phytoplankton photosynthesis. To maximize the benefit of water circulation, aerators can be operated daily between 14:00 and 16:00. Aerators should be operated throughout the night in intensive ponds when dissolved oxygen levels are undersaturated.

In intensive ponds with a biomass <12 t/ha, half of the paddlewheel aerators should be operated during the day. When shrimp biomass density exceeds 12 t/ha, all paddlewheel aerators should be operated continuously (McIntosh, 2000b). In super-intensive ponds with low water exchange rates, such as those that use biofloc technology, continuous aeration is essential to maintain the biofloc in suspension and acceptable dissolved oxygen concentrations.

Energy use in aeration

Mechanical aeration is a major component of energy use in shrimp farming. Small, floating electric aerators usually consume about 1 kW·h of electricity for each hp·h of operation. Using this assumption, the amount of aerator horsepower and energy use per crop per hectare at an aeration rate of 2.5 hp/t shrimp (400 kg/hp) with aerators operated for an average of 16 h/d during an 80-day grow-out period (Table 9.8) verifies that aeration requires considerable energy. Amounts of electricity (and energy) per

TABLE 9.8 Aeration rate at 2.5 hp/t shrimp at different production goals with aerators operated an average of 16 h/d for 80 days.

Production goal (t/ha)	*Aeration rate (hp/ha)*	*Direct energy use*[2]		
		(kW·h/ha)[1]	*(GJ/ha)*	*(GJ/t shrimp)*
2.5	6.25	8000	28.8	11.5
5	12.50	16,000	57.6	11.5
7.5	18.75	24,000	86.4	11.5
10	25.00	32,000	115.2	11.5
12.5	31.25	40,000	144.0	11.5
15	37.50	48,000	172.8	11.5

Notes: [1] Assumption: the 1 hp motors consumed 1 kW·h of electricity per hour of operation (motor efficiency of ~80% and drive train efficiency of ~5%). [2] 1 hp·h = 0.746 kW·h = 2.685 MJ = 0.002685 GJ; 1 kW·h = 1.341 hp·h = 3.60 MJ = 0.0036 GJ; 1 GJ = 1000 MJ = 277.78 hp = 207.22 kW·h.

hectare increase steadily as the production goal increases, but the amount of energy applied per tonne of shrimp for aeration is more or less constant across all production intensities (Boyd et al., 2017). There are several types of aerators (as discussed earlier) and different methods of powering aerators, and energy efficiencies of motors, engines, and drive trains differ. This section discusses the factors influencing the efficiency of energy use in shrimp pond aeration.

FUELS

Aerators often are powered by electricity supplied by national or regional grids. Electric generators, usually powered by diesel engines located onsite, may also provide electricity for aeration. Fuels differ in energy content (Table 9.9) and energy is also used to produce those fuels. The total energy use incurred for a fuel is its energy content plus the amount of energy required for its production (embodied energy). Total energy use may be estimated by multiplying the amount of each fuel used by the embodied energy factor for that fuel (Table 9.9).

TABLE 9.9 Energy content (lower heating value) of common fuels and factors for including embodied energy in fuel energy calculations.

Fuel	*Energy content*[1]	*Embodied energy factor*
Ethanol (E100)	21.27 MJ/l	1.49[2]
Ethanol (E85)	22.58 MJ/l	1.34[3]
LPG (liquid propane gas)	23.48 MJ/l	1.12[4]
Gasoline (no ethanol)	31.80 MJ/l	1.36[4]
Gasoline (10% ethanol)	31.25 MJ/l	1.37[3]
Biodiesel	33.32 MJ/l	1.69[5]
Diesel (No. 2)	35.80 MJ/l	1.28[4]
Electricity	3.6 MJ/kW·h	2.50[6]

Notes: [1] https://afdc.energy.gov/fuels/fuel_comparison_chart_pdf. [2] http://large.stanford.edu/courses/2014/ph240/dikeou1/docs/ethanolnetenergy.pdf. [3] Calculated. [4] http://www.iea.org/statistics/resources/manuals/. [5] https://afdc.energy.gov/files/pdfs/3229.pdf. [6] Shapouri et al. (2002).

TABLE 9.10 Typical fuel use and energy output for different sizes of diesel-powered generators.

Rated output	*Fuel use at 75% load*	*Energy output at 75% load*		
(kW)	*(l/h)*	*(% input)*	*(kW/l)*	*(MJ/l)*
25	7.00	26.9	2.65	9.65
50	13.25	28.4	2.83	10.19
100	21.95	34.3	3.42	12.31
150	31.79	35.4	3.54	12.74
200	41.64	36.1	3.60	12.96
250	51.48	36.5	3.64	13.10
500	99.92	37.6	3.75	13.50
750	148.75	37.9	3.78	13.61
1000	197.20	38.1	3.80	13.68

Source: https://www.dieselserviceandsupply.com/Diesel_Fuel_Consumption.aspx

Knowledge of the energy content of fuels allows determination of the volumes of each fuel necessary to provide the same amount of energy at a specific motor or engine efficiency. For example, 1.68 l of fuel-grade ethanol (E100) would be required to obtain the same amount of energy provided by 1 l of No. 2 diesel fuel (35.80 MJ/l diesel fuel ÷ 21.27 MJ/l ethanol = 1.68 l). Likewise, 9.94 kW·h of electricity represents the same amount of energy as 1 liter of No. 2 diesel fuel.

Electricity-generating plants convert primary energy in coal and other primary fuels with an efficiency of around 40%, and there is an additional loss of about 7% of the original primary energy in transmission of electricity through the grid – an overall efficiency of about 33% (https://www.mpoweruk.com/energy_efficiency.htm; https://www.energycentral.com/c/ec/grid-efficiency-opportunity-reduce-emissions). Larger generators are more efficient than smaller generators in converting the energy of diesel fuel to electricity (Table 9.10), but larger units are only 35–38% efficient, and additional losses will occur in transmission to aerator motors.

Energy efficiency from primary fuel is roughly equal for electricity from the national or regional grid and from on-site generators. The cost of primary fuels per kilowatt-hour generated is less for large power plants than for smaller diesel generators, and grid electricity usually is less expensive than electricity generated at shrimp farms.

ELECTRIC MOTORS

Synchronous induction, alternating current motors of different types and sizes are used to power aerators. Smaller electric motors (1–3 hp) often are single-phase and operate on 110–120 volts (V) or 208–230 V, while larger motors usually are three-phase and operate on 208–230 V or even 460 V. Motors usually are rated at full load to require a specific current, measured in amperes (A) (Table 9.11); the current in A required doubles between 115 V and 230 V motors, and between 230 V and 460 V motors.

The efficiency of electric motors is calculated as the energy output by the motor shaft (brake power) divided by the energy input to the motor (wire power). The energy use by an electric motor can be calculated by the equation:

$$\text{kW} = \frac{\text{Brake hp} \times 0.746}{\text{Motor efficiency}}$$

The typical efficiencies of small electric motors are provided in Table 9.12.

Motors should be operated at around 75% full load because service life and efficiency are greater than at full load. A 10 hp, 230 V, three-phase motor operated at a brake power of 7.5 hp would use around 6.32 kW, or 6.32 kW·h of electricity per hour.

The current used by a motor is roughly in direct proportion to load between 50% and 100% full load (Fig. 9.8). The current drawn by the aerator motor can be easily measured with an ammeter during motor operation and the approximate load estimated. There is a substantial no-load current by electric motors and they waste electricity when under-loaded. The amount of power required to operate paddlewheel aerators increases with greater paddlewheel depth at any rotation speed (Ahmad and Boyd, 1988). Attention to the depth of paddle tip submergence is important because it can change as the floats for aerators shift during operation. This results in greater or lesser submergence of paddles, which can lead to over- or under-loading of motors.

The rotation speed of electric motors varies with the number of magnetic poles in the motor (Table 9.13). Motor slip under-load reduces output shaft speed by 3–5% in 1–5 hp motors and by around 2–3% in motors of 7.5–20 hp. Motors for paddlewheel aerators usually have a shaft speed of about 1735 rpm.

TABLE 9.11 Full-load current (A) for single-phase and small three-phase electric motors.[1]

Motor power rating	*Single-phase*		*Three-phase*	
(hp)	*115 V*	*230 V*	*230 V*	*460 V*
1	16	8	3.6	1.8
2	24	12	6.8	3.4
3	34	17	9.6	4.8
5	56	28	15.2	7.6
7.5	80	40	22	11
10	100	50	28	14
15	–	–	42	21
20	–	–	54	27

Note: [1] http://mechreps.com/PDF/MRI_Formulas_Conversions.pdf

TABLE 9.12 Efficiencies of small electric motors and small stationary internal combustion engines powered by different fuels.

Electric motors (hp range)	*Efficiency[1] (%)*	*Small (<25 hp) stationary internal combustion engine*	*Efficiency[2] (%)*
1–4	78.8	Ethanol (E100)	20–25
5–9	84.0	Liquid propane gas (LPG)	25–30
10–19	85.5	Gasoline	20–30
20–49	88.5	Diesel	28–32

Sources: [1] https://www.engineering toolbox.com/electrical-motor-efficiency-d_655.html. [2] Ashok et al. (2015); Innocent et al. (2013); Kalra et al. (2014); Lin et al. (2010); McCarthy et al. (2011); Reddy and Raju (2018); Takaishi et al. (2008); Tompkins et al. (2012).

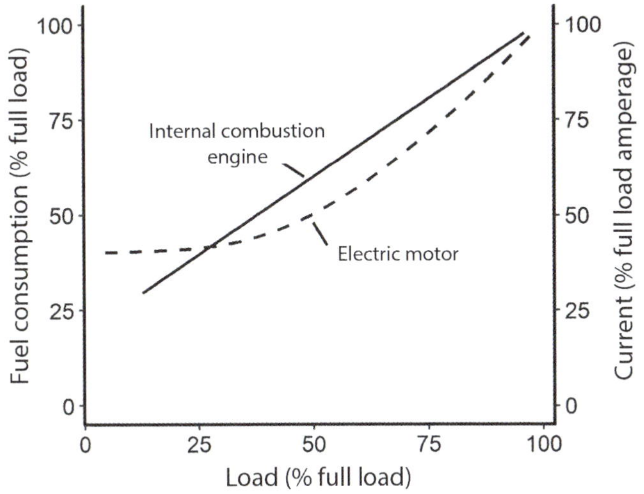

FIGURE 9.8 Load versus fuel use by internal combustion engines and current use by electric motors. Prepared with information from http://dieselserviceandsupply.com/Diesel_Fuel_Consumption.aspx and http://energy.gov/sites/prod/files/2014/04/f15/10097517.pdf

By contrast, paddlewheels, depending on their diameter, transfer oxygen most efficiently at 80–120 rpm (Moore and Boyd, 1992). Motors that turn at such low speeds are more expensive, and speed reducers are placed in the drive train of aerators to allow higher-speed motors to be used.

There also is considerable use of Venturi aerators such as the Aire-O_2® aerator in shrimp ponds. Most of these aerators have two-pole motors that rotate at about 3600 rpm (Table 9.13). This type of aerator does not transfer as much oxygen per unit of energy if it is supplied with a four-pole motor (Ruttanagosrigit et al., 1991). Vertical pump aerators for aquaculture typically are supplied with a four-pole motor.

When electric motors are started up, an inrush of electric current is necessary to start the rotor and accelerate the shaft to full speed. For single-phase motors of 1–10 hp, the inrush current is about three times the full load motor current requirement. For three-phase motors, the inrush current is about twice the full load current. Large motors usually are started with a motor controller (often called a motor starter) with fuses that withstand the large current input at start-up.

TABLE 9.13 Electric motor rotation speed as related to the number of magnetic poles of the motor and the electrical frequency.[1]

Magnetic poles	*Electrical frequency (Hz)*	
	50	*60*
2	3000	3600
4	1500	1800
6	1000	1200
8	750	900
10	600	720
12	500	600
16	375	450
20	300	360

Source: [1] https://www.engineeringtoolbox.com/synchronous-motor-frequency-speed-d_649.html

INTERNAL COMBUSTION ENGINES

Internal combustion engines that run on diesel and other fuels have much lower energy efficiencies than electric motors (Table 9.12), but end-use energy of these two types of power units are derived differently from primary fuels. On average, the efficiency of grid electricity metered at point of use is about 33%. When 1 MJ of primary energy is combusted at an electric power-generating plant, 0.33 MJ can be delivered to a 2 hp motor of an electric aerator at a shrimp farm. The motor of such an aerator is about 79% efficient, and so 0.26 MJ of energy will be available from the aerator motor shaft – an overall efficiency of 26%. This efficiency of end-use electricity is similar to the end-use efficiency from fuels supplied to internal combustion engines.

Internal combustion engines designed to run on different fuels have similar efficiencies (Table 9.12). Nevertheless, fuels differ in price per liter and so a fuel of lower energy content may in some instances be more economical than a fuel of greater energy content.

Internal combustion engines, like electric motors, should be operated at around 75% full load to assure greater service life and efficiency; while engines consume fuel under no load, the percentage fuel use when an engine is idling at no load is lower than the percentage of full-load electricity use by a motor running under no load. The fuel consumption of an engine declines with load in an approximately linear manner, and load can be estimated from fuel use. A typical small diesel engine uses about 0.23 l/h fuel for each horsepower-hour (Yoo and Boyd, 1994).

Output shaft speeds of small engines usually are around 3000 rpm (https://www.hatz-diesel.com). Paddlewheel aerators used in shrimp farming usually operate at speeds of 60–120 rpm. As a result, shaft speeds of motors and engines must be reduced in the drive trains of these aerators.

DRIVE TRAIN

Venturi aerators and vertical pump aerators operate at motor output shaft speed. Only a single coupling between motor and aerator shafts usually is necessary. Paddlewheel aerators are of three main types.

The 1 hp and 2 hp Asian-style floating electric paddlewheel aerators (Fig. 9.3) have motors with a two-directional gear-reducer mechanism mounted on floats with shafts extending from each side of the gear reducer upon which the paddlewheels are mounted. The second type of floating electric paddlewheel aerator that is occasionally used on shrimp farms is a larger (usually 5 hp or 10 hp) aerator which will be called a US catfish pond paddlewheel aerator. The motor of this kind of aerator is mounted on one end of a flotation platform and there is a gear reducer between the motor and the hub (rotor) upon which paddles are welded to form a paddlewheel. In both kinds of aerators, the rotating shafts of the paddlewheels must be supported by bearings or other mechanisms that allow rotation with minimum friction. These aerators usually have a relatively high efficiency of transfer of energy from the motor shaft to the rotating paddlewheel. The energy loss is around 5–10%, or, stated otherwise, there is a 90–95% efficiency of energy transfer from the output shaft of the motor to the aerator shaft.

The other type of paddlewheel aerator is the LAPW aerator popular in Thailand and some other Asian countries (Fig. 9.4). The paddlewheel is mounted on floats in the pond and the engine or motor is mounted on the pond bank. The speed of the engine motor usually is reduced by V-belts or chain-and-sprocket mechanisms in which two shafts are used to reduce the motor output speed to the desired paddlewheel speed. The equation for determining speeds of driven shafts in V-belt speed reductions is:

$$D_1 \times RPM_1 = D_2 \times RPM_2$$

where D_1 and D_2 are the pitch diameters of the driver and driven pulleys, respectively, and RPM_1 and RPM_2 are speeds of driver and driven shafts, respectively.

The equation for speed reduction by chain-and-sprocket mechanisms is similar to that for V-belts and is as follows:

$$T_1/T_2 = RPM_1/RPM_2$$

where T_1 and T_2 are the number of teeth on the driver and driven sprockets, respectively, and RPM_1 and RPM_2 are speeds of the driver and driven shafts, respectively.

A mechanism known as a jackshaft often is used to reduce motor speed when driving a paddlewheel aerator from an engine or motor on a pond embankment. A 15 cm pulley could be placed on the motor shaft (1735 rpm) and a 50 cm pulley could be mounted on a short driven shaft mounted in a metal frame. The driven shaft would turn at 89 rpm [(15 cm × 297 rpm) = (50 cm × RPM_2); RPM_2 = 89], a speed suitable for the paddlewheel. Of course, the two jackshafts would have to be firmly mounted in bearings supported in a metal frame beside the motor shaft.

A shaft must be connected by universal joints between the second driven shaft and a long shaft extended downward at an angle to the aerator shaft and connected to the aerator shaft by another universal joint. On many shrimp farms, directional worm gear mechanisms are used to allow power to be distributed to multiple floating paddlewheels.

The devices used to slow rotation speeds, connect shafts, and change the direction of power transmission in the drive train incur energy losses. Typical efficiencies are 90–95% for V-belt drives (https://www.grainger.com/know-how/equipment-information/kh-types-of-belt-drives-efficiency), 95–98% for roller chain drives, and 85–90% for gear boxes (https://www.machinedesign.com/archive/article/21834659/a-second-look-at-gearbox-efficiencies). Energy loss in couplings that are properly designed and installed is small (~1–2%), but misalignment of couplings and vibration in shafts can result in much larger losses of energy.

The shaft between the jackshaft and the aerator shaft usually is a long, 3.75–5 cm diameter galvanized water pipe that vibrates considerably, and misalignment between shafts and couplings are obvious. To our knowledge, there have been no studies of energy loss in the drive trains of LAPW aerators, but the loss is likely at least 30–40%, and sometimes more, between motor shafts and aerator shafts when only one aeration unit is operated from a motor. Greater losses are no doubt incurred with multiple LAPW aerators powered from a single engine or motor.

Rotating shafts on to which paddlewheel aerators are mounted require support. A rolling bearing supports the load of a rotating shaft by placing rolling elements (balls or short cylinders) between two bearing rings (also called races). Such bearings usually require frequent lubrication with grease, and rolling bearings that do not require frequent lubrication are very expensive. Lubrication of mechanical aerators is difficult because they are mounted on floats and positioned in ponds.

Paddlewheel aerator shafts do not rotate at high speeds; typical speeds are between 60 and 120 rpm. As a result, simple linear shaft supports (or shaft support blocks) that do not require lubrication may be used in place of rolling bearings (roller block bearing). These shaft support blocks often are made by cutting a half circle in the top of a block of high-density plastic that can be firmly attached to the aerator frame. In Asia, such blocks often are made of wood, which induces more friction than is incurred with high-density plastic.

Aerators corrode in many waters, especially in saline water. In the USA, most paddlewheel aerators have been made of mild steel because of the high price of stainless steel. These paddlewheels can be "hot-dipped" galvanized for protection, and some manufacturers attach a piece of sacrificial metal to such aerators to minimize corrosion of the metal working parts of the aerator. It would be greatly beneficial to manufacture plastic paddlewheels according to the aerator design used in US catfish farming. However, there has been very little interest in adopting more efficient aerators of the US design in Asia. Conversely, some of the design features of Asian-style aerators (especially the use of plastics) could be of benefit to users of mechanical aeration in US aquaculture.

POTENTIAL FOR ENERGY CONSERVATION

It was shown earlier that floating electric Asian-style paddlewheel aerators use about 11.5 GJ of energy per tonne of shrimp (Table 9.8). The large floating electric paddlewheel aerators used in US catfish farming are at least 30% more efficient in transferring energy than floating electric paddlewheel aerators used in Asia (Boyd et al., 2018). Based on the rule-of-thumb of 2.5 hp/t shrimp for Asian-style paddlewheel aerators, the aerator rate could be reduced to 1.75 hp/t of shrimp by use of US-style aerators.

Drive-train efficiencies are probably about 70% for LAPW aerators and 90% for Asian-style floating electric paddlewheel aerators. As a result, about 3.22 hp/t shrimp is required for LAPW aerators. By using LAPW aerators, the energy use would be increased to 14.8 GJ/t of shrimp, because the LAPW aerator is less efficient than the floating electric Asian-style paddlewheel aerator by the ratio 2.5 hp/t of shrimp to 3.22 hp/t of shrimp. Likewise, by use of the larger US-style floating paddlewheel aerators, energy use could be reduced to 8.1 GJ/t of shrimp.

According to the aeration schedule used in preparing the data in Table 9.8, aerators would run for 1440 h per crop or 4637 hp·h. For a diesel-powered LAPW aerator with a fuel consumption of 0.23 l/hp·h, 1067 l of diesel fuel would be consumed per tonne of shrimp. This amounts to 38.2 GJ/t of shrimp.

The estimates given above are for direct energy use. Applying embodied energy factors for fuels from Table 9.9 give total energy use as: US-style floating electric paddlewheel aerator, 20.25 GJ/t shrimp; Asian-style floating electric paddlewheel aerator, 28.8 GJ/t shrimp; electric-powered LAPW aerator, 37 GJ/t shrimp; diesel-powered LAPW aerator, 48.9 GJ/t shrimp. These calculations are based on certain assumptions and obviously are subject to uncertainty, but they reveal that the efficiency of shrimp pond aeration could be improved to save a considerable quantity of energy. Assuming that 4 Mt of shrimp are produced in aerated ponds, each 1 GJ/t reduction in energy use would reduce energy use in penaeid shrimp farming by 4 million GJ, or roughly 830 million kW·h (71,367 t of oil or 488,234 barrels of oil equivalent). There would be both energy conservation and farm-level economic benefits to improving the energy efficiency of shrimp pond aeration.

Aerator placement

Placement of aerators to maximize their effect is a complex decision that is a function of production intensity, pond size, pond shape, and the number and type of aerators selected. The goal is to maximize the area with a dissolved oxygen concentration sufficient to support the shrimp crop and minimize the area deficient in oxygen. Thus, aerator placement can affect the bottom surface area suitable for habitation by shrimp. Aerator placement can also affect the distribution and mixing of lime, fertilizers or other chemical supplements, and bacterial amendments added to pond water. There has been relatively little systematic evaluation of the effect of aerator placement in shrimp ponds on dissolved oxygen distribution and shrimp production performance; farmers have taken an empirical or practical approach to aerator placement.

In relatively large (<10 ha) semi-intensive shrimp ponds with modest levels of aeration (1–2 hp/ha), it is not possible to place aerators in such a way as to aerate the entire pond volume. In this context, aerators are most often placed in convenient locations, typically adjacent to pond embankments for diesel-powered paddlewheel aerators. Similarly, the layout of a farm's electricity supply lines and control panels will dictate placement of electric-powered aerators. With modest levels of aeration, aerators can be placed to produce zones of effect that extend from 15–25 m for 1–3 hp paddlewheel aerators. Using aerators of greater power can increase the zone of effect. In a 3.2 ha channel catfish pond with three 10 hp aerators placed side by side, a dissolved oxygen concentration of 3.0 mg/l was measured 45 m from the aerators, and 2.5 mg/l around 95 m from the aerators (Farrelly et al., 2016). In channel catfish ponds, paddlewheel aerators are often placed side by side near the pond bank with the aerator water flow directed toward the pond center, creating a large zone of effect that is sufficient to support the standing crop when dissolved oxygen concentration in the bulk water is at critically low levels, typically late at night and early in the morning. Thus, aerator placement in semi-intensive ponds with modest aeration rates is often based more on convenience than on maximizing the effects of aeration.

One placement option in semi-intensive ponds is to concentrate aerators in 15–20% of the pond area, creating a well-oxygenated aerated zone that is adjacent to a facultative zone that alternates between aerobic and hypoxic conditions over 24 h. Shrimp will swim to seek favorable water quality, especially in response to aerator water currents. The aerated zone is sufficiently large to support the shrimp crop when the facultative zone is hypoxic.

With increasing levels of aeration in large (>10 ha) semi-intensive ponds, one possible technique is to distribute aerators uniformly over the pond surface area, with aerator outflows oriented in the same direction. Each aerator creates a zone of influence such that most of the pond bottom area is habitable by shrimp. It is difficult to establish a circular water flow pattern in ponds >5 ha. Wind-driven water currents can complicate efforts to establish circular flow patterns in such ponds, especially if the prevailing wind direction is in opposition to the direction of aerator water currents.

With increasing use of automatic feeders in semi-intensive, intensive, and super-intensive ponds, aerators can be placed to assure sufficient dissolved oxygen supply to meet the demand of high concentrations of shrimp that may at times inhabit the pond bottom adjacent to feeders. Several aerators can be placed about 25 m from automatic feeders to accomplish aeration of shrimp feeding zones.

Placement of aerators in intensive and super-intensive ponds is more important than placement in semi-intensive ponds. In general, the approach to aerator placement in intensive ponds is to arrange the aerators in such a way so as to generate circular gyres around the pond perimeter that creates a

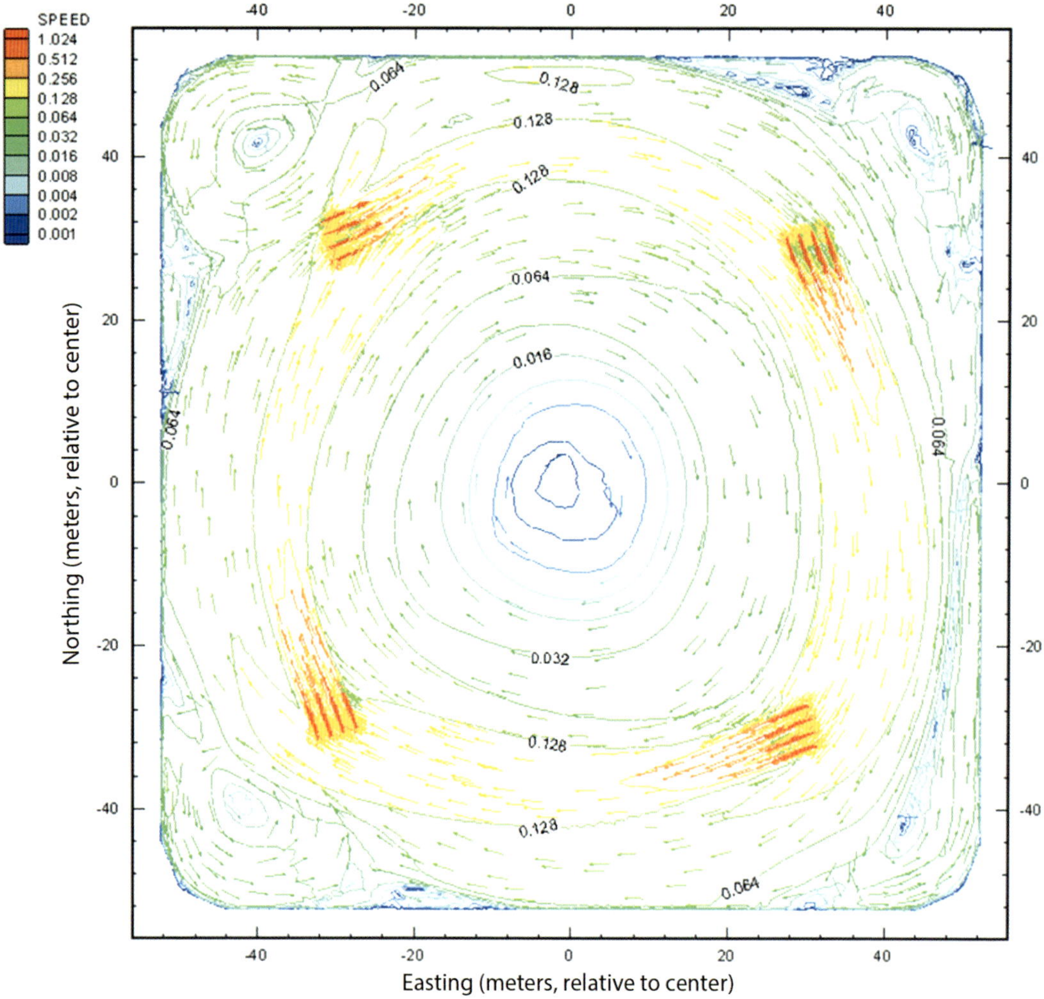

FIGURE 9.9 Simulated surface currents in a 1 ha pond with four 2 hp Asian-style paddlewheel aerators. Velocity vectors are color-coded in m/s. Source: Peterson and Pearson (2000), with permission of Global Aquaculture Advocate.

stagnant quiescent zone near the pond center to facilitate the sedimentation of solids (Fig. 9.9). In earthen ponds, placement of aerators in this way can result in a large mound of sediment scoured from the embankments and the pond bottom near the pond center. In lined ponds, such an arrangement can result in the accumulation of waste solids from feeding and dead phytoplankton and bacteria near the pond center. These solids can be removed through center drains or by siphoning.

In square or rectangular ponds, paddlewheel aerators are often placed in corners to avoid the accumulation of sediment and organic matter in those areas. Depending on the size of intensive pond, two or three concentric rings of paddlewheel aeration current gyres can be put into place, each 10–15 m apart (Fig. 9.10). To accelerate movement of organic particles to the pond center, LAPW aerators should be installed at an angle to the embankment 6–7° less than perpendicular (H. Tung, personal communication,

FIGURE 9.10 Arrangement of 2 hp Asian-style paddlewheel aerators in a 3500 m^2 lined shrimp pond in Guatemala. Note the flow pattern of individual aerators and how aerators are collectively working together to concentrate wastes in the pond center, where they can be removed periodically.

Brisbane, 2020). Regardless of aerator placement, in addition to accumulation of sediment and organic matter at the pond center, small areas of accumulated materials may occur in other areas of the pond as a result of complex water movements. Aerator arrangements can be periodically altered to minimize the size of these smaller areas of accumulated material.

Peterson et al. (2001) evaluated three aerator arrangements of six 2 hp paddlewheel aerators in a 1 ha rectangular pond: (1) in-line, where aerators were arranged in series; (2) parallel, where aerators were placed side by side; and (3) diagonal, where aerators were placed to produce diverting water flows. As indicated by their minimization of the sum of the undesirable scouring area and deposition area, the best arrangements were diagonal or in parallel (Table 9.14).

In general, in intensive and super-intensive ponds, paddlewheel aerators should be placed in such a way that their zones of influence overlap. As water flow rate decays from one aerator, the next aerator downcurrent will impart additional momentum to the water, effecting circulation. In this way, individual aerators are positioned to work together to create an oxygenated zone.

Peterson and Pearson (2000) described what they considered to be the near-optimum arrangement of four 2 hp paddlewheel aerators in a 1 ha square (100 m × 100 m) pond. Each aerator was placed about

TABLE 9.14 Proportion (%) of 1 ha rectangular earthen ponds aerated with four 2 hp paddlewheel aerators and two 2 hp aspirator pump aerators (9 kW/ha) for different aerator arrangements according to benthic shear stress category (after Peterson et al., 2001). Excessive sedimentation occurs at a benthic shear stress of <0.001 N/m^2 (2 cm/s) and excessive erosion occurs at a benthic shear stress >0.03 N/m^2 (12 cm/s).

Aerator arrangement	*Excessive sedimentation*	*Desirable*	*Excessive erosion*
In-line	12	26	62
Parallel	4	35	61
Diagonal	10	39	50

18–20 m from pond embankments and aerator flow was directed toward a downcurrent corner about 80 m away (Fig. 9.9). Aerators were approximately 60 m apart. The direct zone of influence of each aerator was about 25 m, but the prevailing currents established by operation of all aerators created a zone of influence that extended from one aerator to another.

A final note related to aerator placement concerns the lack of importance of the Coriolis effect on the selected direction of circulation (Boyd, 2017). In the northern hemisphere, large water masses are deflected to the right (clockwise), and in the southern hemisphere, they are deflected to the left (counterclockwise). Although the Coriolis effect is important for the rotation of large air and water masses, the size of even large shrimp ponds is too small for the Coriolis effect to have a meaningful effect on water circulation. In shrimp ponds with aeration, the inertial forces of aerator-induced water currents are far greater than Coriolis forces in affecting water circulation.

Aeration hydrodynamics

Although the primary purpose of mechanical aeration in aquaculture is air–water oxygen transfer, aerators also serve a complementary function to circulate water and create currents in ponds. Water circulation is beneficial because it breaks down thermal and chemical stratification, increasing dissolved oxygen concentration at the pond bottom, conserves oxygen produced by photosynthesis in the pond, and generally creates a more homogeneous environment conducive to good shrimp production (Fast and Boyd, 1992; Rogers, 1989; Rogers and Fast, 1988). Although several mixing devices have been tested and their efficacy demonstrated, they have not been widely adopted in commercial shrimp farming. The focus of the following discussion will be on the mixing and water currents provided by conventional aeration equipment, especially paddlewheel aerators.

Circular water currents established by arrays of operating aerators function to concentrate wastes in the pond center and increase the effectiveness of the distribution of important inputs such as feed, chemicals, lime, minerals, and probiotics. Water currents have implications for conditions at the pond bottom and shrimp swimming endurance. Some water movement is beneficial in any shrimp production system, but the importance increases in intensive and super-intensive systems. Optimization of conditions favorable for shrimp growth may require adjustments in management protocols to balance oxygen supply with generation of water currents conducive to good shrimp production.

Movement of paddles through water imparts turbulent kinetic energy, creating pressure in the downstream direction. Paddlewheel aerators deliver about 100 N of horizontal thrust per horsepower

TABLE 9.15 Extent of water currents produced by an 11 hp diesel-powered LAPW aerator measured at the surface and bottom of a shrimp pond as a function of paddlewheel rotation speed (from Taparhudee et al., 2007).

Rotation speed (rpm)	*Limit of surface current (m)*	*Limit of bottom current (m)*
25	15	5
35	15	10
55	20	10
70	25	15
90	25	15

(Peterson, 2000), pushing water away horizontally in the direction of rotation. The momentum of moving water then dissipates away from the aerator by friction with adjacent water layers (eddy viscosity) and the pond bottom.

Asian-style, 1–3 hp paddlewheel aerators produce water currents of 60–80 cm/s immediately adjacent to the aerator (Peterson, 2000). Currents then decay to 20–30 cm/s at 10 m, 10 cm/s at 20 m, 5 cm/s at 30 m, and 2–3 cm/s at 40 m (Kang et al., 2004). The effective distance of aeration by 2 hp Asian-style paddlewheel aerators is considered to be about 25 m (Taparhudee et al., 2007). Of course, more powerful aerators will create a greater zone of influence than Asian-style paddlewheel aerators. The zone of influence of LAPW aerators is a function of their rotation speed (Table 9.15).

The initial thickness of the water layer generated by a paddlewheel aerator is determined by the depth of paddle tip submergence, typically 5 cm. The aerator produces a relatively high-velocity sheet of surface water overlying relatively slower-moving deeper water layers. The water depth affected by paddle rotation expands away from the aerator at a slope of 7:1 (Peterson et al., 2000). Thus, for a pond water depth of 1.4 m, water currents generated by paddlewheel aeration will reach the bottom about 10 m from the aerator. In addition, the water currents at the bottom will be less than surface currents because of friction with the pond bottom and dissipation of turbulent energy.

SOIL EROSION AND DEPOSITION

In intensive and super-intensive shrimp ponds, multiple aerators are deployed in an array and a characteristic flow pattern typically develops. Paddlewheel aerators are often arranged to develop horizontal water currents that move along pond embankments in a circular pattern. Momentum that is lost as water moves away from one aerator is regained by the action of another aerator, thereby maintaining circulation. Although the zone of influence for single paddlewheel aerators in a stagnant pond extends for only about 25 m, in ponds with accelerated flow established by multiple aerators working together, the zone of influence can extend for 80–100 m (Peterson, 2000; Peterson et al., 2000).

The primary circular flow results in the development of a stagnant, low-pressure area near the pond center. The primary flow further leads to a secondary radial flow that extends outward at the pond surface and toward the center at the pond bottom. Surface water currents from aeration of intensive and super-intensive shrimp ponds are about 10–20 cm/s along the pond periphery, with a maximum of 25–30 cm/s, and minimum water currents near the pond center are <5 cm/s (Delgado et al., 2003; Kang et al., 2004; Peterson, 2000; Peterson et al., 2000; Wijesekara et al., 2005).

In earthen ponds, water currents along the pond periphery are sufficient to cause erosion of soils from embankments and the pond bottom. Soil and other types of particles move in spirals progressively toward the pond center in response to circular flows around the perimeter and inward radial flows along the pond bottom (Peterson and Pearson, 2000). The stagnant depositional environment at the pond center leads to accumulation of particles. This so-called "teacup effect" has been observed in intensive aerated earthen or lined shrimp ponds since the 1990s (Briggs and Funge-Smith, 1994; Delgado et al., 2003; Hopkins et al., 1991).

The net effect of the flow pattern in intensive ponds is zonation of the pond into areas that are well oxygenated, with high current speed leading to soil erosion, and areas with low dissolved oxygen concentration, and low current speed leading to particle deposition. There are numerous characteristics of the stagnant center that are undesirable for shrimp culture. Organic solids from feeding (sludge) accumulate and exert a significant oxygen demand for decomposition. The pond soil profile is characterized by a steep gradient in oxidation–reduction potential resulting in the production of potentially toxic reduction products such as hydrogen sulfide. Low dissolved oxygen concentration coupled with high demand for oxygen by decomposing material can result in the absence of a normal thin crust of oxygenated sediment that restricts the diffusion of reduced substances to the overlying water. Shrimp will actively avoid these stagnant areas, which can account for 40–60% of the pond bottom area (Briggs and Funge-Smith, 1994; Delgado et al., 2003), depending on pond size, soil type, and the number and type of aerators. The implication of this phenomenon is that shrimp may be overcrowded in preferred areas at densities that are much greater than the stocking density expressed on a per unit pond area basis.

Turbulent flow at the sediment–water interface can be described in terms of benthic shear stress, which relates the force exerted by water currents to erosion or deposition based on particle density. Lighter particles are suspended at low benthic shear stress and heavier particles are suspended at higher benthic shear stress. Erosion of silt occurs in pond areas with benthic shear stress >0.03 N/m^2 and fine particle deposition occurs in areas with benthic shear stress <0.003 N/m^2 (Peterson, 1999). Based on the benthic shear stress of particle types with a characteristic density, distinct sediment zones can be delineated (Table 9.16). Moderate benthic shear stress between 0.003 and 0.03 N/m^2 (equivalent to a bottom velocity of 4–12 cm/s) is considered optimal, and 0.001–0.01 N/m^2 (2–8 cm/s) is considered ideal (Peterson, 1999; Peterson et al., 2001). The effect of various arrangements of aerators on the distribution of areas with different benthic shear stress is indicated in Table 9.14.

TABLE 9.16 A classification scheme for earthen pond zones based on benthic shear stress (after Peterson, 1999).

Particle type	*Benthic shear stress range (N/m^2)*	*Bottom water velocity range[1] (cm/s)*
Sand	0.1–2.3	30
Silt	0.03–0.1	12–25
Clay	0.01–0.03	8–12
Feed	0.003–0.01	4–8
Algae cells	0.001–0.003	2–4
Dead	0–0.001	0–2

Note: [1] Calculated for a 1.2 m deep water column.

SHRIMP SWIMMING

Swimming is a routine activity used by shrimp for seeking feed or areas of the pond with preferred conditions. Shrimp orient head-first toward water currents (positive rheotaxis). Shrimp swim by undulating strokes of five pairs of pleopods (swimming legs) that are attached to the abdomen. The pleopod beat frequency is directly related to current speed (Zhang et al., 2006b). Swimming demands substantial energy to sustain, and levels of glucose and protein in the hemolymph are quickly depleted if shrimp are forced to swim until they are exhausted (Li et al., 2018; Zhang et al., 2006b). The swimming endurance of Pacific white shrimp depends on temperature and salinity, with maximum endurance at 21°C and 28 ppt (Zhang et al., 2007).

The critical swimming speed (U_{crit}) is an index that estimates the maximum swimming speed that can be sustained by shrimp. It is affected by temperature, salinity, body length, and starvation (Yu et al., 2010). For Pacific white shrimp between 5.5 and 10 cm (~1–5 g), the relative U_{crit} is about 4–5 body lengths/s, with greater relative U_{crit} at smaller body weights. At optimum salinity, the U_{crit} is around 43 cm/s. Pacific white shrimp can maintain their position in water currents up to 11.5 cm/s (Zhang et al., 2006b) but are unable to swim for 200 min at a current velocity of 15 cm/s (Li et al., 2018).

In terms of the hydrodynamics of semi-intensive ponds, shrimp swimming is not limited by water currents because currents are low, with the exception of zones of influence near isolated aerators. However, intensive aeration of intensive and super-intensive ponds can create strong, and in some cases continuous, water currents, depending on the number, type, arrangement, and operation schedule of aerators. In high-current areas, dissolved oxygen concentration is good but shrimp must expend considerable energy to maintain their position. In low-current areas preferred by shrimp, dissolved oxygen concentration may be deficient and crowding may increase susceptibility to the effects of hypoxia and cannibalism. For 10–20 g shrimp, a bottom velocity <5 cm/s is preferred (Wijesekara et al., 2005). Pond areas with high currents (>15 cm/s) and low currents (<3 cm/s) are actively avoided, forcing shrimp to congregate in zones that represent a minor proportion of the total pond bottom area. For shrimp, the trade-off is between preferred swimming speed and dissolved oxygen concentrations that maximize growth potential.

RECENT DEVELOPMENTS

Many of the issues discussed in this section have been addressed through changes in pond design and operation. First, with the trend toward ever-smaller ponds as production has intensified, ponds <0.5 ha are now constructed with plastic liners. This has addressed the problems of soil erosion caused by aerator currents and the need to renovate pond bottoms by removing the central sediment mound or restoring embankment profiles between crop cycles. Second, modern intensive shrimp ponds are constructed with a central settling cone that facilitates removal of material settling in the quiescent central zone of the pond.

Biofloc technology is an alternative approach to address the accumulation of sludge in the central zone of lined ponds. Additional aeration is provided and directed toward the sludge mound to force its resuspension. Such ponds are typically operated with low rates of water exchange and very high aeration rates. Intensive aeration in biofloc systems creates many zones of excessive current velocity, forcing shrimp into low-velocity refuges as they are encountered during cruising.

Another recent technique of aeration and circulation is to combine diffused and LAPW aeration. Lines or grids of diffuser tubing near the pond bottom develop upwelling currents that maintain organic matter in suspension. LAPW aerators create circular currents that move suspended material toward the stagnant center zone, where it can be collected in central cones and pumped away. Upwelling currents created by diffusers can interfere with development of circular currents by paddlewheels.

There are trade-offs in selecting an optimal current speed. The maximum current speed is no longer limited by soil erosion but remains limited by shrimp physiological tolerance (<15 cm/s). A minimum current speed avoids large areas of deposition. Shrimp prefer relatively quiescent, slowly moving water (<5 cm/s), but those pond zones are depositional for organic particles such as bioflocs. In terms of hydrodynamics, the ideal intensive shrimp pond would have a homogenous flow field that would not exceed a current velocity of 5–10 cm/s for maximum benefit.

Overview

Nearly all penaeid shrimp are farmed in earthen ponds in coastal areas of tropical and subtropical climate zones around the world in waters with a broad salinity range, from fresh to hyper-saline. Dissolved oxygen supply is a critical ecosystem service provided by aquaculture ponds to support shrimp production. In all types of fed aquaculture, dissolved oxygen is a fundamental limiting factor affecting production capacity.

The processes that produce and consume oxygen in shrimp ponds are (1) photosynthesis by phytoplankton, (2) respiration by algae and bacteria suspended in the water column, (3) SOD, (4) shrimp respiration, and (5) air–water gas transfer. The relative importance of each varies by production intensity, with photosynthesis and respiration by phytoplankton being the main processes in semi-intensive ponds. The relative importance of shrimp respiration increases with production intensity. Shrimp respiration is affected by body size, activity level, and molt status, as well as temperature and dissolved oxygen and carbon dioxide concentrations,.

Penaeid shrimp have evolved numerous physiological behavioral adaptations to cope with limited oxygen supply, including dendrobranchiate gills and hemocyanin as the oxygen-transport protein, and various behavioral actions that increase the partial pressure of oxygen in the hemolymph. The limiting dissolved oxygen concentration necessary to support routine metabolism is about 2 mg/l. Shrimp growth potential is maximized at a dissolved oxygen concentration of 6 mg/l. Shrimp can tolerate repeated short-term exposure to hypoxia. However, exposure to hypoxia reduces appetite and affects energy allocation to growth and other metabolic processes. Hypoxia also diminishes the function of various parts of the shrimp innate immune system that affect disease resistance. Survival is significantly lower when minimum dissolved oxygen concentration is <2 mg/l. The factors affecting the severity of the response to hypoxia include (1) alterations to gill structure and function, (2) substances (e.g. nitrite, carbon dioxide) that reduce the oxygen-binding capacity of hemocyanin, (3) substances (e.g. sulfide) that reduce the capacity of tissues to utilize oxygen, and (4) conditions that increase shrimp metabolic oxygen demand (e.g. molting).

Shrimp production intensity varies by at least two orders of magnitude, from 0.5 t/ha per crop for extensive systems to more than 50 t/ha for super-intensive systems. In extensive ponds, tidal water exchange is the principal management tool to avoid pond stagnation and episodes of hypoxia. Water exchange using water pumped from external surface sources is the main dissolved oxygen management tool used in semi-intensive shrimp farming. Intensive and super-intensive systems depend on ever greater rates of aeration to support correspondingly greater shrimp production levels. Paddlewheel aerators (1–3 hp) are the most common type of aerator used in shrimp farming. LAPW aerators are now in widespread use in intensive ponds. In super-intensive ponds, combinations of aerator devices are used, including paddlewheel aerators, diffused aerators, and Venturi injectors.

The feed oxygen demand consists of components for shrimp respiration, oxygen needed to oxidize organic matter produced from feeding, and oxygen needed to support nitrification of excreted ammonia. Attempts to calculate aeration requirements based on the total respiration rate are unreliable and do not consider potentially catastrophic events such as phytoplankton die-offs that cause rapid depletion of dissolved oxygen. An empirical relationship between installed aeration capacity and shrimp biomass is widely applied in intensive and super-intensive systems: 1 hp of aeration can support 400–500 kg of shrimp or 1 hp of aeration can support a daily feeding rate increment of 10–15 kg/ha.

Energy used in aeration can vary by aeration efficiency of the device selected, type of power source, fuel type, and drive-train configuration. Placement of aerators affects the proportion of the pond bottom area that can be inhabited by shrimp. Water currents established by aerators have implications for shrimp swimming endurance and erosion and deposition of pond soil and waste organic matter.

Trends and outlook

A clear trend since the transition from growing black tiger shrimp to Pacific white shrimp as the predominant species has been intensification of production, accompanied by mechanical aeration. It is estimated that around 70–80% of farmed shrimp production is derived from aerated ponds (Boyd and McNevin, 2018). More recently, largely in response to concerns about biosecurity, farms have reduced water exchange with external sources of water and increased internal recirculation.

As pond sizes have reduced, farmers have divided the farm into ponds with different functions: some are used to grow shrimp and others are dedicated for waste treatment. Thus, the oxygen demand of organic and nitrogenous wastes can be displaced to treatment ponds. In arranging ponds and water movement this way, more of the oxygen provided by aeration of shrimp production ponds can be used to support the shrimp crop and proportionally less is used to meet the respiratory demand of phytoplankton and bacteria. The oxygen required for the respiratory demand of decomposing organic matter can be met by oxygen provided by photosynthesis in dedicated waste treatment ponds. Ponds designated for settling organic matter collected in central ponds are typically operated with anaerobic conditions. Some treatment ponds are operated as facultative ponds, where the absence of oxygen is desirable to promote denitrification and recovery of alkalinity. Salinity permitting, oxidation ponds often include filter-feeding fish species such as tilapia or mullet to graze undesirable cyanobacteria and stabilize water quality.

Another feature of intensive shrimp ponds is partial coverage with shade cloth to minimize phytoplankton activity and promote bacterial activity. This can increase the stability of water quality by reducing the fluctuation in dissolved oxygen concentration caused by photosynthesis and respiration by phytoplankton. In general, the trend has been to divide the shrimp production system into unit processes of grow-out, solids settling and treatment, nitrification, and oxidation. In this way, water can be reused and good biosecurity maintained.

Modern intensive shrimp farms continue to use Asian-style paddlewheel and LAPW aerators predominantly. Venturi aerators are also used, but typically for specific applications such as the aeration of deepwater zones and to facilitate accumulation of solids at the pond center. Venturi injectors and diffused aeration systems are now increasingly used in hyper-intensive ponds.

It is somewhat perplexing why shrimp farmers have not adopted more efficient aerator designs such as US-made catfish pond aerators that have been manufactured using design criteria developed during research on optimizing oxygen transfer rate and aerator efficiency. Such aerators are more expensive than the Asian-style paddlewheel aerators. However, the investment cost of an aerator is only part of the cost of aeration. Thought should be given to oxygen-transfer efficiency, service life, and maintenance cost of aerators. The long-term benefits of higher quality, greater efficiency, longer service life, and less maintenance are rarely considered and aerator cost seems to be the overriding consideration. Asian-style paddlewheel aerators seem to be viewed as a consumable item rather than capital equipment, with obsolescence understood to be associated with their relatively low cost.

As noted in the section on the ratio of installed aeration capacity to shrimp biomass supported, many shrimp farmers are unable to achieve the 400–500 kg/hp benchmark, suggesting that some ponds are aerated excessively. Most shrimp farmers do not own dissolved oxygen meters to monitor dissolved oxygen concentration in their ponds. Thus, given the high value of a shrimp crop, farmers use their practical experience in making decisions about how much aeration capacity to install and when to operate aeration equipment. The use of automated dissolved oxygen sensors to monitor ponds and operate aeration systems is promising but is not yet affordable for most farmers. Once the production system is standardized, including pond design, stocking density, and management of feeding, then standard protocols for aeration can be developed with the assistance of hand-held or continuous monitoring systems.

Although there is considerable knowledge and experience about managing dissolved oxygen and using aeration in shrimp ponds, there is considerable scope to increase production in existing ponds by applying what is already known. Additional work is needed to understand how to balance oxygen transfer with water circulation, as well as learning more about the most effective aerator placement methods and reducing energy use to increase production efficiency.

References

Ahmad, T. and Boyd, C.E. (1988) Design and performance of paddlewheel aerators. *Aquacultural Engineering* 7, 39–62.

Allan, G.L. and Maguire, G.B. (1991) Lethal levels of low dissolved oxygen and effects of short-term oxygen stress on subsequent growth of juvenile *Penaeus monodon*. *Aquaculture* 94, 27–37.

Allan, G.L., Maguire, G.B., and Hopkins, S.J. (1990) Acute and chronic toxicity of ammonia to penaeid prawns and the influence of low dissolved oxygen levels. *Aquaculture* 91, 265–280.

Ashok, B., Ashok, S. D., and Kumar, C.R. (2015) LPG diesel dual fuel engine – a critical review. *Alexandria Engineering Journal* 54, 105–126.

Barbieri, E., Bondioli, A.C.V., Melo, C.B., and Henriques, M.B. (2014) Nitrite toxicity to *Litopenaeus schmitti* (Burkenroad, 1936, Crustacea) at different salinity levels. *Aquaculture Research* 47, 1260–1268.

Bett, C. and Vinatea, L. (2009) Combined effect of body weight, temperature and salinity on shrimp *Litopenaeus vannamei* oxygen consumption rate. *Brazilian Journal of Oceanography* 57, 305–314.

Boyd, C.E. (1998) Pond water aeration systems. *Aquacultural Engineering* 18, 9–40.

Boyd, C.E. (2009) Estimating mechanical aeration requirement in shrimp ponds from oxygen demand of feed. In: Browdy, C.L. and Jory, D.E. (eds) *The Rising Tide. Proceedings of the Special Session on Sustainable Shrimp Farming.* World Aquaculture Society, Baton Rouge, LA, pp. 230–234.

Boyd, C.E. (2017) Does Coriolis force impact aerator placement in aquaculture ponds? Global Aquaculture Advocate. Available at: https://www.aquaculturealliance.org/advocate/coriolis/ (accessed 4 August 2021).

Boyd, C.E. (2020) *Water Quality: An Introduction*, 3rd ed. Springer Nature Switzerland, Cham, Switzerland.

Boyd, C.E. and McNevin, A.A. (2018) Land use in shrimp aquaculture. *World Aquaculture* 49(1), 28–34.

Boyd, C.E. and McNevin, A.A. (2021) Aerator energy use in shrimp farming and means for improvement. *Journal of the World Aquaculture Society* 51, 6–29.

Boyd, C.E., Torrans, E.L., and Tucker, C.S. (2018) Dissolved oxygen and aeration in Ictalurid catfish aquaculture. *Journal of the World Aquaculture Society* 49, 7–70.

Boyd, C.E., McNevin, A.A., Racine, P., Tinh, H.Q., Minh, H.N., Viriyatum, R., Paungkaew, D., and Engle, C.R. (2017) Resource use assessment of shrimp, *Litopenaeus vannamei* and *Penaeus monodon*, production in Thailand and Vietnam. *Journal of the World Aquaculture Society* 48, 201–226.

Boyd, C.E., D'Abramo, L.R., Glencross, B.D., Huyben, D.C., Juarez, L.M., Lockwood, G.S., McNevin, A.A., Tacon A.G.J., Teletchea F., Tomasso, J.R., Jr., Tucker, C.S., and Valenti W.C. (2019) Achieving sustainable aquaculture: Historical and current perspectives and future needs and challenges. *Journal of the World Aquaculture Society* 51, 578–633.

Broecker, W.S. and Peng, T.H. (1974) Gas exchange rates between air and sea. *Tellus* 26, 21–35.

Briggs, M.R.P. and Funge-Smith, S.J. (1994) A nutrient budget of some intensive marine shrimp ponds in Thailand. *Aquaculture and Fisheries Management* 25, 789–811.

Brouwer, M., Bonaventura, C., and Bonaventura, J. (1978) Analysis of the effect of three different allosteric ligands on oxygen binding by the hemocyanin of the shrimp, *Penaeus setiferus. Biochemistry* 12, 2148–2154.

Browdy, C.L., Ray, A.J., Leffler, J.W., and Avnimelech, Y. (2012) Biofloc-based Aquaculture Systems. In: Tidwell, J. (ed.) *Aquaculture Production Systems*, 1st ed. John Wiley & Sons, Inc., Ames, IA, pp. 278–307.

Burford, M.A. and Longmore, A.R. (2001) High ammonium production from sediments in hypereutrophic shrimp ponds. *Marine Ecology Progress Series* 224, 187–195.

Burford, M.A., Thompson, P.J., McIntosh, R.P., Barman, R.H., and Pearson, D.C. (2003) Nutrient and microbial dynamics in high-intensity, zero-exchange shrimp ponds in Belize. *Aquaculture* 219, 393–411.

Burnett, K.G. and Burnett, L.E. (2015) Respiratory and metabolic impacts of crustacean immunity: are there implications for the insects? *Integrative and Comparative Biology* 55, 856–868.

Castille, F.L., Jr., and Lawrence A.L. (1981) The effect of salinity on the osmotic, sodium and chloride concentrations in the hemolymph of euryhaline shrimp of the genus Penaeus. *Comparative Biochemistry and Physiology Part A: Molecular & Integrative Physiology* 68, 75–80.

Charmantier, G., Soyez, C., and Aquacop, (1994) Effect of molt stage and hypoxia on osmoregulatory capacity in the penaeid shrimp, *P. vannamei. Journal of Experimental Marine Biology and Ecology* 178, 233–246.

Chatvijitkul, S., Boyd, C.E., and Davis, D.A. (2018) Nitrogen, phosphorus, and carbon concentrations in some common aquaculture feeds. *Journal of the World Aquaculture Society* 49, 477–483.

Chatvijitkul, S., Boyd, C.E., Davis, D.A., and McNevin, A.A. (2017). Pollution potential indicators for feed-based fish and shrimp culture. *Aquaculture* 477, 43–49.

Chen, J.C. and Cheng, S.Y. (1995) Hemolymph oxygen content, oxyhemocyanin, protein levels and ammonia excretion in the shrimp *Penaeus monodon* exposed to ambient nitrite. *Journal of Comparative Physiology* 164B, 530–535.

Cheng, S.-Y. and Chen, J.-C. (1999) Hemocyanin oxygen affinity, and the fractionation of oxyhemocyanin and deoxyhemocyanin for *Penaeus monodon* exposed to elevated nitrite. *Aquatic Toxicology* 45, 35–46.

Cheng, S.-Y., Shieh, L.-W., and Chen, J.-C. (2013) Changes in hemolymph oxyhemocyanin, acid–base balance, and electrolytes in *Marsupenaeus japonicus* under combined ammonia and nitrite stress. *Aquatic Toxicology* 130–131, 132–138.

Cheng, W., Liu, C.-H., Yan, D.-F., and Chen, J.-C. (2002) Hemolymph oxyhemocyanin, protein, osmolality and electrolyte levels of whiteleg shrimp *Litopenaeus vannamei* in relation to size and molt stage. *Aquaculture* 211, 325–339.

Clark, J.V. (1986) Inhibition of moulting in *Penaeus semisulcatus* (De Haan) by long-term hypoxia. *Aquaculture* 52, 253–254.

Colt, J. (2012) *Dissolved Gas Concentrations in Water*, 2nd ed. Elsevier, Amsterdam, The Netherlands.

Dall, W. (1986) Estimation of routine metabolic rate in a penaeid prawn, *Penaeus esculentus* Haswell. *Journal of Experimental Marine Biology and Ecology* 96, 57–74.

de Lima, J.P.V., Brito, L.O., Costa, W.M., de Melo Costa, W., and Gálvez, A.O. (2012) Utilização de peróxido de hidrogênio (H_2O_2) no incremento de oxigênio dissolvido em cultivo de *Litopenaeus vannamei* (Boone, 1931). *Pesquisa Agropecuária Pernambucana* 17, 73–77.

Delgado, P.C., Avnimelech, Y., McNeil, R., Bratwold, D., Browdy, C.L., and Sandifer, P. (2003) Physical, chemical, and biological characteristics of distinctive regions in paddlewheel aerated shrimp ponds. *Aquaculture* 217, 235–248.

Ding, S., Wang, F., Dong, S., and Li, Y. (2014) Comparison of the respiratory metabolism of juvenile *Litopenaeus vannamei* cultured in seawater and freshwater. *Journal of Ocean University of China* 13, 331–337.

Engle, C.R., McNevin, A., Racine, P., Boyd, C.E., Paungkaew, D., Viriyatum, R., Tinh, H.Q., and Minh, H.N. (2017) Economics of sustainable intensification of aquaculture: Evidence from shrimp farms in Vietnam and Thailand. *Journal of the World Aquaculture Society* 48, 227–239.

Farrelly, J.C., Chen, Y., and Laskodi, C. (2016) Oxygenation zones in conventional and split earthen catfish ponds in eastern Arkansas. *Aquacultural Engineering* 75, 14–21.

Fast, A.W. and Boyd, C.E. (1992). Water circulation, aeration and other management practices. In: Fast, A.W. and Lester, J.L. (eds) *Marine Shrimp Culture: Principles and Practices*. Elsevier Science, Amsterdam, The Netherlands, pp. 457–496.

Fast, A., Tan, E., Stevens, D., Olson, J., Qin, J., and Barclay, D. (1999) Paddlewheel aerator oxygen transfer efficiencies at three salinities. *Aquacultural Engineering* 19, 99–103.

Felix-Portillo, M., Martinez-Quintana, J.A., Peregrino-Uriarte, A.B., and Yepiz-Plascencia, G. (2014) The metallothionein gene from the white shrimp *Litopenaeus vannamei*: characterization and expression in response to hypoxia. *Marine Environmental Research* 101, 91–100.

Fernández-Prini, R. (1982) Le Chatelier's principle and the prediction of the effect of temperature on solubilities. *Journal of Chemical Education* 59, 551–553.

Forstner, H. and Gnaiger, E. (1983) Calculation of equilibrium oxygen concentration. In: Forstner, H. and Gnaiger, E. (eds.) *Polarographic Oxygen Sensors*. Springer-Verlag, Berlin, Germany, pp. 321–333.

Fregoso-López, M.G., Morales-Covarrubias, M.S., Franco-Nava, M.A., Ramírez-Rochín, J., Fierro-Sañudo, J.F., Ponce-Palafox, J.T., and Páez-Osuna, F. (2017) Histological alterations in gills of shrimp *Litopenaeus vannamei* in low-salinity waters under different stocking densities: potential relationship with nitrogen compounds. *Aquacultural Research* 48, 5854–5863.

Furtado, P.S., Serra, F.P., Poersch, L.H., and Wasielesky, W. (2014) Short communication: Acute toxicity of hydrogen peroxide in juvenile white shrimp *Litopenaeus vannamei* reared in biofloc technology systems. *Aquaculture International* 22, 653–659.

Furtado, P.S. Valenzuela, M.A.J., Badillo, M.A. Gaxiola G., and Wasielesky, W., Jr. (2016a) Effect of dissolved carbon dioxide on oxygen consumption in the Pacific white shrimp, *Litopenaeus vannamei* (Boone 1931). *Marine and Freshwater Behaviour and Physiology* 49, 337–346.

Furtado, P.S., Valenzuela, M.A.J., Rodriguez-Fuentes, G., Campos, B.R., Wasielesky, W., Jr., and Gaxiola, G. (2016b) Chronic effect of nitrite on the rearing of the white shrimp *Litopenaeus vannamei* in two salinities. *Marine and Freshwater Behaviour and Physiology* 49, 201–211.

Garcia, A. and Brune, D.E. (1991) Transport limitation of oxygen in shrimp culture ponds. *Aquacultural Engineering* 10, 269–279.

Gross, A., Abutbui, S., and Zilberg, D. (2004) Acute and chronic effects of nitrite on white shrimp, *Litopenaeus vannamei*, cultured in low-salinity brackish water. *Journal of the World Aquaculture Society* 35, 315–321.

Hopkins, J.S., Hamilton, R.D., Sandifer, P.A., Browdy, C.L., and Stokes, A.D. (1993) Effect of water exchange rate on production, water quality, effluent characteristics and nitrogen budgets of intensive shrimp ponds. *Journal of the World Aquaculture Society* 24, 304–320.

Hopkins, J.S., Sandifer, P.A., and Browdy, C.L. (1995) Effect of two feed protein levels and feed rate combinations on water quality and production of intensive shrimp ponds operated without water exchange. *Journal of the World Aquaculture Society* 26, 93–97.

Hopkins, J.S., Sandifer, P., Browdy, C.L., and Holloway, J.D. (1996) Comparison of exchange and no-exchange water management strategies for the intensive pond culture of marine shrimp. *Journal of Shellfish Research* 15, 441–445.

Hopkins, J.S., Stokes, A.D., Browdy, C.L., and Sandifer, P.A. (1991) The relationship between feeding rate, paddlewheel aeration rate and expected dawn dissolved oxygen in intensive shrimp ponds. *Aquacultural Engineering* 10, 281–290.

Hu, F., Pan, L., and Jing. F. (2009) Effects of hypoxia on dopamine concentration and the immune response of white shrimp (*Litopenaeus vannamei*). *Journal of Ocean University of China* 8, 77–82.

Innocent, D.S., Sylvester, P.O., Yahaya, M.F., Nwadike, I., and Okoro, L.N. (2013) Comparative analysis of biodiesel and petroleum diesel. *International Journal of Education and Research* 1, 1–8.

Jamnongwong, M., Charoenpittaya T., Hongprasith, N., Imai, T., and Painmanakul, P. (2016) Study of liquid film forming apparatus (LFFA) mechanisms in terms of oxygen transfer and bubble hydrodynamic parameters. *Engineering Journal* 20, 77–90.

Jiang, L.-X., Pan, L.-Q., and Bo, F. (2005). Effect of dissolved oxygen on immune parameters of the white shrimp *Litopenaeus vannamei*. *Fish & Shellfish Immunology* 18, 185–188.

Kalra, D., Babu, A.V., and Kumar, M.V. (2014) Effects of LPG on the performance and emission characteristics of SI engine – an overview. *International Journal of Engineering Development and Research* 2, 2997–3003.

Kang, Y.H., Lee, M.O., Choi, S.D., and Sin, Y.-S. (2004) 2-D hydrodynamic model simulating paddlewheel-driven circulation in rectangular shrimp culture ponds. *Aquaculture* 231, 163–179.

Kirk, K.R. (2010) Modeling microbial and nutrient dynamics in zero-discharge aquaculture systems. PhD Dissertation. Clemson University, Clemson, SC, USA.

Kniffin, C.D., Burnett, L.E., and Burnett, K.G. (2014). Recovery from hypoxia and hypercapnic hypoxia: impacts on the transcription of key antioxidants in the shrimp *Litopenaeus vannamei*. *Comparative Biochemistry and Physiology Part B: Biochemistry & Molecular Biology* 170, 43–49.

Le Moullac, G., Soyez, C., Saulnier, D., Ansquer, D., Avarre, C.J., and Levy, P. (1998) Effect of hypoxic stress on the immune response and the resistance to vibriosis of the shrimp *Penaeus stylirostris*. *Fish & Shellfish Immunology* 8, 621–629.

Li, J., Li, W., Zhang, X., and He, P. (2018) Physiological and behavioral responses of different modes of locomotion in the whiteleg shrimp *Litopenaeus vannamei* (Boone, 1931) (Caridea: Penaeidae). *Journal of Crustacean Biology* 38, 79–90

Li, J., Xu, X., Li, W., and Zhang, X. (2019a) Linking energy metabolism and locomotor variation to osmoregulation in Chinese shrimp *Fenneropenaeus chinensis*. *Comparative Biochemistry and Physiology Part B: Biochemistry & Molecular Biology* 234, 58–67.

Li, Y., Wei, L., Cao, J., Qiu, L., Jiang, X., Li, P., Song, Q., Zhou, H., Han, Q., and Diao, X. (2016) Oxidative stress, DNA damage and antioxidant enzyme activities in the Pacific white shrimp (*Litopenaeus vannamei*) when exposed to hypoxia and reoxygenation. *Chemosphere* 144, 234–240.

Li, Z.S., Ma, S., Shan, H.W., Wang, T., and Xiao, W. (2019b) Responses of hemocyanin and energy metabolism to acute nitrite stress in juveniles of the shrimp *Litopenaeus vannamei*. *Ecotoxicology and Environmental Safety* 186, 109753.

Liao, I.C. and Murai, T. (1986) Effects of dissolved oxygen, temperature and salinity on the oxygen consumption of the grass shrimp, *Penaeus monodon*. In: Maclean, J.L., Dizon, L.B. and Hosilos, L.V. (eds) *The First Asian Fisheries Forum*. Asian Fisheries Society, Manila, Philippines, pp. 26–31.

Lin, Y.C. and Chen, J.C. (2003) Acute toxicity of nitrite on *Litopenaeus vannamei* (Boone) juveniles at different salinity levels. *Aquaculture* 224, 193–201.

Lin, W.-Y., Chang, Y.-Y., and Hsieh, Y.-R. (2010) Effect of ethanol–gasoline blends on small engine generator energy efficiency and exhaust emission. *Journal of the Air & Waste Management Association* 60, 142–148.

Madenjian, C.P. (1990) Patterns of oxygen production and consumption in intensively managed marine shrimp ponds. *Aquaculture and Fisheries Management* 28, 407–417.

McCarthy, P., Rasul, M.G., and Moazzem, S. (2011) Comparison of the performance and emissions of different biodiesel blends against petroleum diesel. *International Journal of Low-Carbon Technologies* 6, 255–260.

McGraw, W., Teichert-Coddington, D.R., Rouse, D.B., and Boyd, C.E. (2001) Higher minimum dissolved oxygen concentrations increase penaeid shrimp yields in earthen ponds. *Aquaculture* 199, 311–321.

McIntosh, R.P. (2000a) Changing paradigms in shrimp farming: III. Pond design and operation considerations. *Global Aquaculture Advocate* February, 42–45.

McIntosh, R.P. (2000b) Changing paradigms in shrimp farming: V. Establishment of heterotrophic bacterial communities. *Global Aquaculture Advocate* December, 52–54.

Mohanty, R.K. (2001) Effect of pond aeration on growth and survival of *Penaeus monodon* Fab. *Bangladesh Journal of Fisheries Research* 5, 59–65.

Mohanty, R.K., Ambast, S.K., Panigrahi, P., Thakur, A.K., and Mandal, K.G. (2018) Enhancing water use efficiency in monoculture of *Litopenaeus vannamei*: impacts on pond water quality, waste production, water footprint and production performance. *Aquacultural Engineering* 82, 46–55.

Moore, J. M. and Boyd, C.E. (1992) Design of small paddlewheel aerators. *Aquacultural Engineering* 11, 55–69.

Nonwachai, T., Purivirojku, W., Chuchird, N., and Limsuwan, C. (2011). Effects of dissolved oxygen levels on growth, survival and immune response of juvenile Pacific white shrimp *Litopenaeus vannamei. Kasetsart University Fisheries Bulletin* 35(3), 1–10.

Pérez-Rostro, C.I., Racotta, I.S., and Ibarra, A.M. (2004) Decreased genetic variation in metabolic variables of *Litopenaeus vannamei* shrimp after exposure to acute hypoxia. *Journal of Experimental Marine Biology and Ecology* 302, 189–200.

Peterson, E.L. (1999) Benthic shear stress and sediment condition. *Aquacultural Engineering* 21, 85–111.

Peterson, E.L. (2000) Observations of pond hydrodynamics. *Aquacultural Engineering* 21, 247–269.

Peterson, E.L., Harris, J.A., and Wadhwa, L.C. (2000) CFD modelling pond dynamic processes. *Aquacultural Engineering* 23, 61–93.

Peterson, E.L. and Pearson, D. (2000) Round peg in a square hole: aeration in a square shrimp pond. *Global Aquaculture Advocate* 3(5), 44–46.

Peterson, E.L. and Walker, M.B. (2002) Effect of speed on Taiwanese paddlewheel aeration. *Aquacultural Engineering* 2, 129–147.

Peterson, E.L., Wadhwa, L.C., and Harris, J.A. (2001) Arrangement of aerators in an intensive shrimp growout pond having rectangular shape. *Aquacultural Engineering* 25, 51–65.

Pipoppinyo, S. and Boyd, C.E. (1993) The α-coefficient in pond aeration. *Journal of Aquaculture in the Tropics* 8, 111–119.

Ponce-Palafox, J.T., Ruíz-Luna, A., Gómez, M.G.-U., Esparza-Leal, H.M., Arredondo-Figueroa, J.L., Martinez-Palacios, C.A., and Ross, L.G. (2013) A response-surface analysis of the relative importance of the temperature, salinity and body weight on the respiratory metabolism of the white shrimp *Litopenaeus vannamei* (Boone, 1931). *Marine and Freshwater Behaviour and Physiology* 46, 399–417.

Racotta, I.S., Palacios, E., and Mendez, L. (2002) Metabolic responses to short and long-term exposure to hypoxia in white shrimp (*Penaeus vannamei*). *Marine and Freshwater Behavioral Physiology* 35, 269–275.

Rathburn, C.K., Sharp, N.J., Ryan, J.C., Neely, M.G., Cook, M., Chapman, R.W., Burnett, L.W., and Burnett, K.G. (2013) Transcriptomic responses of juvenile Pacific whiteleg shrimp, *Litopenaeus vannamei*, to hypoxia and hypercapnic hypoxia. *Physiological Genomics* 45, 794 –807.

Re, A.D. and Díaz, F. (2011) Effect of different oxygen concentrations on physiological energetics of blue shrimp, *Litopenaeus stylirostris* (Stimpson). *The Open Zoology Journal* 4, 1–8.

Reddy, E.R.K. and Raju, V.D. (2018) Experimental studies on natural aspirated diesel engine fueled with corn seed oil methyl ester as a bio-diesel. *IOP Conference Series: Materials Science and Engineering* 330, 012104.

Rogers, G.L. (1989) Aeration and circulation for effective aquaculture pond management. *Aquacultural Engineering* 8, 349–355.

Rogers, G.L. (2009) Aeration theory and its application to shrimp culture. In: Browdy, C.L. and Jory, D.E. (eds) *The Rising Tide. Proceedings of the Special Session on Sustainable Shrimp Farming.* World Aquaculture Society, Baton Rouge, LA, pp. 211–219.

Rogers, G. L. and Fast, A.W. (1988) Potential benefits of low energy water circulation in Hawaiian prawn ponds. *Aquacultural Engineering* 8, 155–165.

Rosas, C., Martínez, E., Gaxiola, G., Brito, R., Sánchez, A., and Soto, L.A. (1999) The effect of dissolved oxygen and salinity on oxygen consumption, ammonia excretion and osmotic pressure of *Penaeus setiferus* (Linnaeus) juveniles. *Journal of Experimental Marine Biology and Ecology* 234, 41–57.

Ruttanagosrigit, W., Musig, Y., Boyd, C.E., and Sukhareon L. (1991) Effect of salinity on oxygen transfer by propeller-aspirator-pump and paddlewheel aerators used in shrimp farming. *Aquacultural Engineering* 10, 121–131.

Samocha, T.M. (2019) *Sustainable Biofloc Systems for Marine Shrimp*. Academic Press, Cambridge, MA.

Santa, K.D. and Vinatea, L. (2007) Evaluation of respiration rates and mechanical aeration requirements in semi-intensive shrimp *Litopenaeus vannamei* culture ponds. *Aquacultural Engineering* 36, 73–80.

Seidman, E.R. and Lawrence, A.L. (1985) Growth, feed digestibility, and proximate body composition of juvenile *Penaeus vannamei* and *Penaeus monodon* grown at different dissolved oxygen levels. *Journal of the World Mariculture Society* 16, 333–346.

Shapouri, H., Duffield, J.A., and Wang, N. (2002) The energy balance of corn ethanol: an update. Agricultural Economic Report 814. Office of Energy Policy and New Uses, Washington, DC.

Silbey, R.J., Alberty, R.A., and Bawendi, M.G. (2005) *Physical Chemistry*, 4th ed. John Wiley and Sons, Inc., New York, NY.

Soñanez-Organis, J.G., Peregrino-Uriarte, A.B., Gómez-Jiménez, S., López-Zavala, A., Forman, H.J., and Yepiz-Plascencia, G. (2009) Molecular characterization of hypoxia inducible factor-1 (HIF-1) from the white shrimp *Litopenaeus vannamei* and tissue-specific expression under hypoxia. *Comparative Biochemistry and Physiology Part C: Toxicology & Pharmacology* 150, 395–405.

Steeby, J.A., Hargreaves, J.A., and Tucker, C.S. (2004) Factors affecting sediment oxygen demand in commercial channel catfish ponds. *Journal of the World Aquaculture Society* 35, 322–334.

Suplee, M.W. and Cotner, J.B. (1996) Temporal changes in oxygen demand and bacterial sulfate reduction in inland shrimp ponds. *Aquaculture* 145, 141–158.

Takaishi, T., Numata, A., Nakano, R., and Sakaguchi, K. (2008) Approach to high efficiency diesel and gas engines. *Mitsubishi Heavy Industries Technical Review* 45(1), 21–24.

Taparhudee, W., Benjaprasertsri, M., and Sattiti, B. (2007) Comparative study on paddle-wheel aerators using electric motors and diesel engines in Pacific white shrimp (*Litopenaeus vannamei*) culture ponds. *Kasetsart Journal (Natural Science)* 41, 522–530.

Taw, N. (2014) Shrimp farming in biofloc system: review and recent developments. *World Aquaculture 2014*. Adelaide, South Australia, Australia, pp. 1–33.

Taw, N., Saleh, U., and Slamat, B. (2013) Malaysia shrimp project scales up for production in biosecure biofloc modules. *Global Aquaculture Advocate* January/February, 44–47.

Thulasi, D., Muralidhar, M., and Saraswathy, R. (2020) Effect of sulphide in Pacific white shrimp *Penaeus vannamei* under varying oxygen and pH levels. *Aquaculture Research* 51, 2389–2399.

Tommerdahl, A.P., Burnett, K.G., and Burnett, L.E. (2015) Respiratory properties of hemocyanin from wild and aquacultured penaeid shrimp and the effects of chronic exposure to hypoxia. *The Biological Bulletin* 228, 242–252.

Tompkins, B. T., Song, H., Bittle, J.A., and Jacobs, T.J. (2012) Efficiency considerations for the use of blended biofuel in diesel engines. *Applied Energy* 98, 209–218.

Ulaje, S.A., Ascencio, F., Arreola, L.R., Hernández, P.C., Cota, S.E.L., and Sicard, M.T. (2019) Gene expression and energetic metabolism changes in the whiteleg shrimp (*Litopenaeus vannamei*) in response to short-term hypoxia. *Aquaculture Research* 50, 994–1004.

Vinatea, L. and Carvalho, J. (2007) Influence of water salinity on the SOTR of paddlewheel and propeller-aspirator-pump aerators, its relation to the number of aerators per hectare and electricity costs. *Aquacultural Engineering* 37, 73–78.

Vinatea, L., Muedas, W., and Arantes, R. (2011) The impact of oxygen consumption by shrimp *Litopenaeus vannamei* according to body weight, temperature, salinity and stocking density on pond aeration: a simulation. *Acta Scientiarum. Biological Sciences* 33, 125–132.

Vinatea, L., Gálvez, A.O., Browdy, C.L., Stokes, A., Venero, J., Haveman, J., Lewis, B.L., Lawson, A., Shuler, A., and Leffler, J.W. (2010) Photosynthesis, water respiration and growth performance of *Litopenaeus vannamei* in a super-intensive raceway culture with zero water exchange: Interaction of water quality variables. *Aquacultural Engineering*, 42, 17–24.

Walker, S.J., Neill, W.H., Lawrence, A.L., and Gatlin, D.M., III. (2009) Effect of salinity and body weight on ecophysiological performance of the Pacific white shrimp (*Litopenaeus vannamei*). *Journal of Experimental Marine Biology and Ecology* 380, 119–124.

Walker, S.J., Neill, W.H., Lawrence, A.L., and Gatlin, D.M., III. (2011) Effects of temperature and starvation on ecophysiological performance of the Pacific white shrimp (*Litopenaeus vannamei*). *Aquaculture* 319, 439–445.

Wei, L.-Z., Zhang, X.-M., Li, J., and Huang, G.-Q. (2008) Compensatory growth of Chinese shrimp, *Fenneropenaeus chinensis* following hypoxic exposure. *Aquaculture International* 16, 455–470.

Wei, L.-Z., Zhang, X.-M., Li, J., Huang, G.-Q., and Li, J. (2009) Effects of limited dissolved oxygen supply on the growth and energy allocation of juvenile Chinese shrimp, *Fenneropenaeus chinensis*. *Journal of the World Aquaculture Society* 40, 483–492.

Wijesekara, R.G.S., Nomura, N., and Matsumura, M. (2005) The effect of paddle wheel and jet aerators on bottom velocity, dissolved oxygen levels and shrimp distribution in aquaculture ponds. *Sri Lanka Journal of Aquatic Sciences* 10, 97–114.

Wu, R.S.S., Lam, P.K.S., and Wan, K.L. (2002) Tolerance to, and avoidance of, hypoxia by the penaeid shrimp (*Metapenaeus ensis*). *Environmental Pollution* 118, 351–355.

Yang, H.-C. and Park, S.-K. (2011) Effect of salinity on dissolved oxygen characteristics in an ejector-aerator. *Journal of the Korean Society of Marine Engineering* 35, 640–646.

Yoo, K. H. and Boyd, C.E. (1994*) Hydrology and Water Supply for Pond Aquaculture*. Chapman and Hall, New York, NY.

Yu, X., Zhang, X., Duan, Y., Zhang, P., and Miao, Z. (2010) Effects of temperature, salinity, body length, and starvation on the critical swimming speed of whiteleg shrimp, *Litopenaeus vannamei*. *Comparative Biochemistry and Physiology Part A: Molecular & Integrative Physiology* 157, 392–397.

Zhang, P.D., Zhang, X.M., Li, J., and Huang, G.Q. (2006a) The effects of body weight, temperature, salinity, pH, light intensity and feeding condition on lethal DO levels of whiteleg shrimp, *Litopenaeus vannamei* (Boone, 1931). *Aquaculture* 256, 579–587.

Zhang, P., Zhang, X., Li, J., and Huang, G. (2006b) Swimming ability and physiological response to swimming fatigue in whiteleg shrimp, *Litopenaeus vannamei*. *Comparative Biochemistry and Physiology Part A: Molecular & Integrative Physiology* 145, 26–32.

Zhang, P., Zhang, X., Li, J., and Huang, G. (2007) The effects of temperature and salinity on the swimming ability of whiteleg shrimp, *Litopenaeus vannamei*. *Comparative Biochemistry and Physiology Part A: Molecular & Integrative Physiology* 147, 64–69.

Chapter 10

ZERO WATER EXCHANGE SHRIMP PRODUCTION, INDOOR SYSTEM EVALUATION

MANUEL POULAIN, BARBARA HOSTINS AND
FRANCISCO SARAIVA GOMEZ

Summary

Farming shrimp is a combination of art and science. Farming shrimp also requires passion, which most farmers have in abundance. However, even with dedication and passion, culturing shrimp at a profit is becoming increasingly difficult. Disease prevalence in most production areas is an important challenge. The impact of disease on a country's annual production is huge and it brings about inconsistent outputs. This variability creates uncertainty in all areas of the supply chain and leads to fluctuating prices.

The shrimp culture industry needs to enforce adequate farming practices that will generate enough certainty to attract the required investment. Shrimp farming production forecasts for the upcoming 10 years reflect impressive growth in terms of tonnes produced. It is necessary to bring to the industry any elements that will improve knowledge, thus taking the industry in the right direction.

Fully biosecure production systems exist. Their productivity is high: some say it can reach over 100 Mt/ha/y. These systems are environmentally sustainable and have very low water requirements with reduced discharge. Production systems such as recirculating aquaculture systems (RAS) and biofloc use are some of the techniques used in these high-production, low-discharge systems.

Every system has its advantages and its limitations, both biological and economic. Very different outputs can result from facilities set up in a tropical country in Asia, or those set up in the USA. A RAS shrimp farm business model would be suitable for niche markets in the USA or Europe, whereas a biofloc farm would be best suited for Asia. The success of such operations ultimately depends on the people working in the field. Teamwork and dedication are key. Financial support, especially during the start-up phase, is also critical. Every element counts and must be thoroughly considered if a zero water exchange shrimp production system is to be a reality.

Introduction

Success in shrimp farming relies on the identification and proper management of the different risks encountered in production. A successful risk management strategy will lead to predictable results. The consistency of production is largely impacted by the incidence of disease. Viral diseases were first observed in shrimp production in the 1990s (Taura syndrome, white spot syndrome, yellow head disease, etc.). More complex bacterial diseases have been prevalent since the year 2000 (early mortality syndrome, *Enterocytozoon hepatopenaei* infection, White feces syndrome, etc.). Most of the large production areas have experienced pathogen contamination in their facilities. This is why protecting crops through biosecurity is key for the future development of the industry. Under the pressure of pathogen incidence, shrimp-farming operations have been forced to increase their control over all possible vectors in order to secure a more predictable output.

Biosecurity certainly starts by securing disease-free postlarvae. The animals used for stocking remain the highest contamination risk in farms. Additionally, intensification of production requires more control of environment-to-pond interactions. Intensification also entails an increased demand for water in order to maintain optimal farming conditions. Typically, water is pumped into the pond and then directly discharged back into the environment. This represents a challenge not only in terms of biosecurity but also environmentally, as these effluents are usually loaded with organic matter. The reduction of water exchange remains the most critical point in order to reduce the entry of pathogens into the system.

Some production facilities have built a series of wastewater treatment ponds adjacent to the shrimp grow-out ponds. These units are used to clean and treat the discharged water before it is pumped back into the system. This technique has shown promising results, improving both the biosecurity conditions of the farm and the environmental footprint of the farm's activity. However, the use of discharge water treatment ponds is demanding in terms of land and thus decreases the area that can be used for production at those sites that use this technique. In some cases, the area used for water treatment can be larger than the area used for shrimp production. Most of those operations take place outdoors and are also subject to adverse weather conditions. Control over wastewater treatment is limited, and there is a degree of uncertainty involved.

Other alternatives for shrimp production exist. Farms can operate with very limited water intakes. Some systems focus on engineering and filtration techniques, such as recirculating aquaculture systems (RAS). Other options include bioremediation, using biological processes that take place within the shrimp pond ecology itself to treat and recycle organic waste and eliminate toxic chemical compounds. This system refers mainly to bioflocs, or microbial production systems. RAS and microbial shrimp-production systems are active today in most parts of the world. They have shown positive signs in terms of their productivity. However, there are still very few examples of these types of farms achieving economic sustainability. Several projects have been unviable and have had to close down.

This chapter aims to analyse the benefits and limitations of a zero water exchange approach. The environmental and financial implications have been included in order to provide insight that would allow an assessment of these farming techniques.

Main considerations

POSITION

The main point of reducing water requirements in shrimp farming is to reduce biosecurity issues. The aim of applying these techniques is to progress towards precision farming, incorporating a higher degree of environmental control in the production.

In order to improve day-to-day fluctuations in parameters related to weather changes, zero water exchange operations are usually built indoors. In outdoor conditions, plankton development varies in response to the weather. These changes prevent the system from being highly reliable and consistent. This restriction limits the size of the tanks/ponds used and increases the number of units needed if the necessary volume of production is to be achieved.

Indoor facilities require very large investment in comparison to outdoor farming techniques. High productivities are required if the production unit is to be economically viable. Usual stocking densities range from 150 to 1000 postlarvae/m^2. At these densities, the maintenance of a suitable level of dissolved oxygen in the water becomes the ultimate limiting factor.

Oxygen levels in the water are directly linked to the carrying capacity of the production system. With the regular use of blowers and diffusers, the level of dissolved oxygen could be comparable to those found with a biomass of 5 kg/m^3 if we include a feed loading capacity of around 100 g/m^3 of feed administered to a pond every day of culture. If we consider, for example, a 20 g shrimp that has been fed to 2% of its biomass, the maximum density would be $20 \times 2\% = 0.4$ g of feed per shrimp, given a total carrying capacity of 250 shrimp/m^2 to reach a feed load of 100 g/m^3/d. In this scenario, 250 shrimp/m^2 at a body weight of 20 g will have a total carrying capacity of 250×20 g = 5 kg/m^3. This carrying capacity, in terms of shrimp and feed, is usually considered to be the maximum safe limit for the development of sustainable production. Higher productivities can be achieved, but they involve considerably higher production risks and should be considered very carefully. In start-up operations, a maximum farming density of 250 postlarvae/m^2 is not recommended, in order to reduce risks and create a sustainable foundation for future development.

ZERO WATER EXCHANGE CONSIDERATIONS

As long as there is no water exchange, organic waste and toxic chemical compounds will accumulate in the water. The bacterial load in the system will be high, as a direct consequence of super-intensive farming.

Nitrogen control

The main reason for water exchange in a shrimp production pond is to limit the accumulation of toxic forms of nitrogen in the water. Nitrogen accumulation mainly comes from the feed used in the shrimp-farming process. In super-intensive conditions, it will take an optimal amount of 1.2 kg of feed to produce 1 kg of shrimp of about 20 g weight. An estimated 16% of the protein contents of the feed are identified as nitrogen. In the absence of water renewal, nitrogen compounds accumulate, and the concentrations of toxic compounds such as ammonia or nitrites increase rapidly.

In zero water exchange systems, special attention is given to the enhancement and control of nitrifying biochemical reactions. The most efficient nitrification pathway for nitrogen removal involves the nitrogen cycle, via the activity of autotrophic nitrifying bacteria. Scientific studies have shown that *Nitrosomonas* sp. bacteria are so efficient that a single cell can convert ammonia at a rate that would require up to one million heterotrophs to equal (Nithiya et al., 2016).

In RAS systems, autotrophic nitrification is carried out in external biofilters, or within the shrimp pond itself in the case of microbial systems. In the first case, adequate microbial communities will be enhanced inside a biofilter, where bacteria will attach and develop on the biofiltration medium. In the case of biofloc systems, the biofiltration medium allows the growth of nitrifying bacteria in the tank's biofilms, in the same way as the floc particles. In biofloc systems, the nitrification process is also supported by the heterotrophic pathway, where bacteria will digest nitrogen directly into the biomass (Divya et al., 2015).

Vibrio control

Achieving a suitable balance between 'good' and 'bad' bacterial populations in the pond ecosystem is fundamental. By quorum sensing, opportunistic bacteria such as *Vibrio* sp. can become pathogenic when they reach a certain population density (Defoirdt and Sorgeloos, 2012) (see Chapter 8). It is therefore important to limit the growth of *Vibrio* sp. in the system. The control and maintenance of the right balance of bacterial communities present in the water is key to limiting the risk of opportunistic bacteria becoming pathogenic. In super-intensive conditions, the shrimp biomass must be considered together with the feed being loaded into the water, as these considerably increase the potential risk of vibriosis-related issues. The organic matter in a pond provides a niche for opportunistic bacteria, where *Vibrio* sp. proliferate.

Control of the accumulation of solids in the system, as well as the use of high-quality probiotic bacteria in sufficient quantities, remain the best available tools to reduce this risk. When applying probiotic bacteria, consideration should be given to using a higher number than the estimated *Vibrio* sp. population, to take into account the difference in multiplication time between the two bacterial communities. In zero water exchange, super-intensive production systems, a ratio between probiotic bacteria and *Vibrio* sp. of 10–20 should be considered. In these conditions, the suppression of opportunistic bacteria should focus on the competitive exclusion principle (Callaway et al., 2008).

In RAS, removal of solids is carried out physically through the use of mechanical filters. In biofloc systems, solids are continuously maintained in suspension, acting as activated sludge (Seviour and Nielsen, 2010). In this case, the usual organic waste is recycled within the shrimp tank ecology (Crab, 2012), serving as a medium for beneficial bacterial communities (mainly nitrifiers and probiotics) but also providing food support for the animals (Burford et al., 2004).

Feeding control

Feed is one of the major costs in shrimp production. Feed management also has a major impact on the water ecology. This applies to all shrimp-farming operations but even more so in zero water exchange conditions. The high daily loads of feed introduced into the system linked to super-intensive conditions lead to the need for 24-hour distribution, via the use of automatic feeders. Automatic feeders can be belt feeders for small tanks, or spinner or blower-operated feeders for larger production units.

The distribution of feed throughout the day and night stabilizes and improves nitrification processes and limits the accumulation of feed at the bottom of the tank. This reduced accumulation optimizes pellet accessibility and improves shrimp growth and survival. Feeders are set to feed at specific hours of the day, for a fixed number of times per day. To simplify operations, adjustment of the feeder regulation is done daily, adjusting only the duration of each feeding event.

Systems available

Two main approaches are being used at present in zero water exchange systems, with opposing underlying principles. One approach targets shrimp farming in clear water, based on advanced mechanical engineering, via the use of pumps and filtration equipment, referred to in this chapter as RAS. The second approach focuses on biological processes, or bioremediation, growing shrimp in loaded waters, referred to as biofloc systems.

CLEAR-WATER RECIRCULATING AQUACULTURE SYSTEMS

RAS have been used in aquaculture production for decades. RAS technologies were first used in fish production and were rapidly introduced in shrimp farming. The RAS approach consists of recycling water continuously from the production tank through a series of filtration units to maintain optimal water quality throughout all stages of the farming process. The filtration equipment deals with the main constraints involved in the absence of water renewal, namely nitrogen and bacterial composition.

The microbial control of opportunistic bacteria is carried out by suppression, via mechanical processes. There is a range of equipment available on the market, including equipment specifically applied to shrimp aquaculture, that makes RAS systems a promising approach for the future development of zero water exchange shrimp production systems.

The typical filtration line of a RAS consists of mechanical filtration (sand filter, bead filter, drum filter, etc.), for the removal of solids. Water will then go to a UV sterilizing unit and/or ozone injection unit, to reduce the total bacterial load in the water. Finally, the water will pass through a biofiltration unit, which will limit the accumulation of nitrogen. It is recommended to place the biofiltration unit last in the filtration line, to protect it from possible contamination (i.e. so that water passes through it after sterilization), but also to allow nitrifying bacteria that grow in the biofilter to colonize the plumbing and production tank's biofilms (Thymann, 2013). From stocking to harvest, the daily turnover of water going through the filtration line ranges from 100% at the beginning of the crop up to 600% towards the end.

For the treatment of recirculating water to be effective, tanks are usually preferred to be small in volume. The water level can be low to allow a higher productivity per cubic meter of water. In this case tanks can be stacked one on top of the other, to obtain a considerable increase in production per area.

BIOFLOC SYSTEM

In contrast to RAS, the biofloc, or microbial, systems target shrimp production in loaded waters. Efforts are made to control the pond ecosystem through bioremediation (Divya et al., 2015).

Autotrophic and heterotrophic nitrification pathways are enhanced and promoted to control hazardous nitrogenous waste such as ammonia and nitrites. Due to the relatively slow process of colonization for autotrophic bacterial conditioning, nitrifiers require prior culture in a nitrification bioreactor (Achuthan, 2000). This helps the maturation of ammonia- and nitrite-oxidizing bacterial communities that can be used as an inoculum for the shrimp culture pond. In the presence of nitrogen produced by the shrimp culture process, nitrifying bacteria will then establish in the ecosystem, giving full control over the most toxic nitrogen compounds of nitrogen for shrimp, namely ammonia and nitrites.

Heterotrophic populations are promoted via the continuous addition of high-quality probiotic bacteria at a rate 10–20 times higher than the desired maximum concentration of *Vibrio* sp. When applied in this way, probiotic bacteria will have a varied mode of action. They will serve as a continuous bacterial inoculum to sustain the active degradation of organic waste, and also suppress the growth of *Vibrio* sp. and possible opportunistic bacteria through competitive exclusion.

Both the reduction of food/nitrogen content in the water and the colonization of biofilms promote natural productivity that serves as nutritional support for the animals (Burford et al., 2004). Shrimp such as *Litopenaeus vannamei* are well adjusted to catch suspended particles via the use of their third maxilliped (Alexander et al., 1980).

Through the development of a rich microbial community, biofloc systems do not require water exchange, or mechanical treatments during the grow-out phase. The most important component of the biofloc tanks is the aeration system design. Organic solids produced by the animals and the feed need to be maintained in suspension in order to act as bioflocs or activated sludge and not as a waste/contaminant. Aeration systems work by creating convection currents from the bottom to the top of the pond or tank. Total air flow of 30–50 m^3/h should be supplied to 100 m^3 of culture water. Air is diffused into the water via aeration diffusers, offering the best relation between oxygen transfer and convection current force. Diffusers are placed all across the bottom of the tank, in order to avoid as many accumulation areas as possible. As an example, in the case of aerotube diffusers of 1 inch (2.5 cm) diameter, which are the most commonly used in this application, it is considered that 100 m^3 of culture waters will require the installation of 10–16 m of aerotubes.

The biofloc closed dynamic induces an added value to the monitoring data coming from the laboratory. Fluctuations in environmental parameters within the shrimp pond ecosystem are very important and must be monitored closely. Interactions between pH and alkalinity, as well as dissolved oxygen levels and Imhoff cone evolution, provide accurate indicators of the quality of the husbandry management being carried out. All interactions are part of the shrimp pond ecology and do not include any deviation that could come from filtration lines (in the case of RAS) or water intake, in addition to weather changes, in traditional outdoor farming.

Biofloc systems are preferably installed indoors, and should include natural light shading to allow a reduction of about 70–80% of light. Phytoplankton, as an autotrophic organism, will compete with nitrifiers for space and nutrients (Risgaard-Petersen et al., 2004). Biofloc systems are bacteria based, unlike semi-intensive outdoor shrimp pond systems, which are algae based. Limiting the amount of sunlight reaching the tank water limits phytoplankton development, creating a more stable environment that is not impacted by day/night algae respiration, as well as other weather-related variations.

Environmental implications

ENVIRONMENTAL SUSTAINABILITY

Biosecurity is one of the main issues that has driven the ongoing changes introduced in shrimp culture systems towards indoor zero water exchange practices. Usually, in a conventional culture system, high water exchange rates are used to keep the water quality parameters at levels suitable for shrimp culture. Most conventional shrimp farms occupy extensive areas, with pond sizes ranging between 8 ha and 16 ha (medium- to large-scale farms) (Na nakorn et al., 2017) with one canal for water discharge and water exchange. Additionally, neighbouring farms can be relatively close to one another, sharing the same source of water intake and discharge. Therefore, the risk of spreading pathogenic organisms and/or disease vectors when farms are operated under these systems is substantially higher. As an example, Shinn et al. (2018) reported that shrimp disease-related losses within the Vietnamese Mekong Delta in 2015 were determined to be >US$ 26 million due to acute hepatopancreatic necrosis disease, while the costs of white spot syndrome virus in the same year were >US$ 11 million.

Thus, the shift towards limited or reduced water exchange aims to improve biosecurity measures particularly by 'closing' the system and operating under water recirculation (i.e. RAS systems) conditions, using some of the ponds as reservoirs (recirculation farms), or stimulating microbial biomass production via additional carbon input (biofloc systems). In such systems, nutrients are recycled and/or assimilated, allowing the reuse of water in numerous production cycles.

These practices, and the manner in which biosecurity is driven and managed in indoor, zero water exchange systems, can suggest a link to environmental sustainability. The high rate of water use and discharge in conventional shrimp farming is the most critical issue related to the environmental impact of the shrimp farming industry. Water use in a flow-through-operated system can average 50 m^3 water/kg feed applied daily (Martins et al., 2010), whereas Verdegem et al. (2006) reported that a RAS is operated with 0.1–1 m^3 water/kg feed. Additionally, an intensive zero exchange lined shrimp pond requires only 1–2.26 m^3 water/kg shrimp, whereas a conventional system with regular water exchange may require up to 80 m^3 water/kg shrimp per day (Hargreaves, 2006).

In this context, as well as the need to improve biosecurity in shrimp farms, limited water exchange systems have been developed as a response to increasing and stricter environmental regulations and a scarcity of water and land in an increasing number of countries. An assessment carried out using the internationally standardized life cycle assessment (ISO, 2006) reports a reduction in water dependency of up to 93% (Roque d'Orbcastel et al., 2009). This kind of analysis has been applied in several studies to evaluate the environmental sustainability of aquaculture systems and can provide an overview and an estimate of defined parameters such as global warming potential, eutrophication potential, energy use, and acidification potential in the system.

When water is exchanged in a conventional shrimp pond, the 'used' water is usually discharged into nearby rivers, creeks, or estuaries (Burford and Williams, 2001). The macro- and micronutrients in the discharged water (phosphorus, nitrogen, sulphides) are often present at very high concentrations; in the case of nitrogen, it is estimated that 27–57% of the total nitrogen added to ponds is discharged (Funge-Smith and Briggs, 1998; Preston et al., 2000).

Na nakorn et al. (2017) assessed the effluents produced by 19 different shrimp farms of variable sizes, operated under conventional systems in Thailand, and verified that for farms of any size, the average total suspended solids, organic matter, and ammonia nitrogen of effluents from harvests would not pass the Thai Effluent Standard for Coastal Aquaculture. As mentioned by the authors, environmental regulations for aquaculture systems have been demanding alternatives for the reduction and management of waste during a production cycle. Along these lines, the life cycle assessments carried out in aquaculture systems have highlighted that the feed represents the factor with the highest environmental impact, as a causative agent with a high eutrophication potential. Shrimp feed from early postlarval stages until harvest can contain approximately 35–40% of protein. From the total protein content, cultured shrimp use, on average, only 20–30% of the nitrogen present in the feed for growth. The remaining nitrogen is released into the water, representing one of the major causes of eutrophication in conventional systems. The eutrophication potential is around 20% lower in RAS systems than in flow-through systems due to differences in waste release.

Although in indoor systems discharge of waste can be limited by the limited water exchange, waste management is still a challenge that requires the adoption of adequate treatment, and an appropriate destination, of wastewater. For example, recirculation farms often convert one or more ponds to reservoirs in order to maximize water reutilization. Shrimp ponds operated under this kind of system are equipped with a central drain, from which the excess waste is flushed out of the pond and can be treated accordingly. Alternatives to treat the waste produced in indoor systems have been discussed in several academic publications. Denitrification reactors can be installed in RAS to convert accumulated nitrates (NO_3^-) into nitrogen gas (N_2) by an anaerobic process. These reactors are designed according to the needs of each system and can decrease the need for heating, water, and bicarbonate (Martins et al., 2010). However, nitrogen treatment by nitrification/denitrification can be very energy demanding, it requires expertise, and its costs are high (Hu et al., 2012; Yogev and Gross, 2019). Other options for waste management are the use of settling chambers that concentrate the sludge, followed by drying of the microbial biomass to produce an alternative replacement meal, among other by-products. Nevertheless, all of the techniques mentioned require extra storage facilities, transportation, labour, and disposal fees, making them still limited to commercial upscaling.

There are numerous advances and advantages of indoor zero water exchange systems as alternatives to effectively support the intensification and the changing conditions of shrimp culture systems. However, challenges regarding waste management still need to be overcome and transferred from research findings to practical and commercial application.

MICROBIAL BOTTLENECKS

When water is not exchanged in culture systems, nutrients and potentially toxic nitrogen compounds are recycled in the system by conversion or assimilation. These processes are driven by microbial communities which use these nutrients for metabolic activities.

In RAS, the removal of ammonia nitrogen is achieved through nitrification processes, in which the nitrogen excreted by the cultured shrimp in the form of ammonia is converted into nitrite and nitrate (less toxic forms) by nitrifying bacteria. On the other hand, in biofloc systems, there is a predominance of heterotrophic bacteria groups which, when exposed to a high carbon/nitrogen (C/N) ratio, assimilate ammonia nitrogen and convert it into microbial biomass. These two different microbial processes can occur in zero water exchange systems, and the general pathways used have been well described at experimental scale.

As well as improving the water quality by reducing the level of ammonia in the system, it is reported that the microbial community established in bioflocs can act as a nutritional supplement (Wasielesky et al., 2006) and increase the immune response of shrimp (Ekasari et al., 2014). Hostins et al. (2019) verified that when cultured in biofloc systems, with management of the C/N ratio and the microbial community, *L. vannamei* is protected against *Vibrio* challenges.

The mechanisms of protection and disease defence by which the microbial community can confer greater immunity to shrimp is still unclear and currently under study. However, some hypotheses have been explored at laboratory scale. It has been reported that the microbial biomass present in zero water exchange systems is able to disrupt quorum sensing of *Vibrio* sp. (Defoirdt et al., 2011), which is the mechanism used by these bacteria to signal and induce cell proliferation.

The main challenge and bottlenecks of such systems that are microbially driven when implemented at a commercial scale are the many variables present that can influence the different processes. Currently, limited information on the interactions of microbial communities in aquaculture systems is available. This lack of knowledge highlights the challenge of finding tools to improve and optimize the management of these systems, to improve their consistency and reliability. Emparanza (2009) assessed the main problems affecting nitrification in RAS, and pointed out that management issues such as variations in pH, temperature, and oxygen, high stocking density, and variation in daily feeding are the factors affecting nitrification at a commercial scale.

All the factors mentioned above can affect the microbial community directly, and a better understanding of bacterial community structures is essential to provide tools, products, and the possibility to steer microbial communities in indoor zero water exchange systems.

PHYSIOLOGICAL BOTTLENECKS

Indoor culture systems will have lower phytoplankton growth due to limited light availability and photosynthetic activity. However, these systems are often microbe based, which allows the development of microbial aggregates that can represent a complementary nutritional source for the cultured shrimp.

In the conventional way of culturing shrimp used a few decades ago, the development of microbes in the ponds was considered a negative factor, and high volumes of water exchange took place in order to eliminate any microbial community that could grow in the system. However, recent studies have shown that the way in which these microbes are managed within systems can be of benefit to the production cycle. For example, Wasielesky et al. (2006) confirmed the benefits of natural productivity in a zero water exchange system based on bioflocs. The authors observed better growth performance when shrimp fed on the microbial flocs. The same observations were reported by Burford et al. (2004), suggesting that the natural biota contributes substantially to the nutrition of *L. vannamei*. Besides bioflocs, microorganisms that constitute the biofilms or periphyton can also be an important nutritional supplement (Martínez-Córdova et al., 2015).

The way in which microbial communities are managed (in terms of C/N ratio) will have an influence on the predominant bacterial groups found in microbial-based systems, and will also influence their nutritional contribution to shrimp performance. In a high C/N ratio environment, heterotrophic biomass is the main bacterial group present in biofloc systems. By contrast, the C/N ratio in RAS is lower, and autotrophic bacteria are predominantly established. In these systems, the nutritional contribution of microbial aggregates is minimal or absent. In this sense, Tierney and Ray (2018) attributed better growth performance of shrimp cultured in bioflocs compared with those grown in RAS.

Feed composition is one of the main challenges for indoor zero water exchange systems, especially for RAS. The lack of natural production in these systems, and the production of solids composed of lost and uneaten feed, require the formulation of diets specifically designed for these systems. Martins et al. (2010) highlighted the need for improved RAS diets formulated with ingredients that can boost digestibility and produce faeces that are easily removed from the water, as well as producing fewer fine solids that could be easily fermented by the microbial community. In order to achieve these targets at a commercial scale, more research on feed composition and characteristics, and shrimp gut ecology, is required.

Financial implications

SCENARIOS

To assess the financial implications of zero water exchange systems, investment and operation figures have been set to the standard parameter of 100 t of *L. vannamei* shrimp of average weight 20 g produced per year.

To achieve this production, the environmental conditions of the selected site are a determinant factor in terms of cost. Production in tropical areas will require a different investment structure from production in colder climates. Elements such as heating requirements, cost of utilities, and cost of labour will greatly impact the figures. Together, these factors mean that the development of zero water exchange projects in Asia will be very different from that of systems implemented in Europe. This point is valid not only in relation to infrastructure, but also at an operational and sales level.

In the context of the assessment carried out, biometric performance in terms of stocking density, growth, and survival are assumed to be similar for RAS and biofloc systems. A productivity of 5 kg/m^3 of shrimp per crop was considered. This carrying capacity was considered to be representative of the average capability of most production systems. These set levels of annual production, harvest size targeted, and carrying capacity serve only for comparison purposes for the financial implications study. Elements such as shrimp genetics, the quality of the postlarvae, feed, and additives used, as well as the environmental conditions, robustness of infrastructure and engineering, and, ultimately, labour performance, will all impact the results.

In temperate climates, a heating system and building insulation are required for the indoor production of *L. vannamei*. Additionally, such projects usually target niche markets, or high-end distribution/catering lines. This market will absorb only a certain quantity of fresh product daily. This is why smaller tanks are preferable, in order to allow full harvest practices and to limit partial harvest manipulations, which are labour intensive and stressful for the animals.

In tropical countries, temperatures are suitable for year-round shrimp farming. Simple building solutions can be used, allowing installations to be covered by tunnels or greenhouse structures. This type of building solution usually offers a wider range of free span, giving access to larger shrimp tank construction options. There are also no restrictions in the volume of what the shrimp market can usually process, being able to absorb large quantities of product at any point in time. Due to these factors, when setting up a facility in a tropical country one can consider larger production pond units than in a temperate area. As a reference, in the financial exercise, tanks of 100 m^2 were considered for temperate-climate developments, whereas tanks of 500 m^2 were considered for tropical locations.

Other indicators were also taken into consideration. Due to the maturity of the shrimp farming industry in tropical areas, farmers have much better access to high-quality genetic material and high postlarvae quality than in those operating temperate-climate sites. This increases the average biometric performance of a facility established in a tropical site.

SYSTEM COMPARISON

Under the scenario set, comparing standard outputs, the capital investment required for the operations of a RAS and a biofloc system have been estimated. The unit cost for construction and operation refers to the actual average cost for both areas described: developing tropical countries, such as most of the Asian countries, and countries with a more temperate climate, such as those in Europe or the USA.

Fig. 10.1 summarizes the estimated total investment required for the construction of a 100 Mt/y production farm, together with the estimated average production cost per kg of 50 count shrimp produced.

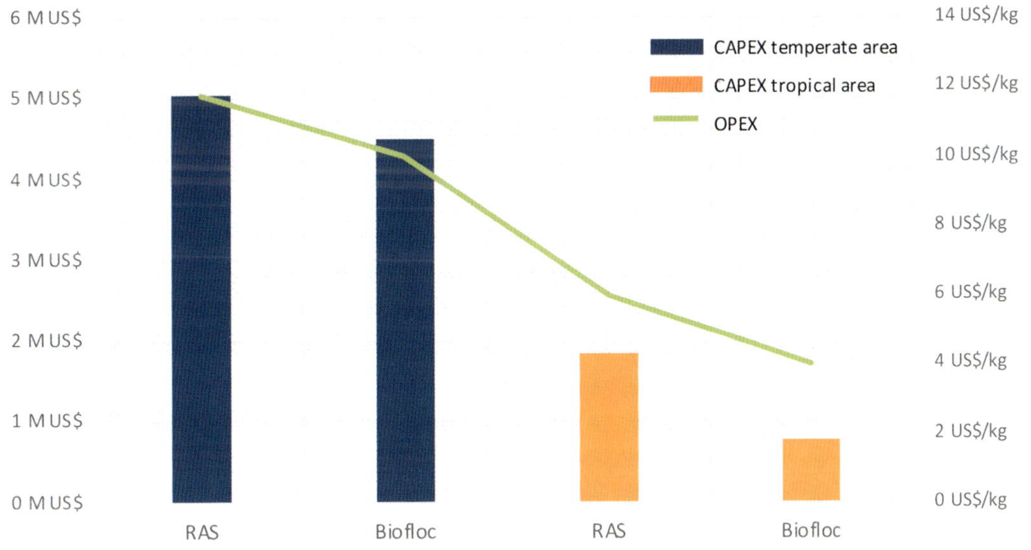

FIGURE 10.1 Capital expenditure (CAPEX)/operating expenditure (OPEX) estimation for zero water exchange production.

Capital investment requirements

The capital investment required ranges from US$5 million for a RAS installed in a developed temperate area, down to around US$800,000 for the development of a biofloc system in a tropical developing country. The trends for production costs follow the same evolution; be it a RAS or a biofloc system, the costs are also considerably lower if the operation has been set up in a tropical country.

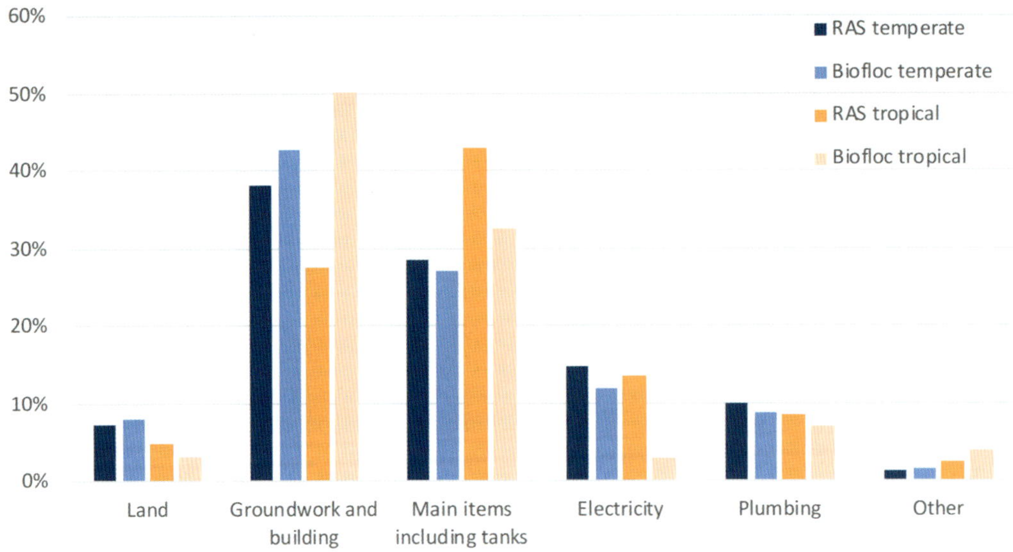

FIGURE 10.2 Capital investment repartition estimates for zero water exchange system.

The spread of the investment costs per system and climate is presented in Fig. 10.2. The major items impacting costs are related to the construction of the facilities. Construction costs include groundwork, concrete structures, and buildings. These costs can represent up to 50% of the investment in the case of a biofloc system installed in a tropical area.

The second highest cost demand relates to the machinery required. This includes the geomembrane used for tanks, all the pumps and filtration equipment, and also the fresh shrimp holding equipment required (climate-controlled room for shrimp selection, ice machine and cold room, etc.). The rest of the capital required is for the purchase of the land to develop the project, electrical and plumbing works, as well as other small equipment used for production and laboratory work.

Operational expenses

Production costs can be three times higher for a RAS production system set up in a temperate area than for a biofloc system established in a tropical country. The trend of cost efficiency follows the same path as the capital investment required, making biofloc systems less expensive than RAS. Fig. 10.3 presents the production cost structure to deploy each system in the different production sites.

The production cost figures include all consumables, such as postlarvae, feed, additives, laboratory requirements, and so on. They also include power, water (mostly used to process animals), and heating. Natural gas was considered as the main energy source for heating.

Production costs in tropical areas are evidently lower than those of an installation built in a colder climate. There is an evident link of cost per unit for most of the items considered. For example, postlarvae can cost around US$3.50–4.50/1000 postlarvae in Asia, while in Europe and the USA, the cost is well over US$10/1000 postlarvae at the farm gate. The cost of labour also makes a major difference when the areas are compared. It is important to note that labour costs are higher than feed costs when operating a farm

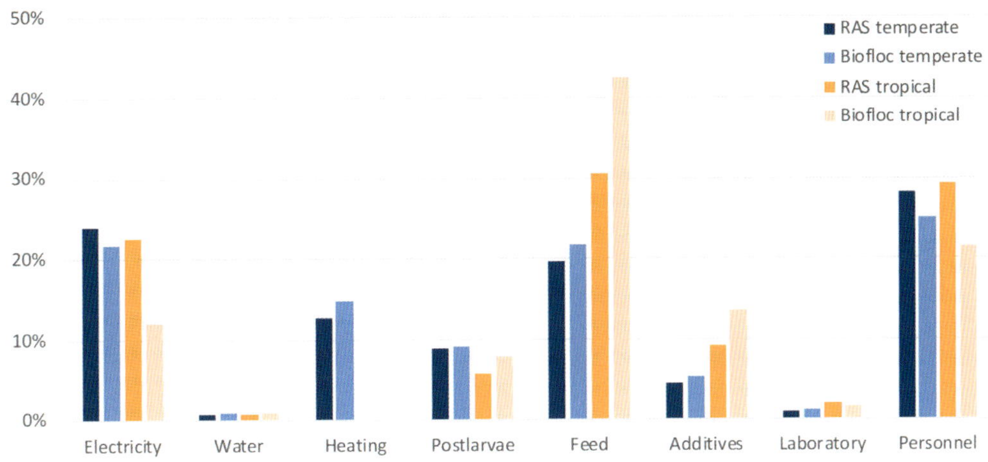

FIGURE 10.3 Production cost structure for zero water exchange system.

of this capacity in a developed country in a temperate area. All the different aspects together considerably increase the production cost of zero water exchange systems in locations such as the USA or Europe.

MARKET IMPLICATIONS

The assessment carried out on the financial implications of a zero water exchange system clearly demonstrates the high disparity of investment and operational costs inherent to the different scenarios analysed.

The high cost of shrimp production in developed countries, such as those in Europe, dictates the need for high sale prices. Average production cost figures estimated for the chosen comparative scenarios are too high to be able to aim for the commodity market. This kind of project requires considerable effort in marketing and subsequently branding of the product. The market targeted is that of high-end seafood retailing units, such as hotels and restaurants. The growing concerns of consumers regarding food quality represents an opportunity for this kind of project, if the operation has been set up near large markets.

In tropical developing areas, the usually high level of local shrimp production in farms is aimed at the high-sale-price markets in Europe or the USA. The interest of the projects in these countries mainly lies in achieving consistent production, with predictable business operations. Zero water exchange systems enforce optimal biosecurity measures, leading to a much lower incidence of environmental and weather-related impacts.

A zero water exchange farm in Europe will target high-end markets, aiming towards batch production, where shrimp will be stocked monthly, and harvested almost daily, all year round. Profitability resides in the efficiency of the tank's rotation together with the marketing/branding carried out to achieve reasonable resale prices exceeding US$15/kg in a short supply circuit (local consumption).

In Asia, the same project will mainly concentrate on the biosecurity aspect, to achieve consistent and predictable production. Marketing/branding impacts are less evident, and it is likely that a minimum volume will be required in order to be able to target higher resale prices.

For a production of 100 Mt/y, in Europe or the USA, branding is key to achieve higher resale prices in order to be profitable enough with a high-end consumption market to be economically viable. This applies regardless of the system used for shrimp production. In Asia, the focus is on biosecurity, to reduce contamination risk factors. The main consideration relates to infrastructure design and application in order to avoid tank leakage, underground leaching, labour movement, or any other possible vectors of contamination.

In developing tropical countries, the set-up of a RAS for shrimp production does not appear to be the best strategy. Production costs remain quite high in comparison to the costs of setting up biofloc systems. The difference in cost comes from the accumulation of higher power costs (to operate higher-level machinery), higher estimated feed costs due to the higher nutrient requirements (because of the low natural productivity), and higher requirements for personnel to monitor and maintain the elements of the system.

Other considerations

The following sections intend to present and discuss additional limiting factors that should be considered. These factors have been identified as important for the development of zero water exchange technologies.

POSTLARVAE QUALITY AND GENETICS

The quality of shrimp genetic material available in the location of the set-up is important to consider. Different strains can have very different outputs in a given situation. A good – and stable – quality of larvae supplies is as important as the genetic quality of the larvae. Bad postlarvae quality, and/or poor genetics, will lead to poor performance in the grow-out phase, regardless of the management or system used.

It is important to consistently check the batches of postlarvae arriving at the facility. The average weight of the population, variation in size, and survival of salinity stress tests are good indicators of quality. The variation in these characteristics, together with the farm's performance record, will be of great use to estimate the necessary practices to apply (stocking density, feeding, treatment, etc.). For example, size variation in a population is a useful indicator for farmers. In farming, high size variation is usually representative of an environmental issue and/or a feeding deviation (either not enough food or too much food, as well as the possible environmental issue). If the postlarvae stocked had high size variation but this was not checked when they arrived at the facility, the farmer will never know whether this problem has arisen from the pond management or the batch of postlarvae, thus losing a valuable management indicator.

In terms of genetics, several large global companies have invested in genomic tools. This step forward for shrimp genetic selection programs, together with enough variability in shrimp strains, opens the possibility to develop strains of animals adapted to super-intensive production in a reasonable number of generations. The focus should be on animal robustness, increased shrimp resistance to environmental stresses such as low dissolved oxygen concentrations or high bacterial load, as well as on optimal growth.

Attention should be given to the feeding regime, which is linked to animal growth, to ensure that daily feeding practices do not exceed the system's carrying capacity. In this sense, strains with higher survival abilities are preferable to fast-growing strains. Even though production rotation will be faster with fast-growing strains, the large amount of feed required to express this genetic high-growth capability may compromise the water treatment systems in place. This point is of even greater importance in the RAS approach, as removal of solids is carried out mechanically in such systems.

HUMAN FACTORS

Human factors are and will always remain at the core of any aquaculture farm success. Competence and experience, as well as motivation, communication, and teamwork, are key components. Labour costs represent the largest cost component in terms of operating expenses, especially in the case of establishing an operation in countries such as those in Europe or in the USA. For this reason, farmers will always prioritize simple facilities and simple protocols over complexity. Facilities need to be conceived with a high level of ergonomics, allowing optimal conditions for the workforce, over time. On-site training, company participation, and career opportunities are key to secure motivated staff who will perform under any circumstances.

Project management is as fundamental as the team members. Skilled, experienced, and motivated managers are key to success. A project manager needs to act as a leader, to maintain personnel motivation at its highest, sharing passion and knowledge. The management also has the very important task of the accurate and rapid communication of relevant information to the company's board, to help ensure the right decisions are made to solve potential problems in an effective manner.

Opportunities

ECONOMIES OF SCALE

Like many other biotech farming processes, the financial implications of a zero water exchange project are impacted by its scale. A pilot project run with commercial-sized tank units is fundamental to minimize the initial risk of setting up the project. This phase aims to ensure attainability of the forecasted production output, and to validate the business model strategy.

The project needs to contemplate a production volume that is large enough to reach sustainable profitability. As suggested in the estimates discussed in the 'Financial implications' section of this chapter, a production of 100 Mt/y will achieve a return on investment over 10 years in the West. Larger production operations will automatically lead to reasonable economies of scale. This is mainly achieved through distribution of management costs over a higher production volume and improvements in the efficiency of utilities consumption (e.g. electricity, fuel) with larger systems, as well as increased bargaining power for consumable purchases such as feed, postlarvae, and additives.

These economies of scale seem to be considerably important in temperate developed countries, where labour costs remain the largest operational cost. Production cost optimization allows a reduction in sale price, opening new distribution channels and allowing entry to the commodity market, which is not accessible for lower production volumes. As an example, the simulation of a biofloc system established in Europe suggests a reduction by half of the time needed to achieve a return on investment for a production reaching 500 Mt/y compared with one of 100 Mt/y.

TECHNOLOGY

Following the assessment carried out, the most efficient farming strategy for zero water exchange production appears to be the biofloc system, using bioremediation as a water treatment pathway. Figures for RAS suggest that the cost of establishing the production unit is similar to that of a biofloc farm, but additional funds are required for filtration machinery acquisition. RAS also require higher levels of energy to operate, as well as higher feed quality to supplement the lack of natural productivity of the system.

Elements such as footprint reduction using a stacked tank system for RAS can greatly reduce the building cost. This system also relies mainly on mechanical engineering, and so carries opportunities for direct automation. With this in mind, RAS seem to be a more competitive solution for application in developed countries. RAS farming systems have a higher chance of success near niche markets, with multiple set-ups rather than one large operation.

Biofloc systems are easier to operate and require less machinery, leading to lower energy requirements compared with RAS. The main constraints of biofloc systems pertain to the limited knowledge available today regarding the complex microbial interactions at play in the bioremediation principles. In this context, some operators will feel more comfortable dealing with machines than with biological processes.

In any case, the future for shrimp production will be indoor and super-intensive. The evolution that is taking place will combine both biology and engineering. As the industry progresses, it clearly appears that both RAS and biofloc systems have much to offer.

INVESTMENT

The rapidly increasing demand for quality protein, driven by global population growth, requires considerable investment. Given that there is limited land and resources available on Earth, production systems will have to focus on super-intensive production, with the lowest possible environmental impacts.

Super intensive production systems such as the zero water exchange models discussed in this chapter seem very promising for future development, and they seem particularly well suited to respond to the main challenges to come. However, at present the development of these types of systems as a commercial investment remains anecdotal. Many claims have been made in favour of both RAS and biofloc systems, but there is still a lack of visibility of actual production units of these kinds with viable economic figures.

The principles behind these systems, in terms of consistency and predictability, generate investor interest, with investors viewing these models as real opportunities for development in the near future. On the other hand, the lack of successful examples of the implementation of these systems dampens investment interest. Additionally, as has been suggested by the financial implications of this type of project, the scale of implementation needs to be large enough as to generate sufficient profit.

Zero water exchange production requires a certain degree of know-how, both for RAS and for biofloc systems. Projects need to be designed taking into consideration the fact that production will take some time to ramp up to full-scale figures. Technology and know-how will need to be acquired and assimilated, and the workforce will need to be trained. Markets need to be created, especially for projects aimed at fresh shrimp retail. All the elements included in setting up operations, from site selection to facility design ergonomics, production protocols and their application, processing options, and distribution lines, need to be working in synergy to maximize production output and minimize implementation risks.

It seems important to consider the added cost of the start-up period in the initial financial package, in order to secure cash flow during this time. A grace period of roughly 2 years should be considered, as this is the time needed to acquire and assimilate the hands-on experience required for the farming process. This will be long enough to produce several crops and select the right strategy to adopt during the ramp-up phase of the project. This period of time can vary from 1 to 3 years depending on the farm's design (modular production or not), the partnership possibilities (technological assets), the project manager, and the capacity of the team.

Conclusion

This chapter intends to synthesize most of the elements that should be taken into consideration when looking at developing zero water exchange projects for shrimp production. The clear-water RAS and microbial (biofloc) technologies are two options that can be considered for zero water exchange production facilities. Their development across the world will allow the supply of high-quality shrimp, produced sustainably. The rapidly growing concerns of consumers related to food safety dictates the development of farming techniques with a higher degree of control, which is the core element of zero water exchange production systems.

The level of maturity of the technologies involved in zero water exchange farming reinforces the economic viability of such a project, whether a RAS or a biofloc system is selected. The assessment carried out clearly demonstrates the requirement to control every aspect of a shrimp-farming production unit in order for it to be successful. Elements such as postlarvae supply, genetics, and quality, as well as management, workforce knowledge, and experience, together with end-product positioning and distribution lines, are all crucial to the success of the project. None of these elements should be underestimated, as they impact the whole business model.

Another important aspect of a zero water exchange system is the need for it to be operated at a large enough scale to obtain enough profits to motivate investors. A project of this kind, producing 50 Mt/y of shrimp, will barely generate an income. A project producing 500 Mt/y or more will have a much better cost structure, opening considerable opportunities for development.

It is believed that zero water exchange technologies will play a big role in the development of future shrimp production techniques. The recent advances in shrimp breeding techniques, together with the forthcoming combination of RAS technology with the microbial know-how used in biofloc systems, provides a positive outlook for the growth of zero water exchange systems around the globe in the years to come. The increased consistency that is brought about by zero water exchange facilities opens the door to the use of automation and artificial intelligence decision-making units. This level of precision in shrimp farming is believed to be required in order to achieve the production levels necessary to feed the planet's growing population in a sustainable manner and to help secure the future protein supply of generations to come. Zero water exchange technologies will play an important role in the changes that the industry requires.

References

Achuthan, C. (2000) Development of bioreactors for nitrifying water in closed system hatcheries of penaeid and non-penaeid prawns. PhD thesis. School of Environmental Studies, Cochin University of Science and Technology, Kerala, India.

Alexander, C.G., Hindley, J.P.R., and Jones, S.G. (1980) Structure and function of the third maxillipeds of the banana prawn *Penaeus merguiensis*. *Marine Biology* 58, 245–249.

Burford, M.A., Thompson, P.J., McIntosh, R.P., Baumand, R.H., and Pearson, D.C. (2004) The contribution of flocculated material to shrimp (*Litopenaeus vannamei*) nutrition in a high-intensity, zero-exchange system. *Aquaculture* 232, 525–537.

Burford, M.A. and Williams, K.C. (2001) The fate of nitrogenous waste from shrimp feeding. *Aquaculture* 198, 79–93.

Defoirdt, T. and Sorgeloos, P. (2012) Monitoring of *Vibrio harveyi* quorum sensing activity in real time during infection of brine shrimp larvae. *The ISME Journal* 6, 2314–2319.

Defoirdt, T., Sorgeloos, P., and Bossier, P. (2011) Alternatives to antibiotics for the control of bacterial disease in aquaculture. *Current Opinion in Microbiology* 14, 251–258.

Divya, M., Aanand, S., Srinivasan, A., and Ahilan, B. (2015) Bioremediation – an eco-friendly tool for effluent treatment: a review. *International Journal of Applied Research* 1, 530–537.

Callaway, T.R., Edrington, T.S., Anderson, R.C., Harvey, R.B., Genovese, K.J., Kennedy, C.C., Venn, D.W., and Nisbet, D.J. (2008) Probiotics, prebiotics and competitive exclusion for prophylaxis against bacterial disease. Cambridge University, December 2008, 217-225.

Crab, R., Defoirdt, T., Bossier, P., Verstraete, W. (2012) Biofloc technology in aquaculture: Beneficial effects and future challenges. Elsevier aquaculture vol. 356-357, 351-356.

Ekasari, J., Azhar, M.H., Surawidjaja, E.H., Nuryati, S., De Schryver, P., and Bossier, P. (2014) Immune response and disease resistance of shrimp fed biofloc grown on different carbon sources. *Fish and Shellfish Immunology* 41, 332–339.

Emparanza, E.J.M. (2009) Problems affecting nitrification in commercial RAS with fixed-bed biofilters for salmonids in Chile. *Aquacultural Engineering* 41, 91–96.

Funge-Smith, S.J. and Briggs, M.R.P. (1998) Nutrient budgets in intensive shrimp ponds: implications for sustainability. *Aquaculture* 164, 117–133.

Hargreaves, J.A. (2006) Photosynthetic suspended-growth systems in aquaculture. *Aquacultural Engineering* 34, 344–363.

Hostins, B., Wasielesky, W., Decamp, O., Bossier, P., and De Schryver, P. (2019) Managing input C/N ratio to reduce the risk of Acute Hepatopancreatic Necrosis Disease (AHPND) outbreaks in biofloc systems – a laboratory study. *Aquaculture* 508, 60–65.

Hu, Z., Lee, J.W., Chandran, K., Kim, S., and Khanal, S.K. (2012) Nitrous oxide (N_2O) emission from aquaculture: a review. *Environmental Science & Technology* 46, 6470–6480.

ISO (2006) ISO 14044:2006. Environmental management – life cycle assessment – requirements and guidelines. ISO, Geneva, Switzerland.

Martínez-Córdova, L.R., Emerenciano, M., Miranda-Baeza, A., and Martínez-Porchas, M. (2015) Microbial-based systems for aquaculture of fish and shrimp: an updated review. *Reviews in Aquaculture* 7, 131–148.

Martins, C.I.M., Eding, E.H., Verdegem, M.C.J., Heinsbroek, L.T.N., Schneider, O., Blancheton, J.P., Roque d'Orbcastel, E., and Verreth, J.A.J. (2010) New developments in recirculating aquaculture systems in Europe: a perspective on environmental sustainability. *Aquacultural Engineering* 43, 83–93.

Na nakorn, A., Chevakidagarn, P., and Danteravanich, S. (2017) Environmental impact of white shrimp culture during 2012–2013 at Bandon Bay, Surat Thani province: a case study investigating farm size. *Agriculture and Natural Resources* 51, 109–116.

Nithiya, A., Rao, P.H., and Kumar, T.S. (2016) Bioremediation of aquaculture water using nitrifying bacteria-microalga consortium with special reference to ammoniacal nitrogen. *International Journal of Current Research and Academic Review* 4(12), 164–177.

Preston, N., Jackson, C., Thompson, P., Austin, M., and Burford, M. (2000). *Prawn Farm Effluent: Composition, Origin and Treatment.* Fishing Research and Development Corporation Final Report 95/162. Fishing Research and Development Corporation, Canberra, Australia.

Risgaard-Petersen, N., Nicolaisen, M.H., Revsbech, N.P., and Lomstein, B.A. (2004) Competition between ammonia-oxidizing bacteria and benthic microalgae. *Applied and Environmental Microbiology* 70, 5528–5537.

Roque d'Orbcastel, E., Blancheton, J.P., and Aubin, J. (2009) Towards environmentally sustainable aquaculture: comparison between two trout farming systems using life cycle assessment. *Aquacultural Engineering* 40, 113–119.

Seviour, R. and Nielsen, P.H. (2010) *Microbial Ecology of Activated Sludge.* IWA Publishing, London, UK.

Shinn, A.P., Pratoomyot, J., Griffiths, D., Trong, T.Q., Vu, N.T., and Briggs, M. (2018) Asian shrimp production and the economic costs of disease. *Asian Fisheries Science* 31S, 29–58.

Thymann, A. (2013) The co-existence of heterotrophs and nitrifiers in nitrifying biofilm. Master's thesis. Roskilde University, Denmark.

Tierney, T.W. and Ray, A.J. (2018) Comparing biofloc, clear-water, and hybrid nursery systems (Part I): shrimp (*Litopenaeus vannamei*) production, water quality, and stable isotope dynamics. *Aquacultural Engineering* 82, 73–79.

Verdegem, M.C.J., Bosma, R.H., and Verreth, J.A.J. (2006) Reducing water use for animal production through aquaculture. *International Journal of Water Resources Development* 22, 101–113.

Wasielesky, W., Atwood, H., Stokes, A., and Browdy, C. (2006) Effect of natural production in a zero exchange suspended microbial floc based super-intensive culture system for white shrimp *Litopenaeus vannamei. Aquaculture* 258, 396–403.

Yogev, U. and Gross, A. (2019). Reducing environmental impact of recirculating aquaculture systems by introducing a novel microaerophilic assimilation reactor: modeling and proof of concept. *Journal of Cleaner Production* 226, 1042–1050.

Chapter 11

IMPORTANT WATER CHEMISTRY VARIABLES FOR SHRIMP PRODUCTION

KATHARINE A. CLAYTON, ROBERT P. ELLIS, ROD W. WILSON AND ALBERT FERRER

Summary

Commercial shrimp production has experienced an exponential increase in growth globally over the past four decades. This expansion has led to the intensification and diversification of traditionally 'extensive' production techniques to now incorporate a range of production settings, levels of productivity and implementation of technological innovation. This development, which has enabled shrimp production to move indoors, has also facilitated ever-increasing production of shrimp in an expanding number of countries worldwide. Nonetheless, irrespective of the production setting or level of technological innovation used, shrimp aquaculture still faces a number of ubiquitous challenges. The drive for increasing production intensity and globalization means that the shrimp aquaculture sector must overcome the challenge of disease associated with suboptimal environmental conditions and subsequent adverse health outcomes, in order to ensure sustainable growth. In this chapter we outline the key environmental parameters that influence shrimp health, indicate the optimal operational ranges for each one and highlight the importance of using organism physiology to guide farm management to optimize productivity in this sector. Moreover, in highlighting the latest capability for pond-side measurement and manipulation of key environmental parameters, the chapter identifies key remaining technology gaps and delineates a pathway for future research, innovation and investment that will significantly enhance global expansion of shrimp production.

Introduction

The Pacific white shrimp, *Penaeus vannamei*, is the most widely cultured crustacean species worldwide. Historically shrimp farming has utilized 'extensive' production techniques where animals are reared in earthen ponds under traditional polyculture techniques, multiple species and multiple phyla often being farmed concurrently at low density. Such production focuses on trying to replicate natural ecosystem functioning, with little manipulation of the production environment and minimal inputs (e.g. feed). The vast majority of shrimp are still produced in such pond environments in Asia and Central America (the top six producing nations being Thailand, China, India, Ecuador, Vietnam and Indonesia), with an estimated area of 2.1 million hectares allocated to shrimp pond production globally in 2015 (Boyd and McNevin, 2018). Nonetheless, production has shifted from extensive to semi-intensive, intensive or even super-intensive production of monoculture crops (Boyd et al., 2018) driven by increasing demand and the globalization of the industry to utilize this unique economic opportunity. Global shrimp production reached almost 4 million tonnes in 2018, representing an increase of 3–5% compared to 2017 (FAO, 2019), whilst the sector is forecast to register a CAGR (compound annual growth rate) of 5.2% over the subsequent 5-year period (2019–2024).

Since the turn of the century global shrimp production has quadrupled, facilitated by local governments aiming to exploit this export opportunity. Contract farming initiatives, in which large exporting firms supply small-scale farmers with the seed, feed and the technology required for intensive production, have also enabled this transition to intensification. Based on conventional outdoor pond production, and thus limited biosecurity, the rapid growth of the sector has been coupled with the emergence of devastating diseases, such as taura syndrome virus (TSV), yellowhead syndrome virus (YSV), white spot syndrome virus (WSSV), necrotizing hepatopancreatitis (NHP) and acute hepatopancreatic necrosis disease (APHND). Imposing US$6 billion in losses annually (World Bank, 2014), disease has resulted in significant losses totalling up to 40% of annual marine and brackish water shrimp harvests (Israngkura and Sae-Hae, 2002). Technological solutions are thus urgently required to overcome this challenge and enable shrimp production to expand in order to match the forecasted increase in global demand.

Classification of shrimp production systems is widely based on productivity (extensive, semi-intensive, intensive or super-intensive systems), on design (outdoor or indoor systems) and on capacity for exchanging water (static, partial exchange or recirculating aquaculture systems – RAS). The concept of culture intensity includes not only the consideration of relative crop yield, but also the broader concept of resource input (Tidwell, 2012; Table 11.1 and Figure 11.1). However, despite key differences between production settings, one feature of commercial production common across all production systems is the

TABLE 11.1 Shrimp production system characteristics. Credit: A. Ferrer.

Parameter category	*Extensive*	*Semi-intensive*	*Intensive*	*Indoor intensive*	*Indoor super-intensive*
Productivity (kg/m^2)	0.1–0.2	0.2–0.5	0.5–3	0.5–3	>3
Stocking rate (ind./m^2)	1–3	5–25	25–150	25–150	>150
Stocking strategy	Direct stocking	2-phase[1]	2–3-phase[2]	2–3-phase	3-phase
Pond shape	Irregular	Irregular	Square/RW[3]	Square/RW	RW/circular
Aeration	No	Yes	Yes/intense[4]	Yes/intense[4]	Yes/intense[4]
Temperature control	No	No	No	Yes	Yes
Water exchange (%)	<5	5–20	20–40	20–40	>100
Rearing system	Natural productivity	Enhanced productivity	Mixotrophic	Nitrifying/RAS	Nitrifying/RAS

[1] 2-phase system (nursery and grow out)

[2] 2–3-phase system (nursery, pre-grow out and grow out)

[3] RW (raceway)

[4] >24 Horsepower/ha

FIGURE 11.1 Different shrimp production systems. Credit: R. Ellis and A. Ferrer.

challenge of adverse environmental conditions, such as water temperature (Huang et al., 2017; Madeira et al., 2015), pH (Han et al., 2018a), calcification-related variables (CO_2, alkalinity, Ca and Mg) (Cai et al., 2017; Lemonnier et al., 2004), dissolved oxygen (DO) (Han et al., 2018b) and salinity (Chen et al., 2016b; Li et al., 2008; McGraw and Scarpa, 2004; Wang et al., 2016), and the subsequent implications for shrimp biology, physiology and the incidence of adverse health outcomes.

This chapter aims to provide an overview of the impact of adverse environmental conditions on the physiology of *P. vannamei*, highlighting the key parameters of concern within a commercial production context. In doing so, the chapter aims to cover background information relating to shrimp physiology and stress indicators that will help the reader understand why shrimp need optimal conditions to maximize health and growth. It will also provide evidence for why monitoring of the aquaculture environment is vital for any production setting, how to interpret environmental data to make informed choices, and how this information can be used to improve or maintain the efficiency of a system.

This remainder of this chapter is split into five further sections. Three of these address the environmental parameters that have the greatest impact on shrimp physiology, and thus disease susceptibility, adverse health outcomes and productivity within the commercial setting, covering:

Section 2: Temperature, salinity and oxygen
Section 3: Calcification-related variables
Section 4: Nitrogenous excretory products (ammonia, nitrite and nitrate)

Sections 2 and 4 outline the effects of these parameters on the growth and productivity, disease and immune function, moulting, metabolism, respiration and reproduction, along with commentary on any other notable information. Section 3 covers calcification-related environmental variables and specifically describes how calcium, alkalinity, CO_2 and magnesium influence post-moult calcification, which ultimately impacts the health of the whole animal and productivity of a system. Subsequently, section 5 (Water quality management) addresses our understanding of these water chemistry parameters within a commercial context, outlining what is understood with respect to operational ranges, measurement capacity or capability, monitoring frequency and identification of key knowledge or technology gaps in relation to environmental monitoring. The key knowledge gaps identified throughout this chapter, in terms of scientific understanding and technological capability, are presented in section 5. It is therefore hoped this chapter will be of a broad interest to both academia and industry, spanning the connection between these two sectors and highlighting how the latest scientific understanding can result in key industry benefits, within a commercially relevant context.

Temperature, salinity and oxygen

BACKGROUND

One of the main reasons *P. vannamei* are produced extensively in aquaculture globally is due to the species' tolerance to a wide range of environmental conditions, in particular salinity, temperature and oxygen. This is an essential quality, particularly for outdoor shrimp farming, as the animals are exposed to large environmental fluctuations as a result of the production environment itself (e.g. water source, stocking density, algal community, farm management practices), as well as natural variability in these parameters that occurs in daily and seasonal cycles.

Temperature

Shrimp are poikilothermic animals, meaning their internal body temperature fluctuates with the external environment. *P. vannamei* has a high temperature coefficient for growth and so environmental temperature is a driving factor governing survival, health and growth in shrimp aquaculture. Whilst a large body of literature has highlighted the importance of understanding the effects of temperature on shrimp biology, key knowledge gaps remain. For example, temperature variation within a production setting has been demonstrated as key for immune function and health in other production species. When exposed to a disease challenge, a number of animal species will preferentially occupy specific temperatures zones, a phenomenon termed behavioural fever (Boltana et al., 2018). This natural response, however, requires a sufficient temperature range to be present in the environment. Understanding of this phenomenon could thus have important implications for the design and management of aquaculture facilities, being hugely beneficial for shrimp production.

Salinity

P. vannamei is described as a euryhaline species known to experience and tolerate a wide salinity range, from ion-poor freshwater to >40 psu (Menz and Blake, 1980). This has resulted in a wide range of environments being deemed suitable for pond culture of this species. However, survival at the extremes of this range varies with life stage due to changes in the organism's capacity to osmoregulate (Chong-Robles et al., 2014).

Oxygen

With respect to dissolved oxygen concentration (DO), *P. vannamei* is exposed to large natural fluctuations, with this variation becoming more extreme in water bodies where water movement, and thus mixing, is reduced. In such static environments, which include many aquaculture settings, DO is a function of the intensity of respiration, which continuously depletes oxygen (particularly when animal and microbial biomass are high), and reciprocally the intensity of photosynthesis during daylight hours when algae and macrophytes often produce more oxygen than respiration removes. Mass-specific oxygen consumption of *P. vannamei* (mg O_2/g/h) is influenced by temperature, salinity, size (ABW) and the interaction of these parameters. For this reason, commercial shrimp production settings often utilize powered aeration systems (e.g. surface paddle wheel aerators, submerged aeration hoses, circulators, diffusers or venturis). Indoor shrimp culture is typically higher intensity, with algae/macrophytes excluded, so it is even more essential to aerate (or even oxygenate using a source of compressed pure O_2) the water to maintain suitable dissolved oxygen levels.

GROWTH AND PRODUCTIVITY

Growth and productivity, as well as survival, feed conversation ratio and body mass gain, are key performance indicators of profitability across global aquaculture. These metrics are regularly monitored to ensure that the production environment is optimal. Here we describe how environmental temperature, salinity and oxygen impact the performance of shrimp, providing evidence for why these parameters should be constantly monitored for shrimp culture optimization.

Temperature

In terms of temperature, growth rate universally increases with rising temperature, at least until the animal reaches its upper thermal limit when growth starts to decline shortly before reaching a critical temperature at which mass mortality occurs (Hartnoll, 2001). *P. vannamei* growth rates are size specific, as for all animals, but it has been proposed that for juveniles (initial mass 3.9 g), sub-adults (initial mass 10.8 g) and adults (initial mass 16 g), reduced feeding and growth occurs at colder extremes, in ponds where the temperature falls below 23 °C. When grown at 23 °C, the specific growth rate (SGR) was reduced in all size categories/production stages when compared to conspecifics grown at 26.2 °C (SGR at the higher temperature was 67, 43 and 56% higher in juveniles, sub-adults and adults, respectively). Similarly, growth was also reduced at warmer extremes, in shrimp maintained >30 °C, but this effect was noted only in adult shrimp (SGR 87% of the control treatment, 26.2 °C) (Wyban et al., 1995). In post-larval shrimp (initial mass ~0.02 g), the SGR at 20 °C was 47% of the highest SGR at 25–35 °C treatment (Ponce-Palafox et al., 1997). Despite the wide thermal tolerance of this species, optimal growth rates have therefore been proposed to occur in a relatively narrow thermal window of between 28 and 30 °C (Box 11.1).

BOX 11.1. SGR

Throughout this chapter we have reported SGR measurements where possible. These have all to be calculated in the same way (see below). However, it is important to note that throughout the literature SGR has been calculated in different ways. Nevertheless, we believe that the various calculation methods produce results that are reliable regarding the percentage change in SGR relative to the control treatment within a given study. The latter tells us the magnitude of a growth rate difference and is straightforward to interpret, and hence we have included these percentages where possible too (see below for the calculations used). For these calculations the final and initial body masses, and duration of the growth period, are required. Finally, when considering growth rate in the literature, it is important to consider the initial body mass from each study, as SGR is faster in smaller animals and thus we have included these where possible to provide context, as otherwise direct comparisons between studies using different life stages can be misleading.

SGR:

$$SGR\ (\%/day) = 100 \times \frac{\ln\ (final\ mass) - \ln\ (initial\ mass)}{Duration\ of\ Growth\ Period\ (days)}$$

Percentage of the control:

$$Relative\ Treatment\ Effect\ on\ SGR\ (\%) = 100 \times \left(\frac{Treatment\ SGR}{Control\ SGR}\right)$$

Salinity

The *P. vannamei* has an isosmotic point of 25 psu (718 mOsm/kg), defined as where the internal (haemolymph) and external osmolality are the same. In theory this is the point at which the animal spends less energy on osmoregulation. However, counterintuitively, the optimum salinity for growth in shrimp is not necessarily the isosmotic point, with literature describing a wide range of salinity for optimum growth. For example, contrasting salinity ranges of 5–25 psu (Bray et al., 1994; Rosas et al., 2001) as well as 33–40 psu (Ponce-Palafox et al., 1997) have been suggested. Ponce-Palafox et al. (1997) concluded that "good growth can be obtained between 25 and 45 psu". The growth of post-larvae (~0.1 g) at very low salinities (2 and 3 psu) after a 35-day growth trial was only 57 and 79%, respectively, that of conspecifics maintained at 30 psu (Laramore et al., 2001). It is also important to note that growth differences occurred only after 18 d in the study of Laramore et al. (2001), with the authors also reporting no difference in growth and survival observed at 4 and 30 psu. Such a trend is further supported by two additional studies. Walker et al. (2009) showed that growth rate was significantly less at 2 psu than at 10 or 28 psu after 211 days (final weights were 20.2, 24.7 and 25.7 g for 2, 10 and 28 psu respectively), whilst Wang et al. (2004) demonstrated that at 1–16 psu the whiteleg shrimp's SGR, food consumption and food conversion efficiency all generally increased with rising salinity.

To take a different approach, in the wild, *P. vannamei* do not inhabit an environment with stable conditions. Experimentally induced salinity fluctuations of between 5 and 10 psu (occurring every 4 days), around a mean of 20 psu, have been shown to increase growth rates in *P. vannamei* compared to a constant salinity treatment of 20 psu. The greatest increase in growth rate was demonstrated with fluctuations of 10 psu every 4 days, where the SGR was 31% higher than under stable conditions (Su et al., 2010). This trend has been repeatedly demonstrated and is proposed to be as a result of greater feed intake, enhanced feed assimilation and reduced faecal energy loss under fluctuating conditions during the 48-day feeding trial (Feng et al., 2008; Su et al., 2010).

Oxygen

Dissolved oxygen has also been shown to significantly influence the growth of whiteleg shrimp. In one study, juvenile shrimp (initial mass 0.6 mg) had significantly decreased growth at the lowest DO treatment (1 mg/l) compared to the higher DO treatments of 2, 3 and 4 mg/l DO at an average temperature of 28 °C after 16 days. SGR of shrimp in the 1 mg/l DO treatment was just 68% of that of shrimp maintained under higher DO treatments (Seidman and Lawrence, 1985). However, these authors found no differences in moisture, lipid content or protein content in shrimp grown at different dissolved oxygen levels (Seidman and Lawrence, 1985).

DISEASE AND IMMUNOLOGY

Immune function and disease resistance of marine invertebrates are experimentally demonstrated to be influenced by temperature, salinity and oxygen (Ellis et al., 2011). Here we discuss how these environmental parameters can influence disease susceptibility in *P. vannamei* production.

Temperature

WSSV, TSV, infectious hypodermal and haematopoietic necrosis virus (IHHNV), AHPND and *Enterocytozoon hepatopenaei* (EHP) are among the most costly diseases in shrimp aquaculture (Walker and Mohan, 2009). These viral agents all have a temperature range that exceeds that of their host, and thus shrimp will always be susceptible to these infections (Millard et al., 2020). Certain temperature regimes, however, have been shown to impact shrimp survival. For example, survival was increased from zero (no survivors) at 25.8 °C to 80% at 32.3 °C in juvenile shrimp infected with WSSV, suggesting a beneficial effect of hyperthermia for this disease (Vidal et al., 2001). This is consistent with results from IHHNV- and TSV-challenged shrimp, where higher temperatures yielded the highest survival rate (Montgomery-Brock et al., 2002, 2007). The cause of this improved survival was suggested to be reduced replication ability of WSSV at elevated temperatures and/or an increase in shrimp antiviral response (Granja et al., 2006; Montgomery-Brock et al., 2007; Moser et al., 2012; Rahman et al., 2006; Vidal et al., 2001).

Elevated temperatures may be of benefit to prevent mortality from three of the main pathogenic viruses (WSSV, TSV and IHHNV), but can cause increased susceptibility to other diseases. For example, the pathogenic bacteria *Vibrio alginolyticus* caused a significant increase in mortality of whiteleg shrimp under high-temperature treatments (32–34 °C) compared to lower-temperature treatments (Cheng et al., 2005). It was also found in this study that whiteleg shrimp have a reduced immune capability when they have been transferred from 27 or 28 °C to 32–34 °C (Cheng et al., 2005).

Where a temperature gradient was used to determine the thermal preference of the whiteleg shrimp, it has been shown that the preferred temperature is 26.2 ± 0.4 °C (González et al., 2010). Although it was not significant, there is a slight trend that when the shrimp are acclimated to higher or lower temperature the preferred temperature is shifted in that direction also (González et al., 2010). Acclimating the shrimp can also increase the critical thermal maxima (temperature with which the shrimp are unable to perform righting behaviour). For example, thermal tolerance was increased by 5.7 °C in shrimp acclimated at 32 compared to those acclimated at 20 °C (González et al., 2010).

Associated with thermal preference is 'behavioural fever'. This is where the animal will raise or lower (usually raise) its body temperature to fight an infection. For homeotherms ('warm-blooded') animals such as mammals, this is achieved through changing their metabolic heat production. For poikilotherms however, this is not possible, but body temperature can sometimes be regulated through behaviour, with the animal seeking out warmer or colder environments purposefully (Evans et al., 2015). This phenomenon has been demonstrated to occur naturally in a wide range of animals including reptiles, fish and insects, and it is also observed in shrimp in ponds (A. Ferrer, personal observation). In demonstrating behavioural fever, the shrimp would have an ability to partially alleviate, or mitigate against, the effect of disease, reducing the requirement of human intervention in disease treatment, often achieved through the administration of expensive and polluting therapeutics (Thornber et al., 2019). This potentially positive benefit of enabling behavioural fever has already been shown in key production species of fish, for example Atlantic salmon (Boltana et al., 2018).

Salinity

Salinity can also influence immune responses to disease in shrimp. *P. vannamei* grown in low salinities (2.5 and 5 psu) demonstrated changes in immune activity, and when they were challenged with *V. alginolyticus* and WSSV, those reared at lower salinities (2.5 and 5 psu) had a significantly higher mortality rate (30%) when compared to those reared at 15–35 psu (76–86%) (Lin et al., 2012).

Other adverse effects on the host are also known to occur as a result of exposure to a combination of suboptimal environmental conditions and infection. The growth rate of *P. vannamei* was reduced under high-salinity (49 psu) treatments, but was slowed even further when infected with IHHNV (initial mean mass 1.6 and 2.2 g for non-infected and infected populations, respectively). Infected populations grown at 25 psu had an SGR that was 80.8% of the non-infected control, while infected populations grown at 49 psu had an SGR that was 74.5% of the non-infected control, meaning that elevated salinity had a significant effect of slowing growth in infected shrimp (Bray et al., 1994).

Oxygen

Oxygen also influences immune function. For example, all immune parameters tested – including total haemocyte count, phenoloxidase activity, phagocytic activity of haemocytes, and bacteriolytic and antibacterial activity of cell-free haemolymph – were significantly lower at both 3.0 and 1.5 mg/l DO compared to 5.5 mg/l DO in the control treatment, in *P. vannamei* (Fawen et al., 2009). When *P. vannamei* were injected with a sub-lethal dose of a pathogenic strain of the pathogenic bacterium *Vibrio campbellii*, shrimp exposed to hypoxia (PO_2 4 kPa compared to the normoxic control, 19–20.6 kPa) had a higher percentage of pathogenic bacteria that remained culturable within their haemolymph and tissues (Burgents et al., 2005). Shrimp cultured under this low-oxygen treatment also presented with a more widespread distribution of bacteria throughout their tissues, with a higher bacterial burden in organs such as the hepatopancreas and gills (Burgents et al., 2005). This suggests that hypoxia inhibited the shrimps' ability to fight a bacterial infection.

MOULTING AND CALCIFICATION

Moulting is strongly regulated by the endocrine system and is a hormonal phenomenon, with several aspects for which hormonal control is important, including determining when a moult occurs, controlling the onset of terminal anecdysis and determining the timing of puberty moult (Hartnoll, 2001). However, the demonstration of stress-induced disruption of moulting indicates the potential for the production environment to impact moulting in shrimp. Whilst it is highly likely that such stress impacts the endocrine control of this process, little is known about the physiological mechanisms by which different hormones in moulting are impacted by environmental variability. Thus, when considering impacts upon moulting, this section of the chapter focuses on the impact influences of temperature, salinity and DO perturbation on the timing of, and absolute growth achieved during, the moulting process rather than its hormonal control.

Temperature

Although limited data exist on the impact of temperature on the moult cycle in *P. vannamei*, there is evidence of an impact in other decapods, such as the Australian freshwater crayfish *Cherax destructor*. Verhoef et al. (1998+) found that as temperature increased from 22 to 25 or 28°C the intermoult period was shortened, decreasing by 25% at the highest temperature compared to the lowest. Whilst this is a single experimental example, it is consistent with the generally accepted hypothesis that intermoult period and moult increment (increase in size, or growth, per moulting event) is reduced with increasing temperature in crustaceans (Hartnoll, 2001). However, in decapods intermoult period is shown to be independent of moult increment, with both processes being regulated in a coordinated yet separate manner (Freeman, 1990).

Salinity

Following on from the positive effect of fluctuating salinity on shrimp growth rate, Feng et al. (2008) also demonstrated that salinity fluctuations can induce a change in the length of the intermoult period (and thus moulting frequency). When exposed to salinity fluctuations over a 48-day period, the time between moults (intermoult period) in shrimp became shorter (i.e. faster rate of moulting) compared to that of conspecifics maintained under static salinity conditions. As the frequency and amplitude of salinity fluctuations increased, this response became greater (Feng et al., 2008). Moreover, as shown above, this also resulted in significantly larger shrimp (shrimp exposed to salinity fluctuation of 10 psu every 4 days were 42% larger than control conspecifics maintained under static 20-psu conditions) (Feng et al., 2008).

Salinity obviously influences osmoregulation, and the moulting process clearly impacts upon osmoregulation too. Osmoregulatory capacity (OC) can be defined as the difference between the osmotic pressures of haemolymph and external medium at a given salinity. It has been used to identify stress in crustaceans and is also known to be reduced before and after ecdysis, becoming almost isosmotic at 35.5 psu during these stages, suggesting that shrimp lose the ability to maintain large osmotic gradients during these key points in their life cycle (Charmantier et al., 1994). Hyper-OC and hypo-OC, respectively, measured at low salinity and in seawater, were both depressed after 1-2 days exposure to low oxygen tension (PO2).

Calcium is a vital component of calcification of the new exoskeleton after moulting, and crustaceans derive this ion from the external water (see section 3). Ambient concentrations of dissolved calcium will naturally change largely in proportion with salinity, and so any effect of changing salinity on moulting may be in part driven by calcium availability. However, the calcium availability of fresh, brackish and seawater can also vary independently of salinity, especially in artificial settings such as aquaculture (see section 3). Other mitigation strategies may also affect this relationship – for example, where reductions in pH are buffered via the addition of alkaline substances such as calcium oxide the relationship between salinity and calcium will shift, thus potentially influencing calcification. As a result, caution should be taken when inferring the effect of salinity on moulting and an attempt to decouple the impact of calcium concentrations and salinity should be made, particularly where the ionic composition of production water is manipulated artificially, either intentionally or not. As this consideration is largely overlooked in the wider scientific literature, it is difficult to standardize across publications and separate the impact of these two parameters. For more information on the impact of the ionic composition of production water on shrimp biology see below.

Oxygen

When considering the impact of DO on moulting, it is important to note that oxygen consumption varies during the moult cycle in whiteleg shrimp, with higher O_2 consumption being noted immediately post moult at both 0.5 and 30 psu (Ding et al., 2014). These authors hypothesize that this increase in oxygen consumption post moult is due to the increased energy demand for calcification (Ding et al., 2014). This would suggest that the tolerance of shrimp to lower dissolved oxygen concentrations may vary depending on the moult stage of the animal.

RESPIRATORY PHYSIOLOGY AND METABOLISM

Temperature

A major part of homeostasis involves maintaining a stable internal pH, termed acid–base balance, as enzymatic reactions that occur within cells function optimally within a narrow pH window. This is due to the exquisite pH sensitivity of the tertiary structure and therefore functionality of almost all proteins. The internal pH of an organism is directly affected by certain external and internal factors, and crustaceans respond to these challenges by transporting acid–base-relevant molecules across cell membranes to maintain internal pH within, or as close to, the optimal functioning window (Whiteley et al., 2001). Salinity has a key influence on acid–base balance, and evidence suggests that acid–base regulation in decapod crustaceans as a result of salinity is linked to cell volume regulation (Whiteley et al., 2001). For example, low salinity causes metabolic alkalosis while higher salinity results in metabolic acidosis in haemolymph of euryhaline crabs. Also, when changes in environmental parameters force the animal to actively respond, energy is used to maintain homeostasis, which diverts energy away from other processes such as growth and immune function. This is just another example of why maintaining an optimal environment will improve the energetic efficiency and therefore growth of shrimp.

Moving on to monitoring of energy usage in the form of metabolism, oxygen consumption is widely used as a proxy for metabolic rate in a wide range of organisms, due to its relative ease of measurement and its key role in converting energy into a useful form within the cell during aerobic respiration. Subsequently, this section will now mainly discuss how temperature, salinity and dissolved oxygen influence oxygen consumption.

As temperature increases so does the oxygen consumption of *P. vannamei*. Temperature quotient (Q_{10}) values were reported to be <2 with an external temperature change from 20 to 30 °C, giving an average Q_{10} of 1.81, typical of many aquatic ectotherms (Clarke and Johnston, 1999). However, >30 °C the Q_{10} value decreased to 1.14, which is the typical trend as temperatures approach the critical threshold. As shrimp increase in size from 1 to 10 g, their mass-specific oxygen consumption (QO_2) halves, but no significant difference was detected between 10- and 20-g shrimp (Ponce-Palafox et al., 2013) (Box 11.2).

A slightly higher Q_{10} of 2.27 was reported for *P. vannamei* at 20–32 °C, and was deemed consistent with a thermal preference of 26 °C (Villarreal et al., 1994). This was supported by a subsequent study in which behavioural preference was also found to be 26 °C (Gonzalez et al., 2010).

Salinity

The effect of salinity on oxygen consumption is less clear, with some research indicating a marginal effect on oxygen consumption (Villarreal et al., 1994; Walker et al., 2009). Others demonstrate that oxygen consumption increased at salinities <5 psu compared to 30 psu (Ding et al., 2014; Rosas et al., 2001). There is also a large difference in oxygen consumption after feeding in shrimp acclimated to different

BOX 11.2. Q_{10}

For poikilotherms, Q_{10} describes the exponential relationship between metabolism and temperature, specifically the relative increase in metabolic rate for every 10 °C rise in temperature. The metabolic rate increases with increasing temperature, and so the Q_{10} value will be above 1 and typical Q_{10} values of ectothermic animals are around 2; this means that as temperature rises by 10 °C, the metabolic rate will double. To interpret Q_{10}, it is important to understand that the higher the value the steeper the metabolic rate curve in relation to increasing temperature, and vice versa. As temperature exceeds the animal's optimum, metabolism peaks then declines as it approaches its upper lethal limit, and is often used to represent the organism's sensitivity to temperature variation. The metabolic rate can be measured in a number of ways, but typically can be monitored as oxygen consumption because oxygen is essential for aerobic metabolic pathways. Q_{10} is usually calculated using the following equation:

$$Q_{10} = (rate2 \div rate1)^{10 \div (temp2 - temp1)}$$

rate1 = Oxygen consumption rate at temp1
rate2 = Oxygen consumption rate at temp2
temp1 = Lower temperature
temp2 = Higher temperature

salinities: at 5 psu, oxygen consumption increased by 134% (compared to the fasting control acclimated to the same salinity), whereas at 15 and 30 psu a significantly smaller increase, 89 and 51%, respectively, was observed (Rosas et al., 2001). The researchers propose that this pattern was caused by an increased energy requirement for maintaining homeostasis at 5 psu, as shown by an increased oxygen consumption compared to 15 and 30 psu.

Oxygen

Whiteleg shrimp can tolerate moderate hypoxia using physiological adaptations including anaerobic metabolism (for short periods at least) and increased ventilation and oxygen-carrying capacity (Racotta et al., 2002). After exposing shrimp to moderate hypoxia (2–2.6 mg/l DO) for 3 or 14 days, it was found that they mobilized copper from within the hepatopancreas for extra haemocyanin synthesis, which would increase oxygen-carrying capacity compared to shrimp kept under normoxia (mean of 6.1 mg/l DO). These authors also observed an increase in glucose and lactate levels in the haemolymph, which is indicative of anaerobic metabolism (Racotta et al., 2002).

REPRODUCTION

The importance of reproductive quality for aquaculture is ultimately dependent on the intended fate of the shrimp (broodstock animals versus human consumption). Thus, this section has greater relevance when considered in the context of broodstock animals rather than those produced for grow-out and the market for which energy allocated towards reproduction is diverted away from somatic growth and is hence counterproductive. Currently there is a dearth of literature concerning the reproductive effects of salinity and oxygen concentration on reproduction in *P. vannamei*, or indeed in other shrimp species, so we focus on the limited information such as that available in Harlıoğlu et al. (2018) concerning the effects of temperature on *P. vannamei* reproduction.

Temperature

In male *P. vannamei* (48 g) kept under 36 psu salinity, the mean sperm count per compound spermatophore was significantly reduced when kept for 42 days at constant temperatures of 29 °C (0.1 million cells) or 32 °C (no complete sperm cells, only fragments were found) compared to 26 °C (18.6 million cells) (Perez-Velazquez et al., 2001). There was also a lower percentage of sperm abnormalities in the 26 °C treatment (36.7%) compared to the 29 °C treatment (99.7%) (Perez-Velazquez et al., 2001).

CONCLUSION

To summarize, temperature, salinity and oxygen concentration each impose significant impacts on the biology of the whiteleg shrimp and should be carefully considered when farming this species. Regarding temperature, the most common optimum across the available literature appears to be 26 °C. Salinity is more flexible depending on other conditions, but the optimum is proposed to be 20–25 psu. Low dissolved oxygen clearly limits growth and other physiological functions below a critical minimum. From the limited research available, it is thought the optimal dissolved oxygen level is 4–5.5 mg/l or higher. To put this into perspective, the DO of 100% air-saturated water at 26 °C and 25 psu salinity would be 7.15 mg/l. While too much oxygen can become problematic for many aquatic organisms, the upper limit is yet to be determined for whiteleg shrimp in aquaculture.

It is important to note that the research that has been discussed above may not be representative of all life stages and strains of the whiteleg shrimp species. Thermal tolerance in shrimp can be plastic, meaning that optimal temperatures can change depending on their previous acclimation temperature.

Calcification-related variables (CO_2, pH, alkalinity, calcium, magnesium)

BACKGROUND

To allow growth (i.e. a physical increase in size and energy content), all crustaceans must moult. This means shedding their old exoskeleton and producing a new one, which then requires rapid hardening before they have full protection across their outer body surface. In addition, the mouthparts must harden before they become strong enough to resume feeding. While the new exoskeleton remains soft, shrimp (like all crustaceans) are most vulnerable to physical damage from cannibalism and predators, as well as to invasion by disease organisms and parasites, and potentially more susceptible to the uptake of chemical contaminants. In shrimp aquaculture, very high rates of moulting (every few days in the early stages) coupled with the high stocking densities (which maximize the chances of aggressive encounters with conspecifics) further increase the chances of cannibalism, which is a major cause of the losses experienced in shrimp aquaculture (Romano and Zeng, 2017). This section will focus on the water chemistry variables known to have a direct influence on post-moult calcification, and therefore should have potential for management to improve the health and growth of shrimp in aquaculture.

CALCIUM AND ALKALINITY

First let us describe the essentials of the chemical process of calcification. Immediately after moulting, shrimp require large amounts of calcium ions (Ca^{2+}) and carbonate ions (CO_3^{2-}) to rapidly calcify their exoskeleton, i.e. harden it by laying down calcium carbonate ($CaCO_3$) minerals within the cuticle.

$$Ca^{2+} + CO_3^{2-} \rightleftharpoons CaCO_3 \qquad (1)$$

Whilst there is only one form of calcium ion (Ca^{2+}), carbonate ions (CO_3^{2-}) can also be supplied by deprotonation of bicarbonate ions (HCO_3^-), which simultaneously liberates hydrogen ions (H^+):

$$HCO_3^- \rightleftharpoons CO_3^{2-} + H^+ \qquad (2)$$

In marine crustaceans, the majority of the calcium and carbonate ions used for calcification are rapidly taken up via the gills from the external seawater immediately after a moult. In theory, the shrimp can gain carbonate ions for calcification by taking up either bicarbonate ions (HCO_3^-) or carbonate ions (CO_3^{2-}) from the external water, but bicarbonate is usually considered to be the molecule transported across these cell membranes. However, using bicarbonate ions as the main seawater source for calcification therefore requires that protons (generated from the reaction in Eq. 2) must be rapidly excreted by the gills too. Thus, calcification after moulting is associated with a high rate of acid excretion (which has the same effect as base uptake in terms of acid–base chemistry). Within the normal range for seawater pH (e.g. 7.9–8.3) the vast majority of inorganic carbon is in the form of bicarbonate ions, hence gill transport for the purpose of exoskeleton calcification is often measured as 'bicarbonate uptake' or flux. Collectively, bicarbonate and carbonate ion concentrations are known as carbonate alkalinity.

The intensity of moulting in high-density aquaculture systems is so high that it can limit post-moult calcification by removing both calcium and carbonate/bicarbonate ions over relatively short time frames. A further quantitatively important influence on water alkalinity (and therefore shrimp calcification) is from the nitrifying microbes within the biofiltration units of RAS. These microbes form a critical component of the 'life-support systems' that maintain a healthy water quality, but they further deplete seawater alkalinity in the process of converting toxic ammonia excreted by the animals to the much safer nitrate (Eshchar, 2006).

Several studies on crustacean species other than shrimp have demonstrated that the uptake of bicarbonate can be substantially impaired under conditions of low seawater alkalinity (e.g. in European lobster (Middlemiss et al., 2016) and blue crab (Cameron and Wood, 1985)). We have recently confirmed that seawater alkalinity also dramatically influences calcification in whiteleg shrimp, based on a pilot study at the University of Exeter, in a manner that follows Michaelis–Menten kinetics, i.e. bicarbonate

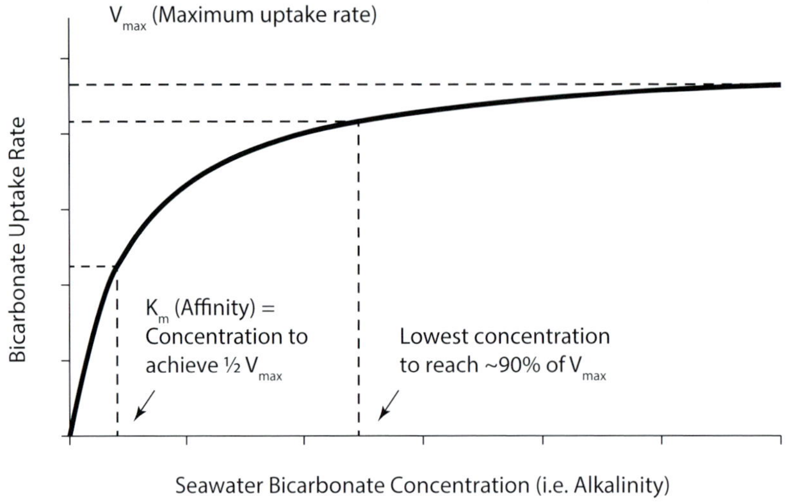

FIGURE 11.2 The relationship between uptake of bicarbonate from, and the concentration of bicarbonate available in, external seawater. This relationship follows the shape of a classic Michaelis–Menten kinetic curve established for ion uptake in aquatic animals: that is, a very steep rise in uptake as concentrations increase from zero, but uptake rapidly approaches a plateau of maximum uptake, with very little further gain as environmental concentrations increase further. The ion transport system is quantitatively defined by two parameters: V_{max} (the maximum uptake rate where the curve plateaus) and K_m (the concentration that achieves 50% V_{max}), which is also the inverse of affinity for the ion (i.e. the lower the K_m the higher the affinity). Credit: R. W. Wilson.

uptake rate (and therefore calcification) increases rapidly as water concentration increases from zero, but then begins to plateau (Figure 11.2). As seawater alkalinity increases further, very little increase in uptake rate and calcification is achieved (Wright, Ellis and Wilson, 2019, unpublished).

This suggests there will be an optimum range for alkalinity (e.g. around 90% of V_{max}) to ensure rapid uptake of bicarbonate from the water, and hence exoskeleton calcification after moulting. The same will also apply to calcium ions (e.g. blue crab; Neufeld and Cameron, 1994). Importantly for the management of shrimp production, at higher seawater concentrations for either ion there will be no advantage to calcification. Indeed, there will likely be problems created as high alkalinity is associated with high pH, which increases the likelihood of ammonia toxicity. Also, high concentrations of bicarbonate and calcium can result in precipitation of unwanted carbonate minerals within the aquaculture system, clogging pipes and requiring increased maintenance of submerged pumps, etc. In addition, increasing alkalinity and calcium by dosing of alkali and calcium salts above the optimum range will have a financial cost with no productivity gains. Thus understanding the precise nature of this curve (particularly the affinity [K_m] and the lowest concentration needed to achieve uptake rates close to the maximum [V_{max}]) for both calcium and alkalinity, for different stages of the shrimp life cycle would provide valuable information to support the evidence-based management of water quality to promote rapid post-moult calcification and shrimp health and productivity.

CARBON DIOXIDE

High densities of animals used in aquaculture inevitably result in a rapid depletion of dissolved oxygen (O_2) and a reciprocal rise in dissolved carbon dioxide (CO_2) as a result of aerobic metabolism (which consumes and generates similar molar amounts of these gases, respectively). This relationship with animal density means that the effect on these dissolved gases is amplified in intensive compared to extensive shrimp culture. The issue is exaggerated even further in recirculating aquaculture systems (RAS) because, in addition to the shrimp themselves, the biofilter microbes carry out considerable aerobic metabolism, and the nitrification process itself consumes O_2 and produces CO_2 in converting toxic ammonia to nitrate (Eshchar et al., 2006). Thus, active gas exchange processes to restore suitable levels of dissolved O_2 and CO_2 are an increasingly important process the more intensive the shrimp culture system.

Oxygen is relatively easy to replace by aeration, and many shrimp culture ponds use powered aeration mechanisms (e.g. surface paddle wheels, submerged aeration hoses, circulators, diffusers or venturis). However, removing CO_2 is much more difficult than replacing O_2, simply due to physical chemistry. Firstly, CO_2 is 30 times more soluble in water than O_2 (Weiss, 1974), so it prefers to stay dissolved rather than enter the atmosphere. Secondly, most of the CO_2 produced by shrimp and biofilter microbes is initially hydrated, producing bicarbonate and hydrogen ions which therefore acidify the water:

$$CO_2 + H_2O \rightleftharpoons HCO_3^- + H^+ \quad (3)$$

However, when trying to remove CO_2, the reverse reaction is very slow relative to the contact time between air and water when using active gas-exchange processes. For example, in water containing low O_2/high CO_2, restoring dissolved oxygen requires only minutes of air equilibration, whereas completely restoring dissolved CO_2 would require several hours (Hoppe et al., 2012). This is true even within the most advanced CO_2 stripping units used in some modern RAS. A further complicating factor is that some intensive RAS oxygenate water using pure oxygen (100% O_2). Because this contains five times as much oxygen per litre of gas than air (21% O_2), it enhances the inward diffusion gradient and therefore the rate at which dissolved O_2 can be restored. However, this also means that fivefold less flow of gas is required to achieve the target DO level. Unfortunately, this also means that potentially fivefold less CO_2 will be removed when using pure O_2 compared to air. The net effect of all these factors is that CO_2 inevitably builds up (especially in RAS using pure O_2), until the gradient between water and air is high enough to achieve a new steady state, which is always at a higher CO_2 level in water than would be expected for equilibrium with atmospheric air.

In 2019, atmospheric CO_2 levels averaged 411 µatm (https://www.co2.earth/), a figure that has been rising exponentially since the Industrial Revolution. This rise is causing concern for aquatic life, with the overall effect on seawater chemistry referred to as 'ocean acidification'. However, all aquatic environments are following the same trend of rising CO_2, including freshwater, so it is perhaps better to refer to 'aquatic acidification' or even 'aquatic carbonation', as many of the biological effects are driven by CO_2 itself rather than the pH of the aquatic environment. Elevated dissolved CO_2 has long been known to directly affect acid–base and ion regulation, respiratory function and aerobic performance in aquatic animals (Brauner et al., 2019; Ellis et al., 2017). The atmospheric CO_2 levels predicted for the end of the century are around 1,000 µatm (Pörtner et al., 2014). However, the CO_2 levels in aquaculture are already well in excess of this, ranging from 2,000 to 45,000 µatm in a wide range of finfish and shellfish systems. The level of CO_2 in shrimp culture systems is rarely reported, but the average level in Bangladesh shrimp ponds was around 17,000 μatm (Sahu et al., 2013; Saksena et al., 2006).

How elevated aquatic CO_2 affects calcification has been studied in many invertebrates. The fundamental principle is that elevated CO_2 simultaneously reduces both pH and carbonate ion concentrations in seawater. Any calcified materials directly exposed to these seawater conditions are therefore at risk for two reasons. Firstly, an increased rate of 'corrosion', i.e. dissolution of calcium carbonate mineral due to the reversal of Eq. 1. Secondly, calcification (the forward reaction in Eq. 1) will be slowed down due to the lack of availability of carbonate ions. However, the reality is somewhat more complex for most calcifying organisms, because they can exert some control over calcification in their exoskeletons and the site of calcification is often internalized, i.e. not in direct contact with the external seawater. Hence, it is the chemistry of their internal body fluids, and in particular those fluids adjacent to the exoskeleton calcification sites, that exert the most influence over post-moult calcification. In addition, the exoskeleton is covered by an organic component that potentially offers some protection against the direct influence of the seawater chemistry. However, we currently know little about any protection these organic layers can provide, or even whether enhanced dissolution in a high-CO_2 environment primarily occurs in the external- or internal-facing parts of the calcified exoskeleton (Melzner et al., 2019). Indeed, whether the organic coatings themselves are affected by elevated CO_2 in crustaceans is also unknown and worthy of study (Melzner et al., 2019).

One important aspect of exposure to high CO_2 that is well understood is the effect on the acid–base balance of aquatic animals. The gills necessarily have high permeability to gases to allow both uptake of O_2 and excretion of CO_2. This means that when environmental CO_2 rises, the CO_2 in the haemolymph rises in parallel until a new steady state is reached with a new gradient that is sufficient to drive the outward diffusion of CO_2 that matches the animal's endogenous production. However, this elevation in haemolymph CO_2 results in an internal acidification (called acidosis) as a result of the reaction in Eq. 3. This happens within minutes of a rise in environmental CO_2. However, given enough time most aquatic animals can regulate haemolymph pH back to normal, by activating processes at the gills that excrete excess H^+ ions and accumulate additional bicarbonate in the extracellular fluid (Brauner et al., 2019; Wilson, 2020). This haemolymph pH compensation is vital for normal functioning and health but takes hours to days, depending on how easily the ion exchange systems in the gill ionocytes can transport these acid–base-relevant ions (Brauner et al., 2019).

Under constant high-CO_2 conditions, crustaceans will therefore have re-established normal haemolymph pH (important for maintaining optimum conditions for all proteins to function), but with a new equilibrium whereby both CO_2 and HCO_3^- are elevated in haemolymph. Thus, in species that are good acid–base regulators, the concentration of bicarbonate available to the exoskeleton calcification sites should actually be greater than during low-environmental CO_2 conditions. This may be part of the explanation for occasional reports of increased calcification in marine invertebrates exposed to elevated CO_2 (Peck et al., 2018; Wood et al., 2008). Indeed, increased calcification in elevated CO_2 conditions (up to 2,900 μatm) has been reported for some crustaceans (e.g. eastern king prawn and blue crabs; Ries et al., 2009), but whether this trend continues at the much higher levels known to occur in intensive shrimp culture (e.g. 17,000 μatm) is still unknown. It is worth noting that European lobster suffered reduced carapace mass and mineral content (calcium and magnesium) when larval stages were cultured in seawater with CO_2 experimentally elevated to only 1,200 μatm (Arnold et al., 2009). This could be linked to an inability to accumulate bicarbonate and regulate haemolymph pH in lobster at temperatures near the upper end of their range (Small et al., 2020). However, how well whiteleg shrimp can regulate their internal acid–base balance in response to high CO_2 conditions, or how temperature influences this ability, are not yet known.

MAGNESIUM

Magnesium is 5.2 times more abundant than calcium in seawater (52 versus 10 mM), and is often incorporated into calcium carbonate minerals, i.e. magnesium ions (Mg^{2+}) simply substitute for the occasional calcium ion (Ca^{2+}) within the crystal lattice (Morse, 2007). Furthermore, high levels of dissolved magnesium are considered to inhibit calcification (Morse, 2007), and the Mg content of the crustacean exoskeleton is thought to play an important role in resistance to bacterial infection (Kunkel et al., 2013). For this reason, monitoring the magnesium content of seawater in shrimp aquaculture may provide a further layer of understanding to shrimp moulting success, health and productivity. Perhaps more specifically, the Mg:Ca ratio in seawater should be related to how much magnesium is incorporated into the exoskeleton. However, the mechanisms underlying the inclusion of magnesium in the exoskeleton, and its role in the biology and health of the shrimp, are currently unrefined.

Ammonia, nitrites, nitrates and phosphates

BACKGROUND

The main inorganic nitrogen compounds found in aquaculture are ammonia/ammonium (NH_3/NH_4^+), nitrites (NO_2^-) and nitrates (NO_3^-). There are two main sources of these nitrogenous compounds: excretory products from the shrimp themselves (primarily ammonia) and the microbial degradation of both uneaten food and the dissolved ammonia excreted by the shrimp (to nitrite and then nitrate). In addition, phosphorus-containing compounds are excreted via faeces by shrimp and released from uneaten food/faeces by microbes. The whiteleg shrimp is ammonotelic, excreting its nitrogenous waste via the gills primarily as soluble ammonia, comprising 62–84% of the total nitrogen excretion of the animal (Jiang et al., 2000). The toxicity of these nitrogen compounds decreases from ammonia to nitrite to nitrate. There is limited literature on the impacts of phosphates on shrimp, and so this will not be covered in this chapter; however, without further data it cannot yet be discarded as a potentially important limiting factor within shrimp aquaculture.

This conversion process from ammonia/ammonium to nitrite, then to nitrate, is known as nitrification. Nitrification is a key component in the nitrogen cycle and is conducted by microbes in a two-step process usually described by the formulae stated below. Each step is conducted by different types of bacteria and archaea, but they usually occur in the same environment which is why nitrate is considered the common end product.

For ammonium:

The oxidation of ammonium to nitrite:

$$2NH_4^+ + 3O_2 \rightarrow 2NO_2^- + 4H^+ + 2H_2O \qquad (4)$$

The oxidation of nitrite to nitrate

$$2NO_2^- + O_2 \rightarrow 2NO_3^- \quad (5)$$

For un-ionized ammonia:

The oxidation of un-ionized ammonia to nitrite

$$NH_3 + O_2 \rightarrow NO_2^- + 3H^+ + 2e^- \quad (6)$$

The oxidation of nitrite to nitrate

$$NO_2^- + H_2O \rightarrow NO_3^- + 2H^+ + 2e^- \quad (7)$$

Total ammonia nitrogen (TAN) is the sum of both ammonium (NH_4^+) and un-ionized ammonia (NH_3) forms. The relative proportion of these two forms within TAN depends on the temperature, salinity and pH of the water. Un-ionized ammonia is the most toxic form because cell membranes have high permeability to this gaseous form, so it enters the body of aquatic animals rapidly when the external concentration is high (Ip and Chew, 2010; Wright and Wood, 2009). However, at high ambient concentrations ammonium ions may disrupt sodium transport mechanisms (Armstrong et al., 1978). The proportion of the NH_3 and NH_4^+ forms changes most dramatically in relation to pH. As pH rises so does the proportion of un-ionized ammonia, with that of ammonium declining. To this end, where sufficient environmental data is presented in the original study, the NH_3-N molarity has been calculated for the examples provided in this section, so all values are directly comparable regardless of environmental conditions. The units of salinity that are reported in this chapter have been standardized to psu also.

Nitrogenous excretion is a dynamic process. Excretion rate (and hence concentrations achieved in the external water), as well as toxicity to shrimp, can be influenced by many factors. Feeding rate has the biggest influence, as ultimately it is the dietary protein (and specifically the amino acids that are absorbed in excess of that required for protein synthesis) that is the source of most excreted ammonia. Other influential factors include environmental conditions, developmental stage and moult cycle stage (Diaz et al., 2001; Jiang et al., 2000; Lin and Chen, 2001; Wajsbrot et al., 1990; Wickins, 1985). Subsequently, the composition and concentration of nitrogenous products is shown to vary within intensive shrimp production water across a range of temporal scales, from minutes to days (Jackson et al., 2003).

This section will discuss the impacts and adverse health outcomes on shrimp of elevated concentrations of nutrient excretory products such as ammonia, nitrite and nitrate. Where literature concerning the effects of nitrogenous compounds is lacking for the whiteleg shrimp, we have attempted to use examples from the wider literature on closely related species, where deemed relevant. It is also important to note that the literature is dominated by studies that use juvenile shrimp, which may not be representative of the impacts across the entire life cycle.

GROWTH AND PRODUCTIVITY

Growth is a key indicator of health and productivity within any aquaculture system, and a build-up of these nutrient excretory products in the environment has been repeatedly shown to limit growth. Here we discuss a number of examples of growth impairment that have been demonstrated in shrimp in response to elevated ammonia, nitrite and nitrate.

When aquaculture tanks populated with whiteleg shrimp were left to accumulate ammonia and nitrite without water replacement or filtration, the ammonia and nitrite concentration both reached ~1.2 mg/l (equivalent to ~0.05 mg/l of NH_3-N at pH 7.5–8.2) after 33 days. The survival rate, gain in mass and length and SGR were all significantly lower than the control (in which water replacement and filtration maintained a consistently low ambient level of TAN <0.038 mg/l and nitrite <0.032 mg/l, respectively; Han et al., 2017). In terms of magnitude, the SGR was significantly reduced in juveniles (20% lower) in tanks without filtration (Han et al., 2017). A similar result was shown by Rostami et al. (2019), albeit at much higher concentrations of ammonia, where in shrimp (mean initial mass 0.3 g) exposed to 13 or 19 mg/l of TAN (equivalent to 0.49 or 0.72 mg/l of NH_3-N at pH 7.8) over 45 days, SGR was reduced to 85 and 79% of the control (0 mg/l TAN).

In zero water exchange systems, higher nitrite levels have been shown to reduce growth in juvenile whiteleg shrimp (mean initial mass 8 g), at 24 psu salinity, whereby mass gained after 30 days declined almost linearly within increasing ambient nitrite. Overall SGR from the 10 mg/l treatment in this study was 65% of the control (NO_2^- = 0.0 mg/l) (Furtado et al., 2016a). Nitrate levels (220 mg/l nitrate-N) at a salinity of 11 psu have been shown to reduce survival, growth and biomass in the whiteleg shrimp (Kuhn et al., 2010). The growth rate of juvenile whiteleg shrimp (initial mass ~1.3 g) exposed to nitrate concentrations of 300 mg/l was significantly lower than the control (75 mg/l nitrate) conditions, whereby the SGR of the treatment was 85% of the control (Furtado et al., 2016a). Apparent feed conversion doubled also, from 1.50 to 3.00 in the control and 300 mg/l treatment, respectively (Furtado et al., 2016a).

In another penaeid, *Penaeus monodon*, which also shows reduced growth when exposed to elevated TAN, re-establishment of control conditions was shown to result in the recovery of growth rate comparable to the control after a period of 30 days (Chen and Lin, 1992). This suggests that, providing high ammonia concentrations are discovered and mitigated against quickly enough, long-term effects can be minimized.

To conclude, according to the current literature safe levels in terms of mitigating loss of productivity from growth rates would be:

- ammonia: <1.2 mg/l TAN (equivalent to 0.05 mg/l of NH_3-N at pH 7.5–8.2)
- nitrite: <10 mg/l N-NO_2
- nitrate: <220 mg/l nitrate-N.

DISEASE AND IMMUNOLOGY

Shrimp susceptibility to diseases can be increased by elevation of nitrogenous products through induction of cell apoptosis and other harmful responses that damage tissues and organs, thus disrupting normal function. Ammonia and nitrite stress – proposed to occur at a level of 20 mg/l TAN (equating to 0.75 mg/l NH_3-N at pH 8.3) and 20 mg/l nitrite-N – have been shown to cause damage to the intestines of whiteleg shrimp, reducing the bacterial diversity of their intestines and disrupting intestine mucosal function and mucus immune components, as a result impairing intestine barrier function (Duan et al., 2018).

There is evidence also that elevated ammonia concentrations can induce gill abnormalities that can result in increased susceptibility of shrimp to pathogens such as WSSV. The gills are integral to ammonia excretion (among other functions) and are a main target organ of WSSV (Escobedo-Bonilla et al., 2007). To support this, whiteleg shrimp exposed to WSSV and TAN concentrations of 10 mg/l (equivalent to 0.26 mg/l NH_3-N at pH 8.3) show that ammonia toxicity inhibits the immune function, which led to an increase in infection severity (Lu et al., 2019). *P. monodon* also demonstrated increased susceptibility to WSSV when exposed to TAN levels as low as 1.1 mg/l (equivalent to 0.12 mg/l NH_3-N at pH 8.35) (Fouzi et al., 2010). Elevated nitrate-N exposure (42 days of exposure to 300 mg/l) has also been shown to cause histopathological damage in the gills of the whiteleg shrimp (Furtado et al., 2014).

Similar to growth performance, healing and recovery from gill damage caused as a result of exposure to sublethal ammonia-N concentrations have been demonstrated in other decapod crustaceans, such as the blue swimmer crab, when the animals were placed in clean water following exposure (Romano and Zeng, 2010). It has also been documented that elevated nitrate concentrations (>220 mg/l) caused lesions in the hepatopancreas, and occasional occurrences of gill fouling were also found in the whiteleg shrimp (Kuhn et al., 2010).

Suppressing immune function is another way in which nitrogenous excretory products can increase susceptibility of shrimp to disease. Haemocytes are integral components of an invertebrate's immune response (Ellis et al., 2011). In *Litopenaeus schmiti*, TAN concentrations of 5 ± 1 mg/l (equivalent to 0.27 mg/l NH_3-N at pH 8.0) caused a decrease in haemocyte count of 66% (Rodríguez-Ramos et al., 2008). In whiteleg shrimp exposed to TAN levels of 20 mg/l (equivalent to 0.28 mg/l NH_3-N at pH 7.8), haemocyte apoptosis was increased, which would probably reduce shrimp immunocompetence, increasing susceptibility to pathogenic infection (Liu et al., 2020).

Another key immune parameter that is regularly used as an indicator of stress in ammonia-challenged whiteleg shrimp is haemolymph coagulation. Haemolymph coagulation both reduces loss of further haemolymph and prevents entry of bacteria or other microorganisms, via open wounds (Söderhäll and Cerenius, 1992). It has been shown that TAN concentrations of 5 and 10 mg/l (equivalent to 0.27 and 0.43 mg/l NH_3-N at pH 7.9) are enough to delay coagulation time by 166 and 218%, respectively, compared to shrimp from control conditions (0 mg/l) (Chang et al., 2015).

Whiteleg shrimp challenged with *Vibrio alginolyticus* (a common pathogenic bacterium) were more susceptible to disease, experiencing higher mortality and suppression of immune function when they were placed in water containing concentrations of TAN >5.24 mg/l (equivalent to 0.25 mg/l NH_3-N at pH 8.0) (Liu and Chen, 2004). Following 168 hours of exposure to significantly elevated TAN concentrations (21.6 mg/l, equivalent to 1.04 mg/l NH_3-N at pH 8.0) alone, no shrimp mortality was experienced while cumulative mortality in shrimp exposed to bacteria in isolation (1×10^6 cfu) under control TAN conditions was 47%. Conversely, shrimp exposed to the combined bacterial and TAN (5.24 mg/l, equivalent to 0.25 mg/l NH_3-N at pH 8.0) treatment experienced a cumulative mortality of 67% after the same 168-hour period (Liu and Chen, 2004). Similarly, after 168 hours when challenged with the same bacteria and subsequently exposed to nitrite-N levels of 5.15 mg/l, cumulative whiteleg shrimp mortality was significantly higher (70%) compared to the bacteria-only control treatment (43%) (Tseng and Chen, 2004). There was also a universal depression noted in immune function observed in shrimp exposed to nitrite-N treatments >4.94 mg/l after 96 hours of exposure (Tseng and Chen, 2004).

Safe levels, in terms of mitigating loss of productivity as a result of disease, are difficult to determine because most of the evidence comes from experiments undertaken at higher concentrations than would typically be experienced within intense-production settings or predicted from other species. From the literature currently available, proposed safe levels in terms of immune function and disease resistance would appear to be:

- ammonia: <1.1 mg/l TAN (equivalent to 0.12 mg/l NH_3-N at pH 8.35)
- nitrite: <5.15 mg/l nitrate
- nitrate: <177 mg/l nitrate-N
- TAN: <1.1 mg/l.

MOULTING AND CALCIFICATION

In order to grow, all crustaceans must moult and, once they have shed their old, hard exoskeleton, a soft shell remains which needs to harden to avoid predation and infection. In the whiteleg shrimp and other penaeids, increasing ammonia concentration has been shown to directly increase moulting frequency by reducing the period between the first and second intermoult, by up to 50% (Chen and Kou, 1992; Chen and Lin, 1992; Rostami et al., 2019). The average intermoult periods of post-larval whiteleg shrimp, in 35 psu salinity, exposed to 0, 6, 13 and 19 mg/l TAN (equivalent to 0, 0.23, 0.49 and 0.72 mg/l NH_3-N at pH 7.8) concentrations were reduced from 11.5 to 10.8, 9.4 and 8.7 days, respectively (Rostami et al., 2019). However, even though the moulting frequency was increased by higher TAN, the body mass gained over the experimental period was significantly higher in the control treatment (4.0 g) compared to the 19 mg/l (equivalent to 0.72 mg/l NH_3-N at pH 7.8) treatment (2.2 g) (Rostami et al., 2019). This must mean that the growth per moult cycle is less under high ammonia conditions, which cannot be of benefit to the animals.

Moult death syndrome (MDS) occurs in decapod crustaceans, including the whiteleg shrimp, and is described as a mortality resulting from an incomplete moult (e.g. Gong et al., 2004), whereby the animal dies before it can shed the old exoskeleton. Nitrite stress (3 mg/l $NaNO_2$-N) has been shown to cause mortality by MDS in the blue swimmer crab (Romano and Zeng, 2009). Although evidence is currently lacking for whiteleg shrimp, it is feasible that a similar outcome may occur in this species due to them both being decapod crustaceans, a possibility that requires further investigation.

ACID–BASE PHYSIOLOGY, METABOLISM AND RESPIRATION

How an organism's physiology responds to a change in the environment can influence its energy budget for other activities such as growth and reproduction. Predicting the effects of ammonia, nitrite and nitrate on homeostatic process such as acid–base physiology, metabolism and respiration can help to mitigate against lowered production efficiency, and so we present a selection of studies that illustrate the homeostatic responses found in shrimp.

As mentioned earlier, un-ionized ammonia diffuses rapidly through cell membranes, including in the gills, and thus when concentrations increase in the production water this results in a comparable increase in tissue concentration within the shrimp. Shrimp typically respond to the build-up of ammonia in their tissues and haemolymph by reducing their feeding rate, which is thought to help by reducing tissue production and therefore internal accumulation of metabolic ammonia (Barbieri, 2010; Colt and Armstrong, 1981; Frias-Espericueta et al., 2000).

Oxygen consumption was reported to be significantly higher in juvenile whiteleg shrimp exposed to TAN concentrations of 2.14 mmol/l (equivalent to 0.07 mmol/l NH_3-N at pH 7.8; ~7 µmol O_2/g/h), compared to 1.07 mmol/l (equivalent to 0.03 mmol/l NH_3-N at pH 7.8; ~5.5 µmol $O_2/g/h^-$), after 24 hours (Racotta and Hernández-Herrera, 2000).

Ammonia can also induce oxidative stress. Exposure to 20 mg/l of TAN (equivalent to 0.49 mg/l NH_3-N at pH 7.9) in the whiteleg shrimp led to apoptosis in the hepatopancreas, in addition to a reduction in superoxide dismutase (SOD) activity, a key antioxidant enzyme (Liang et al., 2016). Other studies have concluded that ammonia stress (20 mg/l, equivalent to 0.28 mg/l NH_3-N at pH 7.5) can disrupt amino acid and nucleotide metabolism and lipid metabolites in the haemocytes, which could decrease haemocyte functionality or haemocyte count by inducing apoptosis in the whiteleg shrimp (Liu et al., 2020).

REPRODUCTION

The literature is scarce on the impacts of nitrogenous compounds on the reproductive success of the whiteleg shrimp. However, research on another penaeid, *Penaeus paulensis*, suggests that combined exposure to 2.62 mg/l TAN (equivalent to 0.07 mg/l NH_3-N at pH 7.77) and 1.50 mg/l NO_2^- had no measurable impact on adult reproductive performance after a 46-day exposure (Cavalli et al., 1998). In whiteleg shrimp it is reported that optimal reproductive potential occurs around 12 months for both females and males, but as this study spanned only 46 days, long-term experiments over the first year of the animal's life could be important for understanding reproductive effects (Ceballos-Vázquez et al., 2003, 2010).

INTERACTIVE EFFECTS OF SALINITY AND NITROGENOUS EXCRETORY PRODUCTS

Salinity is a key factor determining the toxicity of nitrogenous excretory products to whiteleg shrimp. Here we have reviewed the scientific literature to demonstrate the importance of maintaining salinity within an optimal range with regard to ammonia/nitrite/nitrate toxicity. The iso-osmotic point of the whiteleg shrimp (i.e. when osmolality is the same in the internal body fluids and the external environment) has been reported to be 25 psu – this is therefore the salinity that is widely believed to be optimal for osmoregulation in this species (Chong-Robles et al., 2014).

Higher salinities (>25 psu) increased the LC_{50} values of ammonia-N in both juvenile and post-larval whiteleg shrimp, meaning that ammonia tolerance is improved at higher salinities (Lin and Chen, 2001; Rostami et al., 2019). The proposed safe levels are estimated to be 2.44, 3.55 and 3.95 mg/l for TAN and 0.12, 0.16 and 0.16 mg/l for NH_3-N (at pH 8.05) in 15, 25 and 35 psu, respectively (Lin and Chen, 2001). In support of these findings, a separate study showed that low salinity had a negative impact on TAN tolerance, with an increase in mortality at 3 psu compared to 17 and 32 psu when juvenile whiteleg shrimp were exposed to TAN concentrations of 12 mg/l (equivalent to 1.53, 1.40 and 1.27 mg/l NH_3-N at pH 8.30) and above (Li et al., 2007). The impact of lower salinity on TAN tolerance in whiteleg shrimp supports this trend too, where the 48-h LC_{50} of TAN was higher for shrimp maintained at 18 psu (42.6 mg/l, equivalent to 1.58 mg/l NH_3-N at pH 7.8) compared to 10 psu (39.7 mg/l, equivalent to 1.56 mg/l NH_3-N at pH 7.80) (Schuler et al., 2010).

Nitrite toxicity in relation to ambient salinity has also been investigated using the same experimental method as Lin and Chen (2001), but using nitrite rather than ammonia. Results suggest that higher salinity (25 psu and above) increased the LC_{50} values of nitrite-N in juvenile whiteleg shrimp, meaning that nitrite tolerance is improved under higher salinity similar to the effect of salinity on ammonia toxicity (Lin and Chen, 2003). The proposed safe levels of nitrite are estimated at 6.1, 15.2 and 25.7 mg/l of nitrite-N in 15, 25 and 35 psu, respectively (Lin and Chen, 2003). A later study was more conservative in its conclusion, recommending that at salinities of 8 and 24 psu, nitrite (NO_2^--N) concentrations remain <2.5 and 10 mg/l, respectively, to avoid growth impairment and mortality (Furtado et al., 2016a).

The combined toxic effects of ammonia and nitrite should also be considered. Confusingly, both synergistic and antagonistic effects of nitrite (NO_2^--N) and TAN have been demonstrated, suggesting a lack of a consistent impact (Schuler et al., 2010). In one example, when nitrite levels were adjusted to the lowest observed effect concentration (LOEC) (133 mg/l NO_2^--N), the LC_{50} for TAN was also reduced (39.7 mg/l, equivalent to 1.56 mg/l NH_3-N at pH 7.8) without LOEC nitrite, and 28.2 mg/l (equivalent to 1.1 mg/l NH_3-N at pH 7.8) with LOEC nitrite). This means that nitrite at LOEC reduced TAN tolerance in whiteleg shrimp, with this combination having a synergistic impact on survival because elevated nitrite may weaken the shrimp. Conversely, when TAN was reduced to LOEC (30 mg/l), nitrite LC_{50} was increased (154 mg/l without LOEC TAN and 187 mg/l with LOEC TAN) and hence has an antagonistic impact. The reasoning behind these findings was not explored within the study, and thus caution should be exercised in their interpretation or application of findings.

IMPACTS OF LIFE STAGE ON AMMONIA TOXICITY

Shrimp morphology, physiology and ability to tolerate stress all change throughout the life cycle. Most of the available scientific research, and thus the research presented in this chapter, uses juvenile shrimp; however, here we present the few examples from earlier life stages. For example, at the zoea (2–5 days post hatching), mysis (6–9.5 days post hatching) and post-larval 1 (11–12 days post hatching) stages of development, the safe recommended un-ionized ammonia (NH_3-N) concentrations are 0.15, 0.26 and 0.19 mg/l (at pH 8.5), respectively (De Lourdes Cobo et al., 2012). For post-larvae 12 (12 days post-larvae), it is recommended that ammonia-N concentration should not exceed 1.22 mg/l (equivalent to 0.048 mg/l NH_3-N at pH 7.92) (Frias-Espericueta et al., 2000).

CONCLUSION

To optimize production efficiency, it is important to minimize the influence of nitrogenous excretory products on all aspects of shrimp physiology. To this end, the conservative safe levels of ammonia, nitrites and nitrates for rearing whiteleg shrimp are estimated to be 1.2 (equivalent to 0.048 mg/l NH_3-N at pH

7.92), 2.5 and 177 mg/l (Frias-Espericueta et al., 2000; Furtado et al., 2014, 2016a; Han et al., 2017). As discussed, the toxicity of these nitrogenous compounds varies depending on a multitude of other factors that must be considered when optimizing shrimp production through altered pond management.

Water quality management

INTRODUCTION

Effective monitoring and control of temperature, salinity, dissolved oxygen, pH, calcification-related variables and nitrogen excretory products is one of the key challenges to ensure productivity and sustained growth when *P. vannamei* is cultured in different production systems. Therefore, in this section we review how various productions systems have evolved and the technological advances that have recently emerged, to provide a comprehensive understanding on how to monitor, control and manage water quality across those systems.

IMPORTANCE OF WATER QUALITY IN SHRIMP FARMING

Maintaining optimal water quality parameters during shrimp farming is crucial to avoidance of stress during the culture period. High levels of stress in shrimp affect feeding behaviour and growth, reduce immunocompetence and facilitate the spread of disease. Therefore, understanding the ecological and environmental conditions within shrimp ponds/tanks, and how these parameters deviate from optimal conditions, will help farmers to anticipate potential disease outbreaks and mitigate losses (Ferreira et al., 2011).

For commercial production, maintaining cultures under optimal conditions is required to maximize physiological performance, survival and productivity of shrimp, which, as discussed in previous sections, is specific to different life stages. Below we summarize this information, enabling the reader to identify maximum and minimum thresholds required for maintaining optimal productivity and to identify when corrective actions may be required.

Temperature

As discussed in detail in section 2, temperature is a critical variable for *P. vannamei* production (see Table 11.2), with several studies outlining direct impacts of temperature on the growth, physiological performance and survival of the animals (e.g. He et al., 2018; Huang et al., 2017; Zhang et al., 2019).

P. vannamei is cultured mostly in open ponds in tropical and semi-tropical climates where temperature changes are usually minimal (2–3 °C). Temperature in these environments is mainly controlled by ambient air temperatures and solar radiation. While *P. vannamei* tolerates, and performs well, in temperature ranges of 26–30 °C, small temporal variations in temperature caused by day–night cycles, seasonal temperature fluctuations or sudden environmental changes such as storms can affect performance greatly. Therefore pond location, design and management strategy are critical considerations in mitigating against the adverse impacts of temperature fluctuation on shrimp productivity, as far as possible. For example, in

TABLE 11.2 Recommended optimal levels of temperature, salinity and dissolved oxygen for the rearing of *P. vannamei* at different life stages. Credit: A. Ferrer.

Variable	*Life stage*	*End point*	*Unit*	*Reference(s)*
Temperature	Post-larval, 0.02 g	SGR affected	<26 ºC >31 ºC	Ponce-Palafox et al., 1997
Temperature	*P. vannamei*	Thermal preference	26 º C	González et al., 2010
Temperature	Post-larval, PL16 Juvenile, 10.9 g	Lethal LC_{50}	<12.9 ºC <11.7 ºC	Pontinha et al., 2017
Temperature	Juvenile, 3.9 g Sub-adult, 10.8 g Adult, 6 g	SGR affected	<28 ºC >30 ºC	Wyban et al., 1995
Temperature	*P. vannamei*	Susceptibility to WSSV	<32.3ºC	Vidal et al., 2001
Temperature	*P. vannamei*	Susceptibility to *V. algynolicticus*	>32 ºC	Cheng et al., 2005
Temperature	Male adult, 48 g	Reduced sperm count and abnormalities	>26 ºC, 36 psu	Perez-Velazquez et al., 2001
Salinity	*P. vannamei*	Isosmotic point	24.7 psu	Castille and Lawrence, 1981
Salinity	Post-larval, 0.1 g	SGR affected	<3 psu	Laramore et al., 2001
Salinity	*P. vannamei*	Performance and survival affected	<5 psu	Diaz et al., 2001; Li et al., 2007
Salinity	Post-larval, PL16 Sub-adult, 10.9 g	Lethal LC_{50}	<1.80 psu <2.40 psu	Pontinha et al., 2017
Salinity	*P. vannamei*	Susceptibility to WSSV and *V. alginolyticus*	<15 psu	Lin et al., 2012
Salinity	Juvenile, 1.6 and 2.2 g	SGR affected	>49 psu	Bray et al., 1994
Dissolved oxygen	Juvenile, 6 and 8 g	SGR affected	<4 mg/l	Nonwachai et al., 2011
Dissolved oxygen	Juvenile, 6 and 8 g	Survival affected	<2 mg/l	Nonwachai et al., 2011
Dissolved oxygen	*P. vannamei*	Immune function affected	<3 mg/l	Fawen et al., 2009
Dissolved oxygen	Larval, PL6 Juvenile, 5 cm Adult, 11.5 cm Broodstock, 13.5 cm	Lethal LC_{50}	1.23 mg/l 0.57 mg/l 0.59 mg/l 0.65 mg/l	Zhou et al., 2014

sub-tropical climates, production cycles are seasonally coordinated to avoid stocking post-larvae during the coldest months of the year (October–February in the Northern Hemisphere, April–August in the Southern Hemisphere). Instead, farmers use this fallow period to undertake key management activities for sanitation (e.g. pond dry-out) or preparation activities for the next grow-out season. The use of enclosed nurseries to produce juveniles in temperature-controlled settings allows farmers to stock larger shrimp in open ponds, when environment temperatures are optimal, considerably increasing yield while simultaneously reducing production risk.

FIGURE 11.3 Enclosed nursery system in Nicaragua. Credit: A. Ferrer.

In enclosed systems (indoor) the temperature is predominantly determined by facility design and, specifically, system components selected to regulate the temperature (shading, insulation, heating and cooling) (Figure 11.3). The advantage of indoor settings is that temperatures can be maintained, and thus production optimized, year round. Moreover, enclosed systems enable production at latitudes typically deemed unsuitable for shrimp culture (outside tropical or sub-tropical climates), and where outdoor cultures would not be feasible. The main consideration here is the cost of energy and its impact on production costs. Specifically, energy cost is determined by the power consumption of temperature control infrastructure, and the differential between the temperature of source water and optimal culturing temperature for shrimp.

Dissolved oxygen

Dissolved oxygen concentration is one of the most important environmental stress factors in shrimp culture. Its concentration in the pond/tank water is not only affected by *P. vannamei* respiration itself, but also by many environmental factors such as algal biomass and productivity (e.g. blooms or die-off), zooplankton productivity and the microbial decomposition of organic matter, which includes biological waste (faeces and carcasses) and non-consumed feed. Any deviation in dissolved oxygen concentration severely affects survival and growth performance (Allan and Maguire 1991; Seidman and Lawrence, 1985), respiration (Patterson and Thorne, 1995; Wu et al., 2002) and haemolymph osmotic pressure (Charmantier et al., 1994; Cheng et al., 2003), as well as contributing to enhanced sensitivity to pathogens (Chang et al., 2002; Le Moullac et al., 1998) and elevated mortality in shrimp, as discussed in section 2.

In outdoor ponds, hypoxia (low oxygen) can have severe consequences for aquatic organisms, even leading to death (Vaquer-Sunyer and Duarte, 2008). This is especially critical in ponds that have no additional means of aeration, and thus where shrimp can be exposed to hypoxia with little likelihood of remediation (Chantal et al., 2008). The bottom layer of shrimp ponds, where shrimp spend most of their time, may even become anoxic (zero oxygen) at night, due to respiration of all organisms (shrimp, zooplankton and algae) and microbial decomposition of accumulated organic matter (Cheng et al., 2003).

The biological carrying capacity of an un-aerated pond is proposed to be approximately 0.15 kg/m^2 or 1,500 kg/ha. In aerated ponds, where oxygen production does not rely solely on primary productivity (restricted to daylight hours) or passive addition of oxygen via diffusion from the atmosphere, carrying capacity can be significantly higher (Tidwell, 2012). Generally, it is assumed that 1 horsepower of mechanical aeration can contribute to an additional 300–500 kg/ha of biomass under optimal temperature and salinity conditions.

When dissolved oxygen approaches suboptimal levels (see Table 11.2), widely proposed to be <3.0 mg/l, mitigation action is critical. Strategies adopted by farmers to recover optimal DO levels in open ponds are focused on reducing the organic load dissolved in production water and/or accumulated on the bottom of the pond (microalgae, bacteria, detritus, uneaten feeds and wastes), mainly achieved by exchanging a portion of water. Alternatively, farmers can install, or increase the use of, equipment for mechanical aeration to help circulate and aerate the water column, reducing stratification or even manipulating the pond depth to accentuate the influence of wind on the water's surface (however, this strategy could also affect pond temperature).

In enclosed systems (indoors) where biomass can reach >10 kg/m^2 (equivalent to >100,000 kg/ha), maintenance of dissolved oxygen concentrations above the proposed optima for this species, and often above 5 mg/l, is regularly employed. Such levels are maintained through the enhanced use of aeration devices (i.e. regenerative blowers) where the air is distributed by PVC pipes to ponds/tanks and enters the water as small bubbles via microporous hoses or air diffusers. Additional strategies to cope with periods of low dissolved oxygen must account for feeding rate where automatic feeders are used, and include removal of leftover feed and settled solids through siphoning, use of clarifiers, protein skimmers or cyclone filters for suspended solids, partial harvests to reduce shrimp biomass and, lastly, water exchange.

The amount of air or pure oxygen required to maintain normoxia will depend on the production type and carrying capacity of the production system. As a rule of thumb, Tidwell (2012) proposes the below equation for systems relying on TAN fixation by nitrifying and heterotrophic bacteria:

FIGURE 11.4 Emergency mechanical aeration system (Naqua) in a semi-intensive pond, Saudi Arabia. Credit: A. Ferrer.

Production term for oxygen (P_{oxygen}):

P_{oxygen} = 0.25 kg O_2 per kg feed consumed by shrimp
0.12 kg O_2 per kg feed consumed by nitrifiers
0.13 kg O_2 per kg feed consumed by heterotrophs (up to 0.5)
Total P_{oxygen} = 0.5 kg O_2 per kg of feed for system

Oxygen production terms are negative, due to each process removing oxygen from the system as opposed to adding it to the water. If management of solids is not appropriate, a safe value of P_{oxygen} would be close to 1 kg of O_2 per 1 kg of feed added.

Salinity

Changes in environmental salinity affect the growth performance and physiology of *P. vannamei* (Li et al., 2007, 2008), as discussed in detail in section 2. *P. vannamei* is euryhaline, meaning that it has the ability to osmoregulate in a broad range of salinities (1–50 psu; Pante, 1990). However, at low salinity (<5 psu) its performance and survival are greatly inhibited (Diaz et al., 2001; Li et al., 2007), with low stress tolerance (Li et al., 2007, 2008; Lin and Chen, 2001, 2003), muscle cramp, lethargy and disoriented spinning demonstrated in this species (Davis et al., 2002).

In open ponds (outdoors), salinity is mainly controlled by the natural processes of evaporation and rainfall. Generally, in production ponds evaporation reduces the water level by 2–10 mm per day (i.e. removal of 20–100 m^3/ha/day), depending on seasonal temperature and humidity (Boyd and Gross, 2000; Yoo and Boyd, 1994). Strategies used to maintain salinity within optimal range (see Table 11.2) are mainly focused on designing the pond to allow rapid water exchange whenever this is required. These include surface water exchange after heavy rains, mechanical pumping to regulate salinity during high/low tides or even maintaining a high operative water level to avoid a drastic reduction in salinity after heavy rains.

In closed (indoor) systems, evaporation is the primary challenge. Evaporation in these systems depends on water temperature, water vapour pressure (humidity) and the intensity of aeration. Generally, in systems with very low water exchange (<1%) the salinity tends to increase rapidly to levels exceeding optimal values (>30 psu). Subsequently, optimal values can be re-established through the addition of fresh water. Samocha et al. (2017) proposed the approach below to restore salinity to within the desired values:

$$V2 = \left(\frac{C1 * V1}{C2}\right) - \mathrm{V1} \tag{8}$$

C1 = current salinity (psu)
V1 = water volume (m^3)
C2 = target salinity (psu)
V2 = volume of fresh water to add (m^3)

Where salinity falls below optimal or target levels, the addition of commercial salts is widely employed in commercial shrimp farming. Each commercial sea salt brand results in water of slightly different ionic composition, nutrient level, pH and alkalinity. Subsequently, the chemical and nutrient profile of each brand must be accounted for both in terms of target values for production water and in the shrimp feed, to maintain the required ionic balance for optimal performance and survival.

MOULTING AND CALCIFICATION-RELATED VARIABLES

Environmental pH is an important factor affecting crustaceans (Pan et al., 2007). In aquaculture ponds, pH values can fluctuate from 6.6 to 10.2 with this variation in pH predominantly mediated by respiration of the shrimp, diverse biogeochemical processes including photosynthesis/respiration of phytoplankton and zooplankton, and microbial decomposition of organic residues, which may give rise to stress in *P. vannamei* (Boyd, 1990; Chen et al., 2018; Datta, 2012). After sunset, DO concentrations decline as photosynthesis stops and all plants and animals in the pond continue to consume oxygen (respiration). In heavily stocked ponds, CO_2 concentrations can become high as a result of respiration. The free CO_2 released during respiration reacts with water, producing carbonic acid (H_2CO_3) which rapidly dissociates to form hydrogen ions (H^+, which causes the reduction in pH) and bicarbonate ions (HCO_3^-).
When environmental pH falls outside the optimal range (see Table 11.3) shrimp may suffer physiological perturbation, induced through disruption of their internal ionic balance, which can subsequently result in slow growth, abnormal behaviour and increased mortality (Han et al., 2018), suppressed immune activity (Pan and Jiang, 2002) or oxidative damage (Wang et al., 2012).

Total alkalinity is the acid-buffering capacity of the water, being the ability to maintain a stable pH for any given addition of acid – for example, most commonly caused by elevation in CO_2. In most natural waters (fresh and saline) the vast majority of total alkalinity is derived from bicarbonate and carbonate ions, collectively referred to as carbonate alkalinity. Quantitatively, this is calculated as:

$$\text{Carbonate alkalinity} = [HCO_3^-] + 2[CO_3^{2-}] \quad \text{– Eq. 9}$$

This is because each carbonate ion (CO_3^{2-}) can buffer the addition of two H^+ ions before being converted to CO_2, whereas each bicarbonate ion (HCO^-) reacts with only one H^+ ion during conversion to CO_2. In seawater, total alkalinity also includes smaller quantities of other acid-buffering components such as borate, hydroxide, phosphates, silicates and minor organic acids and bases exuded from phytoplankton, the last of these contributing up to 10% of total alkalinity (Ferrer, A., unpublished results). In aquaculture settings, alkalinity is commonly reported in mg/l as calcium carbonate (mg/l $CaCO_3$, or mg $CaCO_3$/l), which is based on the assumption that total alkalinity is derived from carbonate alkalinity alone and that all of this is derived from the theoretical dissolution of solid pure calcium carbonate. According to Van Wyk and Scarpa (1999), alkalinity >100 mg $CaCO_3$/l is ideal for *P. vannamei* development, with higher values reducing the magnitude of daily pH fluctuations. Timmons and Ebeling (2013) demonstrated enhanced performance in *P. vannamei* under $CaCO_3$ levels ranging from 140 to 180 mg/l, which may relate to improved availability of carbonate for rapid post-moult calcification (see 'Calcium and alkalinity' above).

In shrimp ponds pH is stabilized by addition of an alkali. This is done using various carbonate, bicarbonate, oxide or hydroxide compounds to raise the pH of the production water, which also results in an elevation of buffer capacity by raising carbonate alkalinity. Some of the different alkali components used raise carbonate alkalinity directly, e.g. sodium bicarbonate ($NaHCO_3$), potassium bicarbonate ($KHCO_3$), sodium carbonate (Na_2CO_3), potassium carbonate (K_2CO_3) and calcium carbonate ($CaCO_3$). However,

TABLE 11.3 Recommended optimal levels of pH, alkalinity and CO_2 for the rearing of *P. vannamei*..

Variable	*Life stage*	*End point*	*Unit*	*Reference*
CO_2	Juvenile (1.76g)	96-h LC_{50}	<5.9 mg/l	Furtado et al., 2017
CO_2	Sub-adult (12.1g)	Oxygen consumption	<5 mg/l	Furtado et al., 2016a,b
CO_2	Sub-adult (6 g)	Weight gain Productivity Biomass Feed conversion rate (FCR)	<5.38 mg/l	Furtado et al., 2011
pH	Juvenile (1.2g)	Survival Growth (weight and length)	>pH 6.7 <pH 9.7	Han et al., 2018
pH	Juvenile (3.2g)	Digestive morphology Immune function Antioxidant defence	>pH 6.7 <pH 9.7	Duan et al., 2019
pH	Sub-adult (7–8g)	Reactive oxygen species (ROS) Catalase expression Malondialdehyde (MDA) concentration	>pH 6.7 <pH 9.3	Wang et al., 2012
pH	Post-larval (PL10)	Acute (96-h) LC_{50}	>pH 4.04 <pH 9.58	Furtado et al., 2015a
pH	Sub-adult (10.2 g)	Susceptibility to *V. alginolyticus* Immune response	>pH 6.5 <pH 10.1	Li and Chen, 2008
pH	Sub-adult (6 g)	Weight gain Productivity Biomass FCR	>pH 7.5	Furtado et al., 2011
Alkalinity	Juvenile (0.2g)	Nitrification process (in biofloc) Growth	≥150 mg $CaCO_3$/l	Furtado et al., 2015b
Alkalinity	Sub-adult (6 g)	Weight gain Productivity Biomass FCR	>100 mg $CaCO_3$/l	Furtado et al., 2011

other alkalis that are used raise pH directly but only secondarily increase carbonate alkalinity as a result of the change in pH converting existing dissolved CO_2 to HCO_3^- and CO_3^{2-} ions (e.g. sodium hydroxide (NaOH), calcium hydroxide ($Ca(OH)_2$), calcium oxide (CaO) and magnesium oxide (MgO)). The alkalis used have different properties in terms of solubility, strength of reaction (effective pH change) and safety in handling, which must be considered for use within intensive aquaculture settings (Wasielesky et al., 2015). Moreover, the availability and cost of different alkali compounds varies regionally, and thus utility of various forms for application in specific settings must be factored in a context-specific manner when determining best practice for pH and alkalinity control at the farm level.

Carbon dioxide is a critically important, but largely overlooked, water quality-limiting parameter, which is largely due to the technical challenge of measuring it in solution and thus the lack of robust, precise, reliable and affordable equipment to measure this parameter in situ in a commercial setting. Furthermore, this challenge is compounded by the fact that when CO_2 dissolves in water it reacts to first form carbonic acid, before dissociating to form H^+ and bicarbonate ions. This reaction also makes alleviating the challenge of elevated CO_2 far harder than replenishing O_2 (see 'Carbon dioxide' above and Ellis et al., 2017), despite the directly proportional relationship between O_2 removal and CO_2 production by respiration.

In 'Calcification related variables', the key issue is explained of why the physico-chemical properties of the two respiratory gases means that removal of CO_2 is so much slower than replacement of oxygen. This includes the reason why the use of pure O_2 (common practice in RAS to reduce the risk of deoxygenation) exacerbates the issue of CO_2 accumulating in shrimp culture. This misunderstood and poorly appreciated parameter can have dramatic and often misdiagnosed impacts within an intensive setting. Moreover, as biomass increases, the challenge of elevated CO_2 increases proportionally. In extensive or semi-intensive systems, the level of aeration is often enough to remove sufficient CO_2 from the water and control this parameter to levels close to atmospheric equilibrium. However, in super-intensive systems where shrimp biomass can exceed 15 kg/m^2 and where removal of nitrogenous waste utilizes heterotrophic bacteria and nitrifiers for ammonia and nitrite oxidation (AOB and NOB, respectively), additional degassing units are required for its control.

NITROGENOUS EXCRETORY PRODUCTS (AMMONIA, NITRITE AND NITRATE)

Ammonia, nitrites and, of lesser importance, nitrates are nitrogen reactive species that are considered the major environmental pollutants in the aquatic system and are harmful to aquatic organisms if not maintained within optimal ranges (see Table 11.4). Ammonia is the main end product of protein catabolism in the nitrogen cycle while nitrite is an intermediate product in the nitrification process of conversion of ammonia to nitrate. Accumulation of both ammonia and nitrite, and to a lesser extent nitrate, leads to a deterioration in water quality of shrimp ponds/tanks, as discussed in 'Calcification related variables', and is the main cause of growth reduction, immune suppression and pathogen susceptibility in shrimp.

Therefore, ammonia concentration (TAN) must ideally be maintained near to optimal values (<1.1 mg/l, equivalent to ~0.12 mg/l NH_3-N at pH 7.5–8.2). In outdoor ponds (semi-intensive or intensive) this value can vary due to the feeding intensity of the shrimp, but normally it is controlled by photoautotrophs. In indoor systems the ammonia is controlled mainly by heterotrophs and nitrifiers, depending on the rearing characteristics of the system.

There are several methodologies available to control ammonia and consequently to control the build-up of nitrites and nitrates, as these processes are coupled and mediated by ammonia-oxidizing bacteria (AOB) and nitrite-oxidizing bacteria (NOB). Strategies used to manage these nitrogenous products are mainly focused on feed management. Feeds typically contain large quantity of protein, with specific feed composition and protein content dependent on the production stage of shrimp (i.e. juveniles require feeds with a protein content near to 50% while sub-adults require feeds with a protein content of 35%). When ammonia exceeds optimal values, and there are no means for managing it, water exchange is one of the most common and rapid procedures used by farmers, achieved by flushing a percentage of water (30–50%) from the pond, topping up with fresh water and reducing the ammonia concentration by dilution.

Nitrite is oxidized and converted to nitrate by nitrification, a process that is microbially mediated. The rate of this reaction is influenced by changes in temperature, salinity and pH. However, in open shrimp ponds the microbial community present in the environment (sediment and substrates) readily,

TABLE 11.4 Recommended optimal levels of nitrogenous products for the rearing of *P. vannamei* at different life stages and salinities. Credit: K. Clayton.

Nitrogenous product	*Life stage*	*Salinity (psu)*	*End point*	*Optimal TAN (mg/l)*	*Optimal NH_3-N (mg/l) (pH in parentheses)*	*Reference*
TAN	Juvenile, 0.99 and 3.8 g	34	Acute toxicity	6.52 and 7.09	0.29 and 0.13 (8.08)	Frias-Espericueta et al., 1999
TAN	Post-larval, 9.25 ± 4 mg	34	Acute toxicity	1.22	0.048 (7.92)	Frias-Espericueta et al., 2000
Ammonia + nitrite	Juvenile, 0.90 ± 0.02 g	30–31	Survival, weight and length gain, SGR	<1.2	<0.05 (median 7.85)	Han et al., 2017
TAN	Adult, 20 ± 0.5 g	20	Susceptibility to pathogenic bacteria	<5.24	<0.25 (8)	Liu and Chen, 2004
TAN or NH_3-N	Adult, 20 ± 0.5 g	20	Haemocyte coagulation	<5	<0.357 (8)	Chang et al., 2015
TAN	Juvenile, 10.29 ± 1.1 g	39	Metabolic demand	2.14	0.02 mmol/l (7.8)	Racotta and Hernández-Herrera, 2000
TAN	Post-larval	35	Moulting frequency	<6	<0.23 (7.8)	Rostami et al., 2019
TAN	Zoea1	34	Acute toxicity	0.42	0.05 (8.5)	De Lourdes Cobo et al., 2014
Nitrite (NO_2-N)	Juvenile, 8.0 ± 0.50 g	8	Growth and survival	<2.5		Furtado et al., 2016,a,b
Nitrite (NO_2-N)	Juvenile, 8.0 ± 0.50 g	24	Growth and survival	<10		Furtado et al., 2016,a,b
Nitrite-N	Juvenile, 8.4–14.7 g	35	Susceptibility to pathogens and mortality	<5.15		Tseng and Chen, 2004
Nitrite-N	Juvenile, 3.96 ± 0.42 g	15	Acute toxicity	6.1		Lin and Chen, 2003
Nitrite-N	Juvenile, 3.96 ± 0.42 g	25	Acute toxicity	15.2		Lin and Chen, 2003
Nitrite-N	Juvenile, 3.96 ± 0.42 g	35	Acute toxicity	25.7		Lin and Chen, 2003
Nitrate-N	Juveniles, 1.30 ± 0.31 g	23	Growth and survival	177		Furtado et al., 2016,a,b
Nitrate-N	Juvenile, <2.5 g	11	Survival, growth, antenna length and pathology	220		Kuhn et al., 2010

and rapidly, converts nitrite to nitrate, and thus nitrite is often not monitored closely in these settings. In indoor systems conversely, closer monitoring of nitrite is often required, especially where the microbial community is established, and then maintained artificially. In such systems it is recommended that nitrite levels do not exceed 5.15 mg/l, as this indicates insufficient microbial oxidation of nitrite to nitrate and possibly perturbation of the microbial community. As soon as this parameter approaches, and exceeds, optimal levels the measurement frequency for nitrite should be increased, with corrective actions applied urgently, typically via partial water exchange of sufficient volume to deal with any nitrite peaks via dilution.

Nitrate levels must be maintained <220 mg/l (Kuhn et al., 2010). However, the optimal range is proposed as anything <120 mg/l of NO_3-N, especially for *P. vannamei* broodstock indoor cultures. Prolonged exposure to high nitrate values could be associated with an increase in 'loose shell' prevalence in broodstock tanks exposed at nitrate levels >200 mg/l during long periods (A. Ferrer, personal observation). Loose shell syndrome affects the feed conversion efficiency of shrimp, leading to poor meat quality, and affected ponds/tanks can suffer poor survival and chronic mortalities (Raja et al., 2015).

This trend supports the findings of Kuhn et al. (2010), where shrimp exposed to high concentrations of nitrate over a long period of time exhibited shorter antenna length, gill abnormalities and lesions in the hepatopancreas. The hepatopancreas is an organ in shrimp that produces digestive enzymes and is responsible for promoting the normal absorption of digested food. In the presence of elevated nitrate, hepatopancreatic cells were dilated and devoid of epithelial membranes, possibly as a consequence of inadequate nutrition or poor metabolization of feed (Kuhn et al., 2010).

PARAMETER MEASUREMENT AND ASSOCIATED TECHNOLOGY

Water quality parameters are commonly measured manually with the help of various methodologies and instruments in order to determine their status. Procedures for measuring and controlling these parameters, and their methodologies, are widely discussed by several authors (Boyd and Tucker, 1992; Chien, 1992; Feliu et al., 2009), with many measurements practised daily in shrimp farming.

Daily parameter measurements apply to in situ conditions themselves, often reflecting the conditions in a specific locality within a pond/tank and offering a single measurement that reflects only a picture in time without considering other processes that may affect the spatio-temporal distribution of the parameters (Hargreaves and Tucker, 2012). Ponds/tanks are very dynamic environments where water quality changes often occur very quickly and unpredictably. Therefore, careful consideration must be given to the time and position at which measurements are taken in order to provide a representative understanding of pond conditions.

Therefore, the use of robust, reliable methods and well-calibrated measurement equipment is of high importance. Well-planned monitoring programs, ideally performed consistently in the same location in the pond/tank and at the same time(s) each day, allow the detection of temporal trends and help farm managers to anticipate, prevent and correct any adverse water quality deviations to restore optimal conditions in the ponds/tanks as quickly as possible.

The frequency with which environmental parameters are monitored depends primarily on the parameter itself (determined by the technical practicality, time and cost of measurement; see Table 11.5) as well as the production system in which the parameter is being measured, its relevance within the production environment and the life stage of the animals stocked. In extensive and semi-intensive

TABLE 11.5 Recommended monitoring frequency of parameters based on production system characteristics. Credit line: A. Ferrer.

Parameter	*Extensive*	*Semi-intensive*	*Intensive*	*Indoor intensive*	*Indoor super-intensive*
Physical					
Temperature	High	High	High	Very high	Very high
Dissolved oxygen	High	High	High	Very high	Very high
Salinity	Low	Low	Medium	Medium	Medium
Calcification-related					
Carbon dioxide	Not relevant	Not relevant	High	Very high	Very high
pH	Low	High	High	High	Very high
Alkalinity	Low	Low	Medium	Medium	Medium
Calcium	Low	Low	Medium	Medium	Medium
Magnesium	Low	Low	Medium	Medium	Medium
Excretory nutrient products					
Ammonia	Low	Low	Medium	Medium	High
Nitrites	Low	Low	Medium	Medium	High
Nitrates	Low	Low	Medium	Medium	Medium

Low, weekly; medium, bi-weekly; high, daily (am/pm); very high, continuous

systems, it is advisable to record many key parameters at least twice a day, specifically early in the morning and late afternoon, to understand the full magnitude of daily changes associated with light/dark periods. Meanwhile, intensive and super-intensive systems require a more continuous monitoring programme due to the high production loads in terms of biomass per unit area, and therefore the speed at which parameters can deteriorate towards suboptimal levels.

Whilst in many instances measurements are made manually using specialized equipment of wet test kits, many probes and sensors are also available that are able to translate the measurement of a physico-chemical parameter of interest and convert this into an electrical signal that can subsequently be reported, recorded or, in some instances, acted upon automatically as part of a control system (e.g. smart farming systems, see 'Water quality management'). These probes and sensors can either be wired or wireless and can be combined within this control network or utilized in isolation. Such technology can measure many of the key water quality parameters in real time to enable fast remedial action. This section will therefore outline the available manual and automated (probe/sensor) techniques for monitoring of water chemistry; the considerations required for their measurement, which are dependent upon the technique employed; and highlight any current parameters where techniques or equipment are not currently available or not yet ideal for robust, reliable and affordable measurement within a commercial setting.

Temperature

In shrimp production ponds it is important to measure water temperature at least daily where monitoring utilizes manual equipment, although twice daily is recommended. Whilst in most production settings it is not possible to manipulate environmental temperature when this deviates beyond the optimal range, frequent monitoring enables adjustment of the feed ratio, reducing or increasing the amount of feed given to a given biomass of shrimp depending on the water temperature. Whilst manual temperature measurement enables an approximate understanding of temporal variability, sensor capability for continuous monitoring of temperature is reliable and cost effective and therefore represents a good option to gain highly temporally resolved temperature data, even for rural farms.

The most common sensor utilized for water monitoring is the RTD (resistance temperature detector). The reason why these are the most used is their wide working range, accuracy and affordability. Temperature sensors are waterproof to enable operation when fully submersed, and are able to operate in temperature ranges of –5 to 110 °C and pressures >10 bar.

Their placement in the pond/tank must permit continuous water exchange over the sensor element, and thus must account for water level drops >75% of the total volume without becoming emersed (exposed to the air). This results in a high likelihood of probe placement occurring in deeper areas of the pond/tank. However, the possibility of significant temperature stratification, and spatial variability in temperature across ponds, needs to be accounted for when monitoring, and subsequent remediation of temperature deviations. Preventive maintenance activities should include weekly inspections for cleaning the sensor and monthly check-ups for any potential damage.

Salinity

Salinity is most commonly measured manually using a refractometer, although continuous measurement of conductivity (directly proportional to salinity) is technically easy, reliable and relatively inexpensive for commercial production. As with temperature, where monitoring is manual, a higher frequency of measurement enables finer salinity control. It is recommended that salinity is measured at least twice a week, with recommended frequency increasing seasonally due to the significant influence of evaporation or rainfall in tropical and subtropical areas during the rainy season. Frequent monitoring facilitates the alleviation of adverse salinity levels through farm management practices that include the manual addition of salts to the production water during freshening events (rainfall), or the addition of freshwater to remediate elevated salinity levels (evaporation).

For continuous measurement, the recommended conductivity sensors are inductive (also called toroidal or electrodeless), considering that the main requirement is to avoid contact between the sensing element (electrode coils) and water. Most sensors are composed of two copper coils, one that generates an electromagnetic field change depending on water conductivity and the second inducing an electric voltage encapsulated in a PVC-C or PTFE (Teflon) body protecting them from adverse effects.

The conductivity probes must have a broad scale of conductivity (e.g. 0–100 mS/cm) and coupled RTD-type temperature sensor. Their placement in ponds/tanks must also be in well-mixed areas, near any current stream, to provide accurate and reliable information for the control of this parameter. Preventive maintenance activities include monthly inspections for cleaning the sensor and inspections for any potential damage. Once conductivity of the water is known, this value can be converted to salinity using the simple calculation:

$$\text{Salinity (psu)} = 0.4665 \times \text{conductivity}^{1.0878}$$

where conductivity units are mS/cm and salinity units are ppt or psu. Most modern conductivity probes will be attached to a metering system that automatically converts the measurement output to salinity.

Dissolved oxygen

Being controlled by biological processes in the pond/tank, it is highly recommended that DO is monitored at least twice a day, as well as immediately before feeding, in manual systems. Installation of continuous monitoring probes is ideal to enable tight control of this environmental parameter, with alarms linked to these probes alerting farmers to deviations beyond optimal operational ranges and indicating potential problems, enabling corrective action to be taken. Whilst fixed DO probes are significantly more expensive than temperature probes, the value of continuous monitoring enabling fine temporal scale resolution is highly advantageous. This is particularly the case due to the severity of biological impacts of low DO, the dynamic and highly variable nature of DO within shrimp ponds, particularly variable across day and night, as well as the technical feasibility to enact remedial action to restore optimal DO levels and thus avoid adverse health outcomes.

Optical probes using fluorescence technology provide a good option for monitoring DO as they provide a relatively long life time (2–3 years; galvanic probes offer a greater operational life time but require significantly greater maintenance – see below), stability and accuracy. Among other advantages, these probes do not require the replacement and maintenance of electrolyte solutions or membranes as part of their regular maintenance, which are required for traditional galvanic DO probes.

Irrespective of the technology used, for use in aquaculture oxygen probes must have good sensitivity at low DO levels, giving ranges from 0.00 to 300% as air-saturation and 0.00 to 20.00 mg/l as oxygen content with automatic temperature compensation and self-cleaning nozzles.

These probes are typically IP68 rated, enabling full submersion, operate under temperature ranges of –5 to 50 °C and tolerate pressures >10 bar, making them robust for commercial operation.

The placement of DO probes within the pond/tank must be near the bottom, or alternatively at points where low dissolved oxygen readings may persist (e.g. close to where the feed is dispensed either automatically (automatic feeders) or manually (side of the pond/tank selected for feed broadcasting)).

Preventive maintenance activities should include weekly inspection for cleaning the sensor in models without the self-cleaning option, and monthly for any potential damage.

pH

pH should be measured at least once a day, with frequency of measurement varying depending on the production intensity and increasing with biomass. pH measurement is achieved by probes either manually or continuously; however, it is critical that probes are calibrated and cleaned very frequently (daily) for accurate, reliable and robust readings, especially where fixed probes are deployed. Most pH probes are glass-bodied, or plastic-bodied with the pH-sensitive membrane (bulb) at the end of the probe being made of glass, in combination with reference electrode, reference electrolyte and the reference union. This makes them rather delicate and at risk of breakage if mishandled. However, new ISFET pH sensors are increasingly available for commercial use that have no fragile glass components and are therefore much more robust, as well as having good precision. However, they are far more expensive and so are not likely to be used extensively in shrimp farming until they become more affordable. Irrespective of technology, pH electrodes typically display measurement ranges of 0.0–14.0 units of pH, and are often coupled with RTD-type temperature sensors due to the temperature dependence of pH measurements.

The placement of pH electrodes in ponds/tanks must be in well-mixed areas to provide proper information on the parameters for their control. Preventive maintenance activities include at least weekly inspections for calibrations, cleaning the sensor and inspections for any potential damage. It is advised to replace the sensor yearly.

Alkalinity

It is suggested that alkalinity should be measured at least twice a week in outdoor ponds. In addition, measurement of alkalinity should be made following any remedial corrective action to restore optimal pH using the addition of alkali compounds. This will ensure that the dosage is applied successfully. In systems where the oxidation of ammonia and nitrites is performed by bacteria (AOB and NOB), the frequency of alkali addition should be increased due to the nature of microbes utilizing alkalinity during this process. This is particularly important during the first weeks of establishing the bacterial culture, to control the addition of alkalinity and ensure appropriate levels of alkalinity for bacteria.

Alkalinity is widely measured using manual colorimetric kits, such as liquid drop tests, or spectrophotometers, which provide greater accuracy and increased detection limits for its control. Alternatively, far greater accuracy can be achieved using titration techniques; however, this approach requires a greater skill level by the operator, as well as specialized equipment, meaning that titration measurement of alkalinity is infrequently employed in a commercial setting. No commercial sensors are currently available for measuring alkalinity.

CO_2

The measurement of CO_2 in aquaculture has not received great attention, and is particularly overlooked in outdoor systems, whilst attempts are more regularly made to quantify this parameter in intensive and super-intensive systems. As previously discussed, the direct measurement of dissolved CO_2 is technically challenging and the equipment required to enable this measurement is prohibitively expensive in the vast majority of commercial production settings. Indirect approximations of dissolved CO_2 concentration can be achieved via the use of colorimetric test kits where carbon dioxide (as carbonic acid) is titrated with a dilute sodium hydroxide solution to a pH of 8.3 (or using phenolphthalein as a colour change indicator for reaching pH 8.3). Nonetheless, such testing kits have low sensitivity with respect to the requirements for dissolved CO_2 within the optimal operational ranges for shrimp. Also, there is a lack of understanding concerning the complex chemical relationship in water with carbonate chemistry (alkalinity), so at present the measurement capability of this colorimetric method is inappropriate in the context of effective farm management. As with pH and DO, if effective measurement equipment was available to provide a robust, reliable and accurate measurement of dissolved CO_2, this would ideally need to be measured at least twice daily, with measurement frequency increasing with increasing production intensity in order to enable corrective measures to be enacted to re-establish optimal values.

However, some instruments for direct measurement of dissolved CO_2 have been developed and used in commercial aquaculture settings. These are generally probes that rely on a CO_2-permeable membrane that allows CO_2 to diffuse into a gas-filled space inside the probe/instrument (sometimes called the 'head space unit' or HSU) that is submerged in the water to be measured. The CO_2 in this gas phase is then measured with the same industry-standard technology used for measurement of atmospheric

CO_2 levels, i.e. a non-dispersive infrared (NDIR) gas analyser that relies on CO_2 absorbing light at a specific wavelength (4.2 μm). Such sensors have been tested in aquaculture research projects (e.g. Pfeiffer et al., 2011) and are used in some of the more high-tech aquaculture settings. However, they remain very expensive and are not yet considered entirely reliable and robust, especially at the lower aquatic CO_2 levels approaching normal atmospheric equilibration. Other, more precise, options have been proposed based on the same principles (e.g. portable CO_2 equilibrator with NDIR; Watson et al., 2017), but are not yet commercially available and cost ~US$4,000 to custom build. More affordable, but still precise, sensor technology may be providing better options for a wider range of aquaculture settings in the future – for example, based on application of a direct CO_2-sensing chemical membrane (Mendes et al., 2019).

Where deployed, and similarly to DO probes, it is recommended to place CO_2 sensors in an area allowing continuous water exchange, and located in areas where elevated CO_2 conditions are projected to exist, adjacent to the pond bottom or where feed is administered to the pond (either automatically or manually). Preventive maintenance should include weekly sensor cleaning, inspections for any potential damage, calibrations and yearly replacement.

Hardness

Hardness represents the combined concentration of both calcium (Ca^{2+}) and magnesium (Mg^{2+}) ions, i.e. the main divalent cations in natural waters. Concentrations of these two ions vary greatly within freshwater, but seawater has relatively constant proportions of these ions (with Mg^{2+} at 52 mM greatly exceeding Ca^{2+} at 10 mM in full-strength seawater, salinity = 35 psu). Similar to alkalinity, hardness is often reported in mg/l as calcium carbonate (mg/l $CaCO_3$), which assumes that all divalent cations are calcium and all derived from dissolution of solid calcium carbonate. However, hardness is also sometimes measured as degrees of 'general hardness' (dGH or °GH; where 1 dGH = 17.9 mg/l $CaCO_3$). Whilst hardness specifies the total quantity of dissolved calcium and magnesium, it does not reveal whether it is mainly made up of calcium or magnesium or a combination of the two. This lack of specific quantification of the divalent ion(s) involved is clearly an important issue in shrimp farming, as it is calcium rather than magnesium that is required for post-moult calcification.

Hardness can be measured using a titration method based on the addition of a standard solution of EDTA, which complexes Ca^{2+} and Mg^{2+} ions. Once these ions are all complexed, any further addition of EDTA results in a colour change when an appropriate indicator is present (e.g. Eriochrome black T), and the quantity of EDTA used is converted to hardness. For more specific information about calcium specifically, water samples can be analysed (in ascending order of precision) using wet chemistry test kits, flame photometry (FP), atomic absorption spectrophotometry (AAS) or ion chromatography systems (ICS). Whilst it provides increasing precision, FP is expensive (thousands of dollars) and AAS and ICS are very expensive (tens of thousands of dollars) and require high-tech laboratory facilities and great analytical experience. Ion selective electrodes (ISE) for measurement of calcium ions are available, but these are also expensive (thousands of dollars), are far more reliable in freshwater than seawater and, to our knowledge, have only been used in research settings to date and not within aquaculture.

Waste nitrogen variables

There still does not exist a reliable and continuous monitoring technology to accurately measure nitrogen products in seawater including NH_3, NO_2^- and NO_3^- at commercial application levels, even where there is keen interest within the scientific community and the aquaculture industry to develop this kind of technology. For example, the continuous and in situ measurement of ammonia, which is seen as critical in intensive operations of shrimp, and especially in RAS applications, remains a challenge for researchers. The technology developed is based on the use of ion selective electrodes (ISE) for the determination of ammonium ions (NH_4^+) rather than ammonia gas itself (NH_3), and becomes more challenging when water ionic strength (salinity) is high.

Ding et al. (2017) discussed the application of potentiometric sensors for the in situ detection of ammonia nitrogen (TAN, ammonia and ammonium ion) in seawater environments. A design was proposed which combined a polymeric membrane ammonium-selective electrode, integrated with a hydrogel buffer film of pH 7.0 and a gas-permeable membrane. The gaseous form of ammonia NH_3 in seawater tends to diffuse through the gas membrane and converts to ammonium NH_4^+ in the hydrogel buffer. This enables potentiometric measurement of ammonium, improving also the sensitivity of TAN detection by the buffer trap effect.

SMART FARMING

The technology for the remote measuring, controlling and managing in situ water quality parameters is gaining greater traction, facilitating real-time farm management to help lower production risks, improve production planning and increase productivity (and profitability). The technology used to achieve this monitoring has been developed with the objective of reducing errors in measurement and data assimilation and optimizing commercial operations. Despite this technological advancement, there remains a human element to farm management that requires farm operators or technicians to maintain equipment in an effective, reliable and ready measurement state, collect and interpret the resulting data appropriately and, finally, to act upon this information in a timely manner to facilitate effective mitigation of adverse environmental conditions. This creates a large discrepancy between shrimp farming analytics and management, especially at the transition between extensive and intensive production techniques. Subsequently, advances in Information and communication technologies, alongside the development of low-cost sensors, have increased the feasibility and affordability of monitoring and controlling numerous environmental parameters, bringing new opportunities for smart farm management platforms (Su et al., 2019), which are applicable across all production settings and help to bridge the technological divide between extensive and super-intensive production models (see Fig. 11.5). Nonetheless, significant challenges remain with respect to environmental parameter monitoring across all levels of shrimp production. Thus, further advances in technological development and application are key, foremost among these at present, as described in this chapter, being the understanding, measurement and remediation of adverse environmental conditions with respect to carbon dioxide, alkalinity and pH, as well as divalent cations (calcium and magnesium).

- IOT – Internet of things
- SDA – Supervisory data acquisition
- ACB – Automation control box

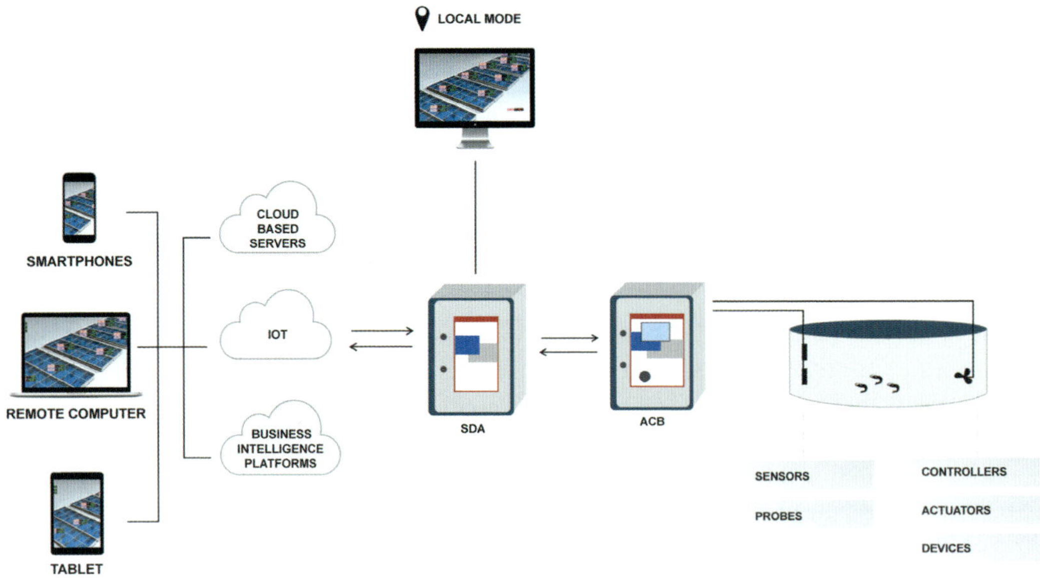

FIGURE 11.5 Typical smart farming architecture. Credit: Ferrer (2020).

Conclusion and future directions

As shrimp production intensity has increased globally, exploiting the huge economic opportunity this sector offers, the importance of maintaining environmental parameters within an optimal range has become unquestionable. In line with the expansion of the sector, an ever-growing body of research has emerged addressing the key production challenges. This has shown that, irrespective of the production system employed, environmental conditions dictate shrimp physiology, immune response (and subsequently susceptibility to disease) and moulting, via both post-moult calcification and growth rate, and as such are vital for the success of production. Moreover, the optimal range for each parameter is determined by the life cycle stage of the shrimp as well as the indicator of performance being measured. Temperature is perhaps the parameter most widely studied with respect to shrimp biology, traditionally being the overriding factor determining where shrimp production is possible globally. Susceptibility to cooler temperatures during early life cycle stages has been widely demonstrated, and an inability to manipulate temperature at the farm level had largely precluded production outside tropical and sub-tropical regions. Nonetheless, a move towards indoor production since 2015 offers an opportunity for greater temperature control in a wide range of climates and thus an avenue with which to expand shrimp farming worldwide. As with temperature, salinity and oxygen are well studied with respect to shrimp health and thus are well understood in the farming context. This has been facilitated by the relative accessibility and affordability of monitoring technology for these parameters commercially, leading to established protocols to maintain these parameters within context-specific (e.g. life cycle stage) optimal ranges, or ease of manipulating salinity and DO when conditions become suboptimal. Conversely, carbonate chemistry parameters such as dissolved CO_2, alkalinity and pH, as well as calcium and magnesium (directly relevant to calcification), are the least studied with respect to shrimp biology and farming. These variables are perhaps where addressing the gap in current understanding offers the greatest opportunity to optimize production environments. As discussed throughout this chapter, the following questions represent key knowledge gaps that are both topical and timely for further investigation:

(1) What are the optimal ranges for post-moult calcification in terms of carbonate chemistry (CO_2, alkalinity, pH) and divalent cations (calcium and magnesium)?
(2) How do these optimal ranges vary with life stage?
(3) What are the options for developing cost-effective and precise real-time sensors for CO_2, alkalinity, calcium and magnesium that can be deployed across all shrimp aquaculture settings?

TABLE 11.6 Glossary of abbreviations. Credit: K. Clayton.

Abbreviation	*Meaning*
LC_{50}	Concentration of substances that will cause 50% mortality
EC_{50}	Concentration of substance that will cause a significant effect in 50% of your samples
WSSV	White spot syndrome virus
TAN	Total ammonia nitrogen: ammonium and un-ionized ammonia
LOEC	Lowest observed effect concentration

TABLE 11.7 Glossary of species names. Credit: K. Clayton.

Scientific name	*Common name*
Penaeus japonicas	Kuruma prawn
Penaeus monodon	Black tiger shrimp
Penaeus stylirostris	Western blue shrimp
Litopenaeus schmitti	Southern white shrimp
Portunus pelagicus	Blue swimmer crab
Penaeus paulensis	São Paulo shrimp
Penaeus vannamei	Whiteleg shrimp

References

Allan, G.L. and Maguire, G.B. (1991) Lethal levels of low dissolved oxygen and effects of short-term oxygen stress on subsequent growth of juvenile *Penaeus monodon*. *Aquaculture* 94, 27e37.

Armstrong, D.A., Chippendale, D., Knight, A.W. and Colt, J.E. (1978). Interaction of ionized and un-ionized ammonia on short-term survival and growth of prawn larvae, *Macrobrachium rosenbergii*. *The Biological Bulletin* 154, 15–31.

Arnold, K.E., Findlay, H.S., Spicer, J.I., Daniels, C.L. and Boothroyd. D. (2009). Effect of CO_2-related acidification on aspects of the larval development of the European lobster, *Homarus gammarus* (L.). *Biogeosciences* 6(8), 1747–1754.

Barbieri, E. (2010) Acute toxicity of ammonia in white shrimp (*Litopenaeus schmitti*) (Burkenroad, 1936, Crustacea) at different salinity levels. *Aquaculture* 306, 329–333.

Barbieri, R.C.J. and Ostrensky, A. (2002). *Camarões Marinhos*. Engorda, Viçosa, Brazil, 370 pp.

Bett, C. and Vinatea, L. (2009). Combined effect of body weight, temperature and salinity on shrimp *Litopenaeus vannamei* oxygen consumption rate. *Brazilian Journal of Oceanography* 57, 305–314.

Boltana, S., Aguilar, A., Sanhueza, N., Donoso, A., Mercado, L., Imarai, M. and Mackenzie, S. (2018). Behavioural fever drives epigenetic modulation of the immune response in fish. *Frontiers in Immunology* 9, 1241.

Boyd, C.E. (1989). *Water Quality Management and Aeration in Shrimp Farming*. Auburn University Press, Alabama, USA.

Boyd, C.E. (1990). *Water Quality in Ponds for Aquaculture*. Auburn University Press, Alabama, USA.

Boyd, C.E. and Gross, A. (2004). Water use and conservation for inland aquaculture ponds. *Fisheries Management and Ecology* 7(1–2), 55–63.

Boyd, C.E. and McNevin, A.A. (2018). Land use in shrimp aquaculture. *World Aquaculture* 49(1), 28–34.

Boyd, C.E., McNevin, A.A., Davis, R.P., Godumala, R. and Mohan, A.B.C. (2018). Production methods and resource use at *Litopenaeus vannamei* and *Penaeus monodon* farms in India compared with previous findings from Thailand and Vietnam. *Journal of the World Aquaculture Society* 49(3), 551–569.

Boyd, C.E. and Tucker, C.S. (1992). *Water Quality and Pond Analyses for Aquaculture*. Agricultural Experiment Station, Alabama, USA.

Brauner, C.J., Shartau, R.B., Damsgaard, C., Esbaugh, A.J., Wilson, R.W. and Grosell, M. (2019). Acid–base physiology and CO_2 homeostasis: Regulation and compensation in response to elevated environmental CO_2. *Fish Physiology* 37, 69–132.

Bray, W.A., Lawrence, A.L. and Leung-Trujillo, J.R. (1994). The effect of salinity on growth and survival of *Penaeus vannamei*, with observations on the interaction of IHHN virus and salinity. *Aquaculture* 122, 133–146.

Burford, M.A., Thompson, P.J., McIntosh, R.P., Bauman, R.H. and Pearson, D.C. (2003). Nutrient and microbial dynamics in high-intensity, zero-exchange shrimp ponds in Belize. *Aquaculture* 219, 393–411.

Burford, M.A. and Williams, K.C. (2001). The fate of nitrogenous waste from shrimp feeding. *Aquaculture* 198, 79–93.

Burgents, J.E., Burnett, K.G. and Burnett, L.E. (2005). Effects of hypoxia and hypercapnic hypoxia on the localization and the elimination of *Vibrio campbellii* in *Litopenaeus vannamei*, the Pacific white shrimp. *The Biological Bulletin* 208, 159–168.

Cai, Y., Chen, T., Ren, C., Huang, W., Jiang, G., Huo, D. and Hu, C. (2017). Molecular charcaterization of Pacific white shrimp (*Litopenaeus vannamei*) sodium bicarbonate cotransporter (NBC) and its role in response to pH stress. *Fish and Shellfish Immunology* 64, 226–233.

Cameron, J.N. and Wood, C.M. (1985). Apparent H^+ excretion and CO_2 dynamics accompanying carapace mineralization in the blue crab (*Callinectes sapidus*) following moulting. *Journal of Experimental Biology* 114(1), 181–196.

Castille, F.L. and Lawrence, A.L. (1981). The effect of salinity on the osmotic, sodium and chloride concentrations in the hemolymph of euryhaline shrimp of the genus *Penaeus*. *Comparative Biochemistry and Physiology* A 68, 75–80.

Cavalli, R O., Peixoto, S.M. and Wasielesky, W. (1998). Performance of *Penaeus paulensis* (Perez-Farfante) broodstock under long-term exposure to ammonia. *Aquaculture Research* 29, 815–822.

Ceballos-Vázquez, B.P., Palacios, E., Aguilar-Villavicencio, J. and Racotta, I.S. (2010). Gonadal development in male and female domesticated whiteleg shrimp, *Litopenaeus vannamei*, in relation to age and weight. *Aquaculture* 308, 116–123.

Ceballos-Vázquez, B.P., Rosas, C. and Racotta, I.S. (2003). Sperm quality in relation to age and weight of white shrimp *Litopenaeus vannamei*. *Aquaculture* 228, 141–151.

Chang, Z.W., Chiang, P.C., Cheng, W. and Chang, C.C. (2015). Impact of ammonia exposure on coagulation in white shrimp, *Litopenaeus vannamei*. *Ecotoxicology and Environmental Safety* 118, 98–102.

Chantal, M., Etienne, Z., Cyrille, G. and Hugues, L. (2008). Combined effect of exposure to ammonia and hypoxia on the blue shrimp *Litopenaeus stylirostris* survival and physiological response in relation to molt stage. *Aquaculture* 274, 398–407.

Charmantier, G., Soyez, C. and Aquacop, K. (1994). Effect of molt stage and hypoxia on osmoregulatory capacity in the peneid shrimp *Penaeus vannamei*. *Journal of Experimental Marine Biology and Ecology* 178, 233e46.

Chen, J.C. and Kou, Y.Z. (1992). Effects of ammonia on growth and molting of *Penaeus japonicus* juveniles. *Aquaculture* 104, 249–260.

Chen, J.C. and Lin, C.Y. (1992). Effects of ammonia on growth and molting of *Penaeus monodon* juveniles. *Comparative Biochemistry and Physiology C* 101, 449–452.

Chen, T., Lin, T., Li, H., Lu, T., Li, J., Huang, W. et al. (2018). Heat shock protein 40 (HSP40) in Pacific white shrimp (*Litopenaeus vannamei*): Molecular cloning, tissue distribution and ontogeny, ersponse to temperature, acidity/alkalinity and salinity stresses, and potential role in ovarian development. *Frontiers in Physiology* 9, 1784.

Cheng, W., Liu, C.H., Hsu, J.P. and Chen, J.C. (2002). Effect of hypoxia on the immune response of giant freshwater prawn *Macrobrachium rosenbergii* and its susceptibility to pathogen *Enterococcus. Fish and Shellfish Immunology* 13, 351e65.

Cheng, W., Liu, C.H. and Kuo, C.M. (2003). Effects of dissolved oxygen on hemolymph parameters of freshwater giant prawn *Macrobrachium rosenbergii* (de Man). *Aquaculture* 220, 843e56.

Cheng, W., Wang, L.U. and Chen, J.C. (2005). Effect of water temperature on the immune response of white shrimp *Litopenaeus vannamei* to *Vibrio alginolyticus. Aquaculture* 250, 592–601.

Chien, Y.H. (1992). Water quality requirements and management for marine shrimp culture. In Wyban, J. (ed.) *Proceedings of the Special Session on Shrimp Farming*, World Aquaculture Society, Baton Rouge, Louisiana, USA, 25 pp.

Chong-Robles, J., Charmantier, G., Boulo, V., Lizárraga-Valdéz, J., Enríquez-Paredes, L.M. and Giffard-Mena, I. (2014). Osmoregulation pattern and salinity tolerance of the white shrimp *Litopenaeus vannamei* (Boone, 1931) during post-embryonic development. *Aquaculture* 422, 261–267.

Clarke, A. and Johnston, N.M. (1999). Scaling of metabolic rate with body mass and temperature in teleost fish. *Journal of Animal Ecology* 68(5), 893–905.

Colt, J.E. and Armstrong, D.A. (1981). *Nitrogen Toxicity to Crustaceans, Fish, and Molluscs* (eds Allen, J.L. and Kinney, E.C.). American Fisheries Society, Northeast Society of Conservation Engineers, Bethesda, Maryland USA.

Datta, S. (2012). *Management of Water Quality in Intensive Aquaculture*. Available at https://www.researchgate.net/publication/259175404_Management_of_Water_Quality_in_Intensive_Aquaculture (accessed 13 July 2021).

Davis, D.A., Saoud, I.P., McGraw, W.J. and Rouse, D.B. (2002). Considerations for *Litopenaeus vannamei* reared in inland low salinity waters. In: Cruz-Suárez, L.E., Ricque-Marie, D., Tapia-Salazar, M., Gaxiola-Cortés, M.G. and Simoes, N. (eds) *Avances en Nutrición Acuícola VI*. Memorias del VI Simposium Internacional de Nutrición Acuícola. 3 al 6 de Septiembre del 2002. Cancún, Quintana Roo, México.

De Lourdes Cobo, M., Sonnenholzner, S., Wille, M. and Sorgeloos, P. (2012). Ammonia tolerance of *Litopenaeus vannamei* (Boone) larvae. *Aquaculture* 45, 470–475.

Diaz, F., Farfan, C., Sierra, E. and Re, A.D. (2001). Effects of temperature and salinity fluctuation on the ammonium excretion and osmoregulation of juveniles of *Penaeus vannamei*, Boone. *Marine and Freshwater Behaviour and Physiology* 34, 93–104.

Ding, L., Ding, J., Ding, B. and Qin, W. (2017). Solid-contact potentiometric sensor for the determination of total ammonia nitrogen in seawater. *International Electrochemical Science* 12, 3296–3308.

Ding, S., Wang, F., Dong, S. and Li, J. (2014). Comparison of the respiratory metabolism of juvenile *Litopenaeus vannamei* culture in seawater and freshwater. *Journal of the Ocean University of China* 13, 331–337.

Duan, Y., Liu, Q., Wang, Y., Zhang, J. and Xiong, D. (2018). Impairment of the intestine barrier function in *Litopenaeus vannamei* exposed to ammonia and nitrite stress. *Fish and Shellfish Immunology* 78, 279–288.

Ellis, R.P., Parry, H., Spicer, J.I., Hutchinson, T.H., Pipe, R.K. and Widdicombe, S. (2011). Immunological function in marine invertebrates: Responses to environmental perturbation. *Fish and Shellfish Immunology* 30, 1209–1222.

Ellis, R.P., Urbina, M.A. and Wilson, R. (2017). Lessons from two high CO_2 worlds – future oceans and intensive aquaculture. *Global Change Biology* 23(6), 2141–2148.

Escobedo-Bonilla, C.M., Wille, M., Alday Sanz, V., Sorgeloos, P., Pensaert, M.B. and Nauwynck, H.J. (2007). Pathogenesis of a Thai strain of white spot syndrome virus (WSSV) in juvenile, specific pathogen-free *Litopenaeus vannamei. Diseases of Aquatic Organisms* 74, 85–94.

Eshchar, M. Lahav, O., Mozes, N., Peduel, A. and Ron, B. (2006). Intensive fish culture at high ammonium and low pH. *Aquaculture* 255(1–4), 301–313.

Evans, S.S., Repasky, E.A. and Fisher, D.T. (2015). Fever and the thermal regulation of immunity: the immune system feels the heat. *Nature Reviews in Immunology* 15, 335–349.

FAO (2019). GLOBEFISH Highlights April 2019 Issue, with Jan.–Dec. 2018 Statistics – A quarterly update on world seafood markets. *Globefish Highlights* no. 2-2019.

Fawen, H., Luqing, P. and Futao, J. (2009). Effects of hypoxia on dopamine concentration and the immune response of white shrimp (*Litopenaeus vannamei*). *Journal of the Ocean University of China Oceanic Coastal and Sea Research* 8, 77–82.

Feliu, V., Rivas, R. and Castillo, F. (2009) Fractional order controller robust to time delay variations for water distribution in an irrigation main canal pool. *Computers and Electronics in Agriculture* 69, 185–197.

Feng, C., Tian, X., Dong, S., Su, Y., Wang, F. and Ma, S. (2008). Effects of frequency and amplitude of salinity fluctuation on the growth and energy budget of juvenile *Litopenaeus vannamei* (Boone). *Aquaculture Research* 39, 1639–1646.

Ferreira, N.C., Bonetti, C. and Seiffert, W.Q. (2011). Hydrological and water quality indices as management tools in marine shrimp culture. *Aquaculture* 318(3–4), 425–433.

Fouzi, M., Shariff, M., Omar, A.R., Yusoff, F. and Tan, S.W. (2010). TaqMan real-time PCR assay for relative quantification of white spot syndrome virus infection in *Penaeus monodon* Fabricius exposed to ammonia. *Journal of Fish Diseases* 33, 931–938.

Freeman, J.A. (1990). Molt increment, molt cycle duration, and tissue growth in *Palaemonetes pugio* Holthuis larvae. *Journal of Experimental Marine Biology and Ecology* 143(1–2), 47–61.

Frías-Espericueta, M.G., Harfush-Melendez, M., Osuna-López, J.I. and Páez-Osuna, F. (1999). Acute toxicity of ammonia to juvenile shrimp *Penaeus vannamei* Boone. *Bulletin of Environmental Contamination and Toxicology* 62, 646–652.

Frias-Espericueta, M.G., Harfush-Melendez, M. and Páez-Osuna, F. (2000). Effects of ammonia on mortality and feeding of postlarvae shrimp *Litopenaeus vannamei. Bulletin of Environmental Contamination and Toxicology* 65, 98–103.

Furtado, P.S., Poersch, L.H. and Wasielesky Jr, W. (2011). Effect of calcium hydroxide, carbonate and sodium bicarbonate on water quality and zootechnical performance of shrimp *Litopenaeus vannamei* reared in bio-flocs technology (BFT) systems. *Aquaculture* 321(1–2), 130–135.

Furtado, P.S., Campos, B.R., Serra, F.P., Klosterhoff, M., Romano, L.A. and Wasielesky, W. (2014). Effects of nitrate toxicity in the Pacific white shrimp, *Litopenaeus vannamei*, reared with biofloc technology (BFT). *Aquaculture International* 23, 315–327.

Furtado, P.S., Fugimura, M.M., Monserrat, J.M., Souza, D.M., Garcia, L.D.O. and Wasielesky, W. (2015a). Acute effects of extreme pH and its influences on the survival and biochemical biomarkers of juvenile White Shrimp, *Litopenaeus vannamei. Marine and Freshwater Behaviour and Physiology* 48(6), 417–429.

Furtado, P.S., Poersch, L.H. and Wasielesky, W. (2015b). The effect of different alkalinity levels on *Litopenaeus vannamei* reared with biofloc technology (BFT). *Aquaculture International* 23(1), 345–358.

Furtado, P.S., Valenzuela, M.A.J., Rodriguez-Fuentes, G., Campos, B.R., Wasielesky, W. and Gaxiola, G. (2016a). Chronic effect of nitrite on the rearing of the white shrimp *Litopenaeus vannamei* in two salinities. *Marine and Freshwater Behaviour and Physiology* 49, 201–211.

Furtado, P.S., Valenzuela, M.A., Badillo, M.A., Gaxiola, G., and Wasielesky Jr, W. (2016b). Effect of dissolved carbon dioxide on oxygen consumption in the Pacific white shrimp, *Litopenaeus vannamei* (Boone 1931). *Marine and Freshwater Behaviour and Physiology* 49(5), 337–346.

Furtado, P.S., Gaona, C.A., Serra, F.P., Poersch, L.H. and Wasielesky Jr, W. (2017). Acute toxicity of carbon dioxide to juvenile marine shrimp *Litopenaeus vannamei* (Boone 1931). *Marine and Freshwater Behaviour and Physiology* 50(4), 293–301.

Gong, H., Jiang, D.H., Lightner, D.V, Collins, C. and Brock, D. (2004). A dietary modification approach to improve the osmoregulatory capacity of *Litopenaeus vannamei* cultured in the Arizona desert. *Aquaculture Nutrition* 10, 227–236.

González, R.A., Díaz, F., Licea, A., Denisse Re, A., Noemí Sánchez, L. and García-Esquivel, Z. (2010). Thermal preference, tolerance and oxygen consumption of adult white shrimp *Litopenaeus vannamei* (Boone) exposed to different acclimation temperatures. *Journal of Thermal Biology* 35, 218–224.

Granja, C., Vidal, O., Parra, G. and Salazar, M. (2006). Hyperthermia reduces viral load of white spot syndrome virus in *Penaeus vannamei. Diseases of Aquatic Organisms* 68, 175–180.

Han, S. Y., Wang B.J., Liu M., Wang M.Q., Jiang K.Y. and Liu, X.W. (2018). Adaptation of the white shrimp *Litopenaeus vannamei* to gradual changes to a low-pH environment. *Ecotoxicology and Environmental Safety* 149, 203–210.

Han, S., Wang, B., Wang, M., Liu, Q., Zhao, W. and Wang, L. (2017). Effects of ammonia and nitrite accumulation on the survival and growth performance of white shrimp *Litopenaeus vannamei. Invertebrate Survival Journal* 14, 221–232.

Hargreaves, J.A. and Tucker, C.S. (2012). Measuring dissolved oxygen concentration in aquaculture. SRAC Publication No. 4601.

Harlıoğlu, M.M., Farhadi, A. and Gür, S. (2018). Determination of sperm quality in decapod crustaceans. *Aquaculture* 490, 185–193.

Hartnoll, R.G. (2001). Growth in Crustacea – twenty years on. *Hydrobiologia* 449, 111–122.

He, P., Wei, P., Zhang, B., Zhao, Y., Li, Q., and Chen, X. (2018) Identification of microRNAs involved in cold adaptation of *Litopenaeus vannamei* by high-throughput sequencing. *Gene* 677, 24–31.

Hoppe, C.J.M., Langer, G., Rokitta, S.D., Wold-Gladrow, D.A. and Rost, B. (2012). Implications of observed inconsistencies in carbonate chemistry measurements for ocean acidification studies. *Biogeosciences* 9(7), 2401–2405.

Huang, W., Ren, C., Li, H., Huo, D., Wang, Y. and Jiang, X. (2017). Transcriptomic analyses on muscle tissues of *Litopenaeus vannamei* provide the first profile insight into the response to low temperature stress. *PLoS ONE* 12, e178604.

Ip, Y.K. and Chew, S.F. (2010). Ammonia production, excretion, toxicity, and defense in fish: a review. *Frontiers in Physiology* 1, 134.

Israngkura, A. and Sae-Hae, S. (2002). A review of economic impacts of aquatic animal disease. In: Arthur, J.R., Phillips, M.J., Subasinghe, R.P., Reantaso, M.B. and McCrae, I.H. (eds) *Primary Aquatic Animal Health Care in Rural, Small-scale Aquaculture Development, Technical Proceedings of the Asia Regional Scoping Workshop*. FAO Fisheries Technical Paper 406, FAO, Rome, pp. 55–61.

Jackson, C., Preston, N., Thompson, P.J. and Burford, M. (2003). Nitrogen budget and effluent nitrogen components at an intensive shrimp farm. *Aquaculture* 218, 397–411.

Jiang, D.H., Lawrence, A.L., Neill, W.H. and Gong, H. (2000). Effects of temperature and salinity on nitrogenous excretion by *Litopenaeus vannamei* juveniles. *Journal of Experimental Marine Biology and Ecology* 253, 193–209.

Jiang, L., Pan, L. and Fang-Bo., T. (2005). Effect of dissolved oxygen on immune parameters of the white shrimp *Litopenaeus vannamei. Fish and Shellfish Immunology* 18(2), 185–188.

Kuhn, D.D., Smith, S.A., Boardman, G.D., Angier, M.W., Marsh, L. and Flick, G.J. (2010). Chronic toxicity of nitrate to Pacific white shrimp, *Litopenaeus vannamei*: Impacts on survival, growth, antennae length, and pathology. *Aquaculture* 309, 109–114.

Kunkel, J.G. and Jercinovic, M.J. (2013). Carbonate apatite formulation in cuticle structure adds resistance to microbial attack for American lobster. *Marine Biology Research* 9(1), 27–34.

Laramore, S., Laramore, C.R. and Scarpa, J. (2001). Effect of low salinity on growth and survival of postlarvae and juvenile *Litopenaeus vannamei. Journal of the World Aquaculture Society* 32, 385–392.

Lemonnier, H., Bernad, E., Boglio, E., Goarant, C. and Cochard, J. (2004). Influence of sediment characteristics on shrimp physiology: pH as principal effect. *Aquaculture* 240(1–4), 297–312.

Le Moullac, G., Soyez, C., Saulnier, D., Ansquer, D., Avarre, C.J. and Levy, P. (1998). Effect of hypoxic stress on the immune response and the resistance to vibriosis of the shrimp *Penaeus stylirostris. Fish and Shellfish Immunology* 8, 621e9.

Li, C.C. and Chen, J.C. (2008). The immune response of white shrimp *Litopenaeus vannamei* and its susceptibility to *Vibrio alginolyticus* under low and high pH stress. *Fish and Shellfish Immunology* 25(6), 701–709.

Li, E., Chen, L., Zeng, C., Chen, X., Yu, N., Lai, Q. and Qin, J.G. (2007). Growth, body composition, respiration and ambient ammonia nitrogen tolerance of the juvenile white shrimp, *Litopenaeus vannamei*, at different salinities. *Aquaculture* 265, 385–390.

Li, E., Chen, L., Zeng, C., Yu, N., Xiong, Z. and Chen, X. (2008). Comparison of digestive and antioxidant enzymes activities, haemolymph oxyhemocyanin contents and hepatopancreas histology of white shrimp, *Litopenaeus vannamei*, at various salinities. *Aquaculture* 274, 80–86.

Liang, Z., Liu, R., Zhao, D., Wang, L., Sun, M., Wang, M. and Song, L. (2016). Ammonia exposure induces oxidative stress, endoplasmic reticulum stress and apoptosis in hepatopancreas of pacific white shrimp (*Litopenaeus vannamei*). *Fish and Shellfish Immunology* 54, 523–528.

Lin, Y.C. and Chen, J.C. (2001). Acute toxicity of ammonia on *Litopenaeus vannamei* Boone juveniles at different salinity levels. *Journal of Experimental Marine Biology and Ecology* 259, 109–119.

Lin, Y.C. and Chen, J.C. (2003). Acute toxicity of nitrite on *Litopenaeus vannamei* (Boone) juveniles at different salinity levels. *Aquaculture* 224, 193–201.

Lin, Y.C., Chen, J.C., Li, C.C., Morni, W.Z., Suhaili, A.S.N. et al. (2012). Modulation of the innate immune system in white shrimp *Litopenaeus vannamei* following long-term low salinity exposure. *Fish and Shellfish Immunology* 33, 324–331.

Liu, C.H. and Chen, J.C. (2004). Effect of ammonia on the immune response of white shrimp *Litopenaeus vannamei* and its susceptibility to *Vibrio alginolyticus. Fish and Shellfish Immunology* 16, 321–334.

Liu, F., Li, S., Yu, Y., Sun, M., Xiang, J. and Li, F. (2020). Effects of ammonia stress on the hemocytes of the Pacific white shrimp *Litopenaeus vannamei. Chemosphere* 239, 124759.

Liu, L., Hu, Z., Dai, X. and Avnimelech, Y. (2014). Effects of addition of maize starch on the yield, water quality and formation of bioflocs in an integrated shrimp culture system. *Aquaculture* 418–419, 79–86.

Llan, G.L. and Maguire, G.B. (1991). Lethal levels of low dissolved oxygen and effects of short-term oxygen stress on subsequent growth of juvenile *Penaeus monodon. Aquaculture* 94, 27e37.

Lu, X., Luan, S., Dai, P., Luo, K., Chen, B., Cao, B., Sun, L., Yan, Y. and Kong, J. (2019). Insights into the molecular basis of immunosuppression and increasing pathogen infection severity of ammonia toxicity by transcriptome analysis in pacific white shrimp *Litopenaeus vannamei. Fish and Shellfish Immunology* 88, 528–539.

Madeira, D., Mendonça, V., Dias, M., Roma, J., Costa, P.M. and Larguinho M. (2015). Physiological, cellular and biochemical thermal stress response of intertidal shrimps with different vertical distributions: *Palaemon elegans* and *Palaemon serratus. Comparative Biochemistry and Physiology A* 183, 107–115.

McGraw, W.J. and Scarpa, J. (2004). Mortality of freshwater-acclimated *Litopenaeus vannamei* associated with acclimation rate, habituation period, and ionic challenge. *Aquaculture* 236(1–4), 285–296.

Melzner, F., Mark, F.C., Seibel, B.A. and Tomanek, L. (2019) Ocean acidification and coastal marine invertebrates: Tracking CO_2 effects from seawater to the cell. *Annual Review of Marine Science* 12, 499–523.

Mendes, J.P., Coelho, L., Kovacs, B., de Almeida, J.M., Pereira, C.M., Jorge, P.A. and Borges, M.T. (2019). Dissolved carbon dioxide sensing platform for freshwater and saline water applications: Characterization and validation in aquaculture environments. *Sensors* 19(24), 5513.

Menz, A. and Blake, B.F. (1980). Experiments on the growth of *Penaeus vannamei* Boone. *Journal of Experimental Marine and Biological Ecology* 48, 99–111.

Middlemiss, K.L., Urbina, M.A. and Wilson, R. (2016). Effects of seawater alkalinity on calcium and acid-base regulation in juvenile European lobster (*Homarus gammarus*) during a moult cycle. *Comparative Biochemistry and Physiology A* 193, 22–28.

Millard, R.S., Ellis, R.P., Bateman, K.S., Bickley, L.K., Tyler, C.R., van Aerle, R. and Santos, E.M. (2020). How do abiotic envrionmetal conditions influence shrimp susceptibilty to disease? A critical analysis focussed on White Spot Disease. *Journal of Invertebrate Pathology* 6 April, 107369.

Montgomery-Brock, D.R., Shimojo, R.Y. and Bullis, R.A. (2002). The reduced impact of Taura syndrome on *Litopenaeus vannamei* held under hyperthermic conditions. Central, *Tropical and Subtropical Aquaculture Aqua Tips* 13, 4–7.

Montgomery-Brock, D., Tacon, A.G.J., Poulos, B. and Lightner, D. (2007). Reduced replication of infectious hypodermal and hematopoietic necrosis virus (IHHNV) in *Litopenaeus vannamei* held in warm water. *Aquaculture* 265, 41–48.

Morse, J.W. et al. (2007). Calcium carbonate formation and dissolution. *Chemical Reviews* 107, 342–381.

Moser, J.R., Álvarez, D.A.G., Cano, F.M., Garcia, T.E., Molina, D.E.C., Clark, G.P., Marques, M.R.F., Barajas, F.J.M. and López, J.H. (2012). Water temperature influences viral load and detection of White Spot Syndrome Virus (WSSV) in *Litopenaeus vannamei* and wild crustaceans. *Aquaculture* 326–329, 9–14.

Neufeld, D. and Cameron, J. (1994). Effect of the external concentration of calcium on the postmoult uptake of calcium in blue crabs (*Callinectes sapidus*). *Journal of Experimental Biology* 188(1), 1.

Pan, L. and Jiang, L. (2002). The effect of sudden changes in salinity and pH on the immune activity of two species of shrimps. *Journal of the Ocean University of Qingdao* 32(6), 903–910.

Pan, L.Q., Zhang, L.J. and Liu, H.Y. (2007). Effects of salinity and pH on ion-transport enzyme activities, survival and growth of *Litopenaeus vannamei* postlarvae. *Aquaculture* 273(4), 711–720.

Patterson, B.D. and Thorne, M.J. (1995). Measurements of oxygen uptake, heart and gill bailer rates of the callianassid burrowing shrimp *Trypaea australiensis* Dana and its responses to low oxygen. *Journal of Experimental Marine Biology and Ecology* 194, 39e52.

Pante, J. R. M. (1990). Influence of environmental stress on the heritability of molting frequency and growth rate of the penaeid shrimp, Penaeus vannamei. University of Houston-Clear Lake: ProQuest Dissertations Publishing.

Peck, V.L., Oakes, R.L., Harper, E.M., Manno, C. and Tarling, G. (2018). Pteropods counter mechanical damage and dissolution through extensive shell repair. *Nature Communications* 9, 264.

Perez-Velazquez, M., Bray, W.A., Lawrence, A.L., Gatlin, D.M. and Gonzalez-Felix, M.L. (2001). Effect of temperature on sperm quality of captive *Litopenaeus vannamei* broodstock. *Aquaculture* 198, 209–218.

Pfeiffer, T.J., Summerfelt, S.T. and Watten, B.J. (2011). Comparative performance of CO_2 measuring methods: Marine aquaculture recirculation system application. *Aquacultural Engineering* 44(1), 1–9.

Ponce-Palafox, J., Martinez-Palacios, C.A. and Ross, L.G. (1997). The effects of salinity and temperature on the growth and survival rates of juvenile white shrimp, *Penaeus vannamei*, Boone, 1931. *Aquaculture* 157, 107–115.

Ponce-Palafox, J., Ruíz-Lunab, A., García-Ulloa Gómez, M., Esparza-Leald H. M., Arredondo-Figueroae, J. L., Martinez-Palaciosf C.A.,& Ross, L.G. (2013) A response-surface analysis of the relative importance of the temperature, salinity and body weight on the respiratory metabolism of the white shrimp Litopeneaus vannamei (Boone, 1931)

Pontinha, V.D.A., Vieira, F.D.N. and Hayashi, L. (2018). Mortality of pacific white shrimp submitted to hypothermic and hyposalinic stress. *Boletim do Instituto de Pesca* 44(2).

Pörtner, H.O., Karl, D.M., Boyd, P.W. et al. (2014) Ocean systems. In: Field, C.B. et al. (eds) *Climate Change 2014: Impacts, Adaptation, and Vulnerability. Part A: Global and Sectoral Aspects. Contribution of Working Group II to the Fifth Assessment Report of the Intergovernmental Panel on Climate Change.* Cambridge University Press, Cambridge, UK and New York, NY, USA, pp. 411–484.

Racotta, I.S. and Hernández-Herrera, R. (2000). Metabolic responses of the white shrimp, *Penaeus vannamei*, to ambient ammonia. *Comparative Biochemistry and Physiology A* 125, 437–443.

Racotta, I., Palacios, E. and Mendez, L. (2002). Metabolic responses to short- and long-term exposure to hypoxia in white shrimp (*Penaeus vannamei*). *Marine and Freshwater Behaviour and Physiology* 35, 269–275.

Rahman, M.M., Escobedo-Bonilla, C.M., Corteel, M., Dantas-Lima, J.J., Wille, M., Sanz, V.A., Pensaert, M.B., Sorgeloos, P. and Nauwynck, H.J. (2006). Effect of high water temperature (33 °C) on the clinical and virological outcome of experimental infections with white spot syndrome virus (WSSV) in specific pathogen-free (SPF) *Litopenaeus vannamei*. *Aquaculture* 261, 842–849.

Raja, K., Gopalakrishnan, A., Singh, R. and Vijayakumar, R. (2015). Loose shell syndrome (LSS) in *Litopenaeus vannamei* grow-out ponds and its effect on growth and production. *Journal of Fish and Aquaculture* 6(4), 1000151.

Ries, J.B., Cohen, A.L. and McCorkle, D.C. (2009) Marine calcifiers exhibit mixed responses to CO_2-induced ocean acidification. *Geology* 37(12), 1131–1134.

Rodríguez-Ramos, T., Espinosa, G., Hernández-López, J., Gollas-Galván, T., Marrero, J., Borrell, Y., Alonso, M.E., Bécquer, U. and Alonso, M. (2008). Effects of *Escherichia coli* lipopolysaccharides and dissolved ammonia on immune response in southern white shrimp *Litopenaeus schmitti*. *Aquaculture* 274, 118–125.

Romano, N. and Zeng, C. (2009). Subchronic exposure to nitrite, potassium and their combination on survival, growth, total haemocyte count and gill structure of juvenile blue swimmer crabs, *Portunus pelagicus*. *Ecotoxicology and Environmental Safety* 72, 1287–1295.

Romano, N. and Zeng, C. (2010). Changes to the histological gill structure and haemolymph composition of early blue swimmer crab *Portunus pelagicus* juveniles during elevated ammonia-N exposure and the post-exposure recovery. *Aquaculture Research* 41, 468–480.

Romano, N. and Zeng, C. (2017). Cannibalism of decapod crustaceans and implications for their aquaculture: A review of its prevalence, influencing factors, and mitigating methods. *Reviews in Fisheries Science and Aquaculture* 25(1), 42–69.

Rosas, C., López, N., Mercado, P. and Martínez, E. (2001). Effect of salinity acclimation on oxygen consumption of juveniles of the white shrimp *Litopenaeus Vannamei*. *Journal of Crustacean Biology* 21, 912–922.

Rostami, F., Davoodi, R., Nafisi Bahabadi, M., Salehi, F. and Nooryazdan, H. (2019). Effects of ammonia on growth and molting of *Litopenaeus vannamei* postlarvae reared under two salinity levels. *Journal of Applied Aquaculture* 31, 1–13.

Sahu, B.C., Adhikari, A.S. and Dey, L. (2013). Carbon, nitrogen, and phosphorus budget in scampi (*Macrobrachium rosenbergii*) culture ponds. *Environmental Monitoring and Assessment* 185(12), 10157–10166.

Saksena, D., Gaidhane, D.M. and Singh, H. (2006). Limnology of Kharland (saline) ponds of Ratnagiri, Maharashtra in relation to prawn culture potential. *Journal of Environmental Biology* 27, 49–53.

Samocha, T.M., Prangnell, D.I., Hanson, T.R., Treece, G.D., Morris, T.C., Castro, L.F. and Staresinic, G. (2017). *Design and Operation of Super-intensive Biofloc-dominated Systems for the Production of Pacific White Shrimp*. World Aquaculture Society, Sorrento, Louisiana, USA.

Schuler, D.J., Boardman, G.D., Kuhn, D.D. and Flick, G.J. (2010). Acute toxicity of ammonia and nitrite to pacific white shrimp, *Litopenaeus vannamei*, at low salinities. *Journal of the World Aquaculture Society* 41, 438–446.

Seidman, E.R. and Lawrence, A.L. (1985). Growth, feed digestibility, and proximate body composition of juvenile Penaeus vannamei and Penaeus monodon grown at different dissolved oxygen levels. Journal of the World Mariculture Society 16, 333–346.

Small, D.P., Calosi, P., Rastrick, S.P.S., Turner, L.M., Widdicombe, S. and Spicer, J.I. (2020). The effects of elevated temperature and PCO_2 on the energetics and haemolymph pH homeostasis of juveniles of the European lobster, *Homarus gammarus*. *Journal of Experimental Biology* 223(8), jeb209221.

Söderhäll, K. and Cerenius, L. (1992). Crustacean immunity. *Annual Review of Fish Diseases* 2, 3–23.

Su, Y., Feng, C. and Ma, S. (2010). Effects of salinity fluctuation on the growth and energy budget of juvenile *Litopenaeus vannamei* at different temperatures. *Journal of Crustacean Biology* 30, 430–434.

Nonwachai, T., Purivirojku, W., Chuchird, N. and Limsuwan, C. (2011). Effects of dissolved oxygen levels on growth, survival and immune response of juvenile pacific white shrimp *Litopenaeus vannamei*. *Journal of Fisheries and the Environment* 35(3), 1–10.

Thornber, K., Verner-Jeffreys, D., Hinchliffe, S., Rahman, M.M., Bass, D. and Tyler, C.R. (2019). Evaluating antimicrobial resistance in the global shrimp industry. *Reviews in Aquaculture* 12(2), 966–986.

Tidwell, J.H. (2012). *Functions and CCharacteristics of All Aquaculture Systems*, 1st edn. John Wiley and Sons, Inc., London.

Timmons, M.B. and Ebeling, J.B. (2013). *Recirculating Aquaculture*, 3rd edn. Ithaca Publishing Company, Ithaca, New York, USA.

Tseng, I.T. and Chen, J.C. (2004). The immune response of white shrimp *Litopenaeus vannamei* and its susceptibility to *Vibrio alginolyticus* under nitrite stress. *Fish and Shellfish Immunology* 17, 325–333.

Van Wyk, P. and Scarpa, J., (1999). Water quality and management. In: Van Wyk, P. et al. (eds) *Farming Marine Shrimp in Recirculating Freshwater Systems*. Florida Department of Agriculture and Consumer Services, Tallahassee, Florida, USA, pp. 128–138.

Vaquer-Sunyer, R. and Duarte, C.M. (2008). Thresholds of hypoxia for marine biodiversity. *Proceedings of the National Academy of Sciences USA* 105, 15452–15457.

Verhoef, G. D., Austin, C. M., Jones, P. L. and Stagnitti, F. (1998). Effect of temperature on molt increment and intermolt period of a juvenile australian fresh-water crayfish, *Cherax destructor*. Journal of Crustacean Biology 18, 673–679.

Vidal, O.M., Granja, C.B., Aranguren, F., Brock, J.A. and Salazar, M. (2001). A profound effect of hyperthermia on survival of *Litopenaeus vannamei* juveniles infected with white spot syndrome virus. *Journal of the World Aquaculture Society* 32, 364–372.

Villarreal, H., Hinojosa, P. and Naranjo, J. (1994). Effect of temperature and salinity on the oxygen consumption of laboratory produced *Penaeus vannamei* postlarvae. *Comparative Biochemistry and Physiology A* 108, 331–336.

Wajsbrot, N., Krom, M.D., Samocha, T.M. and Gasith, A. (1990). Effect of dissolved oxygen and the molt stage on the acute toxicity of ammonia to juvenile green tiger prawn *Penaeus semisulcatus*. *Environmental and Toxicological Chemistry* 9, 497–504.

Walker, P.J. and Mohan, C.V. (2009). Viral disease emergence in shrimp aquaculture: origins, impact and the effectiveness of health management strategies. *Reviews in Aquaculture* 1(2), 125–154.

Walker, S.J., Neill, W.H., Lawrence, A.L. and Gatlin, D.M. (2009). Effect of salinity and body weight on ecophysiological performance of the Pacific white shrimp (*Litopenaeus vannamei*). *Journal of Experimental Marine Biology and Ecology* 380, 119–124.

Wang, W.N., Li, B.S., Liu, J.J., Shi, L., Alam, M.J., Su, S.J. et al. (2012). The respiratory burst activity and expression of catalase in white shrimp, *Litopenaeus vannamei*, during long-term exposure to pH stress. *Ecotoxicology* 21(6), 1609–1616.

Wang, X., Li, E., Chang, X., Qin, J.G., Wang, S. and Chen, X. (2016). Growth, body composition, ammonia tolerance and hepatopancreas histology of white shrimp *Litopenaeus vannamei* fed diets containing different carbohydrate sources at low salinity. *Aquaculture Research* 47, 1932–1943.

Wang, X., Ma, S., Dong, S. and Cao, M. (2004). Effects of salinity and dietary carbohydrate levels on growth and energy budget of juvenile *Litopenaeus vannamei. Journal of Shellfish Research* 23, 231.

Wasielesky, W., Furtado, P., Poersch, L., Gaona, C. and Browdy. C. (2015). Alkalinity, pH and CO_2: Effects and tolerance limits for *Litopenaeus vannamei* superintensive biofloc culture system. An abstract of an oral presentation at Aquaculture America 2015, February 19–22, 2015, New Orleans, Louisiana, USA.

Watson, S.A., Fabricius, K.E. and Munday, P.L. (2017). Quantifying pCO_2 in biological ocean acidification experiments: A comparison of four methods. *PloS ONE* 12(9), e0185469.

Weiss, R.F. (1974). Carbon dioxide in water and seawater: the solubility of a non-ideal gas. *Marine Chemistry* 2(3), 203-215.

Whiteley, N.M., Scott, J.L., Breeze, S.J. and McCann, L. (2001). Effects of water salinity on acid-base balance in decapod crustaceans. *Journal of Experimental Biology* 204, 1003–1011.

Wickins, J.F. (1985). Ammonia production and oxidation during the culture of marine prawns and lobsters in laboratory recirculation systems. *Aquacultural Engineering* 4, 155–174.

Wurts, W.A. and Durborow, R.A. (1992). Interactions of pH, carbon dioxide, alkalinity and hardness in fishponds. SRAC publication.

Wilson, R.W. (2020). The effects of CO_2 and related water chemistry on fish: Climate change v. aquaculture. *Bulletin of the European Association of Fish Pathologists* 40(2), 49–54.

Wood, H.L., Spicer, J.I. and Widdicombe, S. (2008). Ocean acidification may increase calcification rates, but at a cost. *Proceedings of the Royal Society B* 275(1644), 1767–1773.

World Bank (2014). Reducing disease risks in aquaculture. World Bank Report no. 88257-GLB.

Wright, P.A. and Wood, C.M. (2009). A new paradigm for ammonia excretion in aquatic animals: Role of Rhesus (Rh) glycoproteins. *Journal of Experimetnal Biology* 212, 2303–2312.

Wu, R.S.S., Lam, P.K.S. and Wan, K.L. (2002). Tolerance to, and avoidance of, hypoxia by the penaeid shrimp (*Metapenaeus ensis*). *Environmental Pollution* 118(3), 351–355.

Wyban, J., Walsh, W.A. and Godin, D.M. (1995). Temperature effects on growth, feeding rate and feed conversion of the Pacific white shrimp (*Penaeus vannamei*). *Aquaculture* 138, 267–279.

Yoo, K.H. and Boyd, C. (1994). *Hydrology and Water Supply for Pond Aquaculture.* Springer Science + Business Media, Dordrecht, the Netherlands.

Zhang, K., Pan, L., Chen, W. and Wang, C. (2017). Effect of using sodium bicarbonate to adjust the pH to different levels on water quality, the growth and the immune response of shrimp *Litopenaeus vannamei* reared in zero-water exchange biofloc-based culture tanks. *Aquaculture Research* 48(3), 1194–1208.

Zhang, P.D., Zhang, X.M., Li, J. and Gao, T.X. (2009). Effect of salinity on survival, growth, oxygen consumption and ammonia-N excretion of juvenile white leg shrimp, *Litopenaeus vannamei. Aquaculture Research* 40, 1419–1427.

Zhang, W., Chen, B., Niu, C., Yuan, L., Jia, H. and Storey, K. B. (2019) Response of the Chinese soft-shelled turtle to acute heat stress: insights from the systematic antioxidant defense. *Front. Physiol.* 10:710.

Zhang, Z.F., Zhang, W.B. and Liu, L.P. (2012). Status and sustainability problems of white leg shrimp *Litopenaeus vannamei* culture in Zhanjiang [in Chinese with English abstract]. *Guangdong Agricultural Sciences* 14, 158–160.

Zhou, H., Li, Y., Wei, L., Zhang, Z., Huang, H., Diao, X. and Xiang, J. (2014). Sensitivity of larvae and adult and the immunologic characteristics of *Litopenaeus vannamei* under acute hypoxia. *Journal of Chemistry* 2014, 574534.

Chapter 12

SHRIMP FEEDING BEHAVIOUR

GUILLERMO BARDERA, JEAN-BENOÎT DARODES DE TAILLY, MARIA DE FÁTIMA ARRUDA AND CIBELE SOARES PONTES

Summary

This chapter provides an overview of feeding behaviour, which can play an important role in understanding feed efficiency problems in shrimp farming. An integrated ethogram outlines all behavioural units described from various studies, and their relationship to feeding activity is provided. Most behavioural studies in commercial shrimp species have been carried out under laboratory conditions, in order to enhance control over water (e.g. temperature, salinity), environmental (e.g. lighting, conspecifics) and individual conditions (e.g. moult status and feed deprivation). For this purpose, different methodologies have been developed and are outlined in three main sections in this chapter: test arena design, tagging techniques and protocols for recording and analysis of behavioural data. There is also a need for upscaling behavioural observations to commercial conditions, where parameters can considerably differ from those found in laboratory studies. Farmers need more accurate observations of feeding status in their day-to-day operations to further improve feeding efficiency. Therefore, the last part of this chapter focuses on techniques such as telemetry, passive acoustic monitoring and computer vision, which may provide a viable way to measure shrimp feeding behaviour within ponds, potentially overcoming the challenges of commercial conditions.

Introduction

The shrimp farming industry suffers from feeding inefficiencies which represent a cost from both economic and environmental standpoints. Feed can represent as much as half of production costs of commercial farms (Silva et al., 2012; Engle et al., 2017), and excess feeding can result in high feed conversion ratios, poor chemical and microbiological water quality and eventually to the eutrophication of the environment where wastewater is discharged (Smith et al., 2002; Jescovitch et al., 2018; Ullman et al., 2019). A better understanding of shrimp feeding behaviour and its application to feed management will help mitigate those issues.

Economic efficiency depends largely on feeds being quickly recognized and consumed due to the rapid leaching of nutrients. For feed manufacturers, the identification and incorporation of ingredients (i.e. feeding effectors or additives) to commercial feeds is needed to increase ingestion rates. To that end, careful characterization in laboratory conditions of shrimp feeding responses as sequential behavioural descriptors is of importance (Costero and Meyers, 1993; Lee and Meyers, 1996). Interpretation and quantification of behavioural responses in commercial shrimp has been achieved in experimental systems that simulate, under laboratory conditions, the farmed environment characteristics. Although behavioural information in commercial shrimp species is scarce, several studies have already evaluated novel feed effectors (e.g. Montoya-Martínez et al., 2018; Nunes et al., 2006; Suresh et al., 2011), as well as other potential influencers on shrimp feeding behaviour during production, such as circadian rhythms (Pontes et al., 2006), feeding frequencies (Lima et al., 2009) and moulting (Bardera et al., 2019a).

For shrimp farmers, extensive literature from the fish farming industry shows that closer and smarter monitoring and understanding of feeding patterns in culture conditions could help enhance feeding practices and animal welfare (as reviewed in Niu et al., 2018; Saberioon et al., 2017; Zhou et al., 2018; Zion, 2012). State-of-the-art monitoring technologies for underwater observation of animal behaviour are now coming into play and are already shaping the industry, helping farmers adopt feeding protocols that rely less on their own experience and more on quantifiable behaviours from closer and continuous monitoring. This new way of producing, referred to as precision farming, has seen considerable development in the salmon farming industry (e.g. Føre et al., 2011, 2018; Pinkiewicz et al., 2011), and it is now hoped that shrimp farming will follow the same path to reach greater efficiency and sustainability. The present chapter therefore aims at explaining the current knowledge about shrimp feeding behaviour and its future applications for the aquaculture industry.

Describing feeding behaviour

Shrimp can be defined as generalist feeders, because they utilize a wide range of food resources (e.g. live prey, algae, detritus) depending on their availability and on the shrimp life stage (i.e. larvae, juveniles or adults). They are also opportunist, spending large amounts of time scanning the bottom for food (Dall et al., 1990) and interacting with any kind of potential food sources (e.g. organic debris such as shrimp faeces or feed spoilage, cannibalism).

Although shrimp use vision and mechanoreception to identify food sources, their main way of detecting food is chemoreception because aquatic environments are an excellent medium for transmission of chemical cues (You et al., 2006). They respond to primary metabolites (e.g. sugars, proteins, amino acids, etc.) which is the most basic way to detect the presence of food (Hay, 2011). Shrimp possess a large number of chemoreceptors distributed around their body (i.e. cephalothorax, abdomen and telson) and appendages (i.e. antennae, legs, mouthparts) (Derby and Sorensen, 2008). However, the most complex chemoreception is located within the shrimp's antennules. This wide distribution allows shrimp to maximize their ability to locate chemical cues; the antennules are particularly sophisticated in that they allow the shrimp to discriminate chemical cues from food substances from background chemicals and to ascertain the quality and concentration of food (Derby and Steullet, 2001). This is possible thanks to antennule-flicking behaviours, which are characterized by rapid movements of antennules through the surrounding aquatic environment, allowing chemicals to bind repeatedly to receptor sites (Koehl, 2006;

Thiel and Breithaupt, 2011). However, the study of flicking behaviour is quite challenging in shrimp due to the small size of their antennules. In addition, shrimp use information from their mouthparts and pereiopods to locate food sources, by inserting and withdrawing pereiopods continuously into the substrate. Once they are closer to the food source, mechanoreception and visual detection become more relevant and provide additional information about feeding processes (Derby and Sorensen, 2008).

As shrimp feeding behaviour is highly influenced by chemoreception, most behavioural studies have investigated the implications of relevant chemical substances (i.e. feeding effectors or additives) to solve feeding problems in the industry. Lee and Meyers (1996) explained how different feeding effectors can mediate all phases of feeding in shrimp. It is possible to simplify feeding into two main phases: attraction phase (i.e. feed attractability) and consumption phase (i.e. feed palatability). The attraction phase is the time spent by shrimp detecting, orientating and travelling to the feed. In terms of feeding effectors, attractants and arrestants have a positive and negative effect on shrimp attraction, respectively. This phase relies on the accuracy for antennular and pereiopod chemoreception (Derby et al., 2001). The consumption phase is based on the degree of consumption of feed, evaluating the level of feed acceptance or rejection by shrimp. Food consumption is quantified by the time spent feeding and the amount of feed left. Stimulants and deterrents are the feeding effectors in charge of increasing and decreasing, respectively, consumption processes. Feed acceptance or rejection is guided by chemoreceptors and mechanoreceptors located on the dactyls of pereiopods, maxillipeds and the oral cavity (Steiner and Harpaz, 1987). Shrimp use dactyl chemotactic organs as initial feed evaluation before consumption. The feed-handling pattern does not guarantee that shrimp will consume the feed (Harpaz et al., 1987). It is inappropriate to assume that both phases will be directly associated, as shrimp may exhibit a high level of feed attraction but low consumption if feed palatability is poor. In the same way, good feed palatability could be useless if shrimp spend significant amounts of time trying to locate it due to low attractability performance. A summary of behavioural units used by different authors in relation to shrimp feeding behaviour is introduced in Table 12.1.

Marine shrimp exhibit clear changes in their behavioural profile when feed is offered. Although these changes in feeding behaviour may depend on water chemistry (e.g. salinity, temperature), environmental (e.g. photoperiod, stocking density) and individual-level effects (e.g. moult status, feed deprivation) (as reviewed in Bardera et al., 2019b), normally shrimp increase the frequency of behaviours related to detection, orientation and locomotion towards the food source. At the same time, behaviours not related to feeding such as inactivity, burrowing and grooming decrease. Feeding behaviour phases are described following the behavioural units given in Table 12.1.

INACTIVITY

Shrimp remains stationary on the substrate without appendage movements. Inactivity has been observed as the first behaviour to be reduced when food is offered. However, higher frequencies in inactivity are related to low motivation in feeding activity. For example, *Litopenaeus vannamei* shrimp were more inactive during light time (Pontes et al., 2006) and when fed seven times a day (Lima et al., 2009) and this was also negatively related to feeding activity. The same negative correlation between inactivity and feeding was found studying *L. vannamei* feeding behaviour in relation to different densities (Costa et al., 2016), feed deprivation and moult status (Bardera et al., 2019a).

TABLE 12.1 Ethogram of shrimp feeding behaviour

Behaviour	*Description*	*References*
Inactivity	Shrimp remains stationary on the substrate.	Bardera et al., 2019a; Costa et al., 2016; Lima et al., 2009; Pontes et al., 2006; Silva et al., 2012
Detection/ exploration	Shrimp in a steady position uses appendages, such as mouthparts, antennules, antennae and pereiopod dactyls, to search in the substrate.	Bardera et al., 2019a; Costa et al., 2016; Lima et al., 2009; Nunes et al., 2006; Pontes et al., 2006; Silva et al., 2012
Orientation	Shrimp moves to orientate side-on to feed without changing the position of their middle point.	Bardera et al., 2019a; Nunes et al., 2006
Locomotion/ crawling	Shrimp moves forward or backwards with position change.	Bardera et al., 2019a; Costa et al., 2016; Nunes et al., 2006; Silva et al., 2012
Feeding/ingestion	Shrimp uses pereiopod dactyls and maxillipeds to handle and eat food.	Bardera et al., 2019a; Costa et al., 2016; Lima et al., 2009; Nunes et al., 2006; Pontes and Arruda, 2005a,b; Pontes et al., 2008; Silva et al., 2012
Grooming/cleaning	Shrimp rubs pereiopods against carapace; maxilliped beats and antennae wipe.	Bardera et al., 2019a[*1]; Costa et al., 2016 ; Pontes et al., 2006
Attraction-to-feed/ latency-to-feed	Time taken by shrimp to reach feeding area for the first time.	Bardera et al., 2019a; Nunes et al., 2006[*2]; Pontes and Arruda, 2005a; Pontes et al., 2008
Burrowing	Shrimp remains partially or completely buried in the substrate.	Costa et al., 2016; Silva et al., 2012
Swimming/ fleeing	Shrimp moves horizontally or vertically in the water column, with no contact with the bottom of the tank; vigorous movements against the tank wall.	Bardera et al., 2019a; Costa et al., 2016; Lima et al., 2009; Pontes et al., 2006; Silva et al., 2012; Taylor et al., 2004

Notes: [*1]Bardera et al. (2019a) analysed it as the number of times shrimp beat their maxilliped or wiped their antennae during recordings. [*2]Nunes et al. (2006) defined it as locomotion and orientation (loco + orient) time used to reach the feeding area.

DETECTION

Shrimp in a steady position use their pereiopods and maxillipeds to locate chemical substances in the water from a food source. Higher levels of feed motivation are normally exhibited by shrimp striking their pereiopods more rapidly against the ground. However, in behavioural trials a negative relationship in frequency was noted between detection and feeding, as lower detection times are needed to locate and arrive to the food source when shrimp are more attracted to the feed (e.g. Bardera et al., 2019a; Costa et al., 2016; Nunes et al., 2006). After feeding, shrimp increase detection frequency in order to locate new food sources (Silva et al., 2012). In addition, if meals are more dispersed throughout the day (i.e. three meals per day), shrimp spend more time detecting than if meals are consumed closer together (e.g. seven meals per day) (Lima et al., 2009). Some studies have detected no differences in detection frequency between light and dark phases (Pontes et al., 2006), while others explain how detection is more frequent during the morning (Silva et al., 2012).

ORIENTATION AND LOCOMOTION

Orientation has been considered in only a few studies, and is characterized as the time spent by shrimp moving to the sides and becoming oriented to the feed source. Once the shrimp is orientated, locomotion is measured as the time used to search and find the feed. Normally orientation and locomotion behaviours increase as the shrimp senses the chemical stimuli of food in the surrounding water. However, similarly to detection, lower frequencies in locomotion and orientation will be correlated to higher feeding rates. This has been documented in several studies (e.g. Costa et al., 2016; Silva et al., 2012). Food location is related to shrimp size, because larger shrimp have been observed walking more rapidly and scanning larger areas of the ground than smaller individuals. Nunes et al. (2006) used orientation combined with locomotion to calculate the time required by shrimp to reach the feeding area (see attraction-to-feed behaviour). Bardera et al. (2019a) did not find any significant difference in orientation and locomotion when analysing these parameters independently, in relation to feed deprivation and moult status. For that reason, combining orientation and locomotion time for future trials is recommended.

FEEDING

Once a shrimp is close to the food source, it makes contact with the feed and moves towards it with the maxillipeds using the dactyl of the pereiopods. Feed manipulation using dactyls can be characterized by either slower or faster movements, which depend on the shrimp's motivation to feed. Several pellets are usually rejected by the shrimp, even when feeding activity is high and the same feed composition is presented. As previously commented, feed acceptance or rejection is guided by chemoreceptors in the dactyls and maxillipeds, which are used to assess feed quality. Although it depends on pellet size, shrimp use their pereiopods and maxillipeds to hold numerous pellets against the mouthparts due to the slow process of ingestion. Ingestion is indicated by the disappearance of feed from the mouthparts and its appearance in the stomach. Feeding activity can be quantified by the number of pellets eaten by shrimp, using underwater cameras (Bardera et al., 2019a, 2020b), hydrophones (Silva et al., 2019), measurement of feed leftovers (Santos et al., 2016) and by examining the stomach and gut content in vivo (Pontes et al., 2008) or dissecting it after the trial (Soares et al., 2005). Shrimp can identify and discriminate between different diets, with high decision consistency within a population (Nunes et al., 2006). Moreover, shrimp have also been noted to learn how to capture and handle different feed sizes (Nunes and Parsons, 1998) and develop a change in feed acceptability (Steiner and Harpaz, 1987).

GROOMING

After feeding, shrimp present a stereotypic behaviour referred to as antennular grooming (Schmidt and Derby, 2005). This behaviour has been attributed to the activation of binding regions for chemical substances in chemoreceptors through the antennules wiping against maxillipeds and the rubbing movements of maxillipeds against themselves ('autogrooming') (Wroblewska et al., 2002). During a behavioural trial, grooming frequencies are insignificant in comparison to the remainder of behavioural units (Bardera et al., unpublished data; Costa et al., 2016; Pontes et al., 2006). Grooming can also be related to stress responses in shrimp, as studies have discovered that high rates in grooming are correlated to stressful stages such as post-moult (Bardera et al., 2019a) and lower levels of feeding activity (Bardera et al., 2019a; Costa et al., 2016).

ATTRACTION-TO-FEED

This behaviour is specifically used when the time taken by a shrimp to arrive at a feeding area needs to be quantified. A rapid approach to the feeding area does not mean that the shrimp will feed more, as it can reject the feed and leave it immediately. Food rejection is normally observed as shrimp moving around the feed but with little or no interaction happening. On other occasions shrimp can reach the feeding area very quickly due to physiological conditions, the presence of conspecifics and/or good feed attractability. However, if feed palatability is poor, the shrimp will start searching for new sources of food. If any other food source is available, the shrimp will stop eating that feed, which has lost some composition due to the time elapsed since feeding was first initiated. However, Bardera et al. (2019a) and Nunes et al. (2006) have demonstrated that a stable positive correlation between attraction-to-feed and feeding is generally observed, meaning that attraction-to-feed can be used to assess feed attractability in shrimp.

BURROWING

Shrimp use pleiopods to scour the substrate and bury themselves, which helps in saving energy and avoidance of predators. No territorial habits are exhibited by shrimp in relation to burrowing, as they burrow in different places each day. Burrowing behaviour changes depending on shrimp size, age, substrate composition and lighting conditions (Kenyon et al., 1995). Shrimp size and burrowing increase accordingly, with animals of 10–15 cm being buried for longer times. Juvenile shrimp are affected more by substrate composition, while lighting conditions primarily affect the burrowing behaviour of adult shrimp. Most penaeids spend the day buried in the sediment and emerge to feed at night (Dall et al., 1990). However, burrowing habits may differ depending on shrimp species and populations. *Penaeus monodon* populations have been observed to shift their burrowing behaviour during their culture (Primavera and Lebata, 1995), while *L. vannamei* spent only the beginning of the light phase buried until first feed was offered (Costa et al., 2016). Under commercial conditions, a plastic liner is often present at the bottom of the ponds, which does not allow shrimp burrowing. This may present a negative impact on shrimp stress and subsequent growth, especially when shrimp are vulnerable due to ecdysis. The introduction of artificial shelters, distributed around the pond, may be considered as a viable solution to deal with this issue.

SWIMMING

Shrimp move both horizontally and vertically in the water column using the pleiopods. It is a common behaviour exhibited by shrimp population in ponds, more often present during the light phase (Silva et al., 2012) and when high densities are used for the culture (>100 shrimp/m^2) (Costa et al., 2016), as shrimp use more of the water column due to lack of space on the pond bed. However, in behavioural trials, high levels of swimming frequency have been shown to be associated with stress, with shrimp exhibiting erratic swimming behaviour and crashing against the walls of the tank (Bardera et al., 2019a; Taylor et al., 2004), which is sometimes presented as a stereotypical behaviour. Normally when this happens the shrimp do not exhibit any interest in the feed and the trial needs to be stopped.

Measuring feeding behaviour under laboratory conditions

To date, most of the research on shrimp behaviour has been done at a small scale under laboratory conditions. As behaviour is characterized as being a high plastic feature, laboratory conditions allow more control over several important factors (i.e. individual-level, environmental, water chemistry) that may affect shrimp feeding behaviour (Bardera et al., 2019b). Interpretation and quantification of feeding behaviour in commercial shrimp has been assessed in experimental systems that simulate farmed environments within laboratory conditions. However, recommendations taken from these studies need to be considered with caution when applied to commercial conditions.

TEST ARENA DESIGNS

Under laboratory conditions, different test arena designs and methodologies have been developed in relation to the behavioural study of interest. Test arenas are normally made of clear glass or polycarbonate, to provide good visibility of subjects; however, walls can be covered by black vinyl to reduce disruption to shrimp during the trial. Although it is not common to all studies, an acclimation chamber is normally included in the arena. This is essential when shrimp have been moved from another holding tank prior to starting the trial. When holding tanks act as test arenas, experimental shrimp normally go through an acclimation process from 1 to approximately 7 days inside those tanks prior to starting measurement of behaviour. Acclimation periods allow shrimp to adapt to experimental settings (i.e. novel environment, slight differences in lightning and/or water parameters) and to reduce the stress produced by their capture and transport. Acclimation times differ between studies, but 10–20 minutes is deemed appropriate. Acclimation chambers have a removable guillotine door to keep shrimp contained during acclimation periods; once complete, the guillotine door is lifted to release the subjects. It is important that environmental conditions (i.e. lightning) and water parameters are the same as, or as close as possible to, those of the holding tanks, otherwise this may increase stress in experimental shrimp, affecting behavioural data derived from the trials. In addition, in those cases where subjects need to be transferred to test arenas, it is not recommended to use the same subjects for more than two trials per day. On the contrary, if subjects remain in the same tanks for the whole experiment (i.e. holding tanks used as test arena), this recommendation is not needed. In the following, we introduce some of the designs used in shrimp feeding behaviour studies.

Y-MAZE

Test arenas of this design are composed of a rectangular tank divided into two arms at one end that diverge between 45 and 120°, forming a Y-shaped maze. This arena is normally used for testing conditions that may affect animals' learning and memory capabilities. In shrimp aquaculture, these arenas have been used to test dietary choice between different feeding effectors in relation to shrimp feeding behaviour. Experimental feeds are placed in both arms of the Y-maze, and either an individual or a group of shrimp is placed at the bottom of the Y, where the acclimation chamber is located. The Y-shape also allows a considerable distance between feed effectors, avoiding problems associated with mixture of chemical cues at the entrance of both feeding areas. Pittet et al. (1996) designed a Y-maze (base: 30 x 20 x 7.5 cm; arms: 15 x 10 x 7.5 cm (length x width x height)) with a reduced angle (i.e. 10°) between arms to reduce water turbulence. They placed *L. vannamei* within the acclimation chamber (time not specified in publication) and compared the preference and time used to reach feed with different effectors. Nunes et al. (2006), using a Y-tank (base: 100 x 30 x 40 cm; arms: 30 x 15 x 40 cm), placed individual *L. vannamei* within the acclimation chamber for 10 minutes and, after the guillotine door was removed, the time spent arriving at one of the feeds was quantified (Fig. 12.1(a)). Montoya-Martínez et al. (2018) (only the total length of the Y-maze was provided, 145.6 cm) positioned ten *Macrobrachium tenellum* in the acclimation chamber for 60 minutes prior to the start of the experiment, after which time shrimp were observed for 15 minutes and the time required to reach the feeding areas and the number of entrances were measured. All studies concluded that the Y-maze is a valid design for assessment of feeding effectors in shrimp.

Rectangular arena with different chambers

The experimental arena is formed by a rectangular tank, with an acclimation chamber at one end and several chambers (i.e. two to five chambers) at the other. Each chamber presents an opening to allow shrimp to access the feed placed in the chamber. Like Y-mazes, this design is focussed on dietary choice studies. In arenas with more than two feeding chambers there is a potential effect that shrimp will be more likely to choose those located on the sides of the tank. This is due to shrimp often moving along the edges in rectangular arenas, seeking protection (Bardera et al., unpublished data). However, this effect can be sorted if feeds are randomly assigned in the different chambers across recordings. Suresh et al. (2011) assessed the level of attraction to different feed effectors using rectangular tanks (90 x 30 x 30 cm) with three separate chambers in the end part of the arena (30 x 10 x 30 cm; Fig. 12.1(b)). Each feeding chamber had an opening to allow the shrimp to freely access the feed placed in the chamber. Ten *Litopenaeus stylirostris* were introduced to the acclimation chamber for 1 hour before adding feeds (1 g of different experimental feeds in each chamber). An additional 10 minutes were added with feed inside feeding chambers, then the guillotine door was removed and the number of shrimp in each of the feed chambers was counted at 1, 2, 5, 10 and 15 minutes. They found that shrimp needed only 5–10 minutes to discriminate among feeds. Jaime-Ceballos et al. (2007) and Montoya-Martínez et al. (2018) used rectangular arenas (75 x 40 x 40 cm) divided by two internal barriers, with an opening in the upper part of the barriers connecting the three chambers (25 x 40 x 25 each; Fig. 12.1(c)). Ten *Litopenaeus schmitti* (Jaime-Ceballos et al., 2007) or *M. tenellum* (Montoya-Martínez et al., 2018) were introduced in the middle chamber. These behavioural trials, comparing different feed effectors, started without an acclimation period. Two experimental diets were compared, positioning each within one of the two chambers. Jaime-Ceballos et al. (2007) observed feed preference after 5 minutes and Montoya-Martínez et al. (2018) after 15 minutes.

Rectangular arena with one chamber

This arena is also composed of a rectangular tank, with an acclimation chamber at one end and one feeding area only at the other end of the tank. Within the feeding area, one or more types of feed can be added during the behavioural trial. This design allows a more accurate analysis of shrimp feeding behaviour under different physiological or environmental conditions, and their effects on feeding area and/or specific diet type. Costero and Meyers (1993), using this arena design (125 x 25 x 50 cm), tested *L. vannamei* preference in response to two diets, with and without the addition of a commercial chemoattractant. Results showed how the presence of a chemoattractant reduced response times to feed by 50%. In addition, they discovered feed conditioning processes during 6 days of study. A greater number of shrimp arrived at the feeding area, and higher ingestion rates of both diets occurred in the final days of the experiment. Grey et al. (2009) tested feed preference in 55-litre rectangular tanks. These tanks contained two circular feeding areas where two different diets were added (reference diet and test diet). Ten *L. vannamei* were placed in the test arenas and, after 1 hour, the difference between the number of pellets consumed from the test and reference diets was used as the behavioural response. Bardera et al. (2019a) studied the effect of feed deprivation and moult status on *L. vannamei* feeding behaviour (arena 95 x 30 x 15 cm; Fig. 12.1(d)). Shrimp were individually placed in the acclimation chamber (10 x 30 x 15 cm) of the arena for 10 minutes and, after removal of the guillotine door, feeding behaviours were recorded for 20 minutes. Results from this study showed that plastic feeding behaviour appeared to be dependent on the physiological status of shrimp. Shrimp changed their behavioural repertoire, with those under 48-hour feed deprivation and inter-moult status showing the highest interaction rates with feed and reducing the frequency of other behaviours not related to feeding motivation (i.e. inactivity, fleeing).

Circular arenas

Circular open tanks are used to quantify shrimp feeding behaviour, normally without acclimation chambers or any other division within. This type of arena is useful to test behaviour in groups of shrimp and allows analysis of different aspects of behaviour such as feed attraction, feed palatability and interaction between shrimp within groups. The curved walls in this arena also enhance shrimp exploratory activity around the tank. Sanchez et al. (2005) used 5-metre-diameter circular arenas for quantification of *L. vannamei* feeding stimulation. Groups of shrimp were stocked at a density of 2.5, 5.0 and 7.5 shrimp/m^2. To assess feed preference between two commercial feeds, four feeding trays (56 x 30 x 4 cm) were introduced in the arenas (two trays per feed). Behavioural observations were conducted at 1 and 2 hours post-addition of feed in feeding trays. Feed preference was evaluated *via* the number of shrimp on each tray. Suresh et al. (2011) used cylindrical tanks of 50 litres to assess the palatability of feeds in groups of seven *L. stylirostris* in 1-hour observations. At the end of each observation, shrimp were collected from the tank and their mass weighed. Feed leftovers and faeces were also collected and dried in an oven, to calculate the quantity of feed consumed during the trial. Bardera et al. (2021) used circular arenas of 69-cm diameter (75-litre volume) to study *L. vannamei* feeding behaviour with different density groups (6.2, 12.4 and 24.8 shrimp/m^2) (Fig. 12.1(e)). An internal structure was used to form eight acclimation chambers designed to introduce experimental shrimp, one in each, for 10 minutes before starting the recordings. The structure was removed at the start of the trial, releasing all shrimp simultaneously into the whole test arena for each recording. The feeding area (24-cm diameter) was located in the middle of the tank. Feed was added and all acclimation chambers were removed at the same time when recording was started. Shrimp were recorded for 20 minutes, and interactions with feed and among subjects were assessed.

Other arenas

Alternative designs in experimental tanks have been used in order to answer specific questions in relation to shrimp feeding behaviour. Santos et al. (2016) designed rectangular tanks of 40-litre volume with several infrared photocells to study shrimp locomotor and feeding activity in relation to circadian rhythms (Fig. 12.1(f)). Shrimp are not sensitive to infrared wavelengths, which also enables behavioural observations overnight. The feeding area was monitored via an infrared photocell connected to an automatic feeder. This way, each time a shrimp approached the feeding area the sensor was activated, triggering the feed supply. This study demonstrated how shrimp change their feeding activity throughout the circadian cycle and how they can learn to use a self-feeding system. Derby et al. (2016) assessed the level of chemoattraction of krill meal extract (KME) using shrimp behaviour. KME was added to 60-litre tanks as a dilution (i.e. without

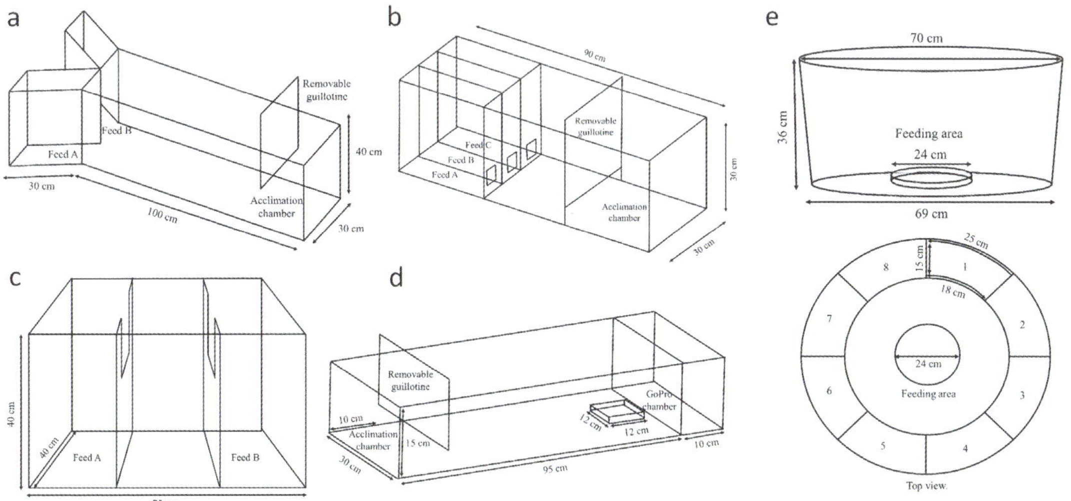

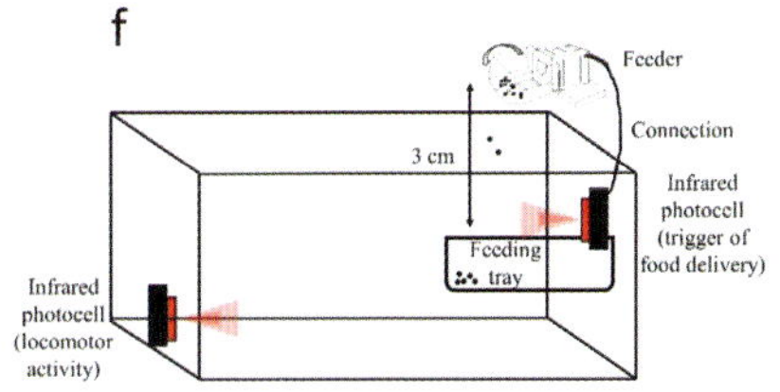

FIGURE 12.1 Experimental arena designs. (a) Y-maze, based on Nunes et al. (2006); (b), (c) rectangular arena with different chambers, based on Suresh et al. (2011) and Jaime-Ceballos et al. (2007); (d) rectangular arena with one chamber, based on Bardera et al. (2019a); (e) circular arena with acclimation chambers, based on Bardera et al. (2021); (f) rectangular arena with infrared photocells. Based on Santos et al. (2016).

feed pellets) with concentrations ranging from 1.33 µg/ml to 13.3 mg/ml. Between 20 and 25 shrimp were introduced to each tank, and different dilutions of KME were added to the tanks using a pump and air-stone. Behavioural responses were monitored for 4 minutes as the number of times shrimp touched or held the air-stone with pereiopods or mouthparts. In addition, instances of when shrimp stopped probing or grabbing the air-stone for >2 seconds were also recorded. Results showed a direct correlation between higher concentrations of KME and higher probing and grabbing events during the trial. However, relatively low KME concentrations (13.3 µg/ml) were enough to elicit a response.

TAGGING SHRIMP

As feeding behaviour can be highly influenced by individual differences, there is often a need to identify individual shrimp within a population. Tagging can be also useful in studies evaluating feeding behaviour in tanks with increased number of individuals, allowing the selection of a given number of shrimp and the extrapolation of behavioural data applicable to the whole population (i.e. focal sampling). The tagging process may disturb and generate stress, as handling of specimens is necessary. Furthermore, careful and speedy handling of individuals is paramount when undertaking such procedures. It is also important to establish before starting the study that tags are not affecting the shrimp's behaviour, and thus an observation period of at least 24 hours prior to starting the experiment is required. In addition, tagging shrimp is a more complex endeavour than tagging fish, as tags need to be designed to remain in the shrimp after ecdysis. Below, some ideas that can be used for shrimp tagging in behavioural trials are presented.

Individual holding tanks

Shrimp can be placed in numbered individual tanks during the experiment. This approach allows individuals to be easily recognized within the entire experimental population throughout the study. However, this is only possible for those behavioural trials seeking to analyse behaviour of individual shrimp. Shrimp can be moved from individual chambers to test arenas every time a behavioural observation is taking place. In addition, having shrimp individually allocated affords the opportunity to monitor the moulting cycle and prevent negative interactions among individuals, which can damage their appendages during the study (see Bardera et al., 2019a) (Fig. 12.2a).

External tags

Visible tags on the outside of the shrimp offer easy visual identification. These kinds of tags are a cheap and quick way to mark shrimp populations and are commonly used in breeding programmes. One example is the use of nylon filament tags which are attached to the thoraco-abdominal muscles and do not interfere with shrimp moulting or growth (Schmalbach et al., 1994). Alternatively, circular rings of various colours or identifiable code can be attached to the shrimp's eyestalk, behind the bulbous portion of the compound eye. These rings are normally made of plastic or metal, similar to those used for tracking birds (Fig. 12.2b). Rings are well tolerated by shrimp and retained through the moulting cycle. In addition, using this tagging method, shrimp do not present any abnormal behaviour during experiments (Bardera et al., 2021; Pontes and Arruda, 2005b).

Internal tags that are visible externally

Some tags may be placed in the subcutis and muscle of shrimp but are still visible to the naked eye. Visible implant tags (VIT) are made of coloured plastic strips (i.e. PVC tags). These tags are normally 2–5 mm long and 1 mm in diameter, and are inserted using forceps through the cuticular articulation membrane into the abdominal musculature under the exoskeleton. Another example is the use of visual implant elastomer (VIE) tags (Fig. 12.2c), which has been gaining popularity in behavioural trials with fish (e.g. Delcourt et al., 2011). This offers a relatively simple and inexpensive way of tagging all sizes of shrimp (i.e. juveniles and adults). VIE tags consist of a two-part, medical-grade liquid mixture which, when mixed, hardens into a pliable solid. The elastomer tagging mix is injected in liquid form and remains that way for several hours before fully curing. VIE tags are available in ten different colours, some with fluorescent material. In studies tagging shrimp using VIE tags, it was found that tag retention was 99.9% in juveniles and 100% in adults after a 10–14-week period. In addition, red-coloured tags were found to be the easiest and lime-coloured the most difficult to identify (Godin et al., 1996). VIE tags have been used only for hatchery operation purposes to identify broodstock in tanks, so their application in behavioural studies needs to be addressed in future studies.

Internal tags

A passive integrated microchip transponder (PIT) tag (Fig. 12.2d) consists of a small integrated microchip and electromagnetic coil encased in glass. These tags are inserted in the abdominal muscle under the exoskeleton using a veterinary syringe, and offer a non-invasive and non-destructive way of identification.

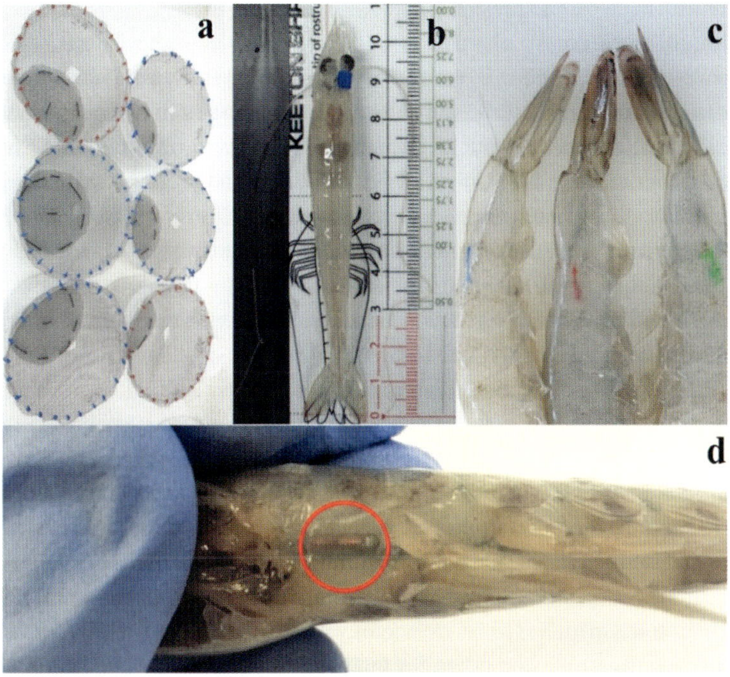

FIGURE 12.2
Different methodologies used for tagging shrimp. (a) Individual holding tanks; (b) eyestalk ring; (c) VIE tags; and (d) PIT tag, location indicated by the circle.

Penaeus monodon populations tagged using PIT tags presented with lower mortality (8%) than when using external eye stalk rings (14%) (Foote et al., 2018). They also need to be smaller than those utilized by fish research; PIT tags 5–7 mm long and 1–2 mm in diameter have been used in shrimp breeding programmes (Caceci et al., 1999; Foote et al., 2018). These tags are dormant until activated inductively by a scanning device, which generates power for the tag to transmit a unique alpha-numeric code (i.e. a barcode). This way, PIT tags may last throughout the entire life of the shrimp and the system allows rapid collection of transmitted data from a large group of tagged individuals. Scanners are available as either portable hand-held readers (10–15 cm) or automated models that are usually used for automated scanning. Automated readers can be either tunnel detectors or detector plates, and can be placed on the bottom of tanks or in a designated area such as feeding stations. These kind of tags can be especially useful when studying shrimp behaviour under commercial conditions, where high water turbidity often impedes stock observation during its development (Smith and Tabrett, 2013). Shrimp movements and distribution around ponds, as well as preference between feeding stations or feeding rhythms, can be assessed this way (see telemetry section).

TECHNIQUES FOR RECORDING AND ANALYSIS OF SHRIMP BEHAVIOUR

Before starting behavioural trials, a pre-test is essential to estimate the duration of observations needed to answer the question of interest. Duration of observation will also depend on the frequency of the analysed behaviour and the sample size used in the study. Behaviour can be monitored at the group or individual level. Alternative means of recording and analysing behaviour are introduced below.

Observer recording

An observer is present during behavioural trials. This method can be used to analyse concise behaviours (e.g. time required by shrimp to arrive at feeding area, time spent feeding), and it is possible to have more control over the trial and allow early termination of the test if the subject presents low interaction rates with the experimental settings. It is important to have shrimp previously acclimated to experimental conditions and habituated to the presence of the observer before data collection (Pontes et al., 2006) to avoid negative interactions, such as startle responses, and any disturbance to the shrimp during the trial. Behaviours must be monitored and recorded in real time during the observation period; good examples for this methodology are presented by Nunes et al. (2006) and Pittet et al. (1996).

Video recording

It may also be possible to utilize video recording equipment to monitor all behaviours, which enables more detailed analysis and minimizes disturbance to shrimp during the trial. There is a growing number of behavioural studies based on video recording methods, as it is a cheaper and more reliable way to analyse behaviour. Cameras can be positioned either above or inside the tanks (i.e. underwater cameras) and record either individuals or groups of shrimp. The observer can leave the room where the trial is taking place, returning when recording is finished. This methodology allows an exact visual record of the behaviour, which can also subsequently be slowed down for analysis of more detailed behaviours.

Video recordings are useful for studying behaviour that is too rapid or complex to analyse in real time (Martin and Bateson, 2007), and data can be stored indefinitely. However, it is necessary to spend more time before starting the experiment to confirm that recording conditions are optimal for video analysis (e.g. angle of cameras, lightning, water reflection, tags recognition). In addition, this methodology does not permit control or early termination of the trial if necessary, as the observer is not present during the recording. There are two options for analysing behaviour using video footage.

Manual observations

Behaviours are analysed from video recordings via a human recorder. While this method still provides accurate and complex behavioural information, it is quite time-consuming. Various observers are needed to analyse videos in order to minimize the effect of observer subjectivity. In addition, it is necessary to take time off between observations in order to prevent observer fatigue, which can negatively influence results derived from the analysis. There are software programmes available to help logging behavioural events in video recordings, one of them being BORIS (Behavioural Observation Research Interactive Software; Friard and Gamba, 2016). This software has already been used to analyse *L. vannamei* feeding behaviour in several studies (Bardera et al., 2019a, 2020b).

Automated observations

Video tracking software may be considered a quicker and more feasible way to analyse behaviour. However, it is more complex than manual observation until the observer is familiar with the software and recording protocols are optimized. It can minimize problems associated with observer subjectivity or fatigue and provide extra behavioural information that cannot be derived from manual observations alone (e.g. animal velocity, distance travelled during the recording). Different software programmes can be used for video tracking and to generate automated responses in relation to feeding behaviour. Videos can be analysed either in real time or from archived video recordings. A good example of video tracking software specialized in animal behaviour is EthoVision XT (Noldus Information Technology, Inc.), for which the analytical reliability of *L. vannamei* feeding behaviour has been validated, with an analysis speed 12 times faster compared to manual observations (Bardera et al., 2020a).

Measuring feeding behaviour under commercial conditions

Today, the assessment of feeding activity in commercial ponds is still heavily reliant on feeding trays, which are lowered onto the bottom of ponds to gain a rough estimate of feed consumption. In addition to being labour intensive, readings from feeding trays remain largely dependent on the observer and provide only past estimates of consumption at a given location (Reis et al., 2019; Smith and Tabrett, 2013). Shrimp are benthic feeders and pond conditions are often highly turbid, which makes observations with the naked eye from the surface difficult or impossible (Smith and Shahriar, 2013; Smith and Tabrett, 2013). However, techniques such as telemetry, passive acoustic monitoring and computer vision can provide in-depth information about movements and current feeding status of animals within ponds.

TELEMETRY

Telemetry is defined as the action of measuring from a distance with electronic devices (Thorstad et al., 2013; Wolcott, 1995), and has been extensively used in aquatic research for fish identification and tracking such as for smolts in rivers (e.g. Dittman et al., 2010; Gerlier and Roche, 1998). Current applications of telemetry go far beyond tracking: this technique is now commonly used to record environmental parameters in tagged individuals (e.g. Davidson and Hussey, 2019) or even noise production in crustaceans (Zenone et al., 2019). However, the use of telemetry techniques for penaeid shrimp can represent a challenge, since they undergo regular moulting processes during which tags can be lost (Freire and González-Gurriarán, 1998) and their small size in commercial conditions does not allow the use of large tags (Wolcott, 1995). For example, active radio and acoustic tracking has proven useful in studying fish movements in the wild or in aquaculture facilities (e.g. Føre et al., 2011; Thorstad et al., 2013). Those techniques enable positioning of individuals using portable antennas or fixed receivers to buoys or to the seafloor (Skerritt et al., 2015; Taylor and Ko, 2011). However, the weight of tag batteries limits their use to large individuals and can negatively affect behaviour (Wolcott, 1995). PIT tags, on the other hand, represent an inexpensive method for tracking a potentially large number of individuals at the same time. Since PIT tags do not have a battery they are small in size (a few millimetres up to 2 centimetres in length) (Thorstad et al., 2013), and the insertion of those devices in penaeid shrimp has already been documented for identification purposes for broodstock in commercial hatcheries (see Caceci et al., 1999 and Foote et al., 2018 for technical considerations) in which good tag retention after moulting was achieved (Caceci et al., 1999). However, the impact of PIT tags on shrimp behaviour has never been accurately assessed. In the mud crab *Scylla serrata*, a combination of PIT tags and T-bar anchor tags has also been documented, the tag therefore being outside the body of the crab (Meynecke et al., 2015). In ponds, PIT tags could be used for tracking a large number of individuals in combination with reading gates (i.e. a fixed copper inductor coil that can be powered by a battery) set at strategic locations – for example, close to feeding stations – to assess daily movement patterns in relation to feeding operations. However, if PIT tags are to be used on a commercial farm, strict procedures should be in place to ensure that no tags are found inside the final product for the consumer (Meynecke et al., 2015) – through the use of a metal detector on the production line, for example.

PASSIVE ACOUSTIC MONITORING

Like many other crustaceans, penaeid shrimp are soniferous species, which means they produce sounds while displaying different behaviours. Contrary to electromagnetic signals, sound propagates well underwater and its propagation is independent from turbidity levels. Penaeid shrimp have the interesting property of producing a specific 'clicking' sound while feeding, characterized on a power spectrum as a 'feeding signature' (Silva et al., 2019; Smith and Tabrett, 2013). This particular sound is caused by the closure of the calcified mandibles while shredding the feed (Silva et al., 2019; Smith and Tabrett, 2013). Feeding signatures of *L. vannamei* and *P. monodon* while consuming feed pellets have been characterized under laboratory conditions, and can be used as proxies of feeding activity (Silva et al., 2019 and Smith and Tabrett, 2013, respectively). In ponds, a change in the number of recorded feeding signatures through a hydrophone after consecutive feeding events can indicate variation in feeding activity (Smith and

Shahriar, 2013), which represents the operating principle of automatic acoustic feeders available on the market. Such systems are now in operation in major shrimp-producing regions, with papers reporting significant improvements in yields compared to timer feeders or hand-feeding protocols, by ensuring that feed is offered at the time shrimp are most likely to consume it (Jescovitch et al., 2018; Ullman et al., 2019). Apart from the practical applications for farmers, acoustic feeders could also be used as tools in studies monitoring feeding behaviour over a large scale in ponds, by providing cues of feeding response over time and across various locations if several feeders and hydrophones are deployed.

COMPUTER VISION

Computer vision, or machine vision, systems have the advantage of being unintrusive and have the potential to detect small changes in behavioural patterns that an observer could not see otherwise (Niu et al., 2018). Camera observations of shrimp in their habitat in aquaculture ponds represent a challenge, since water turbidity can be very high (>100 NTU reported in commercial ponds of *L. vannamei* in Krummenauer et al., 2014), which results in poor visibility and contrast (Schettini and Corchs, 2010). To enhance the quality of underwater footage, defogging and contrast-enhancement algorithms can be applied (e.g. He et al., 2011). Since decapods have been shown not to be sensitive to high light wavelengths (Johnson et al., 2002), near-infrared light sources associated with cameras sensitive to those wavelengths can also be used at the bottom of shrimp ponds to observe animals under low light conditions or even at night (e.g. Hung et al., 2016). Stereo vision (two cameras at different angles recording the same scene) and structured light technologies (the combination of a colour camera and a light projector), such as Microsoft Kinect, can even provide information on the three-dimensional structure of the scene, which allows animal positioning in space and the determination of school structures (e.g. Murphy et al., 2019; Saberioon and Cisar, 2016).

From underwater footage, the automation of observation has taken a large step forward with the emergence of deep learning algorithms for object recognition based on neural networks such as YOLO (You Only Look Once; Redmon et al., 2016). This software and similar ones have already been used for the automatic recognition of fish and shrimp both in their natural environment (e.g. Li et al., 2016; Pedersen et al., 2019) and in aquaculture (e.g. Arvind et al., 2019; Huang et al., 2018). After some training with frames containing the organisms to be observed, deep learning algorithms can achieve good detection accuracy (and speed) and provide coordinates of individuals, thereby allowing the calculation of various metrics such as the number of detected individuals, the distance between them and their respective velocities.

Alternatively, feed detection at the bottom can be automatically achieved through computer vision with image segmentation techniques, thereby providing the farmer with information on feeding activity (as developed for feeding trays in shrimp ponds by Chirdchoo and Cheunta (2019). Differentiating feed from other pixels allows software to assess whether the surface of the feeding tray is still covered by feed within camera range, which has strong commercial applications.

Conclusion

Research in shrimp feeding behaviour can be summarized in two key questions:

- Which factors are influencing feeding behaviour and to what extent?
- How do you efficiently monitor feeding behaviour under commercial conditions and interpret observations for a better feeding management?

Feed often represents the highest production cost on farms, yet today there is a real paradox for the shrimp industry between the economic importance of feeding and the paucity of knowledge surrounding feeding behaviour in aquaculture. Although still in its infancy, laboratory observations have provided most of the results on the influence of physiology and environmental factors on feeding behaviour, which already has strong implications for the industry. Laboratory studies must be based on clearly defined ethograms to allow standardization and repeatability of observations. The future of behavioural bioassays lies in the adoption of tracking software to go further in results interpretation and allow observers to save a considerable amount of time.

On shrimp farms, a major challenge is how to both adapt and adopt pre-existing technologies in a production context to provide a deeper understanding of behaviour in ponds. Assessment of feeding activity on farms should become more of a science based on improved underwater observations and smarter automatic feeders. Some of the techniques presented in this chapter are already in use in most advanced facilities, but others still need to be developed as real decision-support tools for routine feeding operations. Research should also focus on the interpretation of behaviours in order to translate observations into feeding status indicators, which in turn will drive the industry towards better economic performance, animal welfare and sustainability.

References

Arvind, C.S., Prajwal, R., Bhat, P.N, Sreedevi, A. and Prabhudeva, K.N. (2019) Fish detection and tracking in pisciculture environment using deep instance segmentation. In: *TENCON 2019 – 2019 IEEE Region 10 Conference (TENCON).* Presented at the TENCON 2019 – 2019 IEEE Region 10 Conference (TENCON), pp. 778–783.

Bardera, G., Owen, A.G., Pountney, D., Alexander, M.E. and Sloman, K.A. (2019a) The effect of short-term feed-deprivation and moult status on feeding behaviour of the Pacific white shrimp (*Litopenaeus vannamei*). *Aquaculture* 511, 734222.

Bardera, G., Usman, N., Owen, M., Pountney, D., Sloman, K.A. and Alexander, M.E. (2019b) The importance of behaviour in improving the production of shrimp in aquaculture. *Reviews in Aquaculture* 11, 1104–1132.

Bardera, G., Owen, M.A.G., Façanha, F.N., Alcaraz-Calero, J.M., Sloman, K.A. and Alexander, M.E. (2020a) Assessing feed attractability in Pacific white shrimp (*Litopenaeus vannamei*) using an automated tracking software. *Aquaculture* 529, 735692.

Bardera, G., Owen, M.A.G., Façanha, F.N., Sloman, K.A. and Alexander, M.E. (2020b) The influence of sex on feeding behaviour in Pacific white shrimp (*Litopenaeus vannamei*). *Applied Animal Behaviour Science* 224, 104946.

Bardera, G., Owen, M.A.G., Façanha, F.N., Alcaraz-Calero, J.M, Alexander, M.E. and Sloman, K.A. (2021) The influence of density and dominance on Pacific white shrimp (*Litopenaeus vannamei*) feeding behaviour. *Aquaculture* 531, 735949.

Caceci, T., Smith, S.A., Toth, T.E., Duncan, R.B. and Walker, S.C. (1999) Identification of individual prawns with implanted microchip transponders. *Aquaculture* 180, 41–51.

Chirdchoo, N. and Cheunta, W. (2019) Detection of shrimp feed with computer vision. *Journal of Thai Interdisciplinary Research* 14, 13–17.

Costa, F.P., Gomes, B.S.F., Pereira, S.D.N.A. and Arruda, F.M. (2016) Influence of stocking density on the behaviour of juvenile *Litopenaeus vannamei* (Boone, 1931). *Aquaculture Research* 47, 912–924.

Costero, M. and Meyers, S.P. (1993) Evaluation of chemoreception by *Penaeus vannamei* under experimental conditions. *The Progressive Fish Culturist* 55, 57–162.

Dall, W., Hill, B.J., Rothlisberg, P.C. and Staples, D.J. (1990) *The Biology of the Penaeidae. Advances in Marine Biology*. Academic Press, London, UK.

Davidson, E.R. and Hussey, N.E. (2019) Movements of a potential fishery resource, porcupine crab (*Neolithodes grimaldii*) in Northern Davis Strait, Eastern Canadian Arctic. *Deep Sea Research Part I: Oceanographic Research Papers* 154, 103143.

Delcourt, J., Ylieff, M., Bolliet, V., Poncin, P. and Bardonnet, A. (2011) Video tracking in the extreme: A new possibility for tracking nocturnal underwater transparent animals with fluorescent elastomer tags. *Behaviour Research Methods* 43, 590–600.

Derby, C.D. and Steullet, P. (2001) Why do animals have so many receptors? The role of multiple chemosensors in animal perception. *The Biological Bulletin* 200, 211–215.

Derby, C.D. and Sorensen, P.W. (2008) Neural processing, perception, and behavioural responses to natural chemical stimuli by fish and crustaceans. *Journal of Chemical Ecology* 34, 898–914.

Derby, C.D., Elsayed, F.H., Williams, S.A., González, C., Choe, M., Bharadwaj, A.S. and Chamberlain, G.W. (2016) Krill meal enhances performance of feed pellets through concentration-dependent prolongation of consumption by Pacific white shrimp, *Litopenaeus vannamei. Aquaculture* 458, 13–20.

Derby, C.D., Steullet, P., Horner, A.J. and Cate, H.S. (2001) The sensory basis of feeding behaviour in the Caribbean spiny lobster, *Panulirus argus. Marine and Freshwater Research* 52, 1339–1350.

Dittman, A.H., May, D., Larsen, D.A., Moser, M.L., Johnston, M. and Fast, D. (2010) Homing and spawning site selection by supplemented hatchery- and natural-origin Yakima river spring chinook salmon. *Transactions of the American Fisheries Society* 139, 1014–1028.

Engle, C.R., McNevin, A., Racine, P., Boyd, C.E., Paungkaew, D., Viriyatum, R., Tinh, H.Q. and Minh, H.N. (2017) Economics of sustainable intensification of aquaculture: Evidence from shrimp farms in Vietnam and Thailand. *Journal of the World Aquaculture Society* 48, 227–239.

Foote, A.R., Stratford, C.N. and Coman, G.J. (2018) Passive integrated transponder (PIT) tagging black tiger shrimp, *Penaeus monodon*: Applications for breeding programs. *Aquaculture* 491, 321–324.

Føre, M., Alfredsen, J.A. and Gronningsater, A. (2011) Development of two telemetry-based systems for monitoring the feeding behaviour of Atlantic salmon (*Salmo salar* L.) in aquaculture sea-cages. *Computers and Electronics in Agriculture* 76, 240–251.

Føre, M., Frank, K., Norton, T., Svendsen, E., Alfredsen, J. A., Dempster, T., Eguiraun, H., Watson, W., Stahl, A., Sunde, L.M., Schellewald, C., Skøien, K.R., Alver, M.O. and Berckmans, D. (2018) Precision fish farming: A new framework to improve production in aquaculture. *Biosystems Engineering* 173, 176–193.

Freire, J. and González-Gurriarán, E. (1998) New approaches to the behavioural ecology of decapod crustaceans using telemetry and electronic tags. In: Lagardère, J.-P., Anras, M.-L.B., Claireaux, G. (eds) *Advances in Invertebrates and Fish Telemetry*. Springer Netherlands, Dordrecht, the Netherlands, pp. 123–132.

Friard, O. and Gamba, M. (2016) BORIS: a free, versatile open-source event-logging software for video/audio coding and live observations. *Methods in Ecology and Evolution* 7, 1325–1330.

Gerlier, M. and Roche, P. (1998) A radio telemetry study of the migration of Atlantic salmon (*Salmo salar* L.) and sea trout (*Salmo trutta trutta* L.) in the upper Rhine. *Hydrobiologia* 371, 283–293.

Godin, D.M., Carr, W.H., Hagino, G., Segura, F., Sweeney, J.N. and Blankenship, L. (1996) Evaluation of a fluorescent elastomer internal tag in juvenile and adult shrimp *Penaeus vannamei. Aquaculture* 139, 243–248.

Grey, M., Forster, I., Dominy, W., Ako, H. and Giesen, A.F. (2009) Validation of a feeding stimulant bioassay using fish hydrolysates for the Pacific white shrimp, *Litopenaeus vannamei. Journal of the World Aquaculture Society* 40, 547–555.

Harpaz, S., Kahan, D., Galun, R. and Moore, I. (1987) Responses of freshwater prawn, *Macrobrachium rosenbergii*, to chemical attractants. *Journal of Chemical Ecology* 13, 1957–1965.

Hay, M.E. (2011) Crustaceans as powerful models in aquatic chemical ecology. In: Breithaupt, T. and Thiel, M. (eds) *Chemical Communication in Crustaceans*. Springer Science and Business Media, New York, USA, pp. 41–62.

He, K., Sun, J. and Tang, X. (2011) Single image haze removal using dark channel prior. *IEEE Transactions on Pattern Analysis and Machine Intelligence* 33, 2341–2353.

Huang, I.-J., Hung, C.-C., Kuang, S.-R., Chang, Y.-N., Huang, K.-Y., Tsai, C.-R. and Feng, K.-L. (2018) The prototype of a smart underwater surveillance system for shrimp farming. In: *2018 IEEE International Conference on Advanced Manufacturing (ICAM)*. Presented at the 2018 IEEE International Conference on Advanced Manufacturing (ICAM), pp. 177–180.

Hung, C.C., Tsao, S.C., Huang, K.H., Jang, J.P., Chang, H.K. and Dobbs, F.C. (2016) A highly sensitive underwater video system for use in turbid aquaculture ponds. *Scientific Reports* 6, 1–7.

Jaime-Ceballos, B., Cerecedo, R.C., Villarreal, H., López, J.G. and Pérez-Jar, L. (2007) Use of *Spirulina platensis* meal as feed attractant in diets for shrimp *Litopenaeus schmitti*. *Hidrobiológica* 17, 113–117.

Jescovitch, L.N., Ullman, C., Rhodes, M. and Davis, D.A. (2018) Effects of different feed management treatments on water quality for Pacific white shrimp *Litopenaeus vannamei*. *Aquaculture Research* 49, 526–531.

Johnson, M.L., Gaten, E. and Shelton, P.M. (2002) Spectral sensitivities of five marine decapod crustaceans and a review of spectral sensitivity variation in relation to habitat. *Journal of the Marine Biological Association of the United Kingdom* 82, 835–842.

Kenyon, R.A., Loneragan, N.R. and Hughes, J.M. (1995). Habitat type and light affect sheltering behaviour of juvenile tiger prawns (*Penaeus esculentus* Haswell) and success rates of their fish predators. *Journal of Experimental Marine Biology and Ecology* 192, 87–105.

Koehl, M.A.R. (2006) The fluid mechanics of arthropod sniffing in turbulent odor plumes. *Chemical Senses* 31, 93–105.

Krummenauer, D., Samocha, T., Poersch, L., Lara, G. and Wasielesky, W. (2014) The reuse of water on the culture of Pacific white shrimp, *Litopenaeus vannamei*, in BFT System. *Journal of the World Aquaculture Society* 45, 3–14.

Lee, P.G. and Meyers, S.P. (1996) Chemoattraction and feeding stimulation in crustaceans. *Aquaculture Nutrition* 2, 157–164.

Li, X., Shang, M., Hao, J. and Yang, Z. (2016) Accelerating fish detection and recognition by sharing CNNs with objectness learning. In: *OCEANS 2016 – Shanghai*. Presented at the OCEANS 2016 – Shanghai, pp. 1–5.

Lima, P.P., Pontes, C.S. and Arruda, M.F. (2009) Activity pattern of the marine shrimp *Litopenaeus vannamei* (Boone 1931) in laboratory as a function of different feeding frequencies. *Aquaculture Research* 41, 53–60.

Martin, P.and Bateson, P. (2007) *Measuring Behaviour: An Introductory Guide*. Cambridge University Press, Cambridge, UK.

Meynecke, J.-O., Mayze, J. and Alberts-Hubatsch, H. (2015) Performance and physiological responses of combined t-bar and PIT tagged giant mud crabs (*Scylla serrata*). *Fisheries Research* 170, 212–216.

Montoya-Martínez, C., Nolasco-Soria, H., Vega-Villasante, F., Carrillo-Farnés, O., Álvarez-González, A. and Civera-Cerecedo, R. (2018) Attractability and palatability of ingredients in longarm river prawn *Macrobrachium tenellum* feed. *Latin American Journal of Aquatic Research* 46, 615–620.

Murphy, D.W., Olsen, D., Kanagawa, M., King, R., Kawaguchi, S., Osborn, J., Webster, D.R. and Yen, J. (2019) The three-dimensional spatial structure of Antarctic krill schools in the laboratory. *Scientific Reports* 9, 1–12.

Niu, B., Li, G., Peng, F., Wu, J., Zhang, L. and Li, Z. (2018) Survey of fish behavior analysis by computer vision. *Journal of Aquaculture Research and Development* 9, 1–15.

Nunes, A.J.P. and Parsons, G.J. (1998) Food handling efficiency and particle size selectivity by the southern brown shrimp *Penaeus subtilis* fed a dry pelleted feed. *Marine and Freshwater Behaviour and Physiology* 31, 193–213.

Nunes, A.J.P., Sá, M.V.C., Andriola-Neto, F.F. and Lemos, D. (2006) Behavioural response to selected feed attractants and stimulants in Pacific white shrimp, *Litopenaeus vannamei*. *Aquaculture* 260, 244–254.

Pedersen, M., Bruslund Haurum, J., Gade, R. and Moeslund, T.B. (2019) Detection of marine animals in a new underwater dataset with varying visibility. In: *Proceedings of the IEEE. Conference on Computer Vision and Pattern Recognition Workshops*, pp. 18–26.

Pinkiewicz, T.H., Purser, G.J. and Williams, R.N. (2011) A computer vision system to analyse the swimming behaviour of farmed fish in commercial aquaculture facilities: A case study using cage-held Atlantic salmon. *Aquacultural Engineering* 45, 20–27.

Pittet, A.O., Ellis, J.C. and Lee, P.G. (1996) Methodology for the identification and quantitative measurement of chemical stimulants for penaeid shrimp. *Aquaculture Nutrition* 2, 175–182.

Pontes, C.S. and Arruda, M.F. (2005a) Comportamento de *Litopenaeus vannamei* (Boone) (Crustacea, Decapoda, Penaeidae) em função da oferta do alimento artificial nas fases clara e escura do período de 24 horas. *Revista Brasileira de Zoologia* 22, 648–652.

Pontes, C.S., Arruda, M.F. (2005b) Acesso ao alimento artificial e enchimento do trato digestivo de juvenis do camarão marinho *Litopenaeus vannamei* (Boone) (Crustacea, Decapoda, Penaeidae) durante as fases clara e escuro do período de 24 horas. *Revista Brasileira de Zoologia* 22, 1039–1043.

Pontes, C.S., Arruda, M.F., Menezes, A.A. and Lima, P.P. (2006) Daily activity pattern of the marine shrimp *Litopenaeus vannamei* (Boone 1931) juveniles under laboratory conditions. *Aquaculture Research* 37, 1001–1006.

Pontes, C.S., Lima, P.P. and Arruda, M.F. (2008) Feeding responses of juvenile shrimp *Litopenaeus vannamei* (Boone) fed at different frequencies under laboratory conditions. *Aquaculture Research* 39, 1416–1422.

Primavera, J.H. and Lebata, J. (1995) Diel activity patterns *Metapenaeus* and *Penaeus* juveniles. *Hydrobiologia* 295, 295–302.

Redmon, J., Divvala, S., Girshick, R. and Farhadi, A. (2016) You Only Look Once: Unified, real-time object detection. In: *2016 IEEE Conference on Computer Vision and Pattern Recognition (CVPR)*. IEEE Conference on Computer Vision and Pattern Recognition (CVPR), Las Vegas, NV, pp. 779–788.

Reis, J., Novriadi, R., Swanepoel, A., Jingping, G., Rhodes, M. and Davis, D.A. (2019) Optimizing feed automation: Improving timer-feeders and on demand systems in semi-intensive pond culture of shrimp *Litopenaeus vannamei. Aquaculture* 519, 734759.

Saberioon, M.M. and Cisar, P. (2016) Automated multiple fish tracking in three-dimension using a structured light sensor. *Computers and Electronics in Agriculture* 121, 215–221.

Saberioon, M., Gholizadeh, A., Cisar, P., Pautsina, A. and Urban, J. (2017) Application of machine vision systems in aquaculture with emphasis on fish: State-of-the-art and key issues. *Reviews in Aquaculture* 9, 369–387.

Sanchez, D.R., Fox, J.M., Lawrence, A.L., Castille, F.L. and Dunsford, B. (2005) A methodology for evaluation of dietary feeding stimulants for the Pacific white shrimp, *Litopenaeus vannamei. Journal of the World Aquaculture Society* 36, 14–23.

Santos, A.D.A., López-Olmeda, J.F., Sánchez-Vázquez, F.J. and Fortes-Silva, R. (2016) Synchronization to light and mealtime of the circadian rhythms of self-feeding behavior and locomotor activity of white shrimps (*Litopenaeus vannamei*). *Comparative Biochemistry and Physiology Part A: Molecular and Integrative Physiology* 199, 54–61.

Schettini, R. and Corchs, S. (2010) Underwater image processing: state of the art of restoration and image enhancement methods. *EURASIP Journal on Advances in Signal Processing* 2010, 746052.

Schmalbach, A.E., Quackenbush, L.S. and Melinek, R. (1994) A method for tagging the Malaysian prawn *Macrobrachium rosenbergii. Aquaculture* 122, 147–159.

Schmidt, M. and Derby, C.D. (2005) Non-olfactory chemoreceptors in asymmetric setae activate antennular grooming behavior in the Caribbean spiny lobster *Panulirus argus. The Journal of Experimental Biology* 208, 233–248.

Silva, J.F., Hamilton, S., Rocha, J.V., Borie, A., Travassos, P., Soares, R. and Peixoto, S. (2019) Acoustic characterization of feeding activity of *Litopenaeus vannamei* in captivity. *Aquaculture* 501, 76–81.

Silva, P.F., Medeiros, M. de S., Silva, H.P.A. and Arruda, M. de F. (2012) A study of feeding in the shrimp *Farfantepenaeus subtilis* indicates the value of species level behavioral data for optimizing culture management. *Marine and Freshwater Behaviour and Physiology* 45, 121–134.

Skerritt, D., Robertson, P., Mill, A., Polunin, N. and Fitzsimmons, C. (2015) Fine-scale movement, activity patterns and home-ranges of European lobster *Homarus gammarus. Marine Ecology Progress Series* 536, 203–219.

Smith, D.M., Burford, M.A., Tabrett, S.J., Irvin, S.J. and Ward, L. (2002) The effect of feeding frequency on water quality and growth of the black tiger shrimp (*Penaeus monodon*). *Aquaculture* 207, 125–136.

Smith, D.V. and Shahriar, M.S. (2013) A context aware sound classifier applied to prawn feed monitoring and energy disaggregation. *Knowledge-Based Systems* 52, 21–31.

Smith, D.V. and Tabrett, S. (2013) The use of passive acoustics to measure feed consumption by *Penaeus monodon* (giant tiger prawn) in cultured systems. *Aquacultural Engineering* 57, 38–47.

Soares, R., Peixoto, S., Wasielesky, W. and D'Incao, F. (2005) Feeding rhythms and diet of *Farfantepenaeus paulensis* under pen culture in Patos Lagoon estuary, Brazil. *Journal of Experimental Marine Biology and Ecology* 322, 167–176.

Steiner, J.E. and Harpaz, S. (1987) Behavior stereotypes of food acceptance and of the rejection of 'bitter' food in the freshwater prawn *Macrobrachium rosenbergii*. *Chemical Senses* 12, 89–97.

Suresh, A.V., Kumaraguru-Vasagam, K.P. and Nates, S. (2011) Attractability and palatability of protein ingredients of aquatic and terrestrial animal origin, and their practical value for blue shrimp, *Litopenaeus stylirostris* fed diets formulated with high levels of poultry byproduct meal. *Aquaculture* 319, 132–140.

Taylor, J., Vinatea, L., Ozorio, R., Schuweitzer, R. and Andreatta, E.R. (2004) Minimizing the effects of stress during eyestalk ablation of *Litopenaeus vannamei* females with topical anaesthetic and a coagulating agent. *Aquaculture* 233, 173–179.

Taylor, M.D. and Ko, A. (2011) Monitoring acoustically tagged king prawns *Penaeus* (*Melicertus*) *plebejus* in an estuarine lagoon. *Marine Biology* 158, 835–844.

Thiel, M. and Breithaupt, T. (2011) Chemical communication in crustaceans: Research challenges for the twenty-first century. In: Breithaupt, T. and Thiel, M. (eds) *Chemical Communication in Crustaceans*. Springer Science and Business Media, NY, USA, pp. 3–22.

Thorstad, E.B., Rikardsen, A.H., Alp, A. and Økland, F. (2013) The use of electronic tags in fish research–an overview of fish telemetry methods. *Turkish Journal of Fisheries and Aquatic Sciences* 13, 881–896.

Ullman, C., Rhodes, M., Hanson, T., Cline, D. and Davis, D.A. (2019) Effects of four different feeding techniques on the pond culture of Pacific white shrimp, *Litopenaeus vannamei*. *Journal of the World Aquaculture Society* 50, 54–64.

Wolcott, T.G. (1995) New options in physiological and behavioural ecology through multichannel telemetry. *Journal of Experimental Marine Biology and Ecology, Behavioural Ecology of Decapod Crustaceans: An Experimental Approach* 193, 257–275.

Wroblewska, J., Whalley, S., Fischetti, M. and Daniel, P.C. (2002) Identification of chemosensory sensilla activating antennular grooming behaviour in the Caribbean spiny lobster, *Panulirus argus*. *Chemical Senses* 27, 769–778.

You, K., Yang, H., Liu, Y., Liu, S., Zhou, Y. and Zhang, T. (2006) Effects of different light sources and illumination methods on growth and body color of shrimp *Litopenaeus vannamei*. *Aquaculture* 252, 557–565.

Zenone, A., Ceraulo, M., Ciancio, J.E., Buscaino, G., D'Anna, G., Grammauta, R., Mazzola, S. and Giacalone, V.M. (2019) The use of 3-axial accelerometers to evaluate sound production in European spiny lobster, *Palinurus elephas*. *Ecological Indicators* 102, 519–527.

Zhou, C., Xu, D., Lin, K., Sun, C. and Yang, X. (2018) Intelligent feeding control methods in aquaculture with an emphasis on fish: A review. *Reviews in Aquaculture* 10, 975–993.

Zion, B. (2012) The use of computer vision technologies in aquaculture: A review. Computers and Electronics in Agriculture 88, 125–132.

Chapter 13

FEEDING PRACTICES FOR AUTOMATED SYSTEMS IN COMMERCIAL SHRIMP PRODUCTION

JOAO REIS, HERBERT E. QUINTERO AND LUKE A. ROY

Summary

Although shrimp can be reared in different types of system, feed is the most important operating cost in semi-intensive and intensive production systems. Adequate feed management strategies are of utmost importance to achieve optimal growth and higher profitability, and to reduce waste deposition. Effective feed management practices combine quality complete feed with an adequate delivery method. Feed composition is particularly important during the early life stages when nutrient imbalance is more likely to negatively impact physical and physiological development of individuals. Shrimp evolved as benthic grazers with limited storage capacity throughout their digestive tract and rely on frequent ingestion of small quantities of feed. Hence, numerous studies have showed improved growth performance of juvenile shrimp fed multiple daily meals throughout the production cycle. Moreover, dispersion of nutrient input in systems is also beneficial by reducing the likelihood of water quality degradation. Regardless of the advantages, many farms still use traditional hand-feeding techniques or feed trays as their main feed delivery method, making it economically unsustainable to feed multiple meals throughout the day. This is particularly evident in the Americas, where labor cost is substantially higher than other important production regions such as South East Asia. One way of bypassing higher labor costs to provide an increased number of meals to shrimp is the use of automated feed delivery systems. This chapter aims to provide insight into the various aspects of feed management in shrimp production, with particular emphasis on feeding protocols, feed delivery techniques and automated systems. Finally, we aim to explore the effect of different feeding protocols and management regimes on water quality.

Introduction

Feed is the main operating cost of a typical shrimp farm, as well as the primary source of both nutrients and waste. Consequently, good feed management practices have the potential to positively affect animal growth and environmental conditions (Chatvijitkul et al., 2017; Van et al., 2017). Feed management focuses on the combination of high-quality complete feed and the most effective delivery method. Primary productivity within the system can be an important source of supplemental nutrients as well, although its role is reduced as animal density is increased. For a long time, the industry has focused on understanding nutritional requirements of species, which has resulted in the commercial availability of numerous quality complete feeds.

The first step towards understanding the nutritional necessities of a species is to identify nutritional preferences in its natural environment and potential dietary shifts during its life cycle. Shrimp go through larval stages and hence the digestive tract gradually increases its complexity, which ultimately determines substantial variations in nutrient requirements. Extensive research has focused on nutrient requirements and feed sources for shrimp across all metamorphic stages. In general, commercially produced complete feeds for shrimp are regarded as adequate; however, less effort has been directed towards understanding feeding behavior and how to improve feed management practices through more efficient feed delivery methods. Once the juvenile stage is reached, shrimp become benthic grazers with external mastication and very limited capacity to store feed within their digestive tract. Henceforth their natural feeding routine favors slow and continued ingestion of small amounts of feed. Numerous studies have reported improved growth performance with feeding protocols that favor an increased number of meals (Carvalho and Nunes, 2006; Jescovitch et al., 2018; Ullman et al., 2019a,b). However, increasing numbers of meals at the commercial level often comes at a prohibitive increment in labor cost.

Hand feeding, with or without the supplementary use of feed trays, is the primary feeding method for many operations. Nonetheless both are very labor intensive. Many small operations in Asia were able to increase the number of daily meals with limited economic implications, as labor cost in the region is relatively low. Higher labor cost and larger pond size have pushed many shrimp farmers in the Americas towards either maintaining few daily meals or, more recently, the introduction of automated feeding systems.

Optimal water quality is a key factor in the production success of shrimp, as poor water quality conditions are a primary cause of reduced growth, poor survival and increased disease vulnerability. It is of utmost importance to identify and provide optimal water conditions, as well as to understand how system design and management affects these parameters. Feed inputs are an important factor in waste deposition. Water quality degradation in the system can result from both uneaten feed and shrimp metabolites. External mastication and pellet handling by shrimp enhance both loss of water-soluble nutrients and the necessity to monitor water quality. Nutrient leaching and ineffective waste management practices favor bacterial and algal population blooms. This can potentially lead to oxygen depletion, nitrogen waste accumulation and shrimp mortality. Regardless of the production system, it is critical to understand waste management limitations and adjust feed inputs accordingly.

This chapter aims at providing theoretical insight towards the nuances in feed management strategies and planning, as well as explaining how technology has evolved to become one of the driving factors in increasing efficiency in shrimp production. Our ultimate goal is to help readers better understand how feeding practices affect commercial shrimp production biologically (i.e. growth), environmentally and economically.

Artificial feeds and primary productivity

ARTIFICIAL FEEDS

Post-larval, juvenile and grow-out stages

Feed that meets all the dietary requirements of shrimp is paramount for proper feed management. Factors that should be considered when selecting a feed include: life cycle (larval, post-larval, nursery, grow-out, maturation), the production system that it is being used for culturing shrimp (extensive, semi-intensive, intensive systems according to stocking densities), nutrient density of the diet (protein and energy level of the diet) and shrimp size. Other factors that affect nutritional performance include the physical properties and water stability of the diet, transportation and storage, and the feed application method (Tacon et al., 2013).

The shrimp industry has evolved in terms of the feeds used for different life stages. Early stages of shrimp (zoea, mysis, post-larval) depend heavily on algae (e.g. *Chlorella*, *Chaetoceros*) or algal powders (e.g. *Spirulina*), and live, newly hatched *Artemia nauplii*. However, the production of those live foods is restrictive and increases the cost of operation for many hatcheries. Nowadays the use of microparticulate diets, and/or dry/liquid microencapsulated larval feeds, with feeds that are offered in different particle sizes (<100, 100–200, 200–400, 400–600 and 600–850 microns) is widespread across the industry. Once the organisms have reached juvenile stage (>PL30), they rely mainly on formulated feeds that are available in different sizes, and target specific culture units, such as nursery systems followed by grow-out, with feeds offered in particle size ranging from 1 mm (850–1200 microns), 1.5 mm, 2.0 mm and up to 2.4 x 2.5 mm. These feeds have specific dietary protein levels and are typically classified as starter, grower and finisher feeds. Finally, feeds for shrimp broodstock rely on high-quality fresh marine products such as marine polychaetes, squid mantle, *Artemia* biomass, mussels, oysters, shrimp, krill and crabs, as well as formulated dry or semi-moist pelleted feeds (Tacon, 2017). One concern with the use of live feeds and/or processed natural feed items is that they constitute a biosecurity risk due to the potential introduction of pathogens to the cultured shrimp in hatchery or maturation facilities (Tacon, 2017).

In general, shrimp reared in extensive pond production systems rely heavily on natural productivity, while semi-intensive and intensive production systems require a well-balanced diet. In Latin America many producers still rely on extensive production systems using low-protein diets (22–25% crude protein) and fertilizers to stimulate natural productivity, although they may use a high-protein diet (50% crude protein) in the nursery phase. On the other hand, Asian and South East Asian countries rely more on the use of high protein levels, ranging from 36 to 44% crude protein and, in some cases, while culturing black tiger shrimp (*Penaeus monodon*) under semi-intensive and intensive conditions they may use fertilizers and trash fish, but this does not appear to be the case in the culture of whiteleg shrimp (*Penaeus vannamei*) (Hung and Quy, 2013). An additional production system that is becoming more commonplace in the industry is biofloc systems, in which shrimp are produced either in indoor or outdoor tank systems with minimal water exchange and with the availability of microbial flocs as an extra supplemental food source.

Feed inputs depend on water quality considerations and feed consumption. Feed inputs must be adjusted to minimize over- or underfeeding, with feed trays being one of the main tools used by farmers to monitor feed consumption (Figure. 13.1). In Latin America there are two main drivers for the use of feeding trays, one for total feed distribution and the other for the sole purpose of monitoring. Total feed distribution requires people to distribute the feed at 20–25 feed trays per hectare, which requires considerable effort since usually earthen pond size varies from 1 to 5 ha. Feed trays for monitoring are distributed usually as 4–6 per hectare; a fixed amount of feed is placed in the tray, then, after 1–2 hours, the feed trays are checked to determine whether the feed was totally or partially consumed. Those observations are recorded and used to maintain, increase or decrease the feed rate for the next feeding. In South East Asia the number of feed trays is typically 4–10 per hectare (Hung and Quy, 2013), but it is important to consider that pond size is smaller (in the range 0.2–0.5 ha). The amount of feed that is placed in the feed trays varies with shrimp size, with 2.8% body weight of feed distributed if shrimp are in the 5–10 g range, 3.0% of feed in the 10–20 g and 3.3% for shrimp >20 g (Hung and Quy, 2013). While the feed tray method appears objective, there are numerous variables that affect its results, such as the skill and experience of the feeding technician. Variability such as this results in feed conversion ratios (FCRs) that range from 1.1 up to 2.0 or higher among ponds, even from the same commercial farm. The appropriate training of feeding technicians is critical to standardize feed management practices and reduce variability in FCR.

Traditional feed management approaches rely on estimation of the shrimp population (survival rate) to calculate feed inputs. Evaluation of a shrimp population in traditional farming relies on cast netting the pond on a periodic basis, at different locations in the pond, to determine whether the population has changed in the period under evaluation. Traditionally, the number of samples is six cast nets per hectare in extensive and semi-intensive production earthen ponds. Records and maps of the shrimp captured are maintained and compared to historic populations of shrimp in each pond, to estimate population size.

Feed distribution techniques in ponds vary based on the size of shrimp and production stage. During the first weeks of culture, shrimp farmers usually provide a small amount of feed around the edge of the pond to supplement nutrition from primary productivity. As the shrimp begin to grow, feed is subsequently distributed homogeneously in the pond. Feed distribution can be accomplished using several different methods, ranging from small boats or rafts, applying feed following a zig-zag pattern, automated feeders or, as mentioned above, by offering small amounts of feed over feed trays that have been distributed evenly throughout the pond. Some farms also employ land-based blower equipment,

FIGURE 13.1 Indoor shrimp pond with feed tray on dock for use during the first weeks of production.

typically mounted on trailers pulled by tractors or trucks, that can distribute feed out to distances of 15 m from the pond bank (Fox et al., 2001).

Time of application and number of feed offerings vary greatly among farms, but there is a general trend toward feeding a minimum of two and up to four times a day, starting between 06:00 and 08:00 and ending around 16:00–18:00, over an 8–12-hour working day. Due to the difficulty of supervising personnel during night-time hours, most commercial shrimp farms restrict the majority of their feeding to daytime hours. In the past some commercial operations have included night rations; however, there is no clear benefit of such practices and its use has been discontinued due to practical issues and not being cost effective (Fox et al., 2001).

Practical feeding guidelines

The rate of feed offered in commercial shrimp production ponds has been typically calculated based on feed guideline tables that have a targeted/expected weekly growth response (Table 13.1). These guidelines are based on a percentage of the pond shrimp biomass fed as dry weight. That biomass in turn depends on the estimated number of shrimp (survival) and the expected mean weight of the shrimp (growth), which must be monitored on a regular basis. It is common to adjust daily feed allocations based on morning dissolved-oxygen levels, temperature and quantification of feed remaining on feed trays, as well as on past performance of the culture system. These estimations have been found to represent appropriate feed inputs during the first half of the production cycle. However, later in the production cycle, feed inputs tend to rise continually without regard to shrimp growth resulting in higher FCR. Tacon et al. (2002) suggested adjusting feed rates based on shrimp body weight under different water temperatures. Davis et al. (2008) recommended a modified approach that considers the expected FCR and historical data for the farm as the main drivers for determining the ration. Historical data specific for each farm are essential because growth rates vary with species, genetic line and environmental conditions. Using this framework, we can put upper and lower limits on feed management; these will depend on the estimated survival for the specific pond or farm. This theoretical FCR may be used to assist in feed management. When it comes to determining feed inputs there are no absolute answers, so it is critical to look at feed inputs using several tools and to base decisions on the best information available.

Another aspect that is relevant to practical feeding guidelines is the decision on what type of feed is being used on the farm, especially when profits are threatened by price reduction in the world market. Since feed is usually the major driver in production cost, farmers tend to use this input to leverage their operation cost. The current trend is that farmers either reduce the stocking density to reduce the amount of feed and then rely on natural productivity, or reduce the protein content of the feed to reduce feed cost during the production cycle. However, the latter solution may not be the best approach because feed protein content drives growth in shrimp (Kureshy and Davis, 2002; Lee and Lee, 2018). We may consider the following analysis to illustrate the effect of this decision. Current feed price in Panama for 35% crude protein feed is US$0.93/kg, for 28% crude protein feed is US$0.84/kg and for 25% crude protein feed is US$0.81/kg (personal communication). This means that the cost per unit protein is 0.0266, 0.0300 and 0.0324, respectively, which is already worrisome because it represents a higher relative cost for the 25% crude protein feed. This effect is better illustrated if we determine the amount of feed needed to provide, for example, 35 kg of protein. Then, if we have the 35% crude protein feed we would need 100 kg of feed,

TABLE 13.1 Examples of commercial feed tables suggested in different countries.

Shrimp weight (g)	*Biomass (%)*		
	Ecuador	*Colombia*	*Mexico*
0.05	35.0	20.2	
0.10	35.0	17.0	
0.18	35.0	14.7	
0.31	35.0	12.8	
0.54	14.0	11.1	
1.00	10.0	10.1	15.5
2.00	9.0	8.5	12.5
3.00	6.0	7.7	10.1
4.00	4.0	7.1	7.9
5.00	3.0	4.4	6.6
6.00	2.5	4.0	5.6
7.00	2.5	3.6	4.9
8.00	2.5	3.4	4.4
9.00	2.5	3.2	4.0
10.0	2.5	3.0	3.7
11.0	2.0	2.8	3.4
12.0	2.0	2.7	3.2
13.0	1.5	2.6	3.1
14.0	1.5	2.5	3.0
15.0	1.5	2.6	2.9
16.0	1.5	2.5	2.9
17.0	1.5	2.4	2.8
18.0	1.5	2.4	2.8
19.0	1.5	2.3	2.6
20.0	1.5	2.2	2.6

which would cost $93.00; but if we have the 25% crude protein feed we would need 140 kg of feed, which would cost $113.40. Hence, the cost would be 22% higher for the same amount of protein.

Feed supplementation is another trend that has been promoted in the shrimp industry, and consists of the inclusion of additional vitamins and minerals, digestive enzymes (e.g. proteases, amylases, cellulases and lipases), shrimp or fishery products (shrimp paste, hydrolysates and/or silage), probiotics and immunostimulants. These are either included directly in the feed or through 'top-coating' of commercial pelleted feeds to either enhance feed digestibility, stimulate shrimp appetite or improve growth, disease resistance and/or feed consumption (Hung and Quy, 2013; Tacon, 1993, 2002, 2017).

PRIMARY PRODUCTIVITY

The culture system used for growing commercial species, including shrimp, can influence production performance (Tacon et al., 2002). In any system for which shrimp have access to natural foods, this typically provides an additional source of food items to enhance growth (Roy et al., 2012). This is true at the post-larval, juvenile and grow-out stages of shrimp production. Proper feed management must take into consideration food items available in the pond environment in order to minimize FCR and maximize growth of shrimp reared in these systems, as these factors ultimately play a role in the profitability of a farm operation.

A large percentage of the food intake of shrimp reared in ponds can be obtained from primary productivity, particularly early in the production cycle (Hunter et al., 1987). Moss et al. (1992) examined the contribution of select fractions of pond water to enhance shrimp growth in *P. vannamei*. Bombeo-Tuburan et al. (1993) reported on the role of detritus and other natural food items as a food source for pond culture of *P. monodon*. In a study with *P. monodon*, Allan et al. (1995) reported that prepared ponds improved growth performance compared to non-prepared ponds. In a study in which *L. vannamei* were offered anywhere from a 60 to 110% of a daily ration with continuous water flow from a semi-intensive production pond with low-salinity water, there were no differences in growth among treatments, indicating the contribution of pond primary productivity in treatments involving much less feed (Roy et al., 2012). Soares et al. (2005) reported a high percentage of nutrition derived from natural productivity in a study carried out with *Farfentepenaeus paulensis* in pens in Patos Lagoon estuary in Brazil. In that particular study they documented that shrimp consumption of commercial feed increased after shrimp attained a size of 5 g, and the amount of foraging on pond primary productivity decreased as shrimp grew in size. Further work with shrimp and prawns on evaluating the role of additional food items in the pond environment includes studies by Anderson et al. (1987), Reymond and Langardère (1990), Newell et al. (1995), Moriarty (1997), Tidwell et al. (1997), Focken et al. (1998), Moss et al. (2002) and Martinez-Cordova et al. (2003) among others.

Techniques for pond preparation and fertilization are beyond the scope of this chapter, but are reviewed by Boyd et al. (2010) and Green (2015). As such, proper pond preparation to enhance primary productivity prior to stocking of post-larvae is still recommended to accelerate growth, even in commercial ponds receiving shrimp feed during the first few weeks of culture. Furthermore, Ullman et al. (2018, 2019a,b) report on pond preparation and hand feeding techniques used prior to implementing automated feeding systems once shrimp have reached an acceptable size. The use of an additional nursery phase has documented advantages for shrimp prior to stocking in production ponds that have been properly prepared, to enhance primary productivity for grow-out (Samocha, 2010).

Automated systems in shrimp aquaculture

FEED DELIVERY SYSTEMS

Feed delivery protocols in aquaculture have evolved from traditional hand feeding to more sophisticated and efficient methods. A deeper understanding of shrimp behavior in both wild and captive environments allows a better perception on how feeding protocols could be improved. The preponderance of studies indicates that shrimp regularly ingest small quantities of food; this intake is directly related to feed availability. Better growth conditions were achieved in captive shrimp fed numerous meals (Carvalho and Nunes, 2006; Jescovitch et al., 2018; Napaumpaipom et al., 2013; Nunes et al., 2019; Ullman et al., 2019a,b). The repetitive nature of this task led to the adaptation and introduction of automated devices in order to avoid rising labor expenses.

There are generally two types of automatic feeders: timer feeders and on-demand feeders. Timer feeders simply feed a specific preset amount of feed at the time that they are set while on-demand feeders are activated by response to animal feedback (e.g. touch, sound or image). In both cases, workers are required only to refill feeders rather than actively feeding at every meal. If applied properly, feed automation for shrimp farming is designed to improve the efficiency of feed by increasing the number of meals, while reducing the quantity of feed available at each meal ensures that feed is consumed quickly. This in turn reduces feed leaching by reducing the time feed remains in the water. The overall effect of increasing the number of meals results in increased feed intake of higher-quality feed, resulting in higher growth rates (Jescovitch et al., 2018; Reis et al., 2020; Ullman et al., 2019). This improves feed efficiency as well as water quality. The positive impacts on water quality are primarily due to the spreading out of nutrient loads more evenly throughout the day.

Increasing the number of feedings can easily be achieved using a simple timer-based feeding system. However, such systems must be managed as there is no feedback loop. Tying automated systems to measures of demand further improves efficiency. The salmon industry was one of the first to introduce some of these technologies. On-demand feeders for salmon use a diverse range of technologies such as ultrasonic detection and underwater cameras. These feeders detect and monitor animal behavior and feed accordingly. Contrary to salmonids, shrimp are bottom-feeders and grazer animals most commonly reared in outdoor ponds or shallow tanks where turbidity prevents the use of cameras and their small size and proximity to the substrate make ultrasonic detection problematic (Figure 13.2).

One of the most recent technologies developed for shrimp feed management is the use of passive acoustic systems, which allow for the identification and interpretation of clicking sounds produced by the collision and scraping of mandibles during feeding (Smith and Tabrett, 2013) (Figure 13.3). Acoustic patterns for pond-reared *P. monodon* feeding activity were identified, which enabled estimation of food

FIGURE 13.2 Outdoor pond equipped with a timer-feeder located on the far end of the dock.

FIGURE 13.3 Automatic feeder component of an acoustic feedback feeding system in an outdoor pond system.

intake and, more recently, various acoustic parameters described for different size classes of *P. vannamei* (Silva et al., 2019) and diets (Peixoto et al., 2020). A hydrophone is placed inside the pond to detect shrimp activity based on the clicking sound shrimp produce when feeding and disperses feed accordingly. Jescovitch et al. (2018), Napaumpaipom et al. (2013), Reis et al. (2020) and Ullman et al. (2019) reported both enhanced growth performance and higher feed inputs when acoustic feedback feeding systems were used. This technology has increased growth rates, improved FCR and resulted in fewer days to market, hence changing the dynamics of shrimp production.

More recently the technology has focused on integration of all the information provided by feeding and water quality devices to create an intelligent and increasingly autoregulated system. In short, systems are designed to regulate feed inputs based on real-time data for water quality and to activate/deactivate mechanical aeration. The main disadvantages of moving a farm towards automation are the initial investment in technology (sensors, feeders, software, etc.) and adaptation of the farm to integrate new devices (e.g. improved electrical supply, training of employees) on a large scale. Acoustic feedback feeders seem to be the most efficient technology available at present. However, Reis et al. (2020) reported that it is possible to approximate timer-feeder performance to that of an acoustic feedback system by adjusting the feeding schedule and feed amounts. In short, performance can be approximated without the necessity to invest in new devices and software. However, a timer feeding system is still reactive feed management and so, by definition, it is not equivalent to real-time feed management, albeit performance thresholds can be minimized.

Regardless of the automated system put in place, adequate protocols will likely result in better biological production parameters across the board. However, it is paramount to understand that higher feed inputs, larger animals and higher densities (from higher survival or simply stocking density) will increase biological oxygen demand in the system. Hence, oxygenation capacity should also be an important factor taken into consideration for facilities that are considering moving towards more automated protocols.

Water quality measurement and integrated automated systems

Water quality can be described by a multitude of characteristics, but only a handful are usually monitored for their carrying capacity limitation. Oxygen concentration is a major concern and usually the first limiting environmental factor in most types of aquaculture production system. Nitrogenous waste is typically more problematic in intensive and super-intensive systems, with limited to no water exchange as a consequence of both high density and high feeding rates. Hence, farms must monitor these parameters.

Water quality multiparametric sensors are a good tool to measure the primary water quality parameters (dissolved oxygen, temperature, salinity and pH) on an intermittent or continuous basis. Historically, their disadvantage has been the necessity of physically going to the tank or pond to use the device. This problem is amplified by increasing the size of the farm and/or the number of systems, tanks or ponds that need to be monitored. It is common that large farms have a specific person whose task is exclusively to measure and report water quality. As in feeding, the repetitive nature of this task results in development of both underwater sensor devices and automated water quality monitoring systems.

For improved convenience and labor reduction, aquaculture engineering has focused on the development of underwater multiparametric sensors providing real-time water quality data through Wi-Fi connection and software tailored specifically to benefit the aquaculture producer (Figure 13.4).

Introduction of underwater sensors may eliminate the workload associated with measuring water quality, but biofouling is a new issue that must be dealt with. However, technology has evolved sufficiently at this point in time that many sensors require only routine cleaning.

Along with labor reduction, this technology allows farms to maintain historic records of water quality. On a short timescale, having a good perspective on the water quality dynamics of the production system in real time is essential to identify water quality trends (e.g. oxygen crash) and quickly adjust management procedures accordingly. On a longer timescale, keeping a historic record may also be useful to correct mineral deficiencies that negatively affect both animals (e.g. osmoregulation, crustaceans exoskeleton, etc.) and natural productivity (low pH will prevent algae blooms).

FIGURE 13.4 Multiparametric water quality sensor in an outdoor pond system.

Water quality sensors provide software data that may be displayed and analyzed independently by the farmer/technician, or further integrated into intelligent self-managing systems. Self-managing technology integrates data from underwater sensors and manipulates feeders according to the parameters established by the farmer. The objective is to initiate or pause feeding when water conditions fall within a range less than an ideal. Further integration may inclusively trigger the start or pause of mechanical aeration according to pre-established parameters. Although self-managing systems are increasingly available and more efficient, there is still debate about whether management decisions (e.g. aerate, feed) should ultimately rest on a farmer or technician. Considering the wide range of possibilities already available, it is for the farm owner to decide. Self-regulating technologies are available, and have been used and described in the literature as successful.

FARM MANAGEMENT SOFTWARE

The necessity to collect and maintain quality records on multiple aspects of production as a basis for improvement is not a new concept for farmers or companies; in fact, it is not uncommon in larger operations to have personnel assigned the sole responsibility of that task. In a traditional setup, data would be collected and ultimately compiled in a spreadsheet to allow for analysis. Although this is a valuable tool, increasing data volume and diversity may complicate analysis and make the use of spreadsheets difficult. Also, data are usually written down or recorded in a format that may not be easily and quickly integrated in a spreadsheet. Consequently, developers have come up with simpler, and arguably more efficient, software to address this issue.

Following the increasing realization of the necessity to keep records and have as much data available at all times, many farm management software systems have emerged. In some cases, these systems have appeared as a natural and symbiotic technological evolution and complement of water quality monitoring devices or automated feeders. Real-time remote manipulation of devices is particularly important on very large farms. However, a manufacturer does not always develop devices for multiple functions, which results in farms utilizing hardware from different sources. Moreover, different devices from different brands typically use distinct software for data recording or setting manipulation. This can be problematic for farms, as it both requires expertise with different software and retains the problem of not allowing data to be easily centralized in one platform.

Although there is an enormous variety of products on the market, most new software is nothing more than a user-friendly, cloud-based platform that allows recording of data from the various facets of production (e.g. feed usage, water quality data, financial analysis) as well as the ability to graphically display data. Another important aspect has been the customization capacity of the software so that each user (farm/company) can adapt the system to its production setup, and for data to be displayed according to what is more important in a certain sector or specific period of time. Due to the complexity and variety of data included in these platforms, they allow differential (pyramid) access to data. This is particularly helpful for large operations with multiple departments and can be an important tool, by ensuring that employees have access only to the data and functions related to their particular job. For instance, a field technician would have access to pond environment and production data but would not be able to access financial information, which in turn would be restricted to the financial department. Although concentration of data in a single platform presents many advantages, there are a few setbacks: first, the necessity to make sure that production and financial data remain safe and private; and second, compatibility issues between hardware and software can arise. This software can be expensive, and hence acquiring or leasing the necessary equipment may be within the range of only larger farms or companies.

Responding to the necessity of turning raw real-time data into user-friendly integrated platforms, the aquaculture industry has experienced the emergence of cloud-based software that allows both record keeping and real-time data integration from both the entire production facility and individual systems (buildings/units and tanks). Centralizing data in one single platform enables owners to train their staff in one single software package, which is particularly important for farm positions in which workers have limited skills. Albeit these platforms were originally developed from an animal production and water quality monitoring standpoint, in many cases processing information or financial input (e.g. feeds, chemicals, drugs, salaries, etc.) is also possible in order to track both expenses and general company performance.

Feed management and water quality

Feed is the primary source of nutritional input for shrimp raised in semi-intensive and intensive production systems. Physical attributes of feed – such as the presence of fines or dust, water stability and size, as well as feed composition – will determine consumption rate and also potential deleterious effects on water quality. For instance, if there are too many fines or dust on the feed, shrimp will not be able to consume it and feed will be wasted, thus initiating the process of deteriorating pond water quality. Also, improper feeding, especially overfeeding, can have negative effects on the water quality of commercial shrimp production ponds. This is a topic that has been reviewed by authors including Quintero and Roy (2010) and Tacon et al. (2013). Since overfeeding can have such negative consequences on water quality, it can lead to reduced survival, production, sustainability and profit margins for commercial producers. As such, proper feed management can help in the maintenance of water quality that is more conducive to shrimp growth.

Among the negative consequences of overfeeding are excessive enrichment of the pond and the proliferation of harmful algae blooms. Both blue-green and golden algae have been noted to cause problems for shrimp producers (Roy et al., 2018). These harmful algae can be present even when producers are not overfeeding. Nitrogen in excess feed that does not get consumed and is available in the form of nitrogenous waste from culture animals has the potential to further enrich the pond environment. Nitrogenous waste from feed can cause enrichment of the pond, resulting in proliferation of problematic algal blooms.

Farmers must also monitor the levels of dissolved oxygen in their production ponds in order to be good feed managers. If dissolved oxygen levels are too low, shrimp will typically not feed and any feed disbursed in the pond would likely be wasted. Hence, routine monitoring is necessary for managers to be fully aware of current dissolved oxygen levels and observed patterns in shrimp production ponds.

Ammonia and nitrite are additional water quality variables that should be considered, and can be tied to overfeeding and improper feed management. Depending on pond water pH, high levels of ammonia can result in reduced shrimp survival. The level of exposure to toxic levels of ammonia in the pond can be acute or chronic, depending on the situation. If ammonia levels are dangerously high, it may be prudent to reduce or completely limit feeding to allow bacteria in the pond to process ammonia by converting it to nitrite.

PRACTICAL APPROACHES TO MANAGING FEEDING WHEN WATER QUALITY DETERIORATES

Feed management regimes may change temporarily during those periods in which water quality is not optimal in commercial shrimp production ponds. Farmers may choose to interrupt or alter their daily feeding regimes, depending on pond water quality. For instance, if dissolved oxygen is dangerously low one morning and shrimp are piping at the surface, a farmer will likely decide to skip or temporarily postpone feeding in that pond. Under traditional feed management schemes, where the farmer is actively feeding the pond, this is not a big deal. When water quality sensors are integrated, the acoustic-based AQ1 system is typically unaffected during periods of poor water quality as it will self-regulate and not disperse more feed if the shrimp are not actively feeding. However, automated feeders that disburse feed at set times of the day based on programmed timed feedings have the potential to waste feed during periods in which water quality is compromised and shrimp are not feeding. Commercial producers using these types of automated feeder will have to make adjustments to the feeders in real time to avoid wasting feed.

MANAGING WATER QUALITY IN PONDS USING INTENSIVE AUTOMATED FEEDING SYSTEMS

Feed management is a major driving force in pond water quality, particularly in semi-intensive and intensive systems in which an earthen pond is used as the rearing unit. The assimilation of nitrogenous waste from culture animals into the aquatic rearing environment is a product of the culture paradigm and must be managed accordingly. The concept of feed management and water quality management, hence, are very much intertwined for the commercial shrimp producer. The use of automated feeding systems for shrimp production, while documented to increase production and dramatically reduce the amount of time needed to grow a shrimp to market size, will present unique challenges in terms of managing water quality compared to other semi-intensive and intensive production schemes that rely on traditional feed management approaches.

To date, there are limited data on the dynamics of water quality in semi-intensive ponds fed with acoustically triggered automated feeding systems. Under similar experimental conditions, both Jescovitch et al. (2018) and Reis et al. (2020) did not observe a correlation between higher feed inputs and low oxygen occurrences. In these systems, DO is managed through aeration so such observations may be masked. A very large number of studies have evaluated water quality in semi-intensive systems for shrimp production using traditional feed management programs. Using this information, we can surmise the potential impacts based on the altered feeding regimes instituted by the various types of automated feeder.

Traditional feed management programs implemented on commercial shrimp farms rely solely on the decisions of the manager in terms of when the shrimp are fed (timing) and at what amount (feed rate). Automated feeders that are not acoustically based offer a higher degree of control in terms of the number of feedings per day, but since they are not based on demand there is still the possibility for wasted feed. Systems such as the AQ1 alter this approach and rely on shrimp in the pond to make the decision of when and how much to feed, based on real-time response of the shrimp. While the farm manager has control to a certain extent in programming traditional automated feeding devices, if using the AQ1 system shrimp can choose to feed freely at any time of the day or night. These advances in technology have resulted in improvements in growth and FCR using the AQ1 system, but there can be subsequent effects on pond water quality depending on a number of different factors including stocking density, standing crop biomass, water temperature, availability of supplemental aeration, algal blooms, pond history (particularly pond bottom sediment and soil) and the status of other water quality variables in the pond (dissolved oxygen, ammonia, nitrite, etc.), as previously discussed.

Passive acoustic feedback feeding systems for shrimp production have the potential to increase waste loading from an increased amount of feed entering the pond environment. While this has not yet been studied at length formally by the aquaculture research community, the potential for increased waste (due to higher feed amounts over shorter periods of time) entering the pond culture environment based on this altered approach is very likely. It is well documented that excess feed can lead to additional waste in aquaculture ponds (Boyd and Tucker, 2014). Since shrimp are essentially feeding on demand using passive acoustic feedback feeding systems, the amount of feed entering the pond may exceed traditional feeding approaches, particularly in systems having the ability to feed 24 hours per day. Subsequently, this could lead to larger amounts of waste in the pond. These systems have a proven ability to reduce the production period (number of weeks to attain harvest size), and hence larger amounts of feed (more daily feedings) must be assimilated by the pond environment in less time. Jescovitch et al. (2018), in a trial with Pacific white shrimp in which an AQ1 feeder was compared to three other feed management approaches, observed that the AQ1 feeder had superior yields compared to other treatments. However, these superior yields were accompanied by the highest total ammonia nitrogen and nitrite nitrogen levels (albeit within safe ranges for the culture of *P. vannamei*) late in the growing season. Jescovitch et al. (2018) point out that increasing feed rates, while having the advantage of increasing production, will also elevate nutrient loading of nitrogen, phosphorus and organic matter. Consequently, increased nutrients in commercial production ponds leads to a greater oxygen demand and can eventually lead to higher concentrations of toxic metabolites in the water column (Boyd, 2015a,b; Jescovitch et al., 2018). Altogether, pond dissolved oxygen levels and problematic blue-green algae should be tracked on farms implementing acoustic feed systems, as conditions may be favorable for these problems to arise, particularly late in the production cycle when feeding is at its highest peak. The ability to provide sufficient supplemental aeration in the case of low dissolved oxygen will likely be required under certain semi-intensive culture conditions.

Farms implementing acoustic systems such as the AQ1 should have robust protocols and control measures in place to monitor water quality throughout the production cycle. The later weeks of the production season, when larger amounts of feed are entering the pond, will likely be the period in which control measures, such as supplemental aeration, may be necessary. Further controlled studies will likely follow as acoustic systems gain acceptance and are implemented on commercial farms. Additional data should be available in the coming years on the effects of these feeding systems on pond water quality. It will also be important to evaluate the impact of acoustic feeding systems on pond nutrient budgets and soil quality.

Conclusion

Feed management is a primary factor in efficient shrimp farming across most production systems, since it affects multiple facets of both animals and the environment. Feed management has a direct impact on shrimp production, since it is the main source of nutrients, particularly in semi-intensive and intensive production systems. The aquaculture industry has developed a number of feed management techniques to gradually increase carrying capacity – and ultimately the productivity – of aquaculture production systems.

Automation in aquaculture is proving to be a very useful tool to increase production efficiency. Upfront investment in equipment and adaptation of staff and procedures are the main issues associated with transitioning to automated feeding technology. Farms with high-density systems and vast production areas require larger starting and running investments. Although this increases operational costs (feed, salaries, etc.), it also allows larger trade volume for commercial farms. Moving towards production automation will require a very high investment in technology installation, but is also likely to amplify profits in a matter of a few cycles. Various authors have reported efficiency improvements by increasing the number of meals and using acoustic feeding systems. Higher profits are typically attained by increasing overall production efficiency through better feed management, improved water quality and a decrease in labor-associated costs.

For farms with traditional systems willing to move towards automation, the staff must undergo extensive training on the appropriate use of the new technology. In the event of technical issues, staff need to be able to correct problems that arise (e.g. calibration, wiring or electrical issues) or have a backup plan. A backup plan may be feeder substitution via a backup device in case of feeder failure. For aeration control when underwater sensor system failure occurs, it may be simply to bypass the system until a repair is done. While these are important factors to consider, the determining factor in deciding whether a farmer should move towards production automation will always be its business model.

As reported by a number of authors under a variety of culture conditions and different species, there are many advantages associated with the integration of automated systems in aquaculture. There are currently many systems and devices available for different types of culture. Technology engineering for shrimp farming must focus on developing user-friendly systems and devices so that small-scale farmers can solve most of the day-to-day issues, particularly the technology for water quality management. In short, reliable, user-friendly technology, along with trained or qualified workers for technology management and maintenance, will likely reduce or minimize production problems and losses.

It is important to note that as feeding is intensified there will be a shift in the upper limit of feed-associated waste that a shrimp production pond can successfully assimilate. Beyond this point, deterioration in pond water quality will be an eventuality. Hence, increasing feed rates and production performance using automated systems will result in new management issues related to water quality that must be dealt with at both the pond and farm level. While there are indeed excellent and documented advantages to automated systems, farmers must learn to stay within the biological limits of their production systems. This may take some time following the shift from traditional feed management approaches to more modern automated techniques described in this chapter.

As is the case with the development of agriculture technologies in other sectors, technology transfer and the adaptation of this new feeding technology at the farm level may be facilitated by the sharing of information between research institutions, extension and outreach entities and commercial producers. The authors typically advise commercial producers to make a gradual shift to automated systems, and are in the process of working with shrimp farms in the USA on commercial level demonstrations using automated feeding technology. Commercial shrimp producers interested in this technology would be wise to first test automated feeding technology on a small number of their ponds prior to shifting all of their operation to this new approach. This will allow them time to evaluate whether this feeding technology is a good fit for their farm, and provide additional time to troubleshoot potential problems that might arise during the transition from traditional feed management to automated systems.

Aquaculture automation is likely to follow the same trend as most technological innovations, by gradually becoming more accessible (cheaper), user-friendly and reliable. However, the biggest challenge for these technologies will more than likely be convincing farmers to invest and adapt operational procedures to integrate new feeding technology. Widespread commercial implementation of this technology may involve an adaptation of the business model. Integration of automatic systems in aquaculture farms is not an event but a biological, environmental and logistic process.

References

Allan, G. F., Moriarty, D. J. W. and Maguire, G. B. (1995) Effects of pond preparation and feeding rate on production of *Penaeus monodon* Fabricius, water quality, bacteria, and benthos in model farming ponds. *Aquaculture* 130, 329–349.

Anderson, R. K. and Parker, P. L. (1987) A 13C/12C tracer study of the utilization of presented feed by a commercially important shrimp *Penaeus vannamei* in a pond growout system. *Journal of the World Aquaculture Society* 18(3), 148–155.

Bombeo-Tuburan, I. Guanzon Jr, N. G. and Schroeder, G. L. (1993) Production of *Penaeus monodon* (Fabricius) using four natural food types in an extensive system. *Aquaculture* 112, 57–65.

Boyd, C. E. (2015a) *Water Quality: An Introduction* (2nd edn). Springer, New York, NY, USA.

Boyd, C. E. (2015b) Overview of aquaculture feeds: global impacts of ingredient use. In: Davis, D. A. (ed.) *Feed and Feeding Practices in Aquaculture*. Woodhead Publishing, Cambridge, UK. pp. 27–52.

Boyd, C. E. and Tucker, C. S. (1998) *Pond Aquaculture Water Quality Management*. Kluwer Academic Publishers, Boston, MA, USA.

Boyd, C. E. and Tucker, C. S. (2014) *Handbook for Aquaculture Water Quality*. Craftmaster Printers, Auburn, AL, USA.

Boyd, C. E., Boyd, C. A. and Chainark, S. (2010) Shrimp pond soil and water quality management. In: Alday-Sanz, V. (ed.) *The Shrimp Book*. Nottingham University Press, Nottingham, UK, pp. 281–303.

Boyd, C. E., Tucker, C. S., McNevin, A., Bostick, K. and Clay, J. (2007) Indicators of resource use efficiency and environmental performance in aquaculture. *Reviews in Fisheries Science* 15, 327–360.

Carvalho, E.A. and Nunes, A.J. (2006) Effects of feeding frequency on feed leaching loss and grow-out patterns of the white shrimp *Litopenaeus vannamei* fed under a diurnal feeding regime in pond enclosures. *Aquaculture* 252, 494–502.

Chatvijitkul, S., Boyd, C.E., Davis, D.A. and McNevin, A.A. (2017) Embodied resources in fish and shrimp feeds. *Journal of the World Aquaculture Society* 48, 7–19.

Davis, D. A., Roy, L. A. and Sookying, D. (2008) Improving the cost effectiveness of shrimp feeds. In: L. Elizabeth Cruz Suarez, L. E et al. (eds). *Avances en Nutricion Acuicola IX*. IX Simposio Internacional de Nutricion Acuicola. 24–27 Noviembre. Universidad Autonoma de Nuevo Leon, Monterrey, Nuevo Leon, Mexico.

Fox, J., Treece, G. and Sanchez, D. (2001) Shrimp nutrition and feed management. *Methods for Improving Shrimp Farming in Central America*, 65–90.

Focken, U., Groth, A., Coloso, R. M. and Becker, K. (1998) Contribution of natural food and supplemental feed to the gut content of *Penaeus monodon* Fabricius in a semi-intensive pond system in the Philippines. *Aquaculture* 164, 105–116.

Green, B. W. (2015) Fertilizers in aquaculture. In: Davis, D. A. (ed.) *Feed and Feeding Practices in Aquaculture*. Woodhead Publishing, Cambridge, UK. pp. 27–52.

Hung, L. T. and Quy, O. M. (2013) On farm feeding and feed management in whiteleg shrimp (*Litopenaeus vannamei*) farming in Viet Nam. In Hasan, M. R. and New, M. B. (eds). *On-farm Feeding and Feed Management in Aquaculture*. FAO Fisheries and Aquaculture Technical Paper No. 583, FAO, Rome, Italy, pp. 337–357.

Hunter, B., Pruder, G. and Wyban, J. (1987) Biochemical composition of pond biota, shrimp ingesta, and relative growth of *Penaeus vannamei* in earthen ponds. *Journal of the World Aquaculture Society* 18(3), 162–174.

Jescovitch, L. N., Ullman, C., Rhodes, M. and Davis, D. A. (2018) Effects of different management treatments on water quality for Pacific white shrimp, *Litopenaeus vannamei*. *Aquaculture Research* 49, 526–531.

Kureshy, N. and Davis, D. A. (2002) Protein requirement for maintenance and maximum weight gain for the Pacific white shrimp, *Litopenaeus vannamei*. *Aquaculture* 204, 125–143.

Lee, C. and Lee, K. (2018) Dietary protein requirement of Pacific white shrimp *Litopenaeus vannamei* in three different growth stages. *Fisheries and Aquatic Sciences* 21, 30.

Martinez-Cordova, L., Torres, A. and Porchas-Cornejo, M. (2003) Dietary protein level and natural food management in the culture of blue (*Litopenaeus stylirostris*) and white shrimp (*Litopenaeus vannamei*) in microcosms. *Aquaculture Nutrition* 9, 155–160.

Moriarty, D. J. W. (1997) The role of microorganisms in aquaculture ponds. *Aquaculture* 151, 333–349.

Moss, S. M. (2002) Dietary importance of microbes and detritus in Penaeid shrimp aquaculture. In: Lee, C. S. and O'Bryen, P. (eds) *Microbial Approaches to Aquatic Nutrition within Environmentally Sound Aquaculture Production Systems*. World Aquaculture Society, Baton Rouge, LA, USA, pp. 1–18.

Moss, S. M., Pruder, G. D. Leber, K. M. and Wyban, J. A. (1992) The relative enhancement of *Penaeus vannamei* growth by selected fractions of shrimp pond water. *Aquaculture* 101, 229–239.

Napaumpaipom, T., Chuchird, N. and Taparhudee, W. (2013) Study on the efficiency of three different feeding techniques in the culture of Pacific white shrimp (*Litopenaeus vannamei*). *Journal of Fisheries and Environment* 37, 8 16.

Newell, R. I. E., Marshall, N., Sasekumar, A. and Chong, V. C. (1995) Relative importance of benthic microalgae, phytoplankton, and mangroves as sources of nutrition for penaeid prawns and other coastal invertebrates from Malaysia. *Marine Biology* 123, 595–606.

Nunes, A. J., Sabry-Neto, H., da Silva, F. H. P., de Oliveira-Neto, A. R. and Masagounder, K. (2019) Multiple feedings enhance the growth performance and feed efficiency of juvenile *Litopenaeus vannamei* when fed a low-fish meal amino acid-supplemented diet. *Aquaculture International* 27(2), 337–347.

Peixoto, S., Soares, R., Silva, J. F., Hamilton, S., Morey, A. and Davis, D. A. (2020) Acoustic activity of *Litopenaeus vannamei* fed pelleted and extruded diets. *Aquaculture* 525, 735307.

Quintero, H. E. and Roy, L. A. (2010) Practical feed management in semi-intensive systems for shrimp culture. In: Alday-Sanz, V. (ed.) *The Shrimp Book*. Nottingham University Press. Nottingham, UK, pp. 443–454.

Reis, J., Novriadi, R., Swanepoel, A., Guo, J.P., Rhodes, M. and Davis, D.A. (2020) Optimizing feed automation: improving timer-feeders and on demand systems in semi-intensive pond culture of shrimp *Litopenaeus vannamei*. *Aquaculture* 519.

Reymond, H. and Langardère, J. P. (1990) Feeding rhythms and food of *Penaeus japonicus* Bate (Crustacea, Penaeidae) in salt water ponds: role of halophilic entomofauna. *Aquaculture* 81, 125–143.

Roy, L. A., Davis, D. A. and Whitis, G. N. (2012) Effect of feeding rate and pond primary productivity on growth of *Litopenaeus vannamei* reared in inland saline waters of west Alabama. *North American Journal of Aquaculture* 74, 20–26.

Samocha, T. M. (2010) Use of intensive and super-intensive nursery systems. In: Alday-Sanz, V. (ed.) *The Shrimp Book*. Nottingham University Press. Nottingham, UK, pp. 247–280.

Silva, J. F., Hamilton, S., Rocha, J. V., Borie, A., Travassos, P., Soares, R. and Peixoto, S. (2019) Acoustic characterization of feeding activity of *Litopenaeus vannamei* in captivity. *Aquaculture* 501, 76–81.

Smith, D.V. and Tabrett S., (2013) The use of passive acoustics to measure feed consumption by *Penaeus monodon* (giant tiger prawn) in cultured systems. *Aquacultural Engineering* 57, 38–47.

Soares, R., Peixoto, C., Wasielesky, W. and D'Incao, F. (2005) Feeding rhythms and diet of *Farfantepenaeus paulensis* under pen culture in Patos Lagoon estuary, Brazil. *Journal of Experimental Marine Biology and Ecology* 322, 167–176.

Tacon, A. G. J. (1993) Feed ingredients for crustaceans: natural foods and processed feedstuffs. FAO Fisheries Circular No. 866, FAO, Rome, Italy.

Tacon, A. G. J. (2002) Thematic review of feeds and feed management practices in shrimp aquaculture. Report prepared under the World Bank, NACA, WWF and FAO Consortium program on shrimp farming ant the environment. Work in progress for public discussion. Published by the Consortium, 69 pp.

Tacon, A. G. J. (2017) Biosecure shrimp feeds and feeding practices: guidelines for future development. *Journal of the World Aquaculture Society* 48(3), 381–392.

Tacon, A. G. J., Cody, J. J., Coquest, L. D., Divakaran, S., Forster, I. P. and Decamp, O. E. (2002) Effect of culture system on the nutrition and growth performance of Pacific white shrimp *Litopenaeus vannamei* (Boone) fed different diets. *Aquaculture Nutrition* 8, 121–137.

Tacon, A. G. J., Jory, D. and Nunes, A. (2013) Shrimp feed management: issues and perspectives. In: Hasan, M. R. and New, M. B. (eds) *On-farm Feeding and Feed Management in Aquaculture*. FAO Fisheries and Aquaculture Technical Paper No. 583, FAO, Rome, Italy, pp. 481–488.

Tidwell, J. H., Coyle, S. D., Webster, C. D., Sedlacek, J. D., Weston, P. A., Knight, W. L., Hill Jr., S. J., D'Abramo, L. R., Daniels, W. H. and Fuller, M. J. (1997) Relative prawn production and benthic macroinvertebrate densities in unfed, organically fertilized, and fed pond systems. *Aquaculture* 149, 227–242.

Ullman, C., Rhodes, M. A. and Davis, D. A. (2018) Feed management and the use of automatic feeders in the pond production of Pacific white shrimp *Litopenaeus vannaemei*. *Aquaculture* 498, 44–49.

Ullman, C., Rhodes, M. and Davis, D.A. (2019a) The effects of feed leaching on the growth of Pacific white shrimp *Litopenaeus vannamei* in a green-water system. *Aquaculture Research* 50, 3074–3077.

Ullman, C., Rhodes, M., Hanson, T., Cline, D. and Davis, D. A. (2019b) Effects of four different feeding techniques on the pond culture of Pacific white shrimp, *Litopenaeus vannamei*. *Journal of the World Aquaculture Society* 50(1), 54–64.

Van, T.P.T.H., Rhodes, M.A., Zhou, Y. and Davis, D.A. (2017) Feed management for Pacific white shrimp Litopenaeus vannamei under semi-intensive conditions in tanks and ponds. *Journal of Aquaculture Research* 48, 5346–5355.

Chapter 14

MINERAL REQUIREMENTS OF SHRIMP

YANGEN ZHOU, MING HUANG, JIAN GE AND D. ALLEN DAVIS

Summary

This chapter discusses the biological role of macrominerals (calcium, phosphorus, potassium, magnesium, sodium, and chloride) and trace minerals (e.g. copper, iron, manganese, zinc, and selenium) in shrimp nutrition. The mineral nutrition requirements for shrimp and the sources of minerals for shrimp feed are discussed. Symptoms of mineral deficiency are also presented. It is evident that minerals, albeit in small quantities, are an essential part of the shrimp diet.

The importance of mineral elements

Mineral elements are found in every living organism on earth, acting as both intra- and extracellular hinges. They serve as structural components in the formation of hard tissue such as exoskeleton and soft tissue. They also participate in the synthesis of metalloproteins, which are cofactors and activators of many enzymatic systems (Méndez et al., 2001), support metabolism (Frenet and Alliot, 1985), and are involved in the synthesis of nucleic acids and hormones (Moyle and Cech, 2000; Tolonen et al., 1995). In fact, except for osmoregulation (Lovell, 1989), which is maintained by certain soluble mineral elements (e.g., potassium, sodium, calcium, phosphorus, and chloride), the biochemical functions of minerals in aquatic species are identical to those in terrestrial animals.

Mineral elements can broadly be classified into two groups: the seven macrominerals (calcium, chlorine, magnesium, phosphorus, potassium, sodium, and sulfur) and 16 trace minerals (aluminium, arsenic, cobalt, chromium, copper, fluorine, iodine, iron, manganese, molybdenum, nickel, selenium, silicon, tin, vanadium, and zinc). The function of macrominerals, which are defined as minerals found in the diet and body at relatively high concentrations, include electron transfer, acid–base equilibrium regulation, hard tissue formation, membrane potential generation, and osmoregulation. Trace minerals or microminerals, which are typically required in the diet and body at much lower concentrations than the macrominerals, are important components of hormones and enzymes, serve as activators or cofactors for a variety of enzymes, and participate in a wide variety of biochemical process (Lall, 2003; NRC, 2011).

In contrast to terrestrial animals, which absorb minerals only from their diet, aquatic species also obtain minerals from the aquatic medium in which they live. The uptake of numerous minerals from the aquatic medium or diet is mainly accomplished by the gills and intestine, respectively, and excess minerals are excreted in urine, feces as well as the molted exoskeleton. The mechanism and process of osmoregulation has been described specifically by Charmantier (1998) and Lockwood (1962).

In this chapter, we will further elaborate the biochemical functions of mineral elements, determination of nutritional mineral requirements of shrimp, and sources of minerals for shrimp feed development; in addition, we will set out some limitations in our knowledge on the requirements for minerals in shrimp.

Biochemical functions of minerals in shrimp

MACROMINERALS

Sodium and chloride

Sodium and chloride are essential for many physiological processes, including acid–base balance (Lall, 2003). Meanwhile, they are the principal osmotically active solutes in the hemolymph, accounting for 76-94% of hemolymph osmotic pressure (Castille Jr and Lawrence, 1981). As sodium and chloride are typically abundant in water and feedstuffs, the metabolic deficiencies of these minerals have rarely been observed in crustaceans (NRC, 2011).

Owing to recent interest in culturing shrimp, such as Pacific white shrimp *Litopenaeus vannamei*, in low-salinity groundwaters, the effects of dietary supplementation of sodium chloride in this species have been evaluated in low-salinity water. Sodium chloride supplementation at 1 and 2% of diet appeared to promote survival but did not improve weight gain in *L. vannamei* in artificial water with 4% salinity (Roy et al., 2007). In addition, supplementation of sodium chloride in diets to improve the meat quality of *L. vannamei* reared in low-salinity water has also been studied (Zhou et al., 2014).

Potassium

Potassium is essential for maintaining normal physiological activities and growth in shrimp (NRC, 2011; Zhu et al., 2014). It mainly exists in the fluid within the cell and plays an important role in maintaining cellular acid–base balance, ion balance, water balance, and osmotic pressure (Evans et al., 2005; Marshall and Bryson, 1998). In addition, potassium plays important roles in protein synthesis, carbohydrate metabolism, phosphorus metabolic reactions, nutrient transport, and electrical conduction and transmission of nerve impulses in nerve and muscle cells (Black, 1993; Lall, 2003; Luft, 1990).

To date, limited information on deficiency or imbalances of potassium in shrimp feed is not available. Only a few studies note a benefit to potassium supplement, while excess potassium intake in the water may depress the animal's growth (Zhu et al., 2014). In addition, an imbalance between potassium and sodium concentrations in the hemolymph can cause mortality in prawns (Liu et al., 2014).

Calcium

Calcium is a versatile mineral element in shrimp. It primarily functions as a structural component of hard tissue, such as the exoskeleton. In addition, calcium is essential for various forms of physiological regulation. It plays a vital role in the activation of enzymes, including adenosine triphosphatase (ATPase), succinic dehydrogenase, and lipase (Soetan et al., 2010). It is also required for muscle contraction, membrane permeability, osmoregulation, and nerve transmission (Ebashi and Endo, 1968; Huang et al., 2011; NRC, 2011). In addition, calcium ion triggers the molting progress in shrimp (Hou et al., 2016).

Growth performance is inhibited by excessive calcium levels in several species (Maguire and Hume, 1982; Pan et al., 2005; Peñaflorida, 1999). In addition, high levels of calcium result in inhibitory effects on the absorption of phosphorus (Cheng et al., 2006; Davis et al., 1993c; Peñaflorida, 1999).

Dietary deficiency of calcium is generally not noted in shrimp because of the presence of calcium ions in the aquatic environment. Aquatic animals can freely absorb calcium from the environment by drinking water and directly absorbing the ions through the gills and epidermis (Diet, 1978).

Magnesium

Magnesium is the second most abundant cation within the cell, and is mainly bound to nucleic acids and negatively charged phospholipids and proteins (Vormann, 2003). In crustaceans, magnesium occurs both intra- and extracellularly in soft tissues and it is essential for the maintenance of intra- and extracellular homeostasis. Magnesium is essential for respiration and phosphate-transfer reactions involving adenosine tri-, di-, and monophosphate (NRC, 2011). In addition, magnesium is a cofactor in numerous enzymatic and metabolic pathways, including the metabolism of lipids, proteins, and carbohydrates (Cheng et al., 2005; Davis and Gatlin III, 1996). Under low-salinity conditions, magnesium level of the water has been correlated with shrimp survival and osmoregulation (Saoud et al., 2003).

Magnesium is essential for formulating the ATP-Mg^{2+} complex (Robinson and Pratap, 1991), especially Mg^{2+}-dependent Na^+, K^+-ATPase and Mg^{2+}-ATPase, which are both important in osmotic and ionic regulation during acclimation to low-salinity medium (Cheng et al., 2005). In the absence of magnesium, Na^+, K^+-ATPase was found to not hydrate ATP in crustaceans (Furriel et al., 2000).

Phosphorus

Phosphorus has been reported to play multiple roles in shrimp. It is directly involved in all energy-yielding reactions as it is a key component of nucleic acids, phospholipids, phosphoproteins, ATP, and several crucial enzymes (Cheng et al., 2006). In addition, phosphate serves as a buffer to maintain optimal pH in body fluids (Chaitanawisuti et al., 2010). Moreover, phosphorus is related to alkaline phosphatase, which plays a role in response to salinity acclimation and is associated with osmoregulation in crustaceans (Lovett et al., 1994; Pinoni and Mananes, 2004). In association with calcium, phosphorus is mainly found in the exoskeleton of shrimp.

Phosphorus deficiency can occur in most aquatic species because its presence in the water at low levels and its utilization by fish and shrimp are limited. Dietary deficiency of phosphorus impairs intermediary metabolism, resulting in reduced growth and feed conversion (NRC, 2011). In addition, reduced mineralization of hard tissues has been reported following suboptimal phosphorus intake (Sugiura et al., 2004).

TRACE MINERAL ELEMENTS

Copper

The requirement for copper in shrimp is well documented (Zhou et al., 2017a). First, copper acts as a cofactor of proteins and enzymes such as ceruloplasmin, superoxide dismutase, cytochrome *C* oxidase, lysyl oxidase, tyrosinase, and ferroxidase, which are involved in many important biological processes including electron transport, scavenging of free radicals, and collagen synthesis (NRC, 2011; Yuan et al., 2019). Second, a certain amount of dietary copper is necessary for shrimp to achieve optimal growth performance and tissue mineralization (NRC, 2011), and for molting and reproduction (Rao and Anjaneyulu, 2008; Zhou et al., 2017b). In addition, crustaceans utilize hemocyanin, a copper-containing protein, to transport oxygen (Burmester, 2002; Lovell, 1989). It has been estimated that on a fresh-weight basis, 40% of the whole-body copper load in shrimp is found in the hemocyanin (Depledge, 1989). Copper deficiency can lead to several conditions in shrimps, including poor growth; reduced copper levels in the carapace, hepatopancreas, and hemolymph; and enlargement of the heart (Davis et al., 1993a).

Although copper is an essential element, its high levels in the diet and environment may result in toxicity in aquatic animals (Bryan, 1976; Sun et al., 2013). This may be because heavy metals are commonly associated with the formation of reactive oxygen species, leading to oxidative stress and causing cell damage. In addition, dietary copper plays a role in the innate immunity of shrimp species such as *Penaeus vannamei* (Davis et al., 1993a) and *P. monodon* (Lee and Shiau, 2002).

Iron

Iron is a component of cytochromes, catalases, and peroxidases in all aerobic cells and is essential for the synthesis and normal functioning of hemoglobin and myoglobin (Davis et al., 1992a; NRC, 2011). Iron is absorbed and transported in the body in a protein-bound form (Lovell, 1989). In vertebrates, mucosal transferrin binds to Fe^{2+} in the intestinal lumen and transports it across the mucosal border. Within cells, Fe^{3+} is bound to apoferritin forming ferritin (NRC, 2011). Iron-transporting proteins have also been found in the hemolymph of two crab species (Depledge et al., 1986; Ghidalia et al., 1972), which indicates the presence of a regulatory mechanism similar to that in vertebrates. In addition, iron plays a role in the immune system and helps defend against various infections. It appears that its availability to microorganisms affects their virulence (NRC, 2011; Srinivasan et al., 2016).

Although iron has physiological functions in shrimp, its excessive levels may be toxic. Excessive supplementation of iron had a negative effect on the growth of juvenile *P. japonicas* (Deshimaru et al., 1978; Kanazawa et al., 1984). Moreover, iron is a major destabilizing agent in feeds as it increases lipid oxidation and reduces the stability of ascorbic acid (Desjardins et al., 1987; Hilton, 1989; Sutton et al., 2006).

Manganese

Manganese is necessary for normal growth, immunity, reproduction, and bone development in aquatic animals. It is involved in biochemical systems as an enzyme cofactor or active group – for example, in superoxide dismutase (Cai et al., 2017). Furthermore, manganese plays a significant role in improving survival, muscle composition, and stress tolerance in crustaceans (Asaikkutti et al., 2016; Lu et al.,

2018; Xian et al., 2013). Symptoms of manganese deficiency in fish include reduced growth, impaired reproduction, and skeletal abnormalities; however, information on manganese deficiency in shrimps is limited (NRC, 2011).

Zinc

Zinc is required for normal growth, development, and function in animal species (NRC, 1980). It is essential for a wide array of metabolic functions in organisms, such as stabilizing the cellular membranes, tissue components, and organ fluids (Kidd et al., 1996; Thirunavukkarasu et al., 2019; Yamaguchi, 1998). Zinc also regulates cell division, protein synthesis, energy balance, reproduction, the immune system, and stress tolerance, as well as the metabolism of nutrients including vitamin A, lipids, and carbohydrates (Muralisankar et al., 2014; Thirunavukkarasu et al., 2019). Zinc acts as a cofactor in several enzyme systems and is a component of many metalloenzymes, including carbonic anhydrase, carboxypeptidases A and B, glutamic dehydrogenase, alcohol dehydrogenase, lactate dehydrogenase, D-glyceraldehyde-3-phosphate dehydrogenase, malic dehydrogenase, alkaline phosphatase, aldolase, ribonuclease, superoxide dismutase and DNA polymerase (NRC, 2011).

The excess absorption of zinc suppresses reproductive performance, growth, muscle composition, and copper and iron utilization (Muralisankar et al., 2014, 2015; Shiau and Jiang, 2006). Moreover, Muralisankar et al. (2014) have reported the toxic effects of zinc-oxidized nanoparticles, which induce oxidative stress in shrimp.

Muralisankar et al. (2014) have reported that zinc deficiency leads to poor survival, feed intake, and growth in *Macrobrachium rosenbergii*. In addition, zinc deficiency can suppress the functioning of signaling molecules and proteins that are directly involved in DNA replication and repair in an organism; moreover, it can affect the digestibility and metabolism of protein and carbohydrates owing to the reduced activity of carboxypeptidase (Muralisankar et al., 2015).

Selenium

Selenium is a fundamental nutrient in animals, including crustaceans (Kong et al., 2017; NRC, 2011). In general, selenium affects the growth, development, and health of organisms directly (Kong et al., 2017; Sritunyalucksana et al., 2011). It is also an effective exogenous antioxidant that removes and prevents oxidative stress. It is an integral component of glutathione peroxidase (Rotruck et al., 1973), and the function of this enzyme is complementary to that of vitamin E, which protects cell membranes against oxidative damage by implementing reduced glutathione to catalyze reactions necessary for the conversion of fatty acid hydroperoxides and hydrogen peroxide to fatty acid, alcohol, and water (Droge, 2002; Lin and Shiau, 2005). In addition, a certain level of dietary selenium can improve the immune response, disease resistance, stress remission, and nutritional quality (i.e. alter protein and carbohydrate metabolism) in crustaceans (Qin et al., 2016; Wang et al., 2019; Yuan et al., 2018). Moreover, interaction between selenium and mercury has been reported (Bjerregaard et al., 2018; Cuvin-Aralar and Furness, 1991; Turner and Rudd, 1983). Bjerregaard and Christensen (2012) found that selenium supplemented with food can reduce whole-body retention of radiolabeled methylmercury in the brown shrimp *Crangon crangon*. Dietary selenium deficiency or excess can significantly reduce growth, antioxidant ability, and immunity and can induce lipid peroxidation in shrimp (Chiu et al., 2010; Kong et al., 2017).

Chromium

Chromium is an essential element in animals. However, to date, research on the influence of dietary chromium in fish has been rather limited and even less effort has been directed toward crustaceans (NRC, 2011; Soetan et al., 2010). Cai et al. (2016) reported that dietary chromium sources and levels have an effect on the body composition of *L. vannamei* and can regulate the activities of phenoloxidase, alkaline phosphatase, acid phosphatase, and total superoxide dismutase in serum. One of the major metabolic roles of chromium is associated with glucose tolerance factor, an organometallic molecule that stimulates the action of insulin in carbohydrate metabolism (NRC, 2011). In addition, chromium plays a role in maintaining the configuration of the RNA molecule as it has been shown to be particularly effective as a cross-linking agent for collagen (Eastmond et al., 2008).

Other minerals

Several other trace minerals have been deemed essential for vertebrates; however, much less information on their role is available in crustaceans. Nearly every cell in animals contains iodine and, in vertebrates, this mineral is associated with the function of thyroid hormones (NRC, 2011). Cobalt is required as a constituent of vitamin B_{12} and has the same metabolic role (NRC, 2011). In addition to its role in vitamin B_{12}, cobalt is also a cofactor of enzymes involved in DNA biosynthesis and amino acid metabolism (Arinola et al., 2008). Several studies have reported that cobalt supplementation can improve the survival and growth of fish (NRC, 2011). Molybdenum is a component of several metalloenzymes, including xanthine oxidase, aldehyde oxidase, nitrate reductase, and hydrogenase, which play important roles in many physiological functions (Soetan et al., 2010). In contrast, Pierson and Aanes (1958) reported that excess molybdenum may cause copper deficiency. However, there is limited information regarding this in shrimp. Other inorganic elements including arsenic, bromine, barium, strontium, and cadmium are potentially required by the body, but their essentiality can only be established under highly controlled conditions.

Determination of mineral nutritional requirements

Information concerning the mineral nutrition of crustaceans is limited compared with that of most other species (NRC, 2011). In addition, there is much less information on the mineral requirements of aquatic species compared with those of terrestrial animals because of the difficulty involved in conducting research on mineral nutrition of aquatic species (NRC, 2011), as well as the number of species that are cultured. Although research on aquatic animal nutrition is still in its relative infancy, there has been progress regarding mineral requirement data for crustaceans in diverse environments.

METHODS OF ANALYSIS AND METHODOLOGICAL CONSTRAINTS

In the few existing studies on marine crustaceans, the most extensive examinations of mineral requirements have used semi-purified diets under controlled laboratory conditions (Davis et al., 1992b).

However, unlike most terrestrial animals, shrimps have the ability to absorb some inorganic elements not only from their diets but also from their external environment in both freshwater and seawater, which complicates the determination of quantitative dietary requirements (Davis and Gatlin III, 1996). In addition, leaching of minerals from the diet prior to consumption, availability of suitable test diets that have a low concentration of the targeted mineral, and limited data on mineral bioavailability are the constraints in this methodology (NRC, 2011).

Many trace elements are required in such small amounts that it is difficult to formulate purified diets with low mineral content and maintain the water sufficiently free of the test element (Lall, 2003). The most commonly used measure of nutrition status is the level of trace elements in the blood, muscle, and liver. For many essential elements, there is a range of tissue levels compatible with optimal growth and function. Level of mineral intake influences its tissue concentration, which causes a gradual decline in the function of an organ until toxicity and deficiency occur (Fig. 14.1). However, the intermediate stage is difficult to establish, and this method may be unsuitable for the determination of macro elements, such as phosphorus (Peñaflorida, 1999). Moreover, despite advances in the instrumental analysis of trace elements that make lower detection limits possible, there are still many problems associated with accurate measurement in shrimp tissue as well as considerable variability often caused by the molt cycle. Other measures of mineral nutrition status are those that test certain metabolic function of the element rather than analyze a tissue. Nevertheless, well-defined functional tests to assess nutritional status exist only for some elements, such as iodine; for other elements, functional tests are not well established (Lall, 2003). Elucidation of techniques that allow the identification of subclinical and pathological changes in the assessment of nutritional status remains a challenging area for mineral nutrition research.

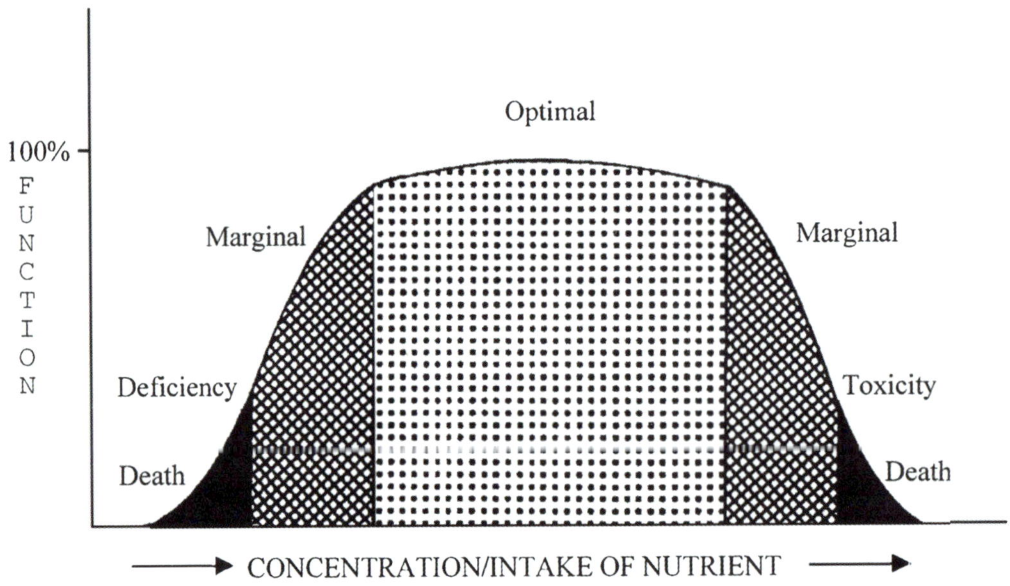

FIGURE 14.1 Biological dose–response curve. Dependence of animal function on the intake of an essential nutrient according to Mertz (1986). Credit: The Academic Press.

REQUIREMENTS OF SHRIMP

The mineral requirement of shrimp is complicated and depends on species, environmental conditions, and the interaction between minerals. Based on the table given in NRC (2011), we summarize the mineral requirements of shrimp by also including data from several recent studies (Tables 14.1 and 14.2). The macrominerals calcium, magnesium, sodium, and chloride are abundant in water and feedstuffs, making the necessity of their dietary supplementation unlikely. However, it is impossible for shrimp to absorb significant amounts of phosphorus from freshwater and seawater as its concentrations in natural water are generally very low. Phosphorus requirement estimates for shrimps in early studies were recommended at 1–2% of the diet (Deshimaru et al., 1978; Kanazawa et al., 1984; Pan et al., 2005). Besides, dietary phosphorus requirements are dependent on the dietary calcium content. Cheng et al. (2006) reported that weight gain in *L. vannamei* was optimal with available phosphorus at a dietary concentration of 0.77% in the absence of supplemental calcium; however, an increase to 1.22% of available phosphorus was needed when 1% supplemental calcium was administered. Similarly, Peñaflorida (1999) demonstrated that, based on weight gain, *P. monodon* required 0.74% dietary phosphorus when the diet was not supplemented with calcium; however, higher phosphorus was needed when the diet was supplemented with calcium.

Potassium nutrition in crustaceans has been studied to an appreciable extent, especially in marine shrimp. Based on growth performance, the dietary requirement of potassium is approximately 1% for *Marsupenaeus japonicus* (Deshimaru et al., 1978; Kanazawa et al., 1984) and at least 1.2% for *P. monodon* (Shiau and Hsieh, 2001). Liu et al. (2014) reported that a diet containing 1.48 K^+ g/100 g diet improved the growth of *L. vannamei* in low-salinity seawater and enhanced the physiological acclimation of the organisms. Potential interaction between potassium and magnesium has been reported by Davis et al. (1992b). In their study, elimination of magnesium from the mineral premix resulted in a considerable reduction in potassium content of the carapace. In contrast, the elimination of potassium from the mineral premix produced a considerable reduction in the magnesium level of the hepatopancreas. Moreover, the addition of potassium to the water can correct imbalances in sodium and chloride and improve growth and survival effectively (Roy et al., 2007; Saoud et al., 2007).

Some trace minerals, including aluminium, arsenic, cobalt, fluorine, molybdenum, nickel, silicon, tin, and vanadium, are typically required in such small amounts that dietary supplementation is not required. The main trace minerals studied in shrimp research are copper, iron, manganese, selenium, and zinc.

Shrimp cannot meet their physiological needs for copper from water, so a dietary source is required for maximum growth and tissue mineralization. Davis et al. (1993a) reported deficiency signs in *L. vannamei* when dietary copper was less than 34 mg/kg. The dietary copper requirement varies in different species based on different assessment indicators. Based on growth and tissue mineralization, the dietary copper requirement was 25.3 mg/kg for the Chinese shrimp *Fenneropenaeus chinensis* (Wang et al., 1997). Based on growth, survival, cytochrome c oxidase activity, and tissue mineralization, Liu et al. (1990) reported a dietary copper requirement of 53 mg Cu/kg diet for *Penaeus orientalis*. Based on weight gain, feed efficiency, and body copper retention, 15–21 mg Cu/kg was required by *P. monodon* (Lee and Shiau, 2002).

Shrimp can absorb iron from water through the gills while some crustaceans, including crayfish (*Procambarus clarkii*) (Miyawaki et al., 1961; Ogura, 1959) and crab (*Cancer irroratus*) (Martin, 1973), have a storage cell containing iron. The hepatopancreas is the organ that is richest in iron. Although dietary iron deficiency has not been reported for shrimp, excessive dietary iron levels may be toxic (NRC, 2011).

TABLE 14.1 Macromineral requirements in shrimp. Credit: National Academies Press.

Macromineral	*Species*	*Dietary protein(s)*	*Requirement (g/100g)*	*Salinity (‰)*	*Reference(s)*
Calcium	Kuruma prawn (*Marsupenaeus japonicus*)	Casein-egg	Dispensable	Seawater	Deshimaru et al. (1978)
		Squid meal	1.2		Kitabayashi et al. (1971)
		Casein	1.0–2.0		Kanazawa et al. (1984)
	Pacific white shrimp (*Litopenaeus vannamei*)	Casein-gelatin	Dispensable	28 ± 1	Davis et al. (1993c)
	Tiger shrimp (*Penaeus monodon*)	Casein-gelatin	Dispensable	Seawater	Peñaflorida (1999)
Phosphorus	*M. japonicus*	Casein-egg	2	Seawater	Deshimaru et al. (1978)
		Casein	1.0–2.0		Kanazawa et al. (1984)
		Squid meal	1		Kitabayashi et al. (1971)
	L. vannamei	Casein-gelatin	0.03% dietary Ca, ≤0.34% P 1% Ca, 0.5–1.0% P 2% Ca, 1.0–2.0% P	28 ± 1	Davis et al. (1993c)
		Casein-gelatin	0.5% Ca, 0.93% total P (EAP = 0.77% P) 1.5% Ca, 2% total P (EAP = 1.22% P)	2	Cheng et al. (2006)
		Fish meal, soybean meal	>1.33%, practical diet	17–20	Pan et al. (2005)
	P. monodon	Casein-gelatin	0.5% (0.74 total) at low Ca	Seawater	Peñaflorida (1999)
Potassium	*L. vannamei*	Casein-gelatin	Unclear	28 ± 1	Davis et al. (1993c)
		Fish meal, corn starch, casein	1.48	4	Liu et al. (2014)
	M. japonicus	Casein-egg	1	Seawater	Deshimaru et al. (1978)
		Casein	0.9		Kanazawa et al. (1984)
	P. monodon	Casein	1.2	19–21	Shiau and Hsieh (2001)
		Casein	1.09	30	Zhu et al. (2006)
Magnesium	*M. japonicus*	Casein-egg	Dispensable	Seawater	Deshimaru et al. (1978)
		Casein	0.3		Kanazawa et al. (1984)
	L. vannamei	Casein-gelatin	0.26–0.35	2	Cheng et al. (2005)

Note: EAP, estimated available phosphorus.

TABLE 14.2 Micromineral requirements in shrimp. Credit: National Academies Press.

Micromineral	*Species*	*Dietary protein(s)*	*Requirement (mg/kg diet)*	*Salinity (‰)*	*Reference(s)*
Copper	*Marsupenaeus japonicus*	Casein	Dispensable		Kanazawa et al. (1984)
	Fleshy prawn (*Penaeus orientalis*)	Fish meal, peanut meal	53		Liu et al. (1990)
	L. vannamei	Casein-gelatin	16–32	30 ± 1	Davis et al. (1993a)
		Casein, fish meal	17.52–20.32	28–32	Zhou et al. (2014a)
	P. monodon	Casein	10–30	19–21	Lee and Shiau (2002)
		Fish meal, corn gluten meal	25–55	20	Xian and Wang (2013)
	Fleshy prawn (*Fenneropenaeus chinensis*)		25.3	22–25	Wang et al. (1997)
	M. japonicus	Casein-egg	Dispensable		Deshimaru et al. (1978)
Iron	*M. japonicus*	Casein	Dispensable		Kanazawa et al. (1984)
	L. vannamei	Casein-gelatin	Dispensable	30 ± 1	Davis et al. (1992a)
	M. japonicus	Casein	Dispensable		Kanazawa et al. (1984)
	Macrobrachium rosenbergii	Fish meal, soybean meal	20 (Fe_2O_3 metal nanoparticles)	Freshwater	Srinivasan et al. (2016)
Manganese	*M. japonicus*	Casein	Dispensable		Kanazawa et al. (1984)
	L. vannamei	Fish meal, soybean meal	150–200 (growth, antioxidant activity)	5	Lu et al. (2018)
		Casein, red fish meal	32.26 (Mn-S) or 23.90 (Mn-met)	28–30	Cai et al. (2017)
Selenium	*L. vannamei*	Casein-gelatin	0.2–0.4		Davis (1990)
	Macrobrachium nipponense	Casein, fish meal	0.58–0.68 (growth, glutathione peroxidase)		Kong et al. (2017)
Zinc	*P. monodon*	Casein	32–34 (growth) 35–48 (immunity)	19–21	Shiau and Jiang (2006)
	L. vannamei	Casein-gelatin	15 (32 total) 200 in the presence of phytate	30 ± 1	Davis et al. (1993b)
	M. rosenbergii	Fish meal, soybean meal	60	Freshwater	Muralisankar et al. (2015)
	M. rosenbergii	Fish meal, soybean meal	60 (zinc nanoparticles)	Freshwater	Thirunavukkarasu et al. (2019)
	M. nipponenseare	Casein	31.90 (growth) 58.62 (carapace zinc deposition)	Freshwater	Guo et al. (2013)

Notes: Mn-s: manganese sulphate; Mn-met: manganese methionine.

Srinivasan et al. (2016) reported that the general health and non-specific immunity of *M. rosenbergii* were improved following up to 20 mg/k of Fe_2O_3 metal nanoparticle supplementation in the diet; however, supplementation beyond this optimal level possibly induced toxicity. Additionally, iron-catalyzed lipid oxidation increases with iron supplementation, which reduces the stability of feed (Desjardins et al., 1987; Sutton et al., 2006) and ascorbic acid (Hilton, 1989). Thus, excessive iron supplementation of crustacean diets should be avoided.

There is a growing body of available information on the manganese requirements of crustaceans. Davis et al. (1992b) reported that the deletion of manganese from the mineral premix resulted in a significant reduction of the manganese content in the carapace of *P. vannamei*, indicating dietary influence. Recently, Lu et al. (2018) reported that a suitable range of manganese supplemental dose for sub-adult *L. vannamei* is 150–200 mg/kg, based on growth performance and antioxidant activity indicators. Kuhn et al. (2017) found that levels of dietary manganese >570 mg/kg significantly impacted the growth of *L. vannamei*. Cai et al. (2017) reported that broken-line regression analysis of weight gain rate indicated that the optimal dietary manganese requirements for juvenile *L. vannamei* were 32.26 mg/kg manganese sulfate and 23.90 mg/kg manganese methionine.

Information on selenium nutrition of shrimp is currently rather limited. Davis (1990) found that *L. vannamei* grew best when fed semi-purified diets supplemented with 0.2–0.4 Se/kg diet. Kong et al. (2017) reported that the recommended minimum dietary selenium level for maximal growth and antioxidant ability of juvenile *Macrobrachium nipponense* was 0.58–0.68 mg/kg diet based on the analyses of weight gain rate and the activities of glutathione peroxidase. Moreover, 1.07 mg Se/kg diet is required to maximize whole-body selenium content.

Zinc nutrition for various shrimp species has also been investigated. Davis et al. (1993b) reported that *L. vannamei* required 33 mg Zn/kg diet to maintain normal tissue mineralization, although growth was not affected by graded levels of dietary zinc. Shiau and Jiang (2006) found that 32-34 mg Zn/kg in the diet was required for *P. monodon* based on weight gain and whole-body zinc retention. In the freshwater prawn *M. rosenbergii*, both Muralisankar et al. (2015) and Thirunavukkarasu et al. (2019) reported that 60 mg Zn/kg can be supplemented with basal feeds to achieve better growth. Guo et al. (2013) reported that the optimum dietary zinc requirements for maximum growth and carapace zinc deposition of *M. nipponense* are 31.90 and 58.62 mg/kg, respectively.

Sources of minerals for shrimp feed development

SOURCE AND FORMS

Among the key components used in practical animal diets, fish meal is the richest in endogenous minerals. Other feedstuffs of animal origin such as meat and bone meal, feather meal, and poultry by-product meal are comparatively rich in minerals. However, as the aquatic animal feed industry reduces the use of animal feedstuffs owing to the shortage of fish meal and the limitations of other animal-based feedstuffs, there is increased use of plant feedstuffs. These feeds often have lower levels of some minerals, contain potential inhibitory factors such as phytase, and consequently lower the bioavailability of minerals.

As the absorption of calcium and phosphorus depends on the solubility at the contact point with the absorbing membranes, mineral sources that are soluble at luminal pH are more likely absorbable.

Generally, the bioavailability of phosphorus is positively correlated with the solubility of the mineral in water and has been indicated in NRC (2011). The bioavailability of phosphorus in marine shrimp appears to be intermediate compared with that in fish, both with and without a true stomach. Apparent phosphorus availability value for *L. vannamei* from calcium phosphate monobasic is 46%, calcium phosphate dibasic is 19%, calcium phosphate tribasic is 10%, potassium phosphate monobasic is 68%, and sodium phosphate monobasic is 68% (Davis and Arnold, 1994).

In addition to the chemical form (solubility, pH) affecting mineral availability, other contents in diets can influence the availability of certain minerals. For example, lactose may increase the permeability of absorptive cells to calcium ions by interacting with them; high intakes of iron, aluminium, and magnesium may inhibit the absorption of phosphorus by forming insoluble phosphates; and fats may interact with calcium, inducing the synthesis of insoluble soaps (Maynard et al., 1979). Therefore, the bioavailability of calcium and phosphorus may be dependent on the mineral form (solubility at intestinal pH), and on the dietary levels of calcium, phosphorus, vitamin D, iron, aluminium, manganese, potassium, magnesium, and fat.

Phosphorus in fish meal exists mainly in the form of hydroxyapatite originating from hard tissues such as bone. Davis and Arnold (1994) have reported the apparent absorption of phosphorus from fish meal to be 46.5% for *L. vannamei*. Phytate phosphorus, which constitutes approximately 67% of the phosphorus in plant feedstuffs, is poorly available to shrimp (Civera and Guillaume, 1990; Davis et al., 1993b). The presence of phytate reportedly inhibits the availability of dietary calcium and phosphorus to *L. vannamei* and *M. japonicus* (Civera and Guillaume, 1990), as well as the availability of 1.5% zinc to *L. vannamei* (Davis et al., 1993b). Specifically, the apparent availability of phytate phosphorus was determined to be 8.4% for *L. vannamei* and 47.3% for *M. japonicus*. Consequently, phytase supplementation of plant feedstuffs or complete diets may limit the availability of specific minerals from this dietary component. In addition, almost all phytases perform optimally between pH 4.0 and 6.0, which may be potential limitations for a dietary supplement for carps and shrimp (Lemos and Tacon, 2017).

TRADITIONAL INORGANIC MINERAL SOURCES AND SYNTHETIC ORGANIC MINERAL PRODUCTS

The bioavailability of minerals chelated to organic molecules has generally been reported to be more efficient than inorganic forms. If an element is chelated by a compound that will release it in the ionic form at the site of absorption, or will be readily absorbed as the intact chelate, this form may greatly enhance the absorption of the element by avoiding its conversion to insoluble chemical compounds in the intestine or by preventing its strong adsorption on insoluble colloids (Scott et al., 1982). Compared with inorganic sources, chelated minerals are also generally less sensitive to the inhibitory action of other compounds (phytate and fibre) and may have higher bioavailability in practical feeds. Another potential benefit of mineral chelates for aquatic species is reduced solubility in water, which reduces the unnecessary loss of materials. However, many mills use mineral chelates regularly even though the cost of most mineral chelates is high.

Several different forms of organic or chelated minerals have been tested for various elements. Amino acid chelates of copper, zinc, manganese, magnesium, and potassium were determined to be more readily absorbable than their inorganic sources, and were determined to potentially enhance the growth performance in *L. vannamei* (Bharadwaj et al., 2014; Katya et al., 2016; Lin et al., 2013; Roy et al., 2009; Saoud et al., 2007). Copper-exchanged montmorillonite is another relatively uncommon form of supplemental copper. When it was included at 1.5 g/kg diet to provide 30 mg of Cu/kg, it increased the growth of Nile tilapia *Oreochromis niloticus* and protected the intestinal mucosa from invasion by pathogenic bacteria (Hu et al., 2007). However, most related studies have rarely focused on crustaceans.

Limitations in our knowledge on the need for mineral in shrimp

Our knowledge on mineral nutrition in aquatic species is considerably limited owing to the infancy of the topic of aquatic animal nutrition and the related difficulty of conducting research on mineral nutrition of aquatic species, as well as the number of species of economic importance. Problems associated with the quantification of mineral requirements include the identification of the potential effects of ion composition in the water (McNevin et al., 2004), leaching of minerals from the diet prior to consumption, availability of suitable test diets that have a low concentration of the targeted mineral, and limited data on mineral bioavailability.

Conclusion

Minerals are essential for the normal metabolism of shrimp. Currently, information concerning the mineral nutrition of shrimp is still inadequate compared to the body of knowledge on terrestrial animals, and more research is required in future decades. We still need to obtain more data on mineral requirements, toxic effects, and mineral deficiencies in shrimp from diverse environments. We also need to develop new approaches to obtain more accurate data to guide future research on mineral nutrition in shrimp.

References

Arinola, O., Nwozo, S., Ajiboye, J. and Oniye, A. (2008) Evaluation of trace elements and total antioxidant status in Nigerian cassava processors. *Pakistan Journal of Nutrition* 7, 770–772.

Asaikkutti, A., Bhavan, P. S., Vimala, K., Karthik, M. and Cheruparambath, P. (2016) Dietary supplementation of green synthesized manganese-oxide nanoparticles and its effect on growth performance, muscle composition and digestive enzyme activities of the giant freshwater prawn *Macrobrachium rosenbergii*. *Journal of Trace Elements in Medicine and Biology* 35, 7–17.

Bharadwaj, A. S., Patnaik, S., Browdy, C. L. and Lawrence, A. L. (2014) Comparative evaluation of an inorganic and a commercial chelated copper source in Pacific white shrimp *Litopenaeus vannamei* (Boone) fed diets containing phytic acid. *Aquaculture* 422–423, 63–68.

Bjerregaard, P. and Christensen, A. (2012) Selenium reduces the retention of methyl mercury in the brown shrimp *Crangon crangon*. *Environmental Science & Technology* 46, 6324–6329.

Bjerregaard, P., John, T. S., Biuki, N. A., Biserova, M. P., Christensen, A. and Pedersen, K. L. (2018) Retention and distribution of methylmercury administered in the food in marine invertebrates: Effect of dietary selenium. *Marine Environmental Research* 138, 76–83.

Black, R. (1993) Acid-base and potassium balance, *Scientific American Medicine*. Scientific American Inc., New York, NY, USA, pp. 23.

Bryan, G. W. (1976) Heavy metal contamination in the sea. *Marine Pollution* 3, 185–302.

Burmester, T. (2002) Origin and evolution of arthropod hemocyanins and related proteins. *Journal of Comparative Physiology B* 172, 95–107.

Cai, H., Tan, B., Yang, Q., Dong, X., Chi, S., Liu, H. and Zhang, S. (2016) Effects of dietary chromium source and supplemental level on growth performance, serum biochemical indices and non-specific immune enzyme activities of juvenile *Litopenaeus vannamei*. *Chinese Journal of Animal Nutrition* 28, 766–779.

Cai, H. R., Yang, Q. H., Tan, B. P., Dong, X. H., Chi, S. Y., Liu, H. Y. and Zhang, S. (2017) Effects of dietary manganese source and supplemental levels on growth performance, antioxidant enzymes activities, tissue Mn concentrations and cytosolic manganese superoxide dismutase (cMnSOD) mRNA expression level of juvenile *Litopenaeus vannamei*. *Aquaculture Nutrition* 23, 603–610.

Castille Jr, F. L. and Lawrence, A. L. (1981) The effect of salinity on the osmotic, sodium and chloride concentrations in the hemolymph of euryhaline shrimp of the genus *Penaeus*. *Comparative Biochemistry and Physiology Part A: Physiology* 68, 75–80.

Chaitanawisuti, N., Sungsirin, T. and Piyatiratitivorakul, S. (2010) Effects of dietary calcium and phosphorus supplementation on the growth performance of juvenile spotted babylon *Babylonia areolata* culture in a recirculating culture system. *Aquaculture International* 18, 303–313.

Charmantier, G. U. Y. (1998) Ontogeny of osmoregulation in crustaceans: a review. *Invertebrate Reproduction & Development* 33, 177–190.

Cheng, K. M., Hu, C. Q., Liu, Y. N., Zheng, S. X. and Qi, X. J. (2006) Effects of dietary calcium, phosphorus and calcium/phosphorus ratio on the growth and tissue mineralization of *Litopenaeus vannamei* reared in low-salinity water. *Aquaculture* 251, 472–483.

Cheng, K. M., Hu, C. Q., Liu, Y. N., Zheng, S. X. and Qi, X. J. (2005) Dietary magnesium requirement and physiological responses of marine shrimp *Litopenaeus vannamei* reared in low salinity water. *Aquaculture Nutrition* 11, 385–393.

Chiu, S.-T., Hsieh, S.-L., Yeh, S.-P., Jian, S.-J., Cheng, W. and Liu, C.-H. (2010) The increase of immunity and disease resistance of the giant freshwater prawn, *Macrobrachium rosenbergii* by feeding with selenium enriched-diet. *Fish & Shellfish Immunology* 29, 623–629.

Civera, R. and Guillaume, J. (1990) Digestive utilization of phytate phosphorus in two species of shrimp: *Penaeus japonicus* and *Penaeus vannamei*. *World Aquaculture Meeting. World Aquaculture Society*, Halifax, Canada.

Cuvin-Aralar, M. L. A. and Furness, R. W. (1991) Mercury and selenium interaction: a review. *Ecotoxicology and Environmental Safety* 21, 348–364.

Davis, D. A. (1990) Dietary mineral requirements of *Penaeus vannamei*: Evaluation of the essentiality for thirteen minerals and the requirements for calcium, phosphorus, copper, iron, zinc, and selenium. *Texas A & M University.* Available at https://oaktrust.library.tamu.edu/handle/1969.1/DISSERTATIONS-1174753 (accessed 17 July 2021).

Davis, D. A. and Arnold, C. (1994) Estimation of apparent phosphorus availability from inorganic phosphorus sources for *Penaeus vannamei*. *Aquaculture* 127, 245–254.

Davis, D. A. and Gatlin III, D. M. (1996) Dietary mineral requirements of fish and marine crustaceans. *Reviews in Fisheries Science* 4, 75–99.

Davis, D. A., Lawrence, A. L. and Gatlin III, D. M. (1992a) Evaluation of the dietary iron requirement of *Penaeus vannamei*. *Journal of the World Aquaculture Society* 23, 15–22.

Davis, D. A., Lawrence, A. L. and Gatlin III, D. M. (1992b) Mineral requirements of *Penaeus vannamei*: a preliminary examination of the dietary essentiality for thirteen minerals. *Journal of the World Aquaculture Society* 23, 8–14.

Davis, D. A., Lawrence, A. L. and Gatlin III, D. M. (1993a) Dietary copper requirement of *Penaeus vannamei*. *Nippon Suisan Gakkaishi* 59, 117–122.

Davis, D. A., Lawrence, A. L. and Gatlin III, D. M. (1993b) Evaluation of the dietary zinc requirement of *Penaeus vannamei* and effects of phytic acid on zinc and phosphorus bioavailability. *Journal of the World Aquaculture Society* 24, 40–47.

Davis, D. A., Lawrence, A. L. and Gatlin III, D. M. (1993c) Response of *Penaeus vannamei* to dietary calcium, phosphorus and calcium: phosphorus ratio. *Journal of the World Aquaculture Society* 24, 504–515.

Depledge, M. (1989) Re-evaluation of metabolic requirements for copper and zinc in decapod crustaceans. *Marine Environmental Research* 27, 115–126.

Depledge, M., Chan, R. and Loh, T. (1986) Iron distribution and transport in *Scylla serrata* (Forskal). *Asian Marine Biology* 3, 101–110.

Deshimaru, O., Kuroki, K., Sakamoto, S. and Yone, Y. (1978) Absorption of labelled calcium-^{45} Ca by prawn from sea water. *Nippon Suisan Gakkai-Shi* 44, 975–977.

Desjardins, L., Hicks, B. and Hilton, J. (1987) Iron catalyzed oxidation of trout diets and its effect on the growth and physiological response of rainbow trout. *Fish Physiology and Biochemistry* 3, 173–182.

Diet, B. (1978) Requirement of prawn for dietary minerals. *Bulletin of the Japanese Society of Scientific Fisheries* 44, 907–910.

Droge, W. (2002) Free radicals in the physiological control of cell function. *Physiological Reviews* 82, 47–95.

Eastmond, D. A., MacGregor, J. T. and Slesinski, R. S. (2008) Trivalent chromium: Assessing the genotoxic risk of an essential trace element and widely used human and animal nutritional supplement. *Critical Reviews in Toxicology* 38, 173–190.

Ebashi, S. and Endo, M. (1968) Calcium and muscle contraction. *Progress in Biophysics and Molecular Biology* 18, 123–183.

Evans, D. H., Piermarini, P. M. and Choe, K. P. (2005) The multifunctional fish gill: dominant site of gas exchange, osmoregulation, acid-base regulation, and excretion of nitrogenous waste. *Physiological Reviews* 85, 97–177.

Frenet, M. and Alliot, A. (1985) Comparative bioaccumulation of metals in *Palaemonetes varians* in polluted and non-polluted environments. *Marine Environmental Research* 17, 19–44.

Furriel, R. P. M., McNamara, J. C. and Leone, F. A. (2000) Characterization of (Na^+, K^+)-ATPase in gill microsomes of the freshwater shrimp *Macrobrachium olfersii*. *Comparative Biochemistry and Physiology B-Biochemistry & Molecular Biology* 126, 303–315.

Ghidalia, W., Fine, J. and Marneux, M. (1972) On the presence of an iron-binding protein in the serum of a decapod crustacean [*Macropipus puber* (Linne)]. *Comparative Biochemistry and Physiology Part B: Comparative Biochemistry* 41, 349–354.

Guo, J., Chen, J., Sun, L., Shen, B. and Pan, Q. (2013) Dietary zinc requirement of juvenile oriental river prawn (*Macrobrachium nipponense*). *Chinese Journal of Animal Nutrition* 25, 661–668.

Hilton, J. W. (1989) The interaction of vitamins, minerals and diet composition in the diet of fish. *Aquaculture* 79, 223–244.

Hou, Y., Shavandi, A., Carne, A., Bekhit, A. A., Ng, T. B., Cheung, R. C. F. and Bekhit, A. E.-D. A. (2016) Marine shells: Potential opportunities for extraction of functional and health-promoting materials. *Critical Reviews in Environmental Science and Technology* 46, 1047–1116.

Hu, C. H., Xu, Y., Xia, M. S., Xiong, L. and Xu, Z. R. (2007) Effects of Cu^{2+}-exchanged montmorillonite on growth performance, microbial ecology and intestinal morphology of Nile tilapia (*Oreochromis niloticus*). *Aquaculture* 270, 200–206.

Huang, G., Ren, L. and Jiang, J. (2011) Purification of a histidine-containing peptide with calcium binding activity from shrimp processing byproducts hydrolysate. *European Food Research and Technology* 232, 281–287.

Kanazawa, A. S., Teixeira, H. and Sasaki, M. (1984) Requirements of the juvenile prawn for calcium, phosphorus, magnesium, potassium, copper, manganese, and iron. *Memoirs of Faculty of Fisheries Kagoshima University* 33, 63–71.

Katya, K., Lee, S., Yun, H., Dagoberto, S., Browdy, C. L., Vazquez-Anon, M. and Bai, S. C. (2016) Efficacy of inorganic and chelated trace minerals (Cu, Zn and Mn) premix sources in Pacific white shrimp, *Litopenaeus vannamei* (Boone) fed plant protein based diets. *Aquaculture* 459, 117–123.

Kidd, M., Ferket, P. and Qureshi, M. (1996) Zinc metabolism with special reference to its role in immunity. *World's Poultry Science Journal* 52, 309–324.

Kong, Y. Q., Ding, Z. L., Zhang, Y. X., Ye, J. Y. and Du, Z. Y. (2017) Dietary selenium requirement of juvenile oriental river prawn *Macrobrachium nipponense*. *Aquaculture* 476, 72–78.

Kuhn, D. D., Lawrence, A. L. and Crockett, J. (2017) Dietary toxicity of manganese to shrimp and its accumulation in bioflocs. *Aquaculture Nutrition* 23, 1121–1127.

Lall, S. P. (2003) The minerals. In: *Fish Nutrition* (ed. Hardy, R.) Elsevier, CA, USA, pp. 259–308.

Lee, M.-H. and Shiau, S.-Y. (2002) Dietary copper requirement of juvenile grass shrimp, *Penaeus monodon*, and effects on non-specific immune responses. *Fish & Shellfish Immunology* 13, 259–270.

Lemos, D. and Tacon, A. G. J. (2017) Use of phytases in fish and shrimp feeds: a review. *Reviews in Aquaculture* 9, 266–282.
Lin, S., Lin, X., Yang, Y., Li, F. and Luo, L. (2013) Comparison of chelated zinc and zinc sulfate as zinc sources for growth and immune response of shrimp (*Litopenaeus vannamei*). *Aquaculture* 406–407, 79–84.
Lin, Y.-H. and Shiau, S.-Y. (2005) Dietary selenium requirements of juvenile grouper, *Epinephelus malabaricus*. *Aquaculture* 250, 356–363.
Liu, F., Liang, D., Sun, F., Li, H. and Lan, X. (1990) Effects of dietary copper on the prawn *Penaeus orientalis*. *Oceanologia et Limnologia Sinica* 21, 404–410.
Liu, H., Zhang, X., Tan, B., Lin, Y., Chi, S., Dong, X. and Yang, Q. (2014) Effect of dietary potassium on growth, nitrogen metabolism, osmoregulation and immunity of pacific white shrimp (*Litopenaeus vannamei*) reared in low salinity seawater. *Journal of Ocean University of China* 13, 311–320.
Lockwood, A. P. M. (1962) The osmoregulation of crustacea. *Biological Reviews* 37, 257–303.
Lovell, T. (1989) *Nutrition and Feeding of Fish*. Springer, New York, NY, USA.
Lovett, D. L., Towle, D. W. and Faris, J. E. (1994) Salinity-sensitive alkaline-phosphatase activity in gills of the blue-crab, *Callinectes sapidus* Rathbun. *Comparative Biochemistry and Physiology B-Biochemistry & Molecular Biology* 109, 163–173.
Lu, Y., Chen, X., Zahng, X., Zheng, P., Wang, D., Xian, J. and Wang, A. (2018) Effects of dietary manganese amount on growth and antioxidant activities of sub-adult *Litopenaeus vannamei*. *Journal of Southern Agriculture* 49, 391–396.
Luft, F. C. (1990) Sodium, chloride, and potassium. In: Brown, M. L. (ed.) *Present Knowledge in Nutrition*, 6th edn. International Life Science Institute Nutrition Foundation, Washinton DC, USA, pp. 233–240.
Maguire, G. B. and Hume, I. (1982) A study of the nutritional requirements of school prawns *Metapenaeus macleayi* (Haswell) in some Australian brackish water farming ponds. *Aquaculture* 29, 261–278.
Marshall, W. and Bryson, S. (1998) Transport mechanisms of seawater teleost chloride cells: an inclusive model of a multifunctional cell. *Comparative Biochemistry and Physiology, Part A: Molecular & Integrative Physiology* 119, 97–106.
Martin, J.-L. (1973) Iron metabolism in *Cancer irroratus* (Crustacea: Decapoda) during the intermoult cycle, with special reference to iron in the gills. *Comparative Biochemistry and Physiology, Part A, Molecular & Integrative Physiology* 46, 123–129.
Maynard, L. A., Loosli, J. K. and Warner, R. G. (1979) *Animal Nutrition*. McGraw-Hill Book Co., New York, NY, USA.
McNevin, A. A., Boyd, C. E., Silapajarn, O. and Silapajarn, K. (2004) Ionic supplementation of pond waters for inland culture of marine shrimp. *Journal of the World Aquaculture Society* 35, 460–467.
Méndez, L., Racotta, I. S., Acosta, B. and Rodríguez-Jaramillo, C. (2001) Mineral concentration in tissues during ovarian development of the white shrimp *Penaeus vannamei* (Decapoda: Penaeidae). *Marine Biology* 138, 687–692.
Mertz, W. (1986) *Trace Elements in Human and Animal Nutrition*, 5th edn. Academic Press, Orlando, FL, USA.
Miyawaki, M., Matsuzaki, M. and Sasaki, N. (1961) Histochemical studies on the hepatopancreas of the crayfish, *Procambarus clarkii*. *Kumamato Journal of Science* 5, 161–169.
Moyle, P. B. and Cech, J. J. (2000) Buoyancy and thermal regulation. In: Moyle, P. B. and Cech, J. J. (eds) *Fishes: An Introduction to Ichthyology*. Prentice Hall, NJ, USA, pp. 69–78.
Muralisankar, T., Bhavan, P. S., Radhakrishnan, S., Seenivasan, C., Manickam, N. and Srinivasan, V. (2014) Dietary supplementation of zinc nanoparticles and its influence on biology, physiology and immune responses of the freshwater prawn, *Macrobrachium rosenbergii*. *Biological Trace Element Research* 160, 56–66.
Muralisankar, T., Bhavan, P. S., Radhakrishnan, S., Seenivasan, C., Srinivasan, V. and Santhanam, P. (2015) Effects of dietary zinc on the growth, digestive enzyme activities, muscle biochemical compositions, and antioxidant status of the giant freshwater prawn *Macrobrachium rosenbergii*. *Aquaculture* 448, 98–104.
NRC (1980) *Mineral Tolerance of Domestic Animals*. National Academies Press, Washington DC, USA.
NRC (2011) *Nutrient Requirements of Fish and Shrimp*. National Academic Press, Washington DC, USA.
Ogura, K. (1959) Midgut gland cells accumulating iron or copper in the crayfish, *Procambarus clarkii*. *Annotationes Zoologicae Japonenses* 32, 133–142.
Pan, Q., Chen, X.-Y., Li, F., Bi, Y.-Z. and Zheng, S.-X. (2005) Response of juvenile *Litopenaeus vannamei* to varying levels of calcium phosphate monobasic supplemented to a practical diet. *Aquaculture* 248, 97–102.

Peñaflorida, V. D. (1999) Interaction between dietary levels of calcium and phosphorus on growth of juvenile shrimp, *Penaeus monodon*. *Aquaculture* 172, 281–289.

Pierson, R. and Aanes, W. (1958) Treatment of chronic copper poisoning in sheep. *Journal of the American Veterinary Medical Association* 133, 307.

Pinoni, S. A. and Mananes, A. A. L. (2004) Alkaline phosphatase activity sensitive to environmental salinity and dopamine in muscle of the euryhaline crab *Cyrtograpsus angulatus*. *Journal of Experimental Marine Biology and Ecology* 307, 35–46.

Qin, F., Shi, M., Yuan, H., Yuan, L., Lu, W., Zhang, J., Tong, J. and Song, X. (2016) Dietary nano-selenium relieves hypoxia stress and, improves immunity and disease resistance in the Chinese mitten crab (*Eriocheir sinensis*). *Fish & Shellfish Immunology* 54, 481–488.

Rao, M. S. and Anjaneyulu, N. (2008) Effect of copper sulfate on molt and reproduction in shrimp *Litopenaeus vannamei*. *International Journal of Biological Chemistry* 2, 35–41.

Robinson, J. D. and Pratap, P. R. (1991) Na^+, K^+-ATPase - modes of inhibition by Mg^{2+}. *Biochimica et Biophysica Acta* 1061, 267–278.

Rotruck, J. T., Pope, A. L., Ganther, H. E., Swanson, A., Hafeman, D. G. and Hoekstra, W. (1973) Selenium: biochemical role as a component of glutathione peroxidase. *Science* 179, 588–590.

Roy, L. A., Davis, D., Saoud, I. and Henry, R. (2007) Supplementation of potassium, magnesium and sodium chloride in practical diets for the Pacific white shrimp, *Litopenaeus vannamei*, reared in low salinity waters. *Aquaculture Nutrition* 13, 104–113.

Roy, L. A., Davis, D. A., Nguyen, T. N. and Saoud, I. P. (2009) Supplementation of chelated magnesium to diets of the Pacific white shrimp, *Litopenaeus vannamei*, reared in low-salinity waters of west Alabama. *Journal of the World Aquaculture Society* 40, 248–254.

Saoud, I. P., Davis, D. A. and Rouse, D. B. (2003) Suitability studies of inland well waters for *Litopenaeus vannamei* culture. *Aquaculture* 217, 373–383.

Saoud, I. P., Roy, L. A. and Davis, D. A. (2007) Chelated potassium and arginine supplementation in diets of Pacific white shrimp reared in low-salinity waters of west Alabama. *North American Journal of Aquaculture* 69, 265–270.

Scott, M., Nesheim, M. and Young, R. (1982) *Nutrition of the Chicken*. ML Scott & Associates, New York, NY, USA.

Shiau, S.-Y. and Hsieh, J.-F. (2001) Dietary potassium requirement of juvenile grass shrimp *Penaeus monodon*. *Fisheries Science* 67, 592–595.

Shiau, S.-Y. and Jiang, L.-C. (2006) Dietary zinc requirements of grass shrimp, *Penaeus monodon*, and effects on immune responses. *Aquaculture* 254, 476–482.

Soetan, K., Olaiya, C. and Oyewole, O. (2010) The importance of mineral elements for humans, domestic animals and plants-A review. *African Journal of Food Science* 4, 200–222.

Srinivasan, V., Bhavan, P. S., Rajkumar, G., Satgurunathan, T. and Muralisankar, T. (2016) Effects of dietary iron oxide nanoparticles on the growth performance, biochemical constituents and physiological stress responses of the giant freshwater prawn *Macrobrachium rosenbergii* post-larvae. *International Journal of Fisheries and Aquatic Studies* 4, 170–182.

Sritunyalucksana, K., Intaraprasong, A., Sa-Nguanrut, P., Filer, K. and Fegan, D.F. (2011) Organic selenium supplementation promotes shrimp growth and disease resistance to Taura syndrome virus. *ScienceAsia* 37, 24–30.

Sugiura, S. H., Hardy, R. W. and Roberts, R. J. (2004) The pathology of phosphorus deficiency in fish - a review. *Journal of Fish Diseases* 27, 255–265.

Sun, S., Qin, J., Yu, N., Ge, X., Jiang, H. and Chen, L. (2013) Effect of dietary copper on the growth performance, non-specific immunity and resistance to *Aeromonas hydrophila* of juvenile Chinese mitten crab, *Eriocheir sinensis*. *Fish & Shellfish Immunology* 34, 1195–1201.

Sutton, J., Balfry, S., Higgs, D., Huang, C.-H. and Skura, B. (2006) Impact of iron-catalyzed dietary lipid peroxidation on growth performance, general health and flesh proximate and fatty acid composition of Atlantic salmon (*Salmo salar L.*) reared in seawater. *Aquaculture* 257, 534–557.

Thirunavukkarasu, M., Periyakali, S. B., Subramanian, R. and Perumal, S. (2019) Influence of two different dietary zinc sources in freshwater prawn *Macrobrachium rosenbergii* post larvae. *Journal of Oceanology and Limnology* 37, 290–299.

Tolonen, J., Grönblad, M., Virri, J., Seitsalo, S., Rytömaa, T. and Karaharju, E. (1995) Basic fibroblast growth factor immunoreactivity in blood vessels and cells of disc herniations. *Spine (Phila Pa 1986)* 20, 271–276.

Turner, M. A. and Rudd, J. W. (1983) The English–Wabigoon River system: III. Selenium in lake enclosures: its geochemistry, bioaccumulation, and ability to reduce mercury bioaccumulation. *Canadian Journal of Fisheries and Aquatic Sciences* 40, 2228–2240.

Vormann, J. (2003) Magnesium: Nutrition and metabolism. *Molecular Aspects of Medicine* 24, 27–37.

Wang, W., Wang, A., Liu, C., Wang, S., Wang, R. and Ma, Z. (1997) Effects of copper concentrations in diets on the growth and copper, zinc and iron contents of *Penaeus chinensis*. *Journal of Fisheries of China* 21, 258–262.

Wang, X., Shen, Z., Wang, C., Li, E., Qin, J. G. and Chen, L. (2019) Dietary supplementation of selenium yeast enhances the antioxidant capacity and immune response of juvenile *Eriocheir Sinensis* under nitrite stress. *Fish & Shellfish Immunology* 87, 22–31.

Xian, J., Chen, X., Wang, A., Ye, J., Guo, H., Miao, Y., Ye, C., Feng, L. and Liao, S. (2013) Effects of dietary manganese levels on growth, antioxidant defense and immune responses of the juvenile white shrimp, *Litopenaeus vannamei*, reared in low salinity water. *Fish & Shellfish Immunology* 6, 1747.

Yamaguchi, M. (1998) Role of zinc in bone formation and bone resorption. *The Journal of Trace Elements in Experimental Medicine* 11, 119–135.

Yuan, L., Zhang, R., Ma, X., Yang, L., Zheng, Q., Chen, D., Li, M., Fan, T., Liu, Y. and Pan, L. (2018) Selenium accumulation, antioxidant enzyme levels, and amino acids composition in Chinese mitten crab (*Eriocheir sinensis*) fed selenium-biofortified corn. *Nutrients* 10, 318.

Yuan, Y., Jin, M., Xiong, J. and Zhou, Q. (2019) Effects of dietary dosage forms of copper supplementation on growth, antioxidant capacity, innate immunity enzyme activities and gene expressions for juvenile *Litopenaeus vannamei*. *Fish & Shellfish Immunology* 84, 1059–1067.

Zhou, X. X., Zhang, J. Y., Liu, S. L. and Ding, Y. T. (2014) Supplementation of sodium chloride in diets to improve the meat quality of Pacific white shrimp, *Litopenaeus vannamei*, reared in low-salinity water. *Aquaculture Research* 45, 1187–1195.

Zhou, Y., Rhodes, M. A., Liu, J. and Davis, D. A. (2017a) Effects of various dietary levels of copper hydroxychloride on growth performance and tissue response in Pacific white shrimp *Litopenaeus vannamei* fed practical diets. *Aquaculture Nutrition* 23, 171–180.

Zhou, Y., Zhang, D., Peatman, E., Rhodes, M. A., Liu, J. and Davis, D. A. (2017b) Effects of various levels of dietary copper supplementation with copper sulfate and copper hydroxychloride on Pacific white shrimp *Litopenaeus vannamei* performance and microbial communities. *Aquaculture* 476, 94–105.

Zhu, W., Liu, M., Chen, C., Wu, F., Yang, J., Tan, Q., Xie, S. and Liang, X. (2014) Quantifying the dietary potassium requirement of juvenile grass carp (*Ctenopharyngodon idellus*). *Aquaculture* 430, 218–223.

Chapter 15

NUTRITION AND SHRIMP HEALTH

WENBING ZHANG, SHUOLI MA, XINXIN LI, YANLIN GUO, HUAICHENG GE, DONG HUANG, FEIYANG CHEN, YANJIAO WU AND KEKE LEI

Summary

The shrimp immune system is characterized by a lack of adaptive mechanisms mediated by antibodies. It is based on innate mechanisms consisting of cellular and humoral components that interplay to recognize and eliminate foreign microorganisms and pathogens. Nutrition from protein (amino acids), lipids (fatty acids), carbohydrates, vitamins and minerals is essential for the health of shrimp. The relation between infection and malnutrition is synergistic. Poor nutrition or poor feeding practices may lead to a reduced immune system response and lower the ability of shrimp to resist diseases. Therefore, optimal quantitative and qualitative nutrition play important roles in shrimp health. At present, nutritionists attach more importance to developing nutritional strategies that improve growth and production, but that do not necessarily optimize for disease resistance.

Non-nutritional immunostimulants are usually recommended for use in feeds that modulate the immune system by increasing the host's resistance against diseases of aquatic organisms. Between 2010 and 2020, the research and development of non-nutritional immunostimulants including polysaccharides, organic acids, single cells, herbs and Biofloc, have increasingly attracted the attention of shrimp health. The immuno-stimulatory effects of glucan, chitin, butyrate and *Lactobacillus* for shrimp have been reported. The most economical and feasible way of administration of immunostimulants to shrimp is by feeding as a feed supplement. Therefore, a review of non-nutritional immunostimulants in shrimp is also presented in this chapter.

Nutrients

PROTEINS AND AMINO ACIDS

Protein and amino acid requirements

The minimum dietary requirement for protein or a balanced mixture of amino acids is a primary concern in aquaculture, since satisfying this requirement is necessary to ensure the adequate growth and health of shrimp, while providing excessive levels is generally uneconomical. Consequently, most studies have been concerned with determining the minimum dietary protein requirements of shrimp. Generally, recommended dietary protein levels vary from 30 to 55% for various species and life stage. The reported estimate of protein requirement must be carefully examined because it is dependent on the quality (e.g. essential amino acid profile and digestibility) of dietary protein, age or physiological state of shrimp, and environmental factors (e.g. salinity, dissolved oxygen, temperature). Proteins are the most important elements of all living cells and, with the exception of water, represent the largest chemical group in the body of living beings. They are essential substances in the nucleus of cells and cellular protoplasm and are responsible for the formation of muscle tissues, organs, and nerves. Protein in the feed must meet the physiological needs of the essential amino acids (EAA), so that each of these carries out its function in the animal's body.

Shrimp require the same ten EAA as fish species. The EAAs including arginine, methionine, valine, threonine, isoleucine, leucine, lysine, histidine, phenylalanine, and tryptophan were initially confirmed by Kanazawa and Teshima (1981). Generally, in contrast to the wide range of quantitative crude protein requirements that have been reported, the EAAs for shrimp are the same.

Protein, amino acids and immune responses

The dietary levels of protein and amino acids can affect shrimp immunity. Shrimp must consume protein to furnish a continuous and well-balanced supply of essential and non-essential amino acids for maintenance and growth. Although growth involves a mixture of soft and hard tissues, the commercial perspective is primarily on protein deposition. If an adequate dietary protein content is not provided, there is a rapid reduction of growth or even a loss of weight because shrimp may withdraw protein from some tissues to maintain the function of more vital tissues. If an excess of protein is supplied, it is broken down to produce energy and proportionally less is deposited as new protein for growth. Consequently, growth rate is the primary driver of protein requirements, also influenced by the quality of the protein used in feed.

Many studies have evaluated the effects of different protein levels and sources on shrimp (Bulbul et al., 2014; Oujifard et al., 2012). For instance, *Litopenaeus stylirostris* (21 g initial weight) were fed six experimental diets for 50 days (Gauquelin et al., 2007). The study found that weight gain (from 21 to 30g) increased with the increase in dietary crude protein content (from 25 to 58%). The result also recorded, on average, 1 kJ shrimp/day or 47 kJ kg live weight/day (22 kJ/0.8 kg/day). In addition, as dietary protein content increased, there was an increase in ammonia production with an O:N ratio indicating that protein was increasingly used as an energy substrate as dietary crude protein increased (Ayisi et al., 2017). Sharawy et al. (2016) recommended that the dietary protein requirements of juvenile Pacific white shrimp, *Peneaus vannamei*, range between 25 and 33.4%.

In recent years, increasing attention has been paid to the effects of dietary protein levels on the immune and antioxidative systems of *P. vannamei*. Xu et al. (2014) reported that the lowest total antioxidative capacity in both plasma and hepatopancreas, as well as the lowest reduced glutathione/oxidized glutathione ratio in plasma, was recorded in shrimp fed with 20% dietary protein. In addition, after 49 days, the highest weight gain (WG) was recorded in shrimp fed diet with 32.9% crude protein (CP). The same group recorded the lowest dietary cost. It is evident that feeding shrimp diets containing 25–33% CP will retard neither growth performance nor feed utilization.

Among the ten EAAs, methionine, lysine, and arginine are usually those most limiting for shrimp in commercial feed formulations, and they are indispensable for normal growth and survival of shrimp (Coloso et al., 1980). *P. vannamei* fed methionine-, lysine-, and arginine-deficient diets had significantly less wet weight gain than those fed the control diet; among the three amino acids, arginine appeared to be least limiting (Mente et al., 2002).

However, arginine is involved in the synthesis of nitric oxide via nitric oxide synthase, which is inducible in shrimp (Teshima et al., 2004). Arginine has been shown to have numerous beneficial effects on T-cell-mediated immunity in various animal models and humans. Thus, its involvement in shrimp health is certainly worthy of further investigation.

LIPIDS AND FATTY ACIDS

Lipid levels

Lipids are an important component of the shrimp diets because they provide a concentrated energy source and supply essential fatty acids such as eicosahexaenoic acid and docosahexaenoic acid. Lipids also serve as precursors for metabolic regulators such as prostaglandins, thromboxans, and prostacyclins – hormones that promote maturation, molting, and growth.

Lipids have the highest energy content among the three major nutritional components, and play important roles in the shrimp life cycle. In addition to providing energy, dietary lipids also provide fatty acids, phospholipid, and cholesterol, which can promote the absorption of fat-soluble vitamins. In addition, lipids contribute to the palatability and attractability of feeds. Though dietary lipids are important nutritionally, the levels and types of lipids used in shrimp feeds are also dictated by economics, constraints of manufacturing procedures, and product quality. According to previous studies, lipid deficiency and excess are proven to be unfavorable to shrimp growth and health. For example, it has been reported that lipid deficiency generally resulted in decreased growth rate and impaired immune function of Pacific white shrimp (*P. vannamei*) (Zhang et al., 2013; Zhu et al., 2010). It has been reported that dietary lipid levels for shrimp ranged from 6 to 7.5%, with a maximum level of 10% being recommended (Akiyama, 1991). Over the last 10 years, researchers have paid more attention to the immune-boosting effects of dietary lipid. Xu et al. (2018) reported that 12% dietary lipid could enhance the immune function of Pacific white shrimp at low salinity. Similar results can be seen from Zhang et al.'s (2013) investigation that a 10–12% dietary lipid level could act as an immunostimulant, benefiting the immune function of Pacific white shrimp. Investigations on ridgetail white prawn (*Exopalaemon carinicauda*) (Suo et al., 2015) showed that optimal lipid levels (5–7%) enhanced antioxidative capacity and immune function.

2.2 Lipid sources

In shrimp culture, fish oil is commonly used in formulated feed owing to its high unsaturated fatty acid (HUFA) content, which is beneficial for animals in regard to improving health functions (Daviglus et al., 1997; Hu et al., 2011). However, with the development of intensive culture and the limited supply of fish oil, evaluation of alternatives to fish oil is imperative. Researchers have tested several substitutes for fish oil, such as soybean oil, soybean lecithin oil, linseed oil, and canola oil. Hu et al. (2011) revealed that there were no significant differences in growth between Pacific white shrimp fed fish meal (34%)-based diets supplemented with fish oil and those partially or totally replaced by soybean oil and soybean lecithin. Recently, the feed industry has seen the introduction of various alternative sources of essential fatty acids. Although these products are relatively new to the market, they appear to be cost-effective and can be an excellent way to supplement some essential fatty acids without adding large amounts of other fatty acids.

In addition to different oils, the quality of oil is also vital to the health of aquatic animals (Chen et al., 2012). Fish oil is susceptible to lipid peroxidation because of the high polyunsaturated fatty acid (PUFA) contents; a previous study on sea bream (*Pagrus major*) showed that oxidized fish oil reduced health status (Gao et al., 2012). Studies on Pacific white shrimp have reported similar results, that dietary oxidized fish oil led to significantly increased oxidative stress (Yang et al., 2015).

Fatty acids

According to previous studies (Kanazawa and Teshima, 1978; Kanazawa et al., 1977, 1979a,b; Lim et al., 1997), four fatty acids are considered essential for shrimp: linoleic (18:2n-6, LA), linolenic (18:3n-3, LNA), eicosapentaenoic (20:5n-3, EPA), and decosahexaenoic (22:6n-3, DHA) acids; among these, the latter two are the most indispensable. González-Félix et al. (2003) confirmed that arachidonic acid (C20:4n-6, ARA) is an n-6 essential fatty acid for Pacific white shrimp. Shrimp, like other crustaceans, have a limited ability to synthesize HUFAs and these are consequently required in the diet.

Numerous studies have demonstrated the important roles of n-3 and n-6 PUFAs in the immune functions of animals. ARA derived from n-6 PUFAs, and eicosapentaenoic acid derived from n-3 PUFAs, led to the generation of eicosanoids such as prostaglandins and leukotrienes, which play vital roles in inflammation in animals (Calder, 2013; Pischon et al., 2003; Wander et al., 1997). Many studies have investigated the effects of HUFAs on immune response of shrimp. Mercier et al. (2009) reported that dietary HUFAs increased the concentration of β-glucan-binding protein in the hemolymph of Pacific white shrimp, which recognized invading microorganisms and amplified the activation of the prophenoloxidase-activating system and, finally, enhanced immune defenses. Aguilar et al. (2012) reported that high ARA levels produced more superoxide and decreased clotting times of hemolymph, which minimized stress response caused by sampling and enhanced immune function of Pacific white shrimp. Zhao et al. (2016) demonstrated that 0.56 mg ARA/g of diet upregulated mRNA levels of lysozyme, Toll receptor, and the immune deficiency of Pacific white shrimp, enhancing immune function. Overall, investigations on the effects of dietary fatty acids on the immune functions of shrimp are limited, which requires investigation in the future.

CARBOHYDRATES

Dietary carbohydrate content and sources

In shrimp the utilization of carbohydrates as an energy source is moderate. Digestibility is dependent on the source or chemical composition of the carbohydrate, as well as the level of cooking. Best results are often attained with wheat starch, and this is commonly the main starch source in shrimp feeds (Cuzon et al., 2000). Carbohydrates are often used as an inexpensive energy source to minimize the catabolism of dietary protein (Cuzon et al., 2004; Guo et al., 2007, 2011; Rosas et al., 2000). Previous studies have suggested that the optimal dietary carbohydrate levels for shrimps are as follows: 26%, *P. chinensis* (Xu et al., 1988); 21%, *P. stylirostris* (Rosas et al., 2000); 10–20%, *P. vannam*ei (Guo et al., 2005).

Carbohydrates of varying molecular complexity have different effects on shrimp. In general, macromolecular starch is suitable for shrimp (Fu et al., 2004; Piedad-Pascual, 1983). Wang et al. (2016) found that the utilization of carbohydrate by shrimp was affected by molecular size, disaccharide bonding, and the spatial conformation of sugar molecules. Macromolecular starch, dextrin, and non-reducing sucrose have the best growth promotion effect, while small molecular glucose has the worst. The most likely reason for this response is that macromolecular starch is slowly released and absorbed after digestion, resulting in slower release of glucose into the blood, which could activate the glycogen synthesis pathway and subsequently increase glycogen content. When the blood glucose level is insufficient to supply energy, glycogen is rapidly decomposed and supplements blood glucose. This could be the reason why the utilization level of dietary corn starch and dextrin was higher than that of dietary glucose (Wang et al., 2016). However, studies have found that sucrose- and glucose-fed groups showed better growth performances compared with those fed corn starch when *P. vannamei* was reared at low salinity (Qiao et al., 2017; Yang et al., 2018).

Carbohydrate and the immune response

Carbohydrate can meet the high-energy requirement of aquatic animals under stress conditions (Tseng et al., 2008). Under the stress of low salinity, 15–20% dietary carbohydrate was found to be optimal for growth and improved the anti-stress ability of *P. vannamei* (Wang et al., 2014). Xiao et al. (2017) found that fructose-enriched artemia can improve the osmoregulation and survival of juvenile *P. vannamei*. During acute low-salinity stress, *P. vannamei* can first use muscle glycogen as energy for osmoregulation, then amino acids in muscle through the gluconeogenesis pathway (Wang et al., 2017).

Intestinal microbiota play important roles in the degradation and fermentation of dietary carbohydrate (Zhang et al., 2014). However, little research is available on the effects of carbohydrates on shrimp intestinal microbiota. Results from an Illumina-based sequencing method indicated that the abundance of opportunistic pathogenic bacteria increased in a corn starch-fed group compared with glucose and sucrose groups, suggesting that the diet exerts selective pressure on the intestinal microbiota (Qiao et al., 2017).

Overall, carbohydrates are a critical component of the diet for feed processing, being required as a non-protein energy source. Types of carbohydrate and processing can clearly influence shrimp performance. Although data are limited, carbohydrate sources also affect the shrimp microbiome, which can influence disease resistance.

VITAMINS

Vitamins are a very small set of compounds which are needed at low levels to maintain essential functions. Vitamins are different from carbohydrates, proteins, and lipids, as they neither provide energy nor are they a structural component of the body. The majority of vitamins need to be provided by the diet as they cannot be synthesized by the organism. There are several known vitamins, including A, B complex,

C, D, E, and K. Vitamins are divided into water- and lipid-soluble types due to their different chemical properties.

Based on important physiological functions of dietary vitamins, their requirement, toxicity, and symptoms caused by deficiency have been investigated thoroughly in shrimp. In recent years, the study of vitamins in the immunity of shrimp has become an area of research focus; the qualitative and quantitative vitamin requirements of shrimp have been evaluated across a number of studies. However, the difficulties in determining vitamin requirements are exemplified by the wide variations in vitamin supplementation utilized by researchers and the commercial industry.

Water-soluble vitamins

The water-soluble vitamins include C (ascorbic acid) and B complex – B_1 (thiamine), B_2 (riboflavin), B_3 (pantothenic acid), B_5 (nicotinamide, nicotinic acid), B_6 (pyridoxine), B_7 (biotin), B_{11} (folic acid), and B_{12} (cobalamin). Recent research on water-soluble vitamins in shrimp mainly involved non-specific immune responses and the sparing effect of multivitamins.

Vitamin C, a famous 'cold' fighter, represents a strong antioxidant known for strengthening the immune system by fighting off infectious diseases. One study in *P. vannamei* showed that oral administration with 0.2 g/kg of vitamin C induced superoxide anion production and had beneficial effects on the immune system. A combination of 1 g/kg beta 1,3/1,6-glucan and 0.2 g/kg vitamin C gave an additive effect on inducing non-specific immune responses in Pacific white shrimp (Wu et al., 2016). Meanwhile, a high level of dietary vitamin C (2000 mg/kg) significantly reduced the content of superoxide anions in hemolymph of Pacific white shrimp under intensive farming (600 and 800 shrimp/m^3), which alleviate the negative effects of intensive culture conditions to a certain extent (Wang et al., 2012). A study in kuruma shrimp, *Marsupenaeus japonicus*, indicated that the hatching and metamorphosis rates of nauplii into zoea I significantly increased with increased levels of either dietary vitamin C or E. In addition, the regeneration effect of vitamin C on vitamin E concentrations and the sparing effect of vitamin E on vitamin C concentrations in the organs of female shrimp were demonstrated (Nguyen et al., 2012).

Research on the vitamin B complex has mainly focused on vitamins B_1 and B_6. Concerning the supplementation of vitamin B_6, this varies under normal growth and low-salinity conditions (3‰). Vitamin B_6 plays an important role in the synthesis of antibodies, production of gastric acid in the digestive system, utilization of fats and proteins, and maintenance of the sodium/potassium balance (stabilization of the nervous system). It is reported that aspartate aminotransferase, alanine aminotransferase, superoxide dismutase, catalase, and lysozyme in the hemolymph of juvenile Pacific white shrimp were significantly influenced by dietary vitamin B_6 levels, which showed that this could affect the health and non-specific immune responses (Cui et al., 2015).

Fat-soluble vitamins

The fat-soluble vitamins include vitamins A, D, E, and K, which are insoluble in water but soluble in fats and organic solvents. They can be stored in large quantities in the body, mainly in the liver. However, recent researches have mainly focused on the effect of vitamin E or its interactions with other nutrients.

It has been found that increasing the dietary concentration of vitamin E from 100 to 300 mg/kg could reduce thiobarbituric acid content in shrimp muscle during frozen storage, and cause higher retention of highly unsaturated fatty acids in the muscle of shrimp fed a high-lipid diet (Ouraji et al., 2011). A high level (400 mg/kg) of vitamin E in diet significantly reduced the expression levels of Toll-like receptor and nuclear factor kappa β in the serum, muscle, and hepatopancreas of *Fenneropenaeus chinensis* (Li et al., 2012).

The combined addition of β-glucan, selenium, and vitamin E to the diet significantly enhanced the anti-oxidative capacity, total hemocyte count, and lysozyme activity in serum of white shrimp (Liu et al., 2013). In addition, the combination of dietary vitamin E and *Schizochytrium* sp. effectively improved the regulation of immune genes and sperm quantity, and promoted spermatophore regeneration in male *P. monodon* (Li et al., 2018). A dietary combination of 400 mg/kg vitamin E and 0.4 mg/kg selenium significantly increased total antioxidative activity, and the activity of phenoloxidase and glutathione peroxidase in the hepatopancreas, which helped balance the antioxidative system in fighting against damage induced by oxygen radicals (Hu et al., 2010).

MINERALS AND SHRIMP HEALTH

Minerals are essential components of bones, teeth, and the exoskeleton. They are necessary for maintenance of osmotic pressure, acid–base balance and thus the regulation of pH in hemolymph, urine, and other body fluids. They are also components of soft tissues, enzymes, certain vitamins, hormones, and respiratory pigments, and are essential for normal muscle contraction and transmission of nerve impulses.

The main functions of minerals in the body include the formation of the skeletal structure, maintenance of colloidal systems (osmotic pressure, viscosity, and diffusion), and regulation of acid–base equilibrium. They are important components of hormones and enzymes, and activators of enzymes. A fixed number of specific trace metals (e.g. Fe, Mn, Cu, Co, Zn, and Se) are firmly associated with a specific protein in metalloenzymes that produce a unique catalytic function. Minerals are essential for the transmission of nerve impulses and muscle contraction.

With the exception of osmoregulation, the maintenance of osmotic balance between body fluids and the water in which the animal lives, the biochemical functions of minerals in aquatic species appear to be similar to those in terrestrial animals. Freshwater species lose ions to the hypotonic environment and therefore suffer from hydration, whereas the reverse is true for marine species.

Unlike terrestrial animals, which are primarily limited to a dietary source of minerals, shrimp may be able to utilize, to some extent, minerals dissolved in the water to meet physiological requirements. Calcium, copper, iron, magnesium, sodium, potassium, selenium, and zinc are generally derived from the water to satisfy part of the requirements of shrimp. Among major minerals, calcium (Ca) and magnesium (Mg) are considered to be very important for molting and new shell formation for cultured shrimp.

Inorganic minerals and shrimp health

Copper

Copper (Cu) is an essential component of numerous oxidation–reduction enzyme systems. For example, it is a component of the enzymes cytochrome oxidase, uricase, tyrosinase, superoxide dismutase, amine oxidase, lysyl oxidase, and caeruloplasmin.

As a component of the enzyme caeruloplasmin (ferroxidase), copper is intimately involved with iron metabolism and, therefore, hemoglobin synthesis and red blood cell production and maintenance.

Copper is also believed to be necessary for the formation of the pigment melanin and consequently skin pigmentation, for the formation of bone and connective tissue, and for maintaining the integrity of the myelin sheath of nerve fibers.

There is a close relationship between the dietary Cu level, activity of enzymes, and cellular and immune functions. Copper deficiency in humans is reflected in the low Cu content in leucocytes. It functions in haematopoiesis and in numerous Cu-dependent enzymes including lysyl oxidase, cytochrome Coxidase, ferroxidase, tyrosinase, and superoxide dismutase. Information on relationships between dietary Cu and immune responses in shrimp is limited. Lee and Shiau (2002) reported on non-specific immune responses in shrimp *P. monodon* fed diets containing various levels of Cu, and found *P. monodon* fed diets containing 10–30 mg Cu/kg diet had the highest total hemocyte count and intracellular superoxide anion production ratio. Collectively, the dietary Cu requirement for maintenance of non-specific immune responses in *P. monodon* reared in brackish water containing 1.53 μg Cu/l water is about 10–30 mg Cu/kg diet. Copper is readily absorbed from the gastrointestinal tract, gills, fins, and skin of fish and crustaceans.

Zinc

The primary functions of zinc are based on its roles as a cofactor in several enzyme systems, and as a component of a large number of metalloenzymes, including carbonic anhydrase, alkaline phosphatase, carboxypeptidase, alcohol dehydrogenase, glutamic dehydrogenase, lactate dehydrogenase, ribonuclease, and DNA polymerase. As an active component or cofactor of many important enzyme systems, zinc plays a vital role in lipid, protein, and carbohydrate metabolism, being particularly active in the synthesis and metabolism of nucleic acids (RNA) and proteins.

As zinc is needed to maintain the normal activity of lymphocytes, making it essential for the integrity of immune system, its dietary supplementation may improve immune function. Information on the relation between dietary zinc and immune responses in shrimp is limited. Shiau and Lee (2006) detected several immune index in *P. monodon* fed graded dietary zinc, and found that both intracellular superoxide anion (O_2^-) production ratios and total hemocyte count were highest in shrimp fed diets with 35 and 48 mg Zn/kg. Furthermore, these results concluded that adequate dietary Zn concentration for better non-specific immune responses in *P. monodon* is about 35–48 mg Zn/kg diet.

Manganese

Manganese is an essential element for most organisms, being an essential component of the enzyme pyruvate carboxylase. As a cofactor or component of several key enzyme systems, manganese is essential for bone formation (re. mucopolysaccharide synthesis), the regeneration of red blood cells, carbohydrate metabolism, and the reproductive cycle.

Manganese is not only a cofactor in enzyme systems, but also plays an important role in modulating growth and metabolism, including in shrimp health (Liu and Lawrence, 1997). In the study of Lu et al. (2018), appropriate manganese supplementation in feed increased the weight gain and specific growth rate of sub-adult *P. vannamei* (4.83 ± 0.39 g), and appropriate manganese supplementation increased total antioxidative capacity by increasing superoxide dismutase activity. However, excessive dietary manganese can compromise the antioxidative activity of sub-adult *P. vannamei* to a certain extent and lead to damage of the hepatopancreas.

Potassium

Potassium is the major cation of intracellular fluid, regulating intracellular osmotic pressure and acid–base balance. Like sodium, potassium has a stimulating effect on muscle irritability. Potassium is also required for glycogen and protein synthesis, and the metabolic breakdown of glucose.

Organic minerals and shrimp health

Many of the inorganic forms of mineral have the potential to react with other nutrients, are easily solubilized in water and consequently leach from the diet and/or may have low biological availability. Chelated minerals are often less reactive, having a lower leaching rate and higher bioavailability. Hence, chelated minerals have attracted considerable attention from feed manufacturers and animal producers as a means of improving performance (Ashmead, 1992; Wang and Lovell, 1997). Mineral sources with higher bioavailability and less reactivity to anti-nutrients, such as phytate, should be considered in feed formulation (Katya et al., 2016). Some studies on chelated minerals have already been conducted in shrimp, and beneficial effects on growth performance or immunity have been reported.

Sritunyalucksana et al. (2011) reported that selenium supplementation and, in particular, supplementation with organic selenium can have a dramatic effect on *P. vannamei* growth, hemocyte number, and survival after taura syndrome virus challenge.

Lin et al. (2013) showed that shrimp (*P. vannamei*) supplemented with zinc methionine had the highest weight gain and immune parameters; those fed diets with organic zinc (zinc methionine, zinc lysine, and zinc glycine) supplementation showed significantly higher growth, survival, and immune parameters than treatment with zinc sulfate. These results also suggested that Zn from zinc methionine was a better source for *P. vannamei*.

Bharadwaj et al. (2014) found that shrimp (*P. vannamei*) required 3–4 times more dietary copper from copper sulfate than from a chelated copper source (chelated to the hydroxy analog of methionine) in regard to promotion of comparable growth. Results from that study suggest that chelated copper is a safe, effective, and highly available source of copper for the Pacific white shrimp.

Katya et al. (2016) used a premix of chelated trace minerals Cu, Zn, and Mn (chelated to 2-hydroxy-4-methylthiobutanoic acid, or the hydroxy analog of methionine) in Pacific white shrimp. Results from feeding trials showed that appropriate chelated premix supplementation in the feed can increase growth performance and Cu/Zn superoxide dismutase activity in the serum and hepatopancreatic tissue.

Liang et al. (2016) showed that 1500–2500 mg/kg iron (Fe) (II)-FPH (hairtail protein ferrous chelating peptide) significantly increased *P. vannamei* survival rate, weight gain, and specific growth rate. Also, serum alkaline phosphatase and lysozyme activity were significantly enhanced by 2000–2500 mg/kg Fe (II)-FPH.

Metal nanoparticles and shrimp health

Recently there has been increasing application of nanoparticles (NPs) in aquaculture. Manufacturers are trying to use nano-based tools to remove the barriers around waterborne food, growth, and reproduction.

Nanoparticles are an important mineral form for antibacterial use and are widely utilized for their strong inhibitory and antibacterial effects. Nanoparticles have higher antibiofilm activity against Gram-positive and -negative microbial pathogens. It was reported that nanoparticles can be used as both bacteriostatic and immunostimulant agents through immersion and dietary administration, thereby enhancing shrimp immunity.

Juarez et al. (2017) showed that both oxygen consumption rate and total hemocyte count remained unaltered after injection of silver nanoparticles compared to polyvinylpyrrolidone alone, which indicated there was no stress. Their results also revealed that the survival rate of white spot syndrome virus (WSSV)-infected shrimp after silver NP administration was 80%, whereas that of untreated shrimp was only 10%, 96 h after infection. These results may increase the potential to explore the potential use of silver NPs as antiviral agents in the treatment of diseases of aquaculture organisms, particularly WSSV in shrimp culture. Bahabadi et al. (2017) found that the activity of silver NPs against *Vibrio harveyi* was rapid, with potential treatment effects against bacterial infection in aquaculture. The antibacterial potential of silver NPs synthesized by tea leaf extract against *V. harveyi*, and its protective effect on juvenile *Feneropenaeus indicus*, were investigated.

Sheikh Asadi et al. (2018) suggested that dietary iron NPs in dietary cumin essential oil (CEO) could improve the growth performance and health status of *P. vannamei*. Glutathione peroxidase activity, along with glucose and cholesterol concentrations, were significantly decreased in shrimp fed with supplementary dietary CEO and CEO NPs compared to those fed control diets.

Tello-Olea et al. (2019) used dietary gold NPs administered (as a single dose) at graded levels to *P. vannamei*. After challenge with *V. parahemolyticus*, survival was higher (80%) and histopathological damages were lower in shrimp treated with the 2 μg/g gold NPs compared to the control groups. The results showed that orally administered gold NPs is proposed as an immunostimulant to protect shrimp against *V. parahemolyticus* infection.

Ishwarya et al. (2018) explored the immunostimulatory potential of zinc oxide (ZnO) NPs in green tiger prawn, *Penaeus semisulcatus*, through dietary administration. Shrimp immune parameters (i.e. total hemocyte count, respiratory bursts, phenoloxidase and superoxide dismutase activity) were significantly improved through ingestion of ZnO NPs. In addition, ZnO NPs also had high antibiofilm activities against both Gram-positive (*Bacillus subtilis*, *Staphylococcus aureus*) and Gram-negative (*Shigella sonnei*, *Pseudomonas aeruginosa*) microbial pathogens.

Non-nutritional immunostimulants

POLYSACCHARIDES

Polysaccharides are biological macromolecules with diverse structures in nature. They consist of repeated structural features that are monosaccharide polymer residues, each bonded by glycosidic linkages. Polysaccharides have significant capacity for biological communication because they have the greatest structural complexity (Ringø and Song, 2016). In aquaculture, most of immunostimulant diets include polysaccharides derived from bacteria, fungi, yeasts, or plants. To date, polysaccharides have been widely used as immunostimulants in shrimp aquaculture (Table 15.1).

TABLE 15.1 Dietary polysaccharides for shrimp.

Family/species	*Polysaccharide/ source(s)*	*Dose (g/kg)*	*Duration of feeding trial (days)*	*Parameters*[a,b]	*Resistance to stresses*	*Reference(s)*
Fungal polysaccharides						
Peneaus vannamei	β-Glucan (*Saccharomyces cerevisiae*)	2	42	SGR, THC, O^{2-}, SOD, PO (↑)	n.a.	Bai et al., 2010
P. vannamei	β-Glucan (*S. cerevisiae*)	2	18	SUR, THC, PO, RB, SOD (↑)	*Vibrio alginolyticus*	Chang et al., 2011
P. vannamei	β-1,3-Glucan (*Laminaria digitata*)	0.5	84	WG, SUR, SGR (↔), SOD, CAT, LYZ, TP, ACP, ALP, RB (↑), CM (↓)	Nitrite stress	Zhao et al., 2011
Fenneropenaeus chinensis	β-Glucan	0.5 or 1	10	THC, proPO, PX, LGBP, CL,	n.a.	Bae et al., 2012
P. vannamei	β-1,3-Glucan (*L. digitata*)	1	14 (recommended)	WG, SUR, CM, CAT, SOD, CAT, LYZ, ALP (↑), ACP (↔), CM (↓)	Nitrite-N stress	Zhao et al., 2012
P. vannamei	MOS (*S. cerevisiae*)	1–8	56	WG, SGR, SUR, DR, VL, PO, SOD (↑)	NH_3 stress	Zhang et al., 2012
Penaeidae monodon	MOS + β-glucan (*S. cerevisiae*)	1–5	60	WG, FCR, SUR, THC, RBA (↑)	WSSV	Andrino et al., 2014
P. monodon	MOS (*S. cerevisiae*)	1–8	63	WG, SGR, THC, RB, Tw/B, Td/B, TmP (↑)	n.a.	Sang et al., 2014
P. vannamei	MOS (*S. cerevisiae*)	3	75	SUR, WG, FCR, SGR (↑), BCC (↔)	n.a.	Aktas et al., 2014
P. vannamei	Yeast β-glucan (*S. cerevisiae*)	1 or 2	35	THC, PO, RBA, SOD, DR (↑)	WSSV	Bai et al., 2014
P. vannamei	β-Glucan (*Ganoderma lucidum* and *Coriolus Versicolor*)	0.5–1	28	THC, DHC, O^{2-}, SOD, PO (↑)	n.a.	Yang et al., 2014
P. vannamei	Crude polysaccharides (*C. versicolor*)	2	52	WG, SUR, FI, PHA, PHI, RBA, PO, ALP, ACP, LYZ, SOD, GSH, ROS, TAC, DR (↑), MDA, GSH-PX (↔)	n.a.	Deng et al., 2015
P. vannamei	β-Glucan	0.5	90	PO, SOD, LGBP, proPO, and SP (↑), LYZ, PE (↔), THC (↓).	Ammonia challenge (48 h)	Wongsasak et al., 2015
P. vannamei	β-1,3,1,6-Glucan (*S. cerevisiae*)	0.001	70	SUR (↑), THC, TP, PO (↔)	Infectious myonecrosis virus	Sabry and Nunes, 2015

P. vannamei	β-Glucan	0.35	90	TP, BUN, TG, CHOL (↔), Glu (↓), Intestinal microbiota, and morphometry,	n.a.	Boonanuntanasarn et al., 2016
P. vannamei	β-1,3,1,6)-Glucan (*G. lucidum*)	1	30	O^{2-}, PO, SOD (↑)	n.a.	Wu et al., 2016
P. vannamei	MOS	2.5, 4 and 10	28	SGR, SUR, TLR1, 2, 3, STAT, crustin, ALF, proPO (↑), CM (↓), intestinal bacteria	*V. alginolyticus* and WSSV	Li et al., 2018
Algal polysaccharides						
Marsupenaeus japonicus	Fucoidan *Undaria pinnatifida*	0–1	56	WG, SGR, SUR, FCR, THC, PO	n.a.	Traifalgar et al., 2010
P. monodon	Fucoidan *Sargassum wightii*	0.1–0.3	45	CM, THC, O^{2-}, PO, SOD, RB, PHA (↑)	WSSV	Immanuel et al., 2012
P. vannamei	Fucoidan	[0.1–0.4 (mg/l)], 0–2	[3 h] 7–21	SUR, proPO, O^{2-} THC, PO, HC, GC, RB, PHA, CE (↑)	*V. alginolyticus*	Kitikiew et al., 2013
P. monodon	Sulfated galactans *Gracilaria fisheri*	0.1 and 0.2 g/ml	7	THC, PO, O^{2-}, SOD, CM (↑)	WSSV	Wongprasert et al., 2014
P. monodon	Fucoidan *S. wightii*	0.1–0.3	60	WG, SGR, THC, PO, RB, SOD, PHA, CE (↑), CM (↓)	*V. parahaemolyticus*	Sivagnanavelmurugan et al., 2014
P. vannamei	Fucoidan (*Fuscus versiculosus*)	2–10	0–6	SUR, HC, GC, THC, PO, RB, SOD, PHA, LGBP, PX, proPO I, II, CE, HHAP (↑)	*V. alginolyticus* and WSSV	Chen et al., 2016
P. vannamei	Sulfated galactans *G. fisheri*	0.5, 1 and 2	7	LGBP, p-NF-KB, IMD, IKKβ, IKKε, proPO-I, II, AP3, PNE-4 (↑), Crustion (↔)	*V. parahaemolyticus* (Immersion, 14 days)	Rudtanatip et al., 2017
P. vannamei	Sulfated polysaccharides *Gracilaria birdiae*	0.1–1.2	32	THC, α2M, PO, O^{2-}, RB (↑)	WSSV	Cantelli et al., 2018
P. vannamei	Trehalose	0.5, 1 and 1.5	84	WG; FE; SUR, PO, SOD, GPx, THC, HC, SGC, GC (↑)	*V. alginolyticus*	Zhu et al., 2018

Notes: [a]ACP: acid phosphatase, ALF: antilipopolysaccharide factor, ALP: alkaline phosphatase, α2M: alpha-2-macroglobulin, BMG: blood urea nitrogen, CAT: catalase, CE: clearance ability, CHOL: cholesterol, CL: c-type lectin, CM: cumulative mortality, DHC: differential hemocyte count, DR: disease resistance, FCR: feed conversion ratio, FI: feed intake, GCs: granular cells, Glu: glucose, GSH: glutathione reduced form, GSH-Px: glutathione peroxidase, HCs: numbers of hyaline cells, HHAP: hemocyte homeostasis-associated protein, LGBP: lipopolysaccharide/β-glucan binding protein, LYZ: lysozyme, MDA: malondialdehyde, MOS: mannan oligosaccharide, O^{2-}: superoxide anion production, PhA: phagocytic activity, PHI: phagocytic index, PO: phenoloxidase, proPO: prophenoloxidase, PX: peroxinectin, RB: respiratory burst, ROS: relative oxygen species, SGCs: semigranular cells, SGR: specific growth rate, STAT: signal transducer and activator of transcription, SOD: superoxide dismutase, SP: serine protease, SUR: survival rate, TAC: total antioxidative capacity, Td/B: dry tail muscle index, TG: triglyceride, THC: total hemocyte count, TLR1, 2, 3: Toll-like receptors 1, 2, and 3, TmP: tail muscle protein, TP: total protein, Tw/B: wet tail muscle index, VL: villi length, WG: weight gain.

n.a.: not available.

[b]Variation in experimental shrimp compared to controls: (↑) significant increase; (↓) significant decrease; (↔) no significant changes.

Fungal polysaccharides

In aquaculture, fungal polysaccharides are widely accepted as a nutritional component that can regulate the growth and health of various aquatic animals. As a type of immune enhancer, in shrimp, fungal polysaccharides can activate the immune system, enhance the ability to fight viruses, bacteria, fungi, and parasites, and increase the ability to resist the stress (Katie, 2007; Lund and Petersen, 2006).

Many studies have indicated that feeding diets with β-glucan will show significantly beneficial effects on the immune parameters of *P. vannamei* (Bae et al., 2012; Deng et al., 2015; Sabry et al., 2015; Wongsasak et al., 2015). However, Bai et al. (2010) believed that the continuous application of β-glucan in the diet caused immunity fatigue in *P. vannamei*. The immune activities of *P. vannamei* were improved by β-1,3-glucan and thereby the resistance against ambient nitrite was increased in this shrimp, the possible reason being that β-1,3-glucan can accelerate energy metabolism and activate the immune system. Taking the immunostimulatory effects and economic aspect together, a feeding schedule of 14 days was recommended for the aquaculture of *P. vannamei* (Zhao et al., 2011, 2012). Secondly, β-glucan and/or mannan oligosaccharide (MOS) not only had beneficial effects on the immune response, but also efficiently protected shrimps (*P. monodon* and *P. vannamei*) against WSSV and *V. alginolyticus* challenge (Andrino et al., 2014; Bai et al., 2014; Chang et al., 2011; Sang et al., 2014; Wu et al., 2016; Yang et al., 2014), the reason possibly being that MOS could promote the growth of beneficial bacteria in the digestive tract or directly activate the innate immune response in these shrimps (Li et al., 2018). Furthermore, β-glucan and MOS both significantly improved growth performance, balanced intestinal microbiota, increased the length of intestinal microvilli (Boonanuntanasarn et al., 2016; Zhang et al., 2012), and enhanced resistance against NH_3 stress in *P. vannamei* (Aktas et al., 2014).

Algal polysaccharides

Fucoidan

Fucoidan is an acid-soluble polysaccharide composed mainly of sulfated-1-fucose sugar and trace amounts of galactose, xylose, rhamnose, and glucuronic acid. In vivo studies indicate that shrimp fed diets containing fucoidan exhibit enhanced growth and immune parameters, as well as resistance against *V. alginolyticus* and/or WSSV challenges (Chen et al., 2016; Immanuel et al., 2012; Kitikiew et al., 2013; Sivagnanavelmurugan et al., 2014). For instance, it has been reported that the fucoidan extracted from *Sargassum wightii* provides potent antibacterial activity against WSSV and *V. parahemolyticus* infection in *P. monodon* by increasing the immune response (Immanuel et al., 2012; Sivagnanavelmurugan et al., 2014). Fucoidan derived from *Fuscus versiculosus* also enhanced immune parameters and clearance efficiency to *V. alginolyticus* and WSSV in *P. vannamei*. (Chen et al., 2016). That in vitro study also indicated that incubation of shrimp hemocytes with fucoidan induced hemocyte degranulation and increased phenol oxidase (PO) activity, release of superoxide anions, clearance efficiency, and resistance against *V. alginolyticus* in white shrimp (Kitikiew et al., 2013).

Sulfated polysaccharides from seaweed

Sulfated polysaccharides are polysaccharides with a sulfate root in the sugar hydroxyl group, including the sulfuric acid derivatives of natural neutral polysaccharides of heparin sulfate extracted from plants and all types of synthetic polysaccharide sulfate (Hao et al., 2005). In shrimp aquaculture, sulfated galactan from *Gracilaria fisheri* was used as a dietary ingredient or immunostimulant to improve *P. vannamei* growth and immunity (Zhu et al., 2018). It has also been proven to show a protective effect against WSSV infection in *P. monodon* (Wongprasert et al., 2014) and against *V. parahemolyticus* infection in *P. vannamei* (Rudtanatip et al., 2017). Cantelli et al. (2018) found that dietary supplementation with sulfated polysaccharides from *Gracilaria birdiae* promoted delayed immunostimulation by enhancing agglutinating capacity and PO activity, and thus increasing the survival of *P. vannamei* challenged with WSSV.

Trehalose

Trehalose is a non-reducing disaccharide composed of two molecules of D-glucose connected by an α,α-1,1-glycosidic linkage (Abuchowski et al., 2010). Trehalose exhibits certain physiological activities, such as inhibiting the inflammatory cascade triggered by lipopolysaccharide (Minutoli et al., 2008) and producing antioxidative activity (Aoki et al., 2010). However, data are limited on the effects of trehalose on the non-specific immunity of shrimp. Zhu et al. (2018) found that trehalose exerted immunostimulatory effects on *P. vannamei* and could be used as a dietary ingredient and immunostimulant for improving its growth and immunity.

Other polysaccharides

Other polysaccharides have been proved to be immunostimulants for shrimp. Chang et al. (2018) found that *Astragalus* polysaccharides (APS) could be used as a safe and effective feed additive in aquaculture of *P. vannamei*, the optimal dose of APS for *P. vannamei* suggested being 0.2 g/kg. Oral *Angelica sinensis* polysaccharide exhibits immunostimulatory effects on *P. vannamei* and may thus be used as an immunostimulant for the enhancement of growth and immunity (Pan et al., 2018). Polysaccharide gel (PG) is a water-soluble pectic polysaccharide containing long-chain polygalacturonic acid with a hairy side chain of neutral sugars. Oral administration of PG in shrimp diets revealed immunostimulating potential and disease resistance in *P. monodon* (Pholdaeng and Pongsamart, 2010). Liu et al. (2011) found that *Panax ginseng* polysaccharide (GSP) could be used as an immunostimulant for shrimp through dietary administration to increase immune enzyme activity and modify the expression of immune genes in *P. vannamei*. Xylo-oligosaccharide (XO) is a type of non-digestible oligosaccharide obtained by xylanase hydrolysis of xylan-rich plants. The addition of XO in the feed improved growth performance, increased non-specific immune antioxidative function, improved intestinal morphology, increased total intestinal bacteria and bifidobacteria, and enhanced resistance to *V. alginolyticus* and WSSV (Chen et al., 2014, 2018; Wang et al., 2010).

ORGANIC ACIDS AND SHRIMP HEALTH

Organic acids and their salts have been widely used as preservatives in diets fed to animals, including fish. These substances have been shown to inhibit microbial growth and the uptake of pathogenic organisms and their metabolites by animals (Freitag, 2007; Malicki et al., 2004; Ng et al., 2009). Organic acids and their salts also reduce pH in the stomach and small intestine, thus inhibiting the growth of disease-producing Gram-negative bacteria (Lückstädt, 2008). By significantly reducing duodenal pH, organic acids and their salts enhance pepsin activities and contribute to improved nitrogen retention and increased nutrient digestibility (Øverland et al., 2000). Also, mineral absorption is enhanced by organic acids and their salts (Ravindran and Kornegay, 1993). Gastric acidity influences the bioavailability of dietary minerals by regulating the chelation and complex formation of the elements, and also by altering the transport mechanisms of minerals (Cross et al., 1990). It has been established that acids and their salts modulate the intestinal microbiota of animals, including fish, and that this modulation can be achieved through a shift in dominant hierarchies of bacteria via lysing Gram-negative bacteria (Hellweg et al., 2006; Zhou et al., 2009). The effects of organic acids and their salts on shrimp are listed in Table 15.2.

TABLE 15.2 Feeding trials in shrimp with organic acids and their salts.

Organic acids and their salts	*Shrimp species*	*Duration of feeding trial*	*Dose*	*Results*	*Reference(s)*
Glycyrrhizin	*P. vannamei*	8 weeks	200 mg/kg	SGR↑; THC↑; PO activity↑; respiratory burst activity↑; SOD activity↑; cumulative mortality after disease challenge↓	Chen et al., 2010
Citric acid	*P. vannamei*	45 days	2 and 3 g/kg	Intestinal protease activity↑; serum phenoloxidase↑; SOD activity↑; LZM↑; accumulative mortality after injection of *V. alginolyticus*↓	Su et al., 2014
Microencapsulated organic acid blend (OAB)	*P. monodon*	22 weeks	20 g/kg	Survival to *V. harveyi* challenge↑; hepatopancreatic damage↓; total viable bacterial and *Vibrio* counts in the hepatopancreas↓	Ng et al., 2015
Microencapsulated organic acid blend (OAB)	*P. vannamei*	50 days	20 g/kg	Growth↑; PO activity↑	Romano et al., 2015
Poly-β-hydroxybutyrate (PHB)	*P. vannamei*	35 days	30 g/kg	FCR↓; intestinal amylase↑; lipase↑; trypsin activity↑ T-AOC↑; inducible nitric oxide synthase↑; expression level of intestinal heat shock protein 70 (HSP70), Toll, immune deficiency (Imd) gene↑; relative mRNA expression levels of mtor signaling-related genes TOR, 4E-BP, eif4e1α, eif4e2↑	Duan et al., 2017a, 2017b
γ-Aminobutyric acid (GABA)	*P. vannamei*	8 weeks	150 mg/kg	Survival↑; weight gain↑; survival after 12, 24, or 36 h of NH_3 stress↑; CAT activity↑; T-AOC↑; SOD activity↑; CAT activity↑; antihydroxyl radical and phenoloxidase activities before and after 36 h NH_3 stress↑	Xie et al., 2017

Organic acids and their salts	*Shrimp species*	*Duration of feeding trial*	*Dose*	*Results*	*Reference(s)*
AviPlus® (AP), a blend of organic acids [citric acid, 25%, sorbic acid, 16.7%] and essential oils [thymol, 1.7%, vanillin, 1.0%]	*P. vannamei*	8 weeks	600 mg/kg	Glutathione and PO activity↑; expression of gut pro-inflammatory genes (TNF-α, LITAF, RAB6A)↓; operational taxonomic unit diversity↑; richness indices in gut microbiota↑	He et al., 2017
ACV® and propionic acid (PA)	*P. vannamei*	60 days	10 and 40 g/kg	Expression of LZM, prophenoloxidase genes↑; expression of penaeidin-3a (Pen-3a) after 30 days↑	Pourmozaffar et al., 2017
Succinic acid (SA)	*P. vannamei*	56 days	2.5, 5, and 10 g/kg	Survival rate of shrimp after ammonia stress for 48 h↑; amylase↑; lipase↑; pepsin activity↑ T-NOS activity↑; proPO and HSP70 gene expression level at 56 day ↑	Duan et al., 2018
				T-AOC↑; PO activity↑; genes (proPO activity, HSP70, Trx, GDH) expression level after exposure to ammonia stress for 48 h↑	
Fulvic acid	*P. vannamei*	60 days	9 and 12 g/kg	Specific growth rates↑; survival rates↑; feed conversion rate↓	Gao et al., 2018
				SOD activity↑; PO activity↑; glutathione content↑; malonaldehyde content↓; HSP70 in hemolymph↓	
Butyrate	*P. vannamei*	47 days	20 g/kg	Feed efficiency↑; nitrogen retention↑; protein efficiency rate↑; survival↑	Silva et al., 2016
Sodium butyrate or polyhydroxybutyrate (PHB)	*P. vannamei*	6 weeks	20 g/kg	Intestinal protease↑; trypsin↑; chymotrypsin↑; amylase↑; lipase↑; digestibility of polysaccharides and lipids↑; length, width, and perimeter of intestinal villi↑	Silva et al., 2016b

SINGLE-CELLED ORGANISMS

Probiotics

Probiotics are defined as live microbial or cultured product feed supplements that beneficially affect the host by producing inhibitory compounds, competing for chemicals and adhesion sites, modulating and stimulating the immune functions, as well as improving the microbial balance. In aquaculture, probiotics can be administered either as a food supplement or an additive to the water, playing a key role in controlling disease, enhancing the immune response, supplementing or – even in some cases – replacing the use of antimicrobial compounds, providing nutrients and enzymatic contributions, and improving water quality. Some of the probiotics used in shrimp come from the gut of terrestrial animals and shrimp (Lakshmi et al., 2013; Liu et al., 2014). The main probiotics commonly used in regulating the nutritional immunity of shrimp are from the genera *Bacillus* (*subtilis, licheniformis, cereus, clostridium*) and *Lactobacillus* (*pentosus, plantarum, pediococcus*) and yeasts (*Rhodotorula, Saccharomyces, Debaryomyces, Phaffia*).

Bacillus

Bacillus is a Gram-positive bacterium with aerobic or facultative anaerobic, which is recognized as a promising probiotic (Nayak, 2010). Previous studies have demonstrated that *Bacillus* strains may be used as potential probiotics for shrimp aquaculture (Hao et al., 2014; Zokaeifar et al., 2014). Studies have shown that *B. subtilis* added to the diet can significantly improve the specific growth rate, food conversion ratio, and digestive enzymatic activity of shrimp in comparison to untreated controls (Zokaeifar et al., 2014). Similarly, Shen et al. (2010) observed that white shrimp fed a diet containing *B. subtilis* had better growth performance, immune response, and antioxidative activity than the control group. Moreover, Zhang et al. (2011) showed that total bacterial counts were significantly increased and *Vibrio* counts were significantly decreased in the intestines of shrimp with supplementation of dietary *B. subtilis*. However, different doses yielded different mortality rates. Zokaeifar et al. (2012) reported on the addition of two different doses of *B. subtilis* in the diets of shrimp: the BM8 group (20%) significantly reduced the cumulative mortality rate while challenged with *V. harveyi* compared with the BM5 group (33.3%) and the control group (63.3%). Shen et al. (2010) suggested that probiotics could be used as an effective immunostimulant, the optimal dose of *B. subtilis* being 5×10^4 colony-forming units (cfu)/kg diet.

Lactobacillus

Lactic acid bacteria (LAB) are the most commonly applied probiotics in terrestrial animal nutrition, their use as probiotics in shrimp species having been reported by several studies (Chen and Wei, 2017; Kongnum et al., 2012; Maeda et al., 2014). Previous studies have demonstrated that *Lactobacillus plantarum* improved the immunity, disease resistance, and survival of shrimp. Similar effects were reported for *Pediococcus pentosaceus*, which enhanced the immunity, stress tolerance, and biological health status in *L. stylirostris* juveniles (Castex et al., 2010) and *P. vannamei* (Leyva-Madrigal et al., 2011). In addition, Kongnum et al. (2012) showed that *L. plantarum* incorporated in feed significantly enhanced growth performance and nutritional utilization in *P. vannamei*. LAB can improve the disease resistance of shrimps. Adel et al. (2017) reported that supplementation of *P. pentosaceus* significantly increased the survival rates of *P. vannamei* when challenged with *V. anguillarum*. Moreover, there is information regarding its efficacy in *L. vannamei* in regard to resistance against the stress of acute low salinity, as indicated by both higher survival rates and higher transcript levels of proPO, SOD and Lys genes (Zheng et al., 2017).

Yeasts

Yeasts are among the probiotics frequently used as dietary supplements in fish and shrimp feeds to increase growth performance, feed intake, survival, and disease resistance, as well as an alternative protein source to fishmeal (Zhao et al., 2017). It was not difficult to find that shrimp fed a diet containing yeast had significantly higher weight gain, specific growth rate, and protein efficiency ratio than those fed the control diet (Qiu et al., 2017; Xiong et al., 2018; Zhao et al., 2017). The reasons for this may be due to the nutritional value of yeast nutrients such as proteins, vitamins, and peptides (Xiong et al., 2018). Mannan oligosaccharides derived from the yeast cell wall have been proven effective in enhancing growth performance, improving gut morphology, modulating intestinal microbiota, and enhancing immunity and stress resistance (Yuan et al., 2017). In addition, yeast culture or extract inhibited the growth of pathogenic microbes colonized in the digestive gut, amended the structure and shape of the intestinal wall, enhanced the activities of Lys and Pro, and attenuated damage by fungal toxin (Deng et al., 2013; Xiong et al., 2018). Biswas et al. (2012) indicated the immunostimulatory effects of nucleotide-rich baker's yeast extract on the kuruma shrimp immune system, supporting its potential use in aquaculture.

SEAWEEDS

Seaweeds are a type of lower unicellular autotrophic plant that live in the ocean and use chlorophyll or other auxiliary pigments for photosynthesis. They can transform inorganic substances in the ocean into organic and pollution-free substances through their photosynthesis and respiration. Although the average protein content of seaweeds is low (~5–30% of dry weight), they are potential feed additives or supplements in aquaculture due to their bioactive compounds and their potential application in several branches of industry. The most common compounds include polysaccharides, terpenes, phenols and polyphenols, minerals, polyunsaturated fatty acids, vitamins, and carotenoids (Miyashita et al., 2013; Pádua et al., 2015; Sanjeewa et al., 2016; Schleder et al., 2017).

The beneficial effects of including microalgae and macroalgae (range 0.05–20%) in the diets of fish and shrimp species have previously been declared in terms of enhanced growth, nutritional performance, health, and coloration (Güroy et al., 2012; Silva-Neto et al., 2012; Teimouri et al., 2013; Zhu et al., 2016). The most common macroalgae include brown seaweeds (*Ascophyllum nodosum*, *Sargassum filipendula*, *Undaria pinnatifida*, *Laminaria japonica*), red algae (*Gracilaria lemaneiformis*, *Gracilaria cervicornis*, *Gracilaria heteroclada*, *Kappaphycus alvarezii*, *Hypnea cercivornis*), and chlorophyta (*Ulva lactuca*, *Caulerpa sertularioides*, *Enteromorpha sp.*). Microalgae include cyanobacteria (*Oscillatoria* sp., *Spirulina platensis*), chlorophyta (*Nannochloropsis oculate*, *Chlorella vulgaris*), and bacillariophyta (*Chaetoceros muelleri*). Nutritional studies in shrimps that have investigated the effects of seaweed meal are summarized in Table 15.3.

TABLE 15.3 Studies on seaweed meal in shrimp feeds.

Shrimp species	*Results*	*Reference(s)*
P. vannamei	*Spirulina* (30 and 60 g/kg) in the diet improved immune response.	Chen et al., 2016
P. vannamei	Inclusion of 4 and 8% of brown and green meal seaweed in commercial shrimp feed supported adequate growth and survival of juvenile *P. vannamei*.	Cárdenas et al., 2015
P. vannamei	*Undaria pinnatifida* (4%) enhanced shrimp immunity and specifically inhibited *Vibrio* spp., while *Sargassum filipendula* increased shrimp resistance to thermal variation.	Schleder et al., 2017
Neocaridina davidi	Inclusion of *S. platensis* meal at a level of ~8–10% improved the growth, survival and, in particular, reproductive performance of red cherry shrimp in commercial aquaria.	Kohal, 2018

Many studies have demonstrated that seaweeds not only contribute to improvement in growth, feed utilization, and carcass quality of cultured fish or shrimp, but also strengthen the immune system against stress and improve the resistance of aquatic organisms to thermal stress.

Macroalgae

The most common compounds found in brown seaweeds, including polysaccharides, terpenes, phenols and polyphenols, minerals, polyunsaturated fatty acids, vitamins, and carotenoids, improve animal resistance against bacterial and viral diseases. The seaweed wakame (*Undaria pinnatifida*) is the most commonly utilized brown alga. Schleder et al. (2017) reported that *U. pinnatifida* enhanced whiteleg shrimp immunity and specifically inhibited *Vibrio* spp., and *S. filipendula* increased shrimp resistance to thermal variation. Similarly, Niu et al. (2015) reported enhancement of phenoloxidase activity in *P. monodon* juveniles fed with diets containing 1–6% of *U. pinnatifida* dry biomass, while superoxide dismutase activity was increased only with 6% treatment. In a previous study, dietary wakame significantly increased intestinal fold area in *P. monodon*, suggesting that dietary wakame improve its nutrient absorptive ability (Niu et al., 2015). These results corroborate the effectiveness of brown seaweed, on a dry biomass basis, in modulating the immune system of shrimp and also the ability to improve the resistance of aquatic organisms to thermal stress (Thanigaivel et al., 2016). In addition, Qiu et al. (2017) found that green seaweed (*Ulva* sp.) could be included in the diet of *P. vannamei* at up to 6.35% to replace 2% fish meal, without resulting in growth depression.

Microalgae

Microalgae, which can absorb excess ammonia, phosphorus, and other substances in water through photosynthesis, while providing food for aquatic animals, play a role in stabilizing and improving water quality. *Spirulina* is rich in protein, vitamins, fatty acids, and minerals. Improved survival rate and enhanced abnormal development in prawn were found by feed containing *Spirulina*. Kohal et al. (2018)

reported that the dietary inclusion of *S. platensis* meal at a level of 8–10% improved the growth, survival, and, in particular, reproductive performance of red cherry shrimp in commercial aquaria at maturity, but only up to the sixth spawning event. Chen et al. (2016) showed that shrimp fed with diets containing *Spirulina* dried powder had increased lysozyme activity and resistance against *V. alginolyticus* infection. Therefore, *Spirulina* can be suggested as the opening bait for *P. vannamei*.

Compound seaweeds

In recent studies, compound seaweeds used as a feed additive can substitute for the deficiency of single microalgal nutrients. It is reported that improvement in the protein efficiency of *P. vannamei* was more effective by the addition of compound algae (*S. filipendula*, *L. japonica*, and *Porphyra* spp., 3%), than by the addition of one or the other to the diet separately (Zhou, 2003). *Spirulina*-okra (4%) compound feed additive was more beneficial in improving the growth performance and survival rate of young shrimp (Wang, 2010). Similarly, it was found that a compound seaweed diet (*L. japonica*, *U. pinnatifida*, and *U. lactuca*) promoted the growth of *Hemicentrotus pulcherrimus*, changed both body and gonadal color, and increased the growth of sea urchin (Chang et al., 1999). In a word, the application of compound algae in shrimp aquaculture has great research potential.

HERBS

Herbs have been widely used as a traditional medicine and immunity booster for humans in China over many centuries. Herbs contain many types of active components, including polysaccharides, alkaloids, and flavonoids. Nowadays, herbs or herb extraction products also play a significant role in the Asian aquaculture industry. They are widely used as feed additives in aquaculture because they can amplify the functions of growth promotion, innate immune response enhancement, and disease resistance in aquatic animals.

Many kinds of herbal medicine have been proved to be able to improve the immunity of shrimp. Guo et al. (2011) found that the combination of *Rheum officinale*, *Galla chinensis*, and *Glycyrrhiza uralensis* in a ratio of 2:2:1 in diets had high antibacterial activity in vitro, and increased disease resistance in *P. vannamei* against *V. harveyi* infection. Qiao et al. (2011) found that the addition of six herbs (Licorice root, *Astragalus membranaceus*, *Poris cocos*, Chinese angelica, *Rhizoma atractylodis* Macrocephalae, *and Codonopsis pilosula*) to *P. vannamei* diets at a dose ratio of 6:30:10:10:30:10 enhanced the immune activity of shrimp. It has been reported that feeding a diet containing 0.4% honeysuckle could improve the growth performance of *P. monodon*, and feeding diet containing 0.2 % honeysuckle diet increased the survival of air exposure-simulated live transportation in shrimp (Chen et al., 2013). The traditional medical herb *Minor bupleurum* decoction was indicated as being able to enhance the non-specific immunity responses of *P. vannamei* (Wu et al., 2017). Chen et al. (2013) found that diets supplemented with zingerone, one of the active components of ginger (*Zingiber officinale*), was able to improve the growth and immunity of *P. vannamei* and enhance disease resistance against the pathogen *V. alginolyticus* (Chang et al., 2012).

In conclusion, immunological parameters including the total hemocyte count, respiratory burst activity, phagocytic activity, hemolymph clotting time, and the activity of lysozyme and phenoloxidase were improved by the addition of Chinese herbs to the diets of shrimp. These herbs included *honeysuckle*

(Chen et al., 2013), zingerone (Chang et al., 2012), *Gynura bicolor* (*roxb. willd*) extract (Wu et al., 2017), *Agathi grandiflora* extract (Bindhu et al., 2014), and *Solanum nigrum* extract (Harikrishnan et al., 2011).

It has been reported that the addition of 0.25% *Rubus coreanus* ethanolic extract to the diet enhanced the growth performance of *P. vannamei* (Subramanian et al., 2013). Lawhavinit et al. (2011) reported that *P. vannamei* fed with diets containing ethanol turmeric extract at 7.5 and 15 g/kg had better growth performance compared to the control group. Some Chinese herb extracts, including *Agathi grandiflora* (Bindhu et al., 2014), *Roxb willd* (Wu et al., 2015), and *Curcuma longa* Linn (Vanichkul et al., 2010) improved shrimp survival after viral and bacterial challenges including *V. alginolyticus*, WSSV, and *V. harveyi*.

Chinese herb extracts enhanced resistance against pathogens by improving the immune activity of shrimp. Harikrishnan et al. (2011) reported that 0.1 and 1.0% *Solanum nigrum* extract added to the diets of *P. monodon* enhanced disease resistance against *V. harveyi*. It was found that ethanol turmeric extract at a dose 15 g/kg added to the diets of *P. vannamei* effectively increased its survival after *V. harveyi* challenge (Lawhavinit et al., 2011).

A extract from nine herbs increased the specific growth rate and food conversion ratio of *P. monodon* when the diets were supplemented with 2.5 ml/kg of this extract; the nine herbs were *Aloe vera*, *Andrographis pariculata*, *Annona squamosa*, *Azadirachta indica*, *Citrus aurantifolia*, *Coriandrum sativum*, *Ocimum sanctum*, *Ollium cepa*, and *Psidium guajava* (Aftabuddin et al., 2017).

It was reported that galangal (*Alpinia galangal* Linn) ethanol extract was capable of promoting resistance to *V. harveyi* infection in *P. vannamei* by upregulating the expression of immune-related genes, such as lysozyme and penaeidin-3 (Aftabuddin et al., 2017). The 0.5% *Rubus coreanus* ethanolic extract added to the diets of *P. vannamei* was able to increase the expression of immune genes, like the gene of catalase, which further improved disease resistance against *V. alginolyticus* in shrimp (Subramanian et al., 2013). In addition to Chinese herbs, there are also some Indian herbs that have been proved to be able to improve the immunity of shrimp. An extract of five Indian herbs (*Acalypha indica*, *Hygrophila spinose*, *Picrorhiza kurooa*, *Tinospora cordifolia*, and *Zingiber officinale*) was able to significantly enhance growth performance and disease resistance against *V. harveyi* in *F. indicus* (Rajeswari et al., 2012).

BIOFLOC

Biofloc is a polymer formed by mutual flocculation of bacterial communities, zooplankton, and organic debris, having the ability to reproduce itself (De Schryver et al., 2008). It comes from the culture system or a biological reactor. In Biofloc culture systems, some nitrogen from feces and uneaten feed is incorporated into bacterial cells that are the main components (Hargreaves, 2013). Typical Biofloc is irregular in shape, tiny, easily compressible, highly porous (up to >99% porosity), permeable to fluids, and has a broad distribution of particle size (Crab et al., 2012). Biofloc is regarded as a food source and immunity booster, facilitating detoxification of systems for cultured organisms and others (Panigrahi et al., 2018).

Biofloc and nutrition

The nutritional quality of Biofloc for cultured animals is good but somewhat variable (Hargreaves, 2013). It is susceptible to microbial composition and community organization. The dry-weight protein content of Biofloc ranges from 25 to 50%, with most estimates at 30–45%. Lipid content ranges from 0.5 to 15%, with most estimates between 1 and 5%. There are conflicting reports about the adequacy of Biofloc in providing the often-limiting amino acids methionine and lysine. Biofloc is a good source of vitamins and minerals, especially phosphorus. It may also have probiotic effects (Hargreaves, 2013).

The Biofloc culture system can provide a direct supplemental source of food protein for shrimp. It can also stimulate the activity of digestive proteinases in shrimp and improve feed utilization and protein retention. The utilization efficiency of proteins in farmed animals can increase from 25 to 45% in Biofloc culture systems (Avnimelech et al., 1994; McIntosh, 2001). Research suggests that 20–30% of shrimp or tilapia growth is derived from the consumption and digestion of microbial protein in the Biofloc culture system. Xu et al. (2012) reported that the dietary protein level of *P. vannamei* juveniles reared in a Biofloc culture system could be reduced to 25% without affecting growth performance. Panigrahi et al. (2018) reported that the growth performance of *P. monodon* in a Biofloc system and fed with a feed containing <24% crude protein was similar to that in shrimp raised under regular intensive culture with a 38% protein diet.

Trials with shrimp fed diets containing up to 30% dried Biofloc proved that dried Biofloc can be used as a substitute protein source and has great potential to reduce the pressure on fishmeal use (Hargreaves, 2013). Shao et al. (2017) reported that the replacement of 15% fishmeal by Biofloc meal in shrimp feed did not have a negative impact on growth performance, digestive enzymes, or the TOR signaling pathway of *P. vannamei*. Kuhn et al. (2016) reported that Biofloc could successfully be used as alternative ingredients for 20% soybean and fishmeal in shrimp diets. Khatoon et al. (2016) reported that when 50% of feed protein was replaced by Biofloc meal in shrimp feed, *P. vannamei* post-larvae had a significantly higher specific growth rate and survival rate. Promthale et al. (2019) reported that the replacement of 100% fishmeal by Biofloc meal did not have negative impacts on growth parameters, including weight gain, specific growth rate, and feed conversion ratio.

Furthermore, Biofloc is a source of further dietary lipids, phospholipids, and essential fatty acids, which are necessary to sustain reproduction and embryonic and larval development (Cardona et al., 2016).

Biofloc and health

Biofloc is not only a source of supplemental nutrition, but also provides substantial probiotic bacteria and bioactive compounds, which play an important role in improving the physiological health of aquatic organisms (Chen et al., 2018). It can regulate the intestinal microecological balance and inhibit the growth and reproduction of pathogenic bacteria by competing with pathogenic bacteria for space and nutrients (Avnimelech, 2007; Burford et al., 2004). Also, *Bacillus*, alkali-producing bacteria, *Pseudomonas*, and other bacteria in the Biofloc can produce poly β-hydroxybutyric acid (PHB) which participates in the carbon metabolism and energy storage of bacteria. When bacteria in the gut of cultured animals die or dissolve, intracellular PHB is released to the outside of the cell and degraded by extracellular PHB depolymerase into β-hydroxybutyric acid, which can inhibit certain pathogens and protect various reproductive factors from bacterial infection (Defoirdt et al., 2006, 2007; Long et al., 2013). Studies showed that Biofloc can reduce the risk of acute hepatopancreatic necrosis disease and enhance the survival of shrimp (Hostins et al., 2019). Promthale et al. (2019) reported that, when the replacement levels of fish meal by Biofloc were 25 and 50%, the survival rates, levels of immune parameters (total hemocyte count, differential hemocyte count, phagocytic activity, and phenoloxidase activity), and the expression of immune genes (proPO-I, PEN-4, and dicer) in shrimp were significantly improved. Moreover, shrimp fed with Biofloc showed notably increased survival rates following *V. parahemolyticus* infection.

The way forward

Great progress has been made in the past decade in the study of shrimp nutrition and health. It has also led to the application of some functional feed additives or functional feeds in shrimp farming. However, with the emergence of new situations in shrimp feed and farming, more in-depth and systematic researches on nutrition and shrimp health are required. First, it is necessary to continue to analyze the mechanisms of nutritional metabolism regulating shrimp immunity, disease, and stress resistance, especially the molecular mechanism. Second, since many fishmeal alternative protein sources are widely used in shrimp feeds, it is necessary to explore in depth the effects of alternative protein sources on shrimp health and their possible mechanisms. Third, because shrimp farming has a variety of culture models and diverse environments, it is necessary to carry out targeted research on the impact of the interaction of environment and nutrient metabolism on shrimp health. Furthermore, development of the feeding strategy for functional feed additives or functional feeds throughout the process of shrimp farming is required.

References

Abasali, H. and Mohamad, S. (2010) Immune response of common carp (*Cyprinus carpio*) fed with herbal iminunostimulants diets. *Asian Journal of Animal and Veterinary Advances* 9, 1839–1847.

Abdel Rahman, S. H., Abdel Razek, F. A., Goda, A. M. S., Ghobashy, A. F. A., Taha, S. M. and Khafagy, A. R. (2010) Partial substitution of dietary fish meal with soybean meal for speckled shrimp, *Metapenaeus monoceros* (Fabricius, 1798) (Decapoda: Penaeidae) juvenile. *Aquaculture Research* 41, e299–e306.

Abuchowski, A., Mccoy, J. R., Palczuk, N. C., Van Es, T., Davis, F. F., Alblewi, F. F., Allohedan, H. A., Rafiquee, M. Z. A., Issa, Z. A. and Altun, S. (2010) Effect of trehalose on protein structure. *Protein Science* 18, 24–36.

Adel, M., Yeganeh, S., Dawood, M. et al. (2017) Effects of *Pediococcus pentosaceus* supplementation on growth performance, intestinal microflora and disease resistance of white shrimp, *Litopenaeus vannamei*. *Aquaculture Nutrition* 23, 1401–1409.

Aftabuddin, S., Siddique, M. A. M., Romkey, S. S. and Shelton, W. L. (2017) Antibacterial function of herbal extracts on growth, survival and immunoprotection in the black tiger shrimp *Penaeus monodon*. *Fish & Shellfish Immunology* 65, 52–58.

Aguilar, V., Racotta, I., Goytortúa, E., Wille, M., Sorgeloos, P., Civera, R. and Palacios, E. (2012) The influence of dietary arachidonic acid on the immune response and performance of Pacific whiteleg shrimp, *Litopenaeus vannamei*, at high stocking density. *Aquaculture Nutrition* 18, 258–271.

Akbarzadeh, A., Pakravan, S., Niroomand, M., Abkenar, K. B., Nimvari, M. E., Karimi, K., Ghazvini, A. and Jalali, S. A. H. (2019) Effect of dietary date seed meal as an alternative carbohydrate source on immune-related gene expression of Pacific white shrimp *Penaeus vannamei*. *Fish & Shellfish Immunology* 86, 994–998.

Akiyama, D. M. (1991) Penaeid shrimp nutrition for the commercial feed Industry. *Revised Proceedings of the Aquaculture Feed Processing and Nutrition Workshop*. American Soybean Association, pp. 80–98.

Aktas, M., Ciger, O., Genc, E., Genc, M. A. and Cavdar, N. (2014) Effects of mannan oligosaccharide and serotonin on molting, growth, body composition and hepatopancreas histology of white leg shrimp *Litopenaeus vannamei* (Boone 1931). *Turkish Journal of Fisheries & Aquatic Sciences* 14, 205–211.

Allan, G. and Fielder, D. (2002) Can aquaculture help address problems with rising salinity in Australia? *Proceedings of Aquaculture 2002*, April 2002, Beijing, China.

Andrino, K. G. S., Apines-Amar, M. J. S., Janeo, R. L. and Corre Jr, V. L. (2014) Effects of dietary mannan oligosaccharide (MOS) and β-glucan on growth, immune response and survival against white spot syndrome virus (WSSV) infection of juvenile tiger shrimp *Penaeus monodon*. *Aquaculture, Aquarium, Conservation & Legislation* 7, 321–332.

Aoki, N., Furukawa, S., Sato, K. and Kurokawa, Y. (2010) Supplementation of the diet of dairy cows with trehalose results in milk with low lipid peroxide and high antioxidant content. *Journal of Dairy Science* 93, 4189–4195.

Ashmead, H. (1992) *The Roles of Amino Acid Chelates in Animal Nutrition*. Noyes Publications, Park Ridge, NJ, USA, pp. 479.

Avnimelech, Y. (2007) Feeding with microbial flocs by tilapia in minimal discharge biofloc technology ponds. *Aquaculture* 264, 140–147.

Avnimelech, Y., Kochva, M. and Diab, S. (1994) Development of controlled intensive aquaculture systems with a limited water exchange and adjusted carbon to nitrogen ratio. *Israeli Journal of Aquaculture-Bamidgeh* 46, 119–131.

Ayisi, C. L., Hua, X., Apraku, A., Afriyie, G. and Kyei, B. A. (2017) Recent studies toward the development of practical diets for shrimp and their nutritional requirements. *HAYATI Journal of Biosciences* 24, 109–117.

Bae, S. H., Kim, B. R., Kang, B. J., Tsutsui, N., Okutsu, T., Shinji, J., Jang, I.K., Han, C. H. and Wilder, M. N. (2012) Molecular cloning of prophenoloxidase and the effects of dietary β-glucan and rutin on immune response in hemocytes of the fleshy shrimp, *Fenneropenaeus chinensis*. *Fish & Shellfish Immunology* 33, 597–604.

Bahabadi, M., Delavar, F., Mirbakhsh, M., Niknam, K. and Johari, S. (2017) Assessment of antibacterial activity of two different sizes of colloidal silver nanoparticle (cAgNPs) against *Vibrio harveyi* isolated from shrimp *Litopenaeus vannamei*. *Aquaculture International* 25, 463–472.

Bai, N., Gu, M., Zhang, W. B., Xu, W. and Mai, K. S. (2014) Effects of β-glucan derivatives on the immunity of white shrimp *Litopenaeus vannamei* and its resistance against white spot syndrome. *Aquaculture* 426, 66–73.

Bai, N., Zhang, W. B., Mai, K. S., Wang, X. J., Xu, W. and Ma, H. M. (2010) Effects of discontinuous administration of β-glucan and glycyrrhizin on the growth and immunity of white shrimp *Litopenaeus vannamei*. *Aquaculture* 306, 218–224.

Bharadwaj, A., Patnaik, S., Browdy, C. and Lawrence, A. (2014) Comparative evaluation of an inorganic and a commercial chelated copper source in Pacific white shrimp *Litopenaeus vannamei* (Boone) fed diets containing phytic acid. *Aquaculture* 422, 63–68.

Bindhu, F., Velmurugan, S., Donio, M. B. S., Michaelbabu, M. and Citarasu, T. (2014) Influence of *Agathi grandiflora* active principles inhibit viral multiplication and stimulate immune system in Indian white shrimp *Fenneropenaeus indicus* against white spot syndrome virus infection. *Fish & Shellfish Immunology* 41, 482–492.

Biswas, G., Korenaga, H., Nagamine, R. et al. (2012) Immune stimulant effects of a nucleotide-rich baker's yeast extract in the kuruma shrimp, *Marsupenaeus japonicus*. *Aquaculture* 366, 40–45.

Boonanuntanasarn, S., Wongsasak, U., Pitaksong, T. and Chaijamrus, S. (2016) Effects of dietary supplementation with β-glucan and synbiotics on growth, hemolymph chemistry, and intestinal microbiota and morphology in the Pacific white shrimp. *Aquaculture Nutrition* 22, 837–845.

Browdy, C., Bharadwaj, A., Venero, J. and Nunes, A. (2012) Supplementation with 2-hydroxy-4-(methylthio) butanoic acid (HMTBa) in low fish meal diets for the white shrimp, *Litopenaeus vannamei*. *Aquaculture Nutrition* 18, 432–440.

Bulbul, M., Kader, M. A., Koshio, S., Ishikawa, M. and Yokoyama, S. (2014) Effect of replacing fishmeal with canola meal on growth and nutrient utilization in kuruma shrimp *M. arsupenaeus japonicus* (Bate). *Aquaculture Research* 45, 848–858.

Burford, M. A., Thompson, P. J., McIntosh, R. P. et al. (2004) The contribution of flocculated material to shrimp (*Litopenaeus vannamei*) nutrition in a high-intensity, zero-exchange system. *Aquaculture* 232, 525–537.

Calder, P. C. (2013) n-3 Fatty acids, inflammation and immunity: new mechanisms to explain old actions. *Proceedings of the Nutrition Society* 72, 326–336.

Cantelli, L., Goncalves, P., Guertler, C., Kayser, M., Pilotto, M. R., Barracco, M. A. and Perazzolo, L. M. (2018) Dietary supplementation with sulfated polysaccharides from *Gracilaria birdiae* promotes a delayed immunostimulation in marine shrimp challenged by the white spot syndrome virus. *Aquaculture International* 27, 349–367.

Cárdenas, J.V., Gálvez, A.O., Brito L.O., Galarza E.G., Pitta D.C., Rubin V.V., 2015. Assessment of different levels of green and brown seaweed meal in experimental diets for whiteleg shrimp (*Litopenaeus vannamei*, Boone) in recirculating aquaculture system. *Aquaculture International* 23, 1491–1504.

Cardona, E., Lorgeoux, B., Chim, L. et al. (2016) Biofloc contribution to antioxidant defence status, lipid nutrition and reproductive performance of broodstock of the shrimp *Litopenaeus stylirostris*: Consequences for the quality of eggs and larvae. *Aquaculture* 452, 252–262.

Castex, M., Lemaire, P., Wabete, N. et al. (2010) Effect of *probiotic Pediococcus acidilactici* on antioxidant defences and oxidative stress of *Litopenaeus stylirostris* under *Vibrio nigripulchritudo* challenge. *Fish & Shellfish Immunology* 28, 622–631.

Chang, J., Zhang, W.B., Mai, K.S., Ma, H.M., Liufu, Z.G., Wang, X.J., Ai, Q.H., Xu, W., (2011) Effects of dietary β-glucan and glycyrrhizin on non-specific immunity and disease resistance of white shrimp, *Litopenaeus vannamei* (Boone) challenged with Vibrio alginolyticus. *Aquaculture Research* 42, 1101-1109.

Chang, Y.P., Liu, C.H., Wu, C.C., Chiang, C.M., Lian, J.L., Hsieh, S.L., (2012) Dietary administration of zingerone to enhance growth, non-specific immune response, and resistance to Vibrio alginolyticus in Pacific white shrimp (*Litopenaeus vannamei*) juveniles. *Fish & Shellfish Immunology* 32, 284-290.

Chang, Y.Q., Wang, Z.C., Wang, G.J., (1999) Effect of temperature and algae on feeding and growth in sea urchin, *strongylocentrotus intermedius. Journal of Fisheries of China* 23(1), 69-76.

Chang, Z.Q., Ge, Q.Q., Sun, M., Wang, Q., Lv, H.Y., Li, J., (2018) Immune responses by dietary supplement with Astragalus polysaccharides in the Pacific white shrimp, *Litopenaeus vannamei. Aquaculture Nutrition* 24, 702-711.

Chen, J.H., Ren, Y.C., Li, Y.Q. et al., (2018) Regulation of growth, intestinal microbiota, non-specific immune response and disease resistance of sea cucumber *Apostichopus japonicus* (Selenka) in Biofloc systems. *Fish and Shellfish Immunology* 77, 175-186.

Chen, N.Z., Wei, W., (2017) Effects of Lactobacillus pentosus on the growth performance, digestive enzyme and disease resistance of white shrimp, *Litopenaeus vannamei* (Boone, 1931). *Aquaculture Research* 48.

Chen, X., Lin, H.Z., Jiang, S.G., Wu, K.C., Liu, Y.J., Tian, L.X., Zhang, Y.Q., Niu, J., (2013) Dietary supplementation of honeysuckle improves the growth, survival and immunity of Penaeus monodon. *Fish Shellfish Immunol* 35, 161-169.

Chen, X., Mai, K., Zhang, W., Wang, X., Ai, Q., Xu, W., Liufu, Z., Ma, H. and Tan, B., (2010) Effects of dietary glycyrrhizin on growth and nonspecific immunity of white shrimp, *Litopenaeus vannamei. Journal of the World Aquaculture Society* 41, 665–674.

Chen, X. M., Cao, J. M., Huang, Y. H., Wang, G. X., Mo, W. Y., Chen, B., Zhao, H. X. and Fu, J. J. (2014) Xylo-oligosaccharide supplementation affects growth performance, non-specific immunity, antioxidant function and anti-white spot syndrome virus capacity of juvenile *Litopenaeus vannamei. Chinese Journal of Animal Nutrition* 2397–2407.

Chen, X. Y., Wang, G. X., P., S. Y., Chen, B., Mo, W. Y., Zhao, H. X., Cao, J. M. and Hhang, Y. H. (2018) Effects of dietary xylo-oligosaccharides on digestive enzyme activities, intestinal morphology and bacterial numbers of juvenile *Litopenaeus vannamei. Chinese Journal of Animal Nutrition* 30, 1522–1529.

Chen, Y. J., Liu, Y. J., Yang, H. J., Yuan, Y., Liu, F. J., Tian, L. X., Liang, G. Y. and Yuan, R. M. (2012) Effect of dietary oxidized fish oil on growth performance, body composition, antioxidant defence mechanism and liver histology of juvenile largemouth bass *Micropterus salmoides. Aquaculture Nu*trition 18, 321–331.

Chen, Y. Y., Chen, J. C., Tayag, C. M. et al. (2016) Spirulina elicits the activation of innate immunity and increases resistance against *Vibrio alginolyticus* in shrimp. *Fish & Shellfish Immunology* 55, 690–698.

Chen, Y. Y., Kitikiew, S., Yeh, S. T. and Chen, J. C. (2016) White shrimp *Litopenaeus vannamei* that have received fucoidan exhibit a defense against *Vibrio alginolyticus* and WSSV despite their recovery of immune parameters to background levels. *Fish & Shellfish Immunology* 59, 414–426.

Chi, S., Tan, B., Mai, K. and Zheng, S. (2009) Growth and feed efficiency of juvenile shrimp *Litopenaeus vannamei* fed formulated diets containing different levels of poultry by-product meal. *Journal of Ocean University of China* 8, 399.

Coloso, R. M. and Cruz, L. J. (1980) Preliminary studies in some aspects of amino acid biosynthesis in juveniles of *Penaeus monodon* Fabricius: I. Incorporation of 14C from (U-14C) acetate into amino acids to precipitable proteins. Available at https://agris.fao.org/agris-search/search.do?recordID=AV2012097972 (accessed 18 July 2021).

Crab, R., Defoirdt, T., Bossier, P. et al. (2012) Biofloc technology in aquaculture: Beneficial effects and future challenges. *Aquaculture* 356, 351–356.

Cross, H., Debiec, H. and Peterlik, M. (1990) Mechanism and regulation of intestinal phosphate absorption. *Minor Electrolyte Metabolism* 16, 115–124.

Cui, P., Zhou, Q. C., Huang, X. L. and Xia, M. H. (2015) Effect of dietary vitamin B6 on growth, feed utilization, health and non-specific immune of juvenile Pacific white shrimp, *Litopenaeus vannamei. Aquaculture Nutrition* 22, 1143–1151.

Cuzon, G., Lawrence, A., Gaxiola, G., Rosas, C. and Guillaume, J. (2004) Nutrition of *Litopenaeus vannamei* reared in tanks or in ponds. *Aquaculture* 235, 513–551.

Cuzon, G., Rosas, C., Gaxiola, G., Taboada, G. and Van Wormhoudt, A. (2000) Utilization of carbohydrates by shrimp. In: *Avances en nutrición acuícola V. Memorias del V simposium internacional de nutrición acuícola*, pp. 19–22.

Daviglus, M. L., Stamler, J., Orencia, A. J., Dyer, A. R., Liu, K., Greenland, P., Walsh, M. K., Morris, D. and Shekelle, R. B. (1997) Fish consumption and the 30-year risk of fatal myocardial infarction. *New England Journal of Medicine* 336, 1046–1053.

Dayal, J. S., Rajaram, V., Ambasankar, K. and Ali, S. A. (2011) Sunflower oil cake as a replacement for fish meal in feeds of Tiger shrimp, *Penaeus monodon* reared in tanks and in net cages. *Indian Journal of Geo-Marine Sciences* 40(3).

De Schryver, P., Crab, R., Defoirdt, T. et al. (2008) The basics of bio-flocs technology: The added value for aquaculture. *Aquaculture* 277, 125–137.

Defoirdt, T., Halet, D., Sorgeloos, P. et al. (2006) Short-chain fatty acids protect gnotobiotic *Artemia franciscana* from pathogenic *Vibrio campbellii. Aquaculture* 261, 804–808.

Defoirdt, T., Halet, D., Vervaeren, H. et al. (2007) The bacterial storage compound poly-β-hydroxybutyrate protects *Artemia franciscana* from pathogenic *Vibrio campbellii. Environmental Microbiology* 9, 445–452.

Deng, B., Wang, Z. P., Tao, W. J., Li, W. F., Wang, C., Wang, M. Q., Ye, S. S., Du, Y. J., Wu, X. X. and Wu, D. (2015) Effects of polysaccharides from mycelia of *Cordyceps sinensison* growth performance, immunity and antioxidant indicators of the white shrimp *Litopenaeus vannamei. Aquaculture Nutrition* 21, 173–179.

Deng, D., Mei, C., Mai, K. et al. (2013) Effects of a yeast-based additive on growth and immune responses of white shrimp, *Litopenaeus vannamei* (Boone, 1931), and aquaculture environment. *Aquaculture Research* 44, 1348–1357.

Duan, Y., Wang, Y., Zhang, J., Sun, Y. and Wang, J. (2018) Dietary effects of succinic acid on the growth, digestive enzymes, immune response and resistance to ammonia stress of *Litopenaeus vannamei. Fish & Shellfish Immunology* 78, 10–17.

Duan, Y., Zhang, Y., Dong, H., Wang, Y. and Zhang, J. (2017a) Effects of dietary poly-β-hydroxybutyrate (PHB) on microbiota composition and the mTOR signaling pathway in the intestines of *Litopenaeus vannamei. Journal of Microbiology* 55, 946–954.

Duan, Y., Zhang, Y., Dong, H., Zheng, X., Wang, Y., Li, H., Liu, Q. and Zhang, J. (2017b) Effect of dietary poly-β-hydroxybutyrate (PHB) on growth performance, intestinal health status and body composition of Pacific white shrimp *Litopenaeus vannamei* (Boone, 1931). *Fish & Shellfish Immunology* 60, 520–528.

Freitag, M. (2007) Organic acids and salts promote performance and health in animal husbandry. *Acidifiers in Animal Nutrition* 31, 131–139.

Fu, S. J. and Xie, X. J. (2004) Nutritional homeostasis in carnivorous southern catfish (*Silurus meridionalis*): is there a mechanism for increased energy expenditure during carbohydrate overfeeding? *Comparative Biochemistry & Physiology Part A Molecular & Integrative Physiology* 139, 359–363.

Gao, J., Koshio, S., Ishikawa, M., Yokoyama, S., Mamauag, R. E. P. and Han, Y. (2012) Effects of dietary oxidized fish oil with vitamin E supplementation on growth performance and reduction of lipid peroxidation in tissues and blood of red sea bream *Pagrus major. Aquaculture* 356, 73–79.

Gao, Y., Zhu, J., Bao, H., Hector, V., Zhao, B. and Chu, Z. (2018) Effect of lignite fulvic acid on growth, antioxidant ability, and HSP70 of Pacific white shrimp, *Litopenaeus vannamei. Aquaculture International* 26, 1519–1530.

Gauquelin, F., Cuzon, G., Gaxiola, G., Rosas, C., Arena, L., Bureau, D. and Cochard, J.-C. (2007) Effect of dietary protein level on growth and energy utilization by *Litopenaeus stylirostris* under laboratory conditions. *Aquaculture* 271, 439–448.

González-Félix, M., Gatlin Iii, D., Lawrence, A. and Perez-Velazquez, M. (2003) Nutritional evaluation of fatty acids for the open thelycum shrimp, *Litopenaeus vannamei*: II. Effect of dietary n-3 and n-6 polyunsaturated and highly unsaturated fatty acids on juvenile shrimp growth, survival, and fatty acid composition. *Aquaculture Nutrition* 9, 115–122.

Goytortúa-Bores, E., Civera-Cerecedo, R., Rocha-Meza, S. and Green-Yee, A. (2006) Partial replacement of red crab (*Pleuroncodes planipes*) meal for fish meal in practical diets for the white shrimp *Litopenaeus vannamei*. Effects on growth and in vivo digestibility. *Aquaculture* 256, 414–422.

Guo, J.-J., Her, B.-Y., Chou, R.-L. and Chen, T.-I. (2011) Screening of modern herbal medicines in white shrimp (*Litopenaeus vannamei*) against *Vibrio harveyi* infection. *The Israeli Journal of Aquaculture Bamidgeh* 63, 558–565.

Guo, R., Liang, G. Y., Liu, Y. J., Tian, L. X., Mai, K. S. and Xiao-Yi, W. U. (2007) Effect of dietary carbohydrate and protein levels on growth performance and digestibility of *Litopenaeus vannamei* juvenile reared in brackish water. *Journal of Fisheries of China* 31.

Guo, R., Liu, Y. J., Tian, L. X., Xia, H. and Wang, J. Q. (2011) Effects of dietary cornstarch levels on fat metabolism of hepatopancreas in *Litopenaeus vannamei. Acta Scientiarum Naturalium Universitatis Sunyatseni* 50, 105–109.

Guo, R., Liu, Y. J., Tian, L. X., Yang, H. J., Liang, G. Y. and Huang, J. W. (2005) The effects of different carbohydrate sources on the growth, survival rate and body composition of *Penaeus vannamei. Acta Scientiarum Naturalium Universitatis Sunyatseni* 44.

Güroy, B., Şahin, İ., Mantoğlu, S. et al. (2012) Spirulina as a natural carotenoid source on growth, pigmentation and reproductive performance of yellow tail cichlid *Pseudotropheus acei. Aquaculture International* 20, 869–878.

Hao, J. G. and Hu, G. T. (2005) Research progress on antiviral effect of polysaccharide sulfate [in Chinese]. *Progress in Veterinary Medicine* 26, 54–57.

Hao, K., Liu, J. Y. et al. (2014) Effects of dietary administration of *Shewanella haliotis* D4, *Bacillus cereus* D7 and *Aeromonas bivalvium* D15, single or combined, on the growth, innate immunity and disease resistance of shrimp, Litopenaeus vannamei. *Aquaculture* 428–429, 141–149.

Hargreaves, J. A. (2013) Biofloc production systems for aquaculture. *Southern Regional Aquaculture Center* 4503, 1–11.

Harikrishnan, R., Balasundaram, C., Jawahar, S. and Heo, M.-S. (2011) *Solanum nigrum* enhancement of the immune response and disease resistance of tiger shrimp, *Penaeus monodon* against *Vibrio harveyi. Aquaculture* 318, 67–73.

He, W., Rahimnejad, S., Wang, L., Song, K., Lu, K. and Zhang, C. (2017) Effects of organic acids and essential oils blend on growth, gut microbiota, immune response and disease resistance of Pacific white shrimp (*Litopenaeus vannamei*) against *Vibrio parahemolyticus. Fish & Shellfish Immunology* 70, 164–173.

Hellweg, P., Tats, D., Männer, K., Vahjen, W. and Zentek, J. (2006) Impact of potassium diformate on gut flora of weaned piglets. *Proceedings of the Society of Nutritional Physiology*, pp. 63.

Hernández, C., Olvera-Novoa, M. A., Aguilar-Vejar, K., González-Rodríguez, B. and de la Parra, I. A. (2008) Partial replacement of fish meal by porcine meat meal in practical diets for Pacific white shrimp (*Litopenaeus vannamei*). *Aquaculture* 277, 244–250.

Hostins, B., Wasielesky, W., Decamp, O. et al. (2019) Managing input C/N ratio to reduce the risk of Acute Hepatopancreatic Necrosis Disease (AHPND) outbreaks in Biofloc systems–A laboratory study. *Aquaculture* 508, 60–65.

Hu, J. and Wang, A. (2010) Effects of dietary vitamin E and selenium on the antioxidant system of shrimp *Litopenaeus vannamei. Oceanologia et Limnologia Sinica* 41, 68–71.

Hu, Y., Tan, B. P., Mai, K. S., Ai, Q. H., Zhang, L. and Zheng, S. X. (2011) Effects of dietary menhaden oil, soybean oil and soybean lecithin oil at different ratios on growth, body composition and blood chemistry of juvenile *Litopenaeus vannamei. Aquaculture International* 19, 459–473.

Immanuel, G., Sivagnanavelmurugan, M., Marudhupandi, T., Radhakrishnan, S. and Palavesam, A. (2012) The effect of fucoidan from brown seaweed *Sargassum wightii* on WSSV resistance and immune activity in shrimp *Penaeus monodon* (Fab). *Fish & Shellfish Immunology* 32, 551–564.

Ishwarya, R., Vaseeharan, B., Subbaiah, S., Nazar, A., Govindarajan, M., Alharbi, N., Kadaikunnan, S., Khaled, J. and Alanbr, M. (2018) Sargassum wightii-synthesized ZnO nanoparticles–from antibacterial and insecticidal activity to immunostimulatory effects on the green tiger shrimp *Penaeus semisulcatus*. *Journal of Photochemistry and Photobiology B: Biology* 183, 318–330.

Juarez, M., Mejía, R., Díaz, F., Reyna, V., Re, A., Vazquez, F., Sánchez, C., Mota, M., Pestryakov, A. and Bogdanchikova, N. (2017) Effect of silver nanoparticles on the metabolic rate, hematological response, and survival of juvenile white shrimp *Litopenaeus vannamei*. *Chemosphere* 169, 716–724.

Kanazawa, A. and Teshima, S. (1978) Effects of eicosapentaenoic acid on growth and fatty acid composition of the prawn, *Penaeus japonicus*. *Memoirs of the Faculty of Fisheries, Kagoshima University* 27, 35–40.

Kanazawa, A. and Teshima, S. (1981) Essential amino acids of the prawn. *Bulletin of the Japanese Society of Scientific Fisheries* 47, 1375–1377.

Kanazawa, A., Teshima, S., Tokiwa, S. and Ceccaldi, H. (1979a) Effects of dietary linoleic and linolenic acids on growth of prawn. *Oceanologica Acta* 2, 41–48.

Kanazawa, A., Teshima, S., Tokiwa, S., Kayama, M. and Hirata, M. (1979b) Essential fatty acids in the diet of prawn. II. Effect of docosahexaenoic acid on growth. *Japanese Society of Fisheries Science* 45, 1151–1153.

Kanazawa, A., Tokiwa, S. and Kayama, M. (1977) Essential fatty acids in the diet of prawn. I. Effects of linoleic and linolenic acids on growth. *Japanese Society of Fisheries Science* 43, 1111–1114.

Katie, B. (2007) Recall shines spotlight on pet foods. *Journal of the American Veterinary Medical Association* 230, 1285–1288.

Katya, K., Lee, S., Yun, H., Dagoberto, S., Browdy, C., Vazquez, A. and Bai, S. (2016) Efficacy of inorganic and chelated trace minerals (Cu, Zn and Mn) premix sources in Pacific white shrimp, *Litopenaeus vannamei* (Boone) fed plant protein based diets. *Aquaculture* 459, 117–123.

Khatoon, H., Banerjee, S., Yuan, G. T. G. et al. (2016) Biofloc as a potential natural feed for shrimp postlarvae. *International Biodeterioration & Biodegradation* 113, 304–309.

Kitikiew, S., Chen, J. C., Putra, D. F., Lin, Y. C., Yeh, S. T. and Liou, C. H. (2013) Fucoidan effectively provokes the innate immunity of white shrimp *Litopenaeus vannamei* and its resistance against experimental *Vibrio alginolyticus infection*. *Fish & Shellfish Immunology* 34, 280–290.

Kohal, M. N., Fereidouni, A. E., Firouzbakhsh, F. et al. (2018) Effects of dietary incorporation of Arthrospira (Spirulina) platensis meal on growth, survival, body composition, and reproductive performance of red cherry shrimp *Neocaridina davidi* (Crustacea, Atyidae) over successive spawnings. *Journal of Applied Phycology* 30, 431–443.

Kongnum, K. and Hongpattarakere, T. (2012) Effect of *Lactobacillus plantarum* isolated from digestive tract of wild shrimp on growth and survival of white shrimp (*Litopenaeus vannamei*) challenged with *Vibrio harveyi*. *Fish & Shellfish Immunology* 32, 170–177.

Kuhn, D. D., Lawrence, A. L., Crockett, J. et al. (2016) Evaluation of Bioflocs derived from confectionary food effluent water as a replacement feed ingredient for fishmeal or soymeal for shrimp. *Aquaculture* 454, 66–71.

Kurnia, A., Muskita, W. H., Astuti, O. and Hakim, A. (2016) Utilization of Telescopium mussel meal as an alternative protein source in the diet of black tiger shrimp, *Penaeus monodon*. *International Journal of Scientific Research* 5, 2006.

Lakshmi, B., Viswanath, B. and Sai Gopal, D. V. (2013) Probiotics as antiviral agents in shrimp aquaculture. *Journal of Pathogens* 2013, 424123.

Lawhavinit, O.-A., Sincharoenpokai, P. and Sunthornandh, P. (2011) Effects of ethanol tumeric (*Curcuma longa* Linn.) extract against shrimp pathogenic *Vibrio* spp. and on growth performance and immune status of white shrimp (*Litopenaeus vannamei*). *Kasetsart Journal (Natural Science)* 45, 70–77.

Lee, M. and Shiau, S. (2002) Dietary copper requirement of juvenile grass shrimp, *Penaeus monodon*, and effects on non-specific immune responses. *Fish & Shellfish Immunology* 13, 259–270.

Leyva-Madrigal, K. Y., Luna-González, A., Escobedo-Bonilla, C. M. et al. (2011) Screening for potential probiotic bacteria to reduce prevalence of WSSV and IHHNV in whiteleg shrimp (*Litopenaeus vannamei*) under experimental conditions. *Aquaculture* 322, 16–22.

Li, E., Yu, N., Chen, L., Zeng, C. and Liu, L. (2010) Dietary vitamin B6 requirement of the Pacific white shrimp, *Litopenaeus vannamei*, at low salinity. *Journal of the World Aquaculture Society* 41, 756–763.

Li, L., Wang, C., Zhu, F. and Wang, H. (2002) Techniques for *Litopenaeus vannamei* culture with inland saline ground water. *Shandong Fish* 19, 6–8 [in Chinese with English abstract].

Li, M., Li, J. and Chen, P. (2012) Effects of vitamin E and Schizochytrium on growth and TLR/NF-kB expression level of *Fenneropenaeus chinensis. Chinese Fishery Quality and Standards* 2, 37–44 [In Chinese].

Li, Y., Liu, H., Dai, X., Li, J. and Ding, F. (2018) Effects of dietary inulin and mannan oligosaccharide on immune related genes expression and disease resistance of Pacific white shrimp, *Litopenaeus vannamei. Fish & Shellfish Immunology* 76, 78–92.

Liang, Y., Lin, H., Shi, Y. and Deng, S., (2016) Effects of hairtail protein ferrous chelating peptide feed on growth, immune performance, and quality characteristics of *Litopenaeus vannamei. Oceanologia et Limnologia Sinica* 47(06), 1257–1262.

Lim, C., Ako, H., Brown, C. L. and Hahn, K. (1997) Growth response and fatty acid composition of juvenile *Penaeus vannamei* fed different sources of dietary lipid. *Aquaculture* 151, 143–153.

Lin, S., Lin, X., Yang, Y., Li, F. and Luo, L. (2013) Comparison of chelated zinc and zinc sulfate as zinc sources for growth and immune response of shrimp (*Litopenaeus vannamei*). *Aquaculture* 406, 79–84.

Liu, D., Huang, J., Zhou, F., Yang, Q., Jiang, S., Yang, L. and Zhu, C. (2018) Synergistic effect of vitamin E and *Schizochytrium* sp. on spermatophore regeneration in *Penaeus monodon. Journal of Guangdong Ocean University* 14, 20–26 [in Chinese].

Liu, F. and Lawrence, A. (1997) Dietary manganese requirement of *P. vannamei. Chinese Journal of Oceanology and Limnology* 15, 163–167.

Liu, H., Li, Z., Tan, B. et al. (2014) Isolation of a putative probiotic strain S12 and its effect on growth performance, non-specific immunity and disease-resistance of white shrimp, *Litopenaeus vannamei. Fish & Shellfish Immunology* 41, 300–307.

Liu, J. (2001) Experiment on raising the survival of Chinese shrimp *Fenneropenaeus chinensis* in saline groundwater. *Shandong Fisheries* 18, 13–14.

Liu, Q., Cao, J., Huang, Y., Wang, G., Wen, Y., Zhou, T. and Sun, Z. (2013) The combined effects of β-glucan with selenium and vitamin E on the growth performance, serum immune and antioxidant indexes, and disease resistance of *Litopenaeus vannamei. Journal of Fishery Sciences of China* 25, 1045–1053 [in Chinese].

Liu, X. L., Xi, Q. Y., Yang, L., Li, H. Y., Jiang, Q. Y., Shu, G., Wang, S. B., Gao, P., Zhu, X. T. and Zhang, Y. L. (2011) The effect of dietary Panax ginseng polysaccharide extract on the immune responses in white shrimp, *Litopenaeus vannamei. Fish & Shellfish Immunology* 30, 495–500.

Long, L., Li, Y., Guan, C. et al. (2013) Research review on the role of Biofloc technology in aquaculture. *Fishery Modernization* 40, 28–33.

Lu Y., Chen X., Zhang X., Zheng P., Wang D., Xian J. and Wang A. (2018) Effects of dietary manganese amount on growth and antioxidant activities of sub-adult *Litopenaeus vannamei. Journal of Southern Agriculture* 49(2), 391–396 [in Chinese with English abstract].

Lückstädt, C. (2008) Effect of organic acid containing additives in worldwide aquaculture – sustainable production the non-antibiotic way. *Acidifiers in Animal Nutrition* 71.

Lund, K. H. and Petersen, J. H. (2006) Migration of formaldehyde and melamine monomers from kitchen- and tableware made of melamine plastic. *Food Additives & Contaminants* 23, 948–955.

Macias-Sancho, J., Poersch, L. H., Bauer, W., Romano, L. A., Wasielesky, W. and Tesser, M. B. (2014) Fishmeal substitution with Arthrospira (*Spirulina platensis*) in a practical diet for *Litopenaeus vannamei*: effects on growth and immunological parameters. *Aquaculture* 426, 120–125.

Maeda, M., Shibata, A., Biswas, G. et al. (2014) Isolation of lactic acid bacteria from Kuruma shrimp (*Marsupenaeus japonicus*) intestine and assessment of immunomodulatory role of a selected strain as probiotic. *Marine Biotechnology* 16, 181–192.

Malicki, A., Zawadzki, W., Bruzewicz, S., Graczyk, S. and Czerski, A. (2004) Effect of formic and propionic acid mixture on *Escherichia coli* in fish meal stored at 12 C. *Pakistan Journal of Nutrition* 3, 353–356.

McIntosh, R. P. (2001) Changing paradigms in shrimp farming: V. Establishment of heterotrophic bacterial communities. *Global Aquaculture Advocate* 4, 53–58.

Mente, E., Coutteau, P., Houlihan, D., Davidson, I. and Sorgeloos, P. (2002) Protein turnover, amino acid profile and amino acid flux in juvenile shrimp *Litopenaeus vannamei*: Effects of dietary protein source. *Journal of Experimental Biology* 205, 3107–3122.

Mercier, L., Racotta, I. S., Yepiz-Plascencia, G., Muhlia-Almazán, A., Civera, R., Quiñones-Arreola, M. F., Wille, M., Sorgeloos, P. and Palacios, E. (2009) Effect of diets containing different levels of highly unsaturated fatty acids on physiological and immune responses in Pacific whiteleg shrimp *Litopenaeus vannamei* (Boone) exposed to handling stress. *Aquaculture Research* 40, 1849–1863.

Minutoli, L., Altavilla, D., Bitto, A., Polito, F., Bellocco, E., Laganà, G., Fiumara, T., Magazù, S., Migliardo, F. and Venuti, F. S. (2008) Trehalose: A biophysics approach to modulate the inflammatory response during endotoxic shock. *European Journal of Pharmacology* 589, 272–280.

Miyashita, K., Mikami, N. and Hosokawa, M. (2013) Chemical and nutritional characteristics of brown seaweed lipids: A review. *Journal of Functional Foods* 5, 1507–1517.

Nayak, S. (2010) Probiotics and immunity: a fish perspective. *Fish & Shellfish Immunology* 29, 2–14.

Ng, W., Koh, C., Sudesh, K. and Siti, Z. (2009) Effects of dietary organic acids on growth, nutrient digestibility and gut microflora of red hybrid tilapia, *Oreochromis* sp., and subsequent survival during a challenge test with *Streptococcus agalactiae*. *Aquaculture Research* 40, 1490–1500.

Ng, W., Koh, C., Teoh, C. and Romano, N. (2015) Farm-raised tiger shrimp, *Penaeus monodon*, fed commercial feeds with added organic acids showed enhanced nutrient utilization, immune response and resistance to *Vibrio harveyi* challenge. *Aquaculture* 449, 69–77.

Nguyen, B. T., Koshio, S., Sakiyama, K., Harakawa, S., Gao, J., Mamauag, R. E., Ishikawa, M. and Yokoyama, S. J. A. (2012) Effects of dietary vitamins C and E and their interactions on reproductive performance, larval quality and tissue vitamin contents in kuruma shrimp, *Marsupenaeus japonicus* Bate. *Aquaculture* 334, 73–81.

Niu, J., Chen, X., Lu, X. et al. (2015) Effects of different levels of dietary wakame (*Undaria pinnatifida*) on growth, immunity and intestinal structure of juvenile *Penaeus monodon*. *Aquaculture* 435, 78–85.

Oujifard, A., Seyfabadi, J., Kenari, A. A. and Rezaei, M. (2012) Growth and apparent digestibility of nutrients, fatty acids and amino acids in Pacific white shrimp, *Litopenaeus vannamei*, fed diets with rice protein concentrate as total and partial replacement of fish meal. *Aquaculture* 342, 56–61.

Ouraji, H., Kenari, A. A., Shabanpour, B., Shabani, A., Sodagar, M. and Jafarpour, S. A. (2011) Growth, survival, and fatty acid composition of Indian white shrimp *Fenneropenaeus indicus* (Milne Edwards) fed diets containing different levels of vitamin E and lipid. *Aquaculture International* 19, 903–916.

Øverland, M., Granli, T., Kjos, N., Fjetland, O., Steien, S. and Stokstad, M. (2000) Effect of dietary formates on growth performance, carcass traits, sensory quality, intestinal microflora, and stomach alterations in growing-finishing pigs. *Journal of Animal Science* 78, 1875–1884.

Pádua, D., Rocha, E., Gargiulo, D. et al. (2015) Bioactive compounds from brown seaweeds: Phloroglucinol, fucoxanthin and fucoidan as promising therapeutic agents against breast cancer. *Phytochemistry Letters* 14, 91–98.

Pan, S., Jiang, L. and Wu, S. (2018) Stimulating effects of polysaccharide from *Angelica sinensis* on the nonspecific immunity of white shrimps (*Litopenaeus vannamei*). *Fish & Shellfish Immunology* 74, 170–174.

Panigrahi, A., Saranya, C., Sundaram, M. et al. (2018) Carbon:nitrogen (C:N) ratio level variation influences microbial community of the system and growth as well as immunity of shrimp (*Litopenaeus vannamei*) in Biofloc based culture system. *Fish & Shellfish Immunology* 81, 329–337.

Pholdaeng, K. and Pongsamart, S. (2010) Studies on the immunomodulatory effect of polysaccharide gel extracted from *Durio zibethinus* in *Penaeus monodon* shrimp against *Vibrio harveyi* and WSSV. *Fish & Shellfish Immunology* 28, 555–561.

Piedad-Pascual, F. (1983) Survival and some histological changes in *Penaeus monodon* Fabricius juveniles fed various carbohydrates. *Aquaculture* 31, 169–180.

Pischon, T., Hankinson, S. E., Hotamisligil, G. K. S., Rifai, N., Willett, W. C. and Rimm, E. B. (2003) Habitual dietary intake of n-3 and n-6 fatty acids in relation to inflammatory markers among US men and women. *Circulation* 108, 155–160.

Pourmozaffar, S., Hajimoradloo, A. and Miandare, H. (2017) Dietary effect of apple cider vinegar and propionic acid on immune related transcriptional responses and growth performance in white shrimp, *Litopenaeus vannamei*. *Fish & Shellfish Immunology* 60, 65–71.

Promthale, P., Pongtippatee, P., Withyachumnarnkul, B. et al. (2019) Bioflocs substituted fishmeal feed stimulates immune response and protects shrimp from *Vibrio parahemolyticus* infection. *Fish & Shellfish Immunology* 93, 1067–1075.

Qiao, F., Liu, Y. K., Sun, Y. H., Wang, X. D., Chen, K., Li, T. Y., Li, E. C. and Zhang, M. L. (2017) Influence of different dietary carbohydrate sources on the growth and intestinal microbiota of *Litopenaeus vannamei* at low salinity. *Aquaculture Nutrition* 23, 444–452.

Qiao, J., Du, Z., Zhang, Y., Du, H., Guo, L., Zhong, M., Cao, J. and Wang, X. (2011) Proteomic identification of the related immune-enhancing proteins in shrimp *Litopenaeus vannamei* stimulated with vitamin C and Chinese herbs. *Fish & Shellfish Immunology* 31, 736–745.

Qiu, X. and Davis, D. (2017) Evaluation of flash dried yeast as a nutritional supplement in plant-based practical diets for Pacific white shrimp *Litopenaeus vannamei. Aquaculture Nutrition* 23, 1244–1253.

Qiu, X. A., Neori, J., Kim, C., Yarish, M., Shpigel, L., Guttman, D., Ben-Ezra, V., Odintsov, D. and Davis, A. (2017) Evaluation of green seaweed *Ulva* sp. as a replacement of fish meal in plant-based practical diets for Pacific white shrimp, *Litopenaeus vannamei. Journal of Applied Phycology* 30, 1350–1316.

Rajeswari, P. R., Velmurugan, S., Babu, M. M., Dhas, S. A., Kesavan, K. and Citarasu, T. (2012) A study on the influence of selected Indian herbal active principles on enhancing the immune system in *Fenneropenaeus indicus* against *Vibrio harveyi* infection. *Aquaculture International* 20, 1009–1020.

Rao, Y. V., Das, B., Jyotyrmayee, P. and Chakrabarti, R. (2006) Effect of *Achyranthes aspera* on the immunity and survival of *Labeo rohita* infected with *Aeromonas hydrophila. Fish & Shellfish Immunology* 20, 263–273.

Ravindran, V. and Kornegay, E. (1993) Acidification of weaner pig diets: a review. *Journal of the Science of Food and Agriculture* 62, 313–322.

Richard, L., Surget, A., Rigolet, V., Kaushik, S. J. and Geurden, I. (2011) Availability of essential amino acids, nutrient utilisation and growth in juvenile black tiger shrimp, *Penaeus monodon*, following fishmeal replacement by plant protein. *Aquaculture* 322, 109–116.

Ringø, E. and Song, S. (2016) Application of dietary supplements (synbiotics and probiotics in combination with plant products and β-glucans) in aquaculture. *Aquaculture Nutrition* 22, 4–24.

Romano, N., Koh, C. and Ng, W. (2015) Dietary microencapsulated organic acids blend enhances growth, phosphorus utilization, immune response, hepatopancreatic integrity and resistance against *Vibrio harveyi* in white shrimp, *Litopenaeus vannamei. Aquaculture* 435, 228–236.

Rosas, C., Cuzon, G., Gaxiola, G., Arena, L., Lemaire, P., Soyez, C. and Wormhoudt, A. V. (2000) Influence of dietary carbohydrate on the metabolism of juvenile *Litopenaeus stylirostris. Journal of Experimental Marine Biology & Ecology* 249, 181–198.

Rudtanatip, T., Boonsri, N., Asuvapongpatana, S., Withyachumnarnkul, B. and Wongprasert, K. (2017) A sulfated galactans supplemented diet enhances the expression of immune genes and protects against *Vibrio parahemolyticus* infection in shrimp. *Fish & Shellfish Immunology* 65, 186–197.

Sabry, N. H. and Nunes, A. J. P. (2015) Performance and immunological resistance of *Litopenaeus vannamei* fed a β-1,3/1,6-glucan-supplemented diet after per os challenge with the Infectious myonecrosis virus (IMNV). *Revista Brasileira de Zootecnia* 44, 165–173.

Sang, H. M., Kien, N. T. and Thanh Thuy, N. T. (2014) Effects of dietary mannan oligosaccharide on growth, survival, physiological, immunological and gut morphological conditions of black tiger prawn (*Penaeus monodon* Fabricius 1798). *Aquaculture Nutrition* 20, 341–348.

Sanjeewa, K. K. A., Kim, E.-A., Son, K.-T. et al. (2016) Bioactive properties and potentials cosmeceutical applications of phlorotannins isolated from brown seaweeds: a review. *Journal of Photochemistry and Photobiology B: Biology* 162, 100–105.

Schleder, D. D., Da Rosa, J. R., Guimarães, A.M. et al. (2017) Brown seaweeds as feed additive for white-leg shrimp: effects on thermal stress resistance, midgut microbiology, and immunology. *Journal of Applied Phycology*. 29, 2471–2477.

Shao, J. C., Liu, M., Wang, B. J. et al. (2017) Evaluation of Biofloc meal as an ingredient in diets for white shrimp *Litopenaeus vannamei* under practical conditions: Effect on growth performance, digestive enzymes and TOR signaling pathway. *Aquaculture* 479, 516–521.

Sharawy, Z., Goda, A. M.-S. and Hassaan, M. S. (2016) Partial or total replacement of fish meal by solid state fermented soybean meal with *Saccharomyces cerevisiae* in diets for Indian prawn shrimp, *Fenneropenaeus indicus*, postlarvae. *Animal Feed Science and Technology* 212, 90–99.

Sheikh Asadi, M., Gharaei, A., Mirdar Harijani, J. and Arshadi, A. (2018) A comparison between dietary effects of *Cuminum cyminum* essential oil and *Cuminum cyminum* essential oil, loaded with iron nanoparticles, on growth performance, immunity and antioxidant indicators of white leg shrimp (*Litopenaeus vannamei*). *Aquaculture Nutrition* 24, 1466–1473.

Shen, W. Y., Fu, L. L., Li, W. F. et al. (2010) Effect of dietary supplementation with *Bacillus subtilis* on the growth, performance, immune response and antioxidant activities of the shrimp (*Litopenaeus vannamei*). *Aquaculture Research* 41, 1691–1698.

Shiau, S. and Lee, C. (2006) Dietary zinc requirements of grass shrimp, *Penaeus monodon*, and effects on immune responses. *Aquaculture* 254, 476–482.

Silva, B., Nolasco, S., Magallón, B., Civera, C., Casillas, H. and Seiffert, W. (2016a) Improved digestion and initial performance of whiteleg shrimp using organic salt supplements. *Aquaculture Nutrition* 22, 997–1005.

Silva, B., Jesus, G., Seiffert, W., Vieira, F., Mouriño, J., Jatobá, A. and Nolasco-Soria, H. (2016b). The effects of dietary supplementation with butyrate and polyhydroxybutyrate on the digestive capacity and intestinal morphology of Pacific White Shrimp (*Litopenaeus vannamei*). *Marine and Freshwater Behaviour and Physiology* 49, 447–458.

Silva, B., Vieira, F., Mouriño, J., Bolivar, N. and Seiffert, W. (2016c) Butyrate and propionate improve the growth performance of *Litopenaeus vannamei. Aquaculture Research* 47, 612–623.

Silva-Neto, J. F., Nunes, A. J., Sabry-Neto, H. et al. (2012) Spirulina meal has acted as a strong feeding attractant for *Litopenaeus vannamei* at a very low dietary inclusion level. *Aquaculture Research* 43, 430–437.

Sivagnanavelmurugan, M., Thaddaeus, B. J., Palavesam, A. and Immanuel, G. (2014) Dietary effect of *Sargassum wightii* fucoidan to enhance growth, prophenoloxidase gene expression of *Penaeus monodon* and immune resistance to *Vibrio parahemolyticus. Fish & Shellfish Immunology* 39, 439–449.

Sritunyalucksana, K., Intaraprasong, A., Sa-Nguanrut, P., Filer, K. and Fegan, D. (2011) Organic selenium supplementation promotes shrimp growth and disease resistance to Taura syndrome virus. *Science Asia* 37, 24–30.

Su, X., Li, X., Leng, X., Tan, C., Liu, B., Chai, X. and Guo, T. (2014) The improvement of growth, digestive enzyme activity and disease resistance of white shrimp by the dietary citric acid. *Aquaculture International* 22, 1823–1835.

Suárez, J., Gaxiola, G., Mendoza, R., Cadavid, S., Garcia, G., Alanis, G., Suárez, A., Faillace, J. and Cuzon, G. (2009) Substitution of fish meal with plant protein sources and energy budget for white shrimp *Litopenaeus vannamei* (Boone, 1931). *Aquaculture* 289, 118–123.

Subramanian, D., Jang, Y.-H., Kim, D.-H., Kang, B.-J. and Heo, M.-S. (2013) Dietary effect of *Rubus coreanus* ethanolic extract on immune gene expression in white leg shrimp, *Penaeus vannamei. Fish & Shellfish Immunology* 35, 808–814.

Suo, S., Li, J., Li, J., Chen, Ge, Q. Q. and Sun, M. (2015) Effects of dietary protein and lipid level on growth performance, non-specific immunity and disease resistance of *Exopalaemon carinicauda. Chinese Fishery Quality and Standards* 5, 40–47.

Tan Heng, S., Yu, H. and Yu, Y. (2003) Growth, feed efficiency of juvenile *L. vannamei* fed practical diets containing different levels of poultry by-product meal. NRA Research Report.

Tan Mai, K., Zheng, S., Zhou, Q., Liu, L. and Yu, Y. (2005) Replacement of fish meal by meat and bone meal in practical diets for the white shrimp *Litopenaeus vannamai* (Boone). *Aquaculture Research* 36, 439–444.

Teimouri, M., Amirkolaie, A. K. and Yeganeh, S. (2013) The effects of *Spirulina platensis* meal as a feed supplement on growth performance and pigmentation of rainbow trout (*Oncorhynchus mykiss*). *Aquaculture* 396, 14–19.

Tello-Olea, M., Rosales-Mendoza, S., Campa-Córdova, A., Palestino, G., Luna-González, A., Reyes-Becerril, M., Velazquez, E., Hernandez-Adame, L. and Angulo, C. (2019) Gold nanoparticles (AuNP) exert immunostimulatory and protective effects in shrimp (*Litopenaeus vannamei*) against *Vibrio parahemolyticus. Fish & Shellfish Immunology* 84, 756–767.

Teshima, S. I., Ishikawa, M., Alam, M. S., Koshio, S. and Michael, F. R. (2004) Supplemental effects and metabolic fate of crystalline arginine in juvenile shrimp *Marsupenaeus japonicus. Comparative Biochemistry & Physiology Part B Biochemistry & Molecular Biology* 137, 210–217.

Thanigaivel, S., Chandrasekaran, N., Mukherjee, A. et al. (2016) Seaweeds as an alternative therapeutic source for aquatic disease management. *Aquaculture* 464, 529–536.

Traifalgar, R. F., Kira, H., Tung, H. T., Michael, F. R., Laining, A., Yokoyama, S., Ishikawa, M., Koshio, S., Serrano, A. E. and Corre, V. (2010) Influence of dietary fucoidan supplementation on growth and immunological response of juvenile *Marsupenaeus japonicus. Journal of the World Aquaculture Society* 41, 235–244.

Tseng, Y. C. and Hwang, P. P. (2008) Some insights into energy metabolism for osmoregulation in fish. *Comparative Biochemistry & Physiology Toxicology & Pharmacology* 148, 419–429.

Vanichkul, K., Areechon, N., Kongkathip, N., Srisapoome, P. and Chuchird, N. (2010) Immunological and bactericidal effects of turmeric (*Curcuma longa* Linn.) extract in pacific white shrimps (*Litopenaeus vannamei* Boone). *Kasetsart Journal (Natural Science)* 44, 850–858.

Wander, R. C., Hall, J. A., Gradin, J. L., Du, S.-H. and Jewell, D. E. (1997) The ratio of dietary (n-6) to (n-3) fatty acids influences immune system function, eicosanoid metabolism, lipid peroxidation and vitamin E status in aged dogs. *Journal of Nutrition* 127, 1198–1205.

Wang, C. and Lovell, R. T. (1997) Organic selenium sources, selenomethionine and selenoyeast, have higher bioavailability than an inorganic selenium source, sodium selenite, in diets for channel catfish (*Ictalurus punctatus*). *Aquaculture* 152, 223–234.

Wang, G. X., Zhou, Y., Huang, W. Q., Huang, Y. H., Liu, X. H. and Dong, S. Z. (2010) Effects of xylooligosaccharide on growth, body composition and non-specific immunity in *Litopenaeus vannamei. Freshwater Fisheries* 40.

Wang, M. X., Guo, R., Xia, H., Li, S., Li, Y. Y., Xie, W., Li, Z. J. and Shen, Q. Z. (2016) Effects of seven kinds of carbohydrate structure on the metabolism of *Litopenaeus vannamei. Journal of Fisheries of China* 4.

Wang, W. (2010) *The Impact of Spirulina-Okra Feed Additive on Growth Performance in Whiteleg Prawn.* Jiangxi Chemical Industry.

Wang, W., Guo, R. and Xia, H. (2012) Effects of high-dose vitamin C on immune factors of *Penaeus vannamei* under crowding stress. *Jiangsu Agricultural Sciences* 40, 202–206 [in Chinese].

Wang, X., Li, E., Qin, J. G., Wang, S., Chen, X., Yan, C., Ke, C., Hou, Y., Na, Y. and Zhang, M. (2014) Growth, body composition, and ammonia tolerance of juvenile white shrimp *Litopenaeus vannamei* fed diets containing different carbohydrate levels at low salinity. *Journal of Shellfish Research* 33, 511–517.

Wang, X., Li, E., Xu, Z., Li, T., Xu, C. and Chen, L. (2017) Molecular response of carbohydrate metabolism to dietary carbohydrate and acute low salinity stress in Pacific white shrimp *Litopenaeus vannamei. Turkish Journal of Fisheries and Aquatic Sciences* 17, 153–169.

Wongprasert, K., Rudtanatip, T. and Praiboon, J. (2014) Immunostimulatory activity of sulfated galactans isolated from the red seaweed *Gracilaria fisheri* and development of resistance against white spot syndrome virus (WSSV) in shrimp. *Fish & Shellfish Immunology* 36, 52–60.

Wongsasak, U., Chaijamrus, S., Kumkhong, S. and Boonanuntanasarn, S. (2015) Effects of dietary supplementation with β-glucan and synbiotics on immune gene expression and immune parameters under ammonia stress in Pacific white shrimp. *Aquaculture* 436, 179–187.

Wu, C.-C., Chang, Y.-P., Wang, J.-J., Liu, C.-H., Wong, S.-L., Jiang, C.-M. and Hsieh, S.-L. (2015) Dietary administration of *Gynura bicolor* (Roxb. Willd.) DC water extract enhances immune response and survival rate against *Vibrio alginolyticus* and white spot syndrome virus in white shrimp *Litopeneaus vannamei. Fish & Shellfish Immunology* 42, 25–33.

Wu, Y.-S., Lee, M.-C., Huang, C.-T., Kung, T.-C., Huang, C.-Y. and Nan, F.-H. (2017) Effects of traditional medical herbs "minor bupleurum decoction" on the non-specific immune responses of white shrimp (*Litopenaeus vannamei*). *Fish & Shellfish Immunology* 64, 218–225.

Wu, Y. S., Liau, S. Y., Huang, C. T. and Nan, F. H. (2016) Beta 1,3/1,6-glucan and vitamin C immunostimulate the non-specific immune response of white shrimp (*Litopenaeus vannamei*). *Fish & Shellfish Immunology* 57, 269–277.

Xiao, Y. C., Chen, J., Xie, C. Y., Peng, T., Liu, Y. and Wang, W. N. (2017) A diet of fructose-enriched Artemia improves the response of juvenile *Litopenaeus vannamei* shrimp to acute low-salinity challenge. *Aquaculture Research* 48, 3935–3949.

Xie, S., Li, Y., Zhou, W., Tian, L., Li, Y., Zeng, S. and Liu, Y. (2017) Effect of γ-aminobutyric acid supplementation on growth performance, endocrine hormone and stress tolerance of juvenile Pacific white shrimp, *Litopenaeus vannamei*, fed low fishmeal diet. *Aquaculture Nutrition* 23, 54–62.

Xiong, J., Jin, M., Yuan, Y. et al. (2018) Dietary nucleotide-rich yeast supplementation improves growth, innate immunity and intestinal morphology of Pacific white shrimp (*Litopenaeus vannamei*). *Aquaculture Nutrition* 24, 1425–1435.

Xu, C., Li, E., Liu, Y., Wang, S., Wang, X., Chen, K., Qin, J.G. and Chen, L. (2018) Effect of dietary lipid level on growth, lipid metabolism and health status of the Pacific white shrimp *Litopenaeus vannamei* at two salinities. *Aquaculture Nutrition* 24, 204–214.

Xu, W.-J. and Pan, L.-Q. (2014) Evaluation of dietary protein level on selected parameters of immune and antioxidant systems, and growth performance of juvenile *Litopenaeus vannamei* reared in zero-water exchange Biofloc-based culture tanks. *Aquaculture* 426, 181–188.

Xu, W. J., Pan, L. Q., Zhao, D. H. et al. (2012) Preliminary investigation into the contribution of Bioflocs on protein nutrition of *Litopenaeus vannamei* fed with different dietary protein levels in zero-water exchange culture tanks. *Aquaculture* 350–353, 147–153.

Xu, X. Z. and Li, A. J. (1988) Study on the suitable content and daily demand of protein, sugar, cellulose and fat in prepared bait of *Penaeus chinensis. Marine Sciences* 12, 1–6.

Yang, C. C., Chen, S. N., Lu, C. L., Chen, S., Lai, K. C. and Liao, W. L. (2014) Effect of mushroom beta glucan (MBG) on immune and haemocyte response in Pacific white shrimp (*Litopenaeus vannamei*). *Journal of Aquaculture Research & Development* 5.

Yang, H.-T., Yang, M.-C., Sun, J.-J., Guo, F., Lan, J.-F., Wang, X.-W., Zhao, X.-F. and Wang, J.-X. (2015) Catalase eliminates reactive oxygen species and influences the intestinal microbiota of shrimp. *Fish & Shellfish Immunology* 47, 63–73.

Yang, P. X., Jia, G. W., Xia, H., Qi, G. S., Li, X. H., Lu, J. J., Zhang, Y. and Guo, R. (2018) Effects of feeding frequency and carbohydrate source composition on carbohydrate metabolism of Pacific white leg shrimp *Litopenaeus vannamei. Journal of Dalian Fisheries University* 5, 35–40.

Yang, Q.-H., Tan, B.-P., Dong, X.-H., Chi, S.-Y. and Liu, H.-Y. (2015) Effects of different levels of *Yucca schidigera* extract on the growth and nonspecific immunity of Pacific white shrimp *Litopenaeus vannamei* and on culture water quality. *Aquaculture* 439, 39–44.

Yang, S. P., Liu, H. L., Wang, C. G., Yang, P., Sun, C. B. and Chan, S. M. (2015) Effect of oxidized fish oil on growth performance and oxidative stress of *Litopenaeus vannamei. Aquaculture Nutrition* 21, 121–127.

Yin, G., Ardó, L., Thompson, K., Adams, A., Jeney, Z. and Jeney, G. (2009) Chinese herbs (*Astragalus radix* and *Ganoderma lucidum*) enhance immune response of carp, *Cyprinus carpio*, and protection against *Aeromonas hydrophila. Fish & Shellfish Immunology* 26, 140–145.

Yuan, X. Y., Liu, W. B., Liang, C. et al. (2017) Effects of partial replacement of fish meal by yeast hydrolysate on complement system and stress resistance in juvenile Jian carp (*Cyprinus carpio* var. Jian). *Fish & Shellfish Immunology* 67, 312–321.

Zhang, J., Liu, Y., Tian, L., Yang, H., Liang, G. and Xu, D. (2012) Effects of dietary mannan oligosaccharide on growth performance, gut morphology and stress tolerance of juvenile Pacific white shrimp, *Litopenaeus vannamei. Fish & Shellfish Immunology* 33, 1027–1032.

Zhang, M. L., Chekan, J. R., Dylan, D., Hong, P. Y., Radlinski, L., Revindran, V., Nair, S. K., Mackie, R. I. and Cann, L. (2014) Xylan utilization in human gut commensal bacteria is orchestrated by unique modular organization of polysaccharide-degrading enzymes. *Proceedings of the National Academy of Sciences USA* 111, E3708–E3717.

Zhang, Q., Tan, B., Mai, K. et al. (2011) Dietary administration of *Bacillus* (*B. licheniformis and B. subtilis*) and isomaltooligosaccharide influences the intestinal microflora, immunological parameters and resistance against *Vibrio alginolyticus* in shrimp, *Penaeus japonicus* (Decapoda: Penaeidae). *Aquaculture Research* 42, 943–952.

Zhang, S. P., Li, J. F., Wu, X. C., Zhong, W. J., Xian, J. A., Liao, S. A., Miao, Y. T. and Wang, A. L. (2013) Effects of different dietary lipid level on the growth, survival and immune-relating genes expression in Pacific white shrimp, *Litopenaeus vannamei. Fish & Shellfish Immunology* 34, 1131–1138.

Zhao, H. X., Cao, J. M., Wang, A. L., Du, Z. Y., Huang, Y. H., Tian, J. X., Li, G. L. and Lan, H. B. (2012) Effect of dietary β-1,3-glucan on the immune response of *Litopenaeus vannamei* exposed to nitrite-N. *Aquaculture Nutrition* 18, 272–280.

Zhao, H. X., Cao, J. M., Wang, A. L., Du, Z. Y., Ye, C. X., Huang, Y. H., Lan, H. B., Zhou, T. T. and Li, G. L. (2011) Effect of long-term administration of dietary β-1,3-glucan on growth, physiological, and immune responses in *Litopenaeus vannamei* (Boone, 1931). *Aquaculture International* 20, 145–158.

Zhao, L., Wang, W., Huang, X. et al. (2017) The effect of replacement of fish meal by yeast extract on the digestibility, growth and muscle composition of the shrimp *Litopenaeus vannamei. Aquaculture Research* 48, 311–320.

Zhao, L. B., Wang, X. L., Huang, X. X., Chen, C. Y. and Li, H. (2016) Effects of dietary arachidonic acid on the immune-related gene expressions and *Vibrio*-resistant ability in *Litopenaeus vannamei. Journal of Fisheries of China* 40, 763–775.

Zheng, X., Duan, Y., Dong, H. et al. (2017) Effects of dietary *Lactobacillus plantarum* in different treatments on growth performance and immune gene expression of white shrimp *Litopenaeus vannamei* under normal condition and stress of acute low salinity. *Fish & Shellfish Immunology* 62, 195–201.

Zhou, Q., 2003. Effects of marine algae powder on the growth of *Penaeus vannamei. Marine Sciences* 27(3), 66–69.

Zhou, Z., Liu, Y., He, S., Shi, P., Gao, X., Yao, B. and Ringø, E. (2009) Effects of dietary potassium diformate (KDF) on growth performance, feed conversion and intestinal bacterial community of hybrid tilapia (*Oreochromis niloticus* ♀ × *O. aureus* ♂). *Aquaculture* 291, 89–94.

Zhu, C., Dong, S., Wang, F. and Zhang, H. (2006) Effects of seawater potassium concentration on the dietary potassium requirement of *Litopenaeus vannamei. Aquaculture* 258, 543–550.

Zhu, D., Wen, X., Xuan, X. et al. (2016) The green alga *Ulva lactuca* as a potential ingredient in diets for juvenile white spotted snapper *Lutjanus stellatus* Akazaki. *Journal of Applied Phycology* 28, 703–711.

Zhu, M., Long, X. H. and Wu, S. J. (2018) Effects of dietary trehalose on the growth performance and nonspecific immunity of white shrimps (*Litopenaeus vannamei*). *Fish & Shellfish Immunology* 78, 127–130.

Zhu, X. Z., Liu, Y. J., Tian, L. X., Mai, K. S., Zheng, S. X., Pan, Q. J., Cai, M. C., Zheng, C. Q., Zhang, Q. H. and Hu, Y. (2010) Effects of dietary protein and lipid levels on growth and energy productive value of pacific white shrimp, *Litopenaeus vannamei*, at different salinities. *Aquaculture Nutrition* 16, 392–399.

Zokaeifar, H., Babaei, N., Saad, C. R. et al. (2014) Administration of *Bacillus subtilis* strains in the rearing water enhances the water quality, growth performance, immune response, and resistance against *Vibrio harveyi* infection in juvenile white shrimp, *Litopenaeus vannamei. Fish & Shellfish Immunology* 36, 68–74.

Zokaeifar, H., Balcázar, J. L., Saad, C. R. et al. (2012) Effects of *Bacillus subtilis* on the growth performance, digestive enzymes, immune gene expression and disease resistance of white shrimp, *Litopenaeus vannamei. Fish & Shellfish Immunology* 33, 683–689.

Chapter 16

VIRAL ACCOMMODATION: A NEW MECHANISM FOR DISEASE CONTROL IN SHRIMP AQUACULTURE

TIMOTHY W. FLEGEL

Summary

Field observations and the results of laboratory experiments prior to 1998 indicate that shrimp are capable of specific, adaptive responses to viral pathogens that could not be explained by the existing knowledge and understanding of their defence against viruses. This was evident from their ability to specifically tolerate single or multiple viruses in persistent infections, without showing signs of disease. The specificity of their responses implied that they possessed an unknown type of memory for tolerated viruses, even though shrimp do not produce antibodies. Tolerance is common to crustaceans and insects, and the words "viral accommodation" were chosen to describe it and to encompass the meanings of viral "lodging and support" in a balanced, co-evolutionary interaction. It was proposed that viral accommodation allowed host animals to harbour one or more viruses in persistent infections (sometimes for a lifetime) with minimal negative effects on health. It allowed for positive evolutionary selection of both reduced pathogen virulence and higher host tolerance over time, such that the host and viral pathogen could eventually reach an evolutionary climactic state of innocuous coexistence. In 2009, based on research since 1998, the viral accommodation hypothesis underwent a major revision in which it was proposed that shrimp had the inherent ability to (1) recognize invading viruses, (2) copy pieces of the viral genetic material into DNA fragments, (3) insert these DNA fragments into their own genome as specific memory elements called endogenous viral elements (EVE) and (4) use EVE for production of anti-viral molecules that could interfere with viral growth and reproduction. The process would specifically reduce the quantity, but not eliminate the originating virus. From 2009 to 2019, research has proven such a mechanism for viral accommodation occurs in insects and I reviewed this technically in the journal *Developmental and Comparative Immunology* in 2020. Here, I will give a brief, non-technical review of the history of viral accommodation as background and then focus on potential applications that will be of interest to shrimp culturists should it be confirmed that the mechanisms discovered in insects are paralleled in shrimp.

Introduction

Penaeid shrimp have only recently been domesticated and subjected to genetic improvement as specific-pathogen-free (SPF) stocks that now provide shrimp farmers with high-quality post-larvae (PL) to stock their rearing ponds. Because of their recent domestication, we know relatively little about shrimp pathogen defence mechanisms when compared to cattle, pigs and chickens, for example. However, it has long been known that shrimp do not have the adaptive response of producing pathogen-specific antibodies (as vertebrates do) when challenged by pathogens. Thus, the strategy of producing traditional vaccines intended to induce long-term protective antibody responses, as in vertebrates, is not possible with shrimp. Indeed, it was previously claimed that insects and shrimp were not capable of a specific adaptive immune response but depended solely on non-adaptive, innate immune responses to defend against viral pathogens. In contrast to this claim, the evidence summarized herein indicates that insects – and probably shrimp, too – have a specific, adaptive response to viral pathogens that involves ribonucleic acid (RNA) and deoxyribonucleic acid (DNA) rather than antibodies. The response can be heritable (not possible for the vertebrate antibody response), if the relevant DNA occurs in the chromosomes of shrimp and insect gametes. Revelation of the detailed mechanisms for these adaptive shrimp responses will make it possible to develop appropriate, safe reagents (vaccines) to induce long-term, heritable defence responses in insects and shrimp.

The field observations and results of laboratory experiments prior to 1998 revealed that shrimp are capable of responding to viral infections in a specific, protective manner that could not be explained by previous knowledge and understanding of their cellular and humoral defences. Characteristics of this response were the ability to tolerate specific single or multiple viral infections without gross or microscopic signs of disease in their tissues. This phenomenon was common to crustaceans and insects, the two groups of animals that constitute the vast majority of arthropods (the dominant animal group on our planet).

The concept of viral accommodation was introduced in 1998 as a testable hypothesis to try to explain the common phenomenon of persistent (sometimes for a lifetime), but innocuous, single to multiple viral infections in shrimp and insects (Flegel and Pasharawipas, 1998). The phenomenon was called "viral accommodation" to encompass the meanings viral "lodging and support" by a host in a manner that allowed viral persistence with minimal negative effects on both host and virus (i.e. high and specific tolerance to infection). In simplified terms, the viral accommodation process allows positive selection for reduced pathogen virulence and for higher host tolerance over time. It was presented as a previously unknown, alternative viral response mechanism to the classical one based on an endless, cyclical battle between host defence mechanisms and pathogen virulence mechanisms.

Viral accommodation (ability to carry innocuous, persistent infections) is a beneficial mechanism for the shrimp industry in terms of shrimp survival in the presence of viral pathogens, but it also means that considerable precautions are required when moving living shrimp from one country to another for the purposes of aquaculture. Uncontrolled transboundary movement of living shrimp aquaculture stocks is hazardous, because grossly healthy animals may be carriers of known or unknown, deadly viruses (Flegel, 2006b; Flegel and Fegan, 2002). Failure to recognize this and take relevant precautions (screening for dangerous viruses before use) has resulted in the rapid spread of devastating shrimp viral diseases that threaten the global shrimp culture industry. Importing grossly healthy, exotic shrimp species and keeping them in quarantine for a specified period of observation to monitor for the development of disease signs (as is done with animals like cats, dogs and horses) does not provide any assurance that imported shrimp are free of dangerous viruses. This problem can be easily minimized by restricting movements for aquaculture to domesticated, SPF stocks (Alday-Sanz et al., 2020).

Viral accommodation may have evolved in shrimp and insects because it more rapidly allows a host and viral pathogen to reach an evolutionary climactic state of innocuous coexistence. The concept was refined in 2001 (Flegel, 2001) and 2007 (Flegel, 2007) to make its proposed mechanisms consistent with accumulating research results. Then, in 2009 (Flegel, 2009a), it underwent a major revision. To understand this major revision, it is necessary to realize that viruses cannot reproduce outside of living host cells. Only when they enter susceptible cells can they utilize the host cellular environment and its biochemical processes for reproduction. Viruses use their host's biochemical processes to make viral proteins, to amplify viral genetic material and to make new viral particles for further viral spread. Unlike humans and shrimp that exclusively use DNA as their genetic material, viruses use either DNA or RNA. However, whether viruses use DNA or RNA as their genomic material, they all need to make a type of RNA called messenger RNA (mRNA) from their genetic material in order to have the host cells produce the proteins and new genetic material needed for their reproduction and spread. This key role of viral mRNA for reproduction in all viruses was central to the major revision of the viral accommodation hypothesis in 2009. I hope to guide readers through the process of discovery that led to that revision, and then to another revision (in 2019).

Early history of viral accommodation

The first observations suggesting that shrimp responded adaptively to viral pathogens were made in the early 1990s, when industrial production of giant tiger shrimp (*Penaeus monodon*) was just beginning in Thailand. PL used to stock shrimp farms were produced by land-based hatcheries that depended on adult broodstock shrimp captured from wild fisheries. Around 1991, without warning, shrimp farms in southeastern Thailand began to experience disease outbreaks characterized by massive shrimp mortality. Affected shrimp were pale in appearance with a distinctly yellow 'head' (carapace) that resulted from an abnormal, yellow hepatopancreas beneath the cuticle. The farmers called this anomaly yellow head disease (YHD). Microscopic tissue analysis of dying shrimp revealed that their lymphoid organ (LO) tissues (likened to human lymph nodes) were massively disorganized and had a 'peppered' appearance, indicating loss of LO function. A similar peppered appearance and consequences were seen in their gill tissue. Rapid methods for presumptive diagnosis were developed using shrimp blood smears and whole-gill mounts for microscopic examination at the farm level (Flegel, 2006a; Flegel et al., 1995).

Subsequent transmission electron microscopy of the LO and gills revealed the presence of unique, previously undescribed, viral particles that we called yellow head virus (YHV) (Boonyaratpalin et al., 1993). Based on morphology and cytoplasmic location, the virus was initially (and mistakenly) likened to DNA granulosis viruses in the family *Baculoviridae*. However, viral purification and nucleic acid analysis eventually revealed that YHV was a single-stranded RNA virus of positive sense (Wongteerasupaya et al., 1995). Eventually it was classified as a member of a new genus *Okavirus* in a new family *Roniviridae* in the order Nidovirales (Cowley et al., 2012; Walker et al., 2005).

Curiously, this serious YHV epidemic ended within about 18 months, after which outbreaks were sporadic. Despite this, we were surprised to find that shrimp in cultivation ponds were often YHV positive by electron microscopy but showed no gross signs of disease (Pasharawipas et al., 1997). In some YHV disease outbreaks we found individual shrimp infected with three viruses (YHV, MBV (monodon baculovirus) and HPV (hepatopancreatic parvovirus)) (Figure 16.1). MBV and HPV were known to be non-lethal. Also, in contrast to bacteria (Figure 16.2), shrimp blood cells (haemocytes) did not surround the areas where the viruses were present (i.e. there was no host inflammatory (swelling) response to the triple virus infection – not even to lethal YHV) (Fig. 16.1).

Another example of acquired tolerance to a viral infection was seen with infectious hypodermal and hematopoietic necrosis virus (IHHNV). Its official scientific name is now *Penstylhamaparvovirus* 1 in the family *Parvoviridae*, sub-family *Hamaparvovirinae* (Pénzes et al., 2020). IHHNV is widespread in wild *P. monodon* in Asia, but was unknown until it jumped from grossly healthy, imported *P. monodon* to fry of the native American shrimp *Penaeus stylirostris* in an American research facility, where it caused massive mortality beginning in the early 1980s (Lightner et al., 1983). Only much later was it realized that imported *P. monodon* was the source of the IHHNV that had caused the disease outbreaks.

IHHNV initially caused high mortality also in *P. stylirostris* stocks being reared in Tahiti for teaching purposes. However, sufficient survivors in each generation allowed for stock continuation. Surprisingly, the successive generations of Tahitian fry that survived to adulthood gave rise to offspring exhibiting progressively improved survival with each generation. Finally, the researchers believed that they had produced a stock resistant to IHHNV. However, when samples of the stock were sent to Dr. Lightner's laboratory in Arizona, they caused massive mortality in his naïve stocks of *P. stylirostris*. The mortality revealed that the grossly normal Tahitian stock was still infected with lethal IHHNV. Thus, the Tahitian stock was tolerant rather than resistant to IHHNV. A subsequent study published more than 10 years

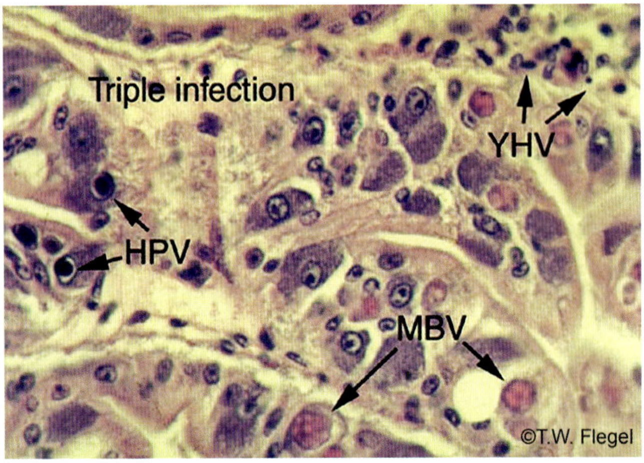

FIGURE 16.1 Photomicrograph showing hepatopancreatic (HP) tissue of a moribund shrimp dying from YHV infection. YHV lesions (dark purple spots giving a 'peppered' appearance) are present in the spaces between the HP tubules. In the HP tubules, reddish lesions of monodon baculovirus (MBV) (now called Penaeus monodon nudivirus) (Yang et al., 2014)) and large purple lesions of hepatopancreatic parvovirus (HPV) (now called *Hepanhamaparvovirus* in the family *Parvoviridae* and sub-family *Hamaparvovirinae* (Pénzes et al., 2020)) can be seen. MBV and HPV are non-lethal for *P. monodon* but can cause retarded growth. Note the absence of masses of shrimp defensive blood cells (i.e. a host inflammatory response) to these three viruses when compared with the massive response to bacterial pathogens shown in Figure 16.2.

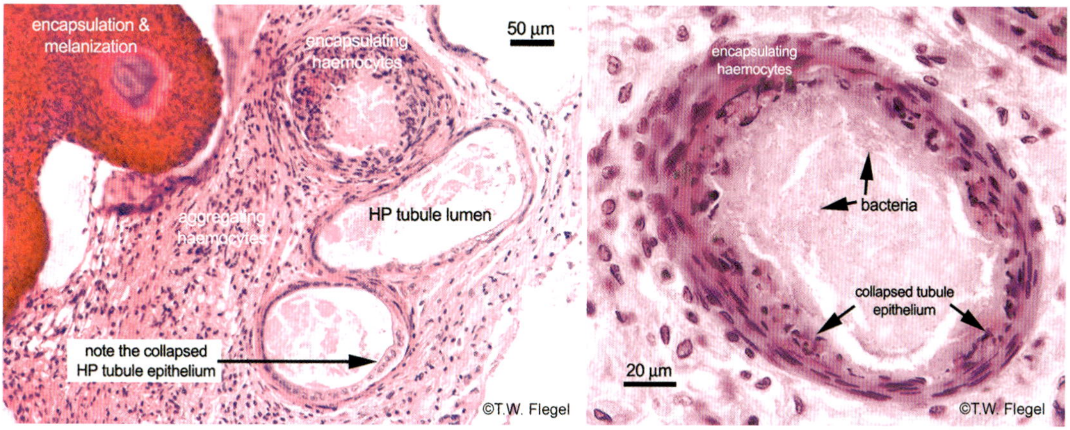

FIGURE 16.2 Photomicrograph of HP tissue infected with bacteria and showing a massive host inflammatory response characterized by aggregation of shrimp blood cells (identified by their darkly stained nuclei) that sometimes encapsulate bacterial colonies in concentric layers and then develop melanin pigment (i.e. granuloma formation followed by melanization). Note that this response to bacterial infection is absent in the viral-infected tissue shown in Fig. 16.1.

later revealed a high prevalence of IHHNV in captured, wild, grossly normal *P. stylirostris* from the Gulf of California (Morales-Covarrubias et al., 1999). This discovery confirmed that IHHNV had been transferred to wild shrimp after its introduction to the Americas and that the wild shrimp population had accommodated it. The authors concluded that "a host–pathogen relationship might have reached a putative equilibrium" in the wild population.

As described above at the beginning of this section, similar accommodation of YHV occurred in Thailand from 1991. It then happened again with WSSV after its introduction to Thailand around 1995. Initially, very severe disease outbreaks occurred but, after 18–24 months, many ponds contained shrimp that tested positive for WSSV but did not exhibit signs of disease. At that time, the primary source of WSSV in shrimp fry (called post-larvae or PL) from hatcheries was grossly healthy parental broodstock captured from the wild but unknowingly infected with WSSV. This transmission pattern is typical for several lethal shrimp viruses.

At the same time, my review of the literature in 2009 (Flegel, 2009b) revealed that despite the severe disease losses from IHHNV, YHV and WSSV in cultivated shrimp in many countries, there was no corresponding decrease in the harvest of susceptible shrimp species from adjacent capture fisheries in the relevant countries. Apparently, during the WSSV epidemic in ponds, similar epidemics did not occur in the wild shrimp populations, suggesting that the infection cycle dynamics and the host–pathogen relationship in the natural environment allowed for relatively rapid acquisition of tolerance.

In succeeding years, selective breeding has allowed for the development of SPF shrimp stocks that may become infected with normally lethal viruses after challenge but show no signs of disease (Alday-Sanz et al., 2020). A good example is SPF shrimp stocks tolerant to infection with Taura syndrome virus (Argue et al., 2002; Moss et al., 2005). However, the mechanistic basis for the viral tolerance in these shrimp is still unclear.

To summarize this background information, Table 16.1 shows what I believe to be fundamental differences between arthropods and vertebrates in their responses to viral pathogens. The viral

TABLE 16.1 A comparison of the difference in response to viral pathogens between shrimp (arthropods) and vertebrates. These are broad comparisons of overall major responses but sufficiently flexible to accommodate 'minor exceptions'. The purpose is to illustrate that the responses of the two groups are fundamentally different.

Shrimp	*Vertebrates*
No inflammatory (swelling) response	Inflammatory (swelling) response
Survivors usually infected (i.e. pathogen persists often for a lifetime)	Survivors usually not infected (i.e. pathogen is cleared)
Survivors infectious for others	Survivors not infectious for others
Tolerance to viruses normal	Tolerance to viruses rare
No antibodies found in serum	Antibodies found in serum
Multiple active infections normal	Multiple active infections rare
Copious virion production	Comparatively low virion production

accommodation hypothesis was formulated to provide a structured framework for the design of experiments to test its predictions regarding the mechanisms behind these differences. The hypothesis could then be modified as necessary based on subsequent experimental results (in a progressive manner) to reveal the underlying mechanisms.

The first formulation of the viral accommodation hypothesis (VAH-1998)

Based on natural and empirical information, it was hypothesized that arthropods have an adaptive process to accommodate (tolerate) viral pathogens and that the first step is characterized by the occurrence of a persistent viral infection arising by an unknown process (Flegel, 2001; Flegel and Pasharawipas, 1998). These persistent infections are active and might last the host's lifetime, during which time the viral pathogen(s) would be transmissible to naïve shrimp for which they might be lethal. Thus, they differ from latent infections in vertebrates, which do not produce infectious virus during latent periods before a disease episode occurs. They also differ from chronic infections, where infectious virus is continuously present after infection but causes periodic disease episodes. It was also hypothesized that tolerance for each virus (i.e. accommodation of each virus) is acquired individually, indicating that specificity is governed by an unknown memory mechanism.

It was proposed that the natural viral accommodation process would go through three evolutionary phases for each new or novel pathogen. Phase 1 would cover the period of exposure to a naïve population, perhaps involving initial catastrophic mortality followed by a relatively rapid decrease in mortality within a few generations. This period would be characterized by the common appearance of tolerated, persistent infections in survivors. Phase 2 would cover a longer term of many generations, characterized by 1) positive selection of host variants (i.e. mutants) less affected by the persistent infections, 2) positive selection of viral variants with a less negative effect on the host and 3) strong negative selection against viral variants with increased virulence. Any changes in the virus or host that simultaneously improved host resistance or tolerance to an unrelated viral pathogen would also be positively selected. Phase 3 would constitute a climactic phase characterized by innocuous host/virus coexistence.

Many tests of the VAH-1998 hypothesis and its implicit predictions were formulated. These included challenge tests to prove that shrimp, insects and insect cell lines that survived infection would be persistently infected but unaffected even by super-challenges with the same virus. Insect cell lines were used for these tests because no immortal cell line for shrimp was available.

The first tests of VAH-1998 using persistently infected insects and insect cells

Tests with densovirus, Dengue virus and Japanese encephalitis virus infections in mosquito cell cultures revealed that persistently infected, grossly normal cell cultures with single, dual and triple infections in individual cells could be produced. The order of challenge for the two or three viruses did not affect the final result of accommodation (Kanthong et al., 2008, 2010a).

During the course of this insect-cell work, it was also discovered that insect cells persistently infected with Dengue virus produced cytokine-like, anionic septapeptides with anti-dengue activity (Kanthong et al., 2010b; Laosutthipong et al., 2013), raising the possibility that similar antiviral cytokines may also occur in shrimp.

In addition to tests with cultured insect cells, trials were carried out with five generations of mosquitoes whose larvae were successively challenged with a stock densovirus. Survivors were then raised to adults to produce larvae for the next generation challenge. Survival of the larvae increased progressively over five generations from 15% to 58% (Roekring et al., 2006). Prevalence of the densovirus infection in surviving mosquito larvae through to adulthood was confirmed by molecular testing (PCR), microscopic tissue analysis (histology), in situ hybridization (ISH) and transmission electron microscopy. Despite the persistent infections, all mosquitoes in the infected populations grew normally to establish each succeeding generation.

In summary, analysis of natural and farmed shrimp specimens and experiments with shrimp, insects and insect cell lines from 1998 to 2007 proved that persistent infections of both RNA and DNA viruses were common in nature and could be easily produced in the laboratory. Essentially, existence of viral accommodation (tolerance to one or more viruses) was confirmed by these experiments. The only remaining questions were related to the nature of the underlying mechanisms. Based on these results from 1998 to 2007, the viral accommodation hypothesis was updated (Flegel, 2007) with the proposal that the predicted memory function proposed in 1998 might be the virus itself in the form of a persistent infection (i.e. 'the virus is the memory'), an idea that was put forward by James A. Brock in a prior dinner discussion. However, subsequent work published in 2006 and thereafter required yet another iteration of the hypothesis with the more interesting possibility that the memory consisted of heritable, viral genome fragments integrated into the shrimp genome. This could be considered a modification of 'the virus is the memory' into 'memory consists of viral genome fragments in the host genome'.

Proposal for heritable, protective viral memory in shrimp

The viral accommodation hypothesis was updated in 2009 (VAH-2009) (Flegel, 2009a), mostly to include a new proposal for the memory mechanism required by the hypothesis. The new proposal was stimulated by a publication in 2006 (Tang and Lightner, 2006) that reported the discovery of incomplete genome fragments of IHHNV-DNA (called 'non-infectious IHHNV') inserted into the host shrimp genome. These fragments had high DNA-sequence identity (around 80%) to existing isolates of IHHNV. Around the same time (2007), one report described bee paralysis virus genome inserts with high sequence homology to the existing bee virus. That report also revealed that the inserts were associated with tolerance to the normally deadly virus (Maori et al., 2007). There had been similar reports of more ancient WSSV-like sequences (very low DNA sequence identity of around 40%) in the genome of *P. monodon* in 2006 (de la Vega, PhD 2006 Thesis, University of Queensland). After that, there were similar reports of ancient WSSV sequences in 2010 in the genome of *P. japonicus* (Koyama et al., 2010) and in 2011 in the genome of *P. monodon* (Huang et al., 2011). However, the low sequence identity of these ancient sequences excluded them as possible candidates for production of negative-sense RNA that could bind with complementary sequences of mRNA from extant viruses. However, in both cases, the discoveries were unexpected because none of the viruses concerned were retroviruses, which had long been known to be scattered widely in eukaryote genomes.

Non-retroviral inserts in the genomes of mammals were discovered around 2010, when they came to be called 'non-retroviral endogenous viral elements'. or just EVE for short (Feschotte and Gilbert, 2012; Katzourakis and Gifford, 2010). There followed a great increase in the number of reports on EVE in vertebrate genomes. Less attention was focused on shrimp, insects and other invertebrates. We subsequently adopted the term 'EVE' to refer to non-retroviral genome fragments found in the shrimp genome. However, it is important to note that most of the EVE reported to date from vertebrates have sequences that have a relatively low percentage of DNA sequence identity to the RNA or DNA sequences of corresponding, existing viruses. This indicates that the vertebrate EVE were fixed in the host genome in ancient times. In contrast, many of the EVE reported from shrimp and insects have very high nucleic acid sequence identities (i.e. 90–100%) to existing viruses. This suggested that they originated very recently in geological time, including the present.

Together with these developments, I recall having earlier asked a colleague, Prof. Anchalee Tasanakachorn from Chulalongkorn University, whether there was any evidence of retroviral infections in her database of expressed genetic sequences (i.e. expressed sequence tags, or EST) from the giant tiger shrimp *P. monodon*. She subsequently informed me that the EST database contained six records for different, putative reverse transcriptase (RT) sequences and seven for different integrase (IN) sequences. The sequences of these EST corresponded to those of well-known endogenous retrotransposon elements in animal species. However, there was no indication of retroviral diseases in the shrimp from which the sequences arose. Thus, the data indicated that the endogenous retroviral sequences (retrotransposons) in the shrimp genome were actively making mRNA transcripts to produce RT and IN enzymes. This was puzzling at the time, because retroviral sequences in vertebrates, in contrast, were known to be mostly silent and of interest only when they caused disease (rarely) by becoming active or being transposed to a new position in the host genome.

While writing a review on shrimp responses to viral pathogens in 2008 (not published until 2011 (Flegel and Sritunyalucksana, 2011), I was seized by the thought that all the data described in the preceding four paragraphs might be connected in a manner that could comprise a memory mechanism for the viral accommodation hypothesis. I reasoned that because non-retroviral EVE did occur in shrimp and insects, they must result from a natural, host genome-insertion mechanism. Shrimp EST with highly conserved sequences for reverse transcriptase and integrase enzymes indicated the presence of enzymes that could serve this purpose. In addition, the high sequence conservation of these enzyme genes suggested that they should have a functional role in shrimp survival. At the same time, the evidence indicated that EVE existed for both RNA and DNA viruses in shrimp, and this, in turn, suggested that the target for host RT of both RNA and DNA viral types would be their mRNA. For both virus types, this was the common vulnerable entity without which the host could not make the viral proteins needed for viral reproduction and spread.

I envisioned a process whereby host RT (arising from endogenous, anciently acquired but actively transcribing retrotransposons) might be able to recognize viral 'foreign' mRNA (either directly or indirectly via molecular pattern receptors) and then make randomly variable viral copy DNA (vcDNA) fragments from it. These vcDNA fragments could then be integrated into the shrimp genome by endogenous IN. There they might give rise to antisense RNA that could activate the RNA interference (RNAi) mechanism, leading to viral tolerance. This process would limit viral replication and lead to persistent viral infections. If the EVE became integrated into germ cells they would become heritable, and those EVE causing viral tolerance would be subjected to positive evolutionary selection. These ideas culminated in the 2009 publication (Flegel, 2009a) of an updated version of the viral accommodation hypothesis (VAH-2009) in a paper entitled 'Hypothesis for heritable, anti-viral immunity in crustaceans and insects'. The hypothesis is summarized in Figure 16.3, as a simplified version of a more complex diagram published in 2017 (Utari et al., 2017).

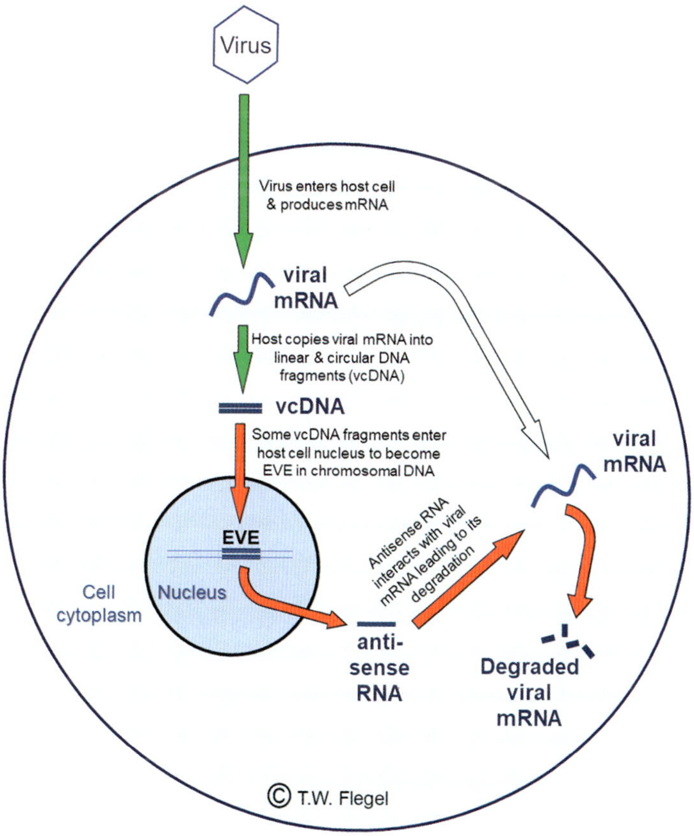

FIGURE 16.3 Simplified illustration of the mechanism for viral accommodation, as updated in 2009.

Basically VAH-2009 proposed that shrimp have a natural mechanism for adaptive tolerance to viral pathogens that relies on nucleic acids (DNA and RNA) for recognition, specific memory and limitation of viral replication. Perhaps this protective, immune mechanism evolved as a substitute for the adaptive antigen–antibody system that occurs in vertebrates but which invertebrates lack. However, the presence of EVE in germ cells gave the viral accommodation mechanism the additional advantage of heritability.

It is important to realize that viral accommodation is an individual cellular response that can occur in both somatic and germ cells. The consequence after exposure to a virus is immense variation in the RNA and DNA extracted from whole individual animals or from millions of cells in a culture dish. This wide variation makes analysis of outputs from the underlying biochemical mechanisms complex. On the other hand, it is interesting that the high variation in germ cells is combined with unusually high numbers of chromosomes in shrimp (e.g. 88 chromosomes or 44 pairs in *P. vannamei*) when compared, for example, to humans (48 chromosomes or 24 pairs). Thus, the possible number of variants in gametes from a single *P. vannamei* animal (based solely on random assortment of chromosomes) is 2^{44} or about 18 trillion (18,000 million) compared to only 2^{24} or 17 million from a single human. As a result, each individual gamete potentially carries a different profile of acquired EVE from new and previously existing viral pathogens. This is combined with massive numbers of eggs per spawn (around 300,000 in *P. vannamei* and around 1,000,000 in *P. monodon*) and with multiple spawns over a shrimp lifetime. The enormous variation in EVE combinations for each generation of a population (driven simply by random assortment of chromosomes) is subject to heavy selection, since a relatively low percentage of offspring end up successfully reproducing. These features may speed up the evolutionary selection of protective EVE.

Tests for EVE in shrimp

Before formulation of VAH-2009, we did not realize that we already had preliminary evidence of EVE in shrimp but did not recognize it at the time it was observed. The first was the discovery of what were called 'latency-related genes' (LRG) of WSSV in *P. monodon* (Khadijah et al., 2003). The second was the occurrence of fragmentary PCR amplicons rather than complete amplicon sets for the whole IHHNV genome in the *P. monodon* that had tested PCR positive for the virus (Chayaburakul, 2005). Subsequent work after 2006 revealed that EVE of high sequence identity to existing IHHNV existed in *P. monodon* and *P. vannamei* (Brock et al., 2013; Jaroenram and Owens, 2014; Saksmerprome et al., 2011) and that the LRG in *P. monodon* were also EVE (Utari et al., 2017). All these putative EVE were variable in number and length in the individual shrimp specimens studied. In addition, RNA production from the LRG-EVE_{366} in *P. monodon* was shown to be variable production (i.e. negative-sense, positive-sense, dual-sense or none). These results supported VAH-2009 in showing that EVE of variable DNA sequences existed for two shrimp viruses. In addition, the work with EVE_{366} revealed the predicted variation in the sense of its RNA transcripts.

One of the key practical ramifications of this work was revelation of the high probability of obtaining false-positive results for shrimp viral infections if the sequence of any particular EVE matched the target sequence of the relevant virus. Please see the accompanying chapter in this book on the problematic consequences that can arise because of this, and how they might be circumvented. This issue would be particularly important in a shrimp breeding programme if a positive PCR test result from a protective EVE was considered an indication of IHHNV infection and thus a reason for culling.

Mendelian inheritance of EVE

The first publication showing Mendelian inheritance of EVE in shrimp was for IHHNV in *P. monodon* (Brock et al., 2013). To confirm EVE heritability for WSSV, EVE_{366} was selected for further study and broodstock pairs from ongoing breeding stock maintenance were screened for its presence (Taengchaiyaphum et al., 2019). In two separate crosses, EVE_{366} was shown to be inherited in a Mendelian fashion. In summary, the results of both studies supported VAH-2009 in revealing that EVE of two DNA viruses occurred in *P. monodon* in a variable manner and were inherited in a Mendelian fashion. What remains to be done with *P. monodon*, or any other shrimp species and its viruses, is to show that some EVE can be protective. Thus, the next step will be to carry out designed crosses that will give offspring with a 50:50 distribution of any particular EVE that shows either antisense or dual-sense expression. This will allow the sibling offspring to be separated into two groups based on the presence or absence of the EVE, in order to determine in challenge tests whether or not a particular EVE is significantly protective.

Recent support for VAH-2009 from insect studies

From 2009 to 2019, several publications revealed the role of EVE and RNAi in protection against viral infections in whole insects and in cultured insect cell lines. These cover the discovery that development of persistent viral infections (viral tolerance) in the fruit fly *Drosophila melanogaster* and mosquitoes is dependent on RT (i.e. production of vcDNA (Goic et al., 2013, 2016; Karlikow et al., 2014, 2016; Poirier et al., 2018; Saleh et al., 2009; Tassetto et al., 2017). Not predicted by VAH-2009, however, was the discovery that the vcDNA produced was of both linear and circular form and was involved not only in EVE formation but also in an immediate, adaptive and systemic RNAi response to viral infection.

With respect to EVE, the publication following the discovery of 472 EVE originating from several families of RNA viruses in a mosquito cell line of *Aedes eagypti* was of particular interest (Whitfield et al., 2017). In that publication and a subsequent publication, (Tassetto et al., 2019), the same research group described how viral EVE are mostly gathered together in genomic PIWI-interacting RNA clusters (piRNA-clusters) that produce long, negative-sense RNA transcripts that are further processed into small RNAs called piRNAs. The piRNAs from viral pathogens interacted specifically with a new type of protein called Piwi4 that used the piRNA as a guide to bind to complementary viral mRNA and cause its degradation. The role of Piwi4 protein interaction with piRNA for specific defence against viral pathogens was not predicted by VAH-2009. In addition, in 2020 it was reported that removal of a natural EVE against cell-fusing agent virus (CFAV) from the genome of the mosquito *A. aegypti* resulted in increased viral replication, showing that the EVE removed had a role in reducing viral replication (Suzuki et al., 2020). However, these recent revelations still support the overall VAH-2009 in showing that the antisense RNA arising from EVE leads to viral mRNA degradation, although the mechanism has turned out to be more complex than originally envisioned. This additional complexity does not alter the consequences for potential applications of EVE for viral disease control in economic insects and shrimp.

The discovery of both linear and circular forms of vcDNA that is capable of inducing an immediate, specific adaptive response to viral infections (Poirier et al., 2018; Tassetto et al., 2017) via the RNA

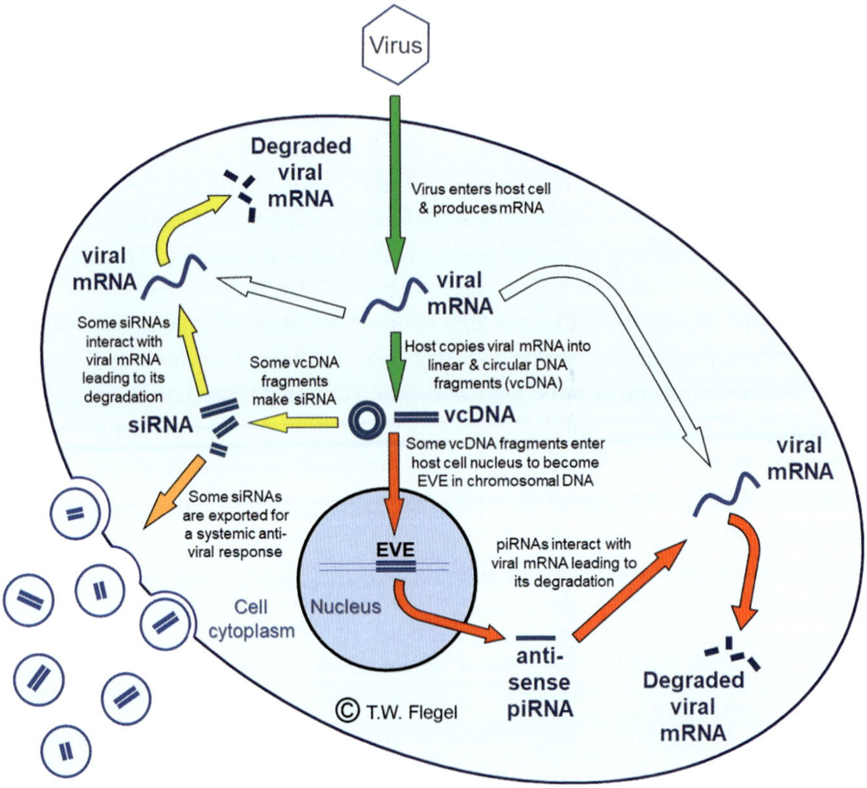

FIGURE 16.4 A simplified diagram of the viral accommodation hypothesis as updated in 2019. This figure shows the same pathways (green and red arrows) as for VAH 2009 in Figure 16.3. The additional pathways indicated by the red and orange arrows were not foreseen in 2009. Specifically, vcDNA was not predicted to occur in both linear and circular form, nor was the occurrence of ELV for systemic dispersal of the RNAi response predicted. These features occur in insects and are now hypothesized to occur also in shrimp.

pathway was not predicted by VAH-2009 either. Nor was the discovery of the systemic nature of this response in *Drosophila* (Poirier et al., 2018) or the disclosure that the response was mediated by blood cells (haemocytes) that could spread the RNAi response throughout the fly body. This is achieved by the release of 'exosome-like vesicles' (ELV) containing protective RNA into the blood serum (haemolymph) for distribution throughout the body to be taken up for the protection of other cells. These discoveries are of particular interest, if a similar process can be shown to occur in shrimp.

As a result of all the above studies, VAH-2009 was updated in 2019 (Flegel, 2020) to include the additional discoveries made with insects and hypothesizing that they also occur in shrimp. A simplified diagram showing the updates is shown in Figure 16.4, where the vcDNA is shown to give rise not only to EVE but also to an immediate and specific adaptive immune response to a viral infection, and where the RNAi response initiated by EVE involves processing by PIWI proteins.

Potential applications for viral accommodation in shrimp

It is very likely that the mechanisms and pathways described above for insects are also present in crustaceans, including shrimp. We can take advantage of the methods and tools from the above insect studies to rapidly determine whether or not this is so. If the answer for the shrimp studies is affirmative, the possibilities for application in shrimp disease control are very exciting.

For example, it should be relatively easy to follow the methods to determine whether circular vcDNA and ELV can be isolated from shrimp and whether they are capable of eliciting a protective response against homologous viruses in both whole animals and primary cell cultures. If so, there are many potential applications. First, we already have good evidence from laboratory tests that injection or feeding of double-strand RNA (dsRNA) to shrimp can be effective in preventing mortality normally caused by homologous viruses (Itsathitphaisarn et al., 2017). However, translation of these results to practical applications has been hindered by both dsRNA production costs and the current lack of practical and cost-effective methods for its delivery by addition to shrimp feed, for example (Itsathitphaisarn et al., 2017). In contrast, a designed form of circular vcDNA would be much superior since it could be produced by simple PCR amplification (i.e. no cell culture needed), and since it would be an additive relatively resistant to the conditions and high temperatures used for preparation of pelleted shrimp feeds. It may become relatively easy to design and prepare tailor-made vcDNA constructs in vitro for any virus and then to screen them for efficacy in protecting shrimp against viral infection by injection. Next, successful constructs could be used in tests of various practical delivery methods for cultivated shrimp. Hopefully this would lead to industrial-scale production of effective viral control reagents for all or most shrimp viruses. At the very least, the injection reagents could be used in shrimp breeding stocks together with antiviral drugs to 'clean up' shrimp that turn out to be infected with a previously unknown virus, or to clean up top performers from grow-out ponds so they may be returned to breeding centres for use in stock development (Itsathitphaisarn et al., 2017).

In addition, we will be able to determine the detailed mechanisms of induction of vcDNA synthesis by endogenous RT and the detailed mechanisms of insertion into the host genome as EVE. With this knowledge, it would be possible to design appropriate nucleic acid reagents for injection into the ovaries and testes of individual broodstock shrimp so that protective EVE would be inserted into the genomes of sperm and eggs at known, effective locations and give rise to tolerant offspring generations. Basically, we would be able to tailor breeding stocks to accommodate almost any virus. The reagents used would mimic molecules already naturally produced in shrimp. The genome insertion would also be accomplished by natural host processes. The development of such stocks would consist of simply steering natural processes that already occur normally in the shrimp immune response to viral pathogens. This screening process would be similar in concept to that used in the development of vertebrate vaccines where pathogen antigens are screened for those capable of inducing the most protective antibody response.

A very recent breakthrough with shrimp

Just prior to reviewing this chapter for publication, we made two important discoveries with shrimp by following the research methods described above for insects. We have discovered that *P. monodon* infected with IHHNV produces a variety of circular DNA transcripts that match the sequence of the infecting virus (Taengchaiyaphum et al., 2021). Since IHHNV is a DNA virus, we suspect that the circular viral copy DNA (cvcDNA) originated from IHHNV mRNA by the action of host RT. We found that the cvcDNA preparation could greatly reduce the replication of IHHNV when injected into naïve *Penaeus vannamei* together with the virus. These results supported our hypothesis that the mechanisms discovered in insects would also be found in shrimp, even for DNA viruses. In addition, we were surprised to find that our cvcDNA preparation contained cvcDNA that arose from the IHHNV-EVE discovered in *P. monodon* in 2006 (Tang & Lightner). This was an exciting new discovery because it revealed that EVE can produce cvcDNA (probably via the RNA they produce). It also revealed that isolation of cvcDNA may be a way of discovering and identifying EVE in existing shrimp breeding stocks. This opens the way for research on identifying and selecting protective EVE in current SPF shrimp breeding stocks.

Natural transgenic organisms reveal that fear of GMO may be unjustified

It is now clear that viral accommodation is common in insects and shrimp and that it occurs by natural processes that can result in autonomous genetic modification (AGMO) that leads to natural transgenic organisms (NTO). This means that millions of species in these groups are NTO and have probably existed long before vertebrates, mammals and humans evolved. Thus, humans have been living together with NTO, growing them, eating them and using their products unknowingly for millennia. A comparable system for AGMO exists (the CRISPR system) in Archaebacteria and Eubacteria (the first life forms on earth), by which they autonomously modify their own DNA to defend against bacterial viruses (called bacteriophages). Thus, these bacteria are also NTO. Plants may be too, due to similar interactions with viral pathogens as occur in insects (da Fonseca et al., 2016). In conclusion, NTO have been with humans since our origin as a species and no apparent harm has resulted. These revelations suggest that the current public fear of GMO may be unfounded, and so it should be possible to convince the public that careful development and testing of NTO can lead to great benefits in terms of reducing the negative environmental impacts of many current pathogen control practices. Realizing these facts should remove any hesitation in pursuing the possibilities of using the natural viral accommodation mechanisms to create natural, transgenic crustaceans, insects and perhaps plants that are tolerant to their viral pathogens.

The fundamental importance of NTO to the biological sciences

In closing, I would like to comment on the significance of the proven existence of NTO, not only for shrimp and insects but also for the biological sciences in general. The CRISPR mechanism and the viral accommodation mechanism have been proven to exist. Thus, we must accept that NTO are widespread and naturally occurring. It has been pointed out that these facts constitute partial vindication of the generally rejected concept of soft inheritance (sometimes called Lamarkism) (Koonin, 2019; Koonin and Wolf, 2009), i.e. the concept that adaptive environmental responses of individual organisms can be passed on to their offspring and contribute to the mainline process of Darwinian evolution.

After the discovery that DNA was the basis for heredity in animals and plants, the soft inheritance concept was essentially dismissed because there was no known or imagined mechanism by which environmental adaptations could be recorded in DNA. Belief that this dismissal was justified was so strong that it became a kind of anti-paradigm, such that any challenges or questions regarding its validity faced total rejection and even disdain or ridicule. Thus, the proven existence of NTO first in bacteria and now in insects revealed that the blanket anti-paradigm was unjustified and may have constituted a kind of scientific taboo that discouraged discussion or even thinking of how heritable adaptations might occur. I now believe this constituted a kind of intellectual blind spot in the biological sciences.

Now that we know that AGMO is a current, natural and widespread process and that NTO are common, it has very important fundamental ramifications. It requires an inherent paradigm shift regarding a former, widely held, belief that there was no possible biochemical route from an animal's response to environmental changes back to its DNA. It opens the way for fertile imaginations regarding other possible evolutionary pathways that may have arisen from the type of underlying molecular mechanisms that result in NTO. These could be environmental responses of individual cells somehow recorded in DNA of somatic or germline cells. For, example, the newly discovered role of DNA and RNA in an immediate, adaptive somatic RNAi defence response in insects was unexpected. Now it seems possible that these mechanisms also occur in crustaceans and perhaps in other invertebrates and plants.

For plants and other eukaryotic organisms like fungi with which methods of somatic (i.e. non-sexual) propagation and cloning are common, many fascinating possibilities for evolution of advantageous, somatic processes might be imagined. A good starting point is the example of vcDNA that leads not only to individual-cell EVE but also to an immediate systemic, anti-viral immune response in host insects (Tassetto et al., 2017). That immediate response involves a somatic shuttle between RNA and DNA, and also involves ELV that can serve as systemic transfer vehicles for vcDNA and siRNA. At the same time, individually variable EVE are generated that can interact with Piwi proteins and their currently known relatives in somatic and germline cells to reduce the negative impact of viral infections. Is it possible, if not probable, that evolution has morphed such fundamental pathways to serve other purposes?

I wonder, for example, whether mechanisms involving currently unknown proteins, small non-coding RNAs and DNA might have morphed in some manner that has allowed DNA (amongst the most stable molecules we know) to serve the somatic purpose of long-term human memory. This idea has flashed through my imaginings at intervals since I began studying biology, but I have always cut the thought line as impossible and a waste of thinking time due to the "anti-Lamarkian" paradigm. Now, however, some of the key features of AGMO seem possibly to be available for such a purpose. These include the potential ability of individual nerve cells to carry out autonomous genetic modifications to generate new piRNA-like genes or to use or modify existing piRNA-like gene clusters via PIWI proteins. Some cells may also have the ability to transfer small RNA and possibly DNA fragments (linear and circular?) among cells via ELV. Indeed, micro RNA (miRNA) is already known to occur in exosomes in cerebrospinal fluid but not in blood serum (Akers et al., 2015; Khasawneh et al., 2018). Currently, these miRNAs are studied mostly in relation to brain dysfunctions. With respect to being a candidate for involvement in long-term memory, a necessary but currently missing feature is some type of protein with nucleic acid binding ability that might be activated by sensations (site, sound, touch, etc.) to initiate a pathway leading to long-term, DNA-based memory. If so, there is potential for enormous variation in miRNA among individual brain cells and even greater variation among individual people. As with EVE, these individual intercellular differences would be difficult to characterize, even by high throughput sequencing. Since these processes might be disruptive if distributed throughout our bodies, is it possible that the blood–brain barrier is there not just to keep some things out but also to keep some things in?

Under sway of the former anti-Lamarkan paradigm, I would have discarded the thoughts above as impossible, wild imaginings and I would have dismissed them out of hand as not worthy of further thought or any kind of investigatory follow-up. Perhaps now, such 'wild imaginings' could be more seriously considered. The possible routes from environmental input back to DNA represent a fascinating new frontier and it is unpredictable where this may ultimately lead. I hope that the upcoming generation of young scientists may have fruitful imaginings. However, even if new imaginings lead nowhere useful, the currently revealed mechanisms will still have dramatic future consequences. I submitted a letter to the journal *Nature* containing the ideas in these closing paragraphs, but it was rejected as too specialized to appeal to a wide audience. I followed up by submitting an opinion paper with similar content to *Biology Direct*, but they rejected the submission as being "unscientific". It was also rejected as a perspective by the journal Science for similar reasons. Given my past experience with interesting ideas, they often find 'new owners' who claim them as their own after they finally sink in. Thus, I would like to go on record here with my 'crazy, speculations about the possibility of other evolutionary permutations of AGMO mechanisms whereby environmental inputs may lead back to DNA in somatic or germline cells, including the possibility of somatic involvement of DNA in human long-term memory.

References

Akers, J. C., Ramakrishnan, V., Kim, R., Phillips, S., Kaimal, V., Mao, Y., Hua, W., Yang, I., Fu, C.-C. and Nolan, J. (2015) miRNA contents of cerebrospinal fluid extracellular vesicles in glioblastoma patients. *Journal of Neuro-oncology* 123, 205–216.

Alday-Sanz, V., Brock, J., Flegel, T. W., Mcintosh, R., Bondad-Reantaso, M. G., Salazar, M. and Subasinghe, R. (2020) Facts, truths and myths about Spf shrimp in aquaculture. *Reviews in Aquaculture* 12, 76–84.

Argue, B. J., Arce, S. M., Lotz, J. M. And Moss, S. M. (2002) Selective breeding of Pacific white shrimp (*Litopenaeus vannamei*) for growth and resistance to Taura syndrome virus. *Aquaculture* 204, 447–460.

Boonyaratpalin, S., Supamattaya, K., Kasornchandra, J., Direkbusaracom, S., Ekpanithanpong, U. and Chantanachooklin, C. (1993) Non-occluded baculo-like virus, the causative agent of yellow head disease in the black tiger shrimp (*Penaeus monodon*). *Fish Pathology* 28, 103–109.

Brock, J. A., Antonio, N. and Argue, B. (2013) Genomic IHHN-related DNA sequences in black tiger shrimp. *Global Aquaculture Advocate* https://www.aquaculturealliance.org/advocate/genomic-ihhn-related-dna-sequences-in-black-tiger-shrimp/ (accessed 19 July 2021).

Chayaburakul, K. (2005) Infection by infectious hyperdermal and hematopoietic necrosis virus in black tiger shrimp *Penaeus monodon*. PhD thesis, Mahidol University.

Cowley, J. A., Walker, P. J., Flegel, T. W., Lightner, D. V., Bonami, J. R., Snijder, E. J. and De Groot, R. J. (2012) Family *Roniviridae*. In: King, A. M. Q., Adams, M. J., Carstens, E. B. and Lefkowitz, E. J. (eds) *Virus Taxonomy: Ninth Report of the International Committee on Taxonomy of Viruses*. Elsevier, New York, USA.

Da Fonseca, G. C., De Oliveira, L. F. V., De Morais, G. L., Abdelnor, R. V., Nepomuceno, A. L., Waterhouse, P. M., Farinelli, L. and Margis, R. (2016) Unusual RNA plant virus integration in the soybean genome leads to the production of small RNAs. *Plant Science* 246, 62–69.

Feschotte, C. and Gilbert, C. (2012) Endogenous viruses: Insights into viral evolution and impact on host biology. *Nature Reviews Genetics* 13, 283–296.

Flegel, T. W. (2001) The shrimp response to viral pathogens. In: Browdy, C. L. and Jory, D. E. (eds) *The New Wave. Proceedings of the Special Session on Sustainable Shrimp Aquaculture, World Aquaculture 2001*, Orlando. World Aquaculture Society, Boca Raton, FL, USA.

Flegel, T. W. (2006a) Detection of major penaeid shrimp viruses in Asia, a historical perspective with emphasis on Thailand. *Aquaculture* 258, 1–33.

Flegel, T. W. (2006b) The special danger of viral pathogens in shrimp translocated for aquaculture. *ScienceAsia* 32, 215–231.

Flegel, T. W. (2007) Update on viral accommodation, a model for host–viral interaction in shrimp and other arthropods. *Developmental and Comparative Immunology* 31, 217–231.

Flegel, T. W. (2009a) Hypothesis for heritable, anti-viral immunity in crustaceans and insects. *Biology Direct* 4, 32.

Flegel, T. W. (2009b) Review of disease transmission risks from prawn products exported for human consumption. *Aquaculture* 290 179–189.

Flegel, T. W. (2020) Research progress on viral accommodation 2009 to 2019. *Developmental and Comparative Immunology*, 103771.

Flegel, T. W. and Fegan, D. F. (2002) Strategies for preventing the spread of fish and shellfish diseases. *Fisheries Science* 68 Supplement 1, 776–788.

Flegel, T. W. and Pasharawipas, T. (1998) Active viral accommodation: a new concept for crustacean response to viral pathogens. In: Flegel, T. W. (ed.) *Advances in Shrimp Biotechnology.* National Center for Genetic Engineering and Biotechnology, Bangkok, Thailand.

Flegel, T. W. and Sritunyalucksana, K. (2011) Shrimp molecular responses to viral pathogens. *Marine Biotechnology* 13, 587–607.

Flegel, T. W., Sriurairatana, S., Wongterrasupaya, C., Boonsaeng, V., Panyim, S. and Withyachumnarnkul, B. (1995) Progress in characterization and control of yellow-head virus of *Penaeus monodon*. In: Browdy, C. L. and Hopkins, J. S. (eds) *Swimming Through Troubled Water: Proceedings of the Special Session on Shrimp Farming.* World Aquaculture Society, Baton Rouge, LA, USA.

Goic, B., Stapleford, K. A., Frangeul, L., Doucet, A. J., Gausson, V., Blanc, H., Schemmel-Jofre, N., Cristofari, G., Lambrechts, L. and Vignuzzi, M. (2016) Virus-derived DNA drives mosquito vector tolerance to arboviral infection. *Nature Communications* 7, 12410.

Goic, B., Vodovar, N., Mondotte, J. A., Monot, C., Frangeul, L., Blanc, H., Gausson, V., Vera-Otarola, J., Cristofari, G. and Saleh, M. C. (2013) RNA-mediated interference and reverse transcription control the persistence of RNA viruses in the insect model *Drosophila*. *Nature Immunology* 14, 396–403.

Huang, S.-W., Lin, Y.-Y., You, E.-M., Liu, T.-T., Shu, H.-Y., Wu, K.-M., Tsai, S.-F., Lo, C.-F., Kou, G.-H. and Ma, G.-C. (2011) Fosmid library end sequencing reveals a rarely known genome structure of marine shrimp *Penaeus monodon*. *BMC Genomics* 12, 242.

Itsathitphaisarn, O., Thitamadee, S., Weerachatyanukul, W. and Sritunyalucksana, K. (2017) Potential of RNAi applications to control viral diseases of farmed shrimp. *Journal of Invertebrate Pathology* 147, 76–85.

Jaroenram, W. and Owens, L. (2014) Separation of endogenous viral elements from infectious *Penaeus stylirostris* densovirus using recombinase polymerase amplification. *Molecular and Cellular Probes* 28, 284–287.

Kanthong, N., Khemnu, N., Sriurairatana, S., Pattanakitsakul, S.-N., Malasit, P. and Flegel, T. W. (2008) Mosquito cells accommodate balanced, persistent co-infections with a densovirus and Dengue virus. *Developmental and Comparative Immunology* 32, 1063–1075.

Kanthong, N., Khemnu, N., Pattanakitsakul, S. N., Malasit, P. and Flegel, T. W. (2010a) Persistent, triple-virus co-infections in mosquito cells. *BMC Microbiology* 10, 14.

Kanthong, N., Laosutthipong, C. and Flegel, T. W. (2010b) Response to Dengue virus infections altered by cytokine-like substances from mosquito cell cultures. *BMC Microbiology* 10, 290.

Karlikow, M., Goic, B., Mongelli, V., Salles, A., Schmitt, C., Bonne, I., Zurzolo, C. and Saleh, M.-C. (2016) *Drosophila* cells use nanotube-like structures to transfer dsRNA and RNAi machinery between cells. *Scientific Reports* 6, 27085.

Karlikow, M., Goic, B. and Saleh, M.-C. (2014) RNAi and antiviral defense in *Drosophila*: Setting up a systemic immune response. *Developmental and Comparative Immunology* 42, 85–92.

Katzourakis, A. and Gifford, R. J. (2010) Endogenous viral elements in animal genomes. *PLoS Genetics* 6, e1001191.

Khadijah, S., Neo, S. Y., Hossain, M. S., Miller, L. D., Mathavan, S. and Kwang, J. (2003) Identification of white spot syndrome virus latency-related genes in specific-pathogen-free shrimps by use of a microarray. *Journal of Virology* 77, 10162–10167.

Khasawneh, A. H., Garling, R. J. and Harris, C. A. (2018) Cerebrospinal fluid circulation: What do we know and how do we know it? *Brain Circulation* 4, 14.

Koonin, E. V. (2019) Lamarckian or not, CRISPR-Cas is an elaborate engine of directed evolution. *Biology and Philosophy* 34, 17.

Koonin, E. V. and Wolf, Y. I. (2009) Is evolution Darwinian or/and Lamarckian? *Biology Direct* 4, 42.

Koyama, T., Asakawa, S., Katagiri, T., Shimizu, A., Fagutao, F.F., Mavichak, R., Santos, M.D., Fuji, K., Sakamoto, T., Kitakado, T., Kondo, H., Shimizu, N., Aoki, T. and Hirono, I. (2010) Hyper-expansion of large DNA segments in the genome of kuruma shrimp, *Marsupenaeus japonicus*. *BMC Genomics* 11.

Laosutthipong, C., Kanthong, N. and Flegel, T. W. (2013) Novel, anionic, antiviral septapeptides from mosquito cells also protect monkey cells against dengue virus. *Antiviral Research* 98, 449–456.

Lightner, D. V., Redman, R. M. and Bell, T. A. (1983) Infectious hypodermal and hematopoietic necrosis, a newly recognized virus disease of penaeid shrimp. *Journal of Invertebrate Pathology* 42, 62–70.

Maori, E., Tanne, E. and Sela, I. (2007) Reciprocal sequence exchange between non-retro viruses and hosts leading to the appearance of new host phenotypes. *Virology* 362, 342–349.

Morales-Covarrubias, M. S., Nunan, L. M., Lightner, D. V., Mota-Urbina, J. C., Garza-Aguirre, M. C. and Chavez-Sanchez, M. C. (1999) Prevalence of infectious hypodermal and hematopoietic necrosis virus (IHHNV) in wild adult blue shrimp *Penaeus stylirostris* from the Northern Gulf of California, Mexico. *Journal of Aquatic Animal Health* 11, 296–301.

Moss, S. M., Doyle, R. W. and Lightner, D. V. (2005) Breeding shrimp for disease resistance: challenges and opportunities for improvement. In: Walker, P. J., Lester, R. G. and Bondad-Reantaso, M. G. (eds) *Diseases in Asian Aquaculture V. Proceedings of the 5th Symposium on Diseases in Asian Aquaculture*. Asian Fisheries Society, Manila, Philippines.

Pasharawipas, T., Flegel, T. W., Sriurairatana, S. and Morrison, D. J. (1997) Latent yellow-head infections in *Penaeus monodon* and implications regarding disease tolerance in crustaceans. In: Menasveta, P., Paisarnrat, S. And Flegel, T. W. (eds) *Shrimp Biotechnology in Thailand*. National Centre for Genetic Engineering and Biotechnology, Bangkok, Thailand.

Pénzes, J. J., Söderlund-Venermo, M., Canuti, M., Eis-Hübinger, A. M., Hughes, J., Cotmore, S. F. and Harrach, B. (2020) Reorganizing the family *Parvoviridae*: a revised taxonomy independent of the canonical approach based on host association. *Archives of Virology* 165, 2133–2146.

Poirier, E. Z., Goic, B., Tomé-Poderti, L., Frangeul, L., Boussier, J., Gausson, V., Blanc, H., Vallet, T., Loyd, H. and Levi, L. I. (2018) Dicer-2-dependent generation of viral DNA from defective genomes of RNA viruses modulates antiviral immunity in insects. *Cell Host and Microbe* 23, 353–365.

Roekring, S., Flegel, T. W., Malasit, P. and Kittayapong, P. (2006) Challenging successive mosquito generations with a densonucleosis virus yields progressive survival improvement but persistent, innocuous infections. *Developmental and Comparative Immunology* 30, 878–892.

Saksmerprome, V., Jitrakorn, S., Chayaburakul, K., Laiphrom, S., Boonsua, K. and Flegel, T. W. (2011) Additional random, single to multiple genome fragments of Penaeus stylirostris densovirus in the giant tiger shrimp genome have implications for viral disease diagnosis. *Virus Research* 160, 180–190.

Saleh, M. C., Tassetto, M., Van Rij, R. P., Goic, B., Gausson, V., Berry, B., Jacquier, C., Antoniewski, C. and Andino, R. (2009) Antiviral immunity in *Drosophila* requires systemic RNA interference spread. *Nature* 458, 346–350.

Suzuki, Y., Baidaliuk, A., Miesen, P., Frangeul, L., Crist, A.B., Merkling, S.H., Fontaine, A., Lequime, S., Moltini-Conclois, I., Blanc, H., Van Rij, R.P., Lambrechts, L. and Saleh, M.-C. (2020) Non-retroviral endogenous viral element limits cognate virus replication in *Aedes aegypti* ovaries. *Current Biology* 30, 1–12.

Taengchaiyaphum, S., Srisala, J., Bunphimpapha, P., Supungul, P., Tassanakajon, A., Chaiyapechara, S., Bowornpinyo, S., Sritunyalucksana, K. and Flegel, T. W. (2019) Mendelian inheritance of endogenous viral elements (EVE) of white spot syndrome virus (WSSV) in shrimp. *Developmental and Comparative Immunology* 96, 144–149.

Taengchaiyaphum, S., Buathongkam, P., Sukthaworn, S., Wongkhaluang, P., Sritunyalucksana, K. and Flegel, T.W. (2021) Shrimp parvovirus circular DNA fragments arise from both endogenous viral elements and the infecting virus. *Frontiers in Immunology* 12. doi.org/10.3389/fimmu.2021.729528.

Tang, K. F. J. and Lightner, D. V. (2006) Infectious hypodermal and hematopoietic necrosis virus (IHHNV)-related sequences in the genome of the black tiger prawn *Penaeus monodon* from Africa and Australia. *Virus Research* 118, 185–191.

Tassetto, M., Kunitomi, M. and Andino, R. (2017) Circulating immune cells mediate a systemic RNAi-based adaptive antiviral response in *Drosophila*. *Cell* 169, 314–325. e13.

Tassetto, M., Kunitomi, M., Whitfield, Z. J., Dolan, P. T., Sánchez-Vargas, I., Garcia-Knight, M., Ribiero, I., Chen, T., Olson, K. E. and Andino, R. (2019) Control of RNA viruses in mosquito cells through the acquisition of vDNA and endogenous viral elements. *eLife* 8.

Utari, H. B., Soowannayan, C., Flegel, T. W., Whityachumnarnkul, B. and Kruatrachue, M. (2017) Variable RNA expression from recently acquired, endogenous viral elements (EVE) of white spot syndrome virus (WSSV) in shrimp. *Developmental and Comparative Immunology* 76, 370–379.

Walker, P. J., Bonami, J. R., Boonsaeng, V., Chang, P. S., Cowley, J. A., Enjuanes, L., Flegel, T. W., Lightner, D. V., Loh, P. C., Snijder, E. J. and Tang, K. (2005) Family *Roniviridae*. In: Fauquet, C. M., Mayo, M. A., Maniloff, J., Desselberger, U. and Ball, L. A. (eds) *Virus Taxonomy, VIIIth Report of the ICTV.* Elsevier/Academic Press, London, UK.

Whitfield, Z. J., Dolan, P. T., Kunitomi, M., Tassetto, M., Seetin, M. G., Oh, S., Heiner, C., Paxinos, E. and Andino, R. (2017) The diversity, structure, and function of heritable adaptive immunity sequences in the *Aedes aegypti* genome. *Current Biology* 27, 3511–3519.

Wongteerasupaya, C., Sriurairatana, S., Vickers, J. E., Anutara, A., Boonsaeng, V., Panyim, S., Tassanakajon, A., Withyachumnarnkul, B. and Flegel, T. W. (1995) Yellow-head virus of *Penaeus monodon* is an RNA virus. *Diseases of Aquatic Organisms* 22, 45–50.

Yang, Y. T., Lee, D. Y., Wang, Y., Hu, J. M., Li, W. H., Leu, J. H., Chang, G. D., Ke, H. M., Kang, S. T., Lin, S. S., Kou, G. H. and Lo, C. F. (2014) The genome and occlusion bodies of marine *Penaeus monodon* nudivirus (PmNV, also known as MBV and PemoNPV) suggest that it should be assigned to a new nudivirus genus that is distinct from the terrestrial nudiviruses. *BMC Genomics* 15, 628.

Chapter 17

THE PROGRESSIVE MANAGEMENT PATHWAY FOR IMPROVING AQUACULTURE BIOSECURITY (PMP/AB): RELEVANCE AND POTENTIAL APPLICATION TO THE SHRIMP AQUACULTURE SECTOR

MELBA G. BONDAD-REANTASO, BRETT MACKINNON, HAO BIN, MELISSA MCLAWS AND HUANG JIE

Summary

An important source of aquatic food protein, the crustacean sector, has also become a significant revenue earner for people engaged in the aquatic food production value chain. This sector generates high-value exports which enable producers to buy lower-value products in the world market – and thus a positive contribution to food security in both producing and exporting countries.

The intensification and globalization through trade of shrimp has led to the emergence and re-emergence of diseases, which have resulted in production and market losses affecting livelihoods, export earnings and the local food supply. Emerging viral and bacterial diseases affecting the shrimp aquaculture sector have been reported since the 1970s and are primarily driven by anthropomorphic factors. At least 15 major diseases of viral, bacterial and parasitic aetiological agents have challenged the sustainability of the sector.

This chapter looks at some of the drivers and pathways of disease emergence in shrimp aquaculture, and the relevance and potential application of a new initiative called the Progressive Management Pathway for Improving Aquaculture Biosecurity (PMP/AB), including the five grades of aquaculture biosecurity at the farm level based on China's experience and their relevance to the PMP/AB stages. The PMP/AB is a four-stage, risk-based and collaborative pathway that addresses the challenges posed by infectious disease agents at the enterprise, national and international levels, and builds on existing frameworks, utilizing public-sector partnership with strong consideration of the cost effectiveness.

The shrimp sector, with its advanced features in terms of technology, industry organization, knowledge of diseases, and farm level and market infrastructure, can take advantage of the benefits to be gained through a systematic and comprehensive approach in dealing with shrimp farming biosecurity challenges where risks and responsibilities are shared with key stakeholders, working together to find common solutions and thus greatly contributing to sector sustainability. It can harness the opportunity of farming shrimp in a sustainable manner by contributing to the establishment of enabling policy environments that this new initiative – the PMP/AB – can offer.

Introduction

The shrimp aquaculture sector has grown exponentially over the past three decades to meet the global demand for seafood. The Food and Agriculture Organization of the United Nations (FAO) reported that, in 2018, total aquaculture shrimp and prawn production reached 6.0 million tonnes, with an average annual growth rate of 8.1% since 1990, compared to only 2.1% for capture shrimp and prawn production (3.5 million tonnes in 2018) over the same period (FAO, 2020a). Of the world total, the People's Republic of China produced 34.2% of the total volume and 41.0% of the total value of shrimp aquaculture production. In 2018, the other nine top producers were Indonesia, Viet Nam, India, Ecuador, Thailand, Mexico, Bangladesh, Brazil and the Philippines.

Shrimps are being farmed around the world in a variety of environments (i.e. fresh water, brackish water and marine), systems (i.e. ponds, recirculating systems, integrated multitrophic systems and polyculture), management strategies (i.e. extensive, semi-intensive and intensive) and sizes of operation (ranging from backyard and subsistence level to small-, medium- and large-scale operations).

Not only an important source of aquatic food protein, the crustacean sector has become a significant source of revenue for people engaged in the aquatic food production value chain. The sector generates high-value exports which enable producers to buy lower-value products in the world market – thus a positive contribution to food security in both producing and exporting countries. For developing countries, production (as food and ornamental) and trade are extremely important (Bondad-Reantaso *et al.*, 2012).

The intensification and globalization through trade of shrimp has led to the emergence and re-emergence of diseases, which have resulted in production and market losses affecting livelihoods, export earnings and local food supplies.

Emerging viral and bacterial diseases affecting the shrimp aquaculture sector have been reported since the 1970s, and are primarily driven by anthropomorphic factors. In the 1970s, the only shrimp viral disease of importance during that period was Monodon baculovirus (MBV) (Lightner and Redman, 1981; Natividad and Lightner, 1992). Going forward to the 1980s, the sector saw the emergence of some of the most significant viral diseases attributed to Hepatopancreatic parvovirus (HPV) (Flegel, 2006), Infectious hypodermal and haematopoietic necrosis virus (IHHNV) (Lightner *et al.*, 1983), Baculovirus penaei (BP) (Couch, 1974) and one bacterial disease, Necrotizing hepatopancreatitis (NHP), caused by *Hepatobacter penaei* (Lightner, 2011). In the 1990s the sector was challenged by three new viral diseases, caused by Yellow head virus (YHV) (Boonyaratpalin *et al.*, 1993), Taura syndrome virus (TSV) (Jimenez, 1992), White spot syndrome virus (WSSV) (Zhan *et al.*, 1998) and the vibriosis group of pathogens (e.g. *V. harveyi*, *V. damsela*, *V. alginolyticus*, *V. vulnificus* and *V. penaeicida*) (Lightner, 2011). In the decade from 2000, while the sector continued to battle with those viral and bacterial problems, new viral pathogens emerged (e.g. Mourilyan virus (MoV), Infectious myonecrosis virus (IMNV), Covert mortality nodavirus

(CMNV), Laem Singh virus (LSNV)) (Flegel, 2006; Lightner *et al.* 2004), and infection with bacterial (Acute hepatopancreatic necrosis disease (AHPND)-causative *V. parahaemolyticus*) and parasitic (*Enterocytozoon hepatopenaei* (EHP)) pathogens, which appeared to be the most significant non-viral diseases of shrimp (Tourtip *et al.*, 2009; Zhang *et al.*, 2012, 2014). The trouble did not end there; recently a new viral disease, Decapod iridescent virus 1 (DIV1), has emerged (Qiu *et al.*, 2017; Srisala *et al.*, 2020; Xu *et al.*, 2016).

This chapter looks at some of drivers and pathways of disease emergence in shrimp aquaculture and the relevance and potential application of the PMP/AB, including the five grades of aquaculture biosecurity at the farm level based on China's experience and their relevant PMP/AB stage.

Disease emergence in shrimp aquaculture: factors, drivers and pathways

The Snieszko circle (Bondad-Reantaso *et al.*, 2001; Snieszko, 1974), also known as the epidemiological triad (Fegan, 2000; van Seventer and Hochberg, 2017), shows the relationship between host, pathogen and the environment in disease development. As applied to the shrimp sector, the triad highlights the interaction between pathogen and susceptible shrimp population in a suitable environment that allows transmission of the pathogen and development of disease in the population (Fig. 17.1). The pathogen must be present for disease to occur, but its presence may not always result in disease. Physiological, reproductive and development stage conditions, stocking density, susceptibility and immunity to diseases are important considerations concerning the host species. In terms of environment, important elements include water and its components (e.g. oxygen, pH, temperature, toxins, wastes, etc.) and the kind of management practices (e.g. handling, drug treatments, transport procedures, etc.). Environment- and population-specific factors influence the level of exposure to a pathogen and vulnerability to infection in a shrimp population, whereas pathogen-specific factors, including infectious dose and survivability, affect the infectivity, pathogenicity or virulence of a pathogen. By managing the environment with good national- and farm-level biosecurity measures (e.g. safe trade, rapid disease detection/response,

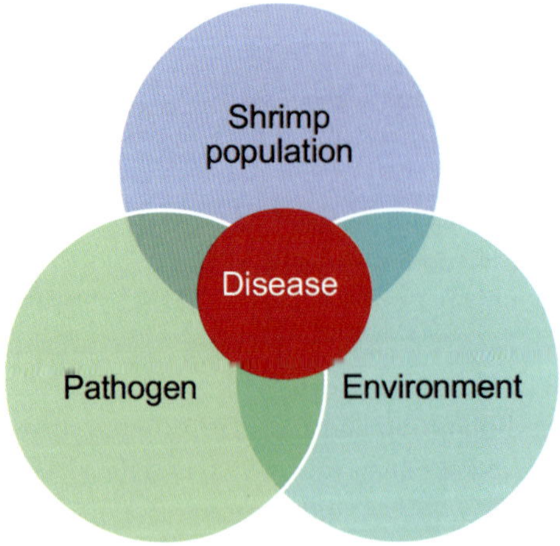

FIGURE 17.1 The Snieszko circle, also known as the epidemiological triad.

disinfection, etc.), exposure to a pathogen and, therefore, prevalence of disease in the population are reduced. The impact of disease can be lessened by enhancing the immunity and disease resistance of the shrimp population through stocking with resistant genetic lines or treating with probiotics, for example.

Several major factors, drivers and/or pathways for disease emergence in aquaculture have been identified, and grouped into four major categories as follows: (1) trade and movement of live animals and products; (2) knowledge of pathogens and their hosts; (3) aquatic animal health management; and (4) ecosystem change (FAO, 2019a,b, 2020b). A thorough understanding of each of these elements may explain the difficulties involved in the prevention or management of diseases affecting shrimp aquaculture production. In this chapter, we tackle the relevance of each of these factors to the shrimp sector.

According to FAO's latest State of World Fisheries and Aquaculture report (SOFIA) (FAO, 2020b), shrimp and prawns have historically been one of the most heavily traded fish products, with the bulk of production taking place in Asia and Latin America (Fig. 17.2) and the major markets located in the USA, the European Union and Japan (Fig. 17.3). In 2017, the total shrimp and prawn trade value was about US$27.32 billion (3.22 million tonnes), which accounted for 17.3% of the world's total trade value of aquatic products (US$158.1 billion). The shrimp and prawn trade value ranked second in all International Standard Statistical Classification of Aquatic Animals and Plans (ISSCAAP)[1] groups, only after the group including salmon, trout and smelt, with a trade value of US$28.06 billion (3.57 million tonnes) in 2017. It should be noted that salmon and trout trade volume already accounted for about 80% of its global total quantity, while the ratio was 35% for shrimp and prawn, which may indicate greater potential of increase in the shrimp and prawn trade.

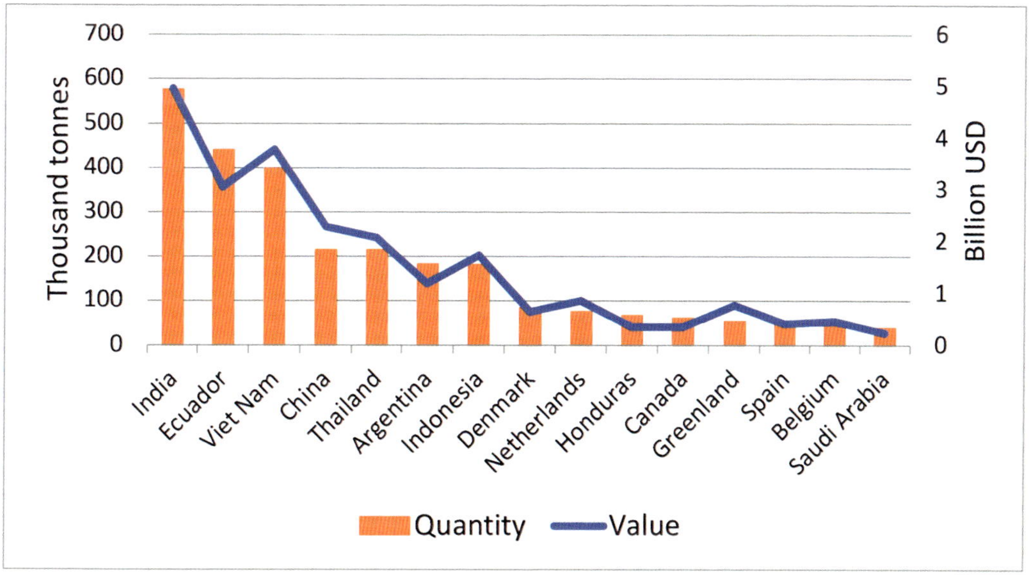

FIGURE 17.2 The top 15 shrimp and prawn exporting countries in 2017 (source: FAO, 2020b).

1 http://www.fao.org/fishery/static/ASFIS/ISSCAAP.pdf

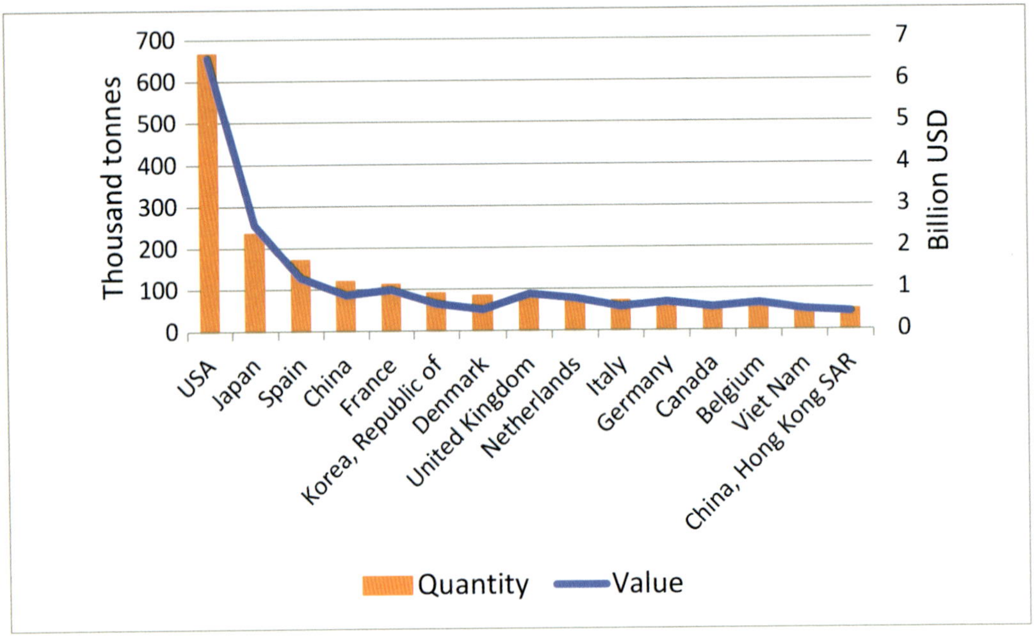

FIGURE 17.3 The top 15 shrimp and prawn importing countries in 2017 (source: FAO, 2020b)

The Chinese market is increasingly becoming an important target for exporters and marketers of shrimp, whereas the scope for further growth in the traditional developed markets is limited. The farmed shrimp sector, which now supplies the majority of volume to the global market, has also suffered from the impact of disease outbreaks. Looking at the experience of the People's Republic of China, the world's biggest farmed shrimp and prawn producer, the range of losses due to diseases officially reported indicates that economic losses are substantial and show an increasing trend. For example, from 2012 to 2016, disease-related losses were approximately US$0.90 million, 0.89 million, 0.96 million, 0.95 million and 0.92 million, respectively. Since then, rapidly increasing losses of about US$1.67 million occurred in 2017 and 2.38 million in 2018[2]. Shrimps and prawns are currently suffering the largest economic losses attributed to various pathogens in China, among which *Penaeus vannamei* had the largest reported loss of all cultured aquatic species in 2018 (US$1.76 million).

Similar problematic conditions may exist in other top shrimp and prawn producing and exporting countries: Viet Nam and Thailand are both among the top five of those. However, Viet Nam saw a sharp decline in production of nearly 20% in 2010, while Thailand had a steep drop in output for two consecutive years in 2013 and 2014, down by 47 and 14%, respectively. Coincidentally, AHPND outbreaks in shrimp farms were first reported in Viet Nam and Thailand in 2010 and 2012, respectively (OIE, 2019). It is possible that the decline in production in these countries was due to AHPND, with major impact on the two countries' local shrimp and prawn industry.

2 Ministry of Agriculture and Rural Affairs of China, 2012–2019, Aquatic Animal Health in China [in Chinese]

A few cases point to movement (introduction and transfer) of infected live shrimp (post-larvae, juveniles and broodstocks) and their products as a major pathway for the introduction and transfer of pathogens from one region to another. A classic example is WSSV, the most serious shrimp viral pathogen, whose current distribution consists of the majority of shrimp-producing countries in almost all regions. The extent of distribution reflects the intense trade of live post-larvae, juveniles and broodstock for aquaculture development. Frozen shrimp is also widely traded around the world.

Because WSSV remains infectious in shrimp tissue even after freezing, frozen shrimp products pose a threat to farmed and wild shrimp populations when used as feed or bait (Hasson *et al.*, 2006). Australia was considered free of WSSV until 2016, when an outbreak was notified to the World Organization for Animal Health (OIE) (OIE, 2016) and Network of Aquaculture Centres in Asia-Pacific (NACA). During the Sixteenth meeting of the NACA's Asia Regional Advisory Group on Aquatic Animal Health (NACA, 2017), it was indicated that bait was considered a likely introductory pathway in Australia during an epidemiological study of the country's first WSSV outbreak in 2016. See Figure 17.4.

Tang *et al.* (2019) reviewed the origin of IMNV, which was first reported in 2002 in populations of farmed *P. vannamei* in the State of Piaui, Brazil (Nunes *et al.*, 2004). Subsequently, it was first detected in Indonesia for the first time in early 2006 (Senapin *et al.*, 2007). In 2016, IMNV was detected in the West Bengal State of India (Sahul Hameed *et al.*, 2017) and spread to Andhra Pradesh State in 2017. Most recently, an IMNV outbreak occurred in the State of Melaka, Malaysia in June 2018 (OIE, 2018). The two

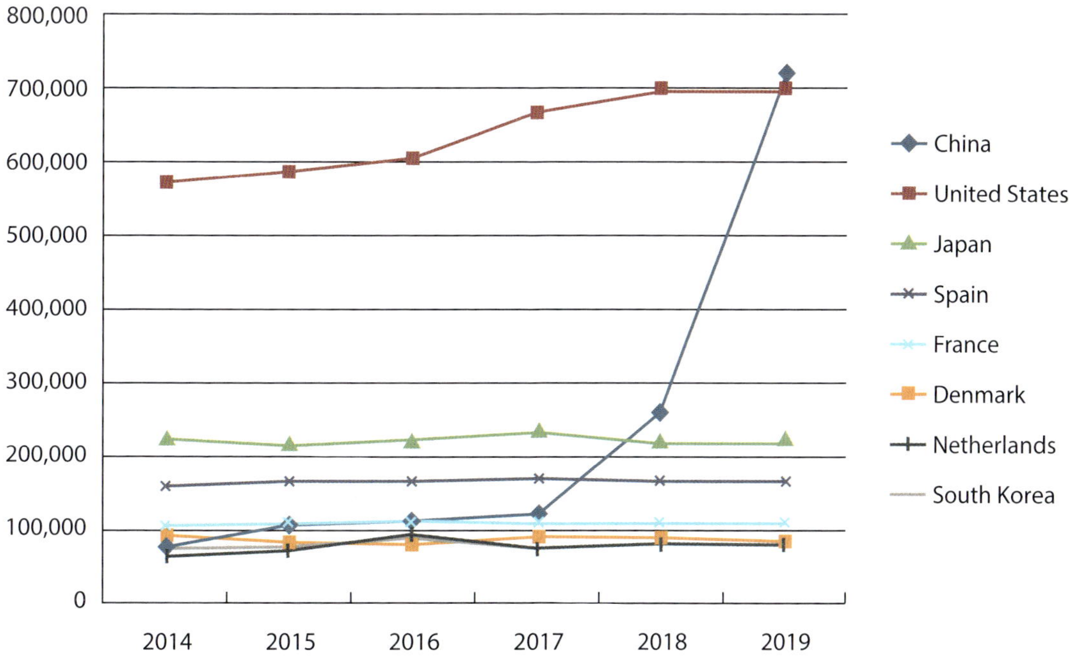

FIGURE 17.4 The top eight shrimp and prawn importing countries, 2014–2019.

latter cases possibly originated from the importation of infected broodstock and post-larvae, or through transmission from subclinically infected *P. monodon* (Sahul Hameed *et al.*, 2017).

The second major driver is knowledge of hosts and pathogens. Although shrimp is one of the most researched commodities (e.g. 710,000 publications in Google Scholar as of 20 August 2020) and shrimp disease is one of the most studied themes in aquaculture (e.g. 22,500 publications in Google Scholar to date), these efforts do not seem to match the significant speed of sector development and the biosecurity needs of the sector.

Penaeus monodon culture dominated the world shrimp production in Southeast and Eastern Asia from the early 1980s to 2002 (Briggs *et al.*, 2005). Its production, however, rapidly dropped the following year when problems with diseases and reduced growth rates became more prevalent. Since then, *P. vannamei* quickly became the preferred species in Asian shrimp culture (Fig. 17.5); the attractiveness of *P. vannamei* stems from several characteristics of this species that include: (1) faster growth, (2) safe stocking at greater density, (3) low-salinity tolerance, (4) cool tolerance, (5) lower protein requirement and (6) breeding and domestication potential (Briggs *et al.*, 2005).

It has been recognized that the transfer of IMNV from Brazil to Indonesia was through the importation of SPF broodstock. Many pathogens affecting the shrimp sector are readily transmitted both vertically and horizontally, and can infect shrimp simultaneously and persist subclinically even at high prevalence of

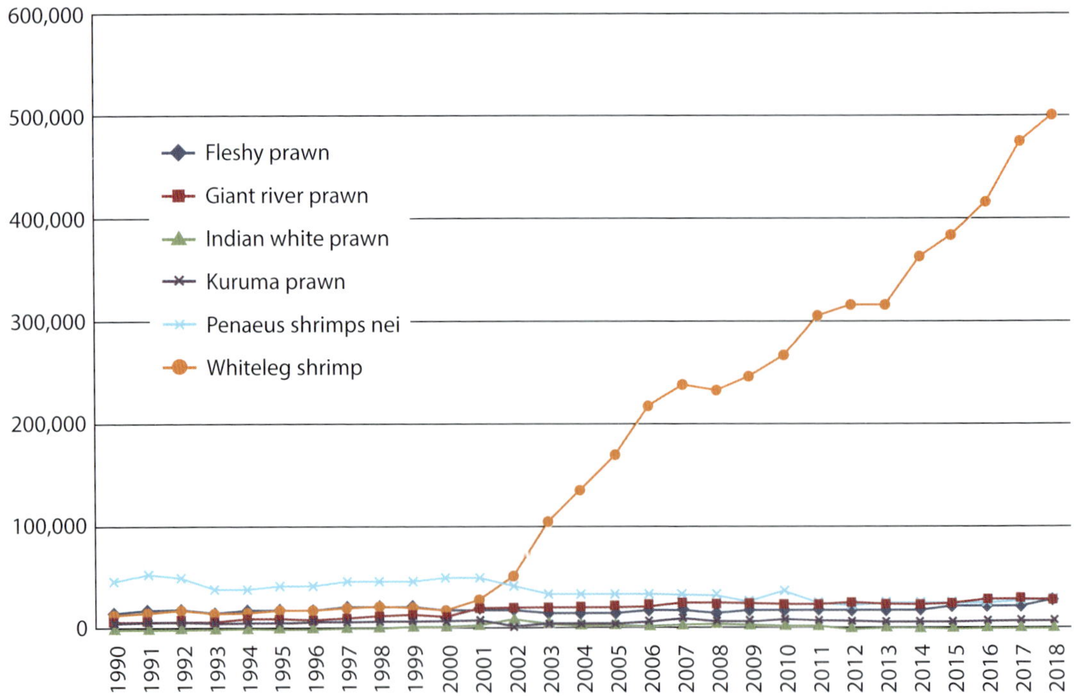

FIGURE 17.5 Production trends for key species of globally farmed shrimp, 1990–2018.

infection (Walker and Winton, 2010); this contributes to the difficulty in controlling outbreaks on farms. Due to an underdeveloped immune system, there is no vaccine available for shrimp.

The third driver is aquatic animal health management and control. Although there is strong awareness on shrimp diseases, there appears to be lack of cooperation or differences in commitment between key stakeholders in the public and private sectors involved in shrimp aquaculture. For example, the capacity and infrastructure of the government required to adequately address the disease issues faced by aquaculture may not develop as quickly as required to support the shrimp sector's rapid growth. As the shrimp sector is highly organized, producers are often the first to note the presence of exotic or emerging diseases; however, low incentive to report diseases to the competent authority is a prevalent problem. If there is a lack of collaboration between academia and governmental authorities regarding surveillance studies for emerging diseases, disagreement in the interpretation of diagnostic test results may result. Another factor may be limited strategic planning for the implementation of biosecurity measures in the shrimp sector; these biosecurity action plans may not be in place, or may be in place but are implemented inconsistently or inadequately to address disease threats. Robust biosecurity systems based on public–private partnerships are essential to support healthy shrimp production, by reducing the likelihood of pathogen introduction and spread at both the farm and national level. Without strategic planning in place, a country may react to serious transboundary aquatic animal diseases in an unsystematic manner, leaving the sector vulnerable to exotic or emerging diseases.

The fourth pathway, as a contributory factor towards disease emergence, is ecosystem change. Aquatic animals are very susceptible to environmental changes as they receive everything they need from water, often requiring very specific conditions. Climate change, therefore, may have major impacts on aquatic ecosystems, including changes in pathogen virulence, transmission and host susceptibility, and local extirpations and migrations. El Niño-related temperature rise may compromise host resistance and increase the occurrence of opportunistic diseases. Increase in the frequency and/or intensity of storms may lead to escapes from damaged facilities, or the introduction of disease or predators into aquaculture facilities during flooding episodes (Bondad-Reantaso *et al.*, 2018); these authors reviewed a few examples that can demonstrate this pathway as described below.

AHPND, caused by a specific strain of *Vibrio parahaemolyticus* (Vp_{AHPND}), is currently the most important non-viral disease threat in cultured shrimp. This pathogenic bacterium is ubiquitous in marine and brackish water environments. The ecology of Vp_{AHPND} suggests that it is affected by temperature, salinity, turbidity and the presence of zooplankton, crustaceans and molluscs. The pathogen can thus be present both in cultured shrimp and in the water, sediments and associated organisms of culture ponds, as well as in the broader aquatic environment. Environmental factors, such as elevated water temperature, salinity >5 ppt and pH >7.0, are believed to promote infection with Vp_{AHPND} (Bondad-Reantaso, 2016). Unconfirmed routes of disease transfer include attachment of flocs to zooplankton that are carried long distances by ocean currents and El Niño phenomena, and attachment on crustaceans and in ballast water of ships (Bondad-Reantaso, 2016; FAO, 2017).

El Niño events in 1987/1988, 1991/1992, 1994/1995 and 1997/1998 have been associated with the emergence of viral diseases including IHHNV, TSV and WSSV, respectively (Bondad-Reantaso *et al.*, 2018).

PMP/AB

The PMP/AB was developed by FAO and partners after seeing a need for strategic planning to further guide and support countries towards achieving sustainable aquaculture biosecurity and health management systems, with strong public and private stakeholder involvement to promote the application of risk management at the sector level as part of the national approach (FAO, 2019a, 1919b, FAO, 2020c, 2020d). The PMP/AB is an extension of FAO's 'Progressive Control Pathways' (PCP) approach, which has been internationally adopted to assist countries as systematic frameworks for planning and monitoring risk mitigation strategies for the reduction or prevention of major livestock and zoonotic diseases (Sumption *et al.*, 2012). However, whereas most PCPs focus on control of single diseases or disease complexes, the PMP/AB focuses on diseases and other biosecurity challenges faced by the aquaculture industry at the commodity and enterprise levels. Given the complex and multifaceted drivers of disease emergence and spread (as described in the previous section), the PMP/AB recognizes that a comprehensive, holistic approach is required to improve aquaculture biosecurity.

As currently defined, in the context of the PMP/AB, biosecurity refers to "the cost-effective management of risks posed by pathogenic agents to aquaculture through a strategic approach at enterprise, national and international levels with shared public–private responsibilities" (FAO, 2020b). Implementation of the PMP/AB is expected to result in the sustainable reduction of burden of disease; improvement in health at farm and national levels; minimization of global spread of diseases; optimization of socioeconomic benefits from aquaculture; attraction of investment opportunities into aquaculture; and achievement of 'One Health' goals.

The PMP/AB is progressive, collaborative and risk based and consists of four stages, designed to be implemented in a stepwise fashion. Countries decide how far and how fast it is appropriate to progress according to their priorities. Due to the wide variation of species being farmed in a single country, aquaculture sectors may advance along the pathway independently at different speeds or with different goals in mind. Strong cooperation between government and industry, such as public–private sector partnership (PPP), is a common theme throughout all four stages of the PMP/AB. This PPP enables the strengths of the public and private sectors to complement each other to achieve a common goal. Key obstacles must be overcome to increase clarity regarding the roles of the public and private sectors, improve capacity and infrastructure and increase awareness among farmers and other key stakeholders along the value chain.

Risk analysis is another key aspect of all stages of the PMP/AB. The foundation of the risk analysis is achieved through a detailed description of the aquaculture stakeholders, sectors and value chains in the country, and identification of high-priority diseases and their descriptive epidemiology. Risk hot spots, areas with the highest concentration of risk along the value chain, are identified and their pathways clearly understood. All the above are expected to lead to the identification of effective and practical biosecurity measures contained in the National Strategy on Aquatic Animal Health (NSAAH) or national aquaculture biosecurity strategy. Such strategies are an essential element of the PMP/AB and consist of several important components supporting aquaculture biosecurity at the international, regional, national and sector (farm) levels (Table 17.1) (FAO, 2007, 2018).

The shrimp sector is characterized by high-value species and export commodities (live and processed products). The sector is relatively well organized: primary producers and many stakeholders are actively engaged in production and provide input and services along the value chain and, at the national, regional

TABLE 17.1 Key components of the NSAAH.

Component (C)	*Title*
C1	Policy, legislation and enforcement
C2	National aquatic pathogen list
C3	Farm-level biosecurity plan
C4	Diagnostics
C5	Surveillance, monitoring and reporting
C6	Zoning and compartmentalization
C7	Border inspection and quarantine
C8	Risk analysis
C9	Emergency preparedness, response and contingency planning
C10	Use of veterinary drugs and avoidance of antimicrobial resistance (AMR)
C11	Information, education and communication (IEC) and aquatic animal health information system (AAHIS)
C12	Research and development, extension and other studies
C13	Human resources and institutional capacity development
C14	Institutional structure (including infrastructure)
C15	Regional and international cooperation

and international levels, there exist various networks. The sector is very knowledgeable and often the primary production segment is the first to know about any change in the disease situation. Although the shrimp industry is an important sector in aquaculture that provides food and livelihoods, it has been devastated by diseases that have resulted in sector collapse.

Farmed-shrimp-producing countries, at whatever stage of industry development (advanced or just starting), will have the opportunity and flexibility to initiate the PMP/AB. For example, one or more of the following scenarios may apply:

- Scenario 1: Country has neither aquaculture biosecurity strategy (AB) nor NSAAH but has aquaculture production or is initiating aquaculture development.
- Scenario 2: Country with NSAAH or other strategies from FAO projects or other assistance projects, at various levels of implementation where it can be investigated how best these strategies can be used, revised and/or expanded to fit the context of the PMP/AB.
- Scenario 3: Country with advanced biosecurity strategies where these strategies can be reviewed and revised/expanded/updated to fit the context of the PMP/AB.
- Scenario 4: Countries sharing water bodies and regions with regional biosecurity strategies.

Table 17.2 describes the four stages of the PMP/AB in terms of objectives, key outcomes and typical activities. We present here the relevance and potential application of the PMP/AB key outcomes to the shrimp sector.

Another tool has been developed, in China, to assess aquaculture biosecurity on farms. Figure 17.6 provides a descriptive classification, including five grades of aquaculture biosecurity at the farm level based on China's experience, and provides the relevant PMP/AB stage. Aquaculture biosecurity grade 1 (ABG1) is based on diagnosis-based treatment, which refers to partial prevention and control under the

TABLE 17.2 Relevance and potential application of the four stages of the PMP/AB to the shrimp sector.

Preliminary key outcomes	Brief overview of what countries need to do
Stage 1 – Biosecurity risks defined *Objective: Define risk-based approaches to support aquaculture health and production*	
Preliminary key outcomes	**Brief overview of what countries need to do**
1 Key stakeholders are identified and production systems, marketing network and associated socio-economic drivers are well described and understood for aquaculture sectors (value chain analysis).	Map stakeholders, including shrimp producers, aquatic animal health specialists, industry partners and government officials, involved in the shrimp sector's value chain. Describe shrimp production systems and movements of live shrimp (i.e. PLs, broodstock) and their products. Form a public–private PMP/AB taskforce, including involvement of shrimp sector associations.
2 Key threats to aquaculture and biosecurity vulnerabilities are identified and described.	Summarize information about the known occurrence and distribution of pathogens in the country that impact wild and farmed shrimp populations. List and prioritize shrimp pathogens present in the country, based on national impact. Describe current biosecurity practices used on farms.
3 Risk hot spots and critical control points are identified through risk analysis.	Conduct preliminary risk assessments via risk-mapping exercises to determine risk hot spots along the shrimp sector's value chain. Conduct a socio-economic analysis. Determine practical and effective risk management measures to mitigate the risk of introduction and spread of pathogens in the shrimp sector (i.e. disinfection, tracing, use of SPF stock).
4 An enabling environment for aquaculture biosecurity is reviewed and developed.	Identify the competent authority for aquatic animal health and/or aquaculture biosecurity (authority and mandate). Assess human and institutional capacity. Review current legal framework. Draft the national aquatic pathogen list (NAPL), including endemic and emerging diseases specific to shrimp (e.g. sources may include the OIE, NACA, existing databases or peer-reviewed literature). Build capacity in laboratory and shrimp health services.
5 Risk-based strategies are developed and endorsed at sector and national levels (*Gateway Pass*).	Develop a National Strategy on Aquatic Animal Health (NSAAH)/national aquaculture biosecurity strategy to address risks based on preliminary risk analysis results. All applicable NSAAH components should be included (refer to Table 17.1). Develop a shrimp biosecurity action plan.
Assuming that key outcomes at Stage 1 are achieved, a country is now ready to move to Stage 2.	
Stage 2 – Biosecurity systems initiated *Objective: Take action to improve biosecurity in aquaculture sectors*	
Preliminary key outcomes	**Brief overview of what countries need to do**
1 Risk-based strategies developed in Stage 1 are implemented by public and private stakeholders.	Implement biosecurity measures, on shrimp farms and at the national level, based on risk analysis results.
2 The management of biosecurity vulnerabilities and occurrence of important pathogens is monitored.	Conduct audits and certification of biosecurity management to ensure compliance of the shrimp sector. Monitor the occurrence and distribution of priority endemic pathogens in farmed shrimp populations via a national surveillance programme.

3 There is evidence that the implementation of risk-based strategies strengthens biosecurity and reduces the impact of pathogens within aquaculture sectors.	Assess whether an increase in shrimp production has occurred on farms. Modify ineffective biosecurity measures in place.
4 Enabling environment is further developed, with enhanced co-operation between public and private sectors.	Adopt the NAPL and strengthen the legal framework. Enhance capacity in laboratory and shrimp health services. Develop a shrimp health information system, which may include disease reports, interventions and animal movements.
5 Risk-based strategies are enhanced and revised, based on evidence (*Gateway Pass*).	Revise the NSAAH to include performance indicators and provisions specific to the shrimp sector for: rapid detection, contingency planning and response activities for emerging and exotic disease incursions; border controls and quarantine; zoning and compartmentalization; import risk analysis; prudent and responsible use of treatments; and ecosystem health.

Assuming that the key outcomes at Stage 2 are achieved, a country is now ready to move to Stage 3. Actions applied in Stage 2 are continuously maintained and/or improved upon.

Stage 3 – Biosecurity and preparedness enhanced

Objective: Enhance biosecurity and preparedness in aquaculture sectors

Preliminary key outcomes	**Brief overview of what countries need to do**
1 Revised risk-based strategies are implemented.	Audit NSAAH implementation to address gaps. Contingency plans are in place for control and/or eradication of exotic, endemic or emerging pathogens of high national concern to the shrimp sector.
2 Exotic, endemic and emerging pathogens under continuous surveillance.	Surveillance is conducted in farmed and/or wild populations to monitor endemic, exotic or emerging pathogens. Trace domestic and international movements of shrimp commodities.
3 Disease incidence and impact are reduced.	Evidence exists to support that reduction or elimination of pathogens is occurring on shrimp farms where management measures are implemented. Reduce the risk of introduction and/or spread of exotic pathogens to the shrimp sector through such risk-management measures.
4 Enabling environment is strengthened.	Strengthen laboratory capacity for rapid detection and diagnosis of all shrimp diseases of high national concern (endemic, exotic, emerging). Strengthen border controls based on import risk analyses. Perform cost–benefit analysis of biosecurity systems. Strengthen legal framework to support full implementation of the NSAAH.
5 Demonstrated commitment, including investment, from public and private stakeholders to safeguard progress (*Gateway Pass*).	Form a national, multi-agency taskforce with capacity for public–private co-regulation to safeguard biosecurity practices in the shrimp sector. Report the performance of the NSAAH. Gain expressed and actual commitment from key stakeholders in the shrimp sector in terms of financial, infrastructure and human resources to safeguard progress.

Assuming that the key outcomes at Stage 3 are achieved, a country is now ready to move to the final stage of the PMP/AB, Stage 4. Actions applied in Stage 3 are continuously maintained and/or improved upon.

TABLE 17.2 *continued*

Stage 4 – Sustainable biosecurity and health management systems established
Objective: Sustainable health management to support the national aquaculture system

Preliminary key outcomes	Brief overview of what countries need to do
1 Risk management activities are sustained and improved based on evidence.	Conduct research on the development or improvement of existing control, management or treatment of diseases of concern to the shrimp sector.
2 Enabling environment is maintained and continuously improved.	Update legal framework and evaluate enforcement as required. Zones and/or compartments are recognized by the World Organization for Animal Health (OIE) (if applicable).
3 Robust socio-economic situation for all (including small-scale producers, food security).	Reduced impact of diseases and/or international confidence in the national biosecurity system has positively impacted the socio-economic situation in the country's shrimp sector.
4 National and international stakeholders have confidence in national aquaculture and ecosystem health.	A One Health approach, considering animal health, human health and ecosystem health, is able to thrive due to the institutions and co-regulation mechanisms developed in Stages 1–3. Support international stakeholders in the development of national shrimp aquaculture biosecurity. Engage in sustainable and safe trade.

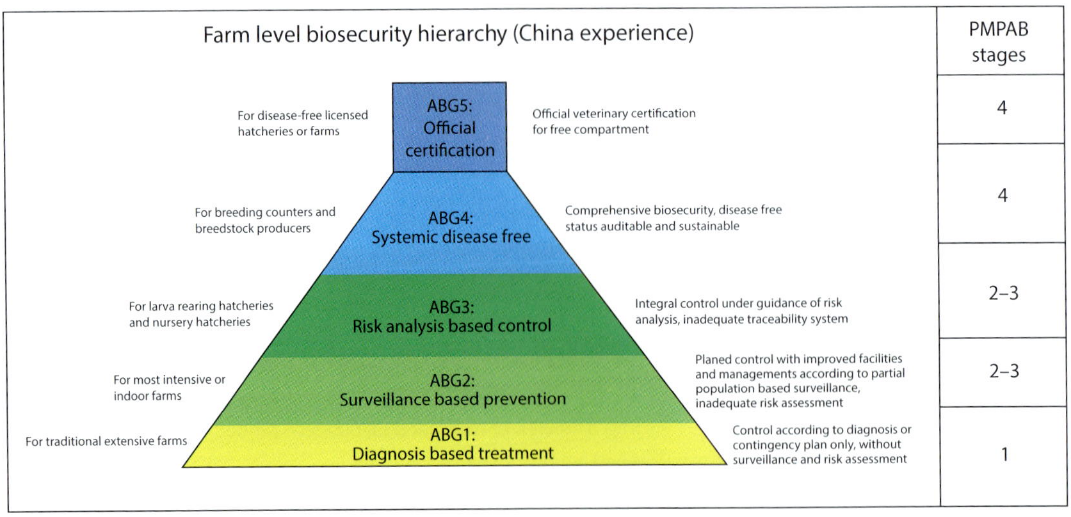

FIGURE 17.6 Descriptive classification, including five grades of aquaculture biosecurity at the farm level, based on experience in China.

guidance of disease diagnosis, mainly including clinical evaluation, disease diagnosis and an emergency response plan, but it lacks disease monitoring and risk assessment.

ABG2 is based on disease prevention, which refers to the implementation of a monitoring plan based on the preliminary risk assessment, and carrying out preventive and control measures through improved facilities and management.

ABG3 is based on risk prevention and control, which refers to the implementation of a relatively complete disease-monitoring plan guided by a more complete risk analysis, thereby forming a more comprehensive and integrated prevention and control strategy whose traceability system is not yet complete.

ABG4 is based on comprehensive security, which refers to the comprehensive implementation of an effective risk analysis and disease-monitoring plan, a complete emergency response plan, the ability to fully implement best management practices (BMP), the comprehensive establishment of a traceability system, and one that will achieve an auditable biosecurity and sustained disease-free state; ABG5 is mainly for disease-free registration based on sustained ABG4 status, which can be officially certificated by the government.

The biosecurity plan needs to consider comprehensive implementation of the detection, diagnosis, prevention and control of different levels of disease, in combination with the corresponding farming management regulation; thus there are intermediate levels. Different farms may formulate different levels of biosecurity plans according to their own conditions and investment capabilities. This scheme is conducive to the wide application of aquaculture biosecurity plans, and will gradually strengthen the implementation of biosecurity during the implementation process.

Conclusion

The PMP/AB, while still a work in progress, is at an advanced stage of development; guidance documents and other toolkits (e.g. pilot testing, governance, development of NSAAH and national aquatic pathogen list within the PMP/AB, risk assessment, surveillance, emergency preparedness, sectoral and farm-level biosecurity action plans, health economics) are all currently in preparation. Farm-level biosecurity is one the PMP/AB tools that will be developed over the course of time. Exploring the potential of China's ABG as a PMP/AB farm-level biosecurity toolkit will be welcomed.

The shrimp sector, with its advanced features in terms of technology, industry organization, knowledge of diseases, and farm-level and market infrastructure, can take advantage of the benefits to be gained through a systematic and comprehensive approach in dealing with shrimp farming biosecurity challenges where risks and responsibilities are shared with key stakeholders, working together to find common solutions and thus greatly contributing to sector sustainability.

The sector can contribute to effective national planning and co-management that can lead to stronger public–private sector partnerships and commitment, and thus wider ownership of the principles, responsibilities and coordination of all relevant activities necessary for biosecurity management. Co-management also means that problems are well recognized and management solutions identified.

As in any food-producing sector, the shrimp sector will continue to face disease challenges. However, it can harness the opportunity to farm shrimp in a sustainable manner by contributing to the establishment of enabling policy environments that this new initiative – the PMP/AB – can offer.

References

Bondad-Reantaso, M. G. (2016) Acute hepatopancreatic necrosis disease (AHPND) of penaeid shrimps: Global perspective. In: Pakingking Jr, P. V., de Jesus-Ayson, E. G. T. and Acosta, B. O. (eds) *Addressing Acute Hepatopancreatic Necrosis Disease (AHPND) and Other Transboundary Diseases for Improved Aquatic Animal Health in Southeast Asia: Proceedings of the ASEAN Regional Technical Consultation on EMS/AHPND and Other Transboundary Diseases for Improved Aquatic Animal Health in Southeast Asia, 22–24 February 2016, Makati City, Philippines* (pp. 16–23). Tigbauan, Iloilo, Philippines: Aquaculture Department, Southeast Asian Fisheries Development Center. (also available at https://repository.seafdec.org.ph/handle/10862/3096, accessed 20 July 2021).

Bondad-Reantaso, M., Garrido-Gamarro and McGladdery, S. E. (2018) Climate change-driven hazards on food safety and aquatic animal health. In: Barange, M., Bahri, T., Beveridge, M. C. M., Cochrane, K. L., Funge-Smith, S. and Poulain, F. (eds) 2018. *Impacts of Climate Change on Fisheries and Aquaculture: Synthesis of Current Knowledge, Adaptation and Mitigation Options*. FAO Fisheries and Aquaculture Technical Paper No. 627. FAO, Rome, pp. 517–534.

Bondad-Reantaso, M. G., McGladdery, S. E., East, I. and Subasinghe, R. P. (2001) Asia diagnostic guide to aquatic animal diseases. *FAO Fisheries Technical Paper*. No. 402, Suppl. 2. FAO, Rome, 240 pp (also available at http://www.fao.org/3/a-y1679e.pdf, accessed 20 July 2021).

Bondad-Reantaso, M. G., Subasinghe, R. P., Josupeit, H., Cai, J. and Zhou, X. (2012) The role of crustacean fisheries and aquaculture in global food security: past, present and future. *Journal of Invertebrate Pathology* 110(2), 158–165 (also available at https://doi.org/10.1016/j.jip.2012.03.010, accessed 21 July 2021).

Boonyaratpalin, S., Supamattaya, K., Kasornchandra, J., Direkbusaracom, S., Aekpanithanpong, U. and Chantanachookin, C. (1993) Non-occluded baculo-like virus, the causative agent of yellow head disease in the black tiger shrimp (*Penaeus monodon*). *Fish Pathology* 28, 103–109.

Briggs, M., Funge-Smith, S., Subasinghe, R. P. and Phillips, M. (2005) Introductions and movement of two penaeid shrimp species in Asia and the Pacific. *FAO Fisheries Technical Paper* No. 476. FAO, Rome, 78 pp.

Couch, J. A. (1974) An enzootic nuclear polyhedrosis virus of pink shrimp: ultrastructure, prevalence and enhancement. *Journal of Invertebrate Pathology* 24, 311–331.

FAO (2007) Aquaculture development. 2. Health management for responsible movement of live aquatic animals. *FAO Technical Guidelines for Responsible Fisheries. No. 5, Suppl.* 2. FAO, Rome, 31 pp.

FAO (2018) Development of a regional aquatic biosecurity strategy for the Southern African Development Community (SADC). FAO Fisheries and Aquaculture Circular No. C1149. FAO, Rome, 344 pp (also available at http://www.fao.org/documents/card/en/c/CA2764EN/, accessed 20 July 2021).

FAO (2019a) Preventing and managing aquatic animal disease risks in aquaculture through a progressive management pathway. Trondheim, Norway, 1–11 pp. (also available at http://www.fao.org/3/na265en/na265en.pdf, accessed 20 July 2021).

FAO (2019b) Report of the FAO/MSU/WB First Multi-Stakeholder Consultation on a Progressive Management Pathway to Improve Aquaculture Biosecurity (PMP/AB), Washington, D.C., USA, 10–12 April 2018. FAO, Rome, Fisheries and Aquaculture Report No. 1254, 76 pp (also available at http://www.fao.org/documents/card/en/c/ca4891en/, accessed 20 July 2020).

FAO (2020a) Fishery and Aquaculture Statistics. Global Aquaculture Production 1950–2018 (FishstatJ). In: FAO Fisheries and Aquaculture Department [online]. FAO, Rome (also available at www.fao.org/fishery/statistics/software/fishstatj/en, accessed 20 July 2021).

FAO (2020b) The State of World Fisheries and Aquaculture 2020. Sustainability in action. FAO, Rome (also available at https://doi.org/10.4060/ca9229en, accessed 20 July 2021).

FAO (2020c) Report of the Progressive Management Pathway for Improving Aquaculture Biosecurity (PMP/AB): First Technical Working Group Meeting, Rome, 20–22 March 2019. FAO Fisheries and Aquaculture Report No. 1322. FAO, Rome (also available at http://www.fao.org/documents/card/en/c/cb0582en, accessed 1 October 2021).

FAO (2020d) Progress Towards Development of the Progressive Management Pathway for Improving Aquaculture Biosecurity (PMP/AB): Highlights of 2019 activities. FAO Fisheries and Aquaculture Circular No. 1211. FAO, Rome (also available at http://www.fao.org/documents/card/en/c/cb0560en, accessed 1 October 2021).

Fegan, D. F. (2000) Evaluation of diagnostic tests: the epidemiological approach. In: Walker, P. and Subasinghe, R. (eds) *DNA-based Molecular Diagnostic Techniques: Research Needs for Standardization and Validation of the Detection of Aquatic Animal Pathogens and Diseases.* Report and proceedings of the Joint FAO/NACA/CSIRO/ACIAR/DFID Expert Workshop. Bangkok, Thailand, 7–9 February 1999. FAO Fisheries Technical Paper. No. 395. FAO, Rome, 93 pp (also available at http://www.fao.org/3/X4946E/x4946e00.htm#Contents, accessed 20 July 2021).

Flegel, T. W. (2006) Detection of major penaeid shrimp viruses in Asia, a historical perspective with emphasis on Thailand. *Aquaculture* 258, 1–33.

Hasson, K. W., Fan, Y., Reizinger, T., Venuti, J. and Varner, P. W. (2006) White-spot syndrome virus (WSSV) introduction into the Gulf of Mexico and Texas freshwater systems through imported, frozen bait-shrimp. *Diseases of Aquatic Organisms* 71(2), 91–100.

Jimenez, R. (1992) Sindrome de Taura (resumen) Acuacultura del Ecuador. In: R Jimenez (ed.) *Revista Especializada de la Camara National de Acuacultura*, pp. 1–16. Guayaquil [in Spanish].

Lightner, D. V. (2011) Virus diseases of farmed shrimp in the Western Hemisphere (the Americas): a review. *Journal of Invertebrate Pathology* 106(1), 110–130.

Lightner, D. V., Pantoja, C. R., Poulos, B. T., Tang, K. F. J., Redman, R. M., Andrade, T. P. D. and Bonami, J. R. (2004) Infectious myonecrosis: new disease in Pacific white shrimp. *Global Aquaculture Advocate* 7(5), 85.

Lightner, D. V. and Redman, R. M. (1981) A baculovirus-caused disease of the penaeid shrimp, *Penaeus monodon. Journal of Invertebrate Pathology* 38, 299–302.

Lightner, D. V., Redman, R. M., Bell, T. A. and Brock, J. A. (1983) Detection of IHHN virus in *P. stylirostris* and *P. vannamei* imported into Hawaii. *Journal of the World Mariculture Society* 14, 212–225.

NACA (2017) *Sixteenth Meeting of the Asia Regional Advisory Group on Aquatic Animal Health: Report of the Meeting.* Bali, Indonesia, pp. 26–27.

Natividad, J. M. and Lightner, D. V. (1992) Susceptibility of the different larval and postlarval stages of black tiger prawn, *Penaeus monodon* Fabricius, to monodon baculovirus (MBV). In Shariff, M., Subasinghe, R.P. and Arthur, J.R. (eds) *Diseases in Asian Aquaculture I*, pp. 111–125. Asian Fisheries Society, Manila, The Philippines.

Nunes, A. J. P., Cunha-Martins, P. and Vasconselos-Gesteira, T. C. (2004) Carcinicultura ameacada. *Review of Panoramic Aquacultre* 83, 37–51.

OIE (2016) *WAHIS, Immediate Notification, White Spot Syndrome Virus, Australia*. https://www.oie.int/wahis_2/public/wahid.php/Reviewreport/Review?reportid=21737

OIE (2018) WAHIS, Weekly Disease Information, Infectious Myonecrosis, Malaysia. https://www.oie.int/wahis_2/public/wahid.php/Reviewreport/Review?page_refer=MapFullEventReportandreportid=27840

OIE (2019) *Manual of Diagnostic Tests for Aquatic Animals* [online]. https://www.oie.int/standard-setting/aquatic-manual/access-online/

Qiu, L., Chen, M. M., Wan, X. Y., Li, C., Zhang, Q. L., Wang, R. Y. and Xiang, J. H. (2017) Characterization of a new member of *Iridoviridae*, Shrimp hemocyte iridescent virus (SHIV), found in white leg shrimp (*Litopenaeus vannamei*). *Scientific Reports* 7(1), 1–13.

Sahul Hameed, A. S., Abdul Majeed, S., Vimal, S., Madan, N., Rajkumar, T., Santhoshkumar, S. and Sivakumar, S. (2017) Studies on the occurrence of infectious myonecrosis virus in pond-reared *Litopenaeus vannamei* (Boone, 1931) in India. *Journal of Fish Diseases* 40(12), 1823–1830.

Senapin, S., Phewsalya, K., Briggs, M. and Flegel, T. W. (2007) Outbreaks of infectious myonecrosis virus (IMNV) in Indonesia confirmed by genome sequencing and use of an alternative RT-PCR detection method. *Aquaculture* 266, 32–38.

Snieszko, S. F. (1974) The effects of environmental stress on outbreaks of infectious diseases of fishes. *Journal of Fish Biology* 6(2), 197–208.

Srisala, J., Sanguanrut, P., Thaiue, D., Laiphrom, S., Siriwattano, J., Khudet, J., Powtongsook, S., Flegel, T. W. and Sritunyalucksana, K. (2020) Urgent warning: Positive PCR detection results for infectious myonecrosis virus (IMNV) and decapod iridescent virus 1 (DIV1) in captured *Penaeus monodon* from the Indian Ocean (available at https://enaca.org/?id=1093, accessed 20 July 2021).

Sumption, K., Domenech, J. and Ferrari, G. (2012) Progressive control of FMD on a global scale. *Veterinary Record* 170(25), 637–639.

Tang, K. F. J., Bondad-Reantaso, M. G. and Arthur, J. R. (2019) Shrimp infectious myonecrosis strategy manual. FAO Fisheries and Aquaculture Circular No. 1187. FAO, Rome (available at http://www.fao.org/3/ca6052en/ca6052en.pdf, accessed 20 July 2021).

Tourtip, S., Wongtripop, S., Stentiford, G. D., Bateman, K. S., Sriurairatana, S., Chavadej, J., Sritunyalucksana, K. and Withyachumnarnkul, B. (2009) *Enterocytozoon hepatopenaei* sp. nov.(Microsporida: Enterocytozoonidae), a parasite of the black tiger shrimp *Penaeus monodon* (Decapoda: *Penaeidae*): Fine structure and phylogenetic relationships. *Journal of Invertebrate Pathology* 102(1), 21–29.

van Seventer, J. M. and Hochberg, N. S. (2017) Principles of infectious diseases: Transmission, diagnosis, prevention, and control. *International Encyclopedia of Public Health* 2017, 22–39.

Walker, P. J. and Winton, J. R. (2010) Emerging viral diseases of fish and shrimp. *Veterinary Research* 41(6), 51.

Xu, L., Wang, T., Li, F. and Yang, F. (2016) Isolation and preliminary characterization of a new pathogenic iridovirus from redclaw crayfish *Cherax quadricarinatus. Diseases of Aquatic Organisms* 120(1), 17–26.

Zhan, W. B., Wang Y. H., Fryer J. L., Yu K. K., Fukuda H. and Meng Q. X. (1998) White spot syndrome infection of cultured shrimp in China. *Journal of Aquatic Animal Health* 10, 405–410.

Zhang, B. C., Liu, F., Bian, H. H., Liu, J., Pan, L. Q. and Huang, J. (2012) Isolation, identification, and pathogenicity analysis of a *Vibrio parahaemolyticus* strain from *Litopenaeus vannamei. Progress in Fisheries Science* 33, 56–62 [in Chinese].

Zhang, Q., Liu, Q., Liu, S., Yang, H., Liu, S., Zhu, L., Yang, B., Jin, J., Ding, L., Wang, X., Liang, Y., Wang, Q. and Huang, J. (2014) A new nodavirus is associated with covert mortality disease of shrimp. *Journal of General Virology* 95, 2700–2709.

Chapter 18

EPIGENETIC MANAGEMENT OF DISEASE RESISTANCE IN FARMED SHRIMP

PARISA NOROUZITALLAB, KARTIK BARUAH, RONALD LULIJWA, PATRICK SORGELOOS, PETER BOSSIER AND DAISY VANROMPAY

Summary

The global demand for high-quality animal protein is expected to increase considerably with the rise in the human population. It is predicted that farmed crustacean shrimps could contribute significantly to this increasing demand, and eventually to the world´s future food protein security and nutrition. Diseases, however, impose a major yield-limiting effect on the farmed production of high-value shrimps, causing devastating socio-economic impacts. To avoid disease-mediated production losses there has been a continuous effort to develop effective and sustainable health management and disease prevention strategies. Lately, epigenetics has attracted considerable attention because there is now compelling evidence indicating that epigenetics alterations are involved in the induction of disease resistance in aquaculture animals, including shrimps, both within and across generations. This chapter brings this information together and discusses the potential of this new concept in the design of new-generation health management and disease preventional strategies in farmed shrimp.

Introduction

In an aquaculture environment, pathogens are generally a part of the natural microflora and, under optimum environmental conditions, they co-exist with the host shrimp without causing a negative impact on the host (de la Vega et al., 2006). However, when the host organism is immunosuppressed or otherwise physiologically stressed due to change in the environmental conditions, some quiescent microbial infections develop into acute diseases, causing significant production losses (Hall and de la Vega, 2004). This suggests that multiple (biotic and abiotic) factors interacting in a complex manner are responsible for causing diseases in shrimp. Considering the threat posed by multiple disease-causing agents to the shrimp, and the limitations in using unsustainable drugs (e.g. antibiotics, disinfectants) due to long-term economic and environmental ramifications, there is a need for a paradigm shift in the development of strategies for managing health and controlling diseases in farmed shrimp. As some of the pathogens are generally a part of the natural microflora of an aquatic environment and are difficult to eradicate from the production environment, one needs to focus on developing an approach to produce robust shrimp that can potentially fight not only against the targeted pathogens but also the pathogens that remain quiescent within the host, and also against environmental stressors. Recently, interest in epigenetics within the aquaculture community has surged as a number of studies have linked epigenetics alterations with the induction of disease resistance, stress tolerance and other economically important traits in farmed aquatic animals, including shrimp.

Epigenetics – the concept

The term 'epigenetics' refers to the long-term or stable regulation of gene expression and function, induced by environmental factors, without changes in the DNA sequence. Because epigenetics has a complex history, the term is therefore often employed loosely and inconsistently and is sometimes used as a synonym for 'epigenetic inheritance'. To avoid misinterpretations, we describe the term epigenetic inheritance in this chapter as mechanisms that permit the stable transmission of parental environmentally induced phenotypes to a subsequent generation or generations without any changes in the genetic sequence (Kelly et al., 2010). The three main types of epigenetic inheritance proposed include: (1) intergenerational or cross-generational effects, such as the impact of *in utero* or paternal exposure to particular environmental cues (e.g. nutrition, stress) on the developing embryo and its germline (only the F1 generation) (Grossniklaus et al., 2013; Radford et al., 2014); (2) multigenerational effects (from the F1 to the F2 generation) (Dias and Ressler, 2014); and (3) transgenerational effects that are found in more than three generations (Anway et al., 2005; Norouzitallab et al., 2016). Numerous mechanisms for epigenetic modifications have been identified in vertebrates and invertebrates, and new mechanisms are likely to emerge. Key mechanisms underlying epigenetic modifications include chromatin remodeling, DNA methylation, histone tail N-terminal modification, and post-translational modification of genes regulation by non-coding/small RNA (ncRNA/sRNA). Collectively, these processes and the components upon which they act constitute the epigenome. Individually or in concert, they play a key role in turning on or off the expression of genes, thus facilitating or inhibiting the production of specific proteins. These epigenetic modifications, which occur primarily during embryonic development, are assumed to facilitate differentiation into specific cell types and can be mitotically and/or meiotically heritable (Daneshfozouna et al., 2015). Epigenetic marks can also be acquired throughout life and be relatively stable, although some are potentially reversible. Therefore, it can be said that epigenetic mechanisms are responsible for mediating the diversified gene expression profiles in a range of cells and tissues. The in-depth description of these mechanistic details is beyond the scope of this chapter and is reviewed elsewhere (see reviews in Norouzitallab et al., 2019; Plunk and Richards, 2020; Sarkies, 2020).

Epigenetics and inheritance of phenotypes

In multicellular organisms, the germline is described as cells that are highly specialized in forming gametes and are responsible for passing the genetic information from parents to the progenies (Hill et al.,

2018). In bisexually reproducing adults, the germline cells develop into haploid gametes by meiotic divisions during the gametogenesis process (Ben Maamar et al., 2018; Hill et al., 2018). These very specialized reproductive cells can subsequently form the most complex tissues and organisms. In bisexual organisms, once the sperm penetrates the egg, a single diploid totipotent zygote cell is formed (Condic, 2014) which is capable of differentiating into all the other possible forms of cells required in an adult organism (Mitalipov and Wolf, 2009). Following fertilization, a series of epigenetic modifications take place in the parental pronuclei. This results in removal of the gamete-specific modifications and enables embryonic development towards birth (Rivera and Ross, 2013). For this purpose the epigenome of the gametic cells is erased through a 'reprogramming' process that brings the cells to a genetic 'blank state' in which new epigenetic marks determine their fate (Figure 18.1). Nevertheless, a very small percentage of genes keep their epigenetic marks during this process and pass unchanged from parent to progeny through the mechanism of a 'self-sustaining feedback loop' that was first described in *Escherichia coli* (Jablonka and Raz, 2009). Dynamic regulation of the oocyte genome is essential for programming the embryo to achieve temporally required developmental landmarks. The embryo's epigenome could also be affected when stem cells are exposed to an environmental cue through parental exposure. The detailed mechanism is beyond the scope of this chapter and is reviewed elsewhere (see review, Norouzitallab et al., 2019).

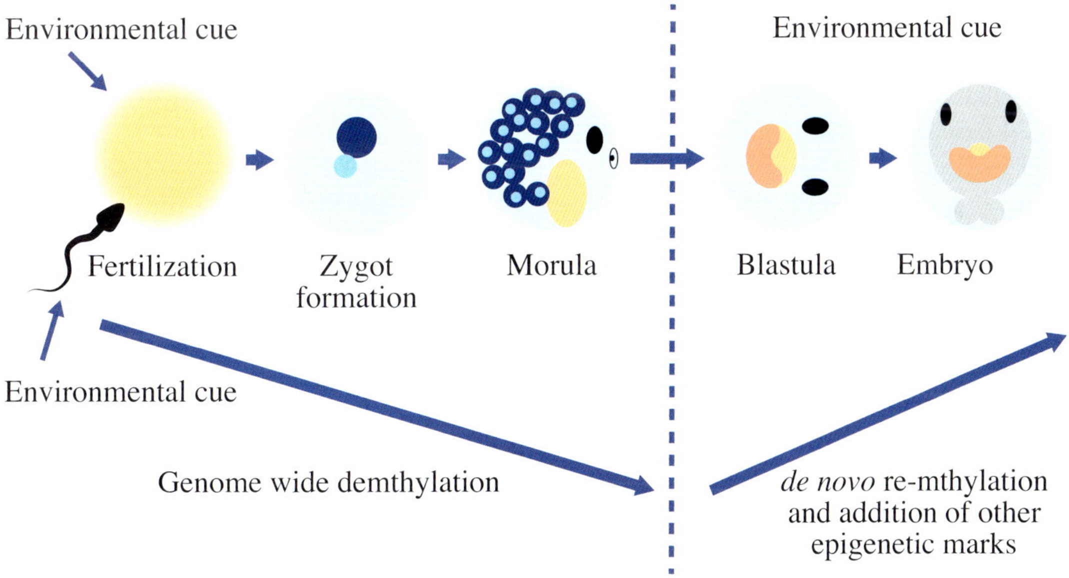

FIGURE 18.1 Hypothetical dynamics of epigenetic marks during embryonic development in shrimp. This model is simulated based on epigenetic marks during vertebrate embryogenesis. Most of the epigenetic marks are erased during early embryonic development while new epigenetic marks are created soon after, during early embryogenesis, after which tissue-specific epigenetic marks are laid down. Parental exposure to environmental cues affects the epigenome of the gametes. Some induced marks are sustained during embyogenesis. New epigenetic modifications may also affect cells if the embryo is exposed to new environmental cues.

Epigenetics and environmental cues

In vertebrate and invertebrate animals, there exist many examples that associate environmental cues with epigenetic modification in organisms. Despite the developmental stability of stem cells, the embryo is highly influenced by the external environment of the parental generation (Spannhoff et al., 2011). There are critical time points during embryogenesis and primordial germ cell specification in which the cells are sensitive to environmental cues and, therefore, environmental changes influence the epigenetic information both within developing individuals and across generations (Bertoldo et al., 2015; Cavalieri and Spinelli, 2017). Apart from the epigenetic modifications that occur during embryonic development at various stages, epigenetic modifications brought about by environmental cues have also been reported to induce development of new phenotypes in fully developed individuals and at various stages in a number of organisms (e.g. *Artemia*, *Daphnia*, mouse). The resultant acquired phenotypic characteristics not only persist throughout the individual's lives but also appear to be transmitted to subsequent generations (Aguilera et al., 2010; Norouzitallab et al., 2014, 2016).

Epigenetic influences have been observed to occur with environmental contaminants (e.g. inorganic contaminants, endocrine disruptors, pesticides or fungicides), nutritional factors, drugs, previous exposure to pathogens and/or immune elicitors (including lipopolysaccharide, heat shock protein, glucan and physiochemical and biochemical environmental variation; for details, see reviews in Skinner, 2015; Norouzitallab et al., 2019).

The impact of environmentally induced epigenetic modifications on an organism's phenotypes can be either detrimental or beneficial. Data collected from several studies indicate that negative environmental stimuli, such as natural climate stressors or sublethal concentrations of chemicals, can generate the accumulation of epigenetic modifications over time, leading to an increased individual's susceptibility to diseases and development of disorders (Mirbahai et al., 2011). Epigenetic changes may also contribute to a more advantageous phenotype. In the case of polyploid organisms, epigenetic changes can provide effective and flexible ways for a neopolyploid to rapidly respond to the enormous change in genetic material, contributing to the survival and potential reproduction of these organisms (Li et al., 2011).

Epigenetics in shrimp health and disease management

Given that environmental cues can influence the epigenome of organisms and subsequently cause concomitant changes in their phenotypes, interest in epigenetics has lately been gaining ground in regard to aquaculture production systems. The epigenome approach can be used to achieve stress and disease tolerance traits or other economically important phenotypic traits (e.g. fast growth, better feed conversion rates) in farmed aquatic organisms, and to facilitate aquaculture productivity for economic benefits while minimizing impacts on wild stocks. The approach to generating environmentally driven phenotypes mediated through epigenetic mechanisms has been considered during the two main periods in the life cycle of farmed animals: (1) embryonic development and early life stages and (2) adult and broodstock. It has been demonstrated that the establishment of epigenetic states of genes related to phenotypic variation occurs mainly during these stages, and exposure to altered environmental conditions during these stages is important for the maintenance of the epigenetic state (Fig. 18.2; Burggren and Blank, 2009;

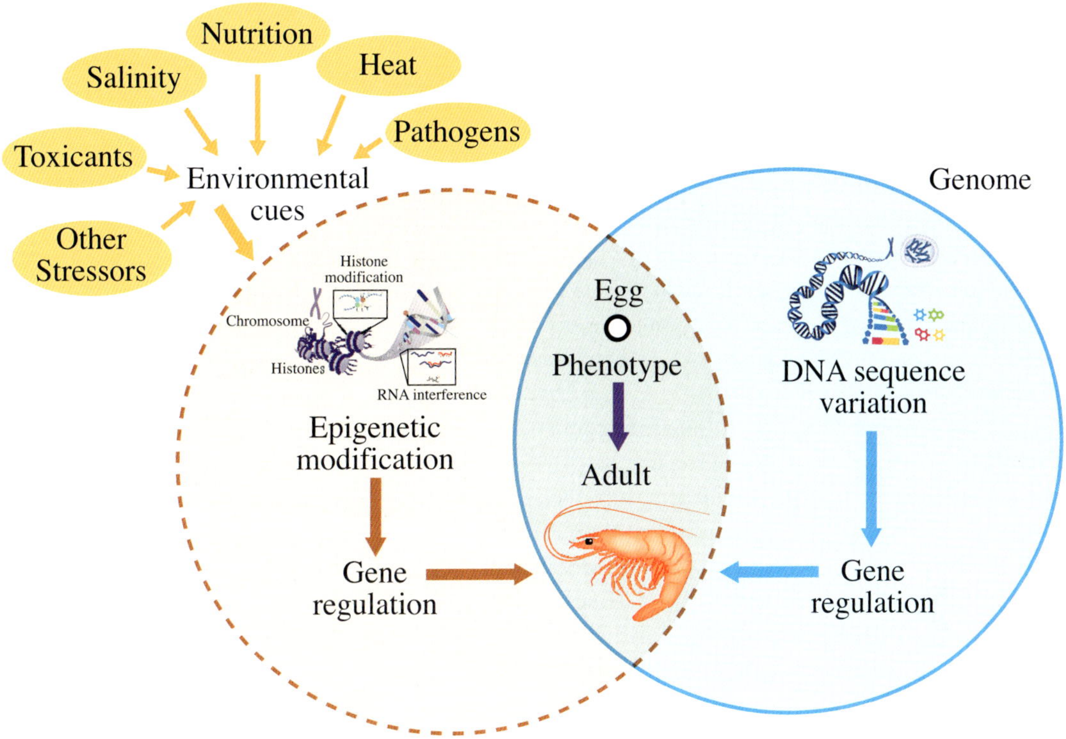

FIGURE 18.2 Environmental cues and epigenome. In organisms the genome determines the phenotype. In addition, environmental cues can also significantly impact the epigenome at various stages of their life. These modifications can result in the development of new phenotypes. Like the genome, epigenetic modifications can be inherited by subsequent generations.

Harris et al., 2012). Many studies, in species ranging from salmonids to oysters and (brine) shrimp, have provided evidence that epigenetic mechanisms are associated with commercially important traits such as metabolism, growth, sex determination, fecundity and behavior (for details, see reviews in Granda et al., 2018 Norouzitallab et al., 2019; Roy et al., 2020).

The study of epigenetics and its applications in health management and disease control in farmed shrimp is very limited. However, studies performed on shellfish animal models revealed evidence linking changes in epigenetic markers to disease resistance and increased tolerance to stressors (see reviews in Fallet et al., 2020; Granada et al., 2018; Norouzitallab et al; 2015, 2016, 2019). For instance, associations between epigenetics mechanisms and temperature-induced tolerance to lethal heat shock and resistance against pathogenic bacteria have been documented in the study of Norouzitallab et al. (2014). In this study, carried out using the brine shrimp *Artemia* (also a model for shrimp), the authors demonstrated an increase in levels of heat shock protein 70, tolerance to lethal heat stress and resistance against the pathogen *Vibrio campbellii*. Parthenogenetic *Artemia* were exposed to daily non-lethal heat shocks for 14 days, starting from day 2 post-hatching. These effects were accompanied by altered levels of global DNA methylation and increased acetylation of histones H3 and H4. Additionally, transgenerational inheritance of acquired traits during three consecutive unstressed generations was observed. Confronted with these interesting findings, the same authors carried out a series of additional studies, under a similar experimental approach, to study whether exposure to other environmental stimuli (e.g. pathogenic bacteria, bioactive food components) can cause similar modulation of the epigenome of *Artemia* resulting in the generation of robustness phenotypes, as described above. The results demonstrated that exposure of the parental generation of *Artemia* at their early life stages to immune challenges with the pathogenic biotic stressor *V. campbellii* caused marked increase in the resistance of the progeny for three successive generations (none of them were immune challenged) against subsequent challenge with the same bacterial strain (Norouzitallab et al., 2015). Furthermore, they showed that robust phenotypes were associated with alteration in the expression of key innate immunity-related genes. Additionally, a stochastic pattern in the acetylation and methylation levels of H4 and H3K4me3 histones, respectively, was also noted in progeny whose ancestors were exposed to an immune challenge, suggesting that epigenetic reprogramming of (selected) innate immune effectors is likely to be involved in the generation of robust phenotypes in *Artemia* (Norouzitallab et al., 2016).

Using a brine shrimp model, the same group also found that exposure of the animals to a plant-derived bioactive compound (phloroglucinol) protected them and their three subsequent non-exposed generations against *V. campbellii* and acute hepatopancreatic necrosis disease-causing *V. parahaemolyticus* challenge (Roy et al., 2019). Transcriptomic analysis revealed that this increased resistance phenotype was associated with altered levels of several innate immune-related genes (DSCAM, LGBP, proPO peroxinectin, Hsp90, Hsp70). Also, it was demonstrated that epigenetic markers, such as DNA methylation, m6A RNA methylation and histone acetylation and methylation (active chromatin markers, i.e. H3K4Me3, H3K4me1, H3K27me1, H3 hyperacetylation, H3K14ac, and repression markers, i.e. H3K27me3, H4 hypoacetylation) were associated with the regulation of gene expression leading to the observed inheritance of resistant phenotypes across successive generations (Roy et al., 2019).

In penaeid shrimp, there is evidence that prior exposure to an immune-stimulating agent or to pathogens has a positive influence on the immune response later in life or on the resistance against subsequent microbial infections, and that this environmental memory can be transmitted to offspring (Amatul-Samahah et al., 2020; Chang et al., 2017; Tanekhy and Fall, 2015; Tassanakajon et al., 2017). Huang and Song (1999) demonstrated that the offspring of black tiger shrimp parents treated with beta-1,3-1,6-glucan possess enhanced protection against white spot syndrome-associated virus, suggesting maternal transmission of environmentally driven protective immune responses in shrimp. The underlying mechanism(s) behind this cross-generational protective immunity in this species are not fully understood. While such an understanding is, arguably, not required to improve productivity in the shrimp farming system, elucidation of the epigenetic mechanisms involved could increase the degree and breadth of ongoing improvements.

Conclusion

In light of current knowledge, outlined above, it is evident that epigenetics has the potential to positively induce disease resistance and stress tolerance phenotypes in farmed shrimp. The traditional approaches to control diseases in an agriculturally important shrimp species are no longer sustainable and environmentally friendly. Here we have proposed the new concept of epigenetics as an alternate approach, and it is anticipated that this new concept will facilitate the design of a new generation of health management strategies to control diseases in penaeid shrimp.

References

Aguilera, O., Fernández, A. F., Muñoz, A. and Fraga, M. F. (2010) Epigenetics and environment: a complex relationship. *Journal of Applied Physiology* 109, 243–251.

Amatul-Samahahab, A. M., Omar, W., Ikhsana, N., Azmai, M., Zamri-Saade, M. and Ina-Salwanyac. M. Y. (2020) Vaccination trials against vibriosis in shrimp: A review. *Aquaculture Reports* 18, 100471.

Anway, M.,D., Cupp, A. S., Uzumcu, M. and Skinner, M. K. (2005) Epigenetic transgenerational actions of endocrine disruptors and male fertility. *Science* 308, 1466–1469.

Ben Maamar, M., Sadler-Riggleman, I., Beck, D. and Skinner, M. (2018) Epigenetic transgenerational inheritance of altered sperm histone retention sites. *Scientific Reports* 8, 5308.

Bertoldo, M. J., Locatelli, Y., O'Neill, C. and Mermillod, P. (2015) Impacts of and interactions between environmental stress and epigenetic programming during early embryo development. *Reproduction, Fertility and Development* 27, 1125–1136.

Burggren, W. and Blank, T. (2009) Physiological study of larval fishes: challenges and opportunities. *Scientia Marina* 73, 99–110.

Chang, Y. H., Kumar, R., Ng, T. H. and Wang, H. C. (2018) What vaccination studies tell us about immunological memory within the innate immune system of cultured shrimp and crayfish. *Developmental and Comparative Immunology* 80, 53–66.

Cavalieri, V. and Spinelli, G. (2017) Environmental epigenetics in zebrafish. *Epigenetics and Chromatin* 10, 46.

Condic, M. L (2014) Totipotency: what it is and what it is not. *Stem Cells Development* 23, 796–812.

Daneshfozouna, H., Panjvinib, F., Ghorbanic, F. and Farahmandd, H. (2015) A review of epigenetic imprints in aquatic animals. *Fisheries and Aquaculture Journal* 6, 119.

de la Vega, E., Hall, M. R., Degnan, B.M. and Wilson, K. J. (2006) Short-term hyperthermic treatment of *Penaeus monodon* increases expression of heat shock protein 70 (HSP70) and reduces replication of gill associated virus (GAV). *Aquaculture* 253, 82–90.

Dias, B. G. and Ressler, K. J. (2014) Parental olfactory experience influences behavior and neural structure in subsequent generations. *Nature Neuroscience* 17, 89–96.

Fallet, M., Luquet, E., David, P. and Cosseau, C. (2020) Epigenetic inheritance and intergenerational effects in mollusks. *Gene* 729, 144166.

Granada, L., Lemos, M. F. L., Cabral, H., Norouzitallab, P., Bossier, P. and Novais, S. C. (2018) Epigenetics in aquaculture – the last frontier. *Reviews in Aquaculture* 10, 994–1013.

Grossniklaus, U., Kelly, W. G., Ferguson-Smith, A. C., Pembrey, M. and Lindquist, S. (2013) Transgenerational epigenetic inheritance: how important is it? *Nature Reviews Genetics* 14, 228–235.

Hall, M. R. and de la Vega, E. (2004) Physiological response to stress and health implications in Crustacea. In: Goarant, C., Harache, Y., Herbland, A. and Mugnier, C. (eds) *Styli 2003: Thirty Years of Shrimp Farming in New Caledonia*. INRA Editions, Plouzane, France, pp. 38–56.

Harris, K. D. M., Bartlett, N. J. and Lloyd, V. K. (2012) *Daphnia* as an emerging epigenetic model organism. *Genetics Research International* 147892.

Hill, P. W. S. et al. (2018) Epigenetic reprogramming enables the transition from primordial germ cell to gonocyte. *Nature* 555, 392–396.

Huang, C. C. and Song, Y. L. (1999) Maternal transmission of immunity to white spot syndrome associated virus (WSSV) in shrimp (*Penaeus monodon*). *Developmental and Comparative Immunology* 23, 545–552.

Jablonka, E. and Raz, G. (2009) Transgenerational epigenetic inheritance: prevalence, mechanisms, and implications for the study of heredity and evolution. *Quarterly Review of Biology* 84, 131–176.

Kelly, T. K., De Carvalho, D. D. and Jones, P. A. (2010) Epigenetic modifications as therapeutic targets. *Nature Biotechnology* 28, 1069–1078.

Li, Z. H., Lu, X., Gao, Y., Liu, S. J., Tao, M., Xiao, H., Qiao, Y. Q., Zhang, Y. P. and Luo, J. (2011) *Chinese Science Bulletin* 56, 245–252.

Mirbahai, L., Yin, G., Bignell, J. P., Li, N., Williams, T. D. and Chipman, J. K. (2011) DNA methylation in liver tumorigenesis in fish from the environment. *Epigenetics* 6, 1319–1333.

Mitalipov, S. and Wolf, D. (2009). Totipotency, pluripotency and nuclear reprogramming. *Advances in Biochemical Engineering and Biotechnology* 114, 185–199.

Norouzitallab, P., Baruah, K., Vandegehuchte, M., Van Stappen, G., Catania, F., Vanden Bussche, J., Vanhaecke, L., Sorgeloos, P. and Bossier, P. (2014) Environmental heat stress induces epigenetictransgenerational inheritance of robustness in parthenogenetic *Artemia* model. *FASEB Journal* 28, 3552–3563.

Norouzitallab, P., Biswas, P., Baruah, K. and Bossier, P. (2015) Multigenerational immune priming in an invertebrate parthenogenetic *Artemia* to a pathogenic *Vibrio campbellii*. *Fish & Shellfish Immunology* 42, 426–429.

Norouzitallab, P., Baruah, K., Biswas, P., Vanrompay, D. and Bossier, P. (2016) Probing the phenomenon of trained immunity in invertebrates during a transgenerational study, using brine shrimp *Artemia* as a model system. *Scientific Reports* 6, 21166.

Norouzitallab, P. Baruah, K., Bossier, P. and Vanrompay, D. (2019) Can epigenetics translate environmental cues into phenotypes? *Science of the Total Environment* 647, 1281–1293.

Pande, G. S. J., Scheie, A. A., Benneche, T., Wille, M., Sorgeloos, P., Bossier, P. and Defoirdt, T. (2014) Quorum sensing-disrupting compounds protect larvae of the giant freshwater prawn *Macrobrachium rosenbergii* from *Vibrio harveyi* infection. *Aquaculture* 406–407, 121–124.

Plunk, E. and Richards, M. (2020) Epigenetic modifications due to environment, ageing, nutrition, and endocrine disrupting chemicals and their effects on the endocrine system. *International Journal of Endocrinology* 2020, 9251980.

Radford, E. J. et al. (2014) In utero undernourishment perturbs the adult sperm methylome and intergenerational metabolism. *Science* 345, 1255903.

Rivera, R. M. and Ross, J. W. (2013) Epigenetics in fertilization and preimplantation embryo development. *Progress in Biophysics and Molecular Biology* 113, 423–432.

Roy, S., Bossier, P., Norouzitallab, P. and Vanrompay, D. (2020). Trained immunity and perspectives for shrimp aquaculture. *Reviews in Aquaculture* 12, 2351–2370.

Roy, S., Kumar, V., Bossier, P., Norouzitallab, P. and Vanrompay, D. (2019) Phloroglucinol treatment induces transgenerational epigenetic inherited resistance against *Vibrio* infections and thermal stress in a brine shrimp (*Artemia franciscana*) model. *Frontiers in Immunology* 27, 2745.

Sarkies, P. (2020) Molecular mechanisms of epigenetic inheritance: Possible evolutionary implications. *Seminars in Cell Developmental Biol*ogy 97, 106–115.

Skinner, M. K. (2015) Environmental epigenetics and a unified theory of the molecular aspects of evolution: a neo-Lamarckian concept that facilitates neo-Darwinian evolution. *Genome Biolology and Evolution* 7, 1296–1302.

Spannhoff, A., Kim, Y. K., Raynal, N. J., Gharibyan, V., Su, M. B., Zhou, Y. Y., Li, J., Castellano, S., Sbardella, G., Issa, J. P. and Bedford, M. T. (2011) Histone deacetylase inhibitor activity in royal jelly might facilitate caste switching in bees. *EMBO Reports* 12, 238–243.

Tanekhy, M. and Fall, J. (2015) Expression of innate immunity genes in kuruma shrimp *Marsupenaeus japonicus* after in vivo stimulation with garlic extract (allicin). *Veterinární Medicína* 60, 39–47.

Tassanakajon, A., Rimphanitchayakit, V., Visetnan, S., Amparyup, P., Somboonwiwat, K., Charoensapsri, W. et al. (2017) Shrimp humoral responses against pathogens: antimicrobial peptides and melanization. *Developmental and Comparative Immunology* 80, 81–93.

Chapter 19

ON SHRIMP PATHOGEN ENTRY: THE CASE OF WSSV AS A GUIDE TO OTHER SHRIMP PATHOGENS

GAËTAN M.A. DE GRYSE, MATHIAS CORTEEL, CESAR ESCOBEDO-BONILLA, THUONG VAN KUONG, PETER BOSSIER, PATRICK SORGELOOS AND HANS J. NAUWYNCK

Summary

Knowledge on where and how pathogens enter the shrimp's body is crucial information in the struggle to find new ways to combat infectious diseases in shrimp farming. Nevertheless, this primary part of pathogenesis research is too often neglected. Here, we relay the story on how the nephrocomplex was identified as a major WSSV entry portal along with salinity shock and moulting as necessary conditions to establish invasion, and why this is pivotal information relevant to scientists, shrimp farmers and practitioners.

Introduction

A profound understanding of disease pathogenesis is one of the most crucial aspects in achieving adequate protection against pathogens, be it prophylactic, in management or curative. Without exception, all infectious diseases start with the very same thing, i.e. pathogen entry and subsequent access to the primary replication site. Thus, knowing where infectious particles enter the animal's body is crucial information. For example, if the entry portal is known, then local defence mechanisms can be studied, upregulated or induced (e.g. local immunity-inducing vaccines, probiotics, etc.). Furthermore, this information provides hints to potential risk factors, while management strategies can be undertaken to cut the chain of transmission (e.g. animal isolation, feed supply, waste management, hygiene, vector control, etc.). Finally, when the method of inoculation mimics the actual, natural way of infection, further pathogenesis studies can be performed with higher accuracy. Based on the general anatomy of shrimp (Young, 1959), we can consider six potential pathogen entry portals: (1) mouth (peroral), (2) anus,

(3) nephropore, (4) integumental gland, (5) genital orifices and (6) the entire surface of the body, when the exoskeleton is somehow compromised (including eyes, gills, etc.). Note that *per ovum* infections are always retraceable to previous generation(s) of shrimp, where the pathogen first succeeded in infiltrating via one of the six aforementioned primary entry portals.

Intrinsic susceptibility of shrimp and first-line defences

Much research on shrimp is based on experiments using intramuscular inoculations. Depending on the subject of investigation, bypassing of the natural entry portal by injecting shrimp can potentially lead to incomplete, incorrect or sometimes missed conclusions. However, these studies do illustrate that shrimp are most susceptible to diseases once the pathogens are inside the body (Escobedo-Bonilla et al., 2005; Khanobdee et al., 2002; Liu et al., 2004). The moment pathogens reach the hemolymph (very often immediately after the first replication cycle in the primary target tissue), pathogens spread quickly and readily throughout the body, reaching the secondary target organs. This swiftness results in a rapid onset of symptoms and, not infrequently, mortality. Even so, not every cell type or organ is susceptible: each pathogen has its own specific set of target cells. White spot syndrome virus (WSSV), for example, is able to infect cells of only mesodermal or ectodermal origin, such as the cuticular epithelium and subcuticular connective tissue of almost every organ: nephrocomplex (antennal gland), stomach, hepatopancreas, gills, cuticular epithelia, heart, lymphoid organ, nervous tissue, some hemocytes, etc. Tubular epithelium cells of the gut and hepatopancreas are not infectible by WSSV (endodermal origin; Chang et al., 1996; Escobedo-Bonilla et al., 2008; Lo et al., 1997; Wu et al., 2013). In contrast, for example, Taura syndrome virus (TSV)'s primary target cells are the cuticular epithelium of mainly foregut, gills, general body cuticle epithelium and appendages (Hasson et al., 1999), while *Hepatobacter penaei* is an intracellular bacterium targeting only hepatopancreacytes (Lightner et al., 1992).

Pathogen susceptibility is determined not only by cell type – species, age and sex are also important. Some pathogens have a narrow host range, like TSV, which is known to be limited to certain penaeid species like *Penaeus vannamei* (mostly at early juvenile stages), whereas *Penaeus stylirostris* is usually resistant to TSV infection. On the contrary, the decapod penstylhamaparvovirus 1, formerly known as IHHNV, is highly virulent to *P. stylirostris*, whereas *P. vannamei* is much less vulnerable (Lightner, 2011; Lightner et al., 1983; Pénzes et al., 2020). In turn, it is suspected that WSSV infects all decapod crustacea species, including shrimp, prawn, lobsters, crabs and crayfish; both cultivated and wild species are at risk (Stentiford et al., 2009). Major economical significant European crustaceans such as *Nephrops norvegicus*, *Homarus gammarus*, *Cancer pagurus* and *Cancer affinis* are also described as highly WSSV susceptible (Bateman et al., 2012; Pradeep et al., 2012; Stentiford et al., 2009). Furthermore, other crustaceans (both saltwater and freshwater) including *Artemia* sp., *Copepoda* (all important food sources in shrimp, crab, lobster and fish aquaculture) and the *Squilla* mantis (mantis shrimp) are also susceptible to the virus. Additionally, WSSV has even been found in some insect larvae (Desrina et al., 2013; Pradeep et al., 2012). Even in non-crustacean species, WSSV had been detected. For a broad enumeration of all sensitive species, see the reviews of Escobedo-Bonilla et al. (2008) and Pradeep et al. (2012). Nonetheless, not all WSSV-susceptible hosts are equally sensitive to white spot disease. In particular, non-penaeid species are considered less likely to succumb to the disease (Stentiford et al., 2009). *Macrobrachium rosenbergii* was reported to be refractive to WSSV infection. Nevertheless, under standardized conditions, it was demonstrated that *M. rosenbergii* does allow WSSV to replicate and subsequently cause mortality, albeit to a lesser extent when compared to the more intensively bred Penaeids. Depending on strain and age of the shrimp or prawn, *P. vannamei* was found to be about 18.6–398 times more susceptible to WSSV infection than the freshwater species *M. rosenbergii* (Corteel et al., 2012; Escobedo-Bonilla et al., 2005).

With so many cell types and species highly susceptible to WSSV and other pathogens, it can almost be termed a miracle that shrimp manage to survive. The answer lies in the shrimp's ability to defend itself against pathogens, particularly in its primary defence system located at the potential pathogen entry points, with the main function of preventing pathogen entry. The first, most evidently apparent, defence system is the shrimp's exoskeleton. This cuticular layer is not only protective against physical threats but is also very effective in fending off other threats such as pathogens and toxins (Corteel et al., 2009; Xiong et al., 2017). The cuticula is made of chitin fibres with protein deposits, reinforced by minerals such as calcium and magnesium. This combination makes the structure strong and resilient, but also very dense. Furthermore, significant parts of the exoskeleton (epicuticle, exocuticle) are formed before the old cuticle is shed, ensuring moult cycle-round protection (see chapter on the integument for a more comprehensive and detailed description of the exoskeleton). This cuticula not only envelops the entire shrimp's external body (including eyes and gills), but the foregut and hindgut are also completely covered by a non-permeable cuticle layer (Thuong et al., 2016b; Young, 1959). In contrast, no cuticle is present to protect the midgut. Instead, a non-cellular, slimy matrix of chitin and proteins, called the peritrophic membrane, ensures local microbial security (Hegedus et al., 2009). It was experimentally demonstrated that no particles of diameter >10 nm were able to pass the peritrophic membrane (Martin et al., 2006). The genital orifices too possess cuticular protection. In open-thelycum species, the female thelycum is a modification of the chitinous sternal plates of the last two pereopoda. Reception of spermatophores occurs only during inter- or pre-moult, when the cuticula is at the height of its protective potential. In so-called closed species, the thelycum is covered by these sternal plates. The male petasma, responsible for depositing the spermatophores, are also a modification of the abdominal appendages, again, covered by a cuticle (Bailey-Brock and Moss, 1992). It is worth noting that differences in the structure and composition of the exoskeleton or cuticle exist, depending on its localization and, thus, does not provide equivalent resistance to noxes. To illustrate, the entire surface of the gills is covered in cuticle, although uncalcified, and at the height of the lamellae (where oxygen exchange takes place) the layer is only 1 μm thick (Bell and Lightner, 1988; Dall et al., 1990).

The only two structures connected to the outside world, and of which the lumen is not covered by an exoskeleton or chitinous layer, are the tegumental glands (Juberthie-Jupeau and Crouau, 1977) and the nephrocomplex (antennal gland; De Gryse et al., 2020). The tegumental glands are clusters of cells located underneath the epidermal cell layer. The secretion of these glands is collected by a central duct, leading towards pores in the shrimp's outer surface (Felgenhaur, 1991). To what extent the integumental glands play a role in shrimp infectious diseases and how, or if, the gland is protected, remain unclear and merit future research. Access to the nephrocomplex (antennal gland) is controlled by the nephropore, situated at the base of the antennal coxipodite. Scanning electron microscopical images of the nephropore reveal a valve-like configuration of the orifice with one valve superimposed over the other, allowing only for efflux of fluid while hindering influx (Fig 19.1). Biodynamic experiments demonstrate the efficiency and robustness of this defence system (De Gryse et al., 2020).

Lastly, an often forgotten first line of defence is the normal microflora living on or inside the entry portals. A healthy, balanced microbiotic community can be very efficient in fending off unwanted pathogens (Holt et al., 2020).

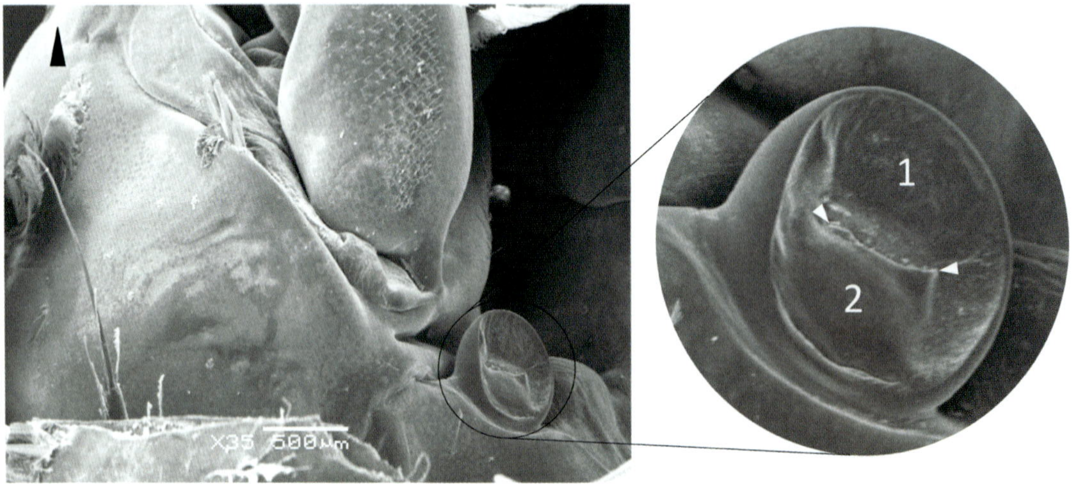

FIGURE 19.1 Scanning electron microscopy and detail of the nephropore at the antennal coxipodite. The rostral valve (1) is superimposed by the caudal valve (2), providing non-return valve-like protection of the nephropore. Biodynamic analysis showed that, under normal conditions, only efflux of fluids is possible through the fissure (indicated by the arrowheads). Credit: De Gryse et al., 2020.

In conclusion, the first lines of defence form an effective, (mostly) mechanical barrier against pathogens trying to invade shrimp via the portals of entry, and ultimately make these animals very resistant against infections.

Risk factors

All potential entry portals are protected by first-line defence mechanisms. In order to bypass these hurdles, certain conditions must be met in order for pathogens to succeed in entering the shrimp and reaching the primary replication sites. Conditions or occurrences that compromise these barriers can generally be termed risk factors. The identification of risk factors is often very useful information for applying management measures to prevent diseases at the farm level; also, it can sometimes lead to the identification of the entry portal. Thus, good pathogenesis research should never neglect possible risk factors. These conditions or occurrences can be identified through statistics-driven epidemiology research, empirical research or sometimes just observational data from shrimp farmers, but should always be confirmed with experiments in a controlled environment, to avoid accepting confounded risk factors as true risk factors.

Epicuticular damage facilitates the entry and nesting of chitinase-producing pathogens on the exocuticle and causing lesions, known as shell disease (Sindermann, 1989). Conditions in which cuticle damage occur more often, e.g. higher stocking density and aggression or inferior feed or water quality, *Vibrio* blooms, directly or indirectly leading to a weaker cuticula, can thus be considered potential risk factors for shell disease. Other possible risk factors include other organisms (pathogenic or non-pathogenic), toxins or the administration of antimicrobials, as is often the case with fungal infections such as *Fusarium* which tends to act as opportunistic secondary pathogens. Risk factors can also include conditions leading to a higher infection pressure or inhibition of cellular and molecular immunity, but these do not fall within the scope of this chapter.

Due to the broad WSSV host range, many risk factors are related to the presence of non-symptomatic virus carriers such as holes in the fences of farms and nets, not filtering pumped water or direct usage of unprocessed seawater (see Tendencia et al. (2011) for a list of articles). Close proximity to the sea (Mohan et al., 2008), use of commercial and fresh feed (Chou et al., 1998; Corsin et al., 2001) and sharing of pond water (Tendencia et al., 2011) are also risk factors, although related more to WSSV gaining entry to pond water than into shrimp. Increased stocking density is also linked to WSSV outbreaks (Tendencia et al., 2011) and can be interpreted in two ways: more shrimp/m^3 water means more infectable and infected shrimp which, in turn, leads to higher virus titres in the pond water and thus to a higher infection pressure. High stocking density also leads to increased cannibalism (eating of infected tissue), aggressiveness (cuticle damage, stress) and an increased need to establish social dominance (higher urination frequency) (Breithaupt and Atema, 2000; Katoh et al., 2008).

Corteel et al. (2009) performed an interesting series of experiments with the goal of assessing moulting stage as a possible risk factor for WSSV. No influence on shrimp susceptibility and mortality was observed when WSSV was injected in shrimp at different moulting stages, meaning that no intrinsic variations exist in sensitivity towards the virus over the different stages. This finding indicated that although immunity parameters can vary according to moulting stage (Liu et al., 2004), the shrimp immune response towards this particular virus remains constant throughout the different stages of the moulting cycle. A different story emerged when the waterborne route was investigated. In an attempt to rule out unwanted cuticle damage caused by hard tank walls as a confounding factor, shrimp were immersed in plastic bags containing WSSV. Post-moult shrimp (stages A and B) were found to be more frequently infected by WSSV than inter- and pre-moult animals (stages C, D1 and D2) during the immersion challenge in polyethylene bags. When one pleiopod was severed, even more shrimp in post-moult stages were infected, whereas shrimp in stages C, D1 and D2 remained uninfected. These series of experiments indicate that, during moulting, first-line defences of entry portals are compromised rather than the shrimp innate immunity. Cuticle damage proved to be of partial effect, but only if the moulting stage had already exerted its effects. A separate study investigating the role of moulting in WSSV infection found 0% mortality when WSSV immersion challenge occurred during pre-moult stages – 53.3% during ecdysis and 26.7% during post-moult (Thuong et al., 2016a).

Another potential risk factor that warranted deeper investigation was the effect of salinity and, more precisely, the transition from one salinity to another. Observations have been made by shrimp farmers that WSSV outbreaks could often be linked to heavy rainfall and after pond water was changed. Large volumes of fresh water could cause a sudden drop in salinity in open-air ponds. Water mismanagement could hypothetically result in similar consequences. Experiments mimicking these acute salinity changes indeed found significant effects on WSSV infection rates. Shrimp acclimatized at 35 g/l were transferred to lower salinities while exposed to WSSV. When salinity was reduced to 10 g/l and lower, the animals became infected with WSSV (Thuong et al., 2016a). Reduction in immunity parameters has also been observed during salinity shock, as well as reactivation in latent WSSV-infected shrimp (Liu et al., 2006). The following questions can be posed: does a salinity shock make shrimp more susceptible to WSSV infection because the immune system is put under stress, or because the first lines of defence are compromised or possibly a combination of both? For instance, it has been demonstrated that shrimp urinate more frequently to regulate their body volume following exposure to lower salinity (Lin et al.,

2000). Every time shrimp urinate, the nephropore valves are opened and could potentially provide a portal for pathogen entry (De Gryse et al., 2020).

Entry portals of WSSV

In the quest to determine how shrimp were becoming infected with the then newly discovered WSSV, several experimental studies found that feeding of WSSV-containing food to shrimp effectively resulted in transmission of the virus (Chang et al., 1996; Rajendran et al., 1999; Wang et al., 1998). These and subsequent studies led to the conclusion that shrimp were perorally infected with WSSV and, thus, the entry portal was somewhere along the digestive tract. Overlooking the fact that the tubular epithelial cells of the digestive system are refractive for WSSV infection, the presence of a chitinous protective barrier layer and an intact microbiome (see section "Intrinsic susceptibility of shrimp"), the gut as the primary entry portal became the leading consensus. Another way in which scientists managed to experimentally infect shrimp (beside intramuscularly) was by water immersion (Chou et al., 1995; Corteel et al., 2009; Kanchanaphum et al., 1998; Witteveldt et al., 2004, 2006). It is important to note that these studies were not performed under the most strictly controlled conditions – even viral titrations of inocula were often not performed. The efficiency of these types of infection was low and sometimes difficult to reproduce (Corteel et al., 2009; Laramore, 2007; Perez et al., 2005). Transovarial infection has also been described (Lo et al., 1997), but this kind of transmission must have originated from ancestor shrimp that became infected through one of the six potential primary entry portals. Based on these observations, together with identification of the risk factors (e.g. salinity shock, moulting and high population density, see section "Risk factors"), several (titration) experiments were conducted in an attempt to identify the WSSV entry portal.

First, in order to rule out cuticle damage, damage was made to the shrimp's general exoskeleton with surgical drills, argon lasers and sandpaper before immersing the shrimp in WSSV-containing seawater, but to no avail – shrimp did not become infected (unpublished data). Outer cuticular damage as facilitator to WSSV invasion is thus ruled out. However, internal damage to chitinous structures such as to the peritrophic membrane was not.

Previous experiments have irrefutably shown that moulting stage influences WSSV entry. During the moulting stage, endogenous chitinases are released that can potentially alter the protective capabilities of the cuticula or peritrophic membrane sufficiently to allow for WSSV entry (Proespraiwong et al., 2010; Tan et al., 2000). This would explain why feeding of WSSV-contaminated feed to shrimp results, although with low efficiency, in WSSV-positive shrimp (most peroral inoculation studies did not account for moulting stage). In insects it was shown that inhibition of this peritrophic membrane led to a higher susceptibility to several pathogens (Rao et al., 2004; Wang and Granados, 2000); and, in shrimp, *Vibrio parahaemolyticus* and *Vibrio harveyi* were shown to use bacterial chitinases to destroy and penetrate the peritrophic membrane in order to colonize the midgut, causing diseases such as acute hepatopancreatic necrosis (Martin et al., 2004; Tran et al., 2013). Moreover, the midgut was previously suggested as a potential entry portal for WSSV by several authors (Arts et al., 2007; Di Leonardo et al., 2005). Thuong et al. (2016b) investigated whether a compromised peritrophic membrane could indeed facilitate WSSV entry at the top of the midgut. They used saltwater to gently flush out the peritrophic membrane before performing titration experiments (Figure 19.2). However, the authors did not retrieve data supporting

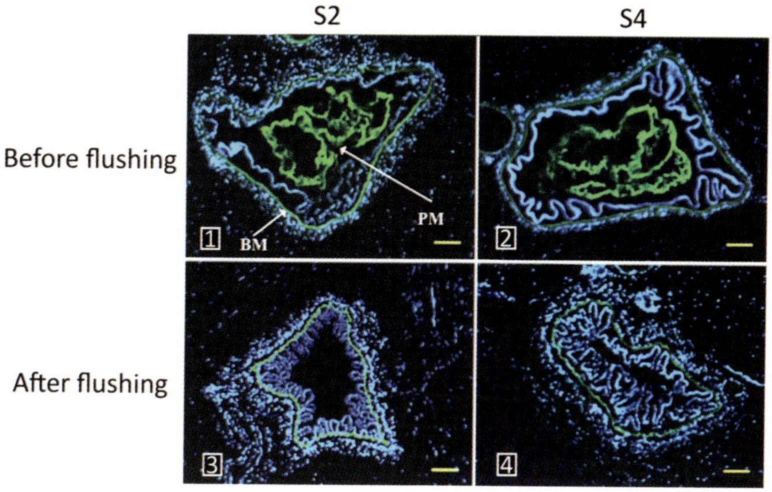

FIGURE 19.2 Flushing of the peritrophic membrane (PM) did not facilitate entry of WSSV. BM, basal membrane; S2 and S4, cross-sections of the midgut at the height of abdominal segments 2 and 4, respectively. Scale 100µm. Credit: adapted from Thuong et al. (2016b).

their claim that removal of this structure significantly facilitates WSSV entry and that the midgut is an entry portal for WSSV. Of course, the tubular epithelia of the midgut are not receptive to WSSV and, together with the basal membrane, provide an effective barrier against WSSV colonization of shrimp even when the peritrophic membrane is compromised. The method of Thuong et al. (2016b) could, however, be useful to test the entry of several other shrimp pathogens via the midgut.

Besides moulting, abrupt changes in saline water concentration have been demonstrated to be a very potent risk factor for WSSV infection. During salinity shock, the opening and closing of the nephropore valve system can potentially provide brief windows of opportunity for pathogen entry (see section "Risk factors"). To test the potential of the nephrocomplex (antennal gland) as an entry portal, titrations with WSSV and *Vibrio campbelli* were performed comparing intramuscular injection to peroral and intra-antennal (intra-bladder, through the nephropore) inoculation (Figure 19.3). Not surprisingly, shrimp were most sensitive to WSSV when injected intramuscularly, but only 56 and 62 times more virus and *V. campbelli*, respectively, was needed when inoculated through the nephropore (intra-bladder). In contrast, when infection of shrimp was attempted perorally, 28.8 x 10^6 and $>10^9$ times more WSSV and *V. campbelli*, respectively, was needed in comparison to intra-bladder inoculation, meaning that far less virus or bacteria is needed to infect shrimp when the pathogen enters through the nephropore than when it does so via the mouth, but only little more than when injected directly into the shrimp body. Infections were confirmed by indirect immunofluorescence (IIF) (De Gryse et al., 2020).

Due to the lack of inner cuticle lining, the role of the nephrocomplex (antennal gland) during a drop in salinity and in the above-mentioned titrations, the nephropore is now considered a prime suspect as a major entry portal for viruses and bacteria. To confirm natural entry of WSSV via the nephropore during the occurrence of acute salinity drop, a WSSV immersion challenge experiment was set up. Shrimp were transferred from 35 g/l saline water to 5 g/l saltwater containing WSSV for 5 hours, after which they were transferred back to 35 g/l seawater. Every 12 hours, urine and hemolymph samples from the same

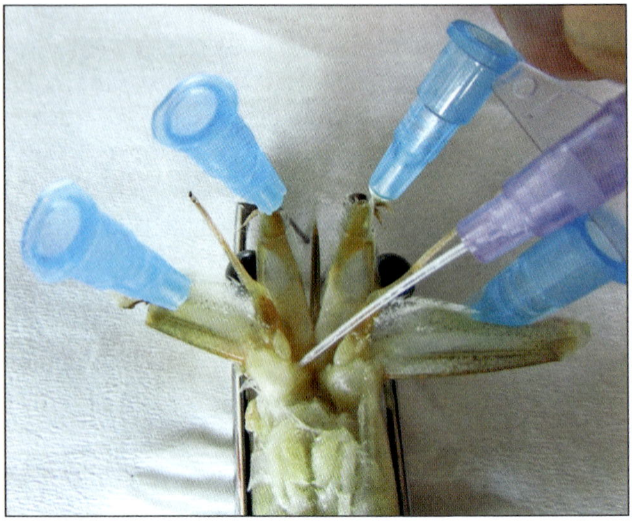

FIGURE 19.3 Demonstration of intra-bladder inoculation (26-gauge, purple catheter). Results showed that 28.8 × 10^6 less WSSV and 10^9 less *V. campbelli* was needed to infect shrimp via intra-bladder inoculation compared to peroral.

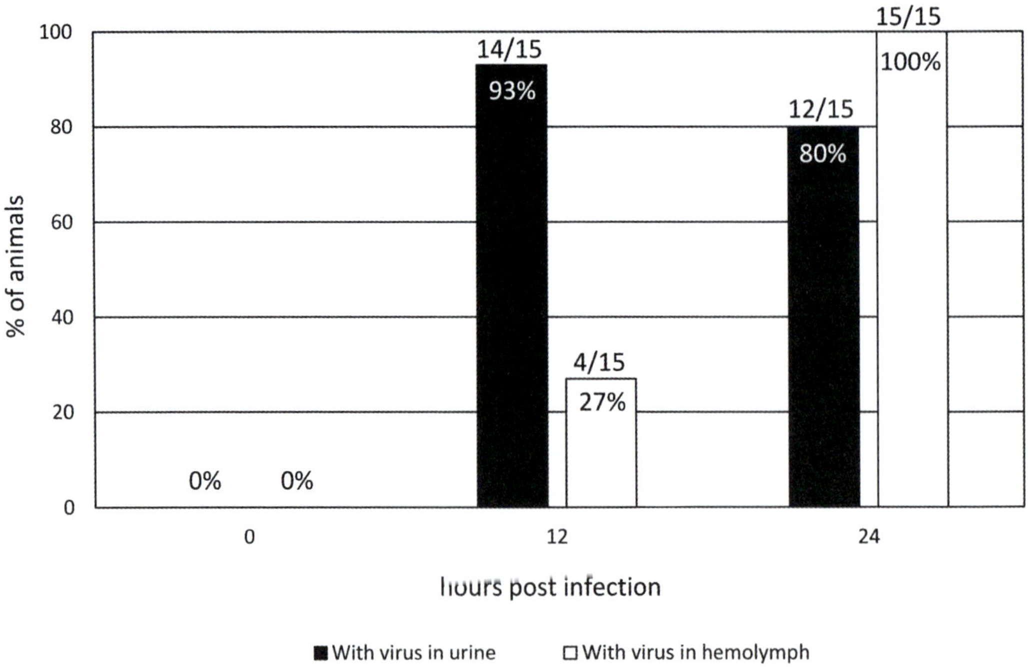

FIGURE 19.4 WSSV immersion challenge to identify the nephropore as the primary entry portal during salinity shock. The presence of WSSV in urine (black) versus hemolymph (white) at different time points post infection. Credit: De Gryse et al. (2020).

animals were collected and checked with qPCR for the presence of WSSV. At 12 hours post infection, WSSV DNA copies were retrieved from shrimp urine while the hemolymph remained mostly negative. At 24 hours post infection, all hemolymph samples tested positive (Figure 19.4). This provides strong evidence that WSSV enters the nephropore during salinity shock, effectively infecting the nephrocomplex (antennal gland) before spreading to the rest of the shrimp's susceptible organ body via the hemolymph. Moreover, viral replication curves were demonstrated in urine and (later) hemolymph, indicating primary replication. An *ex vivo* experiment was set up to investigate potential particle entry during the process of urination. The excised nephropore was mounted on a luer-tip syringe and *mictio* was simulated by pressing the plunger, while the nephropore was inside a container with fluorescent beads (1.0 and 0.2 μm^3). Beads were then retrieved from inside the syringe. This demonstrated that, at the end of the *mictio* process, a window exists in which particles of a size comparable to all viruses and some bacteria succeed in entering the unprotected lumen of the nephrocomplex (antennal gland). This experiment provides the final definitive proof that the nephropore is indeed a key entry portal for WSSV, *V. campbelli* and, quite possibly, other pathogens as well (De Gryse et al., 2020).

Using micro magnetic resonance imaging, a link between moulting and the volume within the nephrocomplex was established. In short, after ecdysis, the nephrocomplex stores a large volume of urine which is gradually replaced by new tissue. Thus, shrimp in post-moult have to frequently urinate small quantities of fluid in order to make room for new muscle and organ mass. Previously, this chapter has established that the mictio process poses opportunities for pathogen entry. These data could provide an explanation for the increased susceptibility of shrimp to WSSV during post-moult (Corteel et al., 2009; De Gryse et al., 2020).

Consequences for pathogenesis, farming and research

It has been demonstrated that shrimp would be highly vulnerable to various pathogens if the various entry portals were not efficiently shielded by first-line defences. These defences are generally provided by some form of chitinous structure such as the exoskeleton, peritrophic membrane or nephropore valves. These barriers not only make shrimp hard to infect, they also impede the search for pathogen entry portals as well. In order for these pathogens to reach the entry points, certain occurrences or conditions capable of compromising these defences, called risk factors, must be present. Concerning WSSV, shrimp that are in post-moulting stages endure harsh salinity shocks or are stocked in high densities are most at risk of contracting this devastating disease. It was demonstrated that these risk factors led to higher urination frequency and thus to greater opportunities for WSSV entry via the nephropore into the nephrocomplex, where primary replication occurs.

Previously, most researchers believed that shrimp were infected with WSSV perorally. This was based on early reports of feeding WSSV-contaminated feed to shrimp, leading to WSSV-infected animals. However, to the best of the authors' knowledge, no accounts exist reporting on WSSV entry via the digestive tract. Furthermore, removal of the peritrophic membrane did not augment the susceptibility of shrimp to WSSV: high viral titres are required to attain WSSV infection through feeding. The contrast here to WSSV entry after immersion via the nephropore could not be higher: intra-bladder inoculation is highly efficient, with proof via qPCR and IIF that primary WSSV replication in the bladder epithelium exists, and particle entry during the urination process has been demonstrated. Peroral inoculation

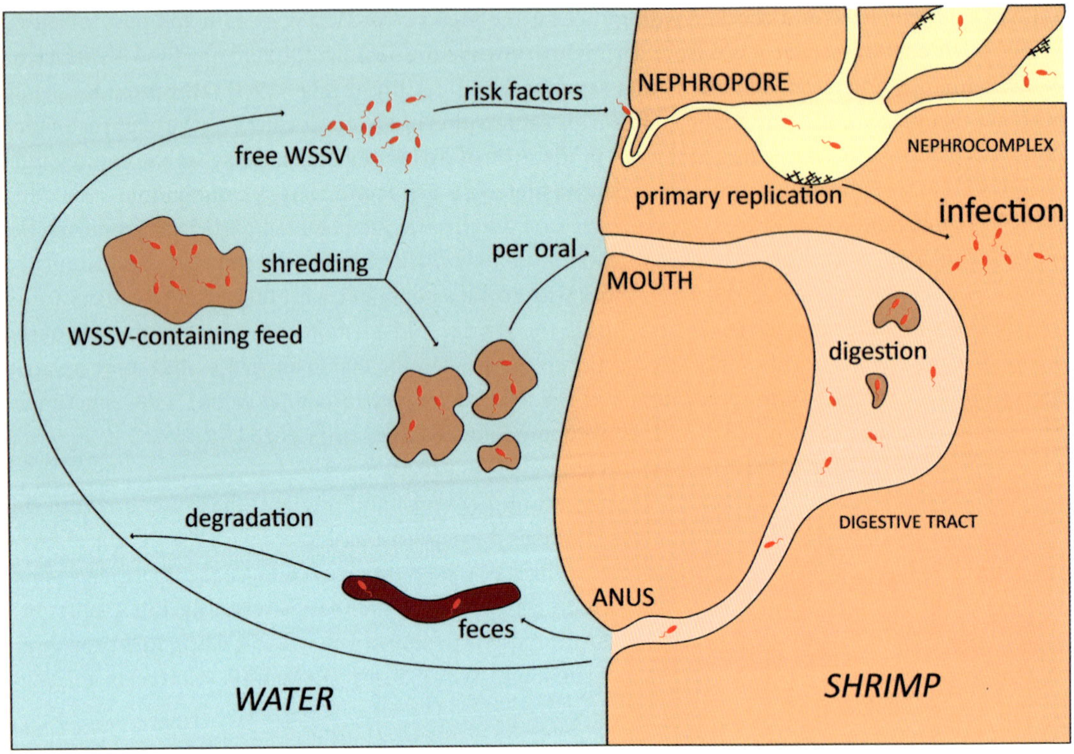

FIGURE 19.5 Hypothetical model of WSSV infection by immersion or feeding.

or a feeding challenge could quite effectively lead to a *de facto* immersion challenge. Shrimp eat very violently, shredding big chunks of feed with their claws and disseminating viral particles throughout the surrounding water. It is worth noting that the nephropores are located very close to the mouth. Besides, when WSSV-containing feed enters the mouth, it is possible that it undergoes a serial passaging of digestive steps, during which WSSV is liberated from it. Consequently, individual WSSV virions could exit the shrimp via the anus or from the degrading bolus, ready to infect shrimp via the nephropore. This does not, however, completely exclude the possibility of WSSV entry via the digestive tract given that certain conditions are met. More research is needed. Figure 19.5 shows a hypothetical model of WSSV infection via immersion versus feeding.

With regard to management, these findings on WSSV pathogenesis provide ample opportunities for direct action at the managemental level concerning biosecurity. First, monitoring of salinity levels and maintaining those levels should be, if not already, a major priority for shrimp farmers. Keeping the salinity up to standard and avoiding sudden fluctuations can be achieved by frequent automated measurements of pond salinity and by avoiding the addition of large volumes of fresh water to the pond (direct rainfall, indirect rainfall, cleaning water, refreshing the pond water, etc.). Next, avoiding aggression and the need to establish social dominance could be helpful. Besides the other effects of stress, feeding at multiple places in the pond as opposed to one central point will reduce fighting and dominant behaviour during the feeding process and thus provide fewer opportunities for WSSV entry.

However, although hardly practical in regard to intensive shrimp farming, one must consider the benefits of adding some type of sand-like soil or other obstacles in which shrimp can hide and escape aggression or other social events, especially during and after the moulting process. Nevertheless, such sand-like soil must be bacteriologically justified and thus it should be possible to clean or regularly replace this soil. Finally, considerations should be made to artificial selection in breeding shrimp to create new generations with optimal growth per moulting cycle, so fewer moulting cycles are required to obtain a similar harvesting weight.

Conclusion

This chapter shows that, with the correct approach and sufficient effort, detection of the entry portal of shrimp pathogens is feasible and is favoured over merely establishing infection protocols. The information given provides insight into the pathogenesis and, consequently, potential managemental measures or other strategies to tackle all types of shrimp diseases.

References

Arts, J. A., Taverne-Thiele, A. J., Savelkoul, H. F. and Rombout, J. H. (2007) Haemocyte reactions in WSSV immersion infected *Penaeus monodon*. *Fish and Shellfish Immunology* 23(1), 164–170.

Bailey-Brock, J. H. and Moss, S. M. (1992) Penaeid taxonomy, biology and zoogeogaphy. In: *Marine Shrimp Culture: Principles and Practises* (eds Fast, A.W. and Lester, L. J.). Elsevier Science, The Netherlands, pp. 9–27.

Bateman, K. S., Tew, I., French, C., Hicks, R. J., Martin, P., Munro, J. and Stentiford, G. D. (2012) Susceptibility to infection and pathogenicity of White Spot Disease (WSD) in non-model crustacean host taxa from temperate regions. *Journal of Invertebrate Pathology* 110(3), 340–351.

Bell, T. A. and Lightner, D. A. (1988) *A Handbook of Normal Penaeid Shrimp Histology* https://agris.fao.org/agris-search/search.do?recordID=US8936124

Breithaupt, T. and Atema, J. (2000) The timing of chemical signaling with urine in dominance fights of male lobsters (*Homarus americanus*). *Behavioral Ecology and Sociobiology* 49, 67–78.

Chang, P. S., Lo, C. F., Wang, Y. C. and Kou, G. H. (1996) Identification of white spot syndrome associated baculovirus (WSBV) target organs in the shrimp *Penaeus monodon* by in situ hybridization. *Diseases of Aquatic Organisms* 27(2), 131–139.

Chou, H. Y., Huang, C. Y., Lo, C. F. and Kou, G. H. (1998) Studies on transmission of white spot syndrome associated baculovirus (WSBV) in *Penaeus monodon* and *P. japonicus* via waterborne contact and oral ingestion. *Aquaculture* 164(1–4), 263–276.

Chou, H. Y., Huang, C. Y., Wang, C. H., Chiang, H. C. and Lo, C. F. (1995) Pathogenicity of a baculovirus infection causing white spot syndrome in cultured penaeid shrimp in Taiwan. *Diseases of Aquatic Organisms* 23, 165–173.

Corsin, F., Turnbull, J. F., Hao, N. V., Mohan, C. V., Phi, T. T., Phuoc, L. H., Tinh, N. T. N. and Morgan, K. L. (2001) Risk factors associated with white spot syndrome virus infection in a Vietnamese rice-shrimp farming system. *Diseases of Aquatic Organisms* 47(1), 1–12.

Corteel, M., Dantas-Lima, J. J., Wille, M., Alday-Sanz, V., Pensaert, M. B., Sorgeloos, P. and Nauwynck, H. J. (2009). Molt stage and cuticle damage influence white spot syndrome virus immersion infection in penaeid shrimp. *Veterinary Microbiology*, 137(3–4), 209–216.

Corteel, M., Dantas-Lima, J. J., Tuan, V. V., Thuong, K. V., Wille, M., Alday-Sanz, V., Pensaert, M. B., Sorgeloos, P. and Nauwynck, H. J. (2012), Susceptibility of juvenile *Macrobrachium rosenbergii* to different doses of high and low virulence strains of white spot syndrome virus (WSSV). *Diseases of Aquatic Organisms* 100(3), 211–218.

Dall, W., Hill, B. J., Rothlisberg, P. C. and Staples, D. J. (1990) The biology of the penaeidae. *Advances in Marine Biology* 27, 1–461.

De Gryse, G. M. A., Khuong, T. V., Descamps, B., Van Den Broeck, W., Vanhove, C, Cornillie, P., Sorgeloos, P., Bossier, P. and Nauwynck, H. J. (2020) The shrimp nephrocomplex serves as a major portal of pathogen entry and is involved in the molting process. *Proceedings of the National Academy of Sciences USA* 117(45), 28374–28383.

Desrina, H., Verreth, J. A. J., Prayitno, S. B., Rombout, J. H. W. M., Vlak, J. M. and Verdegem, M. C. J. (2013) Replication of white spot syndrome virus (WSSV) in the polychaete *Dendronereis* spp. *Journal of Invertebrate Pathology* 114(1), 7–10.

Di Leonardo, V. A., Bonnichon, V., Roch, P., Parrinello, N. and Bonami, J. R. (2005) Comparative WSSV infection routes in the shrimp genera *Marsupenaeus* and *Palaemon*. *Journal of Fish Diseases* 28(9), 565–569.

Escobedo-Bonilla, C. M., Wille, M., Sanz, V. A., Sorgeloos, P., Pensaert, M. B. and Nauwynck, H. J. (2005) *In vivo* titration of white spot syndrome virus (WSSV) in specific pathogen-free *Litopenaeus vannamei* by intramuscular and oral routes. *Diseases of Aquatic Organisms* 66(2), 163–170.

Escobedo-Bonilla, C. M., Alday-Sanz, V., Wille, M., Sorgeloos, P., Pensaert, M. B. and Nauwynck, H. J. (2008) A review on the morphology, molecular characterization, morphogenesis and pathogenesis of white spot syndrome virus. *Journal of Fish Diseases* 31(1), 1–18.

Felgenhaur, B. E. (1991) External anatomy and integumentary structures. In: *Microscopic Anatomy of Invertebrates, Decapod Crustacea* (eds Harrison, F. W. and Humes, A. G.). Wiley-Liss, New York, USA, pp. 7–43.

Hasson, K. W., Lightner, D. V., Mohney, L. L., Redman, R. M., Poulos, B. T. and White, B. M. (1999) Taura syndrome virus (TSV) lesion development and the disease cycle in the Pacific white shrimp *Penaeus vannamei*, *Diseases of Aquatic Organisms* 36(2), 81–93.

Hegedus, D., Erlandson, M., Gillott, C. and Toprak, U. (2009) New insights into peritrophic matrix synthesis, architecture, and function. *Annual Review of Entomology* 54, 285–302.

Holt, C. C., Bass, D., Stentiford, G. D. and van der Giezen, M. (2020) Understanding the role of the shrimp gut microbiome in health and disease. *Journal of Invertebrate Pathology* https://www.sciencedirect.com/science/article/pii/S0022201120300938

Juberthie-Jupeau, L. and Crouau, Y. (1977) The tegumental glands of a troglobitic crustacean. *International Journal of Speteology* 9, 309–319.

Kanchanaphum, P., Wongteerasupaya, C., Sitidilokratana, N., Boonsaeng, V., Panyim, S., Tassanakajon, A., Withyachumnarnkul, B. and Flegel, T. W. (1998) Experimental transmission of white spot syndrome virus (WSSV) from crabs to shrimp *Penaeus monodon*. *Diseases of Aquatic Organisms* 34(1), 1–7.

Katoh, E., Johnson, M. and Breithaupt, T. (2008) Fighting behaviour and the role of urinary signals in dominance assessment of Norway lobsters, *Nephrops norvegicus*. *Behaviour* 145, 1447–1464.

Khanobdee, K., Soowannayan, C., Flegel, T. W., Ubol, S. and Withyachumnarnkul, B. (2002) Evidence for apoptosis correlated with mortality in the giant black tiger shrimp *Penaeus monodon* infected with yellow head virus. *Diseases of Aquatic Organisms* 48(2), 79–90.

Laramore, S. E. (2007) Susceptibility of the peppermint shrimp *Lysmata wurdemanni* to the white spot syndrome virus. *Journal of Shellfish Research* 26(2), 623–627.

Lightner, D. V. (2011) Virus diseases of farmed shrimp in the Western Hemisphere (the Americas): A review. *Journal of Invertebrate Pathology* 106, 110–130.

Lightner, D. V., Redman, R. M. and Bell, T. A. (1983) Infectious hypodermal and haematopoietic necrosis, a newly recognized virus disease of penaeid shrimp. *Journal of Invertebrate Pathology* 42, 62–70.

Lightner, D. V., Redman, R. M. and Bonami, J. (1992) Morphological evidence for a single bacterial etiology in Texas necrotizing hepatopancreatitis in *Penaeus vannamei* (Crustacea: Decapoda). *Diseases of Aquatic Organisms* 13, 235–239.

Lin, S. C., Liou, C. H. and Cheng, J. H. (2000) The role of the antennal glands in ion and body volume regulation of cannulated *Penaeus monodon* reared in various salinity conditions. *Comparative Biochemistry and Physiology Part A: Molecular & Integrative Physiology* 127(2), 121–129.

Liu, B., Yu, Z. M., Song, X. X., Guan, Y. Q., Jian, X. F. and He, J. F. (2006) The effect of acute salinity change on white spot syndrome (WSS) outbreaks in *Fenneropenaeus chinensis*. *Aquaculture* 253(1–4), 163–170.

Liu, C. H., Yeh, S. T., Cheng, S. Y. and Chen, J. C. (2004) The immune response of the white shrimp *Litopenaeus vannamei* and its susceptibility to *Vibrio* infection in relation with the moult cycle. *Fish and Shellfish Immunology* 16(2), 151–161.

Lo, C. F., Ho, C. H., Chen, C. H., Liu, K. F., Chiu, Y. L., Yeh, P. Y., Peng, S. E., Hsu, H. C., Liu, H. C., Chang, C. F., Su, M. S., Wang, C. H. and Kou, G. H. (1997) Detection and tissue tropism of white spot syndrome baculovirus (WSBV) in captured brooders of Penaeus monodon with a special emphasis on reproductive organs. Diseases of Aquatic Organisms 30(1), 53–72.

Martin, G. G., Rebecca, S., Aivan, N. and Amaiak, C. (2006) Peritrophic membrane of the penaeid shrimp *Sicyonia ingentis*: Structure, formation, and permeability. *Biological Bulletin* 211(3), 275–285.

Martin, G. G., Rubin, N. and Swanson, E. (2004), *Vibrio parahaemolyticus and V. harveyi* cause detachment of the epithelium from the midgut trunk of the penaeid shrimp *Sicyonia ingentis. Diseases of Aquatic Organisms* 60(1), 21–29.

Mohan, C. V., Phillips, M., Bhat, B., Umesh, N. and Padiyar, P. A. (2008) Farm-level plans and husbandry measures for aquatic animal disease emergencies. *Revue Scientifique et Technique* 27(1), 161–173.

Pénzes, J. J., Söderlund-Venermo, M., Canuti, M., Eis-Hübinger, A. M., Hughes, J., Cotmore, S. F. and Harrach, B. (2020) Reorganizing the family Parvoviridae: a revised taxonomy independent of the canonical approach based on host association. *Archives of Virology* 165(9), 2133–2146.

Perez, F., Volckaert, F. A. M. and Calderon, J. (2005) Pathogenicity of white spot syndrome virus on postlarvae and juveniles of *Penaeus* (*Litopenaeus*) *vannamei*. *Aquaculture* 250(3–4), 586–591.

Pradeep, B., Rai, P., Mohan, S. A., Shekhar, M. S. and Karunasagar, I. (2012) Biology, host range, pathogenesis and diagnosis of white spot syndrome virus. *Indian Journal of Virology* 23(2), 161–174.

Proespraiwong, P., Tassanakajon, A. and Rimphanitchayakit, V. (2010) Chitinases from the black tiger shrimp *Penaeus monodon*: Phylogenetics, expression and activities. *Comparative Biochemistry and Physiology B: Biochemistry & Molecular Biology* 156(2), 86–96.

Rajendran, K. V., Vijayan, K. K., Santiago, T. C. and Krol, R. M. (1999) Experimental host range and histopathology of white spot syndrome virus (WSSV) infection in shrimp, prawns, crabs and lobsters from India. *Journal of Fish Diseases* 22(3), 183–191.

Rao, R., Fiandra, L., Giordana, B., de Eguileor, M., Congiu, T., Burlini, N., Arciello, S., Corrado, G. and Pennacchio, F. (2004) AcMNPV ChiA protein disrupts the peritrophic membrane and alters midgut physiology of *Bombyx mori* larvae. *Insect Biochemistry and Molecular Biology* 34(11), 1205–1213.

Sindermann, C. J. (1989) The shell disease syndrome in marine crustaceans – a conceptual approach. *Journal of Shellfish Research* 10, 491–494.

Stentiford, G. D., Bonami, J. R. and Alday-Sanz, V. (2009) A critical review of susceptibility of crustaceans to Taura syndrome, Yellowhead disease and White Spot Disease and implications of inclusion of these diseases in European legislation *Aquaculture* 291(1–2), 1–17.

Tan, S. H., Degnan, B. M. and Lehnert, S. A. (2000) The *Penaeus monodon* chitinase 1 gene is differentially expressed in the hepatopancreas during the molt cycle. *Marine Biotechnology* 2(2), 126–135.

Tendencia, E. A., Bosma, R. H. and Verreth, J. A. J. (2011) White spot syndrome virus (WSSV) risk factors associated with shrimp farming practices in polyculture and monoculture farms in the Philippines. *Aquaculture* 311(1–4), 87–93.

Thuong, K. V., Tuan, V. V., Li, W., Sorgeloos, P., Bossier, P. and Nauwynck, H. J. (2016a) Effects of acute change in salinity and molting on the infection of white leg shrimp (*Penaeus vannamei*) with white spot syndrome virus upon immersion challenge. *Journal of Fish Diseases* 39(12), 1403–1412.

Thuong, K. V., Tuan, V. V., Li, W. F., Sorgeloos, P., Bossier, P. and Nauwynck, H. J. (2016b) Per os infectivity of white spot syndrome virus (WSSV) in white-legged shrimp (*Litopenaeus vannamei*) and role of peritrophic membrane. *Veterinary Research* 47, 12.

Tran, L., Nunan, L., Redman, R. M., Mohney, L. L., Pantoja, C. R., Fitzsimmons, K. and Lightner, D. V. (2013) Determination of the infectious nature of the agent of acute hepatopancreatic necrosis syndrome affecting penaeid shrimp. *Diseases of Aquatic Organisms* 105(1), 45–55.

Wang, P. and Granados, R. R. (2000) Calcofluor disrupts the midgut defense system in insects. *Insect Biochemistry and Molecular Biology* 30(2), 135–143.

Wang, Y. C., Lo, C. F., Chang, P. S. and Kou, G. H. (1998) Experimental infection of white spot baculovirus in some cultured and wild decapods in Taiwan. *Aquaculture* 164(1–4), 221–231.

Witteveldt, J., Cifuentes, C. C., Vlak, J. M. and van Hulten, M. C. (2004) Protection of *Penaeus monodon* against white spot syndrome virus by oral vaccination. *Journal of Virology* 78(4), 2057–2061.

Witteveldt, J., Vlak, J. M. and van Hulten, M. C. (2006) Increased tolerance of *Litopenaeus vannamei* to white spot syndrome virus (WSSV) infection after oral application of the viral envelope protein VP28. *Diseases of Aquatic Organisms* 70 (1–2), 167–170.

Wu, W., Wu, B., Ye, T., Huang, H., Dai, C., Yuan, J. and Wang, W. (2013) TCTP is a critical factor in shrimp immune response to virus infection. *PLoS One* 8(9), e74460.

Xiong, H., Pears, C. and Woollard, A. (2017) An enhanced *C. elegans* based platform for toxicity assessment. *Scientific Reports* 7(1), 9839.

Young, J. (1959) Morphology of the white shrimp *Penaeus setiferus* (Linnaeus 175). In: *Fishery Bulletin*. The Fish and Wildlife Service, USA, pp. 145–173.

Chapter 20

WHOLE-GENOME SEQUENCING FOR IDENTIFICATION OF BACTERIAL ISOLATES AFFECTING LOCAL PENAEID SHRIMP FARMS: A NATIONAL SURVEY FOR THE PHILIPPINES

CYNTHIA P. SALOMA, SARAH MAE U. PENIR, LEOBERT D. DE LA PEÑA AND EDGAR C. AMAR

Summary

The shrimp aquaculture industry is a major economic player in the food and livelihood sectors, both globally and in the Philippines. The development of industrialized shrimp farming methods has led to a rapid increase in cultured shrimp production since the 1980s, but the industry has also faced challenges from the introduction of opportunistic pathogens. These pose severe limitations in the success of shrimp culture in both hatcheries and grow-out ponds. We extracted genomic DNA from 423 bacterial isolates from shrimp disease outbreaks in local aquaculture farms throughout the Philippines and subjected them to Illumina paired-end whole-genome sequencing at 50–100-fold coverage. Reads were assembled using a *de novo* approach, and average nucleotide identity measurements were used to establish the identity of the isolates. Outbreak types encountered during the time of sampling include acute hepatopancreatic necrosis disease (AHPND), luminescent vibriosis, white spot disease, and outbreaks of unknown etiology. In both outbreak and non-outbreak sites, members of the genus *Vibrio*, particularly *Vibrio parahaemolyticus*, were observed to be the most abundant. The whole-genome sequence data obtained will aid in the design and implementation of novel shrimp bacterial disease outbreak management and control programs.

Introduction

The shrimp industry is one of the most economically important areas of aquaculture as it is a major player in the food production and employment sectors globally. The shift from wild-captured shrimp production to aquaculture production has led to rapid increases in overall global shrimp production volume, from

72,000 metric tons (MT) in the 1980s to 2.5 million MT in the 2000s (Diana, 2009). In 2017, shrimp or prawn ranked third among the major export commodities in the Philippines, with a total export volume of 8,969 MT and a total contribution of US$51 million to the total export value. Shrimp was also the fourth major species produced by the aquaculture fisheries industry in 2017. Most shrimp are exported in fresh, chilled, or frozen form, accounting for 92% of total shrimp or prawn export revenue for the country. The major destinations of shrimp/prawn products from the Philippines include Japan, the USA, and France (Bureau of Fisheries and Aquatic Resources, 2017).

As a result of exponentially increasing market demands, overintensification of large-scale commercial shrimp farming methods has led to altered ecological balance in breeding environments and encouraged the growth of opportunistic bacteria (Boyd and Clay, 1998; Lavilla-Pitogo and de la Pena, 1998). Disease epizootics due to bacterial pathogens accounted to roughly 22% of worldwide economic losses from shrimp disease in 2001 (Flegel et al., 2008). Varying levels of *Vibrio* species were isolated from the epizootics of luminescent vibriosis in shrimp farms from 11 provinces in the Philippines (de la Pena et al., 2001). The most recent disease to affect the shrimp industry in the Philippines was acute hepatopancreatic necrotic disease (AHPND), which has been described as early mortality syndrome (EMS) due to mass mortalities occurring within 20–30 days after stocking of ponds with post-larvae. The presence of AHPND in *Penaeus vannamei* and *Penaeus monodon* suggests that the disease can also impact late-stage juvenile shrimp (de la Pena et al., 2015). Variability in pathogenicity toward different shrimp species even within the same pathogen underscores the need for accurate identification, typing, and tracking strategies. Traditional phenotypic profiling methods for typing, such as biochemical strip tests, serological techniques, and antimicrobial susceptibility tests, have limited discriminatory power (Cano-Gomez et al., 2009) due to the high degree of phenotypic and genotypic similarity among bacterial strains.

The advent of next-generation sequencing (NGS) has led to a reliable means of resolving taxonomic ambiguities through deep whole-genome sequencing (WGS). The decreasing cost of sequencing and the mounting power of NGS platforms have introduced an era of genomic epidemiology, wherein mutational information from whole genomes of pathogenic isolates collected in disease outbreaks can be used to track the spread of disease (Gardy, 2013). Specifically, the same framework can also be applied in tracking the emergence of aquaculture diseases from an environmental reservoir, and in determining the zoonotic potential and local transmission pattern of a pathogen.

Between 2014 and 2016, the Philippine Shrimp Pathogenomics Program collected, biobanked, and performed WGS on bacterial samples isolated from the hepatopancreas of penaeid shrimp collected from various sites of disease outbreak in the Philippines. With the ultimate goal of managing the emerging shrimp disease in that country, the program biobanked a total 1902 bacterial isolates and sequenced the whole genomes of over 400 isolates. Here, we report the first large-scale, high-throughput analysis of WGS of bacterial species infecting cultured shrimp in local aquaculture farms in the Philippines. Further downstream analysis on the WGS amassed is anticipated to pave the way for the establishment of effective shrimp management interventions, such as the development of fast and affordable detection tools.

Materials and methods

SAMPLE COLLECTION AND BACTERIAL ISOLATION

A total of 14 regions (I, II, III, IV-A, IV-B, V, VI, VII, VIII, IX, X, XI, XII, and XIII) of the country were covered by the sampling program (Figure 20.1). Shrimp sample collections were arranged with the assistance of the Bureau of Fisheries and Aquatic Resources (BFAR) Regional offices, or through direct contact with farms or hatcheries. A site is classified as an outbreak site if any of the following signs of disease are encountered in penaeid shrimp during the time of sampling: weakness, discoloration, tail rot, or other clinical signs. A biased sampling protocol was introduced, wherein animals displaying signs of diseases – usually found near the sides of dikes or under the aerators – were preferentially collected. For each pond, at least 15 animals were collected and transported in oxygenated plastic bags. Metadata pertaining to shrimp samples were gathered by interview with the farm owners or staff.

Bacterial isolation was performed either *in situ* or at the nearest BFAR or private collaborating laboratory. Shrimp hepatopancreas were aseptically dissected and placed in tear-resistant sterile plastic bags for homogenization. Serial dilutions of the homogenate in sterile sea water (10^{-3}, 10^{-4}, and 10^{-5}) were plated on nutrient agar (Pronadisa) with 2% NaCl, thiosulfate citrate bile salts sucrose (TCBS) agar

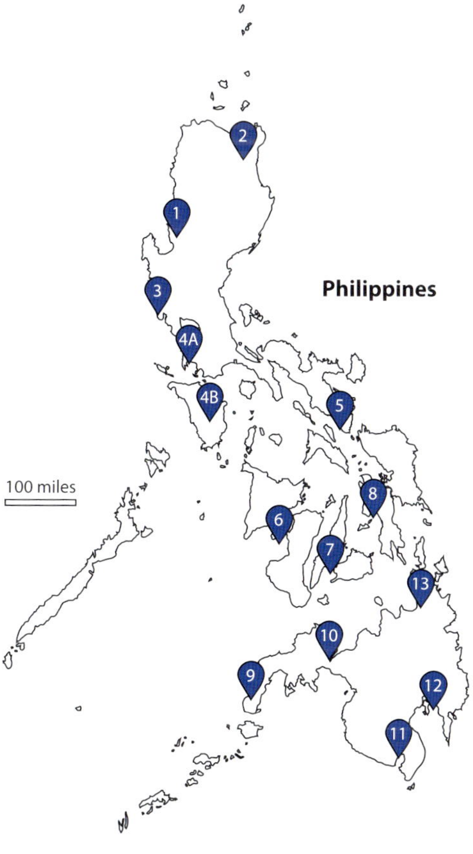

FIGURE 20.1 Map of the Philippines showing the regions covered by the Philippine Shrimp Pathogenomics Program surveillance activity, 2014–2015.

(Pronadisa), and Vibrio chromogenic agar (Pronadisa), in duplicate. After 18–24 h the dominant colonies were isolated and the presence of luminous bacteria was noted. Pure cultures were ascertained before storage with glycerol in a freezer at –80°C.

DNA EXTRACTION AND NGS

The glycerol stock of pure bacterial cultures was inoculated into nutrient broth (Pronadisa) with 1.5% NaCl. Bacterial cultures were incubated in a shaking incubator at 30°C for 18–24 h before genomic DNA extraction. After overnight incubation, bacterial cultures were centrifuged at 3000 x *g* to obtain pellets for automated DNA extraction with the KingFisher Flex System (Thermo Fisher Scientific). The resulting gDNA extracts were incubated with RNAse I (Thermo Fisher Scientific) at 37°C for 10 min, and purified with Agencourt Ampure XP paramagnetic beads (Beckman Coulter) at a gDNA:beads ratio of 0.8:1.0.

Libraries were prepared from the extracted gDNA using the Nextera XT Library Preparation Kit (Illumina). Paired-end libraries were sequenced in an Illumina MiSeq platform (MiSeq Reagent Kit v3, 2 x 300 bp) at the DNA Sequencing Core Facility, UP Philippine Genome Center, at 50–100x coverage.

GENOME ASSEMBLY AND IDENTITY VALIDATION

Raw sequencing reads corrected with Musket (Liu et al., 2012) were assembled *de novo* using SPAdes 3.5.0 (Nurk et al., 2013). Assembly metrics were calculated with assemblathon_stats.pl (Bradnam et al., 2013). *In silico* species identification was performed on the resulting assemblies against reference sequences using calculate_ani.py (Pritchard, 2013).

Results and discussion

VIBRIO SPP. DOMINATE SITES OF DISEASE OUTBREAK IN SHRIMP AQUACULTURE

A total of 423 bacterial samples were sequenced using the Illumina MiSeq platform: these included 360 isolates from shrimp farms with reported disease outbreak, 47 from farms with no reported outbreak, and 17 from farms with unknown data. Farms with recorded outbreaks are located in Aklan, Bohol, Camarines Norte, Camarines Sur, Capiz, Catanduanes, Davao del Sur, Leyte, Marinduque, Masbate, Negros, Oriental Mindoro, Pampanga, and Zamboanga. Farms without outbreak are located in Albay, Bohol, Cagayan Valley, Camarines Sur, Cebu, Davao City, Ilocos Sur, La Union, Masbate, Negros, and Pangasinan.

Among the 423 isolates sequenced, species of the genus *Vibrio* were the most abundant at 72.58%, followed by *Photobacterium* (5.91%) and *Bacillus* (2.60%) (Table 20.1). Other genera detected include *Acinetobacter* (0.95%), *Aeromonas* (0.71%), *Macrococcus* (0.47%), *Staphylococcus* (0.47%), *Enterobacter* (0.24%), and *Providencia* (0.24%). Within the genus *Vibrio* (307 isolates), the most common was *V. parahaemolyticus* at 37.59%, followed by *V. campbellii* (12.77%), *V. alginolyticus* (7.09%), and

V. harveyi (5.44%). Other *Vibrio* species recorded *Vibrio.* sp. (2.6%), *V. vulnificus* (1.89%), *V. sinaloensis* (1.42%), *V. antiquarius* (1.18%), *V. rotiferianus* (0.95%), *V. furnissii* (0.71%), *V. cholerae* (0.47%), and *V. nigripulchritudo* (0.47%).

TABLE 20.1 Overall occurrence of bacterial genera and species in sampled shrimp farms, 2014–2015 (423 isolates).

Species	*Count*	*Frequency (%)*
Vibrio	**307**	**72.58**
Vibrio parahaemolyticus	159	37.59
Vibrio campbellii	54	12.77
Vibrio alginolyticus	30	7.09
Vibrio harveyi	23	5.44
Vibrio sp.	11	2.6
Vibrio vulnificus	8	1.89
Vibrio sinaloensis	6	1.42
Vibrio antiquarius	5	1.18
Vibrio rotiferianus	4	0.95
Vibrio furnissii	3	0.71
Vibrio cholerae	2	0.47
Vibrio nigripulchritudo	2	0.47
Photobacterium	**25**	**5.91**
Photobacterium damselae	23	5.44
Photobacterium ganghwense	2	0.47
Bacillus	**11**	**2.6**
Bacillus megaterium	5	1.18
Bacillus amyloliquefaciens	3	0.71
Bacillus pumilus	2	0.47
Bacillus paralincheniformis	1	0.24
Acinetobacter	**4**	**0.95**
Acinetobacter junii	3	0.71
Acinetobacter johnsonii	1	0.24
Aeromonas	**3**	**0.71**
Aeromonas caviae	1	0.24
Aeromonas schubertii	1	0.24
Aeromonas veronii	1	0.24
Macrococcus	**2**	**0.47**
Macrococcus caseolyticus	2	0.47
Staphylococcus	**2**	**0.47**
Staphylococcus epidermidis	1	0.24
Staphylococcus warneri	1	0.24
Enterobacter	**1**	**0.24**
Enterobacter cloacae	1	0.24
Providencia	**1**	**0.24**
Providencia rettgeri	1	0.24
Unidentified	**67**	**15.84**

Within sites with no outbreaks recorded, 47 isolates were classified into 15 species (Table 20.2). The most abundant species was *V. parahaemolyticus* (23.4%), followed by *V. campbellii* (14.89%). Succeeding *V. parahaemolyticus* and *V. campbellii* were *V. antiquarius* (4.26%), *V. rotiferianus* (4.26%), *Bacillus amyloliquefaciens* (4.26%), *Bacillus megaterium* (4.26%), and *Acinetobacter junii* (4.26%). The following species were also detected in the non-outbreak sites: *V. harveyi* (2.13%), *V. vulnificus* (2.13%), *Bacillus paralincheniformis* (2.13%), *Bacillus pumilus* (2.13%), *Aeromonas veronii* (2.13%), *Macrococcus caseolyticus* (2.13%), *Photobacterium ganghwense* (2.13%), and *Staphylococcus epidermidis* (2.13%).

TABLE 20.2 Overall occurrence of bacterial genera and species in sampled shrimp farms with no recorded outbreaks, 2014–2015 (47 isolates).

Species	*Count*	*Frequency (%)*
Vibrio	**24**	**51.06**
Vibrio parahaemolyticus	11	23.4
Vibrio campbellii	7	14.89
Vibrio antiquarius	2	4.26
Vibrio rotiferianus	2	4.26
Vibrio harveyi	1	2.13
Vibrio vulnificus	1	2.13
Bacillus	**6**	**12.77**
Bacillus amyloliquefaciens	2	4.26
Bacillus megaterium	2	4.26
Bacillus paralincheniformis	1	2.13
Bacillus pumilus	1	2.13
Acinetobacter	**2**	**4.26**
Acinetobacter junii	2	4.26
Aeromonas	**1**	**2.13**
Aeromonas veronii	1	2.13
Macrococcus	**1**	**2.13**
Macrococcus caseolyticus	1	2.13
Photobacterium	**1**	**2.13**
Photobacterium ganghwense	1	2.13
Staphylococcus	**1**	**2.13**
Staphylococcus epidermidis	1	2.13
Unidentified	**11**	**23.4**

A total of 360 isolates were sampled from outbreak sites (Table 20.3). Outbreak types encountered during the time of sampling included AHPND, luminescent vibriosis, outbreak of unknown etiology, and white spot syndrome virus (WSSV). Of the 360 sequenced isolates from outbreak sites, 35% were AHPND, 30% outbreaks of unknown etiology, 21.94% WSSV, and 13.06% luminescent vibriosis.

Similar to the non-outbreak sites, the most abundant species in outbreak sites was *V. parahaemolyticus* (38.89%) followed by *V. campbellii* (13.06%). Apart from *V. parahaemolyticus* and *V. campbellii*, the following species were also observed in outbreak sites: *V. alginolyticus* (8.33%), *Photobacterium damselae* (6.39%), *V. harveyi* (6.11%), *Vibrio* sp. (3.06%), *V. vulnificus* (1.94%), *V. sinaloensis* (1.67%),

TABLE 20.3 Overall occurrence of bacterial genera and species in sampled shrimp farms with recorded outbreaks, 2014–2015 (360 isolates).

Species	*Count*	*Frequency (%)*
Vibrio	**273**	**75.83**
Vibrio parahaemolyticus	140	38.89
Vibrio campbellii	47	13.06
Vibrio alginolyticus	30	8.33
Vibrio harveyi	22	6.11
Vibrio sp.	11	3.06
Vibrio vulnificus	7	1.94
Vibrio sinaloensis	6	1.67
Vibrio antiquarius	3	0.83
Vibrio furnissii	3	0.83
Vibrio rotiferianus	2	0.56
Vibrio cholerae	1	0.28
Vibrio nigripulchritudo	1	0.28
Photobacterium	**24**	**6.67**
Photobacterium damselae	23	6.39
Photobacterium ganghwense	1	0.28
Bacillus	**5**	**1.39**
Bacillus megaterium	3	0.83
Bacillus amyloliquefaciens	1	0.28
Bacillus pumilus	1	0.28
Acinetobacter	**2**	**0.56**
Acinetobacter johnsonii	1	0.28
Acinetobacter junii	1	0.28
Enterobacter	**1**	**0.28**
Enterobacter cloacae	1	0.28
Providencia	**1**	**0.28**
Providencia rettgeri	1	0.28
Staphylococcus	**1**	**0.28**
Staphylococcus warneri	1	0.28
Unidentified	**53**	**14.72**

V. antiquarius (0.83%), *V. furnissii* (0.83%), *B. megaterium* (0.83%), *V. rotiferianus* (0.56%), *V. cholera* (0.28%), *V. nigripulchritudo* (0.28%), *P. ganghwense* (0.28%), *B. amyloliquefaciens* (0.28%), *B. pumilus* (0.28%), *Acinetobacter johnsonii* (0.28%), *A. junii* (0.28%), *Enterobacter cloacae* (0.28%), *Providencia rettgeri* (0.28%), and *Staphylococcus warneri* (0.28%).

Across all outbreak and non-outbreak sites, species of the genus *Vibrio* were observed to be the most abundant. *Vibrio* spp. are Gram-negative, halophilic bacteria naturally found in marine and estuarine ecosystems, as well as in the normal microflora of marine organisms including shrimp (Gomez-Gil et al., 1998). However, they may become opportunistic pathogens from environmental triggers and act as the causative agents of major shrimp bacterial disease outbreaks, such as AHPND – which was confirmed for the first time in 2015 as affecting *P. vannamei* and *P. monodon* cultured in the Philippines (Dabu et al.,

2017; de la Pena et al., 2015) – and luminous vibriosis. The abundance of *Vibrio* in aquatic settings can be explained by their role in nutrient cycling as a consequence of their ability to take up dissolved organic matter (Sherr and Sherr, 2000). They also have the capacity to break down chitin, a homopolymer of N-acetyl-D-glucosamine, one of the largest pools of amino sugars in the ocean (Cottrell and Kirchman, 2000). In addition, they also provide essential polyunsaturated fatty acids to the aquatic food web, which most aquatic organisms cannot intrinsically produce (Cottrell and Kirchman, 2003).

The dominance of *Vibrio* spp. in both outbreak and non-outbreak sites is also manifested in shrimp farms outside the Philippines. Common among the bacterial flora of water, sediment, and shrimp samples of culture environments in India is the high relative abundance of *Vibrio* spp., particularly *V. alginolyticus*, *V. parahaemolyticus*, *V. harveyi*, and *V. vulnificus* (Gopal et al., 2005; Jayasree et al., 2006). Bacterial strains collected from shrimp hatcheries reporting mass mortality in Thailand and Ecuador also largely comprise *Vibrio* spp. with varying degrees of virulence against different larval stages of *P. monodon* (Le Groumellec et al., 1995). *V. alginolyticus* and *V. harveyi* were predominantly present in diseased *Litopenaeus vannamei* larvae, postlarvae, and broodstock sampled from hatcheries in Ecuador and China, as well as from shrimp farms in Mexico (Vandenberghe et al., 1997, 1999).

V. CAMPELLII DOMINATES AT SITES WITH VIBRIOSIS OUTBREAKS

Three (3) *Vibrio* species dominated those sites infected with luminescent vibriosis, namely *V. campbellii* (80.85%), *V. harveyi* (10.64%), and *V. parahaemolyticus* (8.51%) (Table 20.4). WSSV infection was also accompanied by coinfection with various bacterial species. For mortality-causing outbreaks of unknown etiology, species from the genus *Vibrio* were the most dominant at 67.59%, of which *V. parahaemolyticus* comprised 45.37%. Other species observed to occur at sites with outbreaks of unknown etiology were *P. damselae* (7.41%), *P. ganghwense* (0.93%), *B. amyloliquefaciens* (0.93%), *B. megaterium* (0.93%), *B. pumilus* (0.93%), *A. junii* (0.93%), *E. cloacae* (0.93%), and *P. rettgeri* (0.93%) (Table 20.5).

In sites with recorded luminescent vibriosis outbreaks, only three species belonging to the genus *Vibrio* were detected. Luminescent vibriosis, a major disease problem in cultured shrimp worldwide, is caused by bacteria in the family Vibrionaceae (Adams, 1991; Lightner, 1993; Lightner and Lewis, 1975). Infections of *V. harveyi* have caused severe cases of luminous vibriosis leading to mass mortalities in the Philippines (Baticados et al., 1990), India (Karunasagar et al., 1994), Thailand (Jiravanichpaisal et al., 1994), China (Wang et al., 2013), Australia (Pizzuto and Hirst, 1995), and Ecuador (Robertson et al., 1998). Luminescent vibriosis infection can occur during egg, larval, post-larval, and juvenile stages of shrimp growth, and affects major cultured penaeid shrimp species such as *P. monodon*, *P. merguiensis*,

TABLE 20.4 Overall occurrence of bacterial genera and species in sampled shrimp farms reporting luminescent vibriosis outbreaks, 2014–2015 (47 isolates).

Genus	*Count*	*Frequency (%)*
Vibrio	**47**	**100**
Vibrio campbellii	38	80.85
Vibrio harveyi	5	10.64
Vibrio parahaemolyticus	4	8.51

TABLE 20.5 Overall occurrence of bacterial genera and species in sampled shrimp farms with outbreaks of unknown etiology, 2014–2015 (108 isolates).

Genus	*Count*	*Frequency (%)*
Vibrio	**73**	**67.59**
Vibrio parahaemolyticus	49	45.37
Vibrio vulnificus	7	6.48
Vibrio alginolyticus	5	4.63
Vibrio harveyi	4	3.7
Vibrio sp.	4	3.7
Vibrio furnissii	3	2.78
Vibrio rotiferianus	1	0.93
Photobacterium	**9**	**8.33**
Photobacterium damselae	8	7.41
Photobacterium ganghwense	1	0.93
Bacillus	**3**	**2.78**
Bacillus amyloliquefaciens	1	0.93
Bacillus megaterium	1	0.93
Bacillus pumilus	1	0.93
Acinetobacter	**1**	**0.93**
Acinetobacter junii	1	0.93
Enterobacter	**1**	**0.93**
Enterobacter cloacae	1	0.93
Providencia	**1**	**0.93**
Providencia rettgeri	1	0.93
Unidentified	**20**	**18.52**

P. indicus, and *P. vannamei*. Other pathogenic strains from the *Vibrio* family include *V. campbelli*, *V. vulnificus*, *V. alginolyticus*, *V. anguillarum*, *V. parahaemolyticus*, and *V. splendidus*, which cause various vibriosis syndromes, tail necrosis, shell disease, red disease, white gut disease, loose shell syndrome, and AHPND (Jayasree et al., 2006; Tran et al., 2013). Virulence mechanisms of *Vibrio* spp. range from bacterial colonization leading to systemic infections to secretion of toxic extracellular products (ECPs), such as hemolysins, proteases, and lipases that are pathogenic to shrimp (Goarant et al., 2000; Cano-Gomez et al., 2009).

Vibrio harveyi has been cited as the major etiological agent associated with previous reports of heavy mortalities due to luminescent vibriosis in pond-cultured shrimp in the Philippines (Lavilla-Pitogo et al., 1990; Lavilla-Pitogo and de la Pena, 1998; de la Pena et al., 2001). However, with the limitations that come with older methods of species identification, such as morphological, physiological, and biochemical tests, it is possible that older bacterial samples were not confidently identified due to the high degree of genetic and phenotypic similarity among the core of the Harveyi clade (*V. harveyi*, *V. campbellii*, *V. owensii*, *V. jasicida*, and *V. rotiferianus*) (Cano-Gomez et al., 2009). As a result, these likely misidentifications may have overemphasized the importance of *V. harveyi* in prior luminescent vibriosis outbreaks, since recorded misidentifications were usually initially characterized as *V. harveyi* (Lin et al., 2010). Therefore, in order to establish congruence with the observed dominance of *V. campbellii* in Philippine luminescent vibriosis outbreaks from 2014 to 2016 with earlier recorded outbreaks, WGS and species identification through average nucleotide identity analysis may be implemented on archived bacterial samples from older outbreaks.

Except for luminescent vibriosis infection, the most dominant species in both outbreak and non-outbreak sites was *V. parahaemolyticus*, first reported and recognized as a foodborne pathogen in Osaka, Japan in 1951 when the consumption of contaminated sardine led to deaths and hospitalization (Daniels et al., 2000). Under normal conditions, *V. parahaemolyticus* is either free-swimming, attached to underwater surfaces, or associated with a variety of shellfish species in a commensal relationship (McCarter, 1999). *V. parahaemolyticus* may be present all year round in marine areas with water temperatures that do not fall below 15°C (Su and Liu, 2007).

CO-OCCURRENCE OF BACTERIAL SPECIES IN WSSV-INFECTED SHRIMP

WSSV is a bacilliform, non-occluded, enveloped virus of the genus *Whispovirus* and family Nimaviridae (Chou et al., 1995). Shrimp infected by WSSV display apparent white spots or patches of 0.5–3.0 mm diameter in their exoskeleton (Lo et al., 1996). At the pond level, WSSV-infected shrimp gather near the edge of the pond and display clinical signs before the first round of mortality.

In a study by Jang et al. (2014), it was shown that the pacific white shrimp *L. vannamei* died earlier and more frequently with multiple infections (*V. anguillarum* in addition to WSSV) than singly infected shrimp. This suggests that infections by multiple pathogens may induce higher mortality and more serious consequences than those by only one pathogen. Corroborating this trend is a coinfection study which suggests that *V. parahaemolyticus* infection hastens the proliferation of WSSV in *L. vannamei* and vice versa, and that the overall combined accelerated proliferation of both leads to massive mortality (Zhang et al., 2016). As such, it would be interesting to explore in future studies whether the presence of certain species of bacteria contributes to infection by WSSV.

In addition, outbreaks of unknown etiology will require further characterization of the signs and manifestations, at both the clinical and pond level, before the list of causative agents can be narrowed down and identified conclusively. Although *Vibrio* isolates are the most abundant in these sites, it is likely that other factors or elements may be at play regarding shrimp mortality.

Conclusion

Penaeid shrimp samples were collected from 14 of the 17 regions of the Philippines. The genomic DNA of 423 bacterial isolates from hepatopancreas samples was subjected to multiplexed, paired-end WGS on the Illumina MiSeq platform. Genomic relatedness was measured via average nucleotide identity analysis of the resulting assemblies against reference genomes. *Vibrio* spp., which are commonly found as part of the natural microflora of wild and cultured shrimp, were observed to be the most preponderant in both outbreak and non-outbreak sites. Among *Vibrio* isolates, *V. parahaemolyticus* had the highest occurrence with *V. campellii* dominating in sites with luminous vibriosis. Regarding the latter, isolates were exclusively composed of *Vibrio* spp. Data on the occurrence of bacterial species from outbreaks of unknown etiology and WSSV infections were also collected. Whole-genome sequences gathered from this research will contribute to biosurveillance, and to the development of novel methods of disease control of bacterial epizootic outbreaks.

Acknowledgments

The authors acknowledge the Aquaculture Department of the Southeast Asian Fisheries Development Center (SEAFDEC), the Negros Prawn Growers' Cooperative, the Bureau of Fisheries and Aquatic Resources, and the many fish pond operators throughout the country and various research assistants employed by the program for help in the collection of bacterial isolates used in this study; and the UP Philippine Genome Center – DNA Sequencing Core Facility (PGC-DSCF) and Core Facility for Bioinformatics (PGC-CFB) for providing sequencing and bioinformatics analysis services, and Ms Krizelle Alcantara for her assistance in the preparation of this manuscript. This work was made possible through the funding support of the Philippine Council for Agriculture, Aquatic and Natural Resources Research and Development of the Department of Science and Technology (PCAARRD-DOST) for the project "Shrimp Pathogenomics Towards a Competitive and Sustainable Shrimp Industry" under the program "Development of Diagnostic Tools Through Genomics and Establishment of the Philippine Aquatic Pathogen Database Resource for the Philippine Aquaculture Industry". The authors are grateful for the help provided by Drs Adelle Calpe and Dalisay Fernandez of PCAARRD-DOST in the implementation of the program.

References

Adams, A. (1991) Response of penaeid shrimp to exposure to *Vibrio* species. *Fish and Shellfish Immunology* 1, 59–70.

Baticados, M. C., Lavilla-Pitogo, C. R., Cruz-Lacierda, E. R., De La Pena, L. D. and Sunaz, N. A. (1990) Studies on the chemical control of luminous bacteria *Vibrio harveyi* and *V. splendidus* isolated from diseased *Penaeus monodon* larvae and rearing water. *Diseases of Aquatic Organisms* 9(2), 133–139.

Boyd, C. E. and Clay, J. W. (1998) Shrimp aquaculture and the environment. *Scientific American* 278(6), 58–65.

Bradnam, K. R., Fass, J. N., Alexandrov, A., Baranay, P., Bechner, M., Birol, I., Boisvert, S., Chapman, J. A., Chapuis, G., Chikhi, R. and Chitsaz, H. (2013) Assemblathon 2: evaluating de novo methods of genome assembly in three vertebrate species. *GigaScience* 2(1), 10.

Bureau of Fisheries and Aquatic Resources (2017) *Philippine Fisheries Profile 2017*. Department of Agriculture, Quezon City, Philippines.

Cano-Gomez, A, Bourne, D. G., Hall, M. R., Owens, L. and Høj, L. (2009) Molecular identification, typing and tracking of *Vibrio harveyi* in aquaculture systems: current methods and future prospects. *Aquaculture* 287(1–2), 1–10.

Chou, H., Huang, C., Wang, C., Chiang, H. and Lo, C. (1995) Pathogenicity of a baculovirus infection causing white spot syndrome in cultured penaeid shrimp in Taiwan. *Diseases of Aquatic Organisms* 23(3), 165–273.

Cottrell, M. T. and David, K. L. (2003) Contribution of major bacterial groups to bacterial biomass production (thymidine and leucine incorporation) in the Delaware estuary. *Limnology and Oceanography* 48(1), 168–278.

Cottrell, M. T. and Kirchman, D. L. (2000) Community composition of marine bacterioplankton determined by 16S rRNA gene clone libraries and fluorescence in situ hybridization. *Applied Environmental Microbiology* 66(12), 5116–5122.

Dabu, I. M., Lim, J. J., Arabit, P., Orense, S., Tabardillo, J., Corre, V. and Manigas, M. (2017) The first record of acute hepatopancreatic necrosis disease in the Philippines. *Aquaculture Research* 48(3), 792–799 [first published online 19 November 2015].

Daniels, N., Ray, B., Easton, A., Marano, N., Kahn, E., McShan, I., Del Rosario, L., Baldwin, T., Kingsley, M., Puhr, N. and Wells, J. (2000) Emergence of a new *Vibrio parahaemolyticus* serotype in raw oysters: a prevention quandary. *JAMA* 284(12), 1541–1545.

De La Pena, L., Lavilla-Pitogo, C. and Paner, M. (2001) Luminescent *Vibrios* associated with mortality in pond-cultured shrimp *Penaeus monodon* in the Philippines: Species composition. *Fish Pathology* 36(3), 133–138.

De La Pena, L., Cabillon, N., Catedral, D., Amar, E., Usero, R., Monotilla, W., Calpe, A., Fernandez, D. and Saloma, C. (2015) Acute hepatopancreatic necrosis disease (AHPND) outbreaks in *Penaeus vannamei* and *P. monodon* cultured in the Philippines. *Diseases of Aquatic Organisms* (116), 251–254.

Diana, J. S. (2009) Aquaculture production and biodiversity conservation. *Bioscience* 59(1), 27–38.

Flegel, T., Lightner, D., Lo, C, and Owens, L. (2008) Shrimp disease control: past, present and future. In: *Diseases in Asian Aquaculture VI* (eds Bondad-Reantaso, M. G. et al.). Fish Health Section, Asian Fisheries Society, Manila, Philippines, pp. 355–378.

Gardy, J. L. (2013) Investigation of disease outbreaks with genome sequencing. *Lancet Infectious Diseases* 13(2), 101–102.

Goarant, C., Herlin, J., Brizard, R., Marteau, A., Martin, C. and Martin, B. (2000) Toxic factors of *Vibrio* strains pathogenic to shrimp. *Diseases of Aquatic Organisms* 40(2), 101–107.

Gomez-Gil, B., Tron-Mayen, L., Roque, A., Turnbull, J., Inglis, V. and Guerra-Flores, A. (1998) Species of *Vibrio* isolated from hepatopancreas, haemolymph and digestive tract of a population of healthy juvenile *Penaeus vannamei. Aquaculture* 163(1–2), 1–9.

Gopal, S., Otta, S., Kumar, S., Karunasagar, I., Nishibuchi, M. and Karunasagar, I. (2005) The occurrence of *Vibrio* species in tropical shrimp culture environments; implications for food safety. *International Journal of Food Microbiology* 102(2), 151–159.

Han, J., Tang, K., Tran, L. and Lightner, D. V. (2015) Photorhabdus insect-related (Pir) toxin-like genes in a plasmid of *Vibrio parahaemolyticus*, the causative agent of acute hepatopancreatic necrosis disease (AHPND) of shrimp. *Diseases of Aquatic Organisms* 113(1), 33–40.

Jang, I., Qiao, G. and Kim, S. K. (2014) Effect of multiple infections with white spot syndrome virus and *Vibrio anguillarum* on Pacific white shrimp *Litopenaeus vannamei* (L.): mortality and viral replication. *Journal of Fish Diseases* 37(10), 911–920.

Jayasree, L., Janakiram, P. and Madhavi, R. (2006) Characterization of *Vibrio* spp. associated with diseased shrimp from culture ponds of Andhra Pradesh (India). *Journal of the World Aquaculture Society* 37(4), 523–532.

Jiravanichpaisal, P., Miyazaki, T. and Limsuwan, C. (1994) Histopathology, biochemistry, and pathogenicity of *Vibrio harveyi* infecting black tiger prawn *Penaeus monodon. Journal of Aquatic Animal Health* 6(1), 27–35.

Karunasagar, I., Pai, R., Malathi, G. and Karunasagar, I. (1994) Mass mortality of *Penaeus monodon* larvae due to antibiotic-resistant *Vibrio harveyi* infection. *Aquaculture* 128(3–4), 203–209.

Lavilla-Pitogo, C., Baticados, M., Cruz-Lacierda, E. and De La Pena. L. D. (1990) Occurrence of luminous bacterial disease of *Penaeus monodon* larvae in the Philippines. *Aquaculture* 91(1–2), 1–3.

Lavilla-Pitogo, C. and De La Pena, L. D. (1998) Bacterial diseases in shrimp (*Penaeus monodon*) culture in the Philippines. *Fish Pathology* 33(4), 405–411.

Le Groumellec, M., Haffner, P., Martin, B. and Martin, C. (1995) Comparative study of bacterial infections responsible for mass mortality in penaeid shrimp hatcheries of the Pacific zone. In: *Diseases in Asian Aquaculture II* (eds Shariff, M. et al.). Fish Health Section, Asian Fisheries Society, Manila, Philippines, pp. 163–173.

Lightner, D. (1993) Diseases of cultured penaeid shrimp. In: *CRC Handbook of Mariculture* (ed. McVey, J.). CRC Press Inc., FL, USA, pp. 393–486.

Lightner, D. V. and Lewis, D. H. (1975) A septicemic bacterial disease syndrome of penaeid shrimp. *Marine and Fisheries Review* 37(5–6), 25–28.

Lightner, D., Redman, R. et al. (2013) *Documentation of an Emerging Disease (Early Mortality Syndrome or Acute Hepatopancreatic Necrosis Disease) in SE Asia and Mexico.* World Organisation for Animal Health, France.

Lin, B., Wang, Z., Malanoski, A., O'Grady, E., Wimpee, C., Vuddhakul, V., Alves Jr N., Thompson, F., Gomez-Gil, B. and Vora, G. (2010) Comparative genomic analyses identify the *Vibrio harveyi* genome sequenced strains BAA-1116 and HY01 as *Vibrio campbellii. Environmental Microbiology Reports* 2(1), 81–89.

Liu, Y., Schröder, J. and Schmidt, B. (2012) Musket: a multistage k-mer spectrum-based error corrector for Illumina sequence data. *Bioinformatics* 29(3), 308–315.

Lo, C., Leu, J., Ho, C., Chen, C., Peng, S., Chen, Y., Chou, C., Yeh, P., Huang, C., Chou, H. and Wang, C. H. (1996) Detection of baculovirus associated with white spot syndrome (WSBV) in penaeid shrimps using polymerase chain reaction. *Diseases of Aquatic Organisms* 25(1–2), 133–141.

McCarter, L. (1999) The multiple identities of *Vibrio parahaemolyticus. Journal of Molecular Microbiology and Biotechnology* 1(1), 51–57.

Nurk, S., Bankevich, A., Antipov, D., Gurevich, AA., Korobeynikov, A., Lapidus, A., Prjibelski, A.D., Pyshkin, A., Sirotkin, A., Sirotkin, Y., Stepanauskas, R., Clingenpeel, S.R., Woyke, T., McLean, J.S., Lasken, R., Tesler, G., Alekseyev, M.A. and Pevzner, P.A. (2013) Assembling single-cell genomes and mini-metagenomes from chimeric MDA products. *Journal of Computational Biology* 10(1), 714–737.

Pizzutto, M. and Hirst, R. G. (1995) Classification of isolates of *Vibrio harveyi* virulent to *Penaeus monodon* larvae by protein profile analysis and M13 DNA fingerprinting. *Diseases of Aquatic Organisms* 21(1), 61–68.

Pritchard, L. (2013) Retrieved from https://github.com/widdowquinn/scripts/blob/ master/bioinformatics/calculate_ani.py on 14 March 2019.

Robertson, P., Calderon, J., Carrera, L., Stark, J., Zherdmant, M. and Austin, B. (1998) Experimental *Vibrio harveyi* infections in *Penaeus vannamei* larvae. *Diseases of Aquatic Organisms* 32(2), 151–155.

Sherr, E. and Sherr, B. (2000) Marine microbes: An overview. In: *Microbial Ecology of the Oceans* (eds Haddock, B. and Jones, C.). Wiley-Liss, London, UK, pp. 13–46.

Sirikharin, R., Taengchaiyaphum, S., Sanguanrut, P., Chi, T., Mavichak, R., Proespraiwong, P., Nuangsaeng, B., Thitamadee, S., Flegel, T. and Sritunyalucksana, K. (2015) Characterization and PCR detection of binary, Pir-like toxins from *Vibrio parahaemolyticus* isolates that cause acute hepatopancreatic necrosis disease (AHPND) in shrimp. *PloS ONE* 10(5), e0126987.

Su, Y. C. and Liu, C. (2007) *Vibrio parahaemolyticus*: a concern of seafood safety. *Food Microbiology* 24(6), 549–558.

Tran, L., Nunan, L., Redman, R., Mohney, L., Pantoja, C., Fitzsimmons, K. and Lightner, D. V. (2013) Determination of the infectious nature of the agent of acute hepatopancreatic necrosis syndrome affecting penaeid shrimp. *Diseases of Aquatic Organisms* 105(1), 45–55.

Vandenberghe, J., Li, Y., Verdonck, L., Li, J., Sorgeloos, P., Xu, H. and Swings, J. (1997) *Vibrios* associated with *Penaeus chinensis* (Crustacea: Decapoda) larvae in Chinese shrimp hatcheries. *Aquaculture* 169(1–2), 121–132.

Vandenberghe, J., Verdonck, L., Robles-Arozarena, R., Rivera, G., Bolland, A., Balladares, M., Gomez-Gil, B., Calderon, J., Sorgeloos, P. and Swings, J. (1999) *Vibrios* associated with *Litopenaeus vannamei* larvae, postlarvae, broodstock, and hatchery probionts. *Applied Environmental Microbiology* 65(6), 2592–2597.

Vogt, G., Stöcker, W., Storch, V. and Zwilling, R. (1989) Biosynthesis of *Astacus protease*, a digestive enzyme from crayfish. Histochemistry 91(5), 373–381.

Wang, L., Chen, Y., Huang, H., Huang, Z., Chen, H. and Shao, Z. (2015) Isolation and identification of *Vibrio campbellii* as a bacterial pathogen for luminous vibriosis of *Litopenaeus vannamei. Aquaculture Research* 46(2), 395–404.

World Organization for Animal Health (2013) *Acute Hepatopancreatic Necrosis Disease.* World Organization for Animal Health, Paris, France.

Zhang, X., Song, X. and Huang, J. (2016) Impact of *Vibrio parahaemolyticus* and white spot syndrome virus (WSSV) co-infection on survival of penaeid shrimp *Litopenaeus vannamei. Chinese Journal of Oceanology and Limnology* 34(6), 1278–1286.

Chapter 21

EMERGING SHRIMP DISEASES: ACUTE HEPATOPANCREATIC NECROSIS DISEASE (AHPND), VIRAL COVERT MORTALITY DISEASE (VCMD) AND INFECTION WITH DECAPOD IRIDESCENT VIRUS 1 (IDIV1)

QINGLI ZHANG, LIANG QIU, XUAN DONG AND HUANG JIE

Summary

Several emerging shrimp diseases have impacted the shrimp farming industry globally over the past decade. One of the major diseases is a bacterial disease, acute hepatopancreatic necrosis disease (AHPND), which is caused by *Vibrio* spp. carrying the *Photorhabdus* insect-related (Pir) toxins. The two other major diseases are both viral, viral covert mortality disease (VCMD) and infection with decapod iridescent virus 1 (iDIV1), caused by the covert mortality nodavirus (CMNV) and Decapod iridescent virus 1 (iDIV1), respectively. AHPND is caused by specific virulent strains of *Vibrio* (V_{AHPND}) that harbor a 65–74-kb plasmid carrying genes that encode homologues of the Pir toxin, PirA and PirB. CMNV is an icosahedral virus of diameter approximately 20–30 nm. The genome of CMNV contains two segments (RNA1 and RNA2): the 3228-bp RNA1 encodes the RNA-dependent RNA polymerase (RdRp) and the B2 protein, and the 1448-bp RNA2 encoding a capsid protein. DIV1 is a member of the family *Iridoviridae* that exhibits a typical icosahedral structure with a mean diameter of 150 nm. DIV1 has a double-stranded DNA genome of about 166,000 bp; a phylogenetic tree of 27 genes supports the suggestion that DIV1 belongs to a new genus, named *Decapodiridovirus*. The etiological and patho-ecological statistics of the above-mentioned diseases are described and summarized in this chapter. Hopefully, the information in this chapter will be helpful for readers to understand the basic characteristics of these three emerging diseases.

Acute hepatopancreatic necrosis disease (AHPND)

SIGNS OF DISEASE

AHPND was originally referred to as early mortality syndrome (EMS), because mass mortality (70–100%) often occurs within 30–35 days post stocking and progresses extremely rapidly. In some cases, the onset of mortality can start as early as 10 days post stocking (OIE, 2019). Moribund shrimp often exhibit pale and atrophied hepatopancreas (HP), soft shell and empty stomach and midgut (Figure 21.1). The HP of moribund shrimp does not compress easily between the forefinger and thumb. Additionally, affected shrimp often sink to bottom of the pond (NACA, 2012; OIE, 2019). However, these signs are not specific and can only be used for presumptive diagnosis.

PATHOGENIC AGENTS

AHPND is caused by specific virulent strains of *Vibrio* (V_{AHPND}) that harbor a plasmid carrying genes that encode homologs of the *Photorhabdus* insect-related (Pir) binary toxins, PirA and PirB (Dong et al., 2017a; Lee et al., 2015; Liu, L. et al., 2018; Restrepo et al., 2018). This plasmid has been designated as pVA1-type plasmid (Dong et al., 2019a,b; OIE, 2019) with size ranging from 65 to 74 kb. The pVA1-type plasmid has been proved to be responsible for the virulence of AHPND (Lee et al., 2015). Experimental

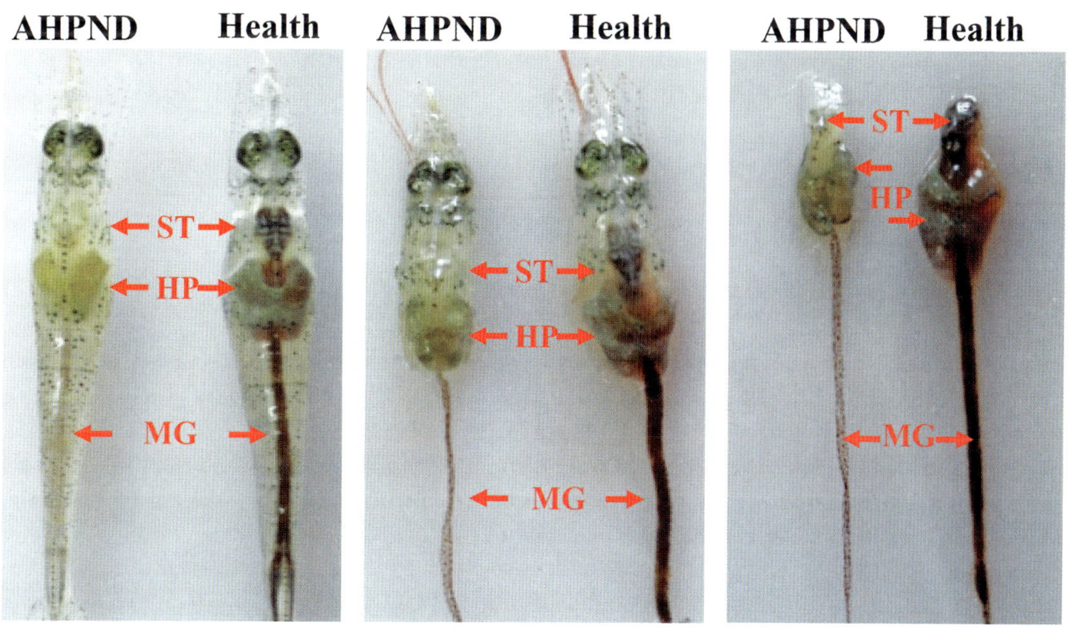

FIGURE 21.1 Signs of acute hepatopancreatic necrosis disease in *Penaeus vannmei*. Three examples of AHPND-affected shrimp (left) compared to healthy shrimp (right). In diseased specimens, empty midgut (MG) and stomach (ST), atrophied and pale hepatopancreas (HP). Healthy shrimp have normal-sized, brownish HP and full stomach and midgut. Credit: Dong et al., 2017a,b.

removal of the pVA1-type plasmid from the AHPND-causing *V. parahaemolyticus* supported the crucial role of this plasmid in AHPND pathogenicity (Lee et al., 2015). More importantly, non-AHPND-causing *V. campbellii* has the ability to cause AHPND in *P. vannamei* after acquiring the pVA1-type plasmid from AHPND-causing *V. parahaemolyticus* by conjugative transfer (Dong et al., 2019b).

The pVA1-type plasmid carries two clusters of conjugative transfer genes, mobilization genes, and a *pndA* PSK system (Lee et al., 2015). It has been shown to have the ability to transfer among certain species of *Vibrio* (Dong et al., 2017a, 2019a,b). To date, strains of five *Vibrio* species – *V. parahaemolyticus*, *V. owensii*, *V. campbellii*, *V. punensis* and *V. harveyi* – have been shown to harbor the pVA1-type plasmid and have the ability to cause AHPND (Dong et al., 2017a; Lee et al., 2015; Liu et al., 2018; Restrepo et al., 2018; Sarmila et al., 2019).

HOST RANGE

It has been demonstrated that *P. vannamei* and *P. monodon* are the susceptible host species of V_{AHPND}. For *P. chinensis*, evidence is incomplete regarding its listing as being susceptible to AHPND. *P. japonicas* has detected positive only using polymerase chain reaction (PCR), but there is a lack of information on active infection in this species (OIE, 2019).

AREAS AFFECTED

The disease was first reported among shrimp farms in China and Vietnam in 2010 (NACA, 2012; Zhang et al., 2012). It was later reported from Malaysia (2011), Thailand (2012), Mexico (2013), Philippines (2014), the USA (2017) and Bangladesh (2017) (de la Pena et al., 2015; Hong et al., 2015; Kondo et al., 2014; Kongrueng et al., 2015b; Nunan et al., 2014; Sperling et al., 2015). It was reported that *V. campbellii* isolates capable of producing PirA and PirB toxins were present in Thailand up to 7 years before AHPND was first reported (Pradit et al., 2018).

PATHOGEN ISOLATION AND IDENTIFICATION

In regard to AHPND-causing *Vibrio*, bacteria can be isolated from gut-associated tissues (HP, stomach or midgut) with either thiosulfate citrate bile salts sucrose (TCBS) plating or marine 2216E agar (5 g tryptone, 1 g yeast extract, 0.01g $FePO_4$, 1 litre filtered seawater, pH adjusted to 7.6) and incubation at 28 °C overnight. Individual colonies can then be selected for subculturing and identification. It is strongly recommended that both green and yellow colonies be tested when using the TCBS plating method, particularly in light of the horizontal transfer ability of the pVA1-type plasmid among *Vibrio* species (Dong et al., 2019a,b).

Bacterial identification can be carried out by use of the API Rapid NE test, detection of *pirAB* genes or pVA1-type plasmid and/or 16S rRNA sequencing from pure cultures. Multilocus sequence analysis (MLSA) can be used to determine taxonomic relationships within *Vibrio* clades using sequences of σ70 factor (rpoD), replication initiator protein (rctB), toxin transcriptional activator (toxR) and 16S rRNA (Dong et al., 2017b; Pascual et al., 2010).

HISTOPATHOLOGY

Histopathology of HP is the key diagnostic technique for AHPND. The HP of healthy shrimp is comprised of four types of epithelial cell: resorptive cells (R-cells), fibrillar cells (F-cells), embryonic cells (E-cells) and blister cells (B-cells). In the early acute phase, the HP tubules begin to degenerate, round up and slough into lumen (Figure 21.2B). Subsequently, the tubules become distorted and gradually degenerate and the epithelial cells become rounded and enter the lumen. The HP in the acute phase of AHPND shows decrease in R-, F- and B-cells, and absence of mitotic activity in the E-cells (Figure 21.2C). The terminal phage is characterized by development of abundant secondary bacterial colonization and hemocytic inflammation (Figure 21.2D) (NACA, 2014; OIE, 2019). Massive sloughing of HP epithelial cells is the only definitive histopathology (NACA, 2014).

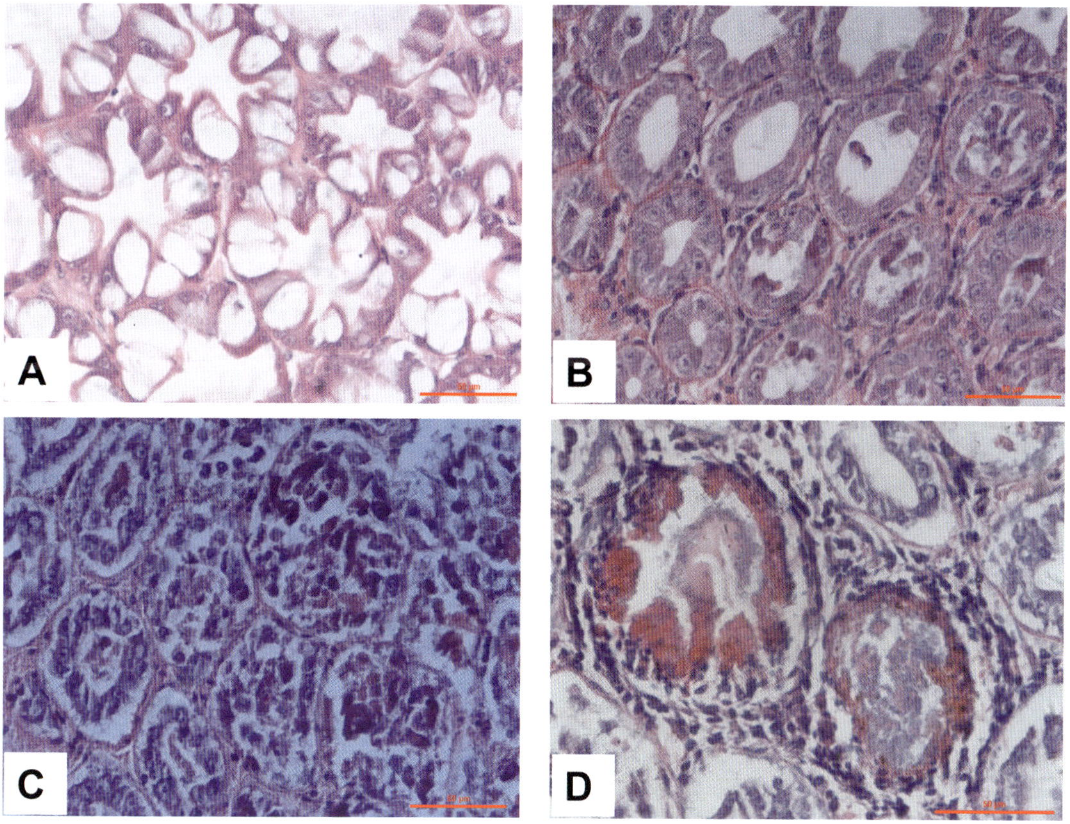

FIGURE 21.2 Histological analysis of AHPND in *Penaeus vannamei* under light microscopy. Scale bars, 50 μm. Credit: X. Dong.

MOLECULAR DIAGNOSTICS

Single-step, nested and real-time PCR and the isothermal loop-mediated amplification protocol (LAMP) have been described (Arunrut et al., 2016; Han et al., 2015a; Kongrueng et al., 2015a; OIE, 2019), while several molecular techniques have been recommended by OIE (OIE, 2019). AP1 primers used for single-step PCR were AP1F (5'-CCT-TGG-GTG-TGC-TTA-GAG-GAT-G-3') and AP1R (5'-GCA-AAC-TAT-CGC-GCA-GAA-CAC-C-3'). AP2 primers used for single-step PCR were AP2F (5'-TCA-CCC-GAA-TGC-TCG-CTT-GTG-G-3') and AP2R (5'-CGT-CGC-TAC-TGT-CTA-GCT-GAA-G-3'). Both of AP1 and AP2 amplified a 700 bp fragment from the non-toxin genes in a pVA1-type plasmid (Flegel and Lo, 2014). AP3 primers used for single-step PCR were AP3-F (5'-ATG-AGT-AAC-AAT-ATA-AAA-CAT-GAA-AC-3') and AP3-R (5'-GTG-GTA-ATA-GAT-TGT-ACA-GAA-3'). PCR amplified a 333-bp fragment from the *pirA* genes in a pVA1-type plasmid (Sirikharin et al., 2015). TUMSAT-Vp3 primers used for single-step PCR were TUMSAT-Vp3F (5'-GTG-TTG-CAT-AAT-TTT-GTG-CA-3') and TUMSAT-Vp3R (5'-TTG-TAC-AGA-AAC-CAC-GAC-TA-3'). PCR amplified a 360-bp fragment from the *pirA* genes in a pVA1-type plasmid (Tinwongger et al., 2014). VpPirA-284 primers used for single-step PCR were VpPirA-284F (5'-TGA-CTA-TTC-TCA-CGA-TTG-GAC-TG-3') and VpPirA-284R (5'-CAC-GAC-TAG-CGC-CAT-TGT-TA-3'). PCR amplified a 284-bp fragment from the *pirA* genes in a pVA1-type plasmid (Han et al., 2015b). VpPirB-392 primers used for single-step PCR were VpPirB-392F (5'-TGA-TGA-AGT-GAT-GGG-TGC-TC-3') and VpPirB-392R (5'-TGT-AAG-CGC-CGT-TTA-ACT-CA-3'). PCR amplified a 392-bp fragment from the *pirB* genes in a pVA1-type plasmid (Han et al., 2015b). Although single-step PCRs including AP1, AP2, AP3, TUMSAT-Vp3, VpPirA-284 and VpPirB-392 have lower sensitivity, nested PCRs such as AP4 are more sensitive and can be used for the direct detection of low levels of V_{AHPND}, without an enrichment step. AP4 primers used for first-step PCR were AP4-F1 (5'-ATG-AGT-AAC-AAT-ATA-AAA-CAT-GAA-AC-3') and AP4-R1 (5'-ACG-ATT-TCG-ACG-TTC-CCC-AA-3'). First-step PCR amplified a 1269-bp fragment from the *pirAB* genes in a pVA1-type plasmid. Primers used for second-step PCR were AP4-F2 (5'-TTG-AGA-ATA-CGG-GAC-GTG-GG-3') and AP4-R2 (5'- GTT-AGT-CAT-GTG-AGC-ACC-TTC-3'). Second-step PCR amplified a 230-bp fragment from the *pirAB* genes in a pVA1-type plasmid (Dangtipa et al., 2015).

Additionally, the TaqMan real-time PCR method has been developed and is more sensitive than conventional PCR methods. VpPirA primers used for real-time PCR were VpPirA-F (5'-TTG-GAC-TGT-CGA-ACC-AAA-CG-3') and VpPirA-R (5'-GCA-CCC-CAT-TGG-TAT-TGA-ATG-3'). The TaqMan probe of VpPirA is (5'-6FAM-AGA-CAG-CAA-ACA-TAC-ACC-TAT-CAT-CCC-GGA-TAMRA-3') (Han et al., 2015a; OIE, 2019). As an alternative, LAMP methods for detection of Vp_{AHPND} have been developed (Koiwai et al., 2016; Kongrueng et al., 2015a). LAMP techniques are recommended for field use because they are easy to perform (only one temperature involved) and rapid (about 60 min).

DISEASE CONTROL

The control and treatment of AHPND is complicated because of the diversity of the pathogenic bacteria, including *V. parahaemolyticus*, *V. owensii*, *V. campbellii*, *V. punensis* and *V. harveyi* (Dong et al., 2017a; Lee et al., 2015; Liu et al., 2018; Restrepo et al., 2018; Sarmila et al., 2019). In addition, some recent studies have reported the horizontal transfer of a pVA1-type plasmid between *Vibrio* species (Dong et al., 2019a,b).

Recent studies have demonstrated that probiotics, phages and other treatments provide significant protection against certain strains of AHPND-causing *V. parahaemolyticus* (Jun et al., 2016, 2018; Santos et al., 2019; Wang et al., 2018). Certain probiotics, such as *Pseudoalteromonas*, *Saccharomyces cerevisiae*, *Lactobacillus casei* and *Rhodopseudomonas palustris*, have been shown to enhance the survival of *P. vannamei* challenged with AHPND-causing *V. parahaemolyticus* (Pinoargote et al., 2018; Santos et al., 2019; Wang et al., 2018). Other studies also focused on bacteriophage prophylaxis and therapy for AHPND, with experimental phage therapy exhibiting significant protection and survival for *P. vannamei* challenged with AHPND-causing *V. parahaemolyticus* (Jun et al., 2016, 2018). Further studies are needed to evaluate the effect of probiotics and phages in additional strains of AHPND-causing *Vibrio* such as *V. owensii*, *V. campbellii*, *V. punensis* and *V. harveyi*. More studies on development of effective treatments for AHPND are needed.

Biosecurity is one of the key ways to prevent and control AHPND. Measures such as the use of high-quality post-larvae (PL) and the application of good aquaculture practices (GAP), screening of PL, improvement in sanitary conditions, disinfection of the farming ponds and considerations with regard to water intake and vector control. Proper disposal of diseased shrimp and decontamination of affected farms are likely to reduce the impacts of AHPND (NACA, 2012). Additionally, the development of AHPND-resistant shrimp lines through selective breeding is also an alternative strategy.

Viral covert mortality disease

Viral covert mortality disease (VCMD) emerged in high-density farming of *P. vannamei* as early as 2002/2003. In the early days the disease was called 'covert mortality disease (CMD)' by local shrimp farming practitioners, due to diseased individuals usually hiding at the bottom of the pond and then dying (Gu, 2012; Song and Zhuang, 2006; Xing, 2004; Xu and Ji, 2009; Zhang, 2004). Covert mortality nodavirus (CMNV) was identified as being the pathogen of the VCMD (Zhang et al., 2014). The horizontal and vertical transmission pathways of CMNV, as well as vectors and reservoirs, have been studied in detail (Liu et al., 2017, 2018). To date, CMNV infection in farming shrimp has been confirmed in several countries (Pooljun et al., 2016; Thitamadee et al., 2016; Wang et al., 2019a; Zhang et al., 2017b).

SIGNS OF DISEASE

General signs

At the pond level, the mortality caused by CMNV infection is significantly different from that by other shrimp viruses. A small percentage of individuals in the diseased shrimp population died every day, with daily mortality occurring until the end of the culture period. Hence, VCMD was also called 'running mortality syndrome' (RMS) in some shrimp-farming areas of Southeast Asian countries. In contrast, significant mass mortality of individuals infected by CMNV occurred in certain cases of sudden change in water temperature or the weather (Zhang et al., 2014, 2017b).

Clinical signs

Atrophy and necrosis of the hepatopancreas and empty stomach and gut are the most common clinical signs of CMNV infection (Figure 21.3). Necrosis of abdominal segment muscle was also observed in typical CMNV-infected individuals. When massive viral replication occurs, this will deplete the host's calcium ions reserves considerably, resulting in calcium deficiency. As a result, CMNV infection usually results in obvious symptoms of softening of the shell, and finally leads to retarded growth of diseased individuals because of molting dysfunction (Zhang et al., 2014, 2017, 2020).

Note: the appearance of typical syndromes of this disease mostly occurs when the viral level in the host is high, but CMNV infection may still occur in the absence of any of the above-mentioned signs, especially in the early phase of infection.

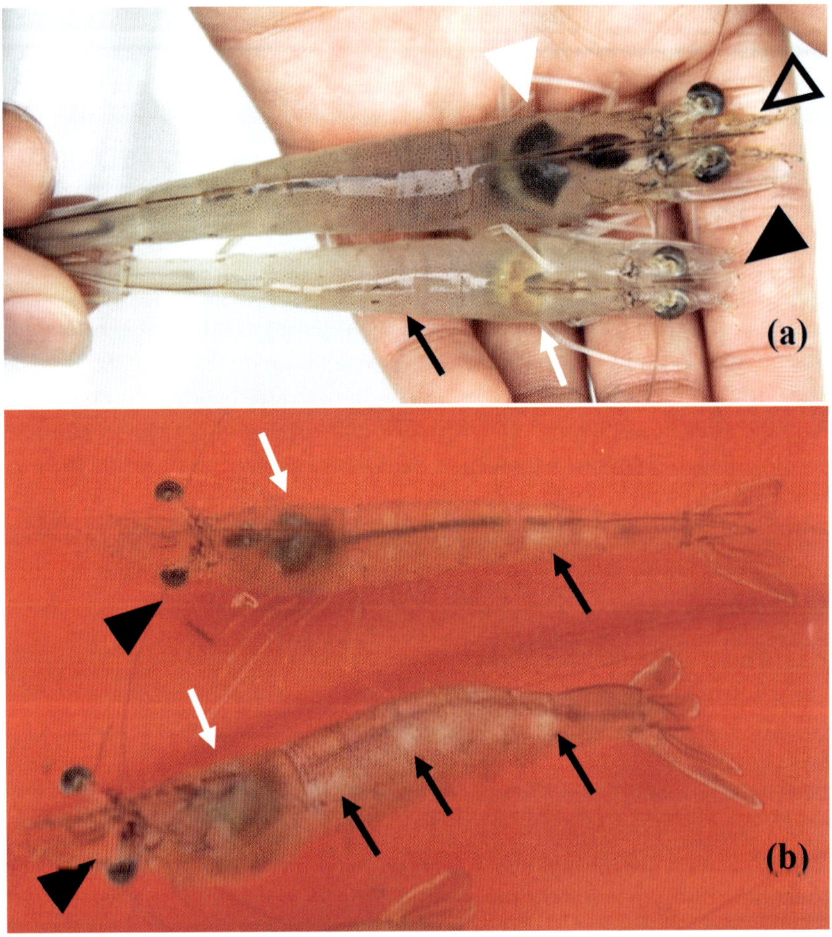

FIGURE 21.3 Clinical signs of VCMD in *P. vannamei*. Figure 21.3A, shrimp are from an artificial infection experiment; Figure 21.3B, shrimp from a farmed pond. Black arrows indicate whitened muscle of the abdominal segment; white arrows indicate hepatopancreatic atrophy and color fading; hollow arrowhead indicates a healthy shrimp; filled black arrowheads indicate shrimp suffering CMD; white arrowhead indicates the hepatopancreas of a healthy shrimp. Credit: Zhang et al., 2014.

PATHOGENIC AGENTS

VCMD is caused by CMNV, a new member of *Alphanodavirus*, Nodaviridae (Liu et al, 2017; Zhang et al., 2014). Similar to other members of Nodaviridae, CMNV is an icosahedral virus without an enveloped structure (Figure 21.4).

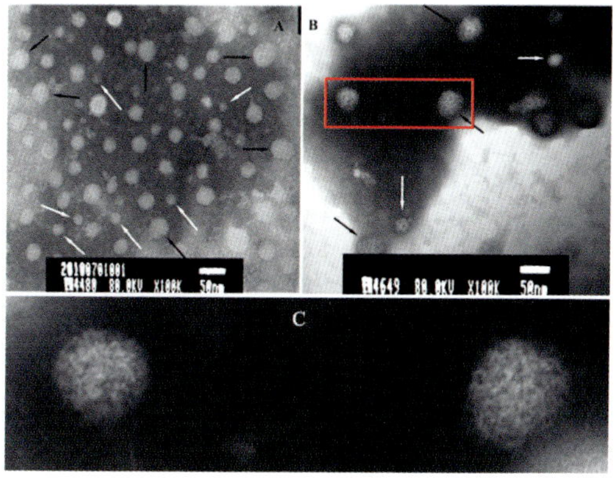

FIGURE 21.4 A, B, Transmission electron micrographs of negative-stained CMNV virions. Virus-like particles collected from sucrose-gradient fraction. The sizes of the virus-like particles are not homogeneous. The larger virus-like particles measure about 32 nm (black arrows) and the smaller ones (white arrows) about 19 nm. C, Magnified micrograph of virus-like particles. Scale bars in A, B, 50 nm. Credit: Zhang et al., 2014.

SUSCEPTIBLE HOSTS

CMNV infection can cause up to 80% mortality in cultured *P. vannamei*, and can also result in high mortality in farmed *P. chinensis*, *P. japonicus*, *P. monodon*, *Procambarus clarkia* and *Macrobrachium rosenbergii* (Zhang et al., 2014, 2017). CMNV was reported as switching its host to fish by cross-species transmission and could infect teleosteans such as *Carssius auratus*, *Paralichthys olivaceus* and *Mugilogobius abei* (Wang et al., 2019a,b; Zhang et al., 2018) (Figure 21.5). In addition, many crustacean species co-inhabiting shrimp farming ponds, including *Ocypode cordimanus*, *Diogenes edwardsii*, *Corophium sinense Zhang*, *Parathemisto gaudichaudi* and *Tubuca arcuate*, were shown to be susceptible to CMNV (ISH positive) (Austin and Newaj-Fyzul, 2017; Liu et al., 2017, 2018; Wang et al., 2019a,b; Zhang et al., 2014, 2017). This was the first shrimp virus to infect fish by cross-species infection, so the high risk of CMNV infecting other fish species, especially freshwater fishes, cannot be ignored.

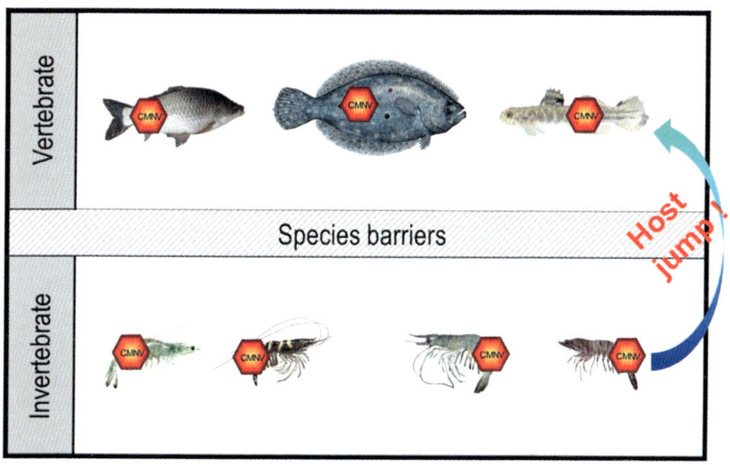

FIGURE 21.5 Cross-species infection by CMNV in fish. Credit: Zhang et al., 2018; Creative Commons license, visit http://creativecommons.org/licenses/by/4.0/.

AFFECTED AREAS

CMNV-positive samples have been reported in most major shrimp-farming areas, including countries in Asia and South America (Li et al., 2019; Pooljun et al., 2016; Thitamadee et al., 2016; Varela, 2016, 2018; Zhang et al., 2014, 2017, 2018) (Figure 21.6).

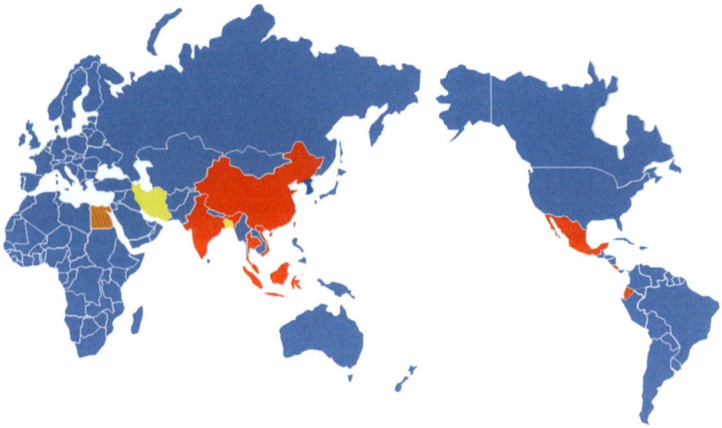

FIGURE 21.6 CMNV prevalence globally. Red indicates the areas where the virus-positive samples have been confirmed; Orange indicates areas that have virus-positive samples yet to be further confirmed; Yellow indicates areas occurred suspecting disease cases but not finished virus-positive samples confirming. Credit: Q. Zhang.

EPIDEMIOLOGY

1. CMNV infection in *P. vannamei* usually occurs within 30–80 days post stocking, and can cause cumulative mortality up to 80%. In serious cases, it occurs within 3–5 days of stocking post-larvae into grow-out ponds. Cases of asymptomatic infection detected with CMNV kits were also found on farms (Zhang et al., 2014, 2017).
2. Horizontal transmission occurs because of the cannibalistic behavior of shrimp, and can be introduced and caused by co-inhabiting vectors in shrimp ponds (Liu et al., 2017, 2018; Zhang et al., 2014).
3. Certain wild crustaceans and fish in ponds are vectors of the disease (Liu et al., 2018; Wang et al., 2019).
4. Migratory birds, aquatic insects, molluscs and humans are likely mechanical vectors of the disease (Liu et al., 2018).

HISTOPATHOLOGY

Figures 21.7–21.9 show histological sections of various tissues from infected shrimp.

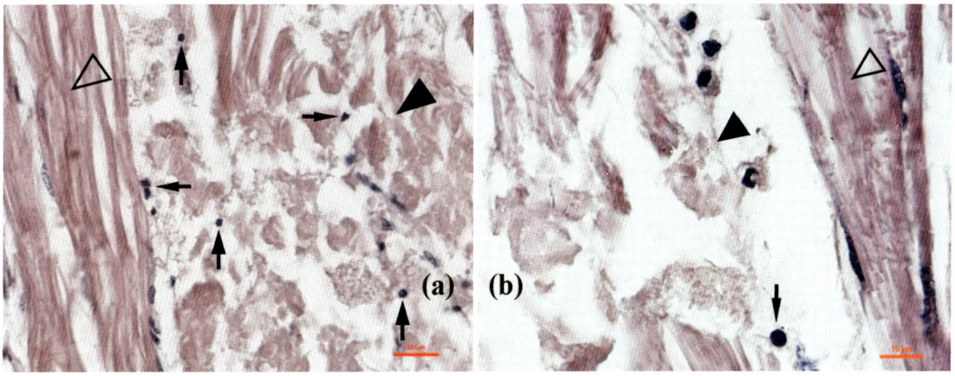

FIGURE 21.7 Histological section of necrotic abdominal muscle from white shrimp. The muscle shows fragmentation tending towards coagulative and dissolving necrosis (filled arrowheads). Hollow arrowheads and arrows indicate normal skeletal muscle and karyopyknotic nuclei, respectively. Scale bars, 20 µm (A), 10 µm (B). Credit: Zhang et al., 2014.

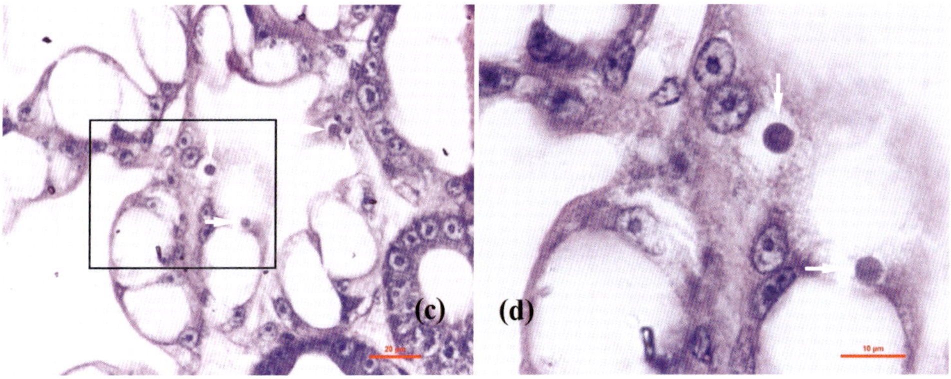

FIGURE 21.8 Histological section of atrophied and necrotic hepatopancreas epithelium from white shrimp. Note that the inclusions in hepatopancreatic tubular epithelium are a typical histopathologic characteristic of CMNV infection. Arrows denote inclusions (light purple). Scale bars, 20 µm (A), 10 µm (B), Credit: Zhang et al., 2014.

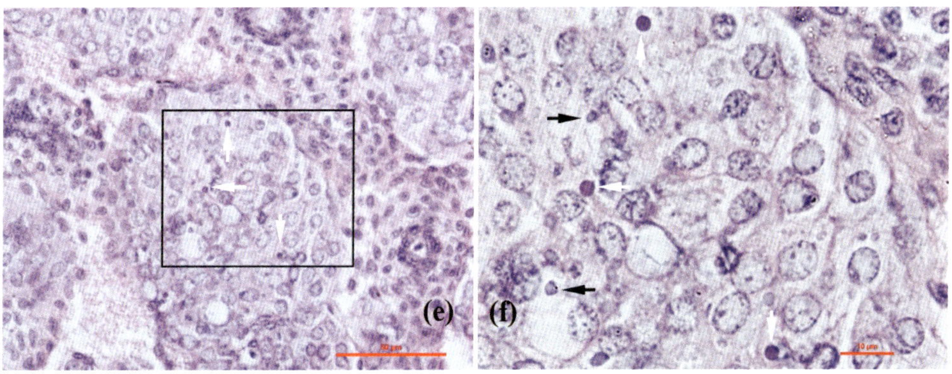

FIGURE 21.9 Histological section of lymphoid organ from white shrimp. Note the inclusions (light purple) (white arrows) and karyopyknotic nuclei (black arrows). Scale bars, 50 µm (A), 10 µm (B). Credit: Zhang et al., 2014.

MOLECULAR DIAGNOSTICS

Reverse transcription amplification-based methods

Various molecular diagnostic methods were established based on the sequence of RNA-dependent RNA polymerase (RdRp). The reported methods include:

- a reverse transcript nest PCR (RT-nPCR) (Zhang et al., 2014)
- a RT–LAMP method (Zhang et al., 2017b)
- a sensitive TaqMan RT–PCR (Pooljun et al., 2016)
- a high-compatibility TaqMan real-time RT–PCR (Li et al., 2018)

Note: The risk of false negative results with CMNV detection using molecular amplification is high (Li et al., 2018) because of the absence of proofreading activity of CMNV RdRp in genome replication. It is strongly recommended that ISH or TEM confirmation should be done in the case of RT-nPCR or RT-qPCR negative results, with particular emphasis on samples associated with mortality and clinical signs (if any) in the host.

Reverse transcription-nested PCR (RT-nPCR)

RT-nPCR is implemented first by performing reverse transcription followed by two separate PCR steps using two pairs of primers targeting the CMNV RdRp gene. In the first PCR step the PCR amplicons are 619 bp and in the second PCR step they are 413 bp. To visualize amplicons, PCR product (5 µl of each) was analyzed in 1.5% agarose gel containing GeneFinder, SYBR Green I or ethidium bromide (Zhang et al., 2014, 2018).

Reverse transcription

1. Denaturation of the sample RNA: prepare denaturing premix tube containing 4 µl of RNase-free water and 1 µl of 10 mmol/l primer Noda-R1. Before reverse transcription, 1 µl of the sample RNA (200 ng/µl) to be tested is added to the abovementioned premix, pre-denatured at 70°C for 5 min and the tube containing the premix placed on ice for 2 min.
2. Prepare reverse transcription premix by mixing the following reagents: 2 µl 5× M-MLV reverse transcriptase buffer, 0.5 µl dNTP (10 mmol/l), 0.25 µl RNase Inhibitor (40 U/µl), 0.5 µl M-MLV Reverse transcriptase (200 U/µl), then add 0.75 µl of RNase-free water to a final volume of 4 µl.
3. Reverse transcription reaction: Mix the reverse transcription premix well by pipetting or vortexing, then add the premix to the denaturing premix tube in (1), gently centrifuge to collect the liquid to the bottom of the tube, then bathe at 42 °C for 1 h, denaturing at 70°C for 5 min. The cDNA template thus obtained should be immediately used for nPCR or stored at –20 °C until use.

PCR, first step

1. The 25-µl PCR reaction mixture contains 2.5 µl of 10× Ex Taq Buffer (Mg^{2+} free), 2 mM $MgCl_2$, 0.2 mM dNTPs, 0.4 µM primers (CMNV-F1: 5′AAA TAC GGC GAT GAC G -3′ and CMNV-R1: 5-ACG AAG TGC CCA CAG AC -3), 0.625 U TaKaRa Ex Taq DNA polymerase (TaKaRa, Dalian, China) and 1 µl of template cDNA.

2. PCR is performed at 95 °C for 3 min, followed by 30 cycles at 95 °C for 30 s, 50 °C for 30 s and 72 °C for 40 s, ending at 72 °C for 5 min.

PCR, second step (nested PCR)

1. PCR mixture as above but with 1 µl of first-step PCR product as template and different primers (CMNV-F2: 5-TCG CGT ATT CGT GGA T -3′and CMNV-R2: 5-TAG GGT CAA AAG GTG TAG T -3).
2. Amplification is performed with the following cycling parameters: denaturation at 94 °C for 3 min; 94 °C for 20 s, 50 °C for 20 s, 72 °C for 30 s, 30 cycles; extension at 72 °C for 7 min.

Figure 21.10 shows an electrophoretogram used for the detection of CMNV.

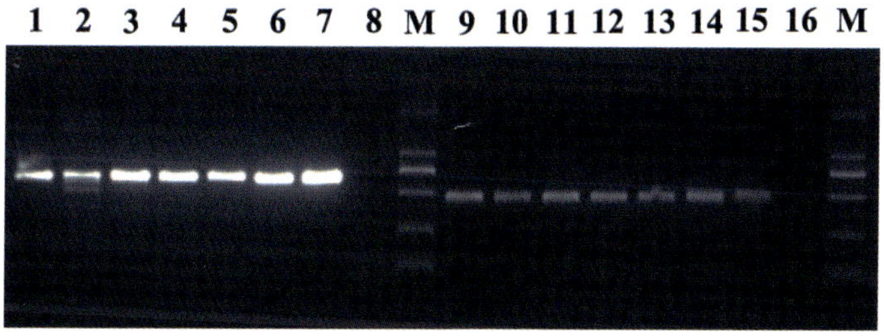

FIGURE 21.10 Electrophoretogram for detection of CMNV in *Mugilogobius abei* from shrimp farming ponds and coastal waters, using RT-nPCR. Lane M, DL2000 molecular weight marker. Lanes 1–8 show the results of the first step of RT-nPCR and lanes 9–16 show the results of the second step. Lanes 1–3 and 9–11 show amplification results of cDNAs from individuals collected from shrimp farming ponds; lanes 4–6 and 12–14 show results for individuals collected from coastal water near the ponds; lanes 7, 8, 15 and 16 show results from the positive and negative controls, respectively. Credit: Zhang et al., 2018; Creative Commons license, visit http://creativecommons.org/licenses/by/4.0/.

Taqman real-time RT-qPCR

The Taqman real-time RT-qPCR assay is conducted using Takara's One Step RNA PCR Kit (TaKaRa, Dalian, China). The sequences of primers targeting a 198-bp amplicon of the ATPase gene are CMNV-Taq-F: 5′-CGA GCT AAT CCA AGC ACT TC-3′), CMNV-Taq-R: 5′-ACC TGT TAG GTA CGC TAC CA-3′. The TaqMan Probe has the sequence: 5′-FAM CGC TCA CGG CTT TGG ATA CCT T TAMRA-3′ (Li et al., 2018). The procedure is as follows (Figure 21.11):

1. Taqman real-time RT-qPCR assay was performed in a 25-µl reaction system consisting of 12.5 µl 2× One Step RT-PCR Buffer III (Takara), 0.2 µM primers, 0.2 µM probe, 0.5 µl Enzyme Mix II, 1 µl RNA template (100 ng/µl) and a moderate volume of RNase-free water.
2. The reverse transcription PCR reaction program: incubate for 15 min at 51 °C for reverse transcription of the RNA template followed by 40 cycles, including denaturation at 94 °C for 10 s and annealing and extension at 52.7 °C for 30 s.

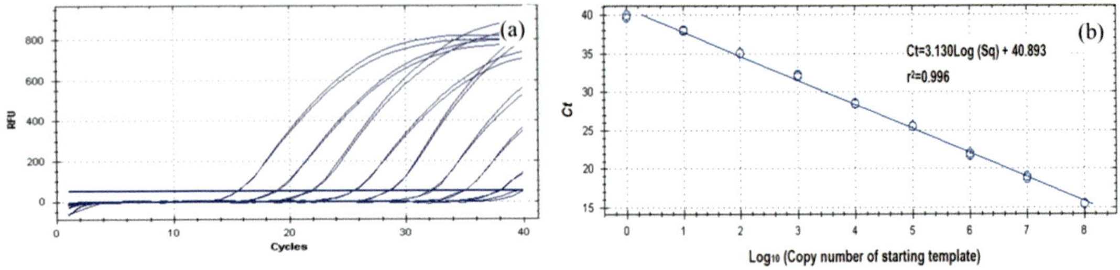

FIGURE 21.11 Amplification plots and standard curve of serial tenfold dilution of plasmids by CMNV-specific TaqMan RT-PCR assay. (A) Plot of TaqMan RT-PCR with serial tenfold dilution of the plasmids. Copies of pMD20-CMNV plasmids (left to right): 5.7×10^8, 5.7×10^7, 5.7×10^6, 5.7×10^5, 5.7×10^4, 5.7×10^3, 5.7×10^2, 5.7×10^1 and 5.7×10^0 copies. (B) Standard curve and standard curve equation for CMNV-specific TaqMan RT-PCR assay generated from amplification plots of the serial tenfold diluted pMD20-CMNV plasmids and *Ct* values. Credit: Li et al., 2018.

Nucleic acid probe-based in situ hybridization (ISH) methods

- An ISH method for CMNV was initially described in 2014 (Zhang et al., 2014).
- An alternative ISH method was subsequently published (Liu et al., 2017; Zhang et al., 2017).

DISEASE CONTROL

1. Disinfection treatment of farming ponds and water. Many species of organisms in sea water and co-inhabiting organisms in shrimp farming ponds were the vectors/carriers of CMNV. Both farming ponds and sea water should be disinfected before shrimp aquaculture.
2. Use seedlings without CMNV. The larval stage is when the risk of CMNV introduction into shrimp farming ponds is highest. Shrimp larvae must be tested before culture to ensure the absence of CMNV.
3. Use fresh bait without CMNV. Fresh bait represents one of the main risks for CMNV introduction into shrimp farming ponds, so fresh bait should be tested before culture to ensure that larvae are free of CMNV.

Infection with Decapod iridescent virus 1

An emerging disease in farmed white-leg shrimp, *P. vannamei* and redclaw crayfish, *Cherax quadricarinatus* caused high mortality in China in 2014 (Qiu et al., 2017; Xu et al., 2016). The new iridescent virus, Decapod iridescent virus 1 (DIV1), was proven to be the etiological agent of this new disease, so the disease was named infection with Decapod iridescent virus 1 (iDIV1). It is also known as 'white head' or 'white spot' disease because of its typical clinical symptoms in the giant freshwater prawn, *Macrobrachium rosenbergii* (Qiu et al., 2019a).

SIGNS OF DISEASE

The following disease signs can be used for presumptive diagnosis of the disease. Infected individuals may show some of the signs below, but disease may still be present in the absence of any signs, especially in the early phase of infection.

Clinical signs of infected individuals include hepatopancreatic atrophy with color fading on the surface, empty stomach and gut and reddish body coloration (Figure 21.12) (Qiu et al., 2017). Diseased *M. rosenbergii* typically exhibit a white area under the carapace at the base of the rostrum, called white head or white spot disease (Figure 21.13) (Qiu, et al., 2019a). Diseased *Exopalaemon carinicauda* also show a slightly whitish area under the carapace at the base of the rostrum (Chen et al., 2019). Moribund individuals sink to the bottom in deep water and dead individuals can be found every day, with cumulative mortality up to 80%.

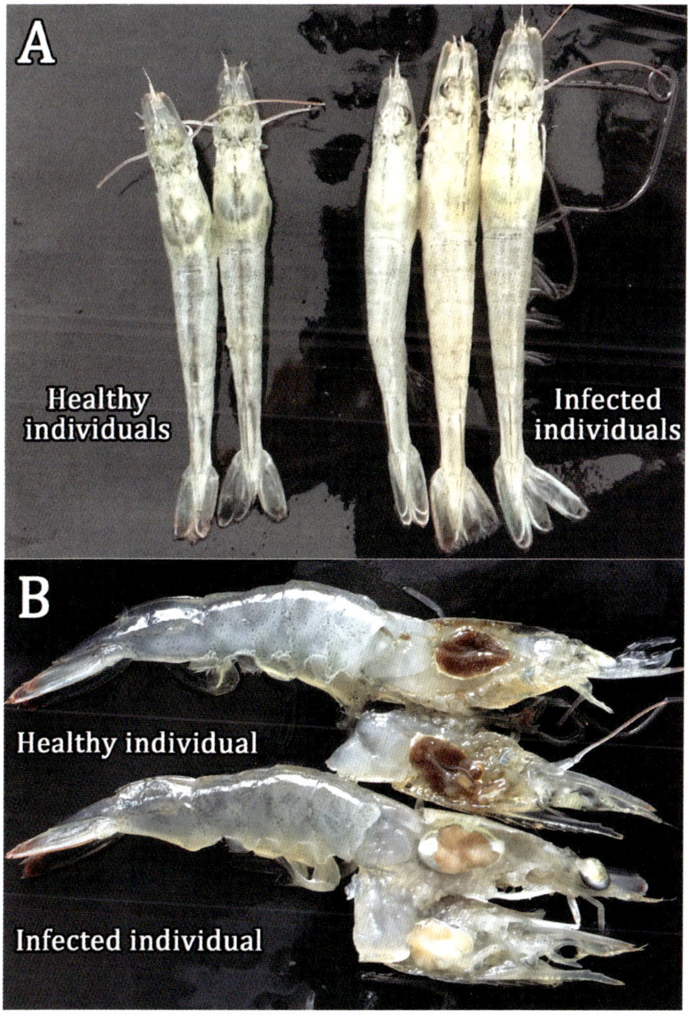

FIGURE 21.12 Clinical signs of *P. vannamei* individuals challenged with DIV1 compared with healthy individuals. (A) External appearance. (B) Section of cephalothorax. Credit: revised from Qiu et al., 2017; Creative Commons license, visit http://creativecommons.org/licenses/by/4.0/

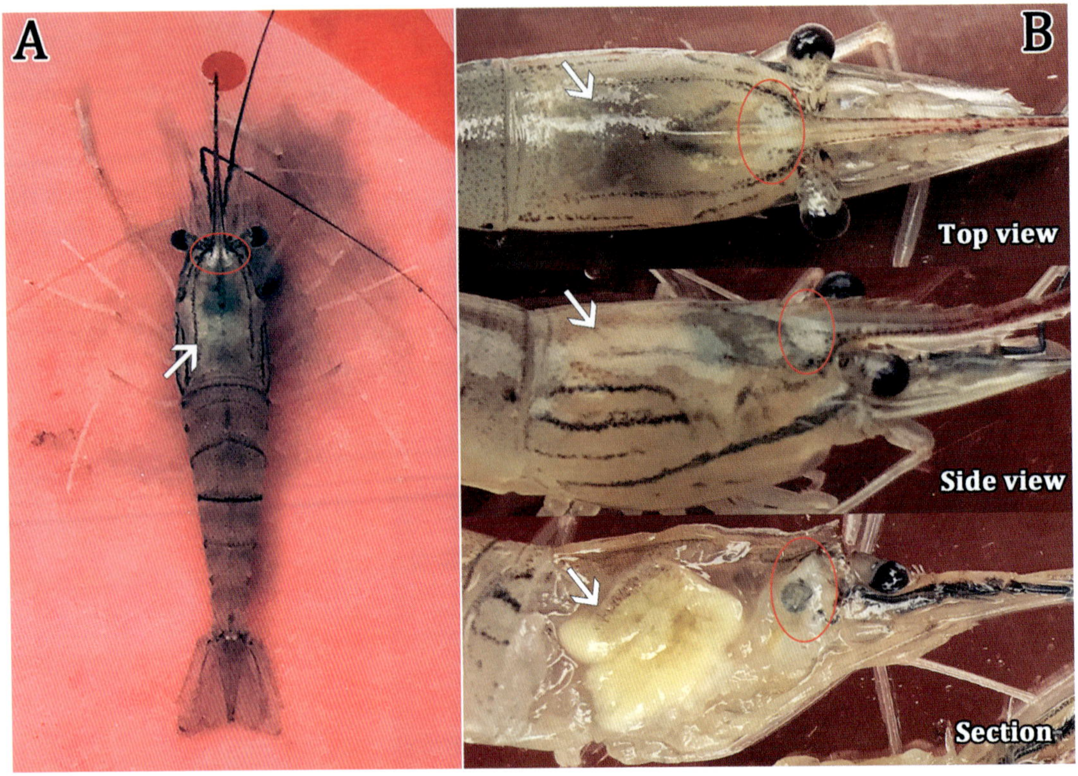

FIGURE 21.13 Clinical signs of *M. rosenbergii* challenged with DIV1. (A) Overall appearance of a diseased prawn in water. (B) Close-up of cephalothorax. Red circles denote white area under the cuticle at the base of the rostrum; white arrows indicate hepatopancreatic atrophy, color fading and yellowing. Credit: revised from Qiu et al., 2019a; Creative Commons license, visit http://creativecommons.org/licenses/by/4.0/

DISEASE AGENTS

DIV1 was assigned by the International Committee on Taxonomy of Viruses (ICTV) as the only member of the genus *Decapodiridovirus* within the family *Iridoviridae*. From July to November 2014, Xu et al. (2016) identified a new iridescent virus in *C. quadricarinatus* samples in Fujian Province using transmission electron microscopy (TEM), sequence analysis of partial major capsid protein gene, virion purification, protein analysis and experimental infections, and temporarily named the new virus *Cherax quadricarinatus* iridovirus (CQIV). In December 2014, Qiu et al. (2017) identified a new iridescent virus in a batch of farmed *P. vannamei* from Zhejiang Province using viral metagenomics analysis, histopathology, TEM, virion purification, experimental infections and *in situ* hybridization, and named it shrimp hemocyte iridescent virus (SHIV) based on the infected organs and susceptible species. SHIV and CQIV both have a typical icosahedral structure with a mean diameter of about 150 nm. Phylogenetic analysis supported that SHIV and CQIV belong to a new genus, which was originally proposed to be named *Xiairidovirus* or *Cheraxvirus*, in the family *Iridoviridae* (Li et al., 2017; Qiu et al., 2017, 2018b). Alignment of the complete genomic sequences revealed that SHIV and CQIV might be different strains

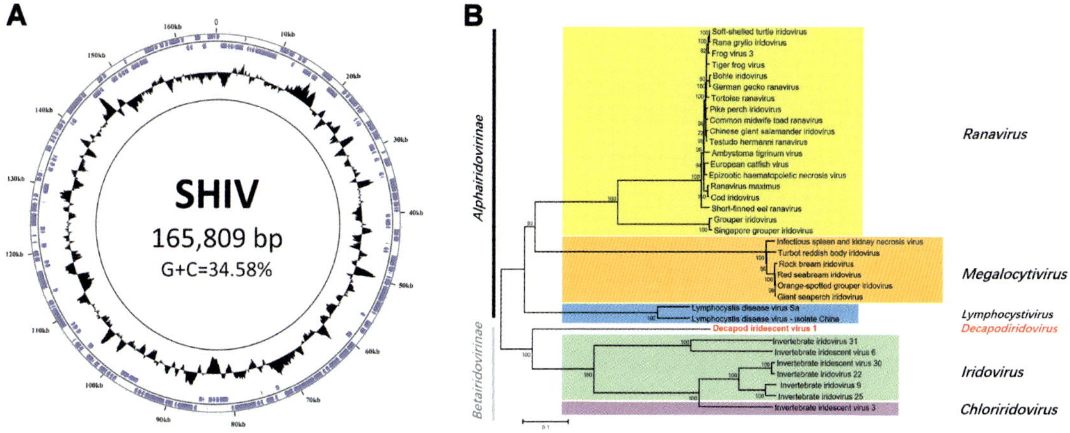

FIGURE 21.14 (A) Circular map of the 165,809-bp DIV1 genome (SHIV 20141215). The outer scale is numbered clockwise in bp. Circles 1 and 2 (from outside to inside) show ORFs on the forward and reverse strand, respectively). Circle 3 represents the G+C content along the genome sequence. Peaks toward the outside indicate G+C content >50%; peaks toward the inside indicate G+C content >50%; the height is proportional to the content). (B) Concatenated phylogenetic tree of family Iridoviridae based on 27 conserved genes. Credit: revised from Qiu et al., 2018b; Creative Commons license, visit http://creativecommons.org/licenses/by/4.0/

or genotypes of the same viral species (Qiu et al., 2018a). In March 2019, ICTV approved the proposal to add a new species, DIV1, in a new genus, *Decapodiridovirus*, with SHIV 20141215 and CQIV CN01 as the original isolates. DIV1 has a double-stranded DNA genome of about 166 Kbp (Figure 21.14) (Li et al., 2017; Qiu et al., 2018b).

HOST RANGE

Currently, known susceptible species include *P. vannamei*, *M. rosenbergii*, *E. carinicauda*, *M. nipponense*, *Procambarus clarkii* and *C. quadricarinatus* (Chen et al., 2019; Xu et al., 2016; Qiu et al., 2017, 2019). Two species of crab, *Eriocheir sinensis* and *Pachygrapsus crassipes*, were infected with DIV1 in experimental challenge by intramuscular injection (Pan et al., 2017), and cannot yet be identified as susceptible species. DIV1 was also detected in *P. chinensis*, *P. japonicus*, *M. superbum*, *Nereis succinea* and some cladocera using only PCR (Qiu et al., 2017, 2018c, 2019a, b).

Symptoms and mortality were observed in infected *P. vannamei* from post-larval to sub-adult stages in the laboratory (Qiu et al., 2017). Target surveillance in China in 2017–2018 revealed that the virus has been detected in shrimp and crayfish, with body length <1 cm to >10 cm; the highest positive rate was in specimens of body length 4–7 cm (Qiu et al., 2018c, 2019b). Other reports have not commented on different levels of mortality by life stage.

GEOGRAPHICAL DISTRIBUTION

iDIV1 has been reported in coastal provinces of China since 2014, including Zhejiang, Guangdong and Hebei (Qiu et al., 2017). Target surveillance in China in 2017–2018 revealed that DIV1 had been detected in 11 of 16 provinces (Qiu et al., 2018c, 2019b). The geographic distribution of DIV1 may be wider than currently assumed (mortality may not have been investigated in other regions).

EPIDEMIOLOGY

Challenge tests in *P. vannamei* and *E. carinicauda per os* and by reverse gavage have demonstrated that direct horizontal transmission is an important route (Chen et al., 2019; Qiu et al., 2017). In addition, polyculture modes with different species of crustaceans increase the risk of outbreaks and spread. In 2019, Qiu et al. (2019a) reported a case of natural infection of *M. rosenbergii*, showing that not only farmed *M. rosenbergii* but also polyculture species of *P. vannamei*, *P. clarkii* and wild *M. nipponense* suffered from different levels of infection. The wild *M. superbum* and some cladocera in the pond were DIV1 PCR positive.

Although there is no evidence of vertical transmission to date, samples from hatcheries have been DIV1 positive (Qiu et al., 2018c).

Target surveillance in China in 2017–2018 revealed that the virus had been detected in shrimp and crayfish at temperatures from 16 to 32 °C, but none at higher temperatures (Qiu et al., 2018c, 2019b).

HISTOPATHOLOGY

The histopathology is characterized by dark eosinophilic inclusions mixed with basophilic staining and karyopyknosis in hematopoietic tissues and hemocytes of the gills, hepatopancreatic sinus, periopods and cuticular epithelium in sections stained by hematoxylin and eosin (H&E) (Figure 21.15) (Chen et al., 2019; Qiu et al., 2017, 2019a).

Positive signals of *in situ* hybridization (ISH) were observed in hematopoietic tissue and hemocytes in gills, hepatopancreas and periopods (Figure 21.16) (Qiu et al., 2017).

Results of *in situ* digoxigenin-labeled, loop-mediated isothermal amplification (ISDL) showed the presence of blue signals in hematopoietic tissue (Figure 21.17A), hemocytes in the hepatopancreatic sinus and in gills (Figure 21.17B, C), some R-cells and myoepithelial fibers of the hepatopancreas (Figure 21.17D), the coelomosac epithelium of the antennal gland (Figure 21.17E), ovarian epithelium (Figure 21.17F) and stomach (Chen et al., 2019; Qiu et al., 2019a).

Visualization with TEM of ultrathin sections of diseased *M. rosenbergii* revealed a large number of icosahedral iridescent particles, both inside and outside the hematopoietic cells (Figure 21.18A). At the cytoplasm edge, many virions were budded from the plasma membrane and, in budding virions, an outer viral envelope was obtained from the plasma membrane (Figure 21.18B). Virion formation occurred in cytoplasmic regions morphologically distinct, termed virogenic stromata, which were electron-lucent areas containing many immature and empty capsids and a few mature virions and were devoid of cellular organelles, with a paracrystalline array of viral particles and budding virions (Figure 21.18C).

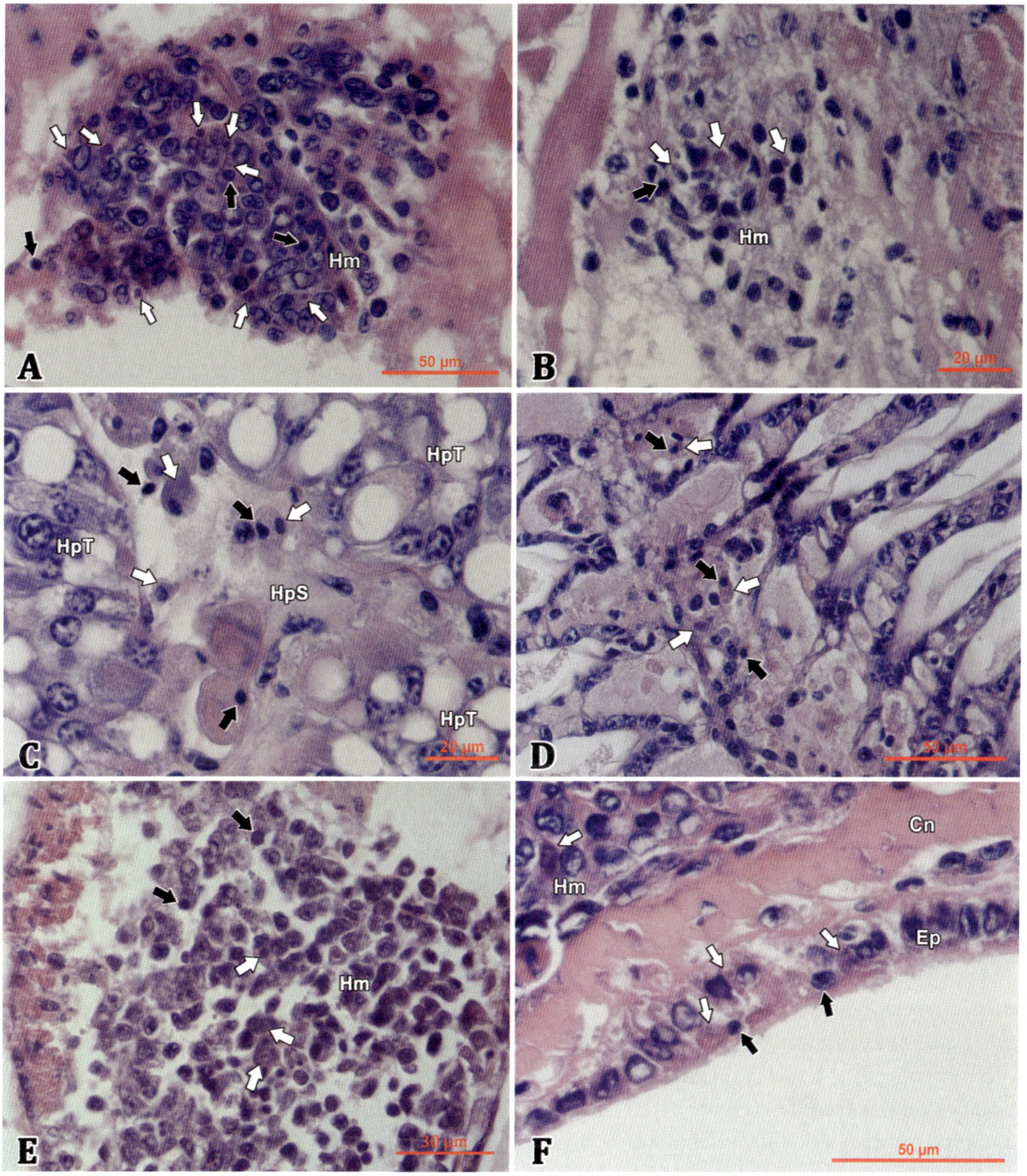

FIGURE 21.15 Histopathological features of diseased shrimp tissues stained using Davidson's alcohol-formalin-acetic acid fixative (DAFA) and H&E. (A, B) Hematopoietic tissue; (C) hepatopancreas; (D) gills; (E) periopods; (F) cuticular epithelium. Black arrows indicate karyopyknosis and white arrows eosinophilic inclusions. HpT: hepatopancreatic tubule; HpS: hepatopancreatic sinus; Hm: hematopoietic tissue; Ep: epithelium; Cn: connective tissue Credit: revised from Chen et al., 2019 and Qiu et al., 2017, 2019a; Creative Commons license, visit http://creativecommons.org/licenses/by/4.0/

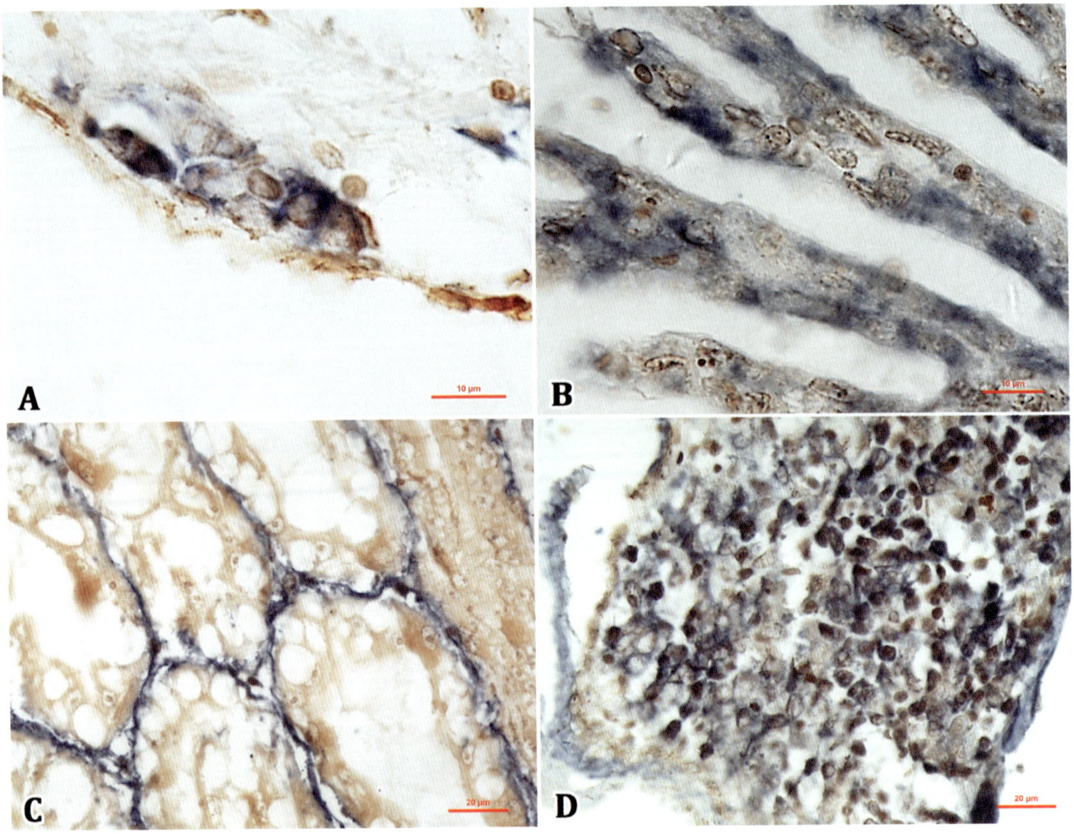

FIGURE 21.16 ISH using a digoxigenin-labeled, 279-bp probe for DIV1 on histological sections of diseased shrimp. (A) hematopoietic tissue; (B) gills; (C) hepatopancreas; (D) periopods. Credit: Qiu et al., 2017; Creative Commons license, visit http://creativecommons.org/licenses/by/4.0/

The nucleocapsid assembly can be described in three stages (Figure 21.18D). In the stage 1, the crescent-shaped structures of early capsid complexes are subsequently assembled into spherical intermediates (Figure 21.18E–G, numbered 1 on Figure 21.18D). In stage 2, icosahedral capsids was formed with a small opening at one vertex (Figure 21.18H, numbered 2 on Figure 21.18D). In stage 3, electron-dense nucleic acid was recruited (Figure 21.18I, numbered 3 on Figure 21.18D). Complete nucleocapsids were observed in a fully filled state (Figure 21.18J, numbered 4 on Figure 21.18D) (Qiu et al., 2019a).

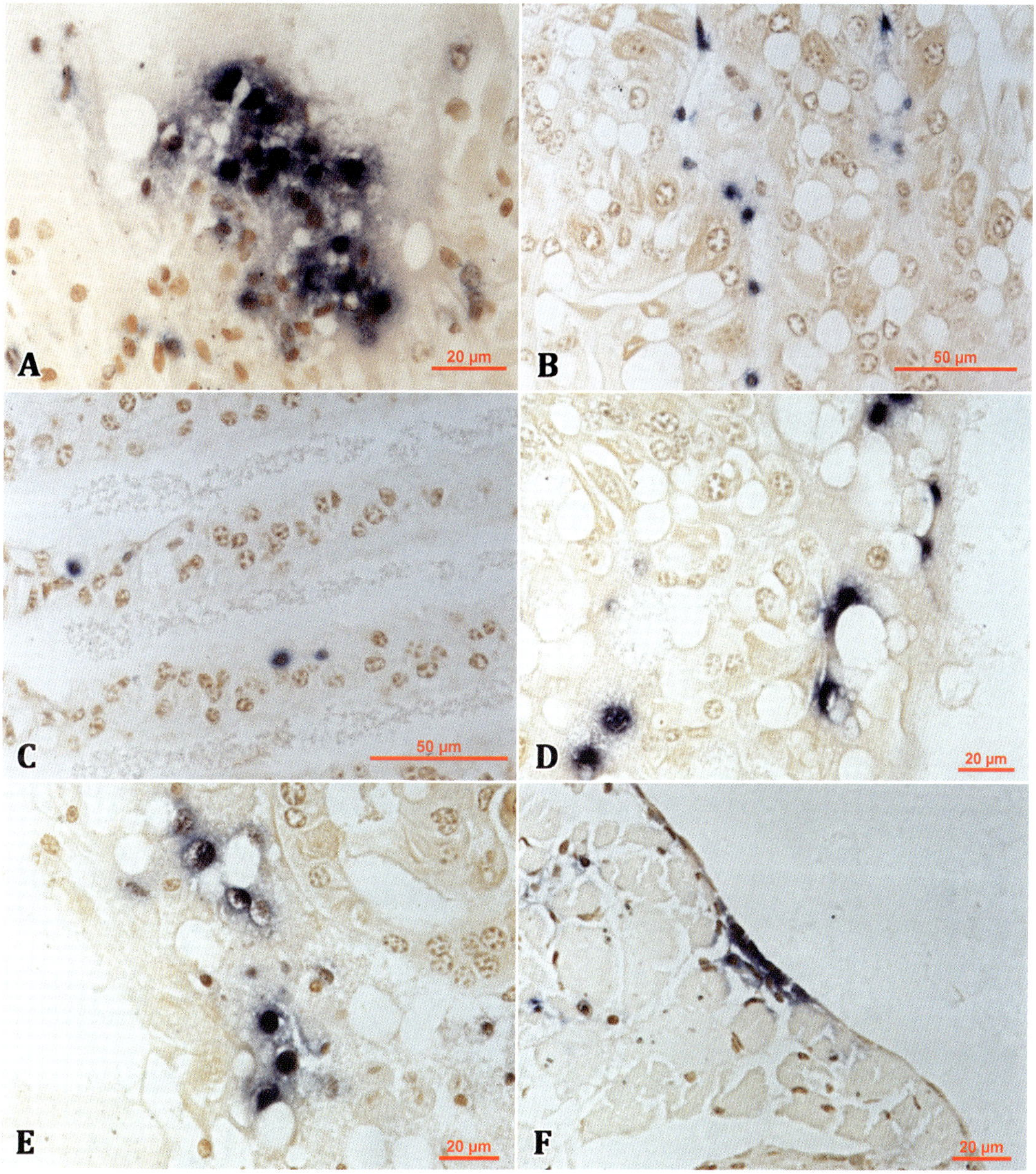

FIGURE 21.17 ISDL targeting the gene of the second largest subunit of DNA-directed RNA polymerase II of DIV1 on histological sections of diseased shrimp. (A) Hematopoietic tissue; (B, D) hepatopancreas; (C) gills; (E) antennal gland; (F) ovaries. In A–C and H, blue signals were observed in hematopoietic tissue, hemocytes in the sinus of the hepatopancreas, and in gills. In D, blue signals were present in some hepatopancreatic R-cells and myoepithelial fibers. In E, blue signals were present in the coelomosac epithelium. In F, blue signals were present in the epithelium. Credit: Qiu et al., 2019a; Creative Commons license, visit http://creativecommons.org/licenses/by/4.0/

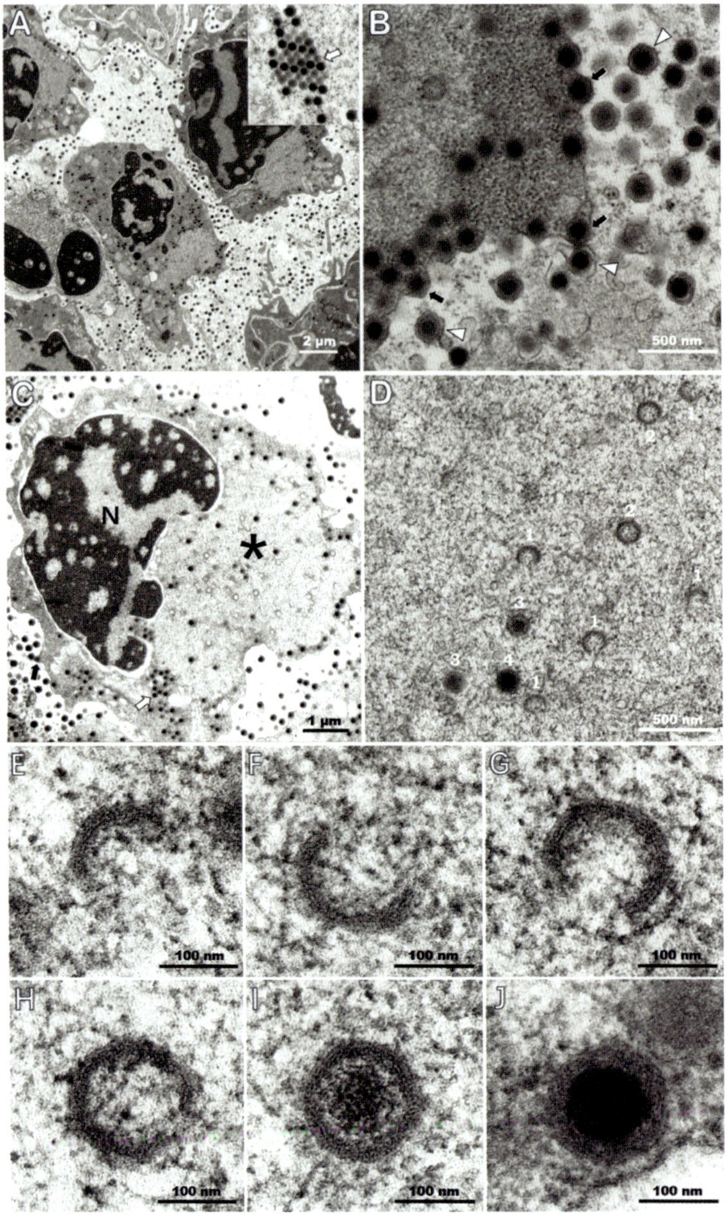

FIGURE 21.18 TEM of hematopoietic tissue of naturally infected *M. rosenbergii* samples. (A) Large number of virions in hematopoietic tissue. (B) DIV1 budded and acquired an envelope from the plasma membrane. (C) DIV1 replication and assembly in hematopoietic cells. (D) The three stages of nucleocapsid assembly, numbered 1–3; a complete nucleocapsid is numbered 4. The capsids at stages 2 and 3 should have a small opening at one vertex but may not be visible in the picture due to the ultrathin section. (E) Crescent-shaped structures. (F–I) As the assembly process continues, the crescent-shaped structure curves to form icosahedral capsids. (J) A mature virion with a dense core is eventually formed. N, nucleus; *, a large electron-lucent virogenic stroma; white arrowheads, paracrystalline array of viral particles; black arrows, budding virions; white triangles, budded virions that acquired an envelope. Credit: Qiu et al., 2019a; Creative Commons license, visit http://creativecommons.org/licenses/by/4.0/

MOLECULAR DIAGNOSTICS

Nested PCR (Qiu et al., 2017)

Nested PCR was carried out in two separate steps using two pairs of primers targeting the ATPase gene of DIV1. In the first step, the PCR amplified a 457-bp amplicon; in the second, a 129-bp amplicon was amplified. To visualize the amplicons, PCR product (5 µl of each) was analyzed in a 1% agarose gel containing GeneFinder (Bio-V, China) (Figure 21.19).

First step

1. The PCR reaction mixture (25 µl) contained 2.5 µl of 10× Ex Taq Buffer (Mg^{2+} free), 2 mM $MgCl_2$, 0.2 mM dNTP, 0.4 µM primers (DIV1-F1: 5′ GGG CGG GAG ATG GTG TTA GAT-3′ and DIV1-R1: 5′-TCG TTT CGG TAC GAA GAT GTA-3′), 0.625 U TaKaRa Ex Taq DNA polymerase (TaKaRa, Dalian, Liaoning, China) and 1 µl of template DNA.
2. PCR was performed at 95 °C for 3 min, followed by 35 cycles of 95 °C for 30 s, 59 °C for 30 s and 72 °C for 30 s, ending at 72 °C for 2 min.

Second step

1. The PCR mixture was as above but with 1 µl of first-step PCR product as template and different primers (DIV1 -F2: 5′-CGG GAA ACG ATT CGT ATT GGG-3′ and DIV1-R2: 5′-TTG CTT GAT CGG CAT CCT TGA-3′).

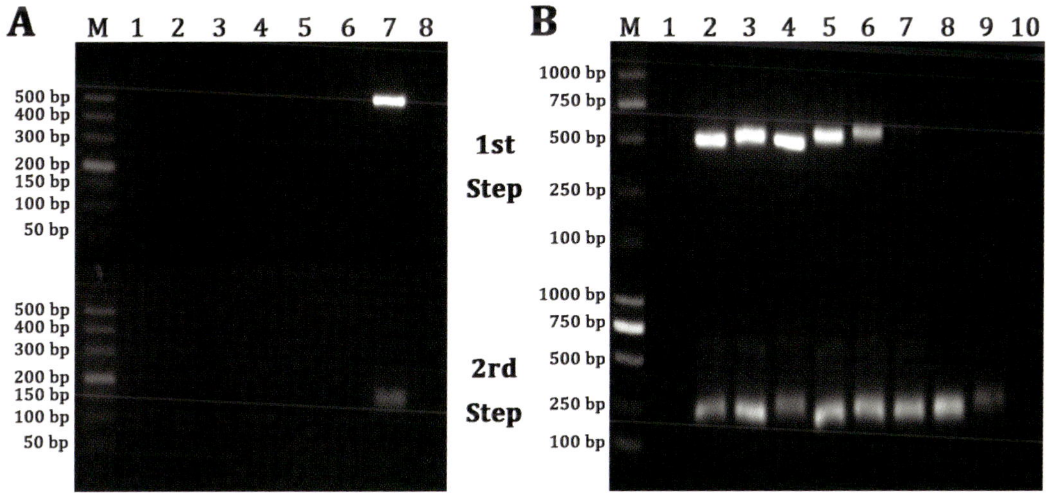

FIGURE 21.19 Nested PCR for DIV1. (A) Specificity analysis of the nested PCR method. M, DL500 molecular mass marker; lanes 1–6, PCR-amplified products with DNA template samples extracted from both healthy *P. vannamei* and *P. vannamei* infected with WSSV, IHHNV, HPV, AHPND and EHP, respectively; lane 7, PCR-amplified products with DNA template sample extracted from shrimp infected with DIV1; lane 8, negative control. (B) Sensitivity testing of the nested PCR method. M, DL1000 molecular mass marker; lane 1: negative control; lanes 2–10, PCR-amplified products with diluted DNA solutions (10^0–10^{-8}) of *P. vannamei* naturally infected with DIV1. Credit : Qiu et al., 2017; Creative Commons license, visit http://creativecommons.org/licenses/by/4.0/

2. Amplification was performed with the following cycling parameters: initial denaturation at 95 °C for 3 min, followed by 35 cycles of 95 °C for 30 s, 59 °C for 30 s and 72 °C for 20 s, with a final extension at 72 °C for 2 min.

Real-time PCR (Qiu et al., 2018a)

TaqMan assay was carried out using the FastStart Essential DNA Probes Master mix (Roche, Mannheim, Germany) (Figure 21.20). The sequences of primers targeting a 188-bp amplicon of the ATPase gene were DIV1-F: 5'-AGG AGA GGG AAA TAA CGG GAA AAC-3', DIV1-R: 5'-CGT CAG CAT TTG GTT CAT CCA TG-3'. The TaqMan Probe had the sequence: 5'-CTG CCC ATC TAA CAC CAT CTC CCG CCC-3'.

1. Real-time PCR amplification was performed in a 20-µl reaction system consisting of 10 µl 2× Master Mix, 500 nM of each primer (DIV1 F/R), 200 nM TaqMan probe and 1 µl DNA template.
2. The PCR profile was one cycle of 95 °C for 10 min for initial denaturation, followed by 40 cycles of 95 °C for 10 s and 60 °C for 1 min.

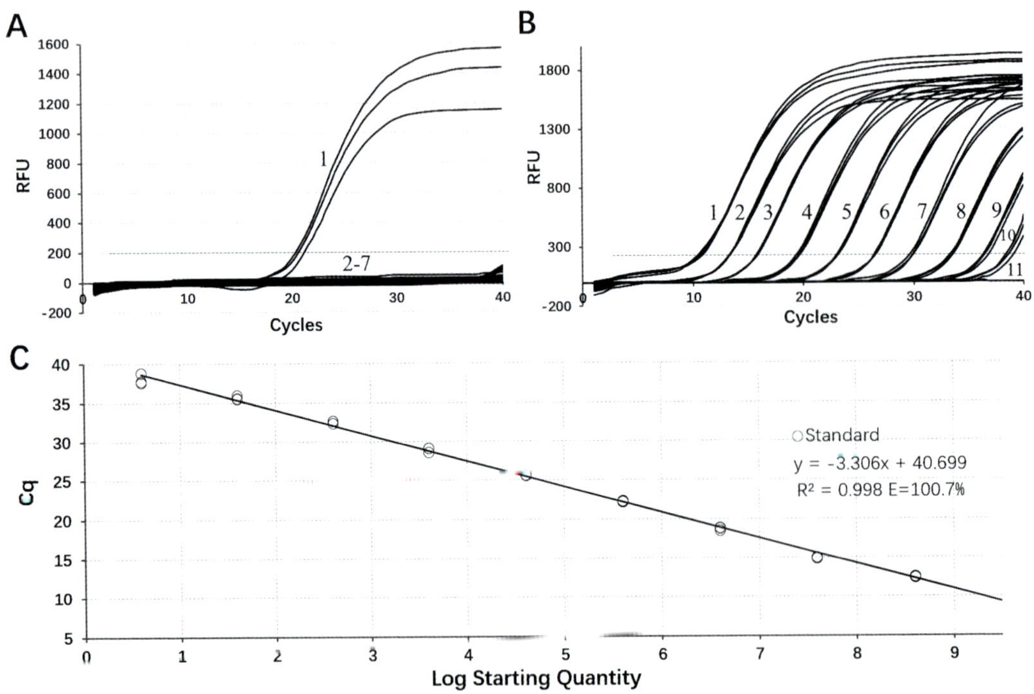

FIGURE 21.20 Taqman-based real-time PCR for DIV1. (A) Specificity of qPCR. Templates for qPCR were the DNA extracted from shrimp infected with 1, DIV1; 2, WSSV; 3, IHHNV; 4, VPAHPND; 5, EHP; 6, HPV; and 7, healthy shrimp. (B) Sensitivity of qPCR. Lanes 1–10, 4 × 10^9 to 4 × 10^0 copies/µl DIV1 DNA; lane 11, water (NTC). (C) Standard curve of qPCR. Credit: Qiu et al., 2018a.

DISEASE CONTROL

Enhancement of the biosecurity system is the key stratagem for gaining control of iDIV1. Disease surveillance plans for farms and quarantine with iDIV1 testing for broodstock and post-larvae need to be implemented. Restrictions on the movement of live crustaceans, and cleaning moribund or dead individuals from farms where the virus is known to occur, will limit the spread of disease. Farms should avoid using polycultures with different crustaceans in the same pond. Hatcheries should avoid the use of live or frozen raw decapods as feed for broodstock, or as feed of polychaetes that are to be fed to broodstock. Generic biosecurity measures to minimize fomite spread via equipment, vehicles or staff (i.e. cleaning and disinfection) should also be implemented (Qiu et al., 2018c).

Conclusions

Spread of disease is facilitated by the use of live or fresh broodstock-infected feeds containing polychaetes and clams. Also, shortages in the supply of SPF broodstock have led some entrepreneurs to utilize PL of imported SPF stocks to produce second-generation broodstock in open shrimp ponds where they became contaminated, and were then used to produce PL for stocking ponds. Such practices have left the whole shrimp industry vulnerable to the rapid spread of the emerging diseases and have resulted in the current crisis in Asian shrimp culture (Thitamadee et al., 2016). Studies revealing the etiology and patho-ecology of emerging diseases will help researchers and farmers to understand the basic characteristics of these diseases, and will also contribute to determining ways and measures to control the transmission of pathogens. Recently, many disease prevention and control measures have been taken by both farmers and governments in major shrimp-farming countries, including the widespread application of rapid, on-site detection kit for seedling quarantine, and the extensive adoption of indoor intensive culture, which should contribute to reduction in the prevalence of these emerging shrimp diseases.

References

Arunrut, N., Kampeera, J., Sirithammajak, S., Sanguanrut, P., Proespraiwong, P., Suebsing, R. et al. (2016) Sensitive visual detection of AHPND bacteria using loop-mediated isothermal amplification combined with DNA-functionalized gold nanoparticles as Probes. PLoS One 11(3), e0151769.

Austin, B. and Newaj-Fyzul, A. (2017) *Diagnosis and Control of Diseases of Fish and Shellfish*. John Wiley & Sons, London, UK, pp. 27–29.

Chen, X., Qiu, L., Wang, H. L., Zou, P. Z., Dong X., Li, F. H. and Huang J. (2019) Susceptibility of *Exopalaemon carinicauda* to the infection with Shrimp hemocyte iridescent virus (SHIV 20141215), a strain of Decapod iridescent virus 1 (DIV1). *Viruses*, 11(4), 387.

Dangtipa, S., Sirikharinb, R., Sanguanrutc, P., Thitamadeed, S., Sritunyalucksanab, K., Taengchaiyaphumb, S. et al. (2015) AP4 method for two-tube nested PCR detection of AHPND isolates of *Vibrio parahaemolyticus*. *Aquaculture* 2, 158–162.

de la Pena, L. D., Cabillon, N. A., Catedral, D. D., Amar, E. C., Usero, R. C., Monotilla, W. D. et al. (2015) Acute hepatopancreatic necrosis disease (AHPND) outbreaks in *Penaeus vannamei* and *P. monodon* cultured in the Philippines. *Diseases of Aquatic Organisms* 116(3), 251–254.

Dong, X., Bi, D., Wang, H., Zou, P., Xie, G., Wan, X. et al. (2017a) *pirAB*^{vp} -bearing *Vibrio parahaemolyticus* and *Vibrio campbellii* pathogens isolated from the same AHPND-affected pond possess highly similar pathogenic plasmids. *Frontiers in Microbiology* 8, 1859.

Dong, X., Wang, H., Xie, G., Zou, P., Guo, C., Liang, Y. et al. (2017b) An isolate of *Vibrio campbellii* carrying the *pir*^{VP} gene causes acute hepatopancreatic necrosis disease. *Emerging Microbes and Infections* 6(1), e2.

Dong, X., Chen, J., Song, J., Wang, H., Wang, W., Ren, Y. et al. (2019a) Evidence of the horizontal transfer of pVA1-type plasmid from AHPND causing *V. campbellii* to non-AHPND *V. owensii*. *Aquaculture* 503, 396–402.

Dong, X., Song, J., Chen, J., Bi, D., Wang, W., Ren, Y. et al. (2019b). Conjugative transfer of the pVA1-Type plasmid carrying the *pirAB*^{vp} genes results in the formation of new AHPND-causing *Vibrio*.. *Frontiers in Cellular and Infection Microbiology* 9, 195.

Executive Committee of the International Committee on Taxonomy of Viruses (ICTV). *One New Genus with One New Species in the Subfamily Betairidovirinae*. Available online: https://talk.ictvonline.org/files/ictv_official_taxonomy_updates_since_the_8th_report/m/animal-dna-viruses-and-retroviruses/8051

FAO (2018) *FAO Yearbook: Fishery and Aquaculture Statistics*.

Flegel, T. W. and Lo, C. F. (2014) Announcement regarding free release of primers for specific detection of bacterial isolates that cause acute hepatopancreatic necrosis disease (AHPND). In: *Network of Aquaculture Centres in Asia-Pacific* (Bangkok, Thailand).

Gu, S. J. (2012) Analysis of causes of the covert mortality disease of Pacific white shrimp and its control strategies. *Scientific Fish Farming* 8, 62–63 [in Chinese].

Han, J. E., Tang, K. F., White, B. L. and Lightner, D. V. (2015a) qPCR assay for detecting and quantifying a virulence plasmid in acute hepatopancreatic necrosis disease (AHPND) due to pathogenic *Vibrio parahaemolyticus*. *Aquaculture* 442, 12–15.

Han, J. E., Tang, K. F., Tran, L. H. and Lightner, D. V. (2015b) *Photorhabdus* insect-related (*Pir*) toxin-like genes in a plasmid of *Vibrio parahaemolyticus*, the causative agent of acute hepatopancreatic necrosis disease (AHPND) of shrimp. *Diseases of Aquatic Organisms* 113(1), 33–40.

Hong, X. P., Lu, L. Q. and Xu, D. (2015) Progress in research on acute hepatopancreatic necrosis disease (AHPND). *Aquaculture International* 24(2), 577–593.

Jun, J. W., Han, J. E., Giri, S. S., Tang, K. F., Zhou, X. H., Aranguren, L. F. et al. (2018) Phage application for the protection from acute hepatopancreatic necrosis disease (AHPND) in *Penaeus vannamei*. *Indian Journal of Microbiology* 58(1), 114–117.

Jun, J. W., Han, J. E., Tang, K. F., Lighter, D. V., Kim, J., Seo, S. W. et al. (2016) Potential application of bacteriophage pVp-1: Agent combating *Vibrio parahaemolyticus* strains associated with acute hepatopancreatic necrosis disease (AHPND) in shrimp. *Aquaculture* 457, 100–103.

Kibenge, F. S. (2019) Emerging viruses in aquaculture. *Current Opinion in Virology* 34, 97–103.

Koiwai, K., Tinwongger, S., Nozaki, R., Kondo, H. and Hirono, I. (2016) Detection of acute hepatopancreatic necrosis disease strain of *Vibrio parahaemolyticus* using loop-mediated isothermal amplification. *Journal of Fish Diseases* 39(5), 603–606.

Kondo, H., Tinwongger, S., Proespraiwong, P., Mavichak, R., Unajak, S., Nozaki, R. et al. (2014) Draft genome sequences of six strains of *Vibrio parahaemolyticus* isolated from early mortality syndrome/acute hepatopancreatic necrosis disease shrimp in Thailand. *Genome Announcements* 2(2), e00221-14.

Kongrueng, J., Tansila, N., Mitraparp-arthorn, P., Nishibuchi, M., Vora, G. J. and Vuddhakul, V. (2015a) LAMP assay to detect *Vibrio parahaemolyticus* causing acute hepatopancreatic necrosis disease in shrimp. *Aquaculture International* 23(5), 1179–1188.

Kongrueng, J., Yingkajorn, M., Bunpa, S., Sermwittayawong, N., Singkhamanan, K. and Vuddhakul, V. (2015b) Characterization of *Vibrio parahaemolyticus* causing acute hepatopancreatic necrosis disease in southern Thailand. *Journal of Fish Diseases* 38(11), 957–966.

Lee, C. T., Chen, I. T., Yang, Y. T., Ko, T. P., Huang, Y. T., Huang, J. Y. et al. (2015) The opportunistic marine pathogen *Vibrio parahaemolyticus* becomes virulent by acquiring a plasmid that expresses a deadly toxin. *Proceedings of the National Academy of Sciences USA* 112(34), 10798–10803.

Li, F., Xu, L. and Yang, F. (2017). Genomic characterization of a novel iridovirus from redclaw crayfish *Cherax quadricarinatus*: evidence for a new genus within the family *Iridoviridae*. *Journal of Genetic Virology* 98(10), 2589–2595.

Li, X. P., Wan, X. Y., Huang, J. et al. (2019) Molecular epidemiological survey on covert mortality nodavirus (CMNV) in cultured crustacean in China in 2016–2017. *Progress in Fishery Sciences* 40(2), 65–73 [in Chinese].

Li, X. P., Wan, X. Y., Xu, T.T. et al. (2018) Development and validation of a Taqman RT-qPCR for the detection of convert mortality nodavirus (CMNV). *Journal of Virology Methods* 262, 65–71.

Liu, L., Xiao, J., Zhang, M., Zhu, W., Xia, X., Dai, X. et al. (2018) A *Vibrio owensii* strain as the causative agent of AHPND in cultured shrimp, *Litopenaeus vannamei. Journal of Invertebrate Pathology* 153, 156–164.

Liu, S, Li, J. T., Tian, Y., Wang, C. et al. (2017) Experimental vertical transmission of Covert mortality nodavirus (CMNV) in *Exopalaemon carinicauda. Journal of General Virology* 8(4), 652–661.

Liu, S, Wang, X. H., Xu, T. T. et al. (2018) Vectors and reservoir hosts of covert mortality nodavirus (CMNV) in shrimp ponds. *Journal of Invertebrate Pathology* 154, 29–36.

NACA (2012) Report of the Asia pacific emergency regional consultation on the emerging shrimp disease: early mortality syndrome (EMS)/acute hepatopancreatic necrosis syndrome (AHPNS). Bangkok, Thailand.

NACA (2014) Diseases of Crustaceans — Acute Hepatopancreatic Necrosis Syndrome (AHPNS). Bangkok, Thailand.

Nunan, L., Lightner, D., Pantoja, C. and Gomez-Jimenez, S. (2014) Detection of acute hepatopancreatic necrosis disease (AHPND) in Mexico. *Diseases of Aquatic Organisms* 111(1), 81–86.

OIE (2019) Acute hepatopancreatic necrosis disease. In: *Manual of Diagnostic Tests for Aquatic Animals.*, Chapter 2.2.1.

Pan, C. K., Yuan, H. F., Wang, T. T., Yang, F. and Chen, J. M. (2017) Study of *Cherax quadricarinatus* iridovirus in two crab. *Journal of Applied Oceanography* 36(1), 82–86 [in Chinese].

Pascual, J., Macian, M. C., Arahal, D. R., Garay, E. and Pujalte, M. J. (2010) Multilocus sequence analysis of the central clade of the genus *Vibrio* by using the 16S rRNA, *recA*, *pyrH*, *rpoD*, *gyrB*, *rctB* and *toxR* genes. *International Journal of Systemic and Evolutionary Microbiology* 60(1), 154–165.

Pinoargote, G., Flores, G., Cooper, K. and Ravishankar, S. (2018) Effects on survival and bacterial community composition of the aquaculture water and gastrointestinal tract of shrimp (*Litopenaeus vannamei*) exposed to probiotic treatments after an induced infection of acute hepatopancreatic necrosis disease. *Aquaculture Research* 49(10), 3270–3288.

Pooljun, C., Direkbusarakom, S., Chotipuntu, P. et al. (2016) Development of a TaqMan real-time RT-PCR assay for detection of covert mortality nodavirus (CMNV) in penaeid shrimp. *Aquaculture* 64, 445–450.

Pradit, W., Siwaporn, L., Suparat, T., Saengchan, S., Paisarn, S. and Parin, C. (2018) PirA & B toxins discovered in archived shrimp pathogenic Vibrio campbellii isolated long before EMS/AHPND outbreaks. *Aquaculture* 497, 494–502.

Qiu, L., Chen, M. M., Wan, X. Y., Li, C., Zhang, Q. L., Wang, R. Y., Cheng, D. Y., Dong, X., Yang, B., Wang, X. H., Xiang, J. H. and Huang, J. (2017) Characterization of a new member of *Iridoviridae*, Shrimp hemocyte iridescent virus (SHIV), found in white leg shrimp (*Litopenaeus vannamei*). *Scientific Reports* 7(1),11834.

Qiu, L., Chen, M. M., Wan, X. Y., Zhang, Q. L, Li, C., Dong, X., Yang, B. and Huang, J. (2018a) Detection and quantification of Shrimp hemocyte iridescent virus by TaqMan probe based real-time PCR. *Journal of Invertebrate Pathology* 154, 95–101.

Qiu, L., Chen, M. M., Wang, R. Y., Wan, X. Y., Li, C., Zhang, Q. L., Dong, X., Yang, B., Xiang, J. H. and Huang, J. (2018b) Complete genome sequence of shrimp hemocyte iridescent virus (SHIV) isolated from white leg shrimp, *Litopenaeus vannamei. Archives of Virology* 163(3), 781–785.

Qiu, L., Dong, X., Wan, X. Y. and Huang, J. (2018c) Analysis of iridescent viral disease of shrimp (SHID) in 2017. In: *Analysis of Important Diseases of Aquatic Animals in China in 2017* [in Chinese]. Fishery and Fishery Administration Bureau under the Ministry of Agriculture and Rural Affairs, National Fishery Technical Extension Center, China Agriculture Press, Beijing, pp. 187–204.

Qiu, L., Chen, X., Zhao, R. H., Li, C., Gao, W., Zhang Q. L. and Huang J. (2019a) Description of a natural infection with Decapod iridescent virus 1 in farmed giant freshwater prawn, *Macrobrachium rosenbergii. Viruses* 11(4), 354.

Qiu, L., Dong, X., Wan, X. Y. and Huang, J. (2019b) Analysis of iridescent viral disease of shrimp (SHID) in 2018. In: *Analysis of Important Diseases of Aquatic Animals in China in 2018* [in Chinese]. Fishery and Fishery Administration Bureau under the Ministry of Agriculture and Rural Affairs, National Fishery Technical Extension Center [in press].

Restrepo, L., Bayot, B., Arciniegas, S., Bajana, L., Betancourt, I., Panchana, F. et al. (2018) PirVP genes causing AHPND identified in a new *Vibrio* species (*Vibrio punensis*) within the commensal Orientalis clade. *Scientific Reports* 8(1), 13080.

Santos, H., Tsai, C. Y., Maquiling, K., Tayo, L., Mariatulqabtiah, A., Lee, C. W. et al. (2019) Diagnosis and potential treatments for acute hepatopancreatic necrosis disease (AHPND): a review. *Aquaculture International* 28, 169–185.

Sarmila, M., Tom, D., Ina-Salwany, M. Y., Fatimah, M. Y., Mohamed, S., Siti, I. I. et al. (2019) *Vibrio parahaemolyticus* and *Vibrio harveyi* causing acute hepatopancreatic necrosis disease (AHPND) in *Penaeus vannamei* (Boone, 1931) isolated from Malaysian shrimp ponds. *Aquaculture* 511, 1–8.

Sirikharin, R., Taengchaiyaphum, S., Sanguanrut, P., Chi, T. D., Mavichak, R., Proespraiwong, P. et al. (2015) Characterization and PCR detection of binary, Pir-like toxins from *Vibrio parahaemolyticus* isolates that cause acute hepatopancreatic necrosis disease (AHPND) in shrimp. *PLoS One* 10(5), e0126987.

Song, S. X. and Zhuang, S. P. (2006) Measures for control of "bottom death" of Pacific white shrimp. *Fish Science & Technology* 6, 36–39 [in Chinese].

Sperling, L., Alter, T. and Huehn, S. (2015) Prevalence and antimicrobial resistance of *Vibrio spp.* in retail and farm shrimps in Ecuador. *Journal of Food Protection* 78(11). 2089–2092.

Thitamadee, S., Prachumwat, A., Srisala, J. et al. (2016) Review of current disease threats for cultivated penaeid shrimp in Asia. *Aquaculture* 452, 69–87.

Tinwongger, S., Proespraiwong, P., Thawonsuwan, J., Sriwanayos, P., Kongkumnerd, J., Chaweepack, T. et al. (2014) Development of PCR diagnosis for shrimp acute hepatopancreatic necrosis disease (AHPND) strain of *Vibrio parahaemolyticus*. Fish Pathology 49(4), 159–164.

Varela, A. (2016) Nodavirus de la mortalidad encubierta (CMNV) en camarones marinos de cultivo. Nota técnica. Rev. Repertorio Científico. *Universidad Estatal a Distancia* 19(1), 33–40. https://revistas.uned.ac.cr/index.php/repertorio/article/view/2533

Varela, A. (2018) Patologías del hepatopáncreas en camarones marinos cultivados en América y su diagnóstico diferencial mediante histopatología. *Revista AquaTIC* 50, 13–30.

Wang, C., Liu, S., Li, X., Hao, J., Tang, K. F. J. and Zhang, Q. (2019a) Infection of covert mortality nodavirus in Japanese flounder reveals host jump of the emerging *alphanodavirus*. *Journal of General Virology* 100(2), 166–175.

Wang, C., Wang, X. H., Liu, S. et al. (2019b) Preliminary study on the natural infection of *Carassius auratus* with covert mortality nodavirus (CMNV). *Progress in Fishery Science* 40(2), 25–32 [in Chinese].

Wang, H., Wang, C., Tang, Y., Sun, B., Huang, J. and Song, X. (2018) *Pseudoalteromonas* probiotics as potential biocontrol agents improve the survival of *Penaeus vannamei* challenged with acute hepatopancreatic necrosis disease (AHPND)-causing *Vibrio parahaemolyticus*. *Aquaculture* 494, 30–36.

Xing, H. (2004) Discussion of the control measures for the "bottom death" (covert mortality disease) of Pacific white shrimp. *China Fisheries* 4, 88–89 [in Chinese].

Xu, L., Wang, T., Li, F. and Yang, F. (2016) Isolation and preliminary characterization of a new pathogenic iridovirus from redclaw crayfish *Cherax quadricarinatus*. *Diseases of Aquatic Organisms* 120(1), 17–26.

Xu, Z. J. and Ji, F. (2009) Comprehensive control of the covert mortality disease of Pacific white shrimp. *Fish Guide to be Rich* 1, 60 [in Chinese].

Zhang, B. C., Liu, F., Bian, H. H., Liu, J., Pan, L. Q. and Huang, J. (2012) Isolation, identification, and pathogenicity analysis of a *Vibrio parahaemolyticus* strain from *Litopenaeus vannamei*. *Progress in Fishery Science* 33(2), 56–62 [in Chinese].

Zhang, Q. H. (2004) To be cautious of "bottom death" in the intensive farming of Pacific white shrimp. *Scientific Fish Farming* 10, 48–49 [in Chinese].

Zhang, Q. L., Liu, Q., Liu, S. et al. (2014) A new nodavirus is associated with covert mortality disease of shrimp. *Journal of General Virology* 5, 2700–2709.

Zhang, Q. L., Liu, S., Li, J., Xu, T. T. et al. (2018) Evidence for cross-species transmission of covert mortality Nodavirus to new host of *Mugilogobius abEi*. *Frontiers in Microbiology* 9, 1447.

Zhang, Q. L., Liu, S., Yang, H. L. et al. (2017a) Reverse transcription loop-mediated isothermal amplification for rapid and quantitative assay of covert mortality nodavirus in shrimp. *Journal of Invertebrate Pathology* 150, 130–135.

Zhang, Q. L., Xu, T. T., Liu, S. et al. (2017b) Prevalence and distribution of covert mortality nodavirus (CMNV) in cultured crustacean. *Virus Research* 2(233), 113–119.

Zhang Q. L. (2020) Investigation and research progress on new incidences of shrimps. National Symposium on Prevention and Control of Aquatic Animal Diseases, 19 November, Fuzhou, China.

Chapter 22

THE MICROSPORIDIAN *ENTEROCYTOZOON HEPATOPENAEI* IN SHRIMP CULTURE

DIVA JANUARY ALDAMA-CANO, DOMINIC WIREDU BOAKYE, KWANTA THANSA, ANUPHAP PRACHUMWAT, SUPARAT TAENGCHAIYAPHUM, ORNCHUMA ITSATHITPHAISARN, THAWATCHAI CHAIJARASPHONG, BRYONY A. P. WILLIAMS, GRANT D. STENTIFORD AND KALLAYA SRITUNYALUCKSANA

Summary

The microsporidian *Enterocytozoon hepatopenaei* (EHP) in shrimp causes a disease called hepatopancreatic microsporidiosis (HPM), resulting in slow growth and size variation in culture ponds. Annual economic losses in the Thai shrimp industry due to EHP infection have been estimated at >US$200 million. Biosecurity programs must be developed rapidly for prevention and control of this disease. Although scientific knowledge on the mechanisms of EHP infection of shrimp is increasing, more research has to be focused on the environmental reservoir(s) or carrier(s) in the hatchery and grow-out pond. This review summarizes the available information on the mechanism of EHP transmission in shrimp, infection mechanism and updated detection methods. Last, the key recommendations for control of EHP in ponds are revealed, especially regarding the use of EHP free-post larvae and utilization of specific-pathogen-free (SPF) stocks regarding EHP.

General information on microsporidians

Microsporidia is a vast phylum comprising approximately 200 genera and >1400 species of unicellular, fungi-related and spore-forming eukaryotic obligate-parasites. They are ubiquitous in nature, which means they are present in terrestrial, marine and freshwater ecosystems and are able to infect hosts from all major animal taxa (Stentiford et al., 2016). Almost half of the known microsporidia genera infect aquatic hosts, from protists to vertebrates, which make them of significant importance for the aquaculture industry (Stentiford et al., 2013b). Until it was found that they possessed a mitosome, a mitochondrial remnant, microsporidia were thought to be primitive eukaryotes (Williams et al., 2002). Microsporidians have experienced reductive evolution, having lost their capacity to produce their own primary metabolites, many rely on 'stealing' them from the host (Dean et al., 2016; Wiredu Boakye et al., 2017).

These organisms are classified according to their ultrastructural characteristics, including spore size (1–10 μm), number of coils of the polar tube (a unique and specific invasive apparatus), their developmental life cycle and the relationship between parasite and host (Vavra and Larsson, 2014; Weiss, 2020; Weiss and Becnell, 2014). Spore morphology is homogenous across species: in the mature stage these are oval or pyriform (pear-shaped), with a resistant wall comprising two layers: an electron-dense, proteinaceous exospore and an electron-lucent endospore composed of chitin and protein, and an inner membrane or plasmalemma that allows it to remain viable outside the host either in an environmental reservoir or a host carrier. Although the life cycle of microsporidian species can be complex, there are three distinctive stages: proliferative (merogeny), spore production (sporogony) and the mature or infective stage, which is the most distinctive as spores are easy to visualize under light microscopy (Vávra and Larsson, 2014). The most relevant stage for microsporidian spread and survival is the mature or infective stage where the spore can germinate, infect potential hosts and propagate. Usually, the infectious process starts through the faecal–oral route, where spore wall proteins from ingested mature spores interact with host cell glycosaminoglycans (GAGs) present in the epithelial cell; if conditions are suitable, this will trigger extrusion of the invasive apparatus (polar tube) at high speed in order to pierce the host cell cytoplasm, inject the infective material and continue the life cycle (Hayman et al., 2005; Jaroenlak et al., 2018; Southern et al., 2007).

ENTEROCYTOZOON HEPATOPENAEI

The microsporidian, *Enterocytozoon hepatopenaei* (EHP) is the causative agent of hepatopancreatic microsporidiosis (HPM). It was first reported in Thailand in 2004 as an unnamed microsporidian in *Penaeus monodon* presenting with growth retardation (Chayaburakul et al., 2004), but it was not fully characterized until 2009 (Tourtip et al., 2009). EHP is confined to the tubule epithelial cells of the hepatopancreas and is expelled through faeces (Jaroenlak et al., 2016). Microsporidia do not need an intermediate host to fulfill its life cycle and can propagate in a pond as the shrimp host cohabit and cannibalize infected carriers. Other sources of infection include shellfish present in the pond, live feed, contaminated water, soil or equipment/machinery (Salachan et al., 2017; Tang et al., 2016; Tangprasittipap et al., 2013; Thitamadee et al., 2016). Although EHP does not cause mortality in infected shrimp, it is associated with severe growth retardation when EHP copy numbers are above 10^3/ng of total DNA extracted from the hepatopancreas (Liu et al., 2015). EHP has been reported from many shrimp-farming countries around the globe including Vietnam, China, Thailand, Indonesia, Malaysia, Brunei and Venezuela (Figure 22.1) (Chayaburakul et al., 2004; Ha et al., 2010; Hudson et al., 2001; Rajendran et al., 2016; Shen et al., 2017; Tang et al., 2017, 2015; Tourtip et al., 2009).

A study on the economic losses caused by an HPM outbreak in Asian aquaculture was carried out in 2018, which indicated that EHP infections in penaeid species had been an important source of productivity

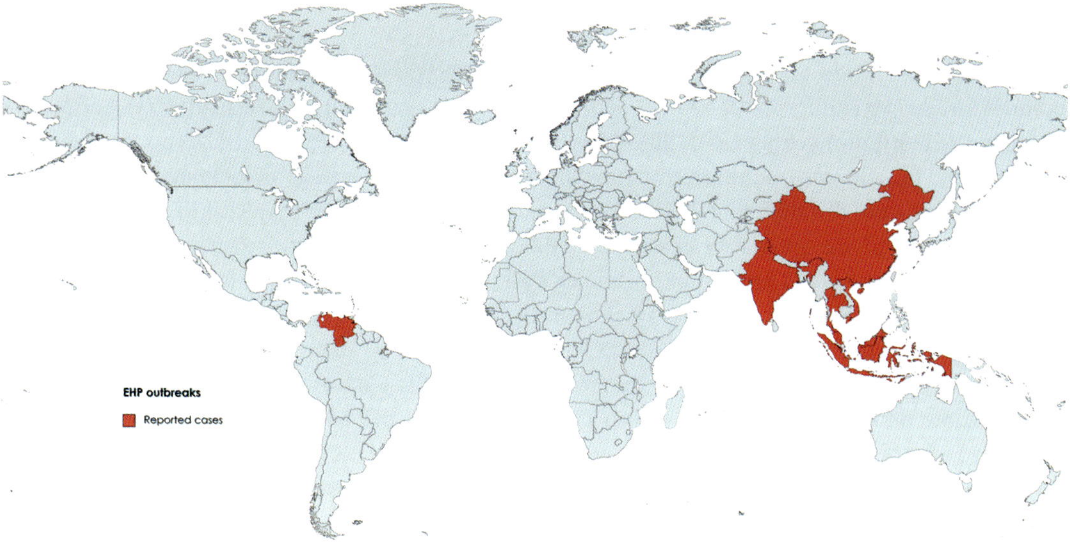

FIGURE 22.1 Global map of countries reporting the presence of EHP in *P. vannamei* and/or *P. monodon* culture (red).

loss to the industry by negatively affecting the survival rate and growth of shrimp in affected ponds (Liu et al., 2015; Shinn et al., 2018). Reports from farmers indicated that slower growth rates, characteristic of heavily infected stock, could force an early harvest. This would mean an estimated economic loss of US$4,500/ha compared to harvesting at the end of the cycle (US$32,000/ha). Early harvesting means lower production yield (9 rather than 12 tonnes/ha under normal harvesting conditions), carried out to cover production costs and resulting in reduced profits. Estimated consequential losses (mostly due to undersized stock) from EHP-related issues in Thailand, which has a 60.7% prevalence of EHP infection, could reach US$180–232 million per annum (FAO, 2020; Sanguanrut et al., 2018; Shinn et al., 2018). Some farmers have estimated their EHP-related economic losses at around US$5000/ha (in India) or US$7538/ha (in Indonesia) (Shinn et al., 2018).

EHP infection causes retardation in shrimp growth

To date, only the horizontal route has been reported for EHP transmission. Horizontal transmission transfers microsporidia from one individual to another by oral ingestion of spores through contaminated food and water, including cannibalism of microsporidia-infected animals or faecal matter (Bandi et al., 2001; Cali and Takvorian, 2004; Didier et al., 2004; Goertz et al., 2007; Tang et al., 2016; Tangprasittipap et al., 2013). It has been challenging to prove that aquatic animals become infected with EHP by vertical transmission, due to the lack of concrete evidence showing that the pathogen is present and transmitted to offspring via the gonads and ovarian cells of infected parents (Chaijarasphong et al., 2020; Tang et al., 2016; Vu-Khac et al., 2018). The horizontal route has been demonstrated using several bioassay models involving co-habitation, oral administration of spores or rearing shrimp in the EHP-contaminated water (Karthikeyan and Sudhakaran, 2019; Mai et al., 2020; Salachan et al., 2017; Tourtip et al., 2009; Tangprasittipap et al., 2013; Tang et al., 2016).

As stated in the Introduction part, although EHP infection is not a direct cause of severe mortality outbreaks in shrimp culture ponds, reduction in growth rate and variability in size in grow-out ponds have been reported (Sritunyalucksana et al., 2015). Evaluation of our in-house recorded data (not published) of EHP-infected PL, and close monitoring of growth performance among five grow-out ponds, revealed that shrimp presenting with a high rate of EHP infection had significant growth retardation. This evaluation was determined using comparative detection methods such as SWP-polymerase chain reaction (PCR) (Jaroenlak et al., 2016) and growth performance evaluation (body weight and average daily growth). The same stock of shrimp postlarvae (PLs) were reared in five different ponds. After one month of culture, shrimp found to be positive SWP-PCR and the pond with 100% shrimp detecting SWP-PCR showed severe growth retardation (reduction in body weight (BW) and average daily growth). The other 4 ponds showed the number of 10–30% of shrimp infected with EHP and the growth performance were found to be better than those in the pond with 100% shrimp infection. Supporting information that EHP infection interferes with growth has also been provided by proteomic and metabolomics studies in EHP-infected *P. vannamei*. Significant changes in several pathways such as energy- production and growth hormone-associated proteins were observed in EHP-infected shrimp that presented with stunted growth. Alteration of several protein markers related to growth hormone disorders, including farnesoic acid o-methyltransferase, juvenile hormone esterase-like carboxylesterase 1 and ecdysteroid-regulated protein, was also identified, which suggests a potentially impaired moulting process in EHP-infected shrimp (Ning et al., 2019).

EHP as a risk factor for other emergent diseases

One of the driving forces in the growth of the aquaculture industry on a global scale is the increasing demand for farmed seafood. This increase in production has been possible due to technological advances and intensification of farming practices. However, intensification brings certain risks, including devastating disease outbreaks that are hard to control due to the dynamics and nature of aquatic ponds, as well as transboundary movement of infected specimens and/or raw materials. As previously indicated, an outbreak of a known pathogen or new emerging disease can cause production and economic losses that may be hard to overcome for small-scale producers.

After the outbreak of shrimp early mortality syndrome (EMS) in Thailand, a comprehensive cohort study of 196 ponds was carried out to check for the other aetiology of EMS, besides acute hepatopancreatic necrosis disease caused by a binary toxin (PirAB)-producing *Vibrio parahaemolyticus* (VP_{AHPND}) (Tran et al., 2013; Joshi et al., 2014; Sirikharin et al., 2015). At the time of the study, it was found that EHP had a 60.7% prevalence and was found to have no apparent significant correlation to the EMS outbreak (Sanguanrut et al., 2018). A study by Aranguren et al. (2017) was carried out under laboratory conditions to study the potential role of EHP as a risk factor for secondary bacterial infections such as AHPND and septic hepatopancreatic necrosis (SHPN). The results obtained indicated that EHP-infected shrimp were more susceptible to diseases caused by opportunistic bacteria, such as *Vibrio sp.*, by disruption of the structure and function of the hepatopancreas, allowing the onset of secondary infections. The results suggest the potential risk of EHP as a gateway for secondary opportunistic pathogens.

The process of microsporidial infection

The important key steps in the overall infective process of microsporidian species are the interaction of microsporidia spore and the host cells, activation of the polar tube extrusion process and replication in, and exit from, infected host cells. However, it is important to note that even though the infective process in microsporidian species shares these steps, variations in specific infection processes and triggers are observed depending on the host cell targets/species (Bandi et al., 2001; Stentiford et al., 2013b). Microsporidian infection in target host cells requires three main parameters: the microsporidian itself, a cellular target and the surrounding factors between microsporidia and host cells. Although the exact mechanism of each step of the infection process is not yet fully understood, this section describes in detail the current understanding regarding the three steps involved. The infection process of microsporidia in host cells uses a unique invasion method that involves the spore wall (SW), polar tube (PT) and sporoplasm (SP) to interact with host intracellular responsiveness during the initial infection step, in order to finally extrude the polar tube from within the spore piercing through the membrane into the cytoplasm of the target cells (Han et al., 2020; Hayman et al., 2001; Jaroenlak et al., 2018; Leitch and Ceballos, 2008; Ning et al., 2019; Peuvel-Fanget et al., 2006; Xu and Weiss, 2005; Yang et al., 2018).

The first step of this process is adherence to the host cell by microsporidian spores, which commonly begins with the interaction between spore wall proteins (SWPs) and host cell surfaces. SWPs are classified as playing a critical role in the attachment process by binding to heparin-binding motifs (HBMs) and host cell surface sulfated glycosaminoglycans (GAGs) (Chen et al., 2017b; Hayman et al., 2001, 2005; Jaroenlak et al., 2018; Li et al., 2009; Southern et al., 2007; Wu et al., 2009). Leonard and Hayman (2017) demonstrated that host cell integrin is also involved in the process of adherence and invasion between the microsporidian *Encephalitozoon intestinalis* and its host. A number of hypothetical proteins predicted to contain the canonical integrin-binding motif arginine-glycine-aspartic acid (RGD), related to the interaction between the extracellular matrix (ECM) proteins and host cell integrins, have been identified in many pathogens (e.g. viruses, bacteria and parasites; Han et al., 2020). Inhibition of spore adherence and the progress of infection by the presence of RGD-peptides and some recombinant integrin proteins, reported in a previous study (Leonard and Hayman, 2017), illustrates the importance of the microsporidian spore attachment process during germination and invasion.

During the invasive process, an inactivated or activated spore enters the cytoplasm of the host cell through endocytosis or phagocytosis (Cali and Takvorian, 2014; Franzen, 2004; Franzen et al., 2005; Han et al., 2019; Takvorian et al., 2005, 2020; Tourtip et al., 2009; Yang et al., 2018). Activated (triggered) microsporidian spores undergo changes to their structure and content. A bulge at the apical end of the spore with narrowing of the endospore coat in that region is initially observed. Next, the apical attachment complex of the polar filament and membranes are rearranged. A collar-like structure formed at the apical complex, called the polar tube, exits the spore wall (Cali and Takvorian, 2014; Takvorian et al., 2005) to serve as a pathway for discharge of sporoplasm into target cells (Cali and Takvorian, 2014; Takvorian et al., 2020). In the case of non-activated spores, phagocytosis by the host cell is a potential reason for the spore to enter the host cell and finally transfer its sporoplasm via PTs into the host cytoplasm (Franzen, 2004; Franzen et al., 2005). A microenvironment called 'invasion synapse' is established after the extruded PT interacts with the host cells. The polar tube protein (PTP), sporoplasm surface proteins and host cell receptor, in association with environmental factors affecting PT germination, play key roles in this invasion synapse. Invasion occurs after the sporoplasm is transferred to the host cell via the PT

(Han et al., 2017). However, the mechanism used to deliver sporoplasm into host cells is still poorly understood. Possibly the PT either punctures the membrane of the host cell in order to transmit the sporoplasm to the target host, or the sporoplasm itself interacts with the host cell membrane during invasion (Han et al., 2017; Takvorian et al., 2013). Some previous reports used the morphological changes of PT (e.g., the invagination compartment in which the PT penetrated and the discharge of PT contents into the invagination of host cell plasmalemma) to confirm the entry of PT (e.g. *Encephalitozoon* spp.) into host cells via phagocytosis (Bigliardi and Sacchi, 2001; Franzen, 2004; Schottelius et al., 2000). It was shown that adding cytochalasin D could inhibit the uptake of *E. cuniculi* spores into host cells via actin-dependent phagocytosis (Couzinet et al., 2000). Moreover, phagocytosed spores of *E. cuniculi* were found escaping from lysosomes and infected the cytoplasm of host cells (Franzen, 2004), suggesting that phagocytosis could be served as another route for microsporidians to infect host cells (Franzen, 2004; Han et al., 2020). Regarding the entry of microsporidian spores into host cells by endocytosis, it was shown that the interaction between the *Encephalitozoon hellem* PTP4 and the host cell transferrin receptor (TfR1) could support the process of microsporidian invasion within the invasive synapse via clathrin-mediated endocytosis pathway (Han et al., 2017), as commonly seen in invasion by viruses and bacteria (Cossart and Helenius, 2014).

In the case of EHP, the existing evidence showed that the invasion synapse between the EHP spore and its infected host cells may be formed by the interaction between the SWP, called EhSWP1, and hepatopancreatic cells. EhSWP1 is the first SWP of EHP found in both layers of the spore wall and is classified into the SWP12 group of microsporidian SWPs (Jaroenlak et al., 2018; Wiredu Boakye et al., 2017). It also carries three HBMs, which facilitate adherence to the host cell for many intracellular parasites, including microsporidia (Kobayashi et al., 2010; Oliveira et al., 2012), in its N-terminal domain and one Bin-amphiphysin-Rvs-2 (BAR2) domain in its C-terminus, together with the presence of heparin-binding activity (Jaroenlak et al., 2018). Moreover, the discovery of heparin and genes involved in the biosynthesis pathway of glycosaminoglycans of the *P. vannamei* hepatopancreas led to the concept that EhSWP1 may support host cell invasion by tethering spores to the cell wall of hepatopancreatic epithelial cells through interaction between HBMs and heparin (Chen et al., 2015). However, no potential method has been proven involving the direct participation of EhSWP1 and host cells during the adherence process, due to the absence of a hepatopancreatic cell culture in penaeid shrimp that could be used as an EHP infection model. Recently, PTP2 has been identified in mature EHP spores and it has been localized on the whole polar tube of germinated EHP spores, suggesting that PTP2 may play a crucial role in the EHP invasion (Wang et al., 2021). There is limited information available on the mechanism of EHP infection. Research on the interaction between other microsporidia and their target host cells may improve our general knowledge of microsporidian biology in turn benefiting knowledge of EHP.

As soon as the invasion synapse is pinched off from the membrane to the host cell cytoplasm, the sporoplasm starts to develop into proliferative meronts, sporonts, sporoblasts, and mature spores, respectively (Cali and Takvorian, 2014). The location of spore development within the host cell depends on the genus of microsporidia, it can be observed either in a parasitophorous vacuole (e.g., genus of Encephalitozoon), in a parasite-secreted amorphous coat (e.g., genus of Pleistophora, Trachipleistophora and Thelohania), in direct contact with the host cytoplasm (e.g., genus of Enterocytozoon and Nosema), or surrounded by host endoplasmic reticulum (e.g., genus of Endoreticulatus and Vittaforma) (Han et al., 2020). Basically, the replication of EHP begins in the host cell cytoplasm by maturation of the sporoplasm via physical growth and nuclear division, thereby providing the branched plasmodium (Tourtip et al.,

2009). With the uniqueness to the Enterocytozoon group microsporidia (EGM), the separation of sporonts from sporogonal plasmodium is controlled by sporogony involved in the production of spore extrusion precursors (e.g., the polar filament and anchoring disk). After separation, sporonts develop to sporoblasts and then mature spores (Stentiford et al., 2013b, 2019; Tourtip et al., 2009).

Until now, only target species for EHP is shrimp. The simple infection cycle of EHP is shown to be different from many microsporidians that have more than one target host species in terms of the production of different types of spores in each type of host cell. Different spore types can infect the same host, whereas others infect only alternative hosts (Stentiford et al., 2013a,b). Transitions between the life stages of these complex cycles of infection probably contain also the haploid (monokaryotic) and diplokaryotic nuclei (Flegel and Pasharawipas, 1995). The most critical issue facing shrimp farmers is whether EHP can infect alternative host species. It is because EHP has been detected (by PCR) in polychaetes and bivalves used to feed shrimp broodstock (Thitamadee et al., 2016). Although there is no proof that these animals have great potential to become reservoirs of EHP that might infect shrimp, it is better not to allow environmental reservoirs for EHP in the shrimp culture pond.

Microsporidia produce energy through glycolysis, but due to the smaller size of mitosomes than typical eukaryotic mitochondria, and to the lack of cristae and their own genome, these organelles do not produce ATP through oxidative phosphorylation (Goldberg et al., 2008; Williams et al., 2002). The source of ATP production by glycolysis in microsporidia ceases completely during intracellular growth and replication within the host, suggesting that microsporidia need their host to provide energy during infection and replication (Dolgikh et al., 2011; Han et al., 2019; Heinz et al., 2014). Such energy salvage from the host cell involves transporter proteins on the extruded sporoplasm plasma membrane (e.g., an ATP-binding cassette (ABC) transporter superfamily protein NoboABCG1.1, nucleotide transport proteins NTT1-4 and microsporidia major facilitator superfamily (MFS) transport proteins), which are able to acquire energy and nucleotides from host cells (Dean et al., 2018; He et al., 2019; Heinz et al., 2014; Katinka et al., 2001; Keeling et al., 2010; Major et al., 2019; Tsaousis et al., 2008). These transporter proteins can interact with host cell mitochondria at some point in order to acquire essential nucleotides for energy and biosynthesis. For example, *E. hellem* sporoplasm surface protein 1 (EhSSP1), commonly localized to the surface of extruded sporoplasm, may support energy acquisition from host cells via its interaction with all voltage-dependent anion selective channels (VDAC1–3) detected on the outer mitochondrial membrane (Han et al., 2019). Moreover, the reorganization of host mitochondria around the parasitophorous vacuole membrane (PVM) found in *Encephalitozoon*-infected cells is proposed to be the strategy used by microsporidia to acquire nucleotides from their host for energy production (Dean et al., 2016; Hacker et al., 2014; Scanlon et al., 2004), which begins in the early stage of infection, as well as the formation of parasitophorous vacuole (PV) after invasion (Scanlon et al., 2004). Direct reaction between mitochondria and PVM is supported by the protein–protein interaction resulting from the disturbance of electron-dense proteinaceous structures to tether mitochondria by protease in *E. cuniculi*-infected cells (Hacker et al., 2014; Scanlon et al., 2004), together with the production of energy for microsporidia replication due to interaction between the mitochondrial VDAC of host cells and the PVM (Hacker et al., 2014; Han et al., 2019). Interaction among transporter proteins found on the surface of extruded sporoplasm of microsporidia, host cell mitochondria and the PV is sufficiently important to warrant further investigation.

Exit of microsporidia from the infected host cell

It is believed that exocytosis is the major mode for microsporidia to leave the healthy host cell, but not many studies on the intracellular pathways which contribute to the exit of microsporidia from host cells have been reported. Szumowski et al. (2014) found that *Nematocida parisii* spores escaped from *Caenorhabditis elegans* intestinal cells to intestinal lumen via host vesicle trafficking system. In this study, the apical recycling endocytosis pathway components RAB-5, RAB-10 and RAB-11 were also identified, and RAB-11 was confirmed to play an important role for exocytosis of *N. parisii* spores. During the process of exocytosis, the actin-5 (ACT-5) coats formed around the membrane compartments containing the exocytosing spores after fusion with the apical membrane. The formation of this actin coat was controlled by RAB-5, RAB-11, cdc-42 and ced-10/Rac1 (Szumowski et al., 2016). For other routes of microsporidia egress from infected host cells, spores can exit the infected host, which dies from body decomposition, cannibalism, scavenging and physical wounds (Cali and Takvorian, 2004). Regarding the exit of EHP from infected hepatopancreatic epithelial cells of penaeid shrimp, mature EHP spores can be released from the host for transmission to other hepatopancreatic cells, either by the rupture of infected host cells or to the environment via faeces (Tourtip et al., 2009).

Environmental factors affecting microsporidian spore germination

To date, the mechanism that controls microsporidian spore germination remains unclear. It is believed that activation of spore germination could involve many processes, such as increased spore osmotic pressure, PT eversion and the transfer of sporoplasm via extruded PT (Han et al., 2020; Weiss and Becnel, 2014). In this section, environmental factors affecting the germination of microsporidian spores will be explained. The induction of intrasporal osmotic pressure is the main factor involved in spore germination. The PT is generally coiled within the mature spore before germination. Once spore germination begins, the PT extrudes through the anterior pole of the spore at the synaptic points, between the PT and host cell membrane, to create a hollow tube (Wittner and Weiss, 1999). Different species of microsporidia have their own spore germination conditions, depending on the target host and the local environment (Han and Weiss, 2017). However, there must be a common factor in terms of intracellular responsiveness, as it was previously reported that external stimuli (e.g. cations, anions and pH) can activate spore germination (Han et al., 2020) through the stimulation of aquaporins (AQPs) or possibly the cleavage of the disaccharide trehalose into glucose and its metabolites, resulting in increased osmotic pressure within the spore, swelling of its polaroplast and posterior vacuole and thus PT extrusion and transfer of sporoplasm into the host cell (Ghosh et al., 2006; Undeen and Frixione, 1990). One study clearly demonstrated that the promotion of *Nosema bombycis* spore germination by an increase of intracellular osmotic pressure is not regulated by the trehalose pathway, due to the absence of significant change in intracellular macromolecules (e.g. trehalose, nucleic acids and proteins) observed before spore germination by single-cell Raman spectroscopy (Miao et al., 2018). Therefore AQPs, among the cell membrane channel proteins of the superfamily of major intrinsic proteins (MIPs) that facilitate osmosis in all living cells (Agre and Kozono, 2003; Chen et al., 2017a; Ghosh et al., 2006; Nehls and Dietz, 2014; Verma et al., 2014; Verkman and Mitra, 2000), are proposed to be the main resource for uptake of water into the spore to increase osmotic pressure due to limitation of water uptake via the lipid bilayers of the spore wall (Frixione et al., 1997; Undeen and Frixione, 1990). This result is supported by the finding that spore germination in *Nosema algerae* is inhibited by the use of mercury salts (Frixione et al., 1997), which suppress AQP activity (Agre and Kozono, 2003; Yang, 2000).

Nevertheless, AQPs may not be the only channel protein affecting intracellular osmotic pressure in microsporidia. Other integral MIPs (e.g. small and basic intrinsic protein (SIP)-like MIPs and aquaglyceroporins (AQGPs)) have been discovered to play roles in microsporidian osmosis (Verma et al., 2014). One research group reported that all SIP-like members, which are small in size and hydrophobic, are found only in the microsporidia of lower fungi. The loop D in these channels is longer than in other MIPs and is highly charged. It is believed that this channel plays key roles in the transport of hydrophobic solutes (Verma et al., 2014). For AQGPs, it has been revealed that these transmembrane channels are capable of transporting a wide variety of solutes (e.g. glycerol, urea, reactive oxygen species hydrogen peroxide, ammonia, carbon dioxide and nitric oxide), whereas transport of water is limited (Borgnia et al., 1999; Hansen et al., 2002; Ishibashi et al., 1994; Mukhopadhyay et al., 2014; Nehls and Dietz, 2014; Pedro et al., 2015). It is no surprise that basic knowledge on external factors affecting the mechanism of EHP spore germination is quite limited, or less advanced. This is, of course, due to the lack of an *in vitro* cell culture model for EHP infection.

The genome of *E. hepatopenaei*

Microsporidia now represent a primary model system for understanding the process of metabolic, cellular and genomic reduction in eukaryotes: their genomes as small as 2.3 Mb (*E. intestinalis*), encoding as few as 1907 predicted genes (Peyretaillade et al., 2012) although others are as large as 51.3 Mb (*Edhazardia aedis*) with 4190 predicted genes (Desjardins et al., 2015). This reduction in gene number probably occurred when a large set of core eukaryotic genes were jettisoned early in the evolutionary history of microsporidia (Nakjang et al., 2013). This left microsporidia without a mitochondrial genome and without the ability to generate ATP via the mechanism of oxidative phosphorylation. Lineages such as *Mitosporidium*, *Paramicrosporidium*, *Nucleophaga* and metchnikovellids, that branch in between the microsporidia and other fungi, variously retain a mitochondrial genome or parts of the electron transport chain or tricarboxylic acid cycle and have less compacted genomes (Corsaro et al., 2014; Galindo et al., 2018; Haag et al., 2014; Williams et al., 2002). Many microsporidia still retain enzymes in the glycolysis pathway capable of producing ATP. The genome sequence of EHP was released in 2016 as part of a broader project to understand the patterns of loss and retention of glycolysis in microsporidia, but provides a valuable database for underpinning future analyses of host–parasite interactions between the microsporidian and its shrimp host (Wiredu Boakye et al., 2017).

The EHP genome assembly, released in 2016, was sequenced using a combination of Illumina and PacBio technology and DNA extracted from spores isolated from infected *Litopenaeus vannamei* from shrimp ponds in Thailand. This resulted in a genome assembly of 62 scaffolds with a total size of 3.25 Mb, which is towards the lower end of microsporidian genome size. Overall, the GC content is 25.5% and there were 2536 predicted protein-coding genes at the time of annotation. These had an average size of 916 bp and occupy 71% of the genome, making the EHP genome relatively compact as shown in Table 22.1. A multiprotein maximum-likelihood phylogeny of 22 conserved proteins showed that EHP branched as the sister group to the closest relative to *E. bieneusi* and *Enterospora canceri*, the edible crab pathogen (Figure 22.2). Transposases and retrotransposons represent a small part of the predicted genome, with just 24 of predicted ORFs predicted to be transposable elements. Just like *E. bieneusi*, EHP has a reduced complement of genes coding for components of the splicing machinery and there is no clear evidence of introns, even in genes where introns are thought to be well conserved across the microsporidia (Whelan et al., 2019). Unlike *Encephalitozoon* spp. and close relatives, EHP encodes proteins required for the RNA interference pathway including Argonaute (AGO2, OQS54083) and putative dicer proteins (EHP00_2492, OQS53451; EHP00_1630, OQS53389). See Table 22.1.

Energy metabolism

The microsporidia jettisoned many metabolic pathways early in their evolutionary history (Nakjang et al., 2013), including those involved in oxidative phosphorylation, meaning that glycolysis is the principal means of generating ATP, supplemented by importation of ATP from the host via ATP transporters acquired by horizontal gene transfer (Dean et al., 2018). The emerging consensus is that glycolysis is active in the spore but less active in the intracellular stage, when parasites have access to host ATP. EHP, like most microsporidia, has lost the ability to produce ATP in mitochondria but, like *E. bieneusi* and other members of the Enterocytozoonidae family, it has also lost genes coding for parts of the glycolytic pathway, meaning that they are unlikely to complete glycolysis. With no obvious intrinsic means of generating energy, this makes the Enterocytozoonidae even more dependent on their hosts for energy than other microsporidia. Although members of the Enterocytozoonidae family have lost the ability to produce ATP via glycolysis, the entire microsporidia phylum, including members of the Enterocytozoonidae family have maintained the enzyme repertoire necessary for converting host-derived partially charged nucleotides such as NDPs, NMPs, dNDPs and dNMPs to fully charged dNTPs and NTPs such as ATP (Wiredu-Boakye et al., 2017) (Purine and Pyrimidine metabolism in Figure 22.3). It is however unlikely for ATP produced in this manner to be used for energy provision as it is energetically expensive. Instead, this nucleotide synthesis pathway in microsporidia is likely to be mainly for dNTP production thereby making DNA precursors readily available for DNA replication during the proliferative merogonial life stage. In addition to the loss of glycolysis, EHP lacks an intact pentose phosphate pathway, trehalose metabolism pathway or pathways involved in fatty acid biosynthesis typically conserved in other microsporidia (Figure 22.3).

TABLE 22.1 Genome statistics.

Assembly size (Mb)	3.26
Contigs	64
Mean coverage	363
Contig N50 (bp)	125008
GC content (%)	25.45
GC content coding (%)	27.81
Coding regions (%)	71.87
Splicing machinery	7/29 genes
Genes*	2536
of which are hypothetical proteins*	1650
of which are TEs*	24

* At time of 2016 annotation

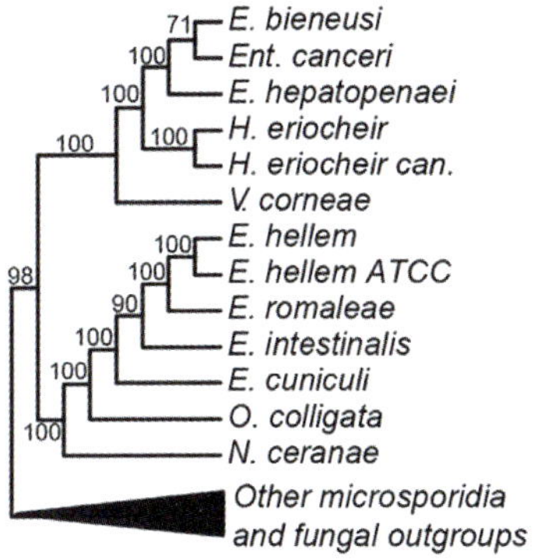

FIGURE 22.2 Cladogram showing the results of a maximum likelihood analysis of sequenced microsporidian genomes with fungal and animal genomes as outgroups. Values on nodes represent boostrap support. Analyses were performed on a concatenated alignment of 21 proteins. Credit: adapted from Wiredu-Boakye et al. (2017).

In terms of interaction with the host, the EHP genome encodes a fairly typical number of predicted transmembrane transport proteins compared to other microsporidia. These transporters are expected to determine what enters and is exported from the cell (Figure 22.3). This analysis of transporter protein complement indicates no clear set of transporters that would compensate for the reduction in other metabolic pathways in EHP.

The genome sequence provides a first step to unravelling the biology of this pathogen, as well as understanding and mitigating the damage caused to its host. Secreted proteins or 'effectors' represent an important mechanism by which parasites can influence the host environment. A SignalP/TMHMM screen of EHP proteins to identify those that predicted secretion signals, but no transmembrane domains that could potentially be leaving the parasite cell to mediate changes in the host, resulted in a list of 230

TABLE 22.2 Putative secreted proteins with similarity to annotated proteins in NCBI.

EHP00_532, OQS53986	Acid phosphatase precursor
EHP00_1308, OQS54455	Basic endochitinase CHB4-like
EHP00_208, OQS54854	Class 3 lipase
EHP00_372, OQS55830	Coatomer subunit beta-3-like
EHP00_841, OQS54198	Coatomer subunit delta
EHP00_266, OQS55147	CTD small phosphatase
EHP00_1845, OQS54264	DNAJ-class molecular chaperone
EHP00_194, OQS54839	Endoplasmic reticulum membrane
EHP00_1094, OQS53676	Glucosamine 6-phosphate N-acetyltransferase
EHP00_1408, OQS54759	Cell cycle serine/threonine-protein kinase CDC5/MSD2
EHP00_2325, OQS54749	RhoGAP protein
EHP00_2050, OQS53543	Leucine-rich repeat protein
EHP00_2091, OQS53545	Leucine-rich repeat protein
EHP00_2692, OQS53554	Leucine-rich repeat protein
EHP00_761, OQS53741	N-acetylglucosaminyl-phosphatidylinositol de-N-acetylase-like
EHP00_1920, OQS53550	Leucine-rich repeat protein
EHP00_2689, OQS53549	Leucine-rich repeat protein
EHP00_2690, OQS53551	Leucine-rich repeat protein
EHP00_453, OQS55715	SCP PR1 domain-containing
EHP00_1097, OQS53673	Serine hydroxymethyltransferase

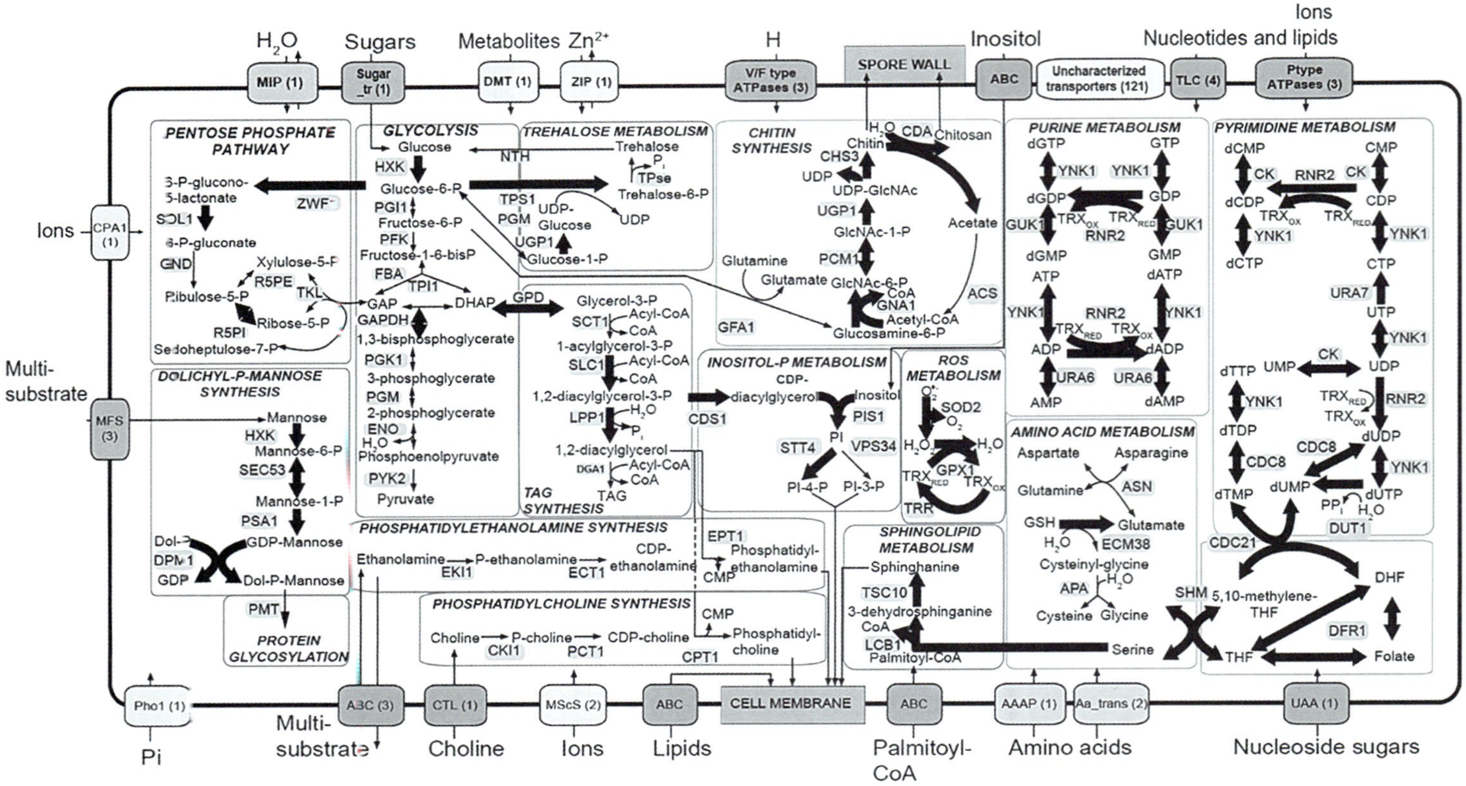

FIGURE 22.3 Summary of metabolic pathways in *E. hepatopenaei*. Within the cell, pathways present in *E. hepatopenaei* are marked with thick lines, whilst those that are absent are thinner. Predicted membrane transporters are marked around the cell, and predicted numbers of each are indicated in brackets. Dol, Ddolichyl; TAG, triacylglycerol; GAP, glyceraldehyde-3-phosphate; DHAP, dihydroxiacetone phosphate; GlcNAc, N-acetlyglucosamine; PI, phosphatidylinositol; GSH, glutathione; THF, tetrahydrofolate; DHF, dihydrofolate; TRX, thioredoxin; HCO_3, anion transport protein; MIP, major intrinsic protein; MFS, major facilitator superfamily; DMT, drug/metabolite transporter; ZIP, zinc/iron permease; SulP, sulfate permease; AAAP, amino acid/auxin permease; V/F typeATPases, vacuolar vesicular proton pump; P type ATPases, E1/E2 ATPases; Sugar_tr, glucose transporter/sugar transporter family; Cation efflux, cation efflux pump; Ca CIC, calcium-dependent chloride channels; MscS, small conductance mechanosensitive ion channel; TLC, ATP/ADP translocase; UAA, UDP-N-acetylglucosamine transporter; ABC, ATP-binding casette; CTL, choline transporter; POT, proton-dependent oligopeptide transporter; Pho1, phosphate transporter. Credit: adapted from Wiredu-Boakye et al. (2017).

TABLE 22.2 *continued*

EHP00_231, OQS54877	Similarity to PUTATIVE LIPASE ATG15
EHP00_2613, OQS53402	Predicted small GTP-binding protein
EHP00_944, OQS55031	Spore wall protein 7
EHP00_1785, OQS53601	Thioredoxin domain-containing protein
EHP00_2597, OQS53382	U-box domain leucine-rich repeat protein
EHP00_2611, OQS53400	U-box domain leucine-rich repeat protein
EHP00_2687, OQS53547	U-box domain leucine-rich repeat protein
EHP00_613, OQS55241	TPR SEL1 subfamily protein
EHP00_151, OQS54653	Zinc finger C2H2 ECU08_0560-like

proteins. Of these, 29 had some sequence similarity to annotated proteins in NCBI (Table 22.2). As in other microsporidia, there are several leucine-rich repeat proteins amongst the list, which may represent a rapidly evolving gene family of effectors (Campbell et al., 2013).

Detection of EHP

In the past decade, several EHP detection methods have been developed to create a more comprehensive diagnosis, either on-site or in the laboratory. These vary from simple light microscopy identification using specific stains to more complex methods such as molecular detection by PCR or histopathological analysis of preserved tissue. Each method has its advantages, most of which can be used with fresh samples (i.e. faeces, hepatopancreatic biopsy) from specimens without the need to sacrifice them; however, it is recommended to use a coupled detection method, meaning more than one at the same time (i.e. PCR and histology, microscopic identification and PCR, etc.) in order to obtain highly specific results that will allow for a better assessment of the particular situation. In the following sections we will describe the EHP detection methods most commonly used in aquaculture.

There are no distinctive gross signs accompanying EHP infection that can be used for presumptive diagnostics, in contrast to other diseases such as WSSV. The most distinctive sign of EHP infection is the effect on shrimp growth during the culture period. However, it has been demonstrated that even though there is a direct correlation between the severity of EHP infection and shrimp growth, in some cases shrimp are able to grow to normal size (Chaijarasphong et al., 2020). It is recommended that pathological examination of suspected animals or ponds is supported by molecular techniques.

MICROSCOPIC IDENTIFICATION

Due to the morphological development of EHP spores during its life cycle, only mature EHP spores are relatively easy to identify in fresh squash mounts from infected hepatopancreatic tissue and/or faeces via microscopic observation, due to their characteristic morphology. Several methods have been published on how to identify spores in several types of samples such as fresh tissue, either from smearing a whole dissected organ or a biopsy taken from suspected infected animals (Cruz-Flores et al., 2019). However, the easiest method is to stain the hepatopancreatic squash mount with a 2% Phloxine B solution (v/v) and observing under 100x magnification. The use of 2% Phloxine B is recommended as it will stain mature spores with a deep magenta colour and will trigger the extrusion of the polar tube, making easier its identification against tissue debris (Aldama-Cano et al., 2018). It is not recommended to use this method as routine, due to the complexity of EHP development and distribution in tissue (Chaijarasphong et al., 2020), which could be misleading when using it on its own. However, it is a farmer-friendly method that can be used for monitoring on site.

MOLECULAR DETECTION BY PCR

In the case of EHP, molecular methods based on 18S rRNA sequence, and spore wall protein 1 gene sequence (Tang et al., 2015; Tangprasittipap et al., 2013; Jaroenlak et al., 2016; Munkongwongsiri et al., 2021), have been developed. The methods were designed for one-step PCR, nested PCR and quantitative PCR. Whereas the one-step and nested PCR are semi-quantitative, the quantitative PCR can quantify the level of a pathogen based on a correlation between the copy numbers of that pathogen's DNA fragment in the template and the number of cycles (CT) required to reach a predetermined threshold by the standard control (Liu et al., 2015; Munkongwongsiri et al., 2021). Loop-mediated isothermal amplification (LAMP) is another alternative, in contrast to PCR and real-time PCR; LAMP results can be visualized in agarose gel or they may also be assessed indirectly by the amount of white magnesium pyrophosphate precipitate formed as a byproduct (Cai et al., 2018; Sathish Kumar et al., 2018; Suebsing et al., 2013). Similarly, the newly developed CRISPR-Cas method uses a polar tube protein (PTP2) as the detection target (Kanitchinda et al., 2020). Both LAMP and CRISPR-Cas are methods of high sensitivity and lower cost compared to standard molecular amplification techniques, because they need only simple equipment.

All of the above techniques have been developed to be highly specific (avoiding false positives from cross-reaction with other closely related microsporidian species) and sensitive (able to detect low infection status) tests. This is a great advantage when monitoring specimens, crops, broodstock, etc., as it allows for sensible decision making. However, it is important to couple molecular detection methods with other tests, such as squash mounting or *in situ* hybridization. This is recommended in order to determine whether the tested sample has an active infective status or if the molecular techniques are detecting the presence of dead spores or their remnants (for example in soil, water, live feed samples, etc.).

HISTOPATHOLOGY

The technique used most commonly to identify the characteristic histopathological features of EHP infection in paraffin-embedded tissue is the standard haematoxylin-eosin (H&E) staining method (Bell and Lightner, 1988). By using H&E, the presence of early life stages such as plasmodium stained in a basophilic colour indicating that the presence of mature spores within the cell cytoplasm or hepatopancreatic tubule epithelial cells and lumen can be observed. It is worth noting that it may be appropriate to carry the observation on under high magnification or an oil-immersion lens as some of the current lesions, life stages or mature spores could be difficult to observe at lower magnification. Other differential staining techniques that have been developed may be useful to differentiate EHP infection sites and tissue, such as acid chromotrope (Moura et al., 1997; Weber et al., 1992) and *in situ* hybridization. Differential stains, such as acid chromotrope, have been developed to identify spores in embedded tissue by staining them in a colour that contrasts to that of shrimp tissue, making their identification easier. However, this technique has the potential of reaching an incorrect diagnosis because it is not a very sensitive technique and can lead to underestimation of the severity of the infection (Chaijarasphong et al.,2020).

In situ hybridization (ISH) is another differential staining technique commonly used for diagnosis. This technique is recommended for its sensitivity and specificity, as it uses a specific DIG-labelled DNA probe. This DNA probe can be constructed using the fragment of choice, either SSU rRNA (Tang et al., 2015; Tangprasittipap et al., 2013) or SWP (Jaroenlak et al., 2016). Even though ISH is a more complex methodology, it is a better tool for estimation of the severity of an EHP infection as the DNA probe will bind to all stages of EHP development. This method is highly recommended when carrying out research and disease diagnosis; however, due to its complexity it is not suitable for on-farm settings or pond site detection.

Recommended EHP control measures with potential field application

The implementation of strong biosecurity protocols, consistent health monitoring and carrying out of preventive measures based on previous recommended guidelines (Sritunyalucksana et al., 2015; Thitamadee et al., 2016) is the approach most recommended to minimize the risk of EHP infection. In the following section we will describe some of the recommended approaches for its potential application on the farm.

In order to control the spread or outbreak of a disease, it is often recommended that we follow prophylactic actions rather than remedial. In the case of EHP, it is recommended for producers and farmers to carry out a screening of PL in hatcheries before stocking the farm, as a preventive measure. Farmers are advised to ensure that the supply of PL received is PCR-negative, and to exercise appropriate biosecurity measures before and after stocking to prevent infection from environmental sources. We also strongly recommend that industry practitioners who currently use wild, captured broodstock check for EHP-free status. Indeed, since domesticated stocks of *P. monodon* SPF for EHP are available, we do not recommend the use of captured wild broodstock for PL production.

Fresh living feed intended for broodstock needs to be tested to confirm that it is free of pathogens. In case the feed cannot be tested, it is recommended that it is treated by either freezing (–20°C to –80°C), for a few days before giving it to living shrimp, to ensure that any potential pathogens have been neutralized, or by using high temperatures (>75°C), for a few minutes under dry heat, autoclaving, etc. (Aldama-Cano et al., 2018; Munkongwongsiri et al., 2021), because EHP spores are highly susceptible to extreme temperature changes.

It is also common practice to use certain products such as calcium oxide or other chemicals (usually disinfectants with high oxidizing properties) between crops and during pond preparation to ensure that the pond, water and equipment to be used will be free from potential pathogens. With this in mind, several *in vitro* models have been developed to understand how to inactivate EHP activity using the above products and, thus, to reduce the risk of transmission and infection. These models use purified active EHP spores and test them under laboratory conditions to screen the effectivity of certain products for their use in the control of this parasite. Results have indicated that strong oxidizing disinfectants, such as potassium permanganate ($KMnO_4$), 65% active chlorine and 70% ethanol, have excellent inhibitory effect on EHP spore activity (Aldama-Cano et al., 2018). Other commercially available products such as biocides (products used to kill living organisms) have shown potential after carrying out *in vitro* testing (Sang-ngam et al., 2020). In both cases, it is strongly recommended that these products are not used when there are living animals in the system. A different approach would be the use of functional additives in the feed or water to reinforce the shrimp defence system against EHP. Functional feed ingredients, such as phytobiotic-based products, have been tested *in vitro* to evaluate their potential to inhibit spore germination (Nuez-ortín and Isern-subich, 2019). Initial results showed that some products have the potential to reduce spore viability and, therefore, could potentially be used as a control tool along with preventive measures.

Conclusion and future research

The microsporidian parasite *E. hepatopenaei* was formally characterized in 2009 and infects the hepatopancreas of its host, causing hepatopancreatic microsporidiosis (HPM), a condition that has been associated with slow growth of the host in aquaculture settings. Unlike other infectious disease agents that have caused economic losses in global shrimp aquaculture, EHP has proven more challenging because not enough is known about its environmental reservoirs and modes of transmission in the industrial shrimp production process. This review summarizes our current knowledge of the EHP life cycle and the molecular strategies it employs as an obligate intracellular parasite. It also provides an analysis of available and new methodologies for diagnosis, since most of the current literature on EHP focuses on that topic. The summary of current knowledge of EHP infection and transmission dynamics and currently recommended, practical control measures to limit its negative impact on shrimp cultivation has been revealed. However, there are major gaps in knowledge that urgently need to be bridged in order to improve control measures.

Acknowledgements

The authors thank the Agricultural Research Development Agency (ARDA), Thailand (to KS and OI) and Thailand Research Fund (TRF) for continuous support of our team in working rapidly on the issue of EHP. GDS was supported by the UK Department for Environment, Food and Rural Affairs (DEFRA) contract nos. FB002 and FX001. We also appreciate financial support from Newton Institutional Links 2017, the Newton prize's Chairman's award and the International Collaborative Award (ICA\R1\180038) from the Royal Society (to GDS, Cefas/UK and KS, BIOTEC/Thailand). In addition, we thank the Faculty of Science, Mahidol University/Thailand and National Science and Technology Development Agency/Thailand for their continuous support of our research.

References

Agre, P. and Kozono, D. (2003) Aquaporin water channels: Molecular mechanisms for human diseases. *FEBS Letters* 555, 72–78.

Aldama-Cano, D. J., Sanguanrut, P., Munkongwongsiri, N., Ibarra-Gámez, J. C., Itsathitphaisarn, O., Vanichviriyakit, R. et al. (2018) Bioassay for spore polar tube extrusion of shrimp *Enterocytozoon hepatopenaei* (EHP). *Aquaculture* 490, 156–161.

Aranguren, L. F., Han, J. E. and Tang, K. F. J. (2017) *Enterocytozoon hepatopenaei* (EHP) is a risk factor for acute hepatopancreatic necrosis disease (AHPND) and septic hepatopancreatic necrosis (SHPN) in the Pacific white shrimp *Penaeus vannamei*. *Aquaculture* 471, 37–42.

Bandi, C., Dunn, A. M., Hurst, G. D. D. and Rigaud, T. (2001) Inherited microorganisms, sex-specific virulence and reproductive parasitism. *Trends in Parasitology* 17, 88–94.

Bell, T. A. and Lightner, D. (1988) *A Handbook of Normal Shrimp Histology*. World Aquaculture Society, Baton Rouge, LA, USA.

Bigliardi, E. and Sacchi, L. (2001) Cell biology and invasion of the microsporidia. *Microbes and Infection* 3, 373–379. doi:10.1016/S1286-4579(01)01393-4.

Borgnia, M., Nielsen, S., Engel, A. and Agre, P. (1999) Celluar and molecular biology of the aquaporin water channels. *Annual Review of Biochemistry* 68, 425–458.

Burri, L., Williams, B. A. P., Bursac, D., Lithgow, T. and Keeling, P. J. (2006) Microsporidian mitosomes retain elements of the general mitochondrial targeting system. *Proceedings of the National Academy of Sciences USA* 103, 15916–15920.

Cai, S. X., Kong, F. De, Xu, S. F. and Yao, C. L. (2018) Real-time loop-mediated isothermal amplification for rapid detection of Enterocytozoon hepatopenaei. *PeerJournal* 6, e5993.

Cali, A. and Takvorian, P. M. (2004) The microsporidia: Pathology in man and occurrence in nature. *Southeast Asian Journal of Tropical Medicine and Public Health* 35, 58–64.

Cali, A. and Takvorian, P. M. (2014) Developmental morphology and life cycles of the microsporidia. In: *Microsporidia: Pathogens of Opportunity*, 1st edn (eds Weiss, L. M. and Becnel, J. J.). Wiley Blackwell Press, NJ, USA, pp. 71–133.

Campbell, S., Williams, T., Yousuf, A., Soanes, D., Paszkiewicz, K. and Williams, B. (2013) The genome of *Spraguea lophii* and the basis of host–microsporidian interactions. *Plos Genetics* 9(8), e1003676.

Chaijarasphong, T., Munkongwongsiri, N., Stentiford, G. D., Aldama-Cano, D. J., Thansa, K., Flegel, T. W., Sritunyalucksana, K. and Itsathitphaisarn, O. (2020) The shrimp microsporidian *Enterocytozoon hepatopenaei* (EHP): Biology, pathology, diagnostics and control. *Journal of Invertebrate Pathology* https://www.sciencedirect.com/science/article/abs/pii/S0022201120301646

Chayaburakul, K., Nash, G., Pratanpipat, P., Sriurairatana, S. and Withyachumnarnkul, B. (2004) Multiple pathogens found in growth-retarded black tiger shrimp *Penaeus monodon* cultivated in Thailand. *Diseases of Aquatic Organisms* 60, 89–96.

Chen, G., Wang, W., Chen, H., Dai, W., Peng, X., Li, X. et al. (2017a) Functional characterization of an aquaporin from a microsporidium, Nosema bombycis. PLoS ONE 12, e0181703.

Chen, K., Li, E., Li, T., Xu, C., Wang, X., Lin, H. et al. (2015) Transcriptome and molecular pathway analysis of the hepatopancreas in the Pacific white shrimp *Litopenaeus vannamei* under chronic low-salinity stress. *PLoS ONE* 10, e0131503.

Chen, L., Li, R., You, Y., Zhang, K. and Zhang, L. (2017b) A novel spore wall protein from *Antonospora locustae* (Microsporidia: Nosematidae) contributes to sporulation. *Journal of Eukaryotic Microbiology* 64, 779–791.

Corsaro, D., Walochnik, J., Venditti, D., Müller, K., Hauröder, B. and Michel, R. (2014) Rediscovery of *Nucleophaga amoebae*, a novel member of the Rozellomycota. *Parasitology Research* 113(12), 4491–4498.

Cossart, P. and Helenius, A. (2014) Endocytosis of viruses and bacteria. *Cold Spring Harbor Perspectives in Biology* 6, a016972.

Couzinet, S., Cejas, E., Schittny, J., Deplazes, P., Weber, R. and Zimmerli, S. (2000) Phagocytic uptake of Encephalitozoon cuniculi by nonprofessional phagocytes. Infection and Immunity 68, 6939–6945. doi:10.1128/IAI.68.12.6939-6945.2000.

Cruz-Flores, R., Mai, H. N., Noble, B. L., Schofield, P. J. and Dhar, A. K. (2019) Detection of *Enterocytozoon hepatopenaei* using an invasive but non-lethal sampling method in shrimp (*Penaeus vannamei*). *Journal of Microbiology Methods* 162, 38–41.

Dean, P., Hirt, R. P. and Embley, T. M. (2016) Microsporidia: Why make nucleotides if you can steal them? *PLoS Pathogens* 12, 1–13.

Dean, P., Sendra, K. M., Williams, T. A., Watson, A. K., Major, P., Nakjang, S. et al. (2018) Transporter gene acquisition and innovation in the evolution of Microsporidia intracellular parasites. *Nature Communications* 9, 1–12.

Desjardins, C., Sanscrainte, N., Goldberg, J., Heiman, D., Young, S., Zeng, Q. et al. (2015) Contrasting host–pathogen interactions and genome evolution in two generalist and specialist microsporidian pathogens of mosquitoes. *Nature Communications* 6(1), 7121.

Didier, E. S., Stovall, M. E., Green, L. C., Brindley, P. J., Sestak, K. and Didier, P. J. (2004) Epidemiology of microsporidiosis: Sources and modes of transmission. *Veterinary Parasitology* 126, 145–166.

Dolgikh, V. V., Senderskiy, I. V., Pavlova, O. A., Naumov, A. M. and Beznoussenko, G. V. (2011) Immunolocalization of an alternative respiratory chain in *Antonospora locustae* (Paranosema) spores: Mitosomes retain their role in microsporidial energy metabolism. *Eukaryotic Cell* 10, 588–593.

FAO (2020) *The State of World Fisheries and Aquaculture 2020. Nature and Resources*. FAO, Rome, https://doi.org/10.4060/ca9229en

Flegel, T. W. and Pasharawipas, T. (1995) A proposal for typical eukaryotic meiosis in microsporidians. *Canadian Journal of Microbiology* 41, 1–11.

Flegel, T. W. (2012) Historic emergence, impact and current status of shrimp pathogens in Asia. *Journal of Invertebrate Pathology* 110, 166–173.

Franzen, C. (2004) Microsporidia: How can they invade other cells? *Trends in Parasitology* 20, 275–279.

Franzen, C., Muller, A., Hartmann, P. and Salzberger, B. (2005) Cell invasion and intracellular fate of *Encephalitozoon cuniculi* (Microsporidia). *Parasitology* 130, 285–292.

Frixione, E., Ruiz, L., Cerbon, J. and Undeen, H. A. (1997) Germination of *Nosema algerae* (Microspora) Spores: Conditional inhibition by D_2O, ethanol and Hg^{2+} suggests dependence of water influx upon membrane hydration and specific transmembrane pathways. *Journal of Eukaryotic Microbiology* 44, 109–116.

Galindo, L., Torruella, G., Moreira, D., Timpano, H., Paskerova, G., Smirnov, A. et al. (2018) Evolutionary genomics of *Metchnikovella incurvata* (Metchnikovellidae): An early branching microsporidium. *Genome Biology and Evolution* 10(10), 2736–2748.

Ghosh, K., Cappiello, C. D., McBride, S. M., Occi, J. L., Cali, A., Takvorian, P. M. et al. (2006) Functional characterization of a putative aquaporin from *Encephalitozoon cuniculi*, a microsporidian pathogenic to humans. *International Journal of Parasitology* 36, 57–62.

Goertz, D., Solter, L. F. and Linde, A. (2007) Horizontal and vertical transmission of a *Nosema* sp. (Microsporidia) from *Lymantria dispar* (L.) (Lepidoptera: Lymantriidae). *Journal of Invertebrate Pathology* 95, 9–16.

Goldberg, A. V., Molik, S., Tsaousis, A. D., Neumann, K., Kuhnke, G., Delbac, F., et al. (2008) Localization and functionality of microsporidian iron-sulphur cluster assembly proteins. *Nature* 452, 624–628.

Ha, N. T., Ha, D. T., Thuy, N. T. and Lien, V. T. K. (2010) Occurrence of microsporidia *Enterocytozoon hepatopenaei* in white feces disease of cultured black tiger shrimp (*Penaeus monodon*) in Vietnam. *Aquatic Animal Diseases* December [in Vietnamese with English abstract].

Haag, K., James, T., Pombert, J., Larsson, R., Schaer, T., Refardt, D. and Ebert, D. (2014) Evolution of a morphological novelty occurred before genome compaction in a lineage of extreme parasites. *Proceedings of The National Academy of Sciences USA* 111(43), 15480–15485.

Hacker, C., Howell, M., Bhella, D. and Lucocq, J. (2014) Strategies for maximizing ATP supply in the microsporidian *Encephalitozoon cuniculi*: Direct binding of mitochondria to the parasitophorous vacuole and clustering of the mitochondrial porin VDAC. *Cellular Microbiology* 16, 565–579.

Han, B. and Weiss, L. M. (2017). Microsporidia: Obligate intracellular pathogens within the fungal kingdom. *Microbiology Spectrum* 5, 10.1128/microbiolspec.

Han, B., Ma, Y., Tu, V., Tomita, T., Mayoral, J., Williams, T. et al. (2019) Microsporidia interact with host cell mitochondria via voltage-dependent anion channels using sporoplasm surface protein 1. *mBio* 10, 1–24

Han, B., Polonais, V., Sugi, T., Yakubu, R., Takvorian, P. M., Cali, A. et al. (2017) The role of microsporidian polar tube protein 4 (PTP4) in host cell infection. *PLoS Pathogens* 13, e1006341.

Han, B., Takvorian, P. M. and Weiss, L. M. (2020) Invasion of host cells by microsporidia. *Frontiers in Microbiology* 11, 1–16.

Hansen, M., Kun, J. F. J., Schultz, J. E. and Beitz, E. (2002) A single, bi-functional aquaglyceroporin in blood-stage *Plasmodium falciparum* malaria parasites. *Journal of Biological Chemistry* 277, 4874–4882.

Hayman, J. R., Hayes, S. F., Amon, J. and Nash, T. E. (2001) Developmental expression of two spore wall proteins during maturation of the microsporidian *Encephalitozoon intestinalis*. *Infection and Immunity* 69, 7057–7066.

Hayman, J. R., Southern, T. R. and Nash, T. E. (2005) Role of sulfated glycans in adherence of the microsporidian *Encephalitozoon intestinalis* to host cells in vitro. *Infection and Immunity* 73, 841–848.

He, Q., Vossbrinck, C. R., Yang, Q., Meng, X. Z., Luo, J., Pan, G. Q. et al. (2019) Evolutionary and functional studies on microsporidian ATP-binding cassettes: Insights into the adaptation of microsporidia to obligated intracellular parasitism. *Infection, Genetics and Evolution* 68, 136–144.

Heinz, E., Hacker, C., Dean, P., Mifsud, J., Goldberg, A. V., Williams, T. A. et al. (2014) Plasma membrane-located purine nucleotide transport proteins are key components for host exploitation by microsporidian intracellular parasites. *PLoS Pathogens* 10, 1–12.

Hudson, D. A., Hudson, N. B. and Pyecroft, S. B. (2001) Mortalities of *Penaeus japonicus* prawns associated with microsporidean infection. *Australian Veterinary Journal* 79, 504–505.

Ishibashi, K., Sasaki, S., Fushimi, K., Uchida, S., Kuwahara, M., Saito, H. et al. (1994) Molecular cloning and expression of a member of the aquaporin family with permeability to glycerol and urea in addition to water expressed at the basolateral membrane of kidney collecting duct cells. *Proceedings of the National Academy of Sciences USA* 91, 6269–6273.

Jaroenlak, P., Boakye, D. W., Vanichviriyakit, R., Williams, B. A. P., Sritunyalucksana, K. and Itsathitphaisarn, O. (2018) Identification, characterization and heparin binding capacity of a spore-wall, virulence protein from the shrimp microsporidian, *Enterocytozoon hepatopenaei* (EHP). *Parasites and Vectors* 11, 1–15.

Jaroenlak, P., Sanguanrut, P., Williams, B.A.P., Stentiford, G.D., Flegel, T.W., Sritunyalucksana, K. and Itsathitphaisarn, O. (2016) A nested PCR assay to avoid false positive detection of the microsporidian enterocytozoon hepatopenaei (EHP) in environmental samples in shrimp farms. *PLoS One* 11. https://doi.org/10.1371/journal.pone.0166320

Kanitchinda, S., Srisala, J., Suebsing, R., Prachumwat, A. and Chaijarasphong, T. (2020) CRISPR-Cas fluorescent cleavage assay coupled with recombinase polymerase amplification for sensitive and specific detection of *Enterocytozoon hepatopenaei. Biotechnology Reports* 27, e00485.

Karthikeyan, K. and Sudhakaran, R. (2019) Experimental horizontal transmission of *Enterocytozoon hepatopenaei* in post-larvae of whiteleg shrimp, *Litopenaeus vannamei. Journal of Fish Diseases* 42, 397–404.

Katinka, M. D., Duprat, S., Cornillott, E., Méténler, G., Thomarat, F., Prensier, G. et al. (2001) Genome sequence and gene compaction of the eukaryote parasite *Encephalitozoon cuniculi. Nature* 414, 450–453.

Keeling, P. J., Corradi, N., Morrison, H. G., Haag, K. L., Ebert, D., Weiss, L. M. et al. (2010) The reduced genome of the parasitic microsporidian *Enterocytozoon bieneusi* lacks genes for core carbon metabolism. *Genome Biology and Evolution* 2, 304–309.

Kobayashi, K., Kato, K., Sugi, T., Takemae, H., Pandey, K., Gong, H. et al. (2010) *Plasmodium falciparum* BAEBL binds to heparan sulfate proteoglycans on the human erythrocyte surface. *Journal of Biological Chemistry* 285, 1716–1725.

Leitch, G. J. and Ceballos, C. (2008) Effects of host temperature and gastric and duodenal environments on microsporidia spore germination and infectivity of intestinal epithelial cells. *Parasitology Research* 104, 35–42.

Leonard, C. A. and Hayman, J. R. (2017) Role of host cell integrins in the microsporidium *Encephalitozoon intestinalis* adherence and infection in vitro. *FEMS Microbiology Letters* 364, 1–8.

Li, Y., Wu, Z., Pan, G., He, W., Zhang, R., Hu, J. et al. (2009) Identification of a novel spore wall protein (SWP26) from microsporidia *Nosema bombycis. International Journal for Parasitology* 39, 391–398.

Liu, Z., Zhang, Q. L., Wan, X. Y. and Huang, J. (2015) Development of real-time PCR assay for detection of microsporidian Enterocytozoon hepatopenaei and detection in shrimp samples under different growth rates. *Fisheries Science* https://doi.org/10.11758/yykxjz.20150512003' https://doi.org/10.11758/yykxjz.20150512003 [in Chinese with English abstract].

Mai, H. N., Cruz-Flores, R., Aranguren Caro, L. F., White, B. N. and Dhar, A. K. (2020) A comparative study of Enterocytozoon hepatopenaei (EHP) challenge methods in Penaeus vannamei. *Journal of Invertebrate Pathology* 171, 107336.

Major, P., Sendra, K. M., Dean, P., Williams, T. A., Watson, A. K., Thwaites, D. T. et al. (2019) A new family of cell surface located purine transporters in microsporidia and related fungal endoparasites. *eLife* 8, 1–23.

Miao, Z., Zhang, Y., Zhang, P., Wang, X., Liu, J. and Wang, G. (2018) Characterization of the germination kinetics of individual *Nosema bombycis* spores using phase contrast microscopy imaging and Raman spectroscopy. In: *Optics in Health Care and Biomedical Optics VIII* (eds Luo, Q., Li, X., Gu, Y., Tang, Y. and Zhu, D.). International Society for Optics and Photonics, Washington DC, USA.

Moura, H., Schwartz, D. A., Bornay-Llinares, F., Sodré, F. C., Wallace, S. and Visvesvara, G. S. (1997) A new and improved 'quick-hot Gram-chromotrope' technique that differentially stains microsporidian spores in clinical samples, including paraffin-embedded tissue sections. *Archives of Pathology & Laboratory Medicine* 121, 888–893.

Mukhopadhyay, R., Bhattacharjee, H. and Rosen, B. P. (2014) Aquaglyceroporins: generalized metalloid channels. *Biochimica et Biophysica Acta* 1840, 1583–1591.

Munkongwongsiri, N., Aldama-Cano, D. J., Suebsing, R., Thaiue, D., Prasartset, T., Itsathitphaisarn, O. and Sritunyalucksana, K. (2021) Microsporidian enterocytozoon hepatopenaei (EHP) spores are inactivated in 1 min at 75 °C. *PubAg* https://doi.org/10.1016/j.aquaculture.2020.736178

Nakjang, S., Williams, T., Heinz, E., Watson, A., Foster, P., Sendra, K. et al. (2013) Reduction and expansion in microsporidian genome evolution: New insights from comparative genomics. *Genome Biology and Evolution* 5(12), 2285–2303.

Nehls, U. and Dietz, S. (2014) Fungal aquaporins: cellular functions and ecophysiological perspectives. *Applied Microbiology and Biotechnology* 98, 8835–8851.

Ning, M., Wei, P., Shen, H., Wan, X., Jin, M., Li, X. et al. (2019) Proteomic and metabolomic responses in hepatopancreas of whiteleg shrimp *Litopenaeus vannamei* infected by microsporidian *Enterocytozoon hepatopenaei. Fish and Shellfish Immunology* 87, 534–545.

Nuez-ortín, B. W. G. and Isern-subich, M. M. (2019) Validating the efficacy of a phytobiotic-based feed additive against white faeces syndrome – their health is your wealth. *Aquaculture Asia Pacific* 15(4).

Oliveira, F. O. R., Alves, C. R., Souza-Silva, F., Calvet, C. M., Côrtes, L. M. C., Gonzalez, M. S. et al. (2012) *Trypanosoma cruzi* heparin-binding proteins mediate the adherence of epimastigotes to the midgut epithelial cells of *Rhodnius prolixus. Parasitology* 139, 735–743.

Pedro, S. L. S., Alves, J. M. P., Barreto, A. S. and De Souza Lima, A. O. (2015) Evidence of positive selection of aquaporins genes from *Pontoporia blainvillei* during the evolutionary process of cetaceans. *PLoS ONE* 10, 1–11.

Peuvel-Fanget, I., Polonais, V., Brosson, D., Texier, C., Kuhn, L., Peyret, P. et al. (2006) EnP1 and EnP2, two proteins associated with the *Encephalitozoon cuniculi* endospore, the chitin-rich inner layer of the microsporidian spore wall. *International Journal for Parasitology* 36, 309–318.

Peyretaillade, E., Parisot, N., Polonais, V., Terrat, S., Denonfoux, J., Dugat-Bony, E. et al. (2012) Annotation of microsporidian genomes using transcriptional signals. *Nature Communications* 3(1), 1137.

Rajendran, K. V., Shivam, S., Ezhil Praveena, P., Joseph Sahaya Rajan, J., Sathish Kumar, T., Avunje, S., Jagadeesan, V., Prasad Babu, S. V. A. N. V., Pande, A., Navaneeth Krishnan, A., Alavandi, S. V. and Vijayan, K. K. (2016) Emergence of *Enterocytozoon hepatopenaei* (EHP) in farmed *Penaeus* (Litopenaeus) *vannamei* in India. *Aquaculture* 454, 272–280.

Salachan, P. V., Jaroenlak, P., Thitamadee, S., Itsathitphaisarn, O. and Sritunyalucksana, K. (2017) Laboratory cohabitation challenge model for shrimp hepatopancreatic microsporidiosis (HPM) caused by *Enterocytozoon hepatopenaei* (EHP). *BMC Veterinary Research* 13, 9.

Sang-ngam, B. K., Aldama-cano, D. J., Itsathitphaisarn, O., Mahl, P., Do, P. and Phan, H. (2020) In vitro evaluation of two commercial biocides on EHP spore extrusion rate. *Global Aquaculture Alliance* https://www.aquaculturealliance.org/advocate/in-vitro-evaluation-of-two-commercial-biocides-on-ehp-spore-extrusion-rate/' https://www.aquaculturealliance.org/advocate/in-vitro-evaluation-of-two-commercial-biocides-on-ehp-spore-extrusion-rate/

Sanguanrut, P., Munkongwongsiri, N., Kongkumnerd, J., Thawonsuwan, J., Thitamadee, S., Boonyawiwat, V., Tanasomwang, V., Flegel, T. W. and Sritunyalucksana, K. (2018) A cohort study of 196 Thai shrimp ponds reveals a complex etiology for early mortality syndrome (EMS). *Aquaculture* 493, 26–36.

Sathish Kumar, T., Navaneeth Krishnan, A., Joseph Sahaya Rajan, J., Makesh, M., Jithendran, K. P. P., Alavandi, S. V. and Vijayan, K. (2018) Visual loop-mediated isothermal amplification (LAMP) for the rapid diagnosis of *Enterocytozoon hepatopenaei* (EHP) infection. *Parasitology Research* 117, 1485–1493.

Scanlon, M., Leitch, G. J. and Visvesvara, G. S. (2004) Relationship between the host cell mitochondria and the parasitophorous vacuole in cells infected with *Encephalitozoon* microsporidia. *Journal of Eukaryotic Microbiology* 51, 81–87.

Schottelius, J., Schmetz, C., Kock, N. P., Schüler, T., Sobottka, I., and Fleischer, B. (2000). Presentation by scanning electron microscopy of the life cycle of microsporidia of the genus Encephalitozoon. Microbes and Infection 2, 1401–1406. doi:10.1016/S1286-4579(00)01293-4.

Shen, H., Jiang, G., Wan, X., Fan, X., Qiao, Y., Shi, W., Li, H. and Wang, L. (2017) Multiple pathogens prevalent in shrimp *Penaeus vannamei* cultured from greenhouse ponds in Jiangsu Province of China. *Journal of Aquaculture Research and Development* 8, https://doi.org/10.4172/2155-9546.1000516

Shinn, A. P., Pratoomyot, J., Griffiths, D., Jiravanichpaisal, J. and Briggs, M. (2018) Asian shrimp production and the economic costs of disease. *Asian Fisheries Science* 31S, 29–58.

Sirikharin, R., Taengchaiyaphum, S., Sanguanrut, P., Chi, T.D., Mavichak, R., Proespraiwong, P., Nuangsaeng, B., Thitamadee, S., Flegel, T.W. and Sritunyalucksana, K. (2015) Characterization and PCR detection of binary, pir-like toxins from vibrio parahaemolyticus isolates that cause acute hepatopancreatic necrosis disease (AHPND) in shrimp. *PLoS One* 10, e0126987. https://doi.org/10.1371/journal.pone.0126987

Southern, T. R., Jolly, C. E., Lester, M. E. and Hayman, J. R. (2007) EnP1, a microsporidian spore wall protein that enables spores to adhere to and infect host cells in vitro. *Eukaryotic Cell* 6, 1354–1362.

Sritunyalucksana, K., Sanguanrut, P., Salachan, P. V., Thitamadee, S. and Flegel, T. W. (2015) Urgent appeal to control spread of the shrimp microsporidian parasite Enterocytozoon hepatopenaei (EHP). *Network of Aquaculture Centres Asia-Pacific (NACA)* https://enaca.org/publications/health/disease-cards/urgent-appeal-to-control-ehp-3.pdf

Stentiford, G. D., Bass, D. and Williams, B. A. P. (2019) Ultimate opportunists—the emergent enterocytozoon group microsporidia. *PLoS Pathogens* 15, 1–6.

Stentiford, G. D., Bateman, K. S., Feist, S. W., Chambers, E. and Stone, D. M. (2013a) Plastic parasites: Extreme dimorphism creates a taxonomic conundrum in the phylum Microsporidia. *International Journal for Parasitology* 43, 339–352.

Stentiford, G. D., Becnel, J .J., Weiss, L. M., Keeling, P. J., Didier, E. S., Williams, B. A. P., Bjornson, S., Kent, M. L., Freeman, M. A., Brown, M. J. F., Troemel, E. R., Roesel, K., Sokolova, Y., Snowden, K. F. and Solter, L. (2016) Microsporidia – emergent pathogens in the global food chain. *Trends in Parasitology* 32, 336–348.

Stentiford, G. D., Feist, S. W., Stone, D. M., Bateman, K. S. and Dunn, A. M. (2013b) Microsporidia: Diverse, dynamic, and emergent pathogens in aquatic systems. *Trends in Parasitology* 29, 567–578.

Suebsing, R., Prombun, P., Srisala, J. and Kiatpathomchai, W. (2013) Loop-mediated isothermal amplification combined with colorimetric nanogold for detection of the microsporidian *Enterocytozoon hepatopenaei* in penaeid shrimp. *Journal of Applied Microbiology* 114, 1254–1263.

Szumowski, S. C., Botts, M. R., Popovich, J. J., Smelkinson, M. G. and Troemel, E. R. (2014) The small GTPase RAB-11 directs polarized exocytosis of the intracellular pathogen N. parisii for fecal-oral transmission from C. elegans. *Proceedings of the National Academy of Sciences USA* 111(22), 8215–8220.

Szumowski, S. C., Estes, K. A., Popovich, J. J., Botts, M. R. and Troemel, E. R. (2016) Small GTPases promote actin coat formation on microsporidian pathogens traversing the apical membrane of *Caenorhabditis elegans* intestinal cells. *Cell Microbiology* 18, 30–45.

Takvorian, P. M., Buttle, K. F., Mankus, D., Mannella, C. A., Weiss, L. M. and Cali, A. (2013) The multilayered interlaced network (MIN) in the sporoplasm of the microsporidium *Anncaliia algerae* is derived from Golgi. *Journal of Eukaryotic Microbiology* 60, 166–178.

Takvorian, P. M., Han, B., Cali, A., Rice, W. J., Gunther, L., Macaluso, F. et al. (2020) An ultrastructural study of the extruded polar tube of *Anncaliia algerae* (Microsporidia). *Journal of Eukaryotic Microbiology* 67, 28–44.

Takvorian, P. M., Weiss, L. M. and Cali, A. (2005) The early events of *Brachiola algerae* (Microsporidia) infection: Spore germination, sporoplasm structure, and development within host cells. *Folia Parasitologica* 52, 118–129.

Tang, K. F. J., Han, J. E., Aranguren, L. F., White-Noble, B., Schmidt, M. M., Piamsomboon, P. et al. (2016) Dense populations of the microsporidian *Enterocytozoon hepatopenaei* (EHP) in feces of *Penaeus vannamei* exhibiting white feces syndrome and pathways of their transmission to healthy shrimp. *Journal of Invertebrate Pathology* 140, 1–7.

Tang, K. F. J., Aranguren, L. F., Piamsomboon, P., Han, J. E., Maskaykina, I. Y. and Schmidt, M. M. (2017) Detection of the microsporidian *Enterocytozoon hepatopenaei* (EHP) and Taura syndrome virus in *Penaeus vannamei* cultured in Venezuela. *Aquaculture* 480, 17–21.

Tang, K. F. J., Pantoja, C. R., Redman, R., Han, J. E., Tran, L. H. and Lightner, D. V. (2015) Development of in situ hybridization and PCR assays for the detection of *Enterocytozoon hepatopenaei* (EHP), a microsporidian parasite infecting penaeid shrimp. *Journal of Invertebrate Pathology* 130, 37–41.

Tangprasittipap, A., Srisala, J., Chouwdee, S., Somboon, M., Chuchird, N., Limsuwan, C. et al. (2013) The microsporidian *Enterocytozoon hepatopenaei* is not the cause of white feces syndrome in whiteleg shrimp *Penaeus* (Litopenaeus) *vannamei. BMC Veterinary Research* 9, 139.

Thitamadee, S., Prachumwat, A., Srisala, J., Jaroenlak, P., Salachan, P. V., Sritunyalucksana, K. et al. (2016). Review of current disease threats for cultivated penaeid shrimp in Asia. *Aquaculture* 452, 69–87.

Tourtip, S., Wongtripop, S., Stentiford, G. D., Bateman, K. S., Sriurairatana, S., Chavadej, J. et al. (2009) *Enterocytozoon hepatopenaei* sp. nov. (Microsporida: Enterocytozoonidae), a parasite of the black tiger shrimp *Penaeus monodon* (Decapoda: Penaeidae): Fine structure and phylogenetic relationships. *Journal of Invertebrate Pathology* 102, 21–29.

Tran, L., Nunan, L., Redman, R. M., Mohney, L.L., Pantoja, C.R., Fitzsimmons, K. and Lightner, D.V. (2013) Determination of the infectious nature of the agent of acute hepatopancreatic necrosis syndrome affecting penaeid shrimp. *Dis Aquat Org* 105, 45–55.

Tsaousis, A. D., Kunji, E. R. S., Goldberg, A. V., Lucocq, J. M., Hirt, R. P. and Embley, T. M. (2008) A novel route for ATP acquisition by the remnant mitochondria of *Encephalitozoon cuniculi. Nature* 453, 553–556.

Undeen, A. H. and Frixione, E. (1990) The role of osmotic pressure in the germination of *Nosema algerae* spores. *Journal of Protozoology* 37, 561–567.

Vavra, J. and Larsson, J. I. R. (2014) Structure of the Microsporidia. In: *Microsporidia, Pathogens of Opportunity* (eds Becnel, L. M. and Weiss, J. J.). Wiley Blackwell, Oxford, UK, pp. 1–70.

Verkman, A. S. and Mitra, A. K. (2000) Structure and function of aquaporin water channels. *American Journal of Physiology. Renal Physiology* 278, F13–F18.

Verma, R. K., Prabh, N. D. and Sankararamakrishnan, R. (2014) New subfamilies of major intrinsic proteins in fungi suggest novel transport properties in fungal channels: Implications for the host-fungal interactions. *BMC Evolutionary Biology* 14, 1–16.

Vu-Khac, H., Thanh, T. N. T., Thu, G. N. T., Le, C. H. and Nguyen, V. D. (2018) Vertical transmission and early diagnosis of the microsporidian *Enterocytozoon hepatonaei* in whiteleg shrimp *Penaeus vannamei. Journal of Pure and Applied Microbiology* 12, 1125–1131.

Wang, L., Lv, Q., Meng, X., Chen, J., Wang, Y., Pan, G., et al. (2021) Identification and characterization polar tube protein 2 (PTP2) from Enterocytozoon hepatopenaei and its potential effect on shrimp microsporidian germination activity evaluation. *Aquaculture* 544, 737062. doi:10.1016/j.aquaculture.2021.737062.

Weber, R., Bryan, R. T., Owen, R. L., Wilcox, C. M., Gorelkin, L. and Visvesvara, G. S. (1992) Improved light-microscopical detection of microsporidia spores in stool and duodenal aspirates. *New England Journal of Medicine* 326, 2014.

Weiss, L. M. (2020) Microsporidiosis. In: *Hunter's Tropical Medicine and Emerging Infectious Diseases* (eds Ryan, E. T. et al.). Elsevier, London, UK, pp. 825–831.

Weiss, L. M. and Becnel, J. J. (2014) *Microsporidia: Pathogens of Opportunity*. Wiley Blackwell Press, Oxford, UK.

Whelan, T. A., Lee, N. T., Lee, R. C. H. and Fast, N. M. (2019) Microsporidian introns retained against a background of genome reduction: Characterization of an unusual set of introns. *Genome Biol. Evol.* 11, 263–269.

Williams, B. A. P., Hirt, R. P., Lucocq, J. M. and Embley, T. M. (2002) A mitochondrial remnant in the microsporidian *Trachipleistophora hominis. Nature* 418, 865–869.

Williams, B. A. P., Haferkamp, I. and Keeling, P. J. (2008) An ADP/ATP-specific mitochondrial carrier protein in the microsporidian *Antonospora locustae. Journal of Molecular Biology* 375, 1249–1257.

Wiredu Boakye, D., Jaroenlak, P., Prachumwat, A., Williams, T. A. P., Bateman, K. S., Itsathitphaisarn, O. et al. (2017) Decay of the glycolytic pathway and adaptation to intranuclear parasitism within Enterocytozoonidae microsporidia. *Environmental Microbiology* 19, 2077–2089.

Wittner, M. and Weiss, L. M. (1999) *The Microsporidia and Microsporidiosis*. ASM Press, Washington, DC, USA.

Wu, Z., Li, Y., Pan, G., Zhou, Z. and Xiang, Z. (2009) SWP25, a novel protein associated with the *Nosema bombycis* endospore. *Journal of Eukaryotic Microbiology* 56, 113–118.

Xu, Y and Weiss, L. M. (2005) The microsporidian polar tube: A highly specialised invasion organelle. *International Journal for Parasitology* 35, 941–953.

Yang, B. (2000) The human aquaporin gene family. *Current Genomics* 1, 91–102.

Yang, D., Pan, L., Chen, Z., Du, H., Luo, B., Luo, J. et al. (2018) The roles of microsporidia spore wall proteins in the spore wall formation and polar tube anchorage to spore wall during development and infection processes. *Experimental Parasitology* 187, 93–100.

Chapter 23

GEOLOGICALLY RECENT SHRIMP ENDOGENOUS VIRAL ELEMENTS AND THEIR IMPLICATIONS FOR SHRIMP DISEASE CONTROL AND TRADE

VANVIMON SAKSMERPROME, SAROCHA JITRAKORN, NARONG ARUNRUT, WANSIKA KIATPATHOMCHAI AND TIMOTHY W. FLEGEL

Summary

Integration of non-retroviral fragments into animal genomes has been found for a few decades as a result of infection and completion of their life cycles. To date, 'endogenous viral elements' (EVE) of infectious hypodermal and hematopoietic necrosis virus (IHHNV) and white spot syndrome virus (WSSV) have been found in penaeid shrimp. For a practical diagnostic approach, this necessitated a change in the routine method of distinguishing between infected shrimp and EVE-containing samples. In the case of IHHNV, because infected shrimp species have developed resistance to the virus it is challenging to separate shrimp carrying both the infectious form of virus and EVE from those that are infected only (i.e. both would be recognized by conventional PCR testing). Discarding of domesticated shrimp breeding stocks based on such false-positive results might have negative consequences, if such inserts are related to shrimp viral disease tolerance according to the 'viral accommodation hypothesis'. Improvement in accuracy is necessary in the diagnosis of IHHNV infection. Multiplex PCR analysis can be developed to amplify the entire IHHNV genome, ensuring an accurate diagnosis, but the technique must be convenient for practical application. More recently, isothermal nucleic acid amplification techniques, such as loop-mediated isothermal amplification (LAMP), involve primers targeting OIE-recommended regions and the 3′ end of the viral genome, for which lower occurrence of EVEs has been reported for the shrimp genome.

The implications of EVE using IHHNV, as the first model involving viruses in the specific-pathogen-free (SPF) list, are also discussed in regard to potential disease control and trading issues.

Introduction

Integration of retroviral fragments into animal genomes has been known for several decades as a result of infection and completion of their life cycles (Feschotte and Gilbert, 2012). More recently, several studies have reported a significant number of non-retroviral sequences that are endogenized in the host genome, including humans, insects and crustaceans. To date, EVE derived from non-retroviruses of IHHNV and WSSV have been found in penaeid shrimp (Saksmerprome et al., 2011; Tang and Lightner, 2006; Tang et al., 2007; Utari et al., 2017). Viral-like sequences have been identified in a Fosmid clone library constructed from the *Penaeus monodon* genome (Huang et al., 2011), suggesting that other endogenous segments of shrimp viruses may be more common than expected.

Certain viral infections lead to life-long, persistent infection with no signs of disease in the host, so-called viral accommodation (Flegel, 2007). Since reverse-transcriptase (RT) and integrase (IN) sequences are commonly found in the genome of shrimp and other arthropods, it is hypothesized that animals use these RT to recognize 'foreign' mRNA of both RNA and DNA viruses, and also use the INs to randomly insert short cDNA sequences into their genomes (Figure 23.1). Host-directed, random insertion of viral sequences may lead to specific, heritable resistance to disease caused by the corresponding virus. Figure 23.2 shows that some EVEs are employed to produce antisense, immunospecific RNA (imRNA) that binds with viral mRNA to produce dsRNA capable of triggering RNAi and suppressing viral propagation (Flegel, 2009).

The occurrence of EVEs in shrimp genome could have implications in disease control and trading in shrimp industry. This chapter will focus on IHHN-EVE in Thai *P. monodon* as an example of random

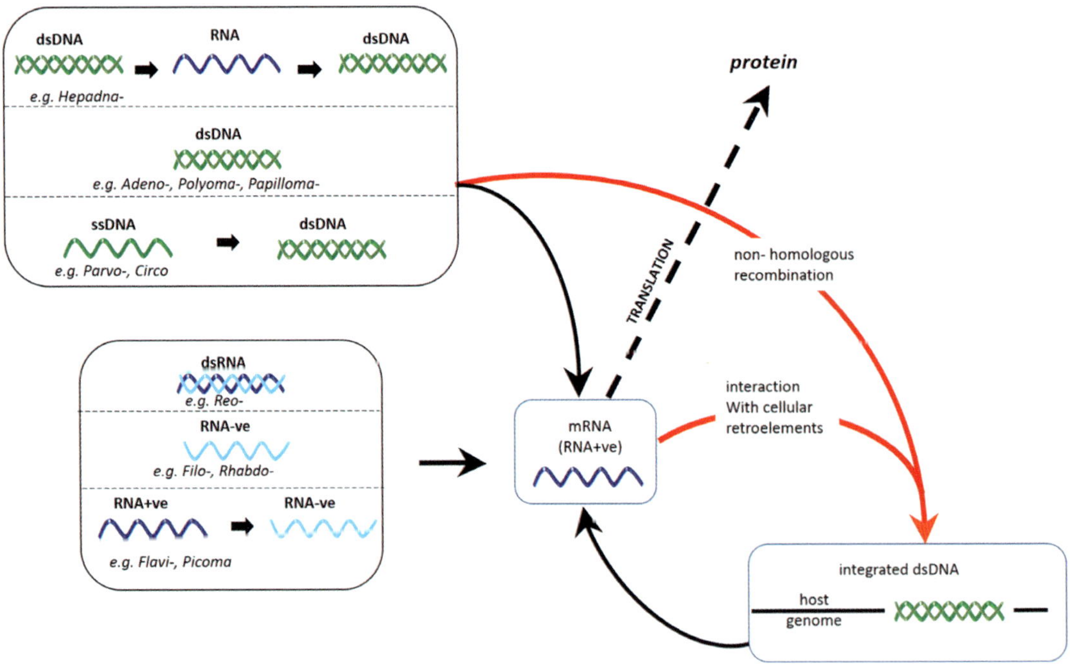

FIGURE 23.1 Occurrence of EVE derived from non-retroviruses. Credit: adapted from Katzourakis and Gifford (2010).

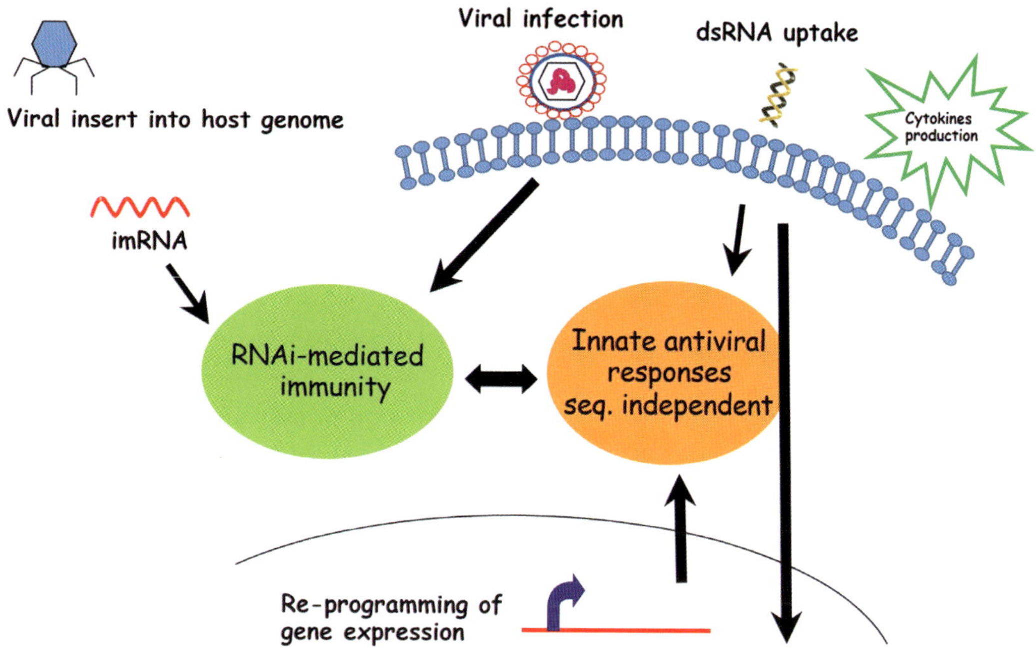

FIGURE 23.2 Hypothesis of the shrimp specific response mechanism for antiviral defense based on EVE. Credit: adapted from Flegel (2009) and Robalino et al. (2007).

and variable inserts in the crustacean genome; challenges in confirming SPF shrimp in EVE-containing shrimp stocks using recommended methods; and the potential to create beneficial IHHNV inserts to produce defined, certifiable, specific-pathogen-resistant (SPR) shrimp stock from SPF stocks.

IHHNV and IHHN-EVE

In 2012–2019 editions of *Manual of Diagnostic Tests for Aquatic Animals*, three genotypes of *Penaeus stylirostris* densovirus (commonly known as IHHNV) were described based on its originally discovered sequence and on sequences of subsequent geographical isolates. Type 1 from the Americas and East Asia and Type 2 from South-East Asia are infectious to penaeid shrimp. IHHN-EVE were Type 3A from East Africa, India and Australia, and Type 3B from the western Indo-Pacific region (Dhar et al., 2014; Tang and Lightner, 2006; Tang et al., 2007). However, regarding evidence of common occurrence of random and variable IHHNV genome inserts (Saksmerprome et al., 2011), it would be more appropriate to assign the infectious form of the virus as 'IHHNV' and the homologous sequence to virus embedded in the shrimp genome as 'IHHN-EVE'.

In regard to a practical diagnostic approach, this necessitated a change in the routine method used to distinguish between infected shrimp and EVE-containing samples. Several infected shrimp species have developed resistance to IHHNV (OIE, 2019), and thus it is a challenge to separate shrimp carrying both the infectious form of virus and EVE from those that are infected only (i.e. both would be recognized by conventional PCR testing). A false-positive result for infectious IHHNV using the recommended methods would certainly have a negative effect on the international seafood trade. In addition, discarding domesticated shrimp breeding stocks based on such false-positive results might also have negative consequences if such inserts are related to shrimp viral disease tolerance according to the above-mentioned viral accommodation hypothesis.

The following sections provide details on findings relating to random number and length of sequence fragments of IHHNV insertions in the giant tiger shrimp genome, and the development of detection methods to reduce false-positive results due to EVE.

Discovery of IHHN-EVE in the shrimp genome

Persistent infection with IHHNV and EVE in the black tiger shrimp, *P. monodon* genome from East Africa and Australia is commonly found without shrimp showing apparent signs of disease (Saksmerprome et al., 2010; Tang and Lightner, 2006). A study by Saksmerprome et al. (2011), using PCR analysis, examined whether giant tiger shrimp from Thailand contained random genome insertions. The study was originally designed to survey the prevalence of IHHNV-infected shrimp from different provinces in Thailand. All shrimp samples used were selected from one Thai shrimp hatchery, because these gave positive test results by PCR with the OIE 309F/R method (Tang et al., 2007) and/or the new IQ2000TM IHHNV Detection and Prevention System for detection of infectious IHHNV. However, using a set of seven overlapping primers to cover most of the IHHNV genome, only 79 out of 99 shrimp had amplicons for all seven sets (Table 23.1), indicating that these contained the template for the complete IHHNV genome and

TABLE 23.1 Positive (+) and negative (–) PCR results obtained with DNA templates of shrimp samples using overlapping primers targeting seven regions covering the whole IHHNV genome. In column 1, the numbers in parentheses indicate the number of shrimp samples with the same pattern of positive and negative results as that for the sample number given. For example, sample B11 had an identical pattern to that of B53. Gray shading indicates the approximate length of the purported IHHNV inserts in the shrimp genome. Credit: Saksmerprome et al. (2011).

Isolate (no.)	*Primer pairs used and positions relative to AF273215*						
	1 158–723	*2 702–1260*	*3 1059–1507*	*4 1451–2027*	*5 2002–2569*	*6 2544–3100*	*7 3031–3625*
Infective IHHNV (total 79 samples)							
B2 (79)	+	+	+	+	+	+	+
One insert (1–2 gaps; total 14 samples)							
B50 (11)	+	+	+	+	+	+	–
B61 (1)	–	+	+	+	+	+	–
B39(2)	–	+	+	+	+	+	+
Two inserts (1–3 gaps; total 6 samples)							
B11, B53	+	+	–	–	+	+	+
B22(1)	+	–	+	+	+	+	+
B58 (2)	+	+	+	–	+	+	–
B63 (1)	–	+	+	–	+	+	–

suggesting that the shrimp from which they had originated carried infectious IHHNV. The remaining 20 shrimp showed from one to three gaps in the IHHNV sequence, suggesting that they were not infected with IHHNV but carried between one and three fragments of the IHHNV genome, probably inserted into their genomic DNA (Table 23.1).

Given the consistency of the PCR results for the seven IHHNV genome regions for 79 of the specimens, the simplest explanation for random failures with the other 20 specimens was that they indicated gaps representing the absence of portions in the IHHNV genome. This was confirmed by subsequent genome walking (see below). There was no obvious pattern in the gaps found from the 20 samples (Table 23.1), but all gave positive results for regions 2 and 5. The target sequence of the 309F/R method for detection of infectious IHHNV (Tang et al., 2007) runs from the middle of our region 4 to the middle of our region 5 (i.e. positions 1719–2027 in GenBank reference sequence AF273215). Single inserts dominated, followed by double inserts. It was not determined whether the double inserts were in the same linkage group or separated in the shrimp genome.

Detailed analysis was carried out with *P. monodon* sample B11 (Table 23.1) by genome walking in the region 4 direction from the putative insert covering IHHNV regions 5–7 (Table 23.1); two-step PCR yielded a 950-bp amplicon (Figure 23.3). When this amplicon band was subjected to cloning and sequencing, a fragment of 587 bases was obtained (Figure 23.3) that contained 313 bases with 99% identity to a matching portion of our region 5 of the IHHNV genome (positions 1889–2201 of the reference sequence GenBank AF273215), separated by 35 bases of unknown identity from a further 239 bases with 97% identity to a portion of a thrombospondin-like expressed sequence tag (EST) of *P. monodon* (GenBank CF805583). All this information, together with the sequencing protocol, suggested that the 239-bp portion of the chimeric fragment was an actual part of the *P. monodon* genome and not a result of vector contamination. This was confirmed by amplification of the chimeric fragment, shown in Figure 23.4, by PCR from the genomic DNA extract of B11 and other shrimp specimens with a gap in our region 4 of the IHHNV genome (see below). With respect to the thrombospondin-like EST of *P. monodon* (GenBank CF805583) that showed 97% identity to 238 bp at the 5′ end of our chimeric fragment from B11, the orientation of the match was plus/minus so that the reading direction for the GenBank thrombospondin element was opposite to that of the IHHNV insert. Therefore, reorientation of the chimeric fragment with respect to the thrombospondin record yielded the sequences shown for 'chimeric fragment 1' in Figure 23.4.

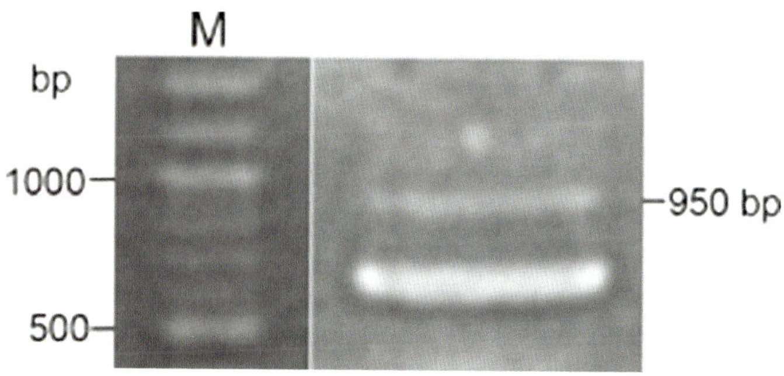

FIGURE 23.3 DNA walking with sample B11. Agarose gel showing an amplified product from PCR-based DNA walking in the B11 DNA library. Intervening lanes in the gel have been removed. Credit: Saksmerprome et al. (2011).

5′ *TTCGCCATTCAGGCTGCG*CAACTGTTGGGAAGGGCGATCGGTGCGGGCCT

CTTCGCTATTACGCCAGCTGGCGAAAGGGGGATGTGCTGCAAGGCGATTAAG

TTGGGTAACGCCAGGGTTTTCCCAGTCACGACGTTGTAAAACGACGGCCAGT

GAATTGTAATACGACTCACTATAGGGCGAATTGGGCCCGACGTCGCATGCTC

CCGGCCGCCATGGCGGCCGCGGGAATTCGATTACTATAGGGCACGCGTGGTC

GACGGCCCGGGCTGGTCTGTCACTAATTACAAACCTGCAGAATATATCAAAA

AGAAAACCAGACTACGACAATATGCAGTGGATAAAATACATGTTAGCCAACA

ACGACATCCGTGTACCAGAAATCTTAGCTTGGATAATCTTCGTAGCAGACAA

AAAACTGGACAAAATAAACACTCTTGTACTCCAAGGACCAACAGGAACAGGA

AAATCTCTTACCATCGGTGCACTACTCGGAAAACTGAACACTGGCCTAGTAA

CAAGAACAGGAGACTCAAACACCTTCCATCTACAAAACC ***TCATCGGAAAGTC***

CTACGCTCTCTTCGAAG ***3′***

FIGURE 23.4 Sequence of B11 chimeric amplicon 1. Nucleotide sequence of the amplicon obtained from the B11 isolate (chimeric fragment 1) by genome walking. The portion of the sequence in bold typeface at the 5′ end of the sequence had 97% identity to a *P. monodon* EST, while the underlined portion at the 3′ end of the sequence had 99% identity to the IHHNV reference sequence (GenBank AF273215) from positions 1998 to 2161. The region outlined in gray gave no significant homology to any sequence in the GenBank database, and the underlined text in bold italic indicates the positions of the primers used for amplification of this chimeric fragment. Credit: Saksmerprome et al. (2011).

Based on the sequence of the chimeric fragment obtained from sample B11, a set of primers was designed for specific amplification of the whole chimeric fragment (587 bp). When used for PCR with DNA templates from B11, B53 and three other samples showing gaps in region 4 (Table 23.1), all gave positive results, yielding amplicons of approximately 587 bp (Figure 23.5). This suggested that all had IHHNV insertions in the same region of the shrimp genome. In addition, sequence analysis of the respective amplicons revealed 100% identity with the chimeric sequence from B11. Also, four out of 24 shrimp samples arbitrarily chosen from the group of 79 that were positive for all seven IHHNV regions also yielded amplicons using the chimeric shrimp/IHHNV primers (Figure 23.4), indicating that the shrimp were not only infected with IHHNV but also had IHHNV inserts in their genome. Surprisingly, specimen B50 also gave a positive result for chimeric amplicon 1 (Figure 23.6), despite the fact that it showed a gap for region 7 only and no gap for region 4. As with the 79 specimens positive for all regions of the IHHNV genome, these tests revealed that multiple IHHNV insertions in a single shrimp specimen would not be revealed by overlapping primer analysis. In addition, the lack of an amplicon in PCR assays with primers for chimeric fragment 1 did not exclude the possibility that different chimeric primer pairs would be able to reveal other insertions masked by overlapping primer analysis. The following section describes further analysis of the phenomenon of multiple insertions, and how to reduce false-positive results using newly developed detection methods.

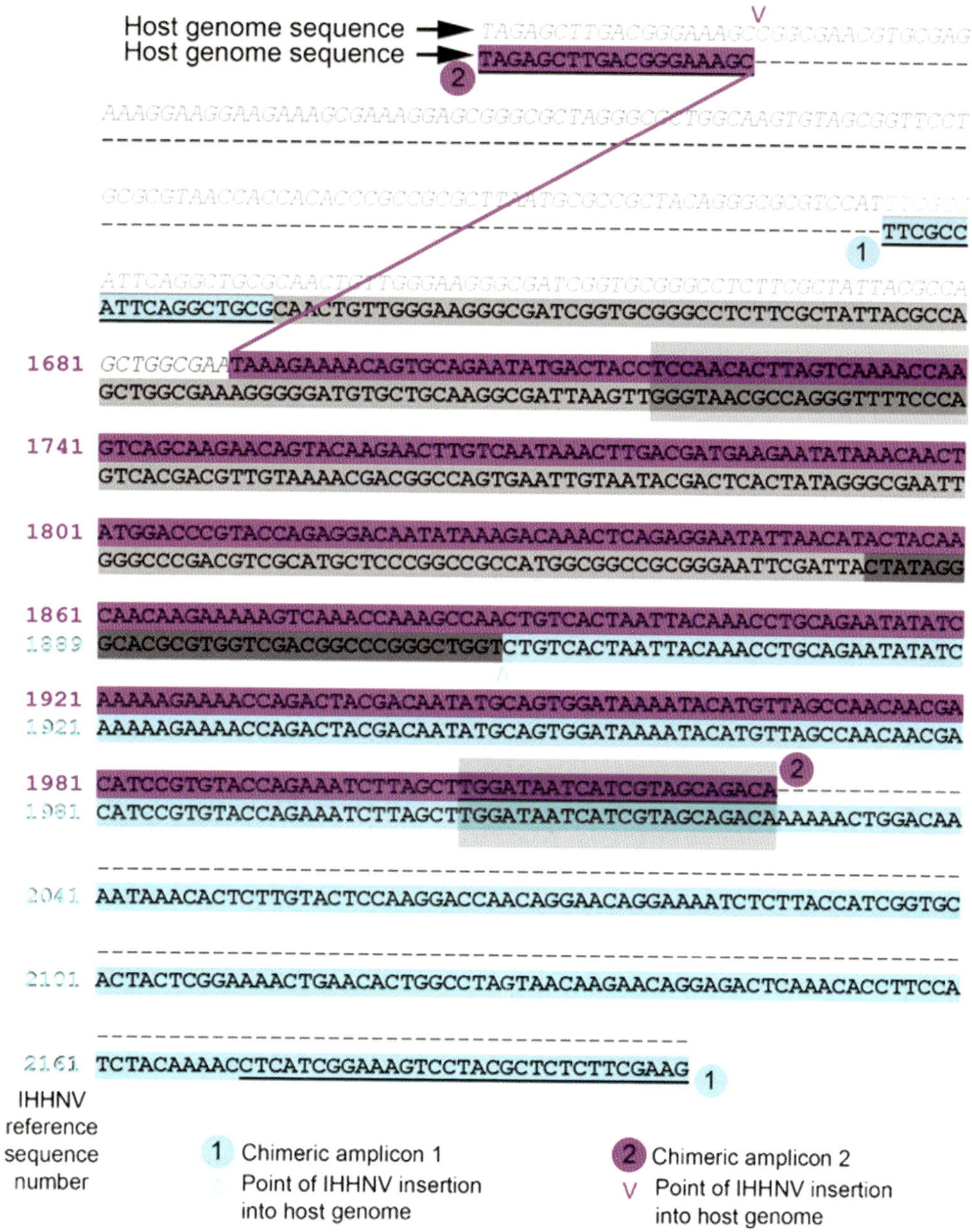

FIGURE 23.5 Relationship between two chimeric amplicon sequences from B11. Diagram showing the relationship between the sequences of the two chimeric amplicons (1 and 2) obtained from sample B53 in relation to the host shrimp sequence and reference sequence (GenBank AF273215). The host genome sequence portion of chimeric amplicon 1 is outlined in light gray, except for the forward primer portion shown together with the IHHNV portion, outlined in blue. All of chimeric amplicon 2 is outlined in magenta, including the host sequence primer portion. The region of the forward and reverse primers for the detection method of Tang et al. (2007) is boxed in gray; this clearly reveals why the insert giving rise to chimeric amplicon 2 can give a positive reaction with the detection method while that for chimeric amplicon 1 cannot. Numbering on the left indicates IHNV genome positions relative to the reference sequence. Credit: Saksmerprome et al. (2011).

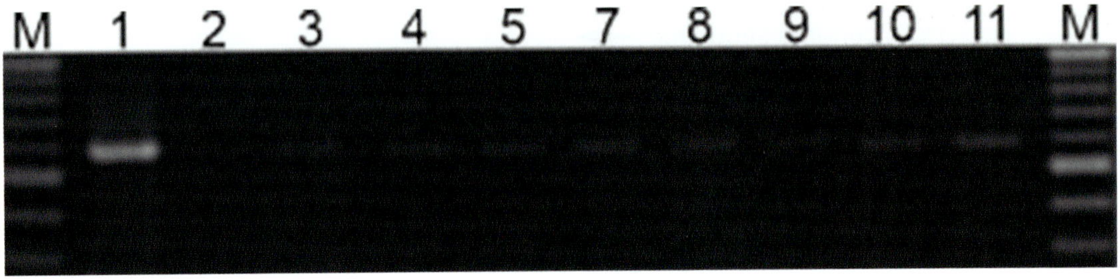

FIGURE 23.6 Chimeric amplicons from a variety of shrimp specimens. Agarose gel showing example bands of PCR amplicons (587 bp) using specific primers for the chimeric shrimp/IHHNV fragment. Lane M, DNA ladder; lane 1, positive control from a plasmid containing B11 fragment; lane 2, negative control; lane 3, sample B50 from IHHNV sequence gap in our region 7; lanes 4–7, samples B51–53 and B57 arbitrarily selected from the B2 group in Table 23.2 showing no gaps; lanes 8–10, samples B58, B63 and 49 with gaps in region 4, like sample B11; lane 11, sample B61 with gap in regions 1 and 7. Credit: Saksmerprome et al. (2011).

Challenges in confirming SPF shrimp among EVE-containing stock

GENERAL PCR METHOD FOR DETECTION OF INFECTIOUS IHHNV

Several PCR methods and a number of commercial PCR kits are available for detection of IHHNV. OIE recommends the PCR method following the methods outlined in Tang et al. (2000) and Nunan et al. (2000). However, recent minor modifications – including the sources of the reagents and the use of an automated DNA extraction instrument – are acceptable. These modifications include a DNA extraction method, primers (Table 23.2) and the volume of reaction. For primers, the OIE manual describes the PCR protocols of primers 389F and 389R, which produce a 389bp fragment from IHHNV-infected samples (Tang et al., 2000), and primers 77012F and 77353R, which produce a 356bp fragment (Nunan et al., 2000). The PCR profile comprises a 3–5-minute step at 95 °C to denature DNA, followed by 35 cycles

TABLE 23.2 Recommended primer sets for one-step PCR detection of IHHNV (OIE Manual of Diagnostic Tests for Aquatic Animals, 28 June 2019).

Primer	*Product*	*Sequence*	*GenBank*
389F	389 bp	5'-CGG-AAC-ACA-ACC-CGA-CTT-TA-3'	AF218266
389R		5'-GGC-CAA-GAC-CAA-AAT-ACG-AA-3'	
77012F	356 bp	5'-ATC-GGT-GCA-CTA-CTC-GGA-3'	AF218266
77353R		5'-TCG-TAC-TGG-CTG-TTC-ATC-3	
392F 3	92 bp	5'-GGG-CGA-ACC-AGA-ATC-ACT-TA-3'	AF218266
392R		5'-ATC-CGG-AGG-AAT-CTG-ATG-TG-3'	
309F	309 bp	5'-TCC-AAC-ACT-TAG-TCA-AAA-CCA-A-3'	AF218266
309R		5'-TGT-CTG-CTA-CGA-TGA-TTA-TCC-A-3'	
MG831F	831 bp	5'-TTG-GGG-ATG-CAG-CAA-TAT-CT-3'	DQ228358
MG831R		5'-GTC-CAT-CCA-CTG-ATC-GGA-CT-3'	

of 95°C for 30 seconds, 60°C for 30 seconds and 72°C for 30 seconds, with a final extension at 72°C for 5 minutes. However, OIE suggests that direct sequencing of amplified products should be performed to confirm the presence of IHHNV or the other primers that detect the IHHNV and do not react with IHHNV-related sequences present in *P. monodon* stocks from Africa or Australia (Tang et al., 2007), or from Thailand (Saksmerprome et al., 2011), to discriminate IHHN-EVE from the actual virus.

METHODS AVAILABLE FOR IDENTIFICATION OF INFECTIOUS IHHNV

Multiplex PCR

Evidence from Saksmerprome et al. (2011) suggested that random insertion of IHHNV fragments may yield false-positive results, particularly if the inserted sequences can be targeted by currently recommended detection methods. Thus, it is necessary to improve the diagnostic accuracy of IHHNV infection. Multiplex PCR analysis has been developed to amplify the entire IHHNV genome, ensuring accurate diagnosis, and the technique is suitable for practical application. Molthathong et al. (2013) developed multiplex PCR reactions in a one-step reaction using a two-tube method. The PCR assay is separated into two reaction mixtures, the first containing three primer pairs – F158-R723 (no. 1), F1451-R2355 (no. 2) and F3031-R3782 (no. 3) – and the second with two primer pairs – F702-R1578 (no. 4) and F2002-R3100 (no. 5). Actin-derived primers are applied in this reaction as an internal control amplification, and actin F/R as internal control. Primer sequences used in the assay are shown in Table 23.3. In the case of samples from actual infections, three and two bands were obtained from the first and second reactions, respectively (Figure 23.7A,B). This is because the infected sample should have its entire viral genome detected by primers with the multiplex method; on the other hand, uninfected specimens (EVE-IHHN) would produce incomplete results. The absence of any five DNA fragments is interpreted as negative to the hypothetical insert by agarose gel electrophoresis. The minimum number of DNA templates providing visible PCR products for both reactions is approximately 20 pg (Figure 23.7). The detection capability of this multiplex PCR system is estimated to be equivalent to 150 copies of the virus. It is possible that shrimp carrying non-infectious inserts are infected by the virus and, if it is

TABLE 23.3 Primers used in multiplex PCR for differentiation of IHHNV-infected and -inserted types. Credit: Molthathong et al. (2013).

Primer (no.)	*Primer (name)*	*Sequences (5′ to 3′)*	*Product size (bp)*
1	F158	ATG GAA GAT ACG AAC AAC CA	566
	R723	GGA CCT GGG GTG AGA AGG CT	
2	F1451	GTT ACC TTT GCT GCC AGA GC	905
	R2355	GGA GGT ACC CAG TAG TCT ATA TC	
3	F3031	CTA AGG AAA CCG ACG TAA CA	752
	R3782	AAG TGA CGG CGG ACA ATA TC	
4	F702	CAA GCC TTC TCA CCC CAG G	877
	R1578	ATG GCG TGG CCA AGA C	
5	F2002	AGC TTG GAT AAT CAT CGT AGC AG	1099
	R3100	GCT GTT GAT TGT ACG GTC ACA AG	

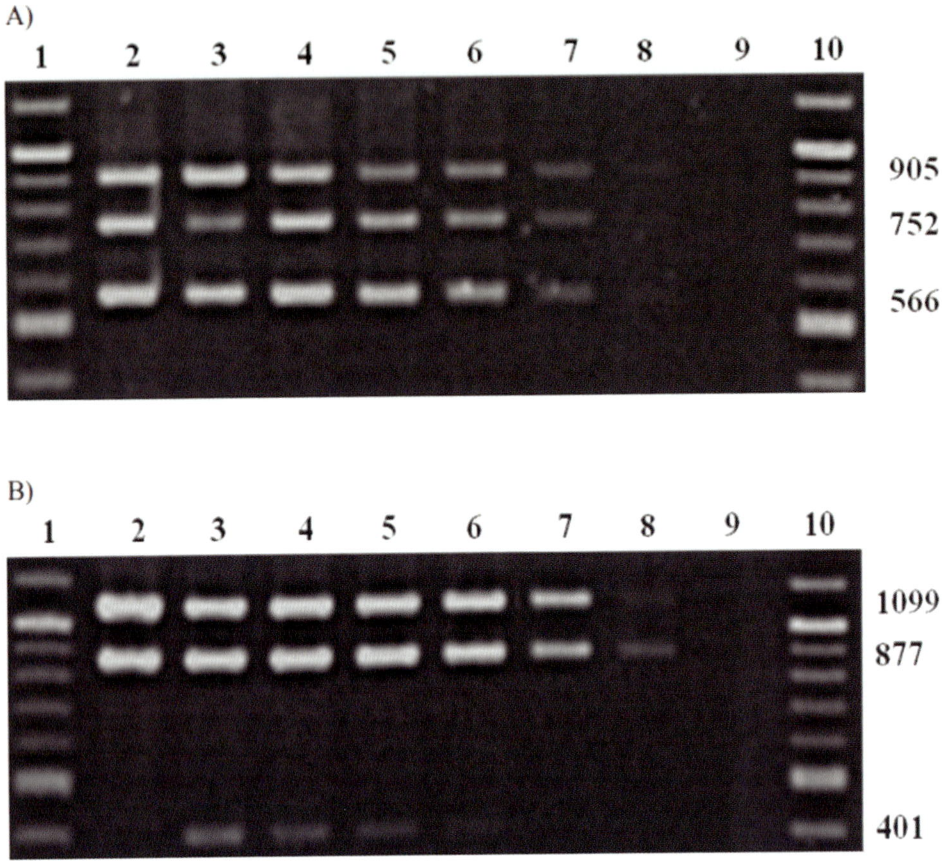

FIGURE 23.7 Sensitivity analysis of the first multiplex reaction (A) and the second (B). Lanes 1 and 10, DNA ladder; lane 2, positive control; lanes 3–8, the amount of DNA template – 200 ng, 20 ng, 2 ng, 200 pg, 20 pg, 2 pg, respectively; lane 9, negative.

the case, multiplex PCR may not be able to separate such specimens from those with IHHNV infection only. Nevertheless, the method developed remains useful for selection of SPF shrimp.

LAMP for IHHNV detection

Currently, several isothermal nucleic acid amplification techniques show promise in the detection of IHHNV, because they can efficiently amplify targets without the expensive thermal cycling equipment required by conventional PCR assay. For example, LAMP has been used for the detection of IHHNV based on sensitivity and specificity (Arunrut et al., 2011; Sun et al., 2006). The LAMP technique is a highly efficient amplification method that allows synthesis of large amounts of DNA in a short time and requires only a controlled thermal instrument, such as a heating block. In addition, recombinase polymerase amplification (RPA) have been developed to detect pathogens using real-time RPA targeting nucleotides 868–1000 of GenBank Accession no. 272315 (Xia et al., 2015) and a lateral flow dipstick (LFD) targeting nucleotides 3094–3214 of GenBank Accession no. 272315 (Jaroenram and Owens, 2014).

However, the above-mentioned methods can detect only one target region of the nucleotide sequence of the IHHNV. False-positive results appear in the case of endogenous viral elements in the shrimp genome (Jaroenram and Owens, 2014; Saksmerprome et al., 2011). Similar to the idea of multiplex PCR, double-LAMP (D-LAMP) was further designed in the detection of two target regions of the IHHNV genome by Arunrut et al. (2019). The D-LAMP comprises two LAMP reactions, the first being an assay targeting the OIE-recommended nucleotide sequence 1740–1968 (GenBank Accession no. 273215) (Arunrut et al., 2011). The second of these assays was developed to target positions 3110–3314 (GenBank Accession no. 273215) at the 3′ end of the viral genome, in which lower occurrence of EVEs has been reported in the shrimp genome (Arunrut et al., 2011). The primer sequences are shown in Table 23.4. This method can detect two target sequences at high sensitivity – 100 copies of plasmid DNA (Figure 23.8) – when using LAMP conditions of 63°C for 45 minutes. In addition, amplification results can also be observed with colorimetric gold nanoparticles probes (AuNPs) for 5 minutes, offering advantages over previously published methods for detection of infectious IHHNV by increasing specificity using D-LAMP reactions, while maintaining the platform's simplicity and convenience. To interpret the results, D-LAMP with *P. monodon* samples giving positive results for both LAMP primer sets will be identified as actual infection with IHHNV.

TABLE 23.4 Primers and probes used in D-LAMP.

Primer name	*Sequence 5′ – 3′*	*Reference*
LAMP 1		Arunrut et al., 2011
R1-F3	AGTCAGCAAGAACAGTACA	
R1-B3	AACATGTATTTTATCCACTGCA	
R1-FIP	TTGTCCTCTGGTACGGGTCCTTTTACTTGTCAATAAACTTGACGAT	
R1-BIP	AACAAGAAAAAGTCAAACCAAAGCCTTTTGGTTTTCTTTTGGATATATTCTGCA	
R1-LF	TAGTTGTTTATATTCTT	
R1-LB	ACTGTCACTAATTACAAA	
R1-probe-thiol	(SH)-A10-TAAAGACAAACTCAGAGGAA	
LAMP 2		Arunrut et al., 2019
R2-F3	TGAAGACCCAATCCAACAA	
R2-B3	TGTATGGTGGTCTGCTGG	
R2-FIP	CCAATTTCTAGTGTCACCATGGATTTTTTGATTCATGGAACAAATGCGA	
R2-BIP	CAACCGGACCCACCTTCATCTTTTAGACTAGGTTTCCAAGGGA	
R2-LF	ATAGGCTTTTCTGTCACC	
R2-LB	GGTGGTCAATTAAAATGGG	
R2-probe-thiol	(SH)-A10-ATGGCGGAGAAATACCA	

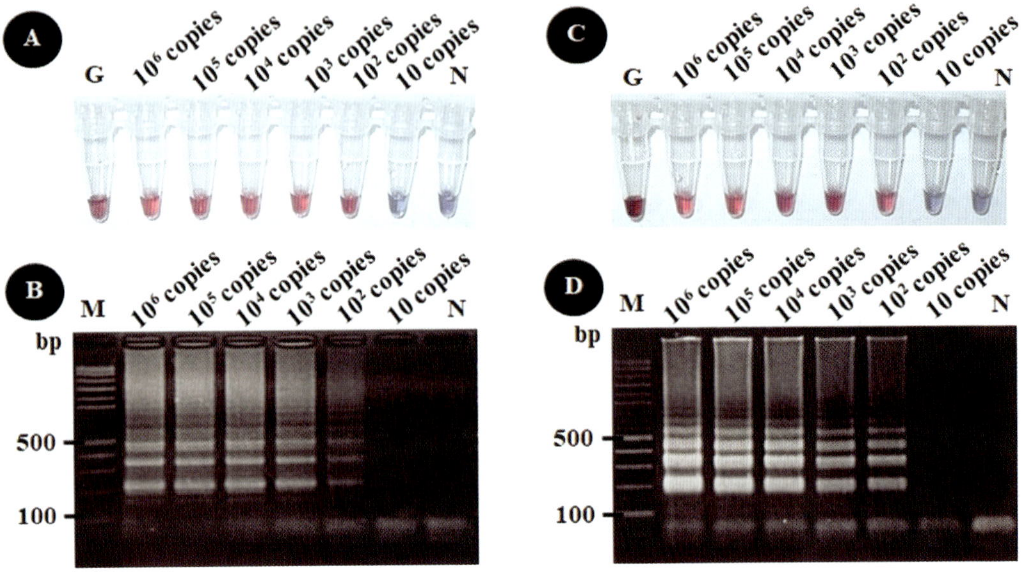

FIGURE 23.8 Sensitivity of D-LAMP-AuNP assay in the detection of IHHNV using DNA plasmid templates of tenfold serial dilutions. (A) Sensitivity of LAMP-1 analysis by AuNP probe assay. (B) agarose gel electrophoresis (AGE) results of LAMP-R1 reactions. (C) Sensitivity of LAMP-2 analysis by AuNP probe assay. (D) AGE results of D-LAMP reactions. Lane G, gold probe; lane M, 2 log DNA marker; lane N, distilled water as negative control. Credit: Arunrut et al. (2019).

TESTING THE EXISTENCE OF FALSE POSITIVES OR POP-UP POSITIVES FROM TRACEABLE BREEDING STOCK

As hypothesized above, there are potential implications related to the selection of disease-resistant shrimp in breeding programs if such inserts turn out to be protective against disease caused by IHHNV. Ideally, 'pop-up positives' from EVE-containing shrimp stocks could be traced from their current descendants to identify crosses that might potentially result as a result of crossover.

For example, male shrimp specimens with a variety of known IHHN inserts could be crossed with female shrimp having no IHHNV inserts, and the offspring from each individual cross could be divided into two groups based on the presence or absence of IHHNV inserts. The two groups of offspring (with and without inserts) could then be challenged with IHHNV and observed for signs of IHHNV disease. Any significant improvement in tolerance to IHHNV (i.e. reduction in signs of disease and reduction in viral load) in the insert groups over the non-insert groups would support the hypothesis. Further analysis could be carried out to determine whether improved tolerance was associated with the production of antisense imRNA from the IHHNV inserts. In addition, dsRNA knockdown against shrimp dicer gene transcription could be carried out to determine whether it blocked any tolerance associated with IHHNV inserts.

PREVENTION OF 'POP-UP' FALSE-POSITIVE PCR RESULTS FOR IHHNV ARISING IN SPF THAI SHRIMP DUE TO RECOMBINATION BETWEEN COMPLEMENTARY EVE

With small but significant frequency, batches of SPF giant tiger shrimp exported from Thailand were rejected by importing countries because some individuals in the batch give positive PCR test results for IHHNV using primers recommended by the World Organization for Animal Health (OIE). However, follow-up tests revealed that these were false-positive results because the relevant shrimp do not give positive PCR test results for other target sequences characteristic of the 4kb IHHNV genome. Our preliminary evidence suggests that these false-positive results arose from relatively low-frequency crossover between two complementary EVE in one of the parents that gave rise to the relevant shrimp specimens. These complementary EVE each contain an incomplete portion of the OIE target sequence for IHHNV, so the source parent is negative for IHHNV using the standard test. However, after crossover between these complementary EVE, the complete target sequence of the OIE method was restored in a proportion of the offspring, resulting in what have been termed pop-up false-positive PCR results for IHHNV. The possibility of this problem is universal for all shrimp viruses in the SPF list of breeding stocks. Thus, we propose the use of IHHNV as the first model to assist Thai SPF shrimp stock producers in preventing pop-up false-positive test results for IHHNV in their exported stocks. Once the methodology has been tested and proven effective, it would be applicable to any virus in the SPF list.

Recommendations for development of international standard detection targets for SPF stocks of shrimp used commercially

A review on the current challenges of EVE regarding identification of SPF stocks is clearly seen in a section entitled 'Problems confirming SPF status', by (Alday-Sanz et al. (2018)). The development of highly accurate and practical detection methods to validate EVE-containing stocks, and calls for scientific agreement on healthy shrimp, should be prioritized to reduce the impact of EVE on the shrimp trade. Meanwhile, if beneficial inserts could be identified, it would be possible to preserve these by monitoring breeding stocks using PCR methods based on chimeric shrimp–viral primer pairs designed for specificity to desired beneficial inserts. This would make it possible to produce defined and certifiable, SPR shrimp stock from SPF stocks. Another benefit of developing stocks with known, monitored IHHNV inserts would be the ability to avoid the previously reported appearance of IHHNV-positive test results in the offspring from IHHNV-negative parental stock. It is likely that this occurred due to recombination between grandparental IHHNV inserts that contained incomplete but complementary portions of the PCR target sequence for the detection method used. Since these methods are defined for controlled diseases such as IHHN, management of inserts could be arranged to eliminate this possibility.

Conclusion

Discovery of IHHN-EVE in Thai *P. monodon* a decade ago has recently gained much attention, because the phenomenon poses challenges in confirming SPF shrimp, especially in EVE-containing shrimp stocks. Rejection of Thai SPF giant tiger shrimp by importing countries, occasionally due to pop-up or false positives, could lead to profit losses for Thai shrimp producers. Screening for EVE in other shrimp viruses, such as WSSV, are encouraged to further prove the concept of variable occurrence of EVE in the shrimp genome. Because of the recent revelation of WSSV-EVE with high sequence identity to matching sequences in extant WSSV, there is the probability that similar EVE will eventually be discovered for other existing and newly emerging viruses. Once additional cases are tested and confirmed, highly accurate and practical detection methods need to be developed and updated accordingly to reduce false positives from EVE when certifying SPF status. IHHNV can be used as the first model for natural occurrence of EVE in the shrimp genome and its implications for the shrimp trade, because it is an economically important shrimp virus on which most research has been done to date. Future studies should include determination of beneficial inserts to produce defined and certifiable SPR shrimp stocks from SPF stocks, as described herein.

References

Alday-Sanz, V., Brock, J., Flegel, T. W., Mcintosh, R., Bonda-Reantaso, M. G., Salazar, M. and Subasinghe, R. (2018) Facts, truths and myths about SPF shrimp in aquaculture. *Reviews in Aquaculture* 12, 76–84.

Arunrut, N., Jitrakorn, S., Saksmerprome, V. and Kiatpathomchai, W. (2019) Double-loop-mediated isothermal amplification (D-LAMP) using colourimetric gold nanoparticle probe for rapid detection of infectious *Penaeus stylirostris* densovirus (PstDNV) with reduced false-positive results from endogenous viral elements. *Aquaculture* 510, 131–137.

Arunrut, N., Prombun, P., Saksmerprome, V., Flegel, T. W. and Kiatpathomchai, W. (2011) Rapid and sensitive detection of infectious hypodermal and hematopoietic necrosis virus by loop-mediated isothermal amplification combined with a lateral flow dipstick. *Journal of Virological Methods* 171, 21–25.

Dhar, A. K., Robles-Sikisaka, R., Saksmerprome, V. and Lakshman, D.K. (2014) Biology, genome organization and evolution of parvoviruses in marine shrimp. *Adv Virus Res* 89, 85–93.

Feschotte, C. and Gilbert, C. (2012) Endogenous viruses: insights into viral evolution and impact on host biology. *Nature Reviews Genetics* 13, 283.

Flegel, T. W. (2007) Update on viral accommodation, a model for host–viral interaction in shrimp and other arthropods. *Developmental & Comparative Immunology* 31, 217–231.

Flegel, T. W. (2009) Current status of viral diseases in Asian shrimp aquaculture. *The Israeli Journal of Aquaculture – Bamidgeh* 61(3), 229–239.

Huang, S.-W., Lin, Y.-Y., You, E.-M., Liu, T.-T., Shu, H.-Y., Wu, K.-M., Tsai, S.-F., Lo, C.-F., Kou, G.-H. and Ma, G.-C. (2011) Fosmid library end sequencing reveals a rarely known genome structure of marine shrimp *Penaeus monodon*. *BMC Genomics* 12, 242.

Jaroenram, W. and Owens, L. (2014) Separation of endogenous viral elements from infectious *Penaeus stylirostris* densovirus using recombinase polymerase amplification. *Molecular and Cellular Probes* 28, 284–287.

Katzourakis, A. and Gifford, R. J. (2010) Endogenous viral elements in animal genomes. *PLoS Genetics* 6, e1001191.

Molthathong, S., Jitrakorn, S., Joyjinda, Y., Boonchird, C., Witchayachamnarnkul, B., Pongtippatee, P., Flegel, T. and Saksmerprome, V. (2013) Persistence of *Penaeus stylirostris* densovirus delays mortality caused by white spot syndrome virus infection in black tiger shrimp (*Penaeus monodon*). *BMC Veterinary Research* 9, 33.

Nunan, L., Poulos, B. and Lightner, D. V. (2000) Use of polymerase chain reaction for the detection of infectious hypodermal and hematopoietic necrosis virus in penaeid shrimp. *Marine Biotechnology* 2, 319–328.

OIE (2019) *Manual of Diagnostic Tests for Aquatic Animals*. OIE, Paris, France.

Robalino, J., Bartlett, T. C., Chapman, R. W., Gross, P. S., Browdy, C. L. and Warr, G. W. (2007) Double-stranded RNA and antiviral immunity in marine shrimp: inducible host mechanisms and evidence for the evolution of viral counter-responses. *Developmental & Comparative Immunology* 31, 539–547.

Saksmerprome, V., Jitrakorn, S., Chayaburakul, K., Laiphrom, S., Boonsua, K. and Flegel, T. W. (2011) Additional random, single to multiple genome fragments of *Penaeus stylirostris* densovirus in the giant tiger shrimp genome have implications for viral disease diagnosis. *Virus Research* 160, 180–190.

Saksmerprome, V., Puiprom, O., Noonin, C. and Flegel, T. W. (2010) Detection of infectious hypodermal and haematopoietic necrosis virus (IHHNV) in farmed Australian *Penaeus monodon* by PCR analysis and DNA sequencing. *Aquaculture* 298, 190–193.

Sun, Z.-F., Hu, C.-Q., Ren, C.-H. and Shen, Q. (2006) Sensitive and rapid detection of infectious hypodermal and hematopoietic necrosis virus (IHHNV) in shrimps by loop-mediated isothermal amplification. *Journal of Virological Methods* 131, 41–46.Tang, K. F. and Lightner, D. V. (2006) Infectious hypodermal and hematopoietic necrosis virus (IHHNV)-related sequences in the genome of the black tiger prawn *Penaeus monodon* from Africa and Australia. *Virus Research* 118, 185–191.

Tang, K. F., Durand, S. V., White, B. L., Redman, R. M., Pantoja, C. R. and Lightner, D. V. (2000) Postlarvae and juveniles of a selected line of *Penaeus stylirostris* are resistant to infectious hypodermal and hematopoietic necrosis virus infection. *Aquaculture* 190, 203–210.

Tang, K. F., Navarro, S. A. and Lightner, D. V. (2007) PCR assay for discriminating between infectious hypodermal and hematopoietic necrosis virus (IHHNV) and virus-related sequences in the genome of *Penaeus monodon*. *Diseases of Aquatic Organisms* 74, 165–170.

Utari, H. B., Soowannayan, C., Flegel, T. W., Whityachumnarnkul, B. and Kruatrachue, M. (2017) Variable RNA expression from recently acquired, endogenous viral elements (EVE) of white spot syndrome virus (WSSV) in shrimp. *Developmental & Comparative Immunology* 76, 370–379.

Xia, X., Yu, Y., Hu, L., Weidmann, M., Pan, Y., Yan, S. and Wang, Y. (2015) Rapid detection of infectious hypodermal and hematopoietic necrosis virus (IHHNV) by real-time, isothermal recombinase polymerase amplification assay. *Archives of Virology* 160, 987–994.

Chapter 24

DISINFECTANT USE IN AQUACULTURE, AN UNDERAPPRECIATED ELEMENT OF BIOSECURITY?

STEPHEN G. NEWMAN

Summary

Aquaculture is the fastest-growing agribusiness globally, with many predicting that the protein derived from aquaculture is essential to feed a burgeoning global population. Biosecurity is a weak link in aquaculture, and many tools are used in an effort to shift the balance in favor of the farmer. Disinfectants are one such groups of tools. In the USA and the EU, their use is tightly regulated. In the 21st century, the vast majority of production of farmed aquatic animals takes place in SE Asia, where regulations are not always enforced to the same extent as in the USA and the EU. Ultimately consumers' concerns about the presence of residues will likely drive stricter enforcement. The regulatory framework in the USA and the EU will be discussed as it relates to imports. The critical role of disinfectants in sustainable production will be highlighted and some conclusions drawn about what lies ahead.

Introduction

Biosecurity is broadly defined as the sum of measures that are undertaken to prevent pathogens from impacting animals (can apply to any organism), both directly and indirectly (Yanong and Erlacher-Reid, 2012). The focus of this chapter is the use of disinfectants in aquaculture, specifically the farming of animals (shrimp and fish) in aquatic environments. Aquaculture has been, and continues to be, plagued by biosecurity failures ensuring that sustainability will remain elusive. Disinfectants are critical tools for tilting things in favor of the animals.

Perhaps the single greatest area that producers can impact in terms of lessening the overall susceptibility of their animals to disease is to rear them under conditions that are designed to minimize stress (Yanong and Erlacher-Reid, 2012). Foremost among those that are manageable include ensuring proper levels of oxygen, the use and maintenance of clean water, freedom from contaminating materials that could be immunosuppressive and/or toxic in other ways, and the use of feeds that are appropriate for the life stages being grown and that meet the nutritional needs of a given species under the conditions of culture. There are also genetic components that cannot be ignored. Proper application of disinfectants should not cause undue levels of stress.

While minimizing stress is important, pathogen exclusion is also critical (Yanong and Erlacher-Reid, 2012). Disease can be managed proactively by controlling pathogens in production systems. A critical element in this is ensuring that farmed animals are derived from stocks determined as being free of pathogens via routine screening, and preferably from established nucleus breeding centers (Lotz, 1997). Appropriate use of disinfectants can further ensure high levels of biosecurity.

Human activities must be monitored and managed to ensure that no inadvertent sources of contamination enter the production facility. The tools used to achieve this are wide ranging (Yanong and Erlacher-Reid, 2012), and this chapter is focused on one specific aspect of that, the use of disinfectants.

The use of compounds sold as disinfectants in aquaculture is ubiquitous. Many, if not most, fish and shrimp hatcheries and farms (ponds, cages, raceways, etc.), as part of their standard operating procedures (SOPs), use a variety of compounds for the purpose of killing bacteria, viruses and other pests that might impact productivity. Depending upon which country the farming is in and the types of compound being used, regulations impacting their use range from being non-existent to onerously complex. Ultimately the decision to use these compounds depends on availability, history of use by staff, perceived usefulness and effectiveness and cost effectiveness of the compounds. Unfortunately, puffery and often outright lies and distortion of facts are also an element of this process. Legality is not, in general, a determining factor in the use of disinfectants in large sectors of the global industry (Newman, S., personal observations).

There are many important constraints affecting the proper use of disinfectants, and it is incumbent on the user to understand that irresponsible use of disinfectants can be hazardous. Improper usage can negatively impact the health of employees, even in some instances causing death. It can also have a deleterious impact on the animals on which they are being used. Improper use can be damaging to the environment. Residues can end up in drinking water and the immediate production environment. The use of disinfectants is complicated and, in some ways, may be a double-edged sword. When used properly they can reduce the risks of disease and improve biosecurity.

What is a disinfectant?

While one would think that it would be simple to define, the reality is that there is not one universal definition. The common definition with which most are familiar is the dictionary definition. The *American Heritage Dictionary* defines a disinfectant as 'an agent, such as heat, radiation, or a chemical, that destroys, neutralizes, or inhibits the growth of disease-carrying microorganisms' (ahdictionary.com).

Disinfectants are often classified into groups, independent of their composition, based on their mechanisms of action. This dictionary definition is the most widely used, and encompasses a wide range of chemical and mechanical agents, a number of which are described briefly in Table 24.1. The reader is referred to several review articles that discuss these in detail (Munakata and Kuo, 2016; Singh and Singh, 2018).

The dictionary definition notwithstanding, each country has their own legislated definitions and approved uses, and these can be much narrower in scope than the dictionary definition. For the most

TABLE 24.1 Summary of different types of disinfectant used in aquaculture.

Type of disinfectant	*Mode of action/comments*
Quaternary ammonium compounds (QACs)	Many dozens of formulations. Typically limited in effectiveness in the presence of organic matter. Not a broadly active disinfectant. Highest degree of activity is against Gram-positive bacteria, with variable activity against Gram-negative bacteria and viruses. Unlike many other disinfectants, they are not corrosive. They can be highly toxic to aquatic animals. They act by binding to both membranes and DNA.
Oxidizing agents	As with QACs, limited effectiveness when organic matter is present. Some are very toxic to aquatic animals although others are not. Common agents in widespread use are chlorine and iodine-based compounds, ozone and peroxygen derivatives.
Biguanides	Not effective in hard or alkaline water and, in general, not as effective against a range of pathogens as other disinfectants. Non-corrosive and commonly used on skin.
Aldehydes	Work well in the presence of organic matter. Act by denaturing proteins. Very effective against a wide range of pathogens. Common agents are formaldehyde and glutaraldehyde. Unlike other disinfectants, they can be used in mists. Toxicity is variable.
Acidity and alkalinity modifiers	pH modifiers work well in the presence of organic matter. These are a versatile group of disinfectants permitted in areas in which other disinfectants are not effective.
UV irradiation	Affected by presence of organic matter. Used widely in water treatment post removal of organics. Different wavelengths have different applications.
Heat	Difficult to apply in a uniform manner. Moist heat typically more effective than dry heat.
Desiccation	Drying out of production systems is one way to kill pathogens. It is common practice to periodically dry out shrimp hatcheries and ponds. Direct sunlight and desiccation are widely used. This is a triple threat to potential pathogens, with UV light, heat and drying out acting together.
Combined methods	Using several different methods can be synergistic, with a greater range of killing activity and thus greater assurance that high levels of pathogen inactivation are achieved.

part, disinfectants that are legal for use in aquaculture are approved for use on surfaces and not for direct use on or in animals. This varies, however, with different countries having different criteria. A few are legal for use via animal contact, although this is highly variable and there is all too often little apparent scientific reasoning behind the approval process.

As is common in the regulation of chemicals and drugs used in aquaculture, how a given material is regulated is often more a matter of who has asked for the approval of a given product and what claims they are asking to be legally allowed to use than what common sense or science would dictate. Given the immense scope of the global aquaculture industry it is not possible to review all of these, but we can offer a broad overview.

Why use disinfectants?

Aquaculture is the fastest-growing agribusiness globally, with an average annual growth rate of 6–10% over the last few decades (FAO, 2018). This trend is certain to continue as there is an insatiable demand for the high-quality protein derived from seafood. Wild fisheries cannot possibly supply what is needed – only aquaculture can. Strong and slowly increasing demand in the USA and the EU, and the burgeoning health-conscious middle-class consumers in India and China, will continue to drive this growth (FAO, 2018).

As with all agribusinesses, aquaculture has biosecurity challenges inherent in the production processes and the specter of pathogen caused diseases is always there. Even in highly controlled and well-managed environments, diseases can and do still impact productivity. Pathogens evolve, and changing production environments drive this evolution. Diseases that are due to the presence of pathogenic bacteria, viruses and fungi are responsible for billions of dollars of losses annually (Lafferty et al., 2015). These losses fluctuate from year to year depending on the species affected and the characteristics of the pathogen. Many, if not most, losses are due to infections with opportunistic pathogens. Pathological processes can be due to the presence of stressors, preexisting health conditions, sublethal infections with other pathogens, etc. A few are obligate pathogens in that their presence is, *per se*, the cause of disease. No mitigating factors need be present.

For many, the tools of choice have been antibiotics for treatment and, typically and ineffectively, for prevention. Antibiotics have been widely abused since their invention. Aquaculture is no exception (Newman, 2007) and, given the nature of aquatic environments, there are serious concerns that resistance can readily spread, impacting the use of critical antibiotics in terrestrial agriculture and human medicine. Resistance is a natural consequence of the use of these types of compounds. Their mechanisms of action are such that there are strong selection pressures for targeted microbes to become tolerant to these (requires a higher dosage for effectiveness) or resistant to (cannot be impacted by any dosage). Disinfectants act via a very different set of mechanisms and resistance, while this does occur, it is simply not an issue in most cases.

Biofilms routinely form in aquatic environments, and they may shield potential pathogens from the impact of short-term disinfectant usage. Much as we wash our hands to prevent potential disease-causing organisms from entering our bodies, the widely suggested first line of defense against obligate pathogens – cleaning and disinfection – is also a critical component of biosecurity for aquaculture (Yanong and Erlacher-Reid, 2012). These are essential steps in minimizing the potential spread of preventable diseases, by killing opportunistic and obligate pathogens. Table 24.2 outlines, for a typical aquaculture operation, most of the areas where the use of disinfectants would be expected to be part of SOPs.

TABLE 24.2 Likely sites for use of disinfectants in aquaculture.

Nets and anything that may come in direct contact with the animals at any life stage	Tanks, ponds, etc., this includes empty structures before and after usage, as well as the water contained within
Pipes used for addition and removal of water	Floors and walls of structures
Air in some cases (many pathogens are airborne)	Vehicles entering the farm (tires)
Hands, feet (boots), etc. of all personnel entering the facility	Any containers used to hold and move animals, whether living, moribund or dead, and any animal products
Various life stages (eggs, nauplii, juveniles, broodstock)	Various live feeds such as Artemia, algae, frozen feeds fed to broodstock

Ideally, disinfectants should be used at any point in the production process where potential pathogens are present. This will vary depending on the production systems, the operating philosophy and the experience of management. Usage is also impacted by the legal framework for their use locally and what legal constraints that there might be for them to be used in products for export.

Regulatory aspects of disinfectant use

Each country has its own set of regulations that impact what is considered to be a disinfectant and how best to use them. The reader is advised to visit specific government websites for the country in which they are producing, as well as for those countries to which they are exporting, to learn what they can and cannot do.

The country that imports (and produces) the greatest volume of farmed seafood is China. The USA and the EU are also major importers, although not on the same scale. Regulatory oversight is inconsistent globally and there can be vast differences in how one gains approval for the use of any given disinfectant. The sale of chemical disinfectants is regulated in the USA by the Environmental Protection Agency (EPA) under the Federal Insecticide, Fungicide and Rodenticide Act (FIFRA) (GPO, 2020). For a detailed description of the EPA's legal responsibilities, consult the US Code of Federal Regulations, Title 40, parts 150–189 (GPO, 2020).

From the US regulatory standpoint, chemical disinfectants are a category of antimicrobial pesticides. These are defined by FIFRA as 'substances or mixtures of substances intended to destroy or suppress the growth of harmful microbiological organisms, and pesticides that protect inanimate objects and surfaces from organisms such as bacteria, viruses, or fungi' (GPO, 2020).

The US Animal and Plant Health Inspection Service (APHIS) has published guidelines that outline the responsible use of disinfectants (USDA, 2020), summarized in Table 24.3. It has been my experience (Newman, S., 2020, personal observations) that these guidelines are rarely followed outside of the USA, even though most users of disinfectants would benefit from following them. Note that these guidelines are for the use of disinfectants on environmental surfaces, and they apply to use in hatcheries and on farms, as well as where there is no intentional animal contact. Ensuring that the immediate environment is clean is a critically important step in ensuring effective pathogen control. Although guidelines will vary as to the compounds being used, in general these are applicable to the use of most disinfectants. Responsible use and disposal of disinfectants is an essential element of sustainable aquaculture, although in many parts of the world there is little oversight.

The types of disinfectant used in aquaculture are considered non-public health products and are used to control the growth of algae, odor-causing bacteria, bacteria causing spoilage, deterioration or fouling of materials and microorganisms infectious only to animals. The regulations are focused on the environment, and disinfectants are licensed solely to be used on hard inanimate surfaces and objects to destroy or irreversibly inactivate infectious fungi and bacteria – but not necessarily their spores and other pathogens. FIFRA details the processes and procedures that must be followed to permit the legal use of a chemical disinfectant in the USA. There are severe penalties for violating the Act, and companies that fail to work with the EPA, regardless of the established history of the disinfectant with which they are working, face fines, inventory seizure and even jail in some cases.

FIFRA requires that the EPA review data submissions from the manufacturers detailing the chemistry, proof of efficacy to support claims, toxicity data and potential environmental impacts in addition to

TABLE 24.3 Guidelines for the responsible use of disinfectants.

Step	*Comments*
Selection of appropriate disinfectant	Following of guidelines outlined by the appropriate regulatory authority in the country where the culture is taking place is essential, as are following the guidelines for the country to which the final product will be exported.
Preparation of disinfectant solution	Follow label instructions. Fresh solutions should be prepared daily at the very least, to ensure maximum potency. Using outdated materials and failure to properly use the products can negatively impact effectiveness, as well as pose a risk to the animals, the environment and the applicator.
Application	Follow label instructions. Each disinfectant will have specific directions for where, when and how to use it. Methods depend on the compounds used but may include wiping, immersing, spraying, fogging, creating a gas, etc.
Safety considerations	The product label will contain specifics regarding the use of the disinfectant and what protection the user must take. This almost always requires the use of gloves, eye protection and masks. Protection of exposed skin may be needed.
Contact time	Disinfectants require sufficient contact time to ensure that they work effectively in killing microorganisms. This varies between compounds.
Rinsing	Once a disinfectant is used, treated areas and surfaces must be rinsed until they are free of any residual disinfectant. Many are harmful to animals, and surfaces should be rinsed with chlorine-containing potable water (although chlorine can also be quite harmful).
Drying of surfaces	A final step in the use of a given disinfectant is to ensure that no residual material is present. Ideally, when animal contact surfaces and materials have been cleaned and disinfected, sufficient time should be allowed to ensure that they have worked as effectively as possible. Thorough drying while keeping animals away from the immediate production area is critical.
Responsible disposal of empty containers or unused product	Many compounds should not be dumped into the environment. Containers need to be cleaned and recycled or destroyed. Products that are not used and outdated need to be destroyed according to the manufacturer's recommendations.

proposed packaging and labeling. Each successfully registered disinfectant has a unique EPA registration number. Labels contain usage directions, and safety precautions are available at the EPAs Pesticide Product and Label System website (EPA, 2020).

Registered disinfectants must be labeled and used according to regulatory mandates. State regulations supersede federal regulations, and state requirements may be more limiting than Federal requirements. This adds an additional layer of complexity and potential confusion to their use, with 50 states and multiple territories each enacting their own legal requirements for the use of chemical disinfectants.

In the USA, when disinfectant compounds are used on living animals, they are considered to be drugs and are regulated by the United States Food and Drug Administration (FDA). If a compound is used specifically for the purpose of treating or preventing a disease in an animal, it is considered a drug. This is a different regulatory arena and falls outside of the scope of this chapter. In other countries this is not the case, and disinfectants that are used involving direct contact of animals are not regulated as drugs.

In general, EPA-approved disinfectants cannot legally be used directly on animals, although there are some exceptions. It is important to recognize that regulations are not necessarily about efficiency and optimum utility of tools. Some compounds that are widely utilized would be toxic at the levels

TABLE 24.4 Selected dictionary-defined disinfectants commonly used in aquaculture.

Compound	*Description*	*Comment*
UV light	Uses lamps generating UV light to kill bacteria, viruses, etc.	Widely used in hatcheries and in RAS
Ozone	O_3; toxic derivative of oxygen	Widely used in hatcheries. Can form toxic residues
Quaternary ammonium compounds	Broad-spectrum chemical compounds known as "quats"	Widely used on farms in SE Asia, in water with animals
Iodophors	Iodine-based compounds; complexed with PVP	Egg disinfectant, foot baths and hand sanitizers
Chlorinated compounds	Contain chlorine as the active ingredient	Foot baths, water sterilization, most very toxic to fish and shrimp
Copper compounds	Copper either in elemental or chelated form	Algicides, residue issues
Calcium hydroxide/oxide	Lime, burnt lime, etc.	Algicides, also used to kill bacteria and viruses in water
Peroxides	Use the killing action of oxygen	Many different types are in use. Algicides, also used to kill bacteria and viruses in water
Dyes	Malachite green and crystal violet kill pathogens	Still in use even though their use is banned, but very effective for killing pathogens

Note: This list is for illustration purposes only. It is not intended to be all inclusive nor is it an endorsement of the use of these compounds.

recommended for surface disinfection if used in direct contact with animals, yet are harmless and even beneficial at much lower levels of use (Hung et al., 2003). They are not legal for these uses in the USA because of the high burdens in complying with the regulations. Fortunately, not all countries look at these regulatory issues in the same manner that the USA does: if they did, much of the world's aquaculture operations would not be profitable.

Table 24.4 shows the diversity of the types of disinfectant employed in global aquaculture, while Table 24.5 expands on this with some specifics.

Table 24.6 lists the few compounds that are approved specifically for aquaculture as disinfectants in the USA. The Federal Joint Subcommittee on Aquaculture Working Group on Quality Assurance in Aquaculture Production has published a 'Guide to Drug, Vaccine, and Pesticide Use in Aquaculture' that can be consulted to get a better idea of what is available for legal use in the USA (JSA, 2007). Consult Table 24.5 in the 'Guide to Drug, Vaccine, and Pesticide Use in Aquaculture' for further specifics.

The EU handles things differently, requiring a substantial amount of testing that the EPA does not require (European Chemicals Agency, 2017). Disinfectants are regulated as biocides. The EU requirements in some respects are much more stringent than those in the USA. Some disinfectants such as copper compounds, while legal for certain uses in the USA are not in the EU. Detailed information on all biocides registered is available (European Chemicals Agency, 2017).

China also regulates disinfectants (Wan, 2012). Their definition is much more comprehensive than that of the USA and the EU: it includes chemicals used to disinfect water, the environment, food (fruits and vegetables), surfaces of objects, air, medical and health consumable goods, sewage and animal skin/mucosa. Their registration requirements are very thorough, mirroring those of the USA and the EU.

TABLE 24.5 Abbreviated list of common disinfectants, usage and hazard or concern.

Disinfectant	*Use*	*Hazard or concern*	*Comments*
Chlorinated halogens			
Calcium hypochlorite	Sanitizer and disinfectant in hatchery	5	Effective at low concentrations with germicidal activity
Chloramine T	Disinfects hatchery tanks, water, eggs	2	As above
Chlorine gas	Disinfects water	5	As above
Sodium hypochlorite	Sanitizer and disinfectant in hatchery, water	5	As above
TCCA (trichloro cyanuric acid)	Disinfects eggs	5	As above
Iodophores (halogens)			
			Bactericide, germicide, remains active in presence of
Polyvinyl pyrrolidone	Disinfects eggs and equipment	4	organic matter
Povidone iodine	Disinfects eggs and equipment	4	As above
Potassium iodide	Disinfects eggs	1	Antifungal agent
Oxidizing agents			
Chlorine dioxide	Antifungal disinfectant, bactericide	4	Treats gill disease, fungal control of eggs
Hydrogen dioxide	Antifungal disinfectant, bactericide	4	As above
Hydrogen peroxide	Antifungal disinfectant, bactericide	4	As above
Percarbonates	Broad-spectrum disinfectant, potent sanitizer	3	Effective against bacteria & viruses, animal friendly, long shelf life
Quarternary ammonia compounds			
Benzalkonium chloride (BKC)	Bactericide, fungicide, egg disinfection	1	Kills parasites and protozoa
Didecyl dimethyl ammonium chloride	Pond disinfectant	1	Fungicide, algicide, microbiocide
	Hatchery disinfectant		
Mixtures	Pond disinfectant	1	Fungicide, algicide, microbiocide
	Hatchery disinfectant		
Lime products			
Calcium carbonate	Disinfects water and pond bottoms	1	Reduction in organic matter, sanitation benefits
Hydrated lime (calcium hydroxide)	Disinfects water and pond bottoms	1	As above
Quicklime (calcium oxide)	Disinfects water and pond bottoms	5	As above

Disinfectant	*Use*	*Hazard or concern*	*Comments*
Pesticides used as disinfectants			
Trichlorfon (organophosphate)	Disinfects water	5	Kills protozoans, small crustaceans
Carbaryl (carbamate)	Disinfects water	5	As above
Aldehydes			
Formalin (diluted formaldehyde)	Antifungal agent & disinfects eggs	5	Bactericide, antifungal agent
Glutaraldehyde	Antifungal agent & disinfects eggs	2	Wide germicidal spectrum
Parasite S (formalin-based product)	Antifungal agent & disinfects eggs	1	Kills parasites and protozoa
Stabilized formaldehyde	Disinfectant, sanitizer for water, equipment, eggs	5	Bactericide, antifungal agent
Acids			
Dilute acetic acid (vinegar)	Disinfectant, sanitizer for water, equipment	3	Limited use as disinfectant
Citric acid	Disinfectant, sanitizer for water, equipment	3	Limited use as disinfectant
Fertilizers			
Anhydrous urea	Water disinfectant	1	Unpredictable in control of predators
Ammonium sulfate/calcium hydroxide	Water disinfectant	1	Unpredictable in control of predators
Other products			
Acriflavine	Egg disinfectant	1	Fungicide and protozoacide
Gentian violet	Disinfect eggs	5	As above
Malachite green	Egg disinfectant, controls fungus	5	As above
Methylene blue	Egg disinfectant	5	As above
Potassium permanganate ($KMNO_4$)	Manage ectoparasites and protozoans	1	Fungicide and protozoacide
Diquat dibromide	Algal and bacterial control in hatcheries	5	Herbicide potential ground water contaminant
Sodium chloride (salt)	Disinfectant for eggs, others	2	
Pyceze (bromopol)	Egg disinfectant	5	Bactericide, biocide
Tribromophenol	Hatchery disinfectant for eggs and tanks	1	Bactericide and germicide

Disinfectant	*Use*	*Hazard or concern*	*Comments*
Ozone (ozonation)	Disinfect water	1	Biocide
UV light	Disinfect water	1	Inactivates bacteria, viruses, other microorganisms

Notes: 1 generally safe; 2 slightly toxic; 3 generally safe at recommended levels, mildly corrosive; 4 corrosive and moderately toxic; 5 extremely toxic, toxic residues, corrosive, worker safety issues, potential environmental issues.

TABLE 24.6 Legally used disinfectants in the USA.

Disinfectant	*Surfaced approved*	*Comments*
Benzalkonium chloride, 500 ppm for 10 min		Non-corrosive and inexpensive; can be highly toxic to aquatic animals at dosages used for surface disinfection. Not labeled for aquatic use, although has been used in baths
Didecyl dimethyl ammonium chloride, 400 ppm for 5 min	Plastics, floors, counter tops	Non-corrosive, not labeled for aquatic use
Phenols, 15 min contact	Hard surfaces	Common household products such as Lysol or Pine Sol. Not labeled for use on field gear. Skin irritant.
Chlorine, 200–500 ppm, 10–60 min	All surfaces except plastics	Works well, cheap, highly corrosive, forms toxic byproducts
Virkon Aquatic, 0.5–1%, 5–30 min	Waders, boots, boats, nets,	Non-corrosive, biodegradable, expensive and with limited proof of efficacy
Ethyl alcohol 70–80% Isopropyl alcohol 60–80%	Hands, tools, counter tops	Readily available, inactivated by sunlight, very volatile and flammable
Iodine 100–250 ppm, 20–30 min	Hands, tools, counter tops	Antiseptic, inexpensive, corrosive and potentially highly toxic to aquatic animals
Hydrogen peroxide 3–5%, 5 min	Field gear and hard surfaces	Can be toxic
Peroxigard 1:16 dilution	Field gear and hard surfaces	Spray
Formaldehyde 1–3%	Field gear and hard surfaces	Highly toxic to personnel

Note that application times and approved concentrations should be followed as per the label specifications. These may vary from what is shown in the tables in this chapter.

At present the USA does not regulate the many compounds that are used outside of the country on shrimp and fish that are exported into and consumed in the USA, and this is not likely to occur in the near future. For most regulators the presence of residues from a given chemical is of concern, as is the presence of metabolites. These are indicative of the use of a given compound. In some instances (malachite green), this can pose a real threat to the consumer (Fallaha and Baran, 2014). It is likely that, as regulatory oversight increases, eventually residues will be found in imports for many compounds currently not being looked for, and this will spur regulatory action to limit the use of specific compounds. Many compounds that are widely used globally have never been adequately tested for residues under specific conditions of usage, and no tolerances have been established.

For now, increased regulatory scrutiny is largely unenforceable. There are several reasons for this. Notably, for animals such as farmed shrimp, huge numbers are produced in any given production environment, making it impossible to screen significantly sized samples for realistic compliance. Probably the major stumbling block is socioeconomic. The most widespread production paradigms for both fish and shrimp farming are largely poverty driven. There are many subsistence farmers who pool their crops, which does not lend itself to regulatory oversight. Many use whatever is available with no thought as to legality, etc. If the entire livelihood of a family is in a few ponds in one's front yard and animals are dying, farmers will often do whatever they have been led to believe can help: it's either that or go broke and starve. This is why chemicals like malachite green still show up in residue tests.

Contraindications for the use of disinfectants

Disinfectants, when used properly and appropriately, have the desired, positive impacts. They can prevent serious disease problems from occurring and even stop them from spreading and limit the impact when they do occur. Disinfectants act by damaging the (micro-) organisms to which they are directed. The mechanisms of action are usually very non-specific. The impacts are broad reaching and can impact aquatic systems at different trophic levels both directly and indirectly. Any treatment modality that opens up – either as a result of unforeseen activity or intentionally – niches that are readily available for other pathogens creates the potential for problems. As the use of disinfectants has become a SOP, far too often these risks are ignored. Science has only recently developed the tools needed to evaluate the microbiome in and on animals. The vast majority of the organisms that comprise any given microbiome cannot be cultured using conventional approaches, so much of the historic work on the impact of the use of disinfectants on the microbiome is not going to be accurate. Even proper usage can have a negative impact on the microbiome (Mohammed and Arias, 2015) and result in more harm than good.

Much as we have developed sensitive tools that allow us to determine the impact of the microbiome, our analytical chemistry tools allow us to identify many compounds at very low levels that a few decades ago would not have been possible. One risk that has been largely ignored is that of disinfectant byproducts DBPs. This is an increasingly serious problem with both approved and unapproved products. It has become apparent that while disinfectants can eliminate heterotrophic bacteria (including nonpathogens), they also react with many elements that are present in the water to produce a variety of toxic compounds. Many of these compounds remain to be characterized. Some of the known compounds are listed in Table 24.7.

As analytical technologies improve, many previously unknown compounds have been identified and more are being discovered. Some of these are extremely toxic and, more than likely, are responsible for some of the observed side effects following the use of disinfectants. These are being found in RAS systems and almost every production system that uses any one of a variety of disinfectants.

Perhaps the most widely used chemical disinfectant is chlorine. To date, several hundred DBPs have been identified and there is mounting suspicion that the widespread use of these products in shrimp and fish hatcheries may be weakening animals (Azhar et al., 2018; Nihemaiti et al., 2014; Shi, 2010; Zhimin et al., 2015). Given the widespread use of chlorine, this potential for harm, while widely ignored, should not be: the potential for impact on both animal health and on consumers is substantial. The presence of these compounds should be regulated and closely monitored.

TABLE 24.7 DBPs from disinfectant usage.

Compound	*Source*	*Comments*	*Reference*
Haloacetic acids (HAA)	Reaction of chlorine with organics in both fresh and sea water	R	Nihemaiti et al., 2014
Trihalomethanes (THM)		R	
Haloacetonitriles (HAN)		NR	
Halonitromethanes (HNM)		NR; nitrogen-containing DBPs are much more toxic than regulated compounds	
Haloacetamides (HAcAms)	Reaction with bromide	NR	

Notes: This is not a comprehensive list. R, regulated; NR, not regulated.

Unapproved use of disinfectants

The greatest risks from the use of disinfectants in aquaculture arise as a result of DBPs and 'unapproved' usage. While there are any number of regulatory frameworks in place that govern the use of disinfectants in aquaculture, the situation is in reality quite chaotic (Newman, S., 2020, personal observations). The use of selected compounds is legal in some countries and not in others. A common practice in a given country could result in a consumer-ready product being declared adulterated if the regulators in an importing country were more vigilant. Even in those cases where the use comes without harm or risk, the user rarely understands that the potential exists for residues in the final product being used by the consumer. This is considered to be irresponsible usage and, while it may escape regulatory action, it nonetheless could result in harm to end users and ultimately in some very serious and economically damaging restrictions.

As analytical methods become increasingly sensitive, the risks become greater. Countries that do not enforce the accepted international standards for the use of these products can rarely afford the economic damage done from the consequences of residues being found. There are many examples, a few of which are addressed below.

Several disinfectants that have been used quite widely historically have been banned. Their use is expressly forbidden, and processing plants are required to test for these compounds and known metabolites as part of GMPs and to meet the legal requirements for importing the final processed products into a given country. As mentioned earlier, however, for all practical purposes much of aquaculture production does not lend itself to adequate oversight and control.

Perhaps one of the best-known examples deals with the use of certain dyes. Some of these are close to what would be described as silver bullets. Their use can have immediate and wide-ranging impacts on pathogen loads. The group of dyes, of which malachite green is a member, function by interfering with DNA synthesis, rapidly killing external parasites; crystal violet is also similar. Malachite green is a powerful disinfectant that is widely used to prevent fungal infections (and often too late, to treat them). Animals are bathed in diluted suspensions of the dye, but it is a presumptive carcinogen and leaves a residue in the animal. It is an excellent fungicide and legally available in many countries for a number of applications that have nothing to do with aquaculture (Fallaha and Baran, 2014).

In the USA, the Delaney Clause of the Federal Food, Drug, and Cosmetic Act (FD&C Act), section 409(c)(3), enacted in the 1950s, bans the use in humans and food animals of food additives that are presumptive carcinogens. Today we have a much better understanding of the nature of carcinogens and their ubiquitous nature (Smith and Perfetti, 2020). It is ironic that, if this clause were widely enforced, there would be little food left that would be legal. Regulators can be forced to ban the use of substances that are not problematic at the levels employed, but that may cause cancer at high doses in laboratory animals. The use of malachite green is banned in aquatic animals. Anybody who imports animals into the USA cannot legally use malachite green on any life stage. This has not stopped its use and, while residues must be tested for, one cannot ensure absolutely that this compound and/or its metabolites do not enter the food chain in the USA. There is little doubt that these dyes and their DBPs do not belong in the human food chain.

This issue is complicated by another factor, relating to traceability. Universal traceability of production lots does not exist. In some countries the output of many small farms is pooled, so there is no way of knowing who they are and if one of them is using malachite green. Enforcement, however, at the local level is at best difficult as there is much less regulatory oversight in many countries than there is in the USA and the EU.

Another example is the broad leaf herbicide, Trifluralin, also known as Treflan. This herbicide is highly toxic to crustaceans and yet it was (and still is) the go-to treatment for *Lagenidium* and *Sirolpidium* spp. infections (Williams et al., 1986). This is a fungal pathogen that thrives on organic matter and readily infects *Nauplii* zooea and mysis and post-larval shrimp. The fungal spores are ubiquitous and virtually impossible to keep out of hatcheries without extreme biosecure measures (sealed buildings with positive airflow). Treflan kills the spores and fungal mycelia at extremely low levels, well below the threshold of reported toxicity to crustaceans. Although its use has been banned, it is and will remain in widespread use. Treflan is used at a fraction of a ppb in environments where it would be rapidly bound to organic matter.

Treflan is widely used in terrestrial agriculture (Williams et al., 1986) because of its broad-spectrum activity, and the risks from its use in shrimp hatcheries are negligible in terms of residues and apparent negative impacts on production. Many hatcheries could not operate without its use. While certain compounds such as Treflan are considered highly toxic for crustaceans, this is not at effective dosage levels in common usage for treatment of the aforementioned infections.

Conclusion

A wide range of compounds are in use in aquaculture as disinfectants. How, when and where they can be used varies greatly from country to country and even within each region within a country. There is no universally agreed approach to their registration and usage. There can be multiple layers of regulatory oversight that impact their use. What is legal in one country may be illegal in another and, while it is not enforced for the most part, the use of any given compound that is not approved for use in a given species at a given dosage, etc. can result in that product being seized and destroyed as adulterated – even in the absence of detectable residues. In theory this means that if a hatchery were to use Trifluralin and they recorded its use, should regulators decide to enforce the illegality of this, even with no detectable residues, the destruction of post-larvae could be enforced and, beyond that, any shrimp that had been produced from these post-larvae. Fortunately, at least so far, this has not been an issue.

The widely disparate nature of the regulatory apparatuses between countries ensures that there is little likelihood that any of this will change in the near future. Chaos reigns and this does little to actually help. Ultimately this approach can undermine the quality of the final product and ensures that sustainability will remain elusive for all too many who see farming of fish or shrimp as a way to break an endless cycle of poverty.

References

American Heritage Dictionary (2020) Available at: https://ahdictionary.com/word/search.html?q=disinfectant&submit.x=41&submit.y=17 (accessed 28 July 2021).

Azhar, A., Khatun, M. S., Bikash S. B., Siddiky, M., Kumer, D. Mondal, S. and Rahman, L. (2018) Effect of chlorination on the production of shrimp (*Penaeus monodon*) in brackish water ponds. *International Journal of Fisheries and Aquatic Research* 3(1), 31–34.

EPA (2020) Available at: https://iaspub.epa.gov/apex/pesticides/f?p=PPLS:1 (accessed 28 July 2021).

European Chemicals Agency (2017) *Biocidal Products Regulation 528/2012. Guidance on biocidal products regulation. Volume II: Efficacy—assessment and evaluation (parts B and C)*. Available at: https://echa.europa.eu/guidance-documents/guidance- on-biocides-legislation (accessed 28 July 2021).

Fallaha, A. A. and Baran A. (2014) Determination of malachite green residues in farmed rainbow trout in Iran. *Food Control* 40, 100–105.

FAO (2018) *State of World Fisheries and Aquaculture*. Available at: http://www.fao.org/3/i9540en/i9540en.pdf (accessed 28 July 2021).

GPO (2020) Available at: https://www.ecfr.gov/cgi-bin/text-idx?tpl=/ecfrbrowse/Title40/40cfrv26_02.tpl (accessed 28 July 2021).

Hung H. S., Ching L. S., Liang C. W., Yuan T. Y. and Liang C. W. (2003) Influence of Timsen™ on *Vibrio* populations of culture pond water and hepatopancreas and on the hemocytic activity of tiger shrimp (*Penaeus monodon*). *Aquaculture* 219, 123–133.

JSA (2007) *Guide to Drug, Vaccine, and Pesticide Use in Aquaculture*. Available at: https://freshwater-aquaculture.extension.org/wp-content/uploads/2019/08/Drug_Guide_7-5-07.pdf (accessed 28 July 2021).

Lafferty, K. D., Harvell, C. D., Conrad, J. M., Friedman, C. S., Kent, M. L., Kuris, A. M., Powell, E. N., Rondeau D. and Saksida, S. M. (2015) Infectious diseases affect marine fisheries and aquaculture economics. *Annual Review of Marine Science* 7, 471–496.

Lieke, T., Meinelt, T., Hoseinifar, S. H., Pan, B., Straus, D. L. and Steinberg, C. E. W. (2019) Sustainable aquaculture requires environmental-friendly treatment strategies for fish diseases. *Reviews in Aquaculture* 12, 943–965.

Lotz, J. M. (1997) Special topic review: Viruses, biosecurity and specific pathogen-free stocks in shrimp aquaculture. *World Journal of Microbiology & Biotechnology* 13, 405–413.

Mohammed, H. H. and Arias, C. R. (2015) Potassium permanganate elicits a shift of the external fish microbiome and increases host susceptibility to columnaris disease. *Veterinary Research* 46, 82.

Munakata, N. and Kuo, J. (2016) Disinfection processes. *Water Environmental Research* 88(10), 1192–1229.

Newman, S. G. (2007) Antibiotics in aquaculture. Is responsible use possible? *Global Aquaculture Advocate* March/April, 36–38.

Nihemaiti, M., LeRoux, J. and Croue, J.-P. (2014) Formation of emerging disinfection byproducts by chlorination/chloramination of seawater impacted by algal organic matter. Conference Paper.

Shi, H. (2010) Analysis and occurrence of disinfection by-products in fresh and saltwater. Ph.D. thesis, Missouri University of Science and Technology, 135 pp.

Singh, B. S., Jain, A. and Mishra, A. (2019) Disinfectants, regulation and requirements – Global Overview. *International Journal of Pharmaceutical Sciences and Research* 10, 4109–4119.

Singh, M. and Singh, P. (2018) Drugs and chemicals applied in aquaculture industry: A review of commercial availability, recommended dosage and mode of application. *Journal of Entomology and Zoology Studies* 6(6), 903–907.

Smith, C. J. and Perfetti, T. A. (2020) Exposure to chemicals from natural processes is ubiquitous. *Toxicology and Research Application* 4, 1–15.

US Code of Federal Regulations, FIFRA. Title 40: Sections 150–189 (accessed 19 September 2020).

USDA (2020) Available at: https://www.aphis.usda.gov/aphis/ourfocus/animalhealth/nvap/NVAP-Reference-Guide/Cleaning-and-Disinfection/Disinfection (accessed 28 July 2021).

Wan, D. (2012) *Disinfectant Registration in China.* Available at: http://www.cirs-reach.com/China_Chemical_Regulation/Disinfectant_Registration_in_China.html. (accessed 28 July 2021).

Williams, R. R., Bell T. A. and Lightner, D. V. (1986) Degradation of trifluralin in seawater when used to control larval mycosis in penaeid shrimp culture. *Journal of the World Aquaculture Society* 17(1–4), 8–12.

Yanong, R. P. E. and Erlacher-Reid, C. (2012) *Biosecurity in Aquaculture, Part 1: An Overview.* SRAC Publication No. 4707.

Zhimin, Q., Zhang, H., Dong, H., Adams, C., Luan, G. and Wang, L. (2015) Formation of disinfection byproducts in a recirculating mariculture system: Emerging concerns. *Environmental Science. Processes & Impacts* 17(2), 471–477.

Chapter 25

GLOBAL SHRIMP PRODUCTION AND TRADE

MARCIO CASTRO DE SOUZA, FELIX DENT, FAHMA AZIZAH, DANDAN LIU AND WEIWEI WANG

Summary

Shrimp is among the most commonly traded and consumed aquatic products worldwide. This chapter analyses the history of global shrimp production, disease outbreaks in shrimp aquaculture and the international shrimp trade. In addition, the chapter will also explore trade measures and dynamics in global shrimp trade, sustainability developments in both wild-captured and farmed shrimp and finally, updates on recent market developments.

Due to overwhelming global demands and limited natural resources, farmed shrimp has taken over the market through its development since the 1970s. The shrimp aquaculture industry has seen many changes, shifting globally from traditional farming methods to modernized technologically enabled intensive methods. Despite increased production with modernized shrimp aquaculture methods, high-density monoculture practices that characterize the bulk of shrimp farming are also highly vulnerable to disease outbreaks. Disease outbreaks in shrimp aquaculture are not only detrimental environmentally, they also have significant negative impacts on international markets, livelihoods, food security and health, especially in developing countries. Shrimp products are some of the most significant income sources in many developing countries. While major developed markets have accounted for the vast majority of global shrimp imports historically, newly emerged markets, particularly in Asia, have been taking over the market at considerable speed. Shrimp products represent the second largest group of exported aquatic species in value terms and they face many international trade measures of different nature. Like other products, international shrimp trade is also affected by global dynamics, including political tensions. While shrimp prices vary among species and markets, price growth, in general, is slower compared with that in the previous century. As sustainability and environmental protection became major factors in current aquaculture and fisheries developments, both wild shrimp and shrimp-farming industries have been the subject of controversy relating to unsustainable practices. As a result, the focus has been on minimizing the environmental impacts in aquaculture and wild-harvested shrimp industries. In the meantime, the global shrimp market is also experiencing some changes as market dynamics shift and new markets emerge. There have also been variations in species that are farmed because most farmers now prefer higher-priced species.

A brief history of global shrimp production

Throughout the growth phase of the shrimp aquaculture industry, beginning in the 1970s and continuing to the present day, global shrimp production has exploded. The combination of unrelenting demand from established and emerging markets with the inability of wild shrimp fisheries to fully meet this demand has served as a powerful catalyst for innovation and expansion of the shrimp aquaculture sector. Traditional extensive farming, originating in East and Southeast Asia, has been steadily replaced by technologically enabled intensive methods throughout the tropical and subtropical regions of Asia and Latin America. In

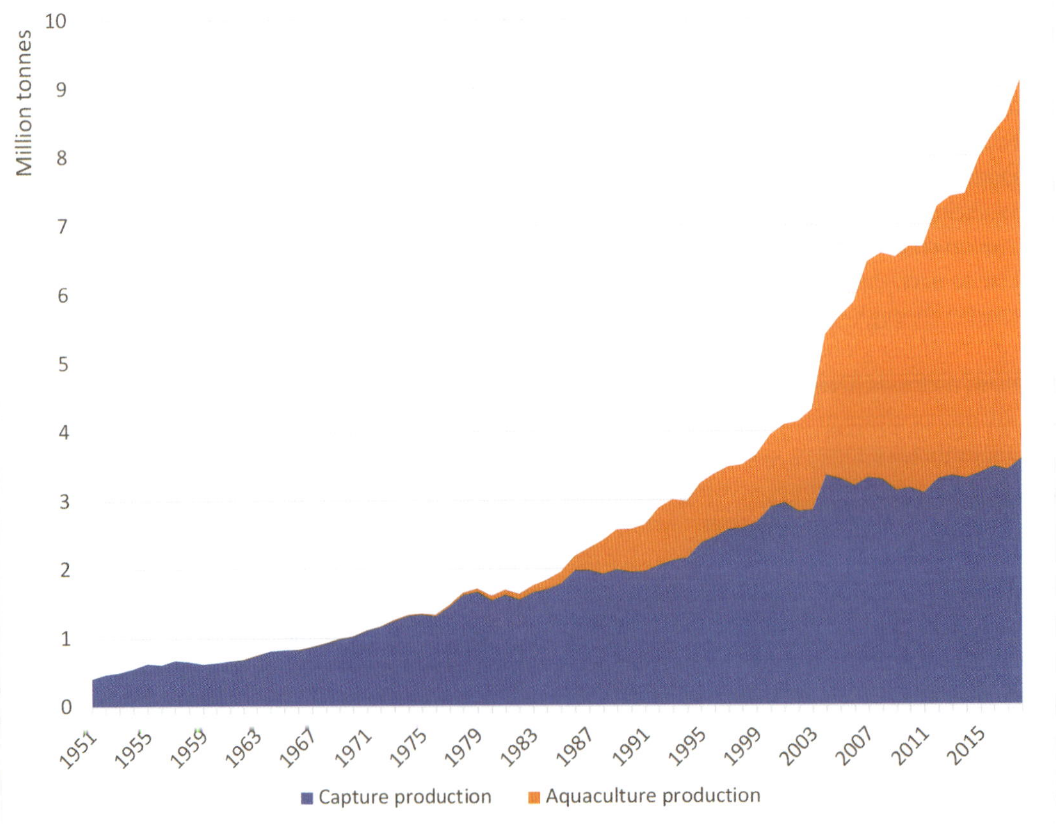

FIGURE 25.1 Global shrimp production, by aquaculture and capture, 1950–2017. Source: FAO (2018a).

2017, total shrimp production, including wild and aquaculture harvests, was 8.7 million tonnes, almost eight times the equivalent figure in 1970 (Figure 25.1). While shrimp aquaculture harvests accounted for less than 1% of total production in 1970, today, they represent around 64% of that total. In contrast, although wild shrimp catches increased from 1.1 million tonnes in 1970 to 3.6 million tonnes in 2017, the rate of production growth has slowed considerably since the early 2000s. In fact, global capture production of shrimp has been relatively stable since 2003, when catches totalled 3.3 million tonnes.

Shrimp species comprising the major proportion of global production fall into three taxonomical families: Penaeidae, Sergestidae and Pandalidae (Figure 25.2). The Penaeidae family includes the two most important species for aquaculture, whiteleg shrimp (*Penaeus vannamei*) and black tiger shrimp (*Penaeus monodon*) and, as a result, this family accounts for an increasingly larger share of total production. A number of Penaeidae shrimp species are also important for capture fisheries, particularly throughout East, Southeast and South Asia. The Sergestidae family includes the Akiami paste shrimp (*Acetes japonicus*), which is caught in large volumes by the Chinese shrimp fleet. The Pandalidae family includes the most commercially important cold-water species, the northern prawn (*Pandalus borealis*). Argentine red shrimp (*Pleoticus muelleri*), the species driving the substantial growth in Argentina's shrimp catches in the 2010s, is classified under the Solenoceridae family.

China is the largest shrimp supplier in the world market and is the top producer of both wild and farmed shrimp. Despite several setbacks due to disease outbreaks, the Chinese shrimp-farming industry has steadily expanded and increased its average stocking density over the last few decades. Since 2000, Chinese shrimp aquaculture production has grown at a constant annual growth rate (CAGR) of 16%. In 2017, China accounted for 35% of global farmed shrimp production with 1.9 million tonnes (Figure 25.3). In Southeast Asia, Indonesia and Viet Nam are now the leading producing countries after the Thai sector experienced a severe contraction following an early mortality syndrome (EMS) outbreak beginning in 2011. Elsewhere, India has been taking on an increasingly important role as a supplier of farmed shrimp in recent years, particularly to the large US market. India has benefited from the decline in Thailand's

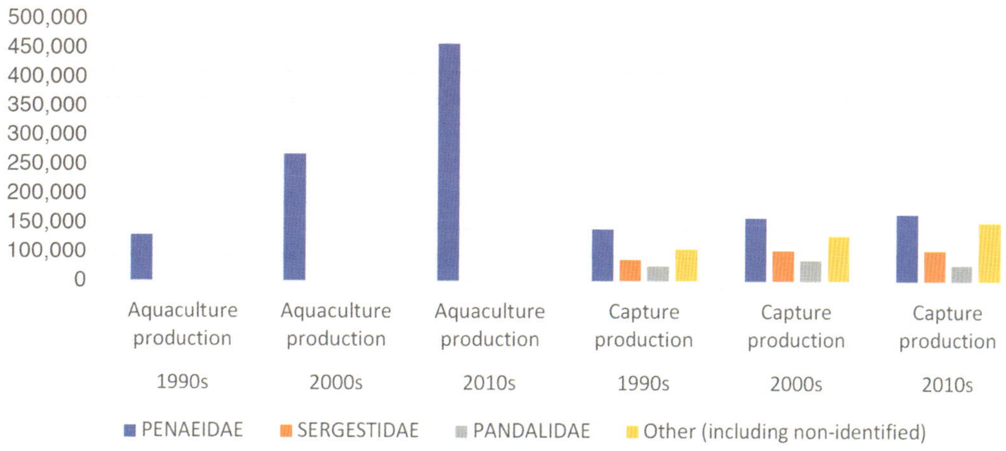

FIGURE 25.2 Global shrimp production, farmed and captured, and family.

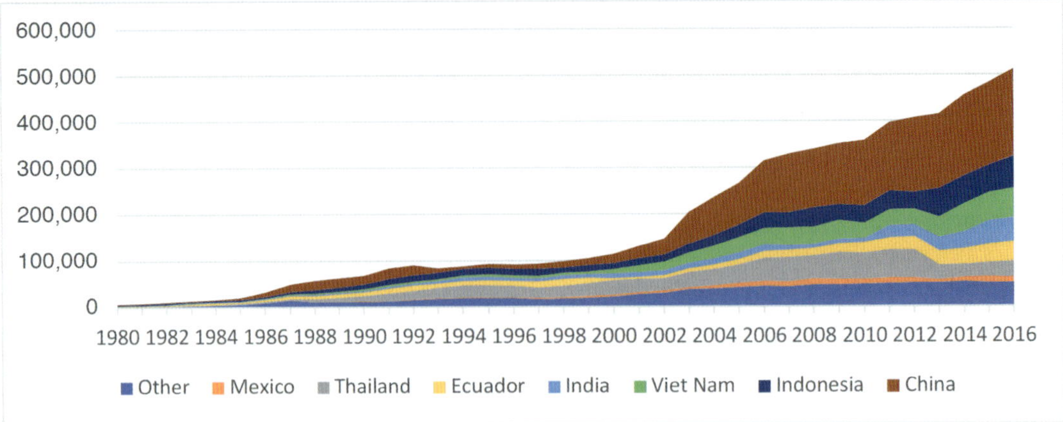

FIGURE 25.3 Contribution of major shrimp-producing countries, 1980–2016.

output and favourable exchange rate developments and the replacement of black tiger shrimp with the faster-growing and more disease-resistant whiteleg shrimp as the primary cultured species.

Bottom trawling is the primary fishing method in most shrimp fisheries, with both industrial fishing fleets and small-scale fisheries contributing significant shares to the total catch. The majority of wild shrimp is caught in the Northwest Pacific and the Indian Ocean. As shown in Figure 25.4A, the Chinese fleet contributed to 32% of total shrimp catches (1.1 million tonnes) in 2017, followed by India with 13% (467,961 tonnes) and Indonesia with 11% (390,694 tonnes). Local artisanal fishers have long exploited shrimp fisheries in these regions, but there has recently been a large-scale shift towards more industrial methods. In the Northwest Atlantic, where the world's most productive cold-water shrimp fisheries are located, the technologically advanced fleets of the USA, Canada and Greenland account for the bulk of catches (Figure 25.4B). Catches in the Northwest Atlantic have been steadily declining since the late 1990s. This production shortfall has been compensated by increased catches of warm-water species in Asia and the Southwest Atlantic, where catches of Argentine red shrimp have been increasing rapidly. Over the last couple of decades, while other shrimp producers have seen minor increases in volumes – and declines in some cases – Argentinian production has increased at 12% CAGR, with most of those gains achieved in the last few years.

A brief history of disease outbreaks in shrimp aquaculture

Disease incidence in shrimp farming tends to be positively correlated with the intensity of cultivation (Bondad-Reantaso et al., 2005). Thus, the high-density monoculture practices that characterize the bulk of the shrimp aquaculture sector today are very vulnerable to disease outbreaks, some of which can devastate 100% of the pond shrimp population in a matter of days (Chamberlain, 2010; Lafferty et al., 2015; Petesch, 2017; Walker and Mohan, 2009).

Shrimp diseases are contagious, and the severity of an outbreak increases with the average geographic proximity of farms and the degree of integration of the industry as a whole. Local transmission occurs through the water itself. The risk is heightened by failure to adopt good management practices and

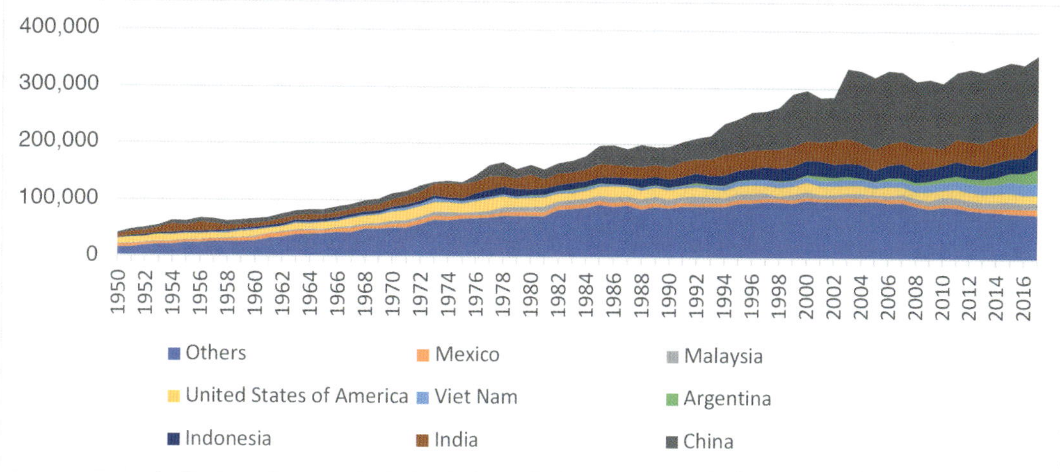

FIGURE 25.4A Shrimp capture by country, 1950–2016.

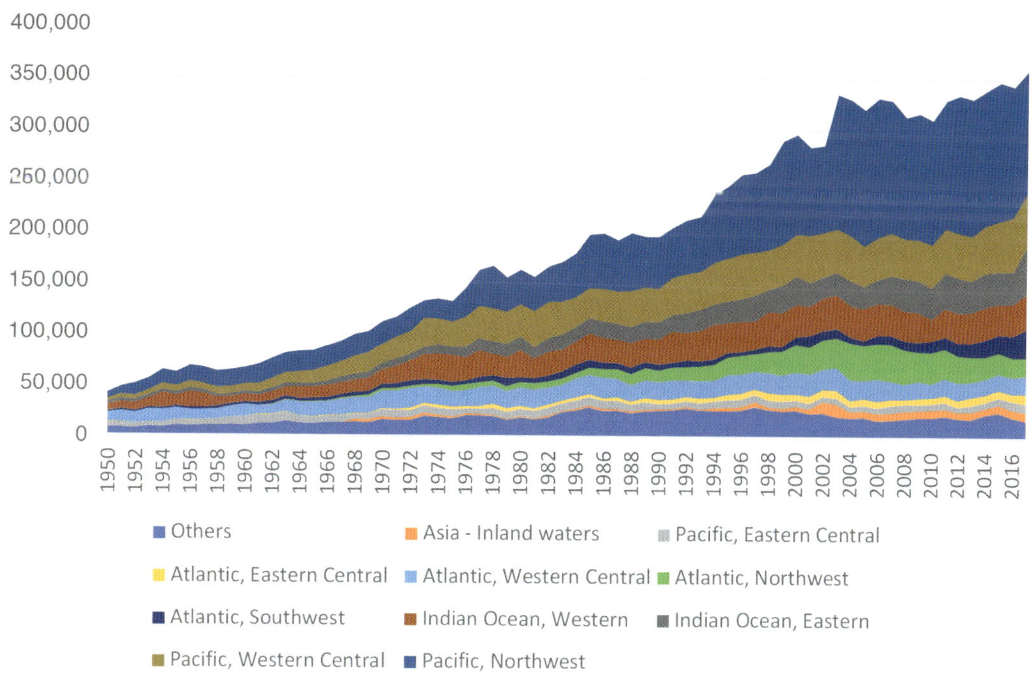

FIGURE 25.4B Shrimp capture by ocean zone, 1950–2016.

implement effective biosecurity measures associated with exogenous elements such as temperature changes or other environmental factors. The process of globalization also facilitates transboundary disease transmission as live infected broodstock, seed shipments and frozen or dead aquatic animals and their processed products are traded between countries (Chamberlain, 2010; Bondad-Reantaso et al., 2005; Leano, 2019). The World Organization for Animal Health (OIE) has listed several pathogens that cause heavy losses in the crustacean sector, including viruses, bacteria and moulds (Table 25.1). Diseases affecting farmed shrimp are primarily caused by viruses and bacteria.

TABLE 25.1 Pathogens of crustaceans according to OIE, and their emergence date.

Year of emergence	*Susceptible hosts*	*Disease*	*Pathogen*	*Distribution*
1860	Freshwater crayfish, Chinese mitten crab	Crayfish plague	*Aphanomyces astaci* (water mold)	Europe, Scandinavia, North America
1977*	Giant tiger prawn (*P. monodon*) and subgenera *P. penaeus*, *P. metapenaeus*, *P. fenneropenaeus* and *P. melicertus*	Spherical baculovirosis	Monodon baculovirus (MBV)	East and South East Asia, Indian subcontinent, Middle East, Australia, New Caledonia, East Africa, Madagascar
1980s**	Penaeid genera or subgenera: *P. litopenaeus*, *P. farfantepenaeus*, *P. fenneropenaeus*, *P. melicertus*, *P. trachypenaeus* and *P. protrachypene*	Tetrahedral baculovirosis	*Baculovirus penaei* (BP)	Americas and Hawaii
1981*	Yellowleg shrimp (*P. californiensis*), giant tiger prawn (*P. monodon*), northern white shrimp (*P. setiferus*), blue shrimp (*P. stylirostris*) and whiteleg shrimp (*P. vannamei*)	Infectious hypodermal and haematopoietic necrosis (IHHN) or runting and deformities disease (RDS)	IHHNV types I and II	Asia-Pacific, Africa, Madagascar, Middle East, Americas
1985*	Whiteleg shrimp (*P. vannamei*)	Necrotising hepatopancreatitis (NHP)	Hepatobacterium penaei	Americas
1990*	Blue shrimp (*P. stylirostris*), dagger blade grass shrimp (*Palaemonetes pugio*), giant tiger prawn (*P. monodon*), jinga shrimp (*Metapenaeus affinis*) and whiteleg shrimp (*P. vannamei*)	Yellow head disease (YHD)	YHV type 1	East and South East Asia, Mexico, Mozambique, Australia
1992*	Greasyback shrimp (*Metapenaeus ensis*), northern brown shrimp (*P. aztecus*), giant tiger prawn (*P. monodon*), northern white shrimp (*P. setiferus*), blue shrimp (*P. stylirostris*) and whiteleg shrimp (*P. vannamei*)	Taura syndrome	TSV	Americas, Middle East, South East Asia

Year of emergence	*Susceptible hosts*	*Disease*	*Pathogen*	*Distribution*
1992*	Decapod crustaceans (e.g. marine, brackish and freshwater shrimp, crab, lobster, crayfish, etc.)	White spot disease (WSD)	WSSV	Asia (East, South East, Middle-East, South), Americas
1995*	Giant freshwater prawn (*Macrobrachium rosenbergii*)	White tail disease	*Macrobrachium rosenbergii* nodavirus	India, China, Taiwan, Thailand, Australia, Caribbean
2002*	Brown tiger prawn (*P. esculentus*), banana prawn (*P. merguiensis*) and whiteleg shrimp (*P. vannamei*)	Infectious myonecrosis (IMN)	IMNV	Brazil, Indonesia, Thailand, China*
2009***	Giant tiger prawn (*P. monodon*) and whiteleg shrimp (*P. vannamei*)	EMS or AHPND	*Vibrio parahaemolyticus* infected by phage	East and South East Asia, Mexico

* Walker and Winton (2010)
** Bondad-Reantaso (2017)
*** Muthukrishna *et al.* (2019)

While disease outbreaks in shrimp farming are very detrimental environmentally, they also have significant negative impacts on international markets, individual livelihoods, food security and health, particularly in developing countries which represent the majority of producers worldwide (World Bank, 2014). Globally, shrimp farmers' losses due to disease were estimated at US$15 billion over the period 1990–2015, with around 80% occurring in Asia (Flegel et al., 2008). Additional costs are also incurred in the wake of severe disease outbreaks, as producers and exporters are typically required to meet new import regulations and production control requirements.

Whitespot syndrome virus (WSSV) is considered to be one of the most commercially damaging pathogens for shrimp farmers due to mortality rates that can reach 80–100% in 5–10 days after the appearance of clinical signs (Lightner et al., 2012; Walker and Mohan, 2009; Walker and Winton, 2010). Since its appearance for the first time in kuruma shrimp (*P. japonica*) culture in China in 1992, the Chinese industry has suffered losses of over 70% of production, from 207,000 to 64,000 tons, equating

TABLE 25.2 Selected estimated losses worldwide due to disease outbreaks. Souce: FAO (2018).

Year(s) of outbreak	*Pathogen*	*Region*	*Estimated industrial loss*
1981	IHHNV	Americas	US$0.5–1 million
1991	YHV	Asia	US$0.5 billion
1991/2	TSV	Americas	US$1–2 billion
1999	TSV	Asia	US$0.5–1 billion
1992/3	WSSV	Asia	US$6 billion
1999	WSSV	Americas	US$1–2 billion
2004	IMNV	Americas	US$100–200 million
2006	IMNV	Asia	US$2 billion
2009–2012	V. parahaemolyticus	Asia	US$1 billion
2013	V. parahaemolyticus	Mexico	US$118 million

to around US$1 billion in financial losses. Over the following years, the disease spread rapidly to many regional Asian countries, with losses totalling US$6 billion (Lightner et al., 2012; Shinn et al., 2018; World Bank, 2014). WSSV was first detected in the western hemisphere in 1995, most likely due to the transport of live animals or import of frozen shrimp products from Asia (Walker and Mohan, 2009), and had spread to Central America and South America by 1999. However, the risk of transmission of WSSV through the import of frozen shrimp intended for human consumption is considered very low (Karunasagar and Ababouch, 2012).

Taura syndrome virus (TSV) first appeared in Ecuador in 1992 and spread throughout North America and South America. Losses in the Americas region were estimated at US$1–2 billion (Walker and Mohan, 2009). TSV spread to Taiwan in 1998 following imports of whiteleg shrimp broodstock from the Americas and, within a few years, appeared throughout East and South East Asia (Walker and Mohan, 2009), leading to total losses estimated at US$0.5–1 billion (Lightner et al., 2012).

The bacterium *Vibrio parahaemolyticus* carrying a virulence plasmid is the cause of early mortality syndrome (EMS) or acute hepatopancreas necrosis disease (AHPND). It was first identified as the cause of a shrimp disease outbreak in Southern China in 2009 (Shinn et al., 2018; World Bank, 2014), before spreading to shrimp farms across South East Asia, including Viet Nam, Malaysia and Thailand around 2011–2012. Thai shrimp production fell by 47% from 2013 to 2014, with total losses reaching US$1 billion in Asia (Zorriehzahra and Banaederakhshan, 2015). The disease was also responsible for lost production worth US$118 million in Mexico (De Schryver et al., 2014) after reaching Latin America within a few years. From 2011 to 2013, US shrimp import prices increased significantly by 17% due to the effects of the EMS disease outbreak on production from Asian farms (FAO, 2018a). It is claimed that transmission of the pathogen in EMS is strongly linked to transportation activity through trading of both living and non-living products (Chamberlain, 2010; Petesch, 2017; Leano, 2019; Lightner et al., 2012; Walker and Mohan, 2009).

A brief history of the global shrimp trade

Rising incomes across the world and the transformative process of globalization have contributed to a substantial increase in global demand for fisheries and aquaculture products over the last few decades and, in turn, to the expansion of international trade of fisheries and aquaculture products. The growth trajectory of shrimp trade has primarily followed this broader trend, with the value of shrimp exports increasing at 7% CAGR from 1976 to 2017, although this rate has slowed to around 4% CAGR since the turn of the millennium. The majority of the largest shrimp exporters are developing countries (Figures 25.5 and 25.6), with this group accounting for 82% (US$22.5 billion) of the total value of shrimp exports in 2017 (US$27.3 billion) and 84% (2.72 million tonnes) of the total exported volume (3.22 million tonnes). Shrimp is a relatively high-priced fish item and the majority of import value has historically been accounted to developed country markets. However, this share has declined as demand for more expensive fisheries and aquaculture products grows amongst the expanding urban middle class in emerging markets (Figures 25.7 and 25.8). In 2017, developed countries reported 1.9 million tonnes of shrimp imports worth US$17.9 billion, representing 75% of the total volume and 81% of the total value.

While China is the leading shrimp producer, a substantial and growing proportion of this production is absorbed by its large domestic market. Indeed, Chinese shrimp exports peaked in 2011 at 305,248

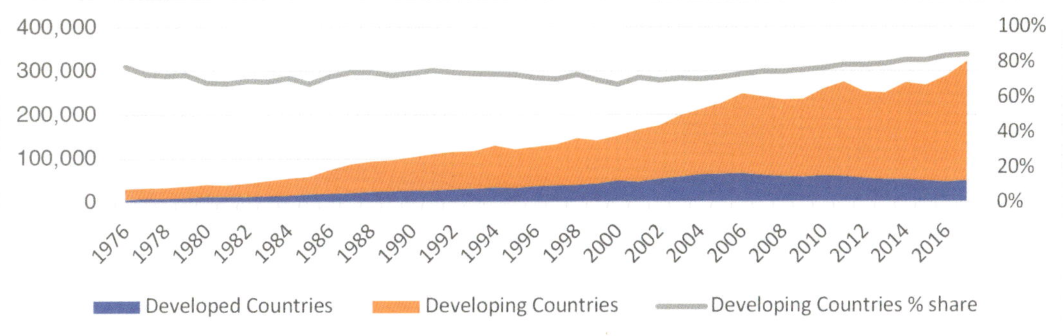

FIGURE 25.5 Contribution of developed and developing countries to global shrimp exports (tonnes), 1976–2016.

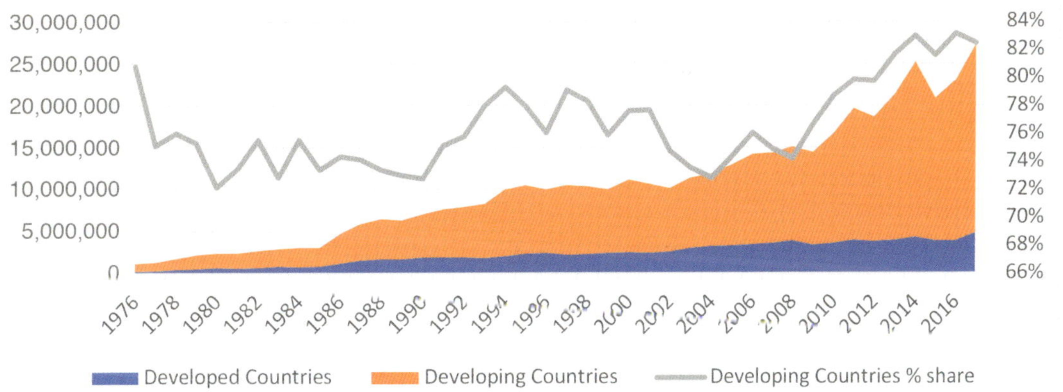

FIGURE 25.6 Contribution of developed and developing countries to global shrimp exports (US$1 000), 1976–2016.

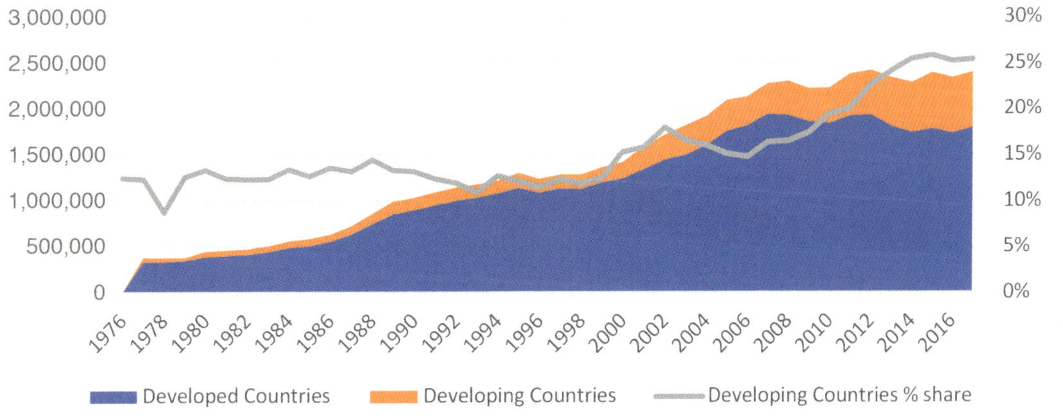

FIGURE 25.7 Value of shrimp imports by developing and developed countries (tonnes), 1976–2016.

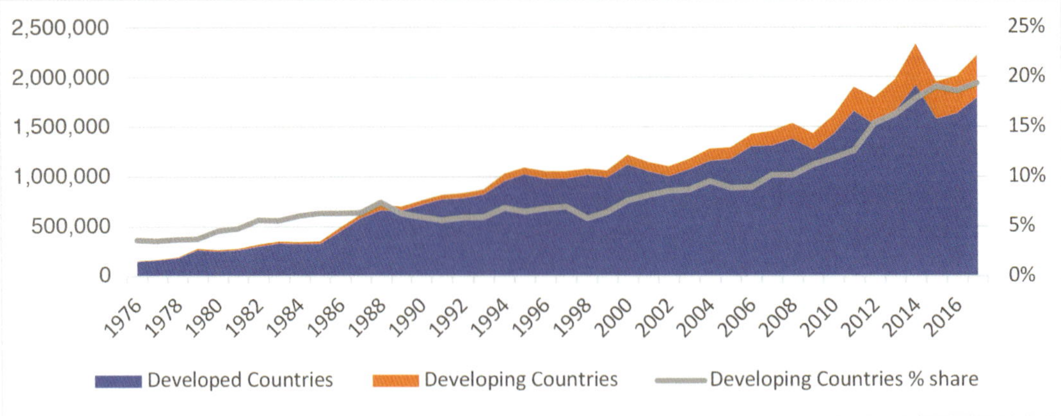

FIGURE 25.8 Value of shrimp imports by developing and developed countries (US$1 000), 1976–2016.

tonnes (US$2.19 billion) and by 2017 had fallen back to 214,954 tonnes (US$2.29 billion), around 7% of the global total (8% of total volume). During the same period, production volume increased by more than 340,000 tonnes, or 20%, while imports more than doubled, indicating a substantive increase in domestic consumption. Chinese shrimp export value over the same period remained approximately stable. However, it reflects an increase in unit values instead. Thai export volumes have primarily followed the industry's production trajectory, dropping steeply in the wake of the EMS outbreak after peaking in 2010 at 427,720 tonnes (US$3.19 billion). In 2017, Thailand exported 214,573 tonnes (US$2.07 billion) of shrimp. Most of the growth in global shrimp exports has come from India, Viet Nam and Ecuador, which have seen annual growth in export volumes of CAGR 18, 14, and 12%, respectively, since 2000 (Figures 25.9 and 25.10).

Historically, the large developed markets of the USA, Japan and the EU have accounted for the vast majority of global shrimp imports. However, the share of these three markets has been declining over

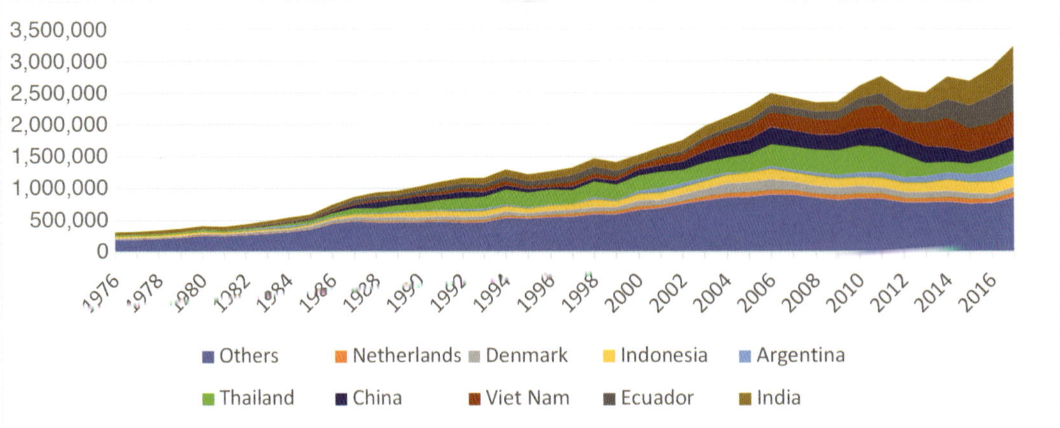

FIGURE 25.9 Value of shrimp exports by major contributors to trade (tonnes), 1976–2016.

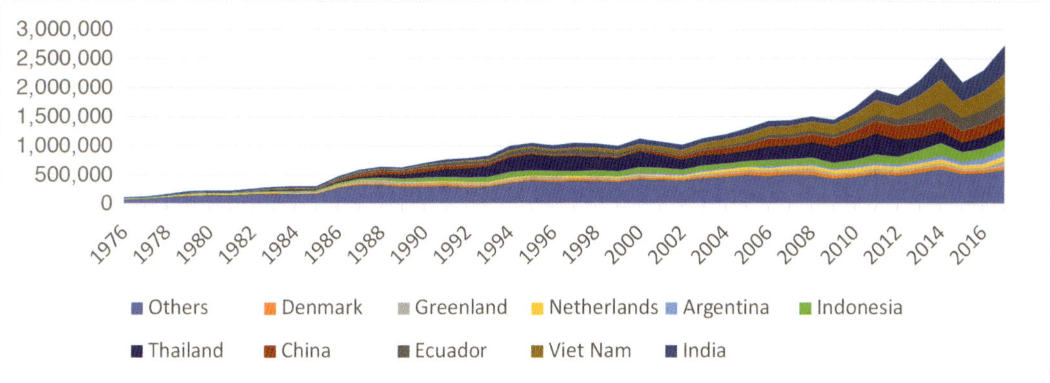

FIGURE 25.10 Value of shrimp exports by major contributors to trade (US$1 000), 1976–2016.

the last few decades as demand from emerging markets, particularly in Asia, has been increasing at a relatively faster rate (Figures 25.11 and 25.12). Another important factor has been sluggish economic growth in Japan and a shift in consumers away from the traditional fish diet towards terrestrial proteins. In 1976 the EU, USA and Japan accounted, respectively, for 22% (15%), 28% (29%) and 33% (47%) of a total of 376,551 tonnes (US$1.58 billion), while in 2017 the respective shares were 32% (32%), 26% (30%) and 9% (12%) of a total of 2.55 million tonnes (US$22.19 billion). Of the other major shrimp importers, China has recorded rapid growth in the last decade, following the financial crisis. As of 2017, Chinese imports represented 5% of the global total in volume terms and 4% by value. The Republic of Korea has realized similar growth during the same period, accounting for 4% of volume and 3% of value in 2017.

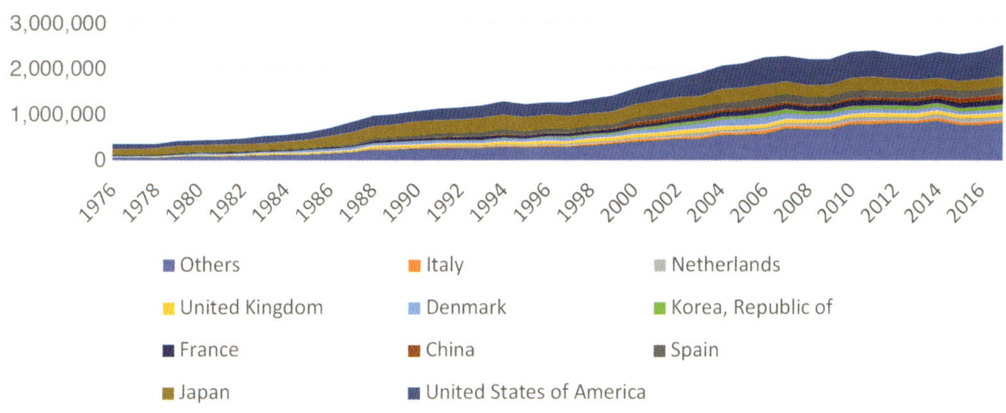

FIGURE 25.11 Value of shrimp imports by major contributors to global trade (tonnes), 1976–2016.

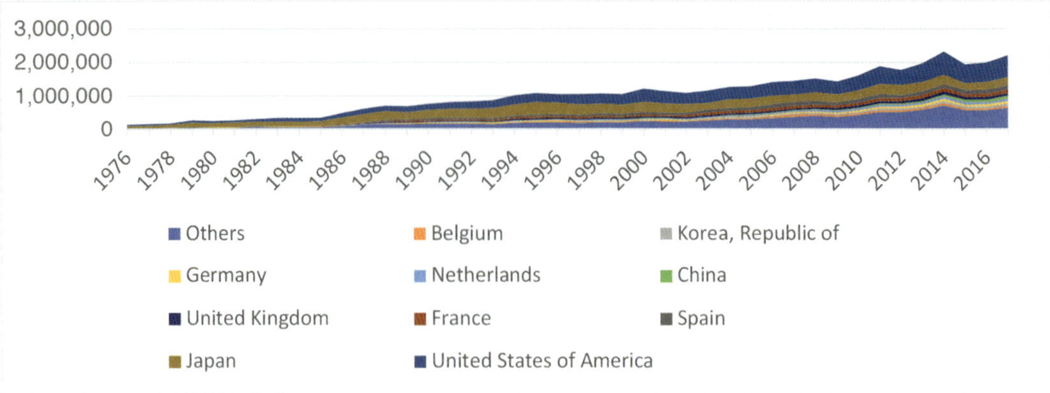

FIGURE 25.12 Value of shrimp imports by major contributors to trade (US$1 000), 1976–2016.

Trade measures in global shrimp trade

It is estimated that 78% of fish and fish products are exposed to international trade competition (FAO, 2018a), and shrimp is no exception. Representing the second largest group of exported aquatic species in value terms, these crustaceans have been heavily affected by trade measures of a tariff and non-tariff nature.

Generally regarded as a high-value product, the top importers of shrimp worldwide are mainly developed countries led by the EU and the USA. The import tariff imposed by these countries on shrimp is relatively low in comparison with agricultural products. Developed countries can impose non-tariff measures permissible under WTO rules (Emran et al., 2019). Among those measures, sanitary and phytosanitary (SPS) regulations and technical barriers to trade (TBT) are particularly relevant for products such as shrimp, where the majority of supply originates in developing countries that might face compliance issues with hygiene and health standards set by the importers. The impact of these measures on developing-country exporters has been confirmed by studies from Indonesia and Bangladesh (Emran et al., 2019; Khaliqi et al., 2018).

Trade tensions between the USA and China and the associated uncertainty have also affected the international shrimp trade. The tariffs introduced on a range of commodities by both China and the USA have inevitably had a visible impact on trade flows of fisheries and aquaculture products between these two major trading partners. In addition, the reluctance of firms to accept uncertainty and risks when the situation can change quickly is another less direct but more widespread consequence of trade tensions. The trade tension scaling between the USA and China has resulted in the setting of a new high 25% import duty on many fisheries and aquaculture products imported into the USA, including shrimp. The USA has sought to make up some of the resulting shortfalls with imports from Thailand, Viet Nam and Indonesia.

In general, however, the overall level of tariff protection applied to shrimp and other fisheries and aquaculture products remains relatively low (Bellmann et al., 2016). In addition, import tariffs and other trade protection measures have been reduced or offset by unilateral or negotiated tariff preferences, including through regional trade agreements (RTAs) or the generalized system of preferences (GSP), which provides preferential trade terms for exporters of fisheries and aquaculture products originating from certain developing countries or least developed countries (LDCs).

Dynamics in global shrimp prices

Movements in global shrimp prices are driven by developments in both supply and demand, although it is typically production changes that are at the heart of most observed volatility. As can be analyzed from FAO's fish price index (FPI), price movements for shrimp and other fisheries and aquaculture products were generally closely correlated throughout the 1990s but have diverged since the start of the new millennium as shrimp price growth has been relatively slower (Figure 25.13).

As the dominant shrimp species in international trade, whiteleg shrimp price movements tend to drive the prevailing trends for shrimp prices overall, particularly in the major markets of the USA and the EU. A notable peak in the shrimp price trend line can be observed for late 2013 due to upward price pressure resulting from poor harvests in the wake of EMS outbreaks in Asia and Latin America. In Thailand, in particular, the ex-farm price of whiteleg shrimp in 2013 increased by 42–50% compared with 2012 (FAO, 2018a). Subsequently, a recovery in production volumes and a weaker EU market during the drawn-out Eurozone crisis combined to reverse the upward trend, and whiteleg shrimp prices had reached record lows by April 2019.

Prices of black tiger shrimp tend to vary significantly across markets. In the US market, import demand has been weak since January 2015 due to the large unsold stocks imported in 2014. Primary demand in 2015 was also poor in Europe and Japan, which resulted in sustained downward pressure on shrimp prices before good seasonal demand during the summer. The weaker US dollar resulted in a rise in shrimp prices in 2017. Good harvests of global shrimp led to some price declines in 2018 but, as of 2019, black tiger shrimp prices have remained firm as a result of increased demand from China and Japan and generally tight supply.

For Argentine red shrimp, prices have been reasonably stable since peaking in 2012, with rapid growth in catch volumes generally inhibiting any sustained upward movement. In contrast, prices for northern

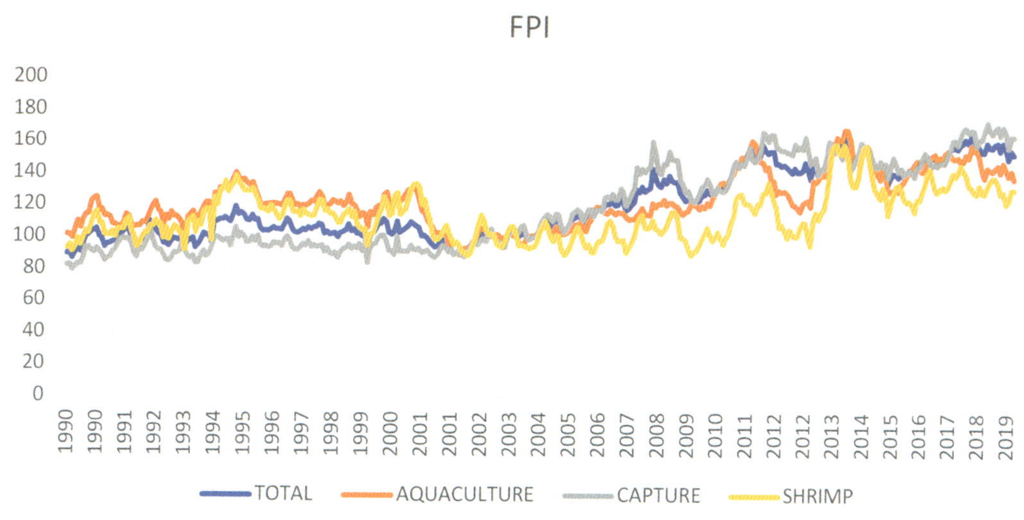

FIGURE 25.13 FPI for farmed and captured shrimp, 1960–2019.

prawns almost doubled from 2010 to 2019, reflecting declining catches and relatively weak correlation with markets for other shrimp species.

Sustainability concerns in shrimp fisheries and aquaculture

The wild shrimp industry and the shrimp aquaculture sector have been the subject of continuous debate related to unsustainable practices.

The widespread use of bottom trawling in shrimp capture fisheries has been shown to contribute to the destruction of marine habitat on the seafloor, particularly coral reefs. Shrimp bottom trawling also involves a rate of by-catch (non-target organisms trapped in the trawl) estimated at around 25% (Gillett, 2008). As a result, an increasing number of countries introduced bans on bottom trawling (Josupeit, 2010).

Traps are a relatively more sustainable fishing method, even though catch quantity per unit of time tends to be lower than trawling. Traps, designed and adapted based on the characteristics of the local shrimp population and their behaviour, are widely used in shrimp fisheries across the world. Research has shown that bottom traps and semi-floating traps are effective means of catching shrimp commercially (Arrasate-Lopez et al., 2012).

Perhaps the most highlighted issue about the sustainability of shrimp farming is the destruction of mangrove forests that are converted into aquaculture ponds. Mangroves are biodiverse habitats that act as breeding and spawning grounds for many aquatic animal species and protect the coastal area from abrasion and disaster damage (Ibharim et al., 2015). Mangroves also facilitate the natural biogeochemical cycle for converting pollutants and purifying water (Ibharim et al., 2015; Paez-Osuna, 2001). However, as most shrimp farming is located near the coast due to seawater accessibility, aquaculture expansion has led to extensive mangrove habitat loss in several countries. In Brazil, for example, more than 50,000 ha of mangroves have been lost to agriculture, and half of that area to shrimp farming precisely (Kauffman et al., 2017). Some scholars estimate that 16% of mangrove-dwelling species are at risk of extinction (Friess et al., 2016).

A second central sustainability issue shrimp farmers face is the effluent produced by aquaculture activities, polluting the surrounding environment and permanently changing the ecosystem. Shrimp aquaculture effluent contains nitrogen- and phosphate-rich compounds from shrimp excretion, uneaten feed, and chemical substances. There is also the risk of transferring pathogens from ponds to wild species. Potential solutions include adopting closed systems to reduce water usage, recycle nitrogen waste and isolate the farming area from the surrounding environment. However, these systems require additional inputs and infrastructure, which translate into other costs.

A third objection levelled at shrimp farming and aquaculture, in general, is that the fish feed used at farms generally comprises several ingredients originating from the ocean, such as fishmeal and fish oil, which are produced from low-value pelagic fish or by-catch species. Although significant effort and investment have been made to develop suitable substitute ingredients to reduce or replace fishmeal and fish oil in fish feed, with an appreciable degree of success, aquaculture still depends on capture fisheries to a large extent which itself can represent a sustainability issue.

Ecolabel schemes are becoming an increasingly helpful tool to address some of these sustainability issues, seeking to promote sustainable development and improved resource management using the power of market forces (Gutierrez et al., 2016). Today, there are many ecolabel standards covering fisheries and aquaculture products from an environmental sustainability angle. Many of them are starting to expand towards addressing social issues, including child labour and other labour rights abuses.

Ultimately, it can be argued that many mass-scale food-producing systems, motivated by the need to produce more foods for an increasing global population, will inevitably have some degree of associated negative environmental impact. The focus has thus been on how to minimize the impact caused by those activities (Martin et al., 2018) rather than to avoid them entirely. For shrimp production, sustainability will be best achieved by incorporating environmentally friendly farming technology, sustainable management practices, and good governance. At the same time, the establishment of national, regional and global communication and coordination platforms is essential. International agencies such as FAO seek to advance this objective through an expressed commitment to achieving the sustainability of food production as part of Sustainable Development Goals.

Market developments

This section is based on GLOBEFISH Highlights, FAO's quarterly market update series for fisheries and aquaculture products (FAO, 2018b). It contains a condensed version of the most recent developments in the global market for shrimp at the time of writing.

PRODUCTION

Global farmed shrimp production in 2018 increased by 3–5% over 2017, which would translate into 5.75 million tonnes of output in 2018 based on FAO's 2017 figures. Production has increased in Viet Nam, Myanmar and other minor producers in South East Asia, while in Latin America, Ecuador's output growth also continues along a sustainable trajectory. Production declined in India, Indonesia and Thailand in 2019 as low whiteleg shrimp prices hit farmer margins. Growth has also been limited in Brazil, Nicaragua and Honduras.

Farmers have noticeably shifted towards aquaculture of black tiger shrimp in Viet Nam, Thailand, Indonesia and Malaysia. Black tiger shrimp accounted for 12–15% share of global farmed shrimp production in 2018, encouraged by relatively stable and high prices compared with whiteleg shrimp.

Argentinian shrimp catches have continued to increase in recent years, breaking records in 2016, 2017 and 2018 before falling back somewhat in 2019. Meanwhile, indications in China suggest that its marine capture production of shrimp has stabilized slightly after successive declines in 2016 and 2017.

TRADE

Exports

Shrimp exports increased in volume for the first five out of the top seven exporters in 2018, but overall growth was slower. Exports earnings declined for all exporters except Ecuador due to a general weakening of prices during the same period. Exporters in Asia have been focusing on more profitable processed shrimp, targeting developed markets.

India was the lead exporter in 2018, with 575,900 tonnes but with a falling growth rate (+7.2%) compared with 2017 (+31%). The leading markets for Indian shrimp were the USA, Viet Nam and the EU. Exports declined to Viet Nam (–15%) and the EU (–13%).

Vietnamese shrimp exports increased by 7% in 2018 compared with 2017, to 530,000 tonnes. The leading markets were the EU (+13.7%), Australia (–4.0%), the USA (+4.6%), Japan (–5.5%) and the Republic of Korea (+23%). Compared with 2017, official exports from Viet Nam to China increased by 300% to 14,000 tonnes in 2018.

Ecuador maintained a stable position in 2018, diverting 61% of its 439,700 tonnes of shrimp exports to Asian markets, but exports to its number one market, Viet Nam, fell by 10% to 202,000 tonnes. However, direct sales to China increased by 512%, reaching 98,000 tonnes in 2018 compared with a mere 16,000 in 2017. Exports also rose to its second-largest market, the EU (+7.8% to 104,300 tonnes).

Imports

The top seven markets imported nearly 2.7 million tonnes of shrimp in 2018, about 31% more than in 2017. This could primarily be attributed to strong Asian markets, particularly China. Imports increased in other emerging markets in East Asia and the Middle East but declined in Viet Nam due to direct imports from China. Among the conventional markets, imports increased marginally in the USA and the EU but declined in Japan and Australia.

US imports increased by 5.1% in 2018 against 2017, but US buyers paid less in 2018 (US$6.2 billion in 2018 compared to US$6.5 billion in 2017) due to record low import prices. These prices had trickled down to end-users by mid-2018, benefitting retail and catering business during 2018 summer and year-end sales. Prices failed to recover in 2019, with the average import price of shrimp in the USA during January–June 2019 averaged 8.5% lower than the same period a year prior. Shrimp remains the most popular fisheries and aquaculture product amongst US consumers. India remained the top supplier to the US market.

Shrimp demand in Japan remained sluggish in 2018. This persistent trend marks the fading appetite for shrimp among Japanese consumers, particularly the younger generation, who increasingly prefer beef, pork, chicken and salmon to shrimp.

China became Asia's leading shrimp importer in 2018. In December 2017, China decreased its import tariff on shrimp from 5 to 2%. Since then, producers worldwide have increased direct exports to China, benefiting farmers in Asia and Latin America. At the same time, China's high surveillance on illegal border trade with Viet Nam reduced unreported shrimp imports to an estimated level of 200,000 tonnes in 2018 compared with 300,000 tonnes in 2017. According to Chinese customs figures, shrimp imports in 2018 doubled to 265,000 tonnes (+117%) compared with 2017, while imports into Viet Nam declined by 18% to 360,000 tonnes in 2018. China produces about one million tonnes of farmed shrimp annually and, considering that less than 20% of this is exported, this makes the country the top consumer market for shrimp in the world.

The EU remains the world's largest shrimp importer, though without much improvement in demand over recent years. Compared with 2017, low shrimp prices and good supplies of tropical shrimp had minimal impact on the consumer market, pointing to sluggish underlying demand in 2018. Total shrimp imports reached 817,700 tonnes (+2.4%) in 2018, with decreasing imports in Spain (–2.4%), France (–1%) and the UK (–1.1%). For the first time in 2018, extra-EU imports exceeded 600,000 tonnes, supplied by Ecuador, Argentina, Viet Nam, India and Greenland.

Elsewhere, increased supplies of farmed shrimp and soft prices supported rising shrimp demand in regional import markets in producing areas of the world. Low export prices also diverted supplies to domestic markets in many producing countries.

PRICES

Shrimp prices in the international market started to weaken in late 2017 after remaining high and stable between 2014 and 2017, with the drop accelerating amidst a price war between the most significant exporter (India) and the largest single market (the USA). Between August and December 2018, prices reached some level of stability but with nominal improvement. In the USA, the average import price fell by 9% to US$8.95/kg in 2018 compared with US$9.83/kg in 2017. Meanwhile, the average export price of Indian shrimp was 13% lower at US$7.58/kg in 2018 (US$8.65/kg in 2017). These price levels largely persisted into 2019, stimulating demand during the high consumption period in the summer months but negatively impacting margins for producers, particularly in India.

References

Arrasate-Lopez, M., Tuset, V. M., Santana, J. I., Garcia-Mederos, A., Ayza, O. and Gonzalez, J. A. (2012) Fishing methods for sustainable shrimp fisheries in the Canary Islands (North-West Africa). *African Journal of Marine Science* 34, 331–339.

Bellmann, C., Tipping, A. and Sumaila, U. R. (2016) Global trade in fisheries and aquaculture products: An overview. *Marine Policy* 69, 181–188.

Bondad-Reantaso, M. G. (2017) FAO efforts on shrimp aquaculture health management including the AHPND International Technical Seminar. *Presentation in International Technical Seminar.*

Bondad-Reantaso, M. G., Subasinghe, R. P., Arthur, J. R., Ogawa, K., Chinabut, S., Adlard, R., Tan, Z. and Shariff, M. (2005) Disease and health management in Asian aquaculture. *Veterinary Parasitology* 132, 249–272.

Brummett, R. E., Alvial, A., Kibenge, F., Forster, J., Burgos, J. M., Ibarra, R., St-Hilaire, S., Chamberlain, G. C., Lightner, D. V., Khoa, L. V. and Hao, N. V. (2014) Reducing disease risk in aquaculture. *The World Bank Report No. 88257*, pp, 1–119.

Chamberlain, G. W. (2010) History of shrimp farming. In: The Shrimp Book (ed. Alday-Sanz, V.). Nottingham University Press, UK, pp. 1–34.

De Schryver, P., Defoirdt, T. and Sorgeloos, P. (2014) Early mortality syndrome outbreaks: a microbial management issue in shrimp farming? *PLoS Pathogens* 10, e1003919.

Emran, S. J., Taslim, Q. N. and Taslim, M. A. (2019) Standards as trade barriers: The case of shrimp export of Bangladesh of EU. *Journal of Bangladesh Studies* 17, 36–45.

FAO (2018a) World Fisheries and Aquaculture. Available at www.fao.org/publications (accessed 6 August 2021).

FAO (2018b) GLOBEFISH Highlights 2018, with Jan.–December. 2018 Statistics – International Markets on Fisheries and Aquaculture Products. Quarterly update.

Flegel, T. W., Lightner, D. V., Lo, C. F. and Owens, L. (2008) Shrimp disease control: past, present and future. Diseases in Asian Aquaculture VI. Fish Health Section. *Asian Fisheries Society* 505, 355–378.

Friess, D. A., Thompson, B. S., Brown, B., Amir, A. A., Cameron, C., Koldeweyn, H. J., Sasmito, S. D. and Sidik, F. (2016) Policy challenges and approaches for the conservation of mangrove forests in Southeast Asia. *Conservation Biology* 30, 933–949.

Gillett, R. (2008) *Global Study of Shrimp Fisheries*, vol. 475. Food and Agriculture Organization of the United Nations, Rome, Italy.

Gutierrez, N. L., Omar, D., Simon, R. B., Doug, S. B., Cathy, A. R. and Andre, E. P. (2016) The current situation and prospects of fisheries certification and ecolabelling. *Fisheries Research* 182, 1–6.

Ibharim, N. A., Mustapha, M. A., Lihan, T. and Mazlan, A. G. (2015) Mapping mangrove changes in the Matang Mangrove Forest using multi temporal satellite imageries. *Ocean & Coastal Management* 114, 64–76.

Josupeit, H. (2010) Shrimp production and trade. In: *The Shrimp Book* (ed. Alday-Sanz., V.). Nottingham University Press, UK, pp. 1–34.

Karunasagar, I. and Ababouch, L. (2012) Shrimp viral diseases, import risk assessment and international trade. *Indian Journal of Virology* 23, 141–148.

Kauffman, J. B., Bernardino, A. F., Ferreira, T. O., Bolton, N. W., Gomes, L. E. and Nobrega, G. N. (2017) Shrimp ponds lead to massive loss of soil carbon and greenhouse gas emissions in northeastern Brazilian mangroves. *Ecology and Evolution* 8, 5530–5540.

Khaliqi, M., Rifin, A. and Kilat Adhi, A. (2018) Trade effect of Sanitary and Phytosanitary (SPS) and Technical Barriers to Trade (TBT) on Indonesia's Shrimp Export. *Indonesian Journal of Agricultural Research* 1, 134–141.

Lafferty, K. D., Harvell, C. D., Conrad, J. M., Friedman, C. S., Kent, M. L., Kuris, A. M., Powell, E. N., Rondeau, D. and Saksida, S. M. (2015) Infectious diseases affect marine fisheries and aquaculture economics. *Annual Review of Marine Science* 7, 471–496.

Leano, E. M. (2019) Transboundary aquatic animal diseases: History and impacts in ASEAN aquaculture. In: *Aquatic Emergency Preparedness and Response Systems for Effective Management of Transboundary Disease Outbreaks in Southeast Asia*. Proceedings of the Asean Regional Technical Consultation. Centara Grand Central Ladprao, Bangkok, Thailand, pp. 72–79.

Lightner, D. V., Redman, R. M., Pantoja, C. R., Tang, K. F. J., Noble, B. L., Schofield, P., Mohney, L. L., Nunan, L. M. and Navarro, S. A. (2012) Historic emergence, impact and current status of shrimp pathogens in the Americas. *Journal of Invertebrate Pathology* 110, 174–183.

Martin, A., Coolsaet, B., Corbera, E., Dawson, N., Fisher, J., Franks, P., Mertz, O., Pascual, U., Rasmussen, L. and Ryan, C. (2018) Land use intensification: The promise of sustainability and the reality of trade-offs. In: *Ecosystem Services and Poverty Alleviation* (eds Schrekenberg, K. et al.). Routledge, London, UK.

Muthukrishnan, S., Defoirdt, T., Ina-Salwany, M. Y., Yusoff, F. M., Shariff, M., Ismail, S. I. and Natrah, I. (2019) *Vibrio parahaemolyticus* and *Vibrio harveyi* causing Acute Hepatopancreatic Necrosis Disease (AHPND) in *Penaeus vannamei* (Boone, 1931) isolated from Malaysian shrimp ponds. *Aquaculture* 511, 734227.

OIE (2019) *Manual of Diagnostic Tests for Aquatic Animals*. Available at: https://www.oie.int/en/standard-setting/aquatic-manual/access-online/

Paez-Osuna, F. (2001) The environmental impact of shrimp aquaculture: Causes, effects, and mitigating alternatives. *Environmental Management* 28, 131–140.

Petesch, T. (2017) Impacts of disease in shrimp aquaculture on U.S. capture fishery price. MSc thesis. Duke University, Durham, NC, USA.

Shinn, A. P., Pratoomyot, J., Griffiths, D., Trong, T. Q., Vu, N. T., Jiravanichpaisal, P. and Briggs, M. (2018) Asian shrimp production and the economic costs of disease. *Asian Fisheries Science* 31S, 29–58.

Walker, P. J. and Mohan, C. V. (2009) Viral disease emergence in shrimp aquaculture: origins, impact and the effectiveness of health management strategies. *Reviews in Aquaculture* 1, 125–115.

Walker, P. J. and Winton, J. R. (2010) Emerging viral diseases of fish and shrimp. *Veterinary Research* 41, 51.

World Bank (2014) Reducing disease risk in aquaculture. World Bank Report Number 88257-GLB, Agriculture And Environmental Services Discussion Paper 09.

Zorriehzahra, M. J. and Banaederakhshan R. (2015) Early Mortality Syndrome (EMS) as new emerging threat in shrimp industry. *Advances in Animal and Veterinary Sciences* 3, 65–72.

Chapter 26

MICROBIOLOGICAL SAFETY OF AQUACULTURED SHRIMP

IDDYA KARUNASAGAR AND INDRANI KARUNASAGAR

Introduction

Epidemiological records show that, in general, shrimp is a safe food. In the USA between 1998 and 2015, 857 seafood-associated outbreaks occurred resulting in 4815 illnesses, 359 hospitalizations and four deaths being recorded (Barret et al., 2017). Of these, the aetiology could be confirmed in 637 outbreaks, of which scombrotoxin accounted for 349 (55%) and ciguatoxin for 227 (36%). Both of these were associated with finfish such as tuna and mahi mahi. The bacterial pathogen *Salmonella* was associated with 978 illness (26%) and 97 hospitalizations (31%), while norovirus was associated with 418 illnesses (11%). There were no shrimp-associated outbreaks or illnesses. Dewaal et al. (2006) noted that in the USA between 1990 and 2003, 63.5% of seafood-borne outbreaks and 32.1% of cases were linked to finfish such as tuna and groupers (histamine and ciguatera poisoning); 15% of outbreaks and 33.9% of cases were linked to molluscan shellfish; 14.3% of outbreaks and 25.8% of cases were linked to seafood dishes such as crab cakes and sushi; and 7% of outbreaks and 8.2% of cases were linked to other seafood including lobster and shrimp. Thus, even among seafood-borne illnesses, shrimp accounts for an extremely small percentage of outbreaks and cases. Alhough *Vibrio* spp. accounted for 9% of seafood-borne outbreaks, 33% of these were associated with molluscan shellfish (Dewaal et al., 2006).

Data from the European Union (EU) show that, in 2018, there were 113 outbreaks associated with fish and shellfish involving 1196 cases (EFSA and ECDC, 2019), but separate data for crustaceans are not available. Table 26.1 summarizes the data from different parts of the world. Although epidemiological data show that shrimp has a very good record of safety, certain pathogens may be associated with it either during primary production in aquaculture systems or introduced as contaminants during handling and processing. This chapter deals with some of the key microbiological hazards associated with shrimp.

TABLE 26.1 Aetiology of foodborne disease outbreaks associated with fish and shellfish.

Causative agent	*Number of outbreaks*		
	USA, 1998–2015[a]	*England and Wales, 1992–2003*[b]	*Australia, 2005–2006*[c]
Bacteria			
Bacillus cereus	3	3	
Campylobacter	1	1	
Clostridium botulinum			
Salmonella	14	-	
Shigella	18	38	
Vibrio cholerae	3	-	
Vibrio parahaemolyticus	3	-	1
Vibrio spp.			
Clostridium perfringens	20	-	
Staphylococcus aureus	1		
		3	
	-		
		1	
	-		
Viruses	18	28	
Toxins			
Ciguatoxin	227		17
Scombrotoxin	349	48	9
Shellfish toxin	11	-	
Parasitic	1	-	
Unknown	220	58	2
Total	857	180	29

Sources: [a] Barret et al. (2017); Lynch et al. (2006); [b] Hughes et al. (2007); [c] OzFoodNet (2006).

Microbial hazards associated with shrimp

VIBRIO SPP.

Vibrio spp. are common inhabitants of the aquatic environment and, of over 80 species included in the genus *Vibrio*, at least 12 are capable of causing human infections (Oliver and Kaper, 2007). Members of this genus are motile, Gram-negative, curved or straight rods with a polar flagellum. *Vibrio* spp. are commonly isolated from estuarine, coastal marine environments (some species like *Vibrio cholerae* are found in fresh water) all over the world, and seafood-borne illnesses are primarily caused by *Vibrio parahaemolyticus*, *V. vulnificus* and *V. cholerae* (FAO/WHO, 2003). Of these, *V. parahaemolyticus* and *V. cholerae* cause gastrointestinal disease while *V. vulnificus* causes septicaemia. *V. vulnificus* infections are mainly associated with consumption of raw oysters by susceptible individuals (those with underlying liver disease, alcoholism, diabetes or other immunocompromised condition), and shrimp are not a documented source of infection.

Vibrio parahaemolyticus

Vibrio parahaemolyticus gastroenteritis has been associated with a variety of fish and shellfish. While oysters are the most common source in recent outbreaks in the USA and South America, there have been reports of involvement of other types of seafood including clam, shrimp, lobster, crayfish, scallop, crab and finfish (Daniels et al., 2000; FAO/WHO, 2011; Oliver and Kaper, 2007). In Japan (Anonymous, 2000), implicated foods include *sashimi* (pieces of raw fish fillet (responsible for 26% of outbreaks)) followed by *sushi* (a vinegary rice ball with pieces of raw fish fillet (23%)), shellfish (16%) and cooked seafood (12%). A large outbreak linked to shrimp occurred in Louisiana, USA in 1978 in which 1133 of the 1700 persons attending a dinner were affected, and this appears to have been caused by cross-contamination between raw and boiled shrimp. Shrimp boiled in the morning were kept in the same wooden seafood box used to transport raw shrimp, were transported 40 miles in an unrefrigerated truck and were held for an additional 7–8 hours before being served for dinner (Oliver and Kaper, 2007).

Vibrio parahaemolyticus has a worldwide distribution and is found in estuarine and coastal environments in tropical to temperate zones (FAO/WHO, 2011; Joseph et al., 1982). This organism is considered to be part of the autochthonous microflora in these environments, and there is no correlation between its presence and faecal contamination of the environment (Joseph et al., 1982; Oliver and Kaper, 2007). *V. parahaemolyticus* has been isolated from seawater, sediment, marine animals, plankton and various fish and shellfish species (FAO/WHO, 2011; Joseph et al., 1982; Karunasagar and Karunasagar, 2018). Thus, *V. parahaemolyticus* is naturally present in areas that grow and harvest shellfish (shrimp and molluscan shellfish). Certain areas may have more favourable environmental conditions that support the establishment, survival and growth of the organism such as temperature, salinity, zooplankton, tidal flushing and dissolved oxygen (FAO/WHO, 2020; Garay et al., 1985; Kaneko and Colwell, 1977; Venkateswaran et al., 1990). In temperate waters, *V. parahaemolyticus* is often detected in warmer months and the organism has been reported to survive in sediment during winter (Kaneko and Colwell, 1977); however, in tropical waters, *V. parahaemolyticus* can be detected throughout the year (Deepanjali et al., 2005; Natarajan et al., 1980). While salinity and temperature are considered important factors influencing the prevalence and levels of *V. parahaemolyticus* in temperate waters (DePaola et al., 2003; Kaneko and Colwell, 1977), salinity may influence levels in tropical waters, low counts being recording during the post-monsoon period (Deepanjali et al., 2005). *V. parahaemolyticus* can grow in sodium chloride concentrations ranging from 0.5 to 10%, with optimum levels 1–3% (Colwell et al., 1984). Adsorption of *V. parahaemolyticus* onto plankton or chitin-containing materials occurs at higher efficiency under conditions of estuarine salinity (Kaneko and Colwell, 1975). In tropical shrimp culture environments, *V. parahaemolyticus* is often present. This organism accounted for 0–27% of the flora in water and sediment of shrimp ponds in India (Gopal et al., 2005; Otta et al., 1999). In shrimp, levels range from undetectable to 10^4/g, high counts being rare (Cann et al., 1981; Karunasagar et al., 1984; Li et al., 2020); in finfish, levels of ~88/g have been reported (Chan et al., 1989).

Previous studies in Japan showed that 96% of clinical strains produce a thermostable direct haemolysin (TDH), while only 1% of environmental strains produce this haemolysin (Joseph et al., 1982). Subsequently, TDH-negative strains from clinical cases were found to produce a TDH-related haemolysin, TRH (Honda et al., 1987). At present, strains producing TDH and TRH are considered pathogenic to man (FAO/WHO, 2011, 2020). Diverse serotypes may be associated with human infections, but recently strains belonging to the O3:K6 serotype and its variants have been found to be the causative agent of several outbreaks in different countries (Nair et al., 2007). Although several publications refer to these strains as "pandemic" strains, Nair et al. (2007) pointed out that this is misleading in the epidemiological sense because outbreaks had not affected an exceptionally high proportion of the population. Nevertheless, strains belonging to this group show clonality in molecular typing methods such as arbitrarily primed (AP) PCR, ribotyping or pulse-field gel electrophoresis (PFGE), and are characterized by the presence of only the *tdh* gene (and not the *trh* gene) and mismatches in nucleotides in the *tox*RS gene, an open reading frame ORF8 derived from the filamentous bacteriophage f237 (Nair et al., 2007). The reported prevalence of pathogenic strains in shrimp culture environments is low. Gopal et al. (2005) noted that only 2/47 isolates from aquaculture sites were positive for the presence of the *tdh* gene. Raghunath et al. (2008) noted the absence of *tdh*-positive strains but the presence of *trh*-positive strains in association with shrimp.

Outbreaks of shrimp mortality at the early stage (<30 days) were reported from China in 2009, which then spread to Vietnam, Thailand and Mexico in subsequent years. The disease was called "early mortality syndrome" (EMS), and a case definition was adopted in 2012 (NACA, 2012). The causative agent was identified as specific types of *V. parahaemolyticus* (Tran et al., 2013) possessing a unique virulence plasmid bearing genes encoding a *Photorhabdus* insect-related (PIR) toxin (Yang et al., 2014). The disease is now called acute hepatopancreatic necrosis disease (AHPND), and all *V. parahaemolyticus* strains causing AHPND tested to date contain the virulence plasmid and are negative for both *tdh* and *trh*. This shows that *V. parahaemolyticus*$_{AHPND}$ is not of food safety concern.

The prevalence and levels of *V. parahaemolyticus* in imported frozen seafood sampled at the port of Osaka (Japan) and imported fresh seafood sampled at Kansai international airport in the period 1998–2000 were reported by Chowdhury et al. (2001). Out of 335 frozen samples examined, 65 (19%) were positive while 234/949 fresh seafood samples (25%) were positive. Tuna had the highest prevalence among several different species of fresh seafood, shrimp had the highest prevalence in frozen seafood while Spanish mackerel had a lower prevalence. Of 1298 *V. parahaemolyticus* strains isolated, two (0.15%) contained the *tdh* gene and 17 (1.3%) contained the *trh* gene. Studies in China showed 15.4% prevalence in tested shrimp, with 7.32% of isolates *tdh*-positive and 21.95% *trh*-positive (Li et al., 2020).

Human volunteer studies indicate that 10^5–10^7 *tdh*-positive *V. parahaemolyticus* cells are required to cause clinical symptoms (Sanyal and Sen, 1974). Considering the low levels found in freshly harvested and frozen shrimp (Cann et al., 1981; Karunasagar et al., 1984; Li et al., 2020), growth of the organism due to mishandling at temperatures permitting growth would be necessary to reach infective doses. The optimum temperature for growth of *V. parahaemolyticus* is 37 °C, and the range at which growth can occur has been reported to be 5–43 °C (ICMSF, 1996). At optimum temperature, the doubling time in shrimp was 9–10 minutes and at 18.3 °C, it was 144 minutes (Katoh, 1965). At 20 °C, the doubling time was 34 minutes in raw shrimp and 28 minutes in cooked shrimp (Liston, 1974). Growth rates in a range of seafoods and tryptic soy broth with 2.5% salt (NaCl) have been recorded and summarized (ICMSF, 1996). These data indicate that moderate populations of 10^2–10^3 organisms/g on seafood can increase to >10^5 organisms/g in 2–3 hours at ambient temperatures between 20 and 35 °C (ICMSF, 1996). A number of studies indicate that *V. parahaemolyticus* dies when exposed to temperatures <5–7 °C, with the highest mortality rate being in the range 0–5 °C (ICMSF, 1996). A drop in numbers of 1–2 $\log_{10}$ occurs during freezing, but the organism can persist in frozen seafood for long periods of time (ICMSF, 1996). Both pathogenic and non-pathogenic strains have been observed to respond similarly to freezing (FDA, 2005). *V. parahaemolyticus* is very sensitive to heat, with a D-value of <1.0 minutes at 65 °C in crab homogenate with an initial inoculum of 10^6 cells (ICMSF, 1996), and hence cooking would greatly reduce the hazard due to this pathogen.

Both conventional microbiological and molecular methods have been used for the detection and enumeration of *V. parahaemolyticus* in various seafoods, including shrimp. FAO/WHO (2016) brought out guidance on selection and application of methods for the detection and enumeration of *Vibrio* spp. in seafood. While conventional methods take several days to complete, enumeration of *V. parahaemolyticus* in 24 hours is possible using colony hybridization (Deepanjali et al., 2005; Raghunath et al., 2008). A DNA probe binding to the thermolabile haemolysin gene (*tlh*) present in all strains of this organism has been used by several investigators for enumeration of total *V. parahaemolyticus* (Deepanjali et al., 2005; DePaola et al., 2003; McCarthy et al., 1999; Raghunath et al., 2008). With the conventional method, TDH-producing strains can be recognized by haemolysin production in a high-salt blood agar medium (Wagatsuma agar; Joseph et al., 1982), but this test requires fresh human or rabbit blood, which is difficult to acquire in many food-testing laboratories. Because there is no phenotypic test for TRH production, molecular methods are widely used for the detection and enumeration of pathogenic strains of *V. parahaemolyticus*. DNA probes binding to *tdh* or *trh* genes have been used (Deepanjali et al., 2005; DePaola et al., 2003; McCarthy et al., 2000; Nordstrom et al., 2006; Raghunath et al., 2007). The US Food and Drug administration (FDA) Bacteriological Analytical Manual (BAM) describes a colony hybridization-based method using alkaline phosphatase (ALP)-labelled probes for enumeration of total *V. parahaemolyticus* and *tdh*-positive *V. parahaemolyticus* (Kaysner et al., 2019). Results obtained with this method and the BAM most probable number (MPN) method were comparable, but the latter was more sensitive (3 MPN/g), compared to direct plating and colony hybridization (10 colony-forming units (cfu)/g), and would be useful for samples with very low *V. parahaemolyticus* density (Ellison et al., 2001; Gooch et al., 2001).

PCR based on amplification of the *tdh* gene for detection of pathogenic *V. parahaemolyticus* has a detection limit of $>10^4$ cfu/g when applied to lysates prepared directly from fish homogenates, which could be improved (<10 cfu/ml) by performing PCR after an 8-hour enrichment in alkaline peptone water (Karunasagar et al., 1996). PCR performed on lysates obtained from seafood homogenates without enrichment using primers amplifying the *tox*R gene yielded a higher number of positives compared to conventional isolation, and PCR was more sensitive than colony hybridization for detection of *tdh*- and *trh*-positive organisms (Raghunath et al., 2008). Positivity increased with duration of enrichment, with PCR performed after 18 hours of enrichment yielding the highest number of positive reactions. TaqMan multiplex real time PCR targeting the genes *tlh*, ORF8, *tdh* and *trh* detected total and pathogenic *V. parahaemolyticus* with detection limits of 1 cfu/g oyster after overnight enrichment (Ward and Bej, 2006). Tyagi et al. (2009) reported SYBR green-based real-time PCR assay targeting the *tdh* gene, with a detection limit of 1 cfu/ml in shrimp homogenates after 6 hours of enrichment.

Vibrio cholerae

Vibrio cholerae is a heterogeneous species consisting of over 220 serotypes, of which only serotypes O1 and O139 are the causative agents of the disease cholera and referred to as choleragenic *V. cholerae* (FAO/WHO, 2005). Strains belonging to non-01/non-0139 serotypes of *V. cholerae* are widely distributed in the aquatic environment and are mostly not pathogenic to humans, though occasionally they may be associated with sporadic cases of gastroenteritis (Kaper et al., 1995; Oliver and Kaper, 2007). The O1 serovar is classified into three antigenic forms: Inaba, Ogawa and Hikojima. *V. cholerae* O1 strains are classified into two biotypes, classical and El Tor, based on their phenotypic characteristics (Kaper et al., 1995). Recent studies have shown that the classical biotype strains are rarely isolated in any part of the world (Sack et al., 2003). The severe form of the disease, termed cholera gravis, is characterized by the passage of voluminous stools of rice water character leading to dehydration, hypovolaemic shock, acidosis and, if appropriate treatment is not initiated, death. However, it has been estimated that only 2% of those infected with the El Tor biotype and 11% of those infected with the classical biotype develop severe disease; 5% of El Tor and 15% of classical infections may result in moderate illness that can be managed in outpatient clinics (Kaper et al., 1995). Symptoms due to serotypes O1 and O139 appear to be identical. The choleragenic El Tor biotype strains of *V. cholerae* are grouped in four major clonal groups: (1) the seventh pandemic, (2) the US Gulf Coast, (3) Australia and (4) Latin America, which seem to reflect broad demographic and epidemiological associations (Wachsmuth et al., 1994). When O139 emerged in the early 1990s and almost replaced the O1 serotype in Southeast Asia it was thought that it might represent a new pandemic, but this strain did not spread beyond Southeast Asia and, even there, cases due to the O139 serotype have declined and the O1 serotype has once more become dominant (Oliver and Kaper, 2007).

Identification of choleragenic *V. cholerae* using antizera has been commonly used, but misidentification can occur due to cross-reaction of some environmental stains with O1 or O139 antizera (FAO/WHO, 2005). The most important virulence factor associated with *V. cholerae* O1 and O139 is the cholera toxin, which has two subunits, A and B. The *ctx* genes (*ctx*A and *ctx*B) encoding production of the cholera toxin are present in a filamentous bacteriophage that is integrated into the genome of *V. cholerae* O1 and O139 (Faruque et al., 1998). Loss of bacteriophage explains the presence of non-toxigenic O1 *V. cholerae* in the environment. Molecular identification methods based on probes or PCR primers binding to *ctx* genes have been widely used to detect toxigenic O1 or O139 *V. cholerae* (FAO/WHO, 2005). Detection of choleragenic *V. cholerae* in fish homogenates containing fewer than 10 cells/ml was possible when PCR was performed after 6 hours of enrichment in alkaline peptone water (Karunasagar et al., 1995). The FDA Bacteriological Analytical Manual recommends *ctx*-based PCR for determining the toxigenicity of *V. cholerae* (Kaysner et al., 2019).

In addition to producing cholera toxin, choleragenic strains of *V. cholerae* possess the ability to adhere to, and colonize, the small intestine, which has been attributed mainly to a toxin-co-regulated pilus (TCP). The pilus colonization factor is also known to act as a receptor for the CTX phage and is encoded by the *tcpA* gene that is part of the *V. cholerae* pathogenicity island (Karaolis et al., 1998). The emergence of the O139 serotype as a choleragenic strain has provided a unique opportunity to study the evolution of new epidemic strains. Molecular epidemiological studies suggest that O139 strains are closely related to O1 El Tor strains, and conversion of the ancestral El Tor strain involved insertion of a large foreign genomic region encoding O139 antigen-specific genes and the deletion of most O1 antigen-specific genes (Faruque et al., 2003). Several investigators have studied the presence of virulence-associated genes in environmental strains of non-O1, non-O139 *V. cholerae* (Chakraborty et al., 2000; Rivera et al., 2001; Saravanan et al., 2007). Although *ctx*-positive non-O1 and non-O139 strains have been found, these often lack the full set of virulence genes found in epidemic strains. Chakraborty et al. (2000) noted the absence of *tcpA* genes in *ctx*-positive strains, while Rivera et al. (2001) noted the absence of genes encoding zonula occludens toxin (*zot*). Although *ctx*-positive non-O1 and non-O139 serovars of *V. cholerae* have been implicated in cholera-like disease, only sporadic cases have been reported (Dalsgaard et al., 2001). Thus, present evidence does not indicate that the emergence of new choleragenic serovars is of serious concern regarding the safety of shrimp (FAO/WHO, 2005).

A variety of fish and fishery products have been involved in outbreaks of cholera in different parts of the world (FAO/WHO, 2005). Crustaceans, molluscs and finfish prepared in a variety of forms have been vectors for the transmission of *V. cholerae*. Transmission of *V. cholerae* by seafood can be acute where fish and shellfish are consumed raw (DePaola, 1981). Seventy five of 336 passengers in an aeroplane were affected in the Americas in 1992, where cold seafood salad was implicated (Eberhart-Phillips et al., 1996). The shellfish most often associated with cholera cases are molluscan shellfish (oysters) and crabs. While oysters are consumed raw in many countries, crabs are generally cooked. However, even after boiling crab for up to 10 minutes or steaming for up to 30 minutes, *V. cholerae* O1 may still retain viability (Blake et al., 1980). A few outbreaks were linked to crustacean shellfish, one of which involved the consumption of raw shrimp in the USA in 1986, where the source was domestic; contaminated crab salad served in an aeroplane flying from Peru to California resulted in 75 cases; an outbreak in Japan in 1978 was associated with lobster imported from Indonesia; and an outbreak was linked to the consumption of raw shrimp in the Philippines in 1962. However, in most cases it is not possible to assess whether *V. cholerae* O1 was naturally present or cross-contaminated after harvesting (FAO/WHO, 2005). Kim et al. (2018) reported three cases of cholera in Korea associated with raw seafood: the source of contamination was traced to be seawater in the area (Kim et al., 2018).

V. cholerae is a component of the autochthonous bacterial flora in estuarine areas and, most often, non-O1 and non-O139 strains are isolated from the environment even in countries where cholera is endemic (Oliver and Kaper, 2007). However, toxigenic *V. cholerae* can persist in the environment for long periods of time in the absence of clinical cases, and this explains the presence of *V. cholerae* O1 in waters of the US Gulf coast and Australia (Oliver and Kaper, 2007). In the aquatic environment, strong association between levels of zooplankton and incidence of *V. cholerae* has been observed. Adhesion to chitin has been shown to strongly influence the ecology of *V. cholerae*. The organism is chitinolytic, and its ability to digest chitin seems to play a role in its persistence in the environment (Colwell et al, 1984). Choleragenic *V. cholerae* has also been reported to attach to the hindgut of crab, and it was noted that the hindgut of crustaceans is an extension of the exoskeleton and is lined with chitin (FAO/WHO, 2005). However, there are very few records of isolation of *V. cholerae* O1 and O139 from shrimp: this could be because shrimp are benthic organisms. Studies from Southeast Asia indicate the absence of *V. cholerae* O1 from raw shrimp (Fonseka, 1990; Karunasagar et al., 1990, 1992; Rattagool et al., 1990). Several studies on shrimp farms in India indicated the absence of choleragenic *V. cholerae* in culture ponds (Gopal et al., 2005; Otta et al, 1999). Dalsgaard et al. (1995a) found *V. cholerae* O1 to be present in 2% (2/107) of water, sediment and shrimp samples collected from a major shrimp culture area in Southeast Asia. However, subsequent testing of the isolates indicated the absence of the *ctx* genes in both O1 strains (Dalsgaard et al., 1995b). Ravi Kiran (1992) and Dalsgaard et al. (1995a) analysed shrimp gut content for the presence of potential human pathogens and noted the absence of *V. cholerae* O1.

Improper hygiene during postharvest handling could result in contamination of shrimp with choleragenic *V. cholerae*. Saravanan et al. (2007) noted the absence of choleragenic *V. cholerae* in shrimp processed under hazard analysis and critical control point (HACCP) conditions in India, but detected the organism in one shrimp sample in a domestic market. The data of DePaola et al. (1993) during the Peruvian cholera epidemic showed high levels of contamination (100%) in a small number of raw seafood samples from street vendors, but only one out of 1011 seafood samples intended for export and processed under HACCP conditions was positive. Chen et al. (2004) isolated *V. cholerae* O1 and O139 from shrimp in a domestic market in Malaysia. FAO/WHO risk assessment for choleragenic *V. cholerae* in warm-water shrimp in international trade looked at the data from testing laboratories in shrimp-importing countries. From a total of 21,857 samples tested, only two originating from an Asian country in 1995 (early stage of HACCP) were positive for choleragenic *V. cholerae* (FAO/WHO, 2005).

The dose response model developed in FAO/WHO risk assessment (FAO/WHO, 2005) indicates that 10^6 choleragenic *V. cholerae* are required to elicit disease. This suggests that if shrimp is contaminated with *V. cholerae*, multiplication of the organism must occur before infective dose is reached. In raw shrimp, *V. cholerae* has to compete with other organisms for growth. The optimum temperature for growth of *V. cholerae* is 37 °C, with a range of 10–43 °C (ICMSF, 1996). Kolvin and Roberts (1982) measured the growth of *V. cholerae* O1 in raw and cooked seafood; no growth was observed in raw prawns, mussels and oysters, but growth occurred in cooked shellfish. Levels of 10^{10} cells/g were reported in cooked prawns and mussels stored at 37 °C. At 22°C, there was a lag phase of 8 hours for the classical biotype and 4 hours for the El Tor biotype. The organism is sensitive to desiccation, with a D-value of 2.65 minutes at 60 °C (ICMSF, 1996). *V. cholerae* survives refrigeration, though some decline in numbers is seen. Viable cells could be recovered from raw shrimp spiked with 10^5 cells/g *V. cholerae* O1 after 4–9 days at 5–10 °C (ICMSF, 1996). Similarly, although freezing causes initial decline in numbers, the organism may survive >180 days in fish (ICMSF, 1996).

FAO/WHO Risk assessment of choleragenic *V. cholerae* in warm-water shrimp in international trade indicated that the risk of transmission of cholera through this commodity is very low (FAO/WHO, 2005). Use of a spreadsheet-based risk assessment tool (Ross and Sumner, 2002) for quantitative risk assessment predicted the likelihood of illness to be 1–2 cases per decade in Japan and the USA, considering the volumes of warm-water shrimp imported and consumed, and one case in 25 years in other shrimp-importing countries (FAO/WHO, 2005). A quantitative approach using a model based on the import-to-consumption pathway (prevalence estimated based on data from testing laboratories in importing countries – two samples in 1995 positive out of 21,857 samples tested between 1995 and 2000) – predicted the illness to be 1–5 cases every 5 years, based on the assumption that 10% of shrimp are consumed raw and 90% after cooking (FAO/WHO, 2005). Thus quantitative risk assessment indicates a very low risk, and there are no epidemiological records of illness linked to imported warm-water shrimp, supporting the very low risk predicted.

SALMONELLA SPP.

The genus *Salmonella* is a member of the family Enterobacteriaceae, and the taxonomy and nomenclature of the members of this genus have been the subject of considerable debate among specialists. Currently, two species are recognized (Tindall et al., 2005): *Salmonella enterica* and *Salmonella bongori*. Six subspecies are recognized in *S. enterica*: subspp. *enterica, salamae, arisone, diarizonae houtenae, indica*. More than 2500 serotypes have been recorded, of which the majority (59%) belong to *S. enterica* subsp. *enterica*, which are also responsible for 99% of *Salmonella* infections in humans and warm-blooded animals (Brenner et al., 2000). Other subspecies may also be associated with cold-blooded animals and the environment, but isolates from both species and all subspecies have occurred in humans (Brenner et al., 2000).

Although the normal habitat of *S. enterica* subsp. *enterica* is the gut of warm-blooded animals, it may survive in the aquatic environment for very long periods of time. Unlike *Escherichia coli*, that has a short survival period outside the host, *Salmonella* can survive and even multiply in the environment – e.g. it can adhere to soil particles and survive and multiply in this ecosystem for at least 1 year (Winfield and Groisman, 2003). In total, 574 isolates of *Salmonella* belonging to 41 serotypes were obtained from Tech River (France) in 1996–1997, some being specific to flood events (Baudart et al., 2000). A 4-year study in coastal waters of Galicia, northwestern Spain, showed a prevalence of 2.4% in molluscs and seawater with *S.* Senftenberg being the most predominant (42%) among 20 different serotypes (Martinez-Urtaza et al., 2004a). The presence of *S.* Senftenberg could not be correlated with environmental parameters, while the presence of other serotypes was associated with wind and rainfall events. *S.* Senftenberg has been very rarely reported in human infections and is halotolerant since it has been isolated from brines with a salt concentration of 30% (Martinez-Urtaza et al., 2004b). Detection of *Salmonella* in 16% of shrimp and 22.1% of mud/water samples in Southeast Asia led Reilly and Twiddy (1992) to suggest that *Salmonella* is part of the normal aquatic flora in tropical environments. However, recent reports of detection of *Salmonella* in fish gut in a natural river system in Texas (Gaertner et al., 2008) suggest that *Salmonella* is more widely present in aquatic systems than previously thought. Between 17 and 33% of fish sampled in San Marcos River, Texas were positive for *Salmonella*, and its presence in fish gut was attributed to ingestion of *Salmonella* present in detritus. *Salmonella* has been isolated from US freshwater catfish ponds with a prevalence of 5% (Wyatt et al., 1979) and from eel culture ponds in Japan with a prevalence of 21% (Saheki et al., 1989). Koonse et al. (2005) reported that, in tropical shrimp aquaculture ponds, the risk of finding *Salmonella* was higher in those with high faecal coliform counts. While this would be understandable, faecal coliform count may not be a suitable indicator for *Salmonella* in aquaculture systems. In US oysters, prevalence was related to season (13.4% positive in summer and 1.6% in winter) and region, but did not correlate with faecal coliform levels (Brands et al., 2005). *Salmonella* can colonize marine mammals including killer whale, bottlenose dolphins, seals, sea lions, elephant seals and porpoises (Higgins, 2000; Stoddard et al., 2005), and the organisms shed by these mammals may contaminate other marine fish. It was found that 21.7% of harbour porpoises in England and Wales were positive for *Salmonella* between 1990 and 2002. At San Miguel Island, California, 33% of fur seal pups and 40% of sea lion pups were positive for *Salmonella* (Higgins, 2000). The presence of *Salmonella* in trout farms in Spain (Gonzalez et al., 1999) and long-term persistence in fish feed plants in Norway has been reported (Nesse et al., 2003). Between 2000 and 2004, 3.78% of environmental samples from Norwegian fish feed production facilities were positive for *Salmonella*; the serovars recovered were mostly *S.* Senftenberg and *S.* Montevedeo, accounting for 2% of human cases in Norway (Lunestad et al., 2007). Byappanahalli et al. (2009) reported that the filamentous alga, *Cladospora*, in Lake Michigan is a reservoir for *Salmonella*, a 3-year study between 2005 and 2007 indicating its presence in 23–72% of samples at densities ranging from 0.16 to 89.46 MPN/g. This alga can be found in both fresh and marine waters.

After reviewing the available evidence, FAO (2010) noted that there are a variety of pathways by which *Salmonella* can enter aquaculture systems including storm water, wild animals, domestic stock, poor sanitation and inappropriate disposal of human and animal waste. Control of such pathways could be a major challenge – e.g. land runoff during rains and control of wild animals such as birds in the farm environment. Noor Uddin et al. (2015) reported that, in Vietnam, *Salmonella* could be detected in both extensive and intensive shrimp culture systems; isolates of *S.* Weltevreden from these systems were clonal in nature. They suggested that S. Weltevreden could survive and possibly even multiply in shrimp culture systems. Studies by Hounmanou et al. (2020) indicated S. Weltvreden as the most common *Salmonella* associated with shrimp culture systems in China, and the isolates were genetically related. Although detection of *Salmonella* in aquaculture products has been reported, this has not led to any major public health problem.

Salmonella is often detected in shrimp and other seafood in markets. In the USA, there were 343 import refusals for fish and fishery products during 2007 due to detection of *Salmonella*. In Malaysia, 25% of raw prawns marketed were positive, the serovars found being S. Blockley, S. Weltevreden and S. Agona (Arumugaswamy et al., 1995); in India, 1% of 500 marketed prawn tested were positive, the serovars being S. Newport and S. Infantis (Prasad and Rao, 1995). Postharvest contamination could additionally contribute to such presence. *Salmonella* including serovar Weltevreden can form biofilms on food contact surfaces and resist sanitizer treatment in biofilms (Joseph et al., 2001). In a study of 353 imported seafood samples in Japan, 2/47 black tiger shrimp were positive, both with S. Weltevreden, and overall contamination levels were <30–40 MPN/100 g (Asai et al., 2008). Analysis of 11,312 imported and 768 domestic seafood samples in the USA during the period 1990–1998 revealed that 10% of imported and 2.8% of domestic raw seafood were positive for *Salmonella*, with an overall incidence of 7.2% and 1.3%, respectively (Heinitz et al., 2000). The most frequent serotypes in imported seafood were S. Weltevreden, S. Senftenberg, S. Lexington and S. Paratyphi B. These common serotypes were rarely (<0.5%) observed in human illnesses in the USA (Helfrick et al., 1997). S. Enteritidis ranked fifth and S. Typhimurium ranked 12th (Heinitz et al., 2000). S. Weltevreden was also the most common serotype isolated from imported food, including seafood, in the USA in 2000 (24/187), followed by S. Thompson (13/187), S. Lexington (12/187) and a number of other serotypes (Zhao et al., 2003). Although S. Weltevreden is rarely associated with human cases in the USA, it is frequently isolated from human cases in Thailand (Bangtrakulnonth et al., 2004).

Although *Salmonella* strains have been isolated from various fish and fishery products including shrimp, fish have been rarely implicated in human cases of salmonellosis. Only 5% of *Salmonella* illness in the USA between 1988 and 1992 was due to seafood (Bean et al., 1997). Out of 273 outbreaks of food-borne salmonellosis between 1980 and 1994 in New York, only four were due to seafood (Wallace et al., 1999). Greig and Ravel (2009) analysed the aetiology of 4093 food-borne outbreaks that occurred globally between 1998 and 2002, and noted that seafood accounted for 4.2% of cases ($n = 991$) due to S. Enteritidis, 4.8% of cases ($n = 270$) due to S. Typhimurium and 2.6% of cases ($n = 657$) due to other *S. enterica* serovars. In USA between 1998 and 2002, 6/488 (1.2%) of seafood-associated outbreaks were due to *Salmonella* (Lynch et al., 2006). In England and Wales, *Salmonella* accounted for 38/180 (21%) of fish- and shellfish-associated outbreaks, with the former accounting for 6% of *Salmonella* outbreaks and desserts accounting for 26%, poultry 25%, red meat 14% and eggs 13% (Hughes et al., 2007). S. Champaign was involved in an outbreak in Japan that affected 330 children who had consumed cuttlefish that had been left to thaw at room temperature for 30 hours and then boiled for a short period (Ogawa et al., 1991). Contaminated well water in a squid-processing plant in Japan was found to be the source of *Salmonella* that affected more than 400 people in 1999 and, in the same year, a cuttlefish snack contaminated with S. Chester was involved in an outbreak that affected more than 1500 people (D'Aoust and Maurer, 2007). S. Livingstone was the cause of an outbreak that occurred in Norway and Sweden in 2001, in which fish gratin manufactured in Sweden was implicated; the egg powder ingredient was suspected to be the source (D'Aoust and Maurer, 2007). One outbreak in which 16 people became ill after a reception in a hotel in the UK in 1981 was attributed to frozen prawns (PHLSCDSC and FHCRL, 1983). Although the implicated food had not been tested, only those who ate prawns were affected and *S.* Bareilly and *S.* Hindmarsh was isolated from those affected. It is not clear whether the prawns were prepared with any other ingredients, which could have been a source of *Salmonella*.

Human food trials indicate that the infectious dose of *Salmonella* is $>10^4$ cells (FAO/WHO, 2002), but outbreak investigations suggest that a low dose (1–10 cells) can cause illness when associated with high-fat foods such as chocolate, cheese and meat (D'Aoust and Maurer, 2007). This has been attributed to protection against gastric acidity afforded to *Salmonella* entrapped in hydrophobic lipid miscelles. However, no data are available for shrimp or seafood, but shrimp do not contain high fat and a low infectious dose is unlikely. Therefore, multiplication of *Salmonella* to an infectious dose would be required for shrimp to be incriminated in illness. *Salmonella* is a mesophilic organism whose growth rate is markedly reduced at temperatures <15 °C, while the growth of most strains is prevented at <7 °C (ICMSF, 1996). Most studies on minimum growth temperature have been done with beef, chicken or eggs using serovars such as Typhimurium or Enteritidis, common in these foods, but these serotypes are not common in seafoods. In raw seafoods containing a variety of bacteria, *Salmonella*, if present has to compete with other flora for growth. *S.* Heidelberg had a generation time of 28 and 31 hours in English sole and sterile crab, respectively, at 8 °C (ICMSF, 1996). In cooked crab inoculated with *Salmonella* and stored at 8–11 °C under modified atmospheres containing low levels of CO_2 (20–50%), proliferation of *Salmonella* has been reported (Ingham et al., 1990). Growth of *Salmonella* is generally inhibited at 3–4% NaCl, but salt tolerance increases with increasing temperature in the range 10–30 °C (D'Aoust and Maurer, 2007) and minimum water activity for growth is 0.94% (ICMSF, 1996). Although the resistance of *Salmonella* to drying varies, this organism may survive for months or even years in dried products and has frequently been isolated from fish meal, meat and bone meal, maize and soy products (Lunestad et al., 2007). Decrease in *Salmonella* numbers occurs during freezing and frozen storage, but this process does not guarantee elimination of salmonellae in foods (ICMSF, 1996). *Salmonella* is heat sensitive, and D-values are influenced by water activity, nature of solutes and pH of the suspending medium (ICMSF, 1996). Typical D-values reported for *Salmonella* are 0.176 minutes in chicken at 70 °C and 0.36 minutes in ground beef at 63 °C (FAO/WHO, 2002). Some strains of *Salmonella*, including *S.* Senftenberg 775W, may show higher heat resistance (ICMSF, 1996). Interestingly, *S.* Senftenberg is a serovar often isolated from fish feed (Lunestad et al., 2007).

Both conventional and molecular methods have been applied in the detection of *Salmonella* in various seafoods, including shrimp. The US FDA BAM culture procedure involves non-selective enrichment, selective enrichment in two broths, isolation on three selective agars and identification of typical isolates using biochemical and serological methods (Andrews et al., 2021). Commercial assay kits based on sandwich hybridization and enzyme-mediated colorimetric detection, as well as PCR, are available. Some of these have undergone AOAC evaluation and been adopted as official methods – e.g. hybridization-based colorimetric screening test AOAC Official Method 990.13, and the PCR-based commercial BAX assay kit adopted as AOAC Official Method 2003.9 for various foods including fish (Mozola, 2006).

Strategies for risk management

Present food safety management systems are based on prevention of contamination at source and the entire food chain, from primary production to consumption, has to be taken into consideration. Therefore, to ensure the safety of aquacultured shrimp, it is necessary to take steps to minimize contamination. Adequate treatment of sewage before discharge into water bodies that may serve as source of water for aquaculture is very important. WHO has recommended a microbial water quality performance target of $\leq 10^4$ *E. coli*/100 ml (WHO, 2006). Faecal coliform/*E. coli* criteria have been successfully used for improvement in the microbial safety of live bivalve molluscs, and these would also be useful in the control of faecal pollution of shrimp aquaculture systems. Control of faecal pollution in aquaculture would be a very important risk management measure for choleragenic *V. cholerae*. Adoption of hygienic practices during postharvest handling and processing would be key, as evidenced by the detection of choleragenic *V. cholerae* in seafood from domestic markets in some developing countries and its absence in shrimp intended for export, which are handled using HACCP-based methods. Table 26.2 summarizes the source and impact of handling on pathogens associated with aquaculture.

Vibrio parahaemolyticus is an autochthonous aquatic organism that cannot be controlled through management of water quality based on faecal coliforms/*E. coli*. The Scientific Committee on Veterinary Measures Relating to Public Health of the European Commission (European Commission, 2001) has opined that judging seafood based on the presence of *V. parahaemolyticus* without considering pathogenic strains should be discontinued. US FDA Guidance level (Compliance Programme 7303.842) for *V. parahaemolyticus* in ready-to-eat fishery products is 10^4/g (Kanagawa positive or negative). Levels of *V. parahaemolyticus* in shrimp at harvest would very rarely be at this level; considering the low numbers of *V. pahaemolyticus* in shrimp at harvest, prevention of multiplication of this organism by cooling shrimp immediately postharvest would be important. The Codex Code Practice for fish and fishery products (CAC/RCP 52-2003) recommends chilling and holding at a temperature as close to 0 °C as possible. The US Interstate Shellfish Sanitation Conference (ISSC) has accepted freezing combined with frozen storage as an acceptable means of postharvest treatment to control *V. parahaemolyticus* and *V. vulnificus*, which should be validated and HACCP compliant according to Code of Federal Regulation 21 CFR 123 (Drake et al., 2007). Cooking would greatly reduce hazard, but it is important to ensure that cooked shrimp is

TABLE 26.2 Summary of hazards associated with aquacultured shrimp (data from various publications cited in text).

Hazard	*Source*	*Growth in product required to cause disease*	*Impact of handling and processing on the hazard*
Vibrio parahaemolyticus	Autochthonous estuarine organism	Yes	Reductions in numbers at temperatures <5°C, during freezing, frozen storage, rapidly killed at temperatures >65°C
Vibrio cholerae	Human faecal contamination, long-term persistence in some areas	Yes	Level of hazard reduced during washing, icing, freezing and cooking
Salmonella	Sewage contamination, animal farm wastes, aquatic birds, marine mammals, reptiles, long-term persistence in aquatic environment	Yes	Growth prevented at <8°C, reduction in numbers during freezing, cooking

not re-contaminated, as happened during the Louisiana outbreak. Therefore, during postharvest handling and processing of shrimp, it is important to adopt HACCP-based measures.

Adequate treatment of sewage before discharge would reduce the risk of *Salmonella* contamination of aquaculture systems. In addition, attention needs to be paid to *Salmonella* derived from animal wastes, aquatic birds, storm water and other sources. The use of HACCP-based methods during postharvest handling and processing is essential to reducing the *Salmonella* hazard in shrimp. Chilling immediately after harvest and holding at a temperature as close to 0 °C as possible, as recommended in Codex Code of Practice for fish and fishery products, would reduce the likelihood of multiplication of any *Salmonella* that might be present as an environmental contaminant in shrimp. Freezing shrimp and cooking before consumption are steps that would reduce the hazard due to *Salmonella* (Table 26.2).

Conclusion

Crustaceans, including aquacultured shrimp, have a very good safety record, particularly in international trade. Very few microbial pathogens are associated with aquaculture systems or have been introduced as contaminants from sewage, storm water or animal wastes. Pathogens may also be introduced during handling and processing if HACCP-based management measures are not used. Therefore, management of shrimp safety requires proper attention to water quality and hygienic practices during primary production, and adoption of HACCP-based measures throughout the food chain.

References

Andrews, W. H., Wang, H., Jacobson, A., Zhang, B.G. and Hammack, T. S. (2021) *Salmonella. Bacteriological Analytical Manual Online.* Available at: https://www.fda.gov/food/laboratory-methods-food/bam-chapter-5-salmonella (accessed 2 September 2021).

Anonymous (2000) Ten-year summary of outbreaks due to *V. parahaemolyticus in* Japan. (1989–1999). The Ministry of Health and Welfare, Japan.

Arumugaswamy, R. K., Rusul, G., Hamid, S. A. and Cheah, C. T. (1995) Prevalence of *Salmonella* in raw and cooked foods in Malaysia. *Food Microbiology* 12, 3–8.

Asai, Y., Kaneko, M., Ohtsuka, K., Morita, Y., Kaneko, S., Noda, H., Furukawa, I., Takatori, K. and Hara-Kudo, Y. (2008) Salmonella prevalence in seafood imported into Japan. *Journal of Food Protection* 71, 1460–1464.

Bangtrakulnonth, A., Pornreongwong, S., Pulsrikarn, C., Sawanpanyalert, P., Hendriksen, R. S., Wong, D. M. L. F. and Aarestrup, F. M. (2004) *Salmonella* serovars from humans and other sources in Thailand, 1993–2002. *Emerging Infectious Diseases* 10, 131.

Barret, K. A., Nakao, J. H., Taylor, A. Y., Eggers, C. and Gould, L. H. (2017) Fish – associated foodborne disease outbreak: United States, 1998–2015. *Foodborne Pathogens and Disease* 14, 537–543.

Baudart, J., Lemarchand, K., Brisabois, A. and Lebaron, P. (2000) Diversity of *Salmonella* strains isolated from the aquatic environment as determined by serotyping and amplification of the ribosomal DNA spacer regions. *Applied and Environmental Microbiology* 66, 1544–1552.

Bean, N. H., Goulding, J. S., Lao, C. and Angulo, F. J. (1997) Surveillance for foodborne disease outbreaks – United States, 1988–1992 *Journal of Food Protection* 60, 1265–1286.

Blake, P. A., Allegra, D. T., Snyder, J. D., Barrett, T. J., McFarland, L., Caraway, C. T., Feeley, J. C., Craig, J. P., Lee, J. V., Puhr, N. D. and Feldman, R. A. (1980) Cholera – a possible endemic focus in the United States. *New England Journal of Medicine* 302, 305–309.

Brands, D. A., Inman, A. E., Gerba, C. P., Maré, C. J., Billington, S. J., Saif, L. A., Levine, J. F. and Joens, L. A. (2005) Prevalence of *Salmonella* spp in oysters in United States. *Applied and Environmental Microbiology* 71, 893–897.

Brenner, F. W., Villar, R. G., Angulo, F. J., Tauxe, R. and Swaminathan, B. (2000) *Salmonella* nomenclature. *Journal of Clinical Microbiology* 38, 2465–2467.

Byappanahalli, M. N., Sawdey, R., Ishii, S., Shively, D. A., Ferguson, J. A., Whitman, R. L. and Sadowsky, M. J. (2009) Seasonal stability of *Cladophora*-associated *Salmonella* in Lake Michigan watersheds *Water Research* 43, 806–814.

Cann, D. C., Taylor, L. Y. and Merican, Z. (1981) A study of the incidence of *Vibrio parahaemolyticus* in Malaysian shrimp undergoing processing for export *Journal of Hygiene (London)* 87, 485–491.

Chakraborty, S., Mukhopadhyay, A. K., Bhadra, R. K., Ghosh, A. N., Mitra, R., Shimada, T., Yamasaki, S., Faruque, S. M., Takeda, Y., Colwell, R. R. and Nair, G. B. (2000) Virulence genes in environmental strains of *Vibrio cholerae*. *Applied and Environmental Microbiology* 66, 4022–4028.

Chan C.H., Woo, M.L., Lam, L.Y. and French, G.L. (1989) *Vibrio parahaemolyticus* and other halophilic vibrios associated with seafood in Hong Kong. *Journal of Applied Bacteriology* 66, 57-64.

Chen, C. H., Shimada, T., Elhadi, N., Radu, S. and Nishibuchi, M. (2004) Phenotypic and genotypic characteristics and epidemiological significance of ctx+ strains of *Vibrio cholerae* isolated from seafood in Malaysia. *Applied and Environmental Microbiology* 70, 1964–1972.

Chowdhury, A., Mori, K, Nakano, Y., Ishibashi, M. and Nishibuchi, M. (2001) Study on the contamination of the imported seafood by *Vibrio parahaemolyticus* using genetic methods *Japanese Journal of the Society for Bacteriology* 56, 323–325 [in Japanese].

Colwell, R. R., West, P. A., Maneval, D., Remmers, E. F., Elliot, E. L. and Carlson, N. E. (1984) Ecology of pathogenic vibrios in Chesapeake Bay. In: *Vibrios in the Environment* (ed. Colwell, R. R.). John Wiley and Sons, New York, USA.

Dalsgaard, A., Huss, H. H., Kithikun, A. and Larsen, J. L. (1995a) Prevalence of *Vibrio cholerae* and *Salmonella* in a major shrimp production area in Thailand. *International Journal of Food Microbiology* 28, 101–113.

Dalsgaard, A., Serichantalergs, O., Shimada, T., Sethabuthr, O. and Echeverria, P. (1995b) Prevalence of *Vibrio cholerae* with heat-stable enterotoxin (NAG-ST) and cholera toxin genes: restriction fragment length polymorphisms (RFLP) of NAG-ST genes among *V. cholerae* O serogroups from a major shrimp production area in Thailand. *Journal of Medical Microbiology* 43, 216–220.

Dalsgaard, A., Serichantalergs, O., Forslund, A., Lin, W., Mekalanos. J., Mintz, E., Shimada, T. and Wells, J. G. (2001) Clinical and environmental isolates of *Vibrio cholerae* serogroups O141 carry the CTX phage and the genes encoding the toxin-coregulated pili. *Journal of Clinical Microbiology* 39, 4086–4092.

Daniels, N. A., MacKinnon, L., Bishop, R., Altekruse, S., Ray, B., Hammond, R. M., Thompson, S., Wilson, S., Bean, N. H., Griffin, P. M. and Slutsker, L. (2000) Vibrio *parahaemolyticus* infections in United States, 1973-1998 *Journal of Infectious Diseases* 181, 1661–1666.

D'Aoust, J. Y. and Maurer, J. (2007) *Salmonella* species. In: *Food Microbiology: Fundamentals and Frontiers* (eds Doyle, M. P. and Beuchat, L.R.) ASM Press, Washington, DC, USA, pp. 187–236.

Deepanjali, A., Kumar, H. S. and Karunasagar, I. (2005) Seasonal variation in abundance of total and pathogenic *Vibrio parahaemolyticus* bacteria in oysters along the Southwest coast of India. *Applied and Environmental Microbiology* 71, 3575–3580.

DePaola, A. (1981) *Vibrio cholerae* in marine foods and environmental waters: A literature review. *Journal of Food Science* 46, 66–70.

DePaola, A., Rivadeneyra, C., Gelli, D. S., Zuazua, H. and Grahn, M. (1993) Peruvian cholera epidemic: Role of seafood. *Proceedings of the 16th Annual Tropical and Subtropical Fisheries Technology Conference of the Americas*, Florida Sea Grant College Programme, Gainsville, FL 32611, pp. 28–33.

DePaola, A., Nordstrom, J. L., Bowers, J. C., Wells, J. G. and Cook, D. W. (2003) Seasonal abundance of total and pathogenic *Vibrio parahaemolyticus* in Alabama oysters. *Applied and Environmental Microbiology* 69, 1521–1526.

Dewaal, C. S., Hicks, G., Barlow, K., Alderton, L. and Vegosen, L. (2006) Foods associated with foodborne illness outbreaks from 1990 through 2003. *Food Protection Trends* 26, 466–473.

Drake, S. L., DePaola, A. and Jaykus, L. A. (2007) An overview of *Vibrio vulnificus* and *Vibrio parahaemolyticus*. *Comprehensive Reviews in Food Science and Food Safety* 6, 120–144.

Eberhart-Phillips, J., Bessser, R. E., Tormey, M. P., Feikin, D., Araneta, M. R., Wells, J., Kilman, L., Rutherford, G. W., Griffin, P. M., Baron, R. and Mascola, L. (1996) An outbreak of cholera from food served on an international aircraft. *Epidemiology and Infection* 116, 9–13.

EFSA and ECDC (2019) The European Union One Health zoonoses report. *EFSA Journal* 17(12), 5926.

Ellison, R. K., Malnati, E., DePaola, A., Bowers, J. and Rodrick, G. E. (2001) Populations of *Vibrio parahaemolyticus* in retail oysters from Florida using two methods. *Journal of Food Protection* 64, 682–686.

European Commission (2001) Opinion of the Scientific Committee on veterinary measures relating to public health on *Vibrio vulnificus* and *Vibrio parahaemolyticus*. Available at: http://ec.europa.eu/food/fs/sc/scv/out45_en.pdf (accessed 30 March 2009).

FAO (2010) The Report of the FAO Expert Workshop on the application of biosecurity measures to control *Salmonella* contamination in sustainable aquaculture. *FAO Fisheries and Aquaculture Report* 937, FAO, Rome, Italy, 30 pp.

FAO/WHO (2002) Risk assessment of *Salmonella* in eggs and broiler chicken. *Microbiological Risk Assessment Series* 2, 1–329.

FAO/WHO (2003) Risk assessment of *Campylobacter* spp. in broiler chickens and *Vibrio* spp. in seafood. *FAO Food and Nutrition Paper* 75, FAO, Rome, Italy, pp. 1–57.

FAO/WHO (2005) Risk assessment of choleragenic *Vibrio cholerae* O1 and O139 in warm water shrimp in international trade. *Microbiological Risk Assessment Series* 9, 1–90.

FAO/WHO (2011) Risk assessment of *Vibrio parahaemolyticus* in seafood. *Microbiological Risk Assessment Series* 16, 111–183.

FAO/WHO (2016) Selection and application of methods for the detection and enumeration of human pathogenic *Vibrio* spp in seafood. *Microbiological Risk Assessment Series* 22, xi–74.

FAO/WHO (2020) Risk assessment tools for *Vibrio parahaemolyticus* and *Vibrio vulnificus* associated with seafood. *Microbiological Risk Assessment Series* 20, ix–69.

Faruque, S. M., Albert, M. J. and Mekalanos, J. J. (1998) Epidemiology, genetics, and ecology of choleragenic *Vibrio cholerae*. *Microbiology and Molecular Biology Reviews* 62, 1301–1314.

Faruque, S. M., Sack, D. A., Sack, R. B., Colwell, R. R., Takeda, Y. and Nair, G. B. (2003) *Proceedings of the National Academy of Sciences USA* 100, 1304–1309.

Fonseka, T. S. G. (1990) Microbial flora of pond cultured prawn (*Penaeus monodon*). *FAO Fisheries Report* 401, 24–31.

Food and Drug Administration (FDA) (2005) Quantitative risk assessment on the public health impact of pathogenic *Vibrio parahaemolyticus* in raw oysters. Available at: https://www.fda.gov/food/cfsan-risk-safety-assessments/quantitative-risk-assessment-public-health-impact-pathogenic-vibrio-parahaemolyticus-raw-oysters (accessed 02 September 2021.

Gaertner, J., Wheeler, P. E., Obafemi, S., Valdez, J., Forstner, M. R., Bonner, T. H. and Hahn, D. (2008) Detection of *Salmonella* in fish in a natural river system. *Journal of Aquatic Animal Health* 20 150–157.

Garay, E., Arnau, A. and Amaro, C. (1985) Incidence of *Vibrio cholerae* and related *Vibrios* in a coastal lagoon and seawater influenced by lake discharges along an annual cycle. *Applied and Environmental Microbiology* 50, 426–430.

Gonzalez, C. J., Lopez-Diaz, T. M., Garcia-Lopez, M. L., Prieto, M. and Otero, A. (1999) Bacterial microflora of wild brown trout (*Salmo trutta*) wild pike (*Esox lucius*) and aquacultured rainbow trout (*Oncorhyncus mykiss*). *Journal of Food Protection* 62, 1270–1277.

Gooch, J. A., DePaola, A., Kaysner, C. A. and Marshall, D. L. (2001) Evaluation of two direct plating methods using nonradioactive probes for enumeration of *Vibrio parahaemolyticus* in oysters. *Applied and Environmental Microbiology* 67, 721–724.

Gopal, S., Otta, S. K., Kumar, S., Karunasagar, I., Nishibuchi, M. and Karunasagar, I. (2005) The occurrence of *Vibrio* spp in tropical aquaculture environments: Implications for food safety. *International Journal of Food Microbiology* 102, 151–159.

Greig, J. D. and Ravel, A. (2009) Analysis of foodborne outbreak data reported internationally for source attribution. *International Journal of Food Microbiology* 130, 77–87.

Heinitz, M. L., Ruble, R. D., Wagner, D. E. and Tatini, S. R. (2000) Incidence of *Salmonella* in fish and seafood. *Journal of Food Protection* 63, 579–592.

Helfrick, D., Bean, N. H., Slutsker, L. and Tauxe, R.V. (1997) *Annual Tabulation Summary 1997. Salmonella Surveillance.* Centres for Disease Control and Prevention, Atlanta, GA, USA.

Higgins, R. (2000) Bacteria and fungi of marine mammals: A review. *Canadian Veterinary Journal* 45, 101–106.

Hondo, S., Goto, I., Minematsu, I., Ikeda, N., Asano, N., Ishibashi, M., Kinoshita, Y., Nishibuchi, N., Honda, T. and Miwatani, T. (1987) Gastroenteritis due to Kanagawa negative *V. parahemolyticu.s Lancet* I, 331–332.

Hounmanou, Y. M., Dalsgaard, A., Sopacua, T. F., Uddin, G. M., Leekitcharoenphon, P., Hendriksen, R. S., Olsen, J. E. and Larsen, M. H. (2020) Molecular characteristics and zoonotic potential of *Salmonella* Weltevreden from cultured shrimp and tilapia in Vietnam and China. *Frontiers in Microbiology* 11, 1985.

Hughes, C., Gillespie, I. A. and O'Brien, S. J. (2007) Foodborne transmission of intestinal infectious disease in England and Wales, 1992–2003. *Food Control* 18, 766–772.

ICMSF (1996) *Microorganisms in Foods 5. Characteristics of Microbial Pathogens* Blackie Academic & Professional, London, UK, pp. 1–513.

Ingham, S. C., Alford, R. A. and McCown, A. P. (1990) Comparative growth rates of *Salmonella typhimurium* and *Pseudomonas fragi* on cooked crab meats stored under air and modified atmosphere. *Journal of Food Protection* 53, 566–567.

Joseph, B., Otta, S. K., Karunasagar, I. and Karunasagar, I. (2001) Biofilm formation by *Salmonella* spp on food contact surfaces and their sensitivity to sanitizers. *International Journal of Food Microbiology* 64, 367–372.

Joseph, S. W., Colwell, R. R. and Kaper, J. B. (1982) *Vibrio parahaemolyticus* and related halophilic vibrios. *CRC Critical Reviews in Microbiology* 10, 77–124.

Kaneko, T. and Colwell, R. R. (1975) Absorption of *Vibrio parahaemolyticus* onto chitin and copepods. *Applied Microbiology* 29, 269–274.

Kaneko, T. and Colwell, R. R. (1977) The annual cycle of *Vibrio parahaemolyticus* in Chesapeake Bay. *Microbial Ecology* 4, 135–155.

Kaper, J. B., Morris, J. G. and Levine, M. M. (1995) Cholera. *Clinical Microbiology Reviews* 8, 48–86.

Karaolis, D. K., Johnson, J. A., Bailey, C. C., Boedeker, E. C., Kaper, J. B. and Reeves, P. R. (1998) A *Vibrio cholerae* pathogenicity island associated with epidemic and pandemic strains. *Proceedings of the National Academy of Sciences USA* 95, 3134–3139.

Karunasagar, I. and Karunasagar, I. (2018) Ecology, virulence factors and global spread of *Vibrio parahaemolyticus. Asian Fisheries Society* 31, 15–28.

Karunasagar, I., Venugopal, M. N. and Karunasagar, I. (1984) Levels of *Vibrio parahaemolyticus* in Indian shrimp undergoing processing for export. *Canadian Journal of Microbiology* 30, 713–715.

Karunasagar, I., Susheela, M., Malathi, G. R. and Karunasagar, I. (1990) Incidence of human pathogenic vibrios in seafoods harvested along coast of Karnataka (India). FAO Fisheries Report 401 (Suppl.), FAO, Rome, Italy, pp. 53–56.

Karunasagar, I., Ismail, S. M. and Amarnath, H. V. (1992) Bacteriology of tropical shrimp and marine sediments. FAO Fisheries Report 470 (Suppl.), FAO, Rome, Italy, pp. 1–8.

Karunasagar, I., Sugumar, G. and Reilley, A. (1995) Rapid detection of *Vibrio cholerae* contamination of seafood by polymerase chain reaction. *Molecular Marine Biology and Biotechnology* 4, 365–368.

Karunasagar, I., Sugumar, G., Karunasagar, I. and Reilly, P. J. A. (1996) Rapid polymerase chain reaction method for detection of Kanagawa positive *Vibrio parahaemolyticus* in seafoods. *International Journal of Food Microbiology* 31, 317–323.

Katoh, H. (1965) Studies on growth rate of various food bacteria. I. On the generation time of *Vibrio parahaemolyticus* Fujino *Japanese Journal of Bacteriology* 20, 94–100.

Kaysner, C. A., DePaola, A. and Jones, J. (2019) Vibrio. *Bacteriological Analytical Manual Online* Available at: https://www.fda.gov/food/laboratory-methods-food/bam-chapter-9-vibrio (accessed 02 September 2021).

Kim, J. H., Lee, J., Hong, S., Lee, S., Na, H. Y., Jeong, Y. I., Choi, E. J., Kim, J., Kawk, H. S. and Cho, E. (2018) Cholera outbreak due to raw seafood consumption in South Korea. *American Journal of Tropical Medicine and Hygiene* 99, 168–170.

Kolvin, J.L., and Roberts, D. (1982) Studies on the growth of *Vibrio cholerae* biotype Eltor and Classical in foods. *Journal of Hygiene* Cambridge 89, 243-252.

Koonse, B., Burkhardt III, W., Chirtel, S. and Hoskin, G. P. (2005) *Salmonella* and the sanitary quality of aquacultured shrimp. *Journal of Food Protection* 68, 2527–2532.

Li, Y., Xie, T., Pang, R., Wu, Q., Zhang, J., Lei, T., Xue, L., Wu, H., Wang, J., Ding, Y. and Chen, M. (2020) Food-Borne *Vibrio parahaemolyticus* in China: Prevalence, antibiotic susceptibility, and genetic characterization. *Frontiers in Microbiology* 11, 1670.

Liston, J. (1974) Influence of U.S. seafood handling procedures on *Vibrio parahaemolyticus*. In: *International Symposium on Vibrio parahaemolyticus* (eds Fujino, T. et al.). Saikon Publishing Company, Tokyo, Japan, pp. 123–128.

Lunestad, B. T., Nesse, L., Lassen, J., Svihus, B., Nesbakken, T., Fossum, K., Rosnes, J. T., Kruse, H. and Yazdankhah, S. (2007) *Salmonella* in fish feed; occurrence and implications for fish and human health in Norway. *Aquaculture* 265, 1–8.

Lynch, M., Painter, J., Woodruff, R. and Braden, C. (2006) Surveillance of foodborne-disease outbreaks – United States, 1998–2002. *Morbidity and Mortality Weekly Reports* 55, 1–34.

Martinez-Urtaza, J., Saco, M., de Novoa, J., Perez-Piñeiro, P., Peiteado, J., Lozano-Leon, A. and Garcia-Martin, O. (2004a) Influence of environmental factors and human activity on the presence of *Salmonella* serovars in the marine environment. *Applied and Environmental Microbiology* 70, 2089–2097.

Martinez-Urtaza, J., Peiteado, J., Lozano-Leon, A. and Garcia-Martin, O. (2004b) Detection of *Salmonella* Seftenberg associated with high saline environments in mussel processing facilities. *Journal of Food Protection* 67, 256–263.

McCarthy, S. A., DePaola, A., Cook, D. W., Kaysner, C. A. and Hill, W. E. (1999) Evaluation of alkaline phosphatase- and digoxigenin-labeled probes for detection of the thermolabile hemolysin (*tlh*) gene of *Vibrio parahaemolyticus*. *Letters in Applied Microbiology* 28, 66–70.

McCarthy, S. A., DePaola, A., Kaysner, C. A., Hill, W. E. and Cook, D. W. (2000) Evaluation of non-isotopic DNA hybridization methods for detection of the *tdh* gene of *Vibrio parahaemolyticus*. *Journal of Food Protection* 63, 1660–1664.

Mozola, M. A. (2006) Genetics-based methods for detection of *Salmonella* spp. in foods. *Journal of AOAC International* 89, 517–529.

NACA (2012) Report of the Asia Pacific emergency regional consultation on the emerging shrimp disease: Early mortality syndrome (EMS)/acute hepatopancreatic necrosis syndrome (AHPNS), 9–10 August 2012. Network of Aquaculture Centres in Asia-Pacific, Bangkok, Thailand.

Nair, G. B., Ramamurthy, T., Bhattacharya, S. K., Dutta, B., Takeda, Y. and Sack, D. A. (2007) Global dissemination of *Vibrio parahaemolyticus* O3:K6 and its serovariants. *Clinical Microbiology Reviews* 20, 39–48.

Natarajan, R., Abraham, M. and Nair, G. B. (1980) Distribution of *Vibrio parahaemolyticus* in Porto Novo environment. *Indian Journal of Medical Research* 71, 679–687.

Nesse, L. L., Nordby, K., Heir, E., Bergsjoe, B., Vardund, T., Nygaard, H. and Holstad, G. (2003) Molecular analysis of *Salmonella enterica* isolates from fish feed factories and fish feed ingredients. *Applied and Environmental Microbiology* 69, 1075–1081.

Noor Uddin, G. M., Larsen, M. H., Barco, L., Minh Phu, T. and Dalsgaard, A. (2015) Clonal occurrence of *Salmonella* Weltevreden in cultured shrimp in the Mekong Delta, Vietnam. *PloS ONE* 10, 0134252.

Nordstrom, J. L., Rangdale, R., Vickery, M. C., Phillips, A. M., Murray, S. L., Wagley, S. and DePaola, A. (2006) Evaluation of an alkaline phosphatase-labelled oligonucleotide probe for detection and enumeration of thermostable direct hemolysin related (TRH) gene of *Vibrio parahaemolyticus*. *Journal of Food Protection* 69, 2770–2772.

Ogawa, H., Tokunou, H., Sasaki, M., Kishimoto, T. and Tamura, K. (1991) An outbreak of bacterial food poisoning caused by roast cuttlefish "Yaki-ika" contaminated with *Salmonella* spp Champaign. *Japanese Journal of Food Microbiology* 7, 151–157.

Oliver, J. D. and Kaper, J. B. (2007) Vibrio spp. In: *Food Microbiology : Fundamentals and Frontiers* 3rd edn (eds Doyle, M. P. and Beuchat, L. R.). ASM Press, Washington, DC, USA, pp. 343–379.

Otta, S. K. and Karunasagar, I. (1999) Bacterial flora associated with shrimp culture ponds growing *Penaeus monodon* in India. *Journal of Aquaculture in the Tropics* 14, 309–318.

OzFoodNet (2006) OzFoodNet Annual Reports. Available at: https://www1.health.gov.au/internet/main/publishing.nsf/Content/cda-cdi3104a.htm (accessed 02 September 2021.

PHLSCDSC and FHCRL (1983) Illness associated with fish and shellfish in England and Wales, 1981–82. *British Medical Journal* 287, 1284–1285.

Prasad, M. M. and Rao, C. C. P. (1995) Occurrence of *Salmonella* infantis and *S.* newport in market prawns. *Journal of Food Science and Technology* 32, 135–137.

Raghunath, P., Pradeep, B., Karunasagar, I. and Karunasagar, I. (2007) Rapid detection and enumeration of trh-carrying *V. parahaemolyticus* with an alkaline phosphatase-labelled oligonucleotide probe *Environmental Microbiology* 9, 266–270.

Raghunath, P., Acharya, S., Bhanumathi, A., Karunasagar, I. and Karunasagar, I. (2008) Detection and molecular characterization of *Vibrio parahaemolyticus* isolated from seafood harvested along southwest coast of India. *Food Microbiology* 25, 824–830.

Rattagool, P., Wongchinda, N., Methatip, P. and Sanghtong, N. (1990) Hygienic processing of shrimp in Thailand. *FAO Fisheries Report* 401 (Suppl.), 32–46.

Ravi Kiran, K. (1992) Potential pathogens and indicator bacteria associated with prawn gut. MFSc thesis, University of Agricultural Sciences, College of Fisheries, Mangalore.

Reilly, A. and Twiddy, D. R. (1992) *Salmonella* and *Vibrio cholerae* in brackish water cultured tropical prawns. *International Journal of Food Microbiology* 16, 293–301.

Rivera, I. N., Chun, J., Huq, A., Sack, R. B. and Colwell, R. R. (2001) Genotypes associated with virulence in environmental isolates of *Vibrio cholerae*. *Applied and Environmental Microbiology* 67, 2421–2429.

Ross, T. and Sumner, J. (2002) A simple, spreadsheet-based, risk assessment tool. *International Journal of Food Microbiology* 77, 39–53.

Sack, R. B., Siddique, A. K., Longini, I. M., Nizam, A., Yunus, M., Islam, M. S., Morris, J. G., Ali, A., Huq, A., Nair, G. B., Qadri, F., Faruque, S. M., Sack, D. A. and Colwell, R. R. (2003) A 4-year study of the epidemiology of *Vibrio cholerae* in four rural areas of Bangladesh. *Journal of Infectious Diseases* 187, 96–101.

Saheki, K., Kobayashi, S. and Kawanishi, T. (1989) *Salmonella* contamination of eel culture ponds. *Nippon Suisan Gakkashi* 55, 675–679.

Sanyal, S. C. and Sen, P. C. (1974) Human volunteer study on pathogenicity of *Vibrio parahaemolyticus*. In: International Symposium on *Vibrio parahaemolyticus* (eds Fujino, T. et al.). Saikon Publishing Co., Tokyo, Japan, pp. 227–230.

Saravanan, V., Kumar, H. S., Karunasagar, I. and Karunasagar, I. (2007) Putative virulence genes of *Vibrio cholerae* from seafoods and coastal environments of Southwest India. *International Journal of Food Microbiology* 119, 329–333.

Stoddard, R. A., Gulland, F. M., Atwill, E. R., Lawrence, J., Jang, S. and Conrad, P. A. (2005) *Salmonella* and *Campylobacter* spp. in northern elephant seals. *Emerging Infectious Diseases* 11, 1967–1969.

Tindall, B. J., Grimont, P. A. D., Garrity, G. M. and Euzeby, J. P. (2005) Nomenclature and taxonomy of the genus *Salmonella*. *International Journal of Systematic Bacteriology* 55, 521–524.

Tran, L., Nunan, L., Redman, R. M., Mohney, L. L., Pantoja, C. R., Fitzsimmons, K. and Lightner, D. V. (2013) Determination of the infectious nature of the agent of acute hepatopancreatic necrosis syndrome affecting penaeid shrimp. *Diseases of Aquatic Organisms* 105, 45–55.

Tyagi, A., Saravanan, V., Karunasagar, I. and Karunasagar, I. (2009) Detection of *Vibrio parahaemolyticus* in tropical shellfish by SYBR green real-time PCR and evaluation of three enrichment media. *International Journal of Food Microbiology* 129, 124–130.

Venkateswaran, K., Kiiyukia, C., Nakanishi, K., Nakano, H., Matsuda, O. and Hashimoto, H. (1990) The role of sinking particles in the overwintering process of *Vibrio parahaemolyticus* in a marine environment. *FEMS Microbiology Ecology* 73, 159–166.

Wachsmuth, K., Blake, P. A. and Olsvik, Ø. (1994) Molecular epidemiology of cholera. In: *Vibrio cholerae and Cholera: Molecular to Global Perspectives* (eds Wachsmuth, I. K. et al.). ASM Press, Washington, DC, USA, pp. 357–370.

Wallace, B. J., Guzewich, J. J., Cambridge, M., Altekruse, S. and Morse, D. L. (1999) Seafood-associated disease outbreaks in New York, 1980–1994. *American Journal of Preventive Medicine* 17, 48–54.

Ward, L. N. and Bej, A. K. (2006) Detection of *Vibrio parahaemolyticus* in shellfish by use of multiplexed real-time with TaqMan fluorescent probes. *Applied and Environmental Microbiology* 72, 2031–2042.

WHO (2000) WHO surveillance programme for control of foodborne infections and intoxications in Europe. 7th Report. Available at: http://www.euro.who.int/foodsafety/Surveillance/20031127_1 (accessed 8 August 2021).

WHO (2006) Guidelines for safe use of waste water, excreta and gray water, Vol 3. Waste water and excreta use in aquaculture. 162 pp Available at: http://whqlibdoc.who.int/publications/2006/9241546840_eng.pdf (accessed 8 August 2021).

Winfield, M. D. and Groisman, E. A. (2003) Role of non-host environments in the lifestyles of *Salmonella* and *Escherichia coli*. *Applied and Environmental Microbiology* 69, 3687–3694.

Wyatt, L. E., Nickelson, R. and Vanderzant, C. (1979) Occurrence and control of *Salmonella* in freshwater catfish. *Journal of Food Science* 44, 1067–1073.

Yang, Y. T., Chen, I. T., Lee, C. T., Chen, C. Y., Lin, S. S., Hor, L. I., Tseng, T. C., Huang, Y. T., Sritunyalucksana, K., Thitamadee, S. and Wang, H. C. (2014) Draft genome sequences of four strains of *Vibrio parahaemolyticus*, three of which cause early mortality syndrome/acute hepatopancreatic necrosis disease in shrimp in China and Thailand. *Genome Announcements* 2(5).

Zhao, S., Datta, A. R., Ayers, S., Friedman, S., Walker, R. D. and White, D. G. (2003) Antibiotic-resistant *Salmonella* serovars isolated from imported foods. *International Journal of Food Microbiology* 84, 87–92.

Chapter 27

IMPORT NOTIFICATIONS OF SHRIMP

ESTHER GARRIDO GAMARRO AND GUILIA LOI

Summary

Shrimp is one of the most internationally traded aquatic products. Exporters often face difficulties in meeting various food safety requirements and standards imposed by different importing countries, including regulations, control procedures and control at borders, where shrimp can be detained, rejected or destroyed, which has an economic impact for exporting countries. This chapter analyses the causes of import notifications by category (chemical, microbiological, histamine, toxins and parasites among others) and explains the import control procedures of Canada, the European Union (EU), Japan and the United States of America (USA). The data used for the development of this chapter have been obtained from various datasets extracted from online portals in which each of the countries makes publicly available data on import notifications. This study evidences major differences among the prevalence of the various reasons behind import notifications in the targeted countries, both in terms of number of import notifications and the typology of the cause. In Canada, for instance, the main reported cause of import notification was due to labelling problems, while in the EU and in the USA veterinary drugs were the leading reported cause. In Japan, antibiotics were the main reported issue at borders. This document throws light on the need to harmonize regulatory requirements and import control procedures and to base actions on food safety risk assessment. Developing a good understanding of import procedures and the causes of import notifications can be very useful for both exporting and importing countries.

Abbreviations

- AIRS: Automated import reference system
- AMR: Antimicrobial resistance
- ASP: Amnesic shellfish poisoning
- CFIA: Canadian Food Inspection Agency

- CFP: Ciguatera fish poisoning
- CFSAN: Centre for Food Safety and Applied Nutrition
- DSP: Diarrhoetic shellfish poisoning
- EFSA: European Food Safety Authority
- FAO: Food and Agriculture Organization
- FDA: Food and Drug Administration
- FIR: Food information regulation
- FSIS: Food Safety Inspection Service
- FSMA: Food Safety Modernization Act
- FSVP: Foreign Supplier Verification Programme
- GDR: Guidance Document Repository
- GMP: Good manufacturing practice
- HACCP: Hazard analysis critical control point
- HS: Harmonized system
- IRR: Import refusal report
- MHLW: Ministry of Health, Labour and Welfare
- NSP: Neurotoxic shellfish poisoning
- OASIS: Operational and administrative import support
- PSP: Paralytic shellfish poisoning
- RASFF: Rapid alert system for food and feed
- TBT: Technical barriers to trade
- USDA: United States Department of Agriculture
- WTO: World Trade Organization

Trade volume and import notifications

Shrimp is one of the most important internationally traded fishery commodities and, in some developing countries such as Argentina, China, Ecuador, India, Indonesia and Thailand, shrimp is the most valuable fishery export. According to the FAO, the main shrimp-importing countries by value (US$ × 1000) in 2017 were the USA (6,556,944), Japan (2,555,460) and Spain (1,269,690). In Canada and in the USA, import notifications of this important commodity represent the 11% of total import notifications for fishery and aquaculture products, whereas in the EE shrimp import notifications comprise the 9% of total rejections for fishery and aquaculture products from 2016 to 2019. In Japan, the percentage is higher, representing 28% of total import notifications for fishery and aquaculture products from 2016 to 2019.

The major producing countries, by tonnes of live weight in 2017, were China (3,069,832), Indonesia (1,124,294) and India (1,111,122). The major exporting countries by value (US$ x 1000) in 2017 were India (4,962,291), Viet Nam (3,792,245) and Ecuador (3,059,195). The species most frequently produced in 2017 were whiteleg shrimp, giant tiger, natantian decapod *nei*, akiami paste shrimp, southern rough shrimp and argentine red shrimp, with most of the notifications corresponding to giant tiger and argentine red shrimp. Most of the exported shrimp is frozen, and this has implications in the profile of notifications, net weight being an important cause of detentions and rejections.

Applicable regulations in Canada, EU, Japan and the USA

The substantial variation among the food safety and quality standards of importing countries constitutes one of the most severe difficulties for exporters in international markets. Despite the existence of the WTO Agreement on the Application of Sanitary and Phytosanitary Measures (SPS Agreement) and the WTO Agreement on TBT aiming to reduce the potential arbitrariness of decisions, differences between national standards and inspection systems continue to exist. For this reason, a good understanding of the procedures, requirements and standards related to food safety and quality is key for shrimp-exporting companies. In addition, traceability requirements to prevent illegal, unregulated and unreported fish from entering the market are of great relevance for certain commodities.

Shrimp can be produced from both marine fishing and aquaculture activities and, depending on the type of fishing or farming, the food safety risk can differ. The main food safety issue with farmed shrimp is the presence of residues of veterinary drugs, especially antimicrobials. These relate to finding residues of banned antimicrobials or residues exceeding their maximum permitted limits.

From the microbiological point of view, although there are differences in the detection of certain pathogens in shrimp produced from wild capture and aquaculture, GMP would prevent any possible contamination of farmed products. *Salmonella* is one of the most common pathogens (Hamilton et al., 2018) causing rejection. The continued reporting of the presence of *Salmonella* in shrimp indicates that the existing practices used by manufacturers or processors are insufficient to eliminate these pathogens at the postharvest stage.

REGULATIONS APPLICABLE IN CANADA

The Facilities Inspection Manual of Canada provides shrimp companies with guidance about establishment registration, quality management programmes and compliance guidelines. This manual provides CFIA inspectors and industry personnel with information on the procedures necessary to determine compliance with Fish Inspection Regulations. Another important guidance document for shrimp exporting companies is the Fish Products Inspection Manual – Policies and Procedures. This serves as a reference for inspection of imported fresh and frozen shrimp and sampling to be carried out at the border. For additional import requirements, shrimp-producing and -processing companies could consult Fish Products Standards and Methods Manual, where special requirements for shrimp can be found in chapter 2 – standard 3 (canned shrimp) and chapter 3 – standard 2 (fresh and frozen shrimp). This manual provides guidance on the documents used to determine good manufacturing practice and compliance, such as Codex Standards; the nomenclature for the product ('The name of the product shall be shrimp, shrimps or prawns'); the forms of product presentation; and type of sampling. In addition, the manual contains a description of defects that could affect shrimp including taint, decomposition, unwholesomeness and failure to meet a standard of identity.

The GDR contains the 'Guide to Additives Permitted in Fish and Fish Products', whereas the Fish Products Standards and Methods Manual in its Appendix 2 provides guidance on microbiological criteria. The majority of shrimp import notifications in Canada are due to labelling issues and, for this reason, it is important for shrimp companies to consult 'Labelling Requirements for Fish and Fish Products' contained within the GDR.

Regulations on residues of veterinary drugs

The Fish Products Standards and Methods Manual provides regulatory information on the use of residues of veterinary drugs in aquaculture, their maximum residue limits and a list of non-permitted veterinary drugs for aquaculture products. Table 27.1 shows the criteria for residues in crustaceans: the name of the residue, the substance, the metabolite, the regulatory use status, the type of species, the tissue to be tested and the action level in ppm and ppb.

TABLE 27.1 CFIA aquaculture therapeutant residue monitoring list.

Class	*Substance*	*Marker residue/ metabolite*	*Regulatory use status*	*Species*	*Tissue*	*Action level (ppm)*	*Action level (ppb)*
Fluoroquinolones	Danofloxacin	N/A	Not accepted for use	Crustaceasn	N/A	0.001	1.0
Macrolides	Erythromycin	N/A	EDR	Crustaceans	Muscle	0.03	30
Nitrofurans	Nitrofurazone	Semicarbazide	Banned	Crustaceans	N/A	DTC	DTC
Quinolones	Flumequine	N/A	Not accepted for use	Crustaceans	N/A	DTC	DTC

Note: N/A, not available.
Source: Fish Products Standards and Methods Manual – Appendix 1.

REGULATIONS APPLICABLE IN THE EU

During recent years, the EU has completed a revision of the legislation governing food hygiene and lain down specific hygiene rules for food of animal origin. 'Regulation (EC) No 178/2002 of the European Parliament and of the Council of 28 January 2002 laying down the general principles and requirements of food law, establishing the European Food Safety Authority and laying down procedures in matters of food safety' is of very broad scope. It establishes the general principles and requirements of food law, lays down procedures on matters of food safety and establishes the structure and role of EFSA. It also covers the basic concepts of equivalence and traceability.

The Regulation applies to all stages of production, processing and distribution of food and animal feed, setting the basic principle of 'the farm to table' approach. It lays down the general principles of food law including risk analysis, the precautionary principle and protection of consumers' interests plus the general obligations of the different bodies in the food chain and their consequent liabilities.

The 'hygiene package' consists of four legislative parts, adopted in April 2004. The first part is the Regulation (EC) No. 852/2004 of the European Parliament and of the Council of 29 April 2004 on the hygiene of foodstuffs, which defines food hygiene as meaning the measures necessary to control any kind of issue and to ensure food safety. The second part is Regulation (EC) No. 853/2004, laying down specific hygiene rules regarding the hygiene of foodstuffs. Then, in Regulation (EC) No. 854/2004 one can find specific rules regarding the organization of official controls on products of animal origin intended for human consumption. The last part is Regulation (EC) No. 882/2004 on official controls performed to ensure the verification of compliance with feed and food law, animal health and animal welfare rules.

The current regulation concerning microbiological and organic criteria applicable to shrimp is Council Regulation (EC) No. 2406/96, laying down common marketing standards for certain fishery products. This regulation sets freshness and size categories for shrimp, as indicated in Table 27.2.

TABLE 27.2 Freshness categories for shrimp under EU regulation.

Requirements	*Criteria*	
	Freshness category	
	Extra	*A*
Minimum requirements	Surface of shell: moist and shiny Shrimp must fall out separately when transferred from one container to another Flesh must be free from any foreign odour Shrimp must be free from sand, mucus and other foreign bodies	The same as for extra category
Appearance of:		
1. Shrimp with shell	Clear reddish-pink in colour with small white flecks; pectoral part of shell predominantly light in colour	Ranging in colour from slightly washed out reddish-pink to bluish-red with white flecks; pectoral part of the shell should be light coloured tending towards grey
2. Deep water prawn	Uniformly pink	Pink with possibility of start of blackening of head
Condition of flesh during and after shelling	Shells easily with only technically unavoidable losses of flesh Firm but not tough	Shells less easily with small losses of flesh Less firm, slightly tough
Fragments	Occasional fragments of shrimp allowed	Small quantity of fragments of shrimp allowed
Smell	Fresh seaweed, slightly sweet smell	Acidulous; no smell of seaweed

Source: Council Regulation (EC) No. 2406/96.

The Commission Regulation (EC) No. 2073/2005 on microbiological criteria for foodstuffs contains specific requirements for crustaceans. Regulation (EC) No. 1333/2008 provides guidance for the use of additives, whereas Commission Regulation (EC) No. 1881/2006 sets the maximum levels for certain contaminants in foodstuffs including dioxins, polychlorinated biphenyls and heavy metals. Regarding labelling and packaging, the EU has two regulations in place. The first is Regulation (EC) No. 1069/2009, laying down health rules as regards animal by-products and derived products not intended for human consumption. The second is Regulation (EU) No. 1379/2013 on the common organization of markets in fishery and aquaculture products. In both, it is possible to find all the requirements regarding labelling and packaging.

Regulations on residues of veterinary drugs

One of the most important regulations on residues of veterinary drugs is Commission Regulation (EU) No. 37/2010 on pharmacologically active substances and their classification regarding maximum residue limits in foodstuffs of animal origin. In its Annex, it sets all the allowed substances classifying them by pharmacologically active substance, marker residue, animal species, maximum residue limits, target tissues, other provisions and therapeutic classification. It is important to highlight that the majority of rejections in the EU are due to residues of veterinary drugs, especially nitrofuran. The use of nitrofurans such as furazolidone, nitrofurazone and nitrofurantoin in animal production has been banned in the EU since 1993.

Another important rule regarding residues of veterinary drugs is Regulation (EC) No. 470/2009, laying down Community procedures for the establishment of residue limits of pharmacologically active substances in foodstuffs of animal origin. It provides information on maximum residue limits, especially on risk assessment and risk management, implementing measures and classification of pharmacologically active substances.

REGULATIONS APPLICABLE IN JAPAN

In Japan, the main important regulatory instruments for food safety regarding fishery and aquaculture products are described in the Handbook for Agricultural and Fishery Products Import Regulations 2009, providing guidance about import examination, inspection of fishery products and microbiological criteria and the Food Sanitation Act. The latter provides guidance about general import, inspection principles and the necessary criteria for labelling and packaging. In addition, this Act provides guidance about import procedures, the obligation for importers to submit import notification and the procedures for import notification of foods and related products. The Act specifies the regulatory requirements for importing fishery products into Japan for commercial purposes as food for direct human consumption. Furthermore, the Food Sanitation Act provides guidance about permitted food additives sold in Japan, their standards and their maximum limits.

REGULATIONS APPLICABLE IN THE USA

In the USA, there are several manuals, regulations and guidance documents regarding food safety for fishery and aquaculture products, including shrimp. One of the most important of these is the FSMA, which aims to ensure that the US food supply is safe by shifting the focus from responding to contamination to preventing it.

Shrimp processors can find assistance in the development of their HACCP systems in the Fish and Fishery Products Hazards and Controls Guidance Fourth Edition. The Guidance provides criteria for microorganisms, additives and heavy metals. Shrimp producers can consult the FSMA FSVP for Importers of Food for Humans and Animals; it requires that importers perform certain risk-based activities to verify that food imported into the USA has been produced in a manner that meets applicable US safety standards. The Food Labelling Guide (Guide for industry) can be useful for shrimp processors and exporters in regard to finding all the requirements on labelling and packaging.

Regulations on residues of veterinary drugs

In the USA, import notifications due to the presence of residues of veterinary drugs are a problem of great importance. For this reason, shrimp-exporting companies could consult the Fish and Fishery Products Hazards and Controls Guidance Fourth Edition, where all approved aquaculture drugs and 'low regulatory priority aquaculture drugs' can be found in chapter 11.

Import notifications in Canada, the EU, Japan and the USA

IMPORT NOTIFICATIONS IN CANADA

The purpose of the AIRS of the CFIA is to provide accurate and timely information on import requirements (FAO, 2018). The application uses a question-and-answer approach to guide the user through a series of questions about the HS codes, origin, destination, end use and miscellaneous qualifiers of the product they want to import. All fish importers must be licensed, which requires the maintenance of all recalls and complaints records and notification to the CFIA if the results of an investigation indicate that the fish may be unsafe or non-compliant with Canadian regulations. When an inspection takes place, the imported products are subjected to microbiological, chemical, bioassay and/or sensory examinations depending on the product type, foreign country standards and Canadian standards.

The tests on random samples are divided into two categories. Standard tests (applied to all products) include:

- Labelling, which involves examination of the label, packaging and code markings to evaluate compliance with Canadian requirements.
- Net content, which involves examination to evaluate conformity with all weight declarations (e.g. net and/or drained weight, including fluid measure where applicable).
- Sensory, which involves examination to evaluate sensory and physical compliance with quality standards.
- Container integrity inspection, to determine the compliance of canned fish with Canadian standards for container integrity.

Specialized tests (applied to those products of safety concern) include:

- Composition analyses to ensure that non-permitted or non-declared ingredients or additives are absent, and to determine whether declared ingredients/additives exceed regulatory guidelines.
- Chemical analyses to determine whether chemical contaminants exceed regulatory limits. This would include toxic elements, pesticides, industrial chemicals and drug residues.
- Natural toxin analyses to determine whether natural toxins exceed regulatory limits – e.g. paralytic shellfish poison, domoic acid and other biotoxins such as ciguatoxin and okadaic acid.
- Bacteriological testing to ensure the absence of pathogenic or indicator organisms – e.g. analyses for organisms such as *Escherichia coli*, *Listeria monocytogenes*, *Salmonella* and *Staphylococcus aureus*.
- Safety parameter testing to ensure that the product is packed in containers that are sealed (FAO, 2018).

Between 2016 and 2018 there were 389 shrimp detentions and rejections in Canada representing 11% of the total import notifications of fish and fishery products (Table 27.3). Most border rejections fall within a broad category called 'others' with 358 cases, representing the 92% of all the causes of detention in shrimp. The number of cases was fairly stable during the three years studied (Figure 27.1), but this was followed by chemical issues with a total of 31 cases – eight in 2016 and 23 in 2017 – whereas none were recorded in 2018.

TABLE 27.3 Import notifications of shrimp in Canada by cause, 2016–2018.

Causes	*2016*	*2017*	*2018*	*Total*
Others	121	111	126	358
Chemical	8	23	0	31
Total	129	134	126	389

Source: CFIA.

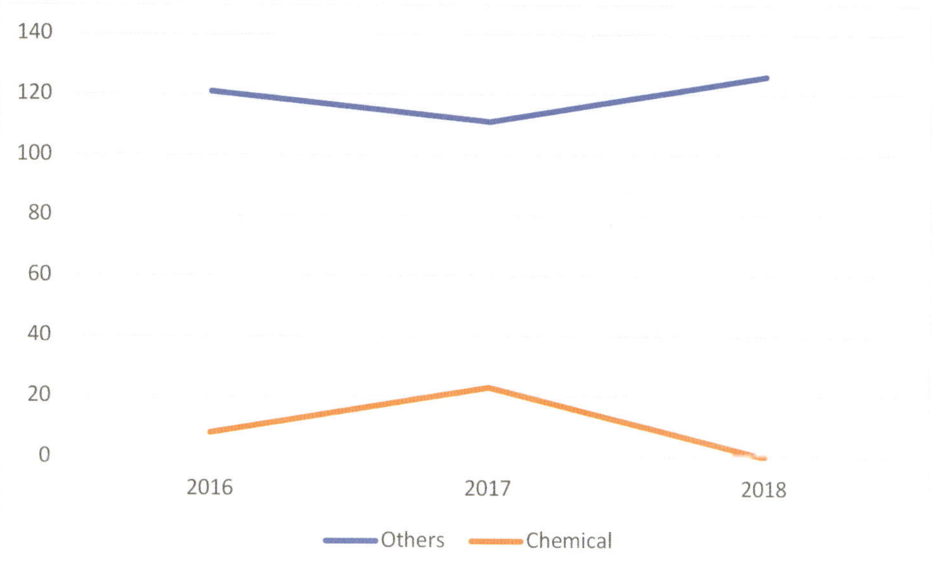

FIGURE 27.1 Import notifications of shrimp in Canada by cause, 2016–2018. Source: Canadian Food Inspection Agency.

The main cause of import notifications in the 'others' category is related to labelling issues, with a total of 316, representing 88% of all causes of shrimp import notifications (Table 27.4). This was followed by lack of net weight determination, with 20 cases – five of these detected in 2016, 12 in 2017 and only three in 2018 (Figure 27.2). Remaining issues under this category were related to food information regulation (FIR) requirements (nine cases), sensory evaluation (seven cases), can coding (five cases) and container integrity (one case).

As shown in Table 27.5, additives represent the most common cause of rejection under the chemical category, with 11 cases, followed by other issues including water activity, fluoroquinolones, nitrofurans, tetracycline, sulphite and tocopherols (Figure 27.3). It is important to note that chemical causes are few in comparison with general issues.

TABLE 27.4 Import notifications of shrimp in Canada, other causes, 2016–2018.

Other causes	*2016*	*2017*	*2018*	*Total*
Labelling	111	91	114	316
Net weight determination	5	12	3	20
FIR requirements	0	5	4	9
Sensory evaluation	5	2	0	7
Can coding	0	0	5	5
Container integrity	0	1	0	1
Total	121	111	126	358

Source: CFIA.

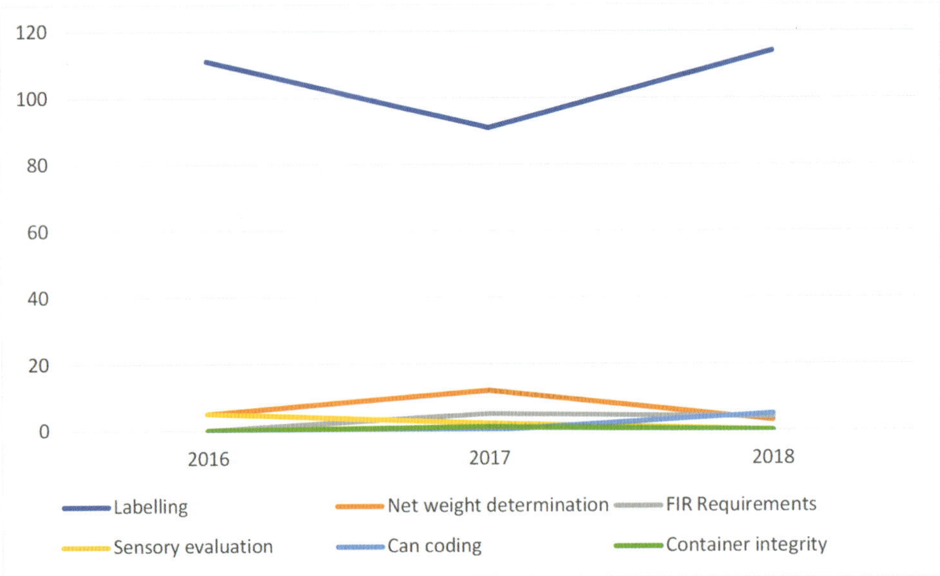

FIGURE 27.2 Import notifications of shrimp in Canada, other causes, 2016–2018. Source: Canadian Food Inspection Agency.

TABLE 27.5 Border rejections of shrimp in Canada, chemical causes, 2016–2018.

Chemical causes	*2016*	*2017*	*2018*	*Total*
Additives	3	8	0	11
Water activity	0	5	0	5
Fluoroquinolones	2	2	0	4
Nitrofurans	1	3	0	4
Tetracycline	1	3	0	4
Sulphite	1	0	0	1
Tocopherols	0	2	0	2
Total	8	23	0	31

Source: CFIA.

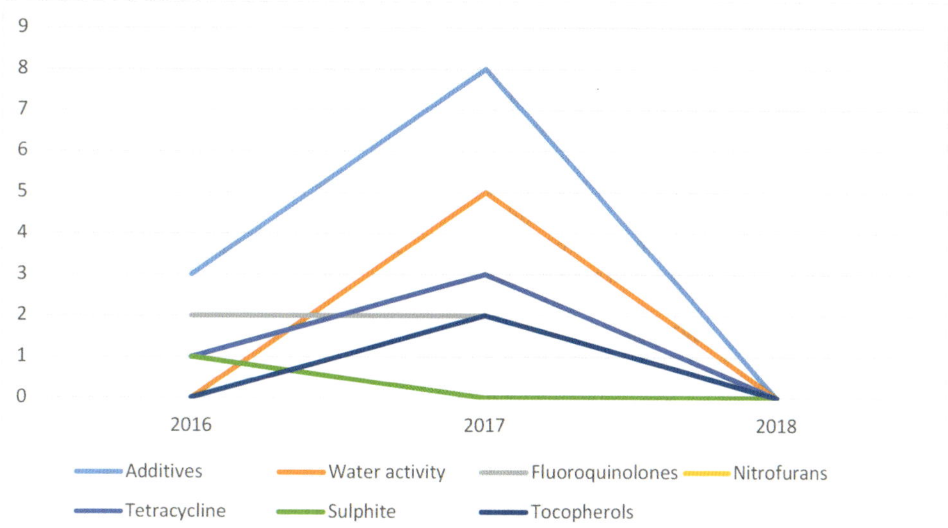

FIGURE 27.3 Import notifications of shrimp in Canada, chemical causes, 2016–2018. Source: Canadian Food Inspection Agency.

IMPORT NOTIFICATIONS IN THE EU

The RASFF is a key tool to ensuring the proper flow of information across borders when public health risks are detected (FAO, 2019a). Whenever a member of this network has any information relating to the existence of a serious risk to human health, or an attempt made to introduce a non-compliant product into the market, this information is immediately notified under RASFF.

Seafood, like all other food or feed, can be subject to:

- Alerts: a notification of a risk that requires rapid action. The product subject to an alert must be withdrawn or recalled from the market.
- Border rejection: a notification about a product that was refused entry into the market for reasons relating to a serious risk to human health.
- Information alert: food can also be subject to information for attention or information for follow-up. These are both notifications that do not require rapid action, either because the risk is not considered serious or the product is not on the market at the time of notification.

The RASFF portal provides a variety of information on products, such as:

- classification of notifications
- date of the case
- notifying country
- subject (e.g. 'mercury in chilled swordfish steaks coming from Spain')

- product category
- product type
- risk decision.

If other details are needed, it is possible to extract information related to:

- action taken
- distribution status
- hazard
- category
- analytical result
- sampling date (FAO, 2019a).

There were 116 shrimp import notifications in the EU between 2016 and 2019, representing 9% of total import notifications of fishery and aquaculture products. Most shrimp import notifications were due to chemical issues, with a total of 63 cases from 2016 to 2019; there was a peak in 2017, with 30 cases recorded through the RASFF (Table 27.6). This was followed by other causes, including issues such as improper health certificates, poor temperature control and packaging problems, with 39 cases of import notifications (Figure 27.4). Other import notifications were due to microbiological issues, with 14 cases.

TABLE 27.6 Import notifications of shrimp in the EU by cause, 2016–2019.

Causes	*2016*	*2017*	*2018*	*2019*	*Total*
Chemical	14	30	15	4	63
Others	6	9	12	12	39
Microbiological	6	3	2	3	14
Total	26	42	29	19	116

Source: RASFF portal.

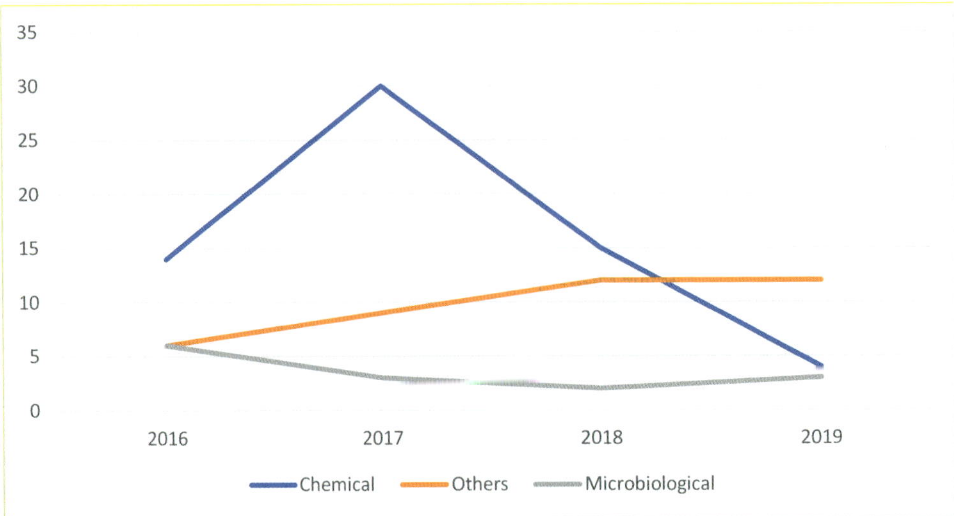

FIGURE 27.4 Import notifications of shrimp in the EU by cause, 2016–2019. Source: RASFF portal.

The main chemical issue detected was nitrofurans, with a total of 29 cases in shrimp, especially frozen product (Table 27.7). The next highest cause of import notification due to chemicals was sulphites, with a total of 18 cases, but this category was found to be decreasing over recent years (Figure 27.5). The third highest cause of chemical issues was cadmium, with eight cases recorded during 2017. Other chemical rejections were due to chloramphenicol (four cases), additives (two cases), sulfadiazine and tetracycline, with one case each.

TABLE 27.7 Import notifications of shrimp in the EU by chemical cause, 2016–2019.

Chemical causes	*2016*	*2017*	*2018*	*2019*	*Total*
Nitrofurans	2	13	11	3	29
Sulphites	9	7	2	0	18
Cadmium	0	8	0	0	8
Chloramphenicol	1	2	0	1	4
Additives	0	0	2	0	2
Sulfadiazine	1	0	0	0	1
Tetracycline	1	0	0	0	1
Total	14	30	15	4	63

Source: RASFF portal.

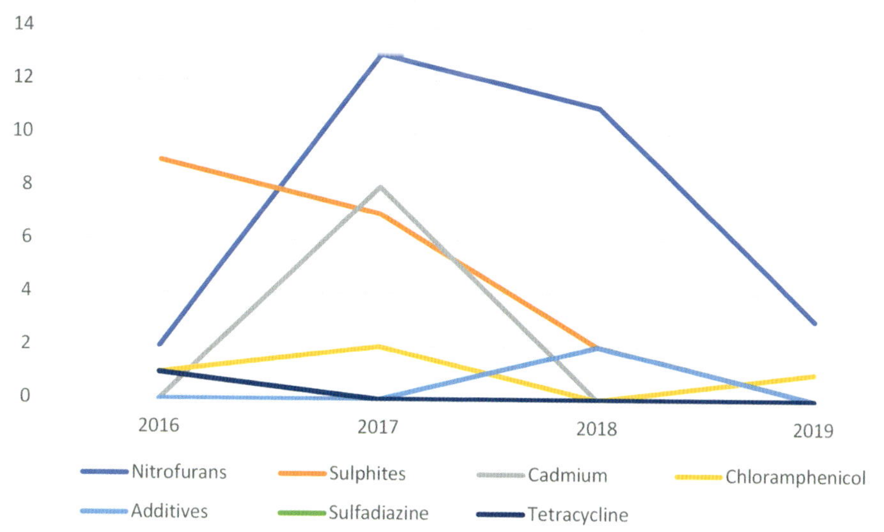

FIGURE 27.5 Import notifications of shrimp in the EU by chemical cause, 2016–2019. Source: RASFF portal.

Among the 39 border import notifications of shrimp due to other issues, the main problem was related to poor temperature control, with a total of 32 cases representing 82% of all other causes (Table 27.8). This was followed by three cases of products unfit for human consumption and two cases each related to allergens and improper health certification (Figure 27.6).

TABLE 27.8 Import notifications of shrimp in the EU, other causes, 2016–2019.

Other causes	*2016*	*2017*	*2018*	*2019*	*Total*
Poor temperature control	5	7	10	10	32
Unfit for human consumption	0	0	1	2	3
Allergens	0	2	0	0	2
Improper health certificate	1	0	1	0	2
Total	6	9	12	12	39

Source: RASFF portal.

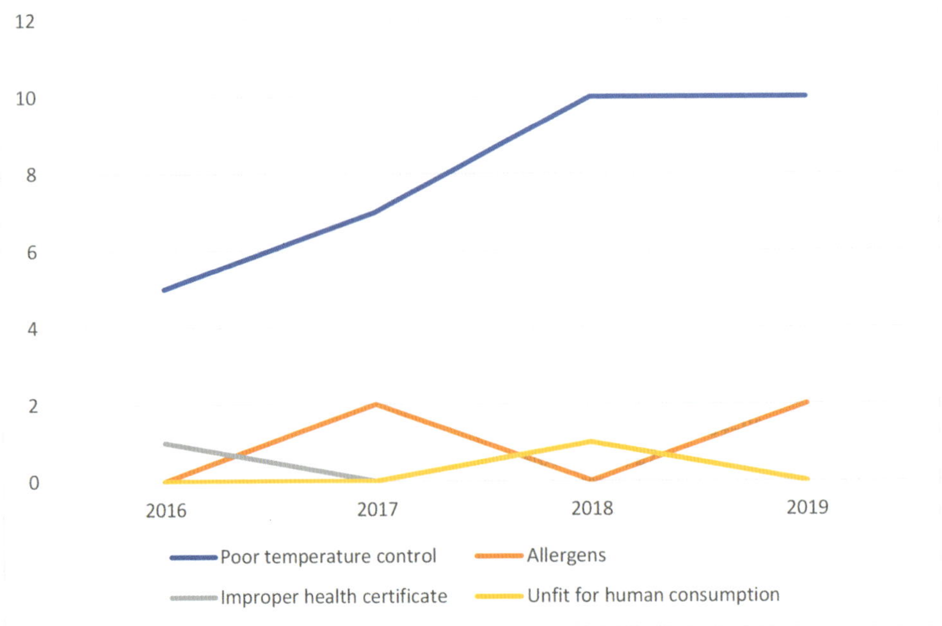

FIGURE 27.6 Import notifications of shrimp in the EU, other causes, 2016–2019. Source: RASFF portal.

The main bacterium detected in shrimp was *Vibrio cholerae*, with seven cases: five during 2016, one in 2017 and one in 2019. This was followed by *Listeria monocytogenes* (four cases), *Salmonella* (two cases) and *Vibrio parahaemolyticus* (one case).

TABLE 27.9 Import notifications of shrimp in the EU, microbiological causes, 2016–2019.

Bacterial species	*2016*	*2017*	*2018*	*2019*	*Total*
Vibrio cholerae	5	1	0	1	7
Listeria monocytogenes	0	0	2	2	4
Salmonella	1	1	0	0	2
Vibrio parahaemolyticus	0	1	0	0	1
Total	6	3	2	3	14

Source: RASFF portal.

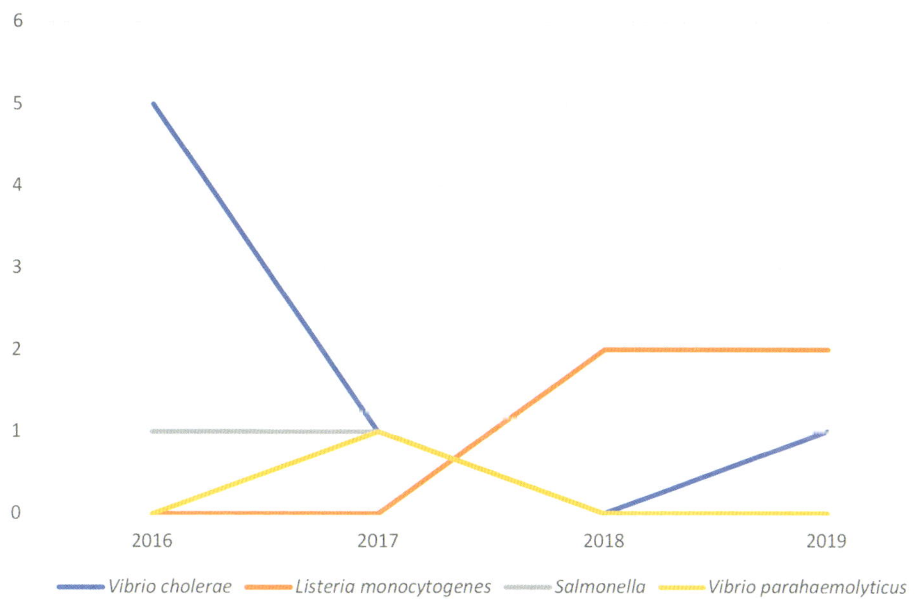

FIGURE 27.7 Import notifications of shrimp in the EU, microbiological causes, 2016–2019. Source: RASFF portal.

IMPORT NOTIFICATIONS IN JAPAN

There are 32 quarantine stations in Japan located at international seaports and airports (FAO, 2019c). At these quarantine stations food inspectors conduct document examination to confirm that the foods comply with the Food Sanitation Law, inspect and monitor imported foods and related products and guide importers on the practical procedures for food importation. During the document examination, the food sanitation inspector validates whether:

- the imported food complies with the manufacturing standards regulated under the Food Sanitation Law
- the use of additives complies with standards

- poisonous or hazardous substances are contained
- the manufacturer or the place of manufacturing has had sanitary issues previously, based on the information reported in the notification form.

The food safety inspectors also examine the notification and attached documents to determine any necessity for on-site, organoleptic, chemical, physical or microbiological examinations. After the document examination, the Ministry of Health, Labour and Welfare (MHLW) carries out an inspection of the imported foods to verify their compliance with the Food Sanitation Act.

When a regulatory violation is detected, proper measures are taken for the relevant products, including disposal or shipping back of the items to the country of origin. The MHLW provides information on its website about importation regulations for all foodstuffs, together with information on the total number of import notifications, as well as on the frequency of inspections and the number of cases of rejected imports. The file provides a variety of information for each case, including the item, the exporting country, the name

TABLE 27.10 Import notifications of shrimp in Japan, 2016–2019.

Causes	*2016*	*2017*	*2018*	*2019*	*Total*
Chemical	29	19	15	15	78
Microbiological	18	16	15	11	60
Others	0	0	0	1	1
Total	47	35	30	27	139

Source: MHLW.

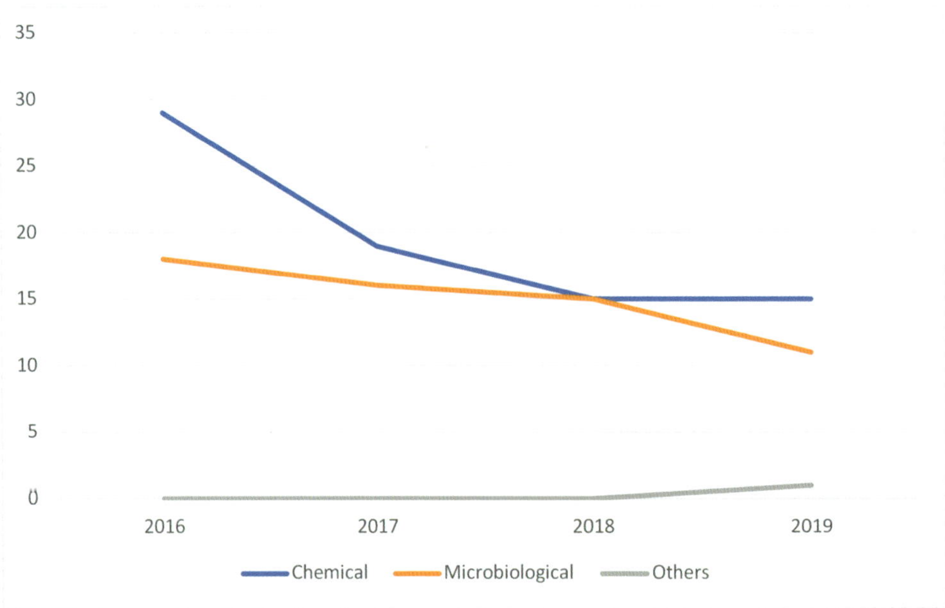

FIGURE 27.8 Import notifications of shrimp in Japan, 2016–2019. Source: MHLW.

of the manufacturer, the shipper, the relevant article of the Food Sanitation Law, the contents of violation, the quarantine station, the cause of violation, the disposal of the cargo, the remark and the publication day.

There were 139 shrimp detentions and rejections in Japan between 2016 and 2019, representing 28% of total import notifications of fishery and aquaculture products. The main causes of import notifications were chemical issues with 78 cases, followed by microbiological issues with 60 cases and by other causes, with only one case due to radiation exposure recorded in 2019 (Table 27.10). In general, import notifications in Japan are decreasing, from 47 cases in 2016 to 27 in 2019 (Figure 27.8).

The most generally reported chemical issues are those represented by antimicrobials, with a total of 70 detentions, representing 90% of total chemical issues in shrimp. The most frequently reported types of antimicrobials were enrofloxacin, furazolidone and sulfadiazine (Table 27.11). Other chemical issues are additives (four cases), sulfur dioxide (three cases) and leucomalachite green (one case). Notifications due to chemical issues decreased from 29 in 2016 to 15 in 2019 (Figure 27.9).

TABLE 27.11 Import notifications of shrimp in Japan, chemical causes, 2016–2019.

Chemical causes	*2016*	*2017*	*2018*	*2019*	*Total*
Enrofloxacin	12	8	7	6	33
Furazolidone	9	9	5	5	28
Sulfadiazine	6	2	1	0	9
Additives	1	0	2	1	4
Sulfur dioxide	0	0	0	3	3
Leucomalachite green	1	0	0	0	1
Total	29	19	15	15	78

Source: MHLW.

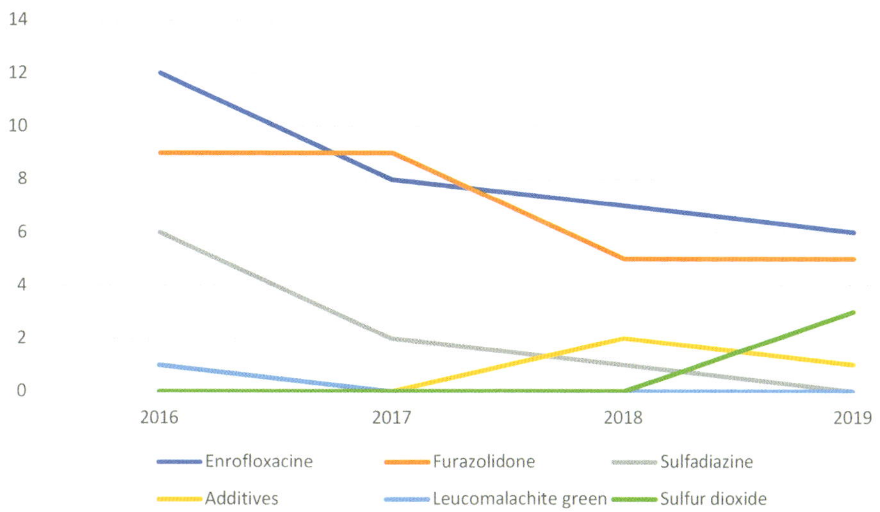

FIGURE 27.9 Import notifications of shrimp in Japan, chemical causes, 2016–2019. Source: MHLW.

Live bacteria represent the main microbiological issue, accounting for 23 cases, followed by *E. coli* in 20 cases (Table 27.12). The third most prevalent cause of import notification was coliform bacteria, with 17 cases (Figure 27.10).

TABLE 27.12 Import notifications of shrimp in Japan, microbiological causes, 2016–2019.

Microbiological causes	*2016*	*2017*	*2018*	*2019*	*Total*
Live bacteria	6	5	7	5	23
Escherichia coli	6	7	4	3	20
Coliforms	6	4	4	3	17
Total	18	16	15	11	60

Source: MHLW.

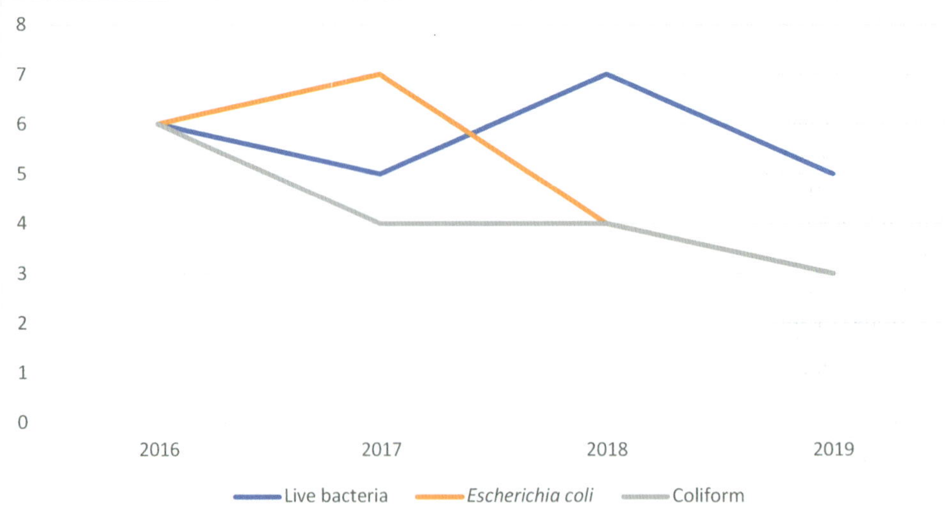

FIGURE 27.10 Import notifications of shrimp in Japan, microbiological causes, 2016–2019. Source: MHLW.

IMPORT NOTIFICATIONS IN THE USA

In the USA, health and hygiene checks are conducted by the FDA of the Department of Health and Human Services and by the FSIS of USDA. A federally mandated seafood rule is the basis for sanitary procedures for processing and importation of fishery products into the country, including good hygienic and manufacturing practices and HACCP. The FSMA encourages companies to participate in a voluntary verification scheme based on periodic certifications, inspections, hazard analysis and biannual preventive control plans. The FDA seeks to prevent entry of unsafe products or to remove them from commerce through advisory, administrative and judicial actions.

The FDA is mandated to examine foodstuffs entering the USA through US Customs and Border Protection, either before entry or after safe delivery to importers. Importers are required to file an application with the US customs authorities for each shipment, and they are also requested to provide copies of customs entry documents together with an invoice of all items listed, to the FDA. The customs authorities notify the FDA of all applications to them for FDA-regulated products. The FDA decides which entries need to be examined and arranges for samples to be collected accordingly. Each year, the programme offices of FDA and the Office of Seafood at CFSAN prepare the compliance programme that directs field inspection and surveillance activities. The programme outlines product areas for emphasis, the types of product to target, the make-up of samples, the types of analyses to conduct on specific products, the analytical methods to be used and the regulatory parameters to determine compliance. If, during the course of the year, concerns about specific products arise, new assignments are written to specify how to conduct inspection and/or sampling in the particular circumstance.

Some of the areas of safety concern in seafood are:

- pathogens: *Salmonella* spp., *Clostridium botulinum*, *Vibrio* spp., *L. monocytogenes*, *Staphylococcus aureus*, *E. coli*
- parasites
- paralytic shellfish toxins causing paralytic shellfish poisoning (PSP), neurotoxic shellfish toxins causing neurotoxic shellfish poisoning (NSP), diarrhoeic shellfish toxins causing diarrhoeic shellfish poisoning (DSP), amnesic shellfish toxins causing amnesic shellfish poisoning (ASP) and ciguatera poisoning (CP)
- histamines
- contaminants and pesticides, including methyl mercury
- aquaculture drugs: unapproved drugs or unapproved applications
- additives: unapproved or improperly declared.

The FDA also inspects products for spoilage decomposition, for being considered filthy, for improper labelling or for showing any forms of economic deception such as short weight or species substitution. All imported seafood is required to meet the same standards as those for domestic goods. The Food, Drug, and Cosmetic Act authorizes the FDA to detain products that appear to be non-compliant with the Act. The FDA district office involved will issue a 'Notice of Detention and Hearing' specifying the nature of the violation, cited to the owner or consignee, who is entitled to an informal hearing to provide testimony and/or documentation on the suitability of the product for import into the USA.

A 'Notice of Detention and Hearing' shows:

- the type and nature of the violations cited by the FDA
- the products, countries and industry involved.

If the owner fails to submit evidence that the product is in compliance, or fails to submit a plan to bring the product into compliance, the FDA will issue a 'Notice of FDA Action' refusing admission of the product. The product then has to be sent back to the exporting country or destroyed within 90 days. Each month, an Import Refusal Report (IRR) is published based on data generated by the FDA's Operational and Administrative Import Support (OASIS) (FAO, 2019b).

There were 677 shrimp detentions and rejections in USA between 2016 and 2019 (FAO, 2019b), representing 11% of total rejections of fishery and aquaculture products (Table 27.13). Figure 27.11 shows that shrimp notificaitons are definitely decreasing, with a total of 251 in 2016 increasing to 138 in 2019. The majority of shrimp detentions were due to chemical issues (312 cases), followed by microbiological causes (187 cases) and 'others' (178 cases).

TABLE 27.13 Import notifications of shrimp in the USA by cause, 2016–2019.

Causes	*2016*	*2017*	*2018*	*2019*	*Total*
Chemical	143	54	50	65	312
Microbiological	52	31	60	44	187
Others	56	56	37	29	178
Total	251	141	147	138	677

Source: FDA

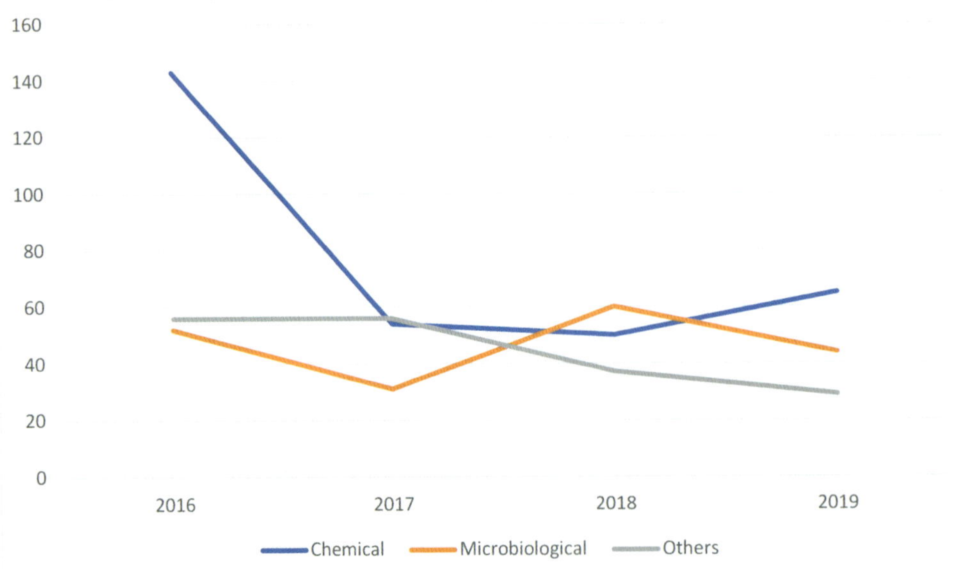

FIGURE 27.11 Import notifications of shrimp in the USA by cause, 2016–2019. Source: FDA.

The main chemical issues were residues of veterinary drugs, with 186 cases, representing 60% of chemical causes in shrimp (Table 27.14). Compared with 2016, there was a decrease in the number of cases of detentions due to the presence of residues of veterinary drugs, from 99 in 2016 to 21 in 2019 (Figure 27.12). This was followed by nitrofurans (90 cases) then chloramphenicol, with 22 cases, which are decreasing. It is important to highlight that although nitrofurans and chloramphenicol are residues of veterinary drugs, they are identified as a separate cause by FDA in this category. Unauthorized additives were the fourth chemical cause, with 13 cases recorded, followed by pesticide with one case recorded in 2019. It is important to highlight that chemical causes in shrimp decreased, from 143 cases in 2016 to 65 in 2019.

TABLE 27.14 Import notifications of shrimp in the USA, chemical causes, 2016–2019.

Chemical causes	*2016*	*2017*	*2018*	*2019*	*Total*
Veterinary drugs	99	35	31	21	186
Nitrofurans	31	2	17	40	90
Chloramphenicol	11	10	1	0	22
Additives	2	7	1	3	13
Pesticides	0	0	0	1	1
Total	143	54	50	65	312

Source: FDA.

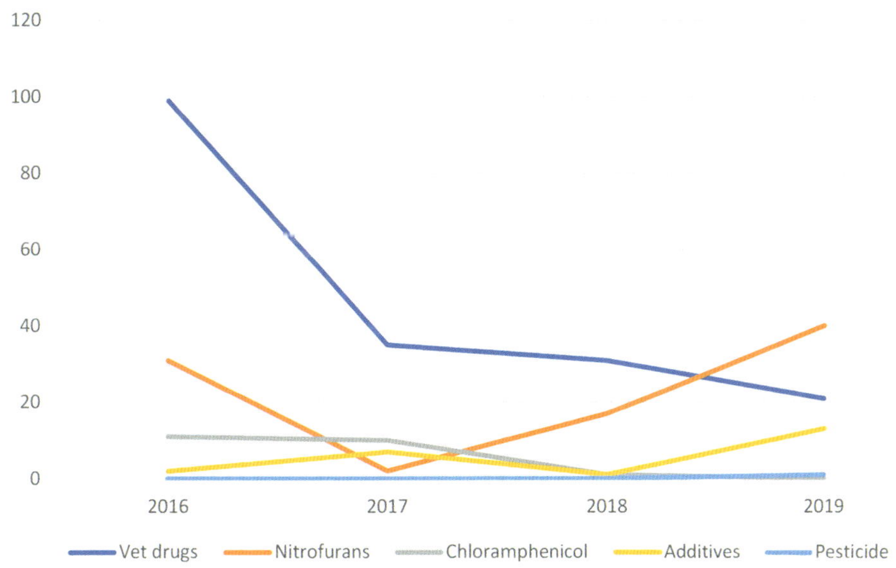

FIGURE 27.12 Import notifications of shrimp in the USA, chemical causes, 2016–2019. Source: FDA.

The main microbiological issue found in shrimp was *Salmonella* with 186 cases of detention, followed by only one case due to *L. monocytogenes*, recorded in 2019 (Table 27.15 and Figure 27.13).

TABLE 27.15 Import notifications of shrimp in the USA, microbiological causes, 2016–2019.

Microbiological causes	*2016*	*2017*	*2018*	*2019*	*Total*
Salmonella	52	31	60	43	186
L. monocytogenes	0	0	0	1	1
Total	52	31	60	44	187

Source: FDA.

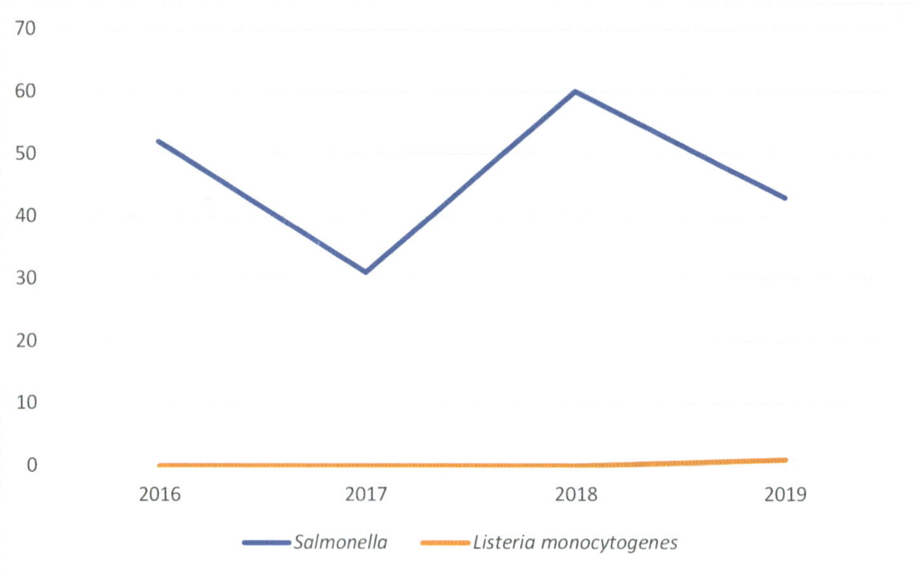

FIGURE 27.13 Import notifications of shrimp in the USA, microbiological causes, 2016–2019. Source: FDA.

Among other causes, most were associated with products found to be unfit for human consumption, decomposed or putrid (Table 27.16). All these issues are categorized as 'filthy', accounting for 141 cases representing 79% of this category. It is important to highlight that cases due to filth decreased, from 48 in 2016 to 16 in 2019. This was followed by misbranding, with nine cases, and adulteration and MfrHACCP,[3] with eight cases each (Figure 27.14).

Other specific reasons for detentions under this category were related to labelling issues, with seven cases, packaging problems with three cases and, last, to allergens, with two cases recorded.

3 The article is subject to refusal of admission pursuant to Section 801(a)(3) of the Act in that it appears to have been prepared, packed or held under insanitary conditions whereby it may have been rendered injurious to health.

TABLE 27.16 Import notifications of shrimp in the USA, other causes, 2016–2019.

Other causes	*2016*	*2017*	*2018*	*2019*	*Total*
Filthy	48	53	24	16	141
Misbranding	6	3	0	0	9
Adulteration	1	0	7	0	8
MfrHACCP	0	0	0	8	8
Labelling	0	0	3	4	7
Packaging	0	0	3	0	3
Allergens	1	0	0	1	2
Total	56	56	37	29	178

Source: FDA.

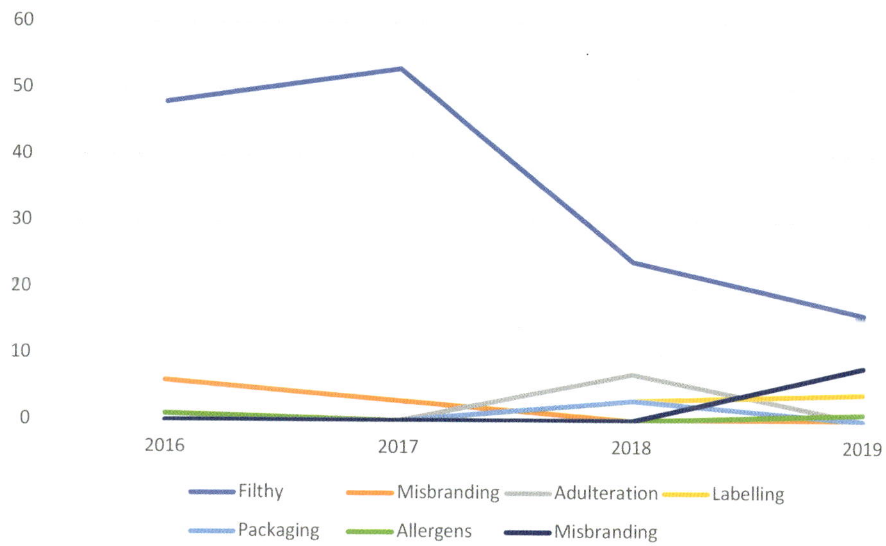

FIGURE 27.14 Import notifications of shrimp in the USA, other causes, 2016–2019. Source: FDA.

Conclusion and recommendations

Shrimp is among the world's most traded commodities. Each country has its own food inspection system at borders to ensure consumer protection. One of the greatest difficulties for shrimp exporters is the variety of inspection frameworks and legal requirements in different exporting markets, especially for veterinary drugs, which seems to be the main food safety concern for this commodity.

The Codex Alimentarius Commission is currently working to promote harmonization of regulatory requirements through different Codex committees, which will be of help for shrimp producers, processors and exporters. Although relevant Codex text was developed through the Codex Committee on Food Import and Export Inspection and Certification Systems, such as the 'Principles and guidelines for the exchange of information between importing and exporting countries to support the trade in food' and 'Guidelines for the Development of Equivalence Agreements Regarding Food Imports and Export Inspection and Certification Systems', important work toward harmonization of maximum residue levels for certain commodities still needs to be done.

References

FAO (2018) *The Automated Import Reference System (AIRS)*. Available at: http://www.fao.org/in-action/globefish/border-rejections/canada/the-airs/en/ (accessed 13 August 2021).

FAO (2019a) *The Rapid Alert System for Food and Feed (RASFF)*. Available at: http://www.fao.org/in-action/globefish/border-rejections/border-rejections-eu/rasff/en/ (accessed 13 August 2021).

FAO (2019b) *FDA's Import Refusal Report (IRR)*. Available at: http://www.fao.org/in-action/globefish/border-rejections/usa/the-irr/en/ (accessed 13 August 2021).

FAO (2019c) *Quarantine Stations*. Available at: http://www.fao.org/in-action/globefish/border-rejections/japan/quarantine-stations/en/ (accessed 13 August 2021).

Hamilton, K. A., Chen, A., Johnson, E. D. G., Gitter, A., Kozak, S., Niquice, C., Zimmer-Faust, A. G., Weir, M. H., Mitchell, J. and Gurian, P. L. (2018) Salmonella risks due to consumption of aquaculture-produced shrimp. *Microbial Risk Analysis* 9, 22–32.

INDEX

B

C

D

I

J

K

L

M

N

Q

R

S

T

U

V

W

Y

Z

5m Books

Our new company 5m Books Ltd was formed in August 2020 and includes the books of 5m Publishing and Nottingham University Press.

Our mission is to publish the highest quality books in veterinary and animal sciences, agriculture, and aquaculture.

Join us at www.5mbooks.com or follow us on Twitter at @5m_Books to be part of our community and to find out more about our books and authors.

If you are thinking of putting together a new book in a subject area in which we publish, we would be delighted to hear from you, please email us: hello@5mbooks.com

www.5mbooks.com

@5mBooks | @5m_Books | @5m_Books
linkedin.com/company/5mbooks

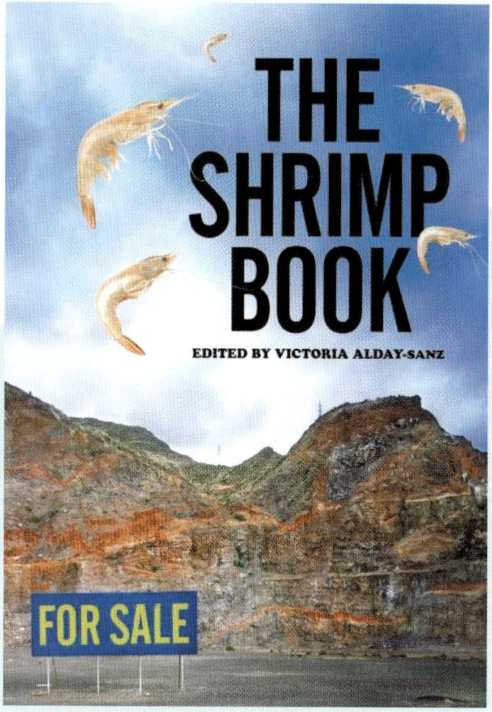

The Shrimp Book

Edited by Victoria Alday-Sanz

This book brings together information often difficult to find, joining the worlds of aquaculture science and industry production, through the leveraging of their synergies and dependencies.

It contains a wealth of scientific knowledge which can only be traced in a fragmented manner in specialised journals. It also includes the description of farming practices and the actual implementation of new technologies, information which is often presented in industry conferences, but rarely reaches publication.

Mar 2010 | ISBN: 9781904761594 £95 | $150 | €115 | 930p HB

Molluscan Shellfish Aquaculture

A Practical Guide

Edited by Sandra Shumway

Molluscan Shellfish Aquaculture is a useable manual for all those interested in an up-to-date introduction to the field. Each of the major cultured species of commercial importance is covered, providing cutting-edge information of practical use to all those involved in shellfish aquaculture.

The book's editor and chapter authors are among the most widely known and respected authorities working in the industry and academia. Species covered include mussels, clams, oysters (including pearl oysters), scallops, cephalopods, abalone and gastropods. *Molluscan Shellfish Aquaculture* contains a huge wealth of information of great use for personnel working in the industry, with chapters covering site selection, hatchery construction, disease, biofouling, best management practice and certification.

Jun 2021 | ISBN: 9781789180107 | £150 | $235 | €180 | 556p HB

5m Books **Aquaculture**

ParaFishControl

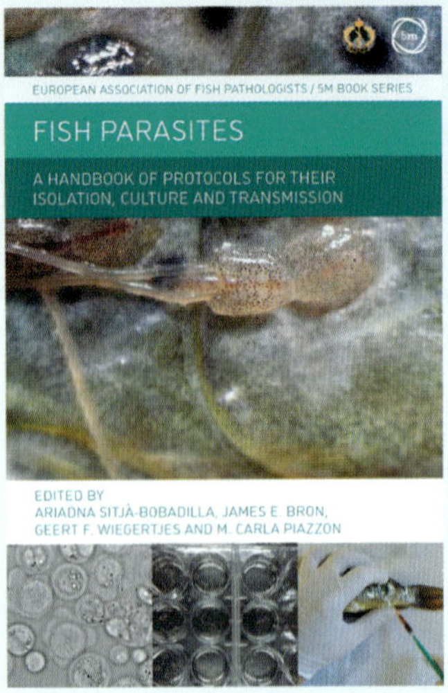

Fish Parasites

A Handbook of Protocols for their Isolation, Culture and Transmission

Edited by Ariadna Sitjà-Bobadilla, James E. Bron, Geert F. Wiegertjes and M. Carla Piazzon

Fish Parasites provides a comprehensive collection of protocols, described in terms of Standard Operating Procedures (SOPs) for parasite isolation, culture and transmission. The book has arisen from the work of The ParaFishControl consortium.

Aug 2021 | ISBN: 9781912178360 | £135 | $210 | €160 | 256p HB

A Handbook of Nutritional Disorders and Diseases of Fish and Shrimp

Albert Tacon and Loc Tran

The field of nutritional pathology is concerned with the study of those health disorders and ailments which result from nutrient deficiencies or dietary imbalances. This exciting new book provides an up-to-date and user-friendly practical guide on nutritional pathology in the main commercially-farmed species of fish and shrimp.

Oct 2022 | ISBN: 9781789181319 | £135 | $210 | €160 | 150p HB

Proliferative Kidney Disease

Edited by Bartolomeo Gorgoglione and Christyn Bailey

As the knowledge pool on Proliferative Kidney Disease deepens, and interest spreads to different fields, appealing to divergent communities, this book offers a comprehensive look on the topic, encompassing a multidisciplinary approach.

Dec 2022 | ISBN: 9781789180527 | £135 | $210 | €160 | 400p HB

The EAFP and 5m Books are pleased to announce a new partnership in the publication of an exciting, definitive and high-quality series of books.

The new EAFP / 5m Books Series will encompass professional-level reference books and practical guides aimed at fish farm operatives and technicians, fish biologists and pathologists, veterinarians, students who are studying aquaculture and fish health, and the many personnel involved in supplying the aquaculture industry with pharmaceuticals, nutraceuticals, feed and equipment.

If you would be interested in putting together a new book to be considered for inclusion into the new EAFP / 5m Books Series, please contact Sean Monaghan s.j.monaghan@stir.ac.uk

5m

EAFP

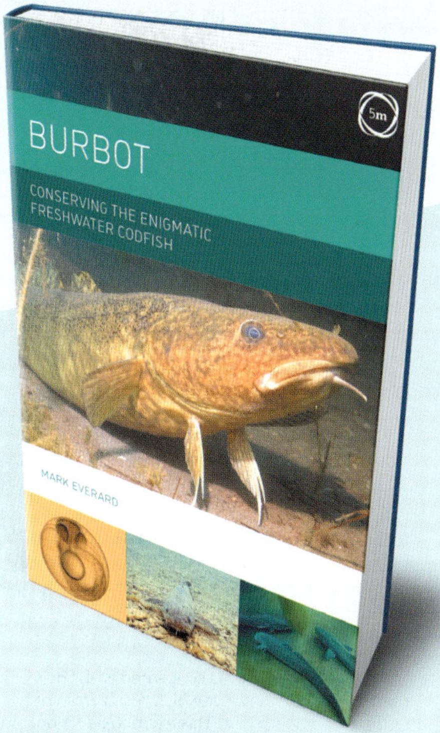

Burbot

Mark Everard

The burbot has a unique ecology as the only member of the order of cod-like fishes found in freshwater. It is the second most widely distributed freshwater fish in the Northern Hemisphere, variously threatened, extinct or thriving across different parts of this wide paleoarctic range.

Burbot were driven to extinction from Britain most probably in the 1970s, the last recorded specimen caught in 1969 in Cambridgeshire. Particularly over the past decade, a large body of work has addressed potential reintroduction of the burbot to Britain.

The burbot's diverse habitat and other needs throughout its life stages also mean that the species is a flagship for a diversity of other wildlife of restored river systems, and of the human benefits that these ecosystems can provide.

Jul 2021 | ISBN: 9781789181456
£29.95 | $49.95 | €35.95 | 176p HB

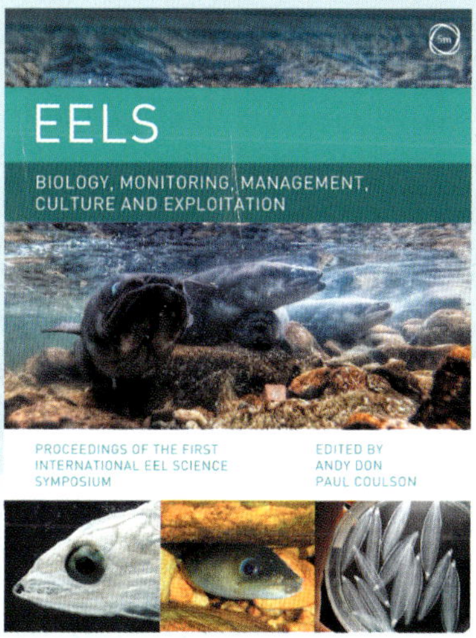

Eels

Biology, Monitoring, Management, Culture and Exploitation

Edited by Andy Don and Paul Coulson

This book, generated from the 1st UK International Eel Science Symposium, is a response to the huge interest worldwide in eels, not only from a biological standpoint, but from fisheries science and management and aquaculture.

Oct 2019 | ISBN: 9781789180695 | £150 | $235 | €180 | 464p HB

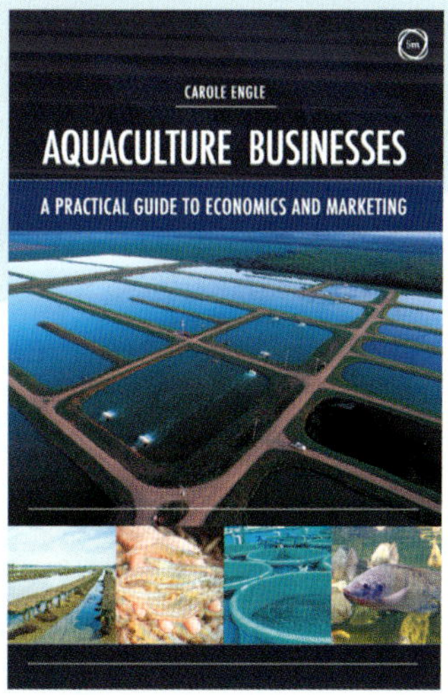

Aquaculture Businesses

A Practical Guide to Economics and Management

Carole Engle

This exciting new book provides practical guidance and advice for individuals who are seeking to manage and develop a successful aquaculture business.

Jan 2020 | ISBN: 9781789180510 | £75 | $115 | €90 | 288p HB

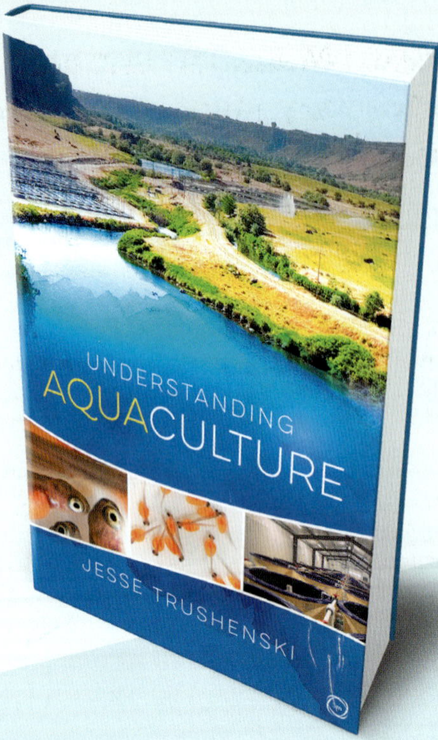

Understanding Aquaculture

Jesse Trushenski

Aquaculture is critical to food security, both now and in the future, and an informed and supportive public is needed to ensure its potential is fully realised. *Understanding Aquaculture* is an introductory guide to Aquaculture, ideal for those studying fisheries, aquaculture, natural resources management, environmental policy and food science, as well as the interested general public. It addresses the common questions associated with aquaculture, such as: Are farmed fish safe to eat? Are wild fish more nutritious? Do fish farms pollute the environment? Is farmed salmon full of antibiotics?

Understanding Aquaculture includes contents and case studies drawn from throughout the world, making it international in scope. It will fulfill the public demand for information about aquaculture product while also being a valuable resource for students and personnel working across all sectors of the aquaculture industry.

Jan 2019 | ISBN: 9781789180114 | £45 | $70 | €55 | 276p HB

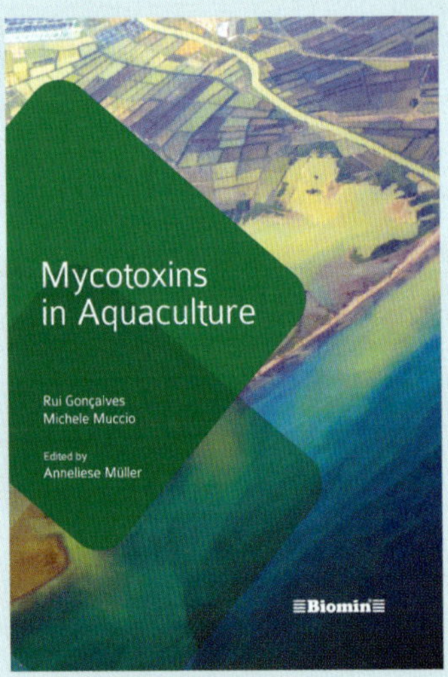

Mycotoxins in Aquaculture

Rui Gonçalves, Michele Muccio, edited by Anneliese Müller

Mycotoxins in Aquaculture aims to help producers in their daily challenges and, at the same time, to contribute to the scientific community, by offering a new tool to whoever is approaching aquaculture in this era of finite resources.

Feb 2019 | ISBN: 9781789180596 | £75 | $115 | €90 | 188p HB

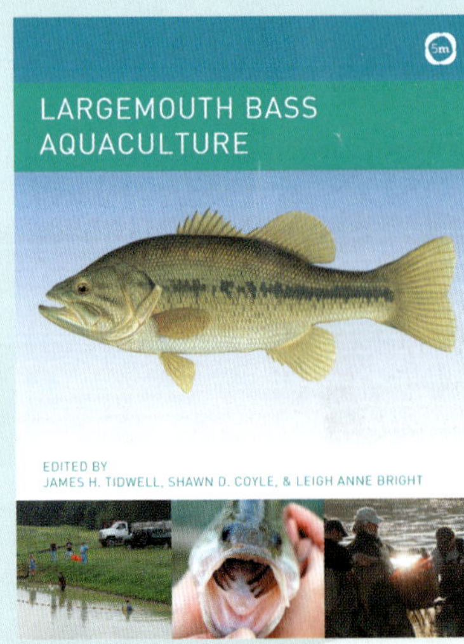

Largemouth Bass Aquaculture

Edited by James Tidwell, Shawn Coyle and Leigh Anne Bright

Largemouth Bass Aquaculture provides comprehensive coverage of all aspects of the farming of largemouth bass, with chapters encompassing all major areas of importance, including: their history, production, environment requirements, reproduction, culture methods, diseases and major markets.

Apr 2019 | ISBN: 9781789180480 | £135 | $210 | €160 | 272p HB

FORTHCOMING 2022 TITLE

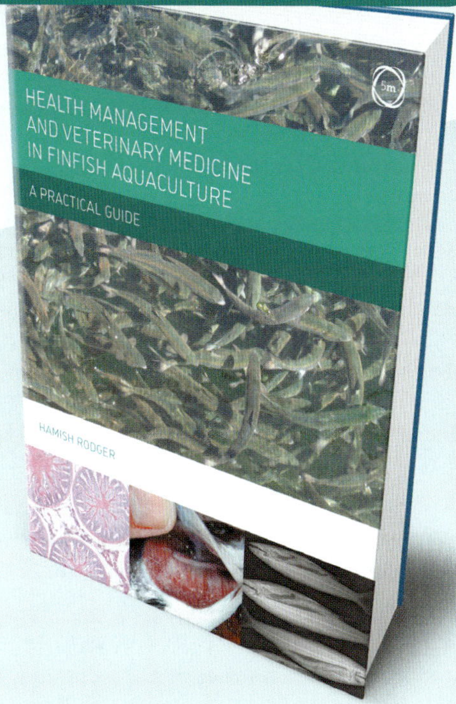

Health Management and Veterinary Medicine in Finfish Aquaculture

Hamish Rodger

This book takes a very practical approach to managing fish diseases and promoting good fish health. It focuses on farmed coldwater fish, mainly salmon and trout, as they are the most widely farmed and economically important species, but also other fish species.

The contents cover the anatomy and physiology of the fish, conditions of the rearing environment, herd level medicine, and feed management. But the main focus is on diseases and disease prevention, with information on the various classes of disease, diseases associated with environmental factors and treatment options.

Dec 2022 | ISBN: 9781789181012 | £85 | $135 | €100 | 320p HB

Cleaner Fish Biology and Aquaculture Applications

Edited by Jim Treasurer

Cleaner Fish Biology and Aquaculture Applications reviews and presents new knowledge on the biology of the utilised cleaner fish species, and provides protocols in cleaner fish rearing, deployment, health and welfare.

Mar 2018 | ISBN: 9781912158218 | £150 | $235 | €180 | 528p HB

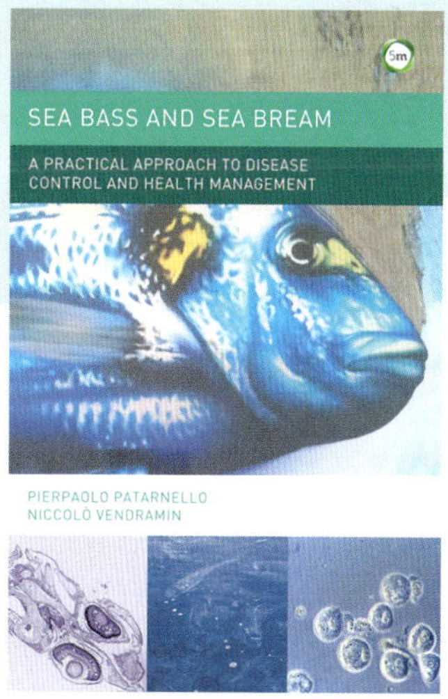

Sea Bass And Sea Bream

A Practical Approach to Disease Control and Health Management

Pierpaolo Patarnello And Niccolò Vendramin

The aim of this book is to provide practical advice and awareness of health management and disease control in sea bass and sea bream, the most widely-farmed fish in the Mediterranean region.

Sep 2017 | ISBN: 9781910455791 | £75 | $115 | €90 | 140p HB